MULTIPHASE FLOW: THE ULTIMATE MEASUREMENT CHALLENGE

MULTIPHASE FLOW: THE ULTIMATE MEASUREMENT CHALLENGE

Proceedings of

The 5th International Symposium on Measurement Techniques for Multiphase Flows (5th ISMTMF)

2nd International Workshop on Process Tomography (IWPT-2) (As a part of ISMTMF)

(5th ISMTMF/IWPT-2, 2006-Macau/Zhuhai)

Macau, China *10 – 13 December 2006*

EDITORS

Xiaoshu Cai

University of Shanghai for Science and Technology, China

Yingxiang Wu

Chinese Academy of Sciences, China

Zhiyao Huang

Zhejiang University, China

Shimin Wang

Southeast University, China

Mi Wang

Leeds University, United Kingdom

All papers have been peer-reviewed

SPONSORING ORGANIZATIONS

The Chinese Society for Measurement
The Japanese Society for Multiphase Flow
National Natural Science Foundation of China
The Chinese Academy of Sciences
The Chinese Society of Engineering Thermophysics
Virtual Center for Industrial Process Tomography

AMERICAN INSTITUTE OF PHYSICS

Melville, New York, 2007

AIP CONFERENCE PROCEEDINGS ■ VOLUME 914

Editors:

Xiaoshu Cai
College of Power Engineering
University of Shanghai for
Science and Technology
516 Jungong Road
Shanghai 200093
China

E-mail: usst_caixs@163.com

Shimin Wang
Institute of Energy and Environment
Southeast University
Nanjing 210096
China

E-mail: smwang@seu.edu.cn

Yingxiang Wu
Institute of Mechanics
Chinese Academy of Sciences
Bei Si Huan Xi Lu 15
Beijing 100080
China

E-mail: yxwu@imec.ac.cn

Mi Wang
IPSE, SPEME, University of Leeds
Woodhouse Lane
Leeds, LS2 9JT
United Kingdom

E-mail: m.wang@leeds.ac.uk

Zhiyao Huang
Institute of Automation Instrumentation
Department of Control Science and Engineering
Zhejiang University
Hangzhou 310027
China

E-mail: zyhuang@iipc.zju.edu.cn

L.C. Catalog Card No. 2007927173
ISBN 978-0-7354-0422-9
ISSN 0094-243X
Printed in the United States of America

CONTENTS

PLENARY PRESENTATION

SESSION A: FUNDAMENTALS OF MULTIPHASE FLOW MEASUREMENT

SESSION B: NOVEL SENSORS AND METHODS

SESSION C: INFORMATION TECHNOLOGY AND INSTRUMENTATION

SESSION D: OPTICAL, NON-INTRUSIVE TECHNIQUES

SESSION E: INDUSTRIAL APPLICATION AND OTHERS

SESSION PT-1: DEVELOPMENTS OF PT TECHNOLOGY

SESSION PT-2: IMAGE RECONSTRUCTION AND APPLICATIONS OF PT TECHNOLOGY

Preface

Multiphase flow is of importance to the design, research and operation for apparatuses of many engineering systems, such as power plants, chemical industry, nuclear energy industry, refrigeratory industry, petroleum industry, metallurgical industry, space industry and environmental engineering industry, etc.

On account of the complication of multiphase flows, their measurements are much more different than that for single phase flows. Up to now, although measurement techniques for multiphase flows have obtained sufficient progress and some new techniques including process tomography have been developed, the accuracy and applied convenience of measurements are still needed to be raised.

The papers presented in this symposium and workshop cover a wide scope of multiphase measurements including measurements for gas-liquid two-phase flows, gas-solid two-phase flows, liquid-solid two-phase flows, liquid-liquid two-phase flows and three phase flows. Advanced methods and recent researching developments in these fields are submitted for academic communication and discussion. It is expected that this symposium and workshop will be a very interesting forum. It will benefit every participant and will promote the development of multiphase measurement techniques.

We greatly appreciate all persons who have given effective help to organize and support this symposium. We also wish to express our grateful thanks to scientists who helped us to invite, collect and review all the submitted papers.

We hope that every participant here will enjoy this fruitful symposium and also enjoy your stay in Macao and Zhuhai.

This collection of this proceedings comprises of 7 parts. All texts of speeches and posts delivered at the symposium and workshop are included. We believe that this collection will be a valuable literature for the measurement techniques of multiphase flows including process tomography.

Zonghu Lin
Chairman of the symposium and workshop

2006 . 11 . 16

ORGANIZERS

- Macau University of Science and Technology
- Southeast University
- University of Shanghai for Science and Technology
- Zhejiang University
- University of Leeds
- OMEC Technology Co.Ltd

SPONSORS

- The Chinese Society for Measurement
- The Japanese Society for Multiphase Flow
- National Natural Science Foundation of China
- The Chinese Academy of Sciences
- The Chinese Society of Engineering Thermophysics
- Virtual Center for Industrial Process Tomography

Symposium Chairman

Zonghu Lin, Chinese Academy of Engineering, China

Co–Chairmen:

Koji Okamoto, University of Tokyo, Japan
Shimin Wang, Southeast University, China
Richard A Williams, University of Leeds, UK
A. A. Xu, Macau University of Science and Technology, China

Advisory Board:

Kefa Cen, Zhejiang University, China
G. Hetsroni, Technion - Israel Institute of Technology, Israel
G. Hewitt, Imperial College, UK
M. Sommerfeld, Martin-Luther-University Halle-Wittenberg, Germany
C. Tropea, Technische Universitat Darmstadt, Germany
Y. Tsuji, Osaka University, Japan
Yiqian Xu, Southeast University, China
F. Yamamoto, Fukui University, Japan

International Scientific Committee for ISMTMF 2006 Macau/Zhuhai

Chairman:

Hongjian Zhang, Zhejiang University, China

Members:

M.X. Francois， Paris 6 University, France
S. Fujikawa, Hokkaido University, Japan
Udo Fritsching, University of Bremen, Germany
Gerard Grehan, Interprofessional Aerothermochemistry Research Complex, France
Liejing.Guo, Xi'an Jiaotong University, China
M. Iguchi, Hokkaido University, Japan
T. Shakouchi, Mie University, Japan
Baosheng Jing, Southeast University, China
Yeon-Won Lee, Pukyong National University, Korea
Haiqing Li, Zhejiang University, China
Jidong.Lu, Huazhong University of Science and Technology, China
Jan Mewis, Kathopieke University Leuven, Belgium
Mingjiang Ni, Zhejiang University, China
Hetang Qiao, Daqing Petroleum Administrative Bureau, China
Huihe Qiu, Hong Kong University of Science and Technology, China,
Kuanfang Ren, Rouen University, France
A. Serizawa, Kyoto University, Japan
T. Takamasa, Tokyo University of Marcantile Marine, Japan
G. Touchard, Universite De Poitiers, France
Naining Wang, University of Shanghai for Science and Technology, China
Yingxiang Wu, Chinese Academy of Sciences, China
Shi Wang, Northeast University, China
Yong Yan, University of Kent, UK
Baofen Zhang, Tsinghua University, China
Zhichu Zheng, Chinese Academy of Sciences, China
Lixing Zhou, Tsinghua University, China

International Scientific Committee for IWPT-2 Macau/Zhuhai

Chairman

Richard A Williams, University of Leeds, UK

Members:

Chad Bennington, theUniversity of British Columbia, Canada
Tom Dyakowski, University of Manchester, UK
Uwe Hampe, FZ Rossendorf e.V., Germany
Bjorn Hjertaker, University of Bergen, Norway
Brian Hoyle, University of Leeds, UK
Zhiyao Huang, Zhejiang University, China
Haibo Jin, Beijing Institute of Petrochemical Technology, China
Jari Kaipio, University of Kuopio, Finland
Apostolos Kantzas, University of Calgary, Canada
Bill Lionheart, University of Manchester, UK
C.L Lin, University of Utah, USA
Shi Liu, Chinese Academy of Sciences, China
Kazuo Maeno, Chiba University, Japan
Rob Morrison, University of Queensland, Australia
Saba Mylvaganam, Telemark University College, Norway
Marchot Pierre, Université de Liège, Belgium
Andrzej Plaskowsk, Mikromax, Poland
Huaxiang Wang, Tianjin University, China
Mi Wang, University of Leeds, UK
Andraw Wilkinson, University of Cape Town, South Africa
Yingxiang Wu, Chinese Academy of Sciences, China
Cheng-gang Xie, Schlumberger Cambridge Research, UK
Wuqiang Yang, University of Manchester, UK
Baofen Zhang, Tsinghua University, China
Zhichu Zheng, Chinese Academy of Sciences, China

International Organizing Committee:

Chairman

Shimin Wang, Southeast University, China

Co- Chairmen

Yassin A. Hassan Texas A & M University, USA
Aliana, Leong Man Wai, Macao University of Science and Technology, China

Members

Gang Chen, Xi'an University of Technology, China
Xiaofang Cheng, University of Science and Technology of China, China
Feng Dong, Tianjin University, China
Qingyu Dong, Better Science Instrumentation Company, China
S. Hattori, Fukui University, Japan
S. Hosokawa, Kobe University Japan
K. Hishida, Keio University, Japan
Zhiyao Huang, Zhejiang University, China
M. Ishikawa, University of Ryukyus, Japan
Y. Kodama, NMRI, Japan
Y. Murai, Hokkaido University, Japan
N. Ninomiya, Utsunomiya University, Japan
K. Primrose, Industrial Tomography Systems Ltd, UK
D. Rhodes, Nexia Solutions/BNFL ,UK
S. Takagi, University of Tokyo, Japan
M. Takei, Nihon University, Japan
F. Takemura, AIST, Japan
R. Tweedie, Malvern Instruments Ltd. UK
Jing Wang, Shanghai Jiaotong University, China
Jianhua Yan, Zhejiang University, China

Changsui Zhao, Southeast University, China
Albert F. Zhang, OMEC Technology Co., Ltd., China
Jianyong Zhang, University of Teesside, UK
Huaichun Zhou, Huazhong University of Science and Technology, China

Paper Committee:

Chairman
Yingxiang Wu, Chinese Academy of Sciences, China

Co- Chairmen
Mi Wang, University of Leeds, UK
Zhiyao Huang, Zhejiang University, China

General Secretary:

Xiaoshu Cai, University of Shanghai for Science and Technology, China
Mi Wang, University of Leeds, UK

PLENARY PRESENTATION

Micro X-ray Tomographic Imaging Of Porous Media

R A Williams[1], X Jia[1] and C Selomulya[2]

[1]*Institute of Particle Science and Engineering (IPSE), School of Process, Environmental and Materials Engineering, University of Leeds, Leeds LS2 9JT, UK. R.A.Williams@leeds.ac.uk*
[2]*Depart of Chemical Engineering, Monash University, Melbourne VIC 3800, Australia. cordelia.selomulya@eng.monash.edu.au*

Abstract. Do not X-ray Microtomography (XMT) is a relatively new technology for measuring and characterising structures of materials at micron scales. It operates on the same principles as hospital CT scanners, but with much higher spatial resolution. The enhanced spatial resolution, portability and affordability, and the non-destructive nature of the measurement technique, are making XMT an increasingly popular means for scientists and engineers to measure or characterise structures of various materials in full 3D with unprecedented level of details. This paper reports on use of two such scanners and their application to porous media imaging.

Keywords: X-ray microtomography, imaging, porous media, flow.

1. INTRODUCTION

X-ray microtomography (XMT) operates on the same basic principles of a medical CT scanner [1], but with a much higher power intensity, pixel resolution (0.5-20 μm/pixel), smaller size and much lower price. The enhanced spatial resolution, portability and affordability, and the non-destructive nature of the measurement technique, are making XMT an increasingly popular means for scientists and engineers to measure or characterise structures of various materials in full 3D with unprecedented level of details [2]. This paper reports on use of two such scanners – Skyscan 1072 and Phoenix Nanotom – and their applications to particle technology. Table 1 below compares their capabilities.

TABLE 1. Comparison of two commercially available XMT scanners

	Skyscan 1072	**Phoenix Nanotom**
Commercially available since	2000 (replaced by 1172 since 2006)	2006
X-ray source design	Closed	Open
Power output	130 kV, 300 μA	160 kV, 250 μA
Spot size	~ 8 μm	< 1 μm
CCD detector size (pixels)	1024x1024	2304x2304
Highest pixel resolution (μm/pixel)	4	0.5
Camera shift mode for large objects	Yes	Yes
Sample manipulation	4 degrees of freedom	6 degrees of freedom
Largest sample size (mm)	35	65
Typical scan time (hours)	2	1
Typical reconstruction time (min)	120 (with dual 2.66 GHz Xeon CPUs)	15 (with a bundled 8 CPU PC-cluster)

A reconstructed image (also known as a tomogram) is a cross-sectional slice of the object, parallel to the X-ray – assuming the X-ray is an idealised fan-shaped beam. Most modern X-ray scanners actually use a cone beam directly to speed up the scanning process, since vertical traversing is in most cases redundant. By stacking the slices, a full 3D digital structure is obtained, which can then be sliced to provide cross-sectional views in any direction. Without

CP914, *Multiphase Flow: The Ultimate Measurement Challenge,* edited by X. Cai, Y. Wu, Z. Huang, S. Wang, and M. Wang

further processing, a reconstructed X-ray tomogram represents the distribution of attenuation coefficients of materials in the cross-section. Therefore, if the materials making up the scanned object are known a prior, distributions of different materials within the object can be ascertained. In an ideal situation where the X-ray is monochromatic and no artefacts are present, the relation between the pixel grey scale value and the material's density is well defined by the Lambert equation. In practice, however, X-ray is polychromatic and artefacts are usually inevitably, it is difficult to relate pixel values with materials densities in an absolute and quantitative manner. One example of the artefacts is due to beam hardening. Beam hardening refers to selective removal of energy from a polychromatic X-ray such that the X-ray becomes more penetrating as it traverses a large object. As a result, the reconstructed attenuation coefficients in the central region of the object appear to be lower than their true values, i.e., an object of uniform density appears to have a darker ring around the edge. Beam hardening can be corrected, but only to a certain extent, especially, for a multi-component structure, as the correction relies on a prior knowledge of the materials distributions. Calibration can help, but uncertainty always exists because, like a normal photograph, the colour or shade of a pixel depends not only "what's there at that point" but also "what's surrounding it". Due to these complications, information obtained from XMT is rarely analysed on its own. The more *a prior* knowledge is available about the scanned object, the better interpretation may be obtained.

(a) (b)

FIGURE 1. Two XMT systems: (a) Skyscan 1072 system, 2000. (b) Phoenix Nanotom system, 2006.

Over the past few years, the authors have used XMT to characterise a wide range of different materials, including:

- Functional materials (material science)
- Biomaterials (medical and bio-engineering)
- Fibrous stuff (textile) (Figure 2a)
- Metal powders (selective laser sintering in rapid prototyping)
- Phase distinction in ores (ore liberation in mining and mineral engineering) (Figure 2b)
- Structural uniformity of particulate products (e.g., catalysts and tablets) (Figure 2c)
- Encapsulated oil in polymer films (cosmetics)
- Agglomerates (micro-mechanics) (Figure 2d)
- Digital structures for simulations

(a) (b) (c) (d)

FIGURE 2. Examples of XMT imaged structures (a) a piece of felt (1 cm); (b) a two component ore (1 cm); (c) part of a detergent tablet (0.5 cm); and (d) a spray-dried agglomerate (0.5 mm).

In many cases, simple statistics and visual inspection of the XMT tomograms are sufficient for engineers to make useful inferences or decisions; in other cases, obtaining the structures is only the first step towards predicting structure-related properties of interest. The remaining part of this paper presents some examples of the latter, namely characterisation of porous structures and their permeability.

METHODOLOGY

Before presenting the examples, it is helpful to briefly introduce the numerical methods that are used together with XMT to produce the results: the DigiPac particle packing model, the Lattice Boltzmann Method (LBM), and LBM assisted extraction of pore network information.

2.1 DigiPac

DigiPac (Structure Vision Ltd, Leeds) is particle packing software developed at Leeds [3]. In DigiPac, shapes are digitised and represented by pixels. After digitisation, a particle is just a coherent collection of pixels, which can be stored in the same format as XMT digitised objects. These pixels are stored and manipulated as integers or as 'bits'. In addition, apart from the initial errors in digitisation, the digital algorithm does not suffer from rounding errors as traditional approaches can. The number of pixels to be used for a shape does not depend on the complexity of shape. Because of these features, DigiPac can handle non-spherical particles more efficiently than traditional vector-based methods. Using the pixel representation also means that the real shapes as imaged using XMT can be used directly by the packing software to generated packing structures. An example is shown in Figure 3 where over 500 pasta pieces of three different shapes are packed. In this example, XMT was used to digitise the pasta pieces, whose external shape, cavities or holes, internal distribution of solids, and surface texture or roughness are all captured. After digitisation, these complex shapes can be handled just as easily as spheres by DigiPac.

FIGURE 3. XMT digitised complex particle shapes (pasta) and an example of their packing.

2.2 Lattice Boltzmann Method

The LBM is a digital equivalent of the convectional computational fluid dynamics (CFD). It is performed on a regular – usually cubic – lattice grid and the calculations are local to each grid cell. This lattice approach and data localisation makes it adept at dealing with complex geometries such as those of porous media. While a fuller treatment of the subject can be found in the literature elsewhere [4], only the basic principles are presented here. The LBM maps physical space onto a regular lattice grid. Associated with each node is a set of mass probability distributions, also referred to as particle distribution functions or simply density. These distributions propagate to neighbouring nodes, as part of mass and momentum transfer, along fixed velocity vectors at each time step, followed by simple collisions designed to conserve mass and momentum. The collision operator is chosen such that the particle velocity distributions at each node obey the Navier-Stokes equations in the low Knudsen number and low Mach number regime. In the LBM, the evolution of the particle distribution functions are described in terms of discrete velocities of fixed directions and magnitude. The time step in a LBM simulation is customarily set to 1, so is the magnitude of the velocity components. As such, everything else (e.g. length scales, velocities and pressure) is in lattice units. The so-called D3Q19 scheme is adopted here for calculating the equilibrium distributions. This scheme operates on a 3D cubic lattice grid with each grid cell interacting with 18 (out of 26) of its closest neighbours. Our code is based on an implementation by Hoffman [5], with added data visualization functionality and graphical user interface for Windows, and is multi-threads enabled for shared memory architecture in dual- or multi-CPU PCs. It is referred to as DigiFlow later.

2.3 Extracting Flow Pathways

Porosity, pore size and tortuosity are commonly used metrics to characterise porous media. However, available definitions of pore size and tortuosity are empirical, and do not lend themselves to direct and independent measurement. The values obtained for pore size and tortuosity depend on the assumptions and equations used to retro-fit measured flow data. Therefore, they are generally only useful for comparison purposes for quality assurance or control where consistency is more important than accuracy.

With XMT, real structures of porous media can now be scanned and reconstructed. Computer models of particle packing structures such as DigiPac are also available to generate realistic packing structures. Both XMT and DigiPac give packing structures in digital volumetric format. In principle, given the digitised structures, it should be possible to calculate any metrics of interest. In practice, however, routine characterisations of these digitised structures have so far been limited to reporting simple statistics such as porosity and its spatial distribution. The difficulty of analysing pore networks directly (based purely on topographical information) may be illustrated with a simple example – random packing of equal spheres. A pore opening on top has at least three channels connected to it. For each of the three channels, there are at least three more channels connected to it. Moving down ten layers of packed spheres, there are at least 3^{10} or ~60,000 channels, all connected to the single starting pore! Given the shear number and geometric complexity of the channels, calculating their properties is by no means a trivial exercise. Further more, structural statistics based purely on geometric information of pore networks can be poor indicators of the actual flow distributions if the collective effect of bottlenecks, dead pores and preferential flow pathways is significant for the porous media. Therefore, it can be argued that analysing the pores which are likely to make a positive contribution to the total flow rate would be of more practical value.

Figure 4 shows a 2D example of LBM simulated flow distribution through a porous structure. Evidently, the majority of the fluid goes through a limited number of pathways and these pathways can be extracted from the LBM calculated flow pattern.

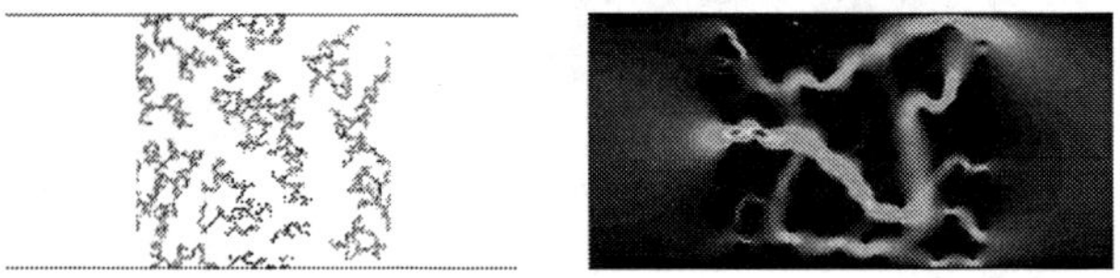

FIGURE 4. A 2D porous structure and LBM simulated flow distribution through it.

Being able to simulate the flow through porous media makes the structural metrics redundant – if the purpose is to estimate, for example, permeability in the first place. However, since these metrics are the commonly quoted for product design, selection and comparison, there is still merit in calculating these metrics based on computer experiments. Also, the metrics calculated in this way are expected to still be more useful than those based on geometry alone even in flow regimes where the current LBM implementation is strictly speaking no longer valid.

3. EXAMPLES

3.1 Characterisation Of Pore Networks

To demonstrate the pore statistics obtained with and without the help of LBM, simulations have been performed for four packing structures [6]:

- PS1 – random packing of equal spheres. The sphere diameter is 17 in lattice units, and the average porosity is 0.38 (Figure 5a).
- PS2 – random packing of ternary mixture of crystalline shapes of identical Carman's diameter (defined as 6V/A where V is particle volume and A particle surface area) to the spheres in PS1. The average porosity is 0.52 (Figure 5b).
- PS3 – random packing of porous agglomerates of crystalline primary particles. The average porosity is 0.77 (Figure 5c).

- PS4 – XMT imaged structure of a piece of foam. The average porosity is 0.73 (Figure 5d).

The volume of flow active pores as a function of the percentage of the total flow rate is shown in Figure 6 for each case. Due to non-uniformity of the flow, the curves bend downward, implying that a high proportion of flow is carried by a small proportion of the flow active pores. For example, it can be seen from Figure 3 that for all cases 50% of the total flow is through only about 25% of the total pore volume. For PS3, there is a steep decrease in the active pore volume near the 100% total flow threshold, indicating the presence of dead pores. This is as expected since flow through internal porosity of agglomerates is negligible. The limiting case is that of a plug flow through a complete void space, shown as the dotted line in Figure 6, where the relation between the active pore volume and the amount of flow is linear. The "closeness" of the actual curve to this limiting line is an indicator of flow uniformity in the porous media.

The two-point probability function $S_2(h)$ defined by Torquato [7] is used here to quantify the correlations of the active pores. $S_2(h)$ is in essence the conditional probability of finding a flow active site of distance h away from a known active site. Figure 7 compares the $S_2(h)$ functions calculated for the total pore volume and the active pores making up 95% of the total flow rate, for the four example structures. With regard to Figure 7, several observations can be made. First, the correlation range of the flow active pore volume is always greater than that of the total pore volume; whereas the correlation values are always lower. This indicates that the active pores structure is more ordered and contains a greater proportion of elongated pores (i.e., flow channels). Secondly, the correlation curve is more likely to oscillate during its decay if the primary particle is of a regular and isotropic shape (cf. Figure 7a for spheres and Figures 7b to 7d for non-spherical shapes). Finally, it is noted that the asymptotic value of $S_2(h)$ is the square of the mean porosity in each case.

Mean empty space [8] and fractal dimensions [9] are sometimes used to characterise pore networks. These have been calculated and are presented in Figure 8. The mean empty space is calculated as the mean spacing between the solids in a network of pores or voids. It is used here as a measure of the average pore size. Fractal dimensions were calculated using a box counting method [10]. Therefore, they can be regarded as mass or volume based fractal dimensions, i.e., the lower the values, the more long branches the structures tend to contain.

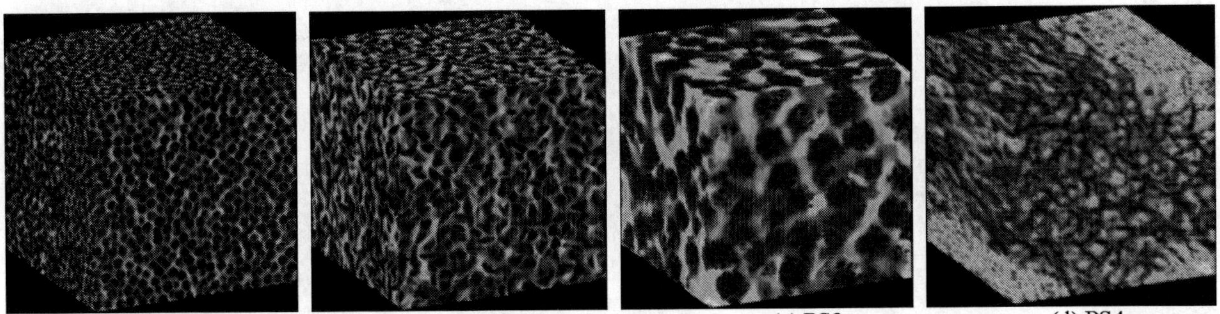

(a) PS1 (b) PS2 (c) PS3 (d) PS4

FIGURE 5. 3D rendering of LBM simulated flow patterns through porous structures. Light areas indicate flow pathways and dark areas where the solids are [6].

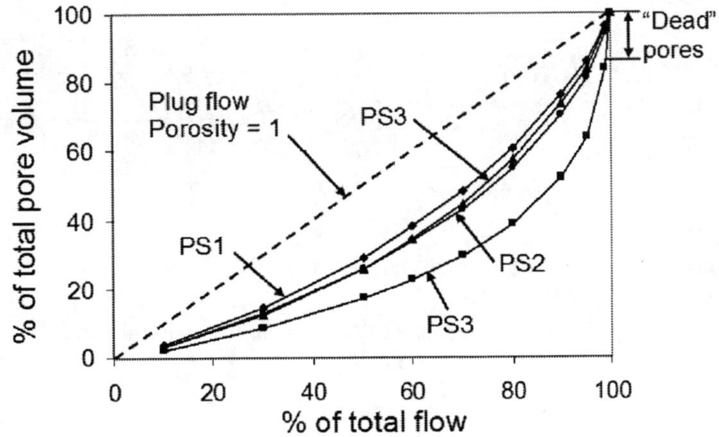

FIGURE 6. Flow active pore volume versus its contribution to the total flow rate [6].

3.2 Permeability

Figure 9 illustrates the concept of "virtual permeametry" – a procedure of assessing filter performance using a combination of XMT, LBM and DigiPac. The clean filter structure is scanned using XMT. LBM is then used to calculate the flow, from which permeability can be estimated. Filter cake can also be scanned using XMT; or it can be simulated using DigiPac. Permeability of the cake on top of the partially fouled filter can again be estimated using LBM.

Figure 10 shows an experimental setup designed to test the above procedure, using model filter cakes made of glass beads. Up to six samples taken from different locations across each filter cake were scanned by XMT and used for calculations. Table 1 summarises and compares the results. For flow calculations, both DigiFlow (our in-house LBM code) and PowerFLOW (a commercial LBM code from EXA) were used. While DigiFlow can make more effective use of memory and hence taking a larger lattice sample (250x250x250), PowerFLOW is more demanding on computing resources and was therefore limited to a lattice sample of 50x50x50 by scaling down the same XMT obtained structures to 1/5 of their original size. This explains the differences in porosity estimated by the two LBM codes. Permeability is normalized against the square of mean particle size, and hence dimensionless. Figure 11 shows a typical cascade from the cake to XMT scan and to LBM calculation. Further details have been described elsewhere by Selomulya et al [11].

Overall, the permeability values are in the same order of magnitude as determined from filtration tests. This suggests that the micro-scale structures in these cases are reasonably representative of the bulk cake structures. The relatively large standard deviation in permeability is partly due to the structural variation between the samples used in fluid simulations. Larger samples would have resulted in smaller deviations, but the sample size is currently limited by computational resources and imaging capability. Nonetheless, the methodology presented here shows the potential for quicker and more accurate assessments, especially for small samples for which measurements are particularly difficult to obtain due to wall effects, i.e. by conducting the entire tests virtually on a computer. Other examples involving sediment of aggregated particles can be found in [12].

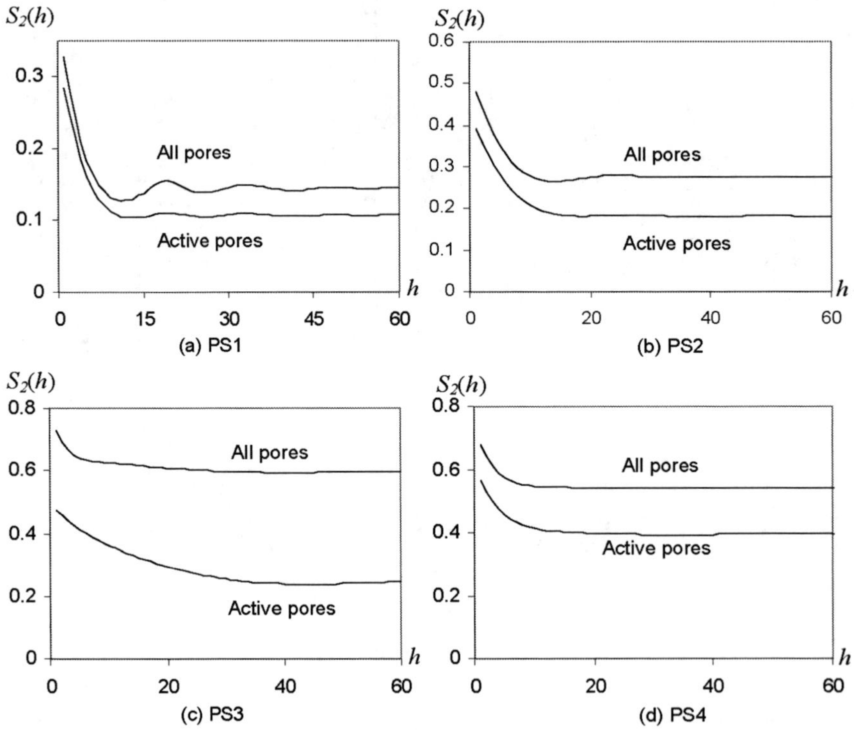

FIGURE 7. Two-point probability functions of the pore structures [6].

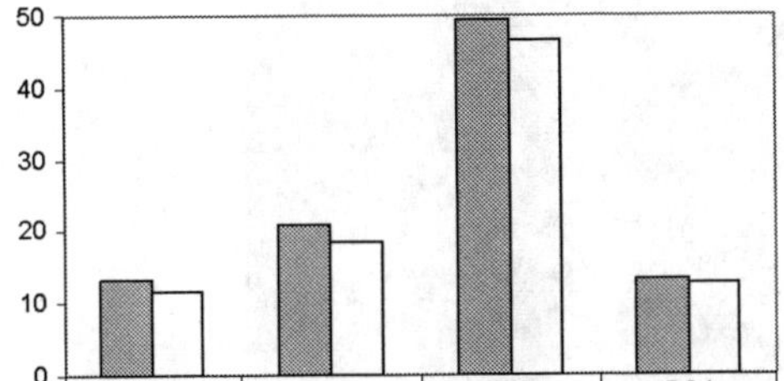

	PS1	PS2	PS3	PS4
All pores	13.32	20.89	49.46	13.31
Active pores	11.51	18.39	46.52	12.66

(a) Mean pore size

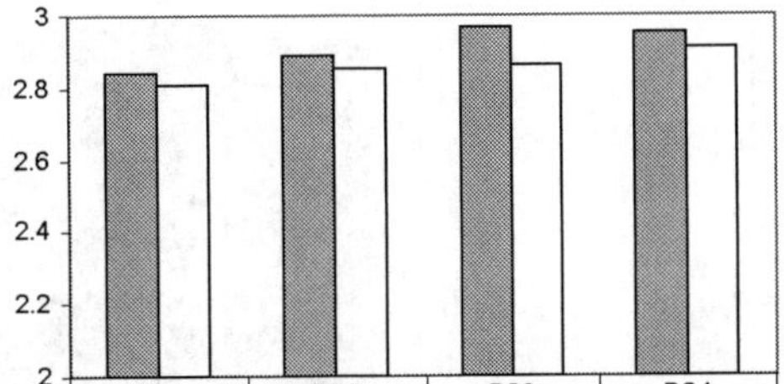

	PS1	PS2	PS3	PS4
All pores	2.84	2.89	2.97	2.95
Active pores	2.81	2.85	2.86	2.91

(b) Fractal dimension

FIGURE 8. Other structural statistics [6].

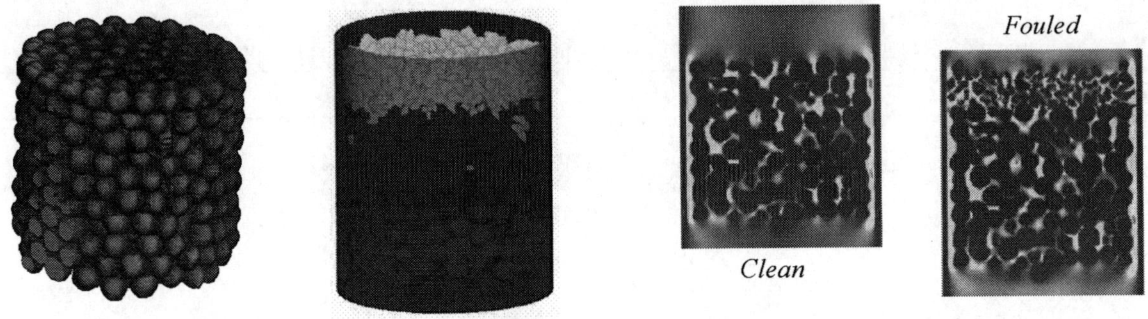

Filter structure obtained by XMT

Partially fouled filter simulated using DigiPac

Cross-sectional views of LBM simulated flow

FIGURE 9. Virtual permeametry based on XMT, LBM and DigiPac.

(a) (b) (c) (d)

FIGURE 10. (a) Experimental setup and an example of the model cakes for XMT scan. (b) SEM of S100. (c) SEM of L100 cake. (d) SEM of L20-S80 cake. [11]

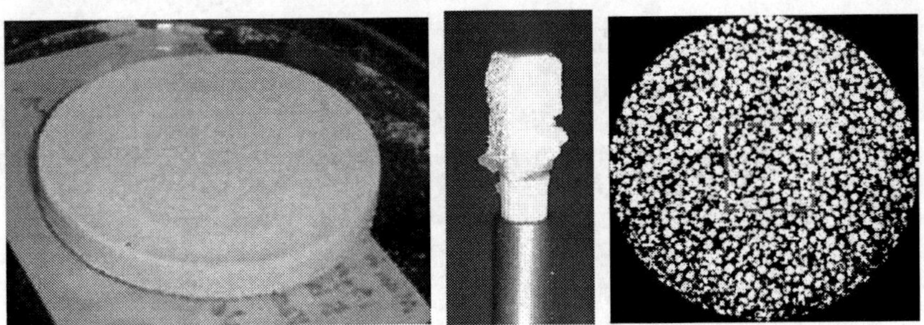

(a) (b) (c)

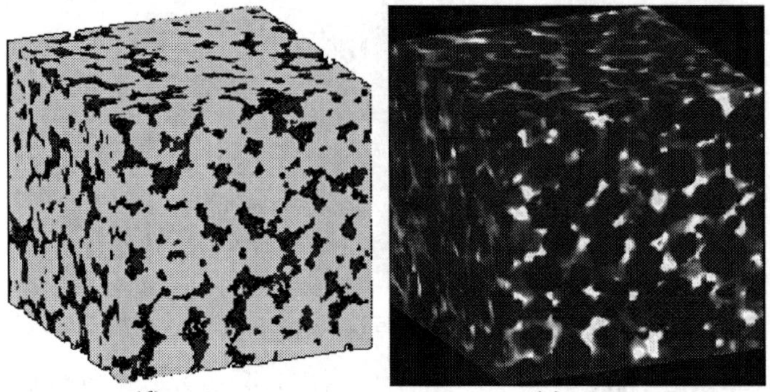

FIGURE 11. From filter cake to LBM estimation of permeability. (a) Filter cake, measuring 5 cm across. (b) A small piece (4mm x 4mm x 8mm) of the filter cake mounted on a XMT sample holder. (c) and (d) A further cut-out from the scanned volume for input to LBM. (e) LBM calculated flow distribution (white indicates high flow rate, black zero flow rate). [11]

TABLE 2. Comparison of measured and LBM estimated permeability values for three model filtration cakes [11]

Sample	S100	L100	L20-S80
Composition	100% of small beads (80 µm mean size)	100% of large beads (531 µm mean size)	20% large, 80% small (178 µm mean size)
Pressure drop (kPa)	10	10	10
Cake thickness (cm)	1.06±0.02	1.08±0.02	1.08±0.04
Measured porosity	0.39±0.02	0.37±0.03	0.41±0.02
Estimated porosity for PowerFLOW	0.34±0.04	0.37±0.01	0.35±0.06
Estimated porosity for DigiFlow	0.31±0.06	0.34±0.05	0.33±0.07
Measured permeability	$2.97\pm0.38 \times 10^{-3}$	$0.17\pm0.04 \times 10^{-3}$	$1.07\pm0.12 \times 10^{-3}$
PowerFlow calculated permeability	$1.51\pm0.56 \times 10^{-3}$	$0.57\pm0.08 \times 10^{-3}$	$0.74\pm0.51 \times 10^{-3}$
DigiFlow calculated permeability	$0.57\pm0.08 \times 10^{-3}$	$0.50\pm0.29 \times 10^{-3}$	$0.85\pm0.26 \times 10^{-3}$

4. CONCLUSIONS

Advances in XMT are enabling porous media to be characterised at micron or finer resolutions with a more representative sample size than allowed by a traditional method such as SEM, on a per scan basis. Meanwhile, advances in commodity computing technologies, coupled with lattice-based numerical techniques, allow measured properties of porous media to be linked directly to their microstructures. Demonstrated here is only one of many possibilities, namely permeability. Other properties, such as effective conductivity, chemical reactions, dissolution, leaching, sintering, etc. can also be calculated based directly on real/realistic microstructures. The insight provided with this type of approach helps to better understand structure-property relationships, hitherto established by and large on empiricism. Such an approach can also be used to provide detailed input to continuum based theoretical models such as finite element analysis, which hitherto mainly uses guesswork or bulk averages for their cell-level property inputs, to perform analyses at a macroscopic scale, thereby bridging the gap between microscopic and macroscopic worlds with more realism than ever before.

REFERENCES

1. A.C. Kak and M. Slaney, Principles of Computerized Tomographic Imaging, IEEE Press, 1988.
2. J. Baruchel, J.Y. Buffiere, E. Marie, P. Merie, G. Peix (Editors), X-ray Tomography in Material Science, Hermes,France, 2001.

3. X. Jia and R.A. Williams, A packing algorithm for particles of arbitrary shapes, Powder Technology, Vol.120, pp.175-186, 2001.
4. S. Succi, The Lattice Boltzmann Equation for Fluid Dynamics and Beyond, Oxford University Press, Oxford, 2001.
5. H.W. Hoffman, A 3D lattice Boltzmann code for modelling flow and multi-component dispersion, Sandia Report, SAND99-0162, 1999.
6. X. Jia, C. Xu and R.A. Williams, Use of lattice Boltzmann method to facilitate characterization of pore networks in porous media, Particulate Systems Analysis 2005, Stratford-upon-Avon, 21-23 September 2005.
7. S. Torquato, Random Heterogeneous Materials: Microstructure and Macroscopic Properties: v.16 (Interdisciplinary Applied Mathematics Series), Springer-Verlag, New York, 2001.
8. M. Zeidan, X. Jia and R.A. Williams, Simulation of aggregation with applications to soot laden lubricating fluids, Particle and Particle Systems Characterization, Vol.21(6), pp.473-482, 2005.
9. P.R. King, The fractal nature of viscous fingering in porous media, J. Phys. A., Vol.20, pp.L529-L534, 1987.
10. G.C. Bushell, Y.D. Yan, D. Woodfield, J. Raperc and R. Amala, On techniques for the measurement of the mass fractal dimension of aggregates, Advances in Colloid and Interface Science, Vol.95, pp. 1-50, 2002.
11. C. Selomulya, T.M. Tran, X. Jia and R.A. Williams RA, An integrated methodology to evaluate permeability from measured microstructures, AIChE Journal Vol.52 (10), pp3394-3400.
12. C. Selomulya, X. Jia X and R.A. Williams, Direct prediction of structure and permeability of flocculated structures and sediments using 3D tomographic imaging, Chem Eng Res Des, Vol.83(A7), pp844-852, 2005.

Recent Work on Flow Boiling and Condensation in a Single Microchannel

Xiaojun Quan, Guodong Wang, Ping Cheng* and Huiying Wu

School of Mechanical and Power Engineering
Shanghai Jiaotong University, 800 Dong Chuan Road, Shanghai, 200240, P.R.China
E-mail address: pingcheng@sjtu.edu.cn

Abstract. Recent visualization and measurements results on flow boiling of water and condensation of steam in a single microchannel, carried out at Shanghai Jiaotong University, is summarized in this paper. For flow boiling of water, experiments were conducted in a single microchannel with a trapezoidal cross-section having a hydraulic diameter of 186 μm and a length of 30 mm. A boiling flow pattern map in terms of heat flux versus mass flux, showing the unstable and stable boiling flow regimes in the microchannel, is obtained. For the investigation of condensation, experiments were carried out for steam condensing inside a single microchannel with a length of 60mm having a hydraulic diameter of 87 μm and 120μm respectively. The location of transition from annular flow to plug/slug flow in a microchannel is found to be dependent on both the dimensionless condensation heat transfer rate as well as the Reynolds number of the steam. The frequency for the occurrence of the injection flow is found to increase with the increasing mass flux.

INTRODUCTION

A great deal of attention has recently been devoted to the study of flow boiling and condensation in microchannels because of their applications to the cooling of microelectronic chips [1,2]. Most of these flow boiling experiments were conducted in parallel microchannels, where interaction of flow from neighboring channels occurred at inlet/outlet headers . For example, Hetsroni et al [3,4] observed flow boiling instability in silicon triangular microchannels. Wu and Cheng [5,6] carried out a simultaneous visualization and measurement investigation in parallel silicon microchannels of trapezoidal cross-sections having hydraulic diameters of 82.8 μm, 158.8 μm, and 185.6 μm respectively. They found that there existed two boiling oscillation flow modes with large fluctuations of wall temperature and pressure in microchannels depending the heat flux and mass flux: Liquid/Two-phase Alternating Flow (LTAF) and Liquid/Two-phase/Vapor Alternating Flow (LTVAF). Qu and Mudawa [7] found flow interaction in flow boiling in parallel microchannels and Li et al. [8] found that reversed flow existed in flow boiling in parallel microchannels. Recently, Hetsroni et al [9] observed the explosive boiling phenomena with periodic wetting and dryout in triangular silicon microchannels, and found that the trigger mechanism of such a regime was the venting of elongated bubble due to rapid expansion. Most recently, Wang et al.[10] have carried out an experiment to investigate unstable and stable flow boiling regimes of water in parallel microchannels under different heat and mass fluxes.

On the other hands, Chen and Cheng [11] performed a visualization study on condensation of steam in parallel microchannels with a hydraulic diameter of 75μm. It was observed that mist flow appeared in upstream and an intermittent flow of vapor and condensate appeared at downstream of the microchannels. Subsequently, Wu and Cheng [12,13] carried out a series of simultaneous visualization and measurement experiments on saturated steam condensing in parallel silicon microchannels, having a trapezoidal cross section with a hydraulic diameter of 82.6μm and a length of 30mm. They observed for the first time the so-called "injection flow", which is the transition process from annular flow to slug/plug flow.

In this paper, we will summary the results of some recent work on simultaneous visualization and measurement studies of flow boiling and condensation in a *single* microchannel. For flow boiling of water, experiments were conducted in a single microchannel with a trapezoidal cross-section having a hydraulic of 186 μm and a length of 30 mm. It was found that reversal flow of vapor also existed in a microchannel under certain conditions of heat to mass

CP914, *Multiphase Flow: The Ultimate Measurement Challenge,* edited by X. Cai, Y. Wu, Z. Huang, S. Wang, and M. Wang

flux ratio, leading to unstable flow boiling mode with large variations of temperature and pressure. Experiments were also carried out for steam condensing in a single microchannel having a hydraulic diameter of 87 μm and a length of 60mm. The influence of the dimensionless condensation heat flux and the Reynolds number on the location of occurrence of injection flow in the condensation of steam in a single microchannel was determined. The influence of the mass flux on the occurrence frequency of the injection flow was also investigated.

FLOW BOILING IN A SINGLE MICROCHANNEL

Experimental Setup

Figure 1 shows the loop of flow boiling experiments, which was similar to those used in our previous experiments [12,13]. The test section was also similar to those in our previous studies [12,13] except that a single microchannel, instead of parallel microchannels, was etched in a <100> silicon substrate. The cross sectional area and the length of the microchannel were also the same: with a hydraulic diameter of 186 μm and a length of 30 mm. Five wall temperatures (Tw1,Tw2,Tw3,Tw4,Tw5), as well as water inlet and outlet temperatures (Ti, To) were measured by seven type-T thermocouples, having a diameter of 0.1 mm and a response time of 0.1s. Two pressure transducers with a response time of 0.001s were used to measure water pressure (Pin and Pout) at the inlet and outlet of the microchannels.

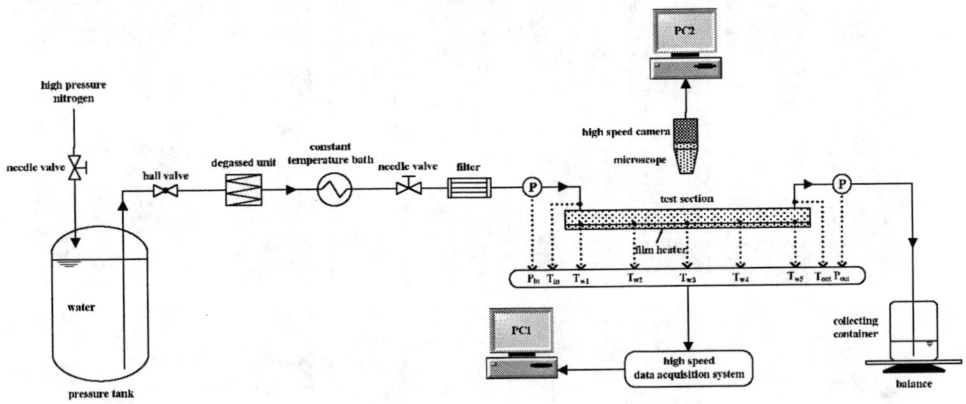

FIGURE 1. Experimental test loop.

Reversed Flow of Vapor in a Single Microchannel

Reversed flow had been found in flow boiling in parallel minichannels and microchannels [7, 8]. The presence of the parallel channels were conducive to the occurrence of reversed flow, which allowed a path of lower flow resistance. The flow and pressure in the other channels compensated the high pressure due to vapor generation to dissipate through the other channels. Interestingly, reversed flow of vapor was also observed in a single microchannel, as is shown in Fig. 2, where the flow direction was from bottom to top. Figure 2(a)-(i) show the progression of the reversed flow of vapor in time sequence. In Fig. 2(a) and (b), it is shown that a bubble nucleated in the channel initially. Fig 2(c) and (e) show the vapor core expanded toward upstream and downstream directions sometime later. The life time of the vapor core expansion, from Fig. 2(c) to (e), was about 0.753ms. This result agreed with those obtained by Hetsroni [13] who concluded that the life time of an elongated bubble did not exceed 0.001s. Figure 2(f) shows the channel in a dry out condition with rising wall temperature because of the rapid evaporation of the liquid film in the microchannel. The life time of the local dryout condition was about 18.981ms, which was much longer than that of the vapor core expansion. Finally, the bottom side of the interface of the vapor continued to move toward the channel exit as is shown in Fig. 2(g)-(i).

Flow Pattern Transition Instability

The flow pattern transition instability in a single microchannel was similar to that in parallel microchannels as discussed in a recent paper by Wang et al.[10]. The trigger mechanism of such instability was the venting vapor core due to very rapid expansion. Figure 3 shows temporal temperature oscillations for q = 297.8 kW/m^2 and inlet water temperature at 35 ℃. A comparison of Fig.3 with those in parallel microchannels having the same hydraulic diameter and length [10] reveals that the corresponding mass flux and oscillation periods of temperature in single microchannel were all larger than those in parallel microchannels, due to the lack of interaction of flow from neighboring channels. For example, the mass flux in a single microchannel varied from 3523.2 kg/m^2s at the beginning to 950.0 kg/m^2s at the end as compared with the corresponding mass flux varying from 308.3 to 142.3 kg/m^2s in parallel microchannels under the same heat flux condition.

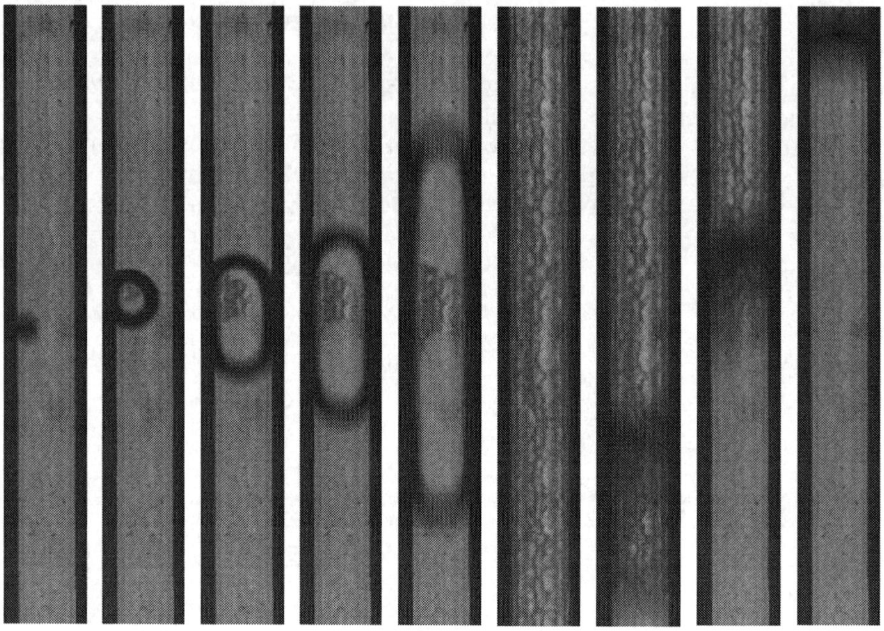

(a) (b) (c) (d) (e) (f) (g) (h) (i)

FIGURE 2. Photos of reversed flow in a single microchannel taken at 6640 frame/s: (a) t= 0 ms; (b) t=1.96 ms; (c) t=2.56 ms; (d) t=2.86 ms; (e) t=3.31 ms; (f) t=19.43 ms (g) t=22.75 ms; (h) t=23.35 ms; (n) t=23.95 ms. Bulk flow direction was from bottom to top with the same location between T_{w1} and T_{w2} at q=84.5 kW/m^2, G=453.9 kg/m^2s, and Tin=35 ℃

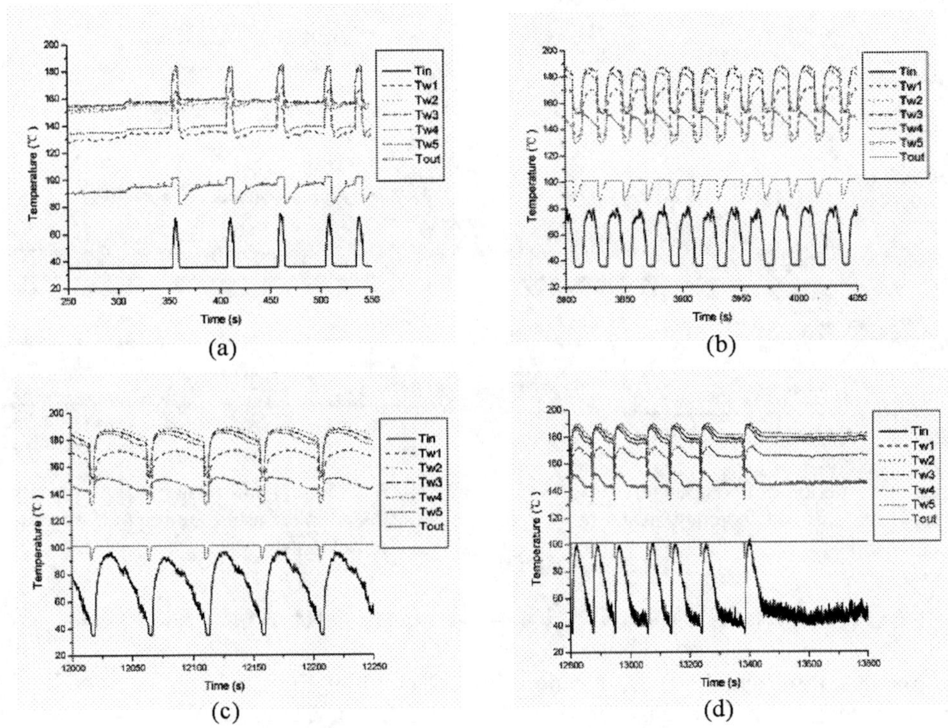

FIGURE 3. Oscillation of fluid and wall temperatures in a single microchannel at q=297.8 kW/m^2, and T_{in}=35 ℃, as G decreasing with time from 3523.2 to 950.0 kg/m^{2}s: (a) 250 s< t < 550 s, G=3523.2 kg/m^{2}s; (b) 3500 s < t < 4050 s, G=2082.5 kg/m^{2}s (c) 12000 s <t <12250 s, G=1048.6 kg/m^{2}s; (d) 12800 s <t <13800 s, G=999.2 kg/m^{2}s.

Figure 4 shows oscillation periods of temperature increased from 10 to 80s when the heat-to-mass flux ratio, q/G, increased from 0.09 to 0.32kJ/kg at different heat fluxes in a single microchannel. It can be seen that oscillation periods of temperature depended on the heat-to-mass flux ratio, but independent of the heat flux, which was the same as those in parallel channels. However, the oscillation periods of temperature in single channel (varying from 10 to 80s) were much longer than those in parallel channels (varying from 2.5 to 9.8s).

Figure 5 is a map showing the stable and unstable flow boiling regimes in a single microchannel. The range of instability was bounded by the lower red line with a slope of 0.09 kJ/kg and the higher red line with a slope of 0.32kJ/kg. A comparison of Fig. 5 with the corresponding figure for parallel microchannels [10] shows that the unstable flow boiling regime in a single microchannel was smaller than that in parallel microchannels with slopes from 0.96 to 2.14 kJ/kg.

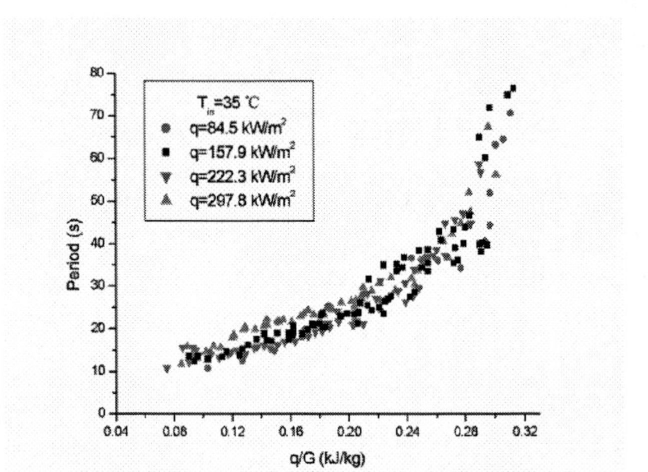

FIGURE 4. Oscillation periods of temperature versus heat-to-mass flux ratio at different heat flux in single microchannel.

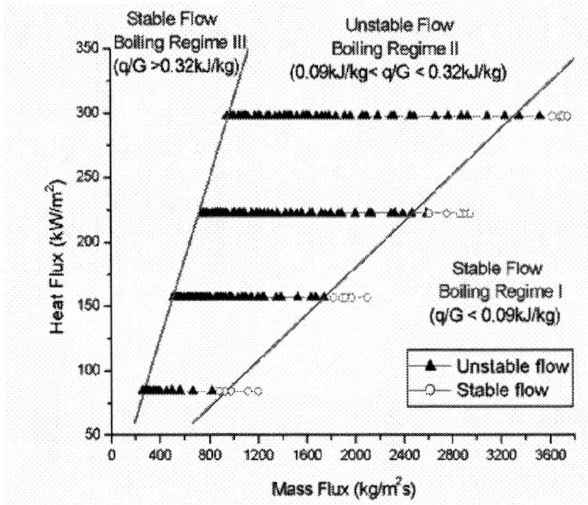

FIGURE 5.. Stable and unstable flow boiling regime of heat flux supplied as a function of the mass flux in a single microchannel.

FLOW CONDENSATION IN A SINGLE MICROCHANNEL

Figure 6 shows that mist flow, annular flow, injection flow, slug/bubbly flow, bubbly flow, and liquid flow exists in the condensation of steam in a microchannel [13]. Recently, Quan, Cheng & Wu [14] have carried out an experimental study on the transition from annular flow to slug/plug flow of saturated steam condensing in a single microchannels, with a hydraulic diameter of 87 μm and 127μm and a length of 60mm. Fig.7 shows the dimensionless location z/L at which the transition from annular flow to slug/plug flow of saturated steam takes place as a function of Reynolds number of steam ($Re= GD_h/\mu_{vin}$) and the dimensionless condensation heat flux ratio ($Co=q_c/Gh_{fg}$). It can be seen that the location where the transition takes place further upstream in the microchannels as the dimensionless heat transfer rate is increased, or as the Reynolds number is decreased.

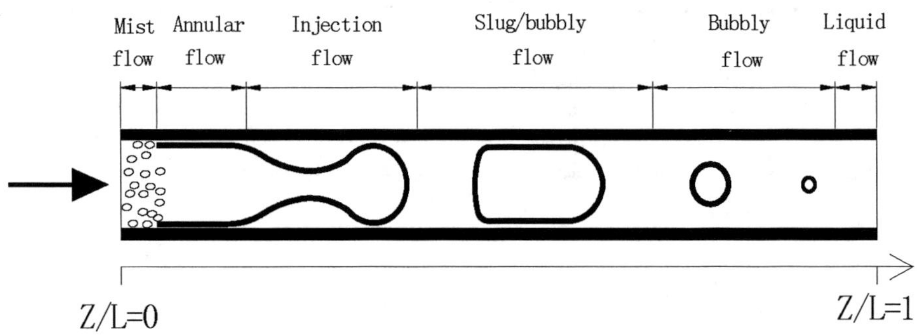

FIGURE 6. Sketch of condensation flow patterns in microchannels

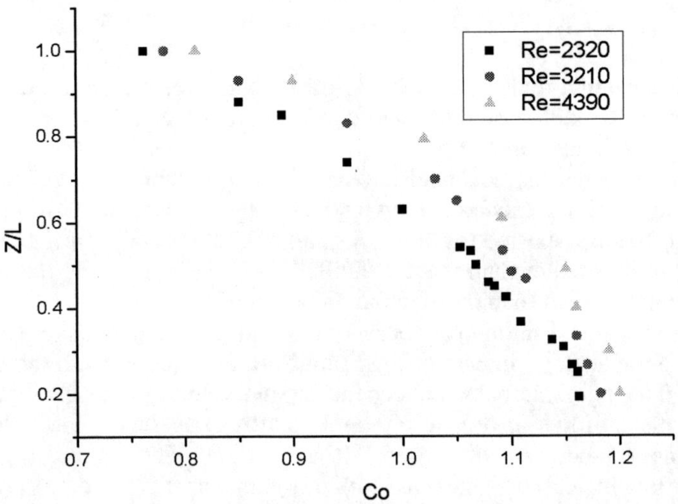

FIGURE 7. Transition locations from annular flow to slug/plug flow of saturated steam Vs. the dimensionless heat flux ratio at different Reynolds number of steam

Fig. 8 shows the effect of mass flux on the occurrence frequency of injection flow in a single trapezoidal microchannel (D_h=120μm) at a constant inlet cooling water temperature of 45°C. It is shown that the occurrence frequency of the injection flow increased with the increase of mass flux. This was because higher mass flux induced instability, leading to higher frequency in the occurrence of injection flow.

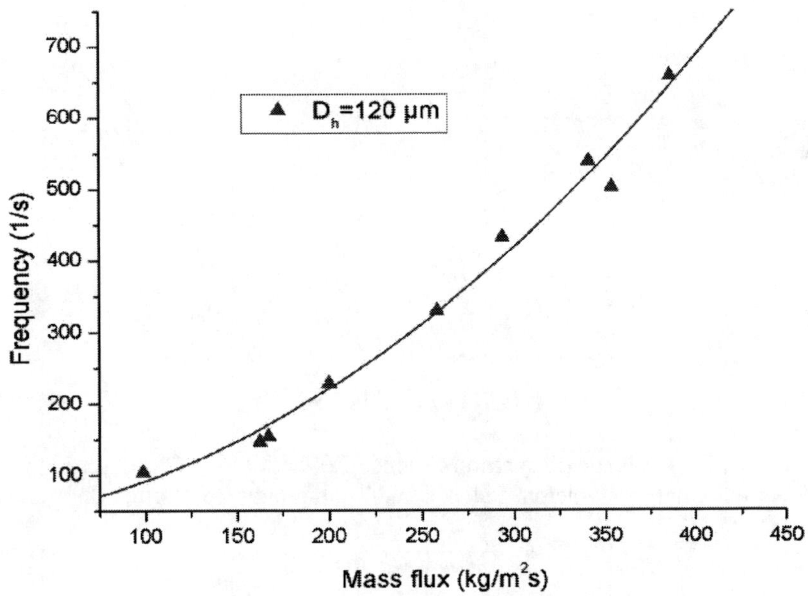

FIGURE 8. The effect of mass flux on the occurrence frequency of injection flow at constant inlet cooling water temperature

CONCLUDING REMARKS

In this paper, simultaneous visualization and measurements of temperature, pressure and mass flux variations have been carried out to investigate flow boiling of water and condensation of steam in a *single* microchannel. The following conclusions can be drawn from this paper:

(1) For flow boiling of water in a single microchannel having a hydraulic diameter of 186 μm and a length of 39mm, it was found that stable flow boiling existed for $q/G < 0.09$kJ/kg and $q/G > 0.32$kJ/kg. On the other hand, unstable flow boiling with large fluctuations existed for $0.09\text{kJ/kg} < q/G < 0.32\text{kJ/kg}$. The unstable flow boiling regime in a single microchannel is found to be smaller than that in parallel microchannels while the oscillation periods of temperature in a single channel is much longer than those in parallel channels.

(2) For condensation of steam in a single microchannel having a hydraulic diameter of 87 μm and a length of 60mm, it is found that mist flow, annular flow, injection flow, plug/slug flow (confined bubbles), and bubbly flow existed in condensation of steam in a single microchannel depending on the local vapor quality. It is found that the occurrence of injection flow took place further upstream in the microchannel as the dimensionless heat transfer rate was increased or as the Reynolds number was decreased. At a constant inlet cooling water temperature, the occurrence frequency of injection flow increased as the mass flux of steam was increased.

FOOTNOTES

h_{fg} latent heat
G mass flux
q heat flux
P pressure
t time
T temperature
Re Reynolds number
L total length of channel
Z transition location
Co dimensionless heat flux ratio
μ viscosity

Subscripts

in inlet
out outlet
w wall
c condensate
v steam

ACKNOWLEDGEMENT

This work was supported by the National Natural Science Foundation of China through Grant Nos. 50536010 and 50476017 as well as by Shanghai Municipal Science and Technology Committee through Grant No. 05JC14025.

REFERENCES

1. P. Cheng and H. Y. Wu, *Advances in Heat Transfer* **39**, 461-563 (2006).
2. P. Cheng, H. Y. Wu and F. J. Hong, *J. Heat Transfer* **129**, 1-7 (2007).
3. G. Hetsroni, A. Mosyak, Z. Segal, *IEEE Transactions on Component and Packing Technologies* **24**, 17-23 (2001).
4. G. Hetsroni, A. Mosyak, Z. Segal, G. Ziskind, *Int. J. Heat Mass Transfer* **45**, 3275-3286 (2002).
5. H.Y. Wu, P. Cheng, *Int. J. Heat Mass Transfer* **46**, 2603-2614 (2003).
6. H.Y. Wu, P. Cheng, *Int. J. Heat Mass Transfer* **47**, 3631-3641 (2004).
7. W. L. Qu, I. Mudarwar, *Int. J. Heat Mass Transfer* **47**, 2045-2059 (2004).
8. H. Y. Li, F. G. Tseng, and C. Pan, *Int. J. of Heat Mass Transfer* **47**, 5591-5601 (2004).
9. G. Hetsroni, A. Mosyak, Pogrenyak, Z. Segal, *Int. J. of Multiphase Flow* **31**, 371-392 (2005).
10. G. D. Wang, P. Cheng, and H. Y. Wu, ASME International Conference on Integration and Commercialization of Micro and

Nano-systems, Jan 10-13, 2007, Sanya, Hainan, China.
11. Y. P. Chen and P. Cheng, *Int. Comm. Heat and Mass Transfer* **32**, 175-183 (2005).
12. H. Y. Wu and P. Cheng. Proceedings of Second International Conference on Microchannels and Minichannels, Rochester, New York, USA, 2004, pp.657-660.
13. H. Y. Wu and P. Cheng, *Int. J. Heat and Mass Transfer* **48**, 2186-2197 (2005).
14. Quan, X. J., Cheng, P. and Wu, H. Y., Transition from Annular to Slug/Bubbly Flow in Condensation in Microchannels, submitted for publication (2007).

Application of Extended Imaging Techniques for Analyzing Elementary Processes in Multiphase Flow

M. Sommerfeld

Mechanische Verfahrenstechnik, Zentrum für Ingenieurwissenschaften
Martin-Luther-Universität Halle-Wittenberg, D-06099 Halle (Saale), Germany
martin.sommerfeld@iw.uni-halle.de

Abstract. For a high-resolution analysis of elementary processes occurring in multiphase flows on the scale of the particles, a planar imaging technique combined with particle-tracking- and particle imaging velocimetry (PTV and PIV) was developed. This technique combines a diffuse background illumination and a single image recording CCD-camera with macro optics, which has a low depth of field. Therefore, the CCD-camera records in-focus as well as out-of-focus blurred objects, namely solid particles, droplets or bubbles. Applying special intensity gradient-based filters all blurred objects are removed from the images. By this combination the imaging plane in the two-phase system of consideration is produced with a thickness being of the dimension of the particles. By adding small tracer particles to the flow, also the flow field around the particles can be obtained by applying PIV. This imaging technique was applied for the analysis of the following elementary processes; droplet evaporation in grid turbulence, collisions of suspension droplets, collisions of non-spherical particles with metal plates and the behavior of ellipsoidal bubbles in bubbly flows.

Keywords: Shadow imaging, particle detection, PTV, PIV, multiphase flows, droplets, solid particles, bubbles.

INTRODUCTION

Imaging techniques are powerful methods for the experimental analysis of multiphase flows, especially when elementary processes occurring on the scale of the particles are of interest, e.g. bubble motion and oscillation, liquid atomization, droplet behavior and particle-wall collisions. Generally, a backlighting illumination is used for producing a planar image of the particles (used as a general term for solid particles, droplets and bubbles) on a recording camera. Applying an appropriate image processing it is possible to estimate the particle contour und from that the area equivalent diameter, the particle shape and the orientation in the flow. When using double image CCD-cameras or high-speed cameras, also the velocity of the particles can be determined by applying a tracking algorithm (PTV).

In the field of atomization and sprays, shadow imaging is a quite common technique for analyzing for example droplet break-up and droplet collisions. Since these phenomena are very fast, mostly high-speed imaging in combination with pulsed high power light sources is needed. Air-assisted atomization of droplets was for example analyzed by Liu and Reitz [1] using a 9 W pulsed copper vapor laser for background illumination and a regular camera having an optics which yields a small depth of field. Hence, a series of events (i.e. individual droplets) were captured on a single image. The results were used for improving modeling of droplet break-up. A similar shadow imaging with a stroboscope was used in the old work of Brazier-Smith et al. [2] for recording the temporal evolution of water droplet collisions on a single image. These results are still used today for deciding on the boundary between grazing collisions and droplet coalescence in connection with numerical predictions of sprays by the Euler/Lagrange approach [3].

In bubbly flows the oscillation and rise behavior of bubbles is of great interest. Deformable and oscillating bubbles have to be observed from two, preferably perpendicular, directions in order to correctly estimate bubbly shape and size. Such a stereo imaging has been successfully applied to single bubbles and bubble chains [4, 5, 6]. In the latter study a special mirror arrangement was used to illuminate and observe the bubbles from two perpendicular

CP914, *Multiphase Flow: The Ultimate Measurement Challenge,* edited by X. Cai, Y. Wu, Z. Huang, S. Wang, and M. Wang

directions. Since the image was divided in two equal parts, images from the two perpendicular viewing directions could be simultaneously recorded with only one CCD-high-speed-camera. In all cases the bubble velocity was obtained with the help of high-speed cameras and a tracking algorithm. A simultaneous measurement of the flow field around the bubbles was not possible with these imaging systems. A high speed PTV-technique based on the application of only one CCD-camera combined with a background illumination of bubbles and tracer particles in a flat bubble columns was proposed by Borchers [7]. The phase separation was performed by sophisticated multi-step object recognition based on a binarisation of the images and the velocities of bubbles and the liquid phase (i.e. tracer particles) were evaluated by a simple tracking algorithm.

Moreover, it is possible to combine an imaging technique with particle image velocimetry (PIV) in order to get the flow field around individual particles. For such an application a combination of laser light sheet and back-lightning or shadow imaging is useful as done for example by Fujiwara et al. [8] and Lindken and Merzkirch [9] for the analysis of the flow field induced by bubble motion. Using fluorescing tracer particles illuminated by a light sheet, one CCD-camera with an appropriate optical filter records the PIV-image with the tracer. The shadow image of bubbles is produced with a LED array having a different wave length and a second CCD-camera only records this image. By overlapping the bubble image and the velocity field of the continuous phase the influence of the bubbles on the flow was analyzed for small bubbles (around 1.0 mm) and very low gas volume fractions, i.e. ≤ 1 % [8]. More recently Sakata et al. [10] analyzed the oscillatory behavior of large single bubbles (between 2.2 and 5 mm) and the temporal evolution of the forces acting on the bubbles as well as the flow field around these bubbles. For a correct determination of the bubble shape, a 90° stereo-shadow imaging was used by illuminating and observing the bubbles from two perpendicular directions. Such methods give detailed information, however requires sophisticated triggering and several CCD-cameras.

In the present contribution a recently developed planar shadow-imaging combined with PTV for particle velocity measurement and PIV for recording the flow field around the particles is described. This system is operating with diffuse back-ground illumination and only one CCD-camera with a lens having low depth of field. Hence, in-focus and out-of-focus blurred tracer and particle images are recorded on the CCD-camera. By applying special gradient based digital filters blurred objects are removed from the image, whereby an imaging plane with defined thickness is created. With this method it is possible to simultaneously measure tracer and particle velocities as well as a projection area equivalent particle size when an appropriate magnification is selected. The applicability and powerfulness of the developed planar imaging velocimetry will be demonstrated for the analysis of the following phenomena; evaporation of multi-component droplets in grid turbulence, collision of suspension droplets, wall collision of non-spherical particles and behavior of bubble swarms.

IMAGE ACQUISITION SYSTEM

The images of the two-phase flow were collected by a double shutter CCD-camera (Type: PCO SensiCam) which allowed to acquire two successive images with a resolution of 1280 by 1024 pixel within a very short time delay. The CCD-camera was equipped with a macro camera lens type Nikkor with focal length: s = 50 mm and f-number: f = 1.2. For the exposure of the images a background illumination mode using a pulsed LED-array consisting of 551 high performance LED's with a total area of 160 mm x 100 mm (Fig. 1) was used. The typical duration of the light pulses was 50 to 70 µs, while the time delay between the two successive images was adjusted to the local velocities. The images were transferred in a digital way from the CCD-camera to the image processing PC. The entire imaging system was mounted on a CNC traversing system which was also controlled by the image processing PC for performing fully automated measurements.

In a first image processing step the images were transformed from the camera's 12 bit format to an 8 bit format grey value image. This step involves an automatic optimization of the contrast and a normalization of the grey-scale values. The histogram of the grey values of each image was analyzed and from this histogram a contrast optimized lookup table for the transformation was created. The online evaluation of the images and all controlling tasks were performed by in-house developed software written in Delphi. By using the MMX (multi-media-extension) technology of the PC an online evaluation of the raw images was possible yielding continuous phase and particle velocity fields as well as other particle properties. Thereby it was avoided to store a huge amount of raw data, which would have been necessary if the entire two-phase system has to be analyzed by moving the imaging system automatically across the facility.

In order to allow measurements of the continuous phase velocity fields the flow was seeded with appropriate tracer particles depending on the two-phase flow system considered. Because of the small depth of field of the macro camera optics it was possible to discriminate between particles and tracer particles inside and outside of the camera's depth of focus by the gradient of grey values. Sharply depicted objects in the focal plane have high gradients, while out of focus blurred objects have low gradients of grey values. Hence, the imaging plane was not produced by a light sheet, but a backlighting combined with a proper selection of the macro lens and digital filter operations. The effective depth of field of course depends on the particle size considered. Preliminary experiments have shown, that the optical configuration described above results in a depth of field smaller than 4 mm for bubble sizes between 2 and 5 mm. The phase discrimination between dispersed phase particles and tracer particles was performed in the evaluation procedure using a series of different digital image filters as will be briefly summarized below [11].

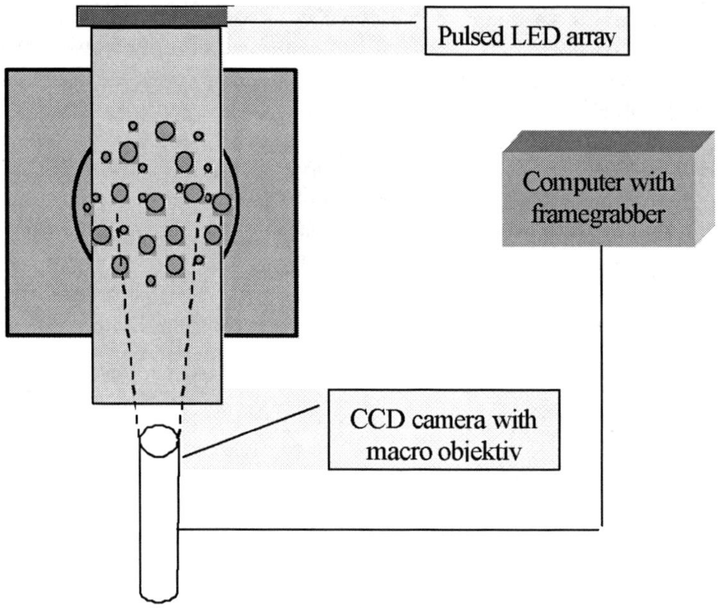

FIGURE 1. Two-phase flow with imaging system (pulsed LED-array and CCD-camera) and data acquisition.

PARTICLE DETECTION

In order to allow for automatic particle detection, the recorded images needed to be prepared by a set of digital filters before object recognition for the particles was applied. The tracer particles and small scale noise objects were erased from the images by a non-linear 5 x 5 median filter (smoothing operation). Then a Sobel filter was applied which provides an image showing the distribution of the intensity gradients. Thereby, it is more reliable to detect the particle contours. Hence, for the particle detection algorithm the gradient of the grey values at the contours of the particles, i.e. the sharpness was used as a criterion for validation. Contours are only accepted if the gradient is above a certain threshold level. This threshold was optimized to yield a maximum particle detection rate. Hence, only particles located in the imaging plane are considered for further evaluation. The particle contour was then calculated by applying a segmentation procedure and a cubic spline interpolation. From closed contours the centroids, the area equivalent particle diameter, the particle shape and orientation was obtained (for details see Bröder and Sommerfeld [11]). Fig. 2 shows an example for the detection of bubble images. It is clearly visible, that out of focus bubbles show a low grey level and hence the contour is not in focus, whereby these bubbles are rejected for further evaluation.

At higher particle volume fractions, however a large number of overlapping images resulting from both in-focus and blurred images are captured on the image, whereby a number of sharply depicted particles did not have a closed contour. Therefore, missing parts of the contour were reconstructed with the help of the spline interpolation. In order to yield high processing speed and a good accuracy in the determination of the area equivalent particle diameter, images were only reconstructed if 85 % of the contour were already detected. In the past, several algorithms were developed to reconstruct missing parts of the contour of overlapping bubbles (or solid particles and droplets in other two-phase flows). A summary of proposed techniques was provided by Honkanen et al. [12]. In their studies an ellipse was fitted to the perimeter belonging to one bubble. A method being applicable to circular images from spherical particles was suggested by Shen et al. [13] using the break-point detection method.

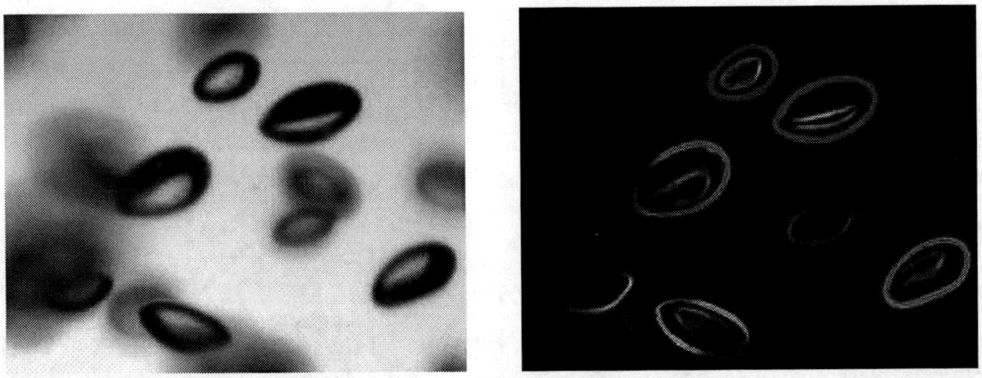

FIGURE 2. Illustration of bubble detection algorithm, a) recorded image (without tracer particles) showing bubbles which are in and out focus, b) validated contours of bubbles which are in focus.

PARTICLE VELOCITY MEASUREMENTS (PTV)

The velocities of the particles were determined by the application of the Particle Tracking Velocimetry (PTV) where the displacement of the particle centroid on two consecutive images was determined. The centroids were obtained by the described particle detection algorithm. Corresponding particle pairs were found by the criteria of overlapping particle contours and the closest neighbor in the estimated direction of the particle motion. In complex flows this was achieved by adjusting the time delay between successive images in such a way that the distance between the two centroids was around one half of the particle area equivalent diameter. In directed flows, the time delay can be larger and the corresponding particles are found from the closest neighbor in the known direction of motion.

CONTINUOUS PHASE VELOCITY MEASUREMENTS (PIV)

The velocity field of the continuous phase was determined by the application of the Particle Image Velocimetry (PIV) to the extracted images of the tracer particles. This extraction process starts with an edge filter proposed by Marr and Hildreth [14], which is often called Laplacian of Gaussian (LoG). The fundamental characteristics of the LoG edge detector are the combination of a smoothing Gaussian filter and an enhancement step which is a Laplacian second derivative in two dimensions. The standard deviation of the LoG (σ_{LoG}) was fitted to the image size of the tracer particles to optimise the performance. This filter operation also ensures that out of focus tracer images were removed from the original image. After the application of the LoG edge detector, the images of the in-focus tracer particles but also the contours of sharply depicted particles were kept inside the images (image B in Fig. 3). Therefore, an additional image operation was used to remove these particle contours. A median filtered image was created which contained only large objects like the dispersed phase particles but no tracer particles (Image C). This median filtered image with the particle contours was subtracted from the image filtered by the LoG edge detector (Image B) and so the resulting image (Image D) was prepared for the PIV evaluation containing only the tracer particles. A typical result for the extraction of the tracer images from the original two-phase image, obtained in a bubbly flow, is also shown in Fig. 3. The regions without tracer particles correspond to the original locations of the bubbles.

The evaluation of the tracer images was performed by PIV with iterative refinement of the interrogation area following the approach suggested by Scarano et al. [15] in order to speed-up the image processing and to improve the sub-pixel accuracy. The determination of the displacement vectors was based on a combination of the Minimum Absolute Difference Method (MAD) and the Minimum Quadratic Difference Method (MQD) developed by Gui et al. [16]. The first two iterative steps were performed with the MAD, since it is about 6-times faster than the MQD. The MAD was applied for an interrogation area size of 128 x 128 pixels for the first step and 64 x 64 pixels for the second step. As the sub-pixel accuracy of the MAD is less than that of the slower MQD only the last iterative step was performed by the more accurate MQD. The interrogation area size for the last step was 40 x 40 pixels. Hence, the results from the larger interrogation area was used as a predictor for the next refinement step, which additionally speeds up the processing and is the basis for the on-line image processing.

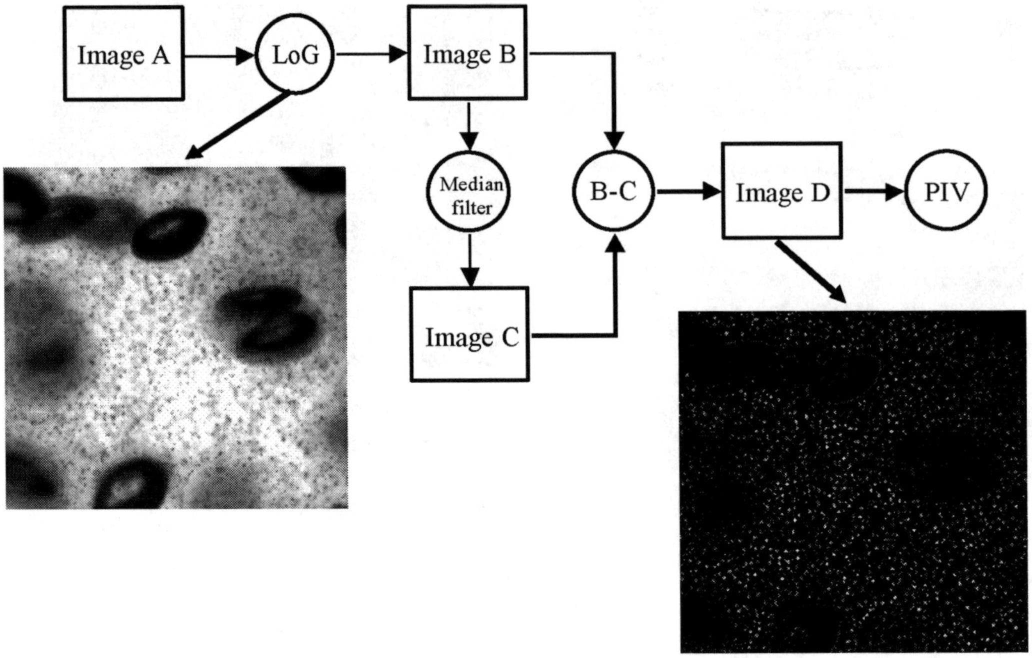

FIGURE 3. Schematic diagram of the filter operations for phase discrimination by extracting the images of tracer particles from the pictures of the two-phase flow (the images are a zoom from the full size image).

EXAMPLES OF APPLICATION

The imaging technique described above was applied to studies of single and multi-component droplet evaporation in grid turbulence, the collision behavior of suspension droplets, the wall collision of non-spherical particles and the motion of bubbles in a bubble column and a loop reactor.

The left image in Fig. 4 shows a multiple-exposed image of droplets evaporating in an upward directed flow downstream of a turbulence grid. Heated air was blown through the grid and the test section with a cross-section of 300 mm x 300 mm at a velocity of 5 m/s [17]. In the present case a single vibrating orifice mounted in the centre of the grid was used in order to produce a droplet chain which dispersed further downstream of the grid. From such images or normally double-pulse images the droplet velocities as well as the droplet sizes were obtained using the particle detection algorithm described above. An imaging technique was applied since the considered droplets with an initial diameter of about 220 µm were found to be non-spherical and oscillating. Therefore, accurate measurements were not possible by applying the standard phase-Doppler anemometry (PDA). The change of the droplet size distribution along the test section was determined by sampling 8000 droplets at each stream-wise measurement location in the centre of the spray with a size of 12.8 mm x 10.24 mm.

The sampled size distributions are shown in Fig. 4 right. Initially, just above the grid, a mono-modal size distribution is measured. Further downstream also larger droplets are found yielding multi-modal size distributions as a result of droplet coalescence in the chain. The evaporation rate was derived from the size reduction of the primary droplet size distribution only. The reduction of droplet surface area for a droplet consisting of a mixture of hexane and decane is shown in Fig. 5. From the slopes one can clearly identify two evaporation regimes. First hexane is evaporated rather fast due to the higher evaporation rate and for heights above 500 mm the remaining decane evaporates much slower. These data were used to develop and validate a multi-component droplet evaporation model

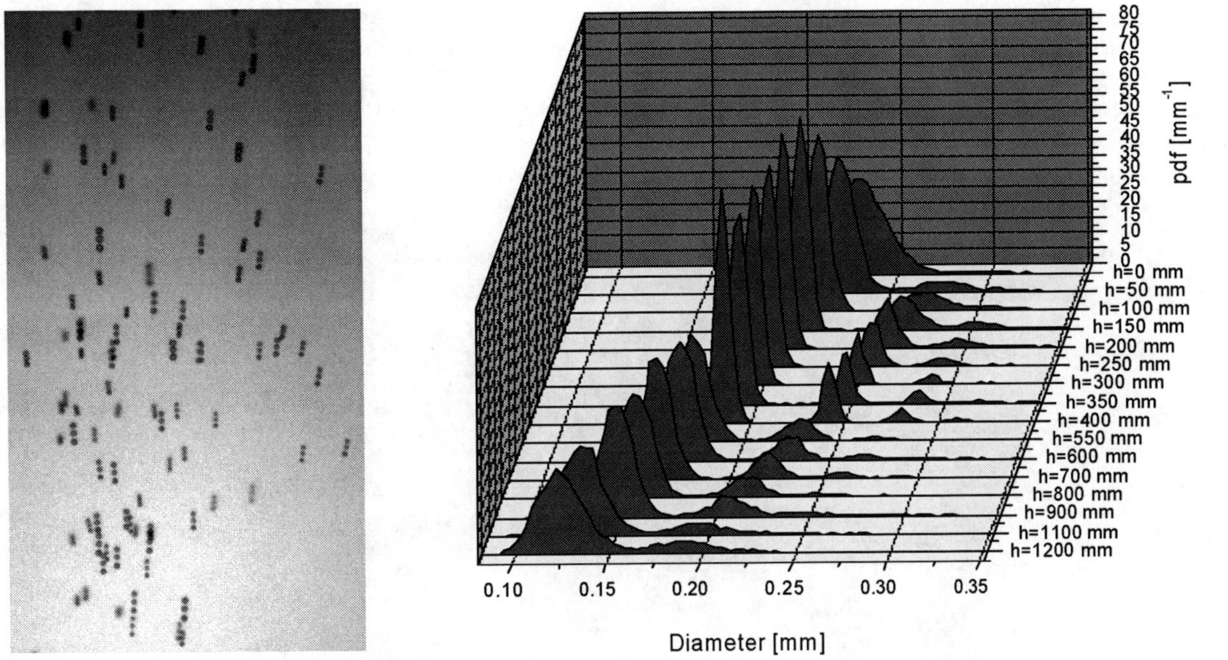

FIGURE 4. Multi-exposure image of evaporating droplets in a grid turbulence (left); chance of droplet size distribution along the test section (right) at a temperature of 200 °C for hexane droplets.

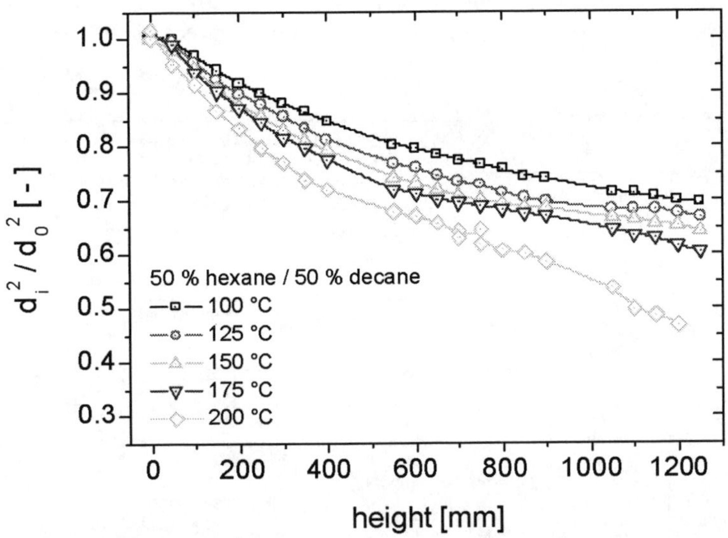

FIGURE 5. Chance of normalized surface area of droplets consisting of a mixture of hexane and decane along the test section for different average air temperatures.

The collision between suspension droplets is very important in spray driers where the production of powders consisting of defined agglomerated particles is anticipated. Since recent years it is tried to optimize the operational conditions of spray dryers by applying methods of computational fluid dynamics (CFD) as for example the Euler/Lagrange approach [18]. However, very little experimental information is available on the collision of suspension droplets which is essential for modeling this elementary process in the frame of numerical computations as demonstrated by Blei and Sommerfeld [19]. Therefore, detailed experiments were performed by producing two droplet chains using piezoelectric generators. These two droplet chains were directed towards each other in order to produce droplet collisions under defined angles. A stereoscopic imaging system was used to view the collision process from two perpendicular directions. The initial droplet sizes as well as the outcome of the collision process, i.e. coalescence or reflective and stretching separation, were determined by applying the object recognition and particle tracking approach described above [20]. Several collision events recorded in this way are summarized in Fig. 6 for the cases of droplet coalescence and droplet separation. For the latter case normally also very small satellite droplets are produced.

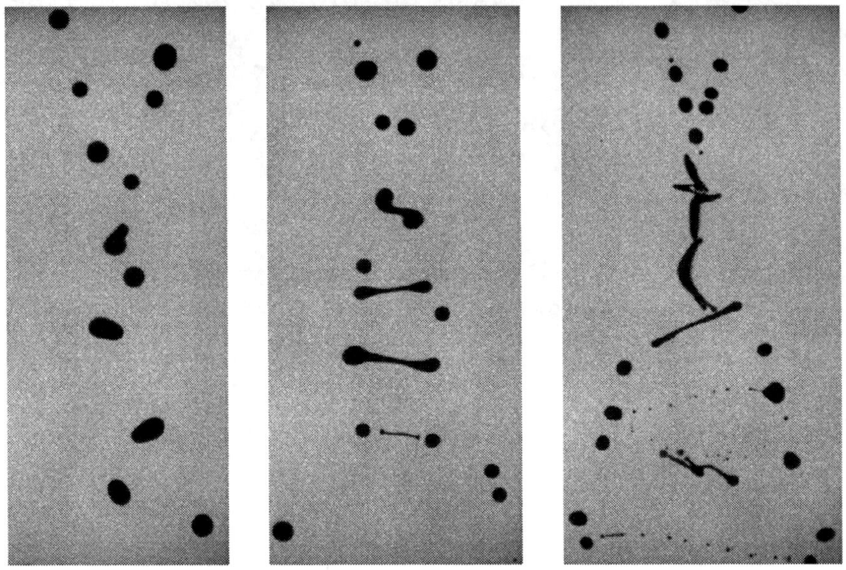

FIGURE 6. Enlargement of droplet collision events showing different collision phenomena (coalescence, reflective separation and stretching separation) for soy based liquid with a solid content of 40% (results from left to right for Weber numbers of 22, 94 and 434).

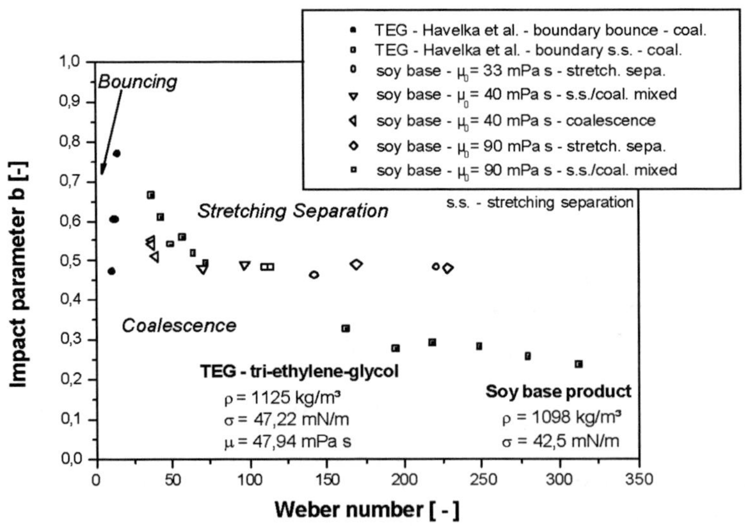

FIGURE 7. Collision map (i.e. impact parameter versus Weber number) showing the border line between coalescence and stretching separation; comparison of the results for soy-based suspension droplets and tri-ethylene-glycol droplets [21].

For modeling droplet collisions two information's are required, namely, under which collision parameters (i.e. collision angle and velocity and droplet sizes) which type of collision occurs and what is the size of the new droplets produced in the case of separation. The observed collision regime is normally characterized by plotting the boundary lines between the different regimes using the impact parameter and the droplet Weber number calculated with the relative velocity. The impact parameter is the non-dimensional separation between the centers of the colliding drops. By averaging a large number of collision events, such a collision map was derived for soy-based suspension droplets (Fig. 7). In this case the border line between coalescence and stretching separation is plotted and compared with the data of Havelka et al. [21] for tri-ethylene-glycol. The present results are very well represented by the correlation proposed by Havelka et al. [21]. Close to the border line (We between 75 and 120) the measurements with soy-based suspension droplets yield both coalescence as well as stretching separation.

Regarding the wall collision process of non-spherical particles almost no experimental information is available. However, in order to allow the derivation of a Lagrangian model for this elementary process detailed experiments using a particle-laden jet which hits a wall specimen with given roughness were performed. The absolute particle impingement velocity was between 10 and 15 m/s. A series of measurements were performed for different types of particles (spherical and non-spherical), varying for each case the impact angle. A multiple-exposed planar image of the trajectory of a cylindrical particle (volume equivalent diameter 480 μm) impinging on a metal plate is shown in Fig. 8. Such an image allows the determination of the particle's translational velocity from the measured spacing between the particle centroids and the specified time delay between subsequent exposures. From the change of the particle velocity normal and parallel to the wall, the restitution and friction coefficient could be derived similar to the work of Sommerfeld and Huber [22]. By avenging around 50 images for each impact angle the velocity ratios were evaluated for different particle types (see Fig. 8 b for the normal component). It is obvious, that the restitution coefficient of non-spherical particles is much lower than that for spherical glass beds, which eventually results in a higher pressure drop in pneumatic conveying [23]. This is the result of the stronger local deformation of particle and wall during impact. Errors in evaluating the particle velocity from planar images may occur due to particle rotation before and after impact, which causes some additional shift of the centroid from one particle image to the next, In the future a stereoscopic imaging will be applied in order to reduce this error and also allow for the determination of all components of the particle rotational velocity.

a)

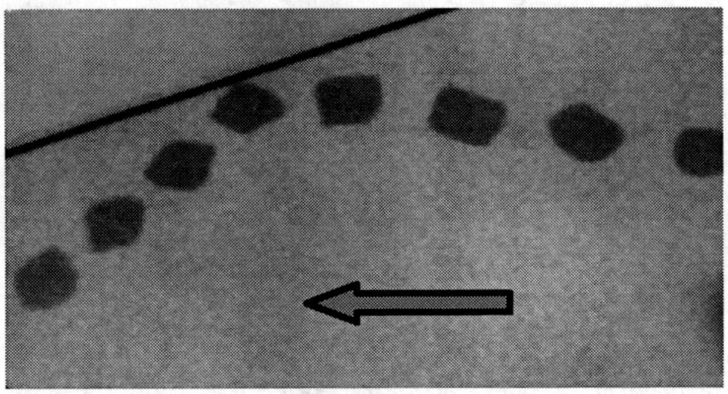

b)

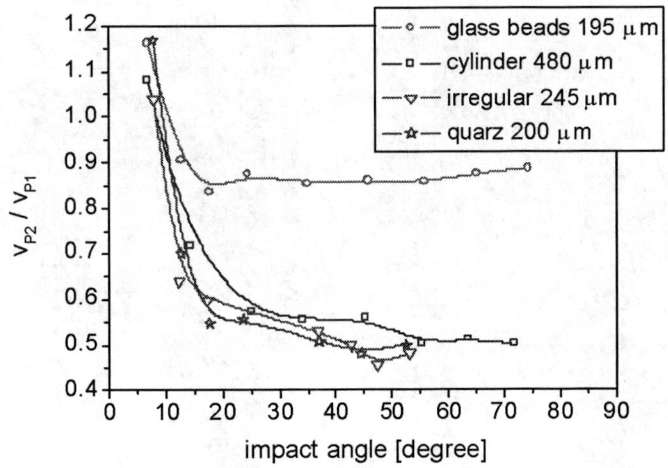

FIGURE 8. Wall collision of non-spherical particles, a) multi-exposed image of a colliding cylindrical particle, b).change of normal velocity component for different types of particles colliding with an almost smooth wall.

The above described particle detection method allows also the measurement of relevant properties in bubbly flows, namely the equivalent bubble diameter based on their cross-section, bubble aspect ratio and bubble orientation. The bubble velocity was obtained using PTV. These studies were performed in a 140 mm diameter bubble column and a special laboratory double loop reactor [11, 24]. A typical result of the eccentricity χ of bubbles (defined as the ration of major to minor axis) as a function of their size obtained in both facilities is shown in Fig. 9. The mean value of χ increases of course with bubble size and reaches a kind of plateau for bubble sizes between 2 and 3 mm with a value of about 1.9. A further increase of bubble size is associated with a decreasing eccentricity approaching a limiting value of about $\chi = 1.6$. The eccentricity of the bubbles measured in the loop facility yields a very similar result than those obtained in the bubble column. The present results are larger than those obtained from single bubble studies [25] and the correlation proposed by Okawa et al. [26]. This is the result of the turbulent flow conditions and bubble interaction effects in the considered bubbly flows with void fractions between 1 and 5 %.

With the same shadow technique also images of tracer particles could be recorded simultaneously in the considered image plane as shown in Fig. 10 a). By applying the described PIV, the continuous phase velocity field in the bubble swarm could be measured. The highest velocities are observed in regions of bubble interaction whereby the turbulence level of the flow increases considerably (Fig. 10 b).

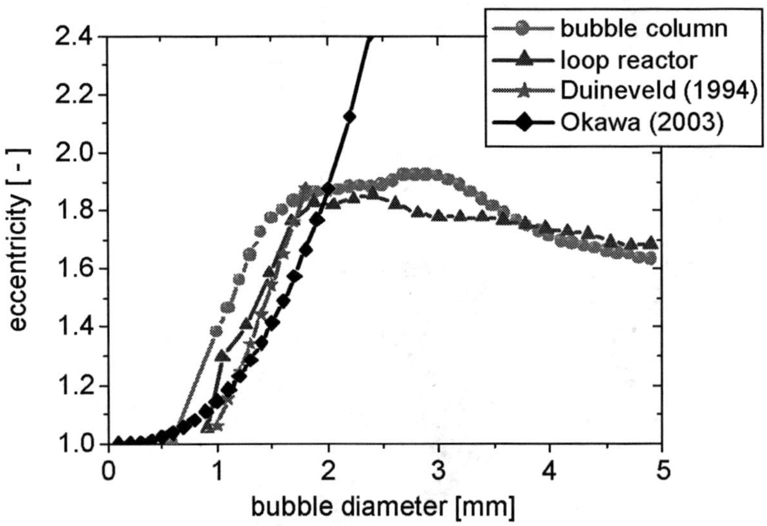

FIGURE 9. Bubble eccentricity as a function of area equivalent bubble diameter in bubble column and loop facility.

a)

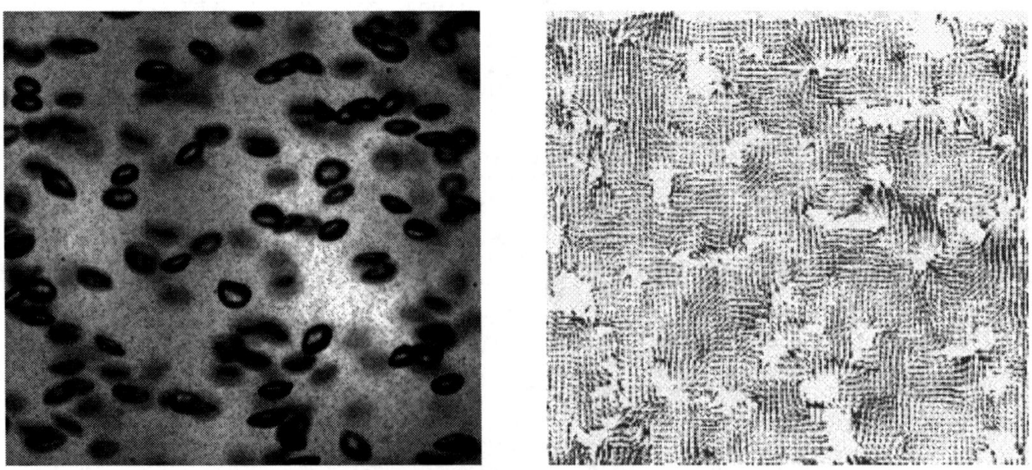

b)

FIGURE 10. Flow structure in a bubble swarm using shadow imaging and pulse light velocimetry, a) Image of bubbles and tracer obtained by back-lightning, b) liquid flow structure within the bubble swarm.

CONCLUSIONS

In the developed planar shadow image velocimetry the imaging plane was created using only one double-image CCD-camera with a macro optics having a low depth of field in combination with digital filter operations which remove out-of-focus objects from the image on the basis of the gradient of the grey scale value. The effective depth of field of course depends on the size of the particles considered and the magnification which was adjusted according to the application. The object detection algorithm incorporated in the in-house image processing software also allowed an accurate determination of the area equivalent particle size, the particle shape and orientation. The particle velocity was determined by PTV while the flow around the particles could be obtained by a multi-grid PIV method when appropriate tracer particles were added. Hence it was also possible to apply the measurement system for multiphase flows with deformable and/or non-spherical particles.

The powerfulness of this imaging technique for the analysis of elementary processes in several multiphase flow systems was demonstrated. Depending on the application, the size of the imaging area was adjusted by using appropriate extension bellows in the range from 12 mm x 10 mm to 60 mm x 40 mm. From the analysis of processes such as multi-component droplet evaporation in grid turbulence, binary collisions of suspension droplets, wall collision of non-spherical particles and behavior of bubbles in turbulent bubbly flows, detailed information could be obtained which are important for modeling these elementary processes in the frame of numerical computations of multiphase flows.

ACKNOWLEDGMENTS

The preset studies were financially supported through several research projects from the Deutsche Forschungsgemeinschaft (contract number SO 204/10, SO 204/19 and SO 204/23) and the European community (contract MAST-B-Liquid No. NNE5-2001-00869). This support is gratefully acknowledged.

REFERENCES

1. Liu, A.B. and Reitz, R.D., *Atomization and Sprays*, **3**, 55-75 (1993)
2. Brasier-Smith, P.R., Jennings, S.G. and Latham, J., *Proc. R. Soc. London*, **A326**, 393-408 (1972)
3. Rüger, M., Hohmann, S., Sommerfeld, M. and Kohnen, G., *Atomization and Sprays*, **10**, 47-81 (2000)
4. Lunde, K. and Perkins, R. J., *Applied Scientific Research*, **58**, 387-408 (1998)
5. Tomiyama, A., Celata, G. P., Hosokawa, S. and Yoshida, S., *Int. J. of Multiphase Flow*, **28**, 1497-1519 (2002)
6. Celata, G. P., Cumo, M., D'Annibale, F., DiMarco, P., Tomiyama, A. and Zovini, C., 5th International Conference on Multiphase Flow, ICMF'04, Yokohama, Japan, Paper No. 477 (2004)
7. Borchers, O. "Zweiphasen-Particle Tracking-Velocimetry (PTV) zur detaillierten Analyse der Hydrodynamik von Blasensäulenreaktoren" Fortschrittsberichte VDI, Reihe 3 Verfahrenstechnik, VDI-Verlag Düsseldorf, Nr. 753 (2002)
8. Fujiwara, A., Takahashi, T., Tokuhiro, A. and Hishida, K., International Conference on Multiphase Flow, New Orleans, U.S.A. (2001)
9. Lindken, R. and Merzkirch, W., *Experiments in Fluids*, **33**, 814-825 (2002).
10. Sakata, Y., Fujiwara, A. and Hishida, K., 5th International Conference on Multiphase Flow, ICMF'04, Yokohama, Japan, May 30 – June 4, 2004, Paper No. 461
11. Bröder, D. and Sommerfeld, M., Submitted to: *Measurement Science Technology* (2007)
12. Honkanen, M., Saarenrinne, P, Stoor, T. and Niinimäki, J., *Meas. Sci. Technol.* **16**, 1760-1770 (2005)
13. Shen, L., Song, X., Iguchi, M. and Yamamoto, F., *Pattern Recognition Letters*, **21**, 21-30 (2000)
14. Marr, D. and Hildreth, E., *Proc. of the Royal Society of London*, **B 207**, 187-217 (1980)
15. Scarano, F. and Riethmuller, M. L., *Experiments in Fluids*, **26**, 513-523 (1999)
16. Gui, L., Lindken, R. and Merzkirch, W., Proceedings of the ASME Fluids Engineering Division Summer Meeting, Paper ASME-FEDSM97-3103 (1997)
17. Horender, S. and Sommerfeld, M., ILASS Europe 2004, Proceedings of the 19th International Conference on Liquid Atomization and Spray Systems, Nottingham, UK, (Ed. B. Azzopardi), pp. 231-236 (2004)
18. Verdurmen, R.E.M.; Menn, P.; Ritzert,J.; Blei, S.; Nhumaio, G.C.S.; Sonne-Sorensen, T.; Gunsing, M.; Straatsma, J.; Verschueren, M.; Sibeijn, M.; Schulte, G.; Fritsching, U.; Bauckhage, K.; Tropea, C.; Sommerfeld, M.; Watkins, A.P.; Yule, A.J. & Schonefeld, H., *Drying Technology*, **22**, 1403-1461 (2004)

19. Blei, S. and Sommerfeld, M., 15th International Drying Symposium (IDS 2006), Budapest, Hungary, August 2006
20. Blei, S., "On the interaction of non-uniform particles during the spray drying process" Shaker Verlag, Aachen 2006
21. Havelka, P., Gotaas, C., Jakobsen, H.A. and Svendsen, H.F., 5th International Conference on Multiphase Flow, ICMF'04, Yokohama, Japan, May 30 – June 4, 2004, Paper No. 123
22. Sommerfeld, M. and Huber, N., *Int. J. of Multiphase Flow*, **25**, 1457-1489 (1999)
23. Lain, S., Sommerfeld, M. and Kussin, J., Proceedings 3rd International Symposium on Two-Phase Flow Modelling and Experimentation, Pisa, September 2004
24. Bröder, D. and Sommerfeld, M., *The Canadian Journal of Chemical Engineering*, **81**, 756-763 (2003)
25. Duineveld, P. C. "Bouncing and coalescence of two bubbles in water" Ph.D. Thesis, University of Twente, Netherlands: ISBN 90-9007541-0 (1994)
26. Okawa, T, Tanaka, T, Kataoka, I. and Mori, M., *Int. J. Heat and Mass Transfer*, **46**, 903-913 (2003)

Multiphase Flow Measurement by Dual Gamma Ray Tomography

Wu Yingxiang, Cui Bin*, Li Donghui, H. Inaki Schlaberg**, Zheng Zhichu, Zhong Xingfu

Institute of Mechanics, Chinese Academy of Sciences, Beijing 100080, China
**Liaohe Oilfield, Panjing 124010, China*
***Institute of Particle Science and Engineering, University of Leeds*

Abstract This paper describes some of our research in three phase flow-rate measurement of oil/gas/water by processing tomography of dual gamma ray, including the instrumental designs on the technique of photons pulse counter, signals of sensor, preamplifier, filter and shaping amplifier, DC base shift correcting circuit, narrow windows of energy spectroscopy, programmable pulse count acquisition system; the FPGA (Field programmable gate array) based data acquisition and processing system for gamma ray tomography; and the oil-water-gas three phase volumetric fraction distributions from experiments on a test flow loop.

Key words: Multiphase flow, flow-rate measurement, gamma ray, process tomography

INTRODUCTION

Multiphase flow measurement, especially oil/gas/water flow-rate measurement is a significant issue in oil industry. Operators in oil companies and designers of oilfield facilities are keen to seek for the multiphase metering technology to simplify the design and improve the control of production facilities. They considered that, unless multiphase metering techniques were mature, it would be virtually impossible to know what was happening in sub-sea production systems or on the unmanned satellite oil production platforms.

Multiphase metering systems used in harsh environments of offshore environment must possess ability to measure accurately flow-rates of each phase in all flow regimes and it should not be affected by changes in fluid properties like density, viscosity, surface tension, and dielectric properties. In addition, they should also be required with non-intrusive property and robust design for convenience in operation and on-line maintenance. In spite of the large number of solutions that have been proposed in recent years, no commercially available three-phase flowmeter yet meets all these requirements.

The multiphase meters can be based on different principles. To some extent a specific principle is used to measure only in a limited range of phase fractions. Some of the available products require a preconditioning of the flow, which is a severe limitation for most applications. Out of the possible measurement principles electromagnetic interaction is promising for cases where the fluid components have distinct dielectric properties. Depending on the application and materials properties one can distinguish resistive, capacitive, radio frequency (RF), microwave, infra-red, optical, ultra-violet, x-ray and gamma-ray sensors[1-4].

Process tomography is another discipline which has seen a significant growth over the last ten years, with laboratory imaging systems having been developed for a number of multiphase applications. It is not yet clear whether such techniques can be used to measure component flowrate any more accurately than existing solutions. However, these techniques may have the potential to enhance the performance of existing commercial flowmeters[5~9].

Gamma ray tomography is a potential method that may be applied to measure the distribution of phases respectively the hold-up of liquid or gas in multiphase transportation pipelines. It is a measurement technique to provide the detailed information on phase distributions of the cross-section of pipeline.

CP914, *Multiphase Flow: The Ultimate Measurement Challenge,* edited by X. Cai, Y. Wu, Z. Huang, S. Wang, and M. Wang

In this paper, we give some research results in three phase flow-rate measurement of oil/gas/water by processing tomography of dual gamma ray, including the electronics design for a dual energy gamma ray, instrumental designs on the technique of photons pulse counter, signals of sensor, preamplifier, filter and shaping amplifier, DC base shift correcting circuit, narrow windows of energy spectroscopy, programmable pulse count acquisition system; the FPGA (Field programmable gate array) based data acquisition and processing system for gamma ray tomography; and the oil-water-gas three phase volumetric fraction distributions from experiments on a test flow loop.

ELECTRONICS DESIGN FOR DUAL ENERGY GAMMA-RAY

In the instrumental design of dual energy gamma ray system, the important parameters to be considered are:

- selection of detector;
- design of preamplifier;
- design of filter and shaping amplifier;
- design of DC base shift correcting circuit;
- design of dual channels analyzer;
- design of programmable pulse counter data acquisition system.

1. Detector

A dual energy γ ray is used as the detecting source of the PT measurement system. It is comprised of two radioactive isotopes of ^{241}Am and ^{137}Cs which have emission energies at 59.5keV and 662keV. NaI (Tl) crystal is selected as the scintillator for its highest detection efficiency and short decay constant to get high count rate and avoid pulse pile up or saturation when the system is operated in count mode. A photomultiplier tube is used for multiplication of the light emitted by scintillator and NaI (Tl), one kind of inorganic scintillators with high density and high atomic number, is adopted to detect gamma rays.

The photo multiplier tube (PMT) consisting of an evacuated glass tube with photocathode at its center and several dynodes in the interior is an integral part of a scintillation counter which amplifies an incident pulse of visible light by a factor of 1000,000 or more in times of 1 ns.

The photons produced in the scintillator enter the PMT and hit the photo cathode. The electron emitted by the photocathode is guided by an electric field towards the successive dynodes. The voltage difference between the two successive dynodes is of the order of 80–120V. Typical commercial photo tubes may have up to 15 dynodes.

2. Amplifier

As the signals from the detector are very weak, usually in mV range, an amplifier is required to increase the amplitude of pulse from detectors by a factor of thousand or more to a few volts to meet the needs of the analyzer that follows the amplifier with the provision for a gain selection and pulse shaping time constant.

The pulse signal from amplifier is a series of negative pulses with a significant positive DC base shift, the amplitude of pulses corresponds to the entrance photon energy.

3. DC base shift correcting circuit

In dual energy gamma ray system, because of the high accuracy requirement of spectrum resolution, the DC base shift is thereby very harmful for pulse altitude measurement, thence a DC base restore circuit is needed to be used to isolate the DC base shift.

In general, a RC differential circuits can be used as DC isolation. If the RC circuit time constant is much shorter than the signal pulse interval, the voltage of capacitance drops to the input signal base rapidly, then the DC base can be restored. As the time constant RC is much shorter than the pulse assert time, the output pulse of RC circuit becomes narrow and sharp. There should be troubles for the behind spectrum analyzer. A improved CD circuit was presented in this study which applied a nonlinear component of diode to replace the resistance in RC circuit. The positive resistance of diode is large when it is not leaded, which will produce a large shift rate than RC circuit.

4. Filter and shaping amplifier

To improve the signal to noise ratio and to shape the pulse into a Gauss-like curve, a filter and a shaping amplifier is engaged in frequency analysis to eliminate some frequencies from the signal. A pole-zero cancellation technique is adopted here.

The output pulses from the shaping amplifier are inverted to positive pulses and are amplified the amplitudes to 0~8V. The pulses are shaped to a Gauss-like curve and are adjusted to only 1~2us wide to avoid pulse pile up for the case of high count.

5. Dual channel analyzer

Dual channel analyzer is an important unit in the gamma ray spectrometer to select the amplitude (and hence the energy) for output pulses from the amplifier. We set two narrow windows for the analyzer to get lower discriminator and upper discriminator level selections. When pulses appear on or out of the shaping amplifier, the lower and upper level discriminators operate, only the photons with energy above the lower threshold energy and below the upper threshold can be counted using the window operation thereby rejecting the unwanted low/high energy pulses.

As the dose of ^{241}Am and ^{137}Cs are 100 mCi and 20 mCi, the peak of ^{241}Am is higher than ^{137}Cs's, and the ^{241}Am uses narrower voltage window than ^{137}Cs, which means that the sensor of scintillator has higher energy resolution and more efficient for ^{241}Am than for ^{137}Cs.

Only the pulse of the amplitude within the low threshold and up threshold can be recorded, lower or upper pulses will be rejected by the discriminator. The discriminator circuit comprised of two high speed voltage comparators and associated logic circuits, two adjustable reference window voltages are provided.

DATA ACQUISITION SYSTEM

The data acquisition system includes two parts, one deals with the detection signals as a programmable pulse counter system, the other communicates with a host computer which sends parameters, such as the acquisition time frame, and sends commands, such as the start signal of an acquisition cycle and collects the acquired results.

1 Programmable pulse counter system

An Intel Pentium industrial control computer is used with eight boards of 32 channel 16 bit pulse counter in its ISA bus as signal acquisition system. The counter's latch time and counter bit length are programmable, the integral time for every channel is programmable as well. The programmable counter board is based on the chip of 8253 which have three channels of programmable interval timer/counter. In order to avoid the pulse deforming (only 100~200ns), digital logic isolation is necessary for the transmission from shaping amplifier to the counting board. An ultra high speed (10MBd) photo coupler 6N137 is employed in the board therefore noise and disturb are eliminated.

Another important constituent in the data acquisition system is the software that enables users to select above control parameters, acquire the data and does further arithmetic.

This is a conventional way of data acquisition based on the technique of photons pulse counter, corresponding to the multi-CPU data acquisition system. It has a simple structure and a low data flow rate from the unclear instruments to computer.

One processor board of the acquisition system is dedicated to perform the communication with the host and interfaces with the rest of the channels of the system. This controller board distributes the commands for the host to each of the channels via a backplane that connects all the boards of the system.

2 FPGA (Field programmable gate array) based data acquisition and processing system

The data acquisition system communicates with a host computer which sends parameters and commands. One processor board of the acquisition system is dedicated to perform the communication with the host and interfaces with the rest of the channels of the system. This controller board distributes the commands for the host to each of the channels via a backplane that connects all the boards of the system.

Each channel of the tomography system can operate independently and report the results to a controlling processor which will communicate with a host computer.

After simple signal conditioning to bring the voltages within the range of the A/D converter specification a dual channel converter chip takes the data of two detectors to the FPGA chip. The FPGA is connected to a DSP which can operate on the sample data or instruct the FPGA to perform operation. A backplane connector provides the bus for communication between the channels and the main controlling board. Although the connector is physically attached to the FPGA, it is the DSP which controls the communication. The A/D converter chip selected operates at 20MSPS with 12 bits resolution. It is pin compatible with 14 bit devices and devices that can sample at rates up to 65MSPS.

The task of the acquisition channel is to count the number of pulses at each energy level and transmit these two values to the host. Combined with the results from other channels these will be used in an image reconstruction algorithm. The procedure for obtaining the information is to first detect the peaks from the data stream of the A/D converter and record its maximum amplitude.

As the data stream enters the FPGA at 20MSPS, the DSP which operates at 200MIPS only has 10 clock cycles to implement the peak detection algorithm between incoming samples. This may prove challenging and also limits the functionality of the algorithm. A further demand on the resources is the fact that each A/D converter has dual inputs and delivers data to the FPGA (and the DSP) from two separate detectors.

The dual data stream may however be compensated by the fact that the DSP has a dual core and can simultaneously operate on two separate data operands albeit with the same instruction. In other words the DSP uses a SIMD (Single instruction, multiple data) architecture that could be exploited by the dual data stream of the A/D converter.

An alternative method to implement the peak detection algorithm makes use of the FPGA. In this way the resources of the device can be employed to operate on the data in parallel rather than the sequential nature of the DSP program. The number of clock cycles can be reduced by employing a pipelined design. However this is at the expense of increased use of internal resources and a longer latency depending on the number of stages in the pipeline. For example, if the design has a five stage pipeline, it would take five cycles for the first data item to emerge. However, a delay of a few clock cycles is not significant in the whole acquisition cycle.

DYNAMIC MEASUREMENT OF VOLUMETRIC FRACTIONS OF OIL/GAS/WATER MIXTURES

1 Dual energy theories

The measurement of component ratios in multiphase flow using γ-ray attenuation was first suggested by Abouelwafa and Kendall (1980), and the technique is used in many current multiphase metering systems. Attenuation of a beam of γ-rays by a material of thickness t is given by:

$$I(E) = I_0(E)B\exp(-\mu(E)t))$$

where I(E) is the transmitted intensity, I0(E) is the incident beam intensity, μ(E) is the linear attenuation coefficient of the attenuating material at energy E and B is the buildup factor, which takes account of radiation scattered into the detector. For oil/water/gas three-phase flow, the attenuation coefficient of the mixture μ(E) is represented by

$$\mu(E) = \alpha\mu(E)_o + \beta\mu(E)_w + \chi\mu(E)_g$$

where μ(E)o, μ(E)w and μ(E)g are the linear attenuation coefficients of the oil, water and gas, and α, β, χ are the respective volume fractions. The transmitted intensity I through a thickness t of a oil/water/gas mixture is therefore

$$I = I_0 \exp[-(\alpha\mu(E)_o + \beta\mu(E)_w + \chi\mu(E)_g)t]$$

if the build-up factor is eliminated through good collimation at the source and detector. After the transmitted flux I1 and I2 at two energies E1 and E2 is measured, the volume fractions can be calculated if the linear attenuation coefficients of the flow components at E1 and E2 are known since

$$\ln(I/I_0)_1 = -[\alpha\mu(E_1)_o + \beta\mu(E_1)_w + \chi\mu(E_1)_g]t$$

$$\ln(I/I_0)_2 = -[\alpha\mu(E_2)_o + \beta\mu(E_2)_w + \chi\mu(E_2)_g]t \quad \text{and} \quad \alpha + \beta + \chi = 1$$

There are three equations with three unknown's volumetric fractions α, β and χ.

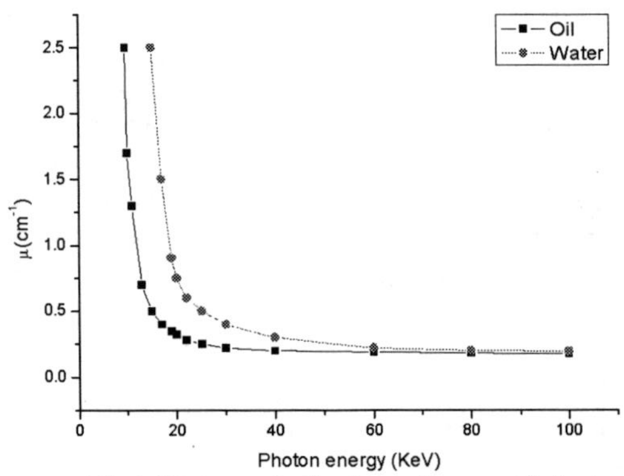

Fig.1 Water and oil attenuation coefficients

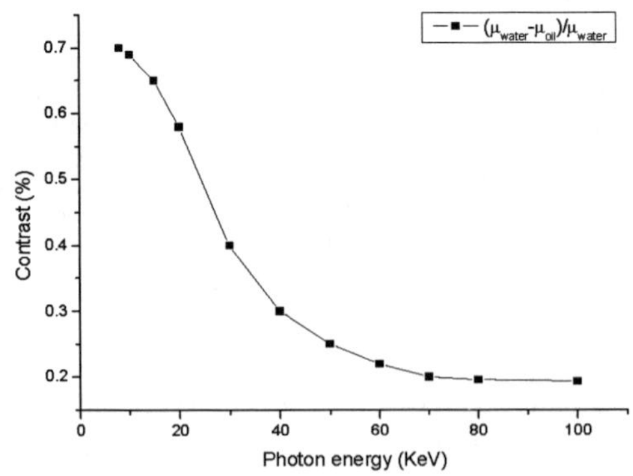

Fig.2 Radiographic contrast between oil and water

The linear attenuation coefficients of water and mineral oil (kerosene) over the same energy range reveals the differences in the photon absorption can be used to distinguish them. Fig.1 shows the photon attenuation of water and oil. Since the photoelectric interaction is a stronger function of photon energy and atomic number than the

Compton interaction, there is greater contrast in the linear attenuation coefficients of oil and water in the region where the photoelectric effect dominates the interaction cross-section of both materials[10].

2 Experimental arrangement

The oil-water-gas three phase flows are very complex and it is impossible to test all the flow rate combinations for each phase media. In order to study the three phase materials attenuation character, an experimental arrangement was designed as shown in Table 1, including total 6 groups of measurements in the experiment. All the flow rate data are superficial flow rate obtained from flow meter of each phase.

Table 1 (Gas1=20SL/min, Gas2=100SL/min)

Oil	Water							
0.75m³/h	2.5m³/h	3.75m³/h	5m³/h	6.25m³/h	7.5m³/h	8.75m³/h	10m³/h	12.5m³/h
1.5m³/h	2.5m³/h	3.75m³/h	5m³/h	6.25m³/h	7.5m³/h	8.75m³/h	10m³/h	12.5m³/h
3.6m³/h	2.5m³/h	3.75m³/h	5m³/h	6.25m³/h	7.5m³/h	8.75m³/h	10m³/h	12.5m³/h

3 Results on three phase flows

Six groups of oil-water-gas three phase flow were surveyed by dual energy γ ray system. The results of three phase volumetric fraction of oil-water-gas multiphase flow are shown in Fig.3.

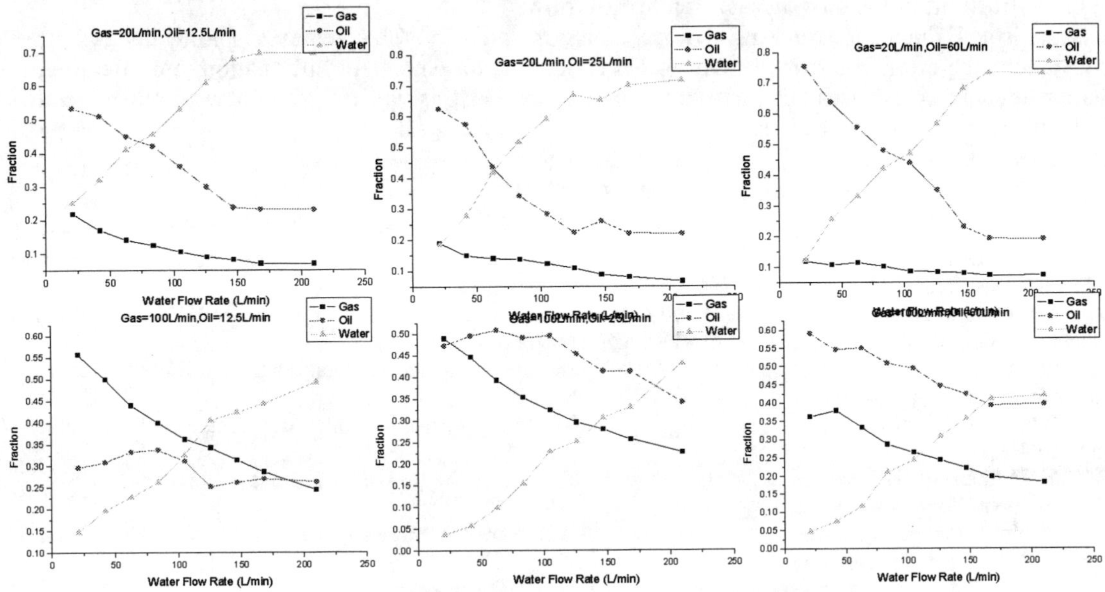

Fig.3 Volumetric fractions of oil-water-gas multiphase flow

In the top row of curves in Fig.3, the gas flow rate is fixed in 20L/min; in bottom row, the gas flow rate is increased to 100L/min. The oil flow rate is changed from 12.5L/min, 25L/min to 60L/min respectively in the diagrams of both rows from left to right in Fig.8. The water flow rate is increased from 21L/min to 210L/min in 9steps as shown on the X axis in each diagram. The sum of gas, oil and water volumetric fraction is equal to 1, and the measurement integral time is 200s in this experiment.

It is easy to find that with the increasing of the oil flow rate as shown in Fig.8 from left to right curves in both rows; the total liquid flow rate increased, and then the gas volumetric fraction decreased. This is because gas is a compressible fluid, when the total liquid fraction increased; the pressure inside the pipe goes up, thus the volume fraction of gas decreased.

The oil volumetric fraction also increased as expected as shown in Fig.8 from left to right curves in both rows. The oil flow rate is changed from 12.5L/min, 25L/min to 60L/min, and the dual energy γ ray system gives out a correct response to the fraction changes. This shows a reasonable measurement accuracy. But there are also some significant error points on these curves; some curves do not look smoothly as they should be.

CONCLUSIONS

A study on oil-water-gas three phase flow measurement system by a dual energy γ ray tomography is presented in this paper. From the study, some conclusions can be drawn that:

(1) There are several key points in the nuclear instrumental designs, including DC base shifting and pulse shaping. An appropriate treatment of above questions will increase the accuracy of the dual energy gamma system.

(2) The system uses dual channel A/D converter chips that operate at 20MSPS and 12 bit resolution, but can be upgraded to 65MSPS and 14bits without hardware modifications other than replacing the pin compatible converter chip. The components used are widely available and result in a low cost per channel. The channel hardware has been successfully tested with a standard signal generator. The data processing for the peak counting has been implemented on the FPGA resulting in a fast and flexible processing of the sensor information.

(3) The algorithm has been simulated on the FPGA platform with data obtained during previous tests on a flow loop and the results show the peaks of energy levels that were expected. Further work on this systems needs to be performed by integrating the data of multiple channels and performing experimental evaluation with the gamma ray sources on the flow loop.

(4) It is clear that the oil-water-gas three phase flows are very complex. Not only there are speed shifts and velocity differences between every phase media but also there are phase volumetric fraction changes in the cross section of the pipe. Until now there are no other more efficient and convenient way that can be used to measure the oil-water-gas multiphase flow. The dual energy γ ray is a promising technique for the purpose of simple and fast estimating the phase fractions of the oil-water-gas multiphase flow.

(5) There is statistical fluctuation in γ ray source emission intensity. This requires a long integral time in measurement to reduce the error, and it is difficult to get short time measurement information, only the long time average results are reliable unless uses large activity γ ray source, but this will bring the new problems related to system cost and radioactive safety protections.

REFERENCES

[1] Miodrag Bolic, Vujo Drndarevic, Digital gamma-ray spectroscopy based on FPGA technology, Nuclear Instruments and Methods in Physics Research A 482 (2002) 761–766

[2] Sergio Garcia Castillo and Krikor B. Ozanyana, Field-programmable data acquisition and processing channel for optical tomography systems, REVIEW OF SCIENTIFIC INSTRUMENTS 76(2005), 095109

[3] Meng Ma, Arjan van Genderen and Peter Beukelman, Developing and Implementing Peak Detection for Real-Time Image Registration, ProRISC, 2005, pp647-652

[4] Bruneau PRP, Mudde RF, van der Hagen THJJ, High-speed flow imaging using gamma-ray computed tomography, 4th WCIPT congress, 2005

[5] H. Toral, S. Cai, E. Akartuna, K. Stothard, A.W. Jamieson, Field Tests of the ESMER Multiphase Flowmeter, 1998, North Sea Flow Measurement Workshop, Gleneagles, Scotland.

[6] C.J.M. Wolff. Required operating envelope of multiphase flow meters for oil well production, 1993, North Sea Flow Measurement Workshop, Bergen, Norway.

[7] Handbook of Multiphase Metering, Norwegian Society of Oil and Gas Measurement

[8] W. Slijkerman, A.W. Jamieson, W.J. Priddy, O. Okland, H. Moestue, Oil companies needs in multiphase flow metering, 1995, North Sea Flow Measurement Workshop, Lillehammer, Norway

[9]N. W. King, Subsea multiphase flow metering-a challenge for the offshore industry? Subsea 90 International Conference, London, England, 1990.

[10]Key, M.J., (1999), Gas Microstructure X-Ray Detectors and Tomographic Multiphase Flow Measurement. Ph.D Thesis, University of Surrey.

Dynamic Non-Obstructed Particle Image Velocimetry Measurements inside a Packed Bed Reactor

Hassan Y., E. Dominguez-Ontiveros, C. Estrada-Perez.

Nuclear Engineering, Texas A&M University, CS, TX, USA, y-hassan@tamu.edu

Abstract. The aim of this work is to present the progress of experimental approaches used to investigate the flow pattern towards mass and heat transfer phenomena within a packed bed. Most of the studies on packed beds consider the porous medium approach to obtain lumped parameters from which bulk characteristics of the flow are determined. This conventional approach neglects the complex flow details within individual pores and constrictions, thus obscuring the physical basis of the model. The experimental approach presented uses Particle Image Velocimetry (PIV) and Matching Refractive Index (MRI) techniques to obtain velocity measurements within a packed bed at the microscopic (pore) level. To completely understand and describe the physical process occurring within the system, local information is required. The main results of velocity measurements for single and two-phase flow conditions are presented under various flow conditions with Reynolds numbers ranging between 50 and 500, based on particle diameter. The results show evidence of various recirculation zones within the pores. Measurements of flow path within a single pore show considerable changes under two-phase flow conditions. Vortex identification at various pore locations exhibits the need for consideration of additional parameters in the models to accurately describe flow structure in a packed bed.

Keywords: Matching Refractive Index, PIV, Packed Bed, Pore level, microscopic.

INTRODUCTION

A typical packed bed is a column where a fluid is flowing through packed grain-like material. A large part of our natural environment as well as substantial fraction of synthesized processes can berepresented as a packed bed. The importance of packed bed reactors in several industries has prompted extensive research on experimental and numerical models for these systems. One important and particular application involves the potential use of this kind of technology in space based applications such as fuel cells, bioregenerative food production, waste treatment, watering and nutrient transport in plant root systems, and filtering processes, among others. Though several studies have been carried out for packed beds under normal gravity; the detailed physics behind this kind of technology is not well understood. Moreover, various studies have shown that under microgravity conditions, these beds may have a different behavior (Motil 2003).

One promising approach to bring insight to the understanding of complex physical phenomena inherent to packed beds is the measurement of parameters at the pore scale level. Many of the macroscopic processes that affect material transport in porous media are manifestations of the flow behavior of the system at the microscopic, individual pore volume, scale. Dispersion, for example, is the result of the cumulative effects of a number of micro-scale phenomena, including the mixing caused by solid obstructions in the flow path, the incomplete connectivity of the medium, eddies in the medium, and recirculation caused by regional pressure gradients. It is common to derive macroscopic transport theory using volume averages of microscale flow. Unfortunately, little is known about micro-scale transport in porous systems because of the difficulty associated with non-invasively measuring interstitial parameters. Lack of information has forced theoreticians to restrict their comparisons to those between theory and available experimental data. Comparisons have been made only in terms of integral or macroscopic results (Northrup et al. 1993).

Multipoint measurements with spatial resolution are necessary to allow visualization of complex flow patterns and to provide data at high spatial density to obtain volume averaging (Northrup et al. 1993). Fluids in porous media, though ubiquitous in nature and modern technologies, are notoriously difficult to study because they are "invisible" to most optical instrumentation. Consequently, theories of flow and transport are largely unverified experimentally (Moroni and Cushman 2001). Recently, optical measuring techniques including qualitative flow visualization, point

CP914, *Multiphase Flow: The Ultimate Measurement Challenge,* edited by X. Cai, Y. Wu, Z. Huang, S. Wang, and M. Wang

measurements with Laser Doppler Velocimetry (LDV), and full field measurements with Particle Image Velocimetry (PIV) have been used due to their improved accuracy and non-intrusive characteristics. One obstacle to the use of these optical methods is the need for optical transparency of the medium under investigation. There are immersion optical methods for studying solid transparent objects, which are based on selecting the refractive index of a liquid medium into which these objects are immersed (Bailey and Yoda 2003, Cui and Adrian 1997, Pokusaev et al 2004). Under stagnant isothermal conditions, the medium and object are immobile and there are no changes in temperature and concentration. However, during experimentation, it is necessary to ensure quality of the refractive indices of the liquid medium and the object by assuring small variation of the refractive indices in case of a change of the medium temperature.

To explore transfer phenomena in multiphase granular beds, it is necessary to account for the selection of the physical properties of the fluid such as density, viscosity, and surface tension. The controlling condition of the immersion is that the refractive index of the liquid should equal the refractive index of the granular bed material. Several investigators have presented tables of available liquids with its respective refractive indices. However, as previously described, the selection of physical properties of the working fluid becomes a key parameter in the experimental design. In this work, an optically transparent packed bed was constructed so the motion of a fluid passing through the bed gaps could be studied by the use of the Particle Tracking Velocimetry (PTV) technique in conjunction with Refractive Index techniques. The measurements obtained were analyzed to provide statistical information from the instantaneous velocity fields. These techniques allowed flow velocity distributions to be measured within the voids formed by neighboring beads under single and two-phase flow (liquid-gas) conditions..

EXPERIMENTAL SET-UP

Experimental apparatus

Experiments were performed using an air/p-cymene system and employ a packing of polymethylmethacrylate (PMMA) beads with an average diameter of 4.7 mm. The column was fabricated out of PMMA material with a square cross section of 3 x 3 cm and height of 30 cm. A thin aluminum grid was placed at the column inlet to serve as a filter in such a way that the first layer of packing material rested over the grid. The fluid was pumped from a reservoir tank through the packed bed using a centrifugal type pump as shown in Figure 1. The inlet flow rate was measured by a propeller type flow meter and controlled by the main valve located in a loop section between the pump and the column inlet. The temperature of the working fluid was controlled by immersing the reservoir tank into a water bath. A type- T thermocouple was submerged into the reservoir tank to monitor the bulk liquid temperature. The gas phase was injected to the system at the column inlet from a compressed air outlet. Gas flow rate was measured with a propeller type flow meter located between the controlling valve and the column air inlet and the test zone was located at a distance, L, such that $L/d_p=53$, from the column inlet. A high speed/ high resolution camera was used to perform the PIV measurements. The laser sheet impinges at a perpendicular angle to one of the column faces as illustrated in Figure 1. In this Figure, the laser sheet comes out of the plane of the paper. The camera was located perpendicular to the laser sheet plane within the test zone. Two different interrogation areas were employed during this work, 7 x 5 mm and 13 x 10 mm. The column's outlet is connected to the closed reservoir tank through a 6 cm pipe that discharges the flow at the upper portion of the tank. The pump flow intake was located in the lower section of the reservoir tank, opposite the discharge pipe. In this way, proper separation of the gas phase was assured at the upper part of the tank and only liquid is flowed through the pump.

Experimental procedure

The reservoir tank was filled up to 70% of its total volume with p-cymene at room temperature. This tank, containing the working fluid was immersed into a water bath at a temperature of 20°C. The working fluid was seeded with Expancel 091 DU particles of an average diameter of 6 μm. A particle-p-cymene solution was prepared in a different container prior to the conduction of the experiment. Optimum mixing of the solution was reached by injection of bubble columns at the bottom of the container and stirring. Pump purging was necessary at the commencement of experiments to replace trapped air inside the pump with pcymene. The liquid was allowed to flow continuously through the loop for at least 20 minutes prior to the experiment in order to remove all the air initially trapped within the bed interstitial spaces and loop piping. The flow was stabilized at the desired flow rate by controlling the main valve lever position. Temperature of the water bath tank was controlled externally.

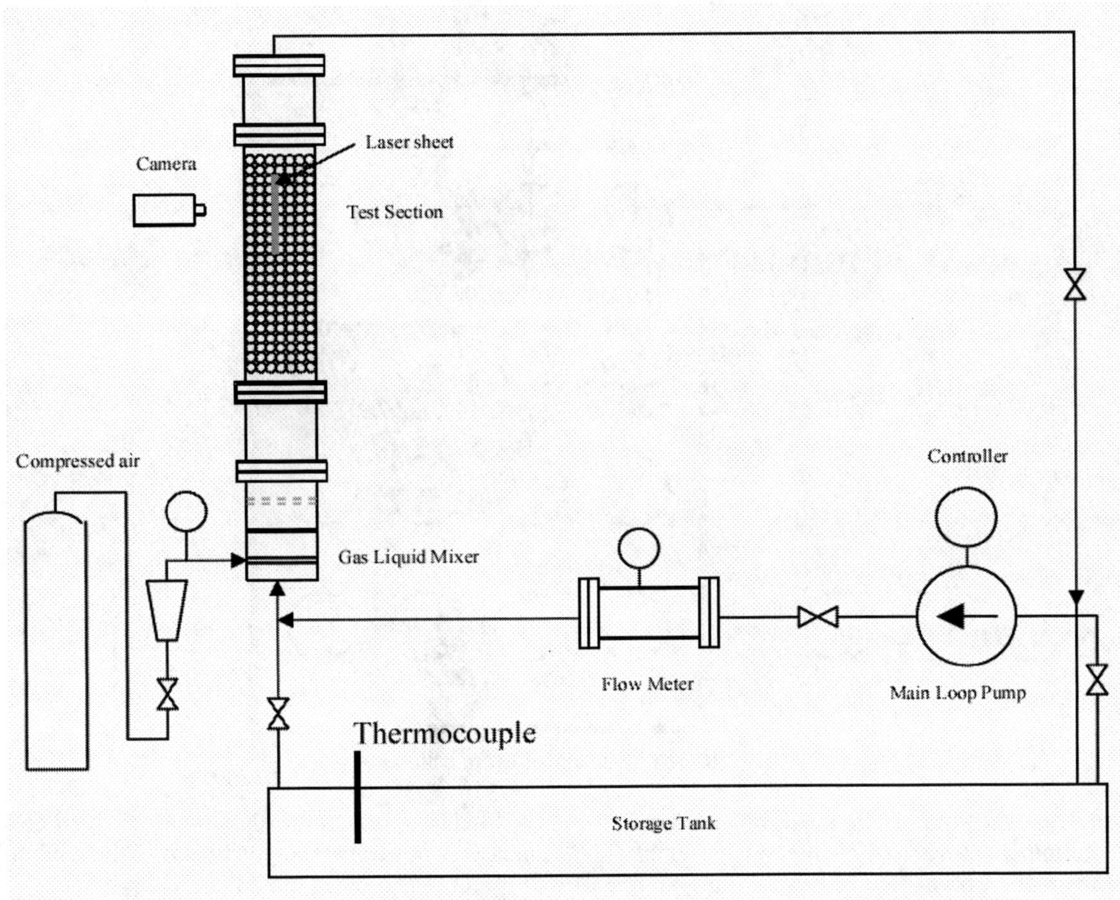

FIGURE 1 . Experimental loop

Calibration

The PIV system was calibrated by using an "in-situ" calibration grid which is located in the core (midplane) of the packed bed near the column's outlet. A thin aluminum plate with wall thickness of 0.7 mm was carved with a circular matrix-like pattern. Distance from center to center of the circular perforations was 2.54 mm in the vertical and horizontal direction with a diameter of 0.9 mm. Figure 2 shows a schematic of the calibration grid.

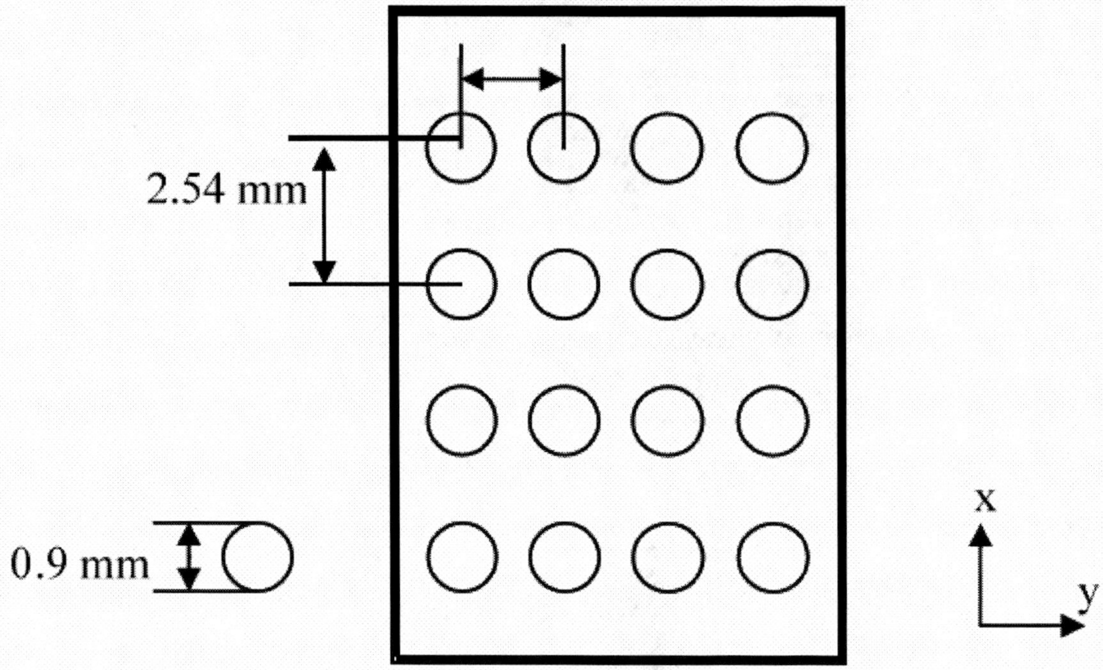

FIGURE 2 . PIV imaging system calibration grid

A high speed/ high resolution camera was attached to a 2-axis motorized translation stage which covered the whole column axial plane (x-axis) and the z-plane. The camera was positioned perpendicular to the calibration grid using the grid as a photographic target. Transparency with the column is only attained when the p-cymene is flowing through it. Therefore, images of the calibration grid are obtained at the front surface of the grid with various f numbers (f/ #) in order to compensate for the depth of field of images. This procedure assures a quantitative proportionality constant between "real world" coordinates and "image" coordinates.

The laser beam was shaped into the form of a thin laser sheet by the use of various, previously noted optical elements. This sheet is directed to coincide with the calibration grid plane, covering the x-y planes of the column. The length of the laser sheet was adjusted, using the appropriate optical elements, in such a way that a portion of the sheet impinges in the calibration grid and extends in the x-plane to the designated interrogation zone. The equivalent viewing area of the camera covered portion of the calibration grid so calibration images are taken at this position. The camera was subsequently moved down to the measurement zone without altering the focal plane by using the precise motorized translation stage. Figure 3 presents a picture of the physical calibration grid within the column. The grid is inserted during the packing procedure and carefully placed at the midplane of the bed. Figure 3 shows the optical accessibility attained when the column is filled with the working fluid. The calibration grid is located in the same plane where the PIV measurements take place.

Data acquisition

The data acquisition process is divided into several data sets. Each data set consisted of 5785 images which correspond to 1.3 sec of elapsed time. The physical conditions for each data set were recorded by the measurement system and by manually recording the readings of the instrumentation. High speed camera recordings were taken at a frame rate of 4700 frames per second (fps). The camera was set to work in a "stand alone" configuration, which sets the internal clock as the base of the timing system. An external trigger synchronized the laser with the camera frame rate. The camera's exposure time was set to 240 μs.

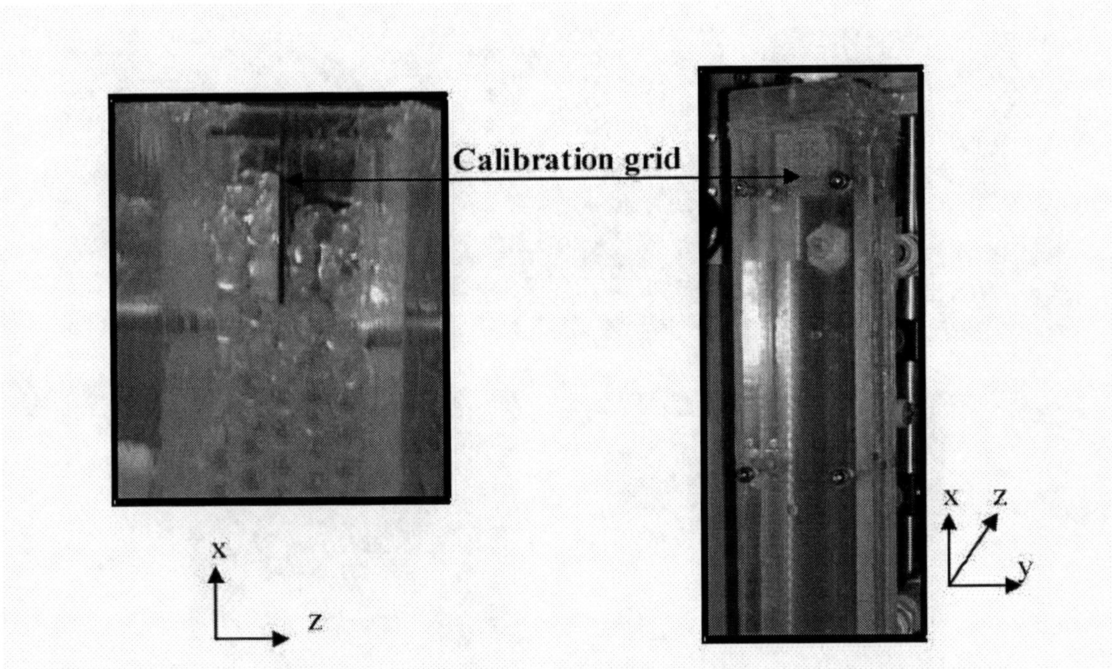

FIGURE 3 . Calibration grid a) Lateral view (the column is filled with air b) 3D view (the column is filled with p-cymene)

RESULTS

This section presents the results and explains the data analysis procedure used. A brief explanation of the problem geometry describes the identified pores in the column. Results of single-phase (only liquid flowing through the column) and two-phase (liquid and gas flowing through the column) flow are presented separately.

Pores

One important parameter in the study of bounded flows is the geometry involved in the problem. In the particular case of packed beds, the interstitial space between the beads is characterized by pores' shapes and sizes. The identification of representative pore geometries near the midplane of the column is an important physical characteristic of the packed bed. In this work, the pore identification is performed by image acquisition at the column's midplane. The use of motorized two-axis position stages makes easy the visual inspection. The geometric distribution of such pores in this experiment can be narrowed to five representative shapes depicted in Figure 4.

In Figure 4 the pore geometry is delimitated by the beads boundaries. It can be observed that the beads do not necessarily have a contact point with other beads in some cases. Small channels of various heights, h are identified within the pore. This variant geometry observed in the pores adds space that enables the fluid (gas or liquid) to flow through it. Inversely, the contact points of neighboring beads represent an obstruction to the fluid which alters its original trajectory. In this work it is intended to show a direct comparison of flow behavior at different pore geometries located in the axial plane of the column.

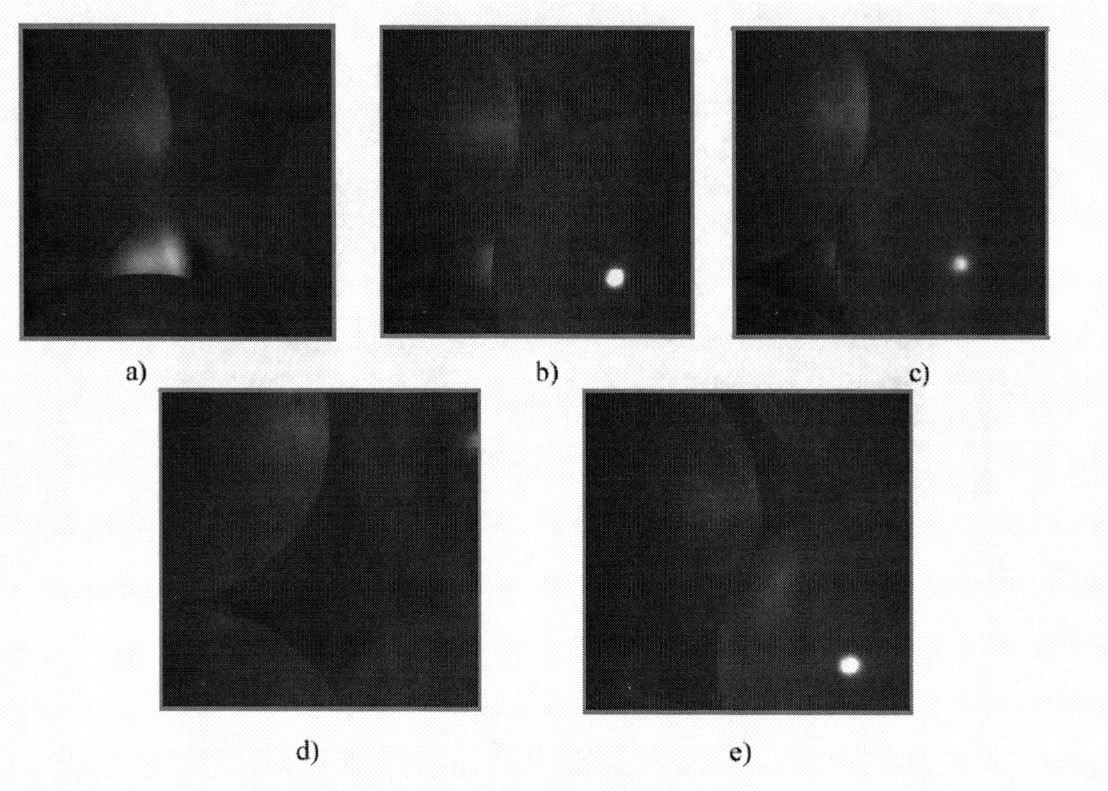

FIGURE 4. Characteristic pore geometry found at the midplane of the packed bed.

3.2 Single phase experiments

The flow characteristics are studied at the selected pore in the midplane of the column. The pore geometry is shown in Figure 5. In this case, the laser sheet plane does not pass through the center of the beads. Figure 5(b) shows a perspective of the selected pore in order to facilitate an appreciation of the detail of the pore geometry.

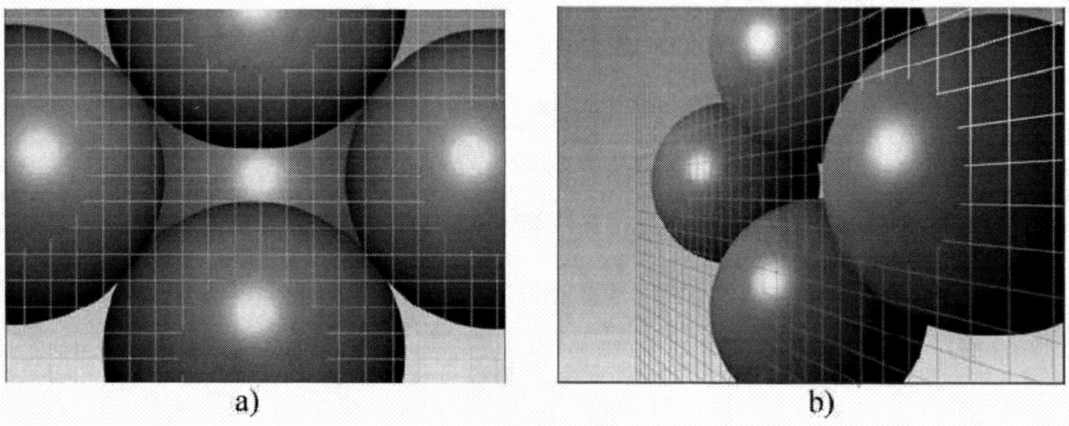

FIGURE 5. Pore geometry at the midplane of the column indicating the cross over of the laser sheet plane a) front view b) perspective.

The obtained data was analyzed by using a PTV algorithm. For this pore, two flow conditions were analyzed, which represent two Reynolds numbers (Re) of 50 and 500. Figure 6 shows the average velocity field in this pore geometry for Re=50. Various streaklines are draw over the velocity vectors to visualize the main flow paths within the pore. It is observed that the average velocity magnitude is higher at the edges of the pore than the magnitude of the vectors within its center. The calculated streak lines show that the flow tends to follow straight lines at the pore edges where the velocity magnitude is higher. The flow smoothly changes direction close to the pore's center.

Figure 7 shows a contour plot of the average magnitude of the velocity vector streamwise component. The maximum measured velocity of 50 mm/s is denoted with a red color. It should be noted that there is a velocity difference of one order of magnitude at the edges of the pore with respect to the center of the pore. Figure 8 shows a contour plot of the spanwise component of the velocity vector as a function of position. The plot shows that this component has positive and negative values of this velocity component. Figure 9 shows the vorticity distribution within the pore. There are higher vorticity values at the edges of the pore than in its center.

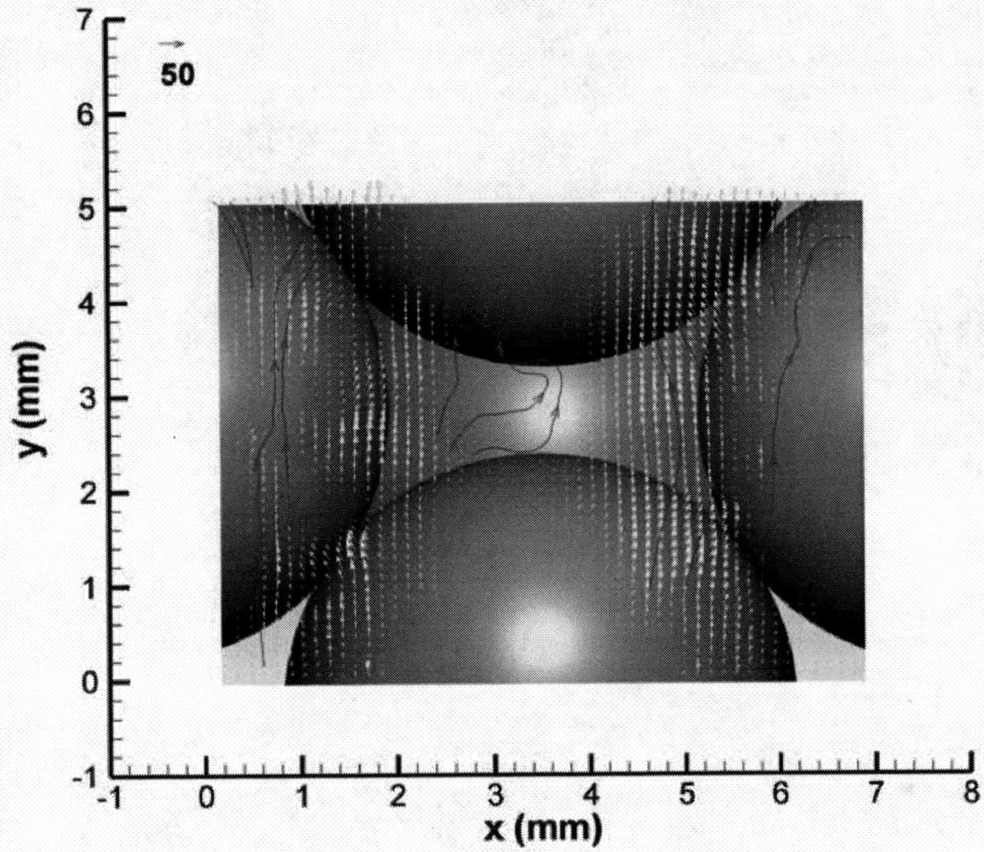

FIGURE 6. Average measured velocity field with streak lines for Re=50. The flow at the edges of the pore tends to follow straight lines; however in the center of the pore where the velocity is decreased the streak lines show a change in the flow direction.

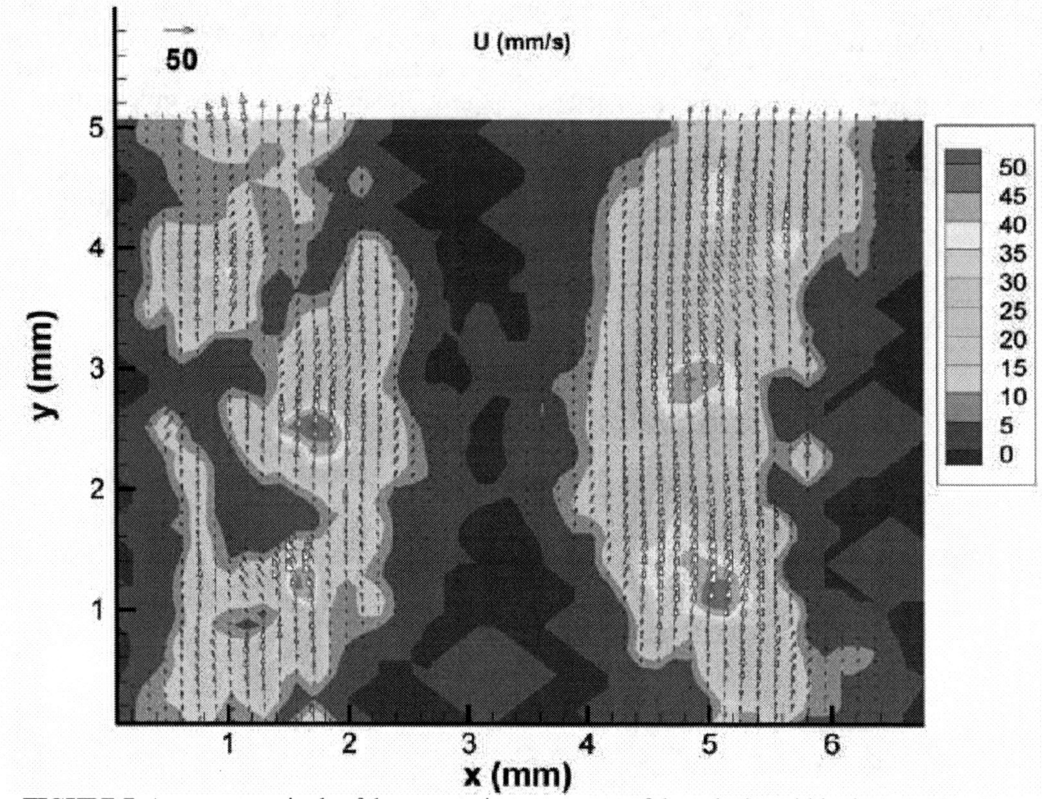

FIGURE 7. Average magnitude of the streamwise component of the velocity within the pore, Re=50.

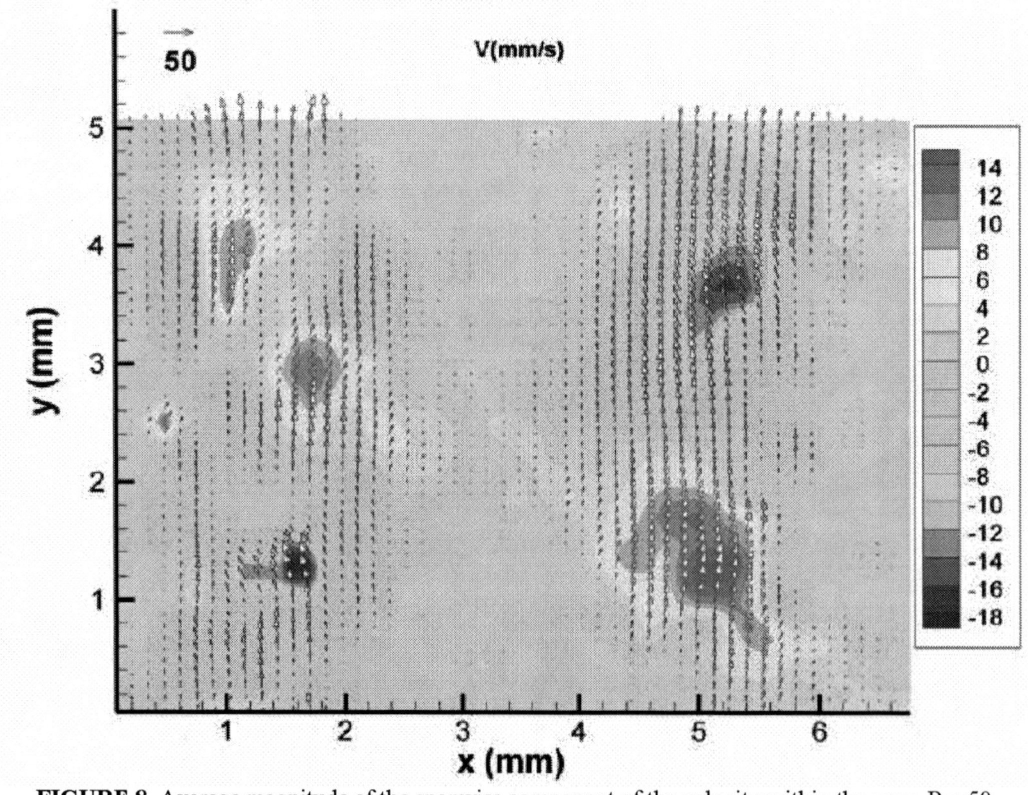

FIGURE 8. Average magnitude of the spanwise component of the velocity within the pore, Re=50.

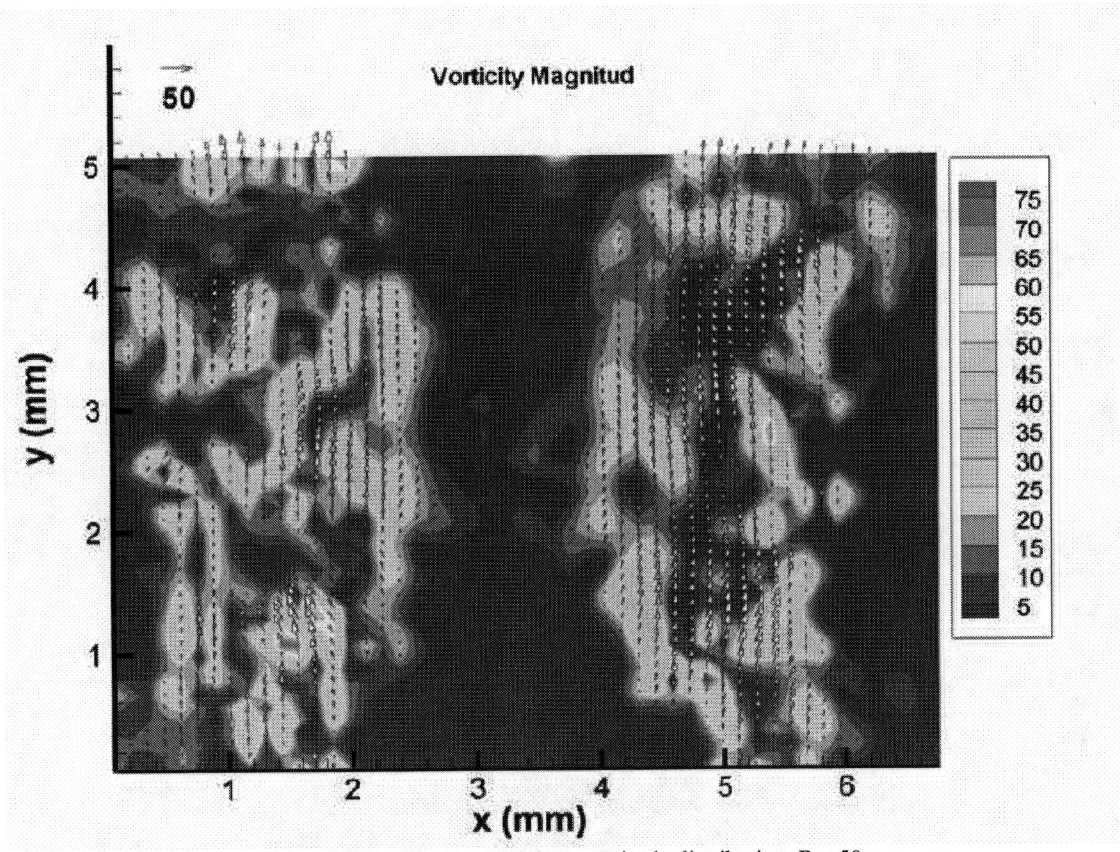

FIGURE 9. Average vorticity magnitude distribution, Re=50.

Figure 10 shows the velocity field for Re= 500. The velocity at the center of the pore is higher by one order of magnitude with respect to the pore edges. The calculated streaklines show drastic changes of direction in this region of the pore. Figure 11 shows evidence of vortex structures at the center of the pore. The velocity gradient increases in this zone and the magnitude of the velocity is decreased by a factor of 100 with respect to the maximum velocity found within the pore. Figure 12 and Figure 13 show the magnitude of the streamwise and spanwise components, respectively, of the velocity vector for Re=500. While, Figure 14 shows the vorticity distribution for Re=500.

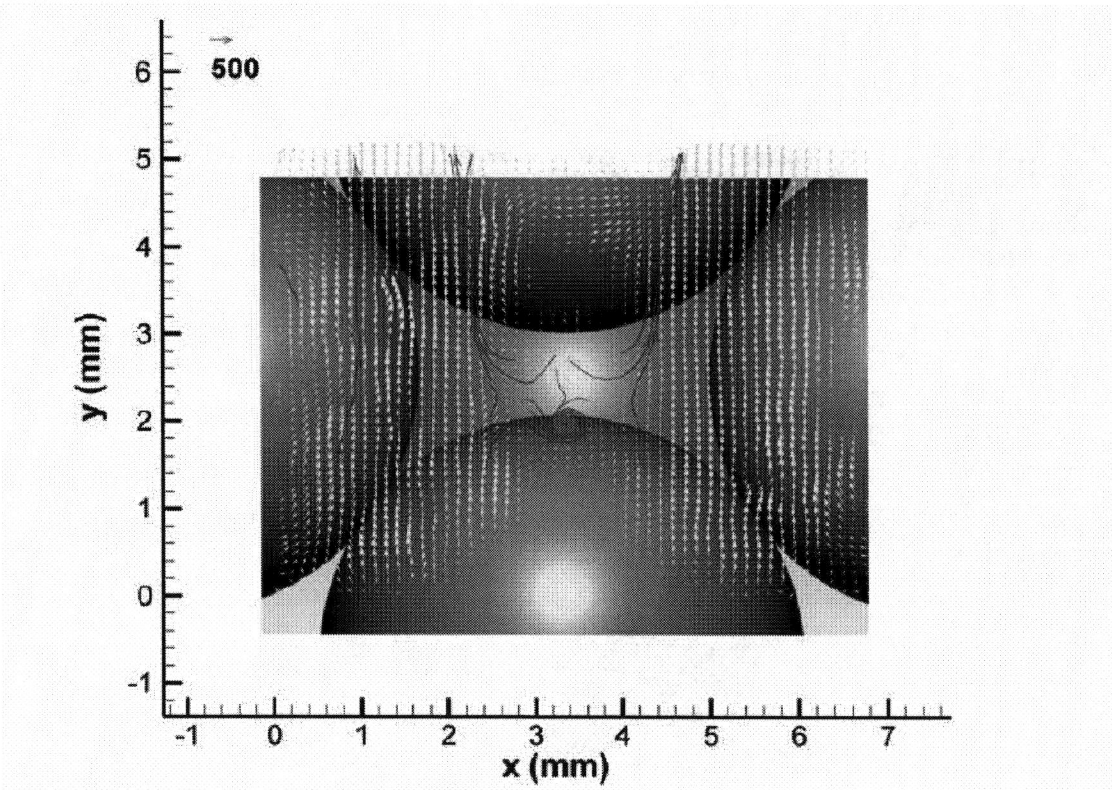

FIGURE 10. Velocity distribution within the pore, Re=500.

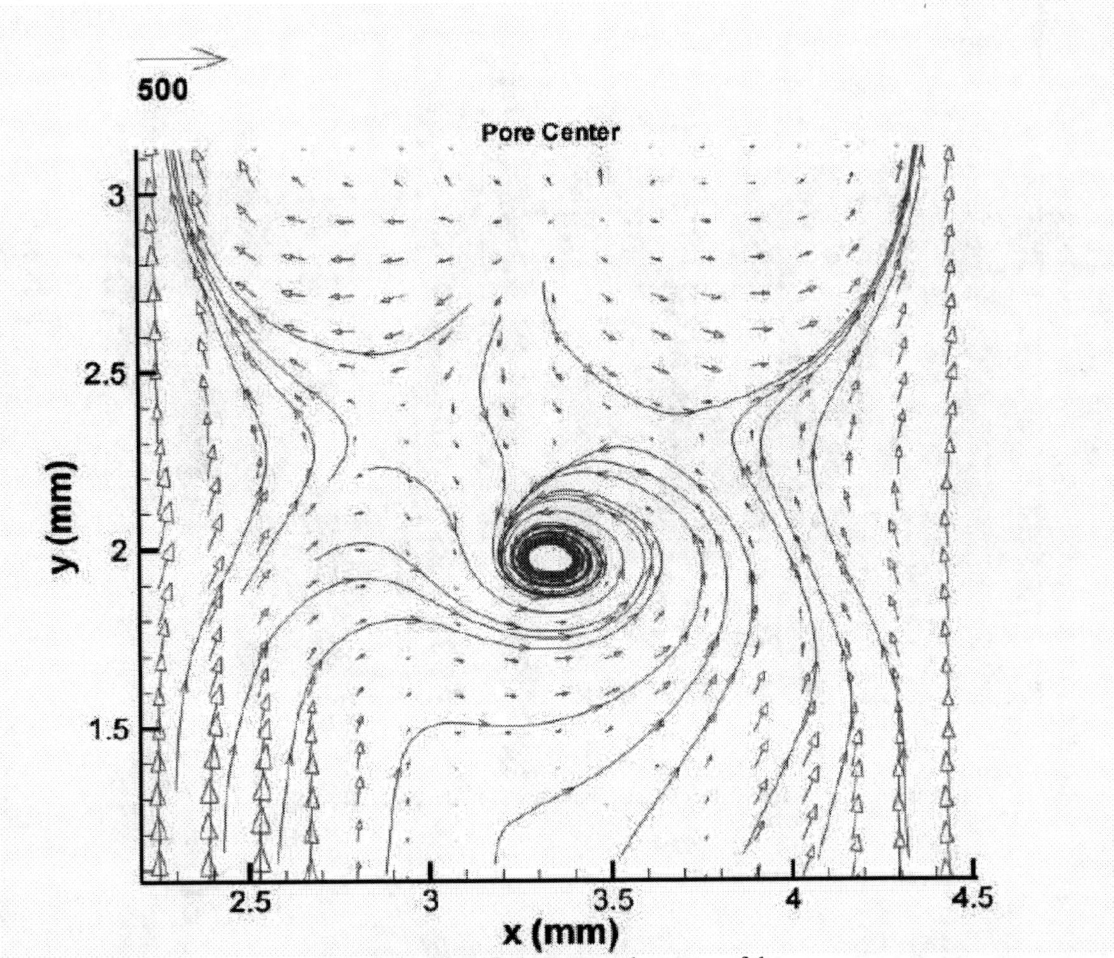

FIGURE 11. Identified vortex at the center of the pore

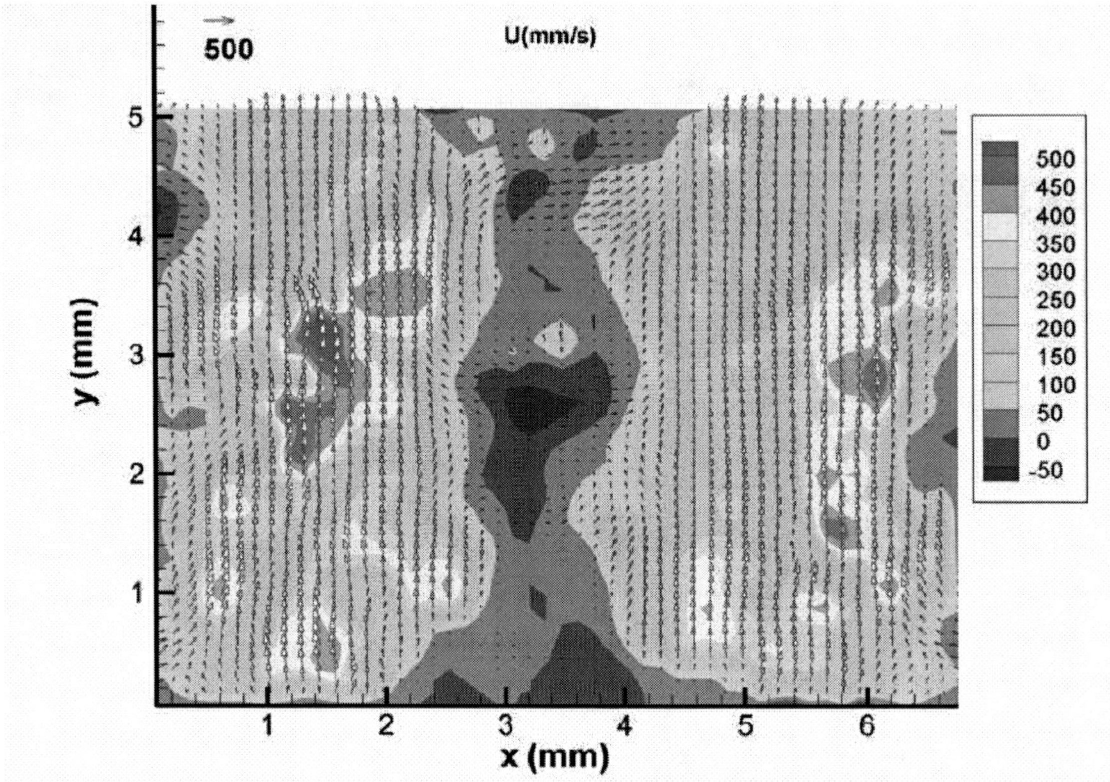

FIGURE 12. Streamwise velocity distribution within the pore, Re=500.

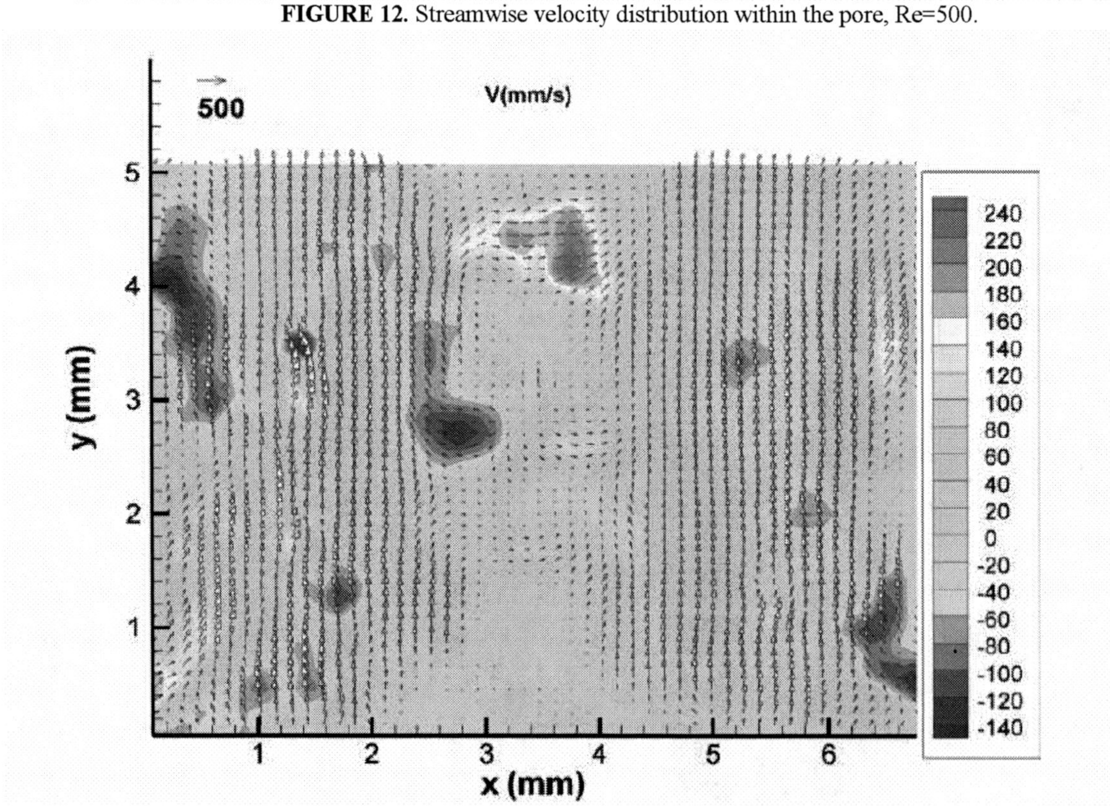

FIGURE 13. Spanwise velocity distribution within the pore, Re=500.

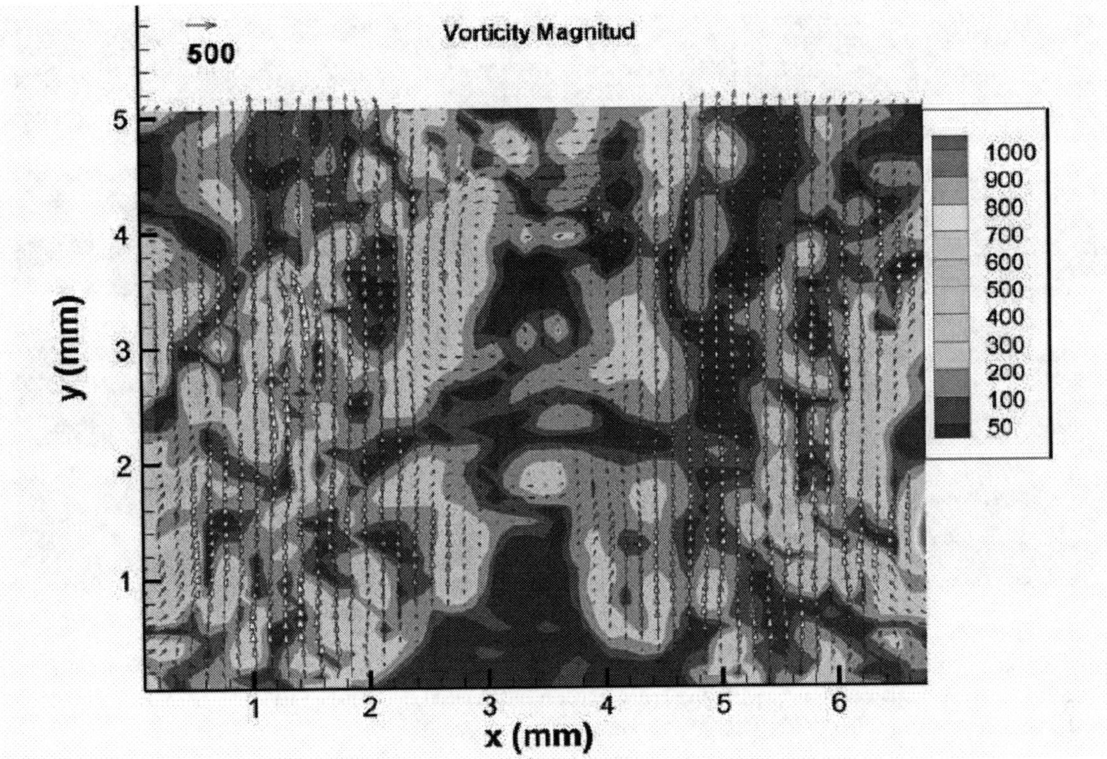

FIGURE 14. Vorticity magnitude distribution within the pore, Re=500.

Two-phase flow experiments

One of the goals of this study was to investigate the extent of the transport of fluid (liquid and gas) material through the porous medium. The effect of the local structure of the void space, at the single pore level, on the local and global transport characteristics of the porous medium is an important parameter. Flow visualizations within the packed bed are presented. The two-phase condition is achieved by injecting air through the air inlet port showed in the experimental setup (Figure 1). For these experiments, the gas rate was constant and the liquid flow rate was varied using the main loop valve. The camera equivalent viewing area was also increased to 13 x 10 mm. This increment was necessary to compensate the apparent decrease in detected vectors due to bubbles covering large areas of the pore. In this way, the probability of detecting valid vectors increased due to the increase in available pores within the image even when large bubbles covered the pore area. Two measurement techniques are employed to measure liquid and gas velocity within pores: Particle Tracking Velocimetry (PTV) and Shadow Particle Image Velocimetry (SPIV). The former is a variant of the PIV technique where the illumination of the target particle is performed by using backlight illumination. The shadow projected by the target into the imaging device is recorded and then tracked using PTV algorithms.

Bubbly flow

The two phase flow experiments were performed with a constant gas flow rate. Various liquid flow conditions were tested. The Reynolds number for the tested flow cases ranges from Re=85 to Re=500.Figure 14 shows an image sequence of typical PIV and SPIV images for a Re=85. It is noticed that large bubbles were formed. The bubble shape adjusted during its journey through the tortuous path caused by the packed bed. Figure 15 shows an enhanced image sequence of typical PIV images for Re=85.

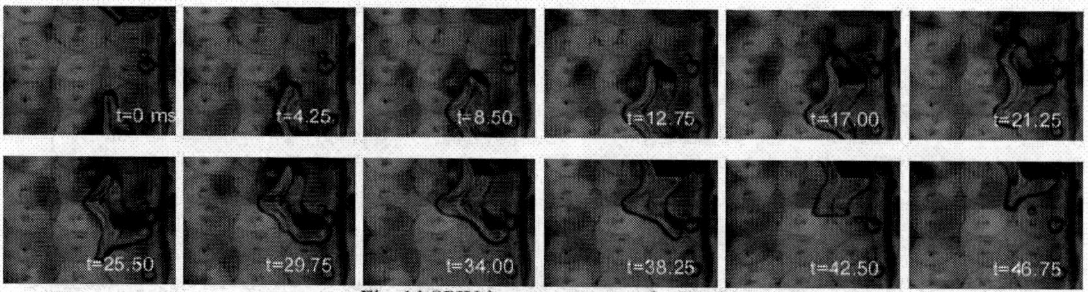

Fig. 14 SPIV images sequence for Re=85

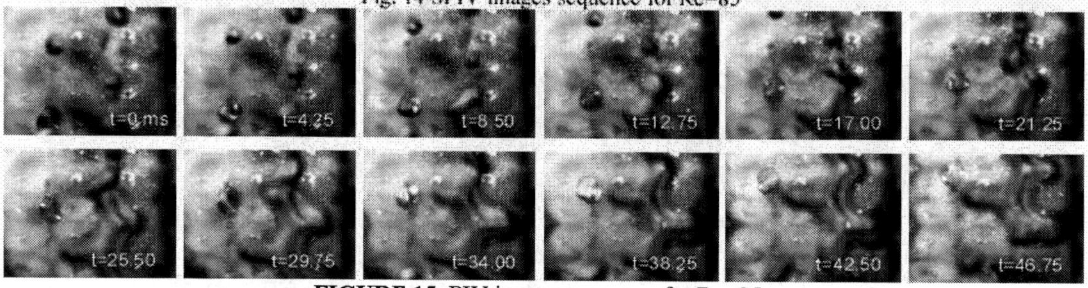

FIGURE 15. PIV images sequence for Re=85

Figure 16 and Figure 17 show typical d PIV techniques respectively for image sequences from SPIV Re=337. The gas flow rate was kept constant, with the only variable being the liquid flow rate. It was observed that the bubble size decrease and bubble coalescence was reduced.

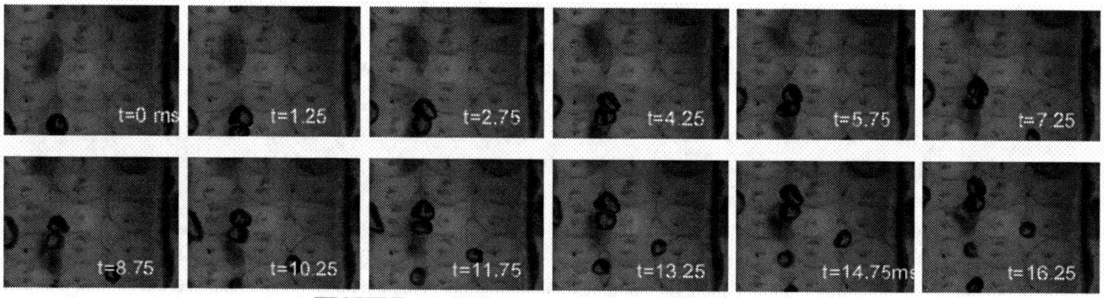

FIGURE 16. SPIV images sequence for Re=337

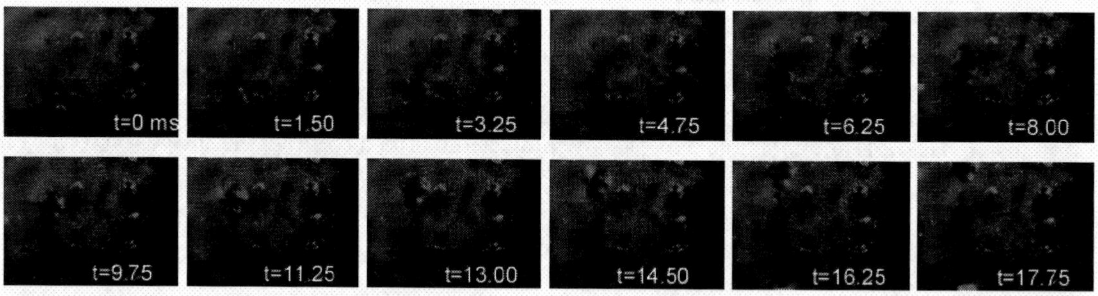

FIGURE 17. PIV images sequence for Re=337

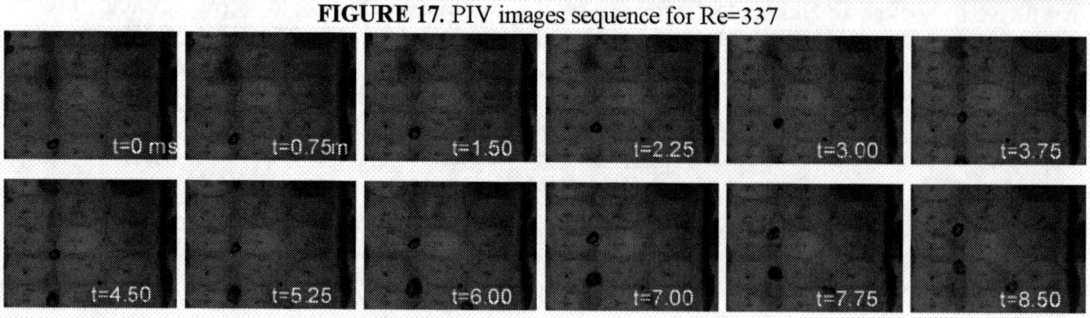

FIGURE 18. SPIV images sequence for Re=494

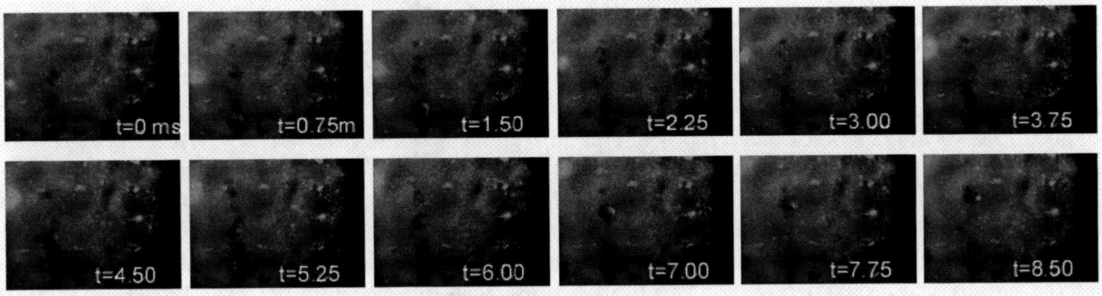

FIGURE 19. PIV image s sequence for Re=494

The velocity magnitude for various flow conditions are shown in Figure 20. In this case, velocity measurements were performed for single and two-phase flow with the following procedure. Liquid flow was stabilized to the desired condition and PIV measurements are performed under this condition. Then, air was introduced to the column at a constant gas flow rate and a data set was acquired with the two-phase flow condition keeping the liquid flow rate constant. This procedure was repeated for each desired flow condition.

The first row in Figure 20 show the measured magnitude of the streamwise component of the velocity vector for th present in the column. It may be observed that channeling is present in the column.The magnitude of the velocity varies from channel to channel and its gradient increase with Reynolds number. The second row in Figure 20 presents the velocity magnitude of the liquid phase for a two-phase flow condition. Here, gas and liquid compete for interstitial space. The average velocity for the liquid phase shows its higher change at a Re=337 with respect to the case of single phase flow. The third row in Figure 20 shows the magnitude of the streamwise component of the velocity for the gas phase in the two-phase flow condition. The gas exhibits comparable velocities to the liquid phase for this particular gas flow rate (0.6 l/min). It is noticed that a higher number of spots with low velocities (dark blue color) are observed for the gas phase. This behavior is due to the constant shape deformation of the bubble when an obstacle is found in its path.

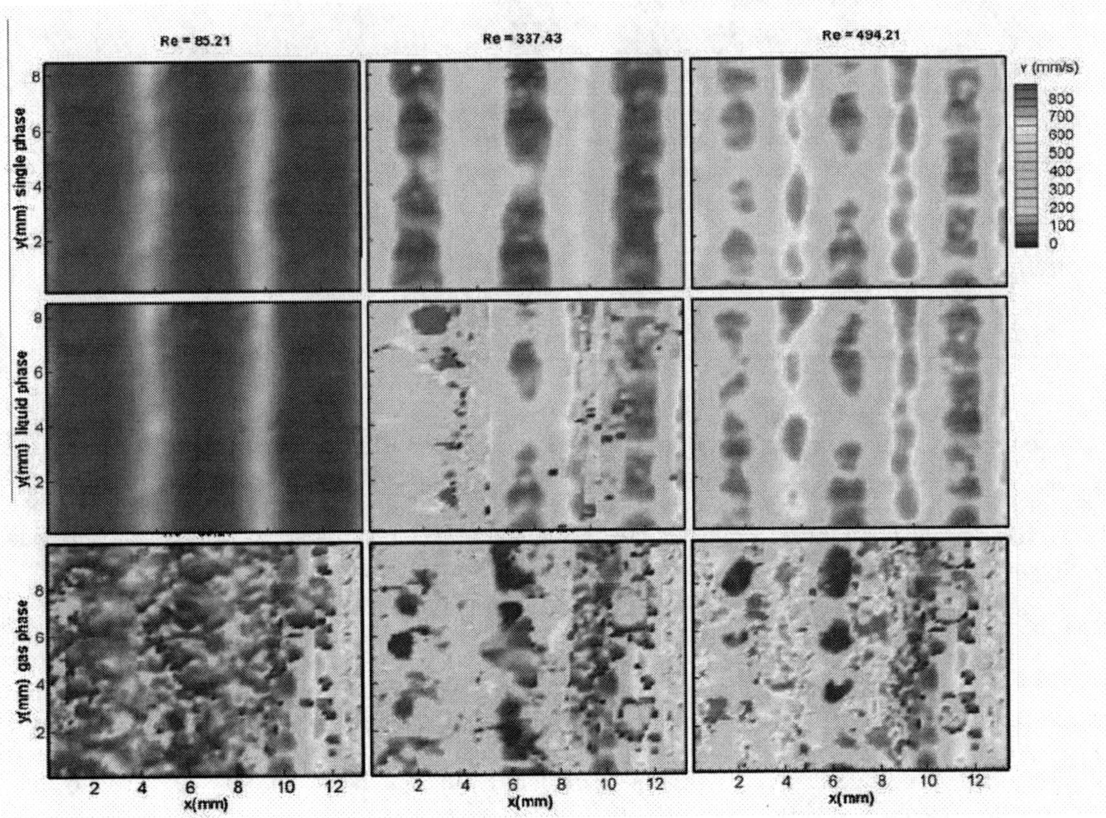

FIGURE 20. Average streamwise velocity contour plots of liquid and gas phase for various Re.

Figure 21 shows the average magnitude of the spanwise component of the velocity for single and two-phase flow at various Reynolds number. It is observed that the liquid phase exhibits a noticeable increase in area of negative magnitude (blue color) with respect to the single phase case. This behavior elucidates the systematic change of direction of the liquid phase when gas is present in the column.. The gas phase shows a higher relative magnitude of this velocity component of the three studied cases. It is also more noticeable for higher Reynolds numbers. The number of locations having velocity magnitudes above 140 mm/s increases (red color) with respect to the liquid phase which elucidates a higher momentum transfer in the y-axis direction of the gas phase.

Figure 22 and 23 shows the average velocity profile for the streamwise component of the velocity vector for the liquid and gas phase under a two-phase flow condition for various Reynolds numbers. The profile exhibits a c cosine shape with maximums at about 3.5 mm, 7.5 mm and 11 mm from the center of the column. This velocity profile illuminates the presence of a channeling effect in the axial direction of the column. The channels are identified for the following x-axis positions: 0-1, 2-5, 6.5 -9 and 10.5-12 mm (see Figure 26). It should be noted that the average velocity in pores between channels can be as high as 100 mm/s. The gas velocity profile shows a decrease in gas velocity in the streamwise direction.

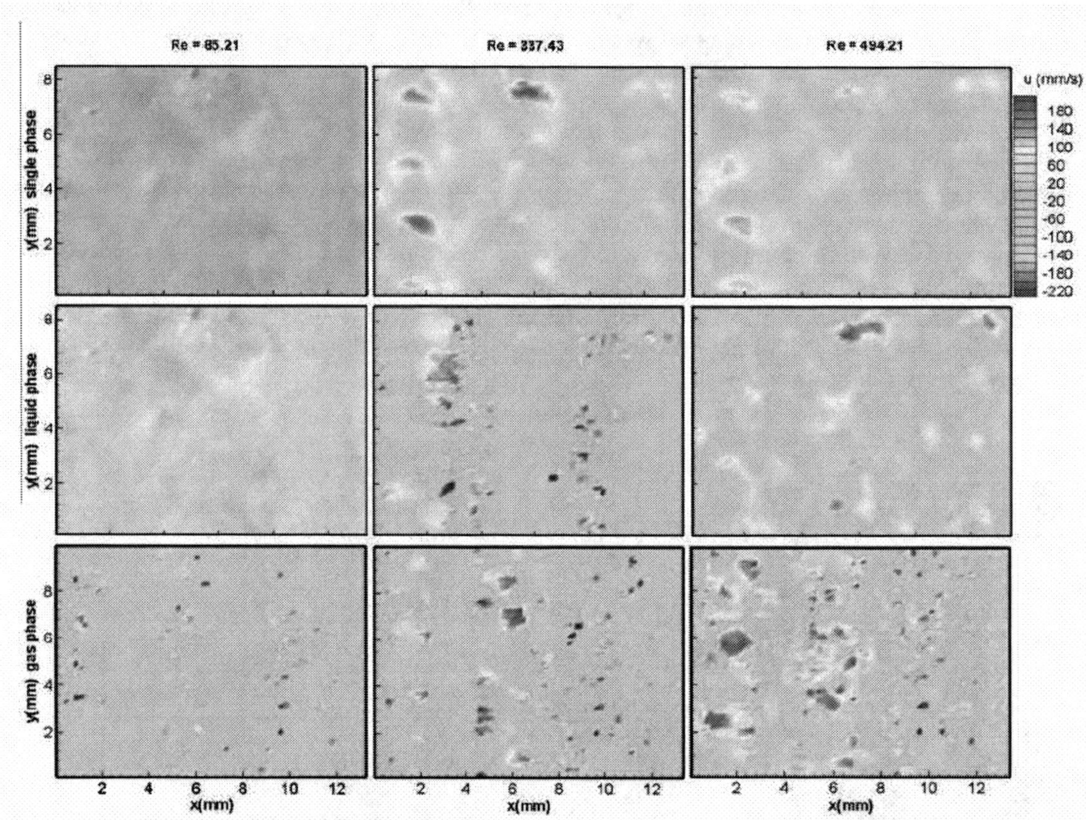

FIGURE 21. Average spanwise velocity contour plots of liquid and gas phase for various Re.

The cosine shape is maintained for low Reynolds numbers. Nevertheless, the velocity profile shape is modified when the gas velocity increases. The velocity of the gas phase in pores between channels exhibit a zero velocity magnitude for Re=337 and Re=494, which contrasts the result of the liquid phase. The reason for this behavior is that gas flow rate was kept constant and only the liquid flow rate was varied during the experiment. The bubble size depends on liquid velocity since at lower flow rates bubble coalescence increases; causing bubble size to increase. This produces larger bubbles at low Reynolds numbers. Bigger bubbles have to deform and change direction once they reach the boundaries of the pore. The change in momentum direction causes the velocity magnitude of the streamwise component to exhibit minimum values. Figure 24 and Figure 25 show the average magnitude of the velocity vector in the spanwise direction for liquid and gas, respectively. The magnitude of the spanwise component in the gas phase is increased for zones between channels and opposes the trend of the streamwise direction.

Figures 26 to 29 show the averaged velocity fluctuations for liquid and gas phase in the streamwise and spanwise directions, respectively. The fluctuating velocities for the gas phase increase dramatically for low Re number in both the streamwise and spanwise directions.

Figure 30 and 31 show the averaged dimensional Reynolds stress for liquid and gas phase respectively. It is observed that for the liquid phase this parameter behaves similar to flow in a channel flow. The Reynolds stress crosses zero and then has a negative magnitude. For a bounded channel flow the Reynolds stress is zero at the center of the channel. This fact may be used to locate the center of the "virtual" channel flow which is identified at a distance of 3.9 mm and 5.7 mm in the x-axis direction. Figure 31 shows a different trend for the gas phase. The magnitude of the stresses increases but it does not show a particular trend except that fact those higher values are located in coincidence with the identified "virtual" channel center. This behavior may be due to the velocity magnitude of the gas phase at the center of the virtual channel where the momentum is in nearly one direction.

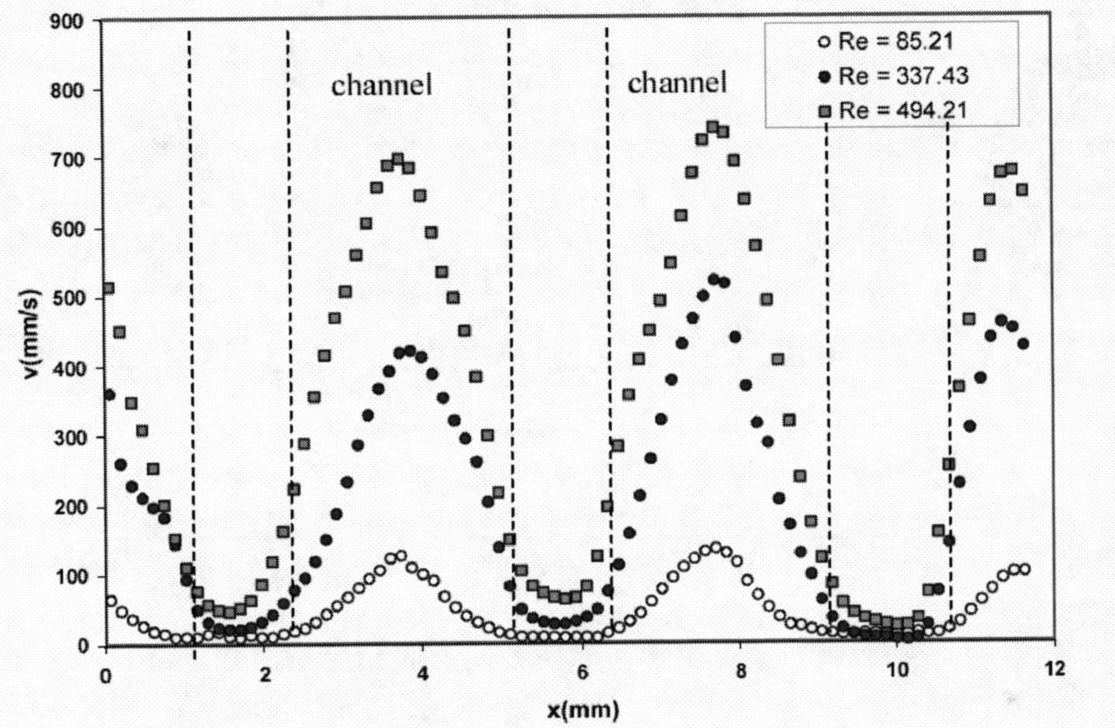

FIGURE 22. Liquid streamwise velocity vs position for various Re

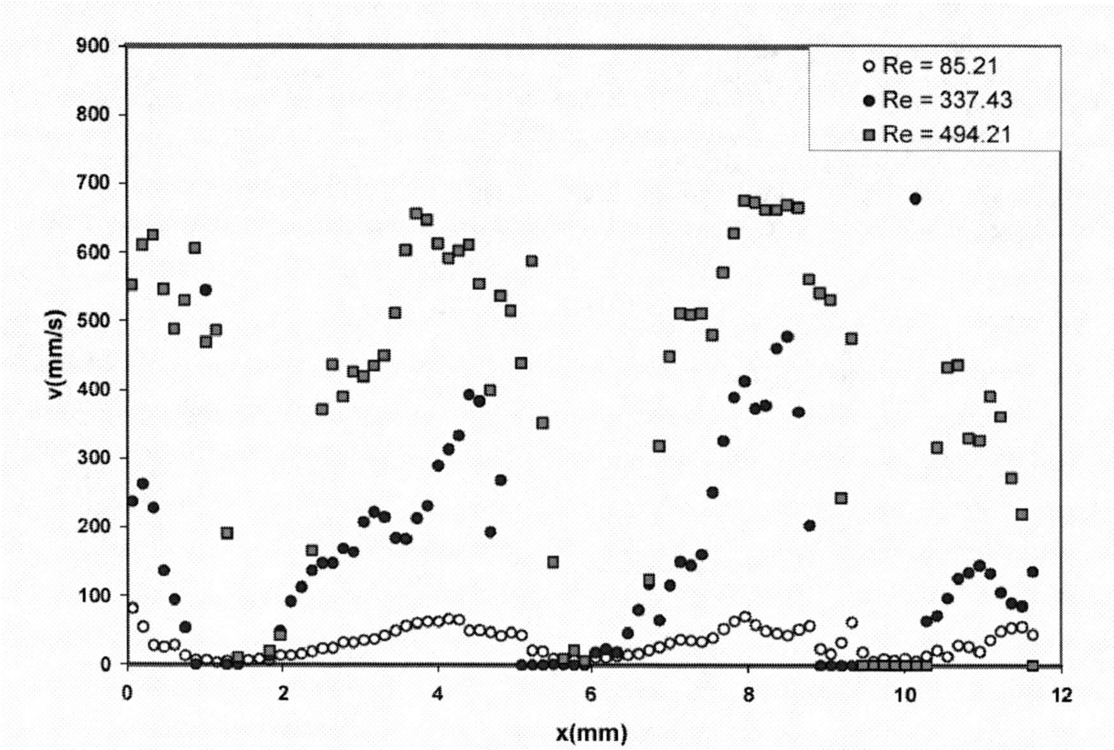

FIGURE 23. Gas streamwise velocity vs position for various Re

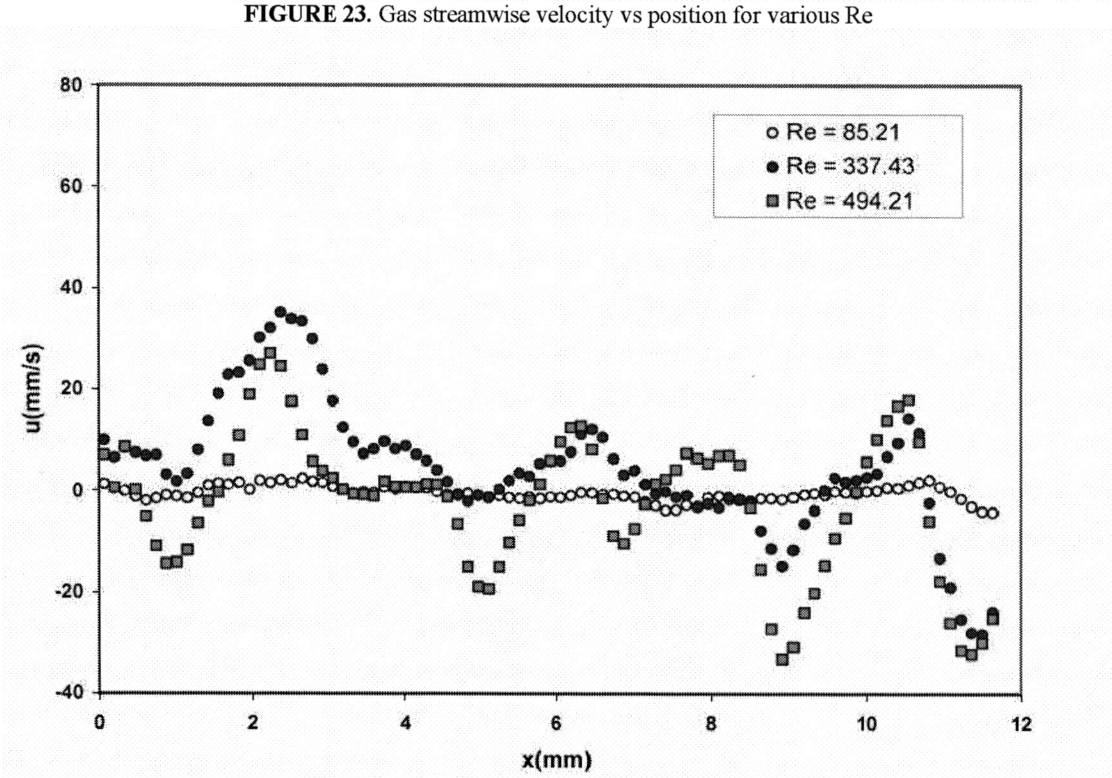

FIGURE 24. Liquid spanwise velocity vs position for various Re

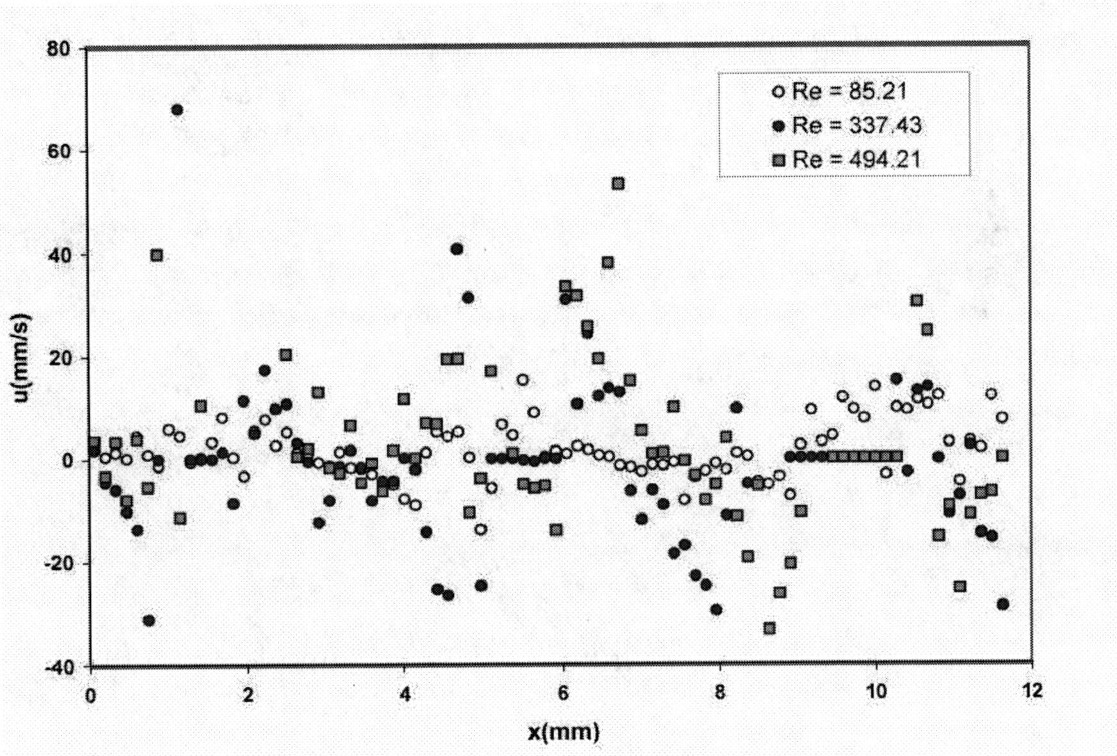

FIGURE 25. Gas spanwise velocity vs position for various Re

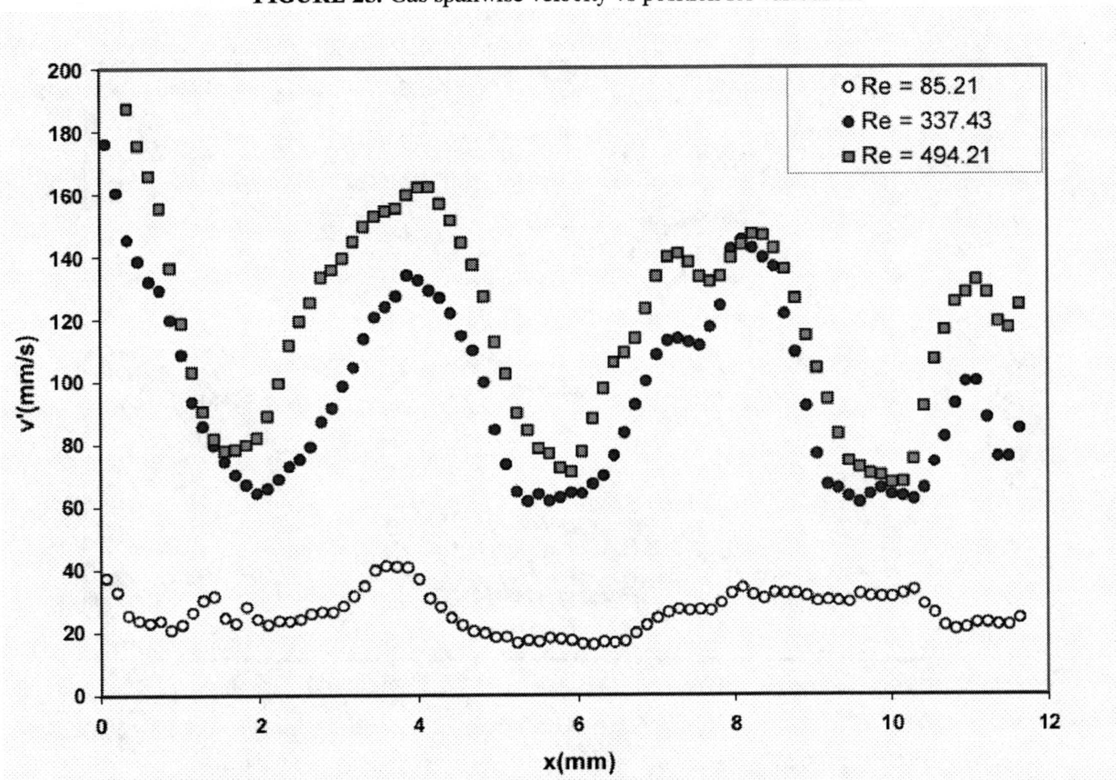

FIGURE 26. ous Re Liquid streamwise fluctuating velocity vs position for vari

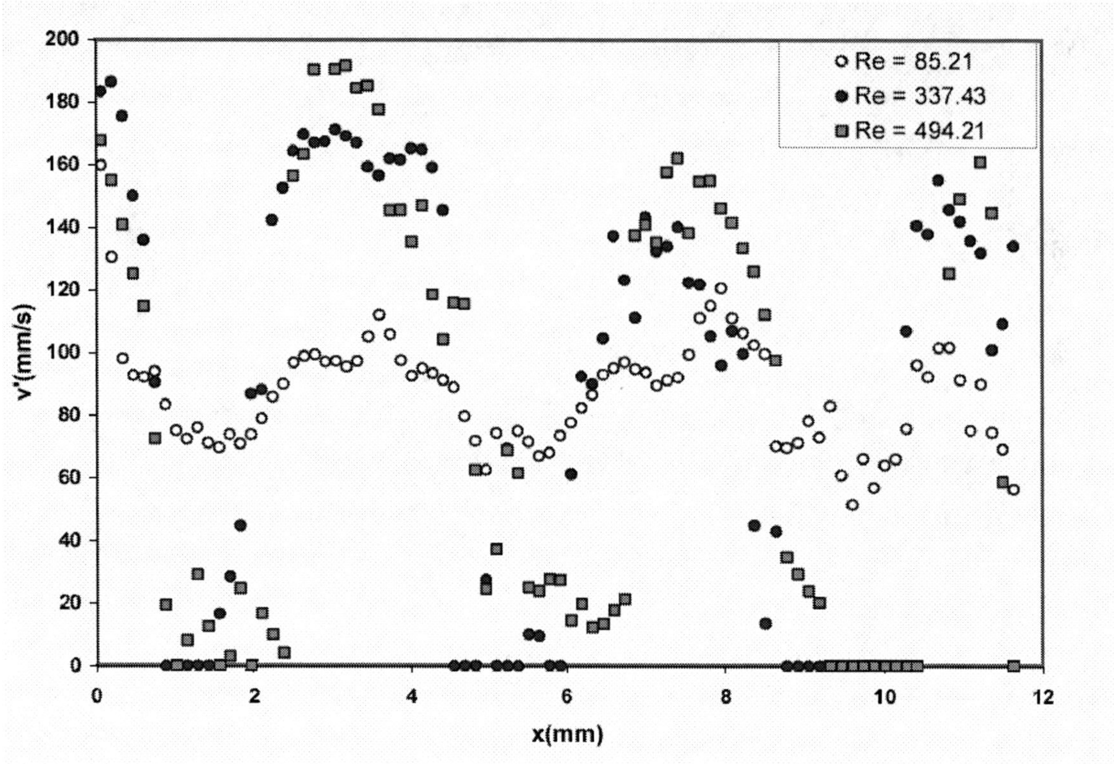

FIGURE 27. Gas streamwise fluctuating velocity vs position for various Re

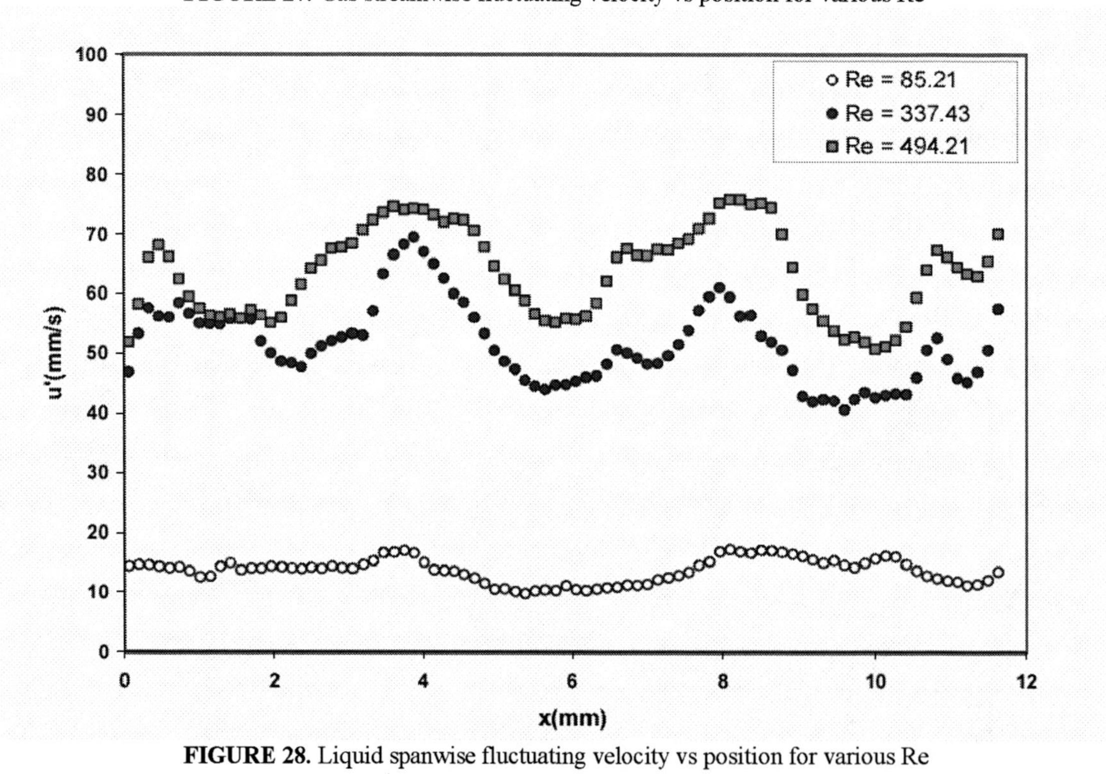

FIGURE 28. Liquid spanwise fluctuating velocity vs position for various Re

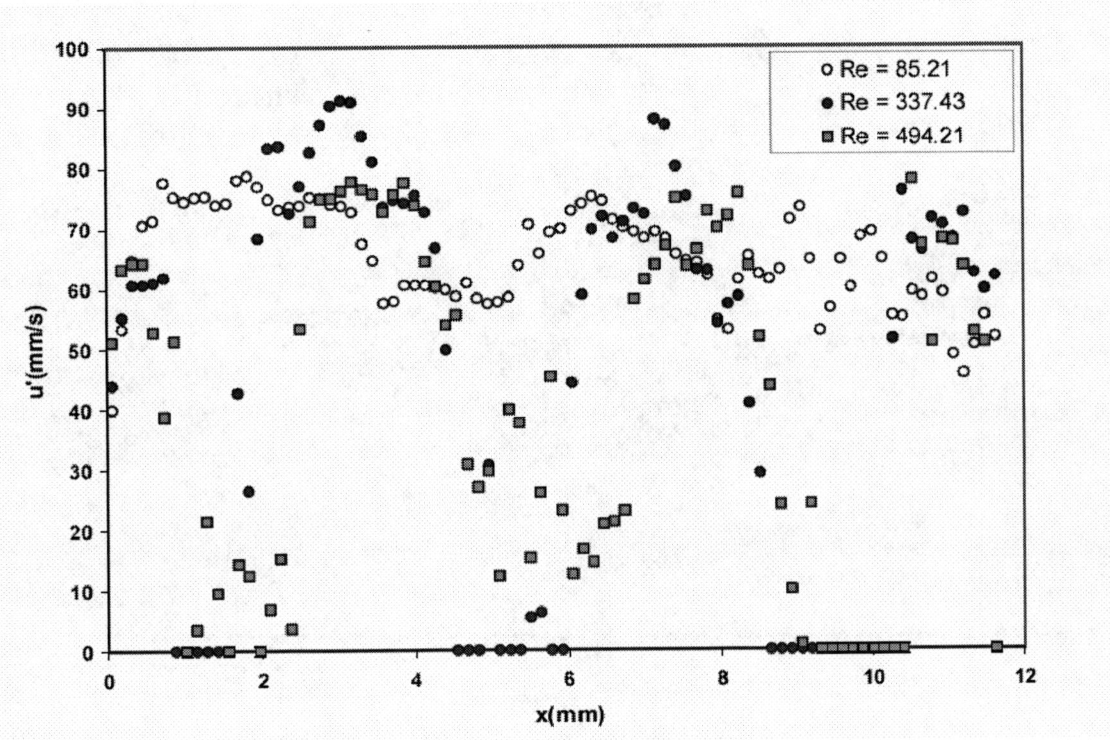

FIGURE 29. Gas spanwise fluctuating velocity vs position for various Re

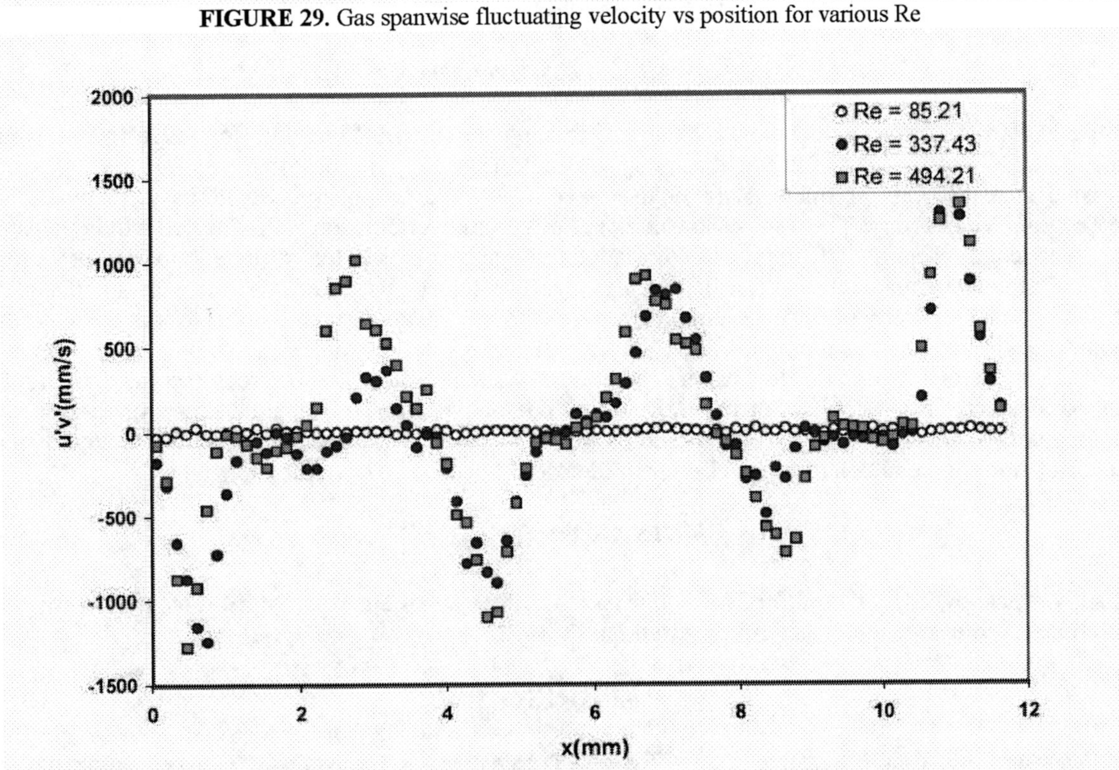

FIGURE 30. Liquid Reynolds stresses vs position for various Re

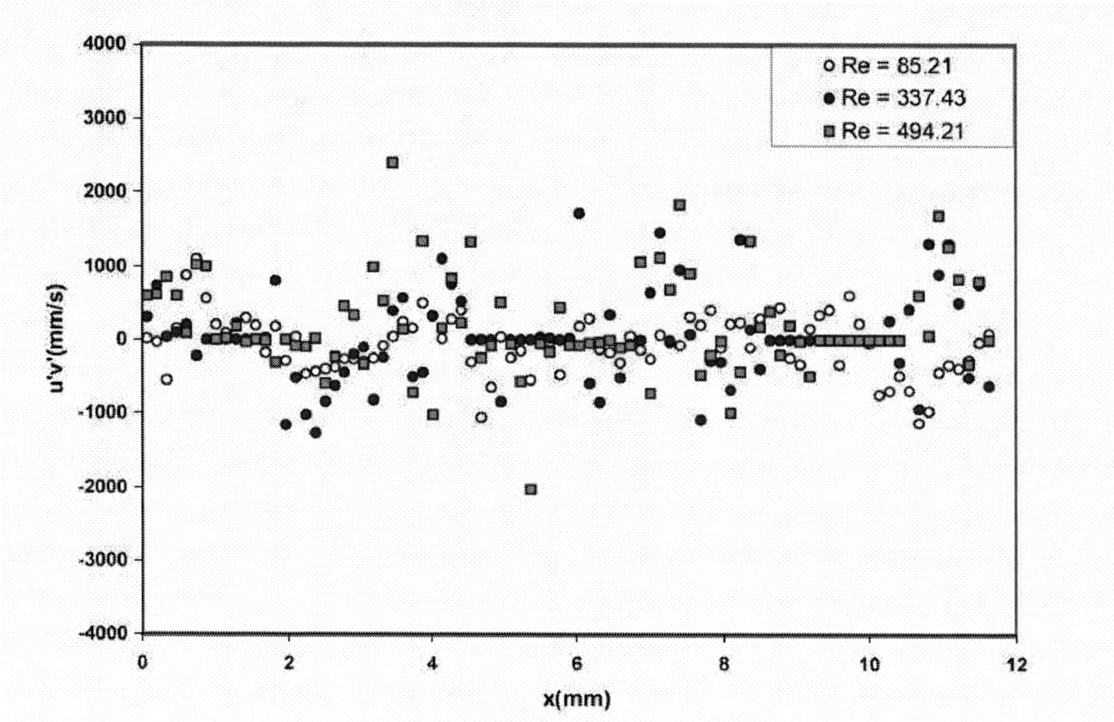

FIGURE 31. Gas Reynolds stresses vs position for various Re

4. CONCLUSIONS

The use of Particle Image Velocimetry techniques, in conjunction with Matching Refractive Index techniques, makes possible the measurement of full field velocity distributions at the pore scale in the midplane of a packed bed. Non-invasive measurements of full-velocity fields under single and two-phase conditions were possible using Particle Tracking Velocimetry. The results show evidence of several recirculation zones located within pores. To the authors' knowledge, there is no empirical correlation that accounts for the recirculation within pores. The measurement of velocity gradients and full field velocity vectors may be used to calculate residence time of a fluid particle. The calculated average residence time for a pore with recirculation, with respect to a pore of similar size and geometry without recirculation, proved to be five times greater on average. Turbulent parameters were calculated for single and two-phase flow cases. There are noticeable differences between the gas and liquid phase turbulent parameters. There is an increment in the magnitude of the average Reynolds stress for the gas phase at specific locations of the bed. The study of presented turbulent parameters may contribute to the improvement of existent correlation for prediction of flow behavior inside packed beds.

ACKNOWLEDGEMENTS

The authors acknowledge the fruitful discussions and the directions of Dr. Brian Motil. This project was founded by the National Aeronautics and Space Administration.

REFERENCES

1. B.C. Bailey and M. Yoda, An Aqueous Low-Viscosity Density- and Refractive-Index Matched Suspension System, Experiments in Fluids, Vol. 35, pp. 1-3, 2003.
2. R. Budwig , Refractive Index Matching Methods for Liquid Flow Investigations, Experiments in Fluids, Vol. 17, pp. 350-355, 1994.
3. R.C Chen,. and L.S. Fan, Particle Image Velocimetry for Characterizing the Flow Structure in Three- Dimensional Gas-Liquid-Solid Fluidized Beds, Chemical Engineering Science, Vol. 47, pp. 3615-3622, 1992.
4. M.M. Cui, and R.J. Adrian, Refractive Index Matching and Marking Methods for Highly Concentrated , Solid-Liquid Flows, Experiments in Fluids, Vol. (22), pp. 261-264, 1997.

5. A.Dybbs and R.V. Edwards, An Index Matched Flow System for Measurement of Flow in Complex Geometries, Proc. Intl. Symp. on the Application of Anemometry to Fluid Mechanics, pp. 171-184, Lisbon, Portugal, 1984.
6. S.J Haam,. R.S. Brodkey, I. Fort, L. Klaboch, M. Placnik and V. Vanecek, Laser Doppler Anemometry Measurements in a Index of Refraction Matched Column in the Presence of Dispersed Beads, Part I, International Journal of Multiphase Flow, Vol. 26, pp. 1401-1418, 2000.
7. F. Hendriks, and A. Aviram, Use of Zinc Iodide Solutions in Flow Research, Rev. Sci. Instrum, Vol.53, pp. 75-78, 1982.
8. A.P. Leis, , S. Schlicher and H.Franke, Optically Transparent Porous Medium for Nondestructive Studies of Microbial Biofilm Architecture and Transport Dynamics, Applied and Environmental Microbiology, Vol. 71, pp. 4801-4808, 2005.
9. T.L. Narrow, , M. Yoda and S.I. Abdel-Khalik, A Simple Method for the Refractive Index of Sodium Iodide Aqueous Solutions, Experiments in Fluids, Vol. 28, pp. 282-283, 2000.
10. T.T. Nguyen, , Y. Biadillah, R. Mongrain, J. Brunette, J.C Tardif and O.F. Bertrand, A Method for Match the Refractive Index and Kinematic Viscosity of a Blood Analog for Flow Visualization in Hydraulic Cardiovascular Models, Journal of Biomedical Engineering, Vol. 126, pp. 529-535, 2004.
11. M.A. Northrup, , T.J. Kulp and S.M. Angel, Fluorescent Particle Image Velocimetry: Application to Flow Measurements in Refractive Index-Matched Porous Media, Applied Optics, Vol. 30, pp. 3034-3040, 1991..
12. M. Maroni and Cushman, J.H. Three Dimensional Particle Tracking Velocimetry Studies of the Transition from Pore Dispersion to Fickian Dispersion for Homogeneous Porous Media, Water Resources Research, Vol. 37 (4), pp. 873-884, 2001.
13. B.J Motil, V. Balakotaiah, and Y. Kamotani, Gas-Liquid Two-Phase Flow Through Packed Beds in Microgavity .
14. B.G. Pokusaev, S.P. Karlov and I. Shreiber, Immersion Tomography of a Gas-Liquid Medium in a Granular Bed, Theoretical Foundations of Chemical Engineering, Vol. 38, pp. 3-8, 2004.
15. C. Stoots, , S. Becker, K. Condie, F. Durst and D. McEligot, A Large-Scale Matched Index of Refraction Flow Facility for LDA Studies Around Complex Geometries, Experiments in Fluids,Vol.30, pp. 391-398, 2001.
16. M. Stohr, , K. Roth and B. Jahne, Measurement of 3D Pore-Scale Flow in Index-Matched Porous Media, Experiments in Fluids, Vol. 35, pp. 159-166, 2003.
17. O. Uzol, Y. C. Chow, J. Katz and C. Meneveau, Unobstructed Particle Image Velocimetry Measurements within an Axial Turbo-Pump using Liquid and Blades with Matched Refractive Indices, Experiments in Fluids, Vol. 33, pp. 909-919, 2002.
18. M.W. Wedd, Determination of PSD Using Laser Diffraction, www.erpt.org/032Q/Wedd-00.htm, 2003.

Flame Measurement and Combustion Diagnoses with Spectrum Analysis

Xiaoshu Cai* Zhihai Cheng* Shimin Wang§

**College of Power Engineering, Shanghai University for Science & Technology*
Shanghai 200093, China
§ The Key Laboratory of Clean Coal Power Generation and Combustion Technology Ministry of Education
Southeast University, Nanjing 210096, China

Abstract. With the spectroscopic technique presented in this paper, a CCD spectrometer is employed to acquire flame emission spectra, from which both the temperature and emissivity of flames can be obtained as well as some characteristic spectral lines correlated with the fuel type and the combustion state. As shown by the experiment results, the shape of flame emission spectrum curve varies with the combustion state and thus may provide additional information for combustion diagnostics.

Key words: flame emission spectra, characteristic spectral line, combustion diagnostics

INTRODUCTION

In recent years, with the rapid economic development and rise in people's living standard, there has been great increase in civil electric power consumption and subsequent massive accession of power sets, among which thermal power sets still predominate in both capacity and amount. There are more varieties of boiler structures and combustion modes with the increase in boiler capacity; in addition, further demands for combustion control arise from more and more strict environmental standards for pollutant emission levels. Therefore, reliable means of detection and corresponding theoretical criterions for combustion status have to come into possession for effective and precise combustion control strategies. However, the probe units commercially available at present can only provide on-off signals, indicating presence or absence of flames. Measuring techniques capable of investigating flame characteristics comprehensively in principle as well as finding industrial application remain unseen in literature[1]. Current devices and methods commonly used for combustion monitoring mainly include: combustion diagnostics based on radiation intensity & flickering frequency analysis [2,3], reconstruction of the temperature field in a combustion chamber based on flame photography [4,5,6,7] & layer analysis [8], flame detectors based on Fourier infrared spectrometers [2,9], flame temperature measurement by multi-wavelength method[9,10,11], measurement of the energy flux injected into the burner & flame radiant energy flux, etc. Acknowledged achievement has been made in improvement of combustion sensing via development of the methods above and it impels gradual evolvement of the functions of combustion detectors into combustion diagnostics. This paper presents a spectroscopic technique, with which flame emission spectra acquired by a CCD spectrometer is analyzed and some development has also been made in seeking the criterion for combustion diagnostics.

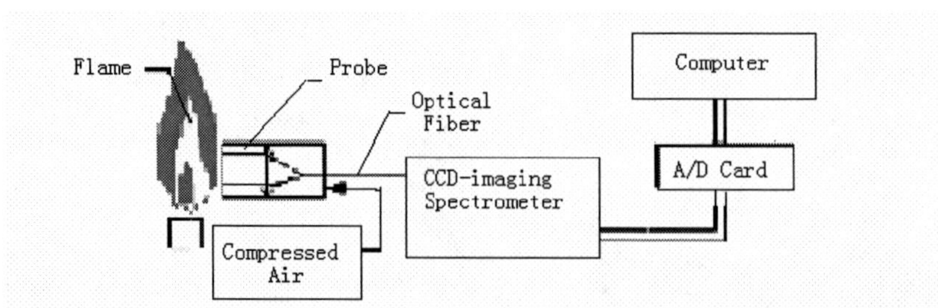

FIGURE 1. Schematic of the apparatus for flame emission spectra acquirement.

CP914, *Multiphase Flow: The Ultimate Measurement Challenge,* edited by X. Cai, Y. Wu, Z. Huang, S. Wang, and M. Wang

APPARATUS FOR FLAME EMISSION SPECTRA ACQUIREMENT

The key component of the whole apparatus is a fiber optic CCD spectrometer with 1100 pixels and a responsive spectral range of 200-1100 nm. The radiation is converged via a lens into a tenuous silica optical fiber, and is then transmitted to the spectrometer where the radiation spectra are captured by the CCD array and converted into electric signals, which are sent to the A/D card whose output is analyzed by the computer. The distance from the lens to the optical fiber can be altered to regulate the measuring scope of the probe. The probe is coated with a metal sleeve and the optical fiber with a flexible tube for protection. The compressed air is introduced to purge the lens of contamination as well as to cool the fore optical components.

Due to the spectral variation in the CCD sensor response, calibration of the spectrometer should be pre-conducted using a black-body furnace as the ideal radiator.

STUDY ON ACTUAL MEASUREMENT OF THE FLAME EMISSIVITY & TEMPERATURE IN A PULVERIZED COAL FIRED BOILER

The radiation-emissivity-temperature relationship is generally described by the Wien formula. In general, the emissivity is wavelength-dependent; however, it can be known by the measuring results of flame emission spectra that the emission characteristic of pulverized coal flames is very near to that of a gray body, i.e. the emissivity can be assumed independent of the emission wavelength. With this assumption, we may fit a curve to the measured spectrum by the least square approach and obtain the apparent emissivity & temperature based on the Wien formula [5,12,13,14]. By means of the above apparatus, the emission spectra of pulverized coal flames in several operating conditions have been measured for a 350MW utility boiler, as shown in Fig. 2. The curve labeled 1, 2, 3, 4 corresponds to the flame emission spectrum measured with 60%, 70%, 80%, 100% of the rated load of the burner, respectively; and the curve labeled 5 corresponds to that measured in the central of the combustion chamber. The radiance over the wavelength range below 550nm is comparatively weak and thus has not been plotted in Fig. 2.

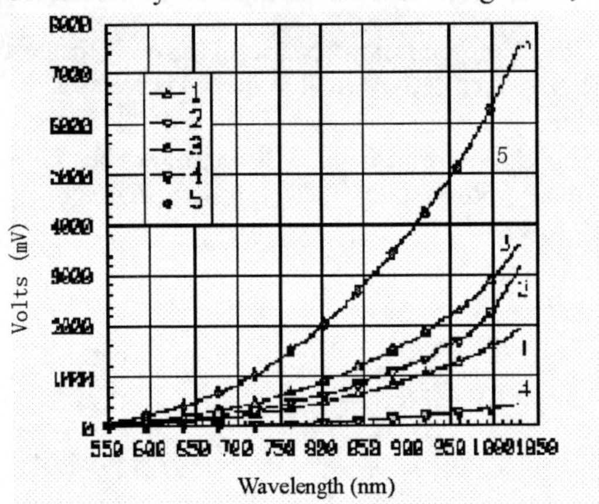

FIGURE 2. Emission spectra of pulverized coal flames measured in several operating conditions for a 350MW utility boiler.

As shown in Fig. 2, the measured emission spectra of pulverized coal flames are smooth and continuous on the whole and their characteristics agree well with gray body radiation, which is alien to the description in earlier literature[18]. This should be ascribed to the presence of large quantities of coal, soot, char & fly ash particles whose emission spectra are all continuous and predominate among the entire pulverized coal flame radiation. In addition, from Fig. 2 one may conclude the followings: (1) the flame radiance increases and the wavelength of peak emission shifts to the ultraviolet end as the burner load goes up, which is consistent with the Wien formula; (2) the radiance at the curve labeled 5 is the highest because the perfect combustion is maintained in the central of the combustion chamber with comparatively high combustion intensity; (3) with the full rated load, the ignition distance is relatively long and thus many coal particles remain unburned in the measuring region, near to the burner exit, resulting in relatively low radiance at the spectrum curve labeled 4.

The least square approach has been employed for fitting curves to the measured spectra. Due to dramatically low responsivity & poor linearity of the CCD sensors above a certain wavelength, the really provided spectral range is from 666nm to 966nm. The data processing results is shown in Tab. 1, where (1) the cumulative relative deviation was calculated with the following expression:

$$\sigma = \left(\int_{666}^{966} \left|e_{\lambda s} - e_{\lambda n}\right| d\lambda \Big/ \int_{666}^{966} e_{\lambda s} d\lambda\right) \times 100\%$$

where $e_{\lambda s}$ and $e_{\lambda n}$ denote the measured and fitted values of monochromatic radiance; (2) the maximum-deviation wavelength refers to the wavelength at which the monochromatic radiance deviation reaches its maximum $Max(\left|e_{\lambda s} - e_{\lambda n}\right|)$; (3) the maximum relative deviation is defined as $(Max(\left|e_{\lambda s} - e_{\lambda n}\right|)/e_{\lambda_m s}) \times 100\%$.

As illustrated in Tab.1, the cumulative relative deviation of each curve is within 4%, manifesting negligible difference between the original and fitted data. This verifies the validity of applying the Wien formula to data processing based on the gray body assumption.

TABLE 1. Results of Spectra Data Posttreatment

Load percent	Center	80%	70%	60%	100%
Temperature (K)	1651	1684	1669	1662	1493
Emissivity	0.482	0.175	0.138	0.116	0.0697
Cumulative relative deviation (%)	1.4	2.3	3.5	2.3	1.6
Maximum-deviation wavelength λ_m (nm)	966	966	966	966	965
Maximum relative deviation (%)	3.7	6.2	8.2	5.1	4.5

Fig. 3 gives the fitted spectrum curves in comparison with the original data. Excellent agreement occurs in the mentioned spectral range. However, as the wavelength goes up above 920nm, there begins a significant increase of the difference between the original and fitted data, thanks to the reduced responsivity and weakened linearity of the CCD sensors.

STUDY ON THE FLAME EMISSION SPECTRA CHARACTERISTICS OF SEVERAL FUELS

The flame emission spectrum varies with the fuel type and thus contains information on fuel characteristics. Ph. NGENDAKUMANA et al. managed combustion diagnostics via analysis of characteristic spectral lines, which may be used for automatic combustion control [16]. In the experiment presented in this section, the flame emission spectra were measured of several fuels in different combustion systems with the fiber optic spectrometer and the possibility of using the measured data for fuel type identification, combustion state monitoring as well as combustion diagnostics was explored.

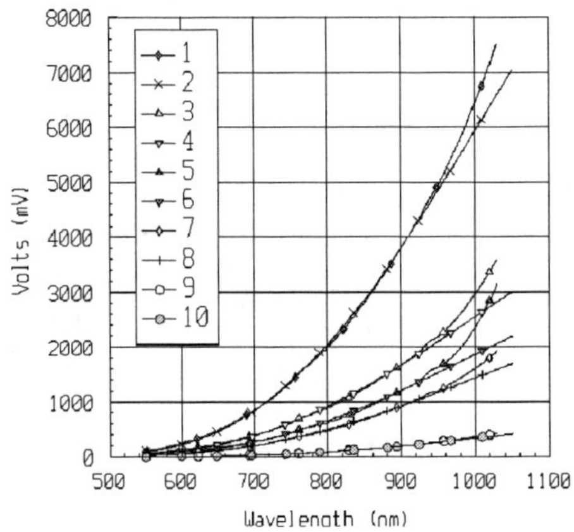

FIGURE 3. Fitted spectrum curves in comparison with the measured data in Fig. 2 (1,2—measured and fit spectrum in the chamber central; 3,4—measured and fit spectrum with 80% load; 5,6—measured and fit spectrum with 70% load; 7,8—measured and fit spectrum with 60% load: 9,10—measured and fit spectrum with 100% load)

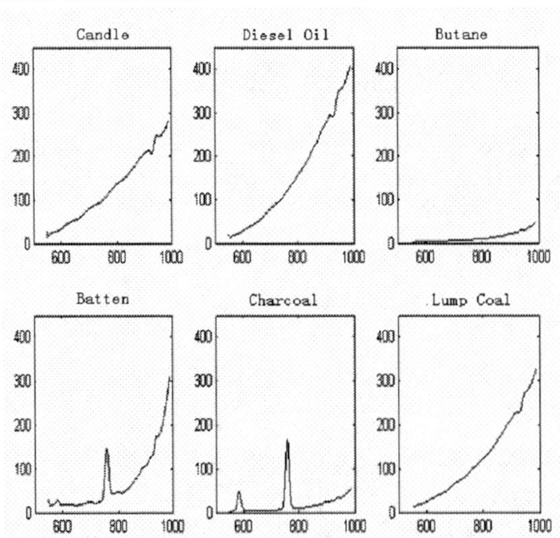

FIGURE 4. The measured flame emission spectra of different fuels

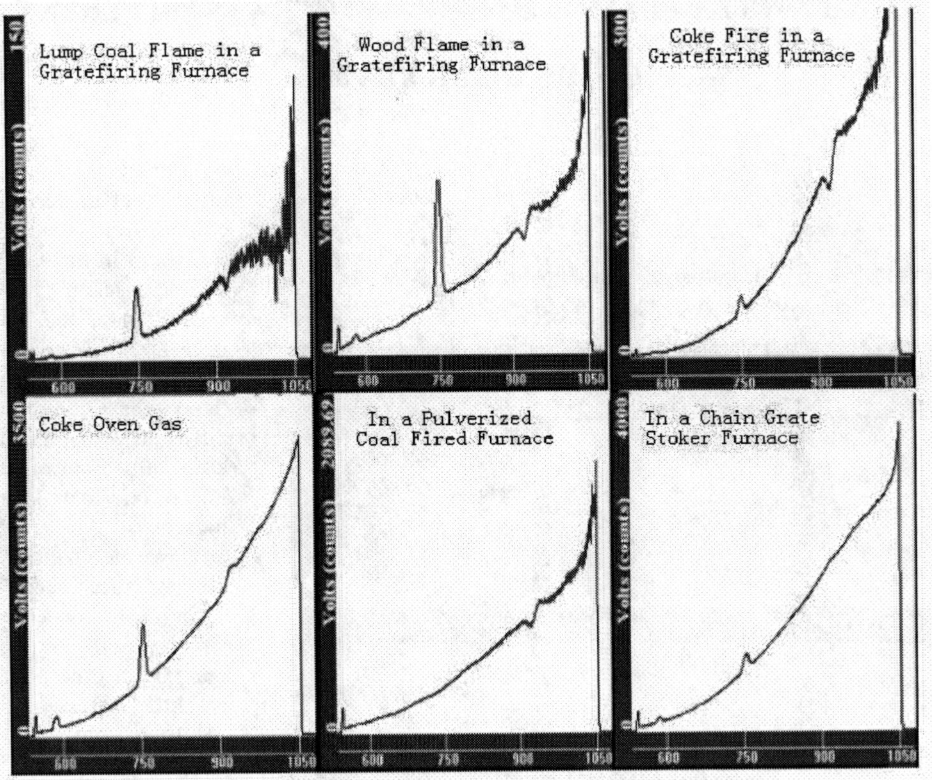

FIGURE 5. Flame emission spectra measured in different combustion systems

Two categories of fuels have been chosen as the measured object, one of which is of alkyl-base fuels including candle, diesel oil, butane, etc., the other is of wood-base fuels including a batten, a paper scroll, charcoal, lump coal, etc. The measured flame emission spectrum of these fuels are shown in Fig. 4 [17].

It can be observed from the figure that the spectrum curves are smooth on the whole except two emission peaks around the wavelengths of 590nm and 766nm and an individual absorption line around 940nm. There are some similarities between the flame emission spectra of alkyl-base fuels such as candle, diesel oil, butane, etc. So are there between those of wood-base fuels such as a batten, a paper scroll, charcoal, lump coal, etc.

By shifting the fiber optic probe in a steady-state flame, the emission spectrum profile can be acquired. The difference between the emission spectra in different positions is found significant. For instance, the absorption line mentioned above exists only in a specific combustion zone, especially for wood-base fuel flames.

The flame emission spectra measured of several wood-base fuels in different combustion systems are shown in Fig. 5. The two emission peaks and the individual absorption line do not exist in all flame emission spectra, correlated with fuel types and combustion systems.

The emission peak around 590nm, that around 760nm, and the absorption line around 940nm have been proved to be the atomic emission line of Sodium (589nm), that of Kalium (766.9nm), and the molecular absorption line of water vapor respectively by the experiments. Emergence of the atomic emission lines of Sodium and Kalium depends mainly upon the flame temperature: in the experiment presented, the atomic emission line of Kalium emerges as the flame temperature reaches 800—820℃ and that of Sodium does as reaches 870—880℃; the two emission lines vanishes as the flame temperature reduces to the same respective levels. Furthermore, the monochromatic intensity at the emission peaks increases as the flame temperature goes up above the critical level. Therefore, the presence and the monochromatic intensity of the characteristic spectral lines may be used for estimation of the flame temperature interval.

The absorption line of vapor around 940nm emerges only under some conditions, which has also direct correlation to the flame temperature. However, the critical flame temperature corresponding to emergence of this line has not been measured due to limited experiment conditions.

STUDY ON THE FLAME EMISSION SPECTRA IN DIFFERENT COMBUSTION STATES

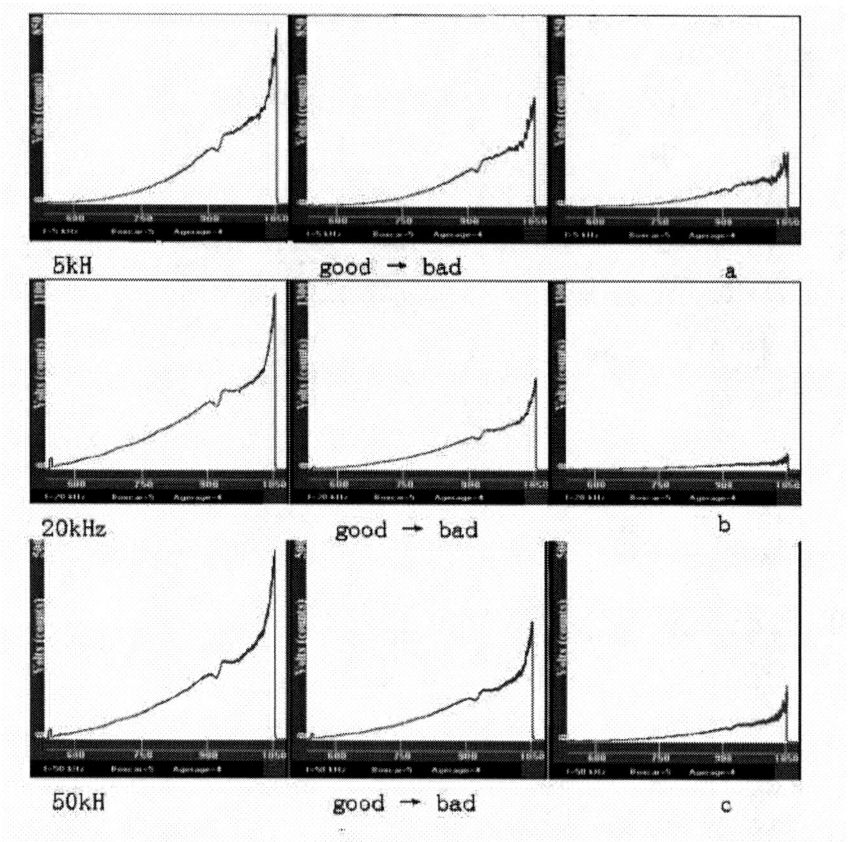

FIGURE 6. Flame Emission Spectra Measured in three Combustion States with Different Scanning Frequencies.

As observed from the experiment, the flame emission spectra bear different characteristics in differ- ent combustion states. The spectra measured in three combustion states with the scanning frequencies of 5 kHz, 20kHz, and 50kHz are shown in Fig. 6. From left to right the combustion state grows poorer and the combustion process transits from diffusion combustion to the kinetically controlled regime. From Fig. 6, one may observe the follow- ings:

(1) the shapes of the spectrum curves measured with different scanning frequencies are quite simi- lar, which implies that the integration time of the spectrometer has little effects on the shape of the spectrum curves, i.e. with the integration time increased, the spectrum curve would not grow smoother while only the absolute signal amplitudes would be changed.

(2) The shape of the flame emission spectrum curves varies with the combustion state. For the diffusion combustion process, the spectrum curves appear quite smooth and the segment in the NIR range fluctuates with a high frequency and a small amplitude; as the combustion process transits to the kinetically controlled regime, the combustion state gets worse and consequently the spectrum curve tends to grow rougher with a reduced fluctuation frequency and an enhanced fluctuation amplitude in the NIR range.

As indicated by the experiments, the shape of the spectrum curve is susceptible to the combustion state, which implies abundant information on the latter. With the dependence relationship of the shape of the spectrum curve with the combustion state obtained, inversion of the combustion state via numerical analysis of the spectrum curve fluctuation may be realized and thus used as the basis of combustion state identification.

STUDY ON THE EMISSION LINES OF KALIUM FOR TEMPERATURE MEASUREMENT

Further study to the emission lines of Kalium finds that the emission line of Kalium is a pair of lines in 766.5 nm and 769.9 nm, respectively, as show in Fig 7. The ratio of intense of these two lines is directly related to flame temperature. Some experiments were done in a black body oven. Fig. 8 gives the measurement results made by a thermal couple and this method.

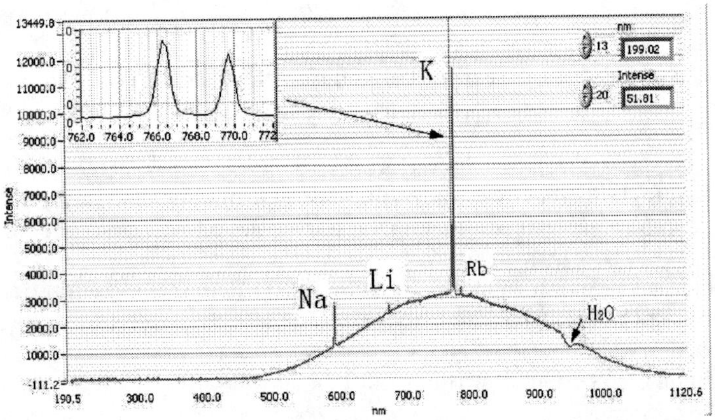

FIGURE 7. Emission lines of Kalium

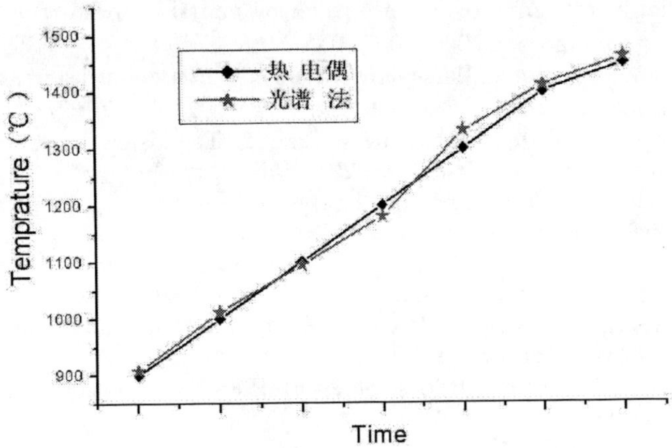

FIGURE 8. Measured temperature by two methods

CONCLUSIONS

The emission spectra of pulverized coal flames approximate very well to that of gray body radiation in the VIS/NIR range, where data posttreatment according to the Wien formula has been proved valid. Thus the micro fiber optic spectrometer with a data-processing computer can be used for in-situ measurement of the apparent flame temperature & emissivity.

The positions & existence of characteristic spectral lines in the flame emission spectrum have dependence relationship with the fuel type and the flame temperature, which implies the possibility of fuel type identification and estimation of the flame temperature interval within certain limits based on characteristic spectral line analysis.

The shape of flame emission spectrum curves has a close relationship with the combustion system, which provides the basis for combustion state identification and further extension to combustion diagnostics.

ACKNOWLEDGEMENT

This project was supported by National Natural Science Foundation of China (NSFC) (No. 50576058).

REFERENCES

1. WANG Fei. Flame temperature field measurement & combustion diagnostics based on computer image processing. PhD Diss., Zhejiang Univ., Hangchou (2000).
2. P.M. Willson, et al. Sensing of Pulverized Coal Flames in Utility Boilers. Boiler Technology (in Chinese), 1991, (2), 21－29
3. Nicolas Docquier, Sébastien Candel. Combustion control and sensors: a review. Progress in Energy and Combustion Science,2002,28,107-150

4. Hanno Scharr1, Bernd J ahne1. Flame front analysis in turbulent combustion. Mustererkennung 2000, 22. DAGM Symposium, 13-15 Sept. 2000,
5. C. PELS LEUSDEN and N. PETERS. Experimental and numerical analysis of the influence of oxygen on soot formation in laminar counterflow flames of acetylene. Proceedings of the Combustion Institute, 2000, Vol. 28, 2619-2625
6. David M. Mosbacher1, Joseph A. Wehrmeyer, and Robert W. Pitz. Experimental and numerical investigation of premixed tubular flames. Submitted to 29th International Symposium on Combustion
7. HUA Yanping, et al. Combustion state estimation based on complexity measure of image gradation. Power Engineering (in Chinese), 2002, 22(2), 1611-1614
8. LU Yonggang. Theoretical and experimental investigation of 3D flame temperature field reconstruction based on optical layered imaging. PhD Diss., Dongnan Univ., Nanking (1999).
9. Norman Chigier. Combustion Measurements. Hemisphere Publishing Corporation, 1991
10. ZHAO Wenhua, et al. Flame temperature measurement in a coal fired boiler. Combustion Science & Technology (in Chinese), 2001, 7(3), 275—277
11. DAI Jingmin. Investigation of a temperature measurement technique with multiple radiation spectra. PhD Diss.,, Harbin Institute of Technology (in Chinese) (1995)
12. CAI Xiaoshu, LUO Wude. Investigation of flame sensing by emission spectroscopy in a pulverized coal fired boiler. Power Engineering (in Chinese), 2000, 20(6), 955—959
13. CAI Xiaoshu, LUO Wude. Investigation of flame temperature & emissivity measurement by emission spectroscopy. Transactions on Engineering Thermal Science (in Chinese), 2000, 21(6), 779—782
14. LUO Wude. Investigation of pulverized coal flame sensing & diagnostics by emission spectroscopy in utility boilers. Master Diss., Univ. of Shanghai for Science & Technology, 1999
15. WAN Zhengang. Investigation of flame temperature measurement & combustion diagnostics in utility boilers. Master Diss., Univ. of Shanghai for Science & Technology, 2002
16. Ph. NGENDAKUMANA. Investigation of fuel oil flames by means of emission spectroscopy. The 20th Task Leaders Meeting of the IEA-Combustion in Ottawa, Canada, July 1998.
17. JI Kun. Investigation of criterion for pulverized coal combustion in emission spectroscopy method. Master Diss., Univ. of Shanghai for Science & Technology, 2003
18. Eckert. Heat & Mass Transfer, Chinese ed. (Peking: Science), 1986

SESSION A

FUNDAMENTALS OF MULTIPHASE FLOW MEASUREMENT

Analysis of Turbulence Modulation in a Sudden-Expansion Flow Laden with Fine Particles

Fei Li,Haiying Qi[1],Changfu YOU

Key laboratory for Thermal Science and Power Engineering of Ministry of Education, Tsinghua University, 100084 Beijing, P.R. China

Abstract. Measurements were conducted in gas-solid tow-phase flow in a vertical sudden expansion pipe with Phase Doppler Anemometry (PDA). Turbulence modulation of dilute two-phase flow with fine particles (55.2 μm) and low Reynolds number (1.98×10^4) was studied. The results imply that the particles reduce the gas turbulence as a whole in the flow field. Before the redeveloping region, the dimensionless root-mean-square (RMS) fluctuation velocities, v, of the gas phase in both the two-phase and single-phase flows are found to be proportional to the dimensionless mean gas velocity gradient, η, in each measured cross-section. The linear relationship between v and η can be represented by $v = -k\cdot\eta+b$. Then the turbulence modulation is proportional to the velocity gradient too by definition. Further analysis indicates that parameters k and b are linearly related with the shear layer thickness. The relationship holds no matter whether adding of particles. However, the particles do delay the development of the sudden expansion flow, which leads to the attenuation of the gas phase turbulence through out the whole jet expansion region. However, turbulence enhancement was observed at the end of the jet expansion region.

INTRODUCTION

Turbulence modulation refers to the change between the gas turbulence intensity in gas-solid two-phase flow and that in single phase flow. To investigate the mechanism of turbulence modulation is of great importance in accurately describing mass and heat transfer process in two-phase flows.

Up to now, a lot of experimental studies have been conducted on homogeneous turbulence, pipe flows and etc, of which there were no free shear layer in the flow fields. The effect of particles on the gas phase turbulence intensity was investigated .It was discovered that fine particles attenuate the gas turbulence intensity and large particles enhance it commonly, while the effect of middle size particles vary according to different position in the

Corresponding author. Tel: +86-10-62796036; Fax: +86-10-62796036; E-mail address: hyqi@mail.tsinghua.edu.cn

CP914, *Multiphase Flow: The Ultimate Measurement Challenge,* edited by X. Cai, Y. Wu, Z. Huang, S. Wang, and M. Wang

flow field[1-4]. Most of the reports are concentrated on the overall effect of particles on the turbulence modulation under different particle properties and inlet conditions. However, the effect of the local space distribution of particles to the turbulence and the analysis of the mechanism of the turbulence modulation are lack of concern, so as the investigation of the turbulence modulation in the flow that contain free shear layer. Because of the limitation of experimental conditions, people seldom measured particle concentration distribution directly.

Therefore, measurements were conducted in a vertical downward round pipe sudden expansion flow. The effect of particles to the shear of the flow and the mechanism of the turbulence modulation were investigated in this article. In the past published paper [5], the turbulence modulation at the inlet section and the overall flow structure were analyzed. It was discovered that the presence of particles delay the development of the sudden expansion flow; the turbulence modulation and velocity gradient at the inlet section took on a linear relationship. Here, further analysis of the turbulence modulation in the whole flow field will be conducted.

EXPERIMENTAL SETUP AND METHOD

Figure 1 shows the experimental setup. The experiment segment and the position of the measured sections are shown in Figs. 2 and 3 respectively. Phase Doppler Anemometry was adopted here to measure the parameters in the flow field [5].

Glass beads with diameter $d_s = 55.2$ μm were used as solid particles, to which glass powder with diameter of 2500 CPSI ($d_s < 6$ μm) were added as trace particles. The density of particle ρ_s is 2450 kg/m^3. The mean velocity of inlet gas U_0 was 5.954 m/s, and the corresponding inlet Reynolds number Re was 19800, with which the flow was turbulence. The maximum velocity (at the axis of the pipe) U_c was 7.486 m/s. The particle mass loading m was 0.146. The turbulence modulation and velocity gradient are represented by dimensionless parameters ξ (Equation 1) and η (Equation 2) respectively.

$$\xi = v_{TP} - v_{SP}, \quad v = u'_{g,rms} / U_0 \tag{1}$$

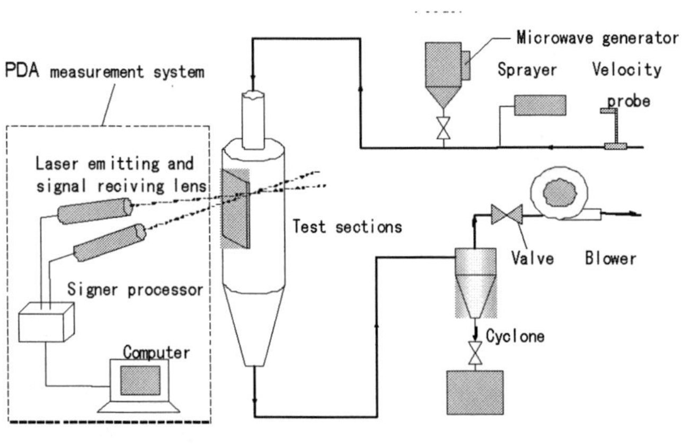

FIGURE 1. Experimental setup

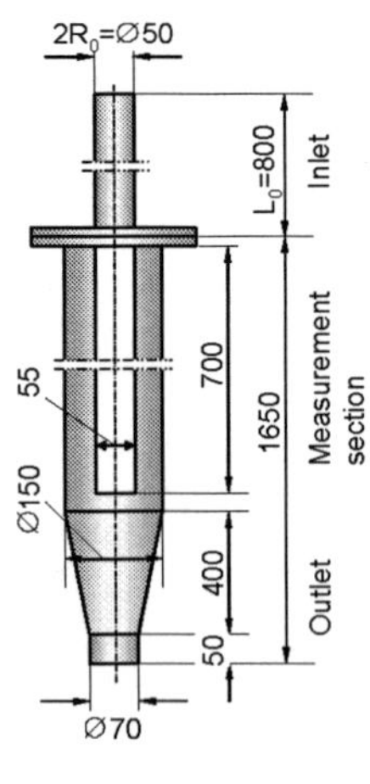

FIGURE 2. Size of experiment segment (mm)

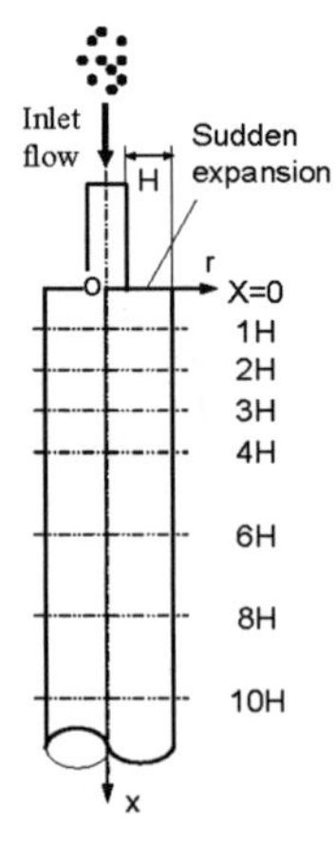

FIGURE 3. Measurement segment and each measured section

$$\eta = \frac{d(u_g/U_0)}{d(r/R)} = \frac{R}{U_0}\frac{du_g}{dr} \tag{2}$$

where the dimensionless fluctuation v is the ratio of the RMS gas fluctuation velocity $u'_{g,rms}$ to the inlet mean velocity U_0, the subscripts TP and SP represent the two-phase and single-phase respectively, R is the radius of the pipe after sudden expansion.

RESULTS AND ANALYSIS

Eight sections were measured in the experiment. Three representative sections of them are selected here to be analyzed. They are section 2, section 5 and section 8 that located at the front, middle and tail of the jet expansion region respectively. Each of them represents local features of the flow field corresponding to the position of the section and also combined they represent the overall flow characteristics of the whole jet expansion region.

Section 2 ($x/H = 1$)

It can be seen from Fig. 4 that, though comparing with single-phase flow, the mean gas velocity distribution of the two-phase flow changes a little, both of them are typical sudden expansion jet flow pattern. The mean velocity distribution can be divided into three parts, the central main flow region A, the shear layer transition region B and the near wall back-flow region C. The boundary of region A and B is the point at where local velocity reaches 99% of the maximum velocity. The boundary of region B and C is the zero velocity streamline.

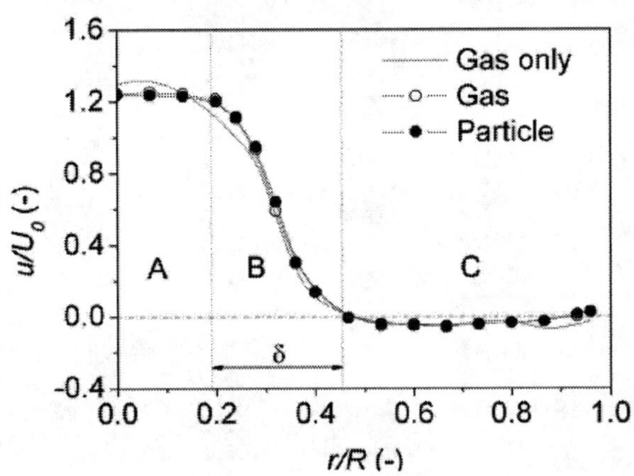

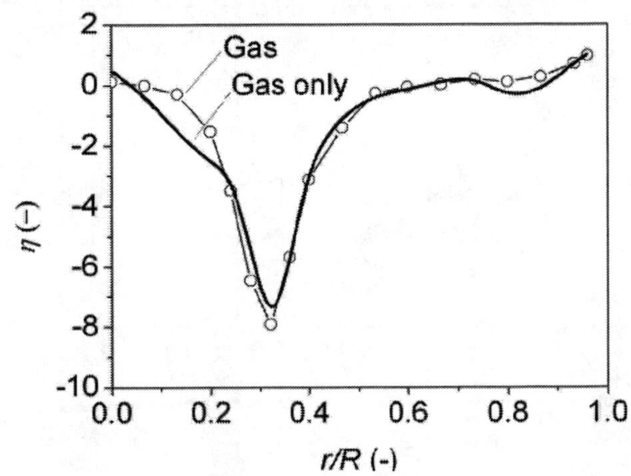

FIGURE 4. Mean axial velocity **FIGURE 5.** Gas velocity gradient

(Notice: "Gas only" represents single-phase flow, "Gas" represents gas-phase in two-phase flow.)

The velocity difference between region A and C is remarkable and the direction of them is opposite. The thickness of region B is δ, and the velocity in which is steeply distributed.

The velocity distribution of the gas-phase and solid-phase are almost the same across the whole section.

Compared with the single-phase flow, the distribution of the gas velocity in the two-phase flow is flatter and the shear layer thickness reduces from 0.38 in single-phase to 0.2. It implies that, with the presence of particles in the vertical downward flow, the inertia of the flow in the main flow region increases and the effect of shear viscosity is further overcome. Because the section is close to the inlet section, the distribution of the mean velocity here inherits the characteristics of that in the inlet section actually [5]. Figure 5 shows the distribution of the mean velocity gradient. The distribution and value of the velocity gradient in both the single-phase and two-phase flows are approximately the same. There exists a negative peak in both cases which is sure to bring intense turbulent fluctuation locally. The distribution of the velocity gradient in the region other than the peak region is much flatter. The velocity gradient of the single-phase flow in the main flow region is larger than that of the two-phase flow.

Not come singly but in pairs, the distribution of the gas phase fluctuation velocity takes on the similar shape to the distribution of the mean velocity gradient (Fig. 6). Unlike the velocity gradient distribution, the position of the maximum fluctuation in the two-phase flow deviates from that in the single-phase flow. The two-phase fluctuation intensity is lower than the single-phase one as a whole. But the contrary was observed in the back flow region, that is , the turbulence is locally enhanced.

According to the above, addition of particles will change the characteristics of the mean flow field and fluctuation field simultaneously, but in the diverse mechanisms. For the former, it is because the inertia is increased, which broadens the velocity distribution that is originally narrow distributed under the effect of the viscous force. The modification of the mean velocity distribution is similar to that caused by the turbulizing of the flow due to the increasing of *Re*. Here since *Re* remains the same, which means the ratio of the inertial force to the viscous force doesn't change, the increasing of flow inertia is apparently contributed to the particle inertia force. And the magnitude of the flow inertia increasing caused by particle inertia exceeds the magnitude of the flow viscosity increasing due to the particles.

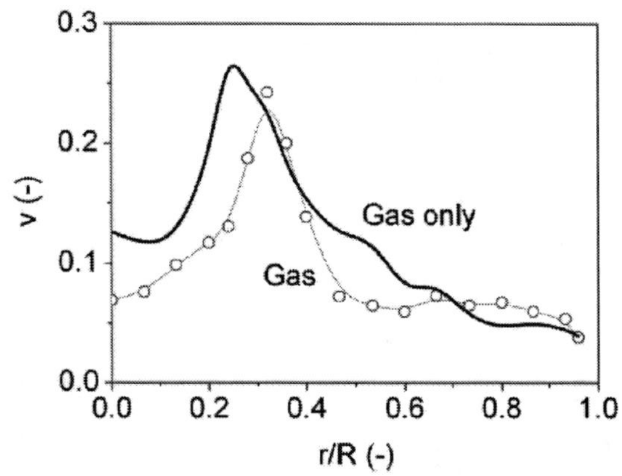

FIGURE 6. Gas fluctuation velocity

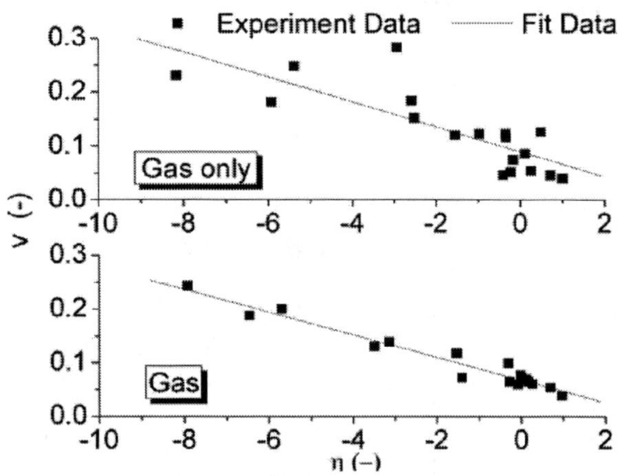

FIGURE 7. Relation between the gas fluctuation and the mean velocity gradient

The modification of the fluctuation field is mainly contributed to the dissipation effect due to the particles, under which the flow tends to laminarize. The result is consistent with the increasing of the flow viscosity due to the presence of particles. Since it is much weaker than the effect caused by particle inertial force, it influences the mean flow field little. Further, it is impossible to explain the tendency of the change of the mean flow distribution just from the view of the fluctuation field modification and flow laminarization.

After all, the effect of particles comes down to the competition and cooperation of the inertial force and viscous force.

It is known from turbulence principle that, the fluctuation velocity of the fluid is proportional to the velocity gradient. And the experimental results from Fig. 5 and Fig. 6 also show the similarity between the two. Then take the two parameters as ordinate and abscissa respectively. It is observed from Fig. 7 that not only in the single-phase flow but also in the two-phase flow, linear relationship exist apparently. After linear fitting, linear equations can be derived (Equation 3). The parameters k and b obtained for the single-phase flow are 0.022 and 0.085 respectively, 0.021 and 0.068 for the two-phase flow. Thus the equation of the turbulence modulation in section 2 can be expressed as Equation 4:

$$v_i = -k\eta + b, \quad i = SP, TP \tag{3}$$

$$\xi = 0.001\eta - 0.017 \tag{4}$$

Section 5

As sudden expansion flow is developing towards downstream, the mean velocity distribution tends to be flatter and milder. Main flow region and back-flow region are shrinking, and the thickness of shear layer increase to 0.53. The maximum velocities (at the axis) of both single-phase and two-phase are still approximately the same, but the single-phase one has reduced compared with that in section 2 (Fig. 8a). Up to now, mean velocity of particle-phase has been larger than gas-phase, because particle velocity reduces slower than gas velocity due to its large inertia, on the other side particle velocity is sure to exceed gas velocity after a certain distance since there exists a settlement velocity of particles. Compared with single-phase flow velocity, distribution of velocity in tow-phase flow holds comparatively high flow inertia, so the shape is little broader. The back-flow region shrinks compared with that in section 2, but it is still broader than that in single-phase flow, which implies that the inertia of particles tend to weaken the transversal diffusion in sudden expansion flow.

It is seen from Fig. 8b that, the distribution of the gas velocity gradient in the section is similar to that in section 2. There exists a negative peak, but the peak value is only a half of that in section 2, however the gradient of both sides of the peak increases commonly.

It is seen from Fig. 8c that, the fluctuation velocity in the two-phase is smaller than that in the single-phase as a whole. The turbulence is still attenuated by the particles. The fluctuation velocity of both sides of the peak increases apparently compared with section 2, which is consistent with the change of the mean velocity gradient. Similar to section 2, the linear relationship still holds between the fluctuation velocity and mean velocity gradient as shown in Fig. 8d. After fitting, the values of k and b can be obtained, which are 0.035 and 0.140 for the single-phase, 0.037 and .115 for the two-phase. Both of the two sets of data are bigger than that in section 2. The turbulence modification is derived as:

$$\xi = -0.002 \cdot \eta - 0.025 \tag{5}$$

Comparison of equation (5) and (4) indicates that, the coefficient of the velocity gradient is negative which means that the turbulence attenuation reduces as the shear increases. Supposing the shear is intensified that the velocity gradient is lower than -12.5, the turbulence will be enhanced. Of course the case will not happen in this section. However, equation (4) indicates that the attenuation of the turbulence increase as the shear increases.

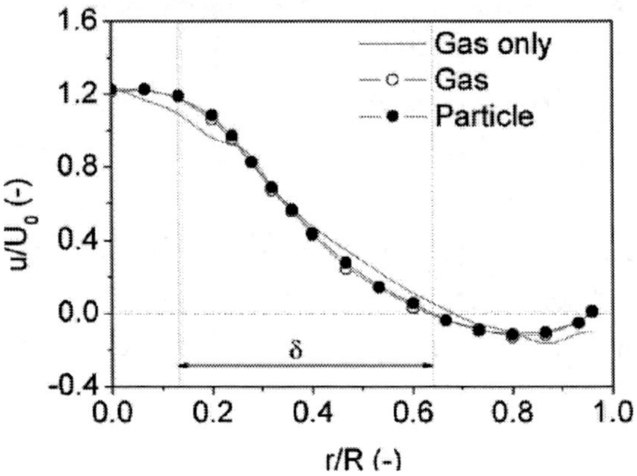

(a) Mean velocity

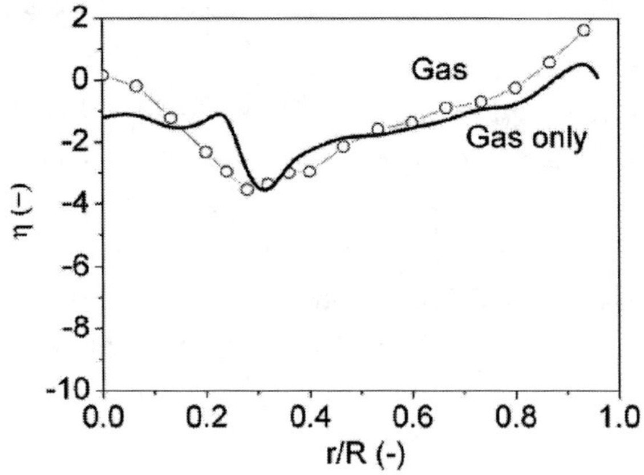

(b) Gas velocity gradient

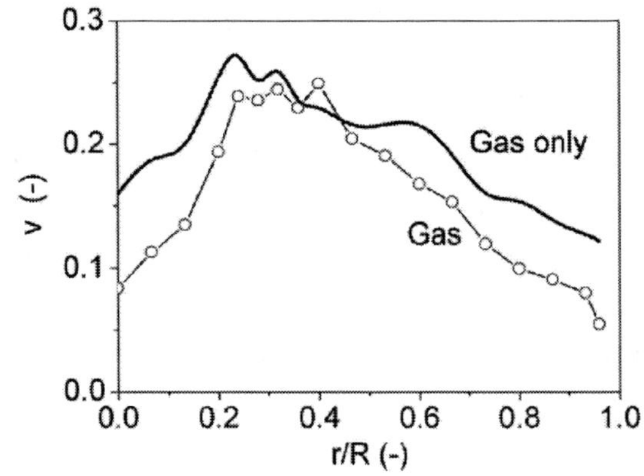

(c) Gas fluctuation velocity

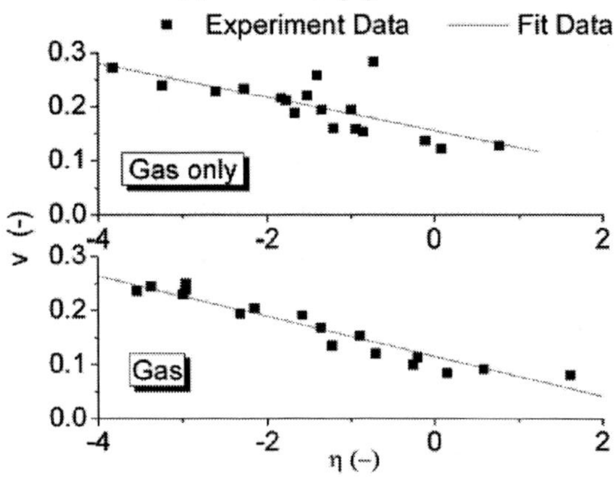

(d) Relation between the gas fluctuation and mean the velocity gradient

FIGURE 8. Distributions of parameters in section 5 (x/H = 4)

Section 8

It is seen from Fig. 9a that, the mean velocity distribution is becoming further flatter, and the maximum velocity at the axis has reduced. The main flow region has almost disappeared. The shear layer is further broadened. The velocity difference between the single-phase and two-phase flows has become remarkable at most places. The lengths of jet expansion region X_R of the single-phase and two-phase flows are 9.63 and 12.64 respectively. Section 8 in the two-phase flow is close to the tail of the jet expansion region while in the single-phase flow, the back-flow region in section 8 has disappeared and the flow has entered the redeveloping region, in which the velocity has changed from reverse direction to forward direction and the magnitude of velocity has exceeded two-phase one.

Corresponding to the mean velocity, the gas velocity gradient in the two-phase flow reduces to just one fourth of that in section 2, but is lower at most place of the section than that in the single-phase flow, which is flat and straight distributed and is close to zero (Fig. 9b). The turbulence modulation is different from the former sections. The fluctuation velocity in the two-phase flow exceeds that in the single-phase at most place in section 8, which implies that the particles enhance the turbulence (Fig. 9c). The linear relationship between the gas fluctuation velocity and velocity gradient is still approximately hold in the two-phase flow, while in the single-phase flow the relationship is not applicable anymore (Fig. 9d). It is considered that, there are two reasons for the enhancement of the turbulence by the particles: one is that in the single-phase flow the gas velocity reduces quickly, and the gradient of which becomes low at the same time, however the velocity gradient in the two-phase flow is comparatively high; the other is that, since the particle velocity is higher than the gas velocity, the particles apply work to the gas and part of which converts to the turbulent kinetic energy of the gas.

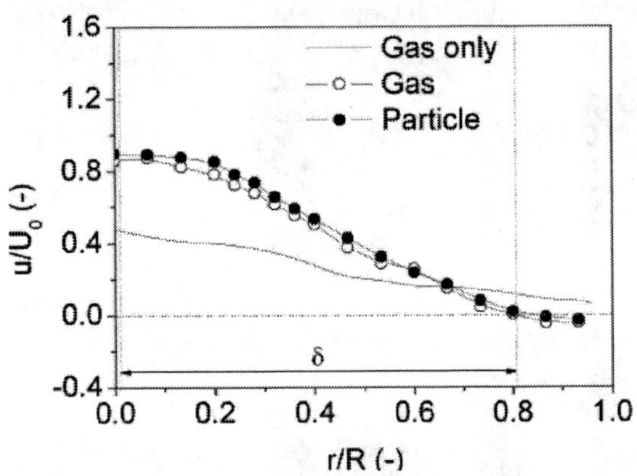

(a) Mean velocity

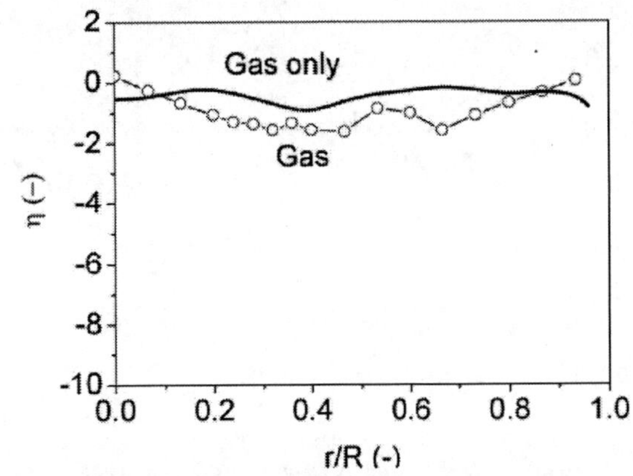

(b) Gas velocity gradient

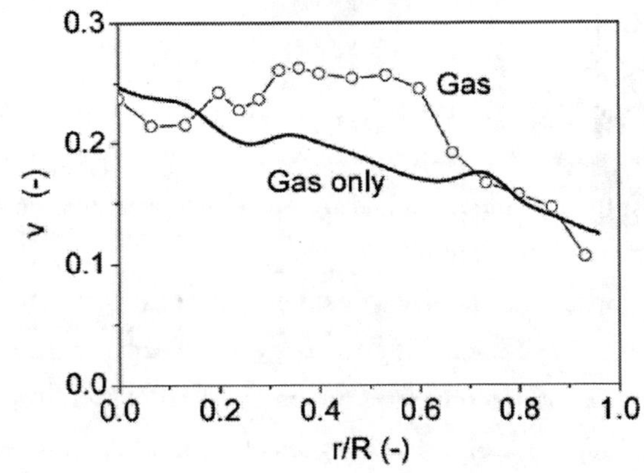

(c) Gas fluctuation velocity

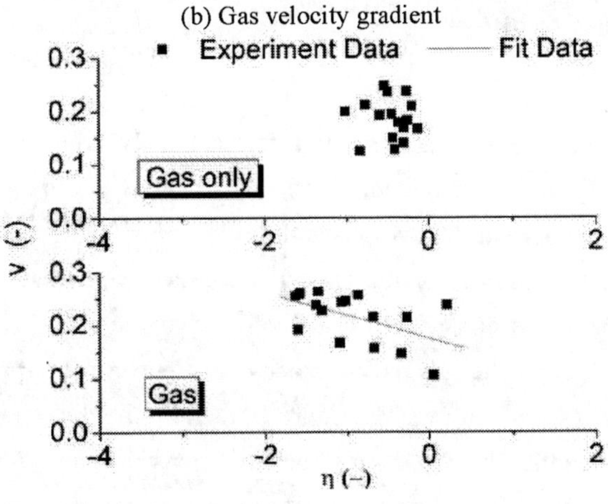

(d) Relation between the gas fluctuation and the mean velocity gradient

FIGURE 9. Distributions of parameters in section 8 (x/H = 10)

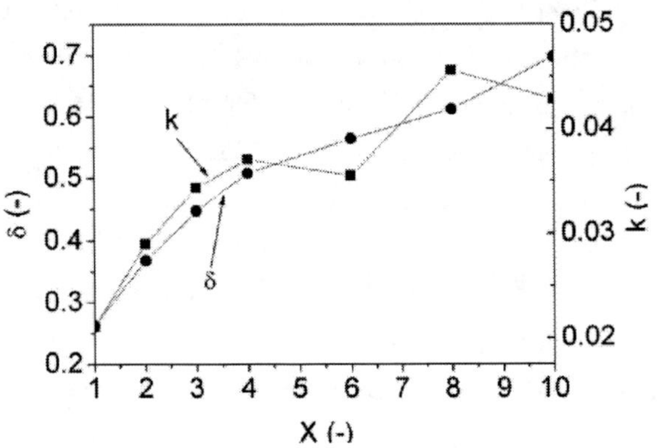

(a) Distributions of k and δ along the flow direction

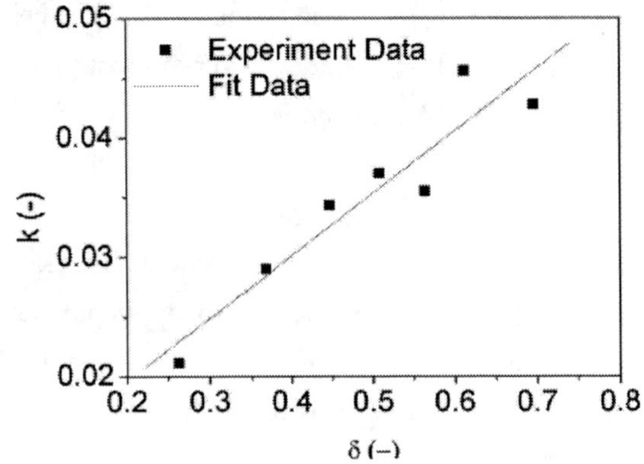

(b) Relation between k and δ

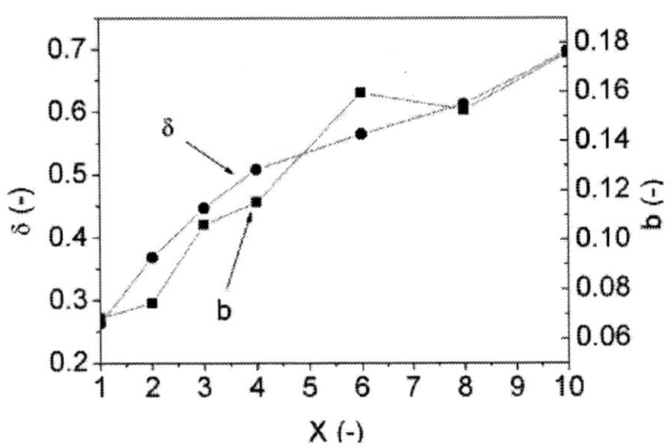

(c) Distributions of b and δ along the flow direction

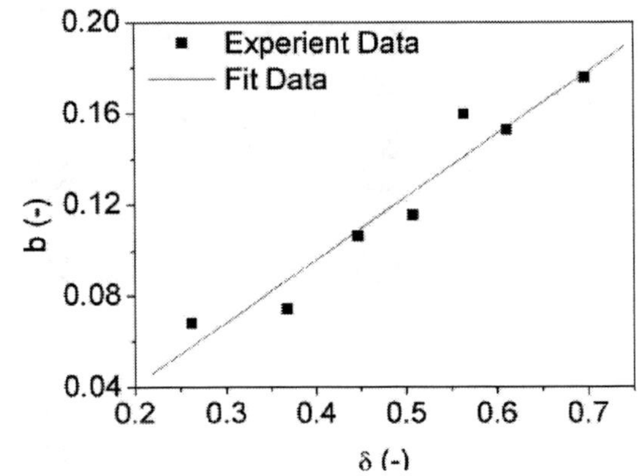

(d) Relation between b and δ

FIGURE 10. Relations of parameters k and b with shear layer thickness δ in two-phase flow

The Laws of *k* and *b*

There are only three sections that are analyzed above, but the linear relationship between the gas fluctuation velocity and the responding velocity gradient exists in all the measured sections, not only in the single-phase flow but also in the two-phase flow. The only changes are just the parameters k and b, the values of which vary as the flow is developing. Thus it is necessary to investigate the laws of k and b.

Take k in consideration at first. It characteristics the proportional relationship between the gas fluctuation velocity and the gas mean velocity gradient. So it is possible to have something to do with the scale of large eddy of turbulent mixing. Figure 10a shows the distributions of k and δ along the flow direction. It is seen from Fig. 10a that the shapes of the distributions of k and δ are similar; both of the two are increasing as the flow is developing. Then the relation of the two can be obtained (Fig. 10b), which shows apparently linear relationship between the two parameters. According to fitting, linear equation is obtained as follow:

$$k_{TP} = 0.052 \cdot \delta_{TP} + 0.009 \qquad (6)$$

Equation 6 implies that parameter k in the two-phase flow is proportional to the thickness of shear layer.

Take the same treatment to parameter b (Fig. 10c and 10d), it is found that there exists another linear equation (Equation 7):

$$b_{TP} = 0.275 \cdot \delta_{TP} - 0.014 \tag{7}$$

Equation 7 implies that the base fluctuation b increases as the shear layer thickness increases. Even if the shear layer thickness equals 0, the base fluctuation still exists which is possible to originate from the systematic fluctuation due to the instability of the flux of blower or the transportation from upstream.

In single-phase flow, the linear relationships between k, b and δ still exist:

$$k_{SP} = 0.052 \cdot \delta_{SP} + 0.009 \tag{8}$$
$$b_{SP} = 0.226 \cdot \delta_{SP} + 0.003 \tag{9}$$

It is seen from equations (6) and (8) that in both the single-phase and two-phase flows, the parameter k

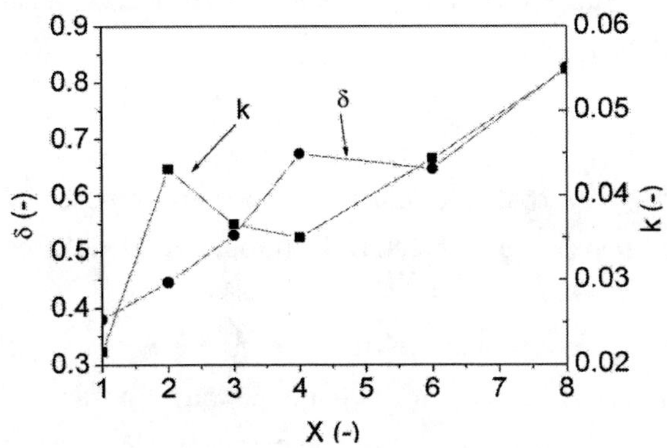

(a) Distributions of k and δ along the flow direction

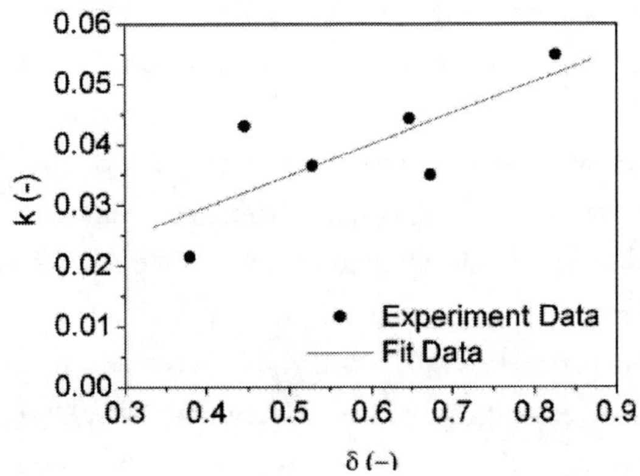

(b) Relation between k and δ

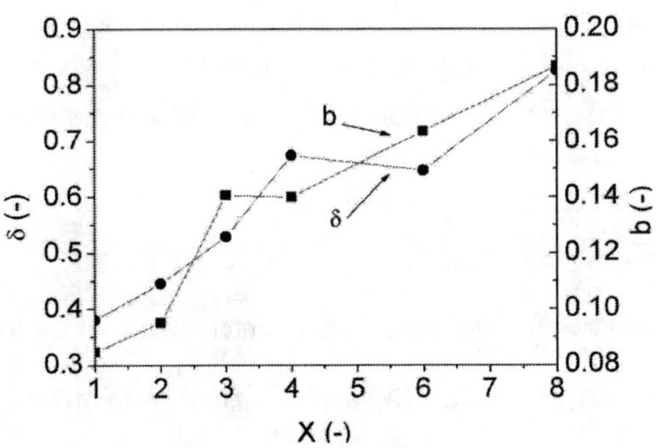

(c) Distributions of b and δ along the flow direction

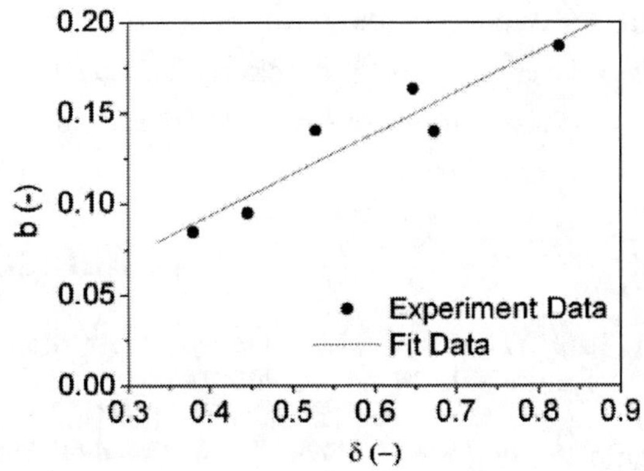

(c) Distributions of b and δ along the flow direction

FIGURE 11. Relations of the parameters k and b with shear layer thickness δ in single-phase flow

changes with the shear layer thickness in the totally same way, or at least the laws of the two are similar in reality. It implies that, under the experiment conditions of this article, the addition of particles has little effect on the law of k vs. shear layer thickness. The main reason for the change of the linear term in the turbulence modulation expressions (4) and (5) is that, the addition of particles modifies the progress of the flow development, thus makes the shear layer thickness of the same section different in the single-phase and two-phase flows, which finally lead to the difference of the linear term in the turbulence modulation expressions. Comparing equation (7) with equation (9), it is found that, the linear part in the expression of the base fluctuation of the two-phase flow is slightly deviate from that of the single-phase flow, but the constant part which represents systematic fluctuation changes comparatively large. The entire same, if the effect of systematic fluctuation is not taken into consideration, the change of the flow development due to the addition of particles would be the main reason for the modification of constant term in the turbulence modulation expressions, just like equations (4) and (5).

CONCLUSION

1) Fine particles attenuated the turbulence as a whole in the jet expansion region. Turbulence enhancement was observed just at the end of the region.
2) Increasing of flow inertia was mainly contributed to the inertial force of the particles; which thus changed the features of the mean flow field, and further influenced the fluctuation velocity.
3) The turbulence enhancement in section 8 is because of the higher velocity gradient in the two-phase flow and energy transfer from the particles to the gas turbulence.
4) In the jet expansion region where there is intense shear, the gas turbulent fluctuation velocity is proportional to the mean velocity gradient in both the single-phase and two-phase flows, thus the turbulence modulation is proportional to the mean velocity gradient too.
5) The parameters k ,b and the shear layer thickness show apparently linear relationship. The expressions of liner relationship between k and the shear layer thickness are the same in both the single-phase and two-phase flows; in the expressions of the base fluctuation, the linear parts are approximately the same in both the single-phase and two-phase flows, while the constant part varies according to different conditions.
6) The addition of particles doesn't changes the laws of the parameters k and b vs. the shear layer thickness, but they do delay the development of sudden expansion and block the transverse diffusion, which makes the shear layer thickness and velocity distribution in the two-phase flow do not agree with those in the single-phase flow in the same section, and further leads to the attenuation of turbulence modulation in the sudden expansion region.

REFERRANCES

1. Y. Tsuji, Y. Morikawa, LDV measurements of an air- solid two-phase flow in a horizontal pipe, J. of Fluid Mechanics, Vol.120, pp.385-409, 1982.
2. Y. Tsuji, Y. Morikawa, H. Shiomi, LDV measurements of an air-solid two-phase flow in a vertical pipe, J. Fluid Mech., Vol.139, pp.417-434, 1984
3. J. Kussin and M. Sommerfeld, Experimental studies on particle behaviour and turbulence modification in horizontal channel flow with different wall roughness, Experiments in Fluids, Vol.33(5): pp.143-159, 2002
4. R. A. Gore and C. T. Crowe, Effect of particle size on modulating turbulent intensity, Int. J. Multiphase Flow, Vol.15, pp.279-285, 1989
5. F. Li , H. Y. Qi, C. F. You, Y. J. Bao, Study on turbulence modulation of gas-solid two-phase flow

in a sudden-expansion tube, J. of Engineering Thermophysics, Vol.27(S1), pp.201-204, 2006
6. Z. S. Zhang, G. X. Cui, C. X. Xu, Theory and modeling of turbulence, Tsinghua University Press, Beijing, 2005

On Characteristics of a Spheroidal Particle by Use of Debye Series

LI Renxian, HAN Xiang'e, JIANG Huifen

School of Sciences, Xidian University, Xi'an, China, 710071

Abstract. We derive the formula of the Debye series decomposition for axially incident plane wave scattering by a spheroidal particle. The Debye series writes each term of the Mie infinite series as another infinite series, and clarifies the physical origins of many effects that occur in electromagnetic scattering, which is of great importance to the study of characteristics of light scattering. The characteristics of a spheroidal particle can be studied within the framework of physical origin by use of the Debye series.

Keywords: Debye series, spheroidal particle, light scattering

INTRODUCTION

For many practical applications such as combustion, environmental control, fluid mechanics, and chemical reaction, we need to measure the properties of particles such as size and refractive index-temperature ratio quickly and precisely. A technique based on light scattering has proved its importance over time.

In fact many particles encountered have spheroidal shapes, so the research on the characteristics of a spheroidal particle is of great importance. Many researchers have been devoted to the research on the light scattering by a spheroid [1~3]. The main method they used is the theory treating the scattering of a linearly polarized plane electromagnetic wave by a dielectric spheroid, known as the Mie scattering formalism. The vector wave equations are separated in the spheroidal coordinates and expanded in terms of the spheroidal wave functions[4~7]. And the boundary conditions are imposed to obtain the unknown coefficients. The Mie scattering formalism is a rigorous solution of the Maxwell equations and contains all the effects that contribute to the scattering. But it gives little clue as to the various physical processes responsible for the scattering.

It turns out that writing each term of the Mie infinite series as another infinite series, known as the Debye series, allows for the decomposition of the global physical process in a series of local interactions, and clarifies the physical origins of many effects that occur in electromagnetic scattering. It is of great importance to the study of characteristics of light scattering. The Debye series for light scattering by a homogeneous[8], coated[9] and multilayered[10] sphere has been developed and employed in the studies of rainbows, glories, coronas and other phenomena.[8~12] The Debye series of diagonally incident plane-wave scattering by an infinite homogeneous cylinder has been developed[13], and been extended to normally incident plane-wave scattering by an infinite multilayered cylinder.[14]

This paper is devoted to the derivation of the Debye series for axially incident plane-wave scattering by a homogeneous spheroid. In the paper, only the TE polarized plane wave is considered, and for the TM mode, the similar method can be employed.

DERIVATION OF THE DEBYE SERIES

Global Scattering Process

The incident wave of unit amplitude polarized in the TE mode can be expressed in terms of infinite series with spheroidal vector wave functions $\vec{M}_{emn}^{r(4)}(c;\eta,\xi,\phi),\vec{N}_{omn}^{r(4)}(c;\eta,\xi,\phi)$ as

CP914, *Multiphase Flow: The Ultimate Measurement Challenge,* edited by X. Cai, Y. Wu, Z. Huang, S. Wang, and M. Wang

$$\vec{E}^{(i)}=\sum_{n=0}^{\infty}\sum_{m=n}^{\infty}i^n\left[g_{mn}(\zeta)\vec{M}_{emn}^{\vec{r}(4)}(c;\eta,\xi,\phi)+if_{mn}(\zeta)\vec{N}_{omn}^{\vec{r}(4)}(c;\eta,\xi,\phi)\right] \tag{1}$$

$$\vec{H}^{(i)}=\frac{k_1}{\omega\mu_1}\sum_{n=0}^{\infty}\sum_{m=n}^{\infty}i^n\left[f_{mn}(\zeta)\vec{M}_{omn}^{\vec{r}(4)}(c;\eta,\xi,\phi)-ig_{mn}(\zeta)\vec{N}_{emn}^{\vec{r}(4)}(c;\eta,\xi,\phi)\right] \tag{2}$$

where, for axially incidence

$$g_{1n}(0)=f_{1n}(0)=2\Lambda_{1n}^{-1}\sum_{r=0,1}^{\infty}{}'d_r^{1n} \tag{3}$$

The internal and scattered waves can be set up in the same form as that of incident wave as follows:

$$\vec{E}^{(s)}=\sum_{n=0}^{\infty}\sum_{m=n}^{\infty}i^n\left[\beta_{mn}(\zeta)\vec{M}_{emn}^{\vec{r}(3)}\left(c^{(1)};\eta,\xi,\phi\right)+i\alpha_{mn}(\zeta)\vec{N}_{omn}^{\vec{r}(3)}\left(c^{(1)};\eta,\xi,\phi\right)\right] \tag{4}$$

$$\vec{H}^{(s)}=\frac{k_1}{\omega\mu_1}\sum_{n=0}^{\infty}\sum_{m=n}^{\infty}i^n\left[\alpha_{mn}(\zeta)\vec{M}_{omn}^{\vec{r}(3)}\left(c^{(1)};\eta,\xi,\phi\right)-i\beta_{mn}(\zeta)\vec{N}_{emn}^{\vec{r}(3)}\left(c^{(1)};\eta,\xi,\phi\right)\right] \tag{5}$$

$$\vec{E}^{(t)}=\sum_{n=0}^{\infty}\sum_{m=n}^{\infty}i^n\left[\delta_{mn}(\zeta)\vec{M}_{emn}^{\vec{r}(1)}\left(c^{(2)};\eta,\xi,\phi\right)+i\gamma_{mn}(\zeta)\vec{N}_{omn}^{\vec{r}(1)}\left(c^{(2)};\eta,\xi,\phi\right)\right] \tag{6}$$

$$\vec{H}^{(t)}=\frac{k_2}{\omega\mu_2}\sum_{n=0}^{\infty}\sum_{m=n}^{\infty}i^n\left[\gamma_{mn}(\zeta)\vec{M}_{omn}^{\vec{r}(1)}\left(c^{(2)};\eta,\xi,\phi\right)-i\delta_{mn}(\zeta)\vec{N}_{emn}^{\vec{r}(1)}\left(c^{(2)};\eta,\xi,\phi\right)\right] \tag{7}$$

Since

$$\vec{M}_{emn}^{\vec{r}(1)}(c;\eta,\xi,\phi)=\frac{1}{2}\left[\vec{M}_{emn}^{\vec{r}(3)}(c;\eta,\xi,\phi)+\vec{M}_{emn}^{\vec{r}(4)}(c;\eta,\xi,\phi)\right] \tag{8}$$

$$\vec{N}_{omn}^{\vec{r}(1)}(c;\eta,\xi,\phi)=\frac{1}{2}\left[\vec{N}_{omn}^{\vec{r}(3)}(c;\eta,\xi,\phi)+\vec{N}_{omn}^{\vec{r}(4)}(c;\eta,\xi,\phi)\right], \tag{9}$$

$\vec{M}_{emn}^{\vec{r}(4)}(c;\eta,\xi,\phi),\vec{N}_{omn}^{\vec{r}(4)}(c;\eta,\xi,\phi)$ are used in Eqs. (1) and (2), then the finally scattering coefficients should be

$$a_{mn}=\frac{1}{2}[1-\alpha_{mn}] \tag{10}$$

$$b_{mn}=\frac{1}{2}[1-\beta_{mn}], \tag{11}$$

where the first term 1/2 of a_{mn} and b_{mn} represents the diffraction of incident waves around the spheroid.

With the help of method by Asano[1], we can obtain the following boundary conditions

$$D[X] = f^{ext} \tag{12}$$

where

$$D = \begin{bmatrix} V_{mn}^{(3),t}(c^{(1)}) & U_{mn}^{(3),t}(c^{(1)}) & -V_{mn}^{(1),t}(c^{(2)}) & -U_{mn}^{(1),t}(c^{(2)}) \\ U_{mn}^{(3),t}(c^{(1)}) & V_{mn}^{(3),t}(c^{(1)}) & -\frac{k_2\mu_1}{k_1\mu_2}U_{mn}^{(1),t}(c^{(2)}) & -\frac{k_2\mu_1}{k_1\mu_2}V_{mn}^{(1),t}(c^{(2)}) \\ Y_{mn}^{(3),t}(c^{(1)}) & X_{mn}^{(3),t}(c^{(1)}) & -Y_{mn}^{(1),t}(c^{(2)}) & -X_{mn}^{(1),t}(c^{(2)}) \\ X_{mn}^{(3),t}(c^{(1)}) & Y_{mn}^{(3),t}(c^{(1)}) & -\frac{k_2\mu_1}{k_1\mu_2}X_{mn}^{(1),t}(c^{(2)}) & -\frac{k_2\mu_1}{k_1\mu_2}Y_{mn}^{(1),t}(c^{(2)}) \end{bmatrix} \tag{13}$$

$$X = \begin{bmatrix} \alpha_{mn} \\ \beta_{mn} \\ \gamma_{mn} \\ \delta_{mn} \end{bmatrix} \tag{14}$$

$$f^{ext} = -\begin{bmatrix} 0 & 0 & V_{mn}^{(4),t}(c^{(1)}) & U_{mn}^{(4),t}(c^{(1)}) \\ 0 & 0 & U_{mn}^{(4),t}(c^{(1)}) & V_{mn}^{(4),t}(c^{(1)}) \\ 0 & 0 & Y_{mn}^{(4),t}(c^{(1)}) & X_{mn}^{(4),t}(c^{(1)}) \\ 0 & 0 & X_{mn}^{(4),t}(c^{(1)}) & Y_{mn}^{(4),t}(c^{(1)}) \end{bmatrix} \begin{bmatrix} 0 \\ 0 \\ f_{mn} \\ g_{mn} \end{bmatrix} \tag{15}$$

The Reflection and Refraction Coefficients

When the incoming wave encounters the interface at $\xi = \xi_0$, a portion of it is transmitted into the spheroid, and another portion is reflected back to the medium. Then the incident, internal and scattered waves can be expressed as

$$\vec{E}^{(i)} = g_{mn}\vec{M}_{emn}^{\vec{r}(4)}\left(c^{(1)};\eta,\xi,\phi\right) + if_{mn}\vec{N}_{omn}^{\vec{r}(4)}\left(c^{(1)};\eta,\xi,\phi\right) \tag{16}$$

$$\vec{H}^{(i)} = \frac{k_1}{\omega\mu_1}\left[f_{mn}\vec{M}_{omn}^{\vec{r}(4)}\left(c^{(1)};\eta,\xi,\phi\right) - ig_{mn}\vec{N}_{emn}^{\vec{r}(4)}\left(c^{(1)};\eta,\xi,\phi\right)\right] \tag{17}$$

$$\vec{E}^{(s)} = R_{mn,TE}^{212}\vec{M}_{emn}^{\vec{r}(3)}\left(c^{(1)};\eta,\xi,\phi\right) + iR_{mn,TM}^{212}\vec{N}_{omn}^{\vec{r}(3)}\left(c^{(1)};\eta,\xi,\phi\right) \tag{18}$$

$$\vec{H}^{(s)} = \frac{k_1}{\omega\mu_1}\left[R_{mn,TM}^{212}\vec{M}_{omn}^{\vec{r}(3)}\left(c^{(1)};\eta,\xi,\phi\right) - iR_{mn,TE}^{212}\vec{N}_{emn}^{\vec{r}(3)}\left(c^{(1)};\eta,\xi,\phi\right)\right] \tag{19}$$

$$\vec{E}^{(t)} = T^{21}_{mn,TE}\vec{M}^{r(4)}_{emn}\left(c^{(2)};\eta,\xi,\phi\right) + iT^{21}_{mn,TM}\vec{N}^{r(4)}_{omn}\left(c^{(2)};\eta,\xi,\phi\right) \tag{20}$$

$$\vec{H}^{(t)} = \frac{k_2}{\omega\mu_2}\left[T^{21}_{mn,TM}\vec{M}^{r(4)}_{omn}\left(c^{(2)};\eta,\xi,\phi\right) - iT^{21}_{mn,TE}\vec{N}^{r(4)}_{emn}\left(c^{(2)};\eta,\xi,\phi\right)\right] \tag{21}$$

Similarly, we can obtain the boundary conditions of matrix form

$$D^{in}\left[X^{in}\right] = f^{in} \tag{22}$$

where

$$D^{in} = \begin{bmatrix} V^{(3),t}_{mn}(c^{(1)}) & U^{(3),t}_{mn}(c^{(1)}) & -V^{(4),t}_{mn}(c^{(2)}) & -U^{(4),t}_{mn}(c^{(2)}) \\ U^{(3),t}_{mn}(c^{(1)}) & V^{(3),t}_{mn}(c^{(1)}) & -\frac{k_2\mu_1}{k_1\mu_2}U^{(4),t}_{mn}(c^{(2)}) & -\frac{k_2\mu_1}{k_1\mu_2}V^{(4),t}_{mn}(c^{(2)}) \\ Y^{(3),t}_{mn}(c^{(1)}) & X^{(3),t}_{mn}(c^{(1)}) & -Y^{(4),t}_{mn}(c^{(2)}) & -X^{(4),t}_{mn}(c^{(2)}) \\ X^{(3),t}_{mn}(c^{(1)}) & Y^{(3),t}_{mn}(c^{(1)}) & -\frac{k_2\mu_1}{k_1\mu_2}X^{(4),t}_{mn}(c^{(2)}) & -\frac{k_2\mu_1}{k_1\mu_2}Y^{(4),t}_{mn}(c^{(2)}) \end{bmatrix} \tag{23}$$

$$X^{in} = \begin{bmatrix} R^{212}_{mn,TM} \\ R^{212}_{mn,TE} \\ T^{21}_{mn,TM} \\ T^{21}_{mn,TE} \end{bmatrix} \tag{24}$$

$$f^{in} = -\begin{bmatrix} 0 & 0 & V^{(4),t}_{mn}(c^{(1)}) & U^{(4),t}_{mn}(c^{(1)}) \\ 0 & 0 & U^{(4),t}_{mn}(c^{(1)}) & V^{(4),t}_{mn}(c^{(1)}) \\ 0 & 0 & Y^{(4),t}_{mn}(c^{(1)}) & X^{(4),t}_{mn}(c^{(1)}) \\ 0 & 0 & X^{(4),t}_{mn}(c^{(1)}) & Y^{(4),t}_{mn}(c^{(1)}) \end{bmatrix}\begin{bmatrix} 0 \\ 0 \\ f_{mn} \\ g_{mn} \end{bmatrix} \tag{25}$$

For the outgoing waves, similar boundary conditions are obtained

$$D^{out}\left[X^{out}\right] = f^{out} \tag{26}$$

where

$$D^{out} = \begin{bmatrix} V_{mn}^{(3),t}(c^{(1)}) & U_{mn}^{(3),t}(c^{(1)}) & -V_{mn}^{(4),t}(c^{(2)}) & -U_{mn}^{(4),t}(c^{(2)}) \\ U_{mn}^{(3),t}(c^{(1)}) & V_{mn}^{(3),t}(c^{(1)}) & -\frac{k_2\mu_1}{k_1\mu_2}U_{mn}^{(4),t}(c^{(2)}) & -\frac{k_2\mu_1}{k_1\mu_2}V_{mn}^{(4),t}(c^{(2)}) \\ Y_{mn}^{(3),t}(c^{(1)}) & X_{mn}^{(3),t}(c^{(1)}) & -Y_{mn}^{(4),t}(c^{(2)}) & -X_{mn}^{(4),t}(c^{(2)}) \\ X_{mn}^{(3),t}(c^{(1)}) & Y_{mn}^{(3),t}(c^{(1)}) & -\frac{k_2\mu_1}{k_1\mu_2}X_{mn}^{(4),t}(c^{(2)}) & -\frac{k_2\mu_1}{k_1\mu_2}Y_{mn}^{(4),t}(c^{(2)}) \end{bmatrix} \tag{27}$$

$$X^{out} = \begin{bmatrix} T_{mn,TM}^{12} \\ T_{mn,TE}^{12} \\ R_{mn,TM}^{121} \\ R_{mn,TE}^{121} \end{bmatrix} \tag{28}$$

$$f^{out} = \begin{bmatrix} 0 & 0 & V_{mn}^{(3),t}(c^{(2)}) & U_{mn}^{(3),t}(c^{(2)}) \\ 0 & 0 & \frac{k_2\mu_1}{k_1\mu_2}U_{mn}^{(3),t}(c^{(2)}) & \frac{k_2\mu_1}{k_1\mu_2}V_{mn}^{(3),t}(c^{(2)}) \\ 0 & 0 & Y_{mn}^{(3),t}(c^{(2)}) & X_{mn}^{(3),t}(c^{(2)}) \\ 0 & 0 & \frac{k_2\mu_1}{k_1\mu_2}X_{mn}^{(3),t}(c^{(2)}) & \frac{k_2\mu_1}{k_1\mu_2}Y_{mn}^{(3),t}(c^{(2)}) \end{bmatrix} \begin{bmatrix} 0 \\ 0 \\ f_{mn} \\ g_{mn} \end{bmatrix} \tag{29}$$

All unknown coefficients in Eqs.(24) and (28) can be easily obtained by using the Cramer rules.

Construction of the Debye Series Expansion

The operator D expressed by Eq.(13) can be split into

$$D = D^L + F^{\text{int}} \tag{30}$$

where

$$D^L = D^{in} = D^{out} \tag{31}$$

and

$$F^{\text{int}} = \begin{bmatrix} 0 & 0 & -V_{mn}^{(3),t}(c^{(2)}) & -U_{mn}^{(3),t}(c^{(2)}) \\ 0 & 0 & -\frac{k_2\mu_1}{k_1\mu_2}U_{mn}^{(3),t}(c^{(2)}) & -\frac{k_2\mu_1}{k_1\mu_2}V_{mn}^{(3),t}(c^{(2)}) \\ 0 & 0 & -Y_{mn}^{(3),t}(c^{(2)}) & -X_{mn}^{(3),t}(c^{(2)}) \\ 0 & 0 & -\frac{k_2\mu_1}{k_1\mu_2}X_{mn}^{(3),t}(c^{(2)}) & -\frac{k_2\mu_1}{k_1\mu_2}Y_{mn}^{(3),t}(c^{(2)}) \end{bmatrix} \tag{32}$$

then D_L is the local interaction and F^{int} is the internal excitation.

If the external interaction is expressed by I^{out}, then

$$D_L\left[I^{out}\right]=f^{ext}=f^{in} \tag{33}$$

with help of Eq.(22), we can obtain

$$I^{out}=X^{in} \tag{34}$$

If the internal interaction is expressed by R^{in}, then

$$D_L\left[R^{in}\right]=-F^{\text{int}} \tag{35}$$

From Eq.(26), we can obtain

$$R^{in}=\begin{bmatrix} T^{12}_{mn,TM} \\ T^{12}_{mn,TE} \\ R^{121}_{mn,TM} \\ R^{121}_{mn,TE} \end{bmatrix}\begin{bmatrix} 0 \\ 0 \\ f_{mn} \\ g_{mn} \end{bmatrix}' \tag{36}$$

Replacing Eq.(35) into Eq.(30) we can write

$$D=D^L+F^{\text{int}}=D^L-D_L\left[R^{in}\right]=D^L\left[I-R^{in}\right] \tag{37}$$

Taking into consideration Eqs.(12),(33) and (37), we can obtain the formula of Debye series expansion

$$[X]=\left[I-R^{in}\right]^{-1}\left[I^{out}\right] \tag{38}$$

Eq.(38) can also expressed as

$$[X]=\sum_{p=0}^{\infty}\left[R^{in}\right]^{p}\left[I^{out}\right] \tag{39}$$

ANALYSIS ON THE CHARACTERISTICS OF A SPHEROIDAL PARTICLE USING DEBYE SERIES

Eqs. (38) and (39) are the Debye series of axially incident plane wave scattering by a spheroid, and the formulas expand the global physical process in a series of local interactions, and clarifies the physical origins of many effects that occur in electromagnetic scattering. When Eq.(39) is employed to the simulation of scattered field, then the corresponding Debye scattered waves consist of the waves that are transmitted into the spheroid, and propagated out the spheroid after p internal reflected. When all the Debye-series terms are considered, then Eq.(39) equals to Eq.(38), and the scattered waves simulated by Eq.(39) or Eq.(38) agree with that by the Mie theory.

CONCLUSION

The Debye series expansion of axially incident plane-wave scattering by a spheroidal particle is derived and employed to the decomposition of the global physical process in a series of local interactions.

REFERENCES

1. Shoji Asano and Giichi Yamamoto, Appl. Opt. 14, 1975, pp.29-49.
2. Shoji Asano, Appl. Opt. 18, 1979, pp.712-723.
3. HAN Yiping, PhD dissertation, Xidian University,China,2000.
4. Le-Wei Li, Xiao-Kang Kang and Mook-Seng Leong, Hohn Wiley & Sons, Inc.,2002.
5. Carson Flammer, Stanford University Press, 1957.
6. Milton Abramowitz, and Irene A. Stegun, National Bureau of Standards Applied Mathematics Series 55, 1964.
7. J. A. Stratton, McGraw-Hill, New York, 1941.
8. E. A. Hovenac and J. A. Lock, J. Opt. Soc. Am. A 9, 1992, pp.781-795.
9. J. A. Lock, J. M. Jamison, and C.-Y. Lin, Appl. Opt. 33, 1994, pp.4677-4690.
10. R. Li, X. Han, H. Jiang and K. F. Ren, Appl. Opt. 45, 2006, pp.1260-1270.
11. P. Laven, J. Quant. Specrtosc. Radiat. Transfer 89,2004, pp.257-269.
12. J. A. Lock, J. Opt. Soc. Am. A 5, 1988, pp. 2032-2044.
13. James A. Lock and C. L. Adler, J. Opt. Soc. Am. A14, 1997, pp. 1316-1328.
14. R. Li, X. Han, H. Jiang and K. F. Ren, Appl. Opt. 45,2006, pp. 6255-6262.

Study on Near Distributed Dynamic Model of a Multifountain-and-multiconfluent Network System of Steam Main-pipeline and Parallel Coursing Units

Pan Lei and Shen Jiong

School of Energy and Environment, Southeast University, Nanjing, China

Abstract. According to the deficiency of the prior modeling methods on main-pipeline system, the paper advances a near distributed modeling way based on subsection, combination and delamination for parallel coursing units and main-pipeline system to fully show its distributed dynamics by an easy-simulated mode of expression. Firstly, a principle is established to plot out the main-pipeline into some short pipe-sections between each pair of fountain/confluent points. Secondly, a near distributed decoupling transfer function matrix model without steady error for each pipe-section is built by rationally approximate deduction and then joined with the dynamic models of the fountain/confluent at the two ends of the pipe-section to form the near distributed model of each subsection. Finally, a smooth arithmetic is adopted to joint all the conterminous subsections into a whole system model. The modeling method above decreases the dimension of matter space and integrates more influence factors on the system dynamic characteristics into the built model, such as pipe length, diameter and the thermodynamics of working fluid, so it more particularly reflects the distributed dynamic characteristics of a multifountain-and-multiconfluent network system of steam main-pipeline and parallel coursing units than before, and can act well as the simulated research object for advanced main-pipeline system control arithmetic and distributed control technology or even validates them. Some simulation experiments have been done and produced good results to prove the validity of the modeling method.

Keywords: parallel coursing unit, dynamic mathematic model, distributed parameter, decoupling, control, multifountain-and-multiconfluent networks

INTRODUCTION

In China, there are more than 2000 power stations for refineries, chemical plants and paper mills themselves are in the unit parallel-coursing mode through main-pipelines, because lots of steam is required by their production process. But few of the control systems configured for these parallel coursing units are put into automation, because their control strategy is concentrated, which may lead to an unsteady control effect especially to long main-pipeline. To solve the problem, some distributed control strategies have been in study for the press regulation, and the basis of such study is to build a mathematic model to fully show the distributed dynamics of the main-pipeline and the parallel coursing unit system by an easy-simulated mode of expression[1][2].

The system has the distributed-parameter attribute for main-pipeline being, thus the distributed parameter model is more precise than concentrated one. An absolutely distributed mathematic model [3-6] is hard to use directly for the simulation of the tuning strategy because the control arithmetic often presents as ODES. By browsing through the literature on the subject, we know that the existing dynamic model of the parallel coursing units are divided into two kinds by its pipeline model, one is concentrated, another is near distributed. The former[1] simply describes the main-pipeline as an integral module, which is the same with the concentrated control simulation and can't satisfy the requirements of the distributed control simulation. The near distributed modeling [7-11] can approach the distributed characterization of pipeline to a great extent and can be in an easily-simulating expression, so it is a feasibly ideal in modeling the control object of main-pipeline control system. The near distributed modeling method for steam main-pipeline system previously [12] [13] regards the main-pipeline as a

CP914, *Multiphase Flow: The Ultimate Measurement Challenge,* edited by X. Cai, Y. Wu, Z. Huang, S. Wang, and M. Wang

distributed object composed of many short pipelines, and treats each section as a concentrated one, and that considers the steam flow between the neighboring sections, so the model established is close to the actual state to a certain extent and can somewhat satisfy the study of the distributed control arithmetic. But it has several intrinsic deficiencies. Firstly, a mutation on each parameter value exists at the border of subsections. Secondly, the influence by the relative location and distance of boilers and turbines in subsection is not considered. Thirdly, the steam pressure at the unit -and-pipeline joint is very important controlled parameter, but here an average of a whole subsection is used. So some error can not be avoided in the method. Dividing the subsections as short as possible can much approach the continuously distributed characteristics of pipeline in theory, but the actual numbers are limited by the need that one turbine or one boiler must be included in a subsection at least.

According to the deficiencies, the paper advances a new near distributed modeling way based on subsection, combination and delamination.

MATHEMATIC MODELLING

The system to be modeled is a multifountain-and -multiconflux network composed of fountain devices like boilers, confluent devices like steam turbines and a single steam main-pipeline. The way to divide the main-pipeline system into subsections leads to modeling ideal. The modeling method in the paper includes such steps: Firstly, a principle is established to plot out the main-pipeline into some short pipe-sections between each pair of fountain/confluent points. Secondly, a near distributed decoupling transfer function matrix model without steady error is built for each basic pipe-section, and then joined with the dynamic models of the fountain/confluent at the two ends of the pipe-section to form a near distributed model of each basic subsection. Finally, a smooth arithmetic is adopted to joint all the conterminous subsections into a whole system model. The basic subsections can be configured into more complex combinations further for easily modeling the larger main-pipeline system by the way of delamination. So the modeling method includes two emphases illuminated as follow.

Compartmentalization and Combination of Main-pipeline

Known from the analysis above, it is necessary to plot out the pipeline section according to the location of the fountain/confluent points. Defining the joints of every two neighboring turbine/boiler as the two end points of a short section can decrease the dimension of influential factors to distributed dynamics and simplify the

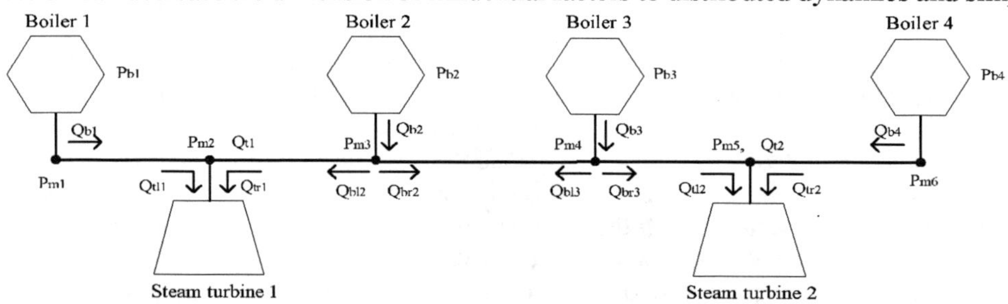

FIGURE 1. Subsections of the main-pipeline system with 4 boilers and 2 turbines

model-solving procedure. Based on the partition principle, the steam main-pipeline system actually in factories is composed of three kinds of subsection that is fountain- conflux, fountain-fountain and conflux-conflux. The first type can be self-balancing; the last two must be connected with other types of subsection.

In the compartmentalization-and-combination modeling, every fountain/conflux together, every fountain and conflux respectively are all numbered 1,2,... from left to right, shown as Fig.1.

Whatever the end type of each pipe-section, the inflow is defined positive and the outflow is negative. When a whole model is combined from subsections, the common joint of every two conterminous subsections must be smooth, that is the parameters at the joint must be satisfied with the smoothly linkage equations which are given by

$$\begin{cases} P_{mli} = P_{mri} = P_{mi}, \text{ when the joint is conflux} \\ Q_{tj} = Q_{tlj} + Q_{trj} \end{cases} \tag{1}$$

or

$$\begin{cases} P_{mli} = P_{mri} = P_{mi}, \text{ when the joint is fountain} \\ Q_{bj} = Q_{blj} + Q_{brj} \end{cases} \tag{2}$$

Where i is the numbers of the fountain/conflux joints on the main-pipeline together, j is the numbers of the fountain or conflux joints respectively, P_{mi} is the steam pressure at $No.i$ fountain/conflux joint, P_{mli} is the $No.i$ fountain/conflux joint pressure calculated from its left subsection, P_{mri} is the $No.i$ fountain/conflux joint pressure calculated from its right subsection. Q_{blj} is the steam flux of the $No.j$ fountain-joint passing in or out its left pipe-section, Q_{brj} is the steam flux of the $No.j$ fountain-joint passing in or out its right pipe-section, Q_{tlj} is the steam flux of the $No.j$ conflux-joint passing in or out its left pipe-section,, Q_{trj} is the steam flux of the $No.j$ conflux-joint passing in or out its right pipe-section. Because the proportion of the two branches of the total flow at each two-subsection-joint is unknown, its balanceable parameters should be solved by iterative arithmetic according to equations (1) or (2).

Near Distributed-parameter Model of the Basic Subsection

The model of each basic subsection is an organic whole of the dynamics of the boiler, turbine and pipe-section. Models of boilers and steam turbines are still the same to power unit's ones actually [12] [14], and the main difference is the model of pipe-section.

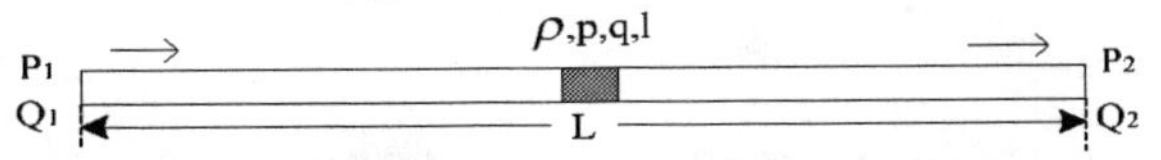

FIGURE 2. A steam pipe-section

The Near Distributed Decoupling Transfer Function Matrix Model of Pipe-section

Fig.2 shows a section of the steam main-pipeline. Inflows are defined positive and outflows are negative. The relationship between flux disturbance and pressure variety at the two fountain/conflux joint of each subsection is the emphasis of modeling the controlled object of parallel coursing units, so each pipe-section is regarded as a 2×2 multivariable object like

$$\begin{pmatrix} \Delta p_1 \\ \Delta p_2 \end{pmatrix} = \begin{pmatrix} W_{11} & W_{12} \\ W_{21} & W_{22} \end{pmatrix} \begin{pmatrix} \Delta Q_1 \\ \Delta Q_2 \end{pmatrix} \tag{3}$$

where

$$\Delta p_1 = p_1 - p_{10}, \Delta p_2 = p_2 - p_{20}, \Delta Q_1 = Q_1 - Q_0, \Delta Q_2 = Q_2 - Q_0 \tag{4}$$

and w_{ij} represents the transfer function from the flow to the pressure at the ends of a pipe-section, here $i, j = 1,2$, '1' denotes the higher-pressure end of a pipe-section, '2' denotes the lower-pressure end of a pipe-section, j belongs to the input and i belongs to the output. P_1 is the higher pressure and P_2 is the lower pressure at the two ends, P_{10} and P_{20} are their corresponding steady-state pressures . Q_0 is the initial steady-state flow, Q_1 and Q_2 are the flows of the higher-pressure end and the lower- pressure end of a pipe-section respectively.

For the least energy loss in flow, the minimal pressure-difference distribution down the main-pipeline is the target pursued in designing, configuring and running a steam main-pipeline, so the steam pressure in the main-pipeline actually varies very flatly, moreover a pipe-section is not very long, accordingly the pressure ratio and temperature ratio of the two ends of each pipe-section are close to 1, and a variety of steam density is ignorable, then a approximate symmetry is obtained that the influence to pressure by flow disturbance at any end or from one end to another is nearly equal to the opposite, that is $w_{11} \approx w_{22}, w_{12} \approx w_{21}$. So equations (3) can be transformed into the decoupling form as:

$$\begin{pmatrix} \Delta p_a \\ \Delta p_s \end{pmatrix} = \begin{pmatrix} W_{11}+W_{12} & 0 \\ 0 & W_{11}-W_{12} \end{pmatrix} \begin{pmatrix} \Delta Q_a \\ \Delta Q_s \end{pmatrix} \tag{5}$$

where

$$\begin{pmatrix} \Delta p_a \\ \Delta p_s \end{pmatrix} = \begin{pmatrix} (\Delta p_1+\Delta p_2)/2 \\ (\Delta p_1-\Delta p_2)/2 \end{pmatrix}, \begin{pmatrix} \Delta Q_a \\ \Delta Q_s \end{pmatrix} = \begin{pmatrix} (\Delta Q_1+\Delta Q_2)/2 \\ (\Delta Q_1-\Delta Q_2)/2 \end{pmatrix} \tag{6}$$

Apply the predigestion and approximation method above to steam main-pipeline, the completely distributed analytic model of pipeline [15] can be transformed by equations (5) and (6) into decoupling matrix for steam pipe-section

$$W_{11}(s)+W_{12}(s) = k_a \cdot \frac{1}{s} + k_b \cdot \sum_{i=1}^{N} \frac{T_{2i}}{T_{2i}\cdot s+1} + \varepsilon_1(s) \tag{7}$$

$$W_{11}(s)-W_{12}(s) = k_b \cdot \sum_{i=1}^{N} \frac{T_{2i-1}}{T_{2i-1}\cdot s+1} + \varepsilon_2(s) \tag{8}$$

where

$$k_a = \frac{\gamma(p_{10}+p_{20})}{V_0 \rho}, k_b = \frac{3\gamma(p_{10}^2-p_{20}^2)^2 k}{2V_0\rho(k-1)}, \quad k = \left(\frac{p_{10}}{p_{20}}\right)^{3/2}, \quad V_0 = \frac{\pi D^2 L}{4}, \quad T_1 = \frac{16L^2\xi Q_0}{(p_{10}^2-p_{20}^2)^2 \gamma\pi D^3} \Big/ \left[\frac{3\pi}{2(k-1)\cdot p_{20}^{3/2}}\right]^2, \quad T_i = \frac{T_1}{i^2} \tag{9}$$

here ρ is the average steam density of a short pipe-section, the pressure ratio and temperature ratio of the two ends of each pipe-section are close to 1。 D represents the diameter of pipeline, L represents the length of pipeline, ξ represents resistance coefficient along the line, γ represents adiabatic index, N denotes the item numbers reserved , $\varepsilon_1(s)$ and $\varepsilon_2(s)$ are the interception errors.

The Near Distributed Pipe-section Model of Finite Item with Zero Steady Error

Try to find the approximate expressions of the errors $\varepsilon_1(s)$ and $\varepsilon_2(s)$, assuming

$$\varepsilon_2(s) = k_b \cdot \frac{T_j}{T_j s+1} \tag{10}$$

The pressure control of steam main pipe-line belongs to the set-value adjustment and has a high-level request for the steady precision of its object model, so the item for compensating the interception error is solved on Laplace Ultimate Theorem. From the Theorem:

$$(\Delta p_1-\Delta p_2)_1 = \lim_{s\to 0}\left\{s\cdot[W_{11}(s)-W_{12}(s)]\cdot\frac{(\Delta Q_1-\Delta Q_2)_0}{s}\right\} \approx \lim_{s\to 0}\left\{s\cdot\left[k_b\cdot\sum_{i=1}^{N}\frac{T_{2i-1}}{T_{2i-1}\cdot s+1}+\varepsilon_2(s)\right]\cdot\frac{(\Delta Q_1-\Delta Q_2)_0}{s}\right\} \tag{11}$$

$$= \lim_{s\to 0}\left\{s\cdot\left[k_b\cdot\left(\sum_{i=1}^{N}\frac{T_{2i-1}}{T_{2i-1}\cdot s+1}+\frac{T_j}{T_j s+1}\right)\right]\frac{(\Delta Q_1-\Delta Q_2)_0}{s}\right\} = k_b\cdot\left(\sum_{i=1}^{N}T_{2i-1}+T_j\right)(\Delta Q_1-\Delta Q_2)_0$$

where, $(\Delta Q_1-\Delta Q_2)_0$ denotes the disturbance increment of flow at the moment of $t=0^+$, $(\Delta p_1-\Delta p_2)_1$ denotes the steady increment of $\Delta p_1-\Delta p_2$ at the ultimate time. From (11) yields:

$$T_j = \frac{(\Delta p_1-\Delta p_2)_1}{(\Delta Q_1-\Delta Q_2)_0\cdot k_b} - \sum_{i=1}^{N} T_{2i-1} \tag{12}$$

The value of $(\Delta p_1-\Delta p_2)_1/(\Delta Q_1-\Delta Q_2)_0$ in expression (12) can be obtained by the field test of little dynamic disturbance on pipeline. The inertia taches in transfer functions (7) and (8) are alternate with the odd and even suffixes, so the solution is

$$\varepsilon_1(s) = k_b \cdot \frac{T_{j+1}}{T_{j+1}s+1} \tag{13}$$

Where

$$T_{j+1} = \frac{T_1}{(j+1)^2}, \quad j \approx \left\lfloor \sqrt{\frac{T_1}{T_j}} \right\rfloor \tag{14}$$

So the transfer functions（7）,（8）and the error compensating expressions（10）,（12）,（13）,（14）constitute the near distributed decoupling pipe-section model of finite-item and Zero steady error.

SIMULATION AND DISCUSSION

The main-pipeline and parallel coursing unit system shown in Fig.1 has been modeled to verify the method. The parameters of chemical units has been adopted in the model, the data is: $\gamma = 1.3$, $\rho = 34.18(kg/m^3)$, the rating of each steam turbine valve is 50 % , the value coefficient $K_t = 60.67(kg/(s\cdot\%))$, the steady drum pressures are: $P_{b1} = 10.5, P_{b2} = 9.24, P_{b3} = 10.46, P_{b4} = 10.5$; the initial pressures of joints on the main-pipeline are P_{m1}=9.1, $P_{m2} = 8.14, P_{m3} = 8.3, P_{m4} = 8.6, P_{m5} = 8.14, P_{m6} = 9.1$, the unit of pressure is MP_a; the steady flow of boilers are $Q_{b1} = 30.5, Q_{b2} = 20.5, Q_{b3} = 40.5, Q_{b4} = 30.5$; the flow of turbines are $Q_{t1} = Q_{t2} = 61$, the unit of flow is kg/s; the length of each pipe-section form left to right is: $L_1 = 50, L_2 = 25, L_3 = 25, L_4 = 25, L_2 = 50$, the diameter of pipeline is 0.5, the unit of length is m。 Do the experiment of combustion rate disturbance。 After 10s running at steady state, the fuel flow M_1 fed into boiler 3 steps from 4.98(kg/s) up to 5.478(kg/s), and the increment is 10%. The influence to all the main-pipeline system is shown in Fig.3.

Curves in Fig.3 reflect the basically running and regulating principles of all main-pipeline systems that are "priority in group "and "priority against distance". The boilers, turbines and pipelines close in geography in main-pipeline systems are often collocated as a balanceable supply-and-demand group in steady state in order to minimize the flow loss in running. In the system shown as Fig.3, boiler 1, pipe-section 1, turbine 1, pipe-section 2 and boiler 2 are the first group, boiler 3, pipe-section 4, turbine 2, pipe-section 5 and boiler 4 are the second group. Known from Fig.7 (a), the increasing rate of the steam pressures at each joint of the main-pipeline in the dynamic response are $P_{m4}, P_{m5}, P_{m6}, P_{m3}, P_{m2}, P_{m1}$ by the order from big to small, which shows that when fuel step in boiler 3 of group 2 occurs, the influence on the main-pipeline pressure P_{m4}, P_{m5}, P_{m6} in the same group is larger than the pressure P_{m3}, P_{m2}, P_{m1} in another group on the whole, and the influence on pipeline pressure in each group is decreasing by its distance to the disturbance spot, viz. P_{m4}, P_{m5}, P_{m6} by the decreasing order of variety rate in group 2 and P_{m3}, P_{m2}, P_{m1} in group 1.

The principle of "priority in group " means that a high aggregation characteristic in group and a low coupling characteristic among groups in every reasonable configured steam main-pipeline system, which helps to the study in control strategy on pressure and load. The principle of "priority against distance" means that the responding rate of the pressure along pipeline is decreasing by its distance to the disturbance spot. The experiment completely validates the two basic principles, so some new control ideas like distributed control strategy can be simulated and validated on such model.

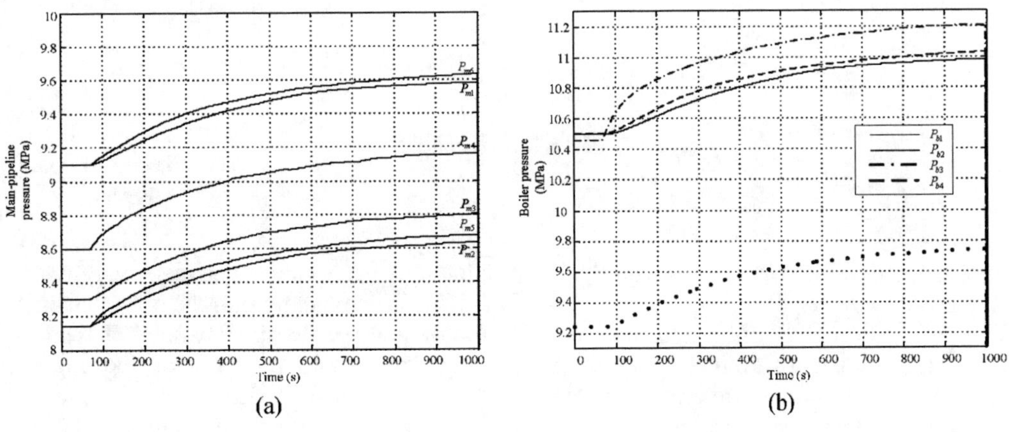

(a) (b)

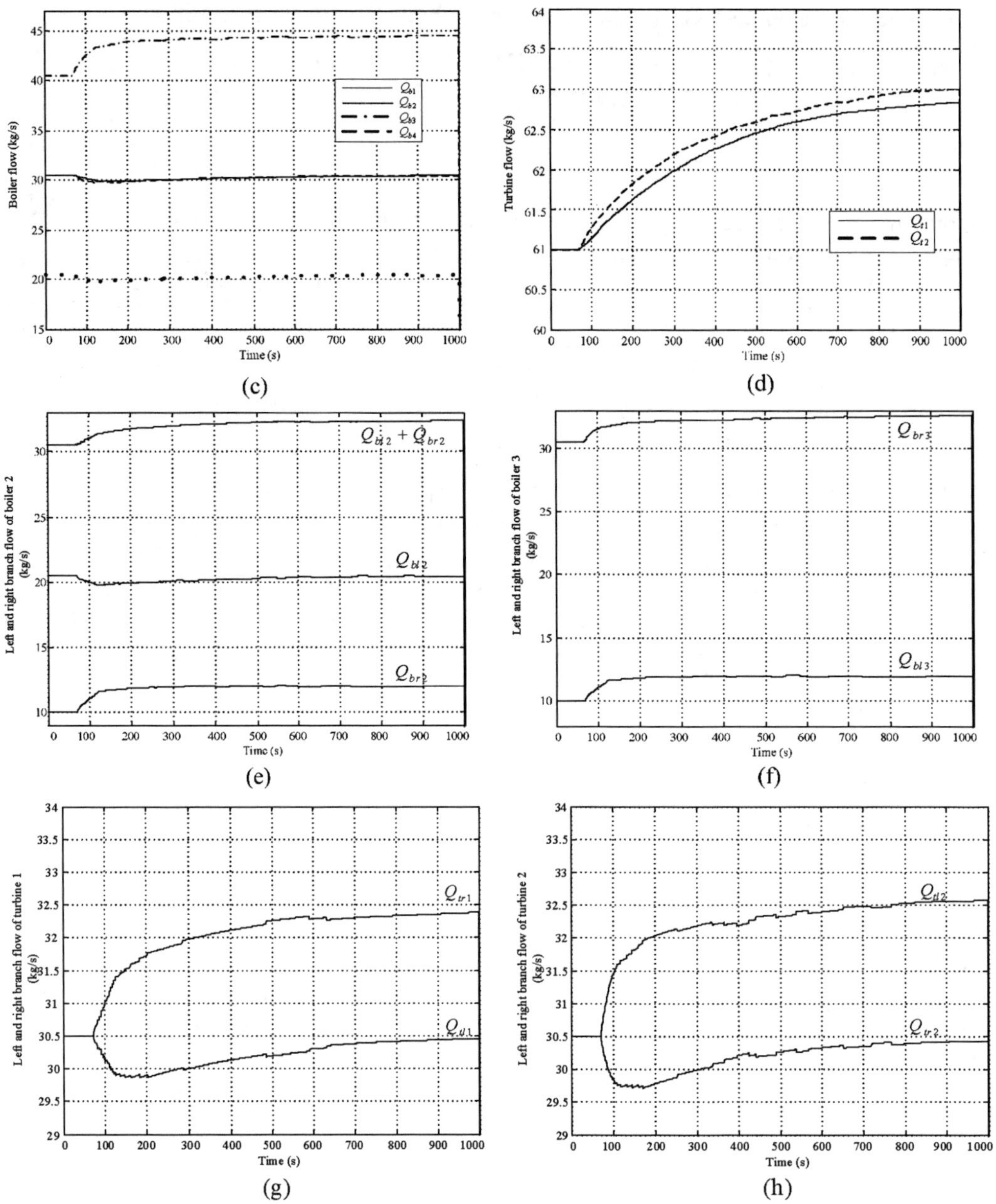

FIGURE 3. Step response curves for 10% disturbance of the fuel fed into a boiler

CONCLUSIONS

The modeling method in the paper has solved the jump in the parameter values between every two conterminous subsections in the prior modeling methods, and integrates more influence factors into the built dynamic model, such as the relative position between boilers and turbines, pipe length and the thermodynamics of working fluid, so it more particularly reflects the distributed dynamic characteristics of a multi fountain/confluent network of steam main-pipeline and parallel coursing units than before. The modeling method decreases the dimension of matter space and makes it easy to model large main-pipeline system by subsection, combination and delamination to the system. The modeling method applies to all the networks composed of the fountains, confluents and the steam-transporting pipeline and the model built is fit well in simulating, studying and verifying the advanced control arithmetic and distributed control technology on main-pipeline system in time domain by its easy-simulated expression mode of transfer function matrix. Some simulation experiments have been done and have well validated the basic principles of "priority in group "and "priority against distance" of main pipe-line systems, which proves the validity of the modeling method.

ACKNOWLEDGEMENTS

The authors are grateful to the support by National Natural Science Foundation of China (50576011), "Study on the pressure control system of parallel coursing units based on immune agent".

REFERENCES

1. Zhao Dongxiao. The pressure control system of parallel coursing boilers based on direct energy balance theory [J]. East China Electric Power, 2002, 10:40-42
2. Pan Weijia. Analysis and study of the combustion control systems of parallel coursing boilers [J]. Industrial instrumentation and automation, 2000, 3: 7-9
3. Wu Chongshi. Methods of Mathematical Physics [M]. Peking university press, 2003:185-291.
4. Oke, Adeyemi ; Mahgerefteh, Haroun; Economou, Ioannis; Rykov, Yuri.A transient outflow model for pipeline puncture[J].Chemical Engineering Science, 2003,58(20): 4591-4604
5. Chen, J.R. A simplified numerical method for transient two-phase pipe flow [J]. Transactions of the Institution of Chemical Engineering, 1993, 71A:304-309
6. Haroun Mahgerefteh. Modelling outflow following rupture in pipeline networks [J]. Chemical Engineering Science, 2006, 61:1811 – 1818
7. Rejowski Jr., R. Scheduling of a multiproduct pipeline system [J]. Computers and Chemical Engineering, 2003, 27(8): 1229-1246
8. Galerkin, Y.B., Danilov, K.A. Gas dynamic design of powerful pipeline compressors not based on model tests[C]. IMechE, Fluid Machinery Group - International Conference on Compressors and their systems 2005, 2005: 243-252
9. Blazic, Saso; Matko, Drago; Geiger, Gerhard. Simple model of a multi-batch driven pipeline [J].Mathematics and Computers in Simulation, 2004, 64(6): 617-630
10. Lindeloff, Niels; Krejbjerg, Kristian.A compositional model simulating wax deposition in pipeline systems [J].Energy and Fuels, 2002, 16(4): 887-891
11. Zhu, G.-Y. Henson, M.A.; Megan, L.Dynamic modeling and linear model predictive control of gas pipeline networks[J]. Journal of Process Control, 2001, 11(2): 129-148
12. Hou Yiwen. A study on the control system of parallel coursing boiler' mail-pipeline pressure based on multiple-agent system [D]. A Dissertation submitted to Southeast University for academic degree of master, Nanjing: Department of power engineering, 2004.2
13. Xie Yingjun, Chen Ying. Dynamic analysis of the pipelines of hydraulic pressure balancing system [J]. Hydraulics pnenmatics and seals, 1998, 1:18-21.
14. Zeng Deliang, Zhao Zheng, Chen Yanqiao,et al. A practical 500MW boiler dynamic model analysis [J]. Proceedings of the CSEE.2003, 23(5): 149-152
15. Ni Weidou. Study on some questions in modeling of thermodynamics system [M]. Science press, 1996:275-331

Experimental Study on Surface Wave and Film Breakdown of Falling Liquid Film Flow

Tang Zhiwei*, Yan Xiaoke+, Jiang Zhangyan*3

*College of Environmental and Energy Engineering, Beijing University of Sciences, Beijing, China, 100022, tangzhiw@263.net
+National Institute of Metrology, Beijing, China, 100013, yanxk@nim.ac.cn
*3Institute of Engineering Thermophysics, Chinese Academy of Sciences, Beijing, China, 100080, zhangyanj@hotmail.com

Abstract. In order to evaluate characteristics of the liquid film flow and their influences on heat and mass transfer, measurements of the instantaneous film thickness using a capacitance method and observation of film breakdown are performed. Experimental results are reported in the paper. Experiments are carried out at Re =250-10000, T_{in}=20℃ - 50℃ and three axial positions of vertically falling liquid films for film thickness measurements. Instantaneous surface waveshapes are given by the interpretation of the test data using the cubic spline method. The correlation of the mean film thickness *vs* the film Reynolds number is also given by fitting the test data. It is revealed that the surface wave has nonlinear behavior. Observation of film breakdown is performed at Re =1.40×10^3-1.75×10^4 and T_{in}=85℃ - 95℃ . From experimental results, the correlation of the film breakdown criterion can be obtained as follows: $Bd = 1.567\times10^{-6}\,Re^{1.183}$

Keywords: two phase flow, falling films, film thickness measurements, capacitance method, film breakdown

INTRODUCTION

Annular two phase and freely falling liquid film flow are important two phase flow patterns. For gravity heat pipes (closed two-phase thermosyphons), the thin liquid film established in the condenser persists into evaporator which have complex hydrodynamic and heat transfer phenomena from a subcooled to a saturated state. These processes influence obviously on design and operation situations of heat pipes. Besides, These flow patterns are also some enhanced heat and mass transfer technologies used extensively for many industry areas, such as chemical and petrochemical industry, food engineering, pharmaceutical plants and advanced containment systems for next generation nuclear power plants[1], etc. Many theoretical and experimental studies have been reported in literatures[2-5] in order to evaluate quantitatively characteristics of the liquid film flow and their influences on heat and mass transfer. Especially, using a multiple-channel conductance method, measurements of surface wave for the freely-falling liquid film flow at the ambient temperature were carried out and nonlinear evolution have been revealed by the experiments[6]. However, knowledge on some interested phenomena of this flow pattern (such as the surface wave evolution at the temperature higher than the ambient temperature and the film breakdown , etc.) is still rather poor, and a better understanding of this field is needed.

The present study is aimed at this direction experimentally. For measuring the film thickness at heated film flow conditions, a capacitance measurement system is at first developed. It is well known that up to now, there are a variety of methods for measuring the film thickness[7], among theme, the conductance method and the capacitance method are quite sensitive, accurate and cheaper, however, the former can not be used for a heated test section. Measuring surface waves of falling film flow are performed at Re =250-10000, T_{in}=20℃ - 50℃ and three axial positions of vertically falling liquid films. Then, observation of film breakdown is carried out at Re =1.40×10^3-1.75×10^4 and T_{in}=85℃ - 95℃. From the interpretation of the test data, the correlation of the mean film thickness and the film breakdown criterion number *vs* the film Reynolds number are given respectively.

CP914, *Multiphase Flow: The Ultimate Measurement Challenge,* edited by X. Cai, Y. Wu, Z. Huang, S. Wang, and M. Wang

CAPACITANCE SENSOR AND EXPERIMENTS

As above, experiments are performed at a heated test section, thus, it is necessary to develop a measurement system of film thickness using a capacitance method. There are two schemes for reaching the goal. The one is measuring voltage variation across two capacitor plates after they are polarized; the other is measuring capacitance variation for the capacitor connected to a bridge circuit or detecting frequency variation of a resonance circuit controlled by the capacitor. The later is selected for the present study as it is lower power

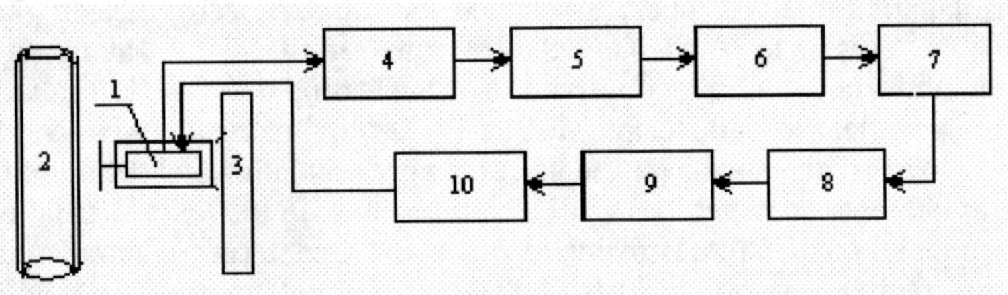

FIGURE 1. Measurement system of film thickness using a capacitance method
1. Measurement probe/oscillator (capacitance sensor) 2.Test section 3.Position fixing tube 4.Frequency converter/amplitude limiter 5.Discriminator 6.Amplifier 7. Frequency / voltage Converter 8.Linearizer 9.A/D converter 10.Microcomputer

dissipation and polarization of the experimental surface can be overcome by means of high frequency developed from the capacitance probe. The block diagram of the capacitance measurement system is shown in Fig.1. The capacitance sensor composed of the measurement probe and the variable oscillator is fixed at the position fixing tube and maintained precisely a distance of 4 mm from the surface of the test section. Measurement circuits composed of the other parts are installed in the microcomputer. The measurement probe is a square plate of 8mm x 8mm, which consists one of two capacitor plates. The surface of the test section is another earth capacitor plate. Variation of the film thickness results in capacitance change of the probe relative to earth and then results in variation of the oscillator frequency. This frequency variation is finally detected as voltage outputs which indicate variation of the film thickness after a series of converting are carried out through measurement circuits. It is well known that the capacitance is given as follows:

$$C=A/\{[(H-h)/\varepsilon_1]+(h/\varepsilon_2)\}\approx A/(H-h) \quad (1)$$

Thanks to $\varepsilon_1=1$ for air and $\varepsilon_2>>1$ for water. The capacitance variation of the probe is about 0.5 pF from calculation according to the above equation, thus, it is paid attention to adjusts of the measurement circuits, such as influence of the lead distribution capacitance, the frequency of the variable oscillator and the frequency converter, etc..

Calibration of the capacitance measurement system is carried out on still water films measured precisely using mechanical methods and a PY-15 digital voltmeter before surface wave experiments of falling film flow are performed. Calibration data is plotted in Fig.2. It is seen that three good linear relations are maintained between the voltage output and the film thickness.

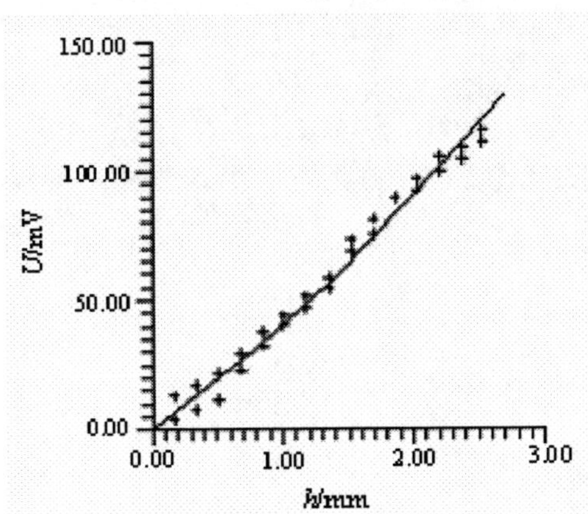

FIGURE 2. voltage output *vs* the film thickness

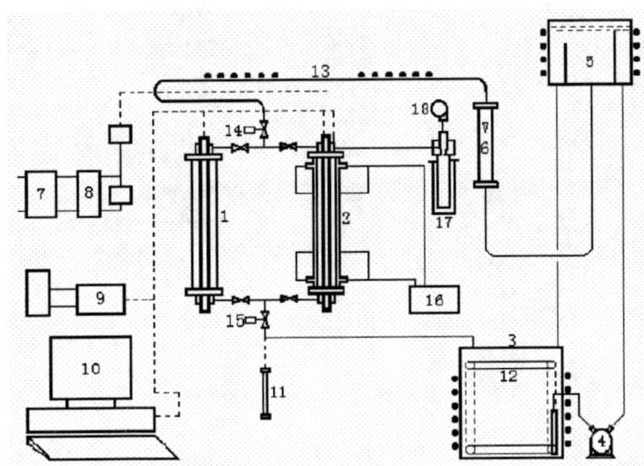

FIGURE 3. Schematic diagram of the test setup

The test fluid circulation loop (as shown in Fig.3) consists of following parts: two test sections 1and 2, a main tank 3, a swirling pump 4, a high-constant level tank 5, a rotameter 6, a plastic coil cooler12, a preheating section 13 made from some graphite tubes wrapped electric heating belts, a water bath 16, a jet pump-condenser 17, a miniature centrifugal ventilator 18, and some plastic joint piping among them. The test section 2 used for present experiments is a stainless steel tube of 20 mm outside diameter and 1.1 m length (the heated effective length is 843 mm, which origin is located at 140 mm from the film origin), and is installed concentrically inside of a jacket shell made from two pyrex tubes with closed head ends. There are 18 pairs of thermocouples embedded in the tube flush with its outside surface and located at two generatrices apart each other 180^0 in order to measure the surface temperatures. Before installation, the test tube 2 assembled the tubular electric heater (the later is inserted in the former and its maximum power is up to 10 kW) together is immersed in a distill water bath and turned on the power supply to inspect heating uniformity based on observation of nucleating bubbles. After that, the test tube is finished again by fine sandpapers and cleaned with aceton, distilled water and alcohol. Twice-distilled water is used as the test liquid. It is unnecessary to go into details for the power supply regulation and the measuring system.

In surface wave measurements of heated films, all electric heaters in the main circulation loop are turned on the power supply and the film inlet temperature is maintained at some predetermined temperature after the test tube 2 is perfectly wetted, if necessary, by means of increasing up and decreasing down the water level of the pool maintained at the bottom of the shell.. Meantime, constant temperature water kept at the same temperature by the water bath 16 is circulated in the annular channel of the jacket shell to maintain constant temperature atmosphere of the test section. Then the flow rate is regulated to a predetermined value and the data acquisition of the film thickness is started while the operation regime of the main circulation loop is steadily maintained 20 min. In film breakdown experiments, the 10 kW tubular electric heater inserted in the test tube 2 is also turned on the power supply except above heaters. The experimental procedure of film breakdown is similar to the above. However, the flow rate is at first regulated to a predetermined value and the heating power of 10 kW tubular electric heater is gradually increased at certain small increments. Data acquisition is started when some predicted heating power value of film breakdown may be approach and the operation regime of the main circulation loop is steadily maintained 20 min. After that, the heating power is continuously increased until film breakdown appeared. Steady operation of the experimental system is maintained as 5 min as possible at this condition and data acquisition is carried out again. The above procedures are repeated several times according to the film flow rate which is in turn varied from smaller to greater and then from greater to smaller.

EXPERIMENTAL RESULTS AND DISCUSSIONS

Nonlinear wave evolution of falling liquid film flow have been verified in theory and experiments[6,8,9], these phenomena are further revealed by present study. Using the capacitance measurement system, data of the instantaneous film thickness at some different locations and Reynolds numbers are obtained and interpreted by the cubic spline method. Instantaneous characteristics of film thickness are shown in Fig.4(b), for comparison, film surface wave measuring from a multiple-point conductance sensor (i.e. the test section 1 in Fig.3) is also given in Fig.4(a). It is seen that the instantaneous variation of film thickness obtained by the capacitance sensor

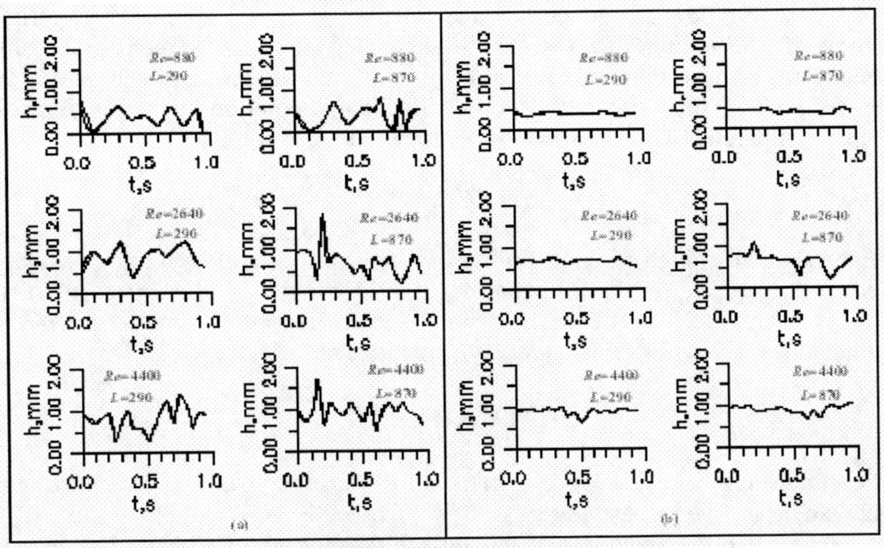

FIGURE 4. Film thickness traces
(a) Results obtained by the multiple-point conductance sensor (b) Results obtained by the capacitance sensor

behaves nonlinearly although the variation altitude is much lower by comparison with that obtained by the multiple-point conductance sensor. The reason may be that the surface area of the test section covered by the plate of the capacitance sensor is much higher than the electrode area of the conductance sensor. Besides, the surface tension of the film is reduced as the temperature of the film is increased.

From measuring results, the time-location-averaged film thickness can be expressed as follows:

$$h= \frac{1}{n}\sum_{i=1}^{n} h_{tai} \tag{2}$$

Thus, the variation of the mean film thickness with the film Reynolds number can be obtained from calculation results, which is plotted in Fig.5. Fitting the test data, following correlation can be written as

$$h=0.295Re^{0.458}(\nu^2/g)^{1/3} \tag{3}$$

It is interested that this correlation is similar to the result interpreted by measuring data from the conductance sensor.

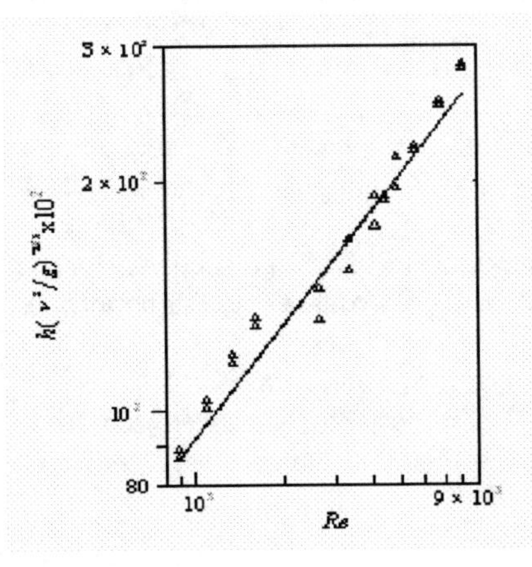

FIGURE 5. Correlation of mean film thickness data

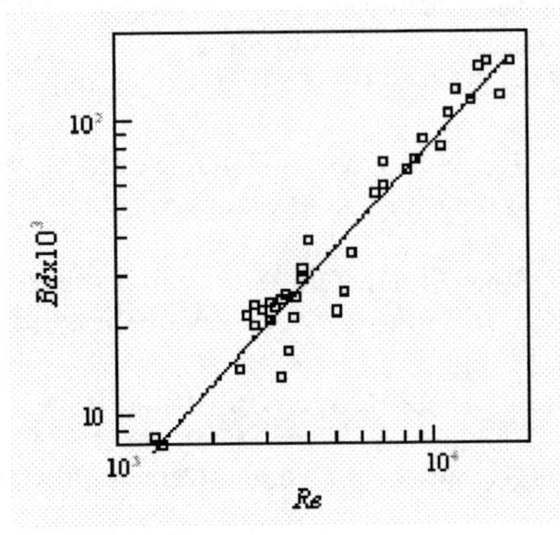

FIGURE 6. Variation of *Bd* with *Re*

Film breakdown of falling film flow is an important and complex process[8], up to now, experimental study are still rather poor. Especially, Film breakdown at the low subcooled condition is often abrupt, observation is very difficult. Interpretation of test data is also a subject to be further explored. According to the opinion[9]suggested by the authors, following correlation of breakdown criterion number *vs* Reynolds number can be obtained by the interpretation of test data which is plotted in Fig.6:

$$Bd = 1.567\times10^{-6}Re^{1.183} \tag{4}$$

It is speculated that breakdown of high subcooled film may be caused by combined role of solitary wave and surface tension difference. However, for low subcooled (or near saturated) films, it seems that this film breakdown mechanism is not suitable as solitary wave and surface tension difference are not remarkable. Perhaps, incipient boiling of low subcooled films may play a role in film breakdown.

SUMMARY

The capacitance measurement system developed is suitable for measuring the instantaneous film thickness at the temperature higher than the ambient temperature.

Measuring results indicate that the amplitude of the surface wave is not notable as that measured by the conductance method, thus, further improvement of the capacitance measurement system is still needed. It is interested that the correlation of mean film thickness is similar for both methods.

Breakdown of low subcooled films is often abrupt. The correlation of breakdown criterion number *vs* Reynolds number is given according to the interpretation of test data. It is speculated that incipient boiling of films may play a role in film breakdown.

REFERENCES

1. International Atomic Energy Agency, Status of Advanced Containment Systems for Next Generation Water Reactors, IAEA-TECDOC-752, IAEA, Vienna, pp.38-40, 1994.
2. Jiang Zhangyan, Ma Tongze, et al., Hydrodynamics and heat transfer in thin water films falling down the outside of a vertical tube, Chinese Journal of Engineering Thermophysics, , Vol.1, pp. 85-93, 1989.
3. S. G. Bankoff, Significant Questions in Thin Liquid Film Heat Transfer, Journal of Heat Transfer, Vol.116, pp. 10-16, 1994.
4. I.Mudawar and R. A.Houpt, Measurement of Mass and Momentum Transport in Wavy-Laminar Falling Liquid Films, International Journal of Heat and Mass Transfer, Vol. 36, pp. 4151-4162, 1993.
5. L.Q.Yu, F.K.Wasden, A.E. Dukler and V. Balakotaiah, Nonlinear Evolution of Waves on Falling Films at High Reynolds Numbers, Physics of Fluids, Vol.7, pp.1886-1902,1995.
6. Jiang Zhangyan, Tao Zhengwen and Yan Weiping, Measurements on Nonlinear Wave Evolution of Annular Two-Phase and Falling Liquid Film Flow, in Development of Measuring Technology for Multiphase Flows (Proceedings of ISMTMF'98)(Edited by Xu Yiqian and Zhang Baofen), pp.306-312,Standards Press of China, Beijing, 1998.
7. Jiang Zhangyan, Recent Progress on Parameter Measurements of Annular Two-phase Flow. Proceedings of the 3rd National Symposium on Measuring Techniques for Multiphase Flow, Shanghai, pp.219-229, 1990 (in Chinese).
8. Jiang Zhangyan, Song Jintian, Qin Wei, et al., Film Breakdown Phenomena and Its Simplified Analysis of Freely-Falling Liquid Films, Proceedings of the 9th International Heat Pipe Conference, Albuquerque, New Mexico, USA, 1995.
9. Tang Zhiwei, Ma Chongfang, Song Jintian and Jiang Zhangyan, Solitary Wave and Film Breakdown of Freely-Falling Film Flow and Their Effects on Operation of Gravity Heat Pipes, Proceedings of the 9th National Heat Pipe Conference, China, HP0403, 2004 (in Chinese).

NOMENCLATURE

A	area of the plate of the capacitance sensor
Bd	breakdown criterion number defined by the authors, $Bd=\beta_T q/\sigma\Gamma c_{pl}$
C	capacitance
c	specific heat
g	gravity acceleration
H	distance between the plate of the capacitance probe and the surface of the test section
h	film thickness

q heat flux of heated test section
Re film Reynolds number, $Re = 4\,\Gamma / \rho\nu$
β temperature coefficient of surface tension
Γ unit perimeter flow rate
ε dielectric coefficient
σ surface tension

Subscripts

in film inlet
l film
p at constant pressure
T varied as temperature
tai time-averaged at a given location
1 for air
2 for liquid film

Experimental Study on The Two-Phase Flow Characteristics Using Conductivity Probes And Laser Doppler Anemometry In A Vertical Pipe

Chiva S*., Mendez S.†, Muñoz-Cobo JL†, Julia J.E*., Hernandez L* ¶

*Department of Mechanical Engineering and Construction, Universitat Jaume I, Campus Riu Sec, E-12071, Castellón de la Plana, Spain

†Department of Chemistry and Nuclear Engineering, Polytechnic University of Valencia, P.O. Box 22 012, E-46071 Valencia, Spain

Abstract. An upward isothermal co-current air-water flow in a vertical pipe (50.2 mm inner diameter) has been experimental investigated. Local measurements of void fraction, interfacial area concentration (IAC), interfacial velocity and Sauter mean diameter were measured using a double sensor conductivity probe. Liquid velocity and turbulence intensity were measured using Laser Doppler Anemometry (LDA). Different air-water flow configurations was investigated for a liquid flow rate ranged from 0.491 m/s to 0.981 m/s and a void fraction up to 10 % .For each two-phase flow configuration twenty five radial position and three axial locations were measured by the conductivity probe methodology, and several radial profiles was measured with LDA at different axial positions.

INTRODUCTION

One of the most important key in the two-phase flow is the existence of multidimensional interfaces between both phases. The correct behaviour prediction of this interfaces, and its quantification is one of the frontiers in the theoretical and experimental studies of this kind of flow, and up to now, there is not an effective technique and methodology for the multidimensional interface characterization in two-phase flow measurement [1] . Then, experimental works play a very important role in the development of new theoretical models and it is essential for design, analysis and behaviour analysis of existing and futures scenarios. In this work a detailed experiments are performed in order to have a good data base about the main parameters in two-phase flow characterization in upward co-flow in vertical pipes.

The basic structure of a two-phase flow can be characterized by three fundamental parameters. These are the void fraction, Sauter mean diameter (for bubbly flow mainly) and interfacial area concentration (IAC) [2]. The void fraction expresses the phase distribution and is a required parameter for hydrodynamic and thermal design in various industrial processes. On the other hand, the interfacial area describes available area for the interfacial transfer of mass, momentum and energy, and it is a required parameter for a two-fluid model formulation. Various transfer mechanisms between phases depend on the two-phase interfacial structures. Therefore, an accurate knowledge of these parameters is necessary for any two-phase flow analyses.

The void fraction measurement is not so complicated using intrusive methods like conductivity probes with only two sensors. But IAC, defined by the quantity of interface between both phases, measured in area units per each unit of mixture volume, is quite complex itself, and with a large difficulties to be measured. Ishii [3] defined the local time-averaged IAC at a fixed position in space by:

$$\overline{a}^{t} = \frac{1}{\Omega}\sum_{l}\frac{1}{\left|\vec{v}_{il}\cdot\hat{n}_{il}\right|} \tag{1}$$

Where $\Omega, \vec{v}_{il}, \hat{n}_{il}$ are time interval, velocity vector of the l-th interphase and the surface normal unit vector of that interphase. That definition opens the door to measure the interfacial area using the velocity of the interphase , if

CP914, *Multiphase Flow: The Ultimate Measurement Challenge,* edited by X. Cai, Y. Wu, Z. Huang, S. Wang, and M. Wang

it is measured with enough detail together its vector components. The theoretical fundaments for this technique was originally proposed by Kataoka [4], and during the last fifteen years numerous researcher have used and improved that technique using intrusive probes of two, four and five sensors.

In this work we have used a double-sensor probe in order to characterize the two-phase flow parameters. This measurement technique is useful for 1D flow when spherical bubble can be assumed, which could be easily supposed in flows trough vertical pipes with relative small diameter and with gas moderated flow. The IAC and interface velocity can be calculated using the distance between the probe tips, but a calibration factor, based in a Mote-Carlo methodology, is proposed to relate the real magnitude to that measured from the experimental signal. It is necessary to introduce a calibration factor, based on statistical methods, to be able to take on account the possible deviation from the vertical direction and the presence of missing bubbles, bubbles that produces bad signals or bubble that only produce a signal in only one of the tips. Interfacial velocity, void fraction, Sauter mean diameter and IAC were measured using double-sensor probe. Also in this work, we include a detailed liquid velocity and turbulence intensity measurements using LDA which it is not usual in other research works where it is measured using hot-wire anemometer (HWA) . LDA is a not intrusive method and it give a more precise results and easier to interpret than HWA when bubbles are present in a liquid flow. Accurate magnitude information about liquid is basic for a good results interpretation.

EXPERIMENTAL FACILITY AND INSTRUMENTATION

The experimental work was performed using a thermo-hydraulic loop placed at the Energy Engineering Institute in Polytechnic University of Valencia (Spain). The loop is schematically illustrated in Figure 1. It consists of a test section, a round transparent tube made by Plexiglas with constant section, an upper plenum and a lower plenum where air and water are mixed. The test section has a 52 mm inner diameter and a length of 3340 mm. The water was circulated by two centrifugal pumps controlled by a frequency controller. The air was supply by a compressor, and it was introduced to the test section through a porous sinter element with a average pore size of 10 mm installed below the mix chamber. The air and water temperature was kept constant during the test assay. The air mass flow rate was measured with a thermal mass flow meter and controller (Bronkhorst ® , EL-FLOW model), liquid flow rate with a Electromagnetic flow meter (Badger Meter ®), and the pressure with a precision pressure transmitter, DRUCK ® PTX 600 series.(INF position: range [0..1] bar, MED, SUP position range [0..250] mbar (Prec. 0.5% F.S.)

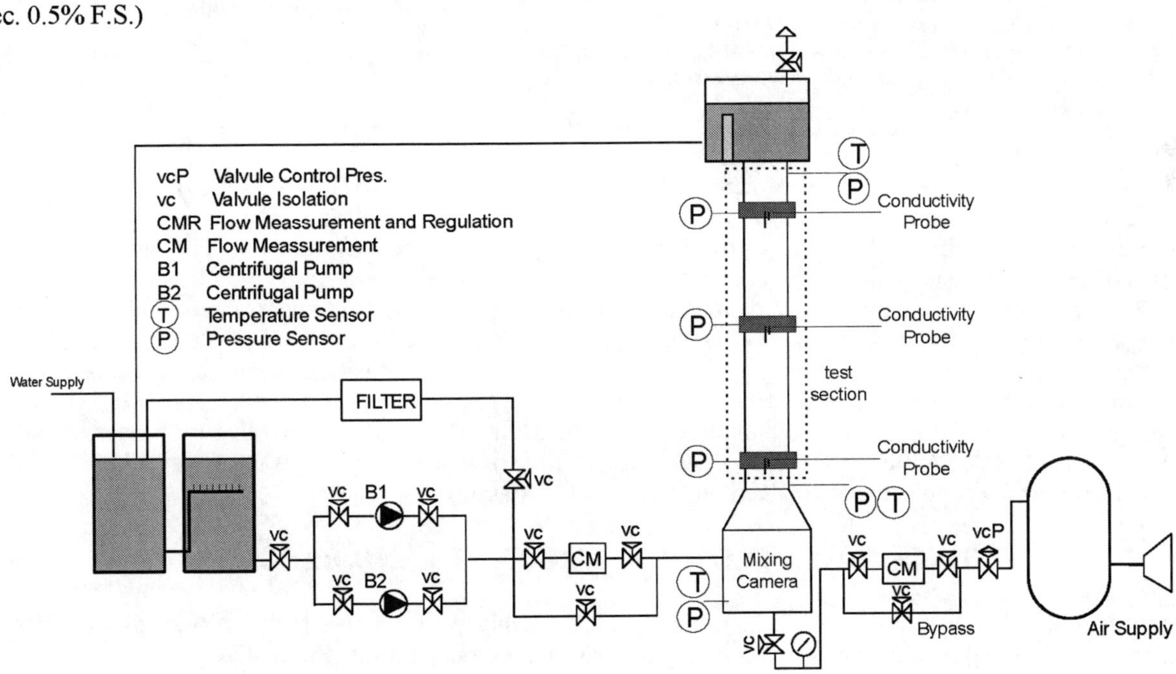

FIGURE 1. Schematic diagram of experimental loop

A double-sensor conductivity probe was used to measure local two-phase flow parameters and Laser Doppler Anemometry to measure liquid velocity and its turbulence intensity. The local flow measurements using the double sensor was performed at three axial localizations, L/D = 3 , 32 , 56, measured from the lower part of the test section. The liquid velocity and its turbulence intensity were measured using, at Z/D = 5 and 56. The sectional differential pressure along the axial direction was recorded using differential pressure gauges and bourdon manometers por auxiliary measurement points at air flow meter and in the bottom of the column...

The LDA equipment consists of a 0.5W Ar+ Ominichrome laser, Dantec Fiberflow beam separator, Dantec FVA 58N40 processor and a PC using the software Floware for data acquisition. A lens of 0.125 m focal length was used and the system was operated in backscattering mode. The vertical component was determined with green (λ=514.5 nm) beams and the horizontal component with blue (λ=488 nm) beams. A preshift frequency of 500 kHz was used. The flow was seeded with hollow particles which were neutrally buoyant with a 10 μm mean size (Dantec HGS-10); therefore, only the liquid velocity was measured. The time series obtained were 60 seconds long

DOUBLE-SENSOR CONDUCTIVITY PROBE METHODOLOGY

The double sensor probe is basically a phase identifier. It consists of two sensors made of stainless steel, coated with gold, with a diameter of 0.22 mm. The vertical distance between both tips was about 1.5 mm . Each sensor is insulated using a insulating varnish except its tips. If the probe is connected to a power supply with a fixed voltage, due to the large difference in conductivity between the liquid phase and the gas phase, the impedance signal acquired rises sharply when a bubble passes through one of the sensor tips, obtaining a more or less square signal. With a suitable signal processing methodology is possible extract precise information from that raw signal

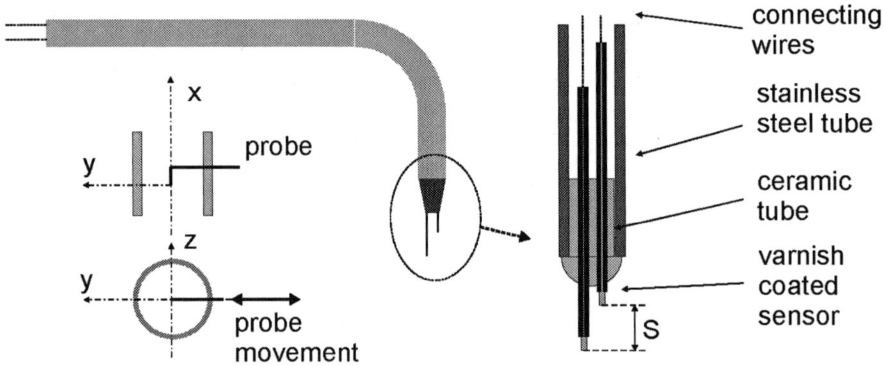

FIGURE 2. Schematic diagram of double sensor conductivity probe

In this work we have chosen the new methodology carried out recently by Muñoz-Cobo et al. [5] to determine the calibration factor. In his study the effects of bubble motions, and probe spacing have been included. The value measured is related to the local bubble interface velocity in the surface normal direction. The probability density function of each variable is identified with the suitable coordinate transformation. Two theoretical calibration factors are defined to relate the mean measurable parameter to the interfacial area concentrations obtained and the measured bubbles, including the missed bubbles. These calibration factors have been obtained through analytical and numerical method, using a Monte Carlo approach. The results of these calculations show that the total interfacial area correction factor is very close to 2, and depends very weakly on the velocity fluctuation, and the relative distance between tips. For the velocity calibration factor, the Monte-Carlo results show that for moderate values of the relative bubble velocity fluctuation and values of the relative distance between tips not too small, the correction velocity factor for the bubble sensor conductivity probe is close to unity.

TWO-PAHSE FLOW EXPERIMNETS AND RESULTS

As a first step in our setup process of our experimental facility we have chosen a matrix of experiments very close to that performed by Hibiki et al. [6] in order to have experimental results from other prestigious researchers to check the methodology and data produced. On the other hand we have checked the reproducibility of the results. We have introduced a more accurate experimental technique to measure the liquid velocity and turbulence intensity using LDA techniques.

TABLE 1. Flow conditions in this experiment

		G01 (●) jg (m/ s) Void Fraction 5%		**G02** (▲) jg (m/ s) Void Fraction 10%	
		Our Work	*Hibiki*	*Our Work*	*Hibiki*
	jf (m/s)	*(solid line)*	*(-)*	*(solid line)*	*(-)*
F01	0.491	0.0226	0.0275	0.0414	0.0556
F02	0.986	0.036	0.0473	0.112	0.113

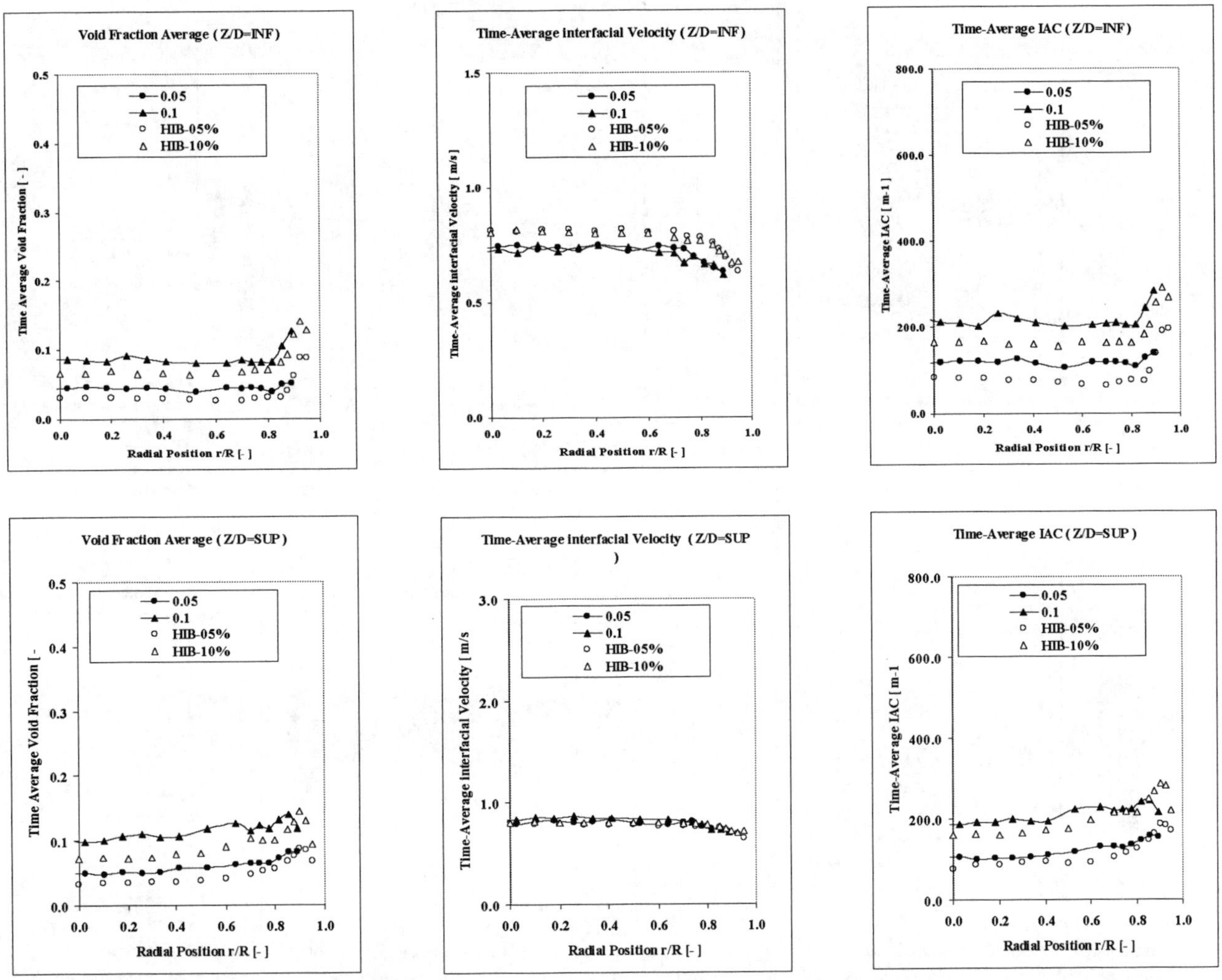

FIGURE 3. Local void fraction, interface velocity and IAC profiles for F01 set (jf = 0.491 m/s) at two position, z/D = 3 (INF) and z/D = 52 (SUP) for present results and for Hibiki data. (0.05) data series shows a 5% void fraction average, (G01), and (0.1) a 10% void fraction average (G02). (HIB) show the Hibiki data for the same void fraction average.

We have performed two sets of experiments, one for each liquid velocity, in each set the gas flow was injected to maintained a void fraction about 5 and 10 % for each case respectively. The bubble diameter at the inlet depends on the gas flow and liquid flow. That initial diameter was not controlled in this experiment. With liquid velocities at the inlet below 2 m/s the turbulence does not produce bubble disintegration (6), then the initial bubble determinates the developing profile. The bubble diameter at the inlet was in the range [2 – 4] mm . Figure 3 shows the void fraction behaviour for, F01, j_f = 0.491 m/s, and it indicate that for G01 and G02 a well-developed wall peaking profile was observed, as in the work of Hibiki, showing a good agreement with his results. The local void fraction profile have a typical evolution of the pressure reduction, it match with lack of bubble break-up o coalescence due to the reduced liquid velocity and the moderated void fraction. Figure. 4 shows the F02 set behaviour, where all the profiles show wall-peak behaviour, with bubble diameters less than 3.6 mm . At the superior position, the peaks in the void fraction profile match quite good with the peaks position described in the literature. As in the former set, the peaks were located more or less at one bubble diameter from the wall. Those peaks were shifted toward the tube centre as the flow develops.

The IAC profile shows a similar behaviour of its respective void fraction profile, since the Sauter diameters are quite uniform in all the profile, and as it is expected for bubbly flow without bubble cups. As it is displayed in Figure 3 and 5 the IAC profiles show a good agreement with the published data

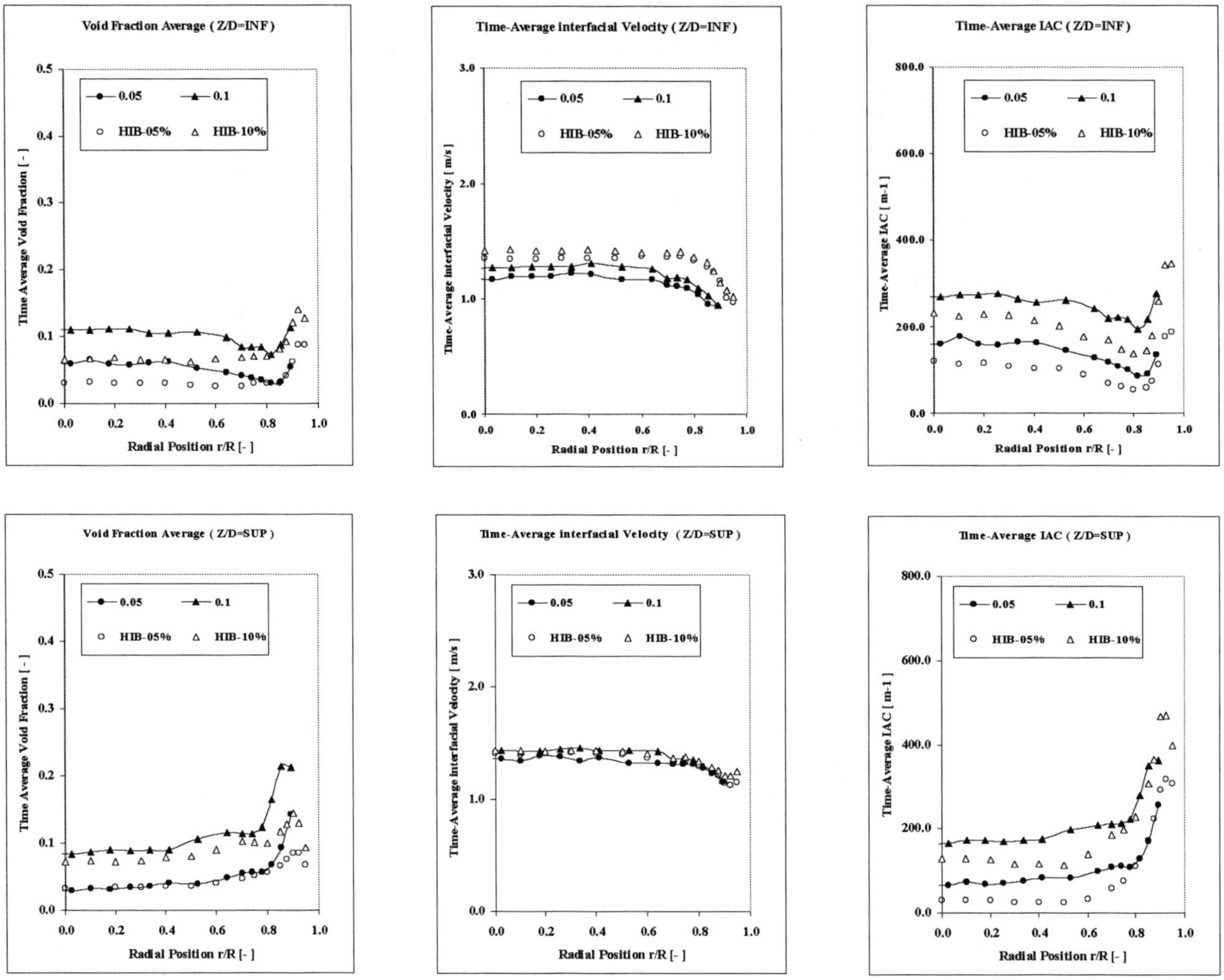

FIGURE 4. Local void fraction, interface velocity and IAC profiles for F02 set (jf = 0.986 m/s) at two position, z/D = 3 (INF) and z/D = 52 (SUP) for present results and for Hibiki data. (0.05) data series shows a 5% void fraction average, and (0.1) a 10% void fraction average. (HIB) show the Hibiki data for the same void fraction average.

The liquid velocity and its turbulence intensity were measured using LDA, , profiles are displayed at Figure 5 for F01 and F02. Those profiles do not match with that measured by HWA, but maintain more or less the same tendency. In general, the agitation increases with the bubble injection to the fluid, but locally, bubbles can reduce the turbulence intensity as you can see at Figure 5. This behaviour matches with that predicted by the literature when the liquid velocity is equal or bigger than 1 m/s. The wall-peak profile in the turbulence intensity can be explained by the large gradient of the velocity and shear stress distribution near the wall, mainly in the lower part of the test section, where the flow is developing.

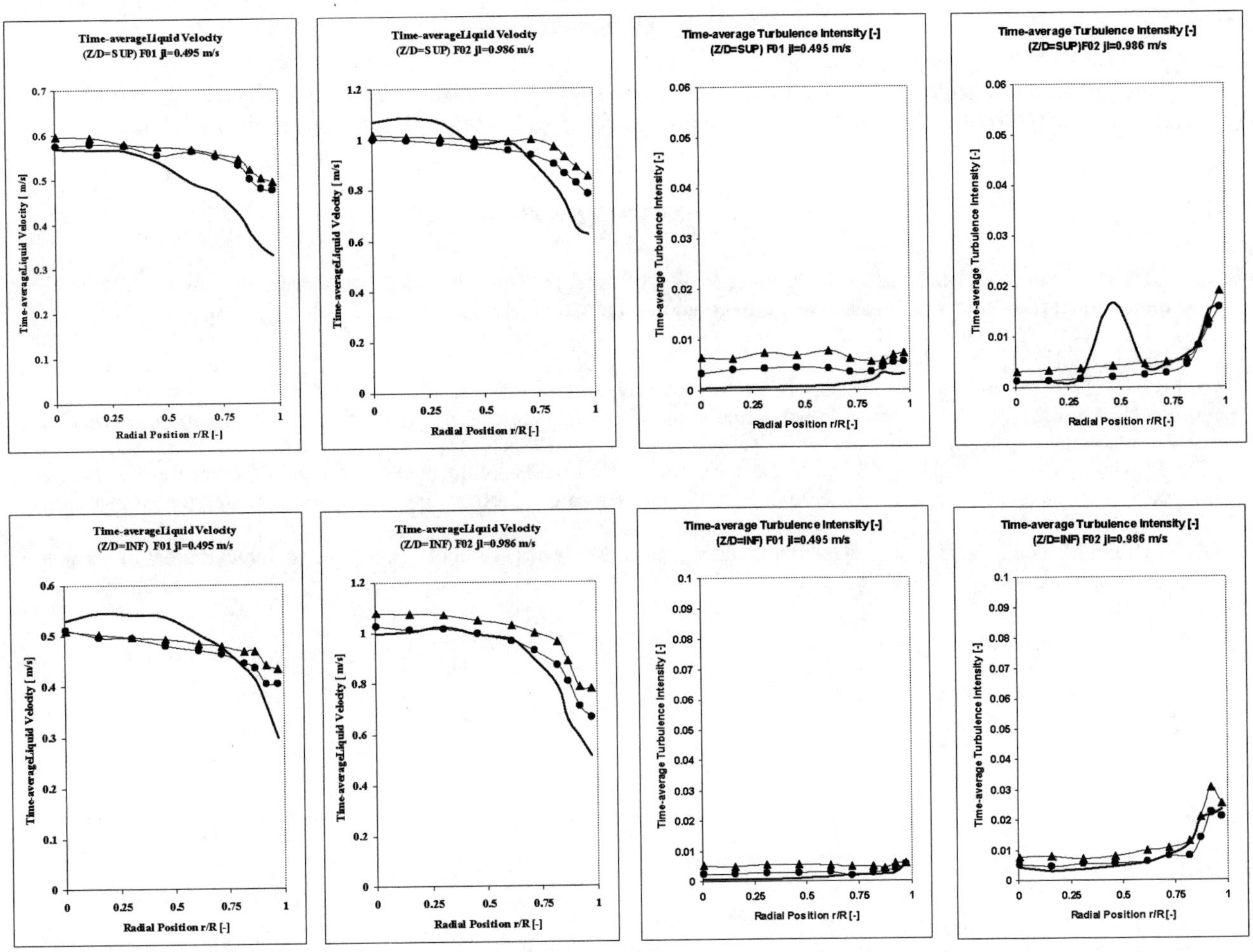

FIGURE 5. Liquid velocity and Turbulence intensity profiles for F01 and F02 set at two position, z/D = 3 (INF) and z/D = 53 (SUP) for our results . The solid line is for single phase velocity (jf = 0 m/s).

The liquid flow rate calculated from the velocity profile was compared to the liquid flow rate measured with the flowmeter with very good results. The liquid velocity profile , in general, is more flatted when bubbles appear, but this effect is quickly saturated when the quantity of bubble increase, as you can see in Figure 5 , it is due mainly to the wall-peak void fraction profile At the lower part of the test section, the presence of bubble can reduce the liquid velocity as it is displayed in Figure 5. It could be due to the agitation effect of the bubble, producing a more flatted profile with a lower velocity in the core.

CONCLUSIONS

As it is showed in the paper, the results maintain a good agreement with those published by Hibiki and co-workers, with a very good prediction of the peaks in the profiles and its tendencies. The liquid velocity and its turbulence intensity were measured using LDA, with more accurate results. Then, the experimental methodology developed to build the double-sensor conductivity probe and to process the data set obtained, give good and repeatable results about the two-phase flow parameters. This methodology could be considered a suitable tool to study two-phase flow sceneries together the use of LDA to measure the liquid magnitudes.

ACKNOWLEDGMENTS

This research was supported by the Plan Nacional de I + D + I. Project MESBITERMO ENE2004-06978-C02-01 and Project ENE2004-06978-C02-02.and from Fundació Caixa Castelló - Bancaixa (project P1 1A2003-17)

.

REFERENCES

1. X. Shen, Y. Saito, K. Mishima, H. Nakamura, Methodological improvement of an intrusive four-sensor probe for the multi-dimensional two-phase flow measurement, International Journal of Multiphase Flow, Vol. 31, pp. 593-617, 2005
2. J.M. Delhaye, P. Bricard, Interfacial area in bubbly flow: experimental data and correlations, Journal of Nuclear Engineering Design, Vol. 151, pp. 65-77, 1994
3. M. Ishii, Thermo-fluid Dynamic Theory of Two-phase flow , Eyrolles Paris, 1975
4. I. Kataoka, M. Ishii, A. Serizawa, Local formulation and measurement of interfacial area concentration in two-phase flow, International Journal of Multiphase Flow, Vol. 12, pp. 505-529, 1986.
5. J.L. Muñoz-Cobo, J. Peña, S. Chiva, S. Mendez, Monte-Carlo calculation of the calibration factors for the interfacial area concentration and the velocity of the bubbles for double sensor conductivity probe, Nuclear Engineering and Design, (article in press), 2006
6. T. Hibiki,M. Ishii, Z. Xiao, Axial interfacial area transport of vertical bubbly flow, International Journal of Heat and Mass Transfer, Vol. 44, pp. 1869-1888, 2001

Particle Velocity and Pressure Drop in Long-distance Dense-phase Pneumatic Conveying of Fly Ash

Zongming Liu[1], Hua Yi[1], Guangbin Duan[1], Zhengyi Fu[2]

[1] School of Materials Science and Engineering, University of Jinan, Jinan, 250022, China
[2] School of Materials Science and Engineering, Wuhan Science and Technology University, Wuhan, 430070, China

Abstract. In process of long-distance dense-phase pneumatic conveying, decrease of pressure along pipeline and compressibility of gas induce the increase of particle velocity. Meanwhile, the disparity of pressure drop over pipeline must be paid attention to. In this paper, with compressed gas being conveying medium and fly ash being conveying material, experiment was employed on a circulating experimental bench which has a pipeline with 202.87m long and the diameter of 80mm. Groups of GP/DP(Gauge Pressure/Differential Pressure) transmitters were assembled along pipeline to achieve experimental data. As a result, particle velocity equation along pipeline was modified well, and pressure drop equation was fitted. Both equations were found to have good agreement with experimental data.

Keywords: dense-phase pneumatic conveying; particle velocity; pressure drop; compressibility

INTRODUCTION

With the development of environmental protection and the comprehensive utilization of fly ash, the study on pneumatic conveying system for transport fly ash has been paid more attention to[1]. To keep the conveying stable in low energy consumption, particle velocity and pressure drop during conveying process must be mentioned. But in most research, compressibility of gas has not been well considered. In fact, compared with static pressure, when pressure drop was not small enough, compressibility of gas can't be ignored[2]. For this reason, particle velocity and pressure drop were investigated in this paper.

INTRODUCTION OF EXPERIMENT

Experiment System

The pneumatic conveying system in this paper is a circulating experimental bench with long-distance pipeline. The transporting distance is 202.87m and the pipe diameter is 80mm. The pneumatic conveying system consists of an air compressor, a feeder, conveying pipeline, measurement and control system, as shown in Fig.1.

Experimental Material and Medium

The fly ash used in the experiment is of an average diameter of d_s=58.73μm , a true density of 2078 kg/m^3 and a bulk density of 771 kg/m^3. The conveying medium of this system was compressed air, clean and dry.

Main Measuring Apparatus

8410-Weight balance: to measure the discharge rate of fly ash into the feeding bin. WT-2000GP/DP transmitter: to measure the pressure of the feeder, setting points along pipeline and pressure drop of test segment. Gas mass flowmeter: to measure the mass flow ratio of the compressed gas.

CP914, *Multiphase Flow: The Ultimate Measurement Challenge,* edited by X. Cai, Y. Wu, Z. Huang, S. Wang, and M. Wang

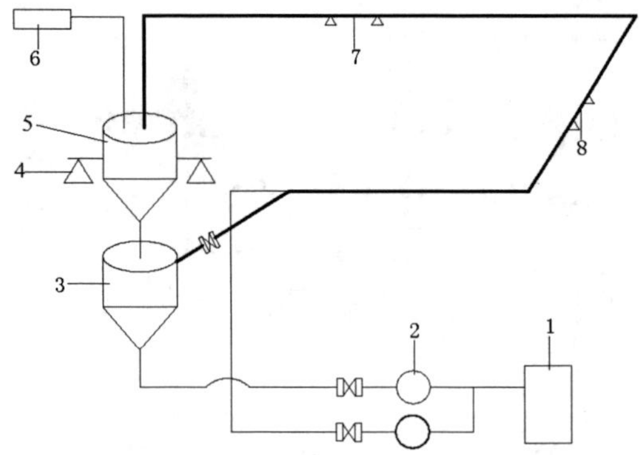

FIGURE 1. Schematic diagram of the experiment system
1. Blower 2. Flowmeter 3. Feeder 4. Weight balance
5. Collecting bin 6. Dust collector 7.8. Pressure test segment

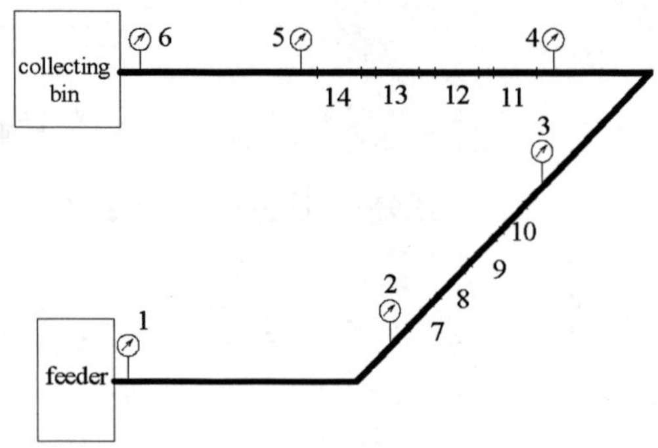

FIGURE 2. Schematic diagram of test segment assembled
1~6. Gauge pressure transmitter
7~14. Differential pressure transmitter

Experimental Procedure

After being put into the feeder, fly ash will be fluidized by compressed air. Then at a predetermined transporting pressure in the feeder, fly ash will pass through the conveying pipeline and reach the collecting bin finally.

Experimental Scheme

Two test segments along horizontal pipeline were employed to analysis the tendency of pressure drop. On each segment, four differential pressure transmitters were assembled with each differential pressure transmitter spanning 1 meter, and the distance between 2 transmitters being 0.2m (7~10,11~14). Gauge pressure transmitters were installed at the point of the initial and terminal of each test segment (2,3,4,5), terminal of pipeline (6) as well as the vent of feeder (1). The layout of transmitters was shown in Fig.2.

Feed pressure was set from 0.26MPa to 0.42MPa with each 0.02MPa interval. Therefore, there are 9 operating conditions in total and several experiments were carried out in each operating condition.

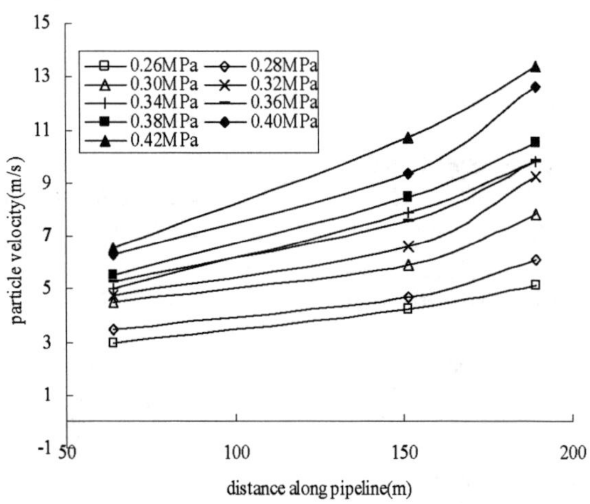

FIGURE 3. Experiment data of particle velocity along pipeline

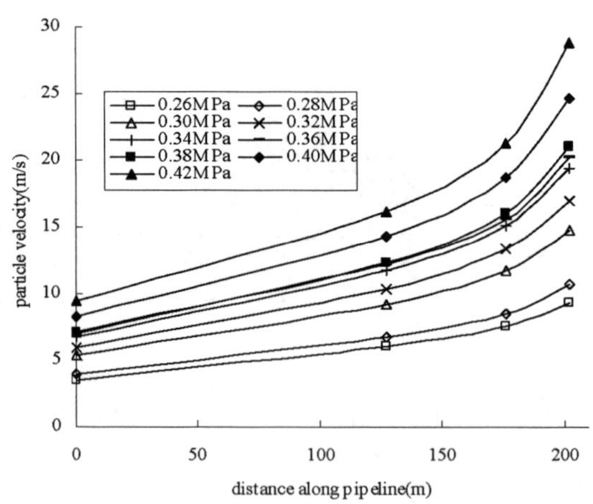

FIGURE 4. Calculated data of particle velocity along pipeline.

PARTICLE VELOCITY ALONG PIPELINE

Experimental Result

As pressure in feeder reaching set value (feed pressure), pneumatic conveying began, and signals from each pressure transmitter were transmitted to MultiF R4000 recorder. The average particle velocities between every two gauge pressure transmitter can be calculated with Eq. (3-1).

Based on time interval of signal emergence of every pressure transmitter and distance between different transmitters, particle average velocities were achieved as follows:

$$u_{ij} = \frac{L_{ij}}{t_{ij}} \tag{3-1}$$

The relationship between particle velocity and distance along pipeline was shown in Fig. 3 based on the results calculated with Eq. (3-1). From Fig. 3, it can be seen that, particle velocity increase gradually along pipeline because of the continuous decrease of pressure and the expansion of gas. This phenomenon should be paid more attention to especially in long-distance pneumatic conveying.

Theoretical Analysis

If conveying material was transported homogeneously, particle velocity could be presented by using Eq. (3-2) [3]:

$$u_s = \frac{G}{(1-\varepsilon)A\rho_s} \tag{3-2}$$

where

$$\varepsilon = \frac{\rho_s}{\mu\rho_g + \rho_s} \tag{3-3}$$

In the condition of this experiment, the gas density was calculated by the following correlation:

$$\rho_g = \frac{p}{RT} \tag{3-4}$$

Substituting Eq. (3-3) and (3-4) into Eq. (3-2) results in

$$u_s = \frac{G_s(\mu p + \rho_s RT)}{A\mu\rho_s p} \tag{3-5}$$

From Eq. (3-5), particle velocity could be calculated when the parameters of p and T were known. Based on the measured data of pressure in the experiment, four particle velocities in each operating condition were obtained as shown in Fig. 4.

Comparing the calculated data by using Eq. (3-5) with experiment data shown in Fig. 3, it is obvious that there is significant difference between them, which indicates that the conveying material was not transported homogeneously.

The modified equation of Eq. (3-5) was obtained with further analysis as follows:

$$u_s = 0.583\frac{G_s(\mu p + \rho_s RT)}{A\mu\rho_s p} \tag{3-6}$$

Using Eq. (3-6), particle velocity along pipeline could be calculated conveniently. Comparing the calculated value with experiment value, good accuracy was achieved with the relative error range -9.56%~10.75%.

PRESSURE DROP

Experimental Result

The relationship between pressure drop and distance along pipeline in different feed pressure was obtained based on the measured data from the eight differential pressure transmitters in the experiment, as shown in Fig.5.

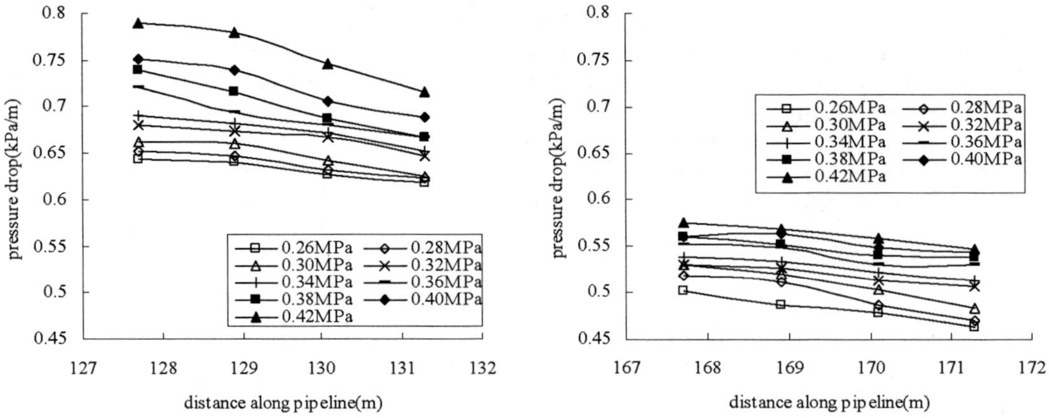

(a) (b)

FIGURE 5. Relationship between pressure drop and distance along pipeline

Note that pressure test segment in Fig.5 (a) is 127~132 m, while Fig.5 (b) is 167~172 m.

From Fig.5, it could be seen that pressure drop decreased gradually with the increase of the distance along pipeline, while on the same point, high feeding pressure induce high pressure drop.

Theoretical Analysis

Provided that the conveying material was transported homogeneously, it means there is no slippage between gas-solid phases[4]. Pressure drop of two-phase flow could be in the presence of research on resistance characteristic of single-phase. In fact, the conveying material was not transported homogeneously as analysised previous, slippage between gas-solid phases existed. So the equation calculating pressure drop could be derived:

$$\frac{\Delta p}{L} = \frac{\lambda \rho_b u_s^2}{2D} \tag{4-1}$$

where

$$\rho_b = \rho_s(1-\varepsilon) + \rho_g \varepsilon \tag{4-2}$$

In process of pneumatic conveying, resistance of friction involves drags between gas and pipeline, gas and solid, solid and pipeline etc[6]. As mentioned previous, influencing factors of resistance coefficient λ are ρ_g, ρ_s, μ_g, D, ε, u_g, u_s, $u_g - u_s$ etc. Suppose the expression of λ as follows:

$$\lambda = f(\rho_s, \rho_g, d_s, D, \mu_g, \varepsilon, u_g, u_g - u_s) = \xi \rho_s^a \rho_g^b d_s^c D^d \mu_g^e u_s^f u_g^g (u_g - u_s)^h \varepsilon^i \tag{4-3}$$

Based on the principle of similitude and dimensional analysis, resistance coefficient can be expressed by the follow equation:

$$\lambda = \xi Y^{c_1} \mathrm{Re}_s^{c_2} \phi^{c_3} (1-\phi)^{c_4} \varepsilon^{c_5} \psi^{c_6} \tag{4-4}$$

where

$$\mathrm{Re}_s = \frac{\rho_g (u_g - u_s) d_s}{\mu_g};\ Y = \rho_s / \rho_g;\ \phi = u_s / u_g;\ \psi = D / d_s$$

In this paper, ψ was a constant, which results in ψ^{c_6} could be involved in ξ. So the undetermined coefficients changed into ξ, c_1, c_2, c_3, c_4, c_5.

To investigate the expression of λ, 18 groups of experiments were carried out in all with eight values of pressure drop in every group of experiment. Therefore, there are 144 groups of data (include λ, Y, ϕ, $(1-\phi)$, Re_s, ε) total. Substituting these data into Eq. (4-4) and adopting linear regression analysis, regression equation of resistance coefficient λ was obtained as follows:

$$\lambda = 1.485 \times 10^{10} Y^{-1.022} \mathrm{Re}_s^{-1.74} (1-\phi)^{1.74} \phi^{-1.817} \varepsilon^{31.144} \tag{4-5}$$

Therefore, resistance equation of horizontal pipe was gained.

$$\frac{\Delta p}{L} = 1.485 \times 10^{10} Y^{-1.022} \mathrm{Re}_s^{-1.74} (1-\phi)^{1.74} \phi^{-1.817} \varepsilon^{31.144} \frac{\rho_b u_s^2}{2D} \tag{4-6}$$

Comparing the calculated value with experiment data, the relative error ranged -7.61%~3.10%. It can be seen that the agreement between predicted and experimental data is quite good. Therefore, it can be used to instruct the engineering application.

CONCLUSIONS

1) For long-distance dense-phase pneumatic conveying, the compressibility of gas induces increasing of particle velocity along pipeline. Particle velocity can be calculated with Eq. (3-6).

2) The distribution of pressure drop along pipeline was obtained experimentally, and it shows pressure drop reduces gradually with the expansion of gas along the pipeline.

3) The pressure drop can be calculated and forecasted by using Eq. (4-6)which was fitted with good accuracy.

ACKNOWLEDGEMENTS

This work was supported by the Natural Scinence Foundation of Shandong Province(grant no.Y2006F04) and Jinan Scinence & Technology Young Star Project(grant no.07103)

REFERENCES

1. Peter W. Wypych, Jianglin Yi. Minimum transport boundary for horizontal dense phase pneumatic conveying of granular materials[J]. Powder Technology, 2003, (119), pp.111-121.
2. Artur J. Jaworski, Tomasz Dyakowski. Investigations of flow instabilities within the dense pneumatic conveying system[J]. Powder Technology, 2002(125), pp.279-291.
3. R.Pan, P. W. Wypych. Pressure drop and slug velocity in low-velocity pneumatic conveying of bulk solids [J]. Powder Technology, 1997(34), pp.123-132.
4. Liu Zongming, Yue yunlong, Lu Haidong. An experimental study on fly ash removal by dense phase pneumatic conveying. Energy and the Environment-Proceeding of the International Conference on Energy and Environment, 2003, pp.1288-1291.
5. Jun Zhao, Zongming Liu, Shougen Hu. Study on the Dense-phase Pneumatic Conveying of Fly Ash and Process Control. The Sixth International Conference on Measurement and Control of Granular Materials (MCGM'2003), Shanghai, China, August, 2003.
6. F. J. Wang, J.-X. Zhu, J.M. Beeckmans. Pressure gradient and particle adhesion in the pneumatic transport of cohesive fine powders. International Journal of Multiphase Flow 26 (2000), pp.245-265.

NOMENCLATURE

ε	void fraction
μ	solid loading ratio
G	conveying capacity of solid (mass flowrate of solid), kg/s
A	cross-sectional area of pipeline, m^2
D	internal diameter of pipe, m
L	length of pipeline, m
t	time, s
ρ_s	particle density of fly ash, kg/m^3
ρ_g	gas density , kg/m^3

u_s	particle velocity, m/s
u_g	gas velocity, m/s
t_{ij}	time interval of signal emergence between two gauge pressure transmitters, s
L_{ij}	distance between the two gauge pressure transmitters, m
λ	resistance coefficient
ρ_b	density of gas-solid mixture, kg/m^3
Re_s	Reynolds number of particle
Y	density ratio of solid to gas
ϕ	velocity ratio of particle to gas
ψ	ratio of pipe diameter to particle diameter

Experimental Study of Influence on The Moving of Sublimation Interface by Precooling Rate and Drying Temperature During Freeze-drying

Xin Xiao and Le-Ren Tao

Institute of Cryomedical Technology and Food Refrigeration, University of Shanghai for Science and Technology, No.516 Jun-Gong Road, 200093, Shanghai, China

Abstract. For complex heat and mass transfer during freeze-drying, the shape and the moving rate of sublimation interface have not been clearly recognized yet. In this paper, Micro-CT scanner was used to follow the moving interface during sublimation. Apple slices cut into 16mm in diameter and 8mm in thickness were used as experimental samples, they were scanned every two hours during sublimation. The scanning images were analyzed and measured, then variation curves of grey value and curves of sublimation rate in two directions were obtained. The moving rates of sublimation interface under various precooling rates and primary drying temperatures were compared. The results show that, heat and mass transfer happens both at the upper and the under surface of the sample. Also it happens at the side surface to some extent. The interface shows as a three-dimensional moving mode, contracts to the geometric centre of the sample and presents an approximately spheral shape. Apple samples frozen at low rate sublimated more quickly than those by high rate. While drying temperature was higher, the sublimation interface moved more quickly. Under slow precooling condition, the sublimation rate rose quickly near the end of sublimation not only in vertical direction, but in horizontal direction.

Keywords: freeze-drying; precooling rate; primary drying temperature; sublimation interface; computed tomography

INTRODUCTION

Freeze-drying is one of the promising methods for food processing and cell storage[1-2]. But the application is limit for high energy input and long drying time. How to increase sublimation rate and shorten drying time are always the keystone study of freeze-drying process. The drying process containing seepage, diffusion, capillary effect and phase transformation is typically a heat and mass transfer coupling process with phase transformation[3].

During freeze-drying, different textures of frozen samples depending on the solidification parameters together with diffusion coefficient for vapour transporting out of drying area have great influence on drying time[4]. Ice crystals mushroom with quick cooling, the growth rate of main branch is high correspondingly, whereas the growth rate is low with slow cooling. Thin and dense branches are formed under high cooling rate, and branches are thick under low cooling rate[5]. As samples are under frozen state firstly, so how to keep them well is the premise although high temperature is better for heat transfer. Therefore, it is necessary to discuss the influence of precooling rate and drying temperature on drying process.

Theoretical simulation is a powerful technique for the study of freeze-drying. Lots of models were acquired yet, from the simple Uniformly Retreating Ice Front Model[6] to the moving sublimation front model using MacCormack predictor-corrector method[7] and Arbitrary Lagrangian-Eulerian method[8]. Liu et al[9] developed a dynamic model to describe the unstable state with heat and mass transfer couplinged.

In another side, experimental methods are useful and practical for freeze-drying study. For it's lossless detecting, X-ray attracts more and more researchers' attention. Steppe et al[10] mensurated the inner structure of wood using Micro-CT technique. Schelenz et al[11] took X-ray photographs at different points of time during primary drying and found the phenomenon that sublimation front proceeded unsymmetrically from all surfaces that contradicted the

CP914, *Multiphase Flow: The Ultimate Measurement Challenge,* edited by X. Cai, Y. Wu, Z. Huang, S. Wang, and M. Wang

previous assumption. Also the phase transformations during the freeze-drying cycle were monitored in real time by X-ray diffractometer (XRD) yet[12].

In this study, according to various precooling rates and primary drying temperatures, the different images showing intuitionistic sublimation interface during sublimation were scanned and analyzed. Grey values under different conditions were calculated and compared. Sublimation rate were also mensurated. The results can provide fundamental informations for recognizing freeze-drying process.

EXPERIMENTAL MATERIALS AND METHODS

Micro-CT Introduction

The experimental Micro-CT Scanner produced in Belgium is SKYSCAN 1074HR. The imaging principle is shown in Fig.1. The X-ray detector is 768×576 pixels 8-bit X-ray camera in on-chip integration mode with lens coupling to scintillator. The maximum object that can be scanned is columniform about 16mm in diameter and 12mm in height.

A shadow image has a size of 736×512 pixels with 16-bit uncompress TIFF format. 16-bit uncompress square BMP grey cross-section images with 736×736 pixels can be reconstructed from those shadow images. In all grey images, the pixel size is about 22 μ m, grey value 0 is black while 255 expresses white.

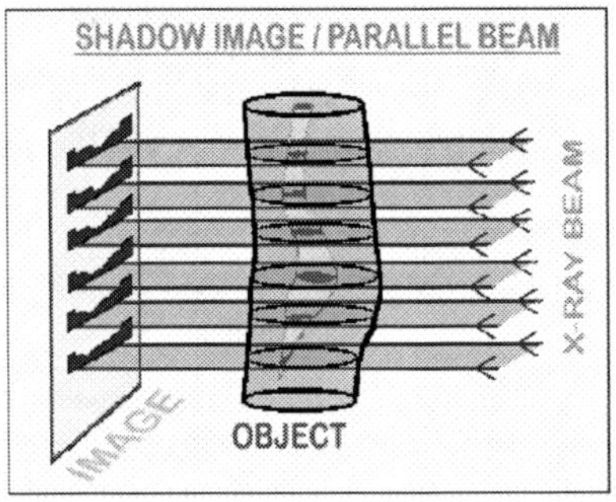

FIGURE 1. Imaging principle of Micro-CT

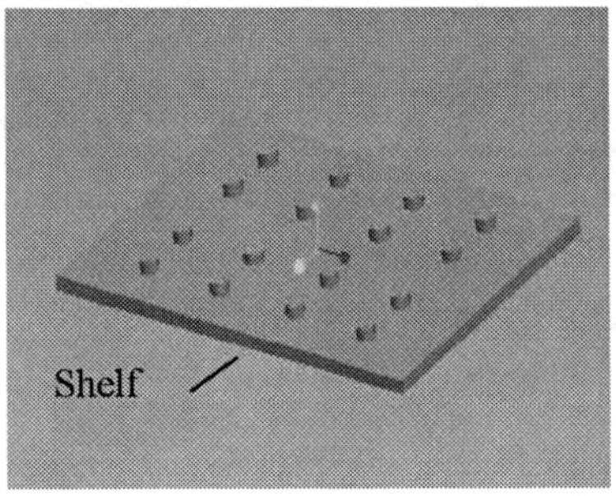

FIGURE 2. Schematic of samples in the shelf

Materials And Methods

Apple slices cut into 16mm in diameter and 8mm in thickness were used as experimental samples. During each experiment, one group samples were frozen on the shelf of lyophilizer, others were frozen in programmable freezer. Then all were moved to lyophilizer for drying. One uniform sample of each group sublimating was taken out for scanning every hour, a series of images showing sublimation interface could be obtained ultimately. Figure 2 shows us the placed position of samples in the shelf.

Freeze-drying Process

Different Precooling Rate

The temperature falling mode is shown in Fig.3, the rate of quick precooling is -14.97 ℃/min and that of slow precooling is just -1.19℃/min. Samples in lyophilizer were frozen to -25℃.

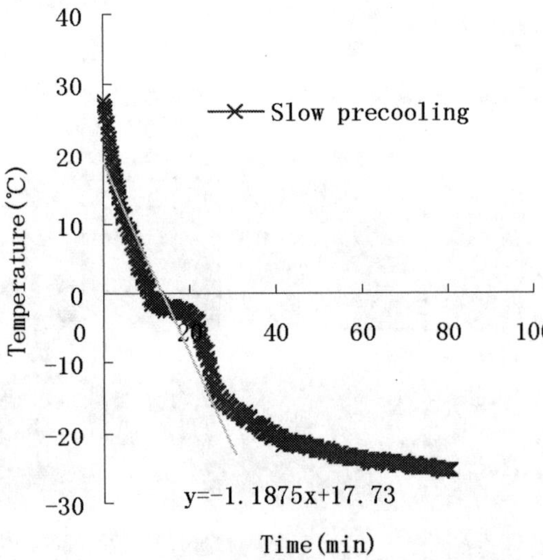

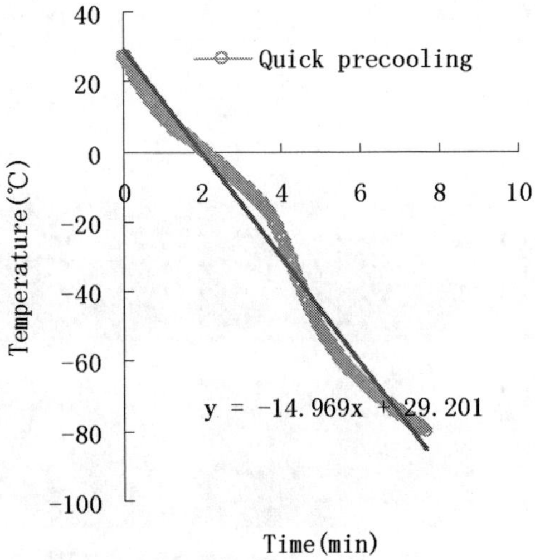

FIGURE 3. The curves of temperature falling of two different precooling rate

Drying Process

Three experiments were done here. The shelf temperature was controlled as Table 1. As we just took the movement of ice crystal during sublimation into account, samples only went to the end of primary drying process.

TABLE 1. Freeze-drying parameters of three experiments

Experiment	Precooling rate (℃/min)		Primary drying temperature (℃)
A	-14.97	-1.19	-15℃
B	-14.97	-1.19	-10℃
C	-14.97	-1.19	-5℃

Data Analysis With Grey Value And Sublimation Rate

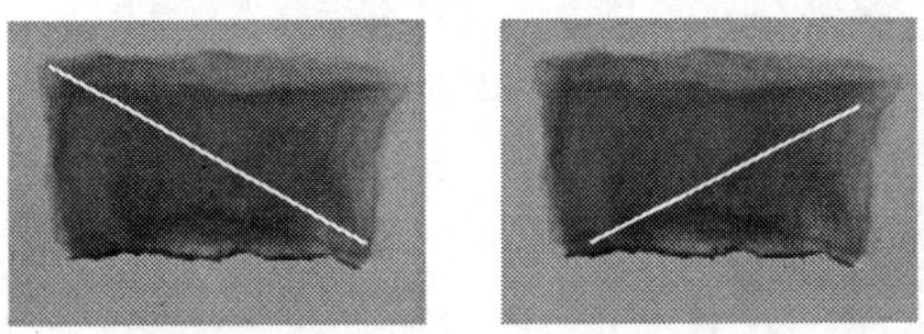

FIGURE 4. Schematic of calculation on mean grey value

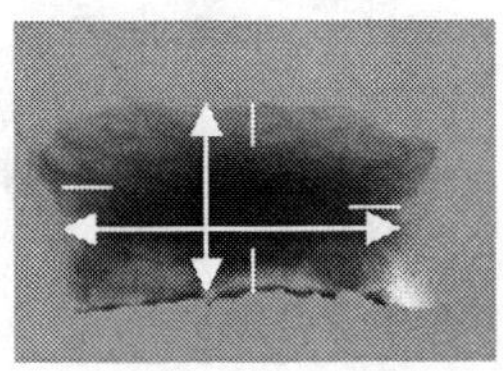

FIGURE 5. Calculation of sublimation rate

During scanning process, X-rays penetrates samples from profiles. Just as shown in Fig.4, to the scanning images obtained every two hours, the mean values of the grey value along two catercorner lines were obtained using Tview software respectively, then the average grey value distribution could be got totally by averaging the two values.

Figure 5 shows us how to calculate sublimation rate. Sublimation process may occur both in the vertical and horizontal direction. To vertical direction, firstly total distance between upper surface and under surface was measured, then dried distance of each surface was measured, so the total ratio meaning dimensionless distance of sublimation indicates the rate. Horizontal sublimation rate is just similar to that.

RESULTS AND DISCUSSION

Direct CT Scanning Images

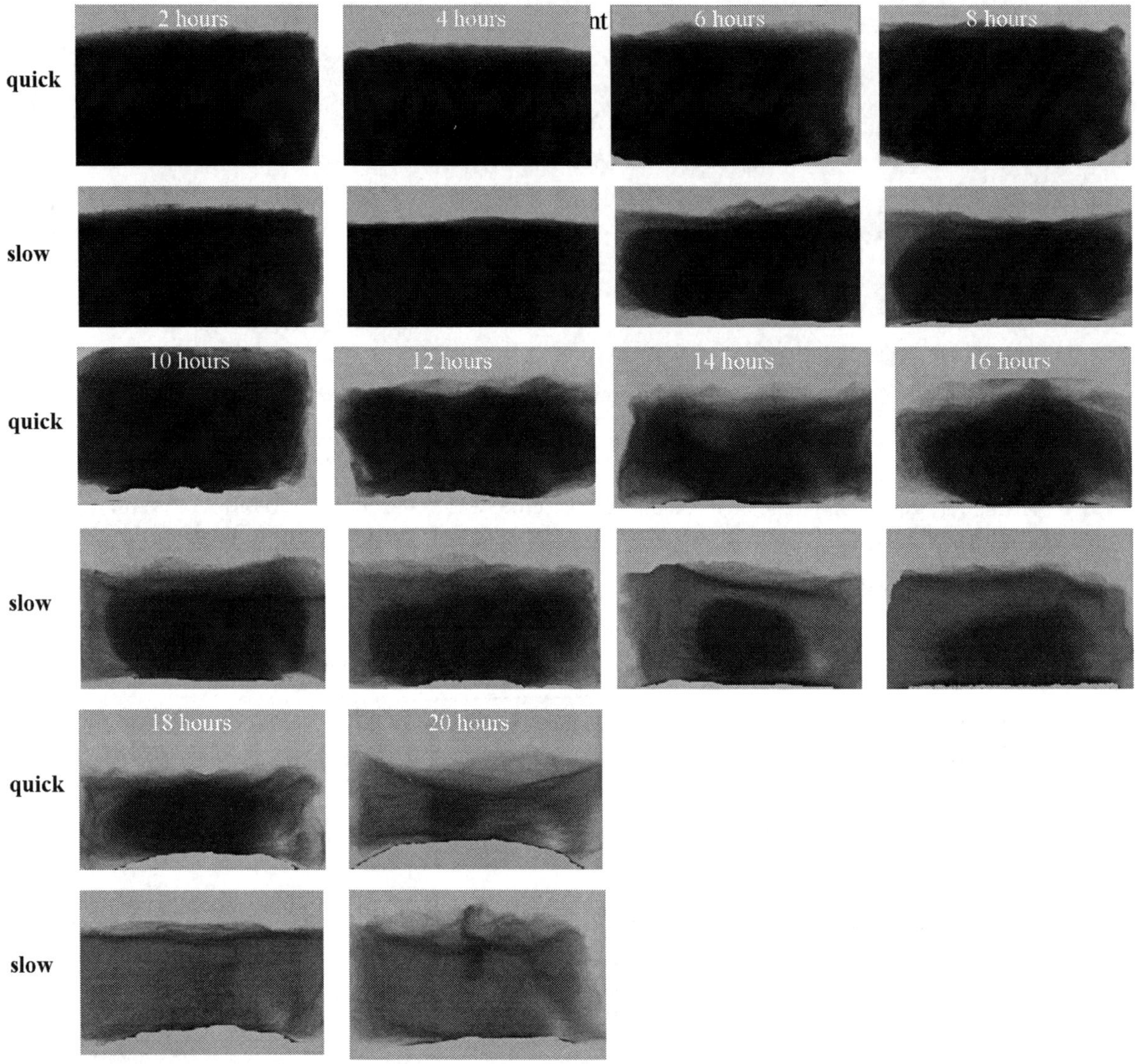

FIGURE 6. Scanning images of samples sublimating under different precooling condition (experiment B)

Figure 6 is the scanning images of apple slices sublimating under quick precooling and slow precooling condition. We just take experiment B for example. Because black pixel represents the absorption of X-ray, when more absorption of X-ray occurs, colour shows deeper. So the dark area indicates the frozen part of sample, and the gray one is the dried part.

At the same time, the drying process of slow precooling is quicker than that of quick precooling. From the images which have been dried for 12 hours, 14 hours, 16 hours and 18 hours, a majority of area has been dried yet of slow precooling, but only the margin gets dried of quick precooling. We can see that under slow precooling condition, samples get dried thoroughly after drying for 16-18 hours. Whereas, under quick precooling condition, samples can't get dried thoroughly after drying for 20 hours, and the frozen area can be seen from the image.

From those images, we can see that the movement of sublimation interface refers to the upper surface、the under surface and the side surface, it contracts to the centre of the sample through the frozen area it wrapped. Under slow precooling condition, an approximately spheral frozen area can be seen clearly from the image after drying for 14 hours, 16 hours. Collapsing phenomena and shrinking of samples are quite distinct under quick precooling condition, but can be avoided well under slow precooling condition.

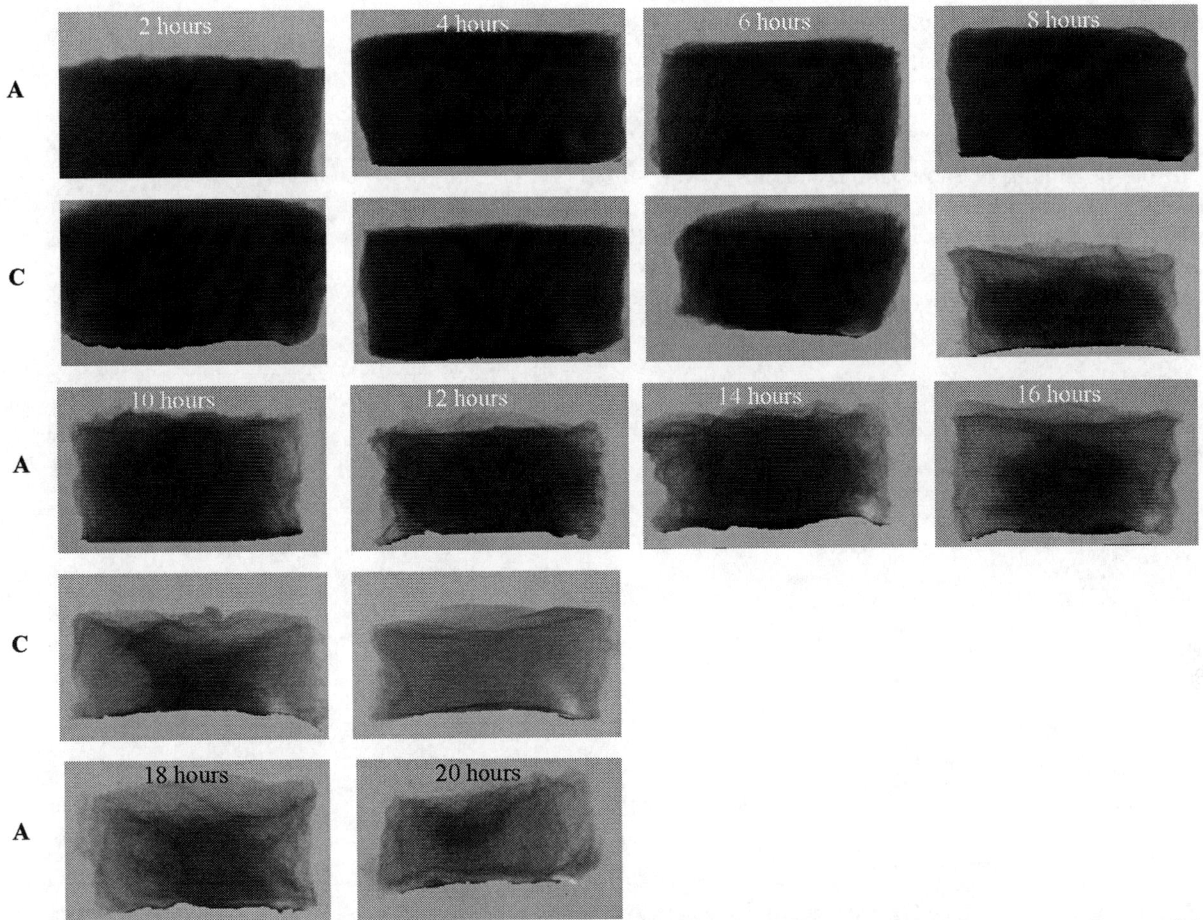

FIGURE 7. Scanning images of samples of experiment A and C during sublimation under slow precooling condition

Figure 7 shows us the scanning images of samples under slow precooling condition. Also we take experiment A and C for example. Comparing the two experiments, we find that the drying process of experiment C with high primary temperature is quick than that of other experiments. For instance, after drying for 8 hours, the samples' margin of experiment C has been dried deep, but the sample of experiment A almost don't start drying.

From those images, we can see that the movement of sublimation interface is three-dimensional, it contracts to the centre of the sample through the frozen area it wrapped. Heat and mass transfer also occur on under surface contacting with shelf and side surface. As a result, an approximately spheral shape frozen area can be seen from the images in Fig.7 after drying for 16 and 18 hours of experiment A, 8 and 10 hours of experiment C.

We can see that after drying for 12 hours of experiment A, samples nearly get dried thoroughly under slow precooling condition. Black frozen area can also be seen after drying for 20 hours of experiment C with primary drying temperature -15℃

Grey Value Analysis

From X-ray absorption characteristic, the absorptivity affects the grey value. A sample with low porosity which means smaller mean pore diameter has high absorptivity, and the grey value is low.

From the curves with grey value, we can see that the grey values of samples under quick precooling condition ascend unconspicuously, and samples dry slowly.

In the beginning of sublimation, samples shrink slightly and some descending points occur in the grey curve. In the middle of sublimation, as porous drying layer with low thermal conductivity has great resistance to heat and mass transfer, a flat area exists in the grey curve.

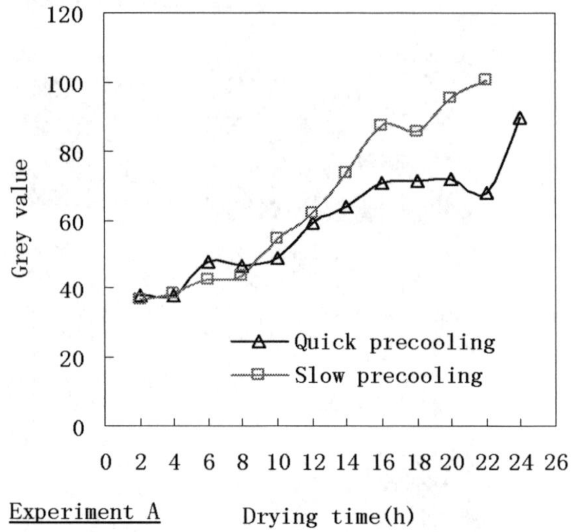

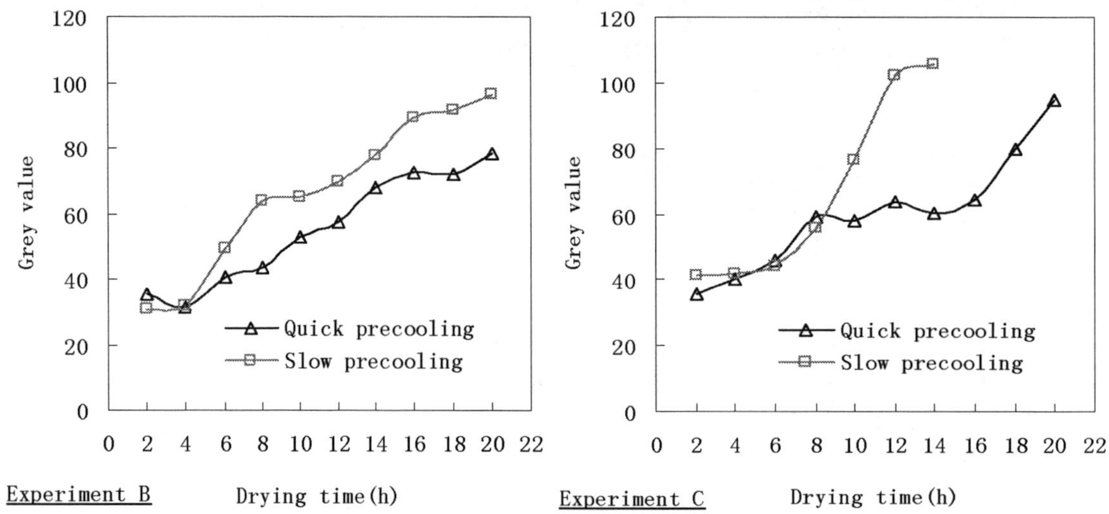

FIGURE 8. The variation curves with grey value of three experiments during sublimation

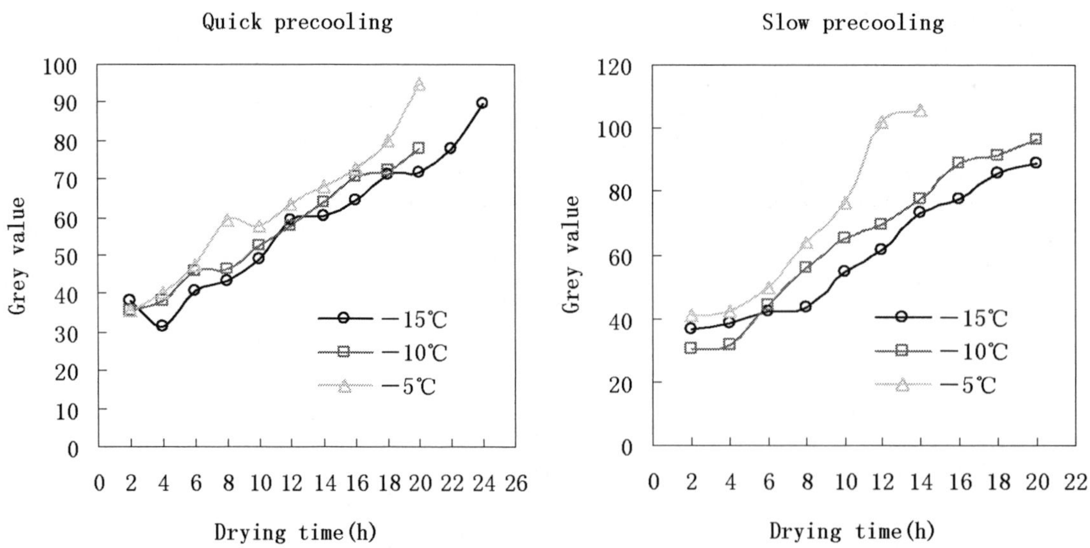

FIGURE 9. The variation curves with grey value of two precooling mode during sublimation

As shown in Fig.9, the difference of grey value is little to various primary drying temperature under quick precooling condition, but that is quite obvious under slow precooling condition. In slow frozen manner, the influence of primary drying temperature on sublimation rate is quite obvious. The reason is that, the porosity of porous dried layer is high under slow precooling condition, then the resistance to ice crystal sublimating is small, so it is the heat dominated process.

Whereas, the process under quick precooling condition is the mass dominated process. That is, the maximum diffusive ability of vapour is less than the ability of heat supply in this state, so it is limited to increase the sublimation rate by the mass transfer of dried area. Therefore, the sublimation process can be quickened obviously via increasing primary drying temperature in slow frozen manner.

Sublimation Rate

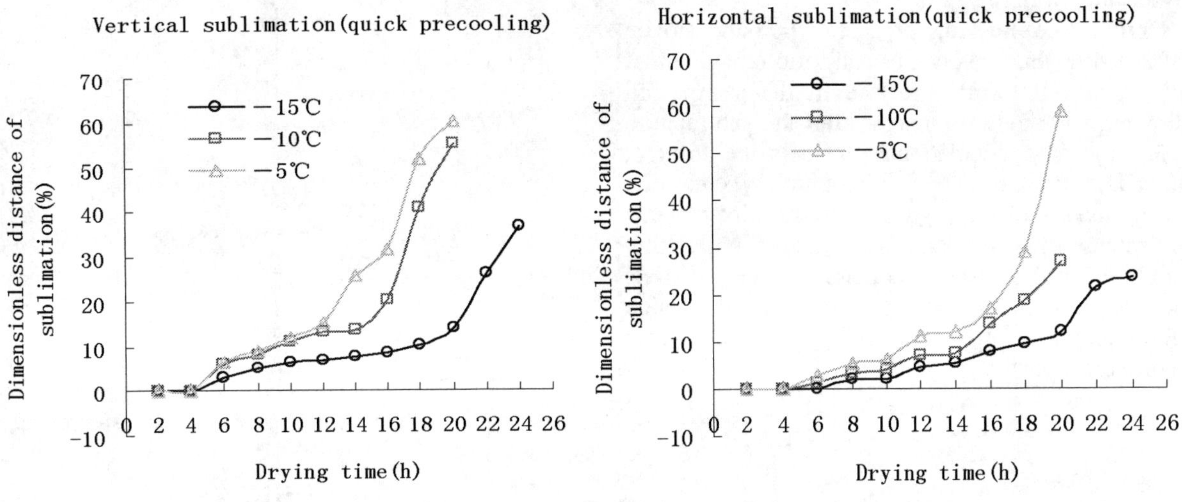

FIGURE 10. The sublimation rate of quick precooling in two direction

We can see in Fig.10 that primary drying temperature influences the horizontal sublimation more obvious than the vertical sublimation under quick precooling condition. The higher temperature, the higher the sublimation rate. On the whole, samples can't be dried completely in this precooling manner, the vertical sublimation rate is higher than the horizontal sublimation one.

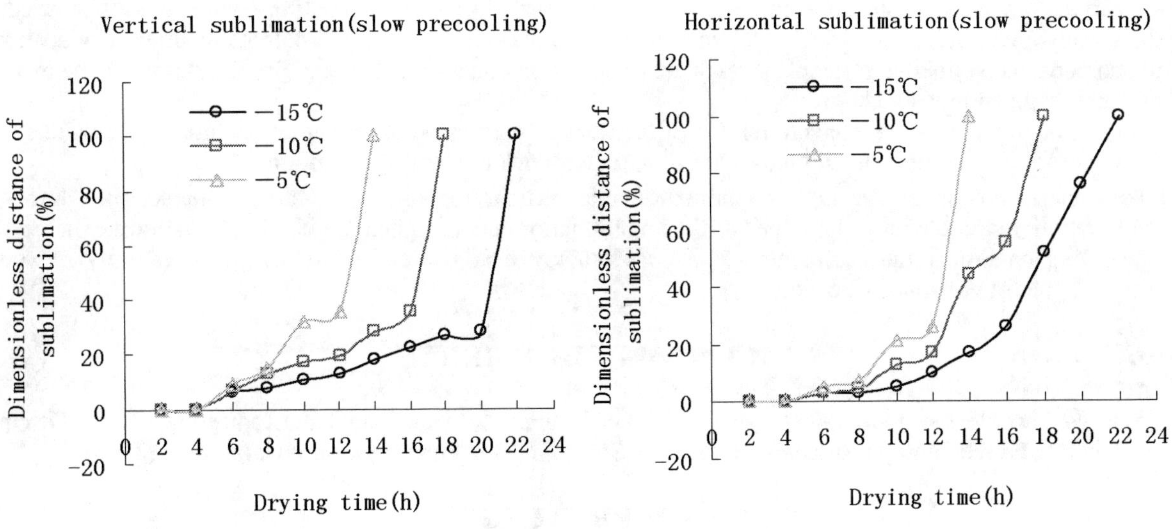

FIGURE 11. The sublimation rate of slow precooling in two direction

From Fig.11, we can see that the sublimation rate rises quickly near the end of sublimation not only in vertical direction, but in horizontal direction, that is, the moving of frozen area accelerates. One reason is that, the sublimation front proceeds from all surfaces, in a spherical fashion. Although porous drying layer with low thermal conductivity has great resistance to heat and mass transfer, total outside surface area decreases rapidly along with sublimation process, so the velocity of frozen area moving to center quickens in the last period. Another one is that, a certain humidity grads exists in porous dried layer, the phenomenon of seepage and capillary effect happens during sublimation.

Mechanism Discuss

We can understand the relationship between the rate of frozen area moving to center and total outside surface area from Fig.12.

During freeze-drying process, when the frozen mode is slow, big ice crystals can form reticulational ice crystal framework structure in the sample. In following subliamtion process, after the sublimation of ice crystals, reticulational channels are formed, which is benefit for vapour escaping. Whereas the frozen mode is quick, a great many slim ice crystals are formed instantly, channels for vapour sublimating can't be formed, resistance to mass transfer increases. Therefore, sublimation process is slow under quick precooling condition, which can be seen clearly from experimental results.

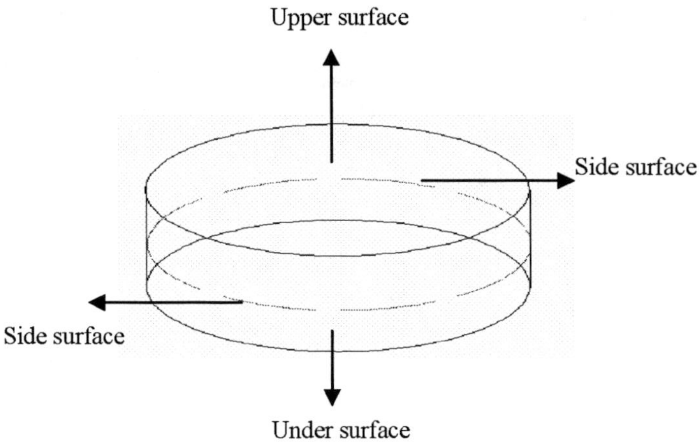

FIGURE 12. Schematic of heat and mass transfer during sublimation

CONCLUSIONS

Experimental observation and analysis on sublimation process during freeze-drying by micro-CT scanning, grey value analysis technique and sublimation rate calculation were carried out in this study, some conclusions can be obtained:

1) Not only with quick precooling but with slow precooling, sublimation will occur on all surfaces of the sample, the sublimation interface contracts to the centre of the sample gradually.
2) The drying process with slow precooling is quicker than that with quick precooling. Samples can't be dried thoroughly under quick precooling condition, but can get dried under slow precooling condition. Collapsing phenomena and shrinking of samples which are quite distinct under quick precooling condition can be avoided under slow precooling condition.
3) To various primary drying temperature, the differences of grey value at the same time are quite obvious under slow precooling condition although they are little under quick precooling condition.
4) The higher temperature, the higher sublimation rate. Primary drying temperature influences the horizontal sublimation more obvious than the vertical sublimation under quick precooling condition. Under slow precooling condition, the sublimation rate rises quickly near the end of sublimation not only in vertical direction, but in horizontal direction.

ACKNOWLEDGEMENTS

The authors would like to acknowledge the financial supports from Shanghai Leading Academic Discipline Project (T0503) and the Sunlight Research Program of Shanghai Municipal Education Committee (2003).

REFERENCES

1. A. I. Liapis and M. J. Pikal, *Drying Technology* **66,** 1265-1300(1996).
2. H. H. Xiao and T. Ch. Hua, *Cryo-Letters* **25**, 111-120(2004).

3. R. T. Lin, *Heat and mass transfer theory of porous medium*, Science Press, Beijing, 1995, pp.9-39.
4. M. Kochs and CH. Korber, *Int J. Heat Mass Transfer* **34**, 2395-2408(1991).
5. T. C. Hua and H. S. Ren, *Cryomedical and Cryobiological Technique*, Science Press, Beijing, 1995.
6. C. O. Sandall and C. J. King, *AICHE Journal* **13**, 428-438(1967).
7. J. F. Nastaj, *Int. Comm. Heat Mass Transfer* **28**, 1079-1090(2001).
8. W. J. Mascarenhas and H. U. Akay, *Computer methods in applied mechanics and engineering* **148**, 105-124(1997).
9. Y. Zh. Liu and Y. Y. Guo, *Journal of Xi'an Jiaotong University* **33**, 61-65(1999).
10. S. Kathy and C. Veerle, *Journal of Structural Biology* **148**, 11-21(2004).
11. S. Gerlinde and J. Engel, *International Journal of Pharmaceutics* **113**, 133-140(1995).
12. P. Abira and Ch. Koustuv, *Pharmaceutical Research* **20**, 802-803(2003).

Experimental Investigations of Mixing-Processes in The Wake of A Circular Cylinder in Stratified Flows

Peter Menzel, Frank Hüttmann, Martin Brede and Alfred Leder

Fluid Mechanics Department, University of Rostock, Albert-Einstein-Strasse 2, 18051 Rostock, Germany

Abstract. The existence of oxygen-rich saltwater in the deeper basins of the Baltic Sea is mainly caused by sporadic inflow-events of salty and oxygen-rich saltwater from the North Sea into the Baltic Sea. These inflows take place over the narrow and shallow Drogden Sill into the first basin, the Arkona Sea. Actually different offshore wind farms are planned in this region, which opens a whole string of questions about the ecological influence of offshore wind farms on the mixing of both layers. To answer these questions, numerical simulations of the mixing processes in the wake of wind turbine bases have been carried out. For the evaluation and quantification of these mixing processes a laboratory-experiment with a simplified model of the natural configuration has been realized. For this purpose a new water-channel has been build. This channel allows to simulate the inflow of saltwater in a size-scale of 1:100 to reality by keeping the densimetric Froude-Number. The experimental configuration consists of a long circular cylinder with a diameter of 8 cm in a 10 cm thick saltwater-layer flowing under a stationary fresh-water layer of 30 cm thickness. Focus point of this investigation is the wake of the cylinder in the stratified flow and the mixing-processes in the shear-layer due to the influence of the cylinder. The stratified flow around the cylinder induces the typical Karman-vortex wake, horseshoe-vortices at the bottom and in the shear layer and Kelvin-Helmholtz-instabilities in the shear-layer. Nonintrusive optical measurements were taken with planar laser-induced fluorescence (PLIF) combined with two dimensional particle imaging velocimetry (PIV). The combination of both techniques allows the determination of instantaneous velocity components u and w from PIV-measurements, the salinity s from PLIF-experiments, their variations u', w', s' and the correlations of those like Reynolds-stress terms ($\overline{u'u'}$, $\overline{u'w'}$, $\overline{w'w'}$) and turbulent- or Reynolds-flux terms ($\overline{w's'}$, $\overline{u's'}$). Especially the vertical Reynolds-flux $\overline{w's'}$ is the characteristic parameter to evaluate entrainment-velocity and entrainment-coefficient.

Keywords: turbulent mixing, bottom plume, stratified flow, PIV, LIF, cylinder, flow, ocean
PACS: 47.20.Ib, 47.20.Ma, 47.27.T-, 47.27.wb, 47.27.wj, 47.80.Cb, 92.10.Lq

INTRODUCTION

The Baltic Sea is a small ocean in North Europe. Due to the very narrow and shallow connection at the Danish straits (see Figure 1) there is only a very weak exchange of water between the North Sea and the Baltic Sea. The Baltic Sea Research Institute Warnemuende (IOW) found out that this exchange is very important for the transport of oxygen into in the deeper basins of the Baltic Sea like the Gotland Basin. The oxygen-rich water is a basic requirement for many biological and chemical processes in the Baltic Sea and thus the basis for a well working ecosystem. The oxygen-transport over such long distance is possibly a result of strong separation between the salty and oxygen-rich water plume from the North Sea and the Baltic Sea water. The reason for this separation is the difference in density which is mainly caused by the difference in salinity. The water from the North Sea has a salinity of 19 PSU while the Baltic Sea water has a salinity of nearly 8 PSU. In consequence of this, characteristic profiles of velocity and salinity were observed. Figure 2 shows measurements, performed by the IOW in the Baltic Sea north of Kriegers Flak on February 7th, 2004 [1]. This exchange only takes place during several inflow-events, depending on the weather situation, only a few times within a decade.

CP914, *Multiphase Flow: The Ultimate Measurement Challenge,* edited by X. Cai, Y. Wu, Z. Huang, S. Wang, and M. Wang

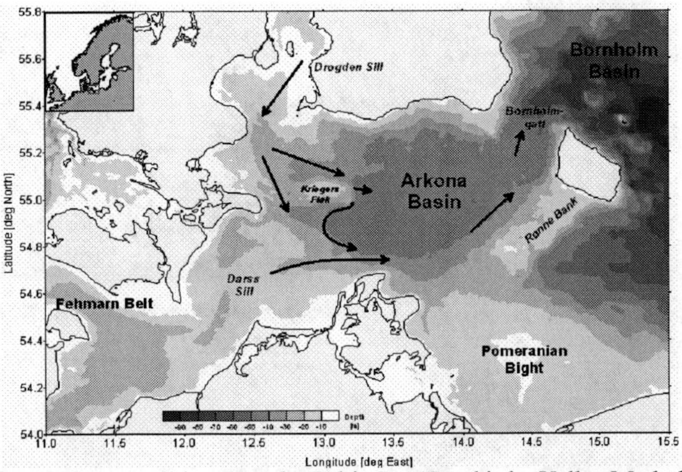

FIGURE 1. Inflow-conditions into the Baltic Sea (Graphic by Volker Morholz, IOW)

Because of the favorable wind-conditions in offshore regions several offshore wind farms are planed in the Baltic Sea. The largest wind farms have been projected in the Arkona Basin. The foundations of the wind turbines are designed as monopiles and tripods. The point of interest now is the impact of these groundings on mixing processes between oxygen-rich saltwater from the North Sea and the brackish water of the Baltic Sea.

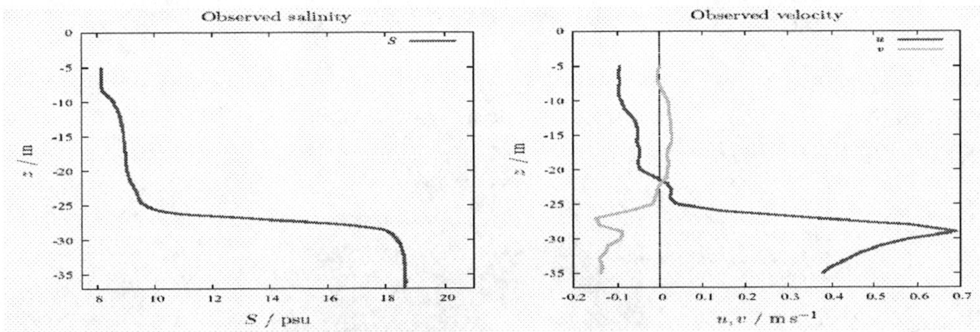

FIGURE 2. Salinity (left) and velocity (right, red southward, green eastward) profile north of Kriegers-Flak during a typical inflow-event, measured by IOW on February 7th , 2004 [1].

EXPERIMENTAL SETUP

The experimental facility consists of a water-channel to create a stratified saltwater flow, a PLIF/PIV-measuring-system and a 3-axis computer-controlled traversing-system.

Description of the water-channel

The water-channel has an overall length of 17 m and a measuring section of 3 m length, 0.8 m width and 0.8 m depth. The measuring section has optical accesses from 5 sides (see Figure 3). The inflow is designed as a bended nozzle followed by a broom as turbulence-generator and a gap of 10 cm height and 30 cm length (see Figure 4). The water-level is controlled by a dam behind the measuring section. Besides this there are two water tanks with a volume of 4 m^3. The upper one is used as reservoir, the lower one as collecting and mixing tank. In this configuration the maximum measuring time at the maximum volume flow is 2 minutes. Moreover it is also possible to run the system in a circulation mode with the collecting tank also used as reservoir. The pump allows a maximum volume flow of 120 m^3/h.

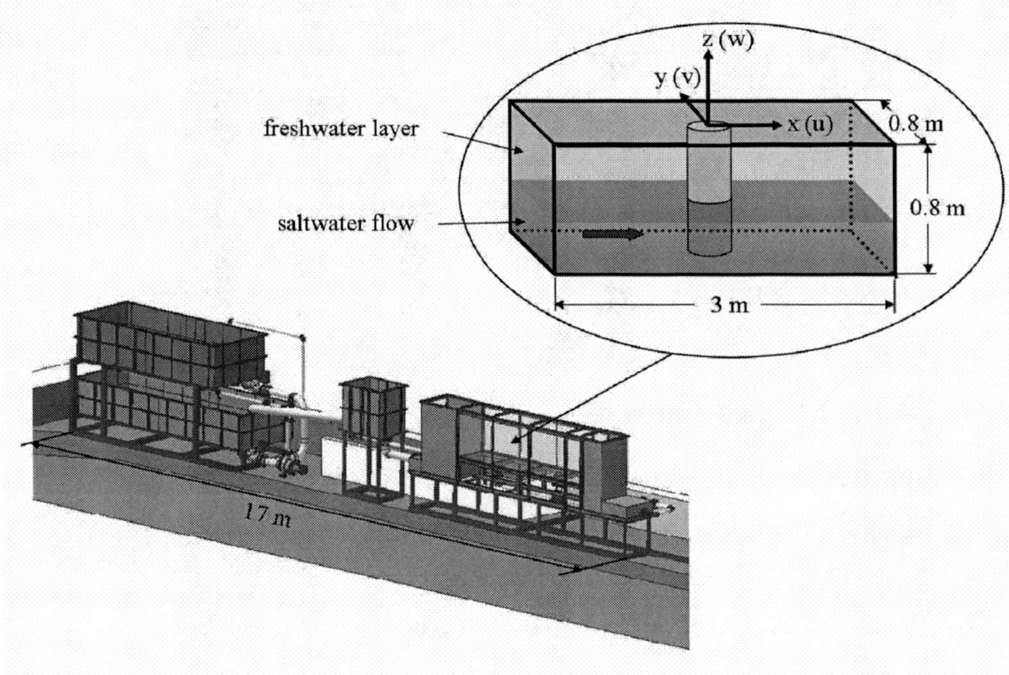

FIGURE 3. Water-channel and definition of the coordinate-system

For measurements the tank first has to be filled with freshwater. Afterwards salt, dye and tracer-particle have to be added to the water in the mixing tank. The saltwater can be pumped into the reservoir for a measurement or be used in circulation mode from the mixing tank.

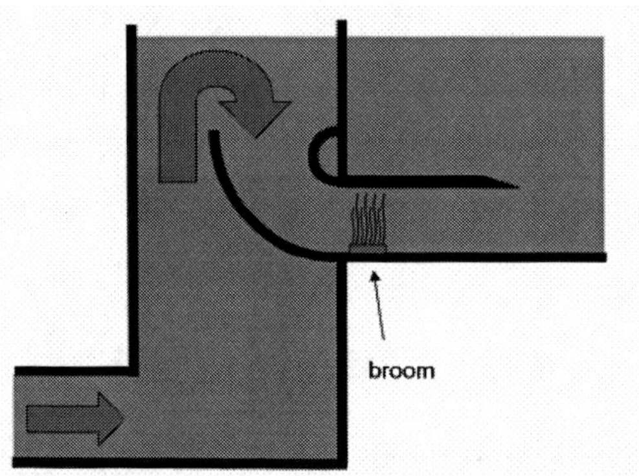

FIGURE 4. Flow-forming facilities at the inflow

The PLIF/PIV measuring-system

For measurements a combination of planar laser-induced fluorescence (PLIF) and digital particle imaging velocimetry (PIV) is used. The system consists of a Nd:YAG double-pulse-laser at a wavelength of 532 nm, two CCD-cameras and the controller (see Table 1 for specifications). One camera is used for 2 component PIV, the other one is used for PLIF with a colour filter in front of the objective. The relation to get the concentration C can be approximated for low concentrations by

$$C = \frac{I_F}{\alpha \cdot I_0} \tag{1}$$

with the intensity of the fluorescence I_F, the intensity of the laser I_0 and a constant α. To calibrate the PLIF-camera the fluorescence of several well known testing-solutions of Rhodamine 6G has to be observed. From these observations the constant α has to be determined as input for the measurements. The two cameras are used with a special camera mount that allows them to observe the same area at the same time. The principle is shown in Figure 5. In that way it is possible to correlate the PIV- and PLIF-Image. As tracer particles VESTOSINT 2157 with a mean diameter of 54 μm and a density of 1016 kg/m³ have been used. The used fluorescence dye has been Rhodamine 6G.

TABLE 1. System features

wavelength	532 nm
pulse-energy	120 mJ
pulse-length	3-5 ns
maximum repetition-rate	15 Hz (double-pulses)
camera	1600 x 1200 pixel
resolution	10 bit
pixel-size	7.4 μm
objective focus-length	85 mm

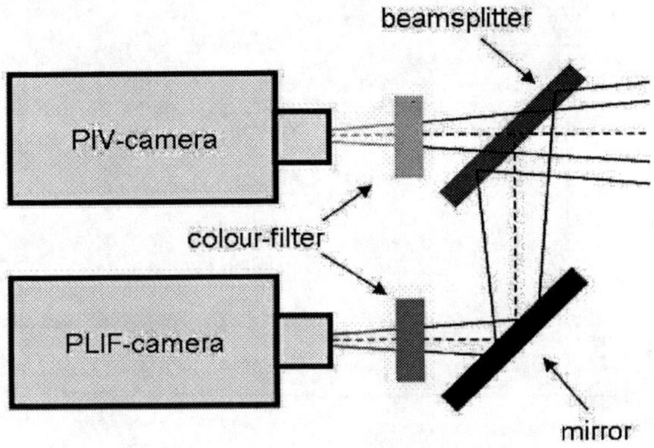

FIGURE 5. Beam-guiding in the camera mount

The results of the PIV-measurements are 2-dimensional, 2-component velocity-vector-fields with the instantaneous velocity components u and w. From the PLIF-Measurement the Rhodamine 6G-concentration c as an indicator for the salinity s is derived. This is possible because the Schmidt-Number of Rhodamine 6G of $Sc_{R6G} = 700$ is nearly the same as for salt which is $Sc_{salt} = 600$ to 1200 [2][3]. So it is possible to calculate the mean quantities and their fluctuations as well as their correlations like Reynolds-stress and Reynolds-flux. Especially the vertical Reynolds-flux-term $\overline{w's'}$ is used for the quantification of mixing processes.

MEASUREMENTS

Flow conditions

In the model-experiments the Arkona Sea with a depth z, a mean salinity S_a and mean density ρ_a is simulated by the depth z_M, the salinity S_{aM} and the density ρ_{aM}. The saltwater flow with the layer-thickness z_b, the mean velocity U_∞, the mean salinity S_b and the density ρ_b is described by the layer-thickness z_{bM}, the velocity $U_{\infty M}$, the salinity S_{bM} and the density ρ_{bM} in the model-experiments. The founding of a wind turbine with a diameter D is approximated as a cylinder with a diameter D_M. For exact values see Table 2. In order to keep the similarity in the experiments, the densimetric Froude number Fr_d has been taken as the characteristic quantity. This Froude number is related to the thickness of the saltwater flow z_b and defined as

$$Fr_d = \frac{U_\infty}{\sqrt{g z_b \dfrac{\rho_b - \rho_a}{\rho_b}}} \tag{2}$$

The Reynolds number Re is related to the diameter of the cylinder D and calculated as

$$Re = \frac{U_\infty D}{\nu} \tag{3}$$

TABLE 2. Characteristic quantities

	parameter	Arkona Sea (original)	water-channel (model)
wind turbine	diameter of cylinder [D /m]	8	0,08
	overall depth [z /m]	40	0,4
upper layer	thickness [z_a /m]	30	0,3
	salinity [S_a /m]	8	0
	density [ρ_a /kgm^{-3}]	1006	998
	velocity [U_∞ /ms^{-1}]	0,0	0,0
saltwater layer	thickness [z_b /m]	10	0,1
	salinity [S_b /PSU]	19	11
	density [ρ_b /kgm^{-3}]	1015	1005
	velocity [U_∞ /ms^{-1}]	0,05 to 0,5	0,043 (max. 0,37)
characteristic numbers	maximum Reynolds number [Re_{max}]	$4\cdot10^6$	$3{,}4\cdot10^3$ (max. $3\cdot10^4$)
	densimetric Froude number [Fr_d]	0,05 to 0,5	0,05 (max. 0,5)

First experimental results

Results of the velocity profile forming-experiments in the entrance region of the measuring section

To achieve a maximum analogy to the conditions in the Baltic Sea the velocity-profile at the entrance of the water-channel measuring section has been formed. For this purpose a nozzle, a broom and a Plexiglas-plate have been installed (see paragraph 2.1). These facilities have been added in order to force a velocity-profile like in the Baltic Sea. Figure 6 displays the velocity profiles measured in the Baltic Sea (left) and in the water-channel (middle) which agree very well.

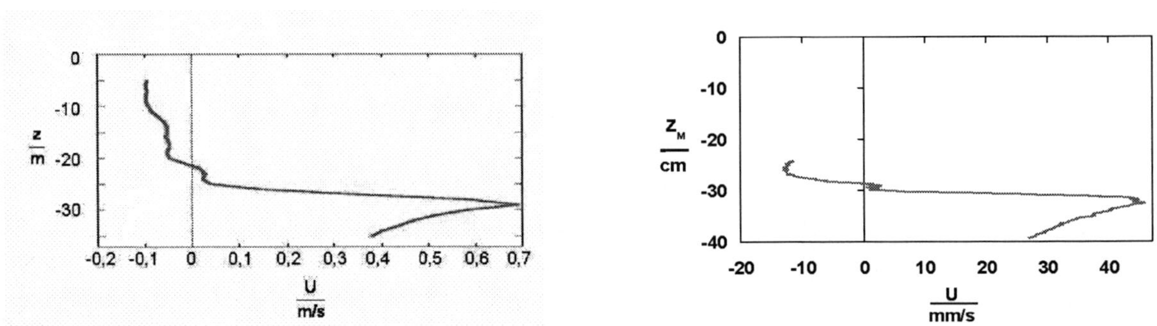

FIGURE 6. Velocity-profiles, measured in the Baltic Sea (left), entrance region of the water-channel measuring section (right)

Combined PLIF and PIV

The combined PLIF and PIV-measurements have been carried out at a double-frame rate of 1 Hertz. In that way 1000 data-fields have been recorded. These measurements have been carried out without a cylinder with the measuring-plane located in the x-z-plane of the system in the centerline of the channel. Figure 7 shows the concentration of Rhodamine 6G and the velocity-field. The left border of the image is located 40 mm downstream the inflow, defined as the end of the Plexiglas-plate (see Figure 4). The red areas indicate the saltwater whereas the purple areas characterize the brackwater. The shear-layer is visualized by the velocity-profile and the strong gradient in concentration between the saltwater and the brackwater. The Reynolds-flux is a characteristic quantity for transport and turbulent mixing. Figure 8 displays the vertical Reynolds-flux $\overline{w'c'}$ from another measurement where 600 PIV- and PLIF-images have been recorded at 1 Hertz. The measuring-plane was located at the end of the measuring section. The red areas in the middle indicate an upward component in the shear-layer region, the interface between freshwater and saltwater.

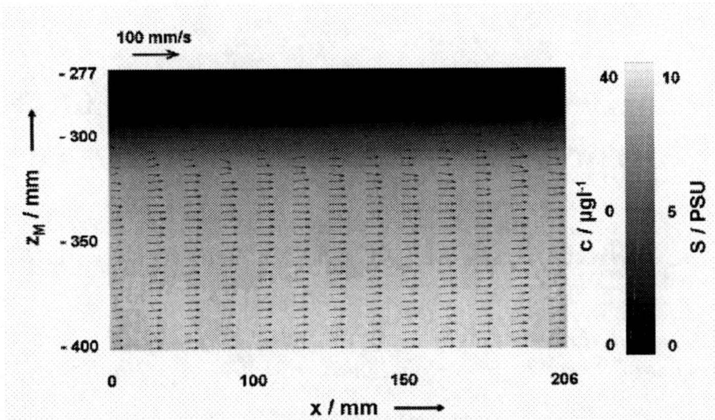

FIGURE 7. Concentration of Rhodamin 6G and velocity-profiles in the entrance region of the measuring section

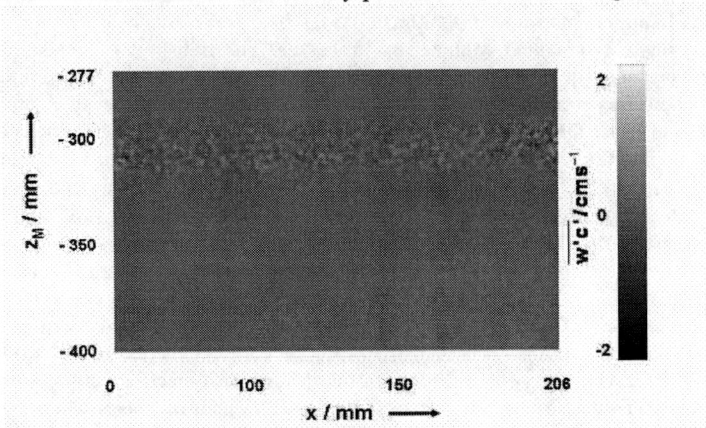

FIGURE 8. Reynolds-flux term $\overline{w'c'}$ / mms^{-1} at the end of the measuring section without cylinder model

PREVIEW

For the near future it is planed to measure velocity- and concentration-fields in the whole measuring section with and without cylinder. From these data the differences in vertical Reynolds-flux, entrainment-velocity and entrainment-coefficient will be estimated to quantify the mixing of both layers due to the influence of the cylinder. Beside this the wake of a cylinder in such stratified flow will be analysed to determine typical vortex-structures which are responsible for the mixing.

ACKNOWLEDGEMENTS

The Project „Quantification of Water Mass Transformations in the Arkona Sea: Impact of Offshore Wind Farms project-part: flow-experiments" (QuantAS-Off) as base of this paper is supported by the Bundesministeriums für Umwelt, Naturschutz und Reaktorsicherheit (BMU) with the project-number 0329957A. The authors like to thank the BMU for supporting this project and the Deutsche Forchungsgemeinschaft (DFG) for financing the participation on the 5th ISMTMF 2006.

REFERENCES

1. H. Burchard, H. U. Lass, V. Mohrholz, L. Umlauf, J. Sellschopp, V. Fiekas, K. Bolding, L. Arneborg, "Dynamics of medium-intensity dense water plumes in the Arkona Basin, Western Baltic Sea" in *Ocean Dynamics*, January 2005
2. C. R. Rehmann, "Effects of stratification and molecular diffusivity on the mixing efficiency of decaying grid turbulence." Ph.D. Thesis, Stanford University, California, 1996
3. C. D. Troy and J. R. Koseff, "The generation and quantitative visualization of internal waves" in *Experiment in Fluids* **38**: pp. 549-562, (2005)

Experimental Research Of Transport Characteristics Of Charged Gas-solid Two-phase Flow

Can Kang*, Zhengping Gao¶ and Tiqian Luo*

*School of Energy and Power Engineering, Jiangsu University,Zhenjiang 212013, China
¶Department of Power Engineering, Southeast University, Nanjing 210096,China

Abstract. A tentative system of charged coagulation was designed and built involving fan, motor, vibratory filler, transparent test part and support platform. Titanium dioxides of certain dimension range were used as solid particles simulating for the tiny dusts in practical engineering processes. The transport of gas-solid flows and dispersed solid particles included was presented in a rectangle-section pipe. Charge-to-mass ratio was measured through grid object method. The quantity of electricity of charged particles was measured. The flow variables under different charge voltages were emphasized in detail. And the advanced flow measurement technique—PIV(Particle Image Velocimetry) system was adopted to study the additional electric field effect on the solid-gas flow. The experimental results manifest that if charged voltage is less than -10kV, the charge-to-mass ratio increases continuously and reversely with the voltage, then reaches a saturation value at charged voltage of -60kV. The coupling between flow and electric field results in the obvious variation of characteristics of flows containing solid particles. With changed voltage, the electric field has linear effect on the gas-solid flows. Through comparison between the data under charged and uncharged conditions, axial attenuation of charged particles is weaker than that of uncharged particles. The vorticity near the pipe's bottom wall reaches its peak value. The vorticity near the boundaries under charged condition is relatively bigger and increases with voltage's increasing. Approaching the heart axis of the pipe, the vorticity increases with voltage's decreasing, as proves that solid particles are less restricted by main flow with the influence of electric field. Turbulent kinetic energy of solid particles keeps declining when the voltage varies from 0kV to -40kV, but increases when the voltage approaches -50kV, then declines again as the voltage varys from -50kV to -70kV.

Keywords: Charged gas-solid two-phase flow, Transport phenomenon, Flow field, PIV.
PACS: 47.57.eb,47.61.Jd,52.80.Wq

INTRODUCTION

Charged gas-solid two-phase flows exist extensively in engineering applications. Electrostatic spray painting[1,2] and electrostatic precipitation[3,4,5,6] are two cases recently focused intensively. If the poisonous tiny solid particles generated in those engineering processes suspending within the main flow are inhaled into human body, lung diseases will be resulted. One of the measures to treat the posed problem is coagulation which makes the coalescence between small particles and engenders big particle agglomerates. It is comparatively simple to collect big-size agglomerates effectively. Charged agglomeration is developed and applied consistently[7,8]. However, the fundamental mechanism between flow's properties and agglomeration has been rarely reported. Particle-particle interactions and particle-fluid interactions are both inevitable and crucial in those flows. Distributions of flow parameters such as velocity, pressure and turbulent kinetic energy have profound effects on agglomeration efficiency. Especially near the boundary, complex interactions among several factors of flows, electricity and boundary enhance relevant researches' difficulty considerably. The coagulation efficiency is affected by transport of charged particles in flow field, as has been approved unanimously.

CP914, *Multiphase Flow: The Ultimate Measurement Challenge,* edited by X. Cai, Y. Wu, Z. Huang, S. Wang, and M. Wang

1. MEASUREMENT OF CHARGE-TO-MASS RATIO

Charge-to-mass ratio embodies some crucial charged characteristics. Titanium dioxide is used here as solid particles simulating tiny dusts in engineering. Charged characteristics are investigated through variation of voltage. Measurement device consists of multi-layer wire netting, exact microampere meter, exact electrical balance and stopwatch. Circuit is made when charged particle groups touches the grounded wire netting and micro-current generated by the electricity in wire netting. The current can be measured by exact microampere meter. The measurement time can be measured by stopwatch and the mass of particles collected by the wire netting can be measured by exact electrical balance, as shown in Fig. 1.

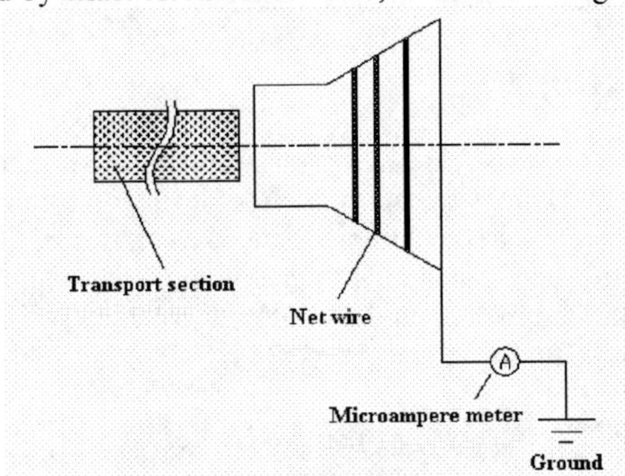

FIGURE 1. Measurement Device for Charge-to-mass Ratio

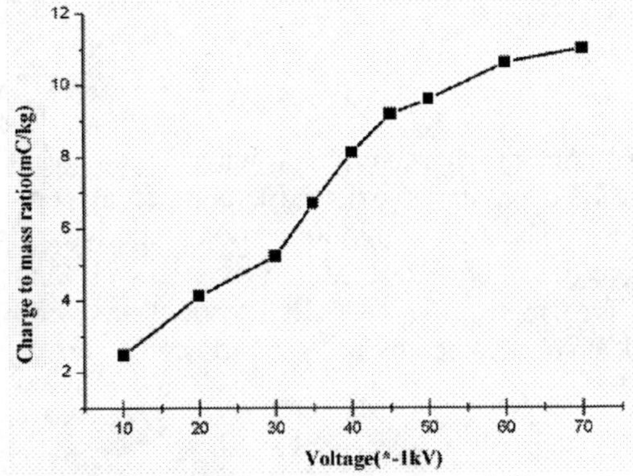

FIGURE 2. Relations between Voltage and Charge-to-mass ratio

2. PARAMETERS MEASUREMENT WITH PIV

2.1 Critical Parameters And Experimental Rig Of PIV System

Continuous progress in PIV techniques[10] have solved the problem of obtaining whole field flow information and high resolution simultaneously and provided new experimental method for investigation of charged two-phase flow. PIV system made by TSI company is employed in the experiment. Operation frequency of the laser generation device is 15Hz , the laser wave length is 532nm and single impulse results in energy of 120mJ. The maximum sampling frequency of the system is 3.75 velocity fields per second. According to the measurement principle of PIV system and solid particles' transport velocity, system parameters are defined as: impulse interval of 90μs, display pixels of 1347×1024 and practical measurement zone of 100mm×200mm. Fig.3 shows the combination of charged particles experimental device and PIV system. Steady delivering pressure of 0.23MPa and transport velocity of 4.8m/s are adopted. The voltage varies from -70kV to 0kV with certain interval of 10kV.

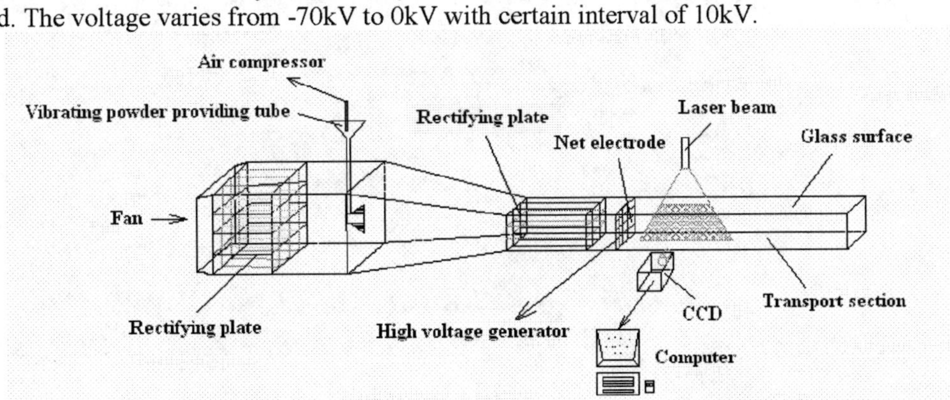

FIGURE 3. Integrated Experimental System of Charged Two-phase Transport Flow

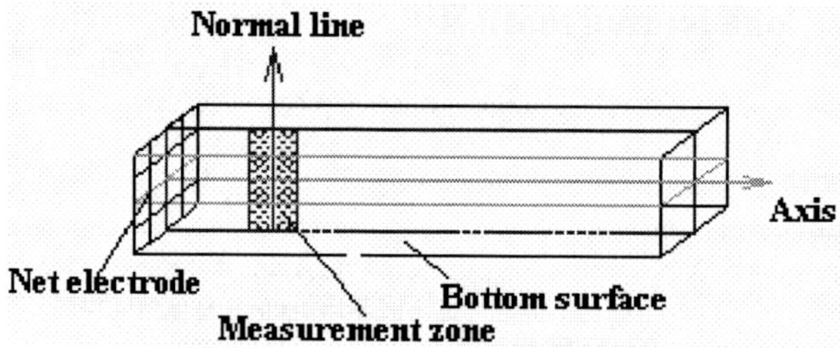

FIGURE 4. Measurement Zone

2.2 Experimental Results

Flow parameters' distributions when charged gas-solid fluid escapes from net electrode are expected to be measured. However, under existing experimental conditions, the whole transport sections can not be measured simultaneously. The critical middle section embodying flow's major characteristics is focused(as shown in Fig.4). In Fig.4, the vertical coordinate of the axis is y=170mm and the horizontal coordinate of the bottom normal is x=150mm. Distributions of velocity, vorticity and turbulent kinetic energy under different voltages(as shown in Fig.5,6 and 7) are obtained by binaryzation cross-correlation image processing arithmetic[11].

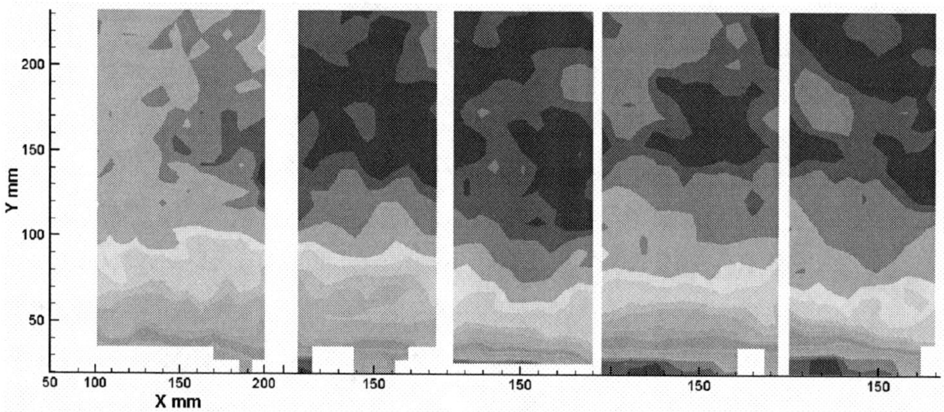

Uncharged -30kV -40kV -50kV -70kV

FIGURE 5. Velocity Contour

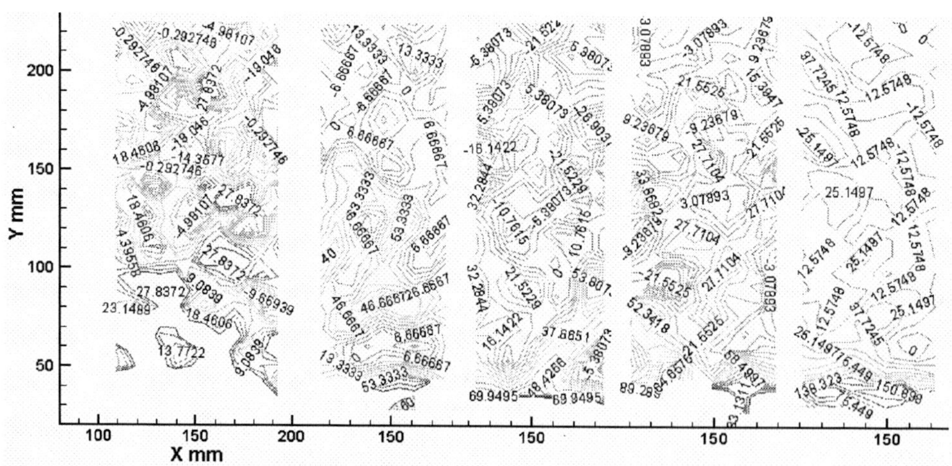

Uncharged -30kV -40kV -50kV -70kV

FIGURE 6. Vorticity Distribution

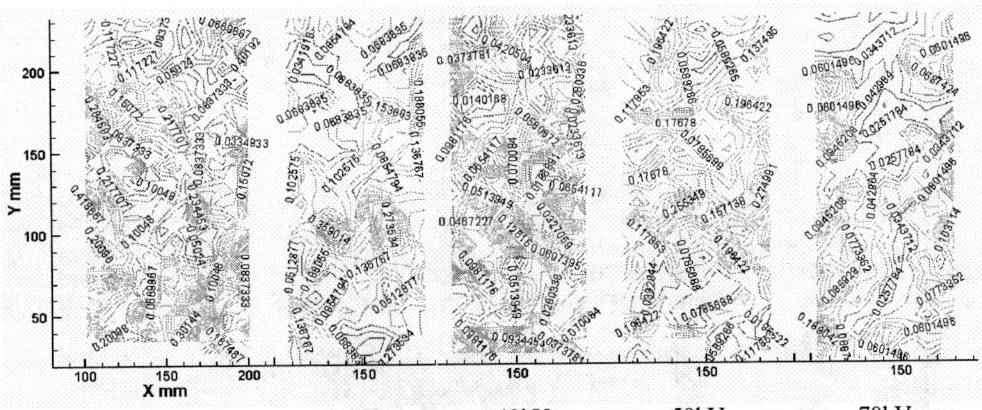

Uncharged -30kV -40kV -50kV -70kV

FIGURE 7. Turbulent Kinetic Energy Distribution

2.3 Analyzation

Under uncharged conditions, as can be seen that evident attenuation of particle velocity in axial direction and velocity's variation along backward bottom normal direction. The coupling between flow and electric fields makes the obvious variation of characteristics of flows containing solid particles. With changed voltage, the electric field has linear effect on the two-phase flows.

Under voltage of -30kV, compared with uncharged condition, the axial particle velocity increases and the attenuation gradient of particle velocity gets greater along backward bottom normal direction. From the whole field view, the turbulent kinetic energy gets decreased, the vortices' dimensions get reduced and the vorticity gets a distinguished enhancement. When the voltage is lowed to -40kV, the axial particle velocity increases. Compared with the flow field under voltage of -30kV, particle velocity has a faster attenuation in backward bottom normal direction, turbulent kinetic energy distribution decreases obviously, vortices' dimensions decrease comparatively, and vorticity gets slightly greater. When the voltage decreases to -50kV, the attenuation of particle velocity along backward bottom normal direction becomes relieved compared with that under voltage of -40kV. The axial particle velocity distribution is analogous with that under -40kV, but the turbulent kinetic energy gets greater. The vortices' dimensions gets increased, however, the vorticity gets decreased. When voltage is lowered down to -70kV, the attenuation gradient of particle velocity along backward bottom normal direction gets bigger, the vortices' dimension gets decreased, vorticity becomes bigger. The turbulent energy distribution shows no obvious variation in terms of that under voltage of -50kV.

3. CONCLUSIONS

1)Through contrast between results under charged and uncharged conditions, axial attenuation of charged particles is weaker than that of uncharged particles.

2)The vorticity near the bottom wall has the maximum value. The vorticity near the boundaries under charged condition is relatively greater and increases with voltage's increasing.

3)Approaching the axis, the vorticity increases with voltage's decreasing, which proves that solid particles are less restricted by main flow with the influence of electric field.

4)Turbulent energy of solid particles keeps decreasing when the voltage varies from 0kV—-40kV, but increases when the voltage reaches -50kV, then decreases again as the voltage's variation from -50kV—-70kV.

ACKNOWLEDGEMENTS

This work was supported by the University Research Project of Jiangsu Province of China under contact of 05KJD470047.

REFERENCES

1. Ying Baosheng, Gao Quanjie and Wang Jiaqing. *Surface Technology* **5**, 12-16(2003).
2. C.U.Bottner and M.Sommerfeld. *Powder Technology* **4**, 206-216(2004).
3. Xu Dexuan. *Environmental Engineeirng* **6**, 25-28(1997).
4. Yu Shui, Li Liguang, Hu Zongjie, Deng Jun and Li Gong. *Transactions of CSICE* **3**, 239-243(2005).
5. Wen Jianlong, Wang Junfeng, Zhang Jun and Luo Tiqian. *Transactions of CSICE* **1**, 31-34(2003).
6. Matthew E Thomas and Roberto Disalvo. *SAE* Paper, No.20123053(2002).
7. Laitinen A , Hautanen J and Keskinen J. Bipolar. *Electrostatics* **38**, 303-315(1996).
8. Stanley B.Grant. *Journal of Colloid and Interface Science* **2**, 238-250(2001).
9. Wen Jianlong, Wang Junfeng, Chen Songshan, Wang Ze and Luo Tiqian. *Drainage and Irrigation Machinery* **19**, 45-47(2000).
10. Fan Jiechuan. *Modern flow visualization.* Publisher,National Defense Industry Press of China,2002,pp 120-167.
11. Wang Canxing. "Theoretical study and application of PIV processing system", Ph.D.Thesis, Zhejiang University, 2000.

Turbulence Modulation Measurements in a Charged Monodispersed Droplet Stream

Wei-Fu Lu, Wei-Hsiang Lai*, Keh-Chin Chang, and Tung-Hsiao Tsai

Department of Aeronautics and Astronautics, National Cheng Kung University, Tainan, 701, Taiwan, R.O.C.

*Corresponding author:whlai@mail.ncku.edu.tw

Abstract. As reported in the study of Chang *et al.* [1], the modulation of the carrier-phase turbulence, that is, the effect of the presence of dispersed phase on the turbulence structure, is heavily dependent on the term of ($<u'_{gi}u'_{pi}>$-2k), where < > denotes the ensemble averaging quantity; k is the turbulent kinetic energy, u'_{gi} and u'_{pi} are the i-th component of fluctuating velocities of carrier and dispersed phases, respectively. Nevertheless, very few experimental data of $<u'_{gi}u'_{pi}>$ can be found in the literature. Wang and Huang [2] made the $<u'_{gi}u'_{pi}>$ measurements by using a phase Doppler anemometer in a planar mixing-layer flow leaded with polydispersed spray. However, the accuracy of the measured $<u'_{gi}u'_{pi}>$ data made by Wang and Huang [2] was in doubt according to the comparison results with the model predictions in the study of Chang *et al.* [1]. Prevost *et al.* [3] preformed the $<u'_{gi}u'_{pi}>$ measurements using a phase Doppler anemometer in the far field of a polydispersed particle-laden jet. Due to unavailability of the thorough measured data that could be used as the inlet conditions required in the theoretical simulation, the experimental works of Prevost *et al.* [3] can not serve as a benchmarked case.

In this study, a monodispersed droplet stream, instead of polydispersed spray, is generated using a monosized atomizer. However, the occurrence of break-up or coalescence due to droplet collisions in the stream downgrades the uniformity of droplet size in the downstream. To avoid the downgrading of the size uniformity in the downstream stream, the electrohydrodynamic (EHD) technique is used. Monodispersed droplets can be made toward rigid body after charging electrons. When two charged droplets move closer, the Coulomb force will help the droplets keep distantly and avoid the occurrence of droplet break-up or coalescence. Measurements of particles' velocity and size in a turbulent stream are made using a Particle Dynamic Analyzer (PDA). Quantities of $<u'_{gi}u'_{pi}>$ and k will be computed using the measured velocities of both carrier and dispersed phases.

INTRODUCTION

Two-phase flow is an interesting subject, and has been paid more attention by many researchers in recent years. Spray or particle-laden jet which is included in this field is widely applied to engineering application, e.g., energy engineering, combustion of liquid fuel, material manufacture, and drying techniques, etc. The turbulence structure of the continuous phase is altered due to dispersed-phase loading, which is attributed to the inability of the dispersed-phase particles to completely follow the fluctuating motion of turbulent eddies [4]. According to Kolmogorov's concept of spectral energy transfer, energy extracted form the low frequency eddies cascades down to the small scale eddies with high frequencies. An expression for the modulation of the turbulence spectrum is formulated. The term "turbulence modulation" is probably due to their first naming for the two-phase interaction [4].

As Chang *et al.* [1] mentioned that turbulence modulation of carrier phase by dispersed phase will depend on the term of ($<u'_{gi}u'_{pi}>$-2k) strongly, where < > denotes the ensemble averaging quantity; k is the turbulent kinetic energy, u'_{gi} and u'_{pi} are the i-th component of fluctuating velocities of carrier and dispersed phases, respectively. But it is hard to find the experimental data in the literature. Wang and Huang [2] made the $<u'_{gi}u'_{pi}>$ measurements by using a phase Doppler anemometer in a planar mixing-layer flow leaded with polydispersed spray. However, the accuracy of the measured $<u'_{gi}u'_{pi}>$ data made by Wang and Huang [2] was in doubt according to the comparison results with the model predictions in the study of Chang *et al.* We estimated the data of $<u'_{gi}u'_{pi}>$ measured by Wang and Huang have been overestimated in the shear layer in the downstream regions. Prevost *et al.* [4] preformed the $<u'_{gi}u'_{pi}>$ measurements using a phase Doppler anemometer in the far field of a polydispersed particle-laden jet. Prevost *et al.* [3] measure fluid-particle correlation in the far field of an axisymmetric jet, but it is to unavailability of the thorough measured data that could be used as the inlet

CP914, *Multiphase Flow: The Ultimate Measurement Challenge,* edited by X. Cai, Y. Wu, Z. Huang, S. Wang, and M. Wang

conditions required in the theoretical simulation.

For getting more experimental data, this study will measure $<u'_{gi}u'_{pi}>$ term in a monosized spray field using phase Doppler anemometer. Therefore, as if Wang and Huang's work, since the velocity was measured by PDA is the instantaneous velocity of a droplet, $u_i^j(t)$, it can be decomposed into mean and fluctuating parts as

$$u_{gi}(t) = U_{gi} + u'_{gi}(t) \quad (1)$$

$$u^j{}_{pi}(t) = U^j{}_{pi} + u^{j'}{}_{pi}(t) \quad (2)$$

where j represents the velocity of j-size drops. Hence, the correlation function can be calculated as follows:

$$\overline{u'_{gi}(t)u^{j'}{}_{pi}(t+\tau)} = \overline{(u_{gi}(t)-U_{gi})(u^j{}_{pi}(t+\tau)-U^j{}_{pi})} = \overline{u_{gi}(t)u^j{}_{pi}(t+\tau)} - U_{gi}U^j{}_{pi} \quad (3)$$

where τ is the delay time between the gas phase and drop phase. When the droplet passes through the measured volume of laser beam, the scattering signal can be detected by the receiver and then processed by the signal processor. The signals detected by the receiver are discrete data and the arrival time for each data is essentially random. The time series of the measured data can be further divided into a number of classes according to the drop size for the purpose of calculating the velocity cross-correlation function. Hence, the velocity cross-correlation function can be expressed in the following equation:

$$\overline{u'_{gi}(t)u^{j'}{}_{pi}(t+\tau)} = \frac{1}{M}\sum_{m=1}^{M}\left\{\frac{1}{N}\sum_{l=1}^{N}u'_{gi}(t_m)u^{j'}{}_{pi}(t_m+\tau)\right\} \quad (4)$$

where N is the quantities of data of time zone at special time t_m, M is the total quantities of calculated number at this point.

Although the velocities of the gas phase and dispersed phase are measured by signals scattering from seeding and drops, simultaneous scattering signals for both phases (i.e., τ=0) is not allowed due to the limitations of PDA measurement. However, since the velocity cross-correlation function, $\overline{u'_{gi}(t)u^{j'}{}_{pi}(t+\tau)}$, is symmetric to τ=0 if it is isotropic in the spray, an interpolation scheme illustrated in Fig. 1 and the value of the velocity cross-correlation function at τ=0 obtained from this figure is used to represent $<u'_{gi}u'_{pi}>$ quantities.

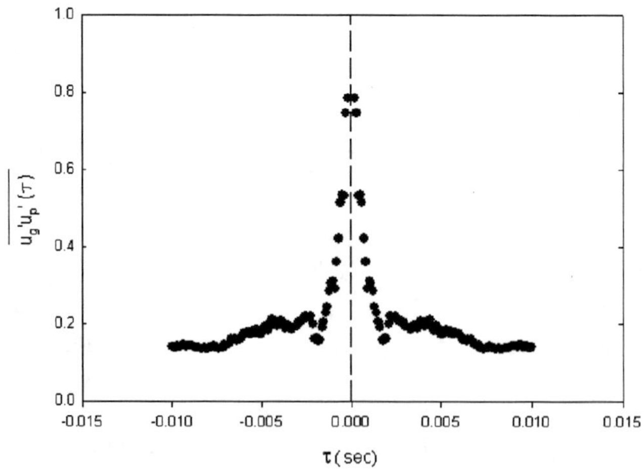

FIGURE 1. Typical cross-correlation function

The $<u'_{gi}u'_{pi}>$ quantities from Wang and Huang's [2] data only could be used in three kinds of particle size classes: 10 μm, 20 μm and 30 μm. And further, it doesn't reach the statistic significance above 30 μm particle size data due to the count is insufficient. In this study, the authors will use monodispersed droplet stream to instead of polydispersed spray for avoiding insufficient data quantities in any size class. However, the occurrence of break-up or coalescence due to droplet collisions in the stream downgrades the uniformity of droplet size in the downstream [5, 6]. To avoid the downgrading of the size uniformity in the downstream stream, the electrohydrodynamic (EHD) technique is used in this study. Monodispersed droplets can be made toward rigid body after charging electrons. When two charged droplets move closer, the Coulomb force will help the droplets keep distantly and avoid the occurrence of droplet break-up or coalescence [7]. Measurements of particles' velocity and size in a turbulent stream are made using a Particle Dynamic Analyzer (PDA). Quantities of $<u'_{gi}u'_{pi}>$ and k will be computed using the measured velocities of both carrier and dispersed phases.

We used a vibrating component to disturb fluid, and pressured on the fluid by nitrogen gas. The fluid would pass through a fixed nozzle to generate monodispersed droplets [8-10]. The fundamental principle of monodispersed droplet generator is built in instability theorem [11]. The vibrating apparatus is manufactured by piezoelectric material mostly. The device can generate monodispersed droplet stream if tune the function generator to make a suitable frequency sine or square wave, and amplify this artificial disturbance to enough intensity. The monodispersed droplet stream has a coherent length where particle size, velocity and droplet space are the same [12]. After this region, the wake of a leading droplet can influence a trailing droplet's behavior to begin collision, coalescence, and even disrupt as smaller droplets [5, 6]. Obviously, the stable region is the requirement we need, and to avoid droplet collision and coalescence is a goal of this study.

We expect to set up monodispersed droplet spray that uniform size distribution will assist us to investigate turbulence modulation. Since original monodispersed droplet stream is not like a spray which owns a spray angle, this characteristic will affect the space range of modulation of dispersed phase. Moreover, the large number of droplet pass the same path as a column will cause seeding cannot be detected by PDA via preliminary experiment. Therefore some modifications for monodispersed droplet stream are necessary. Choi and Delcorio [9] applied a induce electric field on a monodispersed droplet stream in 1990. The neutral liquid droplet passed through a positive electric field and became a dipole by their device; a negative charge accumulates on the surface while a region of positive charge developed in the inner portion of the drop. Our research will keep and modify this idea to achieve our needs, and then get turbulence modulation by measuring the $<u'_{gi}u'_{pi}>$ quantities.

EXPERIMENTAL APPARATUS AND INSTRUMENTS

Monodispersed droplet generator and seeding system

The experimental apparatus is shown in Fig. 2. Constant frequency signals of function generator send to an amplifier, and then drive a vibrating pod which connects the piezoelectric material and disturb the fluid. The working fluid was pressured, issued by nitrogen gas, form a liquid column and then break up. A DC high voltage power supply provides positive electric on the storage though, the grounding connects a charging ring, which is stainless steel and locates nozzle exit about 10 mm. The ring thickness is 2.6 mm, outer diameter is 21.2 mm, and inner diameter is 13.5 mm. The working fluid is pure water, nozzle orifice is 30 μm, gauge pressure is 1.25 kgf/cm^2, volume flow rate is about 0.4 c.c./min. We can generate about 30-70 μm diameter droplet by a 30 μm nozzle. The glass test section is around the monodispersed droplet stream and set a turbulence flow by a fan which Reynolds number is about 7000. The seeding system comprised a humidifier, setting chamber, and annular arrangement elbows. These seeding particles are also composed of pure water as well as monodispersed droplets. The monodispersed droplet is dispersed phase, and the seeding is continuous phase in our experiment. Both these are pure water that can avoid difference of refractive index of two phases while using PDA to measure.

Observation system

The observation system consists of a CCD camera by OSCAR Company, a SZ4045 microscope by Olympus Company and a strobe scope by Monarch Company in this research. The images can be magnified 40-475X and recorded by videocassette recorder, and then output as image file by NI-PCI-1411 frame grabber.

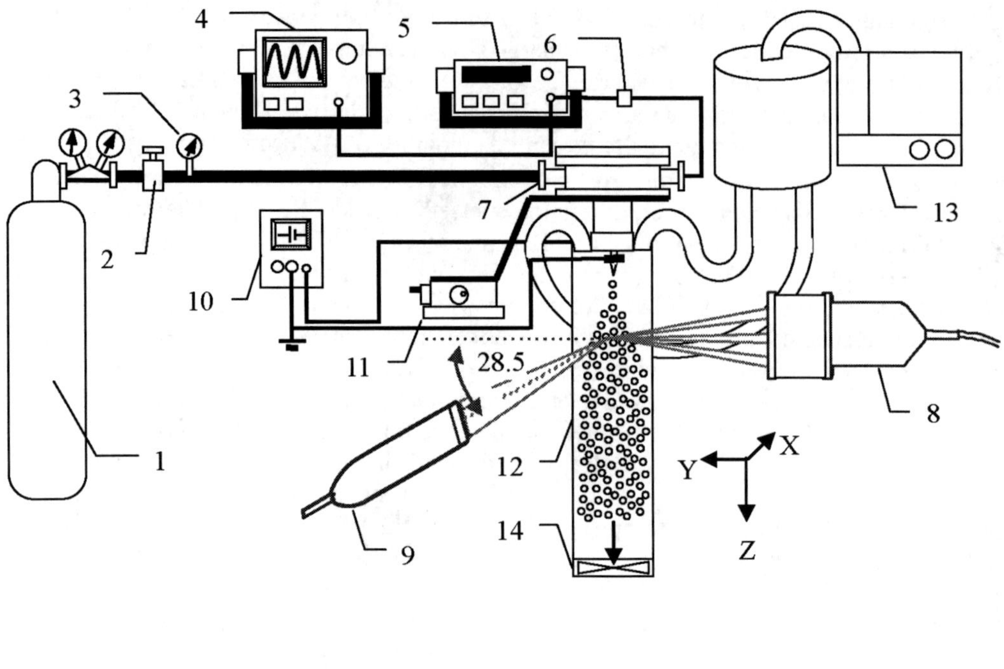

FIGURE 2. Experimental apparatus illustration
1. N_2 gas supply 2. Pressure regulator 3. Pressure gauge 4. Oscilloscope 5. Function generator 6. Amplifier 7. Monodispersed droplet generator ~70 μm(Liquid phase) 8. Transmitter of PDA 9. Receiver of PDA 10. DC HV power Supply 11. 2D traverse 12. Glass test section 13. Humidifier <10 μm(Gas phase) 14. Fan

Particle Dynamic Analyzer

The Particle Dynamics Analyzer, PDA, which is manufactured by Dantec Company, can take particle information at one point of space as Eulerian approach. PDA use scattering light principle to take a large amount of information of particle [13, 14], like as mean size diameter, velocity, fluctuation velocity, concentration, volume flux, etc. PDA has a great accuracy and space resolution, and can supply an instant, non-intruded optic measurement. The PDA setting parameter is as below in this study:

Two Dimensional, the focal length of optical lens of transmitter and receiver are both 600mm, scattering angle is set in 28.5 degree, refraction mode, parallel polarization light. Receiver aperture plate is selected as A type, measurement range of particle size is 0-98.01 μm. Signal validation is set as Yes, -6dB; the maximum phase difference is 15-30 degree, maximum spherical deviation is 10-20%, and PMT high voltage is 1200-1400V. The refractive index of window (glass test section) is 1.51, and 2.36 mm thickness.

According to mass conservation law as Equation (5), the calculated diameter size is 66.43 μm at artificial frequency of 49.4 KHz. The PDA measured diameter size is 68.96 μm at same condition, the error is about 3.8%.

$$D = \left(\frac{6Q}{\pi f}\right)^{\frac{1}{3}} \tag{5}$$

where D is droplet diameter size, Q is volume flow rate, f is artificial disturbed frequency.

RESULT AND DISCUSSION

Electrohydrodynamic technique

It is convenient that use uniformity factor U, which defined as deviation/mean size diameter, to quantitate the monodispersed level. It can be treated as monodispersed droplet if U≦0.1 [15]. The uniformity factor is 0.277 at downstream 60 mm of turbulence field without EHD technique. The high uniformity factor is caused by turbulence flow actually. The average uniformity factor of ten measurements is only 0.0732 at same location but quiescent environment. The uniformity factor turns into 0.118 at downstream 60 mm of turbulence field with an EHD technique. The 3000 DC voltages were selected according to previous experience. The EHD technique has proved can dispersed droplets, and improve the collision and coalesce phenomenon of monodispersed droplet stream efficiently [16].

Turbulence modulation measurement

Figure 3 shows the radial size distributions of the spray in three streamwise stations. Here each value was determined by ten thousands raw data. Due to low data sampling rate at the upstream, i.e. Z=50 and 55 mm, only the measured results around the corn regions are provided. The spray angle is estimated to be about 10° base on geometry principle. This result fits in with the image observation, too. The observed point is at the downstream 60 mm and radial distance 2 mm because the seeding and droplet quantities are almost similar. The image of this location without turbulence interference is shown in Fig. 4. The uniform droplet can be seen clearly from this photo, in addition, the droplet collision and coalesce phenomenon also do not occur.

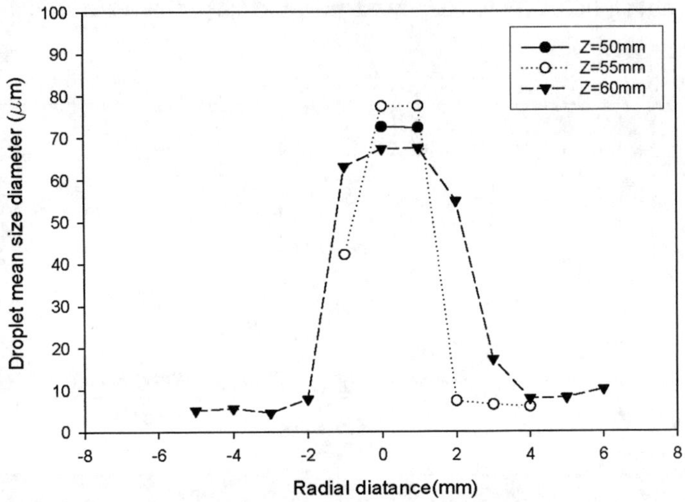

FIGURE 3. Droplet size distributions of different altitude of EHD monodispersed droplet stream

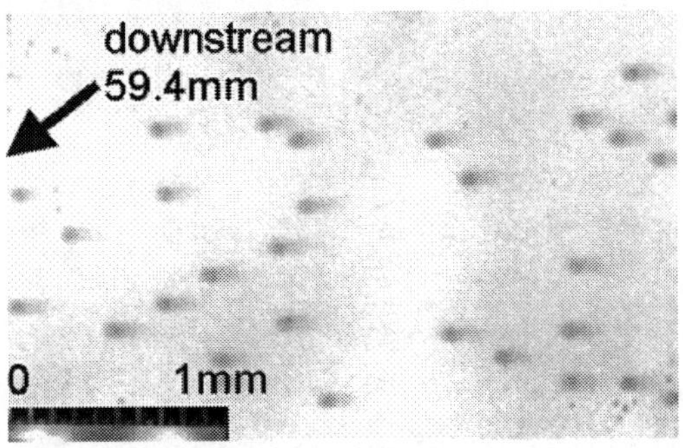

FIGURE 4. Enlarged image of dispersed droplet

When we use PDA measure two-phase flows, a defect will arise because the two phase flows have different refractive index. The software of PDA does not consider this condition and leads some problems. In previous study, the seeding usually is composed of kerosene aerosol or titanium dioxide particle to take account of tiny size. It does not only has a polluted test-section problem in the confined flow, but also leads to the difference of a refractive index of seeding and droplet. However, the most of researchers ignored this issue. When we used the PDA to measure about two-phase flows field, the seeding and droplet are expected to detect using the same setting. However, the scattering angle option is as smaller as possible in refractive mode if we want to measure seeding particles [17]. Unfortunately, the refractive index of particle is very sensitive at this smaller scattering angle. All using pure water to produce droplet and seeding can avoid this problem.

Figure 5 is a redrawing by Wang and Huang's data [2] and Tasi's program [18] that shows a distribution of $\langle u'_{gi}u'_{pi}\rangle$ quantities at downstream 5mm in the mixing layer. The droplet is 30 μm and the seeding is less than 10 μm. The dot represents an interaction of particle to gas phase, $\langle u'_{pi}u'_{gi}\rangle$ quantities, and circle represents an interaction of gas to particle phase, $\langle u'_{gi}u'_{pi}\rangle$ quantities. The graph is asymmetric, not conform to Wang and

Huang's assumption [2]. The assumption of symmetric graph just suits to the smaller droplet, however, 30 μ m diameter size start being unconformable the assumption according to Tsai's analysis [18]. If we set a parameter $T = t + \tau$, the $\overline{u'_{gi}(t)u'_{pi}(t+\tau)}$ will become $\overline{u'_{gi}(T-\tau)u'_{pi}(T)}$. Then compare with $\overline{u'_{pi}(t)u'_{gi}(t+\tau)}$, the particle to gas of $<u'_{pi}u'_{gi}>$ quantities and the gas to particle of $<u'_{gi}u'_{pi}>$ quantities should be a pair of mirror images mathematically if the statistics quantities is enough. The inner region tendency of Fig. 5 is the similar as a pair of mirror images. The cross-correlation function increases 2.06 times suddenly from a lower level at τ =-0.0005 to a higher level at τ=0.0005 because of particle wakes (see the dot).

Figure 6 also shows the difference of cross-correlation function of particle to gas phase and gas to particle phase in this experiment. The droplet is 70 μ m and the seeding is less than 10 μ m. Similarly, the cross-correlation function increases 3.12 times suddenly because of the 70 μ m particle wakes. But the tendency of inner region is not a pair of mirror images any more between particle to gas phase and gas to particle phase. The volume fraction is 0.00262% in this condition, which belong the dilute flow, so it is suitable without considering collision between particles. It is suspected that the larger particle effect causes the asymmetric tendency again.

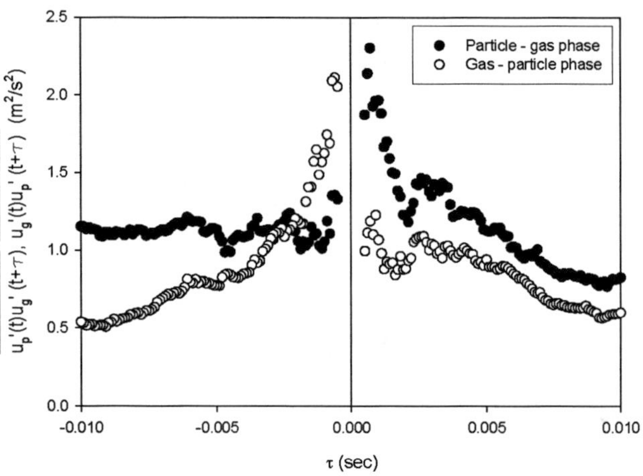

FIGURE 5. Asymmetric cross-correlation function of 30 μ m droplet between particle-gas phase and gas-particle phase in a mixing layer.

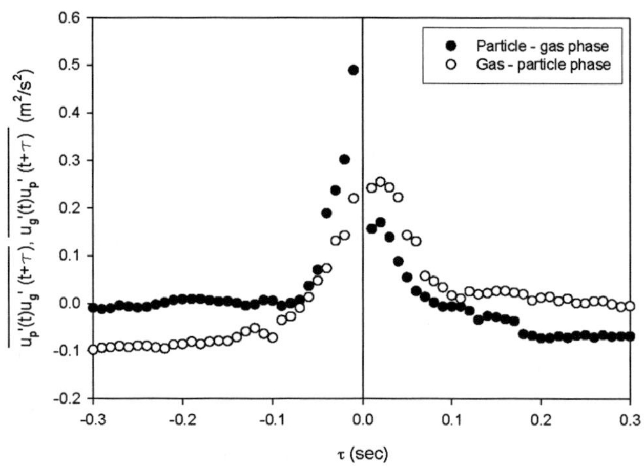

FIGURE 6. The difference of cross-correlation function of 70 μ m droplet between particle-gas phase and gas-particle phase.

This study verifies that our idea can be worked and come up with some problems behind $\overline{u'_{gi}(t)u'_{pi}(t+\tau)}$ and $\overline{u'_{pi}(t)u'_{gi}(t+\tau)}$. The effect of gas to particle phase and particle to gas phase turbulence modulation is different physically. After passing a larger particle, the wake will appear behind the particle but the gas particle will not. Overall, some doubts still need to been solved, but before more experimental data coming, we just only conjecture from these data as possible.

CONCLUSIONS

The study verified the feasibility of turbulence modulation measurement by using an EHD technique and a monodispersed droplet stream. This idea can work successfully and there some conclusions are as below: the EHD technique can assist to improve the collision and coalescence of monodispersed droplet stream. The uniformity factor is down to 0.118 from 0.277 at downstream 60 mm while using DC 3000V. In addition, the seeding and droplet are both composed of pure water; it can reduce the dispute of difference between refractive index of seeding and droplet. Last, the turbulence modulation can be measured indirectly by this method and some doubts came up. The research is a feasibility study and has some space to need to improve. The authors have to say sorry for incompleteness of this paper.

ACKNOWLEDGEMENTS

The authors would like to acknowledge the financial support by the National Science Council of the Taiwan with project no. NSC-87-2218-E-006-068, and to Prof. Jiun-Jih Miau and Muh-Rong Wang who lent us some instruments in the Department of Aeronautics and Astronautics, NCKU, Taiwan, and also to Prof. Jy-Cheng

Chang in the National Defense University, Taiwan, for his suggestion in using PDA.

REFERENCES

1. K. C. Chang, J. C. Yang and M. R. Wang, *AIAA J*, **41-7**, 1292-1303 (2003).
2. M. R. Wang and D. Y. Huang, *Atomization and Sprays*, **5**, 305-328 (1995).
3. F. Prevost, J. Boree, H. J. Nuglisch and G.. Charnay, *Int. J. Multiphase Flow*, **22-4**, 685-701 (1996).
4. A. M. A. Taweel and J. Landau, *Int. J. Multiphase Flow*, **3**, 341-351 (1977).
5. J. C. Chang, D. K. Wang, K. C. Tsai, M. S. Shen, L. C. Lin and K. M. Pang, "The operation characteristic of mono-dispersed droplet generator and its droplet stream structure," in *Proceedings of the 5th National Conference On Combustion Science And Technology*, Kaohsiung, Taiwan, 1995, pp. 489-496.
6. C. S. Cannon and D. Dunn-Rankin, *Atomization and Sprays*, **6**, 485-497 (1996).
7. W. F. Lu, W. H. Lai and K. C. Chang, "Turbulence modulation investigation by a monodispersed droplet stream and electrohydrodynamic method," in *10th International Congress on Liquid Atomization and Spray Systems, ICLASS-2006*, Kyoto, Japan, 2006, A3-02-057.
8. J. M. Schneider and C. D. Hendricks, *Rev. Sci.Instrum.*, **35-10**, 1349-1350 (1964).
9. K. J. Choi and B. Delcorio, *Rev. Sci. Instrum.*, **61-6**, 1689-1693 (1990).
10. C. C. Chen and W. H. Lai, "The study of mono-sized droplet generation," in *Proceedings of the 9th National Conference On Combustion Science And Technology*, Tainan, Taiwan, 1999, pp. 187-191.
11. L. Rayleigh, *Proc. London Math. Soc.*, **10**, 4-13 (1878).
12. K. Anders, N. Roth and A. Frohn, *Part. Part. Syst.Charact.*, **9**, 40-43 (1992).
13. Dantec, *60X series FiberFlow User's Guide*, Denmark.
14. F. Durst and M. Zare, "Laser Doppler measurements in two phase flows, The Accuracy of Flow Measurements by Laser-Doppler Methods," in *Proc. of LDA Symposium*, Copenhagen, Denmark, 1975, pp. 403-429.
15. M. Sato, H. Takahashi, M. Awazu and T. Ohshima, *Journal of Electrostatics*, **46**, 171-176 (1999).
16. W. F. Lu, W. H. Lai and K. C. Chang, "The Effect on the Coalescence of Mono-sized Droplet Streams by EHD Method," in *Proceedings of the 16th National Conference On Combustion Science And Technology*, Keelung, Taiwan, 2006.
17. Dantec, *PDA installation & User's guide*, Denmark, 1994.
18. T. H. Tsai, *On Experimental Determination Of Turbulent Modulation Quantity In a Two-Phase Mixing Layer*, MS thesis, National Cheng Kung University, Tainan, Taiwan, 2003.

Experimental Study on Rising Velocity of Different Nitrogen Bubbles in of Circular Pipe

Hua Zhang, Yiping Liu, and Jing Wang

Institute of Thermal Physics of Engineering, School of Mechanical Engineering, Shanghai Jiaotong University, Shanghai, China

Abstract. Nitrogen bubble rising velocity in bottom of circular pipe has been studied by virtual experiment with high-speed camera. The obtained images are processed systematically. Based on Cole's and Mendelson's empirical formula of bubble rising velocity, fitting formula of nitrogen bubble rising velocity in bottom of circular pipe has been acquired through analyzing nitrogen bubble rising velocity. The treatment results of experimental data show that the rising velocity is related with the diameter of bubbles. And rising velocity of slug bubble is relative to L/D and change law is found. Through this experiment, we research rising velocity's change law, which is the spade work to study propagation mechanism of nitrogen slug bubble in circular pipe.

Keywords: High speed digital camera; Nitrogen bubble; Rising velocity; Image process; Fitting formula.
PACS: 47.55.D-

INTRODUCTION

The development of modern astronavigation brings up new challenge and sky-high demand in application of multi-phase flow theory on cryogenic engineering. In cryogenic engineering, heat leak always exists in conveyor system of cryogenic liquid. So cryogenic two-phase flow is unavoidable[1, 2]. The propagation of two-phase flow makes many problems in conveyor system of cryogenic liquid become complex, such as pressure calculation and stable operation. And higher demand is brought up on safety traffic monitor of cryogenic equipment.

The determination of flow pattern of cryogenic liquid two-phase flow in vertical conveying pipe is an important issue in aerospace engineering. Combination of liquid oxygen and liquid hydrogen or combination of liquid oxygen and kerosene is always used as cryogenic fuel in cryogenic liquid propellant rocket. Then the research of unstable phenomenon of two-phase flow occurring in the process of cryogenic liquid's stockpile and conveying has significant meaning in engine designing. There are many research of bubble rising velocity both here and abroad, which always use normal temperature liquid such as Freon and water, as working medium [3, 4, 5]. And cryogenic liquid is seldom used as working medium in such studies. Their research efforts indicate: bubble rising velocities are different in different normal temperature liquid working medium, and bubble rising velocities are relative to working medium's density, viscosity and surface tension. While cryogenic liquids such as liquid oxygen, liquid nitrogen, liquid hydrogen, all have prodigious compression factor, low density contrast between vapor phase and liquid phase and low latent heat of vaporization, there are prodigious differences between bubble propagation, development in cryogenic two-phase flow and bubble propagation, development in normal temperature two-phase flow. The formula of bubble rising velocity of normal temperature liquids can not be directly used to calculate cryogenic liquid bobble rising velocity.

Aiming at characters of cryogenic liquid, this article studies liquid nitrogen bubble's propagation and rising velocity in bottom of circular pipe through experiment. And high-speed camera is used to acquire experimental dates of liquid nitrogen bubble rising velocity. Based on Cole's and Mendelson's empirical formula of bubble rising velocity, fitting formula of nitrogen bubble rising velocity in bottom of circular pipe has been acquired through analyzing nitrogen bubble rising velocity. Through this experiment, we research rising velocity's change law, which is the spade work to study propagation mechanism of nitrogen slug bubble in circular pipe.

CP914, *Multiphase Flow: The Ultimate Measurement Challenge,* edited by X. Cai, Y. Wu, Z. Huang, S. Wang, and M. Wang

VISUALIZATION EXPERIMENTAL SETUP OF CRYOGENIC LIQUID

Figure 1 is sketch of experimental setup, which includes a liquid nitrogen Dewar, stock tank, cryogenic vertical convey pipe (testing section) and vacuum pump. In order to make a study of flow law of cryogenic vertical convey pipe, aiming at characters of cryogenic liquid, stock tank and cryogenic vertical convey pipe used in the experiment are all made by double layer Pyrex glass, and vacuum interlayer is vacuumized by vacuum pump.

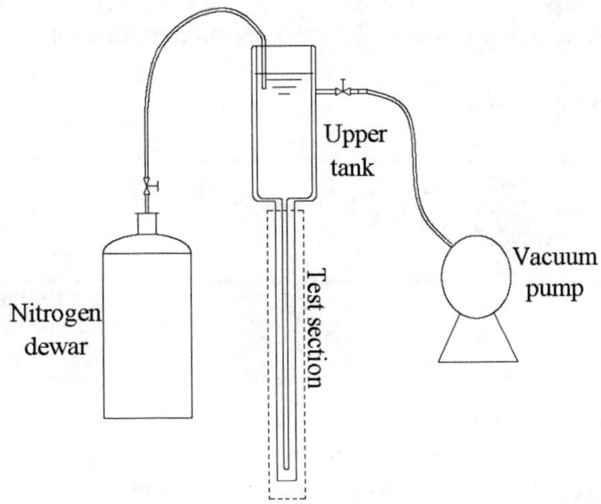

FIGURE 1. Experimental apparatus

The parameter of glass pipe used in experiment and the degree of vacuum in vacuum interlayer after vacuumized are listed in Table 1. High speed digital camera is used to measure and to shot fluid motion condition in pipe.

Figure 2 is the schematic design of image processing system which is used in this experiment. Camera shot devices are composed of high speed digital camera, video camera, monitor, light source and computer. The type of high speed digital camera is Faster-calm super high-speed video camera-10KC. Sampling frequency of high speed DC is 25 frames to 10000 frames per second. Steam bubble motion condition in testing section of cryogenic vertical visualization pipe can be shot and captured through high speed digital camera. The height of pipe that can be shot by digital camera is determined by sampling frequency, frequency is higher, and shooting range is smaller, and definition is higher.

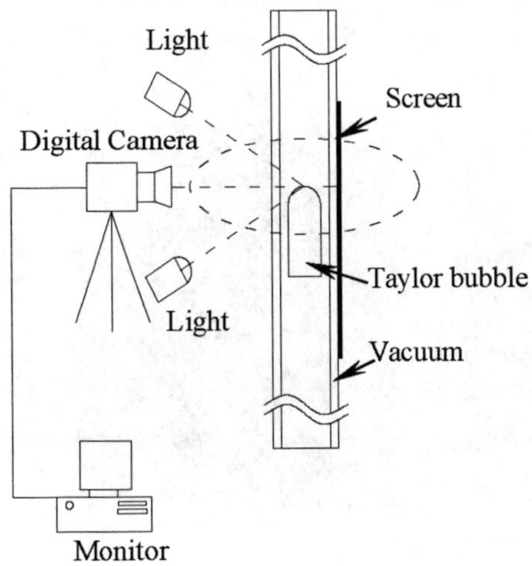

FIGURE 2. The schematic design of Image processing system

In experiment, when nitrogen bubbles are passing by testing section, motion condition is recorded by high speed digital camera, and stored in computer directly or through video collection card. Sampling frequency of high speed DC is 250 frames per second, and photo which is captured by digital camera is 16 bit color picture. Resolution of photo is 25,000 pixels, so it can be used as steam bubble analysis. Two photoflood lamps are used as light source, whose power is 1000w.

Metering zone is from the bottom of pipe, which is 0D, to 4D. In experiment, we use high speed digital camera to shot bubble motion condition in this range.

Table 1 is configuration data of pipe. Double layer conveying pipe is made with transparent Pyrex glass directly. We can use high speed digital camera to shot bubble motion condition in pipe directly.

TABLE 1. Structure of cryogenic transfer pipe

Conveying pipe	Inside diameter of conveying pipe (mm)	length of pipe (m)	Length – diameter ratio	Insulation method
Double layer conveying pipe (vacuum insulation)	16	0.80	50	Vacuum interlayer 15mm, degree of vacuum 6×10^{-2} Pa

VISUALIZATION EXPERIMENTAL RESULTS AND ANALYSES

Rising velocity of bubbles in bottom of tube

Gaining a lot of pictures and data through experiment, we do research on rising velocity of bubbles in 0-4D field of circular pipe. Aiming at the characteristics of cryogenic experiment, in order to reduce measuring error, this article makes use of gas-liquid two-phase visualization experimental method introduced by literature [6, 7], uses image processing software to enlarge the picture before measurement and ensures the essential visualizability and reliability.

For the sampling frequency of high speed digital camera is 250 frames, the interval between two pictures is just 4ms. Because of the short interval, the position of bubbles changes little and the degree of distortion is also small, so we can clearly find the position of bubbles we want to measure in the continual series of pictures. After exposal and gradation processing, using the software to enlarge the picture, we can measure the size and the position of bubbles, and can get average value by arithmetical averaging the date from the pictures, then use the metrical date to calculate the rising velocity.

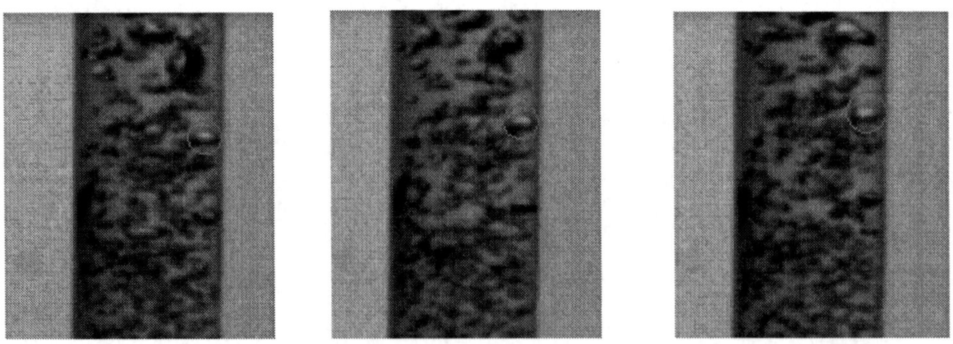

(a) (b) (c)

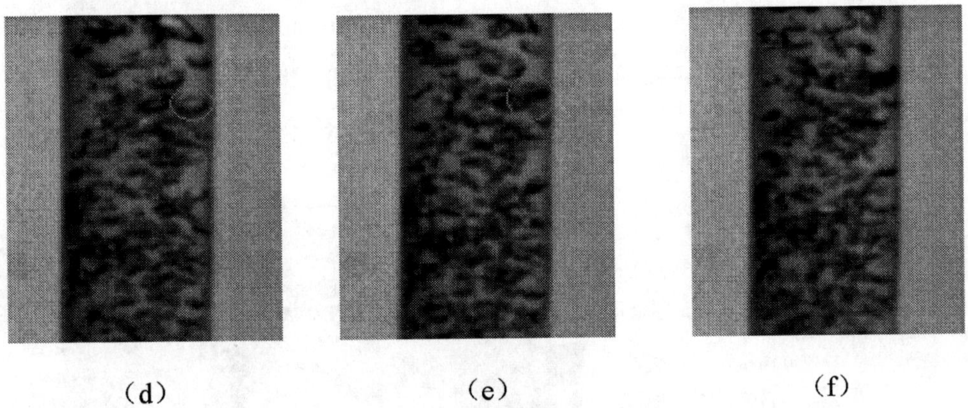

FIGURE 3. An example of a serial of images after processing

Figure 3 is a serial of images of 2D-4Dfield in circular pipe, and it is shot by a high speed digital camera. The pictures in Fig.3 have been handled with exposal and gradation. We can clearly find out that the shape of one bubble doesn't change a lot in deferent picture, so the position of one bubble in deferent picture can be easily fixed.

Measurement and calculation indicate that in 2D-4Dfield the rising velocity of bubbles near the pipe wall is commonly greater than the one near center. The reason is, when bubbles propagate in the pipe, the supply of liquid nitrogen is needed in the bottom of pipe, so the liquid nitrogen below is supplied though the center of pipe from above place, and the natural circulation forms in the pipe. Because of the natural circulation, liquid nitrogen in the center of pipe supplies downwards, liquid nitrogen near the wall moves upwards, so the rising velocity of bubbles in center of pipe is reduced by the downwards moving liquid nitrogen, and the rising velocity of bubbles near wall relatively speeds up; in the meantime, the bubbles move to the wall gradually by the stress from downwards moving liquid nitrogen, and fuse together, so the quantity of small bubbles decreases, and big bubbles formed by small bubbles propagate near wall. Fig.4 is the velocity distributing curve of bubbles rising from the bottom.

Literature [3] introduces the Mendelson's and Harmathy's empirical formula of rising velocity of bubbles as follows:

$$v_b = (2.14\sigma/(\rho_l d_b) + 0.505 g d_b)^{1/2} \tag{1}$$

$$v_b = 1.53(g\sigma/\rho_l)^{0.25} \tag{2}$$

Literature [8] introduces Cole's empirical formula of rising velocity of bubbles as follows:

$$v_b^2 = \frac{4\sigma + g*(\rho_l - \rho_v)*D_b^2}{2*\rho_l*D_b} \tag{3}$$

In Fig.4, we can find there are differences between the experimental data and three diverse empirical formula of rising velocity of bubbles. For the three empirical formula of rising velocity of bubbles which mentioned before are all deduced by gas-liquid two phase flow under natural temperature, the difference may be caused by the different thermo physical property of cryogenic working medium. And Fig.4 indicates that tendency of the data we get from experiment is coherent with tendency of Cole's and Mendelson's empirical formula of rising velocity of bubbles. Based on the empirical formulas, after fitting the experimental data, we obtain the empirical formula of rising velocity of liquid nitrogen bubbles (4):

$$v_b^2 = \frac{1.25\sigma + 0.373*g*(\rho_l - \rho_v)*D_b^2}{1.036*\rho_l*D_b} \tag{4}$$

Where, v_b is the rising velocity of liquid nitrogen bubbles at the bottom of pipe, σ is surface tension of liquid nitrogen, ρ_l is the density of liquid nitrogen, ρ_v is the density of nitrogen gas, D_b is the diameter of pipe.

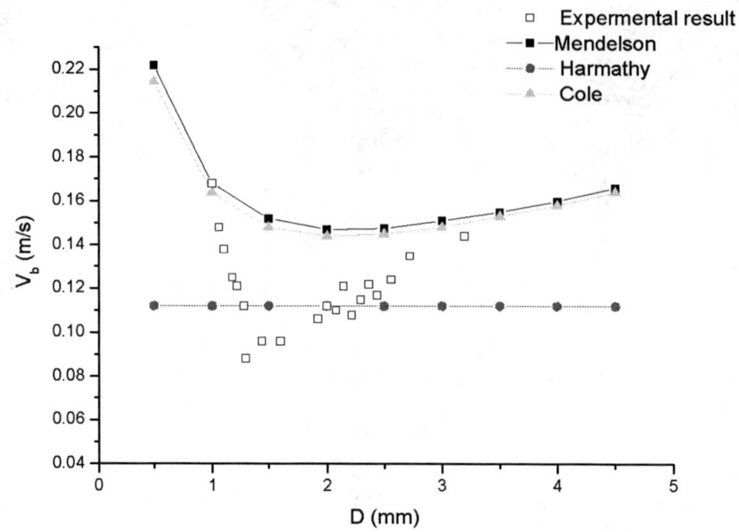

FIGURE 4. D vs. the rising velocity of the bubbles

The experimental result indicates that surface tension is the main factor to affect the shape and rising velocity of bubbles. The bigger the surface tension is, the more round the bubbles are, and the resistance that bubbles get when it is rising will be less, the rising velocity will be larger. The liquid viscosity is mainly effect rising resistance. The bigger the viscosity is, the bigger the resistance will be. Meanwhile, with the density contrast between liquid and gas becoming larger, the rising velocity of bubbles will be greater. So as far as the bottom of the pipe is concerned, it is the place where the bubbles propagate, bubbles have little metamorphose, and bubbles in this field are very close to roundness. So the small bubbles in the bottom of pipe rise faster.

Restricted by the experimental equipment, we can't measure exactly the diameter of bubbles less than 1mm, so there is no metrical data about them.

Rising velocity of slug bubble

Through the pictures (as shown in Fig.5, 6, 7) and data gained from experiment, we measure these pictures and get the rising velocity of slug bubbles. And Fig.8 is gotten through those measured data.

FIGURE 5. The forming slug bubble motion in cryogenic vertical pipe

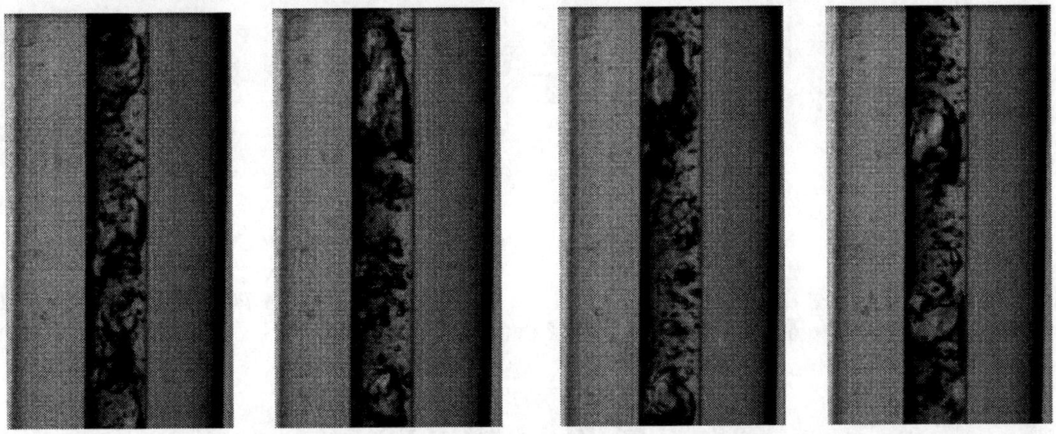

FIGURE 6. the expanding slug bubble motion in cryogenic vertical pipe

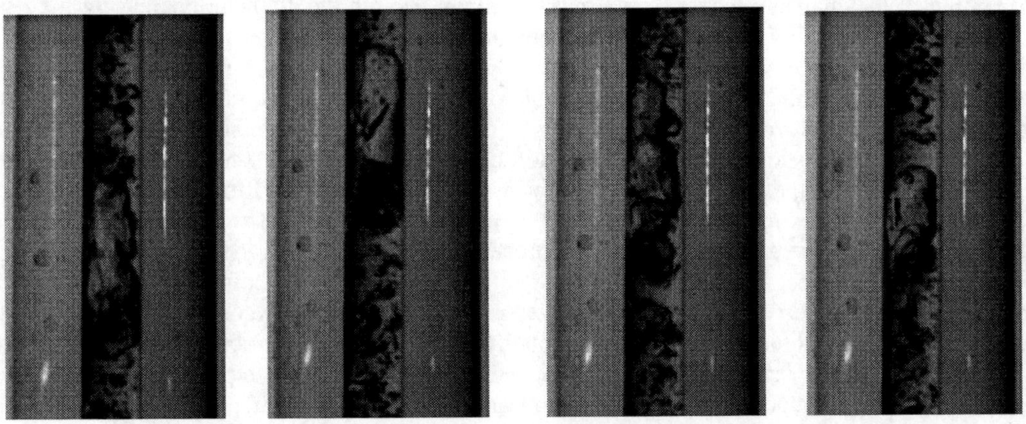

FIGURE 7. the full slug bubble motion in cryogenic vertical pipe

The relationship of rising velocity of slug bubbles in cryogenic vertical pipe and L/D is shown in Fig.8. In order to do some comparison, velocity distribution curves of Moissis and Takatoshi are also given in Fig.8, although water is the working medium in their experiment. In Fig.8, L is the distance in two slug bubbles and D is inside diameter of the pipe. There is shown in Fig.8 that rising velocity of slug bubble increases with L decreasing. Most of all, when L is less than 4D, bubble will affect following bubble and that is so called wake effect. So velocity of following bubble will increase. Once L is greater than 4D, L does little effect on rising velocity distribution and velocity of slug bubble becomes steady. When L is greater than 6D, there is almost no effect between bubbles and velocity won't change any more.

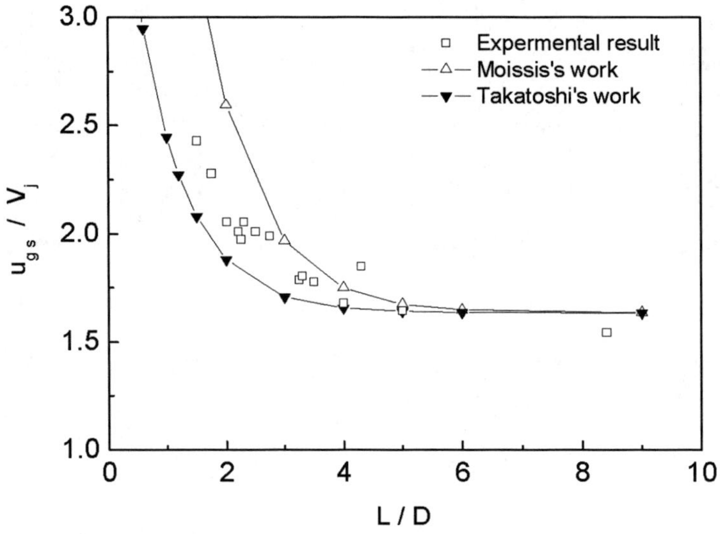

FIGURE 8. L/D vs. the rise velocity of the slug bubble

Restricted by the experimental equipment, the pipe is not long enough and slug bubbles are not fully developed. We can't get enough experimental data that L is greater than 6D in experiment. So slug bubbles in pipe are always in developing state and a few slug bubbles can get the fully developed state.

CONCLUSIONS

Through the experimental research, we get the conclusions as follows: (1) the rising velocity is related with the diameter of bubble. When diameter is less than 1.5mm, companied with the diameter decreasing the rising velocity increases; when diameter is bigger than 1.5mm, companied with the diameter increasing the rising velocity decreases. And based on Cole's equations, this paper gives liquid nitrogen bubbles rising velocity relational expression; (2) there is difference between fitting formula of rising velocity of liquid nitrogen bubbles in the bottom of circular pipe and empirical formula of rising velocity of liquid bubbles in normal temperature, and the difference reflects the physical characteristic of bubbles propagation of cryogenic liquid; (3) small vapor bubbles generate and rise from the bottom of the pipe. The bubbles interfere with each other in performing a release from the system when the population of bubbles reaches the point. At this point the bubbles will swarm and cooperate, then form a single Taylor bubble; (4) As the distance between the preceding slug bubble and the following slug bubble becomes smaller, a strongly wake effect occurs and will destabilize the liquid slug between the slug bubbles and the coalescence of the slug bubbles occurs. With the increase of the distance between two slug bubbles (L>4D), the rising velocity of the slug bubble decreased exponentially. And the rising velocity of the slug bubble becomes steady at last. Through this experiment, it can do advanced research in forecast experiment of cryogenic two-phase flow pattern, and provide reference for doing research of simulation theory calculation and being the spade work to study propagation mechanism of nitrogen slug bubble in circular pipe.

ACKNOWLEDGEMENTS

This work was supported by the National Natural Science Foundation of China (No. 50476015).

REFERENCES

1. Frost, Walter, *Heat Transfer at Low Temperatures*, Publisher, New York, 1975, pp. 1-10.
2. Chen Guobang, Lin Lihe, *Cryoinsulation and Heat Transportation*, Publisher, Hangzhou, 1989, pp. 5-15.
3. Li Xiaoming, Wang Ye, et al, *Journal of Xi'an Jiaotong University* **37**, 971-974 (2002).
4. Kawai T, Yoshino H, Kawabata Y, *Physica: B* **311**, 164-172 (2002).
5. Raymond F, Rosant J M, *Chemical Engineering Science* **55**, 943-955 (2000).
6. L Shemer, D Barnea,, *PhysicoChemical Hydrodynamics* **8**, 243-253 (1987).
7. Takatoshi Takemoto, Mitsuo Matsuzaki, et al, *Journal of Nuclear Science and Technology* **36**, 671-682 (1999).
8. Ruitai Lin, *Boiling and Heat Transfer*, Publisher, Beijing, 1988, pp. 125-130.

Uncertainty of the PIV Testing Results for Gas Flow Field Caused by Tracer Particles and Seeding Method

Shunsen WANG, Jingru MAO, Guanwei LIU

Institute of Turbomachinery Xi'an Jiaotong University, 710049 Xi'an, China

Abstract. The quality of flow visualization is very important for PIV. A lot of testing results indicate that the slip between tracer particles and continuous phase is very severe for high velocity gradient flow field and yield uncertain testing results. In addition, improper seeding method of tracer particles often results in false results. The systematical investigation on the above-mentioned problem has been done based on the theory of dilute phase gas-particle flow. The results indicate that tracer particles must be have the slip with the fluid as long as the velocity gradient presents, even the particle is sufficiently accelerated, and it has severely influence on the reliability of PIV testing results for the flow field with high velocity gradient or (and) larger tracer particles size. The relative error caused by the slip of particles and fluid is in proportion to the particle Stokes time and the particle velocity gradient, but it is in inverse proportion to the correction factor for Stokes-drag force. The above conclusions can be used to judge and correct the measurement error or to select proper tracer particles for the non-uniform flow field. During the study on seeding method of tracer particles, the Reynolds number ratio of dragging gas flow for tracer particles and main gas flow Re_f/Re is considered as the disturbance intensity scale of dragging gas flow on main gas flow. The attenuation process and attenuation distance of disturbance is investigated for different Re_f/Re. The optimal seeding position can be selected based on the data of the attenuation distance for different Re_f/Re.

Keywords: Tracer particles; Slip velocity; Particle image velocimetry (PIV); Measurement errors; Seeding method; Velocity gradient; Reynolds number; Non-uniform flow field.
PACS: 47.80.-v

INTRUDUCTION

Particle-tracking techniques, such as particle image velocimetry (PIV) and laser Doppler velocimetry (LDV), rely on the assumption that particle velocity matches fluid velocity. However, particle-inertia, external forces, and other factors may invalidate this assumption [1]. It is often very difficult to select and get proper tracer particles because of the limit of various insurmountable factors, such as particle density, particle refractive index, laser intensity and etc. A compromise between reducing the particle size to improve flow tracking and increasing the particle size to improve light scattering is, often, necessary [2]. In addition, peak locking error also occurs when the particles images are sampled at a poor spatial resolution (particle image size smaller than two pixels). Therefore, larger particle size is often selected, and many testing results with PIV indicate that the slip between tracer particles and fluid is very severe in high velocity gradient flow field. So, it is necessary to evaluate quantitatively and correct the error caused by particle inertia for improving the reliability of the testing results with PIV.

Errors attributable to particle-inertia effects have been widely studied. Haghgooie et al. [3] investigated LDV techniques in turbulent flows and found that particles of 1 or 2μm in diameter should adequately follow velocity fluctuations in the flow up to 2 or 10 kHz respectively. Samimy and Lele [4] studied the behavior of particles in a compressible shear layer, and recommend that the ratio of the Stokes time to the flow time was kept below 0.05 for accurate flow visualizations. Melling [5] discusses tracer particles and seeding for PIV and finds that a maximum particle size of 1 μm is required to achieve a frequency response of 10 kHz, in accord with the findings of Haghgooie et al. [3]. However, all these research only involved in the method to select proper tracer particles for different flow filed and few studies have reported on how to deal with the condition which the tracer particles is improper. This paper devotes more attention to the PIV measurement on the accelerating flow field and attempts to find out the solution for the error caused by improper tracer particles based on the systematical investigations on

CP914, *Multiphase Flow: The Ultimate Measurement Challenge,* edited by X. Cai, Y. Wu, Z. Huang, S. Wang, and M. Wang

dilute phase gas-particles flow and PIV testing results, and presents a novel method to evaluate quantitatively this error and to correct on the vector field, certainly, it can also be used to select the tracer particles for the non-uniform flow field.

The method and position to introduce tracer particles into the flow are also very critical with PIV, and improper operation can disturb the flow field and result in false results. Solid particles generated by atomization [6] or from powders [7] are preferable for tracer, and oil fog is alternative form, as used for PIV by Stewart et al [8]; Seeding particles should be introduced into the flow in sufficient and stable concentration, with a spatially uniform distribution, while minimizing coagulation and deposition of particles on windows. Frequently a choice must be made between global and local seeding [9, 5]. This paper investigates how to select the proper seeding position for positive pressure flow in section 3.

MEASUREMENT ERROR CAUSED BY PARTICLE INERTIA

Influence of Particle Inertia on PIV Testing Result

The slip velocity between tracer particles and measured fluid $u'=u_p-u_f$ and two phase density ratio $s=\rho_p/\rho_f$ are two important parameters for investigating the particles following ability, namely, the ability of particles to follow the instantaneous motion of the continuous phase, where u_p and ρ_p are the velocity component along x direction and density of tracer particles, respectively, u_f and ρ_f are the velocity component along x direction and density of fluid. Existed theories indicate that tracer particles have worse following ability on gas than on liquid because s is larger for gas-particles flow. In this paper, more attention has been focused on the slip velocity of tracer particles in gas flow field.

For high velocity gradient flow field, many PIV testing results have proved that the particles following ability was very disillusionary and could cause false testing results. The following is an example for PIV measurement on the accelerating flow in a bend passage. Fig.1 shows the profile of passage, and the curve 1 and 2 marked in the figure is the characteristic position in the following research. The inlet and outlet Mach number of passage is 0.2 and 0.98, respectively, and the gas density ρ_f is 1.8kg · m^{-3} in the inlet of passage; tracer particle is hollow glass sphere (ρ_p=2300 kg · m^{-3}) with the Sauter mean diameter of 2.8μm. The particles velocity u_p can be acquired by PIV. This paper employs three-dimensional numerical simulations to obtain the continuous phase velocity u_f in view of computational research program is mature for the single-phase flow field in this research field and its results are proved reliable by engineering practice. The fig.2 shows the value of u_p/u_f with different position of passage, and it should be noticed that u_p is the mean velocity of 1.5~2mm near the curve 1 and 2 (fig.1) here. The corresponding u_p/u_f of two curves is obviously different because of the particles stratification caused by inertial centrifugal force and wall influence on the particles motion. Although the particles size is large for the flow field measurement with PIV, the slip between particles and gas flow is still appalling. Therefore, it is very significant for engineering application and scientific research to investigate the measurement error with PIV caused by the slip of tracer particles and continuous phase.

The PIV combines perfectly the flow visualization and cross correlation technique, so its measurement error also come from the above two aspects. The velocity gradient for two dimension flow field has four components: $\partial u/\partial x$, $\partial u/\partial y$, $\partial v/\partial x$, $\partial v/\partial y$. The gradient $\partial u/\partial y$ and $\partial v/\partial x$ caused by fluid viscosity is along the normal direction of every velocity component, it does often result in inhomogeneous particles motion within an interrogation window. However, it is irrelevant with the slip of particles and fluid. There are specialized investigations on this problem in many literatures [10]-[12], it will not be discussed in this paper. The gradient $\partial u/\partial x$ and $\partial v/\partial y$ caused by pressure gradient is along the tangential direction of every velocity component, it does often cause the slip of particles and fluid, further results in the uncertainty of PIV testing results. The gradients $\partial u/\partial x$ and $\partial v/\partial y$ is default velocity gradients in this paper.

Slip of Tracer Particles and Gas Flow

We consider 1-D flow as the research object of this paper in order to simplify the equation, however, the results is available for 2-D and 3-D flow. For the dilute phase gas-particles flow, the forces acting on the suspended sphere particle can be written as

$$\sum F = F_{SD} + F_{gf} + F_p + F_{vm} + F_B + F_{ma} + F_s \tag{1}$$

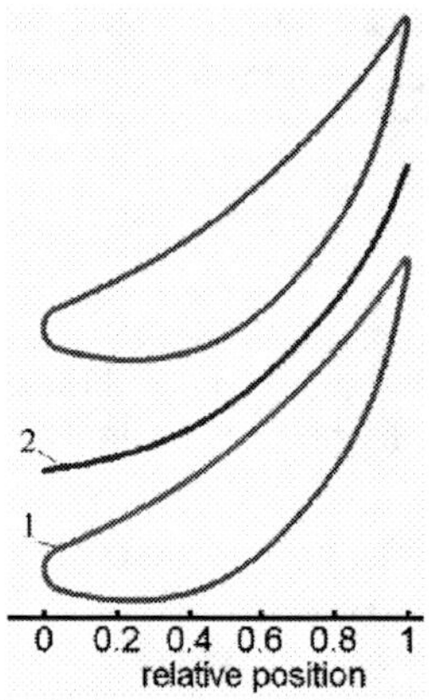

FIGURE 1. Scheme for bend passage and measuring position.

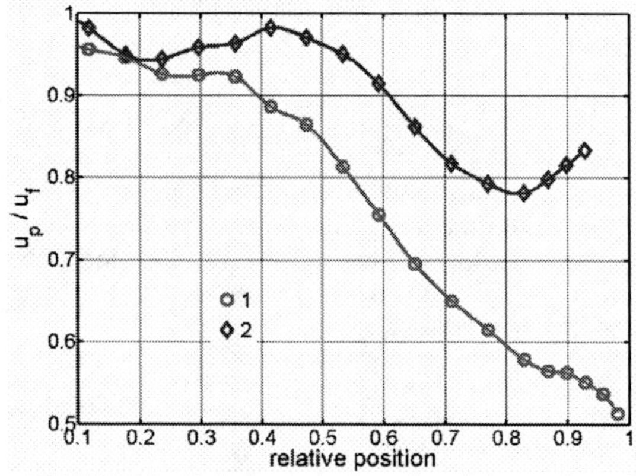

FIGURE 2. Relative velocity of particles and gas flow for different measuring position

Where F_{SD} is the Stokes-drag force, F_{gf} is the combination of gravity and buoyancy, F_p is the pressure gradient force, F_{vm} is the virtual mass force, F_B is the Basset force, F_{ma} is the Magnus force, F_s is the Saffman force. This paper does not involved in temperature gradient, so the thermophoretic force is neglected.

For most experiments it is desirable that seeding particles be non-toxic, non-corrosive, non-abrasive, non-volatile and chemically inert. A wide variety of seeding particles is available for PIV experiments, ceramic materials such as Al_2O_3 (ρ_p= 3970 kg · m^{-3}), TiO_2 (ρ_p= 3500 kg · m^{-3}) and glass bean (ρ_p= 2800 kg · m^{-3}) are favored for seeding gas flow because of the advantage of a high refractive index, and various oil fogs (ρ_p= 900~1000 kg · m^{-3}) is also familiar tracer particles for gas flow on account of its small density.

In order to simplify the equation (1), it is necessary to analyze the magnitude order of various forces based on the characteristic of familiar tracer particles. The particle d_p diameter is in the magnitude order of 10^{-6}m. Compressed air is the most universal gas source, and its density ρ_f and dynamic viscosity coefficient μ are in the magnitude order of 10^0 kg · m^{-3} and 10^{-5} Pa · s, respectively. The flow field measurement on the nozzle flow or jet flow with PIV is a challenge on the particles following ability, and according to the nozzle flow characteristic, pressure gradient $\partial p / \partial x$ is in the range of 10^5~10^6 Pa · m^{-1}, and the slip velocity u' is in the magnitude order of 10^1 m · s^{-1}; corresponding particles maximum rotary speed ω is in the range of 10^4 ~10^5 rad · s^{-1}. Taking above conditions into account, the magnitude order of various forces can be acquired,

$$
\begin{aligned}
F_{SD} &= -3\pi d_p \mu (u_p - u_f) f(\mathrm{Re}_p) && \sim 10^{-9} \\
F_{gf} &= \frac{1}{6}\pi d_p^{\,3} (\rho_p - \rho_f) g && \sim 10^{-14} \\
F_p &= -\frac{1}{6}\pi d_p^{\,3} \frac{\partial p}{\partial x} && \sim 10^{-13}\text{-}10^{-12} \\
F_{vm} &= -\frac{1}{12}\pi \rho_f d_p^{\,3} \frac{d(u_f - u_p)}{dt} && \sim 10^{-13} \\
F_{ma} &= \frac{1}{8}\pi \rho_f d_p^{\,3} \omega \times (u_f - u_p) && \sim 10^{-13}\text{-}10^{-12} \\
F_s &= 1.61(\mu \rho_f)^{0.5} d_p^{\,2} (u_f - u_p) \left| \frac{du_f}{dy} \right|^{0.5} && \sim 10^{-13}\text{-}10^{-12} \\
F_B &= \frac{3}{2} d_p^{\,2} \sqrt{\pi \rho_f \mu} \int_{-\infty}^{t} \frac{d(u_f - u_p)}{\sqrt{t-\tau}} d\tau
\end{aligned}
\tag{2}
$$

It is very difficult to directly analyze the magnitude order of the Basset force, however, we can neglect the Basset force when s>500 according to the literature [13, 14]. The results indicate that Stokes-drag force is the determining factor for particles motion. Because the Stokes-drag force is hundreds times larger than the summation of all other forces, the motion equation (1) for a particle can be reduced to

$$\sum F = F_{SD} \tag{3}$$

In respect that the given conditions for the magnitude order analysis have covered broad flow field, above conclusion can be used to analyze almost all the flow. The motion equation of a particle can be expressed using Newton's Second Law as

$$\sum F = m_p a_p \tag{4}$$

$$a_p = \frac{du_p}{dt} = \frac{du_p}{dx_p}\frac{dx_p}{dt} = \frac{du_p}{dx_p}u_p = \sigma_p u_p \tag{5}$$

Where $\sigma_p = du_p / dx_p$, and it is the velocity gradient of particle in one dimension. The following equation can be obtained from the equation (2) ~ (5)

$$m_p a_p = (\frac{\pi \rho_p d_p^{\ 3}}{6})\sigma_p u_p = -3\pi d_p \mu (u_p - u_f) f(\mathrm{Re}_p)$$

$$-\frac{\rho_p d_p^{\ 2}}{18\mu f(\mathrm{Re}_p)}\sigma_p u_p = u_p - u_f$$

$$u_f = (1 + \frac{\tau_s \sigma_p}{f(\mathrm{Re}_p)})u_p \quad \text{or} \quad \frac{\tau_s}{f(\mathrm{Re}_p)}a_p = u_f - u_p \tag{6}$$

Where $\tau_s = \rho_p d_p^2/(18\mu)$ is the particle Stokes time [1], $f(Re_p)$ is the correction factor for Stokes-drag force, and Re_p is the relative Reynolds number based on the two phases slip velocity. The Re_p is also called the particle Reynolds number and its expression is

$$\mathrm{Re}_p = \frac{\rho_f |u_p - u_f| d_p}{\mu}$$

When $1 \le \mathrm{Re}_p < 50$

$$f(\mathrm{Re}_p) = C_D \frac{\mathrm{Re}_p}{24} = \frac{23.5}{\mathrm{Re}_p^{\ 0.725}}\frac{\mathrm{Re}_p}{24} = 0.979\,\mathrm{Re}_p^{\ 0.275}$$

When $Re_p<1$, $f(Re_p)=1$, so we will reduce the equation (6) to a simpler form

$$u_f = (1 + \tau_s \sigma_p)u_p \quad \text{or} \quad \tau_s a_p = u_f - u_p \tag{7}$$

Above equations indicate that the velocity difference between gas and particles must be present when flow field has velocity gradient, namely, when $\sigma_p \neq 0$.

The velocity of continuous phase and particles can be marked as u_{f0} and u_{p0} when the time t =0; the acceleration of continuous phase is a_f, and it can be considered as the constant if the interrogation window is enough small. Substituting the equation $u_f(t) = u_{f0} + a_f t$ into the equation (7) yields the following equation

$$\frac{du_p(t)}{dt} = -\frac{1}{\tau_s}u_p(t) + \frac{1}{\tau_s}(u_{f0} + a_f t) \tag{8}$$

The following relation can be acquired by solving the inhomogeneous linear ordinary differential equation (8)

$$\frac{u_p(t) - u_f(t) + a_f \tau_s}{u_{p0} - u_{f0} + a_f \tau_s} = e^{-\frac{t}{\tau_s}} \tag{9}$$

When $t \to \infty$, the equation (9) can be reduced to

$$u_f(t) - u_p(t) = a_f \tau_s$$

Above result indicates that the velocity difference of particles and fluid still exist even the particle is sufficiently accelerated, and we call this velocity difference as the terminal slip velocity. The forces acting on the particle is no longer varied with the time and the particle motion is certain after the gas-particle flow arrives at the terminal slip velocity $a_f \tau_s$. We define a parameter to judge the uncertainty of particle slip velocity

$$\varepsilon = \left| \frac{u_f(t) - u_p(t) - a_f \tau_s}{a_f \tau_s} \right| \tag{10}$$

ε is called uncertainty here, and ε=0 means the particle is sufficiently accelerated. $\varepsilon \to 0$ only when $t \to \infty$. In fact, it is not necessary to demand ε=0 for PIV, and the slip velocity of gas-particle flow can be considered as the terminal slip velocity in engineering if ε is smaller than tolerable uncertainty, for instance, ε<2%. It takes about $0.91\tau_s$ and $1.6\tau_s$ seconds to decrease the uncertainty from 5% and 10% to 2% according to the equation (9), respectively.

The accelerating motion of glass bean (ρ_p= 2800 kg · m^{-3}) in gas flow (ρ_f= 1.5 kg · m^{-3}) can be as an example to present us the magnitude order of the terminal slip velocity. Fig.3 shows the terminal slip velocity of tracer particles with different diameter in flow field with different a_f. The range of a_f is from 10^4m/s^2 to ~10^6 m/s^2 and corresponding $\mathrm{Re_p}$ is from 10^{-2} to 10^1. These curves suggest that the slip velocity can not be neglected in the PIV measurement on the flow field of large $a_f$or (and) d_p, and the tolerable range of $a_f$or (and) d_p will be discussed in the following section.

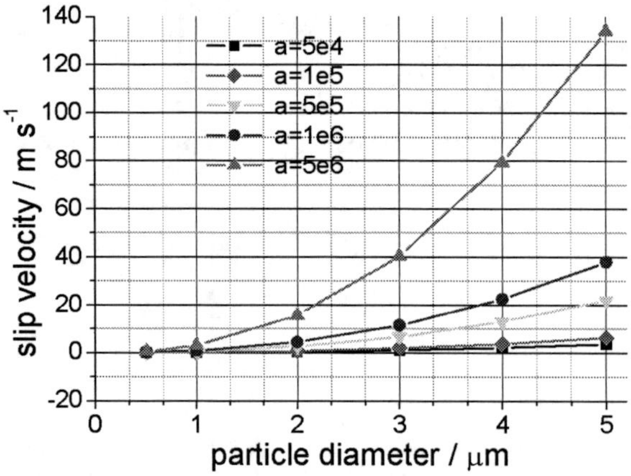

FIGURE 3. Relation of terminal slip velocity and particle diameter for different accelerating flow field.

Measurement Error Caused by Slip Velocity

The measurement error on flow field with PIV caused by the slip of particles and fluid can be obtained from equation (6)

$$\gamma = \frac{u_p - u_f}{u_f} \cong \frac{u_p - u_f}{u_p} = -\frac{\tau_s \sigma_p}{f(\mathrm{Re}_p)} \tag{11}$$

When Re_p<1, it can also be reduced to

$$\gamma = -\tau_s \sigma_p \tag{12}$$

τ_S is a constant for given tracer particles. Getting the velocity gradient among adjacent interrogation window $\sigma_p{}'$ is very convenient for commercial PIV. Although there is some error to substitute $\sigma_p{}'$ for σ_p, it can meet the demand for the error analysis. So the error distribution of measured velocity field can be acquired from equation (12) or (11). The velocity gradient can also be calculated according to the following equation

$$\sigma_{p(i,j)} \cong \left(\frac{u_{(i+1,j)} - u_{(i,j)}}{\Delta x}, \frac{v_{(i,j+1)} - v_{(i,j)}}{\Delta y}\right) \tag{13}$$

The exorbitant mean error implies improper tracer particles selection. We can reselect the tracer particles for demanded maximum relative error according to the following equation

$$d_p < \sqrt{\frac{18\gamma\mu}{\rho_p \sigma_p}} \tag{14}$$

The fig.4 shows the relation of relative error and particles diameter for four kinds different material particles when σ_p=3000 s^{-1}. Particles material are respectively oil, glass (ρ_p=2300 kg · m^{-3}), TiO_2 and Al_2O_3; the gas density ρ_f =1.5 kg · m^{-3} and dynamic viscosity coefficient μ=1.9×10^{-5} Pa·s. The fig.5 provides the tolerable maximum particles diameter for different velocity gradient for the relative error of less 1%. The range of velocity gradient cover from 250 s^{-1} to 64000 s^{-1}, and the other parameters are the same as the fig.4.

The above curves provide the instruction for selecting proper tracer particles for various flow fields. The results indicate that the range of selectable particles is smaller as the increase of velocity gradient.

Correction Method on Error Caused by Slip Velocity

It is very difficult for some experiments to get proper tracer particles because of the limit of many insurmountable factors. Selectable particles diameter is often larger than 1μm, and it can not to ensure the reliability of flow visualization and PIV testing results. Therefore it is very interesting to find out proper method to correct the error.

It is well-known that the results obtained by PIV are the particles velocity u_p, not real gas flow velocity u_f. According to the equation (6), the u_p - u_f can be acquired if a_p or σ_p is known, further the real velocity of measured gas flow can be obtained. This paper attempts to correct the measurement error caused by the slip of tracer particles and fluid based on known velocity gradient among adjacent interrogation window $\sigma_p{}'$ and substituting $\sigma_p{}'$ for σ_p. For the curve 2 of the fig.2, the relative slip of 2%~22% has reduced to 1%~6% by correcting the PIV testing results.

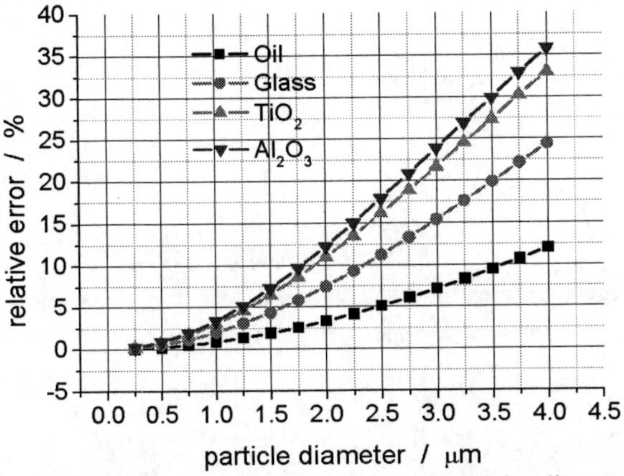

FIGURE 4. The relation of relative error and particles diameter.

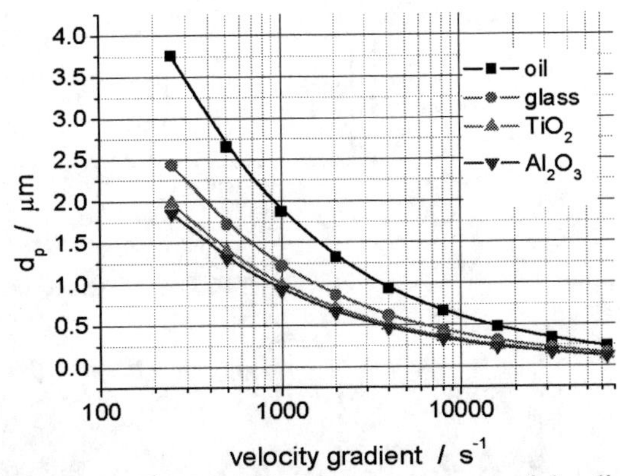

FIGURE 5. The tolerable maximum tracer particles diameter for different velocity gradient.

The residual error can be owed to the combination of cross correlation algorithm error caused by non-uniform tracer particles size and the dilution of precision caused by the difference procedure for computing the σ_p'. Therefore, reliable particles velocity gradient field is the key for improving the error correction.

In recent years, many new methods are presented for improving the accuracy of velocity gradient. Tokumaru and Dimotakis[15] investigated an image analysis technique referred to as correlation image velocimetry (CIV), which is conceived to perform a global optimization of the correlation coefficient in a multidimensional space where not only the displacement component are considered as optimization parameters but also the local spatial derivatives of the displacement. The similar investigation has been done by Hart [16]. However, the multi-dimensional correlation function is computationally very costly, since the normal two-dimensional correlation functions must be calculated for all possible combinations of gradients. Therefore, two alternative methods with a reduced numerical load have been investigated by Nobach and Tropea [17] which recognize the velocity gradients by iteratively maximizing the correlation peak or by using weighting schemes to analyze sub-spaces of the interrogation area. Both techniques iteratively optimize the displacement and deformation towards zero displacement and zero gradients of the residual images. They are alternatives to a multi-dimensional correlation, but with reduced numerical costs. While the multi-dimensional correlation requires the calculation of conventional two dimensional correlation tables for all possible combinations of gradients (four additional dimensions), the number of iterations for the developed algorithms is comparable to conventional iterative PIV methods. Each iteration consists of four two-dimensional correlations for the displacement estimation and another sixteen for the estimations of the effective origins in the case of the interrogation area sub-division or nine two dimensional correlations for the iterative optimization.

There are still many investigations need to be done in this field and they are beyond this paper; however, I believe that accurate velocity gradient would have more excellent performance for correcting the error caused by slip velocity.

SEEDING POSITION FOR TRACER PARTICLES

The purpose to research seeding method and position is to introduce tracer particles into the flow with sufficient, stable concentration and minimum turbulence on the main flow. The dragging gas flow for tracer particles is necessary for positive pressure flow. However the disturbance caused by the dragging gas flow does often determine the particles seeding position because the distance L of testing area and seeding position must be larger than the attenuation distance of the disturbance. However, it is also improper for selecting oversize L, especially for the oil fog. This is because many oil fogs deposited on the wall form the oil film, then it is broken into the coarse secondary oil droplets and influence the measurement accuracy of PIV. This section presents the optimal seeding position by investigating on the attenuation process of the disturbance with the method of numerical simulation.

Fig.6 shows the particles seeding model. The velocity of main gas flow U is along the x direction, and the velocity of dragging gas flow for tracer particles V is along the y direction, and they mix at the seeding inlet. Then, stable gas-particle flow is formed in the pipe which its length L_1 is also the distance of testing area and seeding inlet, and the final velocity is U_1. The Reynolds number of dragging gas flow and main gas flow can be marked as Re_t and

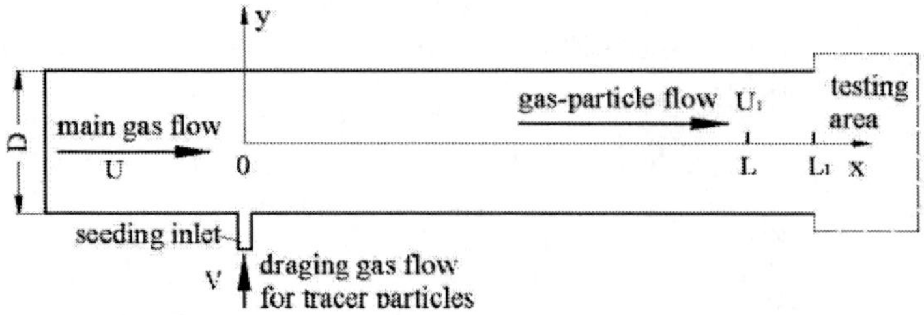

FIGURE 6. Scheme for particles seeding model.

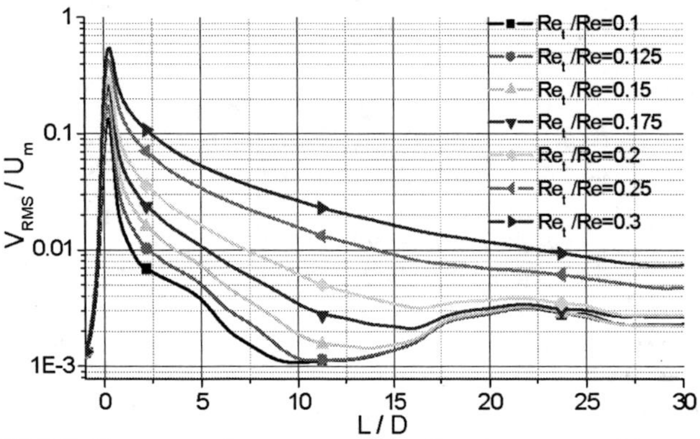

FIGURE 7. The attenuation process of the disturbance caused by the dragging gas flow for tracer particles

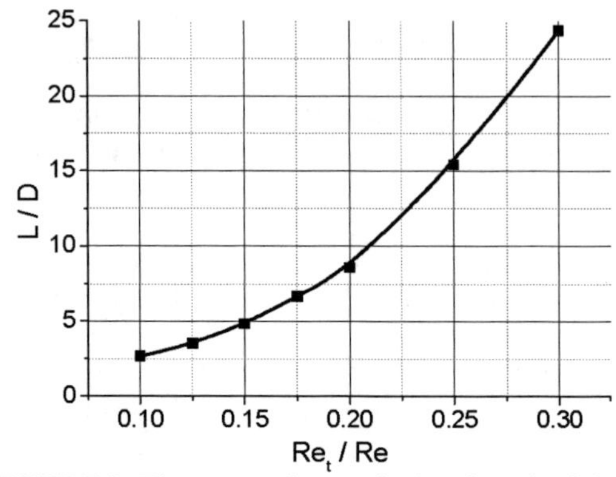

FIGURE 8. The attenuation equivalent length of the disturbance for different Re_t/Re

Re, respectively, and the ratio Re_t/Re implies the disturbance intensity of dragging gas flow on main gas flow. U_m is the mean velocity of gas flow for different *x* position. and the x/D is the equivalent length. V_{RMS} is the root-mean-square of the velocity component *V* and it implies the disturbance component. The value of V_{RMS}/U_m means the relative magnitude of residual disturbance. The disturbance can be neglected in this paper when $V_{RMS}/U_m<0.01$, and we define the distance of seeding inlet and the position x with $V_{RMS}/U_m=0.01$ as the attenuation distance *L* of the disturbance. Fig.7 and fig.8 show the attenuation process and attenuation equivalent length L/D for different Re_t/Re. The results indicate that the larger L/D is necessary for larger Re_t/Re in order to remove the disturbance. The optimal seeding position can be selected according to the attenuation equivalent length for different Re_t/Re. For the seeding such as fig.6, L_1 should be the same or slightly larger than *L*. Some quantitative conclusions can be obtained from fig.8 that the distance of optimal seeding position and testing area is the three times of pipe diameter when $Re_t/Re=0.1$, however, this distance has increased to the twenty five times of pipe diameter when $Re_t/Re=0.3$.

CONCLUSIONS

(1) The results indicated that tracer particles must be have the slip with the fluid as long as the velocity gradient presented even the particle was sufficiently accelerated, and it has severely influence on the reliability of PIV testing results for the flow field with high velocity gradient or (and) larger tracer particles size.
(2) The relative error caused by the slip of particles and fluid is in proportion to the particle Stokes time and the particle velocity gradient, but it is in inverse proportion to the correction factor for Stokes-drag force. The above conclusions can be used to judge and correct the measurement error or to select proper tracer particles for the non-uniform flow field.
(3) During the study on seeding method of tracer particles, the Reynolds number ratio of dragging gas flow for tracer particles and main gas flow Re_t/Re is considered as the disturbance intensity scale of dragging gas flow on main gas flow. The attenuation process and attenuation distance of disturbance is investigated for different Re_t/Re in this paper. The optimal seeding position can be selected according to the attenuation equivalent length for different Re_t/Re. The distance of optimal seeding position and testing area is the three times of pipe diameter when $Re_t/Re=0.1$, however, this distance has increased to the twenty five times of pipe diameter when $Re_t/Re=0.3$.

ACKNOWLEDGMENTS

This project was supported by National Natural Science Foundation of China (NSFC) (No. 50476051; 50323001).

REFERENCES

1. J.M.Bergthorson and P.E.Dimotakis, *Exp. Fluids* 41, 255-263(2006).
2. A.Melling and J.H.Whitelaw, *DISA Information* 15, 5–14(1973).
3. M.Haghgooie, J.C.Kent and R.J.Tabaczynski, *Exp. Fluids* 4, 27–32(1986).
4. M.Samimy and S.K.Lele, *Phys. Fluids* 3, 1915–1923(1991).
5. A.Melling, *Meas. Sci. Technol.* 15, 1406–1418(1997).
6. N.Paone, G.M.Revel and E.Nino, Proc. 8th Int. Symp. on Applications of Laser Techniques to Fluid Mechanics, Lisbon paper 3.4, 1996.
7. D.J.Anderson, C.A.Greated, J.D.C.Jones, G.Nimmo and S.Wiseall, Proc. 8th Int. Symp. on Applications of Laser Techniques to Fluid Mechanics, Lisbon, paper 18.4, 1996.
8. J.N.Stewart, Q.Wang, R.P.Moseley, P.W.Bearman and J.K.Harvey, Proc. 8th Int. Symp. on Applications of Laser Techniques to Fluid Mechanics, Lisbon, paper 18.5, 1996.
9. G.H.Smith, I.Grant, A.Liu, D.Infield and T.Eich, Proc. 6th Int. Symp. on Applications of Laser Techniques to Fluid Mechanics, Lisbon, paper 13.5, 1992.
10. H.T.Huang, H.E.Fiedler and J.J.Wang, *Exp. Fluids* 15, 263–273(1993).
11. P.Meunier and T.Leweke, *Exp. Fluids* 35, 408-421(2003).
12. H.Nobach and C.Tropea, *Exp. Fluids* 39, 614-622(2005).
13. E.E.Michaelides, *Phys. Fluid A* 4, 1579-1582(1992).
14. D.J.Vojir and E.E.Michaelides, *Int. J. Multiphase flow* 20, 547-556(1994).
15. P. T. Tokumaru and P. E. Dimotakis, *Exp. Fluids* 19, 1-15(1995).
16. D. P. Hart, Proceedings of the 10th international symposium on flow visualization, Kyoto, Japan, Paper F0374, 2002.
17. H. Nobach and C. Tropea, *Exp. Fluids* 39, 612–620(2005).

SESSION B

NOVEL SENSORS AND METHODS

Study on a Novel Sensor of Wetness Measurement in Water/Air Flows

Deliang Ning, Changqi Yan, Puzhen Gao

College of Nuclear Science and Technology, Harbin Engineering University, Harbin, 150001, P.R. China

Abstract. In view of the character that wetness measurement for steam is difficult and inaccurate, a novel sensor of wetness measurement for flowing wet steam is developed by making use of capacitance method based on the fact that the water and steam have great different permittivities. The sensor can be installed in the steam pipe directly, so the steam will flow through it with full flow rate. Therefore the error due to drawing steam sample inaccurately is eliminated. By means of this sensor, on-line measurement of the wetness of wet steam can be realized. Based on the fact that air and steam have almost same permittivities, the flowing wet steam is simulated by spraying water into air flow in the experiment. Experiment with the novel sensor of wetness measurement in water/air flows shows that the frequency output decreases as the absolute humidity of flowing wet air increases. The relationship between the absolute humidity of air and the relative frequency deviation are obtained from the experimental data and this result accords with the theoretic conclusion. The sensor can endure high temperature and high pressure and has preferable precision, so it can be used to measure not only the wetness of flowing wet steam in nuclear power plants but also the water content of solid materials on-line after precise calibration.

Keywords: Wetness; Measurement; Capacitance method; Permittivity; Frequency

INTRODUCTION

Wetness of steam is an important parameter for the steam turbine that operates with saturated vapor. It is significant to measure the wetness of wet steam accurately for the safe operation of steam turbine in nuclear power plants. But it is difficult and inaccurate to measure the wetness of flowing wet steam at present. In general, the methods for measuring the wetness of flowing wet steam include heating method, optical method and tracer determination method[1]. The heating method has been studied carefully by Y.F. Li et al. (1997) and W.J. Wang et al. (2001) [2, 3]. This method requires to draw steam sample from main flow of wet steam, and then to heat and measure the sample. Its measuring period is long and result is rough. The optical method for measuring the wetness of flowing wet steam was studied by X.S. Cai et al. (1992) [4] and some achievements have been obtained. With optical method, on-line measurement is realized and drawing steam sample are not needed, but there are some disadvantages, the randomicity of measuring results is notable, the measuring condition demand is very strict, and the price of measuring device is very high. The tracer determination method was investigated by Z.L. Liu et al. (2003) [5]. With this method, there are many disadvantages too, such as longer measuring period, larger error, costly equipment, complicated process of measuring and treating data.

In this paper, a novel sensor of wetness measurement for flowing wet steam is developed by making use of capacitance method based on the fact that water and steam have great different permittivities and the equivalent permittivity of two-phase admixture will be varied with the mass proportions of water and steam. With the sensor, on-line measurement can be realized, and there are lots of advantages, fast response, easy operation, simple configuration and low cost, and it is immune to radioactivity, so it can be used in the places where radiation exists.

PRINCIPLE OF CAPACITANCE METHOD MEASURING WETNESS OF FLOWING WET STEAM

From electromagnetics knowledge, it is well known that the capacitance only relates to the permittivity of

CP914, *Multiphase Flow: The Ultimate Measurement Challenge,* edited by X. Cai, Y. Wu, Z. Huang, S. Wang, and M. Wang

dielectric inside capacitor in the condition that the area and distance of electrodes of capacitor are settled [6]. Based on this truth, many scientists and engineers had measured moisture content in solid, e.g. monitoring sandy soil water content with capacitive sensor was studied by Y.J. Liu et al. (2002) [7]. X.Y. Li et al. (2002) studied the sediment concentration in water current [8], and Y.S. Dai et al. (1994) investigated the water content in oil [9], and A. Jaworek et al. (2004) researched the void fraction in two-phase flow with capacitance method [10]. But current capacitance sensors are restricted by the temperature and pressure of the measured dielectric, and can not to measure the wetness of flowing wet steam with high temperature and high pressure. For this reason, we design a novel sensor for measuring the wetness of steam based on capacitance method. Water and steam have great different permittivities when they are at same temperature and pressure, e.g. the relative permittivity of saturated water at 20℃ ε_f is 80.22, while the relative permittivity of saturated steam ε_v is 1.0002. As water content in vapor varies, the equivalent permittivity of wet steam will change. Consequently, when the wet steam flows through capacitance sensor, the variation of the steam wetness will make the equivalent permittivity change, and then induce the capacitance of sensor to change.

CONFIGURATION OF CAPACITANCE SENSOR

Based on the operational principle of cylindrical capacitance, we develop a coaxial cylindrical capacitance sensor which can be used to measure the wetness of wet steam flowing in pipe on-line. The configuration of the sensor is shown schematically in Fig.1.

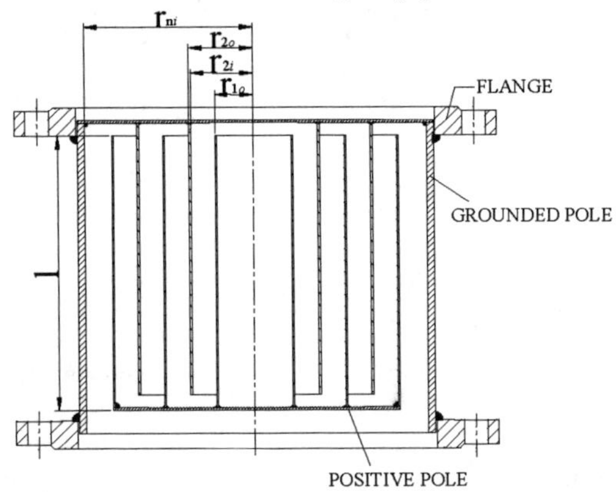

FIGURE 1. Schematic configuration of capacitance sensor

The capacitance sensor presented in this paper is composed of many coaxial cylindrical electrodes, and there are two flanges welded on both sides of the outmost electrode, in order that the capacitance sensor can be connected with steam pipe directly. The wet steam in pipe can flow through the sensor with full flow rate, therefore the measuring error due to inaccurate drawing steam sample can be eliminated. All the electrodes of the capacitance sensor are made up of stainless steel to resist rusting, and the outmost electrode is made of stainless steel seamless pipe with thicker wall for enduring high pressure. The electrodes are connected with wires with parallel connection method, and the outmost electrode serves as grounded pole simultaneously. A low voltage electrical source DC is used to supply power for the sensor, so the dielectric between electrodes can not be destroyed by electric field. The distances between neighbor electrodes are equal and the electrodes are fixed with bakelite boards which have good insulation property and high toughness. Based on the theory of electric field, as ignored the fringe effect, the capacitance C_k of coaxial cylindrical capacitor is given by the formula:

$$C_k = 2\pi\varepsilon_r\varepsilon_0 l / \ln\left[R_i / r_o\right] \quad (1)$$

where ε_r is the relative permittivity of dielectric in capacitor, and ε_0 is the permittivity of vacuum ($\varepsilon_0 = 8.85415\times10^{-12}$F/m), l is the effective length of cylindrical electrodes, R_i and r_o are the inner radius of the outside electrode and the outer radius of the inside electrode respectively. Because the capacitance sensor (Fig.1) is made up of several coaxial cylindrical capacitors connected in parallel with each other, so the equivalent capacitance of the sensor could be given by the formula:

$$C=\sum_{k=1}^{n-1}C_k=\sum_{k=1}^{n-1}2\pi\varepsilon_r\varepsilon_0 l/\ln\left[r_{(k+1)i}/r_{ko}\right] \quad (2)$$

where n is the number of electrodes.

MEASURING CIRCUIT

The capacitance variation of the sensor caused by the wetness change of wet steam is measured by frequency method [11] (F.G. Pang, 1991). The electrical scheme of the measuring circuit is shown in Fig.2.

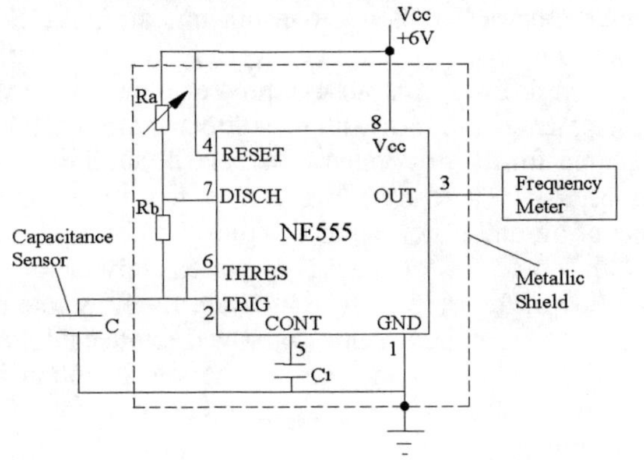

FIGURE 2. Electrical scheme of the measuring circuit

The measuring circuit is made up of an NE555 compositive timer, an invariable resistance R_b (500 Ω), a rheostat R_a (0~20kΩ), a filter capacitance C_1 (0.02 μF) and the capacitance sensor C designed. There is a metallic shield covering the measuring circuit to shield the measuring signal from the environment. As the permittivity of dielectric in the capacitance sensor changes, the capacitance will change, the signal can be captured and transformed to frequency by the measuring circuit and displayed on the frequency meter. If the wetness corresponded to the frequency can be known successfully, then the correlation of the frequency and wetness can be predicted. A low voltage electrical source DC (V_{cc}= +6V) is used to supply power for the sensor and NE555 compositive timer, the discharge and charge processes of the capacitance sensor take place between the threshold voltage ($\approx 0.67V_{cc}$) and the trigger voltage ($\approx 0.33V_{cc}$). The impulse frequency is given by the formula

$$f=\frac{1}{T}=\frac{1.443}{(R_a+2R_b)C} \quad (3)$$

Resistance value of R_a is variable, so the initial frequency output can be adjusted by changing the resistance of R_a. In this way, we can expand the measurement range and improve the precision of results.

EXPERIMENTAL METHOD AND APPARATUS

Different from the humidity of air, the wetness of wet steam is defined as the ratio of droplets mass to the total wet steam mass in unit time, that is

$$Y=\frac{m_f}{m_f+m_v} \quad (4)$$

where m_f and m_v are the mass flow rate of droplets and steam flowing through the capacitance sensor ,respectively. A difficulty for wetness measurement of flowing wet steam at present is that, there is not any calibration apparatus to produce wet steam with exactly known wetness to calibrate the developed or invented instrument for measuring steam wetness. The method as explain subsequently can overcome the difficulty in a certain extent, and calibrate the capacitance sensor.

The relative permittivity of air is very close to that of saturated steam, e.g. at 20℃ and 1atm, the relative permittivity of air ε_a is 1.00054 and that of saturated steam ε_v is 1.0002. Even at high temperature and high pressure, the relative permittivity of saturated steam does not change remarkably, e.g. at 300℃ and 5MPa, ε_v is1.1556, which is yet close to the relative permittivity of air. So we adopt the method of spraying water into air flow to simulate the flowing wet steam to carry out calibration experiment on the capacitance sensor in water/air flows. The experimental stand achieving calibration experiment of capacitance sensor in water/air flows is schematically shown in Fig.3.

In the experimental stand, the volumetric fan blew environmental air into pipe, and then the air entered the wet air injector. The wet air injector was a device in which the blowing air mixed with spraying water. It consisted of 6 air-blast nozzles, mixing chamber and diffuser. The air compressor blew compressed air into 6 air-blast nozzles in which kinetic energy of compressed air was used for water suction and atomization. The flow rate of compressed air in calibration experiment was ignored for its little value. The centrifugal pump pumped water from the water tank into the air-blast nozzles simultaneously. There were two rubber tubes connected with air compressor and water pump, respectively. The flow rate of air and water could be controlled by valves installed on the tubes, as a result, the absolute humidity of wet air could be changed. A vortex flowmeter manufactured by Germanic Co. Endress+Hauser, and a float-type flowmeter were used to measure the flow rate of flowing air and spraying water, respectively. The capacitance sensor in the experimental stand was connected with the mixing chamber of the wet air injector vertically. The measuring circuit would measure the signal output as the two-phase flow of air and droplets flowed through the sensor with full flow rate. To reduce environment influence, cable with shield was used as the signal wire for the measuring circuit.

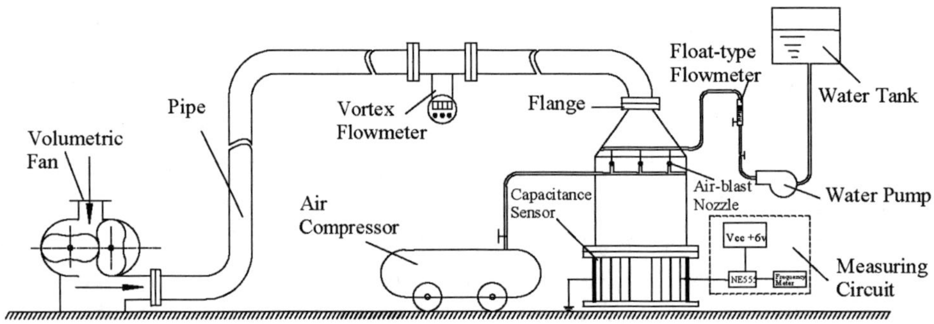

FIGURE 3. Scheme of the calibration experimental stand for the capacitance sensor

The cylindrical capacitance sensor calibrated in this experiment consisted of 9 stainless steel electrodes, so it was made up of 8 coaxial cylindrical capacitors in parallel. Thickness of inner electrodes is 1.5mm and the outmost electrode 5mm. The distance between neighbor electrodes is 20mm, and the effective length of capacitance sensor l is 200mm. The radius of the innermost electrode $r_{1o} = 30\text{mm}$ and the outmost electrode $r_{9o} = 190\text{mm}$.

Supposing that two-phase flow of air and droplets is uniform, then the equivalent relative permittivity of the admixture can be given by the expression below:

$$\varepsilon_e = \varepsilon_a \left[1 + 3 \frac{\varepsilon_f - \varepsilon_a}{\varepsilon_f + 2\varepsilon_a} \frac{1}{1 + \rho_f (1/Y - 1)/\rho_a} \right] \tag{5}$$

where ε_e and ε_a are the equivalent relative permittivity of air and droplets two-phase flow and the relative permittivity of air, ρ_f and ρ_a are the densities of spraying water and air, respectively. In this experiment, Y can be expressed as:

$$Y = \frac{Q_f \rho_f}{Q_f \rho_f + Q_a \rho_a} \tag{6}$$

where Q_f and Q_a are the volume flow rate of spraying water and blowing air, respectively.

ANALYSIS OF EXPERIMENTAL RESULTS

The resistance R_a should be adjusted to an appropriate value before experiment to make the frequency meter display a proper frequency. Then we began the calibration experiment with operating the volumetric fan, the air compressor and the water pump. The absolute humidity of flowing wet air could be controlled by adjusting the valves installed in the experimental stand, and calculated with the formula (6). The relationship between frequency output f and blowing air Q_a before spraying is shown in Fig.4. It is shown in Fig.4 that the frequency output maintains a constant nearly as the flow rate of flowing air increases. That is to say that the operation of capacitance sensor is steady enough.

As the two-phase flow of blowing air and spraying water flows through the capacitance sensor, the frequency output will decrease. The relationship between frequency output and spraying water flow rate is shown in Fig.5. The data points in Fig.5 are obtained in the condition that blowing air flow rate Q_a maintains $250\text{m}^3/\text{h}$. In Fig.5, it is shown that the frequency output decreases as spraying water flow rate increases, i.e. absolute humidity of flowing wet air increases. As the absolute humidity of flowing wet air increases, the equivalent relative permittivity ε_e of two-phase flow of blowing air and spraying water will increase, so the capacitance C of the sensor will increase, and the frequency output f decrease consequently. The correlation of the flowing wet air absolute humidity Y and the relative frequency deviation （$\delta f / f_a$） is expressed with a quadratic curve by means of least square method distinctly, as shown in Fig.6.

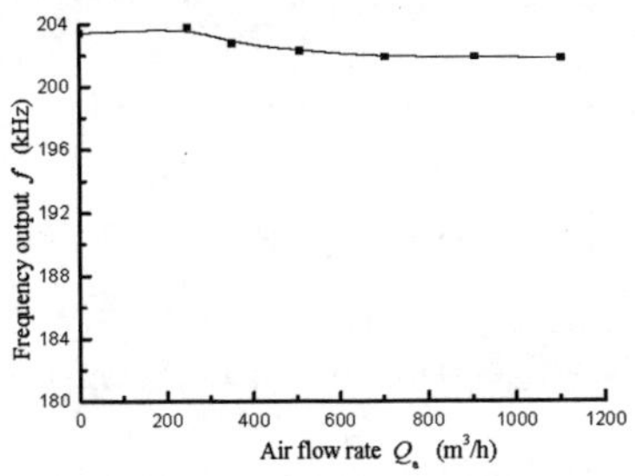

FIGURE 4. Relationship between frequency output and air flow rate

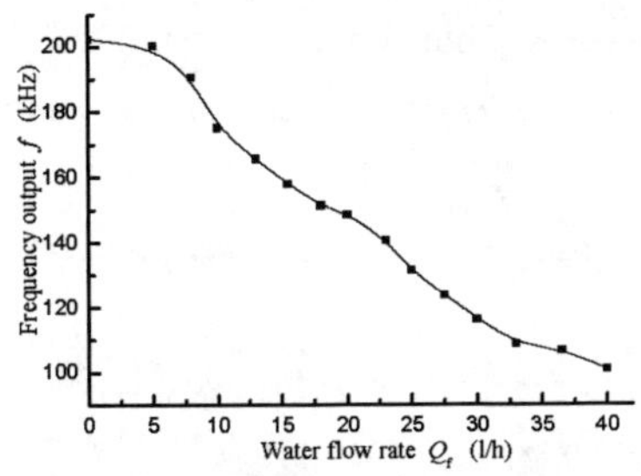

FIGURE 5. Relationship between frequency output and water flow rate

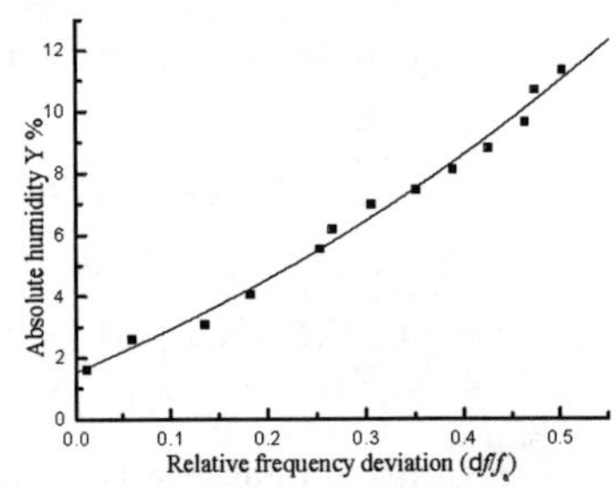

FIGURE 6. Relative frequency deviation vs. absolute humidity

The relative frequency deviation can be defined as

$$F = \delta f / f_a = (f_a - f) / f_a \tag{7}$$

where f_a and f are the steady frequency outputs before spraying water and in spraying. The correlation of the absolute humidity of flowing wet air Y and the relative frequency deviation F can be expressed as a quadratic equation:

$$Y = 12.83F^2 + 12.5F + 1.534 . \qquad (0 < Y < 12) \tag{8}$$

The root-mean-square error of the correlation $R = 0.98989$, standard error $SD = 0.34422$. These errors show that the degree of fitting between the experimental points and the matched curve are well. Based on the calibration experiment results, it is shown that the relationship of relative frequency deviation and absolute humidity of flowing wet air is nonlinear. It accords with the conclusion that obtained by A. Jaworek et al. (2004) [12].

CONCLUSIONS

A method of wetness measurement of flowing wet steam on-line based on measurement of frequency output generated by resonance circuit with a novel capacitance sensor was presented, and verified effectively by calibration experiment operated with air/water flows in this paper. The capacitance sensor developed in this paper had many advantages, such as accurate measuring result, fast response, easy operation, simple configuration and low cost. It can be used to measure not only the wetness of flowing wet steam but also the moisture content of the solid on-line.

The measuring circuit has excellent degree of integration, low cost, and can be manufactured easily. The method that spraying water into air flow to simulate the flowing wet steam is presented in this paper, and the experimental results in water/air flows show that the relationship of relative frequency deviation and absolute humidity of flowing wet air is nonlinear. After more serious calibration, the capacitance sensor can be used in nuclear power plants and other fields.

REFERENCES

1. Y.F. Li and X.J. Wang, *Comparison and analysis of wetness measurement methods for flowing wet steam in turbine*, Turbine Technology, Vol.42, No.3, 2000, pp.156-161.
2. Y.F. Li, X.J. Wang, T.X. Xu, *Heating measuring method of moisture content in flowing wet steam and its device*, Electric Power, Vol.30, No.10, 1997, pp.16-19.
3. Y.F. Li and X.J. Wang, *Experiment technology of heating method for measuring wetness of flowing wet steam*, Journal of Engineering for Thermal Energy and Power, Vol.16, No.2, 2001, pp.153-156.
4. X.S. Cai, N.N. Wang, J.M. Wei, *Measurement of the wetness and droplet size distribution of foggy wet steam flow using the optical method*, Journal of Engineering Thermophysics, Vol.13, No.2, 1992, pp.212-216.
5. Z.L. Liu, G.S. Geng, Z.C. Gou, *Application of non-radioactive tracer determination in determination of primary steam humidity, Heilongjiang Electric Power*, Vol.25, No.3, 2003, pp.168-171.
6. S.C. Fan, *Technology and application of sensors*, China Press, Beijing, 2004, pp.103-108.
7. Y.J. Liu and Z.G. Huang, *Laboratory test of capacitive sensor for monitoring sandy soil water content*, Journal of Shijiazhuang Railway Institute, Vol.15, No.2, 2002, pp.10-12.
8. X.Y. Li, T.W. Lei, W. Wang, *Capacitance sensor measuring sediment concentration in water current*, Acta Pedologica Sinica, Vol.39, No.3, 2002, pp.429-435.
9. Y.S. Dai, G.J. Lei, B.H. Mao, *Development of the compensative capacity water in oil analyzer*, Journal of The University of Petroleum, China Vol.18, No.6, 1994, pp.112-115.
10. A. Jaworek and A. Krupa, *Gas/liquid ratio measurements by rf resonance capacitance sensor*, Sensors and Actuators A, Vol.113, 2004, pp.133-139.
11. F.G. Pang, *Experimental investigation of measurement of void fraction in two-phase flow with capacitance method*, Nuclear Power Engineering, Vol.12, No.6, 1991, pp.78-83.
12. A. Jaworek, A. Krupa, M. Trela, *Capacitance sensor for void fraction measurement in water/steam flows*, Flow Measurement and Instrumentation, Vol.15, 2004, pp.317-324.

Multiphase Flow Measurements by Full Stream Batch Sampling

Wang Dong   Liang Fa-Chun   Zhang Xiu-Gang   Lin Zong-Hu

State Key Lab. of Multiphase Flow in Power Engineering, Xi'an Jiaotong University, Xi'an, 710049, China, Wangdong@mail.xjtu.edu.cn

Abstract. A new method of multiphase flow measurements that based on the "extracting and separating method" was proposed in this paper. An extraction flow (or sample) is diverted in a manner of time sharing or full stream batch sampling, then the sample is separated into single phase flows and metered with conventional flow meters, the total flow rates of each phase are determined according to the metered values and the extraction ratios. Because the full stream is conducted to the extraction loop during sampling, no matter what the flow regime of multiphase flow is, the extracted stream (or sample) is always the representative of the total stream and the extraction ratio can keep stable. Experiments were conducted in an air-water-oil flow test loop. the inside pipe diameter of the test loop was 50 mm, and the superficial gas velocity varied from 4.5 m/s ~22m/s, the liquid superficial velocity was in the range of 0.02~0.3m/s, the oil concentration was in the range of 0.0%~60% by volume. The flow pattern occurring during the experiments included stratified flow ,wave flow and annular flow. The experimental results shown that the full stream batch sample method is feasible to measure the multiphase flow rate, and the average error of flow rates measurements for each phase was less than 3.82%.

Keywords: Multiphase flow; Flow measurement; Sample, Time sharing
PACS:47.61Jd, 47.80.-v, 47.80.cb

INTRODUCTION

Multiphase flows are widely encountered in industrial applications, but the measurements of their flow rates are still a difficult problem in industry. Traditional solution to the problem is to separate the multiphase mixture into single phase flow first, and then measure the flow rate of each phase with single-phase flow meters. Usually a large separator vessel is required to complete the separation of multiphase flow and both the capital and operating costs of equipments tend to be high. Also, in practice the measurements derived from this process are subject to many errors. The ESM(extracting and separating method)[1] is a new method of measurement of multiphase flow rate, in which only a small fraction (1%~20%)of the total flow is extracted and separated, and also the single phase flow meters are employed to meter the separated flows respectively , after the metering ,these flows were returned back to the main stream again, the total flow rates of each phase are determined according to the metered values of single phase flow meters and the extraction ratios. Because only a small portion of the total flow is separated, the separator can be reduced in size greatly compared to the conventional separation method. In order to assure that the extracted stream (or sample) is the representative of total stream or the extraction ratio is stable at any flow regimes, special measures of extracting or sampling should be taken. Andreussi.[2]proposed a sample system in which the two-phase mixture was mixed with a nozzle before a sample was drawn through 3 tubes that fixed in the flow channel. The objective of the present work was to develop a new extraction method and corresponding devices.

THE PRINCIPLE OF EXTRACTION

CP914, *Multiphase Flow: The Ultimate Measurement Challenge,* edited by X. Cai, Y. Wu, Z. Huang, S. Wang, and M. Wang

As shown in Fig.1, the extraction flow or sample of the total flow is diverted in a manner of time sharing or full stream batch sampling, i.e. at an interval of one period, the full stream is alternatively conducted to the main loop and the extraction loop, as long as the frequency of flow switching is high enough and the extraction process is under isokinetic condition a nearly continuous extraction flow can be obtained, and the extraction ratio can easily be controlled by the time fraction during which the full stream is flowing into the extraction loop. Because the full stream is conducted to the extraction loop during sampling, no matter what the flow regime of multiphase flow is, the extracted stream (or sample) is always the representative of the total stream and the extraction ratio can keep stable.

From Fig.1 we can see that when the full stream is switched to the extraction loop, it is similar to acquiring the data of flow rates. As shown in Fig.2 , Q_j represents an arbitrary phase (gas or liquid) flow rate, T is the period of

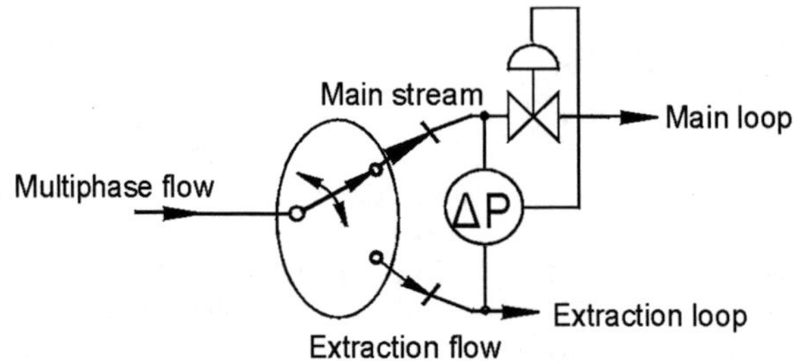

FIGURE 1. Schematic representation of sample system

flow switching, Δt is the duration of sampling in a period, so the total mass of each phase sampled in a period (the shadow area in Fig.2a) is

$$\Delta t \cdot Q_j = T \cdot q_j \tag{1}$$

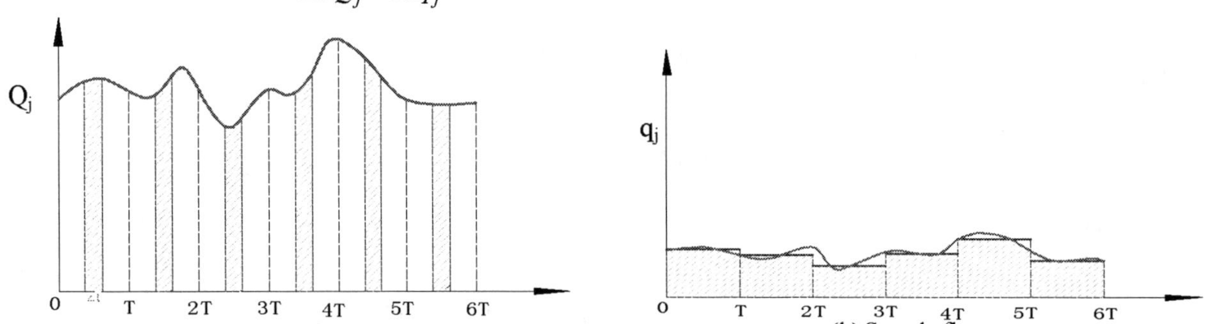

(a) Total flow rate (b) Sample flow

FIGURE 2. Total flow and the corresponding Sample flow

where q_j is the corresponding phase average flow rate in extraction loop, rearranging Equation (1) we can get

$$Q_j = q_j \cdot \frac{T}{\Delta t} \tag{2}$$

Defining the ratio of extraction time as

$$\mu = \frac{\Delta t}{T} \tag{3}$$

Equation (2) can be rewritten as

$$Q_j = \frac{q_j}{\mu} \tag{4}$$

or

$$\mu = \frac{q_j}{Q_j} = \frac{\Delta t}{T} \tag{5}$$

Equation (5) means that the ratio of extraction time is equal to the ratio of flow rates between the extraction flow and the total flow (extraction ratio). If the ratio of extraction time is chosen, the total flow rate can be determined according to Equation (4). The extraction flow q_j is able to represent the total flow Q_j as long as the sample frequency is lager than the double value of the total flow frequency (data acquisition theorem[3]),and this condition is easy to meet.

THE EXTRACTION DEVICE

An actual extraction device (or sampler) is schematically shown in Fig.3. It includes tow parts, a wheel and the fluid rooms. The fluid room is divided into two ports with clapboards, one port is fluidly connected to the main loop, and the other is fluidly connected to the extraction loop. The wheel can rotate about its central axis, and can be motor driven or flow driven, in the present study it is flow driven. The flow entrance of the extraction device is located at the center of the wheel and from there, three paths of flow extend to the edge of the wheel (only one path of flow and one fluid room is shown in Fig.3 for clarity) . As multiphase mixture passes through the wheel, the wheel will be forced to run at a high speed around its axis by the fluid. With the running of wheel, the full stream of multiphase mixture is conducted alternatively into main loop and extraction loop through the fluid room respectively. From Fig.3 we can see that the process of flow switching is smooth and under the condition of isokinetic sampling, the duration of sampling in one period of rotation will be

$$\Delta t = \frac{L}{R\omega} \tag{6}$$

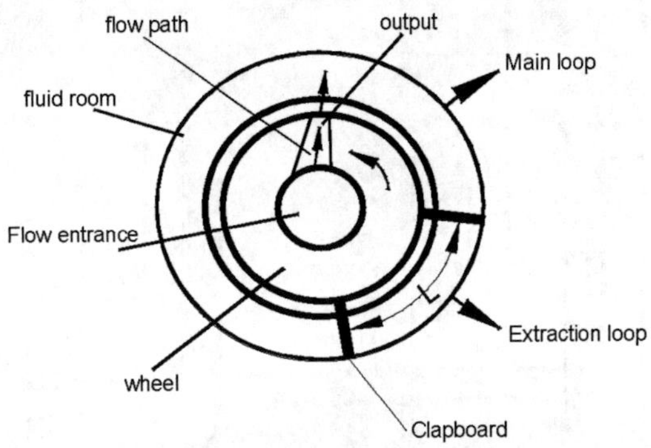

FIGURE 3. Extraction device

where Δt is the duration of sampling in a period of rotation; L is the width of the fluid room connected to the extraction loop; R is the radius of the wheel and the ω is the angular speed of the wheel. The period of the rotation is

$$T = \frac{2\pi}{\omega} \tag{7}$$

substituting Equation (6)and Equation (7) into Equation (3) , the ratio of extraction time will become

$$\mu = \frac{\Delta t}{T} = \frac{L}{R\omega} \cdot \frac{\omega}{2\pi} = \frac{L}{2\pi R} \tag{8}$$

Equation (8) means that the ratio of extraction time depends only on the structure parameters and is independent of the speed of rotation and fluid properties. When flow rate changes , the period of rotation and duration of sampling will also change, but the ratio of extraction time will remain unchanged .We can change the value of the ratio of extraction time by changing the value of L. In the present study, L=15.71mm,R=50mm,according to Equation (8)and Equation (5), the theoretical value of ratio of extraction time and extraction ratio was 0.05. The real values of extraction ratio should be calibrated by experiments.

EXPERIMENTAL SETUP

The extraction device and extraction loop was vertically installed in the downwards side of an inverted U-shape tube as shown in Fig.4. The inside diameter of the tube was 50 mm, and made from Perspex pipe (a transparent pipe) for flow pattern observation. Air, Tap water and lubrication oil (density 871 kg/m^3,viscosity 115mPa • s) were used in the present experiments. The gas flow meter used in the extraction loop was a calibrated venturi meter, and the liquid flow rate was measured using a YOKOGAWA electromagnetic flow meter when the liquid was water, but when the liquid was water-oil mixture, a MILTON ROY meter pump with a accuracy of $\pm 1\%$ was used to meter and pump the liquid. There was 100 pipe diameters length of straight pipe section upstream of inverted U-shape tube. The total air flow rate was metered using a YOKOGAWA vortex flow meter. The water flow rate was measured with two orifice flow meters, one for lower flow rates and the other for higher flow rates. The oil flow rate was metered with a oval gear positive displacement meter. The superficial gas velocity varied from 4.5 m/s ~22m/s, the liquid superficial velocity was in the range of 0.02~0.32m/s, the oil concentration was in the range of 0.0%~60% by volume. The flow pattern occurring during the experiments included stratified flow ,wave flow and annular flow. When the liquid consisted of oil and water, no separation of oil and water could be observed in the pipe, the mixture of water and oil was a fully mixed emulsion, but stronger waves and vibration could observed when oil was added.

RESULTS

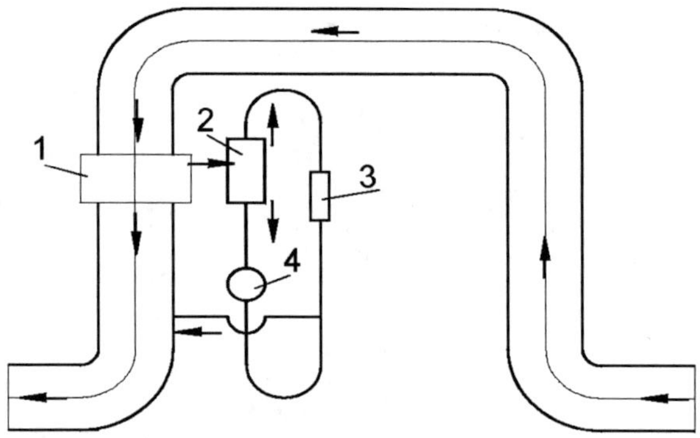

FIGURE 4. Experimental setup
1.Extraction device 2.separator 3.gas flow meter 4. liquid flow meter

Figure 5 shows the comparisons of gas qualities in the extraction loop and in the entrance of the extraction device. Where,X_3 represents the gas quality in extraction loop,X_1 is the gas qualities in the entrance of extraction device. The experiments were made with air and water. From Fig.5 it can be seen that the values of X_3/X_1 were

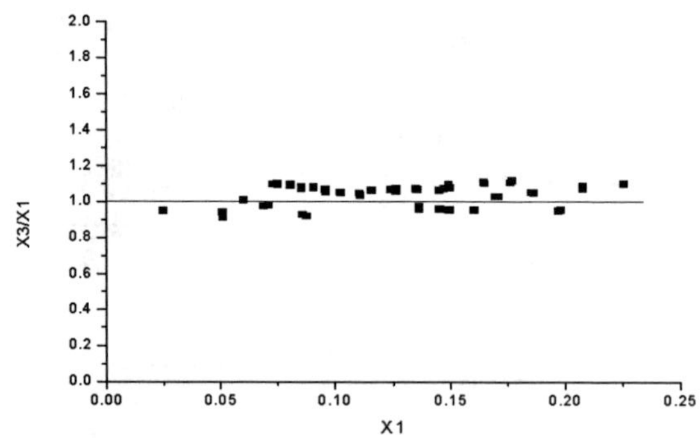

FIGURE 5. Comparisons of gas qualities in extraction loop and entrance of the extraction device (air-water data)

nearly equal to 1,the average difference between $X_3$and $X_1$was less than 2%. This illustrates that the extracted stream was a good representative of the multiphase flow.

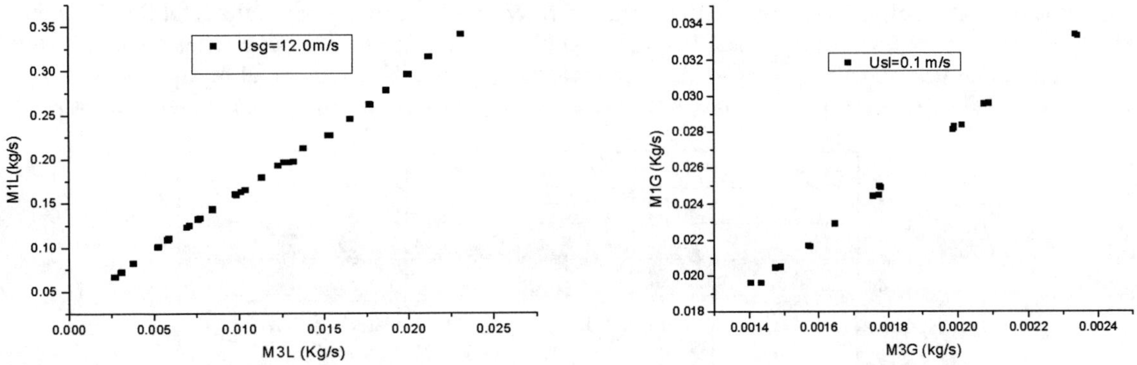

(a) Liquid flow (b) Gas flow

FIGURE 6. Relationship of flow rates between total flow and sample flow (air-water data)

Figure 6 shows the relationships of flow rates between the extraction flow and the corresponding total flow, Where $M_{1L}$and M_{3L} represents the total water flow rate and that in the extraction loop respectively. M_{1G}.and M_{3G} is the total and extracted gas flow rates respectively. U_{sg} and U_{sl} is the superficial velocity of gas and liquid respectively. It can be seen from Fig. 6 that the extracted flow rates M_{3L} and M_{3G} varies linearly with the total flow rates M_{1L} and M_{1G} respectively, these relations can be expressed as

$$M_{3L} = a + bM_{1L} \tag{9}$$

$$M_{3G} = c + dM_{1G} \tag{10}$$

where ,a, b, c and d are constants. Rearranging Equation (9) and Equation (10) the extraction ratios can be obtained

$$K_L = \frac{M_{3L}}{M_{1L}} = \frac{a}{M_{1L}} + b \tag{11}$$

$$K_G = \frac{M_{3G}}{M_{1G}} = \frac{c}{M_{1G}} + d \tag{12}$$

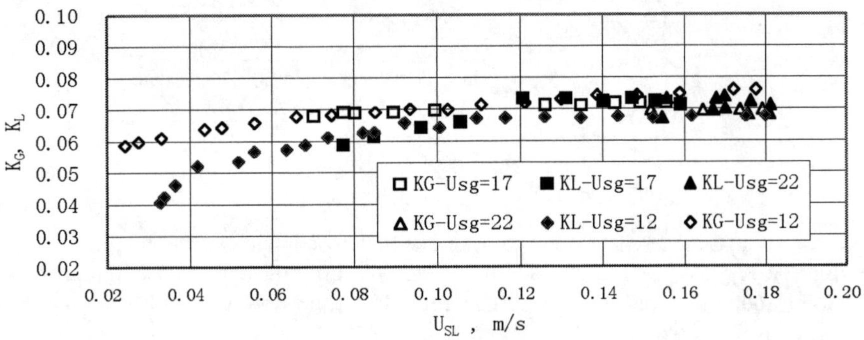

FIGURE 7. Extraction ratio (air-water data)

where K_L and K_G is the liquid and gas extraction ratio respectively. Comparing to Equation (5),it can be seen that in perfect conditions, the constants a and c should be zero, but in actual conditions they are not zero because there is a gap between the wheel and fluid rooms, which has an effect on the process of flow distribution. From Equation (11)and Equation (12) it can also be seen that $K_L \approx b$, $K_G \approx d$, as flow rate M_{1L} and M_{1G} becomes larger, in this case the extraction ratio will be constant and the extracted flow will be proportional to the total flow. Figure 7 shows the actual values of extraction ratios K_L and K_G, they are plotted as a function of superficial liquid velocity U_{SL} for individual superficial gas velocity U_{SG}. From the Fig. it can be seen that K_L and K_G will be equal and take a constant value as the liquid flow rate and gas flow rate become larger ($U_{SL} \geq 0.10$ m/s, $U_{SG} \geq 17$m/s). Figure 8 shows the effect

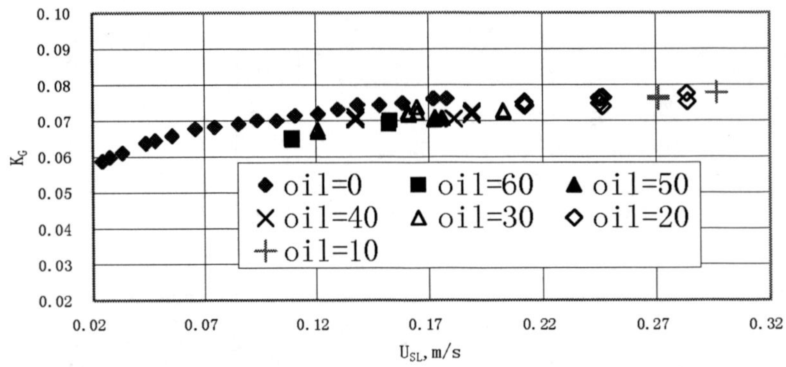

FIGURE 8. The effect of oil concentration on extraction ratio

of oil concentration on the gas extraction ratio K_G, from Fig.8 it can be found that the oil will make the value of K_G drop a little (about 10%) as oil concentration is greater than 50%, it will have little effect upon K_G as oil concentration is lower than 50%. Figure 9 shown the effect of oil concentration on the relationship of liquid flow rates between extraction flow and the total flow. It can be seen that all data nearly fall on the same line except the date of oil concentration of 60% (7% below the line).

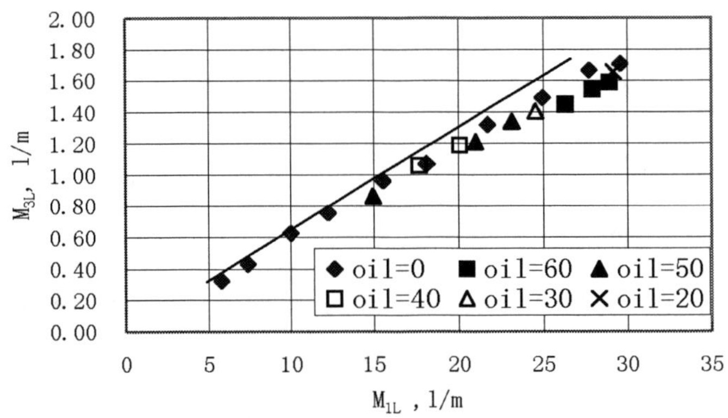

FIGURE 9. The effect of oil concentration on liquid extraction

The calibrated data of K_L and K_G can be used to determine the total flow rate M_{1L} and M_{1G} after the extracted flow rate M_{3L} and M_{3G} has been metered. Experiments shown that the average error of total flow rates measurements for each phase was less than ±3.82%.

CONCLUSIONS

Based on the analysis and experimental work, the following conclusions can be drawn.

(1) Time sharing or full stream batch sampling is a reliable method of extraction , which can extract a good sample of multiphase flow under different flow regime. It is similar to a data acquisition system.

(2) The flow rate of extraction flow always varies linearly with the corresponding total flow rate, and when the gas and liquid flow rates become lager, the values of extraction ratio K_L and K_G will be equal and take a constant., and the flow rates of extraction flow will be proportional to the total flow rate. Total flow can be determined from extraction flow, the average error of this method is less than 4%.

(3) Oil phase has little effect upon the extraction ratio K_L and K_G as oil concentration less than 50%.

ACKNOWLEDGEMENTS

The authors express their great thanks to National Natural Science Foundation of China for financial support (No.50376051)

REFERENCES

1. Wang Dong, Lin Zonghu, Gas-liquid Two-phase Flow Measurement Using ESM, Experimental Thermal and Fluid Science, Vol.26(6-7), 2002.9, pp827-832
2. Paplo Andreussi. Method for measuring the flow rates of the single phases in a multiphase fluid stream and relevant apparatus.US6546809BI.2003.4
3. Jin xi，Yang Er-bin，Zhao ling，The principle of signal process and application, Beijing：Tsinghua university Press，2004
4. Wang Dong, Zhang Xiu-gang, Liang Fa-chun , Lin Zong-hu. Proportional multiphase flow division by time sharing. Journal of Engineering thermophysics,Vol.26(s),2005.6 101-104

Electronics Design for Dual Energy Gamma-Ray Multiphase Flow Meter

Donghui Li, Yingxiang Wu and Keren Wang

Institute of Mechanic, Chinese Academy of Sciences

Abstract. There are two ways to design nuclear electronics for a dual energy γ ray multiphase flow meter. The advanced way is based on high speed A/D data acquisition and DSP digital signal processing, but it involves a complicated system structure and a huge data flow; the conventional way is based on the technique of photon pulse counter, which is complicated in electrical design, but simple in system structure due to its low data flow. Both systems are studied in the Multiphase Flow Laboratory, Institute of Mechanics, Chinese Academy of Sciences. In this paper, the instrumental designs on the technique of photon pulse counter will be mainly discussed, including the γ ray sensor's characteristics, signals of sensor, preamplifier, filter and shaping amplifier, DC base shift correcting circuit, narrow windows of energy spectroscopy, programmable pulse count acquisition system

Keywords: Dual energy γ ray, Multiphase flow

INTRODUCTION

Oil-water-gas multiphase mixed transportation in oil pipe is extensively used in ocean oil industries. How to measure the flow rate and multiphase volumetric fraction are important in this area of studies. Related research work started since the 1980s. The measurement of component ratios in multiphase flows using γ ray attenuation was first suggested by Abouelwafa and Kendall (1980)[1], and the technique has been used in many current commercial multiphase metering systems. Single energy γ ray technique working as a densimeter is satisfactory in two components measurement of gas/liquid or oil/water pipe-flow if not considering the concentration distribution of the different components in the cross-section. But in oil-water-gas three phase flow, the situation is different. Dual energy γ ray techniques have developed rapidly in the last decade, the different attenuation properties of the three phase's media in the oil pipe are used to deduce the components fraction information. Dual energy γ ray technique has been applied to cases with three basis materials having distribution of linear attenuation coefficients with an obvious variance. It is proved to be a very promising technique for the purpose of simple and fast estimation of the volumetric fraction of oil-water-gas multiphase flow, and it becomes the constituent part of radial base multiphase meter[2].

In this paper, the nuclear instrumental design for a dual energy γ ray system is presented for oil-water-air three-phase pipe flow.

DESIGN PROCEDURES

In the instrumental design of dual energy gamma ray system, the important parameters to be considered are

- selection of detector;
- design of preamplifier;
- design of filter and shaping amplifier;
- design of DC base shift correcting circuit;
- design of dual channels analyzer;
- design of programmable pulse counter data acquisition system.

CP914, *Multiphase Flow: The Ultimate Measurement Challenge,* edited by X. Cai, Y. Wu, Z. Huang, S. Wang, and M. Wang

Selection of Detectors

The dual energy γ ray technique is based on the measurement of materials' attenuation coefficients with respect to two radioactive isotopes of emission energies 59.5keV and 662keV. An important step in the γ ray system design is the selection of scintillation detectors. Two parameters are important while selecting the scintillation detectors: detection efficiency and decay constant. High detection efficiency is required to reduce the source strength and short decay constant is required for high count rate to avoid pulse pile up or saturation when the system is operated in count mode. NaI (TI) crystal is the most commonly used scintillator for its highest detection efficiency. Various scintillator properties are tabulated in Table 1.

TABLE 1. The properties of Some inorganic scintillator crystals

Scintillator	Decay constant (us)	Density	Efficiency (%)
NaI (TI)	0.23	3.67	100
CsI (Na)	0.63	4.51	85
CsI (TI)	1.0	4.51	45
CsF	0.005	4.11	3
LiI (Eu)	1.4	4.08	35
CaF2 (Eu)	0.9	3.19	50
BaF2	0.63	4.88	10
KI (TI)	0.24/0.25	3.13	24
Ps	0.002	1.05	5

The function of the detector is to produce a signal for every particle it has recieved. Every detector works on the interaction of particle with matter. A scintillation detector consists of a scintillator and a photo multiplier tube (PMT). Scintillators may be materials of solids, liquids, gases, which produce spark or scintillate light when the ionizing radiation passes through them. The light produced in the scintillator is weak and hence must be amplified before being output as a pulse. A photomultiplier tube is used for multiplication of the light emitted by scintillator. Inorganic and organic scintillators are commonly used for scintillator detectors. Inorganic scintillators, however, are found to be superior to organic scintillators with respective response times of 10 ns and 1 ms. Inorganic scintillators respond much more quickly. NaI (TI) is the most commonly used inorganic scintillator for gamma rays. It has relatively high density and high atomic number, which makes it an extremely efficient gamma detector. Its light conversion efficiency is the highest of all the inorganic scintillators (100%).

The photo multiplier tube or photo tube is an integral part of a scintillation counter. Without the amplification produced by the PMT, a scintillator is not useful for radiation detection. A PMT is essentially a fast multiplier that amplifies an incident pulse of visible light by a factor of 1000,000 or more in 1 ns.

A photo-tube consists of an evacuated glass tube with photocathode at its center and several dynodes in the interior (Fig. 1).

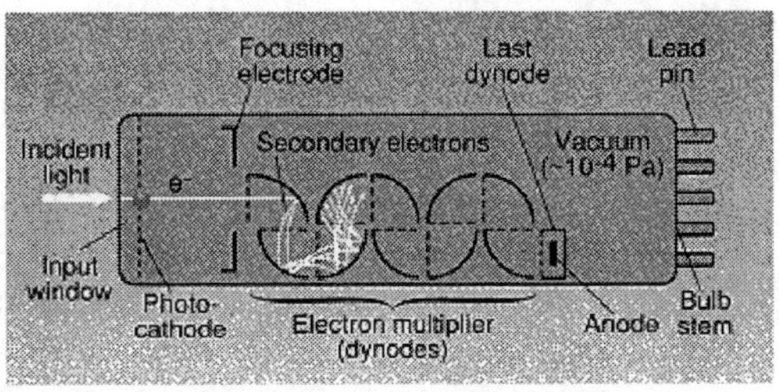

FIGURE 1. Schematic of a photomultiplier tube

The photons produced in the scintillator enter the photo tube and hit the photo cathode, which is made of a material of Cs–Sb compound (used as photocathode in most of the scintillation detectors because its surface has maximum sensitivity for electron emission)and emits electrons when light strikes it. The electrons emitted by the photocathode are guided by an electric field towards the first dynode, which is coated with a substance that emits secondary electrons, if an electron strikes it. The secondary emission rate depends on the applied voltage and on the type of surface. The secondary electrons from the first dynode move towards the second dynode, from there towards the third dynode and so on. The acceleration of electrons from one dynode to another is achieved by applying a successively increasing positive high voltage to each dynode. The voltage difference between two successive dynodes is of the order of 80–120V. Typical commercial photo tubes may have up to 15 dynodes.

The Amplifier

The main purpose of the amplifier is to amplify the signal which comes out of the detector. The signals from the detector are very weak, in mV range. Before it can be recorded, it should be amplified by a factor of thousand or more which is done by the amplifier. In short, the amplifier increases the amplitude of pulse from detectors to a few volts as required by the analyzer that follows the amplifier with the provision for a gain selection and pulse shaping time constant.

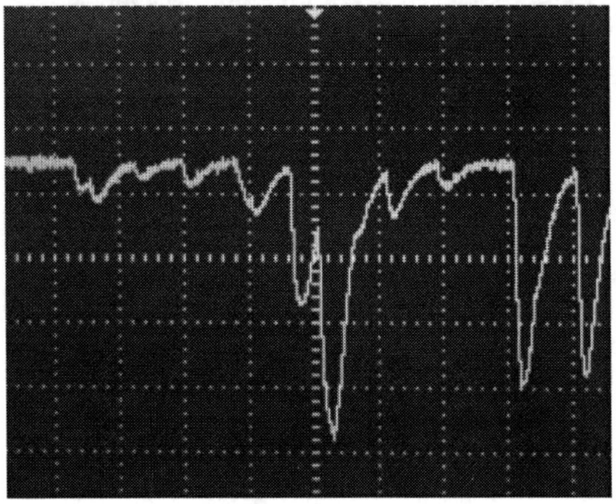

FIGURE 2. The pulse wave form

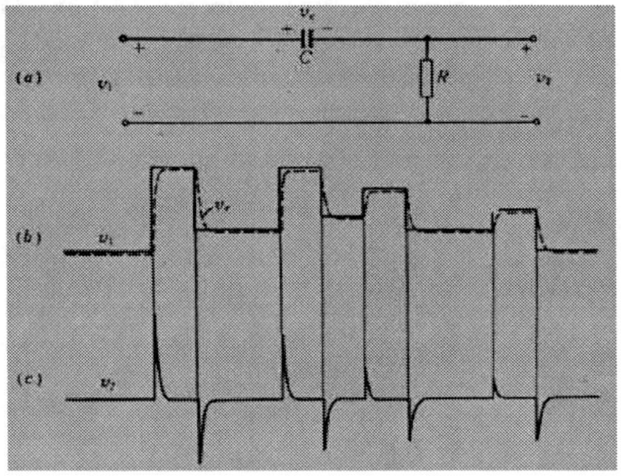

FIGURE 3. RC differential circuit

The pulse signal from amplifier is a series of negative pulses with a significant positive DC base shift, where the amplitude of pulses corresponds to the entered photon energy. The pulse wave form is shown in Fig.2.

DC Base Shift Correcting Circuit

As shown in Fig.3, the positive DC base shift ranges from 1 to 2 volts and takes an indefinite value, perhaps caused by pulse trail pile up, where the shift rate is proportional to the total pulse count rate, a high pulse count rate will cause higher shift voltage at the out of amplifier. In the dual energy gamma ray system, because of the high accuracy requirement of spectrum resolution, the DC base shift is thereby very harmful for pulse measurement, thence a DC base restoring circuit is needed to avoid the DC base shift.

In general, a differential circuit is used as DC isolation. A most common differential circuit is RC circuit, as shown in Fig.3. The circuit time constant RC is chosen to be much shorter than the signal pulse interval. The voltage of capacitance drops to the input signal base rapidly, then the DC base can be restored. Because the time constant RC is much shorter than the pulse assert time, the output pulse of RC circuit becomes narrow and sharp, Fig.3(c), which will make the later spectrum analyzer difficult. An improved CD circuit is proposed as in Fig.4(a). In the CD circuit, a nonlinear component of diode replaces the resistance in RC circuit, its working principle and working wave are shown in Fig.4. The positive resistance of diode is large when it is not leaded, which will produce a larger shift rate than RC circuit, Fig.4(c). In order to solve this problem, an operation amplifier is employed to reduce the inner resistance of the diode,Fig.4(d).

Filter and Shaping Amplifier

The main purpose of the filter and shaping amplifier is to improve the signal to noise ratio and to shape the pulse into a Gauss-like curve. Anything that the filter and shaping does here is actually the same thing in the frequency analysis that they are filtered some frequencies out from the signal. A pole-zero cancellation technique is used. The schematic is shown in Fig.5.

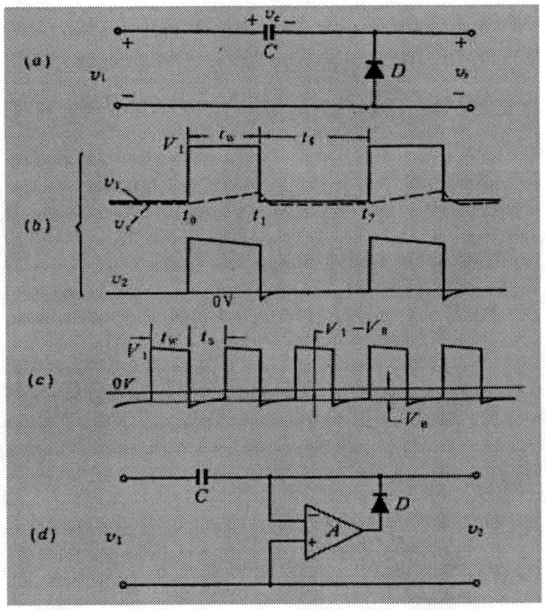

FIGURE 4. CD differential circuit

The output pulses from the shaping amplifier are inverted to positive pulses and the amplitudes are amplified to 0~8V. The pulses are shaped to a Gauss-like curve and are adjusted to only 1~2us wide to avoid pulse pile up for the case of high counts. (Fig.6.)

The pulse amplitude which ranges from 0 to 8V denotes the distribution of the energy spectroscopy as shown in Fig.6.

Dual Channels Analyzer

This is an important unit in the gamma ray spectrometer and does the job of amplitude selection (and hence energy selection) for output pulses from the amplifier. Analyzers have two narrow windows with window selection or lower limit and upper limit level selections. When the pulse goes out of the shaping amplifier, the lower and upper level selections operate, only the photons with energy above the lower threshold energy and below the upper threshold can be counted, thereby rejecting the unwanted low/high energy pulses. The present dual energy system is composed of two radioactive isotopes with emission energies of 59.5keV and 662keV. The dual energy spectroscopy is show in Fig.8.

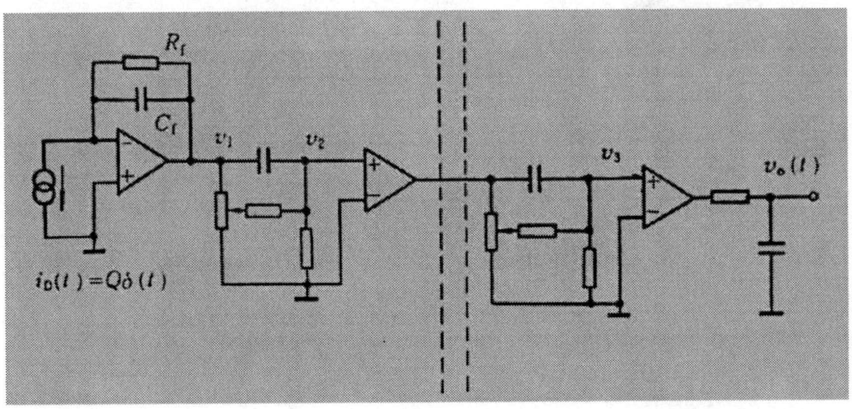

FIGURE 5. A pole-zero cancellation filter and shaping amplifier

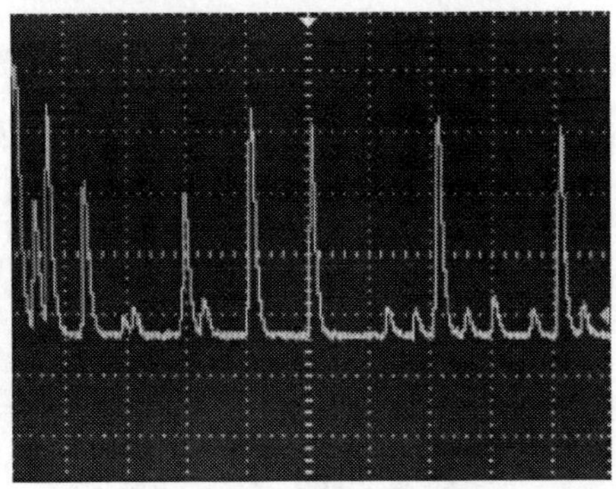

FIGURE 6. Pulses after shaping

With the doses of 100 mCi and 20 mCi, the peak of ^{241}Am is higher than that of ^{137}Cs , and the voltage is narrower , which means that the sensor of scintillator has higher energy resolution and is more efficient for ^{241}Am .

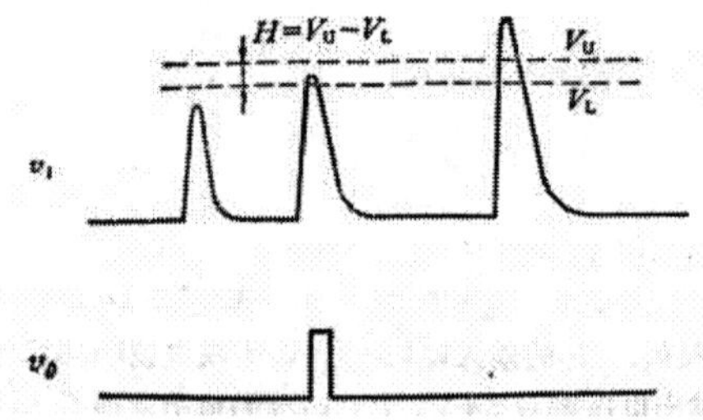

FIGURE 7. Principle of limits

The principle of the two window limits is shown in Fig.7. Only the pulse with amplitude between the low threshold and up threshold can get through, lower or upper pulses will be rejected. The limit circuit comprised of two high speed voltage comparators and associated logic circuits, two adjustable reference window voltages are provided.

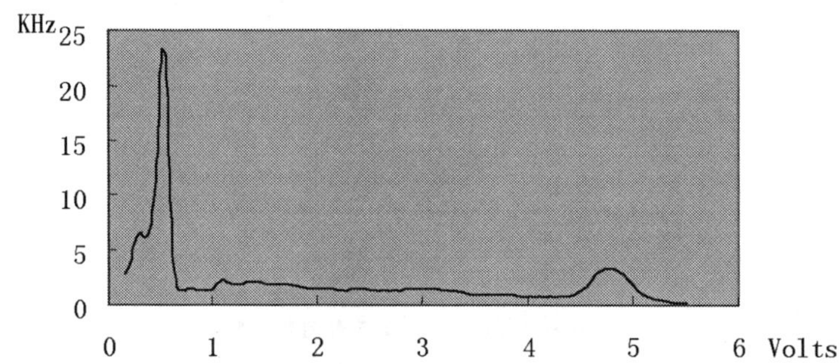

FIGURE 8. Am and Cs dual energy spectroscopy

Data Acquisition System

The data acquisition system is a programmable pulse counter system. Using an Intel Pentium industrial computer, it is equipped with eight boards of 32 channel 16 bit pulse counter in its ISA bus. Total 256 channels of counter are available for the data acquisition system. The counter latch time and counter bit length are programmable; the integral time for every channel is programmable as well. The programmable counter board is based on the chip of 8253, which have three channels of programmable interval timer/counter. It is a commonly used chip in this field, cheap and stable. In order to avoid the pulse deformation (only 100~200ns), digital logic isolation is necessary for the transmission from shaping amplifier to the counting board. An ultra high speed (10MBd) photo coupler 6N137 is employed in the board, therefore, noise and disturbance are eliminated.

FIGURE 9. Data acquisition system

Another important part in the data acquisition system is the software that enables users to select above control parameters, acquire the data and makes further calculations.

This is a conventional way of data acquisition based on the technique of photons pulse counter, corresponding to the multi-CPU data acquisition system. It has a simple structure and a low data flow rate from instruments to computer.

TEST CURVES

Some experiments were carried out on multiphase flow loop in IOM. The test curves on oil-water-gas three phase flow were obtained. A test curve of slug flow is shown in Fig 10. The dispersed area is bubble, and the continuous area is liquid phase. The left diagram is the signal of ^{241}Am ; the right diagram comes from ^{137}Cs channel.

When the bubble passes through the test section, the material attenuation is reduced; then the count rate rises, when the liquid fills the whole test section of the pipe, the material attenuation reaches its maximum, then the count rate drops down. In the ^{137}Cs channel, the signal characteristic is the same as ^{241}Am channel but its total attenuation is less, because of higher emission energy.

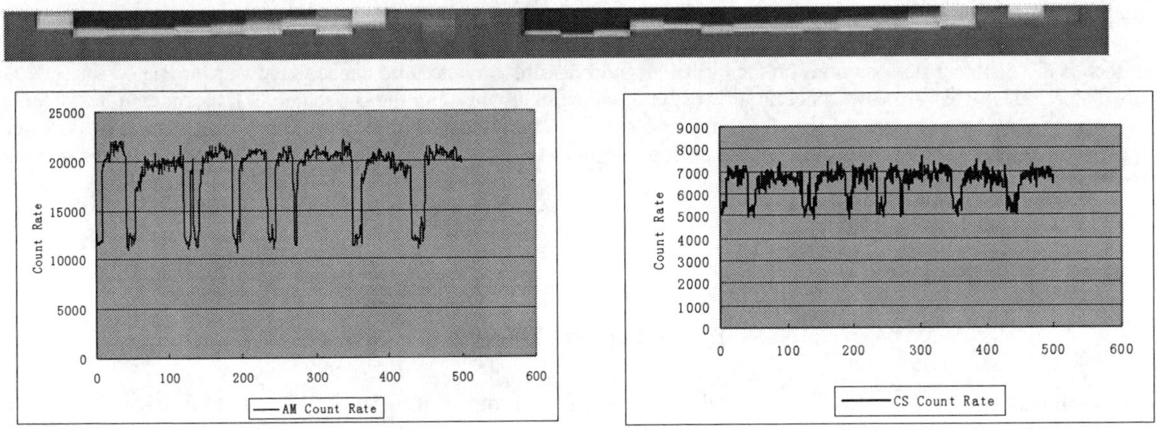

FIGURE 10. Attenuation curve

CONCLUSIONS

In case of a pulse-type dual energy gamma ray system, the output consists of voltage pulses; one pulse per particle is detected. It consists of instruments including scintillator detector, associated amplifiers, dual channel analyzer and data acquisition system.

There are several key points in the nuclear instrumental designs, including DC base shifting and pulse shaping. An appropriate treatment of above questions will increase the accuracy of the dual energy gamma system.

In general, the pulse-type design is a conventional way based on the technique of photons pulse counter; it is simple in system structure and has low data flow

REFERENCES

1. M.S.A.Abouelwafa and E.J.M.Kendall, "The Measurement of Component Ratios in Multiphase System Using γ-ray Attenuation", *J. Phy. E: Sci. Instrum.* **13**, 341(1980)
2. T. Grassler and K.E. Wirth, "Dual-Energy X-Ray Tomography in Process Engineering–A Non-Intrusive Technique to Characterize Vertical Multiphase Flows" 2nd World Congress on Industrial Process Tomography, Hannover, Germany 2001.
3. U. Parasu Veera, *Chemical Engineering Journal* **81,** 251–260 (2001)
4. W.JJ, Nuclear Electronics. Qinghua University

Capacitance Sensor Design For Refrigerant Two-Phase Flow Characterization

Hugo Canière*, Christophe T'Joen*, Arnout Willockx*, and Michel De Paepe*

*Department of Flow, Heat and Combustion Mechanics, Ghent University – UGent
St.-Pietersnieuwstraat 41, 9000 Gent, Belgium

Abstract. A capacitance sensor configuration has been designed to characterize horizontal two-phase flow of refrigerants, based on the difference in electrical permittivity of the vapour and liquid phase. A concave electrode design with active guarding was chosen. Finite element simulations were made for optimizing the design. The thickness of the dielectric layer separating the two-phase flow from the electrodes is an important design parameter and was investigated in detail. The effect of the different flow regimes on the electrical field distribution and total capacitance was simulated for several inscribed electrode angles. An electronic circuit was built to accurately measure the capacitance. Experimental test results with air-water flow in transparent 9mm tubes are presented. Experimental results with refrigerants are necessary for further development of objective and accurate flow regime transition and two-phase flow pattern maps during evaporation or condensation of refrigerants.

Keywords: Capacitance sensor, Two-Phase Flow, Refrigerants, Flow Pattern Detection

INTRODUCTION

During evaporation or condensation of refrigerants in air-conditioning units, horizontal two-phase flow is encountered in smooth or microfinned tubes. Pressure drops and heat transfer coefficients of two-phase flow are strongly dependent of the flow pattern of the two-phase flow. Therefore, objective and accurate flow regime prediction of refrigerants in air-conditioning units is of great importance for designing the circuit layout of the evaporator/condenser.

In order to refine and expand current flow pattern maps, capacitance measurements methods can be used for flow pattern identification because the capacitance value is strongly dependent of the vapour quality and distribution in the cross section of the tubes. These methods are under development for a long time, most of them focusing on air-water or oil-water flows: e.g. Elkow and Rezkallah [1], Tollefsen and Hammer [2] and Jaworek and Krupa [3]. Many types of electrodes were investigated in the past. Helical and double helical electrodes, as well as concave configurations, were developed, mainly for volumetric (void fraction) measurements, e.g. Sami et al. [4] and Geraets and Borst [5]. Concave electrodes were chosen for this study because of simplicity of construction and because of the possibility to acquire local (i.e. almost cross-sectional) measurements. This fits in with the approach of local and flow pattern based heat transfer and pressure drop measurements and models for refrigerants (Thome [6]).

This study investigates the necessary adaptations of capacitance measurements methods for use with refrigerant two-phase flows. A sensor design was made, based on literature data, finite element simulations and experimental tests results. The sensor is able to characterize two-phase flow regimes. The local and dynamic capacitance signals can now be used to further develop objective flow pattern identification, using statistical signal processing tools like probability density estimations and power spectral density analyses.

CP914, *Multiphase Flow: The Ultimate Measurement Challenge,* edited by X. Cai, Y. Wu, Z. Huang, S. Wang, and M. Wang

SENSOR DESIGN CONCEPT

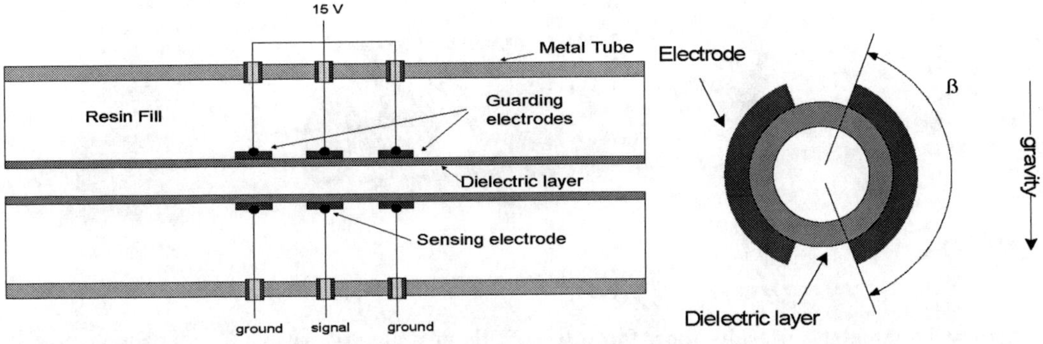

FIGURE 1. Sensor scheme with cross section

A concave electrode configuration was chosen because of the possibility to acquire local capacitance measurements. For that, the width of the electrodes along the flow direction may not be larger than the local flow characteristics. The sensor must be able to track plugs and slugs and to detect smaller bubbles of vapour. On the other hand, smaller electrode width results in lower capacitance. A compromise has to be made. The inner tube diameter is fixed at 9mm as a design specification.

To ensure a proper field line distribution in the direction of the flow, an active guarding technique is used (Reinecke and Mewes [7]). As shown on Fig. 1, three pairs of concave electrodes are placed in a row. The middle pair is the sensing pair, the outer pairs are guards. These guards are connected at one end to the source (which is a 15V block signal) and at the other end to the ground. The space between the electrodes is again equal to the inner tube diameter. To eliminate influences from the environment, the outer metal tube is grounded as well.

The sensor should be carefully mounted with the electrodes positioned one at each side of the tube as shown on the right of Fig. 1. To prevent contact with the fluid, a dielectric layer should be placed between the electrodes and the fluid, to form a tube wall. The thickness of this layer is an important parameter in the design. A series of finite element simulations were made to determine optimal inscribed electrode angle β, to account for phase distributions, to determine the effects of fluid properties and the influence of the thickness of the dielectric layer.

FINITE ELEMENT SIMULATIONS

To optimize the design a finite element method [8] for solving the Poisson's equation (3-1) was used. The electrical potential, field distribution and capacitance between the electrodes can be determined for the geometry of the sensor and the properties of the fluids and materials using equations (3-2) and (3-2),

$$-\nabla\cdot\left(\varepsilon(\vec{x})\nabla u(\vec{x})\right)=\rho(\vec{x}) \tag{1}$$

$$\vec{E}(\vec{x})=-\nabla u(\vec{x}) \tag{2}$$

$$C=\frac{1}{U}\int_S \varepsilon_0\varepsilon_r\left(\vec{E}\cdot\vec{n}\right)dS \tag{3}$$

with ε_0 = 8.8542 x 10^{-12} [F/m] the dielectric constant in the free space. The dielectrical constants used in the simulations were ε_r = 77.5 for water, ε_r = 6.1 for R22(liquid) and ε_r = 1 for air and R22(vapour). The voltage between the source and detector electrode U was set at 15V. The distribution of liquid and vapour in the cross section of the tube was simulated with the flow patterns of Fig. 2. The thickness and permittivity of the dielectric layer was varied as well as the inscribed electrode angle, to investigate their effect on electrical field distribution and capacitance.

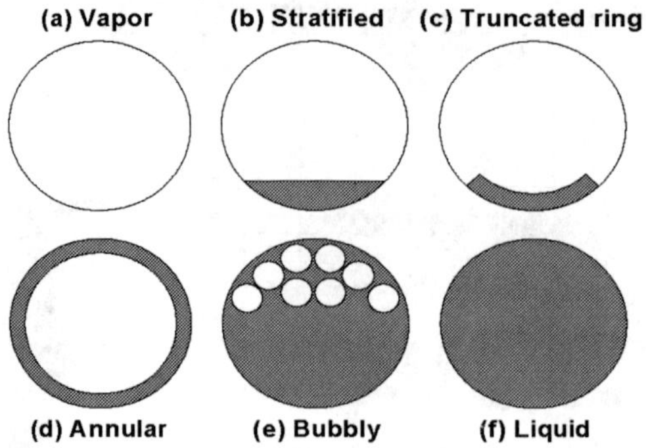

FIGURE 2. Flow structures

Figure 3 shows the simulation results for a 9mm tube without dielectric layer. The capacitance of the vapour-liquid mixture is plotted against the cross sectional void fraction. This void fraction is defined by the ratio of the cross section of the tube occupied by gas, to the total cross section of the tube. A similar effect to the flow patterns is obtained for air-water (left) and the refrigerant R22 (right). Because of the difference in dielectrical constant, the total capacitance of R22 is one order smaller than air-water.

By decreasing the inscribed electrode angle β, the maximum capacitance for air-water drops from 12pF for $\beta=160°$ to 8pF for $\beta=120°$. For this, β should be as high as constructively possible. On the other hand, more linearity is obtained for $\beta=120°$. But concerning flow pattern detection, the difference between the different flow patterns should be as big as possible, which finally favors β as high as possible. When interpreting Fig. 3, there should be mentioned that stratified flows at high void fraction and annular flow at low void fraction, are no realistic flow patterns.

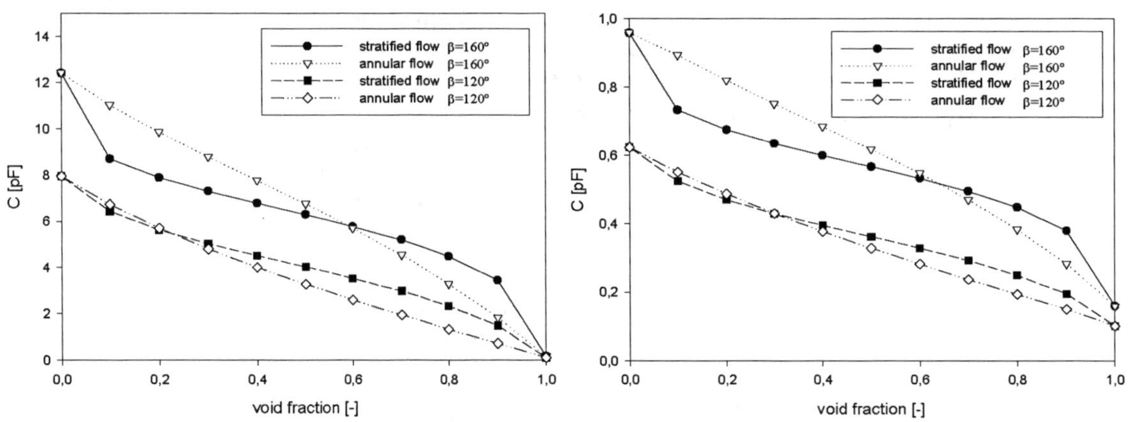

FIGURE 3. Capacitance for different flow patterns (left) air-water (right) R22

For the clearness of the figures, the bubbly and truncated ring patterns are not indicated. Most of the results fall within the curves for annular and stratified flow. This makes the deviation between annular and stratified flow an indicator for optimal flow pattern detection.

Furthermore, it is known that the thickness and relative permittivity of the dielectric layer between the fluid and the electrodes are very important design parameters, especially for media with a high relative permittivity such as water [5]. Connecting capacitances in series, the smallest capacitance will be dominant. The total capacity will be of the order of the smallest capacitance in the series and this smallest capacitance will get the biggest drop in electric potential. For that, the capacitance of the dielectric layer may not be too small, otherwise the sensor won't be able to see the inside of the tube and would only measure the capacitance of the dielectric layer.

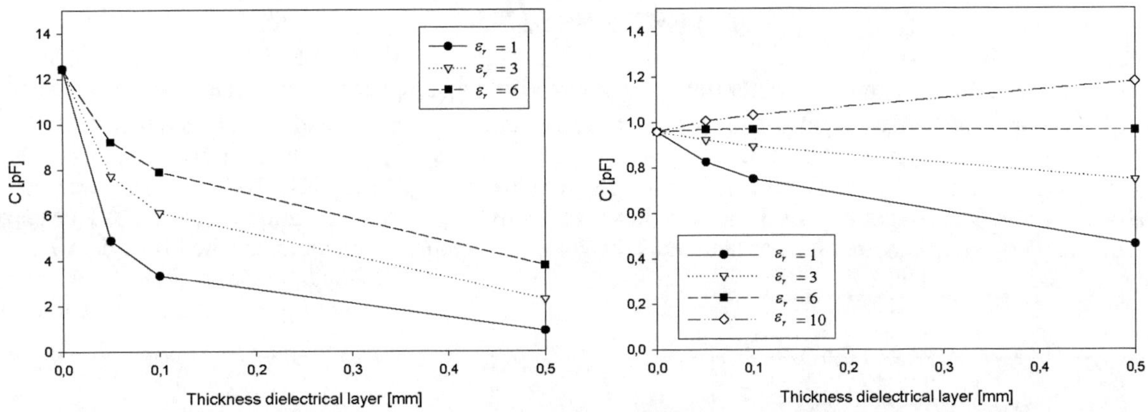

FIGURE 4. Influence of dielectric layer thickness for β=160° (left) air/water (right) R22

Figure 4 shows the simulation results for the influence of the thickness of the dielectric layer for air/water flow and R22, with β=160°. The dielectric constant of the dielectric layer was set at $\varepsilon_r = 3$. Especially for air-water, the thickness of the dielectric layer should be as thin as possible. For R22, and for refrigerants in general, the dielectrical properties of both the fluid and the layer are of the same order of magnitude. As a result the thickness is not as critical as for air/water. For that, the same sensor design can be used for both air-water flow and refrigerant flow.

Figure 5 shows an electrical field density plot with equipotential lines of a 'bad match' between dielectric layer and fluid properties at the left, and the 'final design' at the right. The inscribed electrode angle for both simulations is 120° and the source signal is 15V.

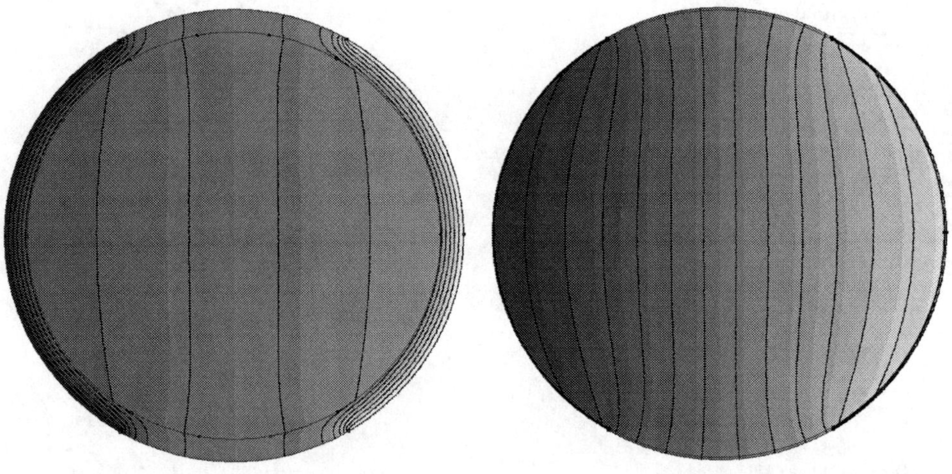

FIGURE 5. Density plot (20colors) and equipotential lines (1V) (left) a 'bad match' – configuration and (right) the 'final design'

ELECTRONIC CIRCUIT

To measure the capacitance, an electronic circuit is needed to obtain a voltage output for the DAQ system. Measured capacitances are small [pF] and amplification is recommended. The circuit of Yang and Yang [9] was chosen and built for that purpose. It is a four-phase charge/discharge measuring circuit. It has an adjustable offset, good stray-immunity and high stability. Measurements can be made at 2 frequencies, 1MHz and 100kHz, resulting in two measurement ranges: 0-2pF for refrigerants and 0-20pF for air-water respectively. A block signal of 15V is connected to the three electrodes, to charge and discharge them. The middle electrode of Fig. 1 is the sensing electrode and is connected to this measurement circuit. Measurements are taken at a sample frequency of 1kHz.

TEST RESULTS

The sensor was tested for horizontal air-water flow. Air coming from a compressor is expanded over a valve and mixed with water pumped out of a tank into a closed loop. Flow rates were controlled as well as temperature and pressure. On Fig. 6, time signals of four different flow regimes are plotted. On the ordinate a dimensionless volt signal, V*, is shown. This is the measured volt signal divided by the voltage of liquid-only-flow at the reference temperature of 25°C. The temperature of the air was lower than this reference temperature (T_{air}=13°C). This makes it possible that the V*-signal becomes greater than unity, because the mixture temperature can be lower than T_{ref}, resulting in a higher capacitance.

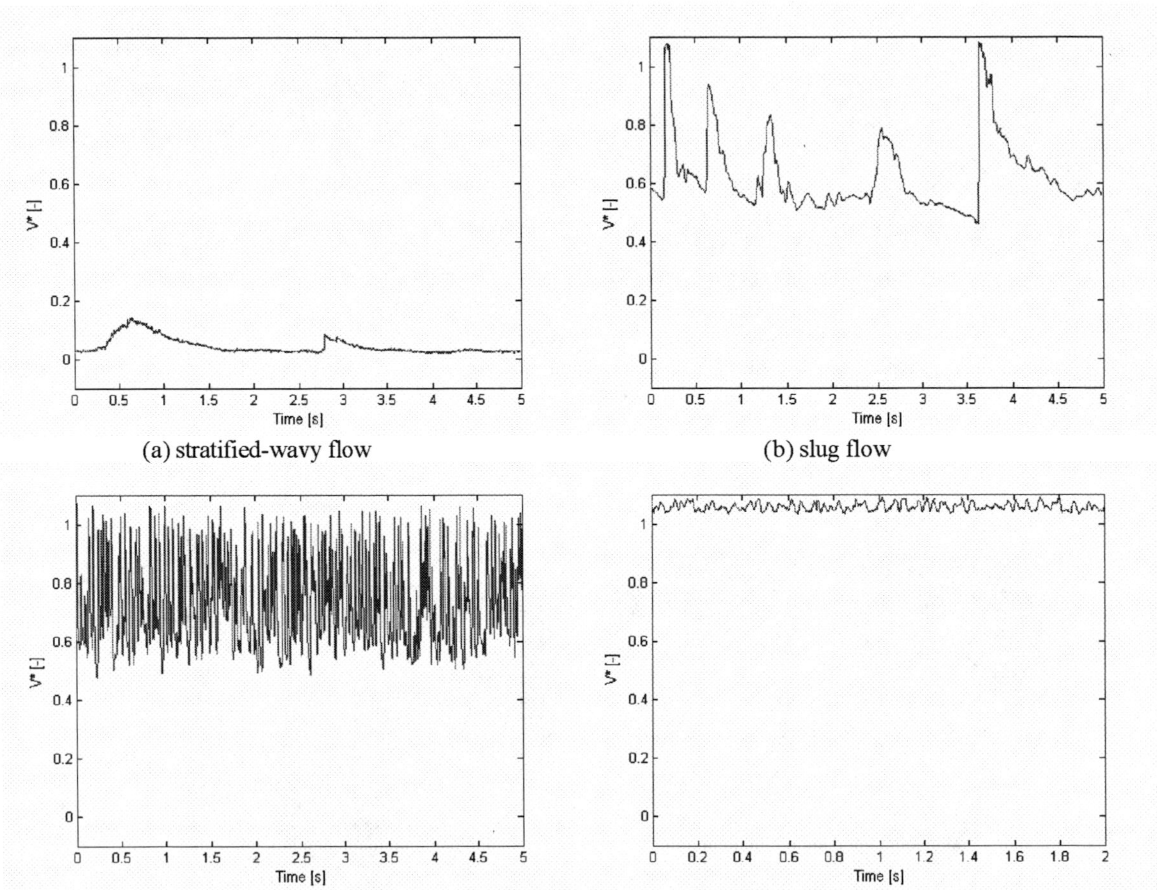

(a) stratified-wavy flow (b) slug flow

(c) annular flow (d) bubbly flow

FIGURE 6. Time plot of sensor signals

Figure 6(a) shows a stratified-wavy flow at near atmospheric pressure (G_{water} = 22.8 kg/m²s and G_{air} = 51.8 kg/m²s). A slow varying liquid level is clearly visible. On Fig. 6(b) the signal of a slug flow is presented (G_{water} = 22.8 kg/m²s and G_{air} = 2.9 kg/m²s). This flow regime is characterized by liquid slugs frequently washing the top of the tube. Annular flow (G_{water} = 59.1 kg/m²s and G_{air} = 47.9 kg/m²s), shown on Fig. 6(c) is characterized by a high V*-signal with disturbances in the highly turbulent liquid film. Finally, Fig. 6(d) shows a bubbly flow (G_{water} = 2230 kg/m²s and G_{air} = 3.1 kg/m²s). The sensor is able to detect the different horizontal two-phase flow regimes. Further analyses of the signal with statistical signal processing tools are under development to determine objective flow pattern transitions.

ACKNOWLEDGMENTS

The authors would like to express gratitude to the BOF fund (B/06634) of the Ghent University-UGent which provided support for this study.

REFERENCES

1. K. Elkow and K. Rezkallah, "Void fraction measurements in gas-liquid flows using capacitance sensors", *Measurements Science & Technology* **7**, 1153-1163 (1996).
2. J. Tollefsen and E. Hammer, "Capacitance sensor design for reducing errors in phase concentration measurements", *Flow Measurement and Instrumentation* **9**, 25-32 (1998).
3. A. Jaworek and A. Krupa, "Gas/liquid ratio measurements by rf resonance capacitance sensors", *Sensors and Actuators* **113**, 133-139 (2004).
4. M. Sami, A. Abouelwafa, E. John and M. Kendall, "The use of capacitance sensors for phase percentage determination in multiphase pipelines", *IEEE Transactions on Instrumentation and Measurement* **29**, 24-27 (1980).
5. J.J.M. Geraets and J.C. Borst, "A capacitance sensor for 2-phase void fraction measurement and flow pattern identification", *International Journal of Multiphase Flow* **14**, 305-320 (1988).
6. J.R. Thome, "Update on advances in flow pattern based two-phase heat transfer models", *Experimental Thermal and Fluid Science* **29**, 341-349 (2005).
7. N. Reinecke and D. Mewes, "Multielectrode capacitance sensors for the visualization of transient two-phase flows", *Experimental Thermal and Fluid Science* **29**, 341-349 (2005).
8. Finite Element Method Magnetics, Version 4.0, 2004, David Meeker, dmeeker@ieee.org
9. S.X. Yang and W.Q. Yang, "A portable stray-immune capacitance meter", *Review of Scientific Instruments* **73**, 1958-1961 (2002).

Experimental Investigation on Electrostatic Spray of Twin-fluid Atomization

Junfeng Wang[+ *,1], Wook Ryol Hwang[*], Tiqian Luo[+]

[+]*School of Energy and Power Engineering, Jiangsu University, 212-013, Zhenjiang, China*
[*]*Research Center for Aerospace Parts Technology, School of Mechanical and Aerospace Engineering, Gyeongsang National University, 660-701, Jinju, South Korea*

Abstract. A kind of electrostatic atomizer of twin-fluid atomization with combined needle and annular electrode was developed, from which a relative high speed electrostatic spray are obtained. The flow filed of charged droplets from the electrostatic spray under different charging voltage was measured by PIV technique. The characteristics of atomization and motion of droplets of uncharged and charged spray was investigated systemically and analyzed by PIV and PDA systems. The flow field of the twin-fluid jet of electrostatic spray and droplet size distribution is visualized by PIV and PDA measurement technique. The results indicate that the flow field of spray is affected apparently by electrical field. The velocity in the spray center become higher and the flow field around the spray shows a change of fluctuation with the increase of charging voltage. The diameter of spray droplets on the centerline increased when the spray was charged and keep the trend of gradually increase from the center point to the edge of the spray. The spray expanded when it was charged and the spray angle become difficult to define since more particles distributed around the spray core.

Key words: Electrostatic atomization, Spray, PIV, PDA, charged multi-phase flow

INTRODUCTION

The charged multi-phase flow is a kind of complex flow phenomenon that exists in the industrial application fields of electrostatic atomization. Since electrostatic atomization was discovered over one hundred years ago, a large quantity of theoretical and practical research on the phenomenon has been reported [1-5]. Compared with conventional atomization techniques of liquids, the electrostatic atomization has some advantages as follows: easy to get ultra fine particles and easy handling of droplets' motion by changing the charging electrical field and avoiding coalescence of droplets due to electric charge of the same polarity on the droplets, a narrow size distribution of generated droplets, more deposition on the target and so on. Therefore, this technique has been applied in pesticide spray, disinfections, spraying coating, nano-particles or micro particles and thin film preparation and so on. The electrostatic atomization was also recently applied in the field of biology, for example, is applied in its

[1]Corresponding author: wangjunfeng@ujs.edu.cn

CP914, *Multiphase Flow: The Ultimate Measurement Challenge,* edited by X. Cai, Y. Wu, Z. Huang, S. Wang, and M. Wang

uniform-dispersion mode to prepare immobilized enzyme gel particles with small size [6]. What's more, electrostatic atomization is a potential technology in the field of spray combustion because it provides a novel way to improve the spray quality for fuel combustion so as to decrease the emission pollutant. But until now people still don't know exactly and clearly the affect of electrical field on the flow field of spray due to its complexity. With the spread of the application of the electrostatic spray, the requirement of more deeply understanding of the charged multi-phase flow has been put forward to the researchers.

Twin-fluid sprayer is widely applied for the atomization of high viscosity fluid in industrial field such as the atomization of heavy oil and coal water mixture. It is also sometime applied in agricultural spray due to high viscosity of the pesticide mixed with sugar to increase the deposition. Since any probe in the flow filed of charged spray will disturb the electrical filed and spray filed, almost at all the time, optical experimental measurement is a main method to discover the spray and motion characteristics of charged droplets in the charged multi-phase flow. Laser diagnostic technology such as LDA, PIV has been already applied in the former research of charged multi-phase flow [7,8]. Presently, PDA is also another more powerful technique to do more deeply research on the charged multi-phase flow. It could obtain not only the accurate velocity of atomized droplets but also the size of the corresponding droplets at the same time. In this paper, a Gas-liquid twin-fluid sprayer was used to generate electrostatic spray under combined annular and needled electrode. The flow field of charged droplets and spray characteristics was investigated by PIV and PDA. The measurement results and research conclusion represented in the paper will support a base of electrostatic spray application for the high viscosity fluid under twin-fluid atomization mode.

EXPERIMENTAL SYSTEM AND TECHNIQUES

In this paper, a kind of twin-fluid atomization nozzle was developed (see Fig.1.a) which was applied in the combustion of coal water mixture. Compared with simple pressure swirl atomization, the spray speed of

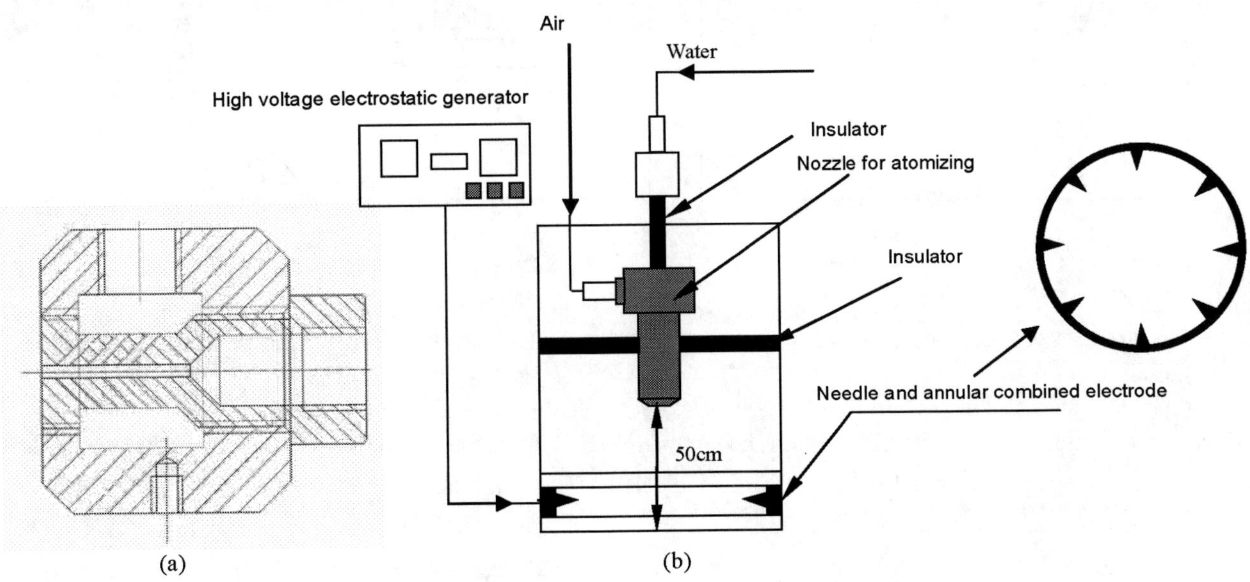

FIGURE 1. Schematic of (a) double fluid atomizer and (b) needle and annular combined charging sprayer

double-fluid atomization is relative higher especially in the spray core area because in the nozzles an additional fluid, compressed air, supplies most of the energy required for dispersing the liquid flow fed to the nozzle into droplets. Air assisted atomizers are especially effective when viscous liquids need to be atomized. To make the charging of the droplet more efficiently we applied the annular electrode combined with needles which can cause a strong electrical filed and make high speed fluid be charged relatively easily (see Fig.1.b). Such kind of atomizer is usually applied in the atomization of high viscosity fluid for example pesticide, heavy oil, coal water slurry et al. The distance between nozzle and electrode is main factor effecting the charging of the spray. We adjust the distance between nozzle and electrode of the sprayer in the present experiment to 50cm to get a satisfying charge to mass ration which could be measure by the equipments as shown in Fig.2 which is similar to the Faraday cage [11].

TABLE 1. Properties of fluid in the experiments.

Surface tension (N/m)	Electrical conductivity (S/m)	Electrical permittivity (C^2/N·M)	Dynamic viscosity (kg/m.s)	Density (Kg/m^3)
7.4×10^{-2}	4.5×10^{-5}	7.10×10^{-10}	9.0×10^{-2}	998

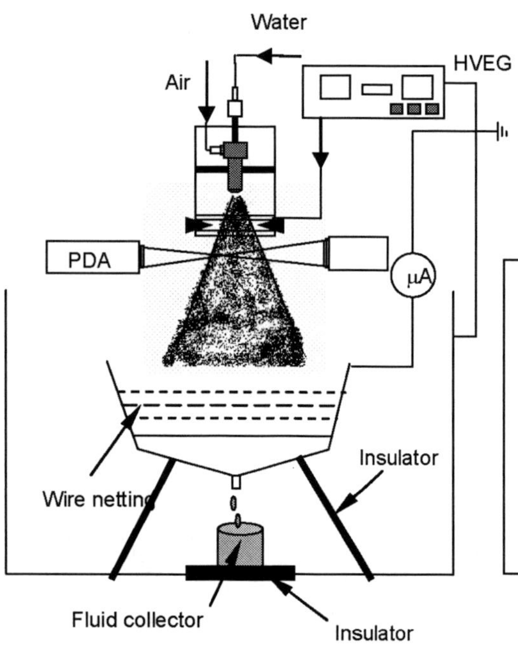

FIGURE 2. Sketch of Faraday Cage

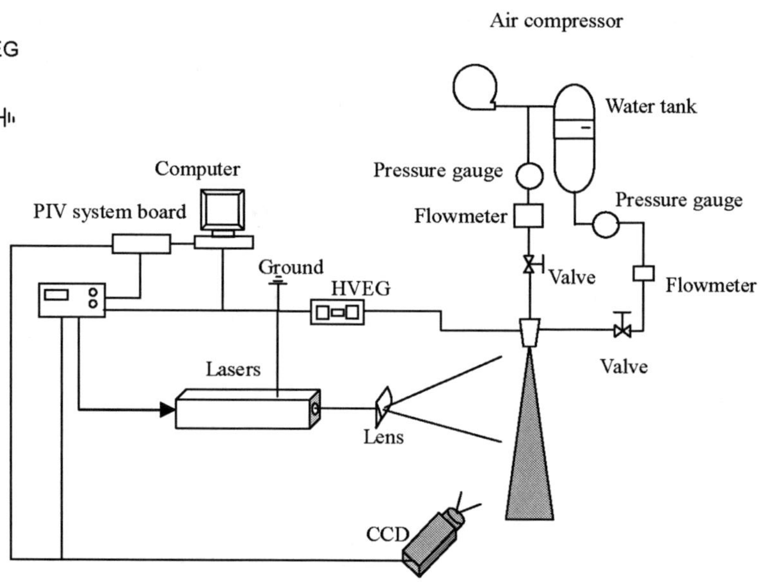

FIGURE 3. Sketch of charged spray and PIV measurement system

The charging system could reach more than 30kV on the electrode of needle and annular combination which is connected with high voltage electrostatic generator (HVEG). A tap water used as a test liquid has the following electrical and physical properties as shown in table 1. By changing the flow rate of water and gas, a stable spray could be obtained and in such situation the water pressure, the flow rate of water and the flow rate of gas are kept at 0.4Mpa, 2L/min and 2L/min, respectively. Fig.2 shows the PIV measuring system which used double-pulsed Yag laser to emit a pulse beam with a wavelength of 532 nm, 120mJ energy per pulse. The type of the CCD camera and Insight software are the same as Ref [11] has mentioned. The CCD camera has a total of 1280×1048 pixels and a Nikon 60mmf/2.8D Micro Nikker lens was used to accompany with the camera. The insight 6.0 software of TSI are applied in the PIV system, in which a direct cross-correlation interrogation method was used to analyses the spray image [5-6]. We obtain 120 photos which supports 60 flow fields for every spray case. The quantitative study of droplets and velocity was also carried out through Dantec 3D PDA system (Fig.2). The characteristic dimensions of the measurement system are given in Table 2. A t a certain flow rate and pressure, the flow filed of charged spray in case of uncharged, charging voltage of 5 kV, 10 kV, 15kV, 20kV, 25kV and 30 kV are measured by PIV and PDA system. For PIV measurement, the interrogation window size is 36×36 pixels and an observation domain of approximately 160×180mm^2 around the electrode in the spray direction is selected. Considering the symmetry of the round spray, the position of the reservation domain deflects to one side of the central axial in order to obtain more information of the flow around the spray.

TABLE 2. PDA system characteristics.

Description	Laser power	Light wavelength	Focal length	Off axis angle	Software
Value	5 W	457.9～514.5 nm	300 mm	60 °	BSA V2.1

RESULTS AND DISCUSSION

Spray Structure

Fig.4 shows the results of uncharged spray filed measure by PIV system. It can be found clearly that the velocity in the spray core is higher than the boundary part of the spray. The spray absorbs the fluid around it and makes it move along the downstream direction of the spray. Different with the pressure swirl spray atomization, the fluid in

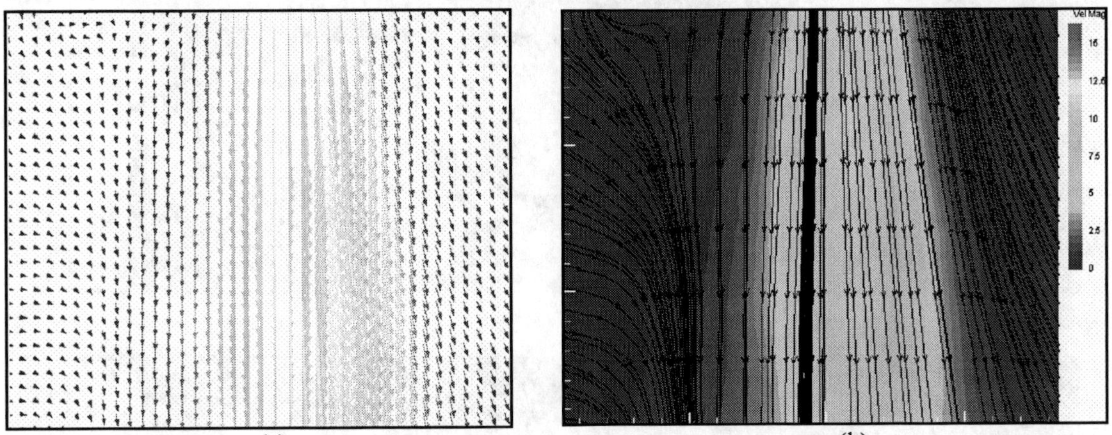

(a) (b)

FIGURE 4. Flow field of uncharged spray: (a) distribution of velocity vector; (b) velocity contour and streamline

the double-fluid atomizer is atomized mainly by the strong shear force of high speed gas and the maximum velocity appears in the center of the spray. The phenomena of fluid absorption by the spray and strong shear vortex are not very clear in the present reservation domain since it is more than 50mm far from the nozzle. The velocity contour and streamline of charged spray under different charging voltage rang from 5kV to 30kV are displayed in Fig.5 shows. The electrode exposed in the air begun to discharge when the charging voltage is more than 32kV. From Fig.

5 we could find that that spray was accelerated greatly when it was charged and the velocity in the center area increase linearly with the increase of charging voltage. The difference with swirl charged spray [11] is that the velocity distribution in the center of the spray didn't fluctuate and almost keep the same during the increase of charging voltage from 10kV to 30kV. The reason for the above mentioned phenomenon is probably that the velocity of double-fluid atomization in the spray center is much higher, which is decided mainly by the gas flow rate. The decreasing velocity distribution in the spray center along axial and horizontal can be seen clearly from Fig.5.

It is interesting that the flow of fluid around the spray show a fluctuating change. For analysis, a schematic flow structure of uncharged and charged spray is shown in Fig.6. When the spray was charged, the flow filed can be divided and named by A, B, C and D four zones. The zone A is the boundary area which represents the boundary area of spray core. When the spray was charged by 5 kV, the zone D and C appear apparently and B almost didn't exist in the flow structure. In the case of 10kV, the zone B appears together with the zone C and D. When the voltage increase to 15kV, the zone B expanded and the zone C disappeared in the observation area. When the voltage increase continuously to more than 15kV, the zone B and C appears again and keep almost the same situation with the increase of charging voltage.

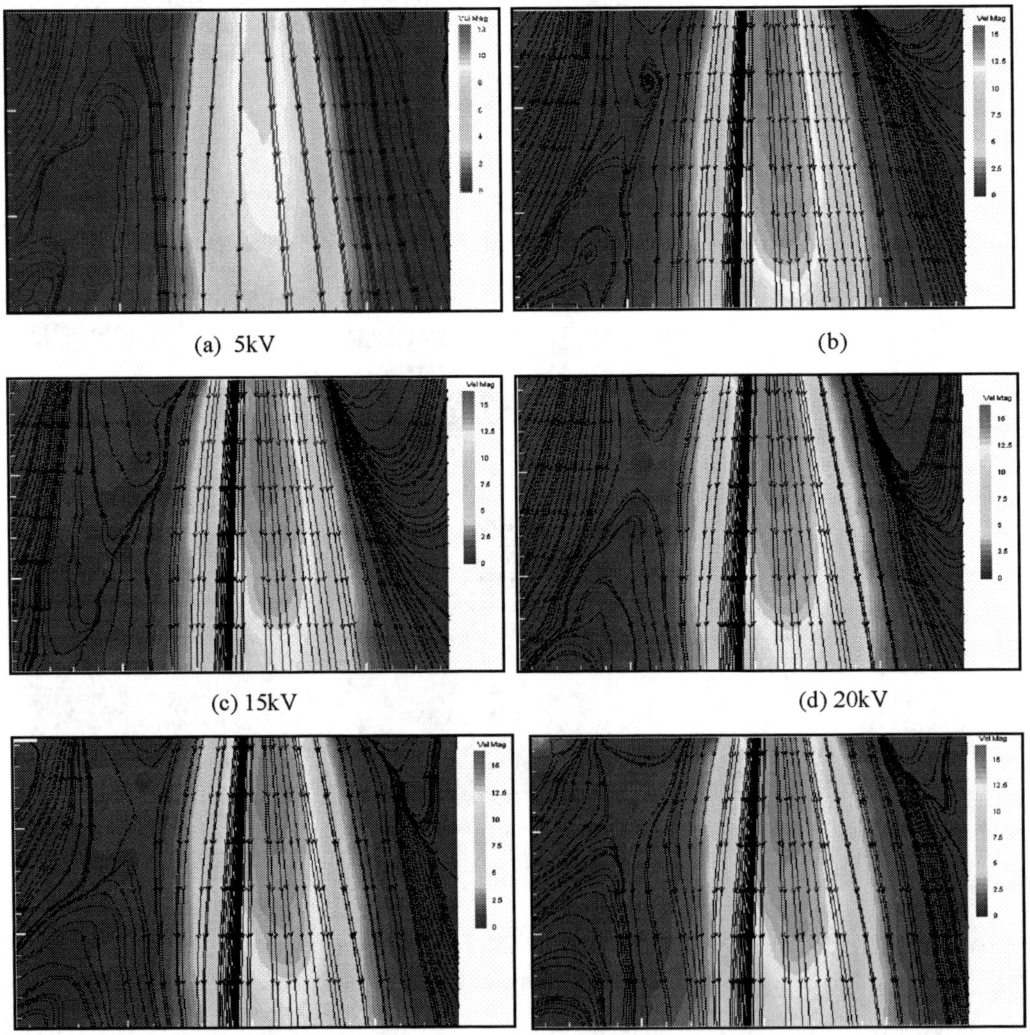

(a) 5kV (b)

(c) 15kV (d) 20kV

(e) 25kV (f) 30kV

FIGURE 5. Velocity contour and streamline under different charging voltage

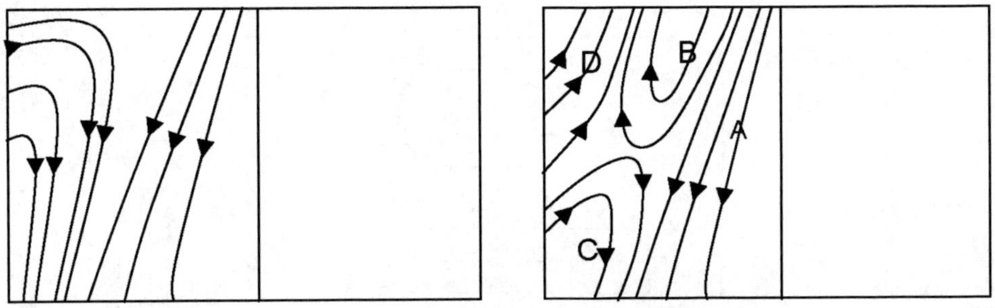

(a) Uncharged spray (b) charged spray

FIGURE 6. Sketch of the spray structure

Droplets Distribution

The Diameter distribution of uncharged and charged spray under different voltage was measure by PAD system. Here we only present several samples. Fig.7 shows the diameter frequency and the change of mean diameter with charging voltage of a point on the centerline with a distance of 70mm to the nozzle. It shows that the amount of small size droplets increase when the spray was charged. Fig.8 shows the diameter distribution in the spray center of uncharged and charged cases. It can be seen that the diameter of droplets on the centerline is the smallest and increase gradually along the horizontal axial direction. When the spray was charged, this trend still exists. But the droplet on the center line becomes larger and shows a trend of uniform distribution from the center to the edge of the

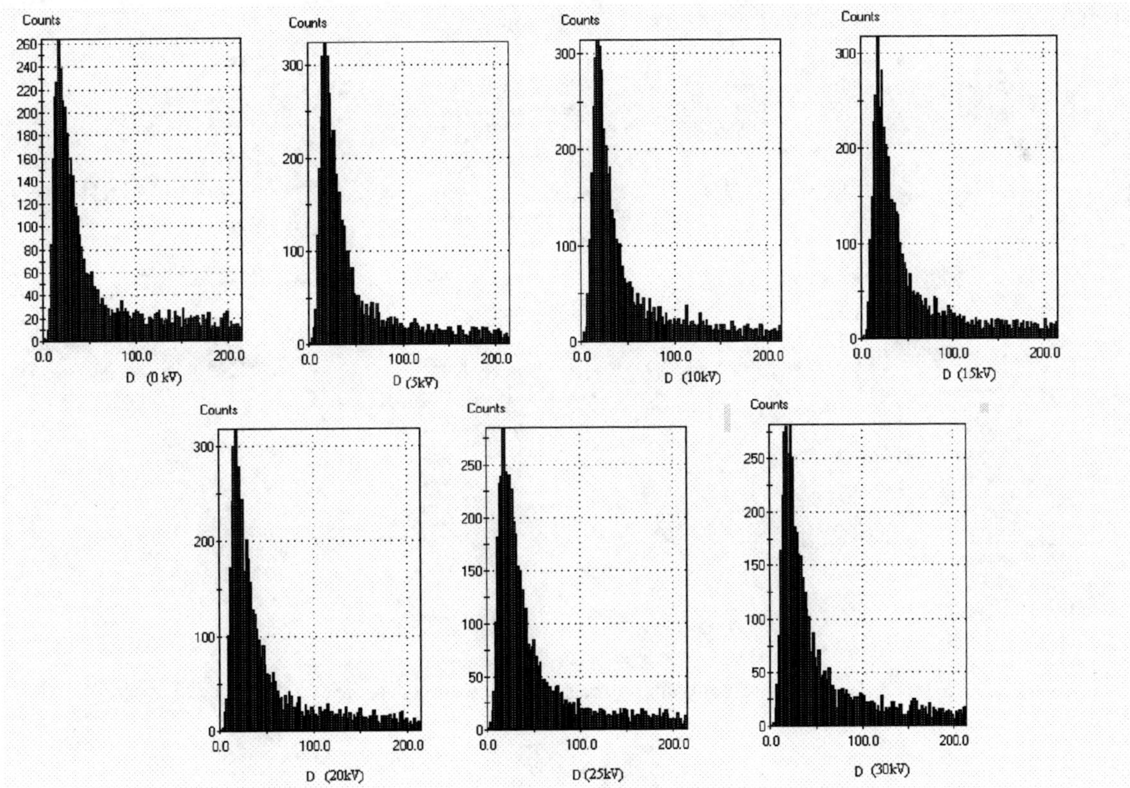

(a)

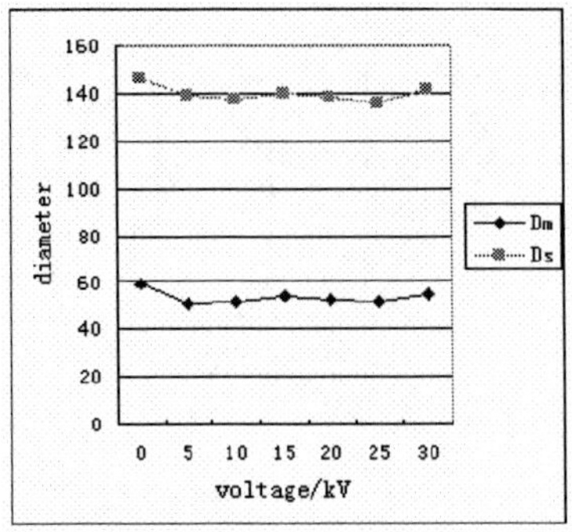

(b)

FIGURE 7. (a) Drop diameter distribution at the centerline (Z= 90mm). (b) Arithmetic Mean diameter (Dm) and SMD (Ds) versus voltage.

spray. It is interesting that the mean diameter of whole spray center increase when the spray was charged which is different with the case of one point on the centerline.

During the increase process of charging voltage, it could be found that more and more droplets began to move to the electrode and adsorbed on it because we found the quantity of dropping water from the annular electrode increase gradually. Whether these droplets flying to the electrode are small size particles? Is it also the reason why the mean diameter in the core of the spray increased when the spray was charged? Since the number of the droplets around the spray is too less to measure by the current PDA system, in the future we need develop some technique to measure the droplet diameter of dilute scattered particle around the field of electrostatic spray to explain the above phenomenon.

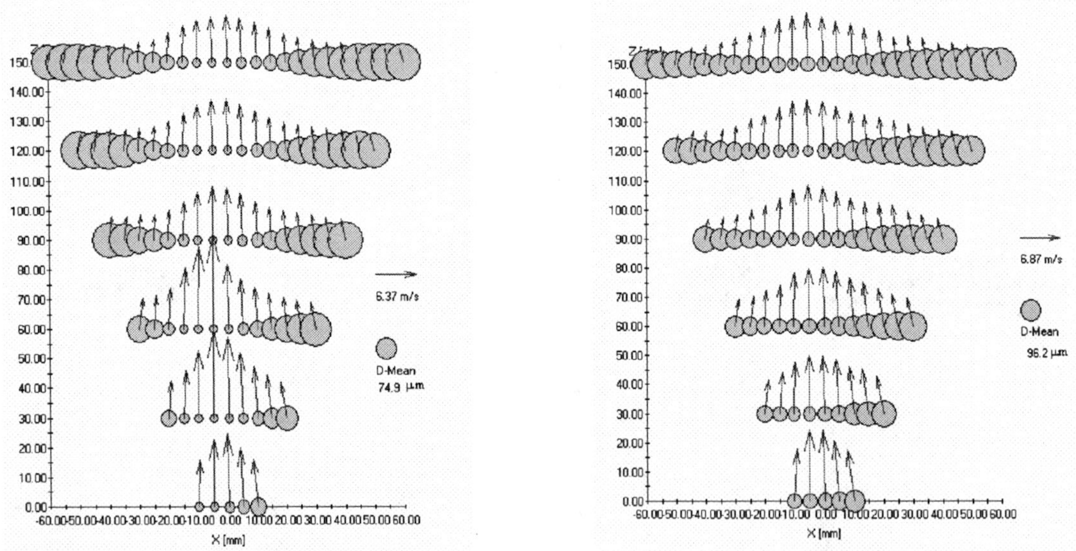

FIGURE 8. Droplets diameter and velocity distribution of (a): uncharged spray and (b): charged spray of 10kV.

CONCLUSIONS

The uncharged and charged spray of double-fluid atomization with needle and annular combined electrode was investigated by the measurement of PIV and PDA. The distributions of flow velocity under different charging voltage and droplet diameter distribution of uncharged and charged spray in the spray center are presented in this paper. When the spray generated by twin fluid atomization is charged, the spray expanded and more droplets distributed around the area of the spray core. So it is difficult to define the spray angle under such condition since the boundary of the spray become more illegible. The measurement and visualization indicate that, the flow field of spray is affected apparently by the electrical field. The velocity in the spray core become higher and the flow around the spray area shows a change of fluctuation with the increase of charging voltage. The diameter of spray droplets on the centerline increased when the spray was charged and keep the trend of gradually increase from the center point to the edge of the spray. The mechanism of the above phenomena needs to be investigated more deeply through more measurement and analysis.

ACKNOWLEDGEMENTS

This work was funded by the National Development and Innovation Committee of China (2003-2140) and Jiangsu Province Natural Research Foundation, China (No.03KJD4300064). Korea Research Foundation Grant funded by the Korean Government (KRF-2005-005-J09902) also supported the research partially.

REFERENCES

[1] W. Balachandran, P. Miao, P. Xiao. Electrospray of fine droplets of ceramic suspensions for thin film preparation. Journal of Electrostatics 2001;50: 249-263.

[2] S.A. Colberta, R.A. Cairncrossb. A discrete droplet transport model for predicting spray coating patterns of an electrostatic rotary atomizer. Journal of Electrostatics 2006; 64: 234-246.

[3] S. Edward Law. Agricultural electrostatic spray application: a review of significant research and development during the 20th century. Journal of Electrostatics 2001; 51-52: 25-42.

[4] H.Okuda, Amold J.Kelly. Electrostatic atomization-experiment, theory and industrial application. Phys. Plasmas3(5),1996

[5] L. F. Gaunta, John F Hughesa, Neale M Harrisonb. Electrostatic deposition of charged insecticide sprays on electrically isolated insects. Journal of Electrostatics 2003; 57:35-47.

[6] H. Watanabea, T. Matsuyama, H. Yamamoto. Experimental study on electrostatic atomization of highly viscous liquids. Journal of Electrostatics 2003; 57:183-197.

[7] J.F. Wang. A fundamental investigation of electrostatic spray combustion. PhD thesis, Jiangsu University 2002; China.

[8] H.H.Jin, J.F. Wang, Z. wang. LDA Research of Charged Particles. J Jiangsu University (in Chinese) 1999; 20(5):47-49.

[9] Cheng Jiusheng, Zhou Xiaonan, Zhen Jiuren,etc. Investigation of electrostatic spray for fuel ethanol. J Combustion Science and Technology 1999; 5(1):70-75.

[10]R.P.A. Hartman, J.-P. Borral, D.J. Brunner,etc. The evolution of electrohydrodynamic sprays produced in the cone-jet mode, a physical model. Journal of Electrostatics 1999; 47:143-152.

[11]J.F.Wang, W.R.Hawang, T.Q.Luo. PIV experimental study on electrostatic spray of swirl atomizer with annular electrode. Proceeding of the 5th International Symposium on Measurement Techniques for Multiphase Flows Dec. 10-13, 2006, Macau, China.533-539.

Electrostatic Introduction Theory Based Spatial Filtering Method for Solid Particle Velocity Measurement

Xu Chuanlong* Tang guanghua* zhou Bin* Yang Daoye*
Zhang Jianyong** Wang Shimin*

*Key Laboratory of Clean Coal Power Generation and Combustion Technology of Ministry of Education, Southeast University, NanJing, 210096, China.
**School of Science and Technology, University of Teesside, Middlesbrough, UK

Abstract. Electrostatic induction theory based spatial filtering method for particle velocity measurement has the advantages of the simplicity of measurement system and of the convenience of data processing. In this paper, the relationship between solid particle velocity and the power spectrum of the output signal of the electrostatic senor was derived theoretically. And the effects of the length of the electrode, the thickness of the dielectric pipe and its length on the spatial filtering characteristics of the electrostatic sensor were investigated numerically using finite element method. Additionally, as for the roughness and the difficult determination of the peak frequency f_{max} of the power spectrum characteristics curve of the output signal, a wavelet analysis based filtering method was adopted to smooth the curve, which can determine peak frequency f_{max} accurately. Finally, the velocity measurement method was applied in a dense phase pneumatic conveying system under high pressure, and the experimental results show that the system repeatability is within ±4% over the gas superficial velocity range of 8.63-18.62 m/s for particle concentration range 0.067-0.130 m^3/m^3.

Keywords: particle velocity; spatial filtering method; electrostatic sensor; wavelet analysis; gas-solid two phases flows

INTRODUCTION

Particle velocity is one of the important parameters describing gas-solid flow processes, and the continuous, on-line measurement of particle velocity in pneumatic pipelines is essential for the metering, economizing energy and control of solid particle transport process. At present, many non-contact velocity measurement methods based on different sensing theories have been investigated and developed, such as Doppler, cross-correlation, optical spatial filtering, NMR and trace method etc [1]. The most promising have been the cross correlation based velocity measurement methods, which correlate the signals derived from a pair of axially spaced sensors to make particle velocity measurement [2]. The non-obstructive measurement method has the advantages of broad measurement range and of good adaptability to hostile industrial environment. A wide range of sensing techniques including capacitive [3], electrostatic [4], ultrasonic [5] and radiometric [6] etc have been utilized. Some useful results have been obtained from the measurement systems based on these sensors, and a number of successful applications for the non-contact, on-line continuous measurement of solid particle velocity in pneumatic transport system have been achieved [7]. Although these methods have been widely applied to industrial processes for the velocity measurement, their applications need to satisfy the conditions that solid particle flow in pipeline is in steady state, particles are distributed homogeneously as possible and particle flow map passing from upstream sensor to downstream sensor should be frozen, the purpose of which assures that the cross correlation function has distinctive peak. But sometimes it is very difficult that these conditions can be satisfied during practical measurement [2]. In recent years, electrostatic sensing technique has created a major interest in the solid particle flow measurement, and its advantages over the other available correlation measurement techniques are well repeatable, reliable and low in cost [8]. Unfortunately the steady flow conditions and "frozen pattern map" should also be satisfied when using the technique [7], which limits its application in industry to some extent.

A mean particle velocity method based on the spatial filtering effect of a ring-shaped electrostatic sensor was proposed in the literatures [9-11]. Based on the frequency analysis and the determination of the peak

CP914, *Multiphase Flow: The Ultimate Measurement Challenge,* edited by X. Cai, Y. Wu, Z. Huang, S. Wang, and M. Wang

frequency of the output signal of the sensor, the new method realizes particle velocity measurement only using a single electrostatic sensor, which reduces the complexity of the measurement system and signal processing, and lowers the dependence of particle velocity measurement accuracy on "frozen pattern" flow conditions. But in these literatures, the Fast Fourier Transform (FFT) algorithm was directly used to analyze the sampled and digitized input signal and to determine the peak frequency f_{max}. Actually, the randomness of particles flow distribution in the pipelines would cause the random distribution of electrical charge carried by the moving particles (termed particle electrostatic flow noise), and therefore the output signal of electrostatic sensor is also a complicated random signal. Obviously, for a random signal, it is not correct to analyze its frequency spectrum directly using FFT. From the statistic characteristics of particle electrostatic flow noise and the frequency response property of electrostatic sensor, the statistic frequency spectrum of output signal of electrostatic sensor was investigated theoretically and further the relationship between the peak frequency f_{max} of the power spectrum and particle velocity was obtained, which provides the theoretical basis for mean particle velocity measurement. Electrostatic sensor acts as a low pass filter in the spatial frequency domain, and the geometric sizes of the sensor affect its spatial filtering characteristics, and thereby the resolution of particle velocity measurement. In this paper, the effects of the length of the electrode, the thickness of the dielectric pipe and its length on the spatial filtering characteristics of the electrostatic sensor were investigated numerically using finite element method in order to provide the theoretical guidance for the design of the electrostatic sensor. Additionally, as for the roughness of the power spectrum characteristics curve and difficult determination of the peak frequency f_{max}, an extraction method for the smooth trend of the power spectrum based on Wavelet analysis was introduced. Finally, the performance of electrostatic induction theory based velocity sensor was tested in a dense pneumatic conveying system under high pressure, and the experimental results are presented and discussed.

MEASUREMENT PRINCIPLE OF THE SPATIAL FILTERING METHOD

During the pneumatic transport process of powder particles, the charging of the particles occurs due to the collisions between particles, impacts between particles and pipe wall, and friction between particles and airflow. An electrostatic sensor can be used to detect the charge carried by the moving particles. The sensing electrode of electrostatic sensor is enclosed in the measuring head, which is illustrated in Fig. 1. The electrode is the main and most important part of electrostatic sensor, whose construction is also described in detail in the former works by J.B. Gajewski [12, 13]. As shown in Fig. 1, the length of the electrode is denoted by W_e, the radius of the screen by R_3, the inner radius of the pipeline by R_1 and the outer radius by R_2, the length of the dielectric pipe by l and its relative permittivity by ε_i. When the charged particles are distributed within the pipelines, the charge and potential are induced on the electrode owning to electrostatic induction. The total amount of the induced charge q on it can be calculated using the following equation due to the axis-symmetry of the measuring head [11]

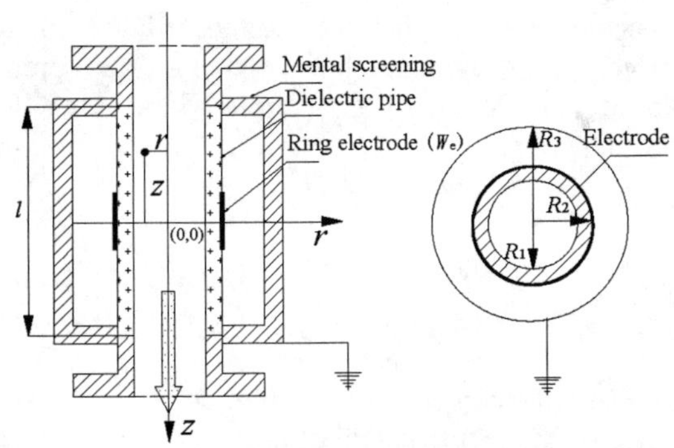

FIGURE 1. Measuring head cross-sections of electrostatic sensor

$$q = \iint i(z,r)\cdot s(z,r)dzdr \tag{1}$$

Where $i(z,r)$ is the surface charge density in the sensing zone of the sensor; $s(z,r)$ is the spatial sensitivity distribution function, which is defined as the induced charge on the electrode when a unity point charge is positioned differently at (z,r) in the sensing zone of the sensor.

In the practical measurement process, the fluctuation of the quasi-electrostatic field caused by the moving charged particles will lead to the change of the induced charge with time on the electrode. If the particles move only along z direction at the velocity v, the induced charge $q(t)$ can then be determined by

$$q(t) = \iint i(z+vt, r)\cdot s(z,r)dzdr \tag{2}$$

The "electrostatic flow noise" $i(z+vt,r)$ produced by the moving charged particles is a random function because of the spatial and temporal random distribution of the particles in the sensing zone of the electrode. From equation

(2), the charge $q(t)$ is an average one of the random signal $i(z+vt,r)$ with the weight function $s(z, r)$ and hence $q(t)$ is also a complicated random signal.

The autocorrelation function of $q(t)$ is defined as

$$\phi_q(\tau) = E[q(t)q(t+\tau)] \tag{3}$$

Then from equation (2), we have

$$\begin{aligned}\phi_q(\tau) &= E\left[\iint i(z+vt+v\tau,r)s(z,r)dzdr \cdot \iint i(\alpha+vt,\beta)s(\alpha,\beta)d\alpha d\beta\right] \\ &= \iint\iint \phi_i(z-\alpha+v\tau, r-\beta)s(z,r)s(\alpha,\beta)dzdrd\alpha d\beta\end{aligned} \tag{4}$$

Where $\phi_i(z,r)$ is the autocorrelation function of electrostatic flow noise $i(z,r)$. According to Wiener-Khintchine theorem, the power spectrum of $q(t)$ is expressed as

$$\begin{aligned}S_q &= \int_{-\infty}^{+\infty}\phi_q(\tau)\exp(-j2\pi f\tau)d\tau \\ &= \iint\iint\int S_i(f_z,f_r)\exp(j2\pi f_z(-\alpha+v\tau)-j2\pi f_r\beta)S^*(f_z,f_r)s(\alpha,\beta)\exp(-j2\pi f\tau)df_z df_r d\alpha d\beta d\tau \\ &= \iint\int S_i(f_z,f_r)\exp(j2\pi f_z v\tau)S^*(f_z,f_r)S(f_z,f_r)\exp(-j2\pi f\tau)df_z df_r d\tau \\ &= \iint S_i(f_z,f_r)|S(f_z,f_r)|^2\delta(f_z v-f)df_z df_r \\ &= \frac{1}{v}\int S_i(\frac{f}{v},f_r)|S(\frac{f}{v},f_r)|^2 df_r\end{aligned} \tag{5}$$

Where $S_i(f_z, f_r)$ is the power spectrum of electrostatic flow noise $i(z, r)$, $S(f_z, f_r)$ is the Fourier transform of the spatial sensitivity function $s(z, r)$, and $S^*(f_z, f_r)$ denotes the complex conjugate of $S(f_z, f_r)$. They are given respectively by

$$S_i(f_z,f_r) = \iint \phi_i(z,r)\exp(-j2\pi(f_z z+f_r r))dzdr \tag{6}$$

$$S(f_z,f_r) = \iint s(z,r)\exp(-j2\pi(f_z z+f_r r))dzdr \tag{7}$$

$$S^*(f_z,f_r) = S(-f_z,-f_r) \tag{8}$$

The electrode has to be connected to a preamplifier to amplify the induced charge $q(t)$. Electrostatic sensor is simplified as an equivalent circuit shown in Fig. 2. Where C_e and R_e are the equivalent capacitance and insulation resistance of the electrode, respectively; C_i and R_i are the equivalent input capacitance and input resistance of the preamplifier, respectively; $u_i(t)$ is the input voltage of the preamplifier. According to Kirchhoff's current law, the following expression applies

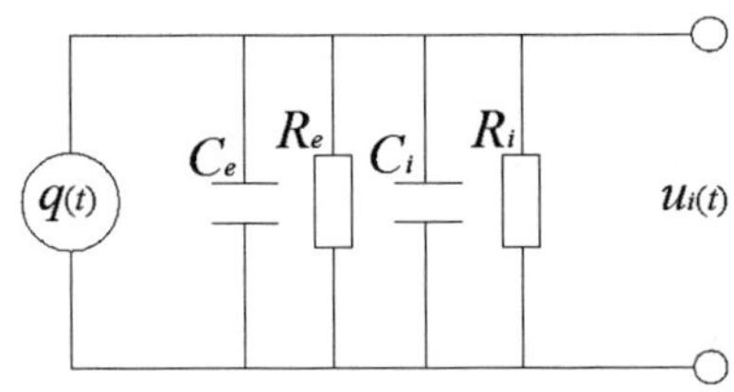

FIGURE 2. Equivalent circuit of electrostatic sensor

$$\frac{dq(t)}{dt} = C\frac{du_i(t)}{dt} + \frac{u_i(t)}{R} \tag{9}$$

Where $R=(R_e \cdot R_i)/(R_e+R_i)$; $C=C_e+C_i$. The application of the Laplace transform to equation (9) gives

$$U_i(s)\Big/Q(s) = \frac{sR}{1+sRC} \tag{10}$$

Where $U_i(s)$ denotes the Laplace transform of $u_i(t)$, and $Q(s)$ is the Laplace transform of $q(t)$. Generally, the frequency of the output signal of the electrostatic sensor is low and less than 2000 Hz. If the preamplifier is designed correctly, the condition $|sRC|<<1$ can be satisfied. Then equation (11) can be obtained

$$U_i(jw)\Big/Q(jw) = jwR \tag{11}$$

Without loss of generality, it is given that R equals to 1 Ω. From equation (5) and (11), we have the power spectrum of the output signal of the electrostatic sensor

$$S_u = |jw|^2 \cdot S_q = (2\pi f)^2\frac{1}{v}\int S_i(\frac{f}{v},f_r)|S(\frac{f}{v},f_r)|^2 df_r \tag{12}$$

Electrostatic flow noise is a white noise with the defined frequency band so that its power spectrum $S_i(f_z, f_r)$ is flat [14]. If the charged particles move along the axial direction at a given radial position r, then an approximation of equation (12) is

$$S_u \approx k_0(2\pi f)^2\frac{1}{v}|S(\frac{f}{v})|^2 \tag{13}$$

Where k_0 is constant. Theoretically, the spatial sensitivity function at the given radial position r can be expressed using the following equation [15]

$$s(z) = ae^{-bz^2} + ce^{-dz^2} \tag{14}$$

Where a, b, c, d are the fitness constants related to the geometric shape and sizes of the measuring head and the radial position r of the point charge in the sensing zone. Therefore we have the Fourier transform of $s(z)$

$$S(f_z) = \frac{a\sqrt{\pi}}{\sqrt{b}}\exp(-\frac{(\pi f_z)^2}{b}) + \frac{c\sqrt{\pi}}{\sqrt{d}}\exp(-\frac{(\pi f_z)^2}{d}) \tag{15}$$

Combining equation (13) and (15), the power spectrum of $u_i(t)$ can be rearranged as

$$S_u \approx k_0(2\pi f)^2 \frac{1}{v}\left[\frac{a\sqrt{\pi}}{\sqrt{b}}\exp(-\frac{(\pi f)^2}{bv^2}) + \frac{c\sqrt{\pi}}{\sqrt{d}}\exp(-\frac{(\pi f)^2}{dv^2})\right]^2 \tag{16}$$

At the peak of the power spectrum characteristics curve S_u, the first derivative of equation (16) is derived in the following form

$$\frac{a}{\sqrt{b}}\exp(-\frac{(\pi f)^2}{bv^2})\cdot(1-\frac{2(\pi f)^2}{bv^2}) + \frac{c}{\sqrt{d}}\exp(-\frac{(\pi f)^2}{dv^2})\cdot(1-\frac{2(\pi f)^2}{dv^2}) = 0 \tag{17}$$

From equation (17), it can be concluded that f/v is constant at the radial position r. If the constant g_r is introduced, the solution of equation (17) gives

$$v/f_{\max} = g_r \tag{18}$$

Where g_r is defined as geometry characteristic constant, which is related to the geometric sizes of the measuring head and the radial position r of the point charge in the sensing zone. $f_{\max}$ denotes the peak frequency of the power spectrum characteristics of the output signal of the electrostatic sensor.

The particle distribution and particle velocity profile within the pneumatic transport pipelines are unknown. If the unknown particle distribution, non-uniform particle velocity profile, particle size and flow inhomogeneities are considered and a dimensionless proportionality k is introduced, equation (18) can be rewritten as

$$v_m = kg_0 f_{\max} \tag{19}$$

Where k can be determined by the calibration experiments and the effects of k on the velocity measurement were analyzed in [9-11]. g_0 denotes the geometry characteristic constant on the central axis of the electrode. Equation (19) is the fundamental formula to estimate the mean particle velocity in pneumatic pipeline.

ANALYSIS OF THE SPATIAL FILTERING CHARACTERISTICS OF ELECTROSTATIC SENSOR

The effects of the geometric parameters of the electrostatic sensor on its spatial filtering characteristics are embodied in the fitness constants (a, b, c, d) of equation (14) and further the constants will influence the central geometry characteristic constant g_0 of the mean particle velocity measurement formula. From equation (19), the change of the peak frequency ($\Delta f_{\max}$) would cause the one of the mean particle velocity (Δv_{m})

$$\Delta v_m = kg_0 \Delta f_{\max} \tag{20}$$

If the dimensionless calibration proportionality k is constant, one can see from equation (20) that for a given value of Δv_{m}, the change of the peak frequency ($\Delta f_{\max}$) will be affected by g_0. The larger central geometry characteristic constant leads to the smaller values of $\Delta f_{\max}$ and hence the lower resolution of the velocity measurement. At the same time, the lower frequency resolution (F) of the power spectrum analysis will result in the larger measurement error of $f_{\max}$. Therefore the smaller central geometry characteristic constant g_0 is required using the spatial filtering method. Additionally, considering the unknown particle distribution in the pneumatic pipeline and the existence of geometry characteristic constant distribution, the mean particle velocity can be calculated using equation (19). But it should be noted that g_0 replaces g_r at different radial positions, which may cause the error of the particle velocity measurement. And thus it is also very necessary to reduce the non-uniformity of geometry characteristic constant distribution.

According to the finite element analysis results of the electrostatic sensing electrode [11], the permittivity of the dielectric pipe (ε_i) and the inner radius of the mental screening (R_3) have the effects on the spatial sensitivity distribution over the central cross section of the electrode, but they have little effects on the axial distribution of the sensitivity. However the length of the electrode (W_e), the length of the dielectric pipe (l) and its thickness (R_2-R_1) have important influences on the axial distribution of the sensitivity. To optimize the construction of the electrostatic sensor and determine g_0 correctly, on the basis of the sensor's mathematical and finite element model established in [11], the effects of these factors (l, R_2-R_1, W_e) on central geometry characteristic constant g_0 are investigated numerically using finite element method. In the numerical simulations, other parameters are kept at ε_i=3.75, R_1=5 mm, R_3=18 mm.

Effect of the Length of the Electrode

Figure 3 shows the effect of the length of the electrode (W_e) on g_r. The lengths are 3 mm, 5 mm and 7 mm, respectively. And other parameters are kept at R_2=10 mm and l=100 mm. g_0 decreases with the increase of the radial coordinate r and the shorter electrode contributes to the reduction of g_0 and hence the higher resolution of mean particle velocity. In addition, for W_e =3 mm, the relative discrepancy between g_0 and g_r at r=4 mm is 3.7% and for W_e=5 mm and 7 mm, the relative discrepancies are 3.1% and 2.4% respectively, which indicates that the longer electrode is helpful to the uniformity of geometry characteristic constant. Therefore it is contradictory to optimize g_0 and to reduce the non-uniformity of geometry characteristic constant distribution by the increase or decrease in the length of the electrode.

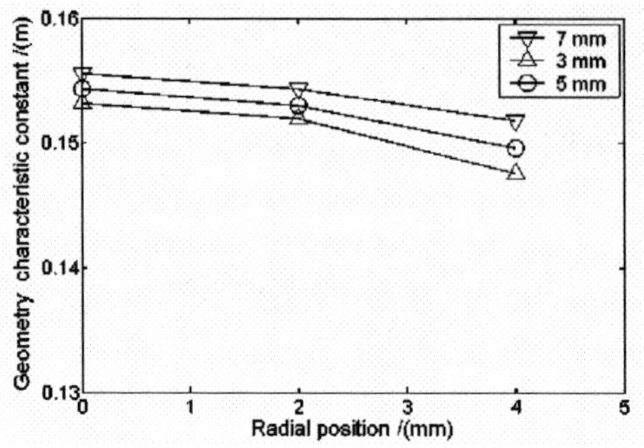

FIGURE 3. Effect of the length of the electrode on g_r

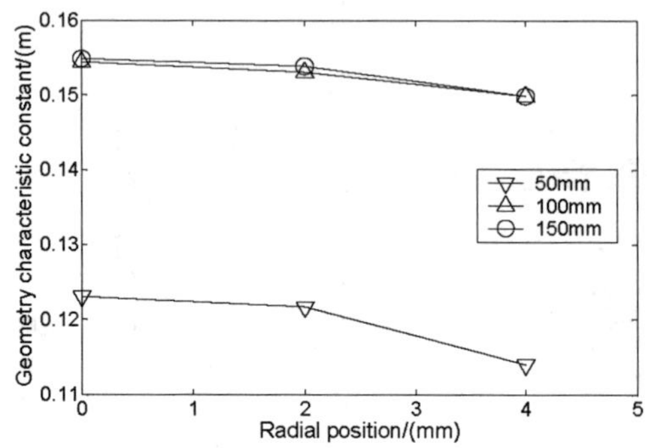

FIGURE 4. Effect of length of dielectric pipe on g_r

Effect of the Length of the Dielectric Pipe

Figure 4 illustrates the effect of the length of the dielectric pipe (l) on g_r. The lengths (l) are 50 mm, 100 mm and 150 mm, respectively. And other parameters are kept at R_2=10 mm and W_e=5 mm. It can be seen from Fig. 4 that the shorter length of the dielectric pipe contributes to the reduction of g_0, but it leads to the larger non-uniformity of g_r. Additionally, with the increase of the length, the values of g_r approach to constant at different radial positions r.

Effect of the Thickness of the Dielectric Pipe

The effect of the thickness of the dielectric pipe on g_r is depicted in Fig. 5. The values of R_2-R_1 are 1 mm, 5 mm and 10 mm, respectively. And other parameters are kept at W_e=5 mm and l=100 mm. For R_2-R_1=1 mm, 5 mm and 10 mm, the corresponding relative discrepancies are 7.3%, 3.1% and 3% respectively, which shows

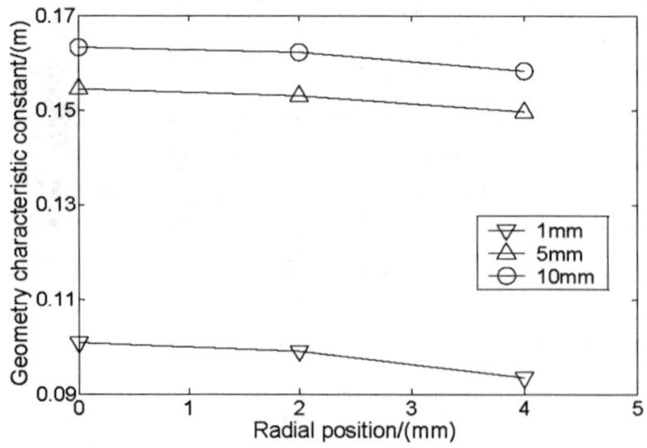

FIGURE 5. Effect of the thickness of the dielectric pipe on g_r

that the increase in the thickness of the dielectric pipe attributes to the uniformity of geometry characteristic constant g_r and results in the increase in the value of g_0 as well.

SIGNAL PROCESSING

Due to the mechanical vibration of the pipeline, pulsation of the particle flow and electrical disturbance, the output signal of the electrostatic sensor contains much noise. The representations of the noise in the power spectrum characteristic curve are that it is not smooth and the indistinctive peak frequency f_{max} may be even submerged by other local peaks, which influences the measurement accuracy of f_{max} and hence the mean particle velocity. A multi-scaling Wavelet analysis based filtering method for the power spectrum was introduced to extract its trend term [16], which effectively overcomes the above interference mentioned and improves the accuracy of the mean particle velocity measurement.

The output signal of the electrostatic sensor is processed in two steps. The power spectrum of the sampled and discrete signal is estimated using Welch method [17]. And then the power spectrum is divided into a smooth trend and a local detail term by wavelet analysis. Finally, the peak frequency can be determined accurately from the extracted trend term. The raw acquired signal and its processing result are shown in Fig. 6. Fig. 6(a) illustrates the raw output signal of the electrostatic sensor with 1 s interval. And obviously, it is very difficult to determine the peak frequency f_{max} from the power spectrum curve with a number of local peaks shown in Fig. 6(b). But the filtered power spectrum curve is smooth with distinctive peak (Fig. 6(c)), the corresponding frequency to which is 36 Hz. In this paper, the 3-order Debechies's orthonormal wavelet basis is used in the filtering method, and the scale function $\varphi_3^D(t)$ and Wavelet function $\psi_3^D(t)$ are given by

$$\varphi_3^D(t)=\frac{1+\sqrt{3}}{4\sqrt{2}}\varphi_3^D(2t)+\frac{3+\sqrt{3}}{4\sqrt{2}}\varphi_3^D(2t-1)+\frac{3-\sqrt{3}}{4\sqrt{2}}\varphi_3^D(2t-2)+\frac{1-\sqrt{3}}{4\sqrt{2}}\varphi_3^D(2t-3) \tag{21}$$

$$\psi_3^D(t)=\frac{1-\sqrt{3}}{4\sqrt{2}}\varphi_3^D(2t+2)-\frac{3-\sqrt{3}}{4\sqrt{2}}\varphi_3^D(2t+1)+\frac{3+\sqrt{3}}{4\sqrt{2}}\varphi_3^D(2t)-\frac{1+\sqrt{3}}{4\sqrt{2}}\varphi_3^D(2t-1) \tag{22}$$

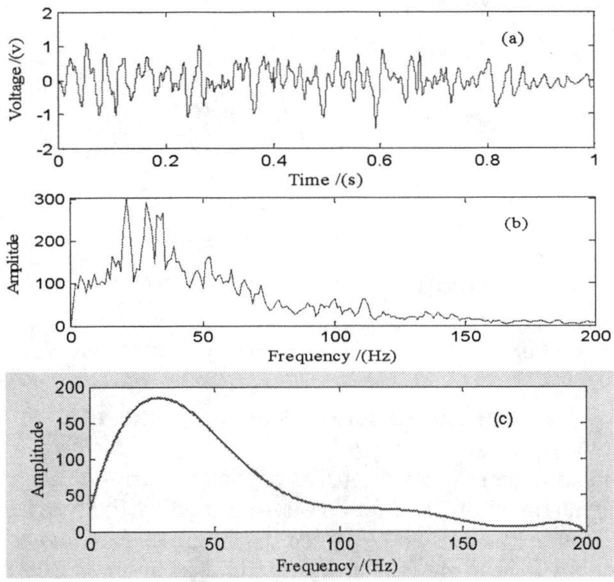

FIGURE 6. Output signal of electrostatic sensor and processing results

EXPERIMENTAL RESULTS AND DISCUSSION

Experimental Apparatus

A plan view of the main apparatus layout is shown in Fig. 7. The constituent components of the apparatus and their functions are as followed

- The volume capacity of the hopper (3) is 0.648 m^3, and by switching the valves in pipelines, the hoppers can be used to feed and receive powder particles, respectively.
- Transport pipeline section consists of 45 m of Φ16x3 mm bore mild steel pipeline.
- Nitrogen cylinder group (8) supplies the transport gas for the dense phase conveyor, and nitrogen exhausted from the pipeline section passes first through the receiving hopper, where the bug dust collector is used to disengage the solid particles from the nitrogen flow. The clean nitrogen then passes through a rotor meter (13) to facilitate the measurement of gas flowrate, before passing through the electrical control valve (1) to atmosphere. The operation pressure of the receiving hopper can be adjusted by the valve.
- A computerized data acquisition system (not shown) allows the signals from all of the measuring sensors to be logged simultaneously and saved for further processing. The sensors include a weigh cell (2) for the measurement of the particle mass in a time interval, 3 rotor meters (10), (11), (12) for the measurement of the transport nitrogen flowrate, 3 pressure transducers and an electrostatic sensor (9).

The high-pressure nitrogen first passes through the gas buffer tank (7) and then is divided into two ways. The first forms fluidizing gas (5), which first fluidizes the powder particles in the feeding hopper, then carries the fluidized particles passing through the transport pipeline, and finally enters the receiving hopper. The second is used to keep the pressure of the feeding hopper (termed keeping pressure gas (6)) and to change the solid gas mass ratio in the transport pipeline (termed supplement gas (4)). Supplement gas is also named secondary gas.

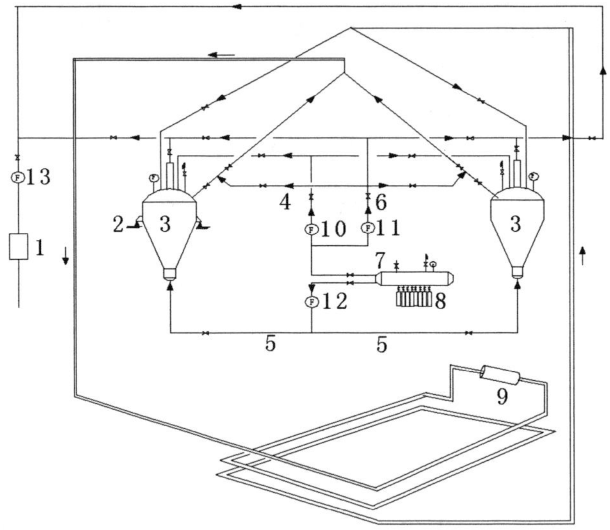

FIGURE 7. Schematic diagram of Dense-Phase Pneumatic Conveying System of Pulverized Coal under High Pressure
1-electrical control valve 2-weigh cell 3-hopper 4-supplement gas 5-fluidizing gas 6-keeping pressure gas 7- gas buffer tank 8-nitrogen cylinder group 9- test pipeline 10~13-rotor meters

Measurement Results and Discussion

The mean particle velocity measurement system consists of the measuring head, a preamplifier and a PC-based data acquisition system. The output signal of the electrode is amplified by the preamplifier, then is sampled, saved and analyzed by the PC. The sampling rate was 5000 Hz and the length of data logged is 5120 points at a time. The geometric parameters of the measuring head are kept at R_1=5 mm, R_2=10 mm, l=100 mm, ε_r=3.75, R_3=18 mm. And the input resistance of the preamplifier R is 2 MΩ. Due to the absence of the calibration instrumentation for gas solid flows measurement, the superficial gas velocity is only a traceable reference with which the measured mean particle velocity is compared. The superficial gas velocity v_a can be calculated as

$$v_a = \frac{Q_a}{A} \tag{23}$$

Where Q_a is the volumetric flowrate of transport nitrogen gas; A is the cross section area of the transport pipeline.

The particle concentration β_s in the pipeline can be estimated using the following equation,

$$\beta_s \approx \frac{Q_s}{Q_a + Q_s} \times 100\% \tag{24}$$

Where Q_s is the volumetric flowrate of coal with the relation $Q_s = M_s/\rho_s$. ρ_s denotes the true particle density and M_s is the coal mass flowrate, which is obtained by weighing the amount of the solid particle collected from the receiving feeder in a suitable time interval. Particle properties are listed in table1.

TABEL 1. Properties of Coal Particle

Material	Mean diameter d_s/(μm)	Density ρ_s/(Kg/m^3)	Shape factor Φ_s
Coal	35.6	1350	0.63

During the experiments, the pressure in gas buffer tank was kept at 2.5 Mpa, the pressure in the feeding hopper was 2.0 Mpa and the flowrate of the fluidizing gas was 0.3 m^3/h. Particle transport experiments were carried out with the system pressure differences being 1.3Mpa, 1.5Mpa and 1.7Mpa, respectively. The corresponding flowrate of the keeping pressure gas were 1.0 m^3/h, 0.8 m^3/h and 0.6 m^3/h. For each pressure difference, the experimental conditions were changed 6 times by gradually increasing the supplement gas flowrate.

Figure 8 shows the comparison between the measured mean particle velocity and superficial gas velocity. The measured velocity is an average one of 10 separate measurements. Over the velocity range of 8.63-18.62 m/s for particle concentration range 0.067-0.130 m^3/m^3, the measured values are slightly higher than the superficial gas velocities. Several reasons for the phenomenon may be found. First, the higher particle concentration over the cross section of the pipeline results in the reduction of the area occupied by the transport gas, and the particles are prone to accumulate within the horizontal pipeline due to their intrinsically adhesive properties and own gravities. As a consequence, if the total particle concentration (β) is considered, the true gas velocities should be estimated using the following equation

$$v_a = \frac{Q_a}{A(1-\beta)} \tag{25}$$

From equation (25), it can be concluded that the true gas velocities are higher than the calculated superficial ones because of the reduced effective cross sectional area of the gas flow. Particle size is an important factor influencing the slip between the particles and the gas flow, and Klinzing's studies showed that the slip velocity is almost proportional to the particle size [18]. In the experiments, the mean size of the pneumatically conveyed particles is 35.6 μm and thus they have better follow characteristics. Therefore it is possible that particle velocities are higher than the superficial ones. Second, the velocity distribution in the lower section of the horizontal pneumatic pipeline is relatively flat due to the higher particle concentration, but the particle concentration and velocity distribution are exactly reverse in the upper section of the pipeline. The output signal of the electrostatic sensor is sensitive to the

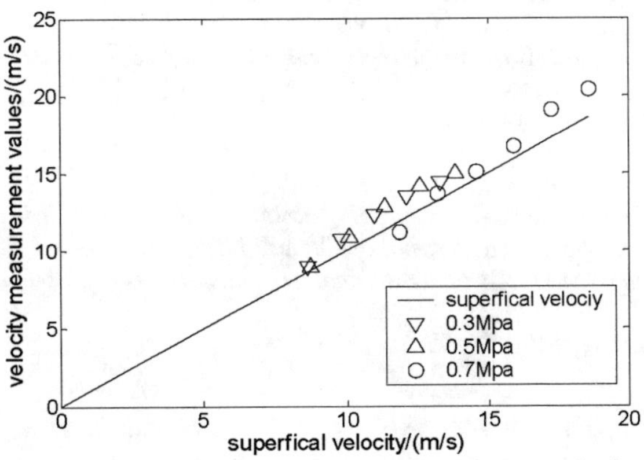

FIGURE 8. Comparison between the measured mean velocity and v_a

particle closest to the pipe wall. Hence the upper particle velocity leads to the increase of the peak frequency of the power spectrum, which attributes to the higher measured mean particle velocity. Finally, for a given geometric sizesof the electrostatic sensor, the geometry characteristic constant gr is related to radial position r while the mean particle velocity is calculated using the substitution g0 for gr for convenience, which also should be responsible for the higher measured velocity. Though the accuracy of the particle measurement velocity using the spatial filtering method cannot be determined due to the absence of the calibration instrumentation and equipment, its repeatability error within $\pm 4\%$ is verified by experiments under the given experimental conditions.

CONCLUSION

The paper presents an electrostatic induction theory based spatial filtering method for the mean particle velocity measurement in pneumatic pipeline. In the view of random signal analysis, a novel mathematical model for the mean particle velocity measurement is proposed using a single electrostatic sensor. And the corresponding measurement system has advantages of simple construction, no contact, low cost and good adaptability to harsh industrial process. Compared with correlation based velocity measurement method, the spatial filtering method has little dependence on the particle flow conditions in the pipeline.

Experimental results in a dense phase pneumatic conveying system of pulverized coal under high pressure indicate that the electrostatic induction theory based spatial filtering method is capable of making mean particle velocity measurement with repeatability better than ±4% over the velocity range of 8.63-18.62 m/s for particle concentration range 0.067-0.130 m^3/m^3. Theoretical analysis and experiments verify the validity of the spatial filtering method, which may be developed as a simple and easy of use mean particle velocity measurement system for pneumatic pipeline.

ACKNOWLEDGEMENTS

This research was supported by Special Funds for Major State Basic Research Projects (2004CB17702-04) and TangShan Application Basic Research Projects (06234501A-1). The authors wish to express their gratitude.

REFERENCES

1. Yan Y., *Meas. Sci. Tehnol.* **7**, 1687-1706 (1996).
2. Xu ling'an, *Cross-correlation Flow Measurement Technique*, Tianjin: Tianjin University Press, 1988, pp. 20-35.
3. Green, R. G, "Capacitance flow transducers for multiphase systems", PH.D. Thesis, university of Bradford, U.K, 1981.
4. Gajewski J. B., *Journal of Electrostatics*, 37, 261-276 (1996).
5. Sheen S H and Raptis A C., *ISA Trans.*,24, 53–58 (1985).
6. Linn J M and Sample D G. Radiometric mass flow measurement of solids/gas stream using radiometric technique, Proc. Symp. Instrumentation and Control of Fossil Energy Processes, Houston, 1982, 12–20.
7. Yan Y. B Byrne, S Woodhead, *Meas. Sci. Technol.*, 16, 515-537 (1995).
8. J. Ma, Y. Yan., *Flow Measurement and Instrumentation*, 11, 195–204 (2000).
9. Xu chuanlong, Zhou Bin, Wang Shimin et al, *ESA/IEEE-IAS/IEJ/SFE Conference on Electrostatics*, 2006.
10. Xu Chuanlong, Huang Jian, Wang Shimin et al., *Journal of Chemical Industry and Engineering (China)*. (To be printed).
11. Xu chuanlong. "Gas-solid two-phase flow parameters and electrostatic measurement", PhD. dissertation, Southeast University, China, 2006.
12. J.B. Gajewski, *J. Electrostat.*, 45, 255–264 (1999).
13. J.B. Gajewski, *J. Electrostat.*, 48, 49–64 (1999).
14. Jianyong Zhang, John Coulthard., *J. Electrostat.*, 63, 1133-1149 (2005).
15. Xu Chuanlong, Zhao Yanjun, Wang shimin et al., *Journal of southeast university (China)*, 36, 556-561 (2006).
16. Zhou Zhe, Zhuang Liangjie, Xiong Zhengnan, *Journal of Chinese inertial technology*, 7, 58-60 (1999).
17. Huang Wenmei. *The application of MATLAB in signal analysis and processing*, Changsha: National University of Defense Press, 2000 pp.201-220.
18. Klinzing G E., *Powder Technol.*, 51, 181-182 (1987).

Electrostatic Charged Two-Phase Flow Equations

Wang Zhentao, Wen Jianlong, Wang Junfeng, Tang Zhihua, Luo Tiqian

School of Energy Resources and Power Engineering, Jiangsu University, Zhenjiang 212013, China

Abstract: Electrostatic charged two-phase flows exit in electrostatic spray、crop-dusting and fuel spray and so on. Electrostatic charged spray applying to FGD scrubber can improve desulfurization efficiency, decrease water usage. For the complexity of two-phase flow's structure in FGD scrubber, and there exit coupled action between non-uniform electric and flow field, also exit phase interaction between charged particles and continuous phase, which makes the flow more complex. So the complete theory has not formed at present. This paper adopts Lagrange and Euler method of combining together and takes the dispersed particle as fluid, and applies the Reynolds transport principle to set up a Reynolds transport equation, which suit electrostatic charged particle and liquid phase. Then based on Reynolds transport equation, equations for the volume average and instantaneous state of the electrostatic charged two-phase flow are obtained. Similar to equations for single phase turbulent flow, this paper applies Reynolds-average method, and develops equations for Reynolds-average equations for electrostatic charged two-phase flow. Finally, according to the model of single phase turbulent flow, equations for electrostatic charged two-phase flows has been closed. So the $k-\varepsilon-k_p$ model is obtained. Contrast of result by PIV and simulation has been finished.

Key words: Electrostatic charged two-phase turbulence, Phases interaction, Reynolds-average equations, $k-\varepsilon-k_p$ model

INTRODUCTION

Electrostatic charged spray research has been abroad developed in crop-dusting、atomization、fuel spray and so on. Electrostatic charged two-phase turbulence flow is very complex. There exist coupled action between space non-uniform electric field and flow field, also exit phase interaction between charged particle phase and continuous phase, which makes the flow more complex. From the point of view of electric field, there exit exterior electric field and derivational electric field by the charged droplets. Ripple of charged droplets could bring the ripple of electric field. So it could influence the force on the droplets, then it could influence the continuous phase through phases interaction. The structure of the charged two-phase turbulence flow of gas and liquid is very complex. This paper mainly discusses the turbulence flow equation in FGD spray scrubber.

ELECTROSTATIC CHARGED TWO-PHASE FLOW BODY AVERAGE CONSERVATION EQUATION GROUP

The Body Average Way

For difference of the diameter of solid particles in the pesticide insufflations or liquid droplets in the spray is very little in the charged two-phase turbulence flow. Furthermore the experiment indicates when the droplets are charged, atomization uniformity observably increase. So, dispersed particles could be regarded as one having the same physics characteristic phase.

According to describing method of two-phase flow, dispersed and continuous phase have been regarded as taking up the same space and having interaction. Choose a control volume dV in the mixture; its surface area is dA. In control volume, the volume of k phase is dV_k, and its surface is dA_k.

Volume average is adopted. Videlicet, in macro flow field, each variable value in liquid tiny unit is the average value in the control volume. The average value of parameter φ could be marked:

CP914, *Multiphase Flow: The Ultimate Measurement Challenge,* edited by X. Cai, Y. Wu, Z. Huang, S. Wang, and M. Wang

$$\langle\varphi\rangle = \frac{1}{V}\int_V \varphi_k dV \tag{1}$$

The sign of < > denotes volume average. $V = V_p + V_c$ is control volume. k denotes particles phase and continuous phase. The volume average has three expressions.

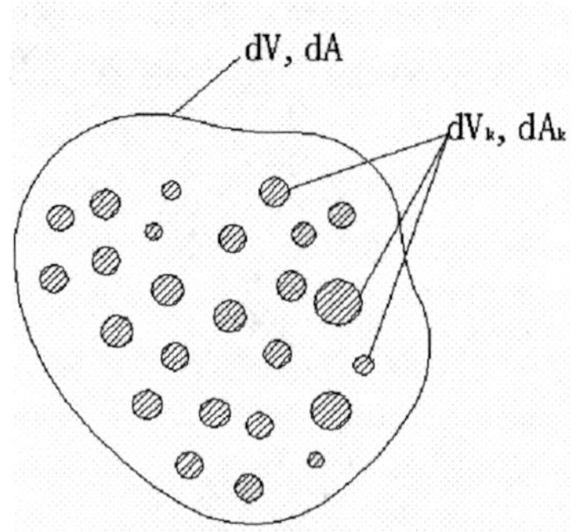

FIGURE 1. The control volume in two-phase flows

1) Gauss formula

$$\int_V \frac{\partial a_j}{\partial x_j} dV = \int_{A_k} a_j n_j dA \tag{2}$$

2) Transport principle

$$\langle\frac{\partial a_i}{\partial t}\rangle = \frac{\partial}{\partial t}\langle a_i\rangle - \frac{1}{V}\int_A a_i v_j n_j dA \tag{3}$$

3) Average principle

$$\langle\frac{\partial a_j}{\partial x_j}\rangle = \frac{\partial}{\partial x_j}\langle a_j\rangle + \frac{1}{V}\int_A a_j n_j dA \tag{4}$$

The Phases Microcosmic Conservation Equation

Dispersed phase would be regarded as liquid, and take the basic equation inside k phase. Adopting apparent density, equation group can be written down.

1) Continuity equation

$$\frac{\partial \rho_k}{\partial t} + \frac{\partial}{\partial x_j}(\rho_k v_{kj}) = 0 \tag{5}$$

2) Momentum equation

$$\frac{\partial}{\partial t}(\rho_k v_{kj}) + \frac{\partial}{\partial x_j}(\rho_k v_{kj} v_{ki}) = \rho_k f_{ki} - \frac{\partial p_k}{\partial x_i} + \frac{\partial \tau_{k,ji}}{\partial x_j} \tag{6}$$

3) Energy equation

$$\frac{\partial}{\partial t}(\rho_k C_{vk} T_k) + \frac{\partial}{\partial x_j}(\rho_k C_{vk} T_k v_{kj}) = -p\frac{\partial v_{kj}}{\partial x_j} + \tau_{k,ij} s_{k,ij} + \frac{\partial}{\partial x_j}(\lambda_k \frac{\partial T_k}{\partial x_j}) + \rho_k q_k \tag{7}$$

ρ_k is k phase apparent density; v_{kj} is k phase velocity; f_{ki} is unit mass force acting on k phase; $\tau_{k,ij}$ is tangency stress tensor; C_{vk} is definite volume specific heat; λ_k is heat exchange coefficient; $s_{k,ij}$ is deformation speed tensor; q_k : the gross of quantity of heat into the k phase.

Volume Average Conservation Equation

Applying the average principle to continuity equation, get:

$$\langle \frac{\partial \rho_k}{\partial t} \rangle = \frac{\partial}{\partial t}\langle \rho_k \rangle - \frac{1}{V}\int_{A_k} \rho_k \vec{v}_s \cdot d\vec{A}$$

$$\langle \nabla \cdot \rho_k \vec{v}_k \rangle = \nabla \cdot \langle \rho_k \vec{v}_k \rangle + \frac{1}{V}\int_{A_k} \rho_k \vec{v}_k \cdot d\vec{A}_k$$

$$\frac{\partial \langle \rho_k \rangle}{\partial t} + \nabla \cdot \langle \rho_k \vec{v}_k \rangle = -\frac{1}{V}\int_{A_k} \rho_k (\vec{v}_k - \vec{v}_s) \cdot d\vec{A}_k$$

According to physics meanings, $\vec{v}_s$ is displacement velocity at phase interface due to phase change. $\vec{v}_k - \vec{v}_s$ is the displacement velocity of *k* phase relative to interface. So, the right side of previous formula is the phase change matter sources in the two-phase flows control volume.

$$-\frac{1}{V}\int_{A_k} \rho_k (\vec{v}_k - \vec{v}_s) \cdot d\vec{A}_k = \frac{1}{V}\int_{A_k} \vec{n} \cdot [\rho_k (\vec{v}_k - \vec{v}_s)] \cdot dA_k$$

$$= \frac{1}{V}\int_{V_K} \nabla \cdot [\rho_k (\vec{v}_k - \vec{v}_s)] \cdot dV_k = \frac{1}{V}\int_{V_K} S_k dV_k = \langle S_k \rangle$$

S_k is the matter sources of volume average in the unit volume due to phase change.

So , *k* phase continuity equation of the volume average is:

$$\frac{\partial \langle \rho_k \rangle}{\partial t} + \nabla \cdot \langle \rho_k \vec{v}_k \rangle = \langle S_k \rangle$$

Removing $< \ >$, macro form *k* phase volume average continuity equation is gained:

$$\frac{\partial \rho_k}{\partial t} + \nabla \cdot (\rho_k \vec{v}_k) = S_k \tag{8}$$

The continuity equations of particle phase and continuous phase are written down apart.

$$\left.\begin{aligned} &\frac{\partial \rho_p}{\partial t} + \frac{\partial}{\partial x_j}(\rho_p v_{pj}) = S \\ &\frac{\partial \rho_c}{\partial t} + \frac{\partial}{\partial x_j}(\rho_c v_{cj}) = -S \end{aligned}\right\} \tag{9}$$

$S = n_p \dfrac{dm_p}{dt}$, n_p is the p phase particle number density; m_p : the mass of one particle.

The momentum and energy equations are also educed by similar method. These equations are gathered:

1) Particle phase

$$\left.\begin{aligned} &\frac{\partial \rho_p}{\partial t} + \frac{\partial}{\partial x_j}(\rho_p v_{pj}) = S \\ &\frac{\partial}{\partial t}(\rho_p v_{pi}) + \frac{\partial}{\partial x_j}(\rho_p v_{pj} v_{pi}) = \rho A_q E_i + \rho_p g_i - \frac{\partial p}{\partial x_i} + \rho_p \frac{v_{pi} - v_{ci}}{\tau_v} + v_{ci} S \\ &\frac{\partial}{\partial t}(\rho_p C_{vp} T_p) + \frac{\partial}{\partial x_j}(\rho_p C_{vp} T_p v_{pj}) = \frac{\partial}{\partial x_j}(\lambda \frac{\partial T_p}{\partial x_j}) + \rho_p q_p - C_{vp} T_p S \end{aligned}\right\} \tag{10}$$

2) Continuous phase

$$
\left.
\begin{aligned}
&\frac{\partial \rho_c}{\partial t}+\frac{\partial}{\partial x_j}(\rho_c v_{cj})=-S\\
&\frac{\partial}{\partial t}(\rho_c v_{ci})+\frac{\partial}{\partial x_j}(\rho_c v_{cj} v_{ci})=\rho_c g_i-\frac{\partial p}{\partial x_i}+\frac{\partial \tau_{ji}}{\partial x_j}-\rho_c\frac{v_{pi}-v_{ci}}{\tau_v}-v_{ci}S\\
&\frac{\partial}{\partial t}(\rho_c C_{vc} T_c)+\frac{\partial}{\partial x_j}(\rho_c C_{vc} T_c v_{cj})=-p\frac{\partial v_{cj}}{\partial x_j}+\tau_{ij}s_{ij}+\frac{\partial}{\partial x_j}(\lambda\frac{\partial T_c}{\partial x_j})+\rho_c q_c+C_{vc}T_p S
\end{aligned}
\right\}\quad (11)
$$

A_q is the charge to mass ratio; $\vec{E}$ is electric field intensity; $v_{ci}S$ is the momentum of phase change matter sources; $C_{vp}T_pS$ and $C_{vc}T_pS$ are separately energy of particle phase and continuous phase in unit volume due to phase change.

CHARGED TWO-PHASE FLOW TIME AVERAGE CONSERVATION EQUATION GROUP

The instantaneous state of the electrostatic charged two-phase flow is obtained previously. For charged two-phase turbulent flow, and similar to equations for single phase turbulence flow, adopting Reynolds-average method, relevant Reynolds-average equations group is obtained.

1) Electrostatic dispersed phase

$$
\left.
\begin{aligned}
&\frac{\partial \rho_p}{\partial t}+\frac{\partial}{\partial x_j}(\rho_p v_{pj})=S\\
&\frac{\partial}{\partial t}(\rho_p v_{pi})+\frac{\partial}{\partial x_j}(\rho_p v_{pj} v_{pi})=\rho A_q E_i+\rho_p g_i-\frac{\partial p}{\partial x_i}+\rho_p\frac{v_{pi}-v_{ci}}{\tau_v}+v_{ci}S+\frac{\partial}{\partial x_j}(\overline{-\rho_p v'_{pj} v'_{pi}})+\overline{\rho_p A'_q E'_i}\\
&\frac{\partial}{\partial t}(\rho_p C_{vp} T_p)+\frac{\partial}{\partial x_j}(\rho_p C_{vp} T_p v_{pj})=\frac{\partial}{\partial x_j}(\lambda\frac{\partial T_p}{\partial x_j})+\rho_p q_p-C_{vp}T_p S-\frac{\partial}{\partial x_j}(\rho_p C_{vp}\overline{T'_p v'_{pj}})
\end{aligned}
\right\}\quad (12)
$$

2) Continuous phase

$$
\left.
\begin{aligned}
&\frac{\partial \rho_c}{\partial t}+\frac{\partial}{\partial x_j}(\rho_c v_{cj})=-S\\
&\frac{\partial}{\partial t}(\rho_c v_{ci})+\frac{\partial}{\partial x_j}(\rho_c v_{cj} v_{ci})=\rho_c g_i-\frac{\partial p}{\partial x_i}+\frac{\partial \tau_{ij}}{\partial x_j}-\rho_c\frac{v_{pi}-v_{ci}}{\tau_v}-v_{ci}S+\frac{\partial}{\partial x_j}(\overline{-\rho_c v'_{cj} v'_{ci}})\\
&\frac{\partial}{\partial t}(\rho_c C_{vc} T_c)+\frac{\partial}{\partial x_j}(\rho_c C_{vc} T_c v_{cj})=-p\frac{\partial v_{cj}}{\partial x_j}+\Phi+\frac{\partial}{\partial x_j}(\lambda\frac{\partial T_c}{\partial x_j})+\rho_c q_c+C_{vc}T_{ps}S-\frac{\partial}{\partial x_j}(\rho_c\overline{C_{vc}T'_c v'_{cj}})+\varepsilon
\end{aligned}
\right\}\quad (13)
$$

Φ is dissipation function; ε is the dissipation of turbulent energy.

THE MODEL OF CHARGED TWO-PHASE TUBULENCE FLOW

Based on time average conservation equations, adopting the supposition that particles are regarded as continuous phase, equation groups are closed.

The Simulation of Continuous Phase

Relevant terms of turbulence fluctuating exists in the equation group are:

$$\overline{-\rho_c v'_{ci} v'_{cj}},\quad \overline{\tau'_{ij} s'_{ij}},\quad \overline{-\rho c_{vc} T'_c v'_{cj}}$$

1) Turbulence stress term $\tau_t = -\rho_c \overline{v'_{ci} v'_{cj}}$ and turbulent energy dissipation term $\overline{\tau'_{ij} s'_{ij}}$ are resolved by the k —ε two equations model of single phase

2) The terms of temperature fluctuating and velocity fluctuating are simulated by the grads.

$$-\overline{T'_c v'_{cj}} = \frac{k_T}{\rho C_v} \frac{\partial T}{\partial x_j} = \frac{\upsilon_t}{\sigma_T} \frac{\partial T}{\partial x_i} \tag{14}$$

$\alpha_T = \dfrac{\upsilon_t}{\sigma_T}$ is named as Prandtl number of turbulence.

The Simulation of Charged Dispersed Phase

Relevant terms of turbulence fluctuating exists in the dispersed equation group are:

$$-\rho_p \overline{v'_{pi} v'_{pj}}, \quad \rho_p \overline{A'_q E'_i}, \quad -\overline{T'_p v'_{pj}}$$

1) Reynolds stress term of charged dispersed phase

$$-\rho_p \overline{v'_{pj} v'_{pi}} = \mu_p \left(\frac{\partial v_{pj}}{\partial x_i} + \frac{\partial v_{pi}}{\partial x_j}\right) - \frac{2}{3} \rho_p k_p \delta_{ij} \tag{15}$$

$\mu_p = \rho_p \upsilon_p$, and that,

$$\upsilon_p = C_{\mu p} \frac{k_p^2}{\varepsilon_p} \tag{16}$$

Equation of resolving dispersed phase turbulent energy k_p is:

$$\frac{\partial k_p}{\partial t} + v_{pj} \frac{\partial k_p}{\partial x_j} = \frac{\partial}{\partial x_j}\left(\frac{\upsilon_p}{\sigma_p} \frac{\partial k_p}{\partial x_j}\right) + \upsilon_p \left(\frac{\partial v_{pi}}{\partial x_j} + \frac{\partial v_{pj}}{\partial x_i}\right) \frac{\partial v_{pi}}{\partial x_j} + \varepsilon_p \tag{17}$$

$$\varepsilon_p = v_{pi} \frac{v_{ci} - v_{pi}}{\tau_r} \tag{18}$$

2) The term of interaction of charged dispersed phase and electric field is:

$$\rho_p \overline{A'_q E'_i} = \frac{\upsilon_E}{\sigma_E} A_q \frac{\partial V}{\partial x_i} \tag{19}$$

V—distributing function of voltage in the electric field; σ_E is Schmidt number of field fluctuating.

3) The terms of temperature fluctuating and velocity fluctuating are:

$$-\overline{T'_p v'_{pj}} = \frac{\upsilon_{pT}}{\sigma_{pT}} \frac{\partial T_p}{\partial x_j} = \alpha_{pT} \frac{\partial T_p}{\partial x_j} \tag{20}$$

COMPARISON OF RESULT BY PIV AND SIMULATION

To verify the result of simulation, the electrostatic charged two-phase turbulence flow field is to be measured by PIV (Particle Image Velocimetry). PIV is modern experiment hydrodynamics improvement. It is a kind of technique for velocity measurement and can gain instantaneous velocity information.

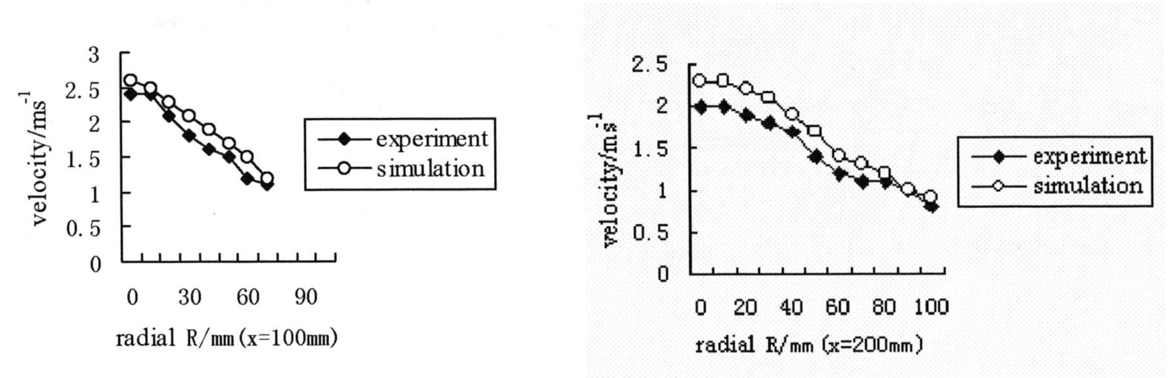

FIGURE 2. Particles phase velocity distributing

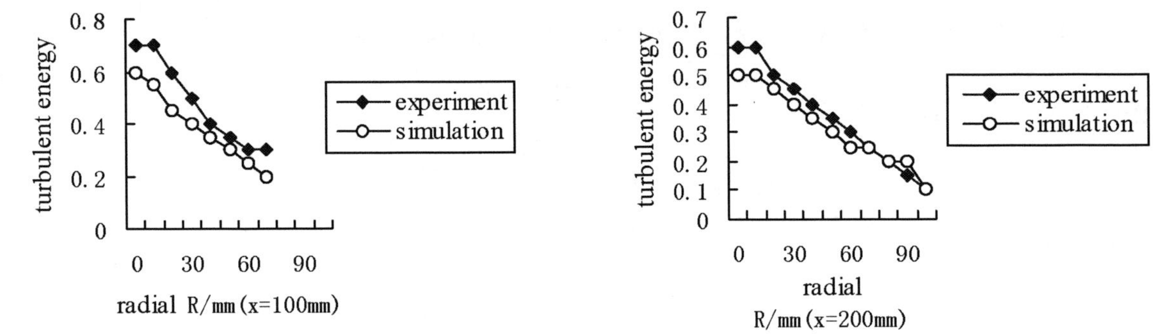

FIGURE 3. Particles phase turbulent energy distributing

The purpose of simulation is to prove that the model of electrostatic charged two-phase flow is right. So, the particles phase velocity and turbulent energy distributing are presented. The result by PIV is compared with simulation (Fig2 and Fig3). The result by PIV shows that the model is correct to simulate electrostatic charged two-phase flow.

CONCLUSIONS

1) Based on analysis of forces in the charged two-phase flow and Reynolds transport equation, adopting the model of supposition that particles phase are regarded as continuous phase, the sequence equation、momentum equation and energy equation for charged dispersed phase and continuous phase are obtained. Three dimensions charged two-phase flow basic equation group is established.

2) Based on the characteristic of charged two-phase turbulence flow, the dispersed phases would be regarded as one phase that has the same physics property. The $k-\varepsilon-k_p$ model for charged two turbulence flow is reasoned. In addition, when setting up the model, the two-phases are apart simulated. So the problem predigests. Above method is attempt to study electrostatic charged two-phase flow. The contrast about flow field results of numerical computation for the $k-\varepsilon-k_p$ model and obtained metrical datum by PIV has proved that the model of the electrostatic charged two-phase flows be available for the engineering simulation. The conclusion is helpful for the development of the electrostatic charged two-phase flows.

REFERENCE

1. Wang Ze, Jin Hanhui. Numerical simulation of charged gas-liquid two-phase jet flows in electrostatic spraying [J]. Chinese Journal of Mechanical Engineering (English Edition), Vol.14, No.3: 266-270. 2001.9. (in Chinese)
2. Wang Junfeng, Wen Jianlong, Luo Tiqian. Numerical Modeling of Charged Two-Phase Jet Flow in Electrostatic Spray[J]. Journal of JiangSu University (Natural Science Edition), 2004，25（1）,13～16. (in Chinese)

3. Wen jianlong. Charged Two-phase Turbulent Flow and Experimental Research on Electro spray Combustion of Diesel Oil [D]. ZhenJiang: Jiang Su University, 2003. (in Chinese)

4. Miller, J. A., Biblarz, O. Electrostatic spray modification in gas turbine combustion. Journal of Propulsion and Power, Vol. 3, No. 2: 187-192.1997.4.

5. Gomez, A. Tang, K. Atomization and dispersion of quasi-monodisperse electrostatic sprays of heptane. Proceedings of the 5th International Conference on Liquid Atomization and Spray Systems. Gaithersburg, USA. 1998.

6. Zhou Lixing. Recent advance in studies on multiphase and reacting flows in China[J]. Acta Mechanica Sinica, 2002,18(2):97~113.

Study of the Gas-Liquid Two-Phase Flow Measuring Method Based on the V-Cone Flow Meter

Dong Feng, Hu Jun

School of Electrical Engineering and Automation, Tianjin University, Tianjin, 300072, China
Email: fdong@tju.edu.cn

Abstract. The paper presents a proper correlation for measuring air-water two-phase flow based on the V-Cone flow meter (β=0.65). The calibrating methods of discharge coefficient and the corrective coefficient θ of V-Cone were introduced. On the basis of theoretical and experimental studies, the Lin Zonghu correlation was studied. The relationship between θ and static pressure (P) or gas-liquid density ratio was obtained. The root-square-errows of gas mass flow and that of liquid mass flow were 2.25% and 2.23% respectively according to the modified Lin Zonghu correlation.

Keywords: gas-liquid two-phase flow; V-Cone; separated flow model; measuring flow

INTRUDUCTION

The flow condition of gas-liquid two-phase flow in pipelines is frequently appeared in industry such as power, chemical industry, nuclear power, refrigeration, petroleum, metallurgy and so on. For example, flow of gas-liquid two-phase flow and the heat conduction problem widely exists in all kinds of ebullition tubes in the nuclear power station and the heat power plant, all kinds of heat interchangers, the chemical reaction equipments, many kinds of condensers, as well as other equipments. Thus all kinds of knowledge about flow and heat transfer of gas-liquid two-phase flow are needed for economically and reliably developing, designing and operating these equipments. The research work of gas-liquid two-phase flow in pipeline began in the 20th century 60's in china. In the recent 30 years, it has made considerable progress along with the development of boiler industry, the nuclear industry, the chemical industry and the oil industry and so on in china. Therefore, it's vital significance to the industry technology and the national economy of china that's developing research work of gas-liquid two-phase flow. It's same as the international situation [1].

Measurement of flow parameters of two-phase flow is a field which is difficult, needed thoroughly to study and explore, its developing tendency and research direction in future roughly might divide into following several aspects:

1) The mature single-phase measurement technology and the measuring instrument apply to parameters measurement of two-phase flow , further explore the method about measuring double parameters of gas-liquid two-phase flow with the single flow instrument, gradually increase its measuring accuracy, reduce the cost;

2) Develops high sensitive, accurate and reliable sensor of two-phase flow and the instrument of parameter measurement by all kinds of development of new technology; measure, analyze and supervise the parameters of two-phase flow by the new signal processing technology and method;

3) Develop model building of parameters measuring system further in two-phase flow process, characteristic parameter extracting, basic theory research to the time denatured auto-adapted function and dynamic track capacity etc, as well as the basic method research of checking and calibrating of multi-phase flow and error analysis etc.

It comes in for extensive research that's the mature single-phase flow measuring technology and the measuring instrument apply to two-phase flow parameters measuring, among, the throttling gear by its own characteristic is widely used to measurement of multi-phase flow. For example, Lin Zonghu uses the venturi tube to measure steam-water two-phase flow [2], Zhang Hongjian uses the orifice to research the air-water two-phase flow [3].

Along with is unceasingly thorough to the research of throttling gear type, a kind of new V-Cone flow meter appeared in the 1980s intermediate stages. The fluid gradually contracts to the pipeline in side wall because of the

CP914, *Multiphase Flow: The Ultimate Measurement Challenge,* edited by X. Cai, Y. Wu, Z. Huang, S. Wang, and M. Wang

V-Cone which s coaxial mounting in the pipeline. The concept of the fluid to contract nearby the pipeline central axis is fundamentally changed. The flow is measured by measuring the differential pressure fore and after of the V-Cone. This kind of V-Cone flow meter had opened a brand-new page for the pressure differential meter. It has been many times tests and the application in the past ten years. It is understood generally and accepted as one kind of more effective flow measuring appliance at present. The practice proves that it is more accurate and effective flow to measure flow by a wider measuring range, however pure or the dirty dirt fluid, under the shorter straight length of pipe condition [4]. It will be bright direction that V-Cone flow meter is applied to multi-phase flow measurement. This paper contains the research about flow measuring of gas/liquid based on V-Cone flow meter for this new kind of throttling gear will utilize in industry technology of china.

THE STRUCTURE AND PRINCIPLE OF THE V-CONE

V-Cone includes the coaxial mounting sharp cone and corresponding pressure port as shown in Fig.1. This measuring tube is precisely processes in advance. The difference presses generates go-and-return of the cone. The positive pressure (P_1) of the difference presses is the static pressure of pipe wall presses before the upstream fluid contraction, the negative pressure (P_2) is pressure of the hole place in the central shaft quarter which faces to the downstream end surface in the cone. This cone apex faces to the fluid orientation. There is a sharp acute angle between the cone and the tail surface. The edge of interface enables the fluid in a smooth transition region before it enters the negative pressure region of downstream [4].

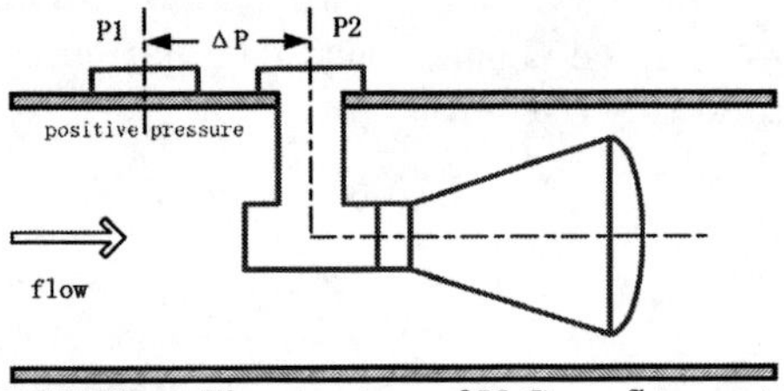

FIGURE. 1 The structure of V-Cone flow meter

THEORETICAL CONSIDERATION

There are three methods analyzing two-phase flow from macroscopic viewpoint: diffusion modeling, finite volume method and average method [5]. Several two-phase flow models are established according to the above analytical method, including homogeneous flow model, separated flow model, drift-flux model and models based on flow patter. There are several modified models of two-phase flow by orifice or the traditional venturi tube which are widely applied in the industry[6], such as: the Murdock correlation [7], the Chisholm correlation [8], Smith & the Leang correlation [9], De the Leeuw correlation [10] as well as Lin Zonghu correlation [11].

Because the structure of Lin Zonghu correlation is simple, its form is flexible, its corrected coefficient only have to do with the gas-liquid density ratio, has nothing to do with the gas or the liquid phase flow, which could be corrected and calibrated by the actual measuring data according to the practical situation. Because measuring principle of the V-Cone is the same as the measuring principle of other throttle gear (for example orifice plate and venture nozzles etc.), therefore this paper promotes Lin Zonghu correlation measuring gas-liquid two-phase flow used in the orifice plates and the venture nozzles flow meter to the new style V-Cone flow meter.

Based on the separated flow model, we assume the following: gas and liquid phased flow separately through a V-Cone; the gas phase is incompressible; the discharge coefficient C is the same for both phases; the pressure drop for each phase is the same as the pressure drop for the two-phase flow in the V-Cone; there is no evaporation during the flow.

Lin Zonghu correlation is:

$$W_{TP} = \frac{\varepsilon CA\sqrt{2\Delta P_{TP}\rho_L}}{\sqrt{1-\beta^4}[(1-x)\theta + x\sqrt{\rho_L/\rho_G}]} \tag{1}$$

We can know that from the deducing of the Lin Zonghu correlation:

$$\alpha = \frac{1}{1+\theta\sqrt{\rho_L/\rho_G}\left[\rho_G/\rho_L\left((1-x)/x\right)\right]} \tag{2}$$

And that we know from the two-phase flow theory:

$$\alpha = \frac{1}{1+S\left[\rho_G/\rho_L\left((1-x)/x\right)\right]} \tag{3}$$

The corrective coefficient θ has some physics meaning, It is the function of velocity ratio S and density ratio ρ_G/ρ_L . As the velocity ratio S is also a function of ρ_G/ρ_L . Therefore θ is the function of density ratio, under given ρ_G/ρ_L , the value of θ is a constant.

According to the modified separated flow model, gas phase is incompressible. So, $\varepsilon=1$, equation (1) becomes:

$$W_{TP} = \frac{CA\sqrt{2\Delta P_{TP}\rho_L}}{\sqrt{1-\beta^4[(1-x)\theta + x\sqrt{\rho_L/\rho_G}]}} \tag{4}$$

The V-Cone discharge coefficient C and corrective coefficient θ could be calibrated by experiment data. Given vapor quality x, total mass flow could be obtained, and then gas mass flow and liquid mass flow could be calculated separately according to the following relations.

The gas mass flow equation:

$$W_G = W_{TP} \cdot \chi \tag{5}$$

The liquid mass flow equation:

$$W_L = W_{TP} \cdot (1-\chi) \tag{6}$$

EXPERIMENTAL APPARATUS AND DATA PROCESSING

Experiment Apparatus

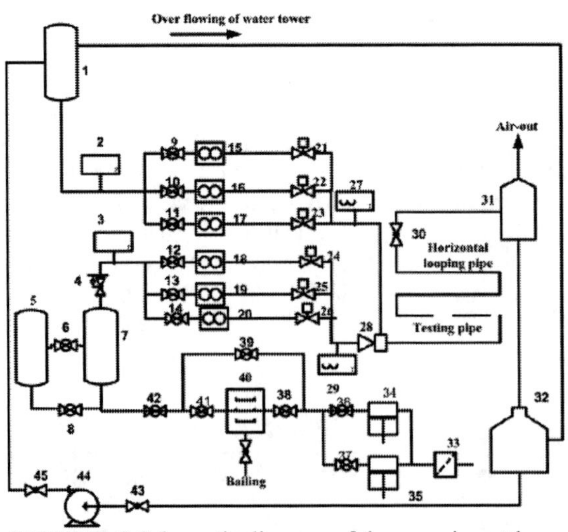

FIGURE. 2 Schematic diagram of the experiment loop

1.water tower 2,3. pressure sensor 4.pressure maintaining valve 5,7. surge tank 6,8,9,10,11,12,13,14,,36,37,38,39,41,42. ball valve 15.badge meter 16. electromagnetic meter 17,19. vortex flow meter 18. roots type flow meter 20. gas flow controller 21,22,23,24,25,26. governor valve 27,29. temperature sensor 28. injector 30,43,45. globe valve 31. gas-liquid separator 32. hydria 33. screen 34,35.compressor 40. freeze drier 44. pump

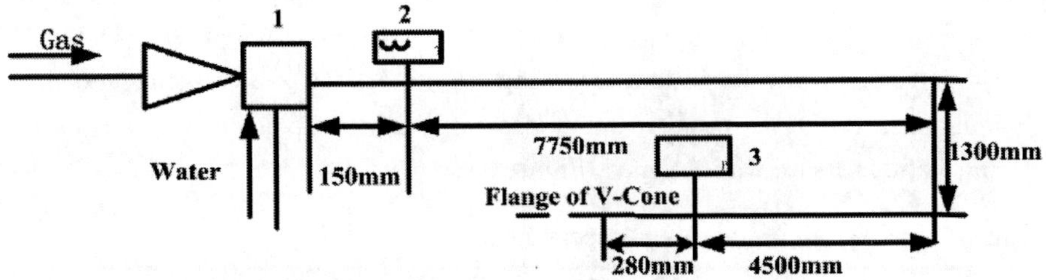

FIGURE. 3 Diagram of horizontal experimental pipe
1. injector 2. temperature sensor 3. pressure sensor

Study of the gas-liquid two-phase flow measuring method based on V-Cone flow meter was carried through in test installation of the Tianjin University automation institute oil/gas/water three-phase flow (Figure.2 and Figure.3); the measuring mediums were air and water, inner diameter of horizontal pipeline was 50mm, the space length of the pressure taps was 56mm, β=0.65, front angle of the V-Cone was 45°, latter angle was 65°.

Water entered the injector after passing through pump, badge flowmeter, electromagnetic flowmeter and vortex flowmeter in turn. Air entered the ejector after passing through compressors, freeze drier, buffer tank and Roots, vortex flowmeter and gas quality controller in turn. Here, the ail-gas two-phase flow rate and vapor quality are known. Then the two-phase flow flowed into the V-Cone. The differential pressure (DP) was measured and changed into 4-20mA current signal, then the current signal was changed into voltage, the result was imported the compute after carrying through proper filtering and analog-digital conversion. So did pressure and temperature.

Experiment Method and Data Processing

Because V-Cone is not the standard throttling gear, its discharge coefficient needs to be calibrated. Considered under gas-liquid two-phase flow of high water ratio condition, generally the liquid phase is the continuous phase, the gas phase is discrete phase, so the liquid should be calibrated first (Figure.4), then the gas phase should be calibrated (Figure.5). As shown in Figure.4 and Figure.5, the discharge coefficient $C_L = 0.866972$, the linearity of discharge coefficient C_L is 1.5%, the discharge coefficient $C_G = 0.855930$, its linearity is 3.7%. So, the mean of C_L and C_G is $C = 0.861451$, namely, the V-Cone discharge coefficient is 0.861451.

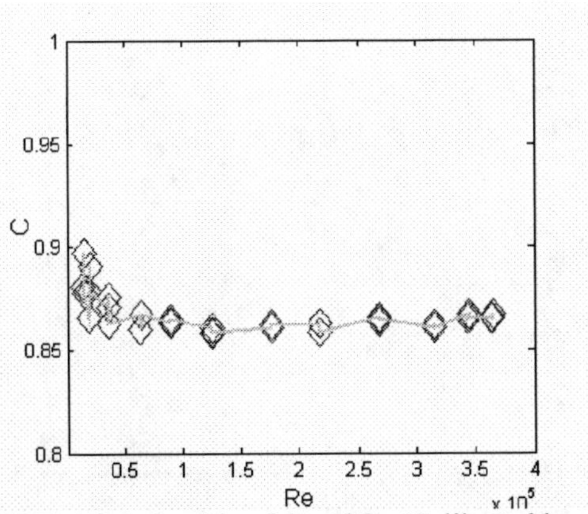

FIGURE. 4 Relationship of C and Re (Calibrated by water)

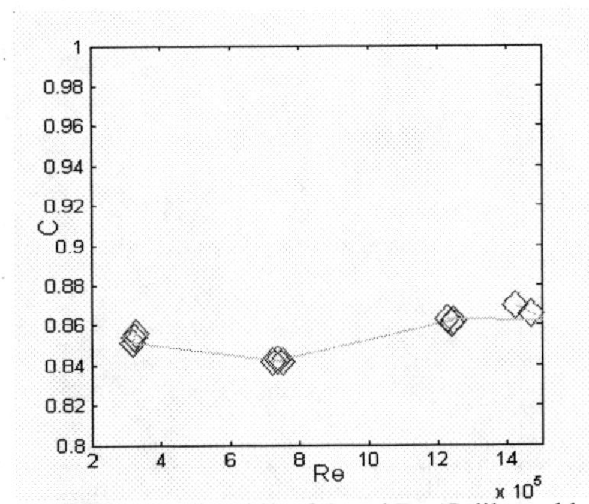

FIGURE. 5 Relationship of C and Re (Calibrated by air)

The corrective coefficient θ should be calibrated according to the following steps:

1) Under the point of static pressure condition, given the different gas phase and liquid phase flow separately, separately obtain ΔP_{TP}, ΔP_G and ΔP_L, which could be reorganized Martinelli parameters

$\sqrt{\Delta P_{TP}/\Delta P_G}$ and $\sqrt{\Delta P_L/\Delta P_G}$. Then fitting linear according to $\theta_1 = \sqrt{\Delta P_{TP}/\Delta P_G} - 1 \Big/ \sqrt{\Delta P_L/\Delta P_G}$ (TABLE 1).

2) The actual corrective coefficient θ_2 could be calculated according to the Lin Zonghu correlation. Obviously, there are some difference between θ_1 and θ_2. Assuming $\theta_3 = b1 * \theta_1 + b0$. The values of $b1$ and $b0$ can be calculated by least squares method ($b1 = -0.0243$ and $b0 = 1.0631$). And the value of θ_3 is approximately as θ_2 .

3) The curve of $\theta = f(\rho_G/\rho_L)$ may be Fitted according to ρ_G/ρ_L as x-coordinate and the value of θ_3 as y-coordinate by obtaining the different ρ_G/ρ_L and its corresponding the value of θ_3 (Figure.6). Samely, the relation between θ and static pressure was obtained (Figure.7).

TABLE 1. Values of air-water experiment(working temperature 30℃)

Sequence number	Liquid flow (kg/h)	Gas flow (kg/h)	x	P (MPa)	ΔP_{TP} (Pa)	ΔP_G (Pa)	ΔP_L (Pa)	θ_1	θ
1	6480.00	55.07	0.008498	0.15386	3841	324	3585	0.73457	1.0447
2	6220.07	36.81	0.005918	0.15752	3670	100	3585	0.84496	1.0437
3	7438.39	55.30	0.007435	0.15823	5172	333	4649	0.78714	1.0436
4	7538.88	48.57	0.006443	0.15900	5258	303	4649	0.80838	1.0434
5	7529.28	27.73	0.003683	0.16321	4932	103	4649	0.88130	1.0425
6	8597.08	48.35	0.005624	0.16442	6290	313	5766	0.81146	1.0423
7	8782.37	55.26	0.006292	0.16522	6784	348	5766	0.83914	1.0422
8	8800.03	20.79	0.002363	0.16671	6084	105	5766	0.89257	1.042
9	8786.55	27.63	0.003145	0.16789	6253	106	5766	0.90590	1.0418
10	8948.51	34.56	0.003862	0.16857	6022	107	5766	0.88579	1.0417
11	9919.79	20.75	0.002092	0.17044	7354	107	7059	0.89763	1.0415
12	9878.45	48.30	0.004889	0.17090	7870	325	7059	0.84125	1.0415
13	9942.86	27.82	0.002798	0.17158	7692	108	7059	0.92013	1.0414
14	10040.82	55.56	0.005533	0.17239	8512	363	7059	0.87143	1.0414
15	9854.96	34.53	0.003504	0.17287	7131	110	7059	0.88042	1.0413
16	11072.96	20.77	0.001875	0.17492	8810	110	8416	0.90904	1.0412
17	11080.00	27.63	0.002494	0.17710	9102	112	8416	0.92482	1.0411
18	11660.00	20.80	0.001784	0.17809	9602	112	9017	0.92064	1.0411
19	11226.33	48.29	0.004302	0.17827	9670	339	8416	0.87118	1.0411
20	11036.10	55.21	0.005003	0.17891	9798	376	8416	0.86750	1.0411
21	11525.10	27.67	0.002401	0.17907	9755	113	9017	0.92824	1.0411
22	11215.87	34.59	0.003084	0.17947	9160	114	8416	0.92700	1.0411
23	11428.00	34.55	0.003023	0.18067	9480	115	9017	0.91266	1.0411

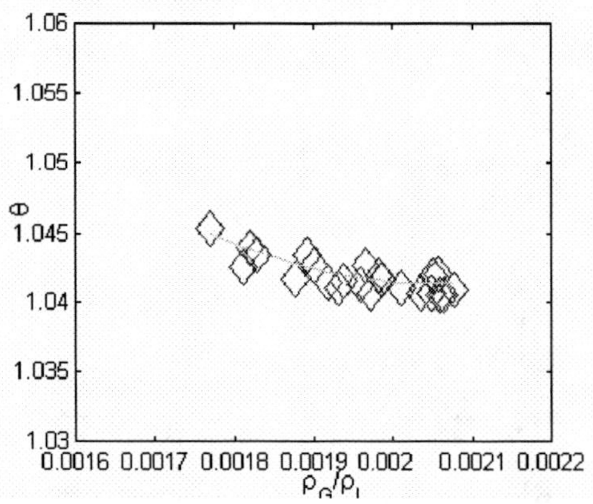

FIGURE. 6 Relationship between θ and ρ_G/ρ_L

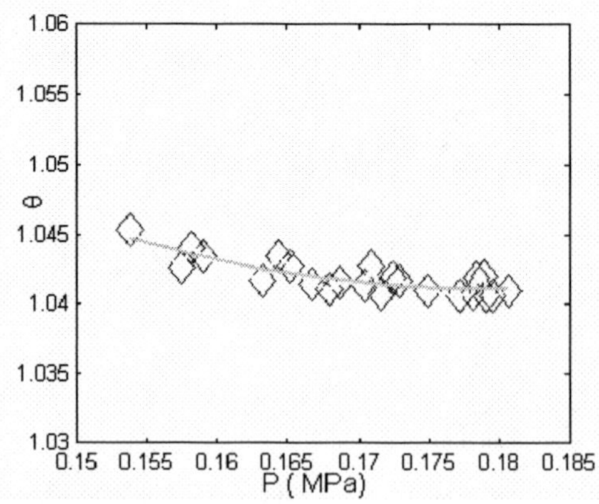

FIGURE. 7 Relationship between θ and P

The curve equation of Figure.6 is:

$$\theta = 42507 \times (\rho_G/\rho_L)^2 - 175 \times (\rho_G/\rho_L) + 1.015 \quad (7)$$

The curve equation of Figure.7 is:

$$\theta = 5.6254 \times P^2 - 2.0166 \times P + 1.2218 \quad (8)$$

Then the value of θ could be obtained according to equation (8) and (2), equation (9) is effective when the working temperature is 30℃.

The total mass flow W_{TP} is obtained according to the Lin Zonghu correlation. The vapor quality x is given, the gas mass flow and liquid mass flow are calculated separately. The range of relative error of gas mass flow are from -2.58% to +5.04% (Figure.8), the range of relative error of liquid mass flow are from -2.58% to+4.96% (Figure.9). The RMS error are 2.25% and 2.23% respectively.

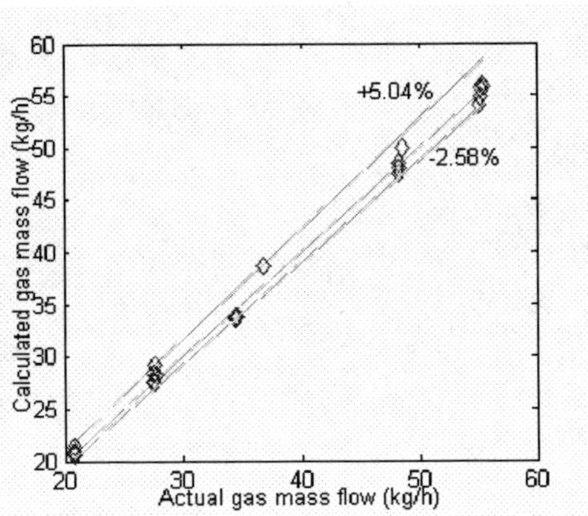

FIGURE. 8 Actual and calculated gas mass flow

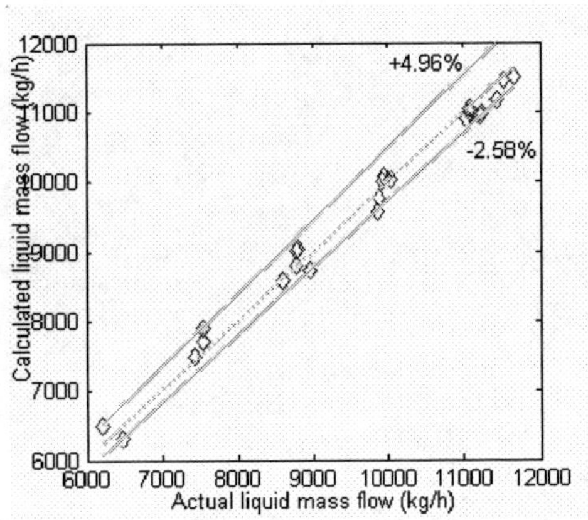

FIGURE.9 Actual and calculated liquid mass flow

CONCLUSION

(1) The discharge coefficient C and the corrective coefficient θ of the V-Cone are calibrated in this paper.

(2) On the basis of theoretical and experimental studies, several correlations, the Lin Zonghu correlation Murdock correlation is tested in air-water two-phase flow based on V-Cone, the corrective coefficient θ of the V-Cone is modified. The relative errors of gas mass flow range from -2.58% to +5.04% (Figure.8), the relative errors of liquid mass flow range from -2.58% to+4.96% (Figure.9). The RMS errors of gas mass flow and that of liquid mass flow are 2.25% and 2.23% respectively.

(3) To improve their general validity, additional experiments with different β, high static pressure, wider range of air phase and water phase flow data are needed. This work may be instructive in further study of air-water two-phase flow measuring using V-Cone.

ACKNOWLEDGEMENTS

The author appreciates the support from National Natural Science Foundation committee of P. R. China (No. 60532020) and Natural Science Foundation committee of Tianjin (No.05YFJMJC11600).

REFERENCES

1. Z. H. Lin, *Characteristics of Gas-Liquid Two-phase Flow in Pipelines and Their Engineering Applications,* Xi'an: Xi'an Jiaotong University Press, 1992.
2. Z. H. Lin, S. Z and D. Wang, *Gas-liquid Two-Phase Flow and Ebullient Heat Transfer*, Xi'an: Xi'an Jiaotong University Press, 2003, pp.182-186.
3. H. J. Zhang, S J Lu, G. Z. Yu, "An Investigation of Two-phase Flow Measurement with Orifices for Low-quality Mixtures", *Int. J. Multiphase. Flow* **18,** 149-155 (1992).
4. Y. Z. Sun, "'V' Shaped inner Conical Flow Meter", *Natural Gas Industry*. **24,** 105-110 (2004).
5. H. Q. Li, *Parameter Measurement of Two-phase Flow and Application*, Zhejiang University Press, 1991.

6. R. N. Steven, "Wet Gas Metering with a Horizontally Mounted Venturi Meter", *Flow. Meas. Instrum*, **12,** 361-372 (2002).
7. J. W. Murdock, "Two-Phase Flow Measurement with Orifices", *J. Basic Eng.* **84,** 419-433 (1962).
8. D. Chisholm, "Research Note: Two-phase Flow through Sharp-edged Orifices", *J. Mech. Eng. Sci.* **19,** No.3, 128-130 (1997).
9. R. V. Smith, J. T. Leang, "Evaluation of Correlations for Two-phase Flow meters Three Current-one New", *J. Eng. Pow.* , 589-594 (1975).
10. H. de Leeuw, "*Wet Gas Flow Measurement by Means of a Venturi Meter and a Tracer Technique*" in *North Sea Flow Measurement Workshop*, Scotland, Oct, 1994.
11. Z. H Lin, "Two-phase Flow Measurements with Sharp-edged Orifices", *Int. J. Multiphas. Flow* **8,** No.6, 683-693 (1982).

New Method of Online Measurement of Oil and Suspended Material Concentration In Flowing Waste Water

Hongwei Liao, Guobing Xu, Xinqiang Xu, Fangde Zhou

Key State Laboratory of Multiphase Flow and Power engineering
Xian Jiaotong University, Xian, China
Email: gbxj@stu.xjtu.edu.cn, FDZhou@mail.xjtu.edu.cn.

Abstract. At present, the most of the measurements of oil and suspended material concentration in waste water measuring are not online surveys. A new method of online measurement of oil and suspended material concentration in flowing waste water is presented. The room experiments and field tests showed that it is suitable to waste water treatment on line. After sampling, It needed to measure immediately the concentration in first time. Then let sample to be in still in 10 – 20 seconds. After that the bulk concentration was measured in second time. Because of the suspended solids having heavy density, they would be dropped from waster water. During ultrasonic operation,, emulsify the oil in waster water, the oil and suspended solid would be depart.. After that the third time measurement was done. In thus way the concentrations of oil and suspended solids can be measured. At present there are two on-site equipments operating in the Changqing oilfield, and the results are pretty well.

Keywords: waste water、oil、suspended material、online surveys、intensity、converter

INTRODUCTION

If water polluted by oil, its color, flavor and soluble oxygen have a great change. The olfactory threshold in drinking water is 5mg/L for light oil, heavy fuel oil to 0.2~10mg/L. Oil is harmful to aquatic organisms, 0.01mg/L oil on the sea having a 24-hour interval is able to fish oil smell. Annual industrial waste water discharged into rivers and seas through the material reached 30,000 tons of oil is one of the main pollutants in. China [1].

According to the survey, major oil-producing areas in China have adopted the oil-water separation technology in waste water treatment plant. But the online measurement equipment of oil and suspended material concentration in waste water t has not been ideal. At present, the main measuring technology principle is optical analytical, Each time it needs to collect samples, a long time testing, online testing can not be consecutive. And also the measurement sensitivity and the measurement range have also been restricted.

National Environmental Protection Departments also attaches great importance to measurement of oil and suspended material concentration of waste waster, and formulated a number of the relevant standard. In addition, oilfield monitored the suspended material concentration on a regular basis. Therefore, the equipment to rapid and accurate detecting oil and suspended material concentration in waste waster is needed . At present, the majority of the suspended particles measuring instrument adopt the "gravimetric method", but it needs a long time's measuring, accuracy is poor, the testing process complicated and cumbersome equipment, and the work efficiency is low. It is not suited to online monitoring, without measuring particle size distribution function, unable to meet the functional and performance requirements.

The measuring methods of the oil concentration currently are including the gravimetric method, batch adsorption method, UV spectrophotometry, fluorescence spectroscopy, infrared spectroscopy and non-dispersive infrared spectrophotometry [2][3].

Gravimetric method is the use of organic acid extraction from the sample of oil extraction plants, solvents evaporated, weighing in terms of oil content. In batch adsorption method, based on petroleum hydrocarbons having good adsorption capacity. it is used for to adsorb oily wastewater. The oil concentration can be obtained according to the weight difference of oily wastewater oil after and before drying,. The measurement of the UV spectrophotometry

CP914, *Multiphase Flow: The Ultimate Measurement Challenge,* edited by X. Cai, Y. Wu, Z. Huang, S. Wang, and M. Wang

[4] uses the UV characteristic absorption bond in the 215~260nm of aromatics compounds containing conjugated double in oil products to measure of oil content. The determination principles of fluorescence spectroscopy is under ultraviolet absorption of organic launched the quantitative fluorescence intensity. Infrared spectrophotometry include non-dispersive infrared spectrometry and infrared spectrophotometry, non-dispersive infrared spectrophotometry use the near-infrared absorption near 34 μm of paraffin oil methyl and methylene, infrared spectrophotometry use the hydrocarbons methyl and methylene and aromatic hydrocarbon three stretching vibration absorption wavelength.

To solve the oily wastewater monitoring; Online monitoring to the scene to resolve issues such as sampling, reducing amount of labor, we have experimented and made a new on- line detection of oil concentration and the concentration of suspended particulate matter concentration .

MEASURING PRINCIPLE

The Oily Wastewater Form

Oil is the raw material for petrochemicals. Oil production in the oil industry, oil refining, oil and petroleum processing, mechanical processing, and regeneration of waste oil, car repair will have a lot of oily wastewater. Oil in the form of the following types of water [5][6]:

(1). Floating oil: put it aside for at least a continuous phase after last film was floating on the water surface;

(2). Scattered oil: the small oil droplets greater than 1 μm diameter suspended and dispersed in water;

(3). Emulsified oil: the insignificant oil droplets in the form of in water - in-oil less than 1μm diameter suspended and dispersed in water;

(4). Dissolved oil: oil dispersed in water oil in the form of molecular methods or chemical methods, oil and water production homogeneous body.

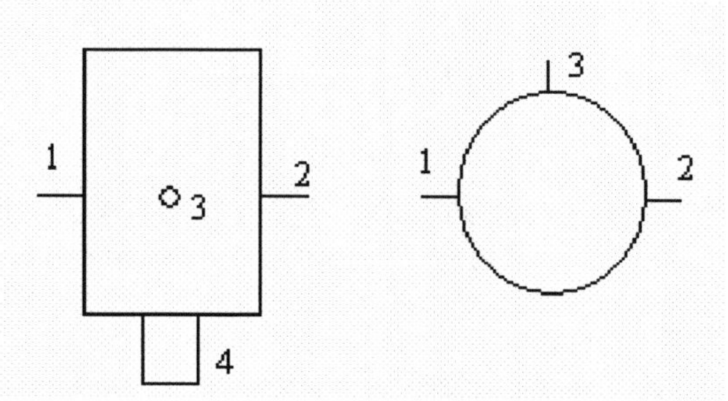

FIGURE 1. Measurement sketch map
1-Light Source 2-T-3DU 3-S-3DU 4-Ultrasound

Light Scattering and Transmission – Ultrasound

On the measuring beam sewage has its characteristic absorption and scattering characteristics. Certain spectrum beam irradiates into the wastewater, it will be scattered by the suspended particulates in the wastewater, the scattering light with a reasonable level of suspended particulates, proportional to the oil concentrations directly .By measuring the amount of light scattering. it can measure the concentration of suspended solids and oil in the waste water. There is a certain intensity index attenuation between the overall amount of light transmission and the concentration of waste water. The attenuation curve present roughly in three times. Light scattering and transmission components and sewage effluent can be a good measurement of oil concentration, the concentration of suspended solids, sediment concentration. To improve the measurement, scattering, optical transmission capacity converted into voltage signals through the photoelectric devices. The correlation of the concentrations of oil and suspended solids would be obtained from the voltage.

Measurement sketch map is showed in figure 1.

The concentration of suspended solids and suspended particles diameter is in a certain relation. There is a certain correlation coefficient between of suspended particles with an average diameter and measurement results. By using ultrasonic cavity effect, the emulsified oil and other particulates from the suspension can to be uniform; the effect of difference size of can be eliminated.

The Online Measurement of Concentration of Suspended Solids and Oil

After sampling, It needed to measure immediately the concentration in first time. Then let sample to be in still in 10 – 20 seconds. After that the bulk concentration was measured in second time. Because of the suspended solids having heavy density, they would be dropped from waster water. During ultrasonic operation,, emulsify the oil in waster water, the oil and suspended solid would be depart.. After that the third time measurement was done. In thus way the concentrations of oil and suspended solids can be measured.

Ultrasonic Wave Cleaning

The oil and suspensions will be attached to the measurement piece if there haven't cleaned, and the measurement results will also be affected. Therefore, it needs to adopt the ultrasonic wave cleaning method to achieve the purpose of online measuring.

This is mainly the use of ultrasonic cavity shock effect. Ultrasonic vibration box issues high frequency and high voltage oscillation signal. Transducer transforms it into vibration and then imports it to the cleaning solution. A continuous radial ultrasonic wave in the medium forward will have thousands of small vacuum bubble. Bubble in a certain pressure in the cleaning surface can form a series of intensive bombing, the constant impact of the surface being cleaned, including through to be cleaned from the other side, and all the media intrusion cavity, blind hole, slit, cleaning the dirt surface to achieve perfect results. Ultrasonic wave also plays an emulsification and neutralization role. The oil can be more effective in preventing re-attached to the object.

Measurement Formula

The relationship equation between scattering-voltage and the concentration of oil and suspended solids:

$$Y = B_0 + B_1 * X \tag{1}$$

Y—the concentration of oil, suspended solids, X—scattering voltage, B_0 , B_1—Parameters of the structure of the equipment, which is only respond to the equipment.

The relationship equation between transmission-voltage and the concentration of oil and suspended solids:

$$Y = A_1 * e^{-X/t1} + A_2 * e^{-X/t2} + A_3 * e^{-X/t3} + A_4 \tag{2}$$

Y—the concentration of oil、suspended solids, X—transmission voltage, A_1、A_2、A_3、A_4、t1、t2、t3—Parameters of the structure of the equipment, which is only respond to the equipment.

Measurement Curves

The correlation curves between scattering-voltage and the concentration of oil and suspended solids, for Figure 2:

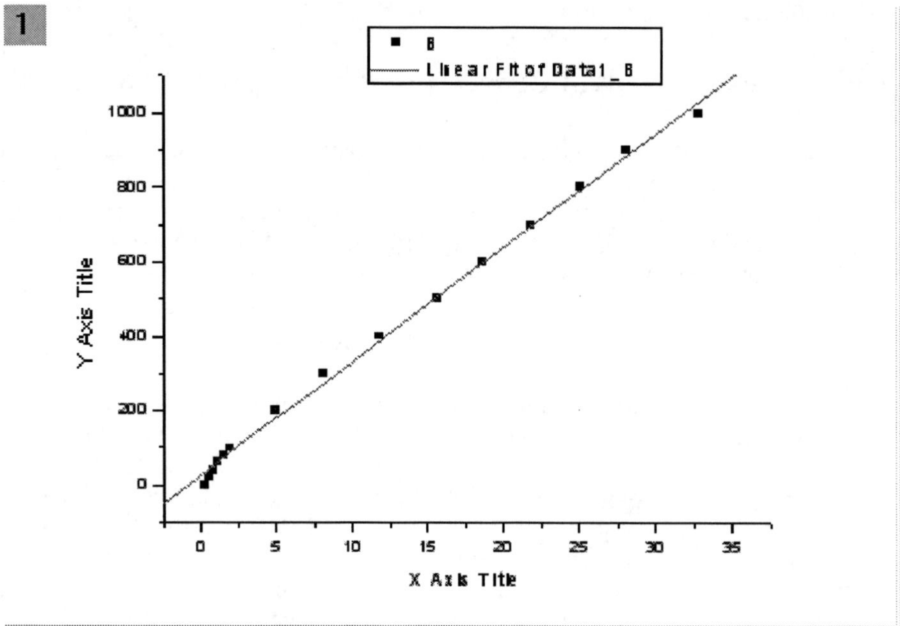

FIGURE 2. The correlation curves between scattering-voltage and the concentration of oil and suspended solids

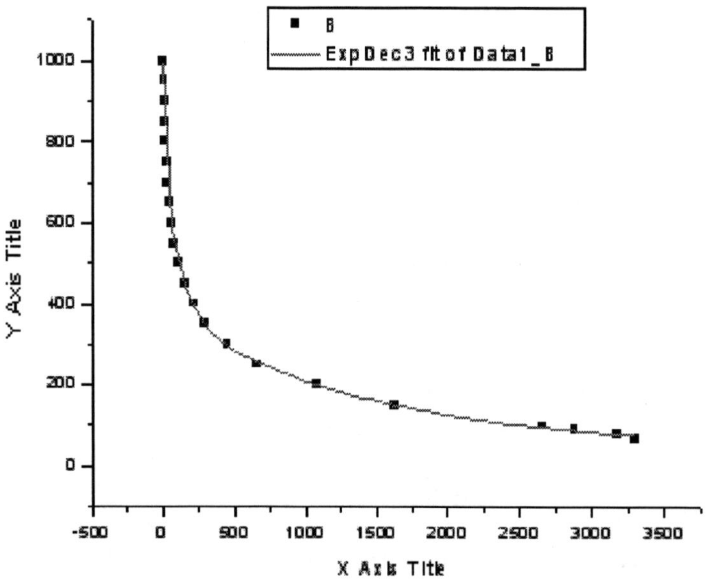

FIGURE 3. The correlation curves between transmission-voltage and the concentration of oil and suspended solids

The correlation curves between transmission-voltage and the concentration of oil and suspended solids, for Figure 3.

EXPERIMENTS

Spectrum Choice Experiment

Through the optical transmission and scattering tests, we concluded that blue and green spectrums accord with the requirements.

Oil Concentration Measurement Test

Through the testing、debugging at the scene, we concluded as below,

A、Oil concentration at entrance of waste water: Source green

The relationship equation between transmission-voltage's difference emulsion around and the concentration of oil and suspended solids:

$$Y = 546.9* e^{-X/28.4} + 424.7 * e^{-X/187.9} + 460.4 * e^{-X/1915.8} + 68.7 \quad (3)$$

Y—the concentration of oil X—transmission voltage

The relationship equation between scattering-voltage's difference emulsion around and the concentration of oil and suspended solids:

$$Y=-2.36+13.76*X \quad (4)$$

Y—the concentration of oil X—scattering voltage

B、YH-500 oil concentration NG at exit: Source blue

The relationship equation between transmission-voltage's difference emulsion around and the concentration of oil and suspended solids:

$$Y = 336.6 * e^{-X/1400.9} + 381.6 * e^{-X/119.7} + 584.6 * e^{-X/15} + 44.5 \quad (5)$$

Y—the concentration of oil X—transmission voltage

The relationship equation between scattering-voltage's difference emulsion around and the concentration of oil and suspended solids:

$$Y=26.5+30.6X \quad (6)$$

Y—the concentration of oil X—scattering voltage

Suspended Solids Concentration Measurement Test

Through the testing、debugging at the scene, we concluded below.

A、oil concentration at entrance of waste water: Source green

The relationship equation between transmission-voltage and the concentration of oil and suspended solids:

$$Y = 818.2* e^{-X/0.75} + 256.0 * e^{-X/7.3} + 649.5 * e^{-X/7.2} -171.0 \quad (7)$$

Y—the concentration of suspended solids X—transmission voltage

The relationship equation between scattering-voltage and the concentration of oil and suspended solids:

$$Y = 0.07537 + 0.000287*X \quad (8)$$

Y—the concentration of suspended solids X—scattering voltage

B、oil concentration at exit of treaded waste water: Source blue

The relationship equation between transmission-voltage and the concentration of oil and suspended solids:

$$Y = 553.2* e^{-X/0.14} + 582.6 * e^{-X/0.14} + 660.6 * e^{-X/1.56} -131.3 \quad (9)$$

Y—the concentration of suspended solids X—transmission voltage

The relationship equation between scattering-voltage and the concentration of oil and suspended solids:

$$Y = -270.1+2353.1*X \quad (10)$$

Y—the concentration of suspended solids X—scattering voltage

Oil and Suspended Solids Mixed Test

Found that the maximum oil droplet transmission angle at 160 degrees, so that we can overlook oil and other impurities, which is capable of measuring the real oil and suspended solids content, as showed in figure 4.

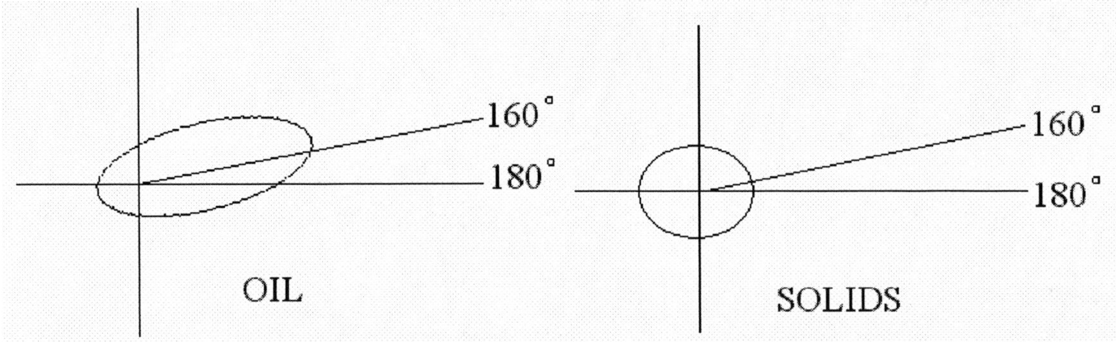

FIGURE 4. Oil droplet's maximum transmission angle

The ultrasound can be emulsified oil even distributed in sewage, and thus the effective measurement of oil content. The ultrasonic cleaning can be effectively conducted the online survey plots.

System Software

The sequential process control equipment used, procedures for the Visual Basic 6.0 development platforms can be achieved oily wastewater concentration, the concentration of suspended solids, sediment concentrations are measured on-line. Statements can be achieved and the automatic generation and printing, and the tendency to see over a period of time.

At present there are two on-site equipments operating in the Changqing oilfield, and the results are pretty well.

CONCLUSION

A new method of online measurement of oil and suspended material concentration in flowing waste water is presented. The experiments and field tests showed that it is suitable to waste water treatment. This kind of oil and suspended material concentration measuring method has stable performance, low detection limit, more sensitive than other equipment, high-precision.

ACKNOWLEDGEMENTS

This research program as the tenth "five years" important plan of China was supported by Nation Science and Technology Department from 2004-2006.

REFERENCES

1. Weiqing Mai, Wei Gu, Duyong Qi. Gravimetric determination of the content of mineral oil to improve sewage. Petrochemical. 1999, volume 28
2. Yuxin Wu. Determination of oil content in the sewage. petrochemical technical analysis tests. 1998.
3. Jiyun Li, Li Li, Chengwu Feng. Soil ultrasonic a UV oil content. University of Petroleum (Natural Science). 23 No. 6, 1999.
4. Dandan Zhang. Spectrophotometric Determination of fluorescence cabin ship oil in wastewater. Industrial Water Treatment. May 2000 Volume 20, No. 5.
5. Zhenmin Yang, Weijin Geng. Oil non-dispersive infrared law on the status of its application in environmental monitoring. Modern scientific instruments. 1999 No. 3
6. Yanliang Liu, Jing Liu, Wenqi Qi, Fusheng Wei. Oil analysis of the water situation. Environmental research. 5, 2000.

A New Measuring Method of Oil-water-gas Three Phase Flowrate by Using of Heat Transfer and Fluid Dynamics

Hongwei Liao, Siyi Zhou, Lei Liu, Fangde Zhou

Key State Laboratory of Multiphase Flow In Power engineering, Xian Jiaotong University, Xian. China.
Email: FDzhou@mail.xjtu.edu.cn

Abstract. It is so difficult to measure individual flowrates of oil-water-gas three phase on-line that very few real three phase flowmeters have been developed in the world. This paper presents a new measuring method of oil-water-gas three phase flowrate by using of heat transfer and fluid dynamics. A cylindrical cyclone is used to separating the most of natural gas from oil-water-gas three phase mixture. A flow mixer is used to homogenize oil and gas flow in a short distance for measurement of bulk flow velocity. A venture-tube is used to measure overall flowrate of three phase mixture. An inverse-U tube is used to measure gas fraction. The heat transfer way is used to measure water fraction. No radiation method is used. So this kind of flowmeter is much more safety, and has much more accuracy. Many room and field experiments have been done. The test results done by the Test Institute of Petroleum Industry of Daqing Oil Field Construction & Design Research Institute were quite well. This kind of three phase flowmeter has been widely used in many oil fields.

Keywords: oil-gas-water, three phase flow, flowrate measurement

INTRUDUCTION

Multi-component and multi-phase flow measurement where any combination of oil, water and air can occurs in many industries, such as in chemical and process industries, in power plant, oil refinery factories, and in particular, in the crude oil and natural gas exploration industry.

The flowrates of crude oil, natural gas and water three-phase mixture in an oil well are the basic parameters in the oil production. This is a basic desire to precisely and continuously measure the flowrates in order to control and predict of an oil well production.

In general natural gas and oil-water are measured separately by using single phase flowmeters, such as a turbo-meter,orifice plate, after heating and separating three-phase mixture into gas and liquid phases. So the measuring equipment is complex and expensive. It needs much fuel to heat fluid. The variation of three-phase flow pattern,phase fraction and liquid viscosity with temperature makes it is impossible to measure multi-phase flowrate by using a single phase flowmeter.

Despite considerable activity in the past. No entirely full multi-phase flowmeter was on the market. Especially in measuring water cut it is very difficult to reach at a precisely level. The one of main difficulties of measuring multi-phase flow is that people knows very few about the flow properties of oil/gas/water mixture. Usually gas and liquid flows in different velocity and regimes. A series of study on flow patterns and phase fraction of oil/gas/water three-phase mixture were done and got useful results.

Since 1988, We have been researched three-phase flowmeter and improved it. We have made a great success.

At first we use a cylindrical cyclone separator to separating the most of natural gas from oil-water-gas three phase mixture. A device of taking liquid sample from oil-water-gas three phase mixture was invented. We use heat transfer way to measuring water fraction in liquid, and use an inverse U-tube to measure phase fraction.

This paper presents the theoretical principle and measurement method for individual flowrates of crude oil, natural gas and water three phase mixtures. It describes the use of heat transfer way to measuring water fraction in liquid, and use an inverse U-tube to measure phase fraction. The phase mixer is used to homogenize the multi-phase mixtures.

CP914, *Multiphase Flow: The Ultimate Measurement Challenge,* edited by X. Cai, Y. Wu, Z. Huang, S. Wang, and M. Wang

MEASURING PRINCIPLE

It is so difficult to measure individual flowrates of oil-water-gas three phase on-line that very few real three phase flowmeters have been developed in the world[5][6]. Since 1988, We have been researched three-phase flowmeter and improved it. We have made a great success. A cylindrical cyclone is used to separating the most of natural gas from oil-water-gas three phase mixture. A flow mixer is used to homogenize oil and gas flow in a short distance for measurement of bulk flow velocity. A venture-tube is used to measure overall flowrate of three phase mixture. An inverse-U tube is used to measure gas fraction. The heat transfer way is used to measure water fraction. No radiation method is used. So this kind of flowmeter is much more safety, and has much more accuracy. From overall flowrate and phase fractions we can individual phase floarates.

Cylindrical Cyclone

A cylindrical cyclone separator is set up at the entrance of measuring device, and used to separating the most of natural gas from oil-water-gas three phase mixture . As strong swirl flow , the most of natural gas is separated and flow via upper pipe, A careful designed cylindrical cyclone can separates over 95% natural gas from whole three phase mixture and no liquid be entrained into gas. The remained overall three phase of low gas fraction is much more convenient to measure. Usually we can use a turbo-meter to measure single gas.

Phase Mixer

We use venture tube to measure the overall flowrate. In actually the phases may not necessary flow at the same velocity and the gas phase may dissolve or evolve form the liquid phase depending on pressure and temperature conditions. In most situations the water and oil are emulsions, oil-dispersed-in water and water-dispersed-in oil. In some cases water may drop out and flow along the bottom of the pipe. It depends on velocity and turbulence. In most gas and oil flow regimes, whether horizontal, vertical or inclined, the gas phase tends to travel faster than that the liquid phase. So We should pursue and empirical analysis to obtain the empirical correlations of phase fractions. However the phase fraction is depended on the flow patterns. Up to now, few data have been reported on relation of phase fraction and floe pattern. Chen [1] and Lin [2] did a series of experiments on the relation of phase fraction and flow pattern taking oil-gas-water three phase mixture to be working fluid.

One way to eliminate the effects of flow patterns on phase velocities is by using a mixer [3]. It is necessary to homogenize oil and gas flow in a short distance for measurement of bulk flow velocity. A mixer such as orifice plate,turbo-meter,nozzle, can take slip ratio between the phases to be zero. King [4] considered the most successful mixer is jet mixer device. In that a portion of the flow is diverted into a bypass loop and through a pump which then reinjects he diverted flow back into the main flow in a high velocity jet. We use a nozzle mixer to homogenize the main flow.

It can simplify the measuring condition, not need long steady straight tube. Experimental data showed that when static flow mixer is installed in front of venturi tube, the accuracy of total multiphase flowrate measurement is increased much higher.

Overall Flowrate

Total volume flowrate of multiphase mixture is measured by use of a venture tube. Here multiphase mixture can be treated as a single-phase fluid with equivalent mixture properties, and standard venture correlation can be applied.

$$Q_t = a\varepsilon\sqrt{\frac{\Delta P}{\rho_m}} \tag{1}$$

a is discharge coefficient, ε is compressibility coefficient, ΔP is differential pressure of venture, ρ_m is density of multiphase mixture.

$$\rho_m = \alpha_0\rho_0 + \alpha_w\rho_w + \alpha_g\rho_g \tag{2}$$

α is phase fraction, subscriptions o, w and g, denote oil, water, and gas

Phase Fraction

Phase Fractions are the main parameters in flowrate measuring. It is defined as the ratio of cross-section area occupied by phase i. A_i and the total cross-section area A_t,

$$h_i = \frac{A_i}{A_t} \tag{3}$$

Whereas volume fraction B is defined as the ratio of volume flowrate I, Q and the total volume flowrate Q,

$$\beta_i = \frac{Q_i}{Q_t} \tag{4}$$

usually they are not equal because of there being slip ratio among phases. For air/water the experiments on phase fraction were done and presented a lot of correlations to predict phase fractions. However very few papers discussed and measured the phase fraction. Minami [5] did an experimental study of two-phase flow to study liquid holdup in wet-gas pipeline. The two-phase mixture were used: kerosene/air, water/air, and water/surfactant/air. So it is necessary to do a serious of experiments and develop an accurate theoretical model for prediction of phase fractions. Chen extended the drift-flux model to three-phase mixture and obtain the correlations of phase fractions [2][4]. Its basic form is,

$$h_i = \frac{\beta_i}{c_o + \frac{<u>}{J}} \tag{5}$$

Where Co is an empirical constant. <u> is weighted average drift velocity, J is average drift velocity, subscript g denotes gas phase. Some results are shown in figure 2. The phase fraction was measured by optical probe.

In practice, we use volume fraction to calculate the individual flowrates, that is,

$$Q_i = K_i u B_i A_i \tag{6}$$

Where K_i is distribution constant of velocity.

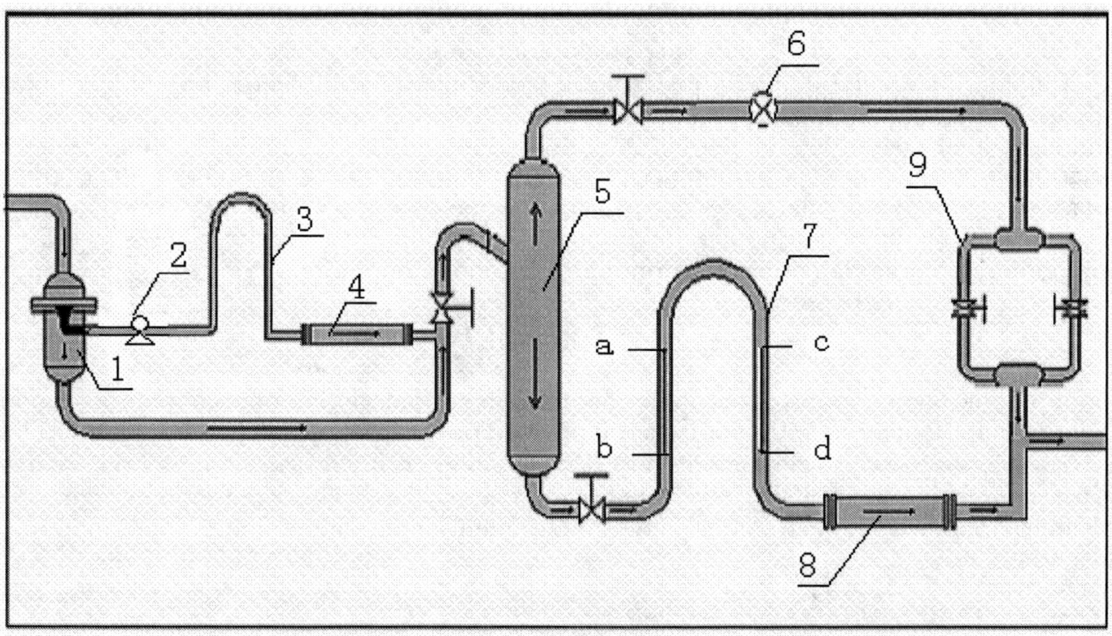

FIGURE 1. A schematic diagram of three phase flowmeter

1. Liquid sample-taking set 2. Gear metering pump 3. Inverse-U tube
4. electrical heater 5. cylindrical cyclone 6. vortex meter
7. Inverse-U tube 8. venture tube 9. pressure controller

As a mixer is set at upstream we take water and oil to be a liquid phase having same velocity after mixing.

As shown in Fig 1, when multiphase mixture flows to inverse-U tube, in section 1.2 and 3.4, there exist pressure drops. These pressure drops are consist of two parts, frictional pressure drop and gravity pressure drop.

$$\Delta P_{a,b} = \Delta P_{f,a,b} + \Delta P_{h,a,b} \quad (7)$$

$$\Delta P_{c,d} = \Delta P_{f,c,d} - \Delta P_{h,c,d} \quad (8)$$

The frictional pressure drop can be eliminated from Eq. (3) and (4), it can be obtain

$$\Delta P_{h,a,b} + \Delta P_{h,c,d} = \Delta P_{a,b} - \Delta P_{c,d} \quad (9)$$

In up flow and down flow, the velocity difference between gas and liquid is not same, as a result of gravity force. In up flow, gas velocity is higher than liquid. In down flow gas velocity is lower than liquid.

The differential pressure drops are measured by special designed transducer. That is no-contact flanged long distance-transmit transducer.

Special ring are installed in front of measuring spots $\Delta P_{,a,}$ ΔP_c and in rear of measuring sports $\Delta P_{,b,}$ ΔP_d in order to eliminate the upstream effect on differential pressure measurement.

Water Fraction

The measurement of water fraction is the most difficult thing in three phase flowmeter.

The heat transfer way is used to measure water fraction. There is no radiation in measuring.. So this kind of flowmeter is much more safety, and has much more accuracy.

A special sample-taking has been designed. This special sample-taking device can take and supply single liquid from three phase mixture. Using a metering gear pump to take a certain fluid can avoid the flowrate accumulation deviation. Even if there is few gas in the liquid, It can be measured by a U-tube device. The water fraction is obtained from thermal equilium.

When oil-water two phase mixture flowed a heated tube. The heat of tube is transfered to fluid. The heat transfer is related to the fluid flowrate and thermal physics properties.

$$Q = G * \rho_n * \Delta T * Cpn \quad (10)$$

$$Cpn = Hw * Cpw + Ho * Cpo \quad (11)$$

$$\rho n = Hw * \rho w + Ho * \rho o \quad (12)$$

Where Q is Heat flux, G is volume flowrate, Cp is specific heat of mixture of oil and water. Subscripts o,w ,n denote oil , water and two-phase . ρis density .

The differences of specific heat of water and oil are much larger. So ,according to the increases of fluid temperature, the water phase fractions are able to be measured. An electricity heater having constant power is used to heat liquid sample. As the difference of specific heat between water and oil is big. According to the heat temperature differences, the water fraction can be obtained. By using heat transfer way The water faction measuring accuracy is much higher than other ways.

TEST RESULTS

So many room and field experiments have been done these years. Through tests We improved the measuring way and device. Recently We did a formal test in daqing oil field by The Test Institute of Petroleum Industry of Daqing Oil Field Construction & Design Research Institute. They use crude oil, natural gas and water to be test medium. The test device has been tested by China Measuring Science Research Institute and has gotten the ceitification. The main technique parameters has reached at , even over the world level.

Measuring ranges are,

liquid flowrate 2.7 – 27 m /h,

gas flowrate 10-100 m /h (working condition)

water fraction 0 – 100 %.

Diameter of flowmeter is 80 mm. Flowmeter is set up DN80mm horizontal line of multiphase flow test device. In front of flowmeter there are 12 m long tube for full developed flow. In rear of flowmeter there are 4 m straight tube.In test process, test time is 10 minutes. The stable time is about 10 minutes.

This test took51 points, 55 date for three phase flowmeter have been done. The accumulation flowrates of Oil,water ,gas flowrates are tested at standard condition. The flowrates errors are relative deviations.

Liquid test error

In 55 test date, there are 44 date of Liquid flowrates is in less than ±5.0 % deviation. It is about 80% of the sum; there are 4 date is in ±5.0 ±8.0 % deviation. It is about 7.2% of the sum; there are 7 date is bigger than ±8.0 % deviation. It is about 12.8% of the sum;

Gas test error

In 55 test date there are 41 date is in ±8.0 % deviation. It is about 74.5% of the sum; there are 6 date is in ± 8-10.0 % deviation. It is about 10.9% of the sum; there are 8 date is bigger than ±10.0 %, deviation. It is about 14.6% of the sum;

Water fraction test error

In 55 test date there are 45 date is in ±2.0 %, deviation. It is about 81.8% of the sum; there are 4 date is in ±2.0-±5.0 % deviation. It is about 7.3% of the sum; there are 6 date is bigger than ±5.0 % deviation. It is about 10.9% of the sum;

The big deviations of some test date were caused by unstable transducers. It can be decreased by implove transducers.

Now, There are over 50 this kind oil-gas-water three phase flowmeter have been used in Dagang, Zhongyuan Victory, Jidong and xingjiang oil fields. There are 6 sets of three phase flowmeters were used at offshore of CNOOC. They operates very well.

SUMMARY

A theoretical principle and the measurement method are proposed for individual flowrates of crude oil,natural gas and wate three phase mixture.

A cylindrical cyclone is used to separating the most of natural gas from oil-water-gas three phase mixture. It is very effective for gas-liquid separating, can separates over 95% natural gas from whole three phase mixture and no liquid be entrained into gas.

The heat transfer way is used to measure water fraction. There is no radiation in measuring.. So this kind of flowmeter is much more safety, and has much more accuracy.

This kind of three phase flowmeter has been widely used in many oil fields. It still needs great effort to improve it and to further raise the measuring accuracy.

ACKNOWLEDGEMENTS

This research program was supported by nation development and reform committee from 1998-2000.

REGERENCES

1. Chen, X. Chen, I. Zhou, F. and Lin, F.: "Grude Oil-Natural Gas-Water Three-Phase Flowmeter," SPE annual Conference and Exhibition. SPE 18219. Houston. (2-5 October 1988).
2. Chen, X. : " The Study of Flow Patterns-Pressure Drop-Phase Fractions of Oil/Air/Water Three-Phase Mixture in Vertical Tube," phD.Thesis. Xi'an Jiaotong University, China. (1990)
3. King.N.W.: "Multi-phase Flow Measurement at NEL," J of Measurement and Control. 21 (1998) 237-239.
4. Chen, X., Zhou, F,; "Determination of the Void Fraction in Helical Coiled Tubes," Alterative Energy Sources 3. Edited by T.N. Veziroglu. Hemisphere Publishing Co, U.S.A. (1986).
5. Pal,R., Rhodes, E.: " Methods for Metering oil and Water production of Wells," 2nd Int. Conference on Multi-phase Flow. 19 1 (March 1980) 24-27.
6. Kinghorn, F .: "Challenging Areas in Flow Measurement," J. Measurement and control. 21 (October 1988) 229-235.

Signal Processing In Transmission Fluctuation Spectrometry With Band-Pass Filter

Jianqi Shen*, Yamin Xu*, Bin Yu*, Xiaoai Guo+ and Ulrich Riebel+

* *Institute of Science, University of Shanghai for Science and Technology, Shanghai 200093, China*
+ *Lehrstuhl Mechanische Verfahrenstechnik, Brandenburgische Technische Universität Cottbus, Cottbus 03046, Germany*

Abstract. Transmission fluctuation spectrometry is a recently-developed method for particle size analysis, in which both the particle size distribution and particle concentration can be obtained simultaneously. This work presents a new technique of data processing of the fluctuating transmission signal. The 1st order band pass filter is employed to improve the resolution of the measurement. The analytical expression is obtained on the basis of the layer model. Numerical results and simulated results are given to make a comparison.

Keywords: transmission fluctuation, particle analysis, particle size distribution, particle concentration
PACS: 42.62.-b, 42.68.-w

INTRODUCTION

Transmission fluctuation spectrometry (TFS) is a new method for particle size analysis. Different from the traditional extinction measurement, the TFS measures the transmitted light with high temporal and spatial resolutions. A narrow beam is used and the signal is received by a detector with a broadband sensing circuit. So the transmission signals show significant fluctuations, which include the complete information on particle size and particle concentration. In the past years, different techniques of the TFS have been developed according to the methods of signal processing. One of those is the TFS with variable spatial averaging, whereby the transmission fluctuation spectrum, expressed in terms of the expectancy of transmission square (ETS), is obtained with a variable beam diameter [1]. Another is the TFS with variable temporal averaging, in which the fluctuating transmission signals are first subjected to a variable temporal averaging and then the non-linear operation to obtain the ETS [2]. Recently, the TFS is realized in the frequency domain, whereby the transmission fluctuation spectrum is obtained with analog signal filtering [3]. This method offers a possibility of real-time measurement. However, it is found that the resolution of the measurement on particle size is low, due to the characteristics of the transfer function of the first low pass filter. In our recent publications [4, 5], we attempted to improve the measurement resolution, by using the high order low pass filters. This work will theoretically present the technique of TFS with the band pass filters.

THEORY

For simplicity, let's consider an extremely thin beam, compared to the particle size. The transmission signal is a binary one (i.e. 0 and 1) according to the presence and/or the absence of the particles (see Fig.1). So, while the particle suspension flows across the beam with a constant velocity, the blockage of the beam contains the information on particle sizes and the average of the fluctuating transmission signal includes the information on particle concentration. In first view of this, we might image that a certain width of the beam blockage corresponds to a certain particle size. However, the case is much complicated. Even for those particles with the same size, the beam blockage is different according to the location of the particles relative to the beam. Thus a certain particle size corresponds to a certain range of the beam blockage or a certain range of frequency in the Fourier space. The above discussion may also apply to the case of a beam whose diameter is narrow (i.e. the beam diameter is smaller than, compared to or a little larger than the particle size).

CP914, *Multiphase Flow: The Ultimate Measurement Challenge,* edited by X. Cai, Y. Wu, Z. Huang, S. Wang, and M. Wang

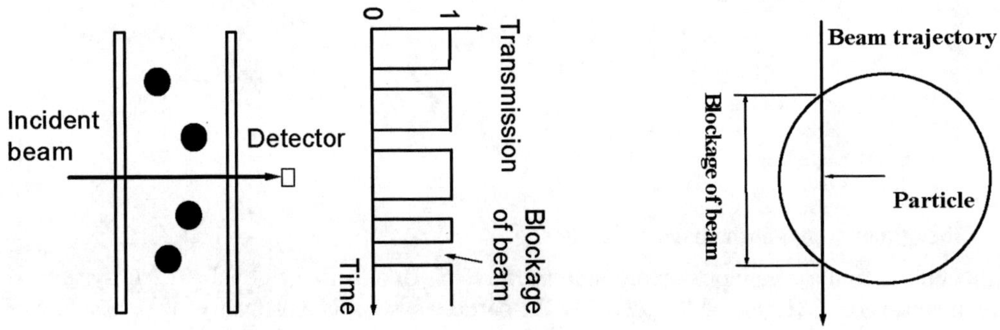

FIGURE 1. The fluctuating transmission signal.

In past, the TFS in the frequency domain employed an analog system of 12 parallel channels, each channel includes a 1st order low-pass filter [3]. The cut-off frequencies are uniformly distributed on a logarithmic scale (see Fig.2). This means the channel with a higher cut-off frequency includes the information which is included in all those channels with lower cut-off frequencies and this tends to reduce the resolution of particle size analysis. A possible improvement might be the use of the band pass filters, which can diminish the overlapping of the signals between different channels.

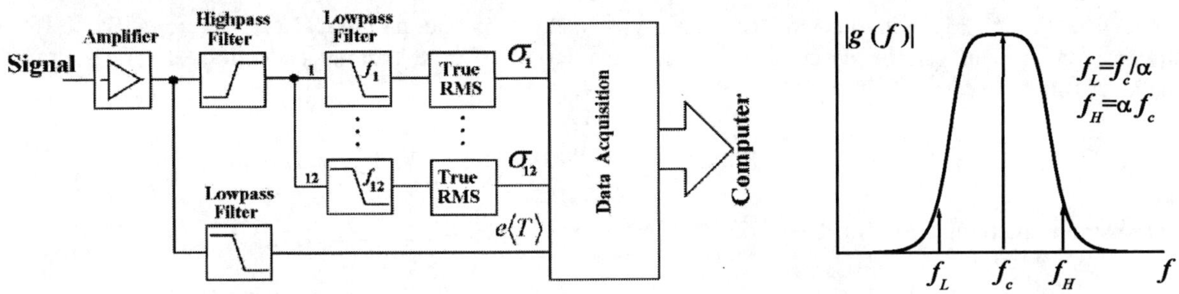

FIGURE 2. Scheme of analog signal processing by electronics.

The analog system is almost same as shown in Fig.2, whilst the low pass filters are replaced by the band pass filters, each containing band pass filter (i.e. a 1st order low pass filter and a 1st order high pass filter in series). The cutoff frequency of the high pass filter is f_L and that of the low pass filter is f_H. So, as shown in Fig.2, the central frequency of the band pass filter is $f_C = (f_L f_H)^{\frac{1}{2}}$ and the band width is determined by a parameter $\alpha = (f_H / f_L)^{\frac{1}{2}} > 1$. We assume that α is same for all channels. The transfer function of the band pass filter is

$$\left|g_{f_L, f_H}\right| = \frac{1}{\sqrt{1+(f/f_H)^2} \cdot \sqrt{1+(f_L/f)^2}} \tag{1}$$

After the fluctuating transmission signal is subjected to such a data processing, the signal is expressed in terms of the ETS, $e\{T^2\}$. Statistically, it can be further described as

$$e\{T^2\} = e\{T\}^2 + \sigma_T^2 \tag{2}$$

whereby $e\{T\}$ is the average transmission and σ_T is the standard deviation, defined as

$$e\{T\} = \lim_{t_s \to \infty} \frac{1}{t_s} \int_0^{t_s} T(t) \mathrm{d}t$$

$$\sigma_T^2 = \lim_{t_s \to \infty} \frac{1}{t_s} \int_0^{t_s} [T_g(t)]^2 \mathrm{d}t \tag{3}$$

here, $T_g(t)$ is the output signal of the band pass filter.

It is very difficult to find an analytical expression of the ETS for a 3-dimensional particle system. However, this can be done on the basis of the layer model, whereby the particle system is reasonably taken as a pile of monolayers [3,6]. For a 3-dimensional mono-dispersed particle system with a pathlength ΔZ, the average transmission is a product of those through the single layers and so is the ETS as long as the particle concentration is low enough.

$$e\{T\} = e\{T_{ML}\}^{N_{ML}}$$

$$e\{T^2\} = e\{T_{ML}^2\}^{N_{ML}} \tag{4}$$

Here, the subscript *ML* is an abbreviation of "monolayer". $N_{ML} = 1.5\Delta Z / Px$ is the number of layers of the particle system. $P \geq 1.5$ is the structure parameter and x is the particle size. Therefore, the ETS of the 3-dimensional system can be described with a single layer. After a tedious deduction, the ETS through the monolayer is obtained as

$$e\{T_{ML}^2\} = 1 - \beta \cdot [2 - \chi(\Lambda, \Psi_C, \alpha)] + \beta^2 \cdot [1 + \varepsilon(\Lambda, \Psi_C, \alpha, \beta)] \tag{5}$$

The transition functions are defined as

$$\chi(\Lambda, \Psi_C, \alpha) = \int_0^\infty F_{\text{spat}} \cdot F_{\text{part}} \cdot F_{\text{filt}} \cdot \mathrm{d}u \tag{6}$$

$$\varepsilon(\Lambda, \Psi_C, \alpha, \beta) = \int_0^\infty F_{\text{spat}} \cdot F_{\text{part}} \cdot F_{\text{filt}} \cdot G_{PY}(u, \beta) \cdot \mathrm{d}u \tag{7}$$

whereby F_{spat} is a factor due to the spatial averaging over the beam cross section ($F_{\text{spat}} = \exp\left[-(u\Lambda/2)^2\right]$ for a Gaussian beam and $F_{\text{spat}} = [2J_1(u\Lambda)/(u\Lambda)]^2$ for a uniform beam). $F_{\text{part}} = 2J_1^2(u)/u$ is a factor of the spherical particle and F_{filt} is the factor from the band pass filter, which is found to be

$$F_{\text{filt}} = \frac{1}{1-\alpha^{-4}} \left[\sqrt{1 + \left(\frac{u}{\alpha \Psi_C}\right)^2} - \sqrt{1 + \left(\frac{\alpha u}{\Psi_C}\right)^2} \right] \tag{8}$$

$G_{PY}(u, \beta)$ is the factor describing the structure of the monolayer. $\Lambda = D/x$ is the ratio of the beam-to-particle diameter and $\Psi_C = \pi x / v$ is the dimensionless frequency expressing the centre of the band. D is the beam diameter and v is the flow velocity. $\beta = P \cdot C_V$ is the monolayer density and C_V is the volume concentration.

For a dilute mono-dispersed monolayer, Eq.(5) may be rewritten as

$$e\{T_{ML}\} = 1 - \beta$$
$$\sigma^2_{T_{ML}} = \beta \cdot \chi(\Lambda, \Psi_C, \alpha) + \beta^2 \cdot \varepsilon(\Lambda, \Psi_C, \alpha, \beta) \approx \beta \cdot \chi(\Lambda, \Psi_C, \alpha) \tag{9}$$

However, for a 3-dimensional particle system, the case is much complicated. Assuming low particle concentration (i.e. $\beta \to 0$) and combining Eq.(5) with Eq.(4), we may find

$$\ln e\{T^2\} \approx -\frac{1.5\Delta Z}{x} \cdot C_V [2 - \chi(\Lambda, \Psi_C, \alpha)] \tag{10}$$

Also, we may obtain the expression of the average transmission.

$$\ln e\{T\} \approx -\frac{1.5\Delta Z}{x} \cdot C_V \tag{11}$$

Finally, we may describe the standard deviation as

$$\sigma_T^2 \approx \frac{1.5\Delta Z}{x} \cdot C_V \cdot \chi(\Lambda, \Psi_C, \alpha) \tag{12}$$

The above results can be extended to the poly-dispersed, dilute particle system.

$$\ln e\{T\} \approx -\sum_i \frac{1.5\Delta Z}{x_i} \cdot C_{V,i} \tag{13}$$

$$\sigma_T^2 \approx \sum_i \frac{1.5\Delta Z}{x_i} \cdot C_{V,i} \cdot \chi(\Lambda_i, \Psi_{C,i}, \alpha) \tag{14}$$

Particle size distribution and particle concentration can be obtained from Eqs.(13-14) with the inversion algorithm.

NUMERICAL CALCULATION AND SIMULATION

To our knowledge, the transition function $\chi(\Lambda, \Psi_C, \alpha)$ can be obtained by numerical calculation only. As an example of this, Fig.3 plots the curves of the transition function for Gaussian beams with variant values of the beam diameter Λ and the filter parameter α.

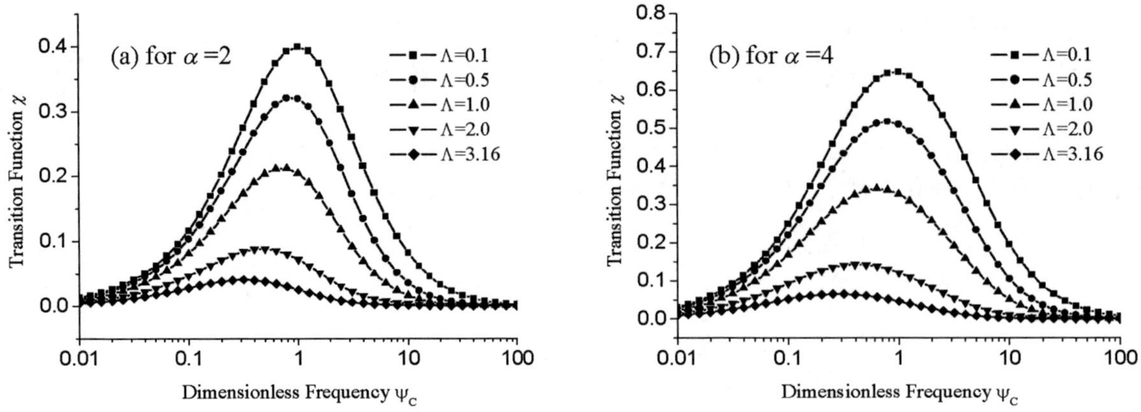

FIGURE 3. Theoretical results of the transition function $\chi(\Lambda, \Psi_C, \alpha)$.

It can be found that $\chi(\Lambda, \Psi_C, \alpha)$ decreases along with the increase of the beam diameter due to the spatial averaging of the fluctuating transmission signal over the cross section of the beam. Each curve has a peak in the vicinity of $\Psi_C = 1$, which includes the information of the particle size. The filter parameter α, denoting the band width of the filter, has effects on the transition function. Comparing Fig.3a and 3b, we may find that the transition function is larger and wide for a bigger α, allowing higher values of the standard deviation but a lower resolution of the measurement. Therefore, this parameter should be chosen carefully.

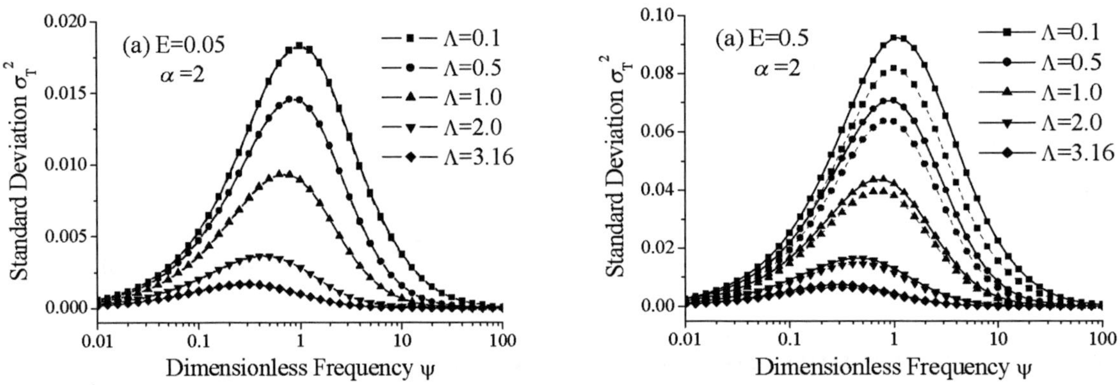

FIGURE 4. Comparison of the standard deviations σ_T^2 obtained from theory (in dash, thin lines) and simulation (in solid, thick lines).

The simulation is performed in a 3-dimensional mono-dispersed particle system. The number of layers is 40 and the particle volume concentrations are 0.0833% and 0.827%, corresponding to the extinction values of E=0.05 and 0.5. The standard deviations by simulation and by theory (calculated with Eq.(12)) are shown in Fig.4 to make a comparison. We may find a complete match between the theory and the simulation for low particle concentrations. However, the difference between theoretical and simulated results occurs and becomes larger as the particle concentration increases.

CONCLUSIONS

A new method of data-processing for the transmission fluctuation measurements is presented in this work, whereby the band pass filter is employed. An analytical expression is obtained on basis of the layer model and the assumption of low particle concentrations. By numerical calculation and simulation, it is found that the theory is capable to match the simulation satisfactorily for dilute mono-dispersed particle systems. This offers a possibility to measure the particle size distribution and particle concentration in dilute two-phase flows.

ACKNOWDGEMENTS

The authors thank for the financial support from DFG (grant Ri 533/7-2), NSFC (grant 50376041) and Shanghai Education Development Foundation (grant 04SG49).

REFERENCES

1. U. Kräuter: Grundlagen zur in-situ Partikelgrößenanalyse mit Licht und Ultraschall in konzentrieten Partikelsystemen. Thesis, Karlsruhe, 1995
2. M. Breitenstein: Grundlagenuntersuchung zur Statistischen Partikelgrößenspektrometrie mittels Auswertung der Transmissionsfluktuationen von Licht in Dispersen Systemen. Thesis, Cottbus, 2000
3. J. Shen: Particle size analysis by transmission fluctuation spectrometry: Fundamentals and case studies, Dissertation, Cuvillier Verlag Göttingen, 2003
4. Y. Xu, J. Shen: Simulation on high concentration effects of transmission fluctuation spectrometry, Chinese Journal of Laser, Vol.33, 2006, pp.253-260 (in Chinese)
5. Y. Xu, B. Yu, L. Liu & J. Shen: Measurement results on transmission fluctuation spectrometry by using the 2nd order low pass filter, ACTA OPTICA SINICA, Vol. 26, 2006, pp.1495-1500 (in Chinese)
6. U. Riebel, U. Kräuter, Extinction of radiations in sterically interacting systems of monodisperse spheres, Part 1: Theory. Particle and Particle System Characterization, Vol.11, 1994, pp.212-221

Measurement Of Multiphase Flow Water Fraction And Water-cut

Cheng-gang Xie

Schlumberger Cambridge Research Limited, Madingley Road, Cambridge CB3 0EL, UK

Abstract. This paper describes a microwave transmission multiphase flow water-cut meter that measures the amplitude attenuation and phase shift across a pipe diameter at multiple frequencies using cavity-backed antennas. The multiphase flow mixture permittivity and conductivity are derived from a unified microwave transmission model for both water- and oil-continuous flows over a wide water-conductivity range; this is far beyond the capability of microwave-resonance-based sensors currently on the market. The water fraction and water cut are derived from a three-component gas-oil-water mixing model using the mixture permittivity or the mixture conductivity and an independently measured mixture density. Water salinity variations caused, for example, by changing formation water or formation/injection water breakthrough can be detected and corrected using an online water-conductivity tracking technique based on the interpretation of the mixture permittivity and conductivity, simultaneously measured by a single-modality microwave sensor.

Keywords: multiphase flow measurement, microwave transmission, water fraction, water-cut, salinity

INTRODUCTION

Multiphase flowmeters (MPFM) are being increasingly used in well testing, well management, and production allocation [1]. For topside applications the main advantage of MPFM is the elimination of bulky test separators and the associated metering equipment and test lines. For subsea wells MPFM have the potential of eliminating the platform. With a meter on each well, hydrocarbon and water production can be monitored continuously. A single meter installed downstream of the test manifold allows well testing to be performed more quickly and frequently than with conventional systems. The need for an MPFM to measure accurately over a wide gas-fraction range requires a robust flow rate measurement for all flow regimes, both water-continuous (high water cut) and oil-continuous flows, and over a wide range of water salinities and fluid viscosities. It is important for an MPFM to have minimal calibration requirements, particularly for permanent and/or subsea applications.

Water cut (or water-liquid-ratio, WLR) is a key measurement in deriving the oil and water flow rates from the measured total liquid flow rate. To improve multiphase flow measurement capability in permanent installations where the water salinity can vary over time (e.g., caused by changing formation water or breakthrough of formation/injected water), it is necessary to detect and correct for such changes. An online water conductivity tracking technique based on interpreting multiphase flow mixture permittivity and conductivity simultaneously measured by a single-modality microwave sensor is described in the section following the Introduction.

A proprietary microwave-transmission system is then described to make WLR measurements for both water- and oil-continuous flows and over water-conductivity and gas-cut ranges that are far beyond the capability of microwave-resonance-based sensors currently on the market. This system measures the amplitude attenuation and phase shift across a pipe diameter at multiple frequencies using cavity-backed antennas, and the multiphase-flow mixture permittivity and conductivity are derived by utilizing a unified microwave transmission model, which further allows tracking of the water conductivity. For three-phase flow WLR determination, a gas-oil-water three-component mixing model is established (with tracked water conductivity as input in changing salinity situations), using mixture permittivity, or mixture conductivity and an independently measured mixture density.

CP914, *Multiphase Flow: The Ultimate Measurement Challenge,* edited by X. Cai, Y. Wu, Z. Huang, S. Wang, and M. Wang

ONLINE WATER CONDUCTIVITY FROM MICROWAVE COMPLEX PERMITTIVITY

Microwave sensors have been developed for the online tracking of multiphase-flow water conductivity by interpreting the simultaneously measured mixture permittivity (ε_m) and mixture conductivity (σ_m) [2]. In gas-liquid flows of high WLR (water-continuous with the gas phase dispersed in the liquid phase), and with a constant water conductivity, the data points in the mixture conductivity and mixture permittivity parameter space fall on a line connecting 100% oil and 100% water points. (See Fig. 1a for measured multiphase flow, ε_m-σ_m, data with water conductivity up to ≈16 S/m.) In these water-rich multiphase flows, the mixture complex permittivity is dominated by the presence of water, whereas the gas and oil (hydrocarbon) have small and similar complex permittivities. The hydrocarbon (gas and oil) volume fraction, sensed over the same sensing volume using a single-modality microwave sensor, influences σ_m and ε_m in a similar way. (A conventional low-frequency system based on a separate electrical capacitance sensor for oil-continuous flow and an inductance sensor for water-continuous flow is incapable of measuring ε_m and σ_m simultaneously and over the same sensing volume.) This characteristic trend (Fig. 1a) is unique to water-rich flows and the slope of the set of measured mixture conductivity versus permittivity data points forms the basis of the online water-conductivity estimate. The water salinity (proportional to density) is determined from the water conductivity, given knowledge of the (dominant) salt species and the measured fluid temperature. This broad concept is applicable to all microwave measurement techniques including transmission, reflection, and resonance methods; a common process exists for estimating the water conductivity from the fundamental conductivity-permittivity parameter space. It is worth noting that the commercially available microwave resonance-based method has a very limited salinity-working range (e.g., NaCl solution with salinity < 30k ppm or < 4.3 S/m at 20°C), because of the absence of a resonance response when the multiphase-flow mixture has a moderately high bulk electrical-conductivity loss.

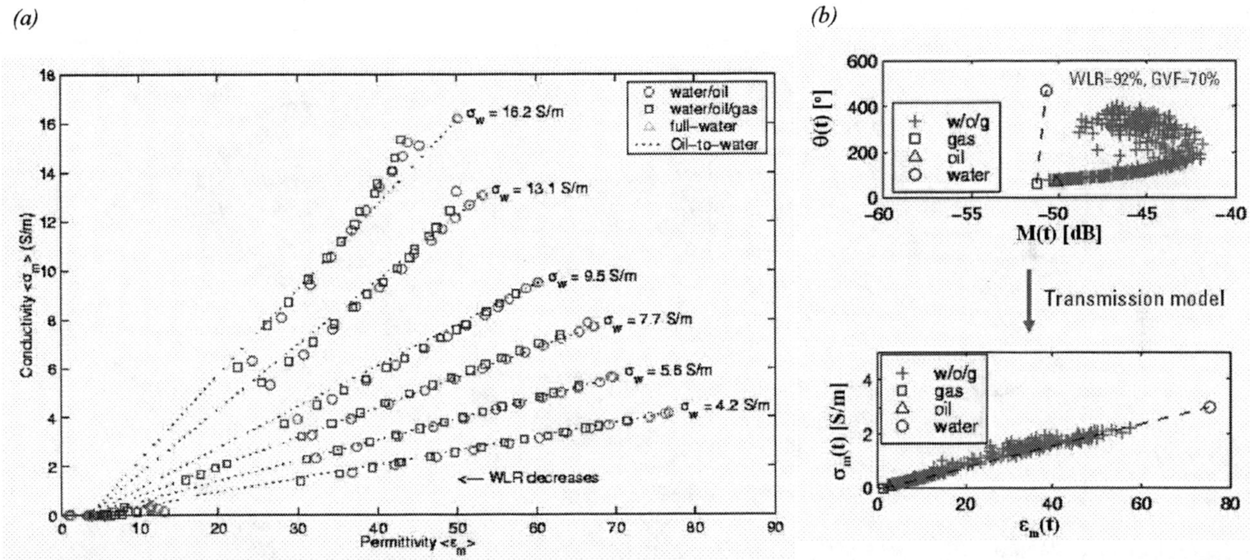

FIGURE 1. *(a)*: Cross plot of time-averaged mixture conductivity σ_m versus permittivity ε_m measured by a microwave open-coaxial probe installed at pipe wall for flows with water conductivity σ_w = 4.2 to 16.2 S/m (≈ 20°C). Oil-continuous (WLR < 0.4) flows are at low σ_m. *(b)*: Illustration of transforming from nonlinear instantaneous microwave transmission magnitude-phase (M,θ) plane to linear mixture (ε_m,σ_m) plane (from the transmission model in Equation (4); (M,θ) data sampled at 50 Hz (shown for 10 s of data) for a water/oil/gas flow with WLR=92%, gas-cut or GVF=70%, salinity = 50 g/L $MgSO_4$, temperature = 37°C.

The key benefit of working in the (ε_m,σ_m) parameter space is the better linearity than e.g. in the transmission amplitude attenuation/phase shift space for a transient gas-liquid flow. This is illustrated in Fig. 1b where there is no unique value of phase-shift θ for a given amplitude-attenuation M, showing the significance of linearizing the instantaneously sampled data by using a transmission model as described in the following section. Another key benefit of working in the (ε_m,σ_m) space is to utilize the fundamentally established relationship between the ε and σ values at the single-phase water end point. The brine permittivity and conductivity, ε_w and σ_w, as a function of temperature T and salinity s, can be correlated as:

$$\varepsilon_w(s,T) = \varepsilon_w(0,T) - K(s,T)\sigma_w(s,T) \tag{1}$$

where $\varepsilon_w(0,T)$ is the well-established temperature correlation of pure water permittivity (s = 0) and $K(s,T)$ is an empirical factor that decreases with increasing temperature and is largely salinity independent [2]. $K(s,T)$ can be calculated, for example, from the Stogryn correlations [3] for NaCl solution low-frequency water permittivity $\varepsilon_w(s,T)$ and water conductivity $\sigma_w(s,T)$. To obtain a water conductivity estimate $\hat{\sigma}_w$, a method is to utilize Equation (1) and demand that the slopes in the ε_m -σ_m space under multiphase and single-phase water conditions are equal. Thus, by requiring $S = (\sigma_m - \sigma_b)/(\varepsilon_m - \varepsilon_b) = (\sigma_w - \sigma_b)/(\varepsilon_w - \varepsilon_b)$, the following relationship is derived (with intercept permittivity $\varepsilon_b \approx \varepsilon_{oil}$ and conductivity $\sigma_b \approx \sigma_{oil} \approx 0$ S/m):

$$\hat{\sigma}_w = [\varepsilon_w(0,T) - \varepsilon_b]S \;/\; (1 + \langle K(T)\rangle\, S) \tag{2}$$

For a formation water with NaCl salt dominant (> 90% by weight of total dissolved solids), the conductivity can be estimated in multiphase-flow conditions to within ±5% of reading, based on the time-averaged mixture conductivity-permittivity measurements and the permittivity-conductivity relation established for pure NaCl solutions. In an intermittent water/oil continuous flow, instantaneous ε_m and σ_m data corresponding to the water continuous regime and recognized by their relatively large ε_m and σ_m values, are selected for time averaging.

MICROWAVE TRANSMISSION FOR OIL-WATER FLOW WATER FRACTION

A research prototype water fraction measurement system based on dual-frequency microwave transmission for sensing amplitude attenuation and phase shift (M,θ) is shown in Fig. 2, with a transmitting antenna, Tx, and at least a receiving antenna, Rx.

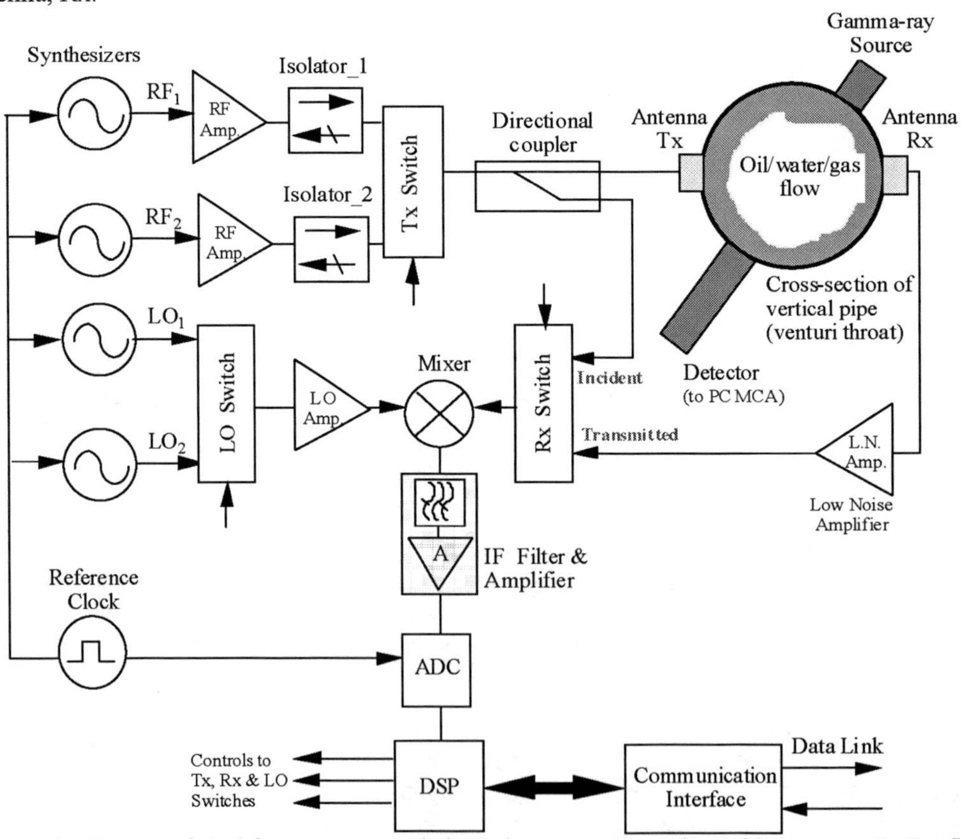

FIGURE 2. Principle diagram of dual-frequency transmission microwave electronics, with antennas Tx-Rx flush mounted across a pipe diameter. Fluid pressure and temperature sensors are not shown.

The local oscillator frequencies (LO_1 and LO_2) are chosen to have the same Intermediate Frequency (IF = $LO_1 - RF_1 = LO_2 - RF_2$). For a desired IF (a few tens of thousands Hz), four synthesizers generate two RF frequencies (e.g., RF_1 = 400 MHz and RF_2 = 800 MHz) and the corresponding LO_1 and LO_2. A "Tx switch" selects RF_1 or RF_2 and the "LO Switch" selects the matching LO_1 or LO_2. For each selected RF frequency, a directional coupler samples a small fraction of the transmitted RF power (as the "incident" signal), while the main fraction is applied to the transmitting antenna Tx. The RF signal, transmitted through a flow mixture, is detected by the receiving antenna Rx to generate the "transmitted" signal. An "Rx switch" selects alternately the incident and the transmitted signals for down-conversion to the respective IF signals by a mixer, which are then low-pass filtered and amplified by an "IF Filter and Amplifier." The incident and the transmitted IF voltage waveforms are digitized by an analog-to-digital converter (ADC). Their amplitude and phase are reconstructed by a digital signal processor (DSP), which also controls the sequencing of all RF switches. The absolute amplitude attenuation and phase shift (M,θ) of the RF transmission signal (the S-parameter S_{21} = transmitted/incident) is then calculated from the amplitude ratio and phase difference of the transmitted signal with respect to the incident.

The microwave-transmission measurement section is installed vertically with upward flow, where oil and water tend to be well mixed. The antennas Tx-Rx are flush mounted (thus nonintrusive) across a pipe diameter. They are the modifications of the cavity-backed antennas used in an electromagnetic propagation tool for downhole logging and evaluated for tomographic flow measurement [4]. The current probe of the antenna is circumferential to the pipe with the equivalent magnetic dipole parallel to the pipe axis. (The TE_{11} mode is preferentially excited in the pipe waveguide when fluid-mixture loss is low and the operating frequency is above the TE_{11} cutoff.) Circular-aperture (diameter ϕ = 15 mm) antennas have been developed and the performance compared with the square-aperture ones (18 × 18 mm) at dual operating frequencies below the TE_{11} cutoff (e.g., 400 and 800 MHz for pipe radius a = 26 mm). The dual-frequency, single-modality microwave transmission measurement provides a robust water fraction determination, under both water- and oil-continuous flow conditions, and over a wide range of water salinities that is far beyond the capability of microwave-resonance-based sensors on the market. It is also more robust against changing conditions than a commercial dual-modality low frequency system based on a separate electrical capacitance sensor for oil-continuous flow and an inductance sensor for water-continuous flow; such a system is incapable of measuring mixture permittivity and conductivity simultaneously over the same sensing volume to track water-conductivity change.

For a lossy dielectric mixture, the complex wave propagation constant γ_m can be expressed as $\gamma_m = \alpha_m + j\beta_m$, with α_m being the attenuation constant and β_m the phase constant. Denoting L as the effective path-length of the Tx-Rx antenna pair, the transmission amplitude attenuation M (dB) and phase shift θ (radian) can be related to α_m and β_m as:

$$\alpha_m = (M - M_o)/(8.686L) = \Delta M/(8.686L) \quad (3a)$$

$$\beta_m = (\theta - \theta_o)/L = \Delta\theta/L \quad (3b)$$

where M_o and θ_o are respectively the amplitude and phase offsets of a Tx-Rx antenna pair (e.g., characterized with empty-pipe measurement). In a metallic pipe of inner radius a, it is well known that $\gamma_m^2 = k_c^2 - k_m^2$; here k_c is the cutoff wavenumber of the TE_{11} dominant mode (k_c =1.84/a), k_m is the wavenumber of the mixture of complex permittivity $\varepsilon_m^* = \varepsilon_m - j\sigma_m/(\varepsilon_o\omega)$, and at angular frequency ω, $k_m^2 = \omega^2\mu\varepsilon_o\varepsilon_m^*$. Equating real and imaginary parts of $\gamma_m^2 = k_c^2 - k_m^2$ leads to the following relations between measurable quantities and the mixture permittivity ε_m and conductivity σ_m:

$$\varepsilon_m = (c_o/\omega)^2 \left(k_c^2 - \alpha_m^2 + \beta_m^2\right) \quad (4a)$$

$$\sigma_m = 2\alpha_m\beta_m/(\omega\mu) \quad (4b)$$

The unified microwave transmission model, given in Equations (4a) and (4b), transforms the measured (M,θ) (as shown in Fig. 3 for vertical water/oil flows) to the fundamental electrical parameters of mixture permittivity and conductivity (ε_m,σ_m) (shown in Fig. 4), without the need of *a priori* knowledge of the existence and properties of any liquid continuous phase. The similarity in the transmission (M,θ) responses for both types of antennas, at multiple water conductivities and frequencies (Fig. 3 and 4), indicates that the unified transmission model is robust. From the derived (ε_m,σ_m), the water fraction (α_w) of vertical water/oil flows (with WLR = α_w) can be determined within ±2.5% by the appropriate use of permittivity and conductivity mixing laws (Fig. 5a), such as the simplified Ramu-Rao and Bruggeman mixing laws [5] (as indicated in Fig. 4a and 4b). Knowledge of the liquid continuous phase is needed to use an appropriate mixing law for water fraction determination; the nature of the continuous liquid can be inferred, e.g., from low values in the measured phase-shift θ (Fig. 3b), or in the transformed mixture-conductivity σ_m (Fig. 4b). To determine water fraction of a water-continuous flow based on the derived mixture conductivity σ_m, the water conductivity (σ_w) is often required as a manual input to a conductivity mixing law (e.g., Ramu-Rao), as is the case for a commercial MPFM based on a low-frequency inductance-sensor incapable of tracking water conductivity. The high-frequency microwave transmission system described here is capable of tracking the change in the water conductivity over a wide salinity range, because of the simultaneous measurement of mixture conductivity and mixture permittivity (see Fig. 4 and the previous section).

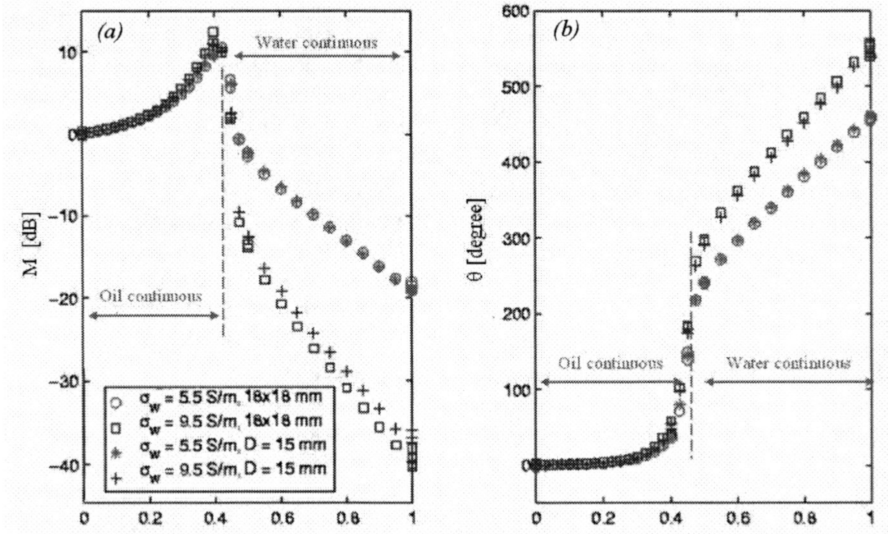

FIGURE 3. Plotted versus WLR for square (18 × 18 mm) and circular-aperture (ϕ=15 mm) antennas, the transmission responses at 800 MHz of *(a)* amplitude M and *(b)* phase-shift θ with vertical oil/water flow.

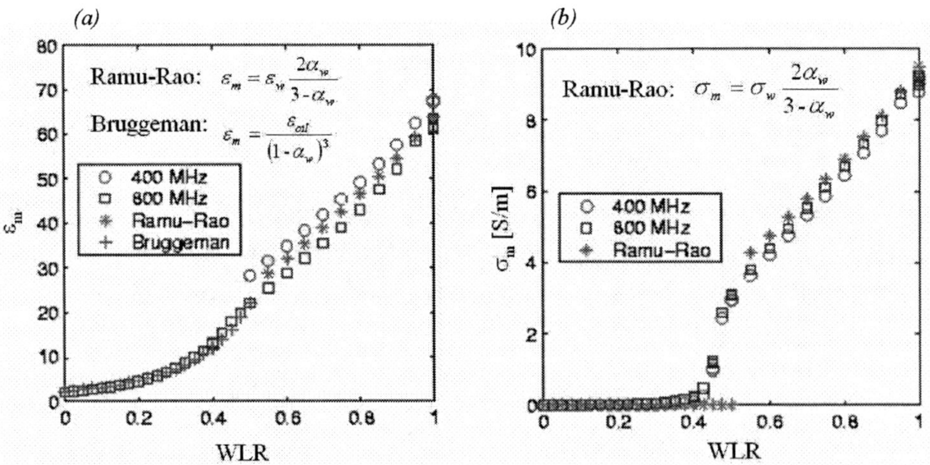

FIGURE 4. For circular-aperture antennas at 400 and 800 MHz, plotted versus WLR for vertical oil/water flows, the transmission model (of Equations 4a and 4b) calculation of *(a)* ε_m and *(b)* σ_m, based on the measured (M,θ) in Fig. 3. The ε_m and σ_m estimated from the Ramu-Rao and Bruggeman mixing laws[5] as indicated (Bruggeman's law used for oil-continuous flow) are also shown (using WLR as the reference water fraction α_w; single-phase fluid parameters σ_w, ε_w, and $\varepsilon_{oil} \approx 2.2$ are known).

Figure 3 shows that M and θ are weakly dependent on the water-conductivity in oil-continuous flows; hence online water conductivity estimation is largely unnecessary for water fraction determination when oil is continuous. It is nevertheless possible by installing a microwave reflection open-coaxial probe (that has a shallow sensitivity depth in the probe-aperture proximity) at a water-rich region (e.g., near pipe wall) to detect water conductivity/salinity change in low WLR flows [2], even at high gas-cut conditions.

Note that the microwave transmission system can also determine independently the water fraction of a water-continuous oil/water flow from the derived mixture-permittivity ε_m, with water permittivity (ε_w) as input to a permittivity mixing law.

MICROWAVE TRANSMISSION FOR THREE-PHASE FLOW WATER CUT WITH DENSITY INPUT

To measure three-phase oil/water/gas flow WLR, a mixture density (ρ_m) measurement is needed to determine the liquid fraction: ($\alpha_L = \alpha_o + \alpha_w$). This can be derived using the relationship $\rho_m = \rho_g(1-\alpha_L)+\rho_L\alpha_L$,

$$WLR = \frac{\alpha_w}{\alpha_L} = \frac{\alpha_w(\rho_o - \rho_g)}{\rho_m - \alpha_w(\rho_w - \rho_o) - \rho_g} \tag{5}$$

The research prototype has a gamma ray densitometer installed at the same pipe cross section as the microwave transmission antennas (Fig. 2). The combination of transmission microwave mixture complex-permittivity and mixture density measurements made at the same pipe cross section (and preferably at the same time) minimizes the effect on the phase fraction and the WLR interpretation of an axially developing multiphase flow.

When the gas phase is present, it can be clearly observed from Fig. 5a that the use of water/oil two-component mixing laws generally leads to an underestimate in the water fraction for both oil- and water-continuous flows. To overcome this water fraction underprediction, caused by the presence of gas giving rise to large scale spatial nonuniformity in the permittivity and conductivity distribution and also affecting the local effective values, a three-component oil-water-gas mixing model, incorporating a mixture density (ρ_m) measurement, is established for vertical multiphase flows. The model has an annular liquid-rich layer near the pipe wall (of area fraction α_{LF} and of an effective permittivity ε_{LF} and conductivity σ_{LF}) and a gas rich core (permittivity $\varepsilon_g \approx 1.0$ and conductivity $\sigma_g = 0$). Inside the liquid rich layer, water and oil are assumed well mixed with negligible entrained gas, and its water fraction (with respect to liquid layer α_{LF}) is denoted as $\alpha_{w,LF}$ (WLR $\approx \alpha_{w,LF}$).

TABLE 1. Three-component gas-liquid annular two-layer mixing models based on simplified Ramu-Rao and Bruggeman mixing law [5], for well-mixed liquid layer [a].

Mixture permittivity: $\varepsilon_m^x = \alpha_{LF}\,\varepsilon_{LF}^x + (1-\alpha_{LF})\,\varepsilon_g^x$ ....(6a)

Mixture conductivity: $\sigma_m^y = \alpha_{LF}\,\sigma_{LF}^y + (1-\alpha_{LF})\,\sigma_g^y = \alpha_{LF}\,\sigma_{LF}^y$ ....(6b)

Mixture density: $\rho_m = (1-\alpha_{LF})\,\rho_g + \alpha_{LF}\,[(1-\alpha_{w,LF})\rho_o + \alpha_{w,LF}\,\rho_w]$ ....(7)

Water-fraction: $\alpha_w = \alpha_{w,LF}\,\alpha_{LF}$ ....(8)

The mixed oil/water flow is a special case of the annular two-layer gas-liquid model (with $\alpha_{LF} \to 1$). Equation (7) is solved by constraint minimization for the example cases below. (Substitute α_{LF} and $\alpha_{w,LF}$ into equation (7) for solving σ_{LF} or ε_{LF}, and then for α_{LF}, $\alpha_{w,LF}$ and α_w.):

Water continuous	*Oil continuous*
From mixture conductivity (Ramu-Rao mixing law):	From mixture permittivity (Bruggeman mixing law):
$\alpha_{w,LF} = 3\sigma_{LF}/(\sigma_{LF}+2\sigma_w)$ ....(9a)	$\alpha_{w,LF} = 1-(\varepsilon_{oil}/\varepsilon_{LF})^{1/3}$ ....(10a)
$\alpha_{LF} = (\sigma_m/\sigma_{LF})^y$ ....(9b)	$\alpha_{LF} = (\varepsilon_m^x - \varepsilon_g^x)/(\varepsilon_{LF}^x - \varepsilon_g^x)$ ....(10b)

[a] Here x and y are empirical exponents. All single-phase permittivity (ε_g, ε_{oil} and ε_w), conductivity σ_w, and densities (ρ_g, ρ_o and ρ_w) are assumed known.

The detailed description of the annular three-component mixing model is summarized in Table 1. From this the water fraction α_w and WLR ($\approx \alpha_{w,LF}$) can be derived based on microwave mixture permittivity ε_m or conductivity σ_m, and mixture density ρ_m measurements with single-phase densities and electrical properties (permittivity and/or conductivity) as inputs. Note that water conductivity σ_w (and permittivity ε_w) inputs can be tracked online (see the section following the Introduction); water density ρ_w can be estimated from the conductivity knowing the dominant salt species. The increase in oil permittivity (ε_{oil}) with oil density (ρ_o) should be taken into account, especially for the WLR determination of oil-continuous flows (e.g., for Oseberge crude [6], ε_{oil} = ~2.19 to ~2.27 for ρ_o = ~850 to ~910 kg/m^3).

By using the three-component mixing model, Fig. 5b illustrates the improvement in the water fraction prediction of three-phase flows (in comparison with Fig. 5a) mostly to within ±2.5% of a gamma ray derived reference. The resulting WLR is largely predicted within ±5% of the reference (Fig. 5c).

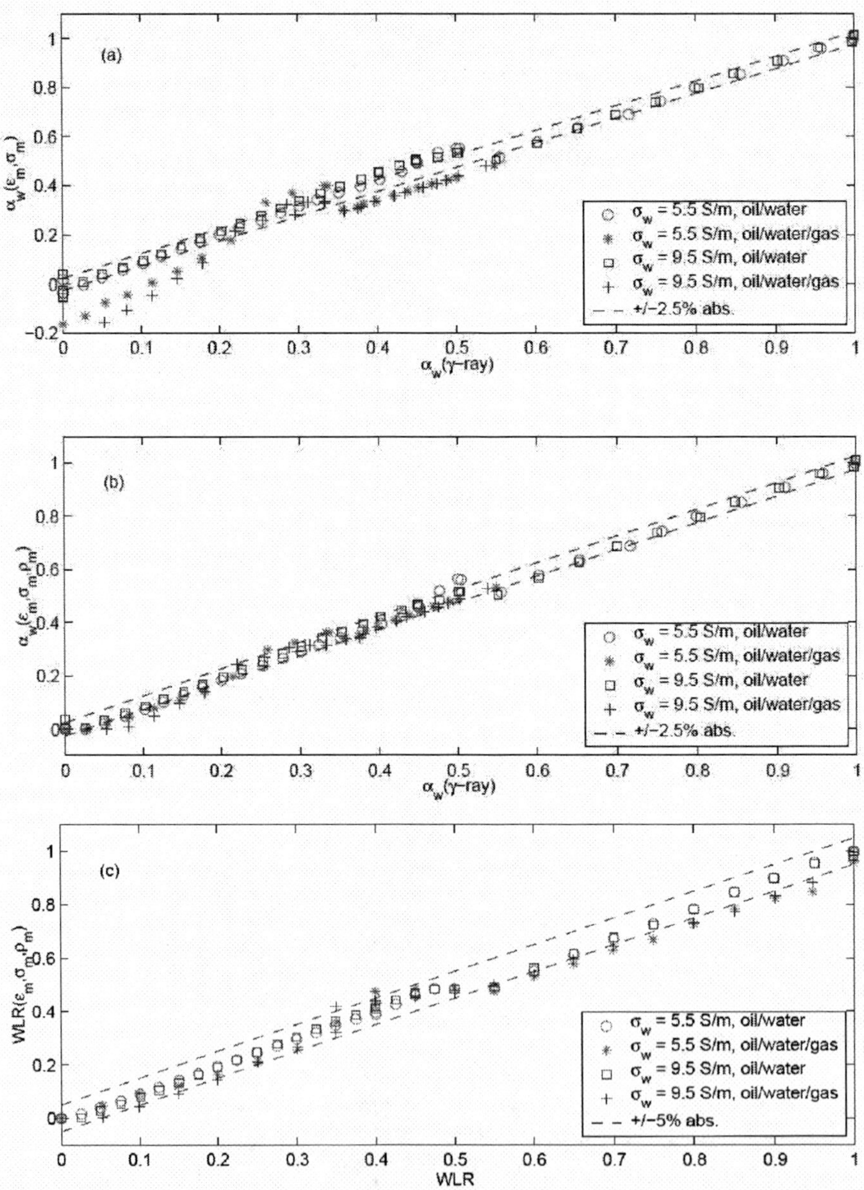

FIGURE 5. For circular-aperture antennas, vertical oil/water and oil/water/gas flows (gas cut tested at 0% and from 45% to 65%), plotted versus a reference gamma ray derived water fraction, *(a)*: water-fraction α_w predictions from the two-component mixing model (Bruggeman's ε_m mixing for oil-continuous at 800 MHz, Ramu-Rao's σ_m mixing for water-continuous at 400 MHz), and *(b)*: α_w from the respective three-component mixing model using gamma ray mixture density ρ_m (Table 1). *(c)*: WLR prediction from the three-component mixing model compared with the reference WLR.

REMARKS

Research in a novel microwave measurement interpretation has demonstrated a potential multiphase flow water-cut measurement system that is robust to changes in water salinity. This reduces the calibration requirements in varying salinity situations caused, for example, by changing formation water or breakthrough of formation/injection water. The potential of a microwave transmission system providing a complementary and high quality water-cut measurement has been further exploited in this work. Vertical flow water fraction and the water cut can be derived from a gas-liquid three-component mixing model, using the microwave mixture permittivity, or mixture conductivity and a mixture-density measurement. The prototype system has been shown to work for both water- and oil-continuous flows over a water conductivity range that is far beyond the capability of microwave-resonance-based sensors currently on the market. The system is also capable of detecting and tracking the change in the water conductivity over a wide salinity range, based on the interpretation of the simultaneously measured mixture conductivity and permittivity. These capabilities may also complement the existing multiphase flow measurements presently deployed in the field based on the Vx technology [1] using dual-energy gamma ray and Venturi combination. Furthermore, the technique developed in this paper makes it more robust against changing fluid properties than a dual-modality low-frequency system based on a separate electrical capacitance sensor for oil-continuous flow and an inductance sensor for water-continuous flow.

ACKNOWLEDGMENTS

The author thanks Schlumberger for the permission to publish this work and acknowledges the collaborations with colleagues at Schlumberger Oilfield Services' research and engineering centers.

REFERENCES

1. I. Atkinson, B. Theuveny, M. Berard, G. Conort, T. Lowe, A. McDiarmid, P. Mehdizadeh, B. Pinguet, G. Smith and K.J. Williamson, "A New Horizon In Multiphase Flow Measurement", *Oilfield Review*, Winter 2004/2005, p.52.
2. C.-G. Xie, G. Segeral, G. Roux and P. Hammond, "Methods and apparatus for estimating on-line water conductivity of multiphase mixtures", US Pat. No. 6831470B2.
3. A. Stogryn, "Equations for calculating the dielectric constant of saline water", *IEEE Transactions on Microwave Theory and Techniques*, August, pp.733-736 (1971).
4. R. Taherian and T. Habashy, "Microwave device and method for measuring multiphase flows", US Pat.No. 5485743
5. E.A. Hammer, "Flow permittivity models and their applications in multiphase meters", Proc. Multiphase Metering, IBC Technical Services, Aberdeen, 12 – 13 March 1997.
6. UK National Engineering Laboratory, "The relative permittivity of oils". Report No.251/99, June 1999.

Measure Technology for Strongly Swirling Fluid Flow in Cyclone Separators: Application and Comparison

Bingtao Zhao*, Jianmin Zhang, and Shouyu Zhang

College of Power Engineering, University of Shanghai Science and Technology, No. 516 Jungong Road, Shanghai 200093, China

Abstract. Three representative measurement technologies, Multiple Holes Pitot Tubes (MHPT), Laser Doppler Velocimetry (LDV) and Particle Image Velocimetry (PIV) on the strongly turbulent and swirling flow in the conventional cyclone separator, are described in the present work. The experimental data are analyzed and compared in order to evaluate the accuracy level of these methods. The comparison results show that these measurement technologies agree with each other well in general for mean velocity in cyclone separators. In addition, the calibrated MHPT method is simple with not much difference in measured quantities. However, as the new and modern methods, LDV and PIV technology can provide the fluid flow characteristics in detail, including the turbulent parameters even gas-particle two phases flows and avoid the effect of intrusion on flow pattern.

Keywords: Cyclone separator, swirling flow, measurement technology
PACS: 47.61.Jd

INTRODUCTION

Cyclone separators, as shown in Fig. 1, are widely used in gas-particle separation in the fields of both industrial application and aerosol sampling. In order to reveal the mechanism of gas-particle separation and pressure drop more clearly, it is essential to acquire the flow pattern in the cyclone separator [1]. However, due to high turbulent, strong anisotropic and three dimensional characteristics, the flow pattern inside cyclone separators is extremely complicated. With the development of model measurement technology for multiphase flow, many methods can be applied to determine the flow pattern in cyclone separators. The representatively experimental observation techniques are divided into the intrusive methods including the Multiple Holes Pitot Tubes (MHPT) and hot wire anemometers, and non-intrusive methods including the Laser Doppler Velocimetry (LDV) and Particle Image Velocimetry (PIV).

In the present work, the detailed analysis, summaries and comparison of velocity profiles in cyclone by the measured methods of MHPT, LDV and PIV, are performed. In addition, their accuracy and availability are also discussed.

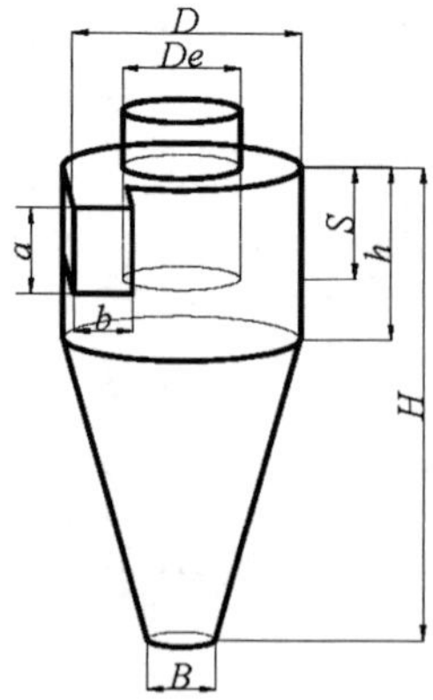

FIGURE 1. Schematic diagram of cyclone separator.

* Corresponding Author. Tel: +86-21-55272704; Fax: +86-21-55272376
E-mail: zhaobingtao@usst.edu.cn

CP914, *Multiphase Flow: The Ultimate Measurement Challenge,* edited by X. Cai, Y. Wu, Z. Huang, S. Wang, and M. Wang

CHARACTERSTICS OF FLOW PATTERNS IN CYCLONE SEPARATORS

Tangential Velocity

In the cyclone separators, the actual tangential velocity of gas phase was obeyed the "Rankine" vortex including the quasi-free vortex in the outer part and the quasi-forced vortex in the inner part, as shown in Fig. 2. Generally, it can be expressed as a function of the velocity at the wall and a vortex exponent for effective separation region:

$$\frac{v_\theta}{v_{\theta w}} = \left(\frac{R_w}{r}\right)^n \tag{1}$$

$$v_\theta r^n = const \tag{2}$$

Where, the vortex exponent was developed by Alexander (1949) [2]. He suggested the vortex exponent as functions of cyclone diameter and temperature:

$$n = 1 - \left(1 - 0.67 D^{0.14}\right)\left(\frac{T}{283}\right)^{0.3} \tag{3}$$

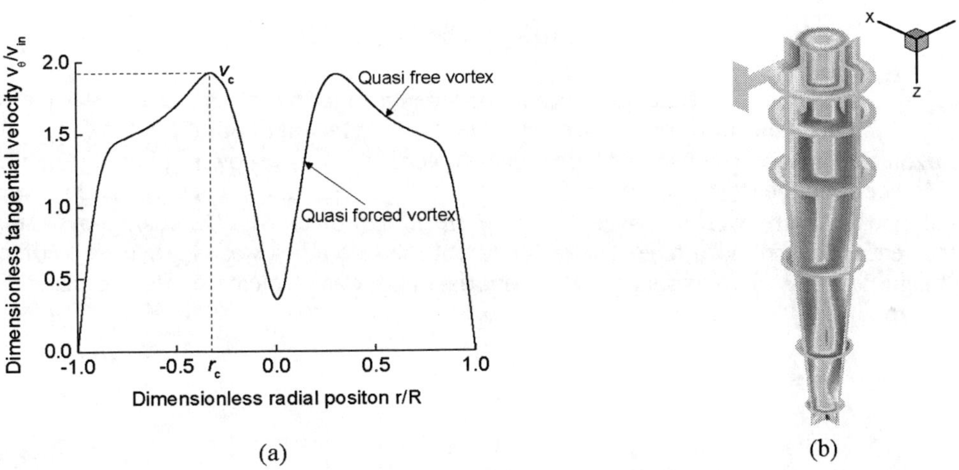

FIGURE 2. Tangential velocity distribution in cyclone separators: (a) 2D and (b) 3D maps.

Axial Velocity

The axial velocity in cyclones is directed downward near the wall and upward in the region around the axis named outer and inner vortex respectively, as describes in Fig. 3. The surface of zero axial velocity between the outer and the inner is approximately coincided with the prolonged surface of vortex finder.

Radial Velocity

The radial velocity is the smallest of the three components compared to the other velocities, and is often assumed constant over the area of zero axial velocity. Usually, the radial velocity is difficult to be measured directly by intrusive methods due to the sensitivity.

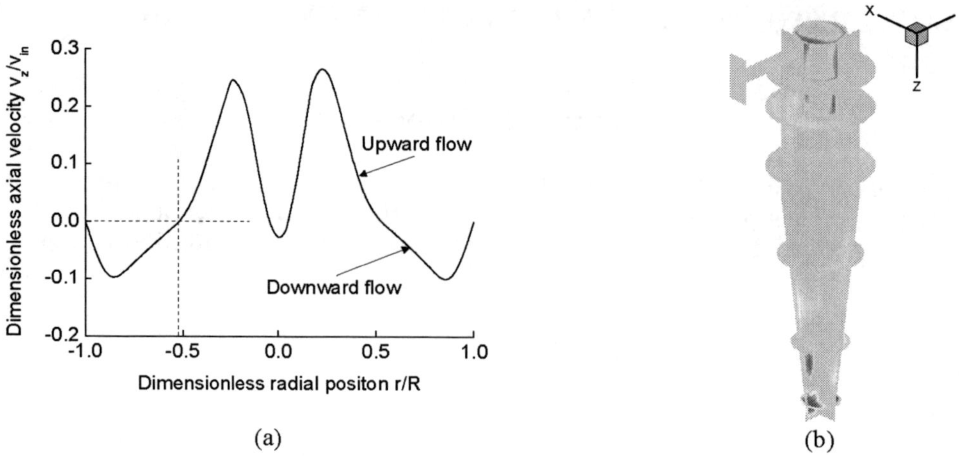

FIGURE 3. Axial velocity distribution in cyclone separators: (a) 2D and (b) 3D maps.

MEASUREMENT TECHNOLOGY FOR FLOW PATTERN IN CYCLONE SEPARATORS

Principles and Methods of MHPT

The multiple holes Pitot tube combines the means for simultaneous measurement of total pressure, dynamic pressure and flow direction by one instrument. It consists of a forward facing Pitot tube at the centre, two chamfered side tubes in horizontal axis and two chamfered side tubes in vertical axis as shown in Fig. 4. The tips of four side tubes were separately chamfered to by means of a special forming designed for this purpose. This was followed by careful stoning after which burrs were removed. The four chamfered tubes were assembled by a small amount of solder around the central square-ended tube. The symmetry of the side tubes was checked by a contour measuring projector using magnification scale. The shape of the probe assembly was formed according to the cranked design.

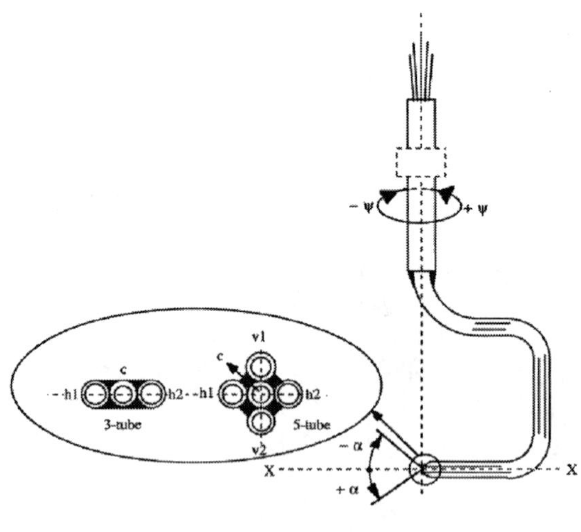

FIGURE 4. The construction of the multiple holes Pitot tube [3].

Principles and Methods of LDV

In LDV shown in Fig. 5, the velocity of the gas is measured as the speed of small seed particles, considered to follow the gas flow faithfully. A laser light beam is split in two, and the two parts are made to cross, which creates an interference pattern (the measuring volume) through which the seed particles fly. The light scattered by the particles is detected as a "Doppler burst", the frequency of which is a measure of the speed with which the particles traverse the measuring volume. In classical LDV, each of the components of the gas velocity is measured consecutively. In newer systems, different velocity components can be measured in the same volume simultaneously using different colored laser light, which is detected separately. If the direction in which the particle crosses the measuring volume is unknown, it can be determined by shifting the frequency of one of the beams to create a moving, rather than a stationary interference pattern.

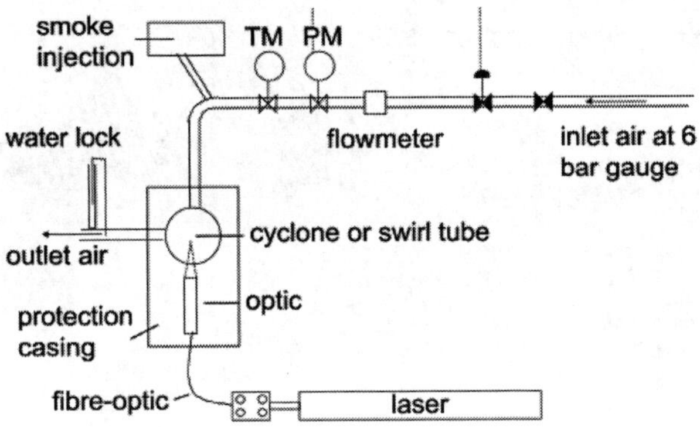

FIGURE 5. The experimental setup of LDV system [4].

Principles and Methods of PIV

Advanced stereoscopic PIV uses two cameras to view a flow field from two perspectives so that the out of plane velocity component can be measured. The two components of velocity nominally perpendicular to the camera optical axis are measured from each camera viewpoint. The pair of two-dimensional velocity vectors for a point in the flow is then combined to yield a three dimensional velocity vector. By combining the vector fields from the two cameras, the three-dimensional velocity field for the plane in the fluid is measured. Among several optical arrangements of the two cameras, the Scheimpflüg configuration was found to be the best layout. The Scheimpflüg configuration requires the object plane, image plane, and lens principle plane to intersect at a common point, as shown in Fig. 6. By tilting the image sensor plane and the lens principle plane to the Scheimpflüg condition, the plane of best focus can be adjusted so that it is aligned with the light sheet.

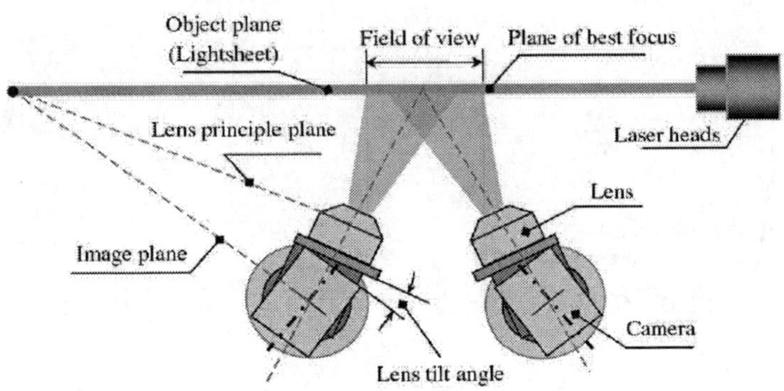

FIGURE 6. The configuration of Stereoscopic PIV system [5].

RESULTS AND DISCUSSION

Comparison of MHPT and LDV Measurements

Fig. 7(a) and 7(b) show the comparison of tangential and axial velocity distribution in cyclone, respectively. The experimental investigation was measured by a calibrated MHPT[6] and LDV system [7, 8], and operated with the Reynolds number of $Re_c = 1.24 \times 10^5$ and in the vertical stations of z/h = 0.412 and z/h = 0.979. The results indicate that the tangential velocity profiles are in good agreement over all redial positions in cyclones. In spite of the difference in the specified region of 0.25 < r/R <0.75 which result from cyclone configurations used, the axial velocity profiles also provide the approximate trends.

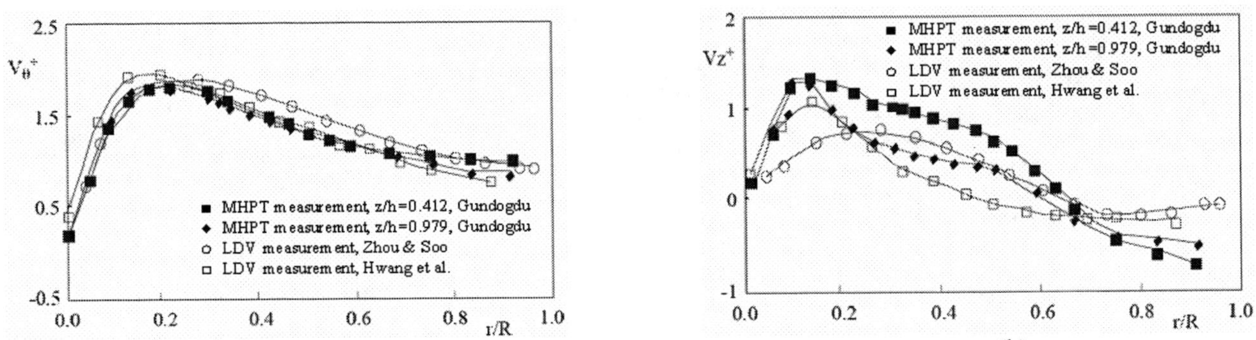

(a) (b)

FIGURE 7. The comparison of dimensionless (a) tangential and (b) axial velocity distribution for MHPT and LDV measurements.

Comparison of MHPT and PIV Measurements

Fig. 8(a) and 8(b) also show the comparison of tangential and axial velocity distribution in cyclone which are tested by a MHPT [9] and PIV system [10]. In these experiments, the flow patterns are measured in different cyclone dimensions and operational conditions. In order to compare the results more conveniently, all experimental data are non-dimensioned to confirm the availability of testing methods according to the similarity characteristics of flow field. As illustrated in Figure, there are some discrepancies for the both maximum velocity with radial position of 0.2 < r/R <0.4 as well as the tangential velocity with left position which may be caused by the instrumental sensitivity (calibrated factor) and cyclone dimensions. However, the good consistencies on the whole are also presented for both methods.

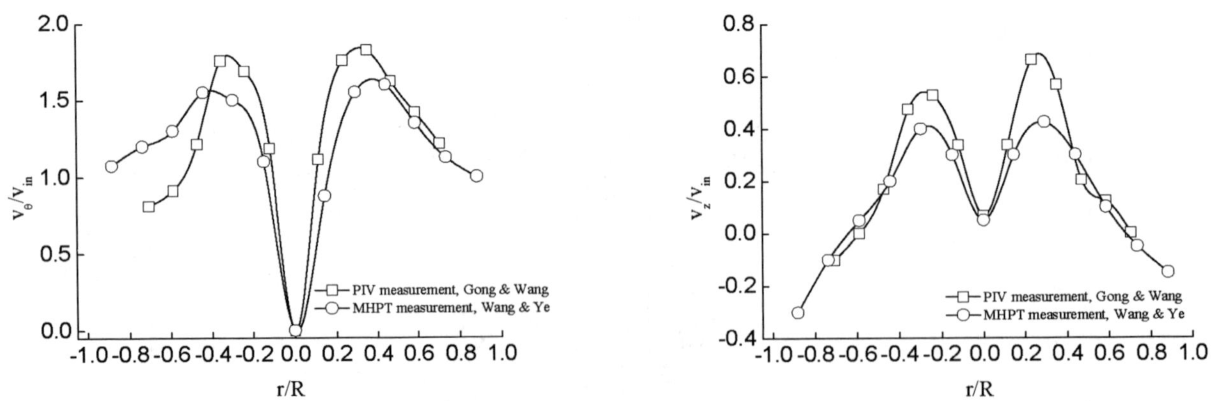

(a) (b)

FIGURE 8. The comparison of dimensionless (a) tangential and (b) axial velocity distribution for MHPT and PIV measurements.

CONCLUSIONS

The measurement of mean flow parameters in turbulently swirling flows was achieved by using the representative technology including MHPT, LDV and PIV, with the calculation scheme and visualization software.

In comparison with LDV and PIV measurements, the MHPT method is simple and cheap with not much difference in measured quantities. However, the difficulty in data reduction from the calibration charts needs further computational work to make the utilization of probes somehow more practical and rapid. On the other hand, as the new and modern methods, LDV and PIV technology need more complicated experimental setups and processes, however, they can provide the fluid flow characteristics in detail including the turbulent parameters even gas-particle two phases flows and avoid the effect of intrusion on flow pattern.

ACKNOWLEDGMENTS

The authors gratefully acknowledge the financial support of this work by the Excellent Young Teachers Fund of Shanghai Colleges and Universities, the Visiting Scholars Fund of Key Laboratory for Ultra-fine Materials of Ministry of Education in ECUST, the Sunlight Research Program of the Shanghai Education Committee (03SG44), the Basic Research Priorities Program of the Shanghai Education Committee (04ZZ23) and the 973 Program (2004CB217705).

REFERENCES

1. B. Zhao, *Chem. Eng. Technol.* **28**, 969-972 (2005).
2. R. M. Alexander, *Proc. Aust. Inst. Min. Met. (New Series)* **152-153**, 202-228 (1949).
3. M. Y. Gundogdu and O.T. Goksel, *J. Eng. Environ. Sci.* **22**, 223-233 (1998).
4. W. Peng, A. C. Hoffmann and P.J.A.J. Boot et al., *Powder Technol.* **127**, 212-222 (2002).
5. Z. Liu, Y. Zheng, L. Jia, J. Jiao and Q. Zhang, *Chem. Eng. Sci.* **61**, 4252-4261 (2006).
6. M. Y. Gundogdu and M. O. Carpinlioglu, *Flow Measure. Instru.* **9**, 243-248 (1998).
7. L. X. Zhou and S. L. Soo, *Powder Technol.* **63**, 45-53 (1990).
8. C. C. Huang, H. Q. Shen, G. Zhu and M. M. Khonsan, *J. Fluids Eng.* **115**, 21-25 (1993).
9. A. L. Gong and L. Z. Wang, *Aerosol Sci. Technol.* **38**, 516-512 (2004).
10. L. Z. Wang and L. Ye, *Aerosol Sci. Technol.* **31**, 187-193 (1999).

On Measurement Property of Vortex Flowmeter in Bubbly Two-Phase Flows

Zhiqiang Sun*[¶], Hongjian Zhang*, and Yinjie Xiang*

*Department of Control Science and Engineering, Zhejiang University, Hangzhou 310027, Zhejiang, China
[¶] School of Energy Science and Engineering, Central South University, Changsha 410083, Hunan, China

Abstract. The measurement behavior of vortex flowmeter was examined under the condition of bubbly two-phase flows, in which the Reynolds number of the two-phase mixture was $4.63 \times 10^4 - 23.45 \times 10^4$ with volume void fraction up to 30%. Regular Karman vortex shedding was found, but its stability was strongly influenced by the introduction of gas-phase component. The frequency of the bubbly vortex shedding was in approximately linear proportion to the two-phase Reynolds number, whereas the volume void fraction has little impact on the frequency. These findings provide a possibility of using a vortex flowmeter for bubbly two-phase flow measurement.

Keywords: gas-liquid two-phase flow; bubbly flow; vortex flowmeter; vortex frequency; flow measurement
PACS: 07.90.+c

INTRODUCTION

Gas-liquid two-phase flow has a wide range of industrial applications. Accurate measurement of its parameters is very important for safety, environmental protection, process control, energy conservation, and quality assurance in industry. Flowrate is one of the most fundamental parameters of gas-liquid two-phase flows, but techniques for measuring them are still under development due to the complex physical nature of two-phase flows. Kinds of flow sensors and methods, whether conventional or novel ones, have been used to investigate the characteristic relationship between two-phase flow parameters and the measurable parameters over the past decades [1–7], however, there is still a long way to go before a universal, accurate and reliable system for the measurement of gas-liquid two-phase flow parameters is available.

Vortex flowmeter, which is based on the phenomenon of Karman vortex street, has gained widely successful use in many industrial and civil fields. One of the unique virtues of vortex flowmeter is that the vortex shedding frequency is not sensitive to the physical properties of the fluid like viscosity and density [8]. For this reason, the vortex flowmeter is suitable for most single-phase fluid measurement including liquid, gas and steam, and it is regarded as a promising substitute device for rotameter and orifice plate [9]. At the same time, it is natural for researchers to attempt to extend range of use of vortex flowmeter to multiphase flow measurement.

The objective of this study is to investigate the measurement property of a vortex flowmeter as it is placed in bubbly gas-liquid two-phase flows, and on the other hand we also try to find a practical method for the measurement of two-phase flowrate with a single vortex flowmeter.

EXPERIMENTAL SET-UPS

Experiments were conducted in a vertical pipe with inner diameter D = 50 mm, and the tested media were air and water both flowing upward. Figure 1(a) shows the layout of the experiment system. The flowrates of air and water were measured respectively by standard flowmeters and regulated by the valves before they entered into the two-phase mixer. The air flowrate at the entrance was 0.2 – 15.5 m^3/h, and the water flowrate Q_L was 5.0 – 19.0 m^3/h. Additionally, pressures of air and water at this part were also measured. After passing through the two-phase mixer, the mixture of air and water reached the test section, in which the experiments were carried out and signals were acquired. This section was composed of the tested vortex flowmeter, the upstream/downstream straight pipes, and a

CP914, *Multiphase Flow: The Ultimate Measurement Challenge,* edited by X. Cai, Y. Wu, Z. Huang, S. Wang, and M. Wang

transparent observation window for visually identifying flow patterns. The upstream and downstream straight pipe were 30*D* and 20*D* long respectively, which could ensure the flow patterns fully developed and the vortices successfully shed. The pressure in this section was also measured, which could be used to calculate the actual tested air flowrate Q_G with the combination of parameters obtained before the entrance of two-phase mixer.

The details of the tested vortex flowmeter are shown in Fig. 1(b). The gas-liquid mixture flowed upward across the vortex flowmeter. A trapezoidal cylinder with a blockage ratio of $b = d/D = 0.28$ (d was the width of the cylinder) was used as the bluff body and was installed at the center of the pipe perpendicular to the flow direction. From the two pressure tags on the pipe wall downstream of the front face of bluff body, the dynamic differential pressure signals were detected by a Honeywell 24PC sensor, and then sampled and processed by a computer. The sampling rate of 1 000 Hz was adopted and 10 000 points were included in each data set.

During each group of experiments the water flowrate was kept constant, while the air flowrate was increased gradually. Correspondingly, the flow pattern in the tested section was transformed from the single flow to bubble, plug, churn or annular flow. It is found that the transformation of flow pattern is dependent on the volume void fraction β and Reynolds number Re_D, where

$$\beta = Q_G / (Q_G + Q_L) \tag{1}$$

$$Re_D = v_{TP} \rho_{TP} D / \mu_{TP} \tag{2}$$

$$v_{TP} = (Q_G + Q_L)/S \tag{3}$$

$$\rho_{TP} = \rho_G \beta + \rho_L (1 - \beta) \tag{4}$$

$$\mu_{TP} = \mu_G \beta + \mu_L (1 - \beta) \tag{5}$$

in which S is the cross-sectional area of the pipe. Signals were acquired only when the bubbly flow pattern was stable.

RESULTS AND DISCUSSION

Phenomenon of Two-Phase Vortex Shedding

Figure 2(a) presents the raw signals obtained at different Reynolds number Re_D. It is obviously seen there exists regular vortex shedding behind the bluff body which oscillates periodically. The shown cases were featured with the same water flowrate, Q_L = 18.66 m^3/h, while the air flowrates of them were increased gradually and so did the volume void fraction β and the Reynolds number Re_D. The amplitude of the fluctuating differential pressure decreased with the increment of β. When β was greater than 15.00%, the signal began to become unstable. It indicates that the introducing of air into liquid phase flow has an influence on the regular Karman vortex shedding.

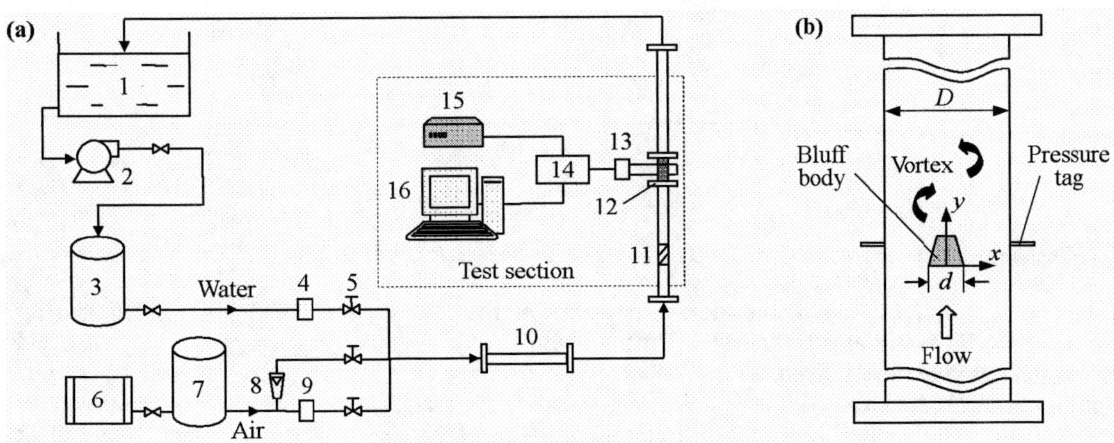

FIGURE 1. Experiment system: (a) Schematic diagram; (b) Details of the tested vortex flowmeter. (1. Storage pool 2. Pump 3. Water surge tank 4. Electromagnetic flowmeter 5. Valve 6. Air compressor 7. Air surge tank 8. Rotameter 9. Standard vortex flowmeter 10. Two-phase mixer 11. Observation window 12. Tested vortex flowmeter 13. Differential pressure sensor 14. Signal conditioning board 15. DC power supply 16. Computer.)

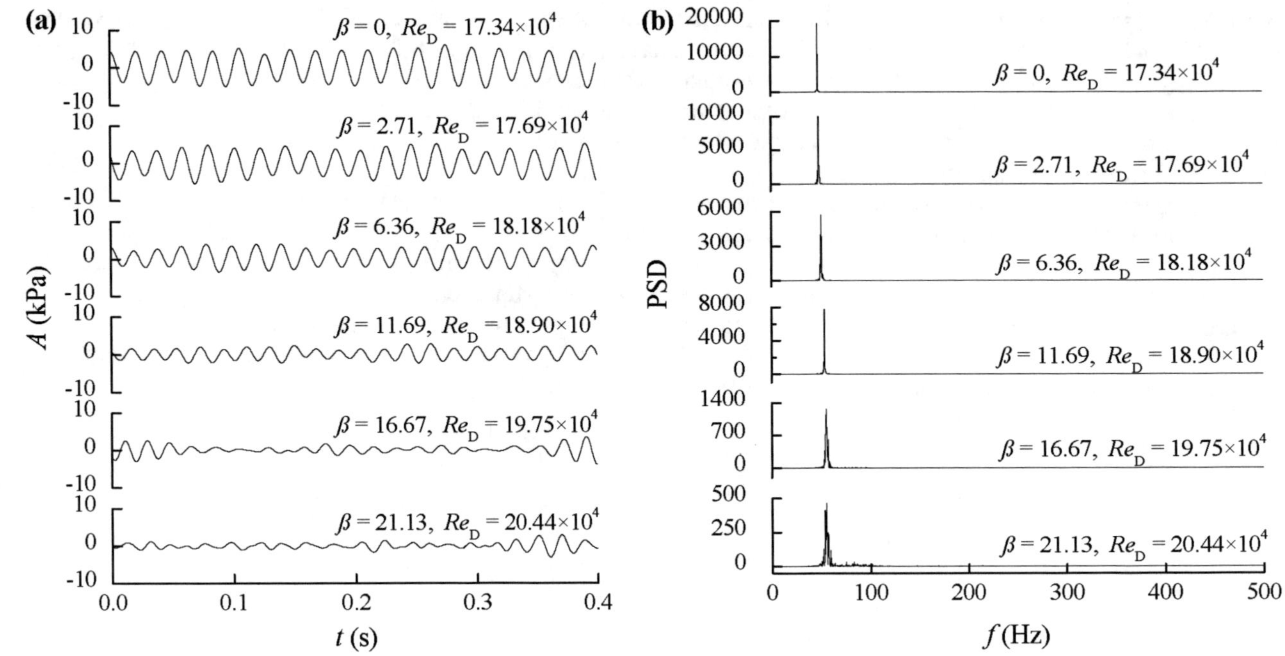

FIGURE 2. Differential pressure signals sampled on the duct wall of the vortex flowmeter: (a) Raw signals obtained at a number of Reynolds numbers studied; (b) The corresponding power spectral densities.

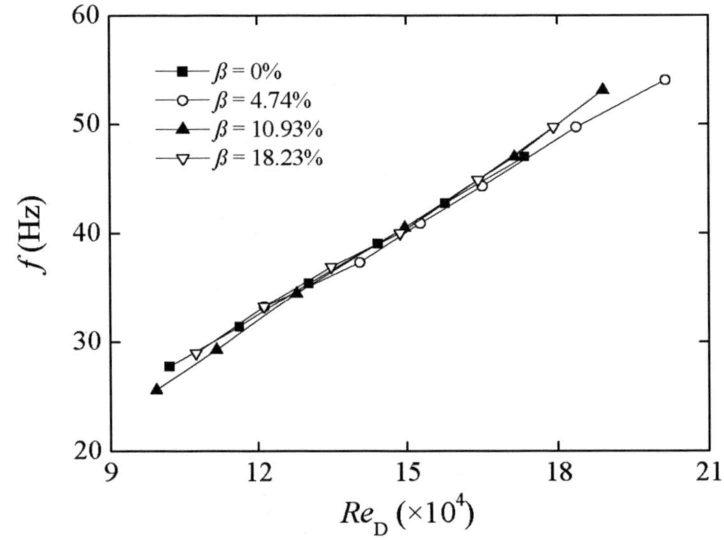

FIGURE 3. Vortex shedding frequency of bubbly two-phase flow versus Reynolds number at the same volume void fraction.

Moreover, the corresponding power spectral densities of the raw signals, which were calculated by the fast Fourier transform (FFT), are also given in Fig. 2(b). Although the amplitude got decreased along with the Reynolds number's increase in the time domain, which differs greatly from the case in single-phase flows, the frequency of the signal still got increased with Re_D little by little, just the same like that of single-phase flows. However, say when β was greater than 30.00%, the fluctuating differential pressure at vortex shedding frequency was not detectable. Consequently, the present study of gas-liquid two-phase vortex shedding was limited to the range of $\beta = 0 - 30\%$, equivalent to $Re_D = 4.63 \times 10^4 - 23.45 \times 10^4$.

Frequency of Vortex Shedding

The frequency f of gas-liquid two-phase vortex shedding was calculated by the FFT of the dynamic differential pressure. Figure 3 shows that f increases with increasing Re_D linearly at various volume void fractions, and the slopes of these lines are nearly of the same value. The experimental results could be approximated by the following Equation (6) according to the least square method, whose accuracy of approximation is within ± 5% for all the experimental data.

$$f = 2.797 Re_{\mathrm{D}} - 1.2607 \tag{6}$$

Note that the validation of Equation (6) is confirmed by the present study under the following conditions: bubbly two-phase flow with β in the range from 0 to 30% and Re_D in the range from 4.63×10^4 to 23.45×10^4.

Measurement of Two-Phase Flowrate

Figure 4 presents the relationship of the relative Strouhal number with β in the bubbly two-phase vortex flows. The relative Strouhal number is defined as the ratio of the two-phase Strouhal number St_{TP} to the single-phase Strouhal number St_{SP}, in which

$$St_{\mathrm{TP}} = \frac{f \cdot d}{v_{\mathrm{TP}}} \tag{7}$$

and St_{SP} is acquired by the calibration of the same vortex flowmeter with single-phase air or water.

As shown in Fig. 4, the relative Strouhal number maintains constant within the tested β range up to 25%, and the values of the relative Strouhal numbers are very close to 1 with less than ± 5% relative errors. When β is between 25% and 30%, there appears to be an unstable zone corresponding to the transition from bubble flow to other flow patterns confirmed by the visual observation. The errors of the relative Strouhal numbers here approach to ± 10%. If β is increased further, the bubbly flow pattern is very difficult to occur.

In the present study, the measuring range the vortex flowmeter for two-phase flows is $\beta = 0 - 25.90\%$ and $Re_D = 9.47 \times 10^4 - 20.44 \times 10^4$. Within these ranges, the following Equation (8) is valid with fixed $K_{TP} = K_{SP} = 9.217 \times 10^3$.

$$Q_{\mathrm{TP}} = f / K_{\mathrm{TP}} \tag{8}$$

which is still a fundamental correlation of vortex flowmeter for the bubbly two-phase flow measurement with less than ± 5% relative error.

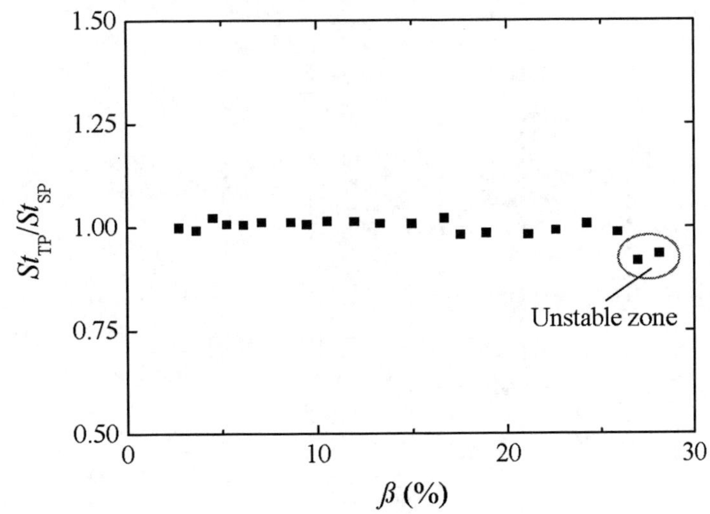

FIGURE 4. Relative Strouhal number of bubbly two-phase flow versus volume void fraction.

CONCLUSIONS

This paper studied the behavior of vortex flowmeter in bubbly gas-liquid two-phase flows. Major findings are summarized as follows.

1) Regular Karman vortex shedding may exist in bubble flows, but its stability is strongly influenced by the introduction of gas-phase component. In the present study, when the volume void fraction β is about 15%, vortices begin to become unstable; and when β is greater than 30%, regular vortex shedding can not be observed any longer.

2) Within the range of β = 0 – 30% and Re_D = 4.63 $\times 10^4$ – 23.45 $\times 10^4$, the frequency of the bubbly vortex shedding is in approximately linear proportion to the two-phase Reynolds number Re_D, whereas the volume void fraction β has little impact on the frequency.

3) If β is restricted to 0 – 25.90% and Re_D is within 9.47 $\times$ 10^4 to 20.44 $\times 10^4$, the two-phase flow Strouhal number keeps constant at the value as in single-phase flows, so the meter coefficient will not vary when the vortex flowmeter is used for bubbly two-phase flow measurement.

ACKNOWLEDGEMENTS

This work was supported by the Special Research Fund for the Doctoral Program of Higher Education of China under the Project No. 20030335058, and the High-tech Research and Development Program of China (863 Program) under the Project No. 2001AA413210.

REFERENCES

1. E. S. Rosa and R. E. M. Morales, *Journal of Fluids Engineering* **126**, 457-467 (2004).
2. W. Wang and Y. Tong, *Flow Measurement and Instrumentation* **6**, 265-270 (1995).
3. T. Sun, H. Zhang and C. Hu, "Identification of Gas-Liquid Two-Phase Flow Regime and Quality" in *Proceedings of IEEE Instrumentation and Measurement Technology Conference* 2, Anchorage, AK, 2002, pp. 1471-1747.
4. H. Asano, N. Takenaka and T. Fujii, et al., *IEEE Transactions on Nuclear Science* **52**, 280-284 (2005).
5. P. P. Bhattacharjee and S. Sen, "Wire-Mesh Tomograph for Gas-Liquid Flow Measurement" in *Proceedings of IEEE Indicon 2005 Conference* 1, Chennai, India, 2005, pp. 427-430.
6. B. Sun, H. Zhang and L. Cheng, et al., *Chinese Journal of Chemical Engineering* **14**, 24-30 (2006).
7. H. Zhang, W. Yue and Z. Huang, *Journal of Zhejiang University (Science)* **6A**, 601-606 (2005).
8. G. L. Pankanin, *Measurement Science and Technology* **16**, R1-R16 (2005).
9. Y. Li, Z. Cai and M. Li, et al., "A Study of Measuring Gas-Liquid Two-Phase Flow Rate and Component with Gas-Liquid Two-Phase Vortex Street Characteristics" in *Development of Measuring Techniques for Multiphase Flows*, edited by Y. Xu et al., Proceedings of the 2nd International Symposium on Measurement Techniques for Multiphase Flows 1, Standards Press of China, Beijing, China, 1998, pp. 100-104.

Analyses of Characteristics of Ring-shaped Electrostatic Meter

Jianyong Zhang*, Chuanlong Xu, Donglai Xu*, John Coulthard* Shimin Wang****

** School of Science and Technology, University of Teesside, Middlesbrough, UK*
*** Key Laboratory of Clean Coal Power Generation and Combustion Technology of Ministry of Education, Southeast University, NanJing, 210096, China.*

Abstract: This paper describes the characteristics of ring-shaped electrostatic pulverised fuel meters and their applications. At the University of Teesside, UK, electrostatic technique has been used to measure pulverised fuel flow concentration, velocity and mass flow rate under lean phase condition. The mathematic model has been developed to express the relationship between the root mean square (rms) value of meter's voltage output and solids concentration and mass flow rate. The effects of solids velocity and particle size on measurements have also been reflected in the model. The model also presents the sensitivity variations over cross the sectional area of meter and along the pipe axis. The paper also introduces the research carried out in the Southeast University in China, where the technique has been extended to measure dense phase flow of pulverised coal, which is common in gasification and blast furnaces.
Key words: Electrostatic, Pulverised fuel, air-solids two-phase flow, measurement

INTRODUCTION

The importance of correct fuel/air ratio in achieving clean and efficient combustion is well demonstrated. However the technology in coal fired power stations is complicated, where the fuel (coal) is pulverised and air-coal mixture is then pneumatically conveyed to the burners. Accurate measurement and control of the pulverised fuel (PF) mass flow rate inside each conveyor leading to burner is of critical importance for improving combustion efficiency, lowering the carbon content of the ash and reducing greenhouse gases and other emissions. The potential environmental and economic benefits from achieving effective PF flow measurement and thus combustion control are extremely significant. In a study by PowerGen, the economics of PF distribution control were analysed, suggesting that improving PF distribution with a combustion optimiser can provide simultaneous benefits in reduction of NOx emissions by 30% with instantaneous improvement of efficiency [1].

The use of coal as auxiliary fuels for injection into iron making blast furnaces is driven primarily by economic factors, especially the benefits derived from reducing the consumption of expensive coals for metallurgical coke making. In order to control the process, measuring pulverised coal flow is essential [2].

In pharmaceutical, food and cement making industries, the measurement of powder flow is essential for safety, economic and operation purposes.

However, achieving accurate and reliable measurement of powder flow in complex pneumatic conveying systems has proven technically challenging. A number of methods, based on different principles have been explored. The ring-shaped electrostatic PF system [3] is among those most promising methods [4].

The principle of the passive electrostatic technique for Pulverised fuel (PF) flow measurement is well known. In pneumatic conveying, electrical charging of the conveyed solids occurs. The primary sources of electrification are frictional contact charging between particles, charge transfer from one particle to another, and friction between particles and the conducting pipe wall.

In a dynamic electrostatic system, the fluctuation level in the net charges carried by solids particles is considered to be proportional to the solids concentration in lean-phase condition. This paper presents the test results which support this assumption. The electrostatic pulverised fuel meters described in this paper measure fluctuation in charges carried by particles to indicate solids concentration and solids mass flow rate.

CP914, *Multiphase Flow: The Ultimate Measurement Challenge,* edited by X. Cai, Y. Wu, Z. Huang, S. Wang, and M. Wang

The density of single phase flow is regarded as constant. However in air-solids two-phase flow, inhomogeneous solids distribution is a common phenomenon, which means that pulverised fuel may not be uniformly distributed in the pipe, often concentrating to form a 'roping flow' so that it becomes highly concentrated locally whilst elsewhere there is very little solid. [5][6]. This non-uniform flow pattern presents real difficulties to achieving accurate measurement as flow profiles varies with the pipe runs and loading. The meter's reading is dependent on flow profiles [7] [8]. To overcome the above problems and achieve accurate measurement of such flow, the metering system with uniform spatial sensitivity is ideal. This paper presents the analysis results of the spatial sensitivity of the ring-shaped electrostatic meter.

The velocity of solids and particle size affect the electrical charges carried by solids due to the mechanism of electrification. Under given solids mass flow rate, the net charge carried by the solids varies with the velocity and particle size. Since the average velocity can be measured independently, e.g. using cross-correlation method, theoretically, this effect can be estimated or compensated. This paper demonstrates the effect of the velocity on the measurement. If the mass flow rate or concentration and velocity are known parameter, the particle size information can be obtained from the charges induced on an electrode. This will also be discussed in this paper.

BASIC MATHEMATIC MODEL OF A RING SHAPED ELECTRODE

The charge induced on a ring-shaped electrode from a single particle having unity charge has been modelled [9], which was based on electrostatic theory and calculated using the finite element method (FEM). As shown in Fig. 1, the charge Q induced on the inner surface of the electrode due to a particle of unity point charge inside the pipe was calculated using following expression;

$$Q = \int_S \sigma \cdot ds \qquad (1)$$

where σ is the surface density of the charge induced on the electrode, and S is entire inner surface area of the electrode.

The FEM analysis results were curve fitted to the following equation [9].

$$Q = Ae^{-k^2x^2} \qquad (2)$$

where A and k are two constants that are functions of the width W of the electrode and the radius r from the central axis of the electrode towards the pipe wall. The variable x denotes the distance along the pipeline from the centre of the electrode cross section as illustrated in Fig. 1. The quantity Q is the charge induced on a metallic electrode ring from a particle carrying unity point charge.

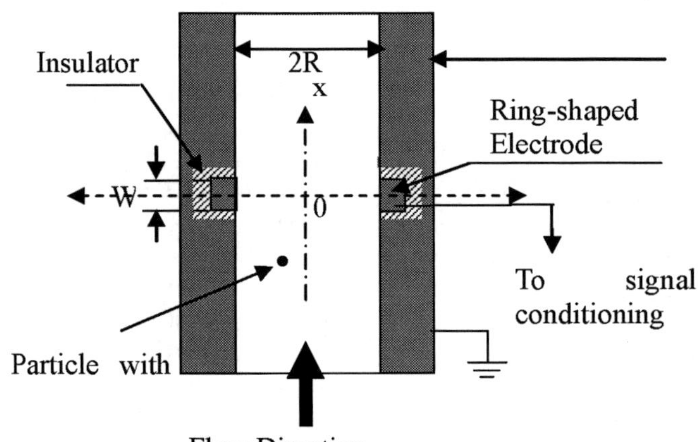

Fig. 1 Schematic of a Ring Shaped Electrostatic Electrode

The sensitivity of a ring-shaped electrode versus the location of a charged particle illustrated in Fig.2 based on equation (2). It can be seen that the sensitivity is a function of the radius r for a given electrode width W. Fig. 3 and Fig. 4 show the spatial sensitivity of a ring-shaped electrode and the relative Ac signal.

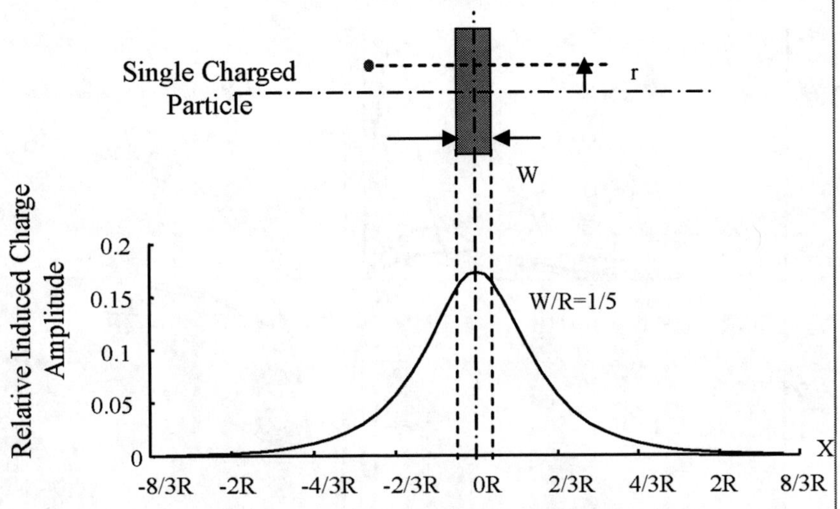

Fig. 2 Variation of Induced Charge on the Electrode

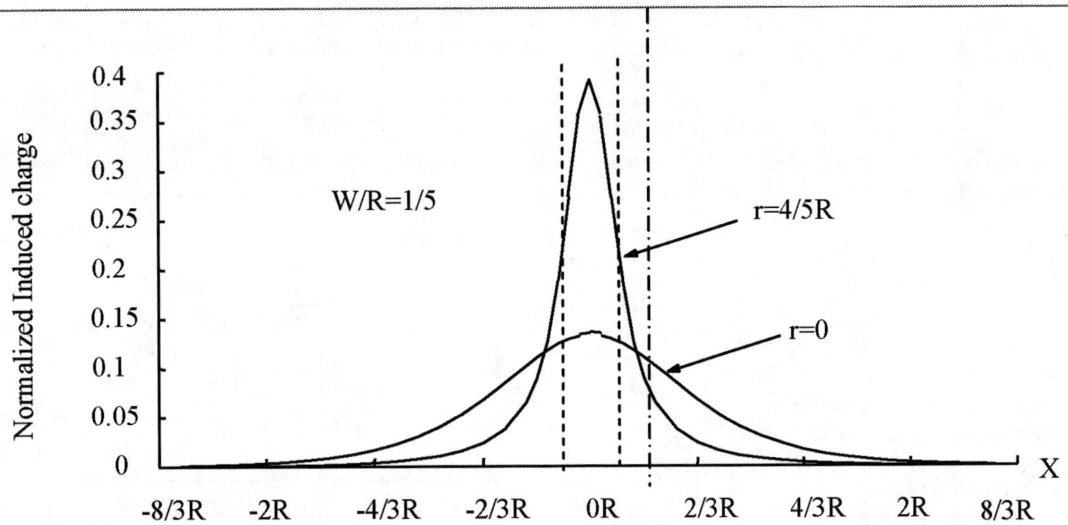

Fig.3 spatial sensitivity of a ring-shaped electrode

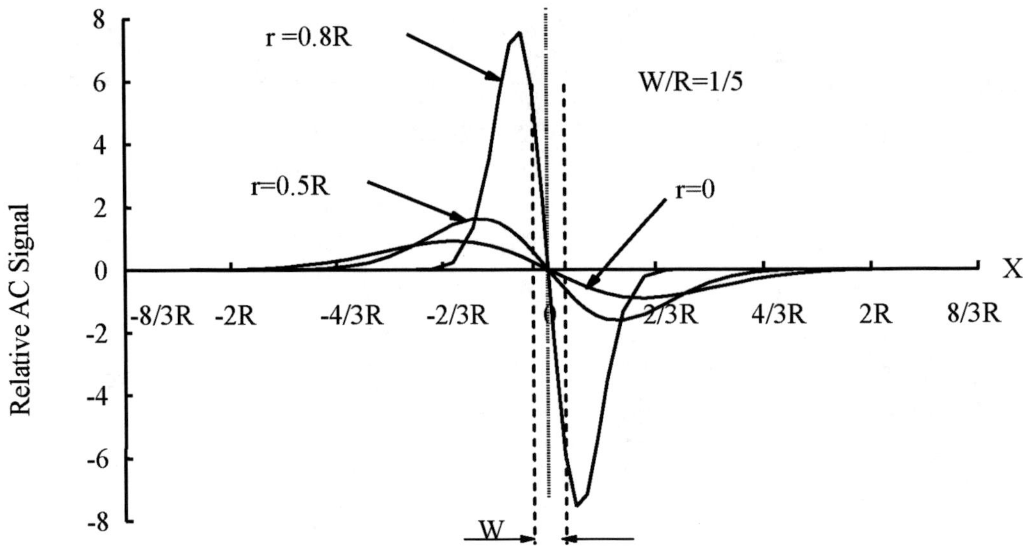

Fig. 4 Relative AC signals

The sensing volume, shown in Fig.5, indicates that the sensor can 'detect' streams of particles moving at different distances on different axes. It can sense particles at greater distances along an axial line closer to the pipe central axis than particles moving closer to the pipe wall. Obviously, the sensing volume is significantly different to the physical volume determined geographically by the electrode width W and the cross sectional area of the electrode.

In the Southeast University, the modelling has results in the following equation,

$$S(z) = ae^{-bz^2} + ce^{-dz^2} \qquad (3)$$

Where S(z) stands for sensitivity along pipe line, z is the co-ordinate along pipe line. Fig. 6 shows the spatial sensitivity based on equation (3). The similarity between these two models is obvious although the electrode is not directly exposed to the follow in the Southeast University design, in which, the inner surface of electrode was lined with a quartz tube.

In equation (3), z=0 is at the edge of the sensing volume, whilst in equation (2), x=0 is at the middle of the sensing length along the pipe axis.

Fig.7 depicts the sensitivity of a ring-shaped electrode against radius r at different position z along pipe line. This is obtained based on the Southeast University Model.

If we take the difference in definitions between x and z into account, it is clear that Fig.5 and Fig. 7 describe the same phenomenon that spatial sensitivity varies with radius r at different position along pipe axis.

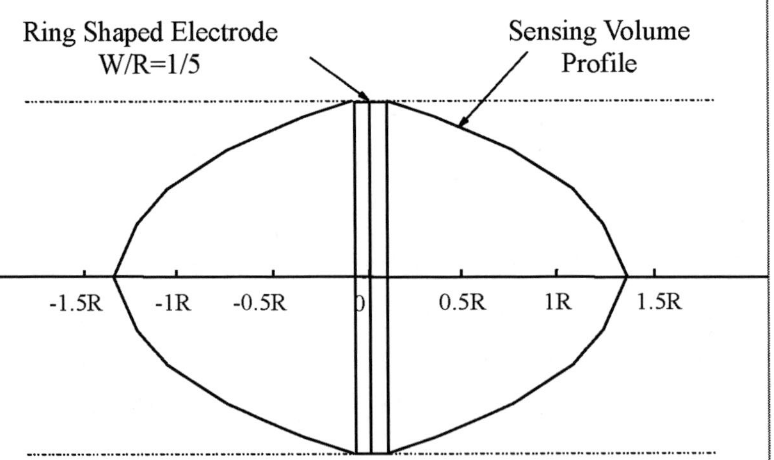

Fig. 5 Sensing Volume

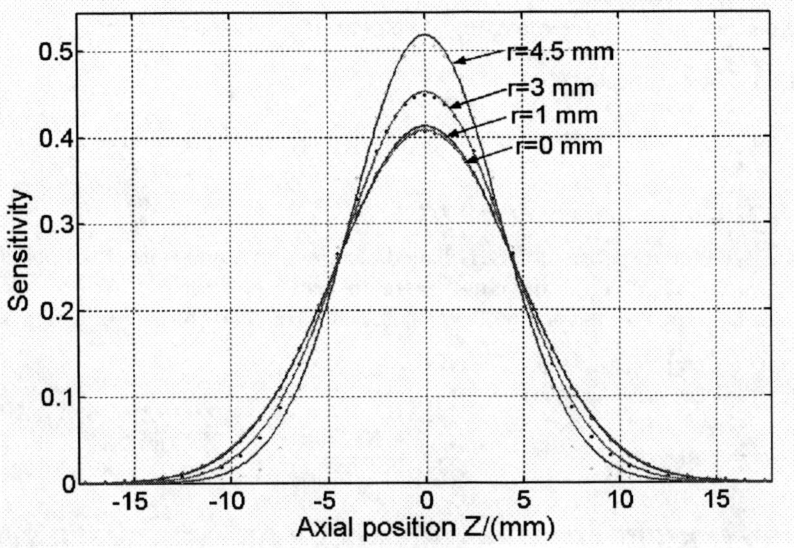

Fig. 6 Spatial sensitivity based on Southeast University Model

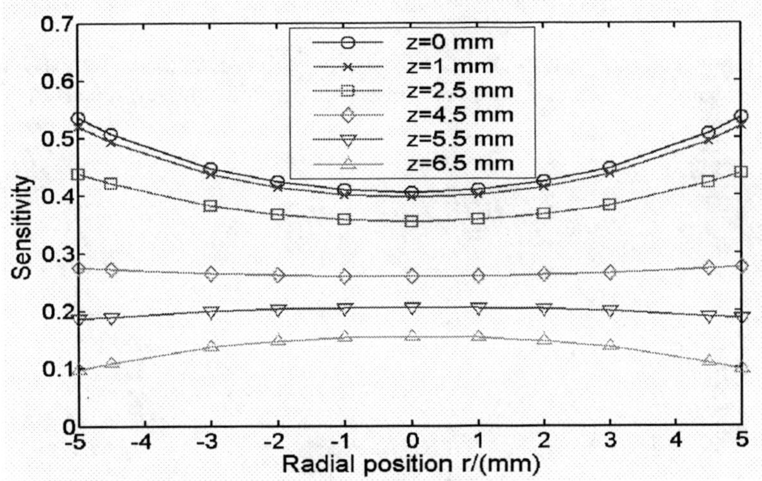

Fig. 7 Sensitivity Versus r at given z

MODEL OF INDUCE SIGNAL TO STREAM OF CHARGED PARTICLES

A ring-shaped stream having inner radius r and outer radius $r + \Delta r$ can represent a roping flow due to the symmetrical property of ring-shaped electrode. Based on the principle of superposition of electrostatic fields, the response of a ring-shaped electrode to this stream can be expressed as follows according to equation (2),

$$Q^{r}_{NET} = \sigma_p \cdot \pi \cdot D_P^2 \cdot N \int_{L_x} A(r,W) e^{-k(r,W)x^2} dx \tag{4}$$

Where A and k in equation (2) are replaced with A(r,W) and k(r,W) to represent their dependences on electrode width and position of the stream with respect to the centre axis [8].

If we consider this roping particle stream wave travelling at velocity V as a delta function, i.e.

$$Q_{NETr}^{P} = \sigma_p \cdot \pi \cdot D_P^2 \cdot N(t) = \delta(t - \frac{x}{V}) \tag{5}$$

At any given time, at only one point along the x direction where Q_{NETr}^{P} exists, this point is x=V*t. Hence Eq. (4) becomes

$$Q_{NET}^{r} = h_{rd}(t) = A(r,W)e^{-k(r,W)V^2t^2} \tag{6}$$

This equation describes the dynamic response of a ring-shaped electrode to a unit impulse of net charge carried by solid particles at radius r. The variable $h_{rd}(t)$ is sensitive to the stream spatial position..

Applying Fourier transformation to Eq. (6), the spatial frequency-response characteristic of a ring-shaped electrode to a unit impulse of net charge carried by particles is,

$$H_r(\varpi) = \int_{-\infty}^{\infty} h_{rd}(t)e^{-j\varpi t}dt = \int_{-\infty}^{\infty} A(r,W)e^{-k(r,W)V^2t^2}e^{-j\varpi t}dt = \frac{A(r,W)\pi^{\frac{1}{2}}}{V\sqrt{k(r,W)}}e^{-\frac{\varpi^2}{4V^2}\frac{1}{k(r,W)}} \tag{7}$$

EFFECTS OF VELOCITY ON MEASUREMENT OF CONCENTRATION AND SOLIDS MASS FLOW RATE

From equation (7), it is evident that the velocity affects the frequency characteristics of the electrode and thus the entire metering system. The detailed analysis of the frequency spatial characteristics of electrode and ring-shaped electrostatic sensors can be found in [8], [9], [10] and [11]. Fig.8 demonstrating that the frequency bandwidth and amplitude peak position of the transfer function of a ring-shaped electrostatic sensor meter increases with velocity [10].

In the Southeast University, the investigations into effects of velocity on the flow measurement as shown in Fig 9 have lead to a novelty method for solids velocity measurement based on signal frequency spectrum analysis [11].

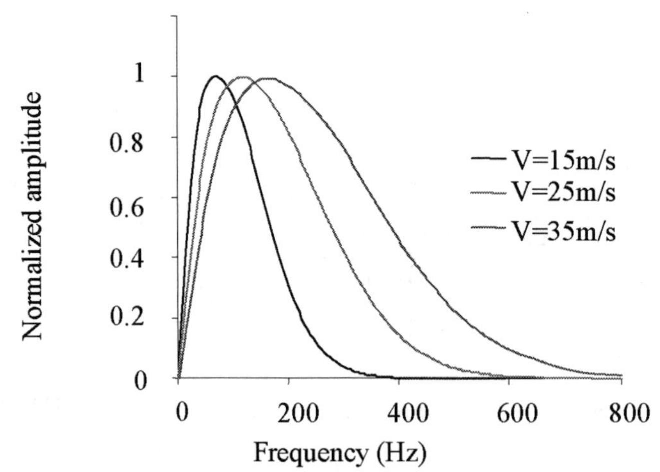

Fig. 8 Effect of Velocity on the Sensor's Frequency Characteristics R=150mm W=40mm

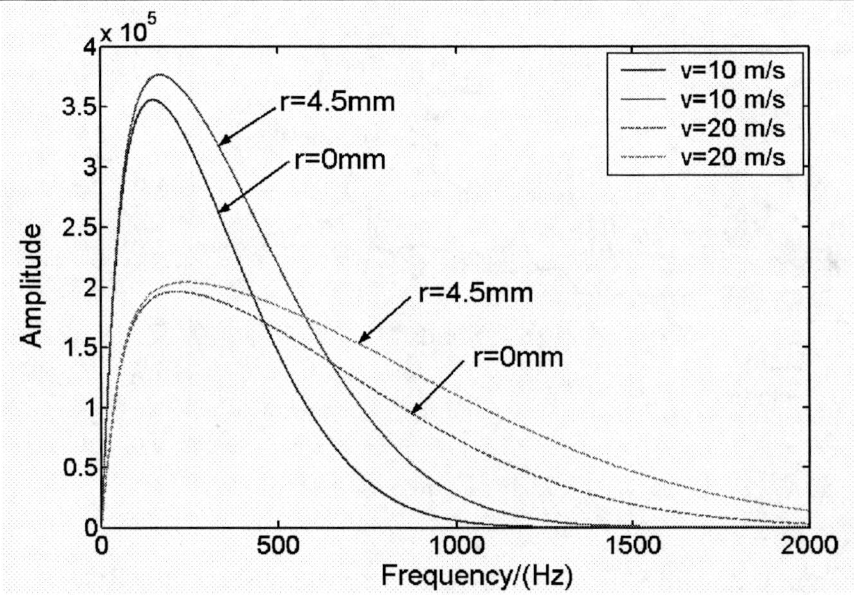

Fig. 9 Spatial sensitivity

A metering system comprises the electrode and the signal conditioning circuit. Thus the transfer function of a sensor can be obtained from the integration of $H_r(\omega)$, i.e. $\int_0^R H_r(\omega)dr$ multiplied by the transfer function of the circuit. For a dynamic measurement system, the root mean square value of meter's output can be related to its power spectral density function according to Parseval's formula. The square value of the system output voltage is given by equation (8) assuming that the solids concentration is proportional to the net charges carried by particles.

$$U_{rms}^2 \propto \frac{con^2}{\varpi_{\max}} V \quad (8)$$

where *Con* is the concentration of flow, V is the solids velocity, ω_{max} is the maximum frequency of band-limited flow noise.

From this equation, it can be seen that the rms value of the output voltage of a sensor is proportional to solids concentration and the square root of the velocity. Thus, in theory, the effect of velocity on concentration measurement can therefore be removed.

It also shows that U_{rms} is linear with solids concentration or solids mass flow rate under a given velocity. Fig.10. and Fig.11 provide laboratory test results supporting the above conclusions.

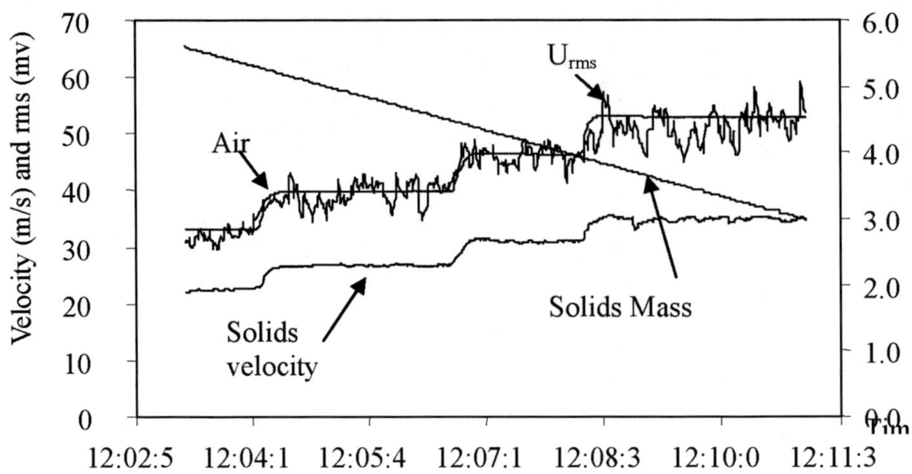

Fig 10. Variation in U_{rms} with Solids Velocity Under constant Mass flow Rate

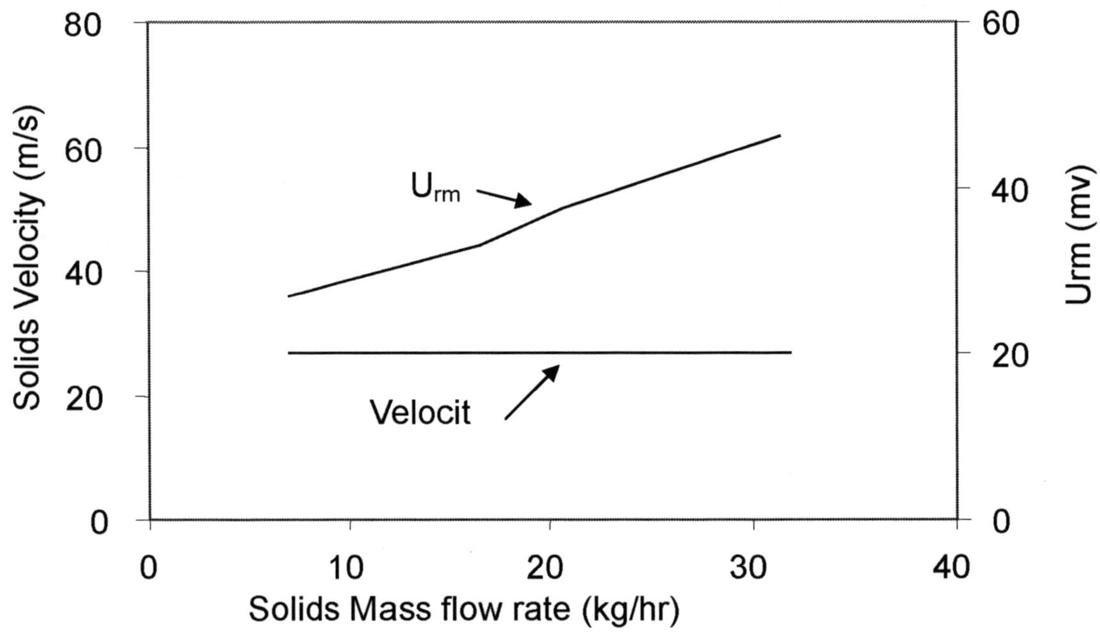

Fig. 11 U_{rms} Variation with Solids Mass flow rate Under constant Velocity

EFFECT OF PARTICLE SIZE ON COMCENTRATION MEASUREMENT

Cheng [9] described the principle of measuring particle size based on signal bandwidths. For a given electrode geometry, the signal waveform and the corresponding signal bandwidth are fundamentally dependent upon the time that particles remain within the sensing volume. Therefore, larger particles remain within the sensing volume for a longer time corresponding to a narrower signal bandwidth, than that produced by smaller particles moving at the same speed. This has been confirmed by modelling carried out at the laboratory of the Southeast University as illustrated in Fig 12.

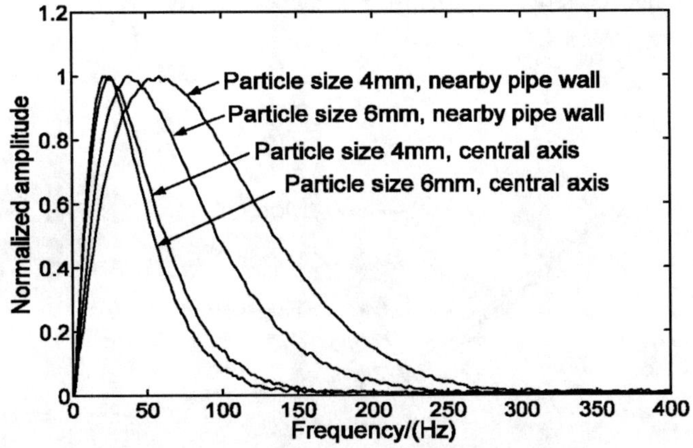

Fig. 12 Effect of particle size on signal frequency spectrum

This has been investigated at the University of Teesside, the project was funded by ECSC (European Coal and Steel community) [12], Fig. 13 shows the autocorrelation curves of signals of sensor output under different particle sizes.

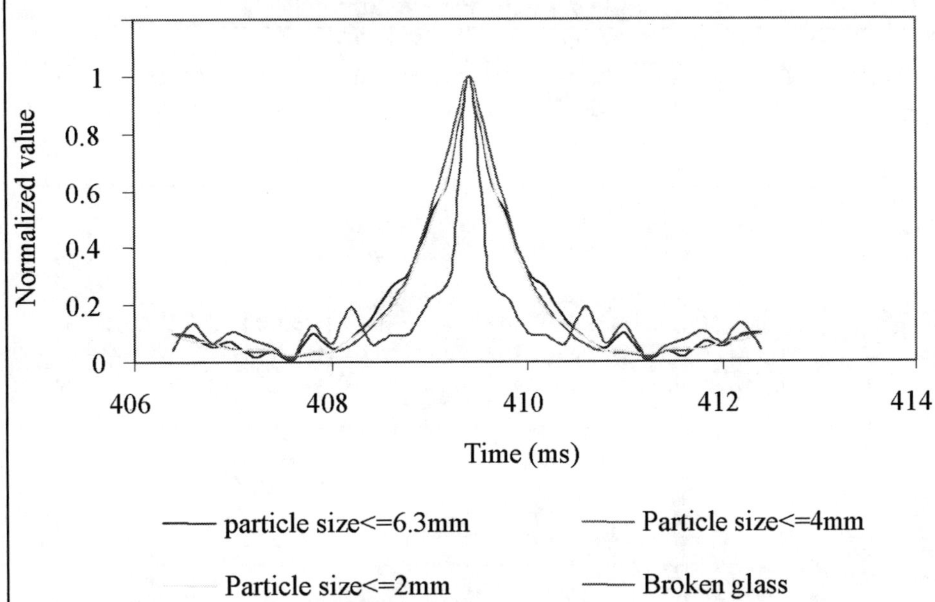

Fig. 13 Auto correlation curves for different size of particles

It can be seen that for the smaller particles, the autocorrelation curve becomes narrower due to higher frequency bandwidth of signal. But the resolution is poor for narrow size distributions. This method may be useful in coal preparation where large and small sizes of coal need to be separated.

Another on-line particle size measurement method investigated is based on the assumption that surface charge density σ_p of particles is constant. It gives that the charge carried by a particle is proportional to its surface or D^2 for spherical particles, where D is particle diameter. [13]

Under the above assumptions, Equation (8) can be modified to include the particle size information, which is

$$U_{rms}^2 \propto \frac{con_{rms}^2}{D^2 \varpi_{\max}} V \tag{9}$$

Fig. 14 shows the test results obtained in a 40mm diameter rig in Casella CRE Energy in Gloucestershire UK. The material was coal powder, the size range covered 75 μm to 1mm. The theoretical signal is also included for comparison.

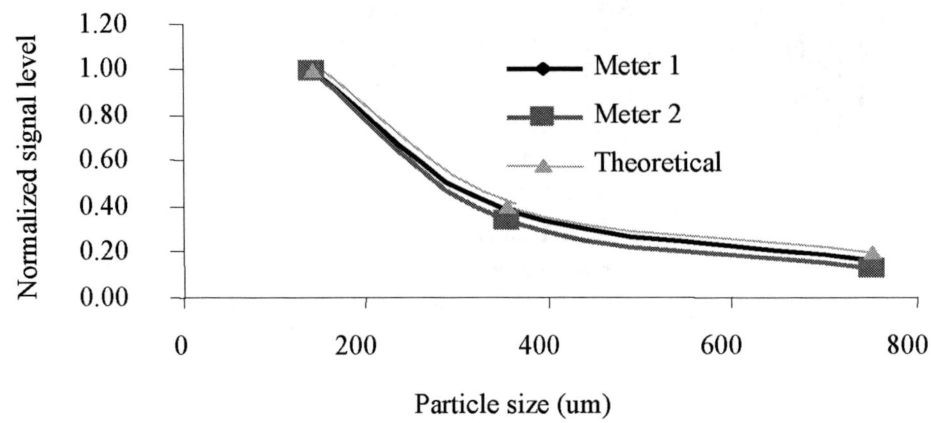

Fig. 14 U_{rms} versus Particle Size

Fig. 15 are based on laboratory tests results at the University of Teesside, the materials was sand, the particle size was in the range of 280 μm to 900 μm. The results again confirmed that the signal rms value decreases with particle size under constant solids mass flow rate and velocity.

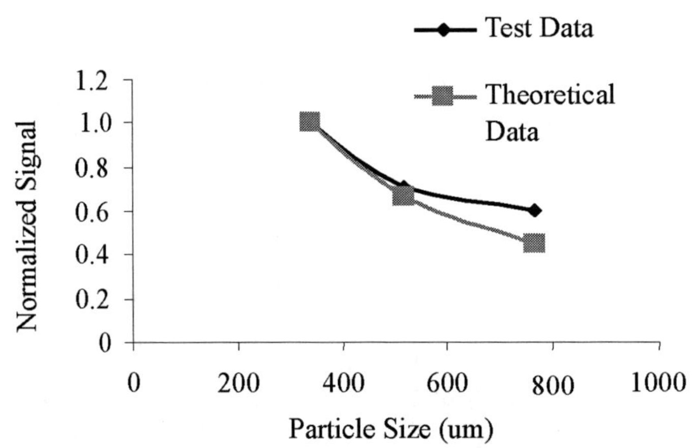

Fig.15 Particle Size Measurement Test Results using Sand

CONCLUSIONS

Although electrostatic meters provide only relative solids mass flow rate and concentration information, it has been successfully used in coal fired power stations for split and velocity measurement and or used as diagnostic tool to identify poor conveying conditions.

Theoretical and experimental investigation into spatial sensitivity, effects of particle size and velocity on flow measurement has played an important role in improving the performance of meters and to extend its applications. For example, in the University of Teesside and the Southeast University, some of research findings have been applied in different aspects such as an innovation method of flow velocity measurement based on frequency spectrum analysis and particle size indication using electrostatic meter, which originally was merely treated as modifying input of mass flow rate measurement.

The research in improving spatial sensitivity and solids mass flow measurement based on Fussy logic are on going activities in these two universities.

The electrostatic meters with ring shaped electrodes are non-intrusive, robust, maintenance free and low cost. The electrostatic method will remain as a practical and promising approach in metering air-solid two-phase flow.

REFERENCES

[1] P. Baimbridge, ECSC Project 7220-PR-050 "Measurement and control Techniques for improving Combustion and Reducing Emissions from coal fired plant", Boiler integration Study, 02/2001.

[2] Angus Thomson, Paul Zulli, Malcolm McCarthy and Keith Horrocks, "Pulverised Coal Injection in Ironmaking Blast Furnaces", http://www.australiancoal.csiro.au/pdfs/Thomson.pdf

[3] Jianyong Zhang, John Coulthard, Ruixue Cheng, Ray Keech, "Measuring and controlling pulverized fuel on coal-fired plant", Journal of clean air, vol 6, No.1 pp 25-35, 2005

[4] Dave Miller et al, "Technology Status Review Report PF Flow Measurement and Control Methods for Utility Boilers", December 2000.

[5]Huber N, Summerfield M, "Characterisation of the Cross-sectional Particle Distribution in Pneumatic Conveying System", Powder Technology Vol.79, 1994

[6]Frank Thomas, Numerical Prediction of Multiphase flows, http://www.imech.tu-chemnitz.de/mpf/mpf_numerik.html.

[7] Wang Shimin, et al, " Advanced Measurement and Diagnosis Technology Applied in Coal-Fired Utility Boilers", Proceedings of 5th International Symposium on Coal Combustion, Nanjing, China, November 23-26, 2003, paper No. 14

[8] Jianyong Zhang, John Coulthard, " Theoretical and Experimental Studies of the Spatial Sensitivity of an Electrostatic Pulverised Fuel meter" Journal of Electrostatics, Vol.63, issue 12, Oct 2005

[9] Cheng R., "A study of electrostatic pulverised fuel meters", Ph.D. thesis, University of Teesside, U.K. 1996.

[10] Jianyong Zhang, " A Study of an Electrostaic Flow Meter", Ph D thesis, April 2002, University of Teesside, U.K.

[11] Xu ChuanLong, "Gas-solids Two phases Flows Parameters and Particle Electrostatic Measurement", Ph D thesis, Souheast University, Nanjing, China, Dec. 2005.

[12]Research Final report (November 2000 to April 2004), "On-line Measurement of Particle size in Fine Coal Transport Systems", project ECSC-7220-PR-105, 31/July 2004.

[13] Jianyong Zhang, John Coulthard, "On-line Indication of Variation of Particle Size Using Electrostatic PF Meters", The proceedings of the 31th International Technical Conference on Coal Utilization and Fuel Systems, May 21-25 Clearwater, Florida, USA

On-line Electromagnetic Wave Based Coal Concentration Monitoring Technique for Pipe Flow

Kefa CEN, Bin Ma, Kunzan QIU, Hao ZHOU*

Zhejiang University, Institute for Thermal Power Engineering,
State Key Laboratory of Clean Energy Utilization, Hangzhou, 310027, P. R. China

Abstract. The on-line coal concentration measuring technique for pipe gas-solid flow is helpful for boiler operation optimization. In this work, a novel electromagnetic wave based coal concentration monitoring technology was developed. The monitoring system consists of an emitting antenna and a receiving antenna. The emitting antenna emits electromagnetic wave, the wave propagates in the coal pipe, and the amplitude of the electromagnetic wave is attenuated by the coal particles. The frequency of the electromagnetic wave can be adjusted to meet the diameter of the coal pipe. The electromagnetic wave motivates current in the receiving antenna, which is amplified by a multiplier and then acquisitioned by the computer. After the digital signal processing, the coal concentration of the gas-solid pipe flow can be achieved. The results show that the electromagnetic wave based coal concentration measuring technology can give the on-line solid concentration information, the output voltage is in linear relationship with the coal concentration. The on-line coal concentration measuring system was employed in the coal pipes of a 660 MWe utility boiler, the filed data illustrated that the system responded quickly with change of the coal concentration, the technique is not sensitive with the coal moisture due the low frequency electromagnetic wave was adapted.

Keywords: Gas-solid flow, electromagnetic wave, concentration measurement

INTRODUCTION

Coal remains the major source of energy resource in the world. Coal fired boiler power stations are now subject to stringent pollution regulations and suffer penalties if the pollution emissions exceeds the acceptable level. Furthermore, coal fired power station efficiency affects the cost of power. The optimization and control of pulverized-coal distribution has assumed greater significance. Balanced PF distribution is helpful for low NOx burner operation to obtain low NOx emissions and high combustion efficiency. Non-uniform distribution of fuel between burner may be caused by the different length of the pipe and /or fuel roping, which results in localized fuel-rich and lean regions in the furnace leading to unacceptably high levels of unburned carbon in ash. The NOx emissions of boiler are also difficult to be reduced in such a case.

At present, extractive sampling techniques are employed to assessing fuel distribution, however, such approaches can only provide data as a spot reading, it can not show the dynamic characteristics of the fuel distribution. There have also been many work and attempts to seek on-line coal concentration measurement techniques, including electrical methods, optical methods, microwave sensors, acoustic/ultrasonic sensors, etc [1]. Some of them have come into commercial application, but most of them were still in the stage of laboratory evaluation.

In this work, a novel electromagnetic wave based coal concentration monitoring technology was developed. The monitoring system consists of an emitting antenna and a receiving antenna. The emitting antenna emits electromagnetic wave, the wave propagates in the coal pipe, and the amplitude of the electromagnetic wave is attenuated by the coal particles. The frequency of the electromagnetic wave can be adjusted to meet the diameter of the coal pipe. The electromagnetic wave motivates current in the receiving antenna, which is amplified by a multiplier and then acquisitioned by the computer. After the digital signal processing, the coal concentration of the

* Corresponding author. Tel: +86-571-87952598; Fax: +86-571-87951616 *E-mail address:* zhouhao@cmee.zju.edu.cn (ZHOU H).

CP914, *Multiphase Flow: The Ultimate Measurement Challenge,* edited by X. Cai, Y. Wu, Z. Huang, S. Wang, and M. Wang

gas-solid pipe flow can be achieved. The results show that the electromagnetic wave based coal concentration measuring technology can give the on-line solid concentration information, the output voltage is in linear relationship with the coal concentration. The on-line coal concentration measuring system was employed in the coal pipes of a 660 MWe utility boiler, the filed data illustrated that the system responded quickly with change of the coal concentration, the technique is not sensitive with the coal moisture due the low frequency electromagnetic wave was adapted.

FUNDAMENTALS OF THE ELECTROMAGNETIC WAVE BASED METHOD

The monitoring system consists of an emitting antenna and a receiving antenna, as illustrated in Fig. 1. The emitting antenna emits electromagnetic wave, the wave propagates in the coal pipe, and the amplitude of the electromagnetic wave is attenuated by the coal particles. The frequency of the electromagnetic wave can be adjusted to meet the diameter of the coal pipe. Information on coal particle concentration in the pipe is obtained from the mean values of the signals with and without particles present.

$$I_p = I_m - I_0 \tag{1}$$

where I, the transmissive wave signal intensity.

The electromagnetic wave are subject to attenuations due to the coal suspension in the pipe. One of the significant problems of electromagnetic wave based method is the disturbance caused by reflected electromagnetic wave in the pipe. The attenuation of electromagnetic wave is small at low loads, which leads to the wave propagates in the pipe system over a long distances and is reflected by the pipe elbows, constrictions, etc. It may result in significant distortions of the accuracy of the measurements.

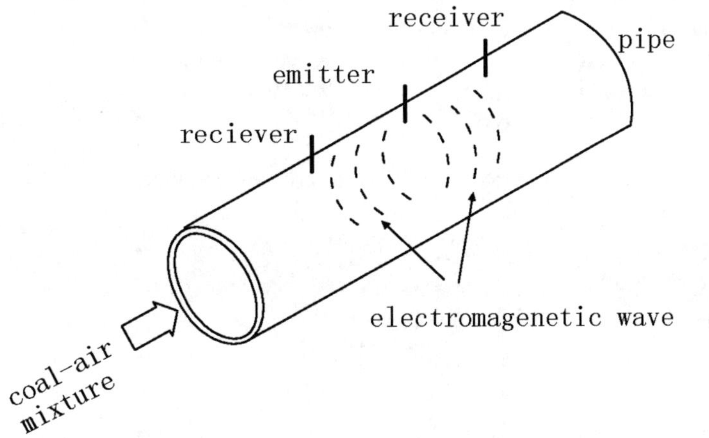

FIGURE 1. Antenna installation on the pipe

Another significant problem about the electromagnetic wave based method is the influence of the moisture content in the pipe. The moisture content has an obvious effect on the attenuation of the electromagnetic wave when the high frequency electromagnetic wave was employed. In this work, the frequency of the electromagnetic wave was chosen as low and the frequency can be adjusted to meet the request of pipes with various diameters.

A laboratory pneumatic conveying test facility consists of a 6-m-long horizontal pipe and a 3.5-m-long vertical pipe (see Fig. 2) was installed. The pipes are made of carbon steel pipes (0.15 m I.D.). A forced-draft fan provided a maximum air velocity of 40 m/s in the pipes. Particles stored in the hopper were discharged into the inlet air stream by means of a screw feeder calibrated for the test solid materials. The metering of the solids flow rate was accomplished by using a variable speed control. A cyclone is used to separate the solid particles from the air. The collected particles are fed to the hopper by an airlock. The remaining fine particles were removed by a fabric filter, and then fed back to the feeder hopper so that continuous operation and measurement were possible. Owing to the continuous circulation of particles in the loop, it is necessary to evaluate the attrition rate of the particles. In a series of experiments, the particles were extracted from the loop every 10 h of operation. Sieve analysis showed no significant attrition for 100 h period covered. After 100 h run of the loop, the particles were renewed.

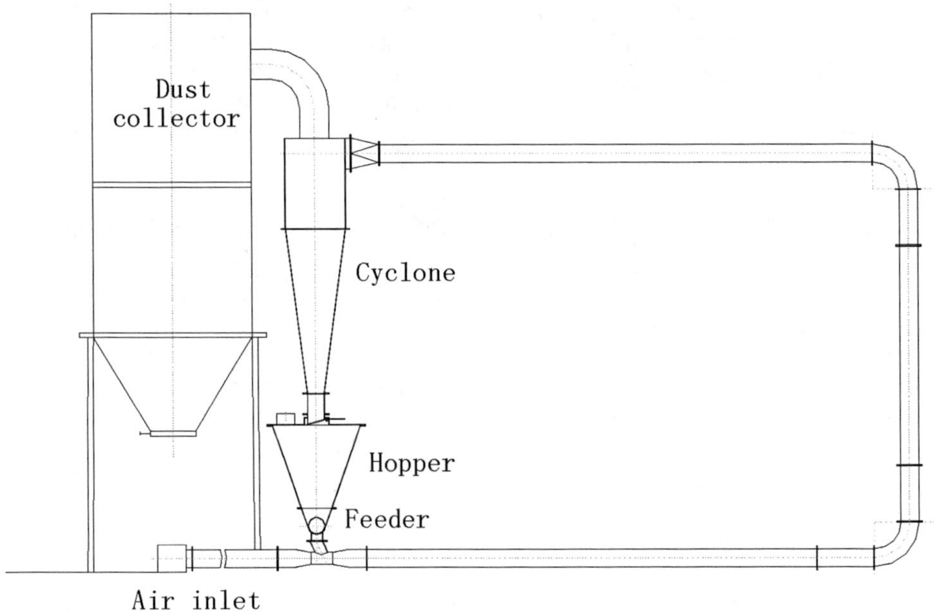

FIGURE 2. Sketch of the pneumatic conveying test facility

The electromagnetic measuring system was calibrated by an isokinetic sampling device in the pneumatic conveying test facility. This sampling device extracted a representative sample of particles to determine the mass flux; the suction velocity at the probe tip could be adjusted so as to be identical to the local gas stream velocity, so that the "isokinetic" condition was achieved. The detailed working principle of the isokinetic sampling probe can be found in Ref [2].

The comparisons showed that the local particle mass fluxes obtained by the two measurement techniques were in good agreement, as illustrated in Fig. 3.

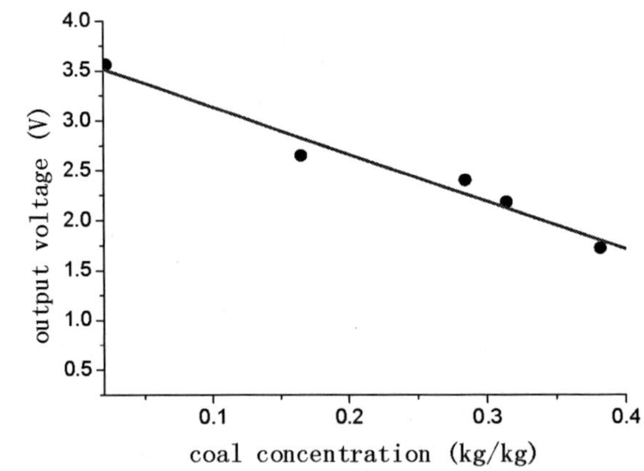

FIGURE 3. The laboratory results

ON-LINE MEASUREMENT RESULTS AND DISCUSSION

The on-line coal concentration measuring system was employed in the coal pipes of a 660 MWe utility boiler, an emitting antenna and a receiving antenna was installed in the pipe, as illustrated in Fig. 4. These two antenna were located at the vertical section of the pipe. A extracting hole was also opened in the pipe to use the isokinetic sampling device to measure the coal concentration in the pipe.

FIGURE 4. Photo of installation of the antenna

Total six cases were carried out in this boiler. The coal properties in various cases are listed in Table 1. There were two comparable cases named cases 2-1 and cases 2-2. The operating parameter such as coal concentration and primary air velocity, etc were all same in these two cases except the coal properties. The moisture content in case 2-1 was larger than that in case 2-2. The monitoring system gave the same output in these two cases, say, the technique is not sensitive with the coal moisture due to the low frequency electromagnetic wave was adapted.

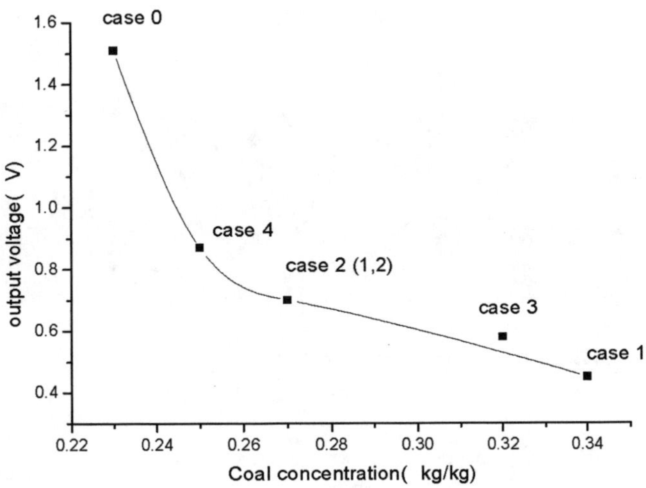

FIGURE 5. Field results

The effect of coal concentration on the monitoring system output is illustrated in Fig. . the monitoring system output decreased with the increase in the coal concentration in the pipe. The filed data illustrated that the system responded quickly with change of the coal concentration, as illustrated in Fig. 6.

TABLE 1. Coal properties

case	Industrial analysis (%)				Ultimate analysis (%)					Heat value
	M_{ad}	A_{ad}	V_{ad}	FC_{ad}	C_{ad}	H_{ad}	O_{ad}	N_{ad}	S_{ad}	(kJ/kg)
0	2.58	16.37	29.78	51.27	69.04	3.68	6.61	1.23	0.49	27489
1	4.96	16.19	27.43	51.42	64.01	2.74	10.68	1.02	0.40	25539
2-1	5.02	16.64	27.24	51.10	64.26	2.67	10.00	1.01	0.40	25717
2-2	4.66	16.27	27.41	51.66	64.00	2.92	10.73	1.00	0.42	25298
3	4.64	17.09	27.46	50.81	63.58	3.00	10.20	1.02	0.47	25285
4	2.89	16.55	29.29	51.27	68.84	3.57	6.42	1.27	0.46	27172

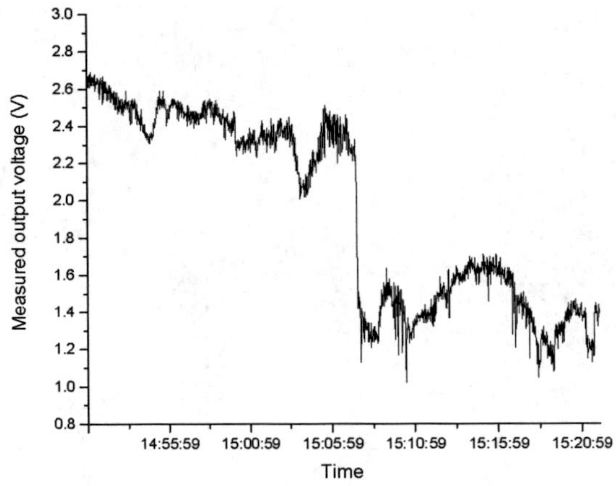

FIGURE 6. Dynamic output of the system

A fiber optic probe [3] was employed to measure the coal concentration distribution in the pipes under various loads. The results showed that the concentration distributions in the pipe cross-section are not uniform. The concentration distributing patterns under various loads are also not same, which implied that the measuring method which just monitors the concentration at one point as the representative in the pipe are not reliable. The electromagnetic wave based method introduced in this work can overcome this problem, because it use surface wave injection, the volume concentration information can be obtained.

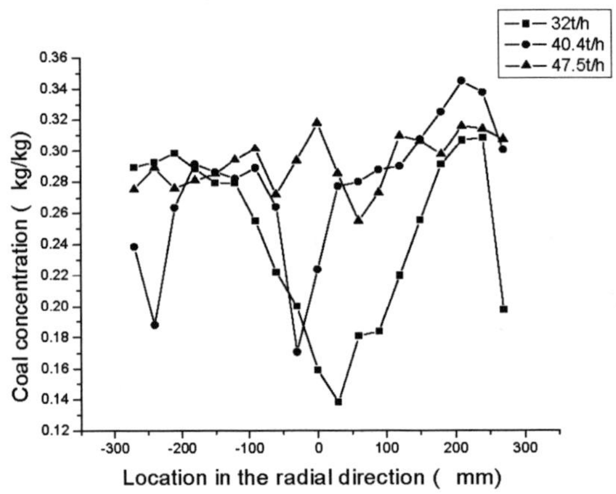

FIGURE 7. coal concentration obtained by optic fiber probe

CONCLUSIONS

A novel electromagnetic wave based coal concentration monitoring technology was developed in this work. The monitoring system consists of an emitting antenna and a receiving antenna. The emitting antenna emits electromagnetic wave, the wave propagates in the coal pipe, and the amplitude of the electromagnetic wave is attenuated by the coal particles. The frequency of the electromagnetic wave can be adjusted to meet the diameter of the coal pipe. The electromagnetic wave motivates current in the receiving antenna, which is amplified by a multiplier and then acquisitioned by the computer. The results show that the electromagnetic wave based coal concentration measuring technology can give the on-line solid concentration information, the filed data illustrated that the system responded quickly with change of the coal concentration, the technique is not sensitive with the coal moisture due the low frequency electromagnetic wave was adapted.

ACKNOWLEDGEMENTS

Supported by National Natural Science Foundation of China(60534030, 50576081), program for Changjiang Scholars and Innovative Research Team in University.

REFERENCE

1. Yan Yong. Mass flow measurement of bulk solids in pneumatic pipelines. Measurement science and Technology, 1996. 7:1687-1706.
2. Nguyen, T., A. Nguyen, and S. Nieh, An improved isokinetic sampling probe for measuring local gas velocity and particle mass flux of gas-solid suspension flows. Powder Technology, 1989. 59: 183-189.
3. Cai, X., et al., In-line measurement of pneumatcially conveyed particles by a light transmission fluctuation method. Flow Measurment and Instrumentation, 2005. 16:315-320.

Concurrent Measurement of 2-D Temperature Distributions and Radiative Properties of Pulverized-coal-fired Flame in a 670 t/h Boiler Furnace

Chun Lou and Huai-chun Zhou*

State Key Laboratory of Coal Combustion, Huazhong University of Science and Technology, Wuhan, 430074 P.R China

Abstract. Both temperature distributions and radiative properties play important roles in boilers and furnaces. Based on radiative imaging models, the 2-D temperature distribution and radiative properties can be simultaneously reconstructed from the boundary radiative information detected by image detectors. In this paper, the 2-D temperature distributions and radiative properties of pulverized-coal-fired flame in three typical cross-sections of a 670 t/h boiler furnace were measured. At the same time, the particles in the three cross-sections of the furnace were sampled and then, the chemical compositions, shapes and size distributions of the particles were given. The experimental results showed that the measured 2-D temperature can be used in combustion diagnosis, and the radiative properties in the burner area of the furnace are larger than those in the outlet of the boiler furnace. The value of measured radiatvie properties is proportional with the carbon content in sampled particles. The radiative properties of smooth ash are smaller while those of porous char are larger.

Keywords: Pulverized-coal-fired flame, 2-D Temperature distribution, Radiative properties, Image Processing
PACS: 44. 40. +a

INTRODUCTION

Pulverized-coal-fired flame in large scale furnaces is a kind of typical gas-solid two-phase flow. Because of high temperature in furnaces, radiative heat transfer plays a key role in these furnaces. Combustion flame in pulverized-coal-fired boiler furnace is made of gases and solid particles such as coal, soot, ash, char. The radiative heat transfer in furnaces is due to both band emission from gases and continuous emission from suspended particles. The radiative properties of particles and gases are estimated by Mie theory [1-2] and narrow-band or wide-band models [3], respectively. In the condition that the radiative properties are given, the two-dimensional (2-D) and three-dimensional (3-D) temperature distributions in furnaces can be reconstructed from the measured radiation information on the boundaries of furnaces based on color image processing [4-6].

However, as of today the knowledge about the chemical compositions of these particles, their shapes and size distributions, and their optical properties are limited. Furthermore, the radiative properties of the particles vary with the change of their temperature, sizes and shapes when they travel through a furnace. Therefore, any prediction of radiative properties from electromagnetic theory would have a certain amount of uncertainty from crude approximations. Thus, more accurate estimation of radiative properties of particulate medium can be expected to be deduced from in-situ measurements, in which the unknown temperature distribution is measured simultaneously with the unknown radiative properties of particulate medium. Only a limited amount of work is available on this topic [7-9]. Zhou et al. have done simulation work on simultaneous estimation of the temperature distribution and radiative properties in one-dimensional (1-D) and 2-D systems [8-9]. Recently, a reconstruction method with good robustness which combines optimization methodology and regularization methodology has been proposed to simultaneously reconstruct the temperature distribution and radiative properties of medium and, the experimental

* Corresponding author: hczhou@mail.hust.edu.cn

CP914, *Multiphase Flow: The Ultimate Measurement Challenge,* edited by X. Cai, Y. Wu, Z. Huang, S. Wang, and M. Wang

investigation on measurements of the radiative properties of particulate medium in pulverized-coal-fired boiler furnaces using the decoupled reconstruction method was conducted and the preliminary experimental results were presented in [10].

In this paper, a detailed concurrent measurement of 2-D temperature distributions and radiative properties of particulate medium in a 670 t/h boiler furnace of a 200MW power generation unit was conducted. Firstly, the measurement principle will be outlined briefly and then, the experimental set-up will be described. After that, the experimental results will be given and analyzed. Finally, some concluding remarks will be discussed.

Measurement Principle

In the narrow wavelength range of visible region, from 380 nm to 780 nm, a pulverized-coal-fired furnace is assumed to be filled by a gray emitting, absorbing, and scattering gas-particle mixture, surrounded by gray water-wall surfaces. For processing a color flame image in the visible region, the radiation emitted, absorbed and scattered by the suspended particles is considered, while the radiation emitted and absorbed by the gaseous components such as CO_2 and H_2O in the flame is neglected, since it is significant only in the infrared region.

Three monochromatic images under the visible wavelengths of red, green, and blue were acquired by a color image detector. A simple calibration procedure and image processing technique was employed to obtain the monochromatic intensity of red, green, and blue and a kind of temperature image $\mathbf{T}_m$ [11].

In order to simultaneously reconstruct the temperature distribution, the radiative properties (absorption coefficient κ and scattering coefficient σ_s) of particulate medium in furnaces, two radiative imaging models were proposed to relate the temperature distribution, the radiative properties in the furnace with the boundary radiative information received by the image detectors [5-6]. In application, the furnace system studied is divided into M spatial meshes for its medium zone and M' surface meshes for its surface area, and the total number of image pixels in the image detectors is set as N. The radiative intensity $\mathbf{I}$ received by image detectors can be calculated as [5-6]

$$\mathbf{I} = \mathbf{AT}, \tag{1}$$

where $\mathbf{T} = (T^4(1)\cdots T^4(i)\cdots T^4(M+M'))^T$ represents the temperature distribution in the furnace, $\mathbf{A} \in R^{N\times(M+M')}$ can be obtained from [5-6], and related with the installation and imaging parameters of the image detectors, and the radiative properties of particulate medium.

Through a radiative temperature model [5-6], a temperature image $\mathbf{T}_m$ is further related to the temperature distribution $\mathbf{T}$ by

$$\mathbf{T}_m = \mathbf{A}'\mathbf{T}, \tag{2}$$

where $\mathbf{T}_m^4 = (T_m^4(1)\cdots T_m^4(j)\cdots T_m^4(N))^T$, and the relationship between matrixes $\mathbf{A}$ and $\mathbf{A}'$ was given in [12]. $\mathbf{T}_m$ holds linear relationship with $\mathbf{T}$.

The reconstruction of temperature and radiative properties can be described as below. The radiative intensity image $\mathbf{I}$ and the radiative temperature image $\mathbf{T}_m$ can be detected by the image detectors and they serve as input data. In the decoupled reconstruction method, a modified Tikhonov regularization method has been used to reconstruct the temperature distribution determined by Eq. (1) from the radiative temperature image $\mathbf{T}_m$ [5-6] while the inverse radiative property problem is formulated as an optimization problem that minimizes the errors between the calculated and measured data of boundary radiative intensity. A Newtonian-type iteration algorithm and the least-squares method are used to estimate the absorption and scattering coefficients of the medium [10]. These two steps are taken alternately till a convergence is reached. It has been seen that the radiative properties converged to stable values after several iterations [10].

Experimental Set-up

The experiment was done in a 670 t/h boiler furnace (four-cornered and tangentially fired) of a 200 MW power generation unit in Qingshan Power Plant, Hubei, P. R. China. The main fuel was coal. The boiler furnace with the horizontal cross-section dimensions of 11.92 m width × 10.8 m depth is shown in Fig. 1.

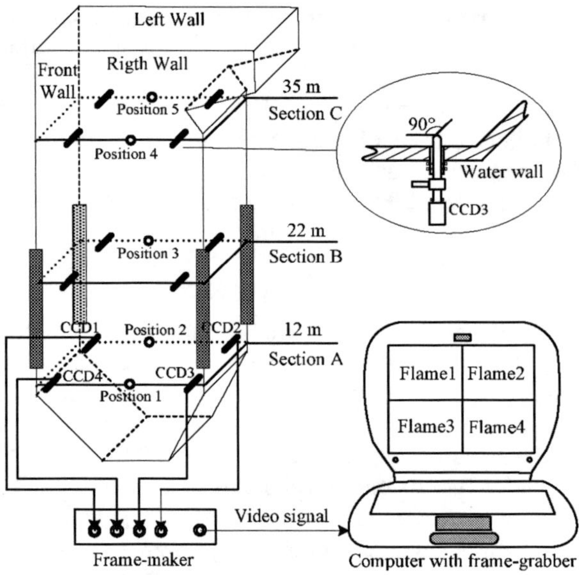

FIGURE 1. Experimental set-up.

The 2-D temperature distribution and radiative properties such as absorption coefficient κ and scattering coefficient σ_s, in three cross-sections were to be measured. As shown in Fig. 1, the three sections are in different position along the height of the furnace. Section A is below the burner area and section B is among the burner area while section C is in the outlet of the furnace. The experimental set-up consisted of four image detectors, a frame-maker, and a microcomputer with a frame-grabber, which has been described in [7-8]. For example, in section A, four flame image detectors were mounted horizontally in the four corners of the cross-section and used to capture flame images in the furnace. By use of a frame-maker and a frame-grabber, the video signals captured by the four detectors were combined into one video signal which was converted into a digital, color image in a computer. Dedicated software was used to process flame images and calculate the 2-D temperature distribution and radiative properties in the section. At the same time, a water-cooled suction probe was used to collect the particles in the three cross-sections of the furnace. As shown in Fig. 1, there were five sampling positions. The chemical compositions, shapes and size distributions of these sampled particles were analyzed in the laboratory.

Each cross-section of the furnace was discrete into $10 \times 10 = 100$ spatial meshes for its medium zone. The temperature distribution of the cross-section was $T(i), i = 1, \cdots, 100$, while the absorption coefficient κ and scattering coefficient σ_s in the cross-section were assumed as uniform. According to the optical imaging process, the image pixels in the central line in each image respond to the radiation from the horizontal cross-section. Consequently, the image pixels in the central line in each image were divided into 90 pixels and totally 360 image-formation pixels were got from four images. During the experimental period, the furnace was configured to burn blending coal with low volatile and the load of the power generation unit was 195 MW.

Experimental results and analysis

The Reconstructed 2-D Temperature Distribution and Radiative Properties

Many flame images were captured during the experiments. Fig. 2 shows the representative flame images captured in the three section of the furnace, respectively. The image pixels indicated in the white line, as shown in Fig. 2 (b), was used to calculate the radiative intensity image and temperature image. The radiation from section B which locates in the burner area is stronger than section A and section C. The shutter speed of the CCD camera of the image detector in section B is 1/2000 s while the shutter speed of the CCD camera in the other two sections is 1/120 s.

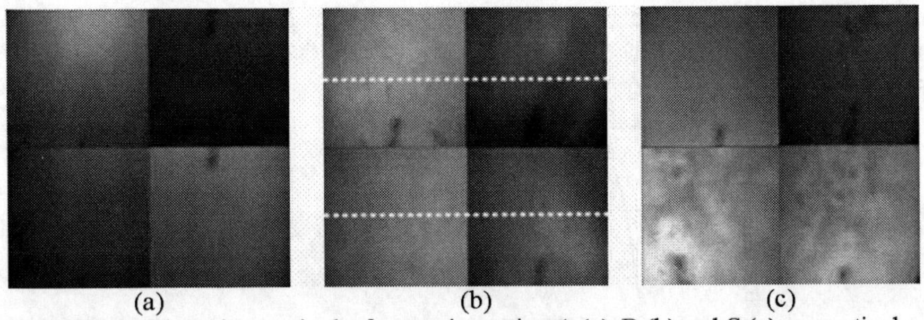

FIGURE 2. Flame images in the furnace in section A (a), B (b) and C (c), respectively.

Table 1 shows the reconstructed uniform absorption and scattering coefficients in the three sections of the furnace, which were calculated from the radiative intensity and temperature images. From Table 1, we can see that the radiative properties in section B are larger than those in section A and C. The reason is that both the concentration of particles in the flue gas and the content of carbon in particles are the highest in the burner zone. So, the radiation is significantly enhanced in the zone. However, near the outlet of the furnace (section C), because the burnout degree increases, the number density of particles decreases and the particles are fly ashes in which the content of carbon is very low. The radaitive properties in section C are the lowest.

TABLE 1. The radiative properties in the three sections of the furnace.

	Section A	Section B	Section C
κ (m^{-1})	0.326	0.468	0.225
σ_s (m^{-1})	0.544	0.693	0.267

The reconstructed 2-D temperature distributions in the three cross-sections are shown in Fig. 3. It can be seen that the temperature distributions typically have a single peak with temperatures higher in the center and lower near the surfaces of the water-wall. The temperatures in the burner zone (section B) are lager than those near the outlet of the furnace (section C) and the bottom of furnace (section A). Besides that, in the section C, the temperatures near the right wall are higher than those near the left wall and the flame center is not in the geometrical center of the furnace. This may be caused by the inhomogeneous fluid field in the furnace and the deviation of the temperature in the outlet of the furnace may cause overheating to the tubes of supper-heaters and re-heaters.

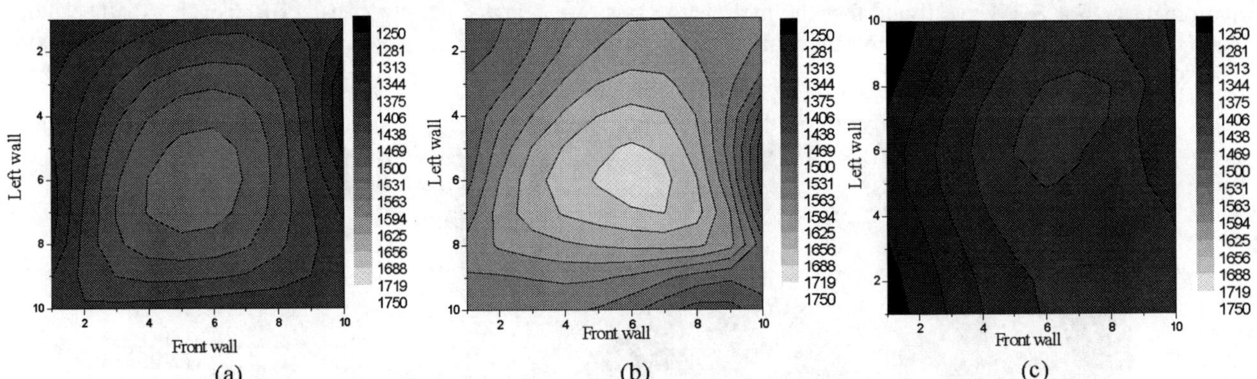

FIGURE 3. Flame images in the furnace in section A (a), B (b) and C (c), respectively.

The analysis of sampled particles

The chemical compositions, shapes and size distributions of the sampled particles were given. The relationship between these physical, chemical parameters and the measured radiative properties were analyzed.

The particle size distributions of three types of pulverized coal are presented in Fig. 4. In the corner fired boiler furnace, the large particles of combustion medium is mainly near the water wall surface, and in the experiment, the fuel is blending coal and the sampled length of the suction probe is only 1.5 m. So the size distributions of all sampled particles are larger. From Fig. 4, we can see that the sizes of particles in sampling position 4 and 5, which both locate near the outlet of the furnace (section C), are smaller, because the burnout degree is higher. The sizes of particles in position 3, which are in the burner area (section B), are moderate, and the sizes of particles in position 1 and 2 (section A) are largest.

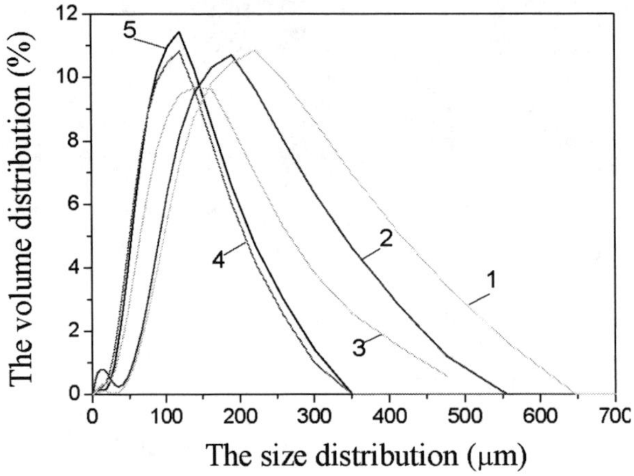

FIGURE 4. Size distributions of the sampled particles.

According to [2], at wavelengths less than 6μm, ash deposits have low emittance. Because the measured spectrum is in the visible region in the experiments, the effect on the reconstructed radiative properties of particulates medium is more related with the content of carbon in particles. Table 2 gives the content of carbon in particles sampled from position 1 to 5. From Table 2, the largest content of carbon in particles is in burner area (position 3 in section B), while the smallest is in the outlet of the furnace (position 4 and 5 in section C) because of the highest degree of burnout.

TABLE 2. The content of carbon of sampled particles in five positions.

Position	1	2	3	4	5
Carbon content (%)	9.71	9.375	39.235	9.395	7.267

The chemical compositions and shapes of sampled particles were given through a Scanning Electron Microscopes (SEM). Figs. 5-9 give some representative SEM graphs and compositions of particles sampled in the five positions. By SEM, we found that the particles in position 4 and 5 mainly consist of smooth ash. The porous char in the position 3 is much than other sampling positions. In position 1 and 2, we found some minerals, which have low content of carbon.

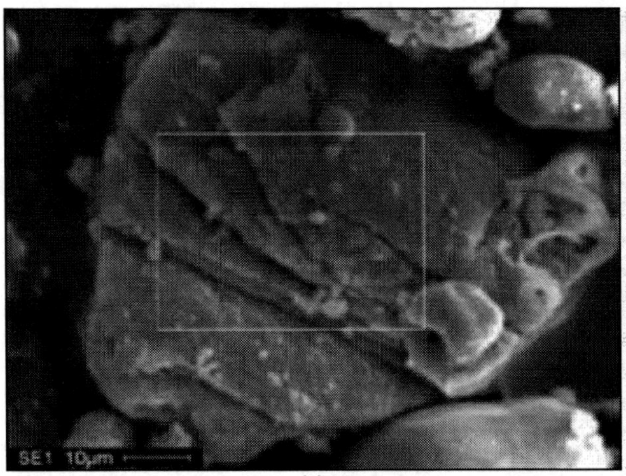

Element	Wt%	At%
CK	10.34	16.93
OK	38.92	47.81
FK	00.24	00.25
MgK	00.24	00.20
AlK	22.29	16.24
SiK	24.27	16.98
ClK	00.37	00.20
KK	00.74	00.37
CaK	00.56	00.27
TiK	00.44	00.18
VK	00.25	00.10
FeK	01.35	00.48
Matrix	Correction	ZAF

(a) (b)

FIGURE 5. The representative SEM graphs (a) and the corresponding compositions (b) of particles sampled in position 1.

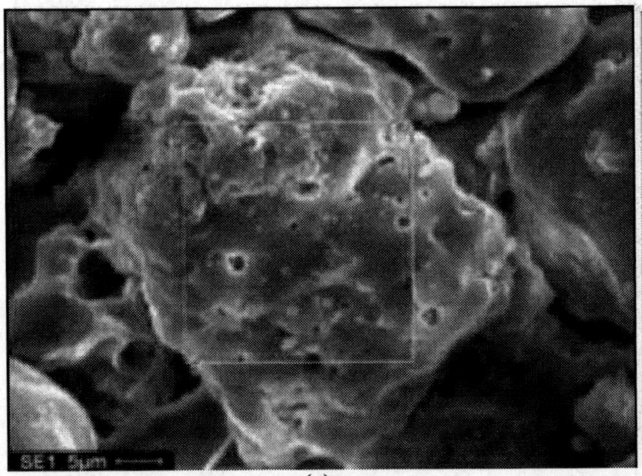

Element	Wt%	At%
CK	12.08	19.94
OK	37.88	46.94
FK	01.79	01.86
MgK	00.45	00.37
AlK	10.95	08.05
SiK	25.93	18.30
ClK	00.50	00.28
KK	02.13	01.08
CaK	00.75	00.37
TiK	01.81	00.75
VK	00.66	00.26
FeK	05.07	01.80
Matrix	Correction	ZAF

(a) (b)

FIGURE 6. The representative SEM graphs (a) and the corresponding compositions (b) of particles sampled in position 2.

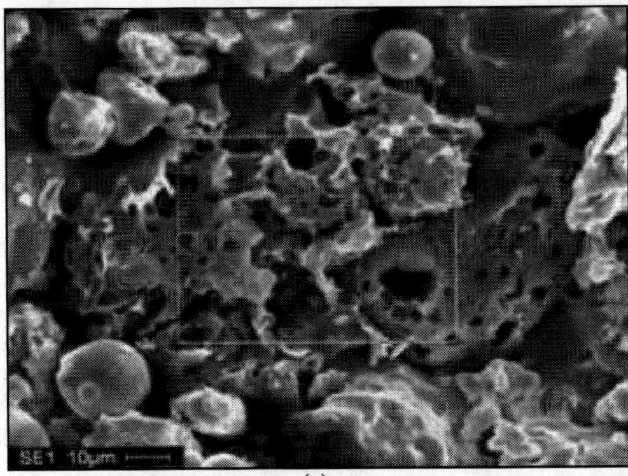

Element	Wt%	At%
CK	88.56	92.66
OK	06.81	05.35
FK	00.16	00.10
MgK	00.13	00.07
AlK	01.81	00.84
SiK	01.75	00.78
ClK	00.00	00.00
KK	00.00	00.00
CaK	00.13	00.04
TiK	00.30	00.08
VK	00.18	00.05
FeK	00.18	00.04
Matrix	Correction	ZAF

(a) (b)

FIGURE 7. The representative SEM graphs (a) and the corresponding compositions (b) of particles sampled in position 3.

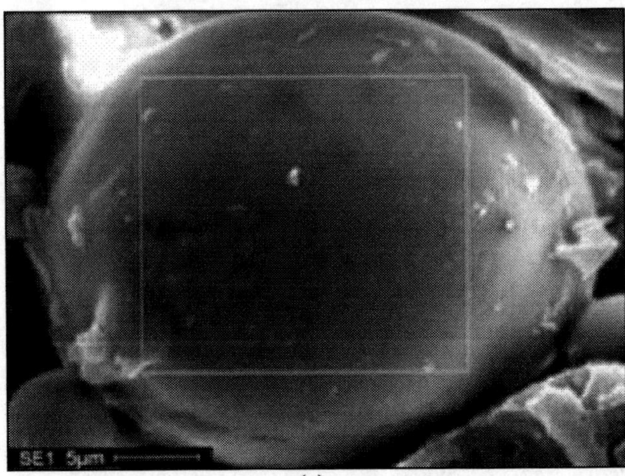

Element	Wt%	At%
CK	16.85	27.00
OK	34.85	41.91
FK	00.53	00.53
MgK	01.31	01.04
AlK	14.27	10.17
SiK	22.28	15.26
ClK	00.16	00.09
KK	03.34	01.64
CaK	00.60	00.29
TiK	00.77	00.31
VK	00.53	00.20
FeK	04.52	01.56
Matrix	Correction	ZAF

(a) (b)

FIGURE 8. The representative SEM graphs (a) and the corresponding compositions (b) of particles sampled in position 4.

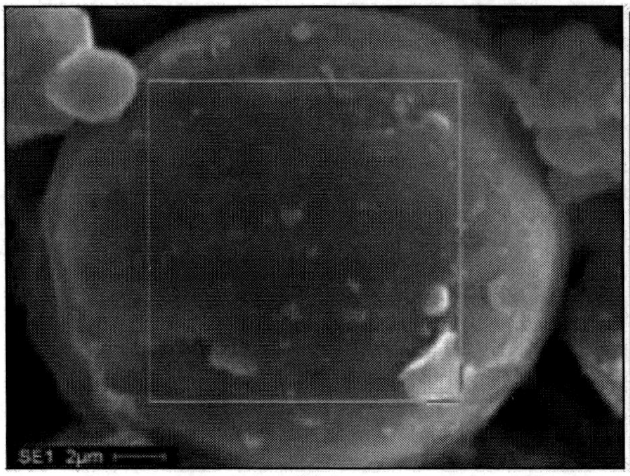

(a)

Element	Wt%	At%
CK	05.72	10.19
OK	34.92	46.71
FK	00.80	00.90
NaK	00.56	00.53
MgK	01.08	00.95
AlK	14.20	11.27
SiK	31.65	24.12
PK	00.36	00.25
SK	00.16	00.10
KK	04.86	02.66
CaK	00.35	00.19
TiK	01.10	00.49
VK	00.35	00.15
FeK	03.89	01.49
Matrix	Correction	ZAF

(b)

FIGURE 9. The representative SEM graphs (a) and the corresponding compositions (b) of particles sampled in position 5.

Compared between the reconstructed radiative properties of particulates medium in pulverized coal flame and the compositions, shapes and size distributions of particles, some conclusions were given as below. The more carbon in particles will enhance the radiation of particles. The radiative properties are also related with the shapes of particles. The porous particles are more absorptive than the smooth particles. The radiative properties in burner area, in which there are high concentration of particles, are larger than those in outlet of furnace.

Conclusion

This paper reported concurrent measurement of 2-D temperature distributions and radiative properties in a 670 t/h pulverized-coal-fired boiler furnace using image processing technique from the color flame images. Besides that, the particles in furnace were sampled by the water-cooled suction probe. The reconstructed results of temperature distributions and radiative properties in three typical cross-sections in the furnace were given. The relationships between the reconstructed radiative properties of particulate medium with the compositions, shapes and size distributions of sampled particles were analyzed. This work is of practical importance in calculating and modeling the radiative heat transfer in combustion computation for pulverized-coal-fired systems. The measured temperature distribution is useful for the combustion diagnosis and control. The measured absorption and scattering coefficients can be used directly in calculation of radiative heat transfer for combustion computation; otherwise, it will be a complicated task to get them, possibly with serious errors.

ACKNOWLEDGMENTS

This work was supported by the National Natural Science Foundation of China (No. 50636010).

REFERENCES

1. C. Kim and N. Lior, *Fuel* 74, 1891-1902 (1995).
2. T.F. Wall, S.P. Bhattacharya, D.K. Zhang, R.P. Gupta and X. He, *Prog. Energy Combust. Sci.* 19, 487-504 (1993).
3. M.F. Modest, *Radiative Heat Transfer*, New York: McGraw Hill, 1993, pp. 323-382.
4. F. Wang, Z.Y. Ma, C.Y. Wei and K.F. Cen, *Proceedings of the CSEE* 20, 40-43 (2000).
5. C. Lou, and H.C. Zhou, *Combust. Flame* 143, 97-105 (2005).
6. H.C. Zhou, C. Lou, Q. Cheng, Z.W. Jiang, J. He, B.Y. Huang, Z.L. Pei and C.X. Lu, *Proceedings of the Combustion Institute* 30, 1699-1706 (2005).
7. L.H. Liu, *Int. Comm. Heat Mass Transfer* 27, 635-643 (2000).
8. H.C. Zhou, Y.B. Hou, D.L. Chen and C.G. Zheng, *J. Quant. Spectrosc. Radiative Transfer* 74, 605-620 (2002).
9. H.C. Zhou and S.D. Han, *Int. J. Heat Mass Transfer* 46, 2645-2653 (2003).
10. C. Lou and H.C. Zhou, P.F. Yu, Z.W. Jiang, *Proceedings of the Combustion Institute*, 31, 2771-2778 (2007).
11. Z.W. Jiang, H.C. Zhou, C. Lou and Z.Q. Yu, *Advances in Multiphase Flows Vol.1-Measurement Technology*, Beijing: International Academic Publishers/Beijing World Publishing Corporation, 2004, pp.361-365.
12. H.C. Zhou, S.D. Han, C. Lou and H. Liu, *Numerical Heat Transfer, Part B: Fundamental* 42, 243-258 (2002).

Development of System and Method for Serial Measurement of Phase Permeability of Sandstone Core Samples During Filtering of Water, Oil and Gas

Anton G. Skripkin[1] ,Yury A. Schemelinin[1], Eugene I. Palchikov[2].

[1] *JSC "TomskNIPIneft VNK", Tomsk, Russia, SkripkinAG@nipineft.tomsk.ru*
[2] *Lavrentyev Institute of Hydrodynamics RAS, Novosibirsk, Russia, Palchikov@hydro.nsc.ru*

Abstract. In this work a novel X-ray method for evaluation of intermediate water and oil saturation in a core sample is presented. Proposed method is considered X-Ray attenuation in the core saturated by doped water\oil and doped oil\kerosene\residual water. Three-phase relative permeabilities can be calculated from two-phase water\oil and gas\oil permeability tests with a help of proposed method. The suggested procedure allows decreasing the time of execution the experiment. The reached results permit to carry out serial measurements of relative phase permeability by steady-state filtration of water\oil\gas in stratal conditions. Tomographic images recovering of water and oil spatial distribution in core samples was used. This image allows classifying core samples, comparing them. It permits to evolve more correct computer simulation model of reservoir flows.

Keywords: X-ray tomography, multiphase fluid, porous media, saturation measurement, partial permeability, oil pool modeling.

INTRODUCTION

Phase permeability of sandstone cores is essential by engineering of oil and gas oilfields. Water-oil-gas relative permeability is especially important for oilfields with solution gas drive or gas cap expansion. With the help of phase permeability for water, oil and gas one can carry out computer simulation technique of processes, which occur in a shelf during oil production, or during pore pressure increase due to the gas injection, during simulation of zones of oil-water and oil-gas boundaries. Joint steady-state flow of three phases is usually simulated in laboratory-like environments with the help of flow systems. Today such systems represent unique and complex devices. A phase permeability (water\oil\gas) determination experiment is realized within 1 or 2 months. Such experiments are of great necessity for oil industry. In our work, one can find the representation of a method and a two-phase X-ray system. The system allows determining three phase relative permeabilities for brine, oil and gas by means of water/oil and oil/gas relative permeabilities.

Today, the X-ray technique is widely used to determine current water saturation in the study of relative phase permeability factors as well as in the visualization of fluid distribution in samples. This technique based on the absorption of X-rays by the mineralized water (brine) saturating the core. In the present paper, results obtained on X-ray system for commercial suppliers from Coretest Systems Inc. are discussed. In the framework of a common industrial technique for the evaluation of the filtration parameters of cores, a novel method to calculate water and oil saturation using X-ray absorption data in saturated core samples is demonstrated. Suggested system was used by tomographic images recovering of water and oil distribution in core samples by X-raying.

EXPERIMENT

In our experiments sodium iodide (NaI) and iodooctane $C_8H_{17}I$ were used as an X-Ray absorbing dopants in the water and oil.

CP914, *Multiphase Flow: The Ultimate Measurement Challenge,* edited by X. Cai, Y. Wu, Z. Huang, S. Wang, and M. Wang

Absorbers, that contained iodine in their structure, were used for best sensitivity and accuracy of the image. As an example of a formation model, in our experiments we used core samples 30 mm in diameter and approximately 50 mm long. Samples with similar porosity and permeability were formed in one column about 300 mm long.

Before the realization of the experiment, oil and water were saturated with accompanying gas in stratal conditions. As a gaseous phase, was used either accompanying gas, saturated with stratal water vapors, or nitrogen.

Investigations were carried out with the help of small-developed ordinary two-phase X-ray system RPXS-841 (Coretest Systems Inc.). The main components of this X-ray system are [1]: two double-plunger pumps, which allow one to control fluid or gas flow within the range of 0.001 ml/min to 50 ml/min with accuracy 0.1%. Fluid or gas rate in steady state flow were measured with a help of pump controllers. At the inlet, it is possible either to inject gas at a fixed pressure through a gas-water separator and doped oil or to inject brine (water, doped by NaI) and oil at a fixed rate by means of pumps.

The rock samples under study are placed into a radiotransparent coreholder consisting of two casings: a rubber one, in which the samples are placed, and an external graphite composite casing. The space between the rubber and the fiberglass casings is filled with pressurized oil. Such an arrangement of coreholder elements makes it possible to model reservoir conditions, at which the core sample is situated: pore pressure that fluids create inside the sample of up to 30 MPa (300 bar) and confining pressure (the pressure of confining the rubber sleeve) of up to 80 MPa (800 bar). The heaters are situated on the graphite composite casing for maintaining reservoir temperature. They allow maintaining the temperature at which the samples are kept, within the accuracy of 0.5 °C in the range between 15 to 150 °C. The system consists of a computer controlled X-Ray source with a capability to operate at user-defined voltages and currents. The range of the preset voltages at the X-ray tube: from 40 kV to 80 kV, currents: from 0,1 to 0,25 mA. The accuracy of maintaining current and voltage of the tube is $\pm$ 0.5%. A collimated X-ray beam, passing through the core saturated with fluid, is partially absorbed in accordance with dimensional distribution of absorbent density. A single spot CsI(Tb) based detector used for measurement of X-Ray attenuation inside 2 mm cross-section. A precision step motor controlled movable slide assembly. The X-Ray detector and X-Ray source are mounted on a moving plate which is positioned along the core sample set within the accuracy of 0,1 mm. The coreholder can be scanned in either vertical or horizontal orientation. Water\oil\gas phase permeabilities of sandstone cores have been derived from constant rate and steady-state flow by measuring of two-phase permeabilities: water-oil and oil-gas. The investigations were realized with samples from Lugineckoye, Zapadno-Moiseevskoye and Igolsko-Talovoe deposits of West Siberia. The porosity of the core samples were 10-14.1%, absolute permeabilities to gas – from $3{,}4\cdot10^{-3}\,\mu m^2$ to $120\cdot10^{-3}\,\mu m^2$.

BRINE\OIL RELATIVE PERMEABILITY

The experiments were carried out using the following scheme: after extraction with centrifuge and drying in oven, the samples under study were put in the coreholder; a confining pressure of 30 MPa and a temperature of 80°C (reservoir conditions) were set. The samples were scanned with a step of 0,8 mm, i.e. rayed by an X-ray beam. The obtained detector signal U_{dry} corresponds to the absorption of X-rays in a dry core with zero water saturation. The absorption value in a dry core is individual for each sample. All other conditions being equal (X-ray tube voltage and current, reservoir temperature and confining pressure), the absorption depends on the mineral composition, structure and geometric dimensions of the core. The magnitude of the signal when scanning a dry sample is the baseline, i.e. the signal to count from. The images were obtained at the voltage of X-ray tube of 75 kV and current of 0,2 mA. Then the samples were vacuumized for 2 hours and then, outgassing water with NaI (brine) was delivered into the samples. After a saturation period of 16 hours the scanning in the same spots was performed.

The absorption dependence in a saturated core transforms into:

$$U_{100} = U_{dry} \exp\left(-\mu_w x_w^{100}\right) \tag{1}$$

where x_w^{100} is the thickness of mineralized water layer in a 100% saturated core.

The obtained value of the signal is the minimal one because the sample is filled with the maximum amount of the absorbing substance. The magnitude of the signal U_{100} is measured for each sample, because it depends on the pore volume and on the absorption efficiency in the dry core U_{dry}.

Furthermore, at a frontal drive of water displaced with oil, the residual saturation was created in the samples. To this end, two pore volumes of pure kerosene and then 3 pore volumes of model reservoir oil were first filtered through the samples. The samples were aged during 5 days at reservoir conditions; scanning was repeated everyday. The brine saturation dependencies are shown on fig. 1.

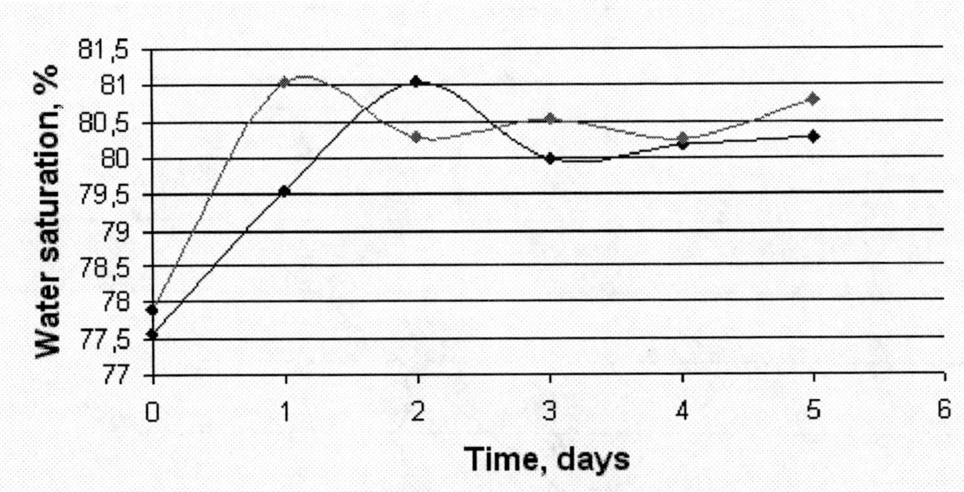

FIGURE 1. Water saturation dependence

It is seen that brine saturation in the core has increased during 2 days. After 2 days the brine saturation becomes constant and filtration can be continued.

Then next scanning of residual water saturated core was performed.

The dependence of the detector output signal *vs.* the amount of absorbing substance will be:

$$U = U_{dry} \exp\left(-\mu_w x_w^{\min} - \mu_k x_k^{\max}\right) \quad (2)$$

where μ_w ,μ_k are mass attenuation coefficients (constant values) of brine and kerosene (viscous oil), U_{dry} is detector signal when raying dry core. x_w^{min} –minimum value of brine layer thickness in the core and x_k^{max} – maximum value of kerosene layer thickness in the core.

Intermediate water saturation in the samples was produced through combined filtration of doped water and oil in proportions: 0:100, 25:75, 50:50, 75:25 and 100:0. The filtration continued until the pressure difference on the sample and detector output signal during scanning stabilized.

Reservoir conditions being constant, the pore volume of the sample does not change, therefore:

$$x_w + x_{oil} = x_w^{100} \quad (3)$$

Considering the fact that in expression (2) the μ_{oil} ,μ_w coefficients are constant and $\mu_{oil}=\mu_k$, we can put $\mu_{oil}=\alpha\mu_w$. Potentiating expressions (1), (2) and substituting x_{oil} from (3) we express x_w in terms of x_w^{100} . Intermediate water saturation in sample S_w is expressed in terms of x_w and x_w^{100}:

$$S_w = \frac{x_w}{x_w^{100}} = \frac{1}{1-\alpha}\left(\frac{\ln U - \ln U_{dry}}{\ln U_w - \ln U_{dry}} - \alpha\right). \quad (4)$$

The experimental measurable values are the readings of the detector U, U_{dry}, U_w, therefore the following expression is convenient as the variable:

$$S^{XR} = \left(\frac{\ln U - \ln U_{sry}}{\ln U_w - \ln U_{dry}}\right). \quad (5)$$

Water saturation S_w in this case is linearly dependent on the measured value S^{XR}. From expression (4) one can see that water saturation becomes zero at a non-zero value of $S^{XR}=\alpha$. This corresponds to the situation when the sample is completely saturated with oil. X-ray absorption in this case is greater than that in the dry core, but the amount of mineralized water in the sample is equal to zero. On the fig.2 the experimental dependence $S_w= S_w(S^{XR})$ is shown. We get it from our experiment.

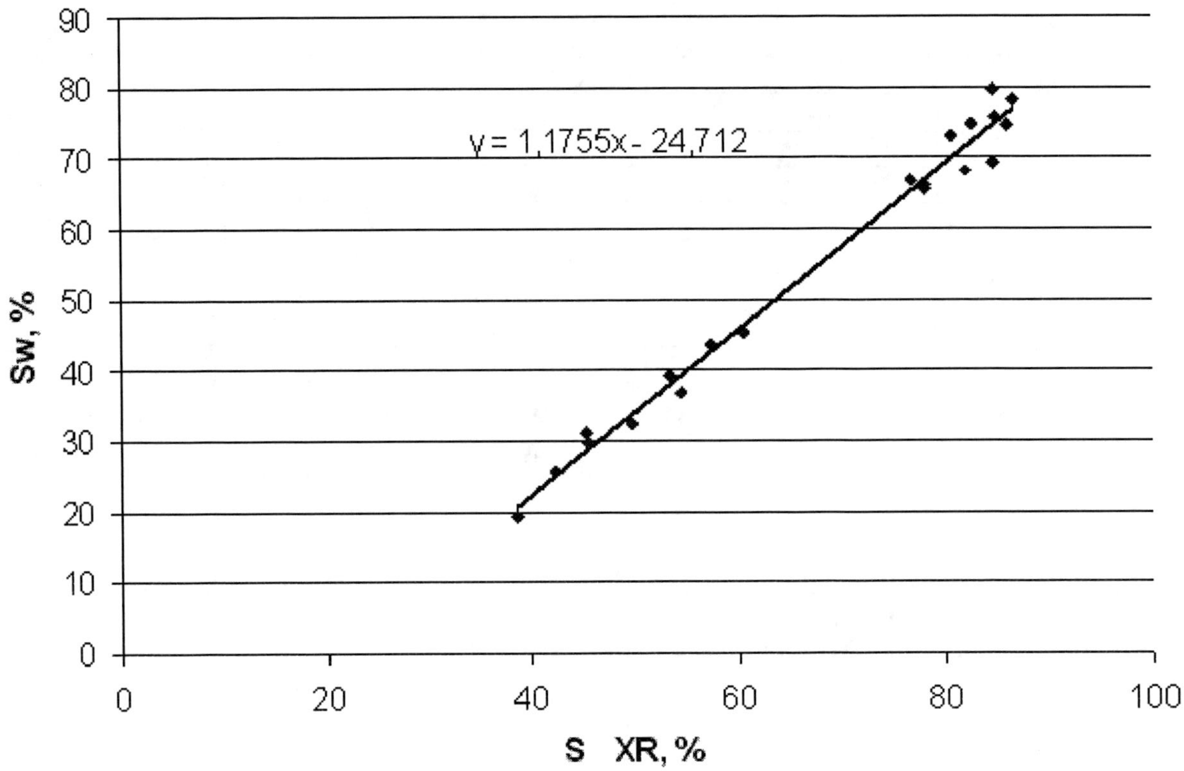

FIGURE 2. Experimental dependence of water saturation vs. the S_w^{XR} parameter

The linear dependence on the fig.2 proves the expression (4). Expression (4) is used in water\oil permeability experiments to calculate water saturation for cylindrical samples with any mineral content and pore geometry. The dispersion of the experimental data around the linear dependence in Figure 1 is first of all due to the non uniform water distribution inside the core. The relative error in this case amounts to about 3%. The overall measurement error was estimated by the round-mean-square dispersion of experimental data around the approximating line. In the water saturation, range between 25 to 80% the relative error was about 4%.

OIL\BRINE TOMOGRAPHY

In order to check a value of non uniform brine distribution a tomography of the core was made. For the tomography a new coreholder was constructed: a rotating plate with a Viton sleeve was situated in front of a digital TV CCD camera. The camera has a light-sensitive chip of 2/3 inch dimension with progressive scanning, pixel size 6.7x6.7 μm, image size 1300x1030 pixels and the Gd_2O_2S converter of X-ray radiation to optical radiation with lens image transfer [2]. The range of camera exposition was from 200 microseconds to 16 seconds, signal-to-noise ratio without cooling >60 dB. The camera was connected to PC via standard interface plate to PCI bus. The camera allows acquiring an accuracy of 12 bit (4096 scales of grey colors) in a single shot. For the raying we used conical X-Ray beam.

The core was vertically situated and squeezed by top and bottom plungers in the coreholder. Brine and oil were filtered through a top plunger and drained through a bottom plunger. For the tomographic data acquisition the core rotated to 360° with a step 7,2°. Raying by conical beam was made for every step. X-Ray scanning was performed for dry, 100% brine saturated and 84% brine saturated core. A median filtering was made for all X-Ray images. Experiment was performed with a core from Sovetskoe oilfield (West Siberia). Gas permeability of the core was $13,2\cdot10^{-3}$ μm^2. 2-D brine distribution inside the core was calculated by means of algebraic reconstruction.

Following classical circuit of ART, the flat field of section is covered by a rectangular grid. Thus it is supposed, that the density of matter in each cell of the grid is a constant (fig.3). The problem of tomography reconstruction is reduced to finding this density in each cell. A path of X-Ray beam L crossing elementary grid is s. A thickness of mineralized water layer in the grid is x. Water saturation inside the grid is x/s.

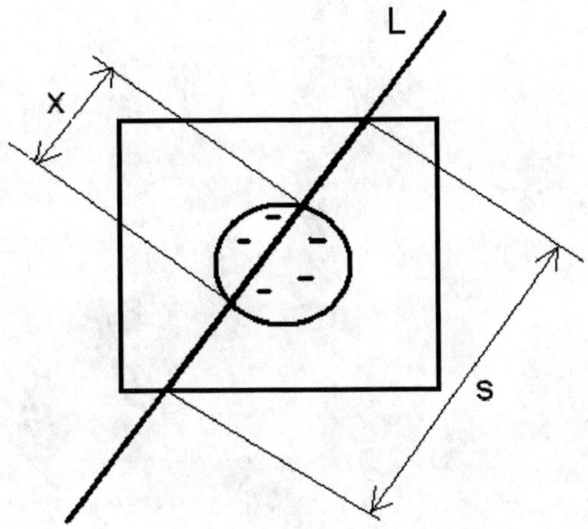

FIGURE 3. Elementary cell with doped water inside

Average water saturation to all N_L grids that crossed by X-Ray beam L is equal to measured water saturation by means of the CCD camera:

$$\frac{1}{N_L}\sum_L \frac{x_i}{s_i} = S_w^L = \frac{1}{1-\alpha}\left(\frac{\ln U - \ln U_{dry}}{\ln U^{100} - \ln U_{dry}} - \alpha\right) \qquad (6)$$

In the expression (6) all grids with water saturation x_i/s_i crossed by X-Ray beam L are summed; S_w^L – measured water saturation for the X-Ray beam L. Path length s of the X-Ray beam for the grid is calculated, x - is unknown required value. Right part of the expression (6) is calculated after scanning of dry, 100% brine saturated and intermediate brine saturated core. Expression (6) is the system of linear equations; water saturation in every grid is calculated with a help of algorithm of algebraic reconstruction. Method of tomographic reconstruction is described in article [2].

The figure 4 shows the tomographic image of cylindrical core. Water saturation is shown by the color on the right of the figure. It is seen non uniform 2D distribution of water saturation; central part of the core is more saturated. In our article [3] we have discussed the problem of non uniform saturation in porous medium by multiphase fluids. Today the non uniform distribution of oil\water is the topical question because of the number of flow dynamic investigations. The figure 5 shows the water saturation histogram with a fitting. The histogram is approximated by Gauss distribution. The sigma of the Gauss distribution in that case σ=0,0641 and mean=0,8331 is characterizes a distribution of saturation inside the core for sells with size 0,01.

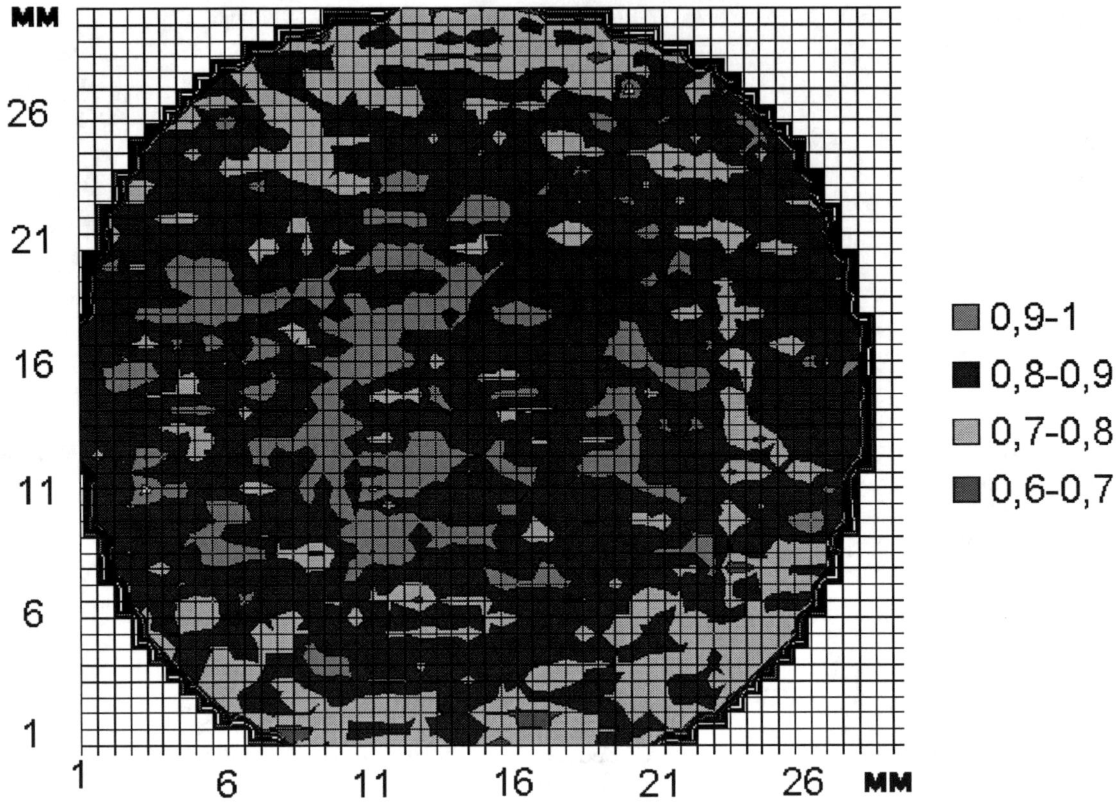

FIGURE 4. Tomographic image of the core (Sovetskoe oilfield)

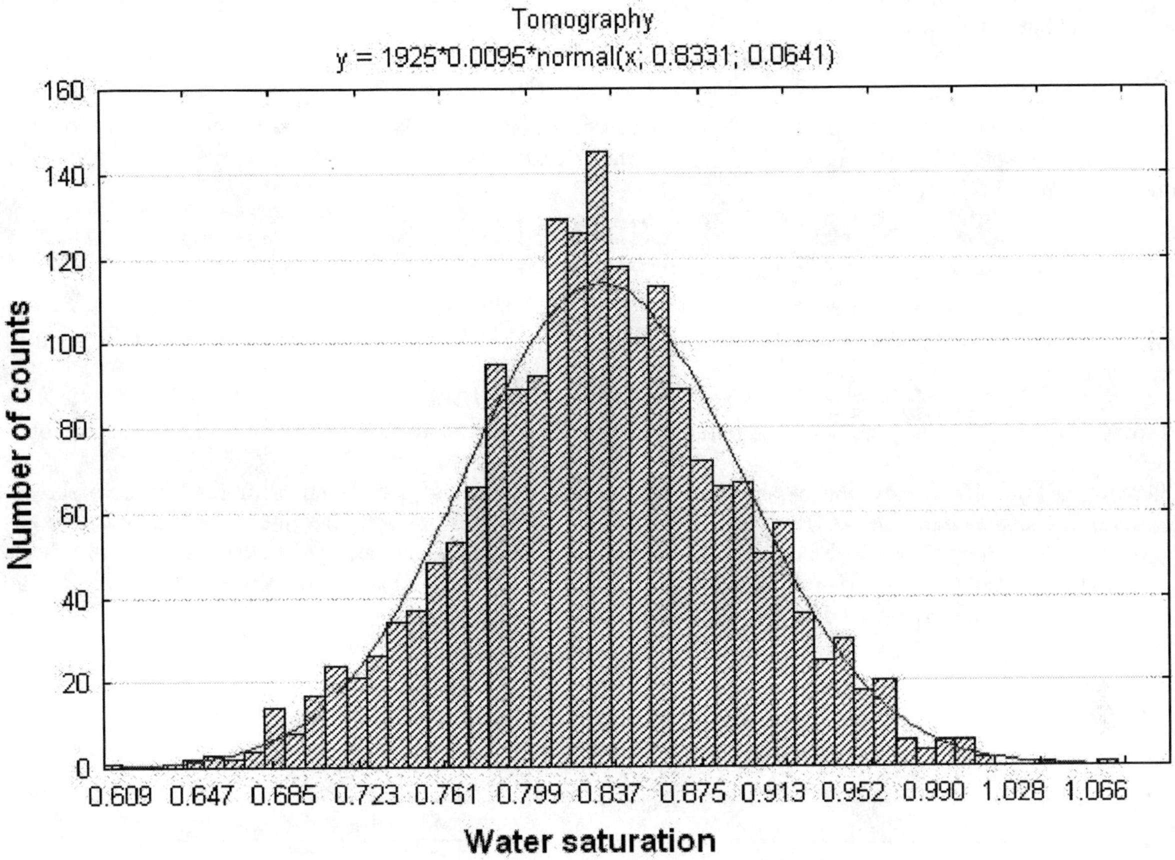

FIGURE 5. Histogram of water saturation

As it is seen water saturation of the core should be calculated to its full extent. Such method of calculation minimizes drawbacks. It is suggested for correct results of water or oil saturation to indicate a characteristic of heterogeneity.

OIL\GAS RELATIVE PERMEABILITY

Next step in three-phase relative permeability is to obtain oil\gas permeability. For the oil\gas relative permeability we used oil dopant $C_8H_{17}I$. Experiments were realized with samples from Lugineckoye, Zapadno-Moiseevskoye and Igolsko-Talovoe oilfields of West Siberia. Before an experiment the cores were extracted and dried, then they were put in the coreholder. A confining pressure of 15 MPa and a temperature of 80°C (reservoir conditions) were set. The residual saturation S_{wr} was created in the samples as written above. Scanning signal in that case:

$$U_S = U_{dry} \exp\left(-\mu_w x_w^{\min} - \mu_k x_k^{\max}\right) \tag{6}$$

where μ_k is mass attenuation coefficient of kerosene or non doped oil. Then kerosene was displaced by doped oil. Scanning signal of the cores filled by doped oil with residual water saturation was:

$$U_S^J = U_{dry} \exp\left(-\mu_w x_w^{\min} - \mu_o x_o^{\max}\right) \tag{7}$$

where μ_o is mass attenuation coefficient of oil with $C_8H_{17}I$ dopant.

Then intermediate oil saturation in the samples was produced through combined filtration of oil and gas in various proportions. In our experiments we obtained 25:75, 50:50, 75:25 and 100:0 gas\oil proportions. The total flow rate of gas and oil was constant; linear velocity of filtration was less then 0,1 m\hour. The filtration continued until the pressure difference on the sample and detector output signal during scanning stabilized. Scanning signal U

in that case was less then U^J_S, because the value of oil layer thickness in the core less then x_o^{max}. Samples were rayed again; the detector signal was:

$$U = U_{dry} \exp\left(-\mu_w x_w^{\min} - \mu_o x_o\right) \quad (8)$$

As written above, we can put $\mu_k = \beta\mu_o$ were β – constant of proportionality between mass attenuation coefficients of doped oil and kerosene. Substituting x_o, x_w^{min}, x_o^{max} from (2), (6), (7), (8) we express intermediate oil saturation in sample S_o in terms of S_{wr} and β:

$$\frac{S_o}{1-S_{wr}} = (1-\beta)\left(\frac{\ln U - \ln U_S}{\ln U_S^J - \ln U_S} + \frac{\beta}{1-\beta}\right) \quad (9)$$

were S_{wr} – residual oil saturation. Ratio of the oil saturation S_o and $(1\text{-}S_{wr})$ in this expression is linearly dependent on the measured value

$$S^{XR} = (\ln U - \ln U_S)/(\ln U_S^J - \ln U_S).$$

From expression (9) one can see that parameter $S_o/(1\text{-}S_{wr})$ is not dependent from mineral content or pore geometry.

There were five oil\gas filtration experiments were made by means of different samples. Figure 6 shows the experimental dependence of $S_o/(1\text{-}S_{wr})$ *vs.* the S^{XR} value calculated from measured values: U – intermediate gas\oil saturation, U_S – scanning signal of the cores with residual water and kerosene and U_S^J – scanning signal of the cores with residual water and doped oil. The figure also shows a linear dependence approximating the data. We have a few data but it is seen that experimental data are agreed with expression (9).

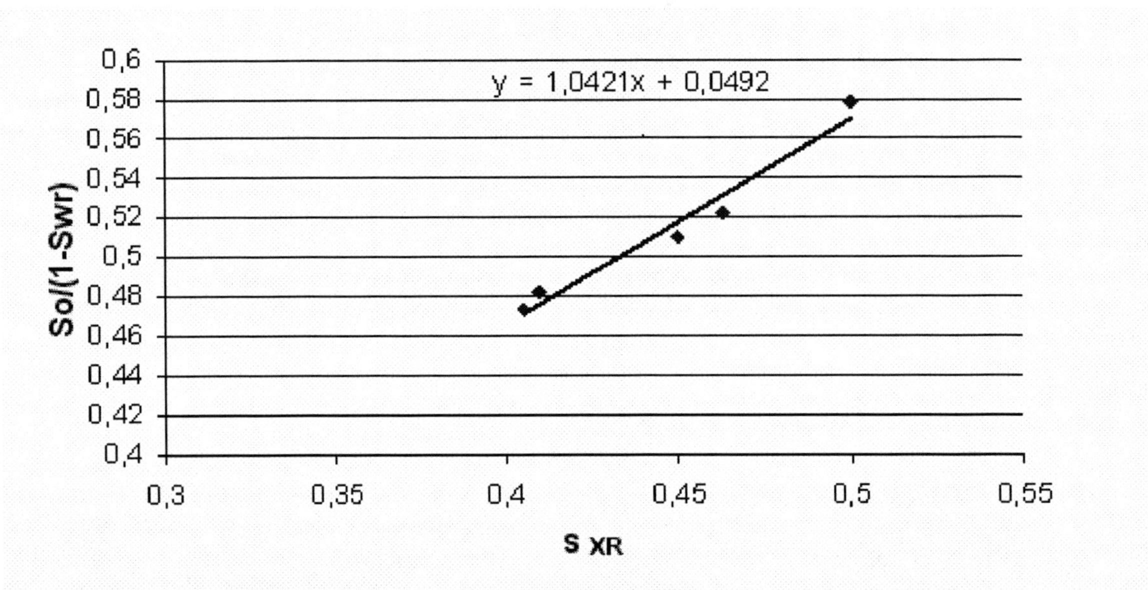

FIGURE 6. Experimental dependence of value $S_o/(1\text{-}S_{wr})$ vs. the S_w^{XR} parameter

In the expression (9) it was assumed that residual water saturation S_{wr} was constant during the gas\oil experiment. To control that assumption water saturation was measured before and after filtration. Before experiment water saturation was measured by means of calibration dependence (fig. 2). After experiment water saturation was measured by an express technique: by vaporizing water from the samples. Water saturation before and after gas\oil experiments were equal.

CONCLUSION

In this work a novel one-stage method for evaluation of intermediate water and oil saturation in a core sample is presented. Proposed method is considered X-Ray attenuation in cores saturated by doped water\oil and doped oil\kerosene\residual water. Three phases relative permeabilities can be calculated from two phases water\oil and gas\oil permeability tests with a help of proposed method. The obtained dependencies substantially improve the accuracy of measuring water and oil saturation in the determination of two phases permeability values. The method allows one to measure phase permeability for water\oil\gas within 1 or 2 weeks while ordinary three phase experiment is realized within 1 or 2 months. Three phase systems are unique and complex devices. In our work we used well-known simple two phase X-Ray system. This system allows simulating reservoir conditions exactly, such as: fluid and gas flow rates, temperature, confining and pore pressures. The proposed method does not require additional operations connected with extraction, processing and loading of samples. The attained accuracy of water saturation determination is about 4%. Suggested modified X-Ray system was used by tomographic images recovering of water and oil distribution in core samples by X-raying [3]. The tomographic image shows non-uniform distribution of water saturation inside the core, hence water saturation of cores should be calculated to its full extent. It is suggested for correct results of water or oil saturation to indicate a characteristic of heterogeneity. This characteristics allows classifying core samples, comparing them. It permits to evolve more correct computer simulation model of reservoir flows.

REFERENCES

1. http://www.coretest.com – RPXS-841 system.

2. A.S. Besov, A. G. Skripkin, Yu. A. Schemelinin, E. I. Palchikov, A. N. Cheremisin, D. Yu. Mekhontsev. Researching the dynamics of movement of gas-liquid mixture in a porous medium by means of low-angle tomography. 5th International Conference on Multiphase Flow, Yokohama, Japan, 2004, Paper No140.

3. Palchikov E.I., Sukhinin S.V., Burlev A.Yu., Bartuli E.R., Mekhontsev D.Yu., Romanjuta M.A., Seleznev K.S. X-ray Tomographic Flow Visualization of Multiphase Fluid Mixture in Porous Media. Seventh Triennial International Symposium on Fluid Control, Measurement and Visualization. Sorrento, Italy, August 25-28, 2003 FLUCOME'03 CD-ROM Proceedings. ISBN 0-9533991-4-1. Paper No 101.

4. E.S.Sprunt, K.P.Desal et al. CT-scan-monitored electrical-resistivity measurements show problems achieving homogeneous saturation. SPE Formation Evaluation, 6, No.3, 134-140 (1991).

The Information Content In Nir Spectral Data For Slurries Of Organic Crystals

Solomon Abebe, Xue Z Wang, Ruifa Li, Kevin J Roberts and Xiaojun Lai

Institute of Particle Science and Engineering, The University of Leeds, Leeds LS2 9JT, UK
Tel: +44 113 343 2427 Fax: +44 113 343 2405 Email: x.z.wang@leeds.ac.uk

Abstract. Despite being widely used in agriculture, food production and environmental monitoring and regarded as on-line chromatograph in petrochemical and biochemical industries, near infrared spectroscopy (NIR) has found difficulties in its application to processes of particle formation through crystallization or precipitation where solids suspended in solutions cause problems in instrumentation as well as distortion of the spectra. The research work reported here was motivated by the speculative idea that the effect of particles on the NIR spectra in effect brings an opportunity instead: the spectra might contain useful information of both the solid and liquid phases. Through carefully designed experiments using both glutamic acid solutions and slurries of varied solid concentrations and particle size and temperature ranges and with the help of chemometric data analysis, it was found that the NIR spectra clearly contain sensitive information about the size, solid concentration, liquid concentration as well as polymorphs of crystals, providing the possibility of using the instrument for simultaneous measurement of the multiple properties of both phases.

Keywords: Near infrared spectroscopy, crystallisation, precipitation, principal component analysis, independent component analysis, process analytical technology.

1. INTRODUCTION

This is t In recent years, there has been a growing interest in applying various process analytical techniques to on-line measurement and characterization of crystallisation or precipitation processes. The techniques include attenuated total reflectance Fourier transform infrared (ATR-FTIR) for in-situ measurement of concentration, X-ray powder diffraction (XRD) for determining crystallographic structure, ultrasound spectroscopy (USS) for particle size distribution, and on-line imaging for shape measurement. On-line spectrometers have yet to overcome some challenges posed by the complexity, harsh conditions and scale of operation in the industrial environment.

NIR has become very popular in agriculture, food manufacture and environmental monitoring, and is known as on-line gas chromatograph (GC) in petrochemical and biochemical industries. However, little work has been reported on its use in monitoring crystallisation and precipitation processes where solids suspended in solutions create difficulties in instrumentation as well as spectra analysis. This is unlike Mid IR where existence of the solid phase does not affect the spectra. It is due to this reason, some people commented that NIR cannot be used in crystallisation monitoring.

The current work was motivated by the speculative idea that the effect of particles on the NIR spectra brings a potential opportunity instead: the spectra might contain information of the solid phase as well as the liquid phase, as long as such information can be effectively extracted. Some research, though still limited, has shined some light on the richness of information such as crystal polymorph form[1,2] and median particle size[3,4] contained in the NIR spectra. This provides the possibility of using NIR for on-line simultaneous measurement of crystal polymorphs, size and solid and solution concentrations. Therefore, it potentially can replace several other techniques. In addition, NIR probe is capable of in-situ observation with flexibility of access to different locations of a reactor and different reactors, because the probe is connected with the spectrometer through fibre optics. The purpose of this study is to investigate the information content in NIR spectra obtained from slurries.

CP914, *Multiphase Flow: The Ultimate Measurement Challenge,* edited by X. Cai, Y. Wu, Z. Huang, S. Wang, and M. Wang

2. EXPERIMENT

The chemical used in this study is L-glutamic acid (LGA) (Figure 1). The solid β-form L-glutamic acid (LGA) crystals were purchased from Van Waters and Rogers (VWR) International Ltd. α-form LGA crystals were obtained through crystallisation from the purchased β-form L-glutamate crystals. The solvent used for the experiments was distilled water.

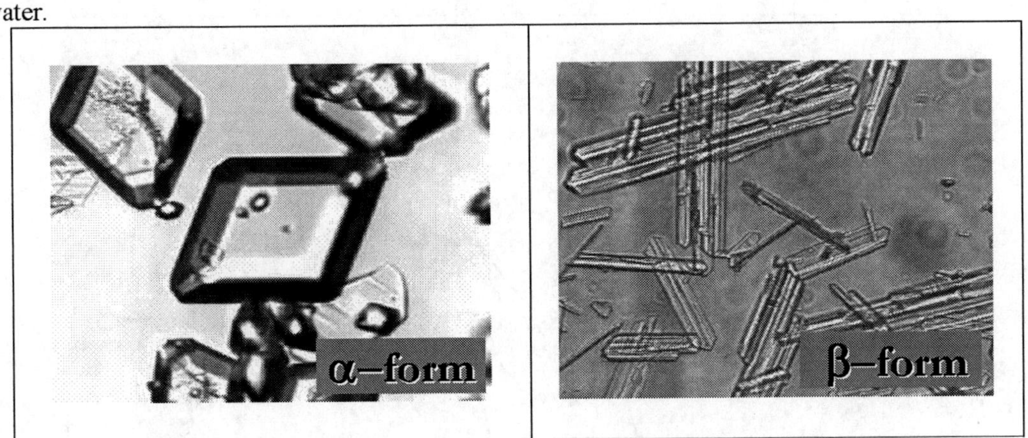

FIGURE 1. L-glutamic acid has two known crystal polymorphs, α-form and β-form.

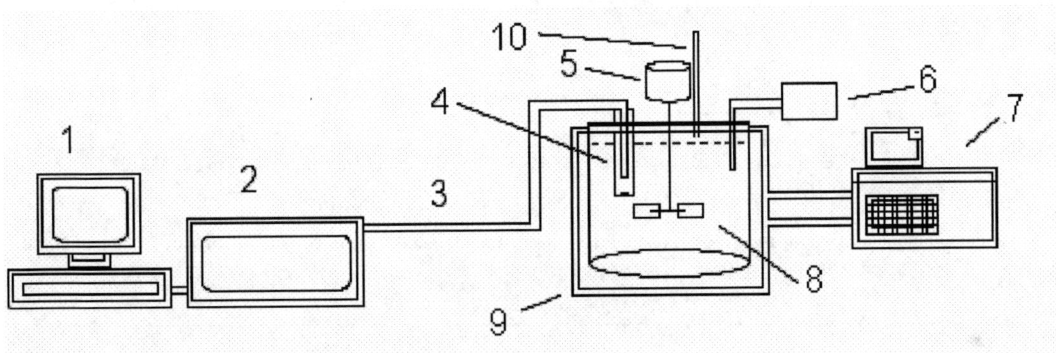

FIGURE 2. Schematic diagram of the experiment set up. 1 Personal computer, 2 NIR instrument, 3 Fibre optics cable, 4 Transflectance probe, 5 Stirrer motor, 6, Temperature probe, 7 Reactor temperature control, 8 Stirrer, 9 glass jacketed reactor (500 ml), 10 Condenser.

The experiments were carried out in a 500 ml jacketed glass reactor (Figure 2) which is equipped with a retreat curve impeller, a temperature control bath, a Platinum resistance thermometers (PT 100), a glass condenser and the NIR transflectance probe with 12 mm optical path length. The signals are transmitted to the FT-NIR ABB Bomen, model SPH 5300G, s/n SPH5305H spectrometer via a 2 meter silica fibre optics cable. The NIR spectra were managed by a software system called GRAMS for display and recording.

Samples were prepared to obtain slurry solutions at six particle size ranges for both polymorphic forms of LGA, α and β. The solutions were prepared according to the solubility of the polymorphs, which is 6 g / 500 ml water for α and 4 g / 500 ml water for β. For β form, the first sample of 4.0 g L-GA was prepared and added to the reactor containing 500 ml distilled water. According to the solubility of β form, 4.0 g of L-GA/500 ml distilled water at 20 ^{0}C should form a saturated solution5. The mixture was stirred at 200 rpm while heated to 30 ^{0}C and held at 30 ^{0}C for an hour to completely dissolve the solids before cooling down to 20 ^{0}C at which NIR spectra were taken in triplicate. Then 1 g of L-GA was added to the saturated solution at 20 ^{0}C. This will form a mixture of slurry of solid concentration, 1 g/500 ml distilled water, for which NIR spectra were taken in triplicate. Following the same procedure, further 2 g, to 10 g of L-GA were successively added to the saturated solution and for each sample triplicate NIR spectra were taken. The same preparation procedure was followed for α form LGA. Solid state samples of six particle size ranges were obtained using a mechanical shaker for both α and β crystals: below 53 μm, 53 – 75, 75 – 100, 100 – 150, 150-200, and above 200 μm as measured by cascade sieves. These samples have been measured using Malvern Mastersizer 2000. The results of mean particle sizes measured are 33, 50, 70, 104, 158,

211 μm respectively for α-form, and 47, 83, 100, 100, 61, 324 μm for β-form. In total, 119 spectra were taken using NIR transflectance probe for all experiments, of which 59 are for α and 60 for β. Although these experiments were conducted for different temperatures, results reported here are only for one temperature 20^{0}C. Repeatability was also tested by examining the variation of the triplicate measurements for saturated solutions and slurries at a fixed particle size range, solid concentration and temperature. It was found that the NIR spectra give satisfactory repeatability.

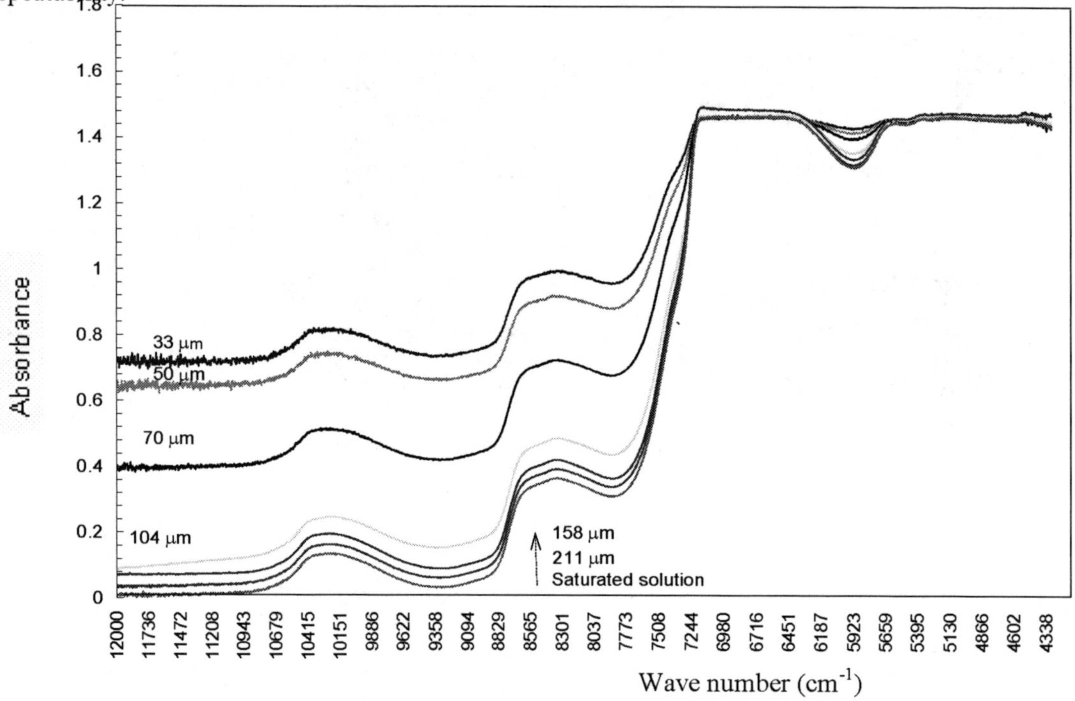

FIGURE 3. The effect of changing particle size on NIR spectra at a fixed solid concentration of 2 g of α form L-glutamate crystals at 20 ^{0}C. Also plotted was the spectrum of saturated solution at 20^{0}C

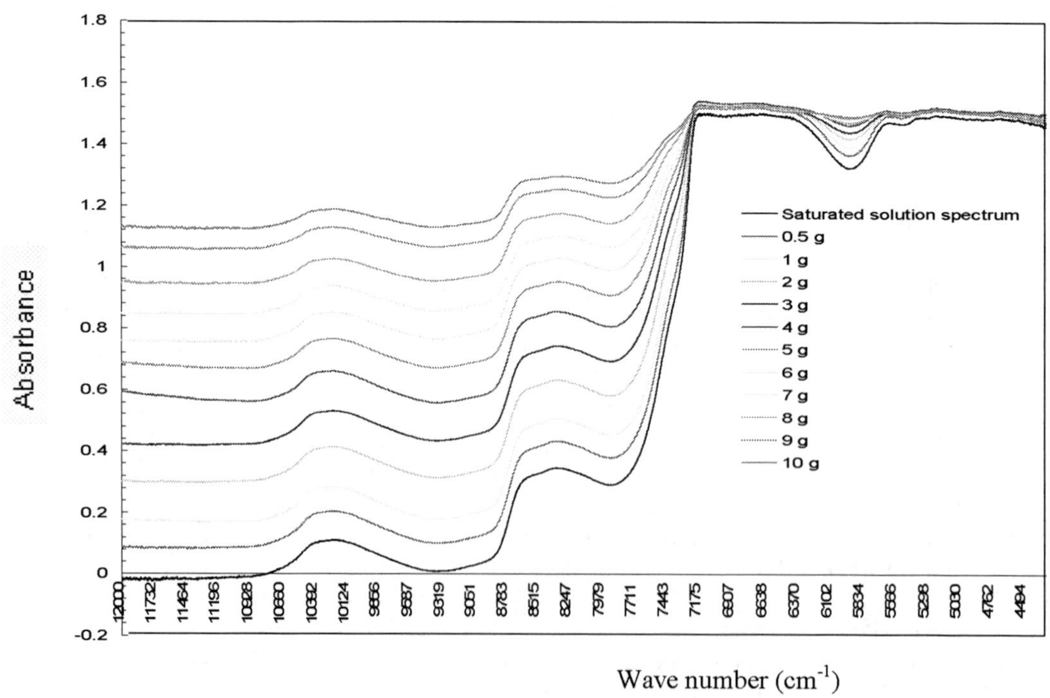

FIGURE 4. Effect of changing solid particle concentrations on NIR spectra at a fixed mean size range of 75-100 μm and a fixed temperature of 20 ^{0}C, for α form L-glutamic acid

3. RESULTS AND DISCUSSION

3.1 Particle Size Information In The NIR Spectra

The effect of particle size on the NIR spectra for α form crystals is shown in Figure 3. For the same solid concentration at 20^{0}C, six different particle size ranges were tested. Also plotted in the figure was the saturated solution without solids at 20^{0}C. It clearly shows that in these spectra the relationship between the baseline elevations and the particle size is inversely proportional.

β form was found to have a similar trend, i.e. generally speaking the smaller the particle size, the larger the base-line shift, except for the mean particle size 83 μm which showed a baseline shift smaller than spectra taken on 100 μm. This was attributed to the shape of the particle and the non-regular size distribution for samples which was reflected in Mastersizer 2000 measurements.

3.2 Solid Concentration Information In The NIR Spectra

Figure 4 shows the NIR absorbance spectra under different solid concentrations of α form LGA for a sample at a fixed size range, 75 – 100 μm and fixed temperature of 20^{0}C. There is clear correlation between the solid quantity and the baseline variation, i.e. the more solids added to the solution the higher the scattering of the radiation hence less light reach to the detector which results in a linear baseline shift across all wave lengths[6]. The same conclusion can be drawn for β form though the results were not shown in here. It is interesting to notice that although in the case of size effect on spectra, β form crystals have slight inconsistency with α crystals which was attributed to the shape of crystals, in the case of solid concentration effect, the qualitative conclusion is the same irrelevant of α or β form.

3.3 Polymorphic Form Identification

For the purpose of polymorphic form identification, principal and independent component analysis (PCA and ICA) was performed. PCA was applied to the raw NIR spectral data of 119 samples, of which 59 were for α, and 60 for β LGA. Only the first two PCs were used in the following analysis. It was found that the first PC, i.e. PC1 mainly contains information about the size and concentration of the solid crystals. Figure 5 shows the plot of PC1 scores vs. solid concentration and particle mean size for α form LGA. It clearly shows that PC1 reflects the changes in size and solid concentration. A similar plot was obtained though not given here due to space limitation for β form crystals though the relationship between PC1 score and size is not fully consistent unlike α form, due to the shape factor described before. In Figure 5, the scores intensity proportionally increases as α solid sample concentration increases. The increase in score intensity was observed as particle size decreases.

To find out what wave numbers are responsible to the size and solid concentration, i.e. to the PC1 scores, contribution plot was used. It was found that the wave numbers in the spectra regions of 12000 – 7278 cm^{-1} and 6088 – 5776 cm^{-1} are higher than the rest regions, indicating that they correlate to the concentration and size variations. These absorbance regions are due to CH bonds 3rd, 2nd and 1st overtone respectively[7].

Figure 6 shows the plot of PC2 scores vs. the samples. It was found the line of PC2 score of zero clearly divides the samples into two parts. Samples above the PC2 zero line, are mainly β form crystals, while below the line are mainly α form. It needs to point out, although on each side of the PC2 zero line, there are six clusters corresponding to six size ranges, it does not mean PC2 has size information. This is because the clusters were formed artificially by plotting x – axis as the number of samples.

Again contribution plot was used to analyze what wave numbers capture the polymorph information. It was found that loading values for the spectra regions 7172 – 6368 cm^{-1} and 5573-4322 cm^{-1} are higher than the rest, indicating that these absorbance regions are mainly responsible for the variation in polymorphs. These regions are OH, NH_2 2nd overtone and COOH, H_2o 1st overtone and the combination band regions[7].

Independent component analysis (ICA) is also performed, which shows that it not only can distinguish α and β form crystals, but can give a better result than PCA. Figure 7 shows the plot of the first independent component, IC1 plot against the sample numbers. In this case, only one sample was misclassified. ICA is a dimension reduction approach, similar to PCA, but has shown to be more effective in some applications associated with multivariate statistical process control[8].

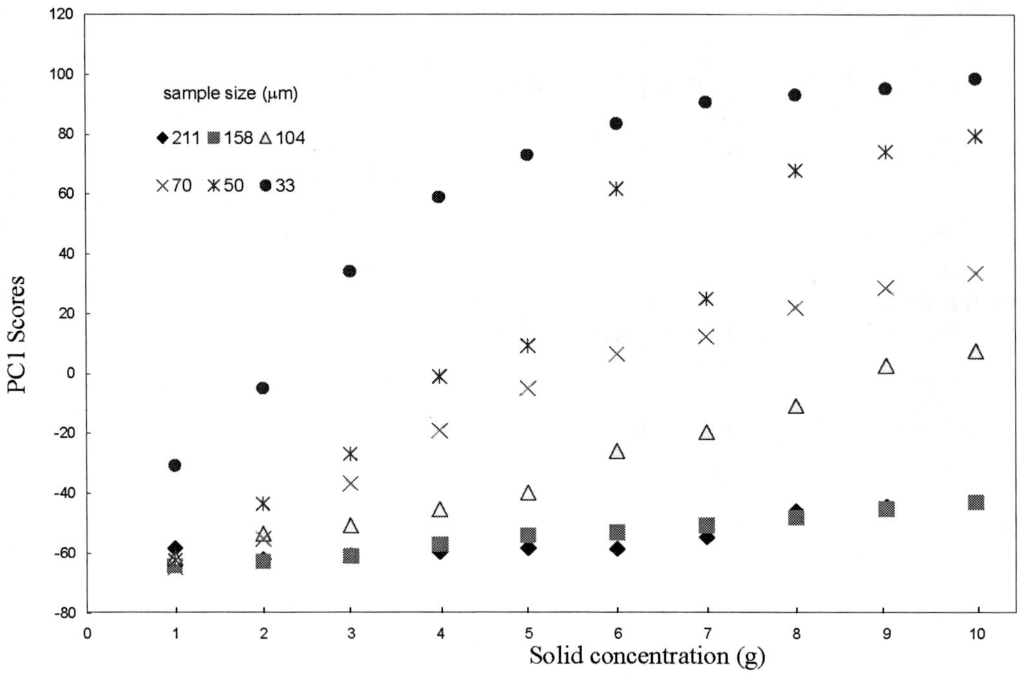

FIGURE 5. PC1 scores vs. solid concentration and particle mean size for α form LGA.

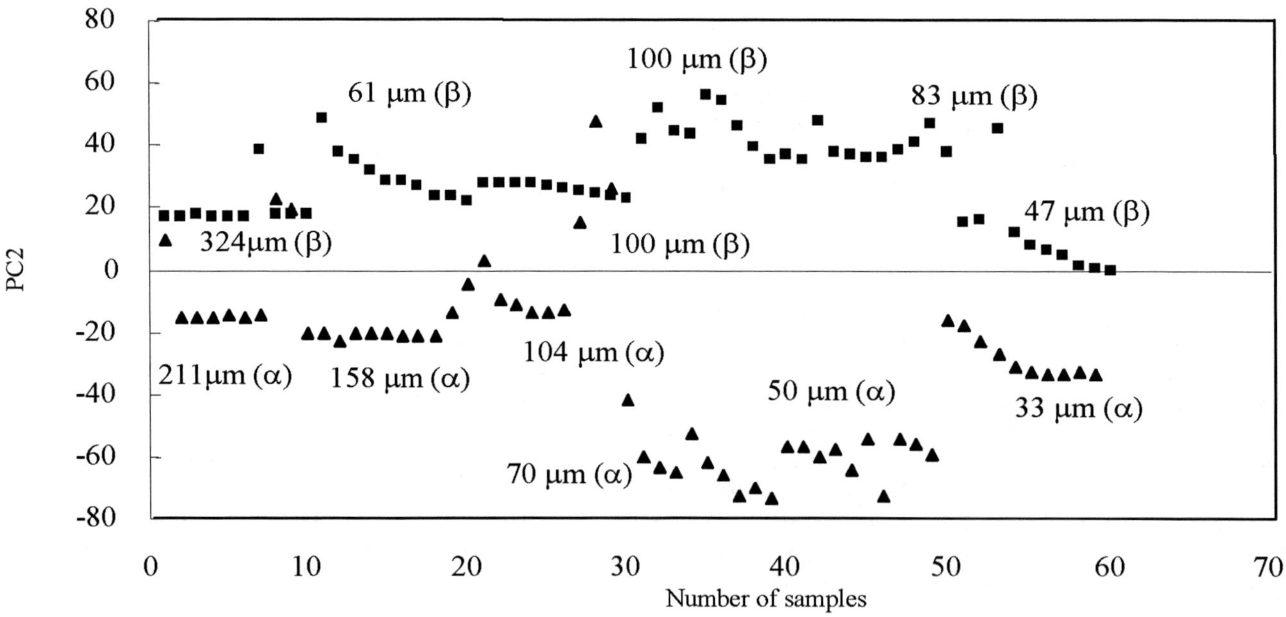

FIGURE 6. PC2 plotted against samples, indicting a clear dividing line between α and β form LGA

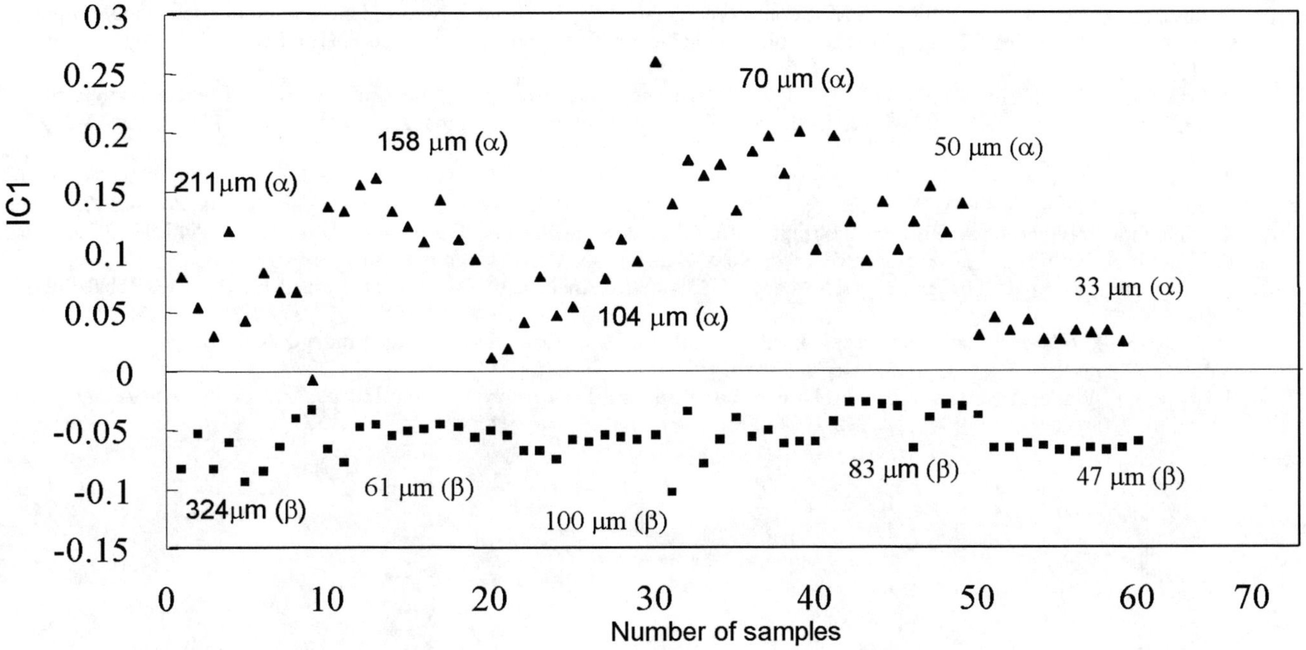

FIGURE 7. Independent component IC1 plotted against samples, indicting a clearer dividing line between α and β form LGA. Above zero line are α forms and below are β forms.

4. CONCLUSIONS

Due to its unrivalled combination of properties of being able to use fibre-optics for rapid data transmission, accurate in measuring both low and high concentrations of solutions, and simplicity, near infrared spectroscopy (NIR) has become popular in agriculture, food production and environmental monitoring, and is known as on-line GC in petrochemical and biochemical industries. However, unlike Mid IR, NIR spectra can be seriously affected by the presence of solid particles in the solution. This study has demonstrated that instead of being a disadvantage, the NIR spectra contain rich information regarding the solid properties, including size, solid concentration and polymorphs. The information is sensitive enough to be extracted for predictive purpose, raising the prospect of using NIR spectrometer for simultaneous measurement of multiple solid and liquid phase properties.

Future work includes developing calibration models for the prediction of the multiple properties of both phases. This will include investigating automatic wave number selections using such as genetic programming and the use of linear (e.g. PLS) and non-linear (e.g. neural networks) models. We also want to investigate independent component analysis to study if the information contents for different properties in the NIR spectra can be blindly separated, i.e. without training or calibration. An earlier study on a simple mixture of three components has demonstrated promising results[9].

ACKNOWLEDGEMENTS

The research is conducted with the support of UK Engineering and Physical Sciences Research Council (Grant reference: EP/C001788/1). Support from Malvern Instruments Ltd for supporting the *IntelliSense* Partnership (http://www.intellisense.org.uk) is also greatly acknowldged. We also would like to thank Clairet Scientific Ltd for providing help in the setup of the NIR system.

REFERENCES

1. M. Blanco, J. Coello, H. Iturriaga, S. Maspoch, C. Perez-Maseda, Determination of Polymorphic Purity by Near Infrared Spectrometry , Analytica Chemica Acta, Vol. 407, pp. 247-254, 2000.

2. E. Rasanen, J. Rantanen, A. Jorgensen, M. Karjalainen, T. Paakkari, J. Yliruusi, Novel Identification of Pseudopolymorphic Changes of Theophylline During Wet Granulation Using Near Infrared Spectroscopy, J. Pharm. Sci., Vol. 90, pp. 389-396, 2001.
3. P. Frake, I. Gill, C.N. Luscombe, D.R. Rudd, J. Waterhouse, U.A. Jayasorriya, Near-infrared Mass Median Particle Size Determination of Lactose Monohydrate, Evaluating Several Chemometric Approaches, Analyst, Vol. 123, pp. 2043-2046, 1998.
4. A.J. O'Neil, R.D. Jee, A.C. Moffat, The Application of Multiple Linear Regression to the Measurement of the Median Particle Size of Drugs and Pharmaceutical Excipients by Near-infrared Spectroscopy, Analyst, Vol. 123, pp. 2297-2302, 1998.
5. M. Kitamura, Polymorphism in the Crystallisation of L-Glutamic Acid, J Crystal Growth, Vol. 96, pp. 541-546, 1989.
6. P. William, K. Norris (Eds.), Near-Infrared Technology, Amer Assoc Cereal Chemists, 2001.
7. M. Claybourn, Infrared Reflectance Spectroscopy of Polymers : Analysis of Films, Surfaces and Interfaces, Global Press, 1998.
8. H. Albazzaz, X.Z. Wang, Statistical Process Control Charts for Batch Operations based on Independent Component Analysis, Ind. Eng. Chem. Res., Vol. 43, pp. 6731-6741(2004).
9. J. Chen, X.Z. Wang, A New Approach to Near Infrared Spectral Data Analysis Using Independent Component Analysis, J. Chem. Inf. Comput. Sci., Vol. 41, pp. 992-1001, 2001.

Visual Observations of Flow and Phase Phenomena in Loop Heat Pipes

Guanghan Wang, Darius Nikanpour

Space Technology, Canadian Space Agency, 6767 route de l'Aeroport, St-Hubert, Québec, Canada
Telephone: 450-926-4800, Fax: 450-926-4695; E-mail: Guanghan.wang@space.gc.ca

Abstract. The Canadian Space Agency is developing loop heat pipe hardware aimed at understanding the thermal performance of two-phase heat transfer devices and in developing numerical simulation techniques using thermo-hydraulic mathematical models, to enable development of novel thermal control technologies. This loop heat pipe consists of a cylindrical evaporator, compensation chamber, condenser along with vapor and liquid lines, which can be easily assembled/disassembled for test purposes. This laboratory setup is especially designed to enable the visualization of fluid flow and phase change phenomena. There are transparent windows in the compensation chamber and transparent lines for vapor line, liquid returning line, and sections of condenser that enable the monitoring of the fluid phase and flowing characteristics along the loop. The setup demonstrated typical loop heat pipe performance. Three working fluids (water, acetone and methanol) were used in the characterization of the loop heat pipe. Visual observations have been made at the compensation chamber, vapor line, liquid line, and the condenser.

ACRONYMS & ABBREVIATIONS

CC Compensation Chamber
CD Condenser
CPL Capillary Pumped Loop
CSA Canadian Space Agency
EV Evaporator
FEP Fluroinated Ethylene Propylene
HL Heat Loop
HP Heat Pipe
id Inner Diameter
LHP Loop Heat Pipe
LL Liquid return Line
MLI Multi Layer Insulation
Ni Nickel
NCG Non-condensed Gas
OD Outer Diameter
SS Stainless Steel
Ti Titanium
VL Vapor Line

INTRODUCTION

It is always desired in the heat transfer community to develop new devices that can transport large amounts of heat at high efficiency with flexibility in operations regarding the varying performances of heat sources and cooling sinks. Spacecraft thermal control systems have even higher operational & life requirements because of the low mass, volume and power budgets and because of the harsh space environment. Advanced thermal controls have the following characteristics:

- Able to handle varying powers (from a few watts to several Kilowatts & high flux densities (ideally up to several hundred watts per cm^2)
- Flexible locations of heat sources and sinks, at distances up to several meters and in different orientations.
- Variable operational regimes: from operating to non-operating, constant rates & transient cycles
- Maintaining stable & accurate thermal control in varying space environments.
- Robust and reliable autonomous operation.
- Low performance degradations during the mission lifetime

CP914, *Multiphase Flow: The Ultimate Measurement Challenge*, edited by X. Cai, Y. Wu, Z. Huang, S. Wang, and M. Wang
2007 American Institute of Physics 978-0-7354-0422-9/07/$23.00

- Multi-system and thermal bus architecture for heat load sharing & optimization in thermal control.

Such ideal devices have been realized by autonomous two-phase fluid systems (with no moving parts) – heat pipes. Capillary-driven heat pipes remove heat by evaporation at evaporators and reject heat by condensation at condensers, and capillary effects drive the fluid flow in micro-sized porous structures. In spacecraft thermal control systems, conventional heat pipes are most popular, in which the wick (grooves) is placed in the inner radius of a tube. The other advanced technology is the capillary-driven Heat Loop (HL). A recent review of the HL technology and its space applications can be found in the reference [1]. Heat Loops cover a number of different design solutions. The most popular ones are Capillary Pumped Loops (CPL) and Loop Heat Pipes (LHP), which differ from each other by virtue of location of the liquid reservoir and compensation chamber(s).

Heat Loop's have been demonstrated to meet the stringent requirements of spacecraft thermal control systems with clear advantages over traditional techniques. They offer a significantly higher thermal control capability (high heat flux densities, wide range of variable heat loads & stable payload temperatures) with a design flexibility & long life operation up to 10 years [2]. Clearly Heat Loops offer novel opportunities in both understanding the two-phase phenomena & in being able to model accurately such devices and explore advanced thermal control technologies and in this context CSA has initiated an LHP R&D activity [1, 3, 4, 5, 6].

LABORATORY SETUPS & VERIFICATIONS

Loop and Setups

A LHP setup has been developed in the Advanced Material and Thermal laboratory at the Canadian Space Agency. This laboratory setup was designed to serve as:

1. A correlation and verification tool for numerical models developed in-house and in industry;
2. A learning platform for better understanding the design-derived requirements for LHP and two-phase heat & mass transfer phenomena;
3. A testing platform for new LHP thermal control technologies.

This LHP uses cylindrical designs for all elements, including the evaporator (EV), the condenser (CD), the liquid compensation chamber (CC), the vapor line (VL), and the liquid line (LL). A schematic sketch is shown in Figure 1 with the evaporator shown in Fig.2. The detailed description has been shown in a previous work [5].

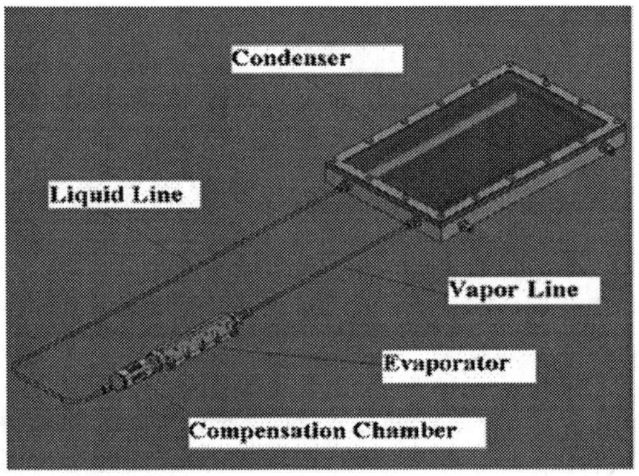

Figure 1: LHP design of the Loop

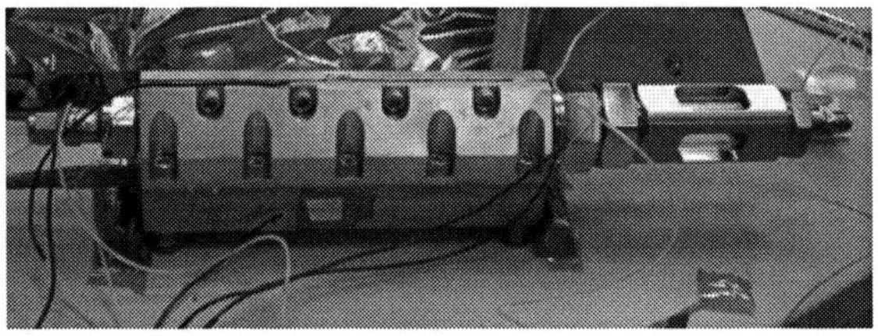

Figure 2: Evaporator and liquid compensation chamber

The VL and LL, as well as the flow paths in CD, are all made of transparent high-temperature FEP tubes in order to view the local fluid flow and phase changes. When visual observation is not required, sections of those tubes can easily be replaced by metal tubes (stainless steel or copper). In the CC, there is one window on each side (four in total) for monitoring the fluid behavior inside the CC. The EV is not designed as see-through or with transparent windows at the present time. A Nikon Coopix 995 Camera was used to record video clips and snapshot photos of the fluid behaviors at different loop locations. The key target element of LHP visualization has been the evaporator such as by Ku et al in 1993 [6], later by Kolos et al in 1996 [7] and recently by Suh et al in 2005 [8]. Since the evaporator is the hottest element of a LHP and any visual windows would affect considerably the EV insulations and hence possibly the thermal performance, this work focuses on the visualization of all other LHP elements instead of EV. Some of the preliminary observations have been presented earlier [5, 9].

Loop Performance Verification

A variety of tests were conducted to verify the actual performance of this LHP setup to assure the designs. Since this LHP setup is also designed to serve as a test platform for new LHP designs, a number of tests have been conducted to determine the influence of design parameters on the loop performance for: Contact conductance between wick & EV case; Heat leakage between EV and CC; Primary wick grooves with varying designs; Primary wick materials; Different working liquids; Fluid charge amount; Gas penetration through the FEP tubes; Local insulations at the LHP elements; Usage of a bayonet tubes; Usage of a secondary wick. By conducting those tests, the performance of this LHP setup has been further improved. The detailed description of all these influences is not included in this work. Some of the test results will be discussed in the next section. A number of tests have been conducted with this LHP setup and typical LHP characteristics were observed. Acetone, water, and methanol have been used as working fluids. The LHP is capable of transferring up to 500 W (water), and the minimum start-up and stable operation power for all the fluids was around 10 W.

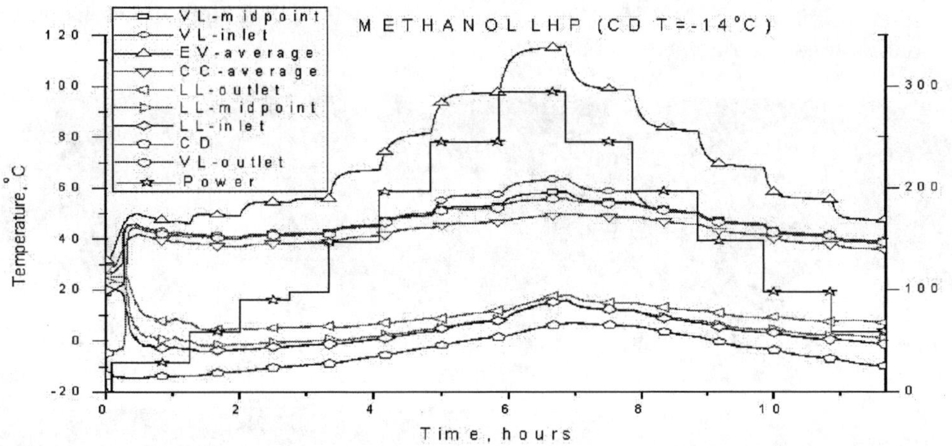

Figure 3: LHP Verification Tests

Figure 3 demonstrates a typical LHP thermal profiles using methanol as working fluid. The temperature difference between heater block and the vapor collected was seen large, which indicates a need of further improvement of thermal contact between the primary wick and the heating source. More figures for other fluids are shown elsewhere [5]. It should also be pointed out that this setup has been found showing a temperature oscillation [5]. The large temperature difference between EV and VL is believed being caused by (1). The non-radial-groove design of the wick; (2). Loose thermal contact between the wick and EV inner wall for which variable LHP performance have been observed.

VISUAL OBSERVATIONS

Flow Patterns in CD & VL

Compared with metal tubing (copper & SS), using FEP tubing in condenser section slows down the heat transfer rate between LHP liquid inside the tube and coolant outside the tube. The process of condensation takes place therefore in longer section length in longer time. For the purpose of visual observation, such elongations can only make better and clearer visualizations, provided with an assumption that the phenomena keep showing the same kind in both metal and plastic tubing. The general two-phase single-component line flow patterns can be found in many references such as Delil [10,11].

For the loop sections other than EV, the following flow patterns have been observed in this LHP:

- Pure superheated-vapor section (in VL with insulation; in CD near entrance);
- Vapor condensation section with small liquid drops mixed (in VL; CD near entrance; CD long vapor sections);
- Very short sections of annular flow and fast transition into long bubble sluggish flow (CD);

- Wavy flow & stratified annular flow (in VL with less thermal insulation);
- Large liquid section forming from small drops (in CD);
- Long vapor bubble trapped between liquid & condensing fast (in CD);
- Sluggish flow with large bubbles trapped between liquid & condensing slowly (in CD, differing from the sluggish flows in Figure 4);
- Sluggish flow with small bubbles trapped in liquid & merging with other bubbles (in CD);
- Sluggish flow with small bubbles trapped in & condensing in sub-cooled liquid (in CD & LL);
- Pure sub-cooled liquid section (in CD & LL).

Liquid Breathing In CC & CD

Not all LHP are designed with a liquid-full CC. It is therefore essential to determine that there is sufficient amount of liquid in CC to grant the continuous liquid supply to EV, and that the fluid status is proper in CC to maintain the operating conditions of the LHP. The four-side windowed CC design allows monitoring the fluid phenomena from all directions left, right, top and bottom.

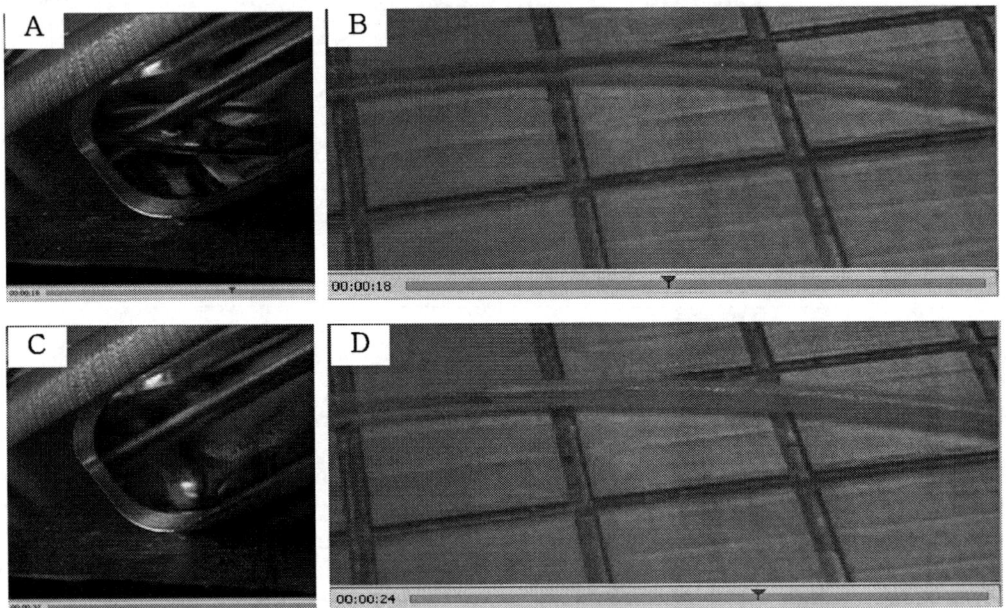

Figure 4: Liquid Breathing in CC: A: CC high; B: CD front while CC high; C: CC low; d: CD front while CC low

During some steady state operations of LHP, the liquid level in CC was observed periodically oscillating (Figure 4). It is called here "Breathing". While the CC liquid was continuing breathing, the fluid in CD also swapped forward and backward correspondingly, as shown in Figure 4. It must be pointed out that due to instrumentation problem, there were no simultaneously taken video clips of both CC and CD. Photos A&C and B&D hence illustrate the same phenomena but do not represent the occurrences at the same moment. The variation of the phase fronts between B and D represents the liquid volume variation in CC. This observation was with water at flow rate 27 Kg/m^{2}s under heat load 150W, vapor temperature from VL at 66°C and coolant 5°C.

All the EV temperatures were observed constant except for the CC entrance (liquid returning from CD) where fluctuation was constantly observed due to the oscillations from CD; the liquid-vapor interface was stably horizontal; no vapor bubbles were observed. A full cycle took about 25 seconds in the current loop. The pressure gauge in the loop could, unfortunately, read only in 10s per data sample and hence could not provide instant readings corresponding to the loop pressure variations during the breathing cycles. It is hence can be described as a half-loop fluid oscillation (CD-LL-CC) resulting only from large-vapor-bubbles condensation in CD. The fluid and temperatures at EV & VL are not affected by this oscillation because of the large volume of CC with sufficient minimum liquid left to continue the evaporation. If CC volume is not large enough, this oscillation can drain all liquid off from CC and discontinue the liquid supply to EV and hence may possibly end up with periodical burnout of LHP. In that case the LHP operates like an oscillation loop. The scale of CC-level breathing depends on the volumes of CC and CD, charged amount of fluid, applied heating power, number of long bubbles trapped in CD, as well as the heat transfer coefficient and total cooling capability of CD.

There has not yet similar phenomenon observed when using thin-wall copper tubes in CD, which may be explained that the condensation heat transfer coefficient of copper tubing was much higher and the vapor bubbles vanish rapidly without a considerable distance of swapping. The same breathing phenomena have however also been observed when EV-CC was inclined in axial angles of 0°, 20°, 45°, 60°. Wider breathing level variation was observed when CC was in higher inclination angles because of the large aspect ratio of CC cylindrical volume (length =150mm, id=25mm). On the other hand, it is possible for similar breathing phenomena to occur in micro-gravity because it is not gravity-assisted. If it happens, the negative effect should be severer than in gravity field, because it will be more difficult to remove the bubbles from the core while easier to onset nucleation in micro-gravity.

A similar breathing phenomenon was recently reported by Cheng [17] in micro-channel boiling.

Vapour & Boiling In CC

For a LHP, the actual loop temperature set-point is by the temperature of the compensation chamber, thus influencing the liquid content, hence the amount of liquid in the rest of the loop and consequently the condenser flooding, as well as the condenser area available for condensation. An opinion generally held among heat pipe specialists is that LHPs are tolerant to bubble generation in the CC. It has however been found in this work that this tolerance depends strongly on the boiling intensity and bubble size. All the instabilities in the LHP operation (poor thermal contact, low power operation, etc.) were connected with a very intense boiling in the CC. The track of small bubbles was also observed during both stable transient and steady state operations (Figure 5A). The speed of bubble evacuation to the top vapor section of the CC was augmented with an increase in input heating power. These bubbles condensed in the liquid in CC, no matter sub-cooled, super-heated or saturated. These small bubbles were not seen affecting the loop performance until their sizes have grown to a certain large and started to disturb the liquid in CC.

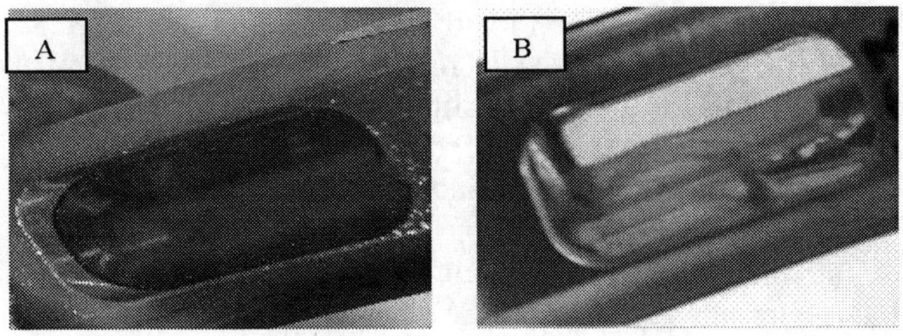

Figure 5: Boiling in CC: A: Small Vapor Bubbles; B: Very Large Vapor Bubbles

On the other hand, there were cases of small bubbles growing into intensive nucleate boiling in CC. Figure 5B is a huge bubble observed when a binary fluid mixture was applied as the LHP working agent. The original small size bubbles grew rapidly in both size and intensity and soon led to violent boiling in CC. This process exaggerated itself and resulted in rapid temperature increases in CC and immediate increase of pressure in the loop. The observed bubbling effects have therefore raised an issue of critical sizes and speeds of bubbles traveling into CC at which nucleate boiling can start in CC and result in unstable LHP performance.

There are three kinds bubbles in CC: (1). Liquid returned to CC can carry some non-condensed bubbles to CC, which are unlikely to onset boiling in CC; (2). Heat leakage from EV generates vapor bubbles in CC, which are possible to grow larger and hence onset boiling; (3). Bubbles generated in EV core can travel to CC by Brownian motion and free convection. Since the liquid temperature in CC is lower than where those bubbles are originally formed, bubbles can both gain heat from and lose heat to the surrounding liquid on its way into CC. Instead of continuously growing, the eventual bubble size can be both larger and smaller than the original size. Computation of the bubble size variation and the speed requires the whole temperature map from EV to CC. This anticipation needs however more experimental data and observations to be verified.

In micro-gravity, due to the coalescence of bubbles [13], it will be easier to form larger size bubbles than in gravity, and hence the bubble occurrences in CC are anticipated to result in severer negative effects than the discussions above.

OBSERVATIONS OF NON-CONDENSABLE GASES

The occurrence of non-condensable gas (NCG) can have a detrimental negative effect on any condensation heat transfer devices, while for LHP, on the boiling element (EV) as well. A small amount (<1%) of NCG blockage in loop can cause significant drop of the overall LHP heat transfer coefficient, by a few percent up to a few times, depending on the locations of NCG occupation. On the other hand, NCG is also a viable tool to control the loop performance. It is therefore extremely important to be able to visually detect the presence of NCG in the loop and hence to confirm the locations.

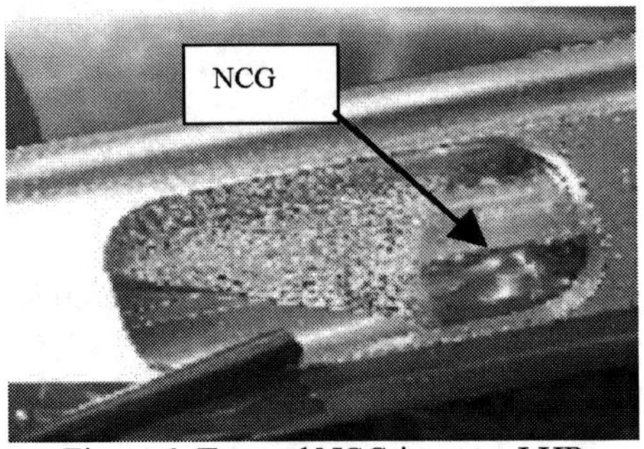

Figure 6: Trapped NCG in water LHP

The influence of NCG to a heat transfer process differs by stationary and moving NCG. In LHP the NCG location is also important. There are mainly four cases for NCG to affect LHP, discussed in following sections.

Case 1: Stationary occupation of evaporation areas

If any NCG is trapped stagnantly in EV, it will not only reduce the area of evaporation in EV, but also increase the thermal conductance between heating source and the fluid. The heat evaporation rate will hence be significantly reduced. The direct detection of the presence of NCG in EV is hence the lower heat transport rate and higher temperature difference between fluid vapor and the heating source. The analytical computation of NCG to LHP evaporation has not been found up to the knowledge of this work. Due the non-transparent design of EV in this setup, it is yet unable for this setup to visually show the traces and performances of NCG in EV.

Case 2: Stationary occupation of condensation areas

The NCG buildup near the condensing film inhibits the diffusion of the vapor from the bulk mixture to the liquid film and reduces the local mass and energy transfer rates. Dead-flow areas are the most possible locations for NCG to be trapped and stay. For condensers with small tubing diameters, stationary NCG will mainly stay in sharply bended sections of CD; for CD with annular flowing areas, NCG will most possibly occupy the corner areas. The immediate effect of stationary NCG is the reduction of condensation areas and hence lower heat transfer rate of CD, and finally increase of outlet temperature from CD. Dube et al [14] investigated the adverse effects of NCG at different locations of CD to a thermosyphon. Similar effect can be expected on a LHP. The mass, momentum and condensation heat transfer with the presence of stationary NCG was investigated in quite a few classical publications such as Sparrow & Eckert [15]. There was not visual detection of NCG in this LHP setup. There have however been some cases of suspecting existence of NCG in coiled copper CD tubes.

Case 3: Moving with flowing fluid along the loop.

Compared to the corresponding stationary NCG, a moving vapor-gas mixture is considered to have a "sweeping" effect, thereby resulting in a lower gas concentration at the interface. While the stationary NCG in vapor-gas mixture can result in significant reduction of heat transfer rate, moving vapor with NCG has been proved showing nearly negligible effects on the flow heat transfer, shown by many investigators [Corradini, 16]. The presence of moving NCG anywhere in the loop is not necessary to be seriously considered for thermal calculation unless it becomes a stationary NCG.

Case 4: Stationary occupation in CC

Due to the larger volume and slower fluid motion in CC, NCG is very possible to be collected in CC and stay there stagnantly. The effect of such stationary NCG in CC is purely a reduction of the compensation volume.

Since NCG is normally small in amount (unless a large amount is purposely introduced in loop), it can hence be considered to have little effect to the heat transfer performance of the loop.

In this laboratory setup, NCG in CC can be collected at the upper right corner by axial inclination of EV-CC. It was observed that some NCG was collected in the CC during a water-LHP operation (Figure 6). The NCG presence had been seen leading to a visible working pressure raise and finally to an EV temperature increase and LHP performance degradation. It was believed from two sources: (1). This NCG was generated in LHP due to the use of aluminum inserts in the preliminary tests; (2). NCG was the gas penetrated through the transparent FEP tube wall by N2 and O2, since the loop was in lower pressure than atmosphere during most of the time. The aluminum inserts were replaced later by brass ones. Some sections of FEP tubes were replaced by copper tubes where no visual observations were necessary latterly. Air leakage into the loop was also suspected a few times but have never been observed. A high-elevation fluid reservoir with valve may be applicable in CC to collect and release NCG in the future design.

MODELING OF CONDENSATIONS IN VL AND CD

The modeling of flow & heat transfer of LHP requires to cover the components of the loop. There are mainly three areas of modeling necessary for an LHP:

- Boiling heat transfer in Evaporator: this is the core element of LHP to conduct the heat transfer duty, including two-phase flow and heat transfer in channels of different cross section geometries (bayonet tube, grooves, spaces between wick and solid surfaces, etc.), two-phase flow and boiling in porous wicks, heat conduction and convection from CC, vapor removing from EV to VL. This is not discussed in this work.
- Fluid Behavior in Compensation chamber: As described in above sections, the phenomena observed in CC include the liquid gathering, heat conduction from EV, bubble formation and phase transfer, pressure oscillation from the condensation in CD. This is not discussed in this work.
- Fluid flow and condensation along the transport lines VL, CD, LL, and the conjunctions (if exist): Although this is to be discussed in the sections below.

The complex phase phenomena observed in the LHP loop are challenging the current numerical predictions of LHP. The numerical modeling and results will de discussed in separate work.

CONCLUSIONS

An experimental LHP setup has been developed, designed and installed in the Material and Thermal Laboratory of the Canadian Space Agency. The setup has two main functions: (1) a correlation and verification tool for numerical models; and (2) a testing platform for new LHP thermal control technologies. This setup has been verified by conducting a variety of thermo-hydraulic tests, and displays typical characteristics of LHPs as a heat transfer device yet the performance of transferring heat needs to be improved.

The design feature with transparent LHP elements makes this setup especially helpful to observe the fluid flow and phase change at designated internal loop locations, and hence aids in better understanding of the physics of LHP's. Several observed cases of flow patterns and fluid phase change in CC, VL and CD are illustrated in photos and discussed. Especially important are the bubble occurrence in CC and the effects on loop performance.

Different cases of NCG presence in the loop were analyzed regarding the effects on the heat transfer performance of the LHP.

This laboratory setup will be further modified to improve performance and to enable flexible controls of transferring heat. The immediate improvements include better thermal contact between primary wick tube and the evaporator wall, a reservoir to collect NCG, and flexible orientation and elevation control of evaporator. More laboratory tests will be conducted to verify the control of bubbles in CC and numerical computation of flow patterns in VL and CD.

REFERENCES

1. Mishkinis D., Wang G., Nikanpour D., MacDonald E., Kaya T., 2005 Advances in Two-Phase Loop with Capillary Pump Technology and Space Applications, SAE paper No 2005-01-2883.

2. Orlov A.A., Goncharov, K., Kotliarov, E.Yu., Tykina, T.A., Ustinov, S.N., and Maydanik Yu.,F. , 1997, The Loop Heat Pipe Experiment on board the GRANAT Spacecraft, Proc. 6th European Symp. on Space Env. Control Systems, Noordwijk, The Netherlands.
3. Mishkinis D., Wang G., Nikanpour D., MacDonald E., Kaya T., 2004 Steady State and Transient Loop Heat Pipe Performance During Periodic Heating Cycles Proc. 13th International Heat Pipe Conference, September 21 – 25, 2004, Shanghai, China.
4. Wang G., Mishkinis D., Nikanpour D., MacDonald E., Kaya T., STAIF 2005, Edited by M S. El-Genk, AIP.
5. Wang G., Mishkinis D., Nikanpour D., Capillary Heat Loop Technology: Space Applications and Recent Canadian Activities, Proc. 5th Minsk International Seminar Heat Pipes, Heat Pumps, Refrigerators, September 12 – 15, 2005, Minsk, Belarus.
6. Mishkinis D., Wang G., Nikanpour D., A Laboratory Setup for Observation of Loop Heat Pipe Characteristics, SAE paper 2006-01-2170.
7. Ku, J., Swanson, T.D., Herold, K., and Kolos, K, 1993, "Flow Visualization within a Capillary Evaporator," in Proceeding of SAE international, Warrendale, PA, pp. 1424-1432.
8. Kolos K.R., Herold K E, Kroliczek E J, Swanson T D, 1996, "Flow Visualization in Capillary Pumped Loop Systems", American Institute of Physics Conference Proceedings, Vol. 361, pp. 731-738.
9. Suh J., Cytrynowicz D., Medis P., Gerner F., Henderson H T, 2005, "Flow Visualization within the Evaporator of Planar Loop Heat Pipe", STAIF 2005, Edited by M S. El-Genk, AIP.
10. Delil A.A.M., Issues of Various Two-Phase Heat Transfer Devices in Gravity Environments Ranging from Micro-Gravity to Super-Gravity, NLR-TP-2000-612, 2000
11. Delil A.A.M., Tutorial on quantification of differences between single- and two-component two-phase flow and heat transfer, NLR-TP-2003-037, 2003
12. Faghri A., Heat Pipe Science and Technology, Taylor & Francis publisher, 1995.
13. J.Kim, J.F. Benton, Highly subcooled pool boiling heat transfer at various gravity levels, Int. J. Heat & Fluid Flow, 23 (2002) 497-508.
14. Dube V., Akbarzadeh A., Andews J., Mochizuki M., The effects of non-condensable gases on the performance of loop thermosyphon heat exchangers, Proc. 7th IHPS, Oct 12-16 2003, Jeju, Korea.
15. E.M. Sparrow and E.G. Eckert, "Effects of Superheated Vapor and Noncondensable Gases on Laminar Film Condensation," AIChE J, 7,3, pp 473-477, 1961.
16. M. L. Corradini, Fundamentals of Multiphase Flow, 1997
17. P. Cheng, "Recent work on boiling and condensation in microchannels", presented at 5th ISMTMF/IWPT2, Dec 10-13 2006, Macau, China.

SESSION C

INFORMATION TECHNOLOGY AND INSTRUMENTATION

Measurement of Solid-Liquid Two-Phase Flow Using a New Robust Normal-Line Hough Transform Method

Xiaoran. Yu*, K. Abe, Y. Hirose, T. Hazuku, T. Takamasa, M. Oshima

**Information System Laboratory/Tokyo University of Marine Science and Technology, Etchujima, Koto District, Tokyo, Japan*

Abstract. In this paper, we derive a technique for the measurement of the solid particle phase in a solid-liquid two-phase flow. The proposed method consists of a two-step algorithm for the localization of particles. In the first step, we devise a Normal-line Hough Transform method (NHT) and utilize it to detect the centre of the particles in the stereo images. In the second step, we validate the existence of the particle position extracted using NHT by comparing with a labeling parameter space that yield using image labeling technique, and this improves the robustness to noise and the precision remarkably. The proposed method provides a high detection rate with particularly low level of false positives, and an evident amelioration in time-consuming as well. We also present extensive experimental results to illustrate the efficiency of the proposed approach.

Keywords: Solid-liquid two-phase flow, image-processing method, normal-line, hough transform, vote.

INTRODUCTION

A high accuracy model of gas-liquid two-phase flow [1], [2] is not only helpful for improving the efficiency of thermal conduction on the ground, such as the normal plants, but also it can play an important role in the designing and development of the space environment as well, such as nuclear reactor design for space environmental use, observation of flow in the fuel pipe, and the heat control design of the living space.

In this research, a gas-liquid two-phase flow is simulated by a solid-liquid two-phase flow which consists of a low density solid particle phase and liquid phase of water, and when it flows in the test tube with an inner diameter of 9 millimeter made of acrylic fiber under microgravity condition [3], [4], images were taken with a digital video camera at z/D = 5, 40 and 60 elevations, respectively. A pair of back-lit images corresponding to two views at right angle to each other and orthogonal to tube axis are taken so that three-dimensional information can be obtained from the stereo images.

To locate the solid particle phase in the solid-liquid two-phase flow is very important for establishing a database, which concerns the interfacial transportation and the space distribution of the solid particle. And for measuring axial development of flow parameters [5] as well, such as void fraction, interfacial area concentration, solid particle phase velocity and solid phase number density. But to detect and to locate the solid particles automatically still remains unsolved, specially under a high void fraction condition which result in superposition of solid particles.

In this paper we present a two-step algorithm for the measurement of particles. In the first step we use a Normal-line Hough Transform (NHT) for the centre detection of the particles in the stereo images, and the second step validates the existence of the detected particle centers by checking the corresponded mapping value in the labeling parameter space, which is parsed using image labeling technique.

The Hough Transform (HT) technique has been widely used in image processing for pattern recognition [6]. When we apply HT, it changes the problem of pattern detection in the image space into the problem of peak detection in the parameter space. In this research, firstly a Canny Edge Detector was adopted to extract the edges of particles in the stereo images. In order to improve the efficiency of HT, such as reduce the time-consuming and memory cost, a Normal-line Hough Transform (NHT) method was devised. It makes use of the property that the normal-line of every edge point on a circle passes through the circle centre consequentially. Voting is carried out on

CP914, *Multiphase Flow: The Ultimate Measurement Challenge,* edited by X. Cai, Y. Wu, Z. Huang, S. Wang, and M. Wang

each candidate center point for each edge point. These voting results are recorded and accumulated in the voting parameter space. The value of the accumulator of the target particle will be the highest one [7].

To improve the robustness to noise and the precision of NHT, a labeling parameter space is utilized. Labeling parameter space is a mapping space on the image using image labeling method, a false means background and a true means foreground labeled object. When the candidate centre point has a true value in the corresponding position of labeling parameter space, this point is voted, otherwise, it will be discarded contrarily.

The rest of the paper is organized as follows. Section *algorithm description* describes our algorithm in details and section *experiment and results* presents experimental results comparing with the NHT and the HT. Section *conclusion and further research* assigned to present conclusion and our plans for future works.

ALGORITHM DESCRIPTION

In this section the major difference between the proposed method and previous method is explained. The HT method is a well known algorithm for extracting geometric primitives from raster images. And it is easily generalized to find more complex features. Usually edge map of the image is calculated then each edge point contributes a circle of radius r to an output accumulator space (Fig. 1a), which we call it voting parameter space. For unknown circle radius, the algorithm should be run for all possible radiuses to form a 4-dimensional voting parameter space, where two dimensions represent the plane coordinates of the center, the third one represents the radius and the fourth for voting values. The voting parameter space has a peak value at where these contributed circles overlap at the center of the original circle. In spite of its popularity owing to its simple theory of operation, the HT has some disadvantages when it works on a discrete image. The large amount of storage and computing power required by the HT are the major disadvantages of utilizing it in real time applications.

To reduce the main deficiency of HT, such as to improve the process speed and reduce the resource cost, we devised and applied a NHT method to locate the particles. In our research, low density solid particles with a same radius are used. As a result, all the circles in an image are of identical known radius, here we suppose that the radius are of length r, and the 4-dimensional voting parameter space can be simplified to a 3-dimensional one. The objective is to detect the coordinates of the centers. Assume that all the circles are of radius r, the center of a circle will certainly fall on the extending line of gradient vectors with a distance of r from the edge pixel, and both sides are possible. Thus, each edge point contributes two points (Fig. 1b) instead of a circle (Fig. 1a).

The Normal-line Hough Transform Algorithm

The following paragraphs describe the major steps of our algorithm.

1. Extract the edges of particles in the stereo images.
2. Calculate the gradient of each pixel on the edges using Equation (1), where *f* is the image function.

$$\theta = \tan^{-1}\left(\frac{\partial f}{\partial y} \Big/ \frac{\partial f}{\partial x}\right) \tag{1}$$

3. Prepare and zero the voting parameter space.
4. Quantize and accumulate the end point of normal-lines to the corresponding coordinates in the voting parameter space.
5. Local maxima in the voting parameter space correspond to the centre point of original circles in the image.
6. Disregard the redundant and fake vote candidates that are adjacent to the peak (centre point) within a fix distance. Here we call the distance as disregard threshold. In our research, the disregard threshold is 3 pixels. If the number of circles detected is less than total circles in the image, then go to Step 5.

A Robust Labeling False Negative Method

To improve the robustness to noise and the precision of NHT, a labeling parameter space is utilized. Figure 2 shows an example of labeling parameter space. Elements of the labeling parameter space are values mapping on the image parsed using image labeling method. A false means background and a true means foreground labeled object. When the candidate centre point has a true value in the corresponding position of labeling parameter space, this point is voted, such as the triangles, otherwise, it will be discarded contrarily, such as the square. In this way, spurious circles can be removed and leave the correct one as result.

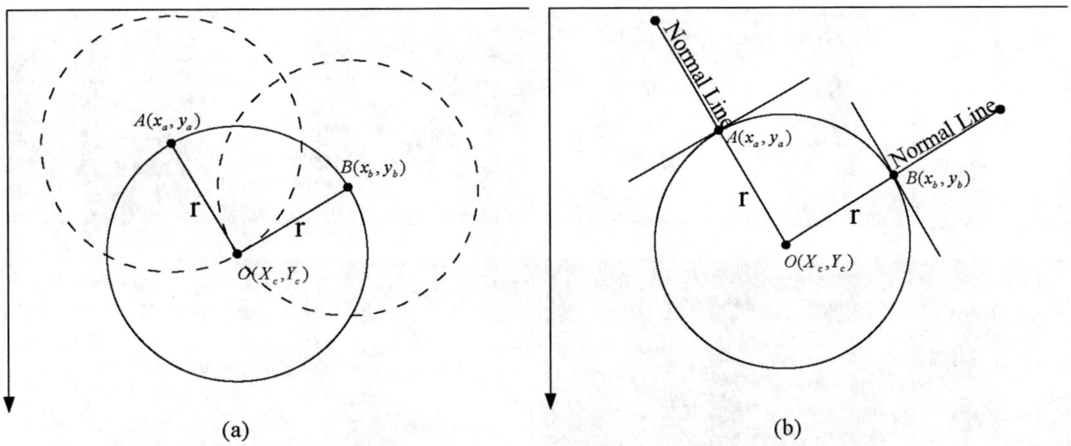

FIGURE 1. (a) Hough Transform. (b)Normal-line Hough Transform.

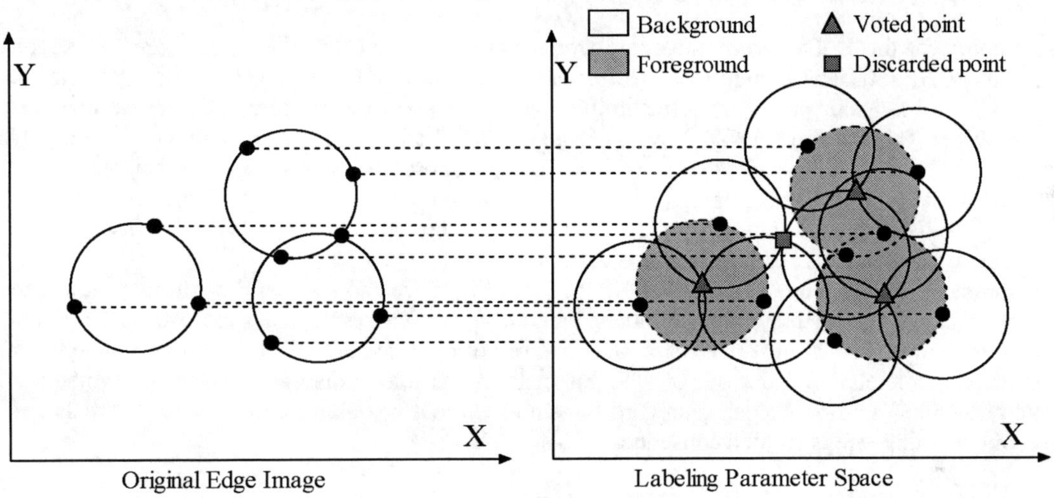

FIGURE 2. Each point in geometric space (left) generates a circle in voting parameter space (right). The circles in voting parameter space intersect at the triangles that are the center in geometric space. Overlapping circles can cause spurious centers to also be found, such as at the square.

EXPERIMENT AND RESULTS

To compare the axial developing flow parameters of a gas-liquid two-phase flow with that of a liquid-liquid two-phase flow, measurement of a liquid-liquid two-phase flow which consists of a low specific gravity silicon phase and liquid phase of water are carried out. However, the acquisition of highly rational axial developing flow parameters is difficult according to the adhesion and the stagnation of the dispersed phase to the test tube wall. As an alternative experiment, in our research, a solid-liquid two-phase flow consists of a low density solid particle phase and water phase flows through tube under normal gravity and microgravity environment, and a pair of images corresponding to two views orthogonal to tube axis are taken so that three-dimensional information is obtained from stereo images.

For measuring axial developments of solid-liquid two-phase flow, we utilize the proposed NHT to process the flow images. Figure 3 shows the result of the NHT algorithm applied to one stereo image. All the particles are extracted effectively in despite of the particle superposition.

To evaluate our algorithm, about 100 images of the solid-liquid two-phase flow were utilized. The examination environments are shown in Table 1. All the particles are of the same radius of 2.38 millimeters.

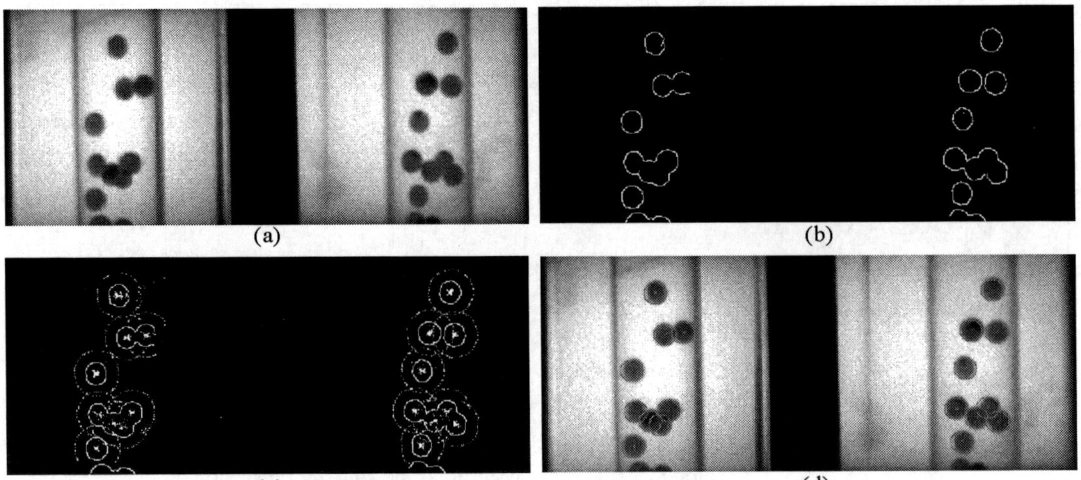

FIGURE 3. (a) Flow image, (b) edge image, (c) voting parameter space image, (d) NHT result image.

We also implement the HT algorithms to compare it with our proposed NHT algorithm. All algorithms are implemented in VC++ 6.0. The experiments were performed using a PC equipped with 1.8 GHz Pentium M processor and 1G RAM. The average time-consuming, average miss extracting rate and average deviation of each algorithm is shown in Table 2. From Table 2, we can find out that the NHT runs at an average about nearly 56 times faster than the NT in processing 640*480 resolution images; moreover, a higher extracting rate of 96.43% is obtained as well. In respect of average deviations, the two methods have approximate deviation of about 0.5 pixels comparing with results of manual observation respectively. As shown in Fig. 4, deviation between our proposed method and manual observation is similar in shape with that of HT.

Voting parameter space example is shown in Fig. 5. From Fig. 5 we can see that the voting accumulator of NHT has a remarkable advantage on reducing redundant candidates, which result in an evident amelioration in time-consuming and memory cost. On the other hand, the peak of voting space of NHT turns to be a sharpen-shaped and discrete one other than a relative wider one of HT, this make it possible to disregard the invalid voted point around the candidate one within a rational small disregard threshold without abandoning those adjacent real ones, thus, an amelioration of extracting rate is yielded consequently.

TABLE 1. Examination environments of solid-liquid two-phase flow

Solid Particle Phase		Liquid Phase	Gravity
Density(k/m^3)	**dia.(mm)**	**j_f(m/s)**	**g(m/s^2)**
908	**2.38**	**1.6**	**1.0**
908	**2.38**	**1.6**	**0**
1140	**2.38**	**1.6**	**1.0**
1140	**2.38**	**1.6**	**0**

TABLE 2. Result comparison between NHT and NT

	Time-Consuming	Miss Extracting Rate	Average Deviation
	(msec)	**(%)**	**(pixel)**
NHT	**284.65**	**3.57**	**0.52**
HT	**15865.70**	**10.71**	**0.49**

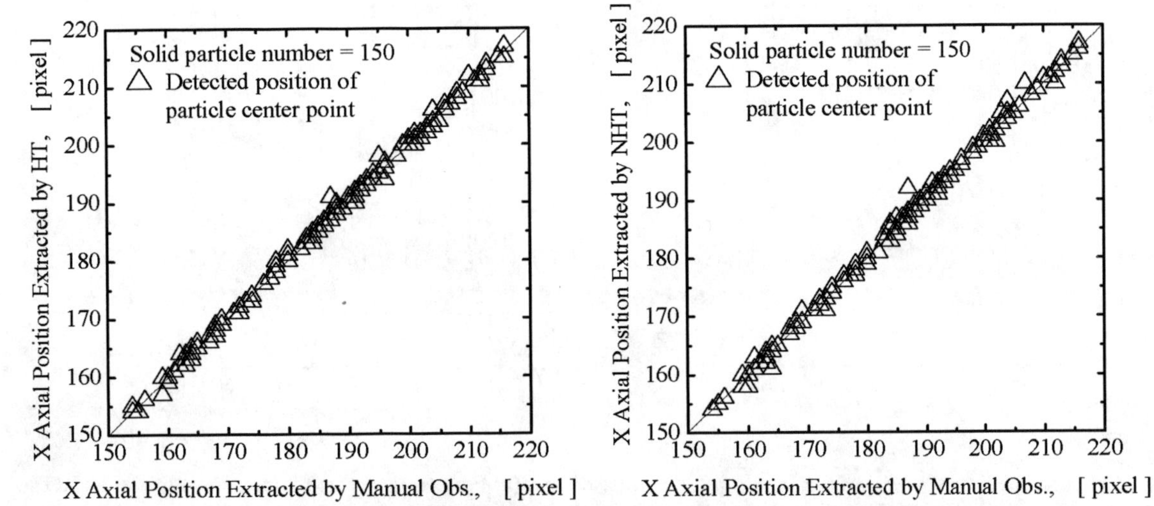

FIGURE 4. Deviation of the two methods compare to manual observation respectively. (a) Deviation of HT compare to manual observation, (b) deviation of NHT compare to manual observation.

Finally, we applied NHT to stereo images of solid-liquid two-phase flow and tried to reconstruct the three-dimensional model of the two-phase flow. A model example is shown in Fig. 6. By further analyzing and evaluating on this model, we can calculate the important physics parameters of the flow, such as the interfacial transportation and the space distribution of the solid particle, as a result.

CONCLUSIONS AND FURTHER RESEARCH

In this paper, we presented a robust NHT method to extract and locate circles. The main idea is based on the fact that the gradient vectors passing through the centre point of circles. The experimental results show that the NHT is nearly 56 times faster than the HT method. In case of 640*480 resolution images, the average time-consuming is about 284.65 micro seconds. Furthermore, the NHT has better resistance against redundant candidates and invalid votes than the HT method as well, which result in an evident improvement in extracting rate.

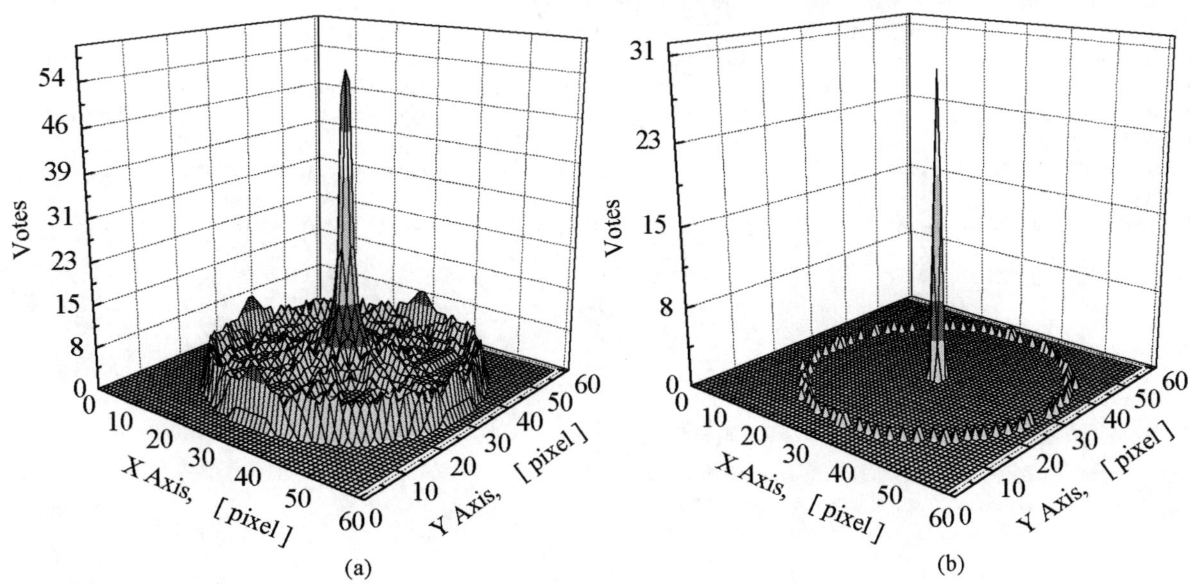

FIGURE 5. (a) Voting parameter spaces of HT, (b) voting parameter space of NHT.

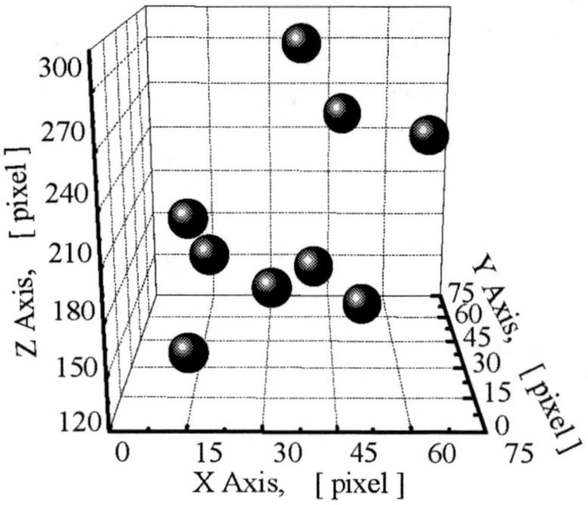

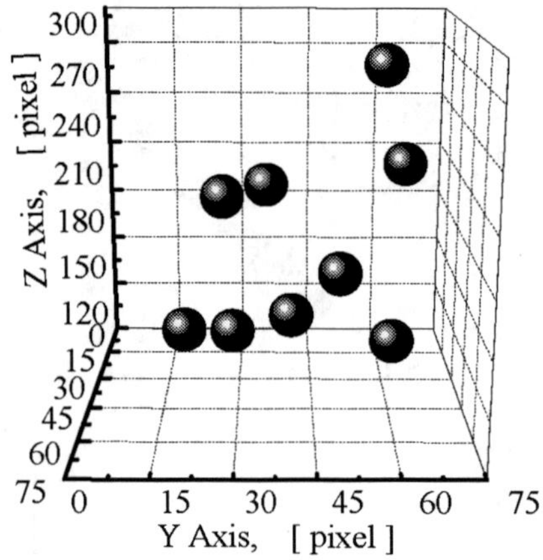

FIGURE 6. Three-dimensional model of the solid-liquid two-phase flow. (a) Projection of three-dimensional model on XOZ coordinates plane, (b) projection of three-dimensional model on YOZ coordinates plane.

Despite of the merit of NHT in processing speed and extracting rate, an average deviation of 0.52 pixels still remains an improving space. For increasing the accuracy of the NHT, one of our further job is to use certain factor for candidate points. These are in terms of an analysis of the peak width in the voting parameter space, and also an analysis of the peak position based on the votes, along with analyzing the distance between each other. Also adjusting gradient algorithm of each pixel on the edge is critical and depends on image features and statistics. Furthermore, by increasing disregard threshold, more candidates will be removed, and this may lead to better result (invalid candidates being removed) or worse result (circle centre point being disregarded). So adjusting these parameters is an art in each application.

REFERENCES

1. T. Takamasa and N. Miyoshi, *Measurements of Bubble Interface Configurations in Vertical Bubbly Flow Using Image-Processing Method,* Transaction of JSME 59, 1993, pp. 64-69 (in Japanese).
2. T. Takamasa, T. Goto, T. Hibiki and M. Ishii, *Experimental Study of Interfacial Area Transport of Bubbly Flow in Small-Diameter Tube*, Int. J. Multiphase Flow 29, 2003, pp. 291-304.
3. T. Takamasa, M. Watarai and K. Kondo, *Gas-Water Bubbly Flow Under Microgravity Condition*, Proc. 31st ASME/ANS/AIAA Natl. Heat Transfer Conf. ASME Symposium. Ser. 310/92, 1996, pp. 180-185.
4. T. Takamasa, T. Hazuku, N. Fukamachi, N. Tamura, T. Hibiki and M. Ishii, *Effect of Gravity on Axial Development of Bubbly Flow at Low Liquid Reynolds Number*, Experiments in Fluids, 37, 2004, pp. 631–644.
5. T. Hibiki, S. Hogsett and M. Ishii, *Local Measurement of Interfacial Area, Interfacial Velocity and Liquid Turbulence in Two-Phase Flow*, Nucl. Eng. Des. 184, 1998, pp. 287–304.
6. R. Duda and P. Hart, *Use of the Hough Transform to Detect Lines and Curves in Pictures*, Communications of the ACM 15, 1975, pp. 11–15.
7. C. Kimme, D. Ballard and J. Sklansky, *Finding Circles by an Array of Accumulators*, Proc. ACM 18, 1975, pp. 120–122.

Two-phase Flow Regime Identification Combining Conductivity Probe Signals and Artificial Neural Network

Leonor Hernández*, José Enrique Juliá*, Sergio Chiva*, Sidharth Paranjape¶, and Mamoru Ishii¶

* *Department of Mechanical Engineering and Construction, Universitat Jaume I, Campus Riu Sec, E-12071, Castellón de la Plana, Spain*

¶ *Thermal-hydraulics and Reactor Safety Laboratory, School of Nuclear Engineering, Purdue Univ., 400 Central Dr., West Lafayette, IN 47907-2017, USA*

Abstract. Important aspects of the hydrodynamics and thus, the correct identification of the flow regime could enhance safety and overall performance in multiphase flow systems. Several works on flow regime identification have been carried out in the past. Most of them consist, in a first stage, on measuring certain flow parameters that can be used as good flow regime indicators and, then, developing a flow regime map using these indicators. In this work, a vertical two-phase flow loop facility was used, whereby local conductivity signals were recorded and utilized for the development of an Artificial Neural Network (ANN) based method for the flow regime classification. The experimental database consists of a total number of 125 test cases covering a wide range of situations in the loop working area. Each experiment flow regime was identified by visual inspection, and classified into bubbly (B), cap-bubbly (CB), slug (S), churn turbulent (CT) or annular (A). The bubble chord length cumulative probability function (CPDF), calculated from the measured conductivity signals was selected as flow regime indicator. Different ANN configurations were designed, trained and optimized. The ANN types and the statistical parameters of the CPFD used as inputs were two of the four parameters varied in the ANN optimization process. The temporal length in the conductivity signal used to calculate the CPDF, was also modified during this study. The range of variation of the temporal signal was from 1 to 60 seconds. Together with this temporal variation, the number of training patterns (increasing with decreasing CPDF processing times or not) was also modified. The modification of all these variables allowed the identification of the ANN configuration that better fit the requirements of the specific study of flow regime classification.

Keywords: Flow regime identification, Artificial Neural Network
PACS: 07.05.Mh, 47.55.Ca, 47.55.dd

INTRODUCTION

Many industrial applications handle systems involving two or more phases, as for example nuclear power plant steam generators, boilers, paper-making industry, processing of chemicals for the production of food, pharmaceuticals, petroleum transportation, etc. The two phases can flow according to several topological configurations called flow patterns or flow regimes, which are determined by the interfacial structure between both phases. Despite the wide usage of multiphase systems, their hydrodynamic and kinematic mechanisms are not completely understood and, in addition, are highly dependent on the flow regime. The correct identification and modelling of the flow regime could enhance safety and overall performance in multiphase flow systems. Vertical two-phase flows are usually classified into four or five basic flow regimes. Considering an upward liquid flow in a vertical pipe with constant flow rate, a gradual increment in the gas flow rate produces, consecutively, the following flow patterns : Bubbly flow (designated as B in the following sections), Cap-Bubbly flow (CB), Slug flow (S), Churn-Turbulent flow (CT) and Annular flow (A).

Several works on the objective flow regime identification have been carried out in the past [1; 2; 3; 4; 5; 6]. Most of them consist, in a first stage, on measuring certain flow parameters that can be used as good flow regime indicators and, then, developing a flow regime map using these indicators.

CP914, *Multiphase Flow: The Ultimate Measurement Challenge,* edited by X. Cai, Y. Wu, Z. Huang, S. Wang, and M. Wang

In this work, a vertical two-phase flow loop facility was used, whereby local conductivity signals were recorded and utilized for the development of an ANN-based method for the flow regime classification. Statistical parameters of the bubble chord length cumulative PFD (CPDF) were selected as flow regime indicators and also used as ANN inputs. Different ANN configurations were designed, trained and optimized throughout this work. The quality of the ANN classification results was evaluated under different conductivity signal temporal lengths in order to analyze the potential of this methodology for fast identification purposes.

The results demonstrated the suitability of the developed ANN-based method for flow regime characterization. The ANN fast prediction responses together with the minimized temporal length in the measured conductivity signal, are combined in this study to create a flow regime identification methodology for fast transient as well as for online flow regime diagnosis.

EXPERIMENTAL SETUP

The experiments have been performed in a vertical air-water flow facility. The experimental facility is designed for simulating adiabatic vertical air-water two-phase flows. The test section consists of an acrylic round pipe with internal diameter (D) of 50.8 mm and height of (L) 3.8 m. The liquid flow is adjusted by valves and measured by an electro-magnetic flow meter with an accuracy of 3 %. Air is supplied by an air compressor which provides a maximum pressure of 60 bar. The air flow rate is adjusted by valves and measured by a rotameter with an accuracy of 8 %. A sparger with porous tip of 10 micron pores mean diameter is employed as a bubble generator. The present injection unit produces nearly uniform bubbles of approximately 1-2 mm in diameter at the inlet of the test section. More technical details about the facility can be found in [7].

Conductivity probes are often used to measure many two-phase flow parameters, such as void fraction, interfacial velocity and interfacial area concentration [8; 9; 10]. A double-sensor conductivity probe was installed at the center radial position and at axial location of 3.4 m (L/D=67) after the inlet of test section for measuring the bubble chord length distribution for each flow condition.

EXPERIMENTAL DATABASE

The experimental database in this study was developed under stationary conditions. A total number of 125 test cases were selected in order to cover a wide range of situations in the loop working area. One important feature of the experimental database from the point of view of ANN was the range of variations of the superficial gas and liquid velocity (jg and jf respectively), as the validity of the predictions is restricted for those values. The ranges of variation covered within this experimental study are from 0.0043 to 9.83 m/s for jg and from 0.0314 to 2.5118 for jf.

Flow regimes were identified by visual inspection, which introduces certain subjectivity to this variable. Based on this observations five flow regimes were identified: bubbly, cap-bubbly, slug, churn-turbulent and annular. The number of identified visual flow regimes also imposed the five different outputs chosen during the ANN development. Acquisition frequency and length of time series of the conductivity signals are important factors for the estimation of statistical properties [11]. For each flow regime, data was acquired with a frequency of 12 kHz during 60 seconds using a 12 bit A/D card. The conductivity signals were post-processed in order to obtain the bubble chord length CPFD, used in turn to calculate certain characteristic parameters that were employed as inputs to the ANN.

Bubble Chord Length CPDF Calculation

The first step in the CPDF calculation consist of a signal pre-processing, where, the noise in the raw signals is removed using a median filter and the signals are normalized with respect to maximum and minimum voltage values. Secondly, the bubble residence time for each of the bubbles is obtained from the front sensor signal using a threshold technique. The last step consists on the chord length calculation, sorting and normalization.

The interfacial velocity is required at this point in order to obtain the chord length distribution. In this work, the cross-correlation between the pre-processed signals of the two sensors together with the known distance between the sensors, are used to calculate the interfacial velocity. This approach presents the advantage of fast calculation time while also maintaining a good accuracy [12].

The CPDFs have been calculated by means of a logarithmic scaling sampling method and, in this way, the bubble chord lengths lower and upper limits are fixed and a constant increment in the logarithmic scale is calculated in order to obtain the required number of CPDF vector components [12; 13]. A minimum number of 150 samples for each CPDF vector have been found to obtain good results in flow regimes identification.

However, 60 seconds is a relatively long temporal signal in case of desiring a tool to instantaneously identify flow regimes in transient flow conditions. An important objective to cover along this work was to analyze the minimum temporal length in the signal so that the corresponding CPDF calculated from it maintains the representative aspects of the flow regime.

ANN FLOW IDENTIFICATION

ANN Inputs and Outputs

Fourteen different CPDF sets were calculated varying the temporal length in the conductivity signals. As the temporal length was reduced, the number of available input patterns calculated for a certain test case was increased. The temporal lengths for the CPDF calculation, together with the multiplying factor of the available patterns (nrep), and the number of the resulting maximum available patterns for each case are shown in Table 1.

As expected, the quality of the calculated CPDF decrease when the temporal length used was reduced: the CPDF shapes were steeper and in some cases, the number of bubbles within the reduced conductivity signal was not enough to properly calculate the CPDF.

Using all the 150 components of the bubble chord length CPDF vectors directly as ANN input could result in a redundancy of information, and thus degrade the ANN training. To overcome this problem, two different procedures were used to select the representative characteristics of the CPDF vectors as ANN inputs: principal component analysis (PCA) and 4 indexes representative of the CPDF (4 IND). The goal of this work was to optimize a tool based on ANN that is able to classify the flow regime starting from certain input data derived from conductivity signals. The identification of the flow regime in each test case was performed by visual inspection, which introduces certain subjectivity to this ANN output. Five different flow regimes were chosen for the classification: bubbly (B), cap-bubbly (CB), slug (S), churn-turbulent (CT) and annular (A).

Special problems arose in the visual identification between churn-turbulent and annular flows. Also, few annular cases were included in the experimental matrix, compared to the other flow regimes. In addition, measured annular cases could not be considered as pure annular in the strict sense, but as a churn-turbulent to annular transition flow regime (pure annular flow is considered to be reached at superficial higher than 10 m/s [14; 15]. These features could complicate the correct ANN classification for annular or churn-turbulent test cases.

ANN Fundamentals

One of the main ANN's advantages is their ability to model complex non-linear relationships between multiple input variables and the required outputs. ANNs are also able to self-identify these complicated relationships, generating non-parametric methods which are capable of non-linear mapping. In the case of flow regime identification, where the traditional classifier systems present some difficulties in the classification [16; 6], this ANN advantage implies an enormous simplification. Another important advantage of the ANN approach is its fast response, which enables it to be included in more complex procedures, such as decision trees and optimization applications. More information about the general features and working principles of ANNs can be found in [17; 18; 19; 20].

There are several fields of applications of the ANN: prediction, classification, data filtering, etc. The specific use of a certain ANN determines its structure and therefore its learning rule. In this work, the flow regime classification was performed by Probabilistic Neural Network (PNN). These ANNs belong to the family of the radial basis networks and are trained by supervised learning. PNN are based on the combination of Bayes decision strategy and Panzen´s method of probability density function (PDF) approximation [21; 22; 23].

TABLE 1. Variations in the temporal length used for CPDF calculation and corresponding available patterns.

Temporal length [s]	1	2	3	4	5	6	7	8	9	10	15	20	30	60
nrep	60	30	20	15	12	10	8	7	6	5	4	3	2	1
Available patterns	7500	3750	2500	1875	1500	1250	1000	875	750	625	500	375	250	125

ANN Training and Evaluation

The selection of the optimized ANN configuration was based on the accuracy of results, following a trial and error procedure. In this work, both ANN implementation and training was developed using the neural network toolbox of Matlab [24].

The characteristic parameters of the CPFD used as inputs (PCA and 4IND) and temporal length in the conductivity signal used to calculate the CPDF were two of the parameters varied in the ANN optimization process. The range of variation of the temporal signal was from 1 to 60 seconds, including 12 intermediate levels (as shown in Table 1). Together with this temporal variation, the number of training patterns (increasing with decreasing CPDF processing times or not) was also modified. The nomenclature used was 1rep if the initial number of training patterns is preserved (that is 125 patterns) and nrep if the number or training patterns increase as the temporal processing lengths decrease). Two important problems were dealt during the ANN configuration selection: to avoid overfitting and to compensate the low number of training patterns. Due to the difficulty and the cost associated to experimental measurements, the available measurement database only contained 125 test cases. These test cases, depending on the temporal length selected for the CPDF calculation, could generate a number of training patterns between 125 and 7500. In order to overcome the problematic mentioned previously, an specific training strategy was established in the study. During the training of each specific ANN configuration, the neural network was trained, using the 90% of the available patterns. The training was repeated 50 times, which resulted in 50 different ANNs. The resulting 50 trained ANN were assembled in an ANN committee, in order to compensate for the differences in the individual networks. The final classification results were obtained averaging the results provided by all the ANN that integrate the ANN committee. The quality parameter was calculated in this case over the whole available patterns, thus evaluating the same matrix of patterns for the different ANN configurations. This quality parameter will be referred to as QP_whole_set and was used to compare the results provided by the different ANN.

ANN Results

Figure 1 compares the efficiency of the PNN when the input data are modified. In the left graph of the figure, the evaluation is performed for the case of increasing the number of training pattern (nrep), and in the right graph, the training patterns are fixed when varying temporal length (1rep). Almost the same results are achieved independent of which input data are used. This fact indicates the consistency of the training procedure and also between the two input sets. The general trend is an ANN classifying efficiency reduction with the temporal length in the conductivity signal. This trend is closer to an asymptotic behaviour for the case of nrep training patterns.

In the case of increasing the number of training patterns when the processing time is decreased (nrep), two opposite effects are affecting the ANN. In one hand, the quality of the CPDF decreases and its representativeness of the corresponding flow regime is lower. In the other hand, the number of training patterns increases, which has a positive effect on the PNN efficiency. These two effects compensate each other from 60 seconds to up to 8 seconds, and from this point, the worsening of the CPDF curves is determinant. On the contrary, in the case of maintaining the same number of training patterns for all the temporal length (1rep), only the negative effect of the CPDF worsening remains.

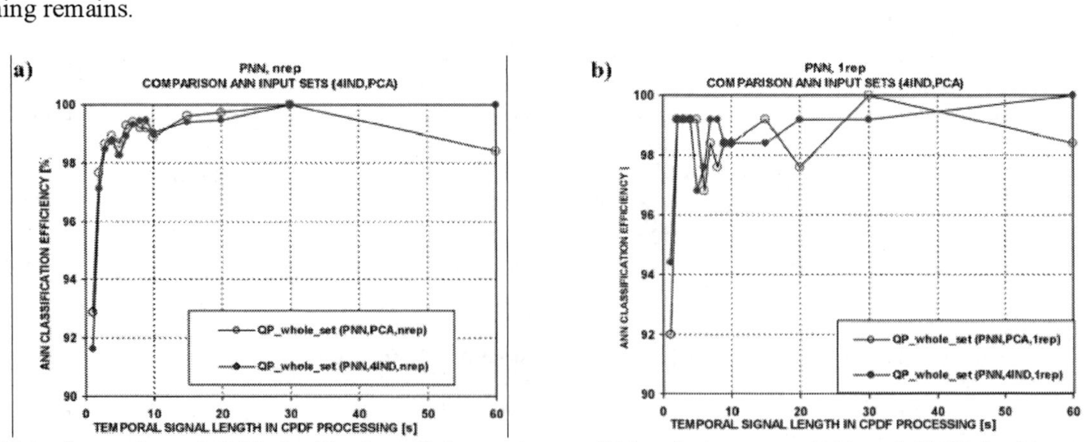

FIGURE 1. Comparison of ANN classification efficiency when modifying the input sets (PCA and 4IND) for (a) nrep training patterns and (b) 1rep training patterns.

As a conclusion of the ANN efficiency results observed along the different configurations it is important to highlight that the difference observed between the two sets of evaluated ANN inputs is not significant and also that the identification results improved when using all the available training patterns (nrep).

The ANN misclassifications are concentrated for most of the cases in adjacent flow regimes. For example, most of the wrongly classified churn turbulent test cases are predicted by the ANN as annular. This is due to the fact that adjacent flow regimes have similar conductivity signals, especially when the temporal lengths are reduced.

CONCLUSIONS

The optimisation of the ANN configuration for the identification of flow regimes in two-phase flow systems was performed. A total number of 125 experimental test cases were obtained using an instrumented test loop and a double sensor conductivity probe. The bubble chord length cumulative PDF was employed as starting information for the ANN inputs. Flow regimes, including bubbly, cap-bubbly, slug, churn-turbulent and annular were identified visually, and used as ANN outputs. The ANN optimisation included the modification of statistical information of CPDF used as ANN inputs (PCA and 4IND) and number of training patterns (nrep and 1rep). The ANN quality was evaluated under the different signal lengths for the CPDF calculation. A specific ANN training methodology was developed in order to compensate the low number of experimental test cases available.

The optimized configuration obtained during this study was PNN architecture, ANN inputs generated by pre-processing with PCA to the CPDF signals, and usage of all the available training parameters (nrep). The selection of the optimized CPDF processing time depends on the specific case: the desired limits in the processing time and in the ANN classification quality should be balanced for each application. The observed results for the ANN optimized configuration demonstrated the suitability of the developed ANN-based method for flow regime identification, as for all the CPDF processing times the ANN quality parameter was higher than 90%.

The ANN misclassifications are concentrated for most of the cases in adjacent flow regimes, and most of them are accumulated for the transition between churn-turbulent and annular. This fact can be explained by the difficult visual classification of these test cases, the similarity between the conductivity signals and maybe the small number of annular test cases. When the CPDF processing time is reduced, the test cases with lowest superficial fluid velocity present high ANN classification failures. The reduction in the probability of bubble hits in the conductivity probe can be responsible of this problem.

The results of this investigation indicate the optimized ANN classifies reasonably well the five different flow regimes starting from the information of a conductivity probe. For 2 seconds in the CPDF processing time, a classification efficiency of 95% was achieved. Future works will be focused in increasing the experimental matrix (including more annular test cases) and also in selecting the conductivity signals in terms of bubble hits instead of a fixed processing times.

ACKNOWLEDGMENTS

This work was performed at Purdue University under the auspices of Mitsubishi Heavy Industries Ltd. through the Institute of Thermal-hydraulics. L. Hernández, J. E. Juliá and S. Chiva gratefully acknowledge the financial support from Fundació Caixa Castelló-Bancaixa (project: P1 1A2003-17) and Ministerio de Educación y Ciencia (project: ENE2004-06978-C02-02).

REFERENCES

1. O.C. Jones and J.M. Delhaye *Int. J. Multiph. Flow* **3**, 389-116 (1976).
2. N.G. Hubbard and A.E. Dukler "The characterization of flow regimes for horizontal two-phase flow" Proc. 1966 Heat Transfer and Fluid Mechanics Institute, Stanford University Press, CA, 1966, pp. 100-121.
3. G. Matsui G *Int. J. Multiph. Flow* **10**, 711-720 (1984).
4. S. Cai , H. Toral, J. Qiu and S. Archer *Can. J. Chem. Eng.* **72**, 440-445 (1994).
5. O.C. Jones and N. Zuber *Int. J. Multiph. Flow* **3**, 273-306 (1975).
6. Y. Mi, M. Ishii and L.H. Tsoukalas *Nucl. Eng. Des.* **184**, 409-420 (1998).
7. H. Goda H, T. Hibiki, S. Kim, M. Ishii and J. Uhle *Int. J. Heat Mass Transf.* **47**, 4101-4118 (2004).
8. I. Kataoka, M. Ishii and A. Serizawa *Int. J. Multiph. Flow* **12**, 505-529 (1986).
9. Q. Wu and M. Ishii *Int. J. Multiph. Flow* **25**, 445-453 (1999).
10. S. Kim, X.Y. Fu, X. Wang and M. Ishii *Int. J. Heat Mass Transf.* **43**, 4101-4118 (2000).

11. T. Xie "Hydrodynamic characteristics of gas/liquid/fiber three-phase flows based on objective and minimally-intrusive pressure fluctuation measurements", Ph.D. Thesis, Georgia Institute of Technology, 2004.
12. J.E. Juliá, S. Paranjabe, Y. Liu and M. Ishii Objective fast local flow regime identification using conductivity probe and neural network techniques Proc. in the 11th Int. Topical Meeting on Nuclear Reactor Thermal-Hydraulics (NURETH-11) Avignon, France, 2005.
13. J.E. Juliá, S. Paranjape, Y. Liu and M. Ishii Local flow regime identification using conductivity probe and neuronal network techniques. Submitted to Exp. Thermal Fluid Science, 2005.
14. T. Taitel, D. Barnea and A.E. Dukler *AICHE. J.* **26**, 345-354 (1980).
15. K. Mishima and M. Ishii *Int. J. Heat Mass Transf.* **27**, 723-737 (1984).
16. K. Sekoguchi, K. Inoue and T. Imasaka *JSME Int. J.* **30**, 1266-1273 (1987).
17. S. Haykin *Neural Networks: a Comprehensive Foundation* , London: Macmillan, 1999.
18. J.M. Zurada *Introduction to Artificial Neural Systems*, Boston: PWS Publishing Company, 1992.
19. A.J. Maren, C.T. Herston and R.M. Pap *Handbook of Neural Computing Applications,* Orlando: Academic Press, 1990.
20. P.D. Wasserman *Advanced Methods in Neural Computing*, New York: Van Nostrand Reinhold, 1993.
21. D.F. Specht *Neural Netw.* **3**, 109-118 (1990).
22. D.F. Specht *Proc. of IEEE Int. Conf. on Neural Networks* **1**, 527-530 (1988).
23. E. Parzen *Annals of Mathematical Statisics* **3**, 1065-1076 (1962).
24. H. Demuth and M. Beale *Neural Network Toolbox for use with Matlab: user's guide version 3*, Massachusetts: The Mathworks Inc., 1998.

Influence of Blocking Flow on Flow Characteristic of Electronic Expansion Valve

Caihua Liang*, Xiaosong Zhang*, Xiuwei Li*

*School of Energy and Environment, Southeast University, Nanjing, 210096, Jiangsu, PRC

Abstract. We study series of flow regime of the refrigerant in electronic expansion valve and make analysis on the flash evaporation and the cavitation erosion phenomenon. Our result shows that the flow of refrigerant in electronic expansion valve is blocking flow. We present the influence of blocking flow on flow rate characteristic of electronic expansion valve and analyze the flow characteristic. We developed a sample machine to conduct some experiments on the valve under the experimental working condition, and the results of the experiments shows that the particular flow rate characteristic of the electronic expansion valve under experimental condition, and found the flow is blocking flow. We present the curves of the flow rate versus the open degree of the valve. The curves expose that there is a linear relationship between the flow rate and the open degree of the valve if the open degree is under a certain value When the open degree is above this value, it will not present the linear relationship any longer, and the current flow rate of the valve is less than that acquired from the theoretical calculation. The difference between the two will be greater as the open degree increases, and the flow rate of the refrigerant will finally not change as the open degree is changed.

Keywords: blocking flow, electronic expansion valve, flow characteristic, refrigeration

PACS: 47.61.Jd Multiphase flows

INTRODUCTION

People pay more attention to comfort and the energy-saving capability of air-conditioner. Thermostatic expansion valve is a wide used throttle component in the field of refrigeration and air condition. However it doesn't fit very well in some occasions, as it has the problem like narrow range of regulation, low precision, slow response and big lag because of the structure and the principle of operation itself. Compared to the thermostatic expansion valve, electronic expansion valve owns the advantage as wide range of regulation, high precision, quick response and wide range of adaptation. Therefore, it has attracted people's great attention and becomes a focus of research in that field. To make better use of electronic expansion valve and fit it well with the refrigeration system, it's important to know more about the flow regime and the flow rate characteristic of electronic expansion valve, since they play a key role in both the design and the application of the valve.

Lately, there's more research abroad about the improvement of the performance of the system using electronic expansion valve [1~3], but less study on the flow characteristic and the controlling method of the valve [4]. The research domestic about the expansion valve starts late and there's few research report about the flow characteristic of electronic expansion valve. Most work concentrate on the acquiring of the pressure, flow rate characteristic and flow rate-opening degree relationship through experimental methods. Based on the hydraulics formula of the throttle valve, Wenbing Weng et al conducted the experimental method to study the influence of the factors on the flow rate of the valve, such as the opening-degree, the pressure-difference before and after the throttle, the pressure ratio, the degree of dryness at the entrance and the pressure before and after the throttle [5]. However, it neglect the fact that the throttle of refrigerant in the electronic expansion valve is a complicated process as the phase transition exists, and that lead to the invalidation of applying the normal the hydraulics formula to describe the flow process and expose the flow characteristic. Founding the physical model of the two phase-flow of the refrigerant in valve, Zhigang Wang et al exposed the pressure field, the velocity field of two phases and the opening degree distributing of the throttle process [6]. But all the above haven't exposed the special flow regime and the flow characteristic caused by the phase transition of the refrigerant during the throttle process. In these studies, it hasn't considered the

CP914, *Multiphase Flow: The Ultimate Measurement Challenge*, edited by X. Cai, Y. Wu, Z. Huang, S. Wang, and M. Wang electronic expansion valve and the refrigerant system as integration under local operation condition, which is necessary to make thorough research on the function of the valve. On that purpose, we analyze the flow regime of the refrigerant in the electronic expansion valve, and expose the flow characteristic of the refrigerant during the throttle process. Conducting experiments on the flow characteristic of the electronic expansion valve under local operation condition, we get the special flow characteristic of the valve while the capacity of the valve doesn't match the system perfectly. Also, we propose some improvements after we analyze the influence of this flow characteristic on the optimization of the electronic expansion valve.

FLOW REGIME ANALYSIS OF THE ELECTRONIC EXPANSION VALVE

Electronic valve takes the effect of throttling and pressure dropping in a refrigeration system, functioning as a throttling organ whose local resistance is variable. Since the process is rapid, we consider equivalently the electronic expansion valves with different opening degree as a series of orifice with different diameters.

The velocity and pressure distribution curves of normal liquid flowing through the orifice are showed in Fig.1.The static pressure of the liquid flowing upstream the orifice is provided as p_1, and $p_1 > p_v$ (p_v is the vapor pressure of the liquid, which is the saturation vapor pressure corresponding to the temperature of the liquid). According to the first law of the thermodynamics, if there's no energy loss because of heat loss or work output at the boundary, the overall energy level of each point remains constant along the certain datum plane of the system. The liquid contracts before the orifice so that it can flow through the orifice. And that lead to the increase of the flow velocity of the liquid, as the velocity is in inverse ratio to the liquid area. Since the sum of the velocity head and the pressure head maintain the same, the pressure head drops down while the velocity head rises up. Downstream the neighbor orifice, it comes to the liquid its minimum area, maximum velocity and the minimum pressure (p_{vc}). If the velocity of the liquid at this position increases sufficiently, the pressure may drop below the saturation vapor pressure, which would cause that liquid flowing through the valve partly boil away, and the bubble appears, too. Downstream away from the position where liquid contracts, the velocity decrease caused by friction resistance, which result in the resurging of the pressure. The liquid would remain boiling away if the resurging pressure still lower than p_v.. In this situation, the gas and liquid phase both exist, and we call this phenomenon flash evaporation. As the flash evaporation phenomenon appearing, the pressure of the liquid after the valve doesn't remain below p_v, but rises sharply above p_v, That will cause the bubble crack and form shock wave with high pressure, and we call this phenomenon cavitation erosion.

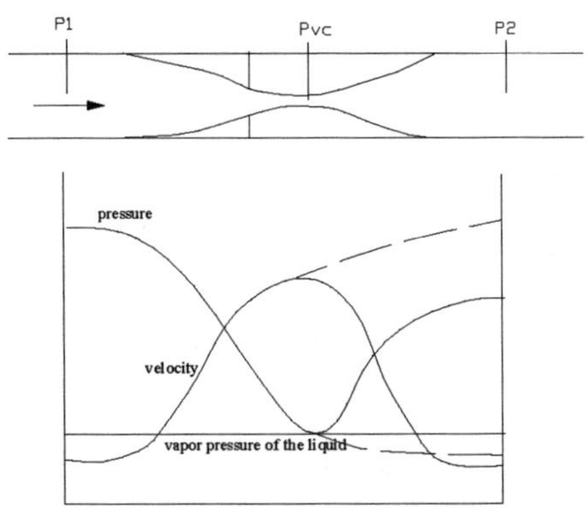

FIGURE 1. Analysis of normal liquid flow regime

Judge of the flash evaporation and the cavitation erosion

The flash evaporation and the cavitation erosion are two interrelated different stages after the liquid boils away. The difference lies on the fact whether the bubble cracks, and the cavitation erosion has a greater degree of damage than the flash evaporation. The judging principal for distinguishing them is as follows [7]:

(1) As the pressure difference of the two ends of the valve: $\Delta p < \Delta p_C$ (Δp_C is the initial flash evaporation pressure difference), there are no flash evaporation and the cavitation erosion.

$$\Delta p_c = K_c(p_1 - p_v) \quad (1)$$

The K_C is the initial coefficient (it could be acquired from the manufacturer by experiment)

(2) While the pressure difference Δp becomes larger and $\Delta p_C \leq \Delta p < \Delta p_S$, the flash evaporation phenomenon appears.

Δp_S is the blocking flow, which means the pressure of the two ends of the valve when the liquid partly boils with the dropping down of the pressure at the contracting position of the valve, and the flow rate of the valve doesn't increase as the pressure difference increasing. And the calculation of the pressure difference is as following formulas:

$$\Delta p_s = F_l^2 (p_1 - F_f p_v) \tag{2}$$

$$F_f = 0.96 - 0.28\sqrt{p_v / p_{cr}} \tag{3}$$

F_l is the restitution coefficient of the pressure of the liquid; F_f is the coefficient of the critical pressure ratio, p_{cr} is the thermodynamics critical pressure.

(3) While $\Delta p \geq \Delta p_S$ and $p_2 \leq p_v$, the flash evaporation comes to be more evident, but there's still no cavitation erosion.

(4) While $\Delta p \geq \Delta p_S$ and $p_2 > p_v$, the cavitation erosion phenomenon appears.

Judge of the blocking flow

In practical design, the checking computation depends on the value of Δp_S. The Δp_S is related to F_l, and the F_l could be acquired by experiment when the valve opens completely. The value of F_l counts on the shape of the valve core, the structure of the valve body and the direction of the liquid flowing through the valve, but has no relationship with the kind of the medium.

It can be seen obviously from the pressure-enthalpy chart that the flash evaporation phenomenon exists as refrigerant flowing through the valve. However, the particular of the flow regime is unknown.

We judge the flow regime of expansion valve using the representative case of the wind- cooling chilling/heating-water unit, which operated steadily under standard working condition of air-condition. Under the standard working condition, the ambient temperature is 35℃, the temperature of the water out of the air-conditioner is 7℃, provided the evaporating temperature is 3℃, the correspondent evaporating pressure is 0.548Mpa, and the condensing temperature is 47℃, the correspondent condensing pressure is 1.812Mpa. The overcooling degree is 5℃ while the correspondent saturation vapor pressure of the refrigerant of this temperature is 1.610Mpa. We can acquire the thermodynamics critical pressure of the R22 as p_{cr}=4.97Mpa. The value of F_l is set as 0.9 according to the structure of the valve. Applying theses parameters in the above formulas, we can obtain that: Δp_S=0.424Mpa; and we have the pressure difference in and out of the valve: $\Delta p = p_1 - p_3$=1.264Mpa. Since p_V=1.610MPa, p_3= 0.548Mpa and $\Delta p \geq \Delta p_S$, $p_2 \leq p_v$, we judge the flow regime of the refrigerant flowing through the electronic expansion valve is blocking flow.

Analysis of flow rate of flash evaporation and blocking flow

When the open degree of the regulating valve remains unchanged, the relationship between the flow rate of the valve and the square pressure difference of the valve on two ends is exposed as Fig.2. It could be seen from Fig.2 that q_v and $\sqrt{\Delta P}$ presents linear relationship within a certain range. As the Δp increases to a certain value, the relationship between the flow rate and $\sqrt{\Delta P}$ will not be linear any longer, and liquid partly flash evaporates. If the pressure at the entrance of the valve drops lower, the amount of the boiling-off liquid increases while the intension of the flash evaporation increases, and the flow capacity decreases further. If the pressure drop is large enough, the flow will be thoroughly blocked so that there will be no increase of the flow rate even the pressure drop further. We call this blocking flow regime. Usually in a flow rate control situation, we always avoid the valve operating while the flow regime is a blocking flow. However, the electronic expansion valve should operate while blocking flow regime when the valve is used as a throttle organ in a refrigeration system. And to a valve, which has a good linear flow rate regulating capability, there will be no proportion relationship between the flow capacity and the flow rate. And the linearization of the open degree and the flow rate of the valve don't mean the linearization of the flow area and the flow rate. When the refrigeration system operates steadily under certain working condition, the pressures in and out the electronic expansion valve depend on the evaporating temperature and the condensing temperature and the difference of the two pressures remain constant.

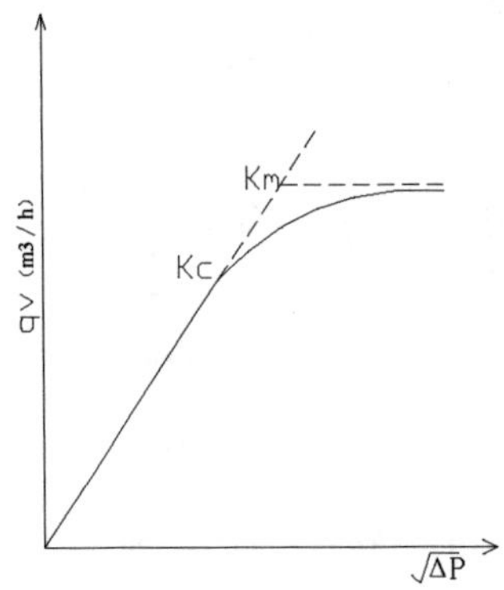

FIGURE 2. Relationship between q_v and $\sqrt{\Delta P}$

During the regulating process of the flow rate, we mainly alternate the open degree to change the flow rate of the refrigeration system, while the pressure difference keeps steady. As the diversity of the refrigeration system, it is impossible that the flow rate of the valve fit the capacity of the system completely under declared working condition. We always consider the flow capacity of the valve (the flow rate when the valve completely open under declared working condition) bigger than the capacity of the system. Therefore, there comes the question that is there still a linear relationship when the declared capacity of the valve is bigger than the system. This is important for an accurate controlling of the valve, and we make research on that through experiments.

EXPERIMENTAL STUDY

The introduction of the experimental system and the electronic expansion valve

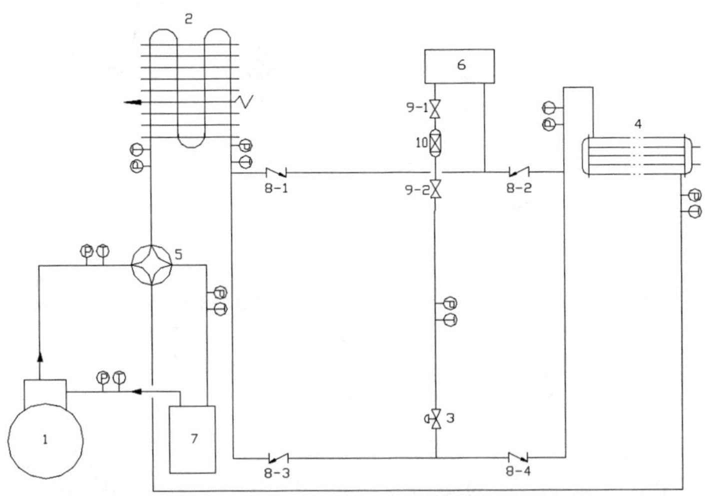

FIGURE 3. Flow chart of the experimental system

1.Screw compressor 2.Heat exchanger at air side 3.Electronic expansion valve 4.Heat exchanger at water side 5. Four-way valve 6.Liquid storage tank 7.Gas-liquid separator 8-1、8-2、8-3、8-4.One-way valve 9-1、9-2.Ball valve 10.Fliter-dryer

The experimental system for studying the flow rate of the electronic expansion valve is described in Fig.3. From Fig.3, we can find that this system is similar to the normal screw heat-pump system of air-source. What is different was that the traditional throttle organ has been replaced by an electronic expansion valve. And there's an additional testing system for temperature and pressure, which is for the purpose to know more about the character when the system operates under steady working condition、starting-up condition and the varied working condition. The locations for temperature and pressure measurement are set at the entrance and the exit of the compressor, the condenser, and the evaporator, too. The location for pressure measurement is of the neighbor to that correspondent temperature measuring position. But there's only pressure measuring at the entrance of the evaporator because of the unsteady temperature there. With these parameters, we can study the operation of the system. And we adopted temperature transmitters of Pt resistance for temperature measurement, and the pressure transmitters for pressure measurement. The signal of the temperature and the pressure is transformed to electronic signal range from 4~20mA.Then it is converted back to the signal of temperature and pressure of that moment by engaging the data acquisition equipment and the computer, and the curve was recorded at that the same time.

The declared capacity of the units is designed to be 160 kW. The electronic expansion valve is selected the EX7-U21 type from company ALCO. It adopts 24V power supply, and a bi-phase bipolar stepper electromotor for driving; the driving frequency is 330 pulses per second while the steps of the valve from completely open to completely close are 1600 pulses. The valve has a capacity of 330kW when it is completely open under declared working condition. According to the curve of the flow rate and the open degree, we can found that the declared open degree is 900 pulses while the capacity of the system is 160kW. The overall performance test of the whole system is taken on a testing system, which is described in detail in another article of the authors whose title is "Experimental Study on Electronic Expansion Valve Appling in a Chilling-Water Unit" [8].

Analysis of the experimental result

We done the experiment, and the experimental condition is set as the standard air-conditioner working condition, i.e. the ambient temperature is 35℃, the temperature of the water in the unit is 12℃, and the water out is 7℃。
We used the electronic expansion valve controller to set the maximum open degree allowed of the valve (The flow rate will not increase even it is not enough if this open degree is reached) as 1000 pulses, 1200 pulses, 1400 pulses and 1600 pulses for the experiment. Actually, in the case of 1000 pulses, we started the system for the experimental operation, and when the system operate steadily we recorded the evaporating and condensing pressure, capacity and the open degree that is displayed on the screen of the controller. The result is showed in Table.1.From Table.1, we could found that under different working condition, the system could operate steadily by setting different maximum open degree allowed of the valve. And there's little difference in these situations of the evaporating pressures (counts on the cooling medium) and the condensing pressures (counts on the ambient temperature) as they depend on the working condition. And there is little difference of the respective capacity, i.e. little difference of the respective flow rate. However, there's big difference of the open degree according to the value of the open degree while the valve operate steadily. The capacity acquired by considering the relationship of the flow rate and the open degree under these working conditions is much bigger than the practical capacity. This is not fit the open-degree and flow rate curve provided by the manufacturer, but why?

TABLE.1 Experimental data of different maximum open degree allowed

Maximum open degree allowed (pulses)	Evaporating pressure (MPa)	Condensing pressure (MPa)	Local capacity (kW)	Set open degree (pulses)	Theory capacity (kW)
1000	0.545	1.809	155.4	985	190
1200	0.548	1.812	157.7	1187	235
1400	0.549	1.818	160.5	1347	275
1600	0.553	1.823	162.3	1536	310

We have judged the flow regime of the valve blocking flow; the relationship between the flow rate and the pressure difference in and out of the valve is exposed in Fig.4. By experiment, we can obtain the relationship between the flow rate and the open degree under constant pressure difference.

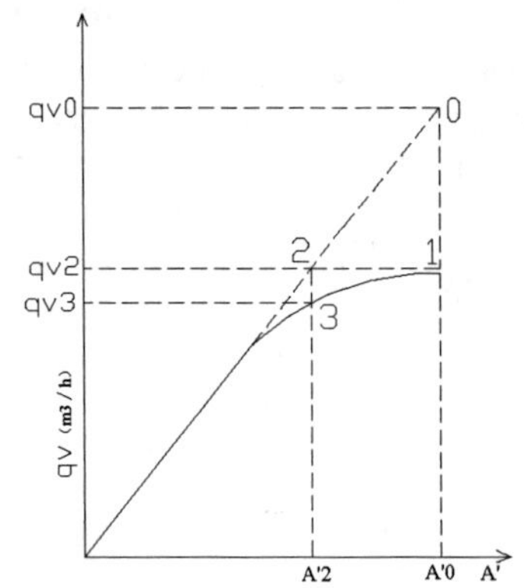

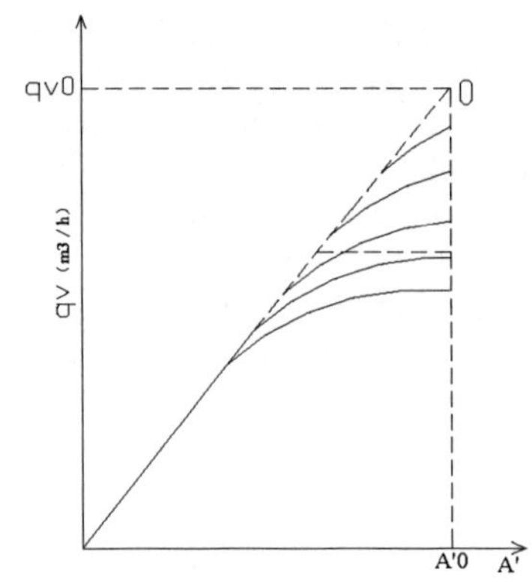

(a) Valve in a certain system (b) Valve in a range of systems

FIGURE4. Relationship between the flow rate and the open degree

Abscissa in Fig.4 is the open degree while the vertical axis is the flow rate. Found from the Fig.4 (a), if the valve and the working condition are set unchanged, i.e. the pressure in and out of the valve is constant. There's a linear relationship between the flow rate and the open degree when the open degree below a certain value (This value related to how the valve fit the whole system). As the open degree exceed this value, the linear relationship will not exist any more; the flow rate of the valve is less than the value we got by calculating theoretically, and the flow rate of the refrigerant will remain unchanged finally. A′0 stands for the maximum of the valve, qv0 stands for the maximum degree calculating theoretically. Since there's no continuum of the flow standard, the maximum declared flow rate is always no smaller than the capacity of the system. qv2 means the maximum capacity designed, the correspondent open degree is A′2 acquired from theoretically calculating. As a matter of fact, when the open degree is A′2, the practical capacity is qv3, and qv3 is smaller than qv2; the difference between the two capacity will rise with the increasing of the open degree.Fig.4 (b) exposes the relationship between the flow rate and the open degree as applying the valve in different system. It could be noticed that the nonlinear range between the flow rate and the open degree shrink as the capacity designed being closer to the declared maximum capacity, which means the valve fit better to the system. If the valve fit the system completely, there will be a completely linear relationship, which is just what the manufacturer provided. Also from Fig.4, $\triangle$A, the increase of the open degree in the linear range differs from the $\triangle$q, which is the flow rate in nonlinear range. Therefore, the nonlinear range brings trouble to the controlling. And we should select the valve whose capacity match the capacity of the system better; or we'll adjust the controlling strategy according to different situation, optimize the arithmetic and improve the effect to make best use a electronic expansion valve.

CONCLUSIONS

(1) We analyze varies of the flow of the liquid flowing through a valve. We judge the flow regime and expose the flow characteristic of the refrigerant.
(2) We study the particular flow characteristic of the liquid flowing through the electronic expansion valve under the practical operation condition in conjunction with the situation that the capacity of the valve is hard to fit the system. And that create the opportunity to optimize the valve.
(3) We got the curves displayed the relationship between the flow rate and the open degree through experiment. While the blocking flow comes out and the pressure in and out of the valve remain constant.
(4) We propose the measures to solve the problems, which are aroused by the particular flow characteristic due to the blocking flow.

ACKNOWLEDGEMENTS

This research is granted by the Natural Science Foundation of China (Contract no: 50376052); and supported by Beijing Municipal Key Laboratory of Heating, Gas Supply, Ventilating and Air Conditioning Engineering.

REFERENCE

1. Aprea Ciro ,Rita Mastrullo. *Applied Thermal Engineering*, **22,** 205-218(2002).
2. Choi J.M., Kim Y..C. *Energy*, **28**, 141-155(2003).
3. Aprea Ciro ,Rita Mastrullo. *Applied Thermal Engineering*, **22**, 161-171(2002).
4. Choi J.M., Kim Y..C. *Energy*, **27**, 391-404(2002).
5. Weng Wenbing, Wang Jinzhu, Jiang Nengzhao. *Fluid Machinery*, **26**, 58-61(1998).
6. Wang Zhigang, Wang Dongqing, Liu Chengwen,et al. *Fluid Machinery*, **27**, 51-55(1999).
7. Liu Hongxia. *Jilin Electric Power*, **149**, 20-22(2000).
8. Liang Caihua, Zhang Xiaosong. *Fluid Machinery*, **32**, 5-8(2004).

Application of a Capacitance Solid Mass Flow Meter in a Dense Phase Pneumatic Conveying System of Pulverized Coal

Xiaolei Guo, Zhenghua Dai, Xin Gong and Zunhong Yu

Institute of Clean Coal Technology, East China University of Science and Technology, 200237 Shanghai, China

Abstract. Capacitance measurement is one of the available methods for measuring the solid mass flow rate in the field of gas-solid flow. Performance of a capacitance solid mass flow meter was tested in a dense phase pneumatic conveying system of pulverized coal. Results have demonstrated that the deviations of mass flow rates in the experimental set-up are almost within ±5% for stable flow patterns while unstable flow patterns and low conveying concentration pose great measurement deviations. On the other hand, the measurement accuracies in a vertical downward flow pipeline and a horizontal pipeline are almost identical and the both are slightly better than that in a vertical upward flow pipeline. Additionally, its performances in a pilot-scale entrained-flow gasification plant with dry feeding are also presented and satisfying results are achieved.

Key words: Capacitance solid mass flow meter; Mass flow rate measurement of solids; Dense phase pneumatic conveying of pulverized coal

INTRODUCTION

Pneumatic conveying technology of pulverized coal is becoming increasingly important in many industries such as gasification, electrical power generation and steel production. To ensure a reliable pneumatic transport operation, online continuous measurement of the mass flow rate of pulverized coal in a pipeline is a common requirement. Although many sensors and instruments based upon different principles have been proposed, few are currently operating in industry [1-5].

With the advantages of the simple structure, low cost, free of maintenance, noninterference of the flow field and a direct electric signal output, solid flow meter employing capacitive sensors was well researched and has been marketed in recent years [5,6], but performances of the capacitance solid flow meter with pulverized coal are few reported. Hu et al [7] employed a novel capacitive system for concentration measurement in a vertical ceramic tube of diameter 100 mm with an upward flow direction and pulverized fuel was used as the solid phase. Their results demonstrated that the system was capable of achieving linearity within ±5% at different temperatures or moisture content. A capacitive solid flow meter developed by Amano et al [5] was tested on a 25 mm vertical tuyere in a pulverized fuel injection system. The achieved measurement error was stated to be within ±5% over the velocity range 5-20m/s. Amano et al also noticed that the measurement accuracy depended upon solids distribution and velocity profiles due to the intrinsically non-uniform spatial sensitivity of the sensors. Mathur et al [6] tested a capacitive mass flow meter manufactured by Auburn International in a 25.4 mm I.D. test loop circulating pulverized coal. Results indicated that variations in the particle size distribution of coal between 30% and 70%—200 mesh and in coal moisture content up to 4.6% had no apparent effects on the flow meter performance. Deviation of data points from the linear regression line was around ±5% to ±15% over the velocity range 5-20m/s.

In this paper, a capacitance mass flow meter manufactured by Ramsey Electronics was tested in a dense phase pneumatic conveying system of pulverized coal so that some instructions may be presented for its application in the field of the basic research of gas-solid flow and online measurement in industry.

CP914, *Multiphase Flow: The Ultimate Measurement Challenge,* edited by X. Cai, Y. Wu, Z. Huang, S. Wang, and M. Wang

EXPERIMENTS

The Capacitance Solid Mass Flow Meter

The solid mass flow meter consists of a capacitive velocity sensor DK13, a capacitive concentration sensor DC13 and a correlator DMK270. The picture of DK13 and DC13 is shown in Fig.1. The concentration sensor DC13 is based upon a simple principle that the presence of solids within a capacitive sensing volume will increase the measured capacitance because of the increased relative permittivity when referenced to gas. This change in the capacitance can be converted into a suitable electrical signal that can be processed by the correlator DMK270. It should especially be pointed out that the concentration sensor DC13 must be calibrated before its application.

The DK13 employs the cross correlation methods to measure the solids velocity in a pipeline [9]. Two identical sensors are installed an axial distance (L) apart from each other. The transit time (T) taken by the particles moving from the upstream sensor to the downstream sensor is measured by cross correlating the two signals using the correlator DMK270. The solids velocity (U_s) is then calculated according to equation (1) from the known sensor spacing L and the transit time (T):

$$U_s = \frac{L}{T} \tag{1}$$

Thus the solid mass flow rate Q_s can be deduced according to the following equation (2) [9] where C is the concentration of solids and A_{sensor} is the area over the sensor's cross section.

$$Q_s = C \times U_s \times A_{sensor} \tag{2}$$

FIGURE 1. Picture of DC13 and DK13 and its installation example

Experimental Set-up

Figure 2 shows the experimental set-up schematically. The test section is a stainless steel pipeline with 20mm inside diameter, and comprises three sections: a riser of length 6.6m, a vertical downcomer of length 5.0m, and a horizontal pipe of length 6.0m. The mass flow meter may be installed in the upward section, downward section, and horizontal section according to test needs, respectively. The measurement positions of the flow meter are also shown in Fig2. When a horizontal conveying test was needed, the riser and downcomer (shown with broken line) were removed.

Load cells are mounted on the receiving hopper to measure the actual mass flow rate. A pressure transducer is used to measure the pressure at the position of the flow meter. The signals of solids velocity and concentration as well as the pressure transducer and load cells readings are all sampled at 1Hz through a data acquisition card and then processed by a computer. The tests were performed with pulverized coal and dried air with superficial gas velocities U_g of 4-15 m/s and solids/gas ratios μ of 50-500 kg/m^3. A feed hopper was operated at a maximum pressure of 0.4 MPa while a receiving hopper was always kept at an atmosphere pressure. The physical properties of pulverized coal are shown in Table 1.

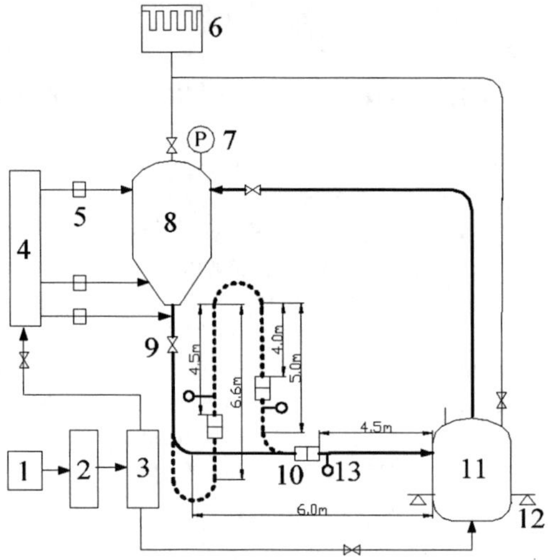

FIGURE 2. Experimental set-up of dense phase pneumatic conveying system
1.Air compressor; 2.Surge tank; 3.Freeze dryer; 4.Air distributor; 5.Gas flow meter 6. Bag dust collector; 7.Pressure gauge; 8.Feed hopper; 9.Ball valve; 10.Solid mass flow meter; 11.Receiving hopper, 12.Load cells. 13.Pressure transducer

TABLE 1. Physical properties of the pulverized coal

Average particle size µm	Bulk density $kg \cdot m^{-3}$	Real density $kg \cdot m^{-3}$	Moisture content %	Angle of repose °
60.67	536	1390	0.63	49.7

The solids/gas ratio μ is obtained by equation (3) where Q_a is the actual solid flow rate obtained from the load cells, V_g is the actual volume flow rate of air through the flow meter. To evaluate the flow meter's performance, a deviation E can be given by equation (4) where Q_m is the average mass flow rate measured by the solid mass flow meter.

$$\mu = \frac{Q_a}{V_g} \tag{3}$$

$$E = \frac{Q_m - Q_a}{Q_a} \times 100\% \tag{4}$$

Calibration procedure of the concentration sensor DC13 is provided as followed: to start up the calibration program of the correlator under a stable gas-solids flowing; at the same time, record an initial reading of the load cells. After several minutes, stop the calibration program and record the reading again. And then input the difference of the load cells' readings into the correlator. Thus a calibration point for this conveying concentration is restored. The solid mass flow meter can restore six calibration points in the correlator corresponding to six different conveying concentrations.

RESULTS AND DISCUSSION

Effects Of Flow Stability On The Measurement

Depending upon the characteristics of solids, gas velocity, and such factors as pipe roughness, pipe size, etc., gas-solids flow patterns can vary from being stable to unstable or an intermediate unstable/stable regime [10]. Generally, with the superficial gas velocity decreases, a transition of flow pattern occurs gradually from a stable fully suspended flow to a strand type conveying, even to an unstable slug flow [11,12]. For stable flows in the horizontal pipeline, typical measuring results of the flow meter are shown in Fig.3 at a superficial gas velocity of 6.9m/s and a solids/gas ratio of 269kg/m^3, with the actual mass flow rate Q_a of 2088kg/h. At the same time, the mass flow rate Q_m measured by the flow meter is 2130kg/h, with a solids velocity of 7.15m/s and a concentration of 263.6kg/m^3. Thus a small mass flow rate deviation E of 2% is achieved. On the other hand, Fig.4 shows the measurement characteristics of an unstable flow in the horizontal pipeline. From Fig.4, it can be seen that the measured values such as the mass flow rate and velocity all fluctuate greatly, and a great deviation E occurs. Thus the unstable flow patterns like slug flow are not suitable to measure by the flow meter; moreover, they are also undesirable in such industries as coal gasification, power generation and steel production.

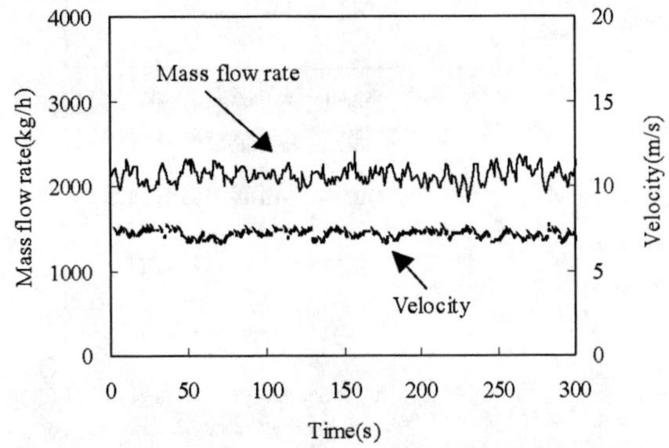

FIGURE 3. Measurement results in a stable flow pattern

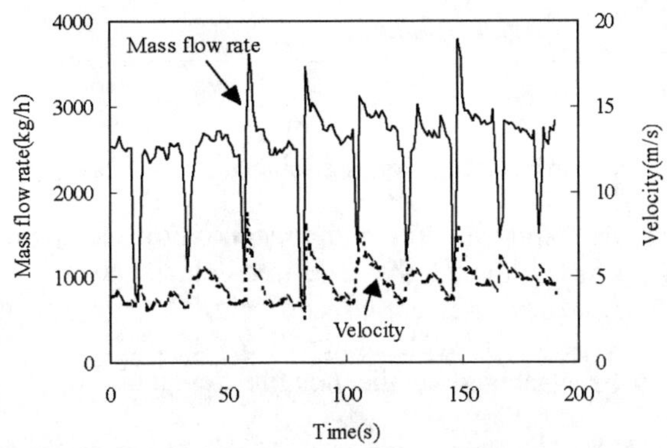

FIGURE 4. Measurement results in an unstable flow pattern

Effects Of Solids/Gas Ratio On The Measurement

With the variations of solids/gas ratio, the distribution of solids over the pipe's cross section can be highly inhomogeneous, which poses some difficulties for the measurement [13]. To reduce the effects of inhomogeneous distribution of solids on the measurement, multiple electrodes are used in the flow meter to produce a rotating sensing field so as to increase the homogeneity of sensing field. Figure 5 shows the effects of solids/gas ratio on the measurement accuracy at the actual mass flow rates of 600-4200 kg/h in the horizontal pipeline, with the corresponding superficial gas velocities of 4-15m/s and the solid/gas ratios of 60-500 kg/m^3. It can be seen that when the solids/gas ratio is more than about 120 kg/m^3, the deviations of mass flow rate are almost within $\pm$5%. However, if the solids/gas ratio decreases to less than 120 kg/m^3, the measured solids mass flow rates are markedly more than the actual values, which can be explained by Fig.6.

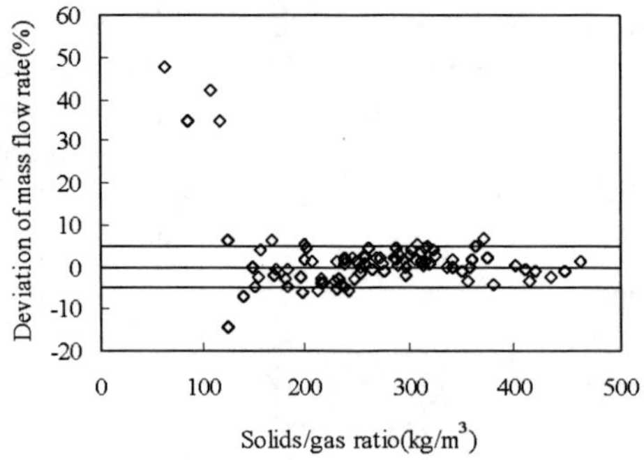

FIGURE 5. Effects of solids/gas ratio on the deviation of mass flow rate in horizontal pipeline

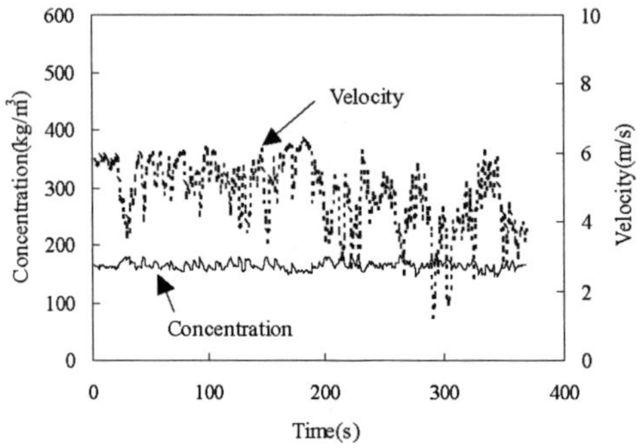

FIGURE 6. Measurement results of solids concentration and velocity at a solids/gas ratio of 64 kg/m^3 in horizontal pipeline

As an example, Fig.6 presents the solids velocity and concentration measured at a solids/gas ratio of 64 kg/m^3 and a superficial gas velocity of 8.4 m/s. Although the measured velocity fluctuates severely, a relatively stable flow pattern can be judged from the stable measurement concentration. On the other hand, the average concentration value of 164kg/m^3 measured by the meter is much higher than the solids/gas ratio of 64 kg/m^3. Thus the notable deviation of mass flow rate can be attributed to the notable deviation of the measured concentration and the abnormal fluctuation of velocity.

As stated in the instruction book of the instrument [9], it would be applicable only to flow conditions under which the conveying concentration in the pipeline is greater than 30kg/m^3; but in the horizontal pipeline, such low concentration as the solids/gas ratios of less than 120kg/m^3 can lead to an abnormal fluctuation of the measured velocity as well as a notable deviation of mass flow rate. Thus the suitable solids/gas ratios are required for the instrument's application.

Effects Of Flow Orientation On The Measurement

The distribution of solids over the cross section of a pipe is closely related to the flow direction. In the vertical upward flow pipeline, tests were performed at superficial gas velocities of 6-12 m/s and solids/gas ratios of 200-400 kg/m^3. As shown in Fig.7, the deviations of mass flow rate are all within $\pm$5% when the solids/gas ratios are less than about 340kg/m^3. However, unstable flows were found when the solids/gas ratios are beyond about 340kg/m^3 and the data points have been marked in Fig.7. The deviation of the mass flow rate grows up rapidly because of the severe fluctuations of the measured parameters. Typical measurement results of the unstable flows are shown in Fig.8 with a mass flow rate deviation of 21.5% when the solids/gas ratio and superficial gas velocity are 371kg/m^3 and 6.7m/s, respectively. On the other hand, the good sensitivity to the flow stability, exhibited by this meter, also implies that it is a suitable instrumentation for the basic research of gas-solids flow.

Tests in the downward flow orientation were performed at superficial gas velocities of 6-15m/s and solids/gas ratios of 150-300 kg/m^3, and the mass flow rate ranges from 1800 to 3700 kg/h. Just as pointed out in the instruction book [9], with the mass flow rate deviations of less than 5%, it is the best flow orientation for the measurement and no unstable flows are found within the ranges studied in this paper.

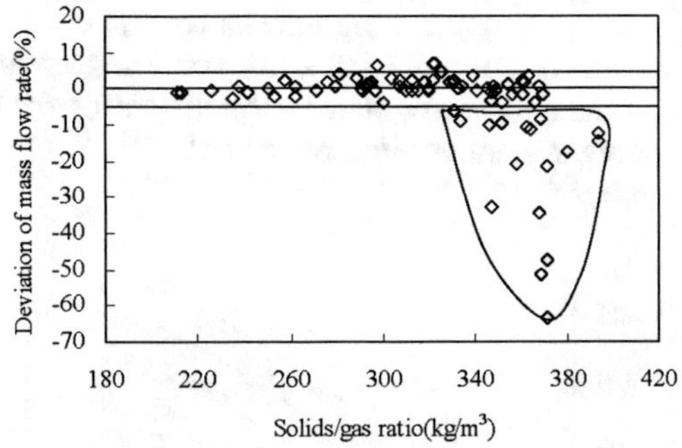

FIGURE 7. Effects of solids/gas ratio on the deviation of mass flow rate in a riser.

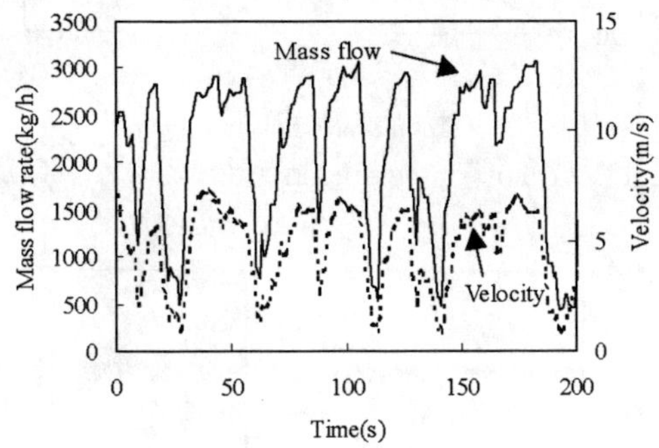

FIGURE 8. Measurement results of mass flow rate and velocity at a solids/gas ratio of 371 kg/m^3 in a riser

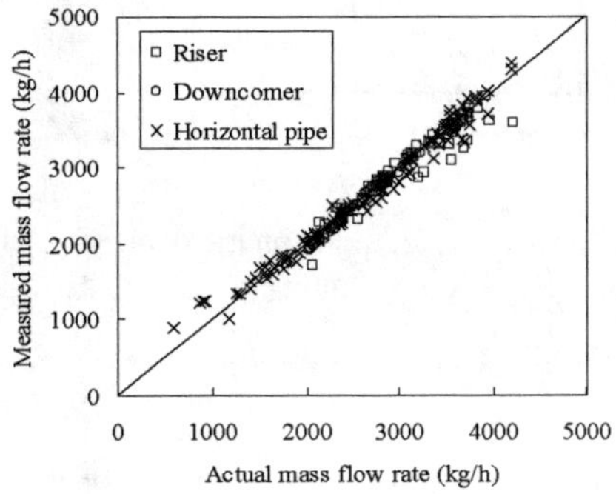

FIGURE 9. Comparison of mass flow rate between actual and measured values in different flow directions

As a comparison, the deviations E of the three different flow orientations in a stable conveying state are shown in Fig.9. From the figure, it can be seen that the measurement accuracy of the flow meter in the vertical downward flow direction and horizontal pipeline is almost identical and slightly better than that in the vertical upward flow pipeline.

Application In A Pilot-scale Gasification Plant Of Pulverized Coal

For an entrained-flow gasifier, it is important to online measure the mass flow rate of pulverized coal so as to ensure a reliable operation. Corresponding to four pneumatic conveying pipelines of pulverized coal, four capacitance solid mass flow meters with 15 mm inside diameter were employed to measure the mass flow rates of pulverized coal in a pilot-scale dry-fed gasification plant that was located at Yankuang Lunan Chemical Fertilizer Plant. Detailed introduction about the pilot-scale entrained-flow gasification plant of pulverized coal can be found in reference [14].

Differing from the above mentioned solid mass flow meter used in the experimental set-up, the model of the correlator used in the pilot-scale plant is Model 2109 instead of DMK270. It can be seen from Fig.10 that the deviations of mass flow rate measured by the meter are within ±9% after a calibration. Fig.11 shows the solid mass flow meter's typical measurement results in 7 hours at a high pressure of about 2.0Mpa. With a measured mass flow rate of about 370kg/h and a conveying concentration of 320kg/m^3, good performances are exhibited even if a sudden flow rate variation is experienced. Thus the solid flow meter is qualified for the online measurement of solid mass flow rate in a high concentration conveying system of pulverized coal.

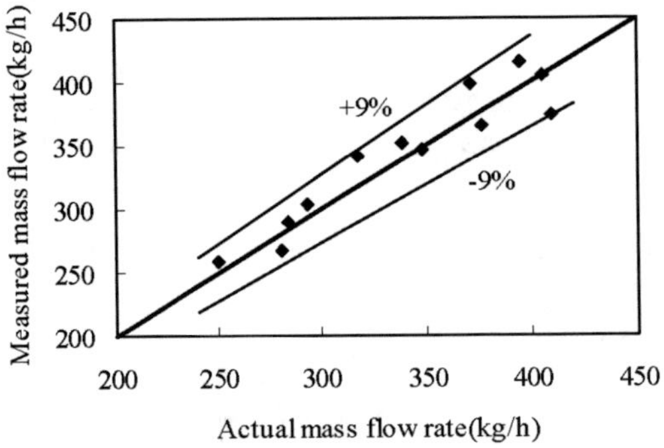

FIGURE 10. Measurement Deviations of the solid mass flow meter with the correlator Model 2109

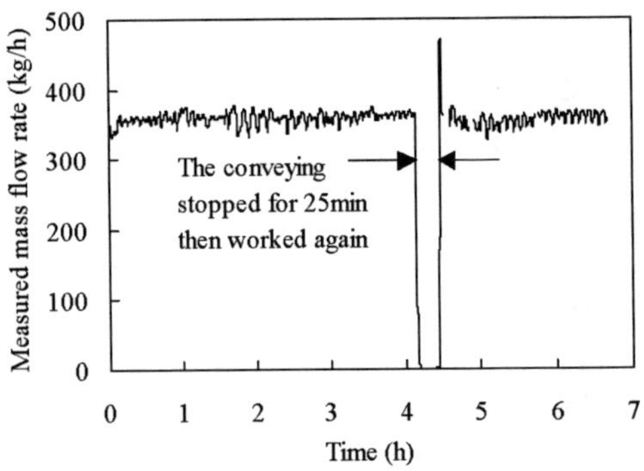

FIGURE 11. Performance of a solid mass flow meter in a pneumatic conveying pipeline to a gasifier (pressure of the gasifer was kept at 2.0MPa)

SOME EXPERIENCES

Many factors may affect the response of the capacitance solid flow meter such as the pipeline orientation, measurement position, phase loading, gas velocity and properties of the solid material [5]. Moreover, unaccountable measurement results occurred more or less in the course of its applications. In any case, some experiences have been achieved from its applications:

The measurement accuracy of mass flow rate is closely related to the calibration such as the concentration range of calibration, distribution of the calibration points and the flow stability.

Performance of the correlator Model 2109 is not the same as DMK270; some problems with Model 2109 occurred such as restricted measurement range and insensitivity (even contrary) to the variation of the conveying concentration. Resultantly, it took supplier of the meters almost one year to solve these problems before the results in Fig.10 were obtained. Thus a suitable design of the correlator is crucial for the measurement.

CONCLUSIONS

Performance of a capacitance solid mass flow meter was investigated in a dense phase pneumatic conveying set-up of pulverized coal. Experimental results have demonstrated that the capacitance solid flow meter is suitable for the gas-solids flow measurements. The deviations of mass flow rate measured by the meter in the experimental set-up are almost within $\pm 5\%$ for stable flow patterns, but unstable flow patterns pose great deviations.

Effects of solids/gas ratio on the measurement tested in a horizontal pipeline show that the flow meter is not qualified for the measurement of low concentration conveying. On the other hand, the measurement accuracy of the flow meter in the vertical downward flow direction and horizontal pipeline is almost identical and slightly better than that in the vertical upward flow pipeline.

Its performances in a high-pressured dense phase pneumatic conveying system of a pilot-scale entrained-flow gasification plant of pulverized coal are also presented and satisfying results are achieved.

REFERENCES

1. R.Thorn, M.S.Beck, and R.G.Green, *J. Phys. E: Sci. Instrum.*, **15**, 1131-1139(1982)
2. M.S.Beck, R.G.Green, and R.Thorn, *J. Phys. E: Sci. Instrum.*, **20**, 835-840(1987)
3. R.A.Williams, C.G.Xie, F.J.Dickin, S.J.R.Simons and M.S.Beck, *Powder Technology*, **66**, 203-224(1991)
4. Y. Yan, *Bulk Solids Handling*, **15**, 447-456(1995)
5. Y. Yan, *Measurement Science and Technology*, **7**,1687-1706(1996)
6. M.P. Mathur and G.E. Klinzing, *Powder Technology*, **40**, 309-321(1984)
7. H.L.Hu, T.M.Xu, S.E.Hui and Q.L.Zhou, *Flow Measurement and Instrumentation*, **17**, 87-92(2006)
8. G.A. Irons and J.S.Chang, *International Journal of Multiphase Flow*, **9**, 289-297(1983)
9. Operating and Service Manual / Model 2109 Correlator, Thermo Ramsey, 1999
10. O.Molerus, *Powder Technology*, **88**, 309-321(1996)
11. Konrad K, *Powder Technology*, **49**,1-35(1986)
12. Y.Yan, B.Byrne, S.Woodhead and J.Coulthard, *Measurement Science and Technology*, **6**, 515-73(1995)
13. Y.Yan, B.Byrne and J. Coulthard, *Flow Measurement and Instrumentation*, **6**, 115-119(1995)
14. X.Gong, X.L. Guo, Z.H.Dai, Z.H. Yu, F.Han, R.T. Zhao, C.L. Lv and W.X. Lu, *Modern Chemical Industry*, **25**, 51-52(2005)

Measurement of Interfacial Area Transport in Oil-water Dispersed Flow by Double-sensor Probe

Dongjian ZHAO, Xiaowei HU, Liejin GUO, Ximin ZHANG

State Key Laboratory of Multiphase Flow, Xi'an Jiaotong University, Xi'an, China, E-mail address: lj-guo@mail.xjtu.edu.cn

Abstract. The double-sensor conductivity probe was designed and applied to measure the distribution and transport characteristics of the oil-water dispersed flow in a vertical upward pipe. Typical radial profiles of interfacial area concentration, oil phase fraction, interfacial velocity, and oil drop Sauter mean diameter were obtained at four axial locations from the test section entrance of z/D =8.75, 33.75, 58.5 and 80.5. The results showed that under most flow conditions, the radial profiles of interfacial area concentration, oil phase fraction and interfacial velocity at the first measuring location were more flat than those at other three locations. This is because the initial size of oil drops were mainly determined by the mixing chamber and the flow of oil-water dispersion were far from fully-development. The interfacial area transport mechanism was also discussed in details.

Keywords: Interfacial area concentration, oil phase fraction, double-sensor probe, oil-water dispersed flow

INTRUDUCTION

Flows of mixture of two immiscible fluids, such as oil and water, are encountered frequently in the design of a variety of industrial processes and equipment, especially in the petroleum and chemical industry. The interest in the transportation of oil-water mixtures has led to a number of works on the flow pattern, averaged holdup and pressure drop[1,2]. However, with the development of more accurate and reliable Computational Fluid Dynamic models of two-phase flow in pipes, a physical understanding of the mechanism involved in the distribution and transport characteristics of the deformable interface is inevitable. The basic geometrical structure of a two-phase flow can be characterized by the local phase fraction and interfacial area concentration. The local phase fraction represents the phase distribution in the two-phase flow, while the interfacial area concentration describes available area for the interfacial transfer of mass, momentum and energy. Over the past ten years, much experimental work has been done on the distribution characteristics of the local phase fraction and interfacial area concentration in gas-liquid two-phase flow[3,4]. However, very few studies have been published on the local flow structure and its transport characteristics of liquid-liquid two-phase flow.

Recently, Farrar and Bruun[5] carried out the pioneering work to measure the distribution characteristics of local flow parameters in a vertical kerosene-water pipe flow. But because of limited flow conditions, they observed only the uniform or wall peaked radial profiles of local oil phase fraction. Lovick and Angeli[6] further studied the size and vertical distribution of drops in dispersed liquid-liquid horizontal flows under dual continuous pattern, where both phases retain their continuity and there is entrainment in the form of drops of one phase into the other. Also under dual continuous pattern of liquid-liquid horizontal flows, Elseth et. al.[7] carried out experimental study on the local distribution of velocity, turbulence and oil phase fraction. However to our knowledge, there is currently no published data available on the transport characteristics of the interfacial area concentration in vertical oil-water two-phase flow.

The present experiment was designed to study the transport characteristics of interfacial area concentration, as well as oil phase fraction, interfacial velocity, and oil drops Sauter mean diameter of an oil-water upward flow in a vertical pipe. Valuable insight will be provided into the physical mechanism of local flow structure in liquid-liquid two-phase flow, which will be different from that of gas-liquid two-phase flow.

CP914, *Multiphase Flow: The Ultimate Measurement Challenge,* edited by X. Cai, Y. Wu, Z. Huang, S. Wang, and M. Wang

EXPERMENTAL FACILITY AND INSTRUMENTATION

Figure 1 shows schematically the experimental system and the corresponding instrumentation. Two fluids used were tap water and 5# white oil. The viscosity of the white oil is 4.1 mPa s at 40 °C, and the density is 824 kg/m^3 at 20 °C. The levels of solubility of water in oil and oil in water are so low that the two fluids can be totally considered as being immiscible. The test section is a vertical round tube made of acrylic resin, with a 40 mm inner diameter and a 3800 mm length.

During the experimentation, water and oil were supplied separately from two storage tanks, and circulated by water pump and oil pump, respectively. Before being introduced into the mixing chamber, the flow rates of oil were measured using a CMF015 Micro Motion mass flow meter with maximum relative deviation of 0.2%, while the flow rates of water were measured by a magnetic flow meter with maximum relative deviation of 0.5%. Flow rates of water and oil can both be well regulated by control valves on the feed and by pass lines. The oil drops were generated by 56 pieces of stainless steel tube located in the mixing chamber. After flowing through the test section, the mixture of oil and water were separated through a horizontal liquid-liquid separator, and then discharged separately into the oil tank and water tank.

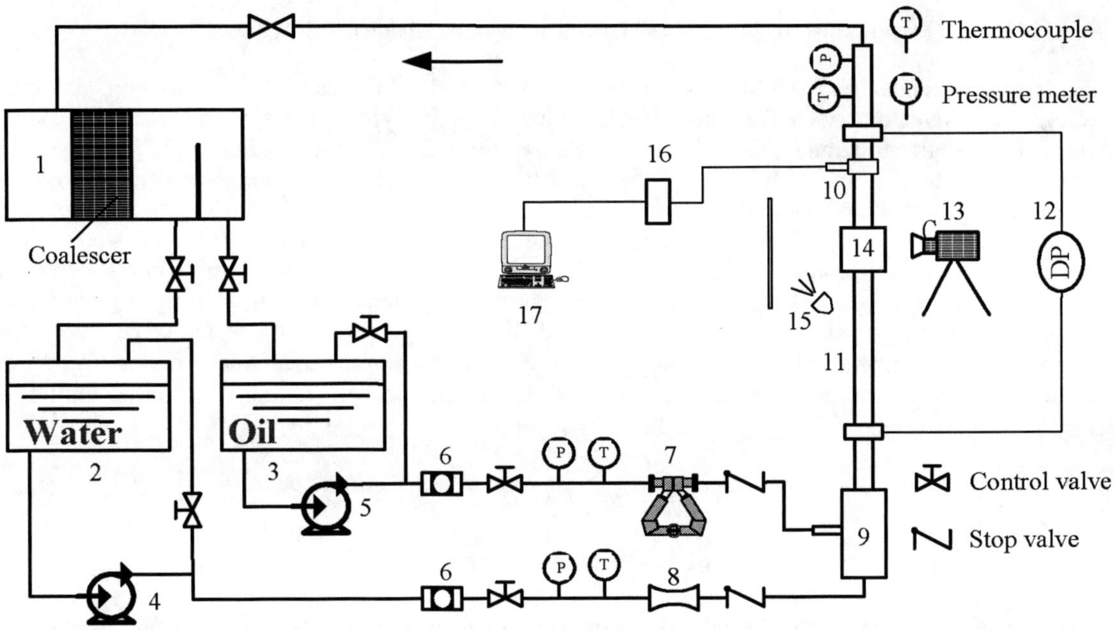

1-Separator, 2-Water tank, 3-Oil tank, 4-Water pump, 5-Oil pump, 6-Filter, 7-Mass flow meter, 8-Magnetic flow meter, 9-Mixing chamber, 10-Mechanical traverser of probe, 11-Test section, 12-differential pressure transducer, 13-Digital high-speed camera system, 14-Light compensation box, 15-Light source and reflection board, 16-Measurement circuit of probe, 17-PC with high-speed acquisition board inside

FIGURE 1. Schematic diagram of experimental loop

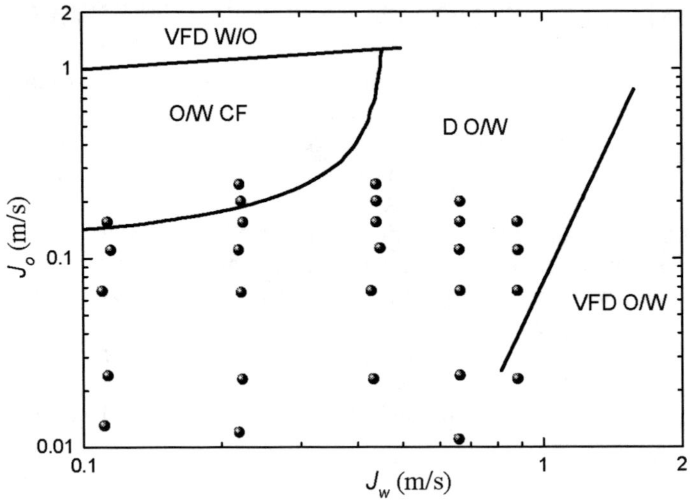

FIGURE 2. Experimental conditions on oil-water two-phase flow regime map proposed by Flores[8]

The double-sensor conductivity probe was designed and applied to measure the distribution and transport characteristics of oil-water dispersed flow in a vertical upward pipe. Typical radial profiles of interfacial area concentration, oil phase fraction, interfacial velocity, and Sauter mean diameter were obtained at four axial locations from the test section entrance of z/D =8.75, 33.75, 58.5 and 80.5. Fifteen radial locations were selected across half of the pipe diameter at each axial location. The water flow rates varied from 0.11 m/s to 0.89m/s, while the oil flow rates ranged from 0.012 m/s to 0.245 m/s. Fig. 2 presents the flow condition in details on the oil-water two-phase flow regime map proposed by Flores[8], where VFD O/W, D O/W and O/W CF denote very fine dispersion oil in water flow, dispersion oil in water flow, and oil in water churn flow, respectively. In all flow conditions, the oil drops dispersed in continuous water were observed. The double-sensor conductivity probe can only be used in current oil drops dispersed in water flow conditions for measuring the local interfacial area concentration. For more detailed information about the experimental facility and the double-sensor conductivity probe methodology, please refer to our early papers[9,10].

RESULTS AND DISCUSSION

Local Flow Parameters

In all flow conditions, the oil drops dispersed in continuous water were observed. Fig. 3 and 4 display typical profiles of local flow parameters at different measuring locations for two different oil flow rates. We can see that under most flow conditions, the profiles of interfacial area concentration, oil phase fraction and interfacial velocity at the first measuring location were more flat than those at other three measuring locations. This is because at the first measuring location, the initial size of oil drops were mainly determined by the mixing chamber and the flow of oil-water dispersion were far from fully-development.

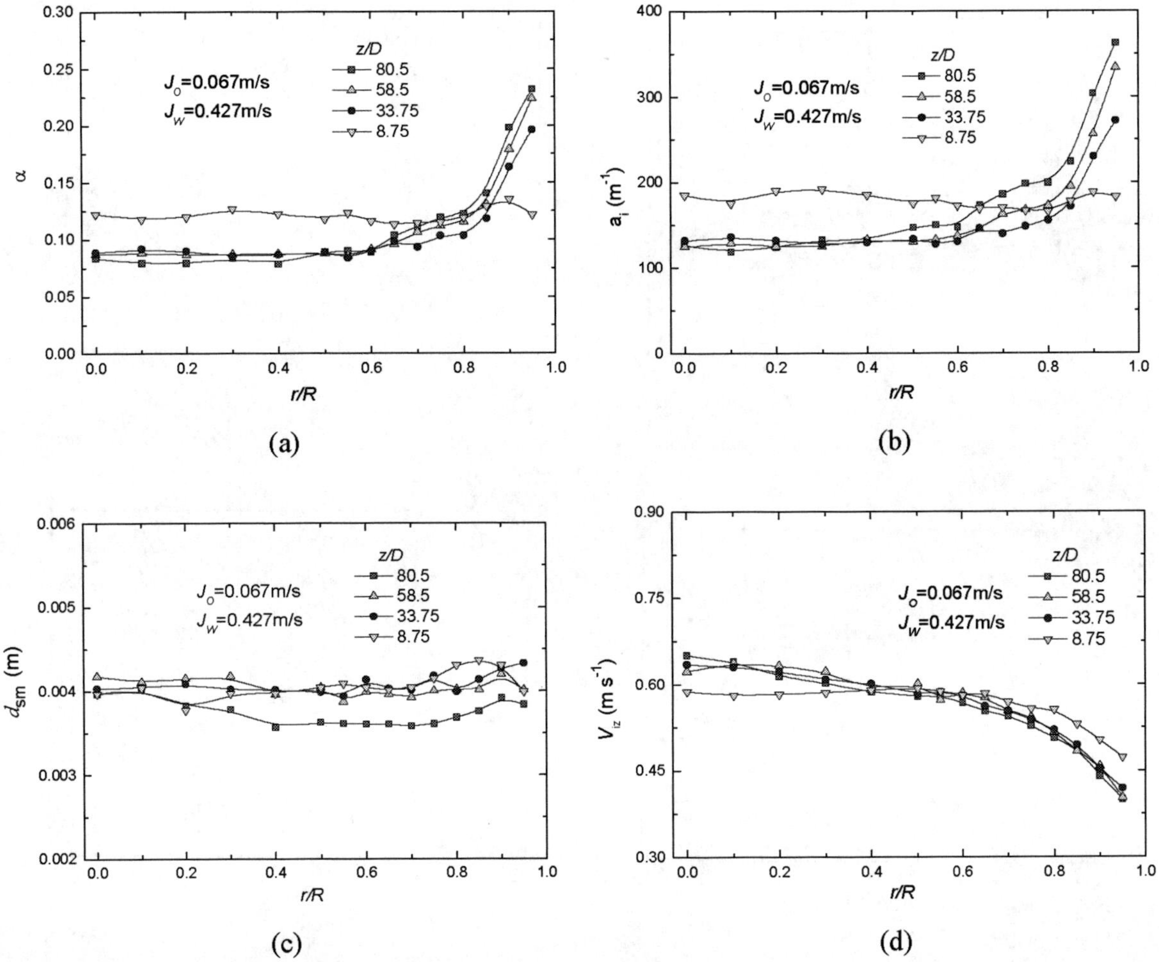

FIGURE 3. Profiles of local flow parameters at different measuring locations for oil flow rate of 0.067 m/s: (a) oil phase fraction, (b) interfacial area concentration, (c) oil drop Sauter mean diameter, and (d) interfacial velocity.

With the development of the oil-water dispersed flow, the steady fully developed flow conditions were achieved gradually at the second or third measuring locations. The interfacial velocity reaches its terminal velocity. And, the oil drop Sauter mean diameters are uniform along the radius of the test section with some increase in size near the wall. The change of the profiles of local flow parameters seemed to be insignificant along flow direction. Since the oil drops collect near the wall of the test section, the profiles of oil phase fraction and interfacial area concentration for oil flow rate of 0.067 m/s have obvious peaks near the wall, as shown in Fig. 3 (a) and (b). When oil flow rate goes as high as 0.245 m/s, the wall peaks of oil phase fraction and interfacial area concentration disappear as illustrated in Fig. 4 (a) and (b).

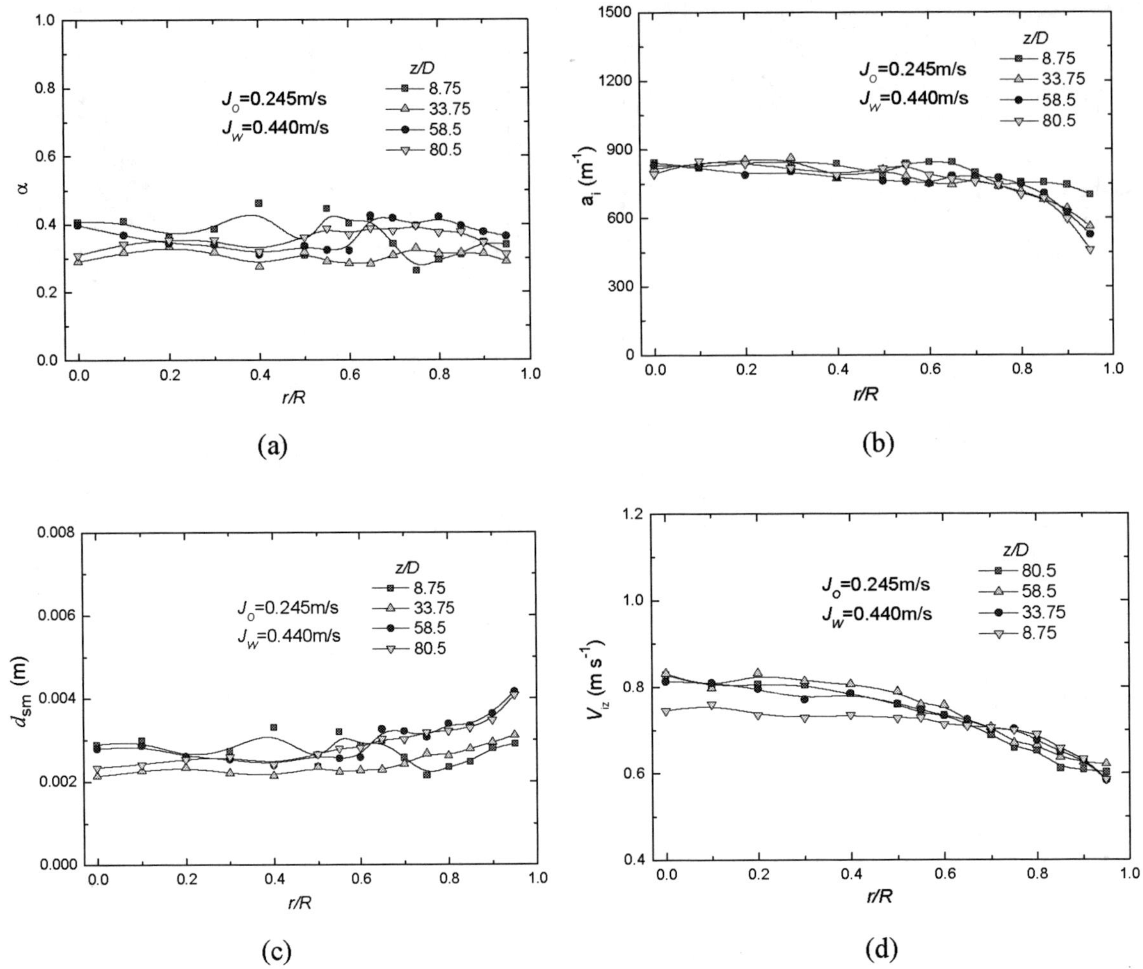

FIGURE 4. Profiles of local flow parameters at different measuring locations for oil flow rate of 0.245 m/s: (a) oil phase fraction, (b) interfacial area concentration, (c) oil drop Sauter mean diameter, and (d) interfacial velocity.

Area-averaged Interfacial Area Transport

In order to understand the mechanisms of interfacial area transport characteristics of oil-water two-phase dispersed flow, the area-averaged interfacial area concentrations under different flow conditions are plotted against z/D in Fig. 5. Because the dispersed oil drops are totally incompressible, the change of interfacial area concentration along flow direction was mainly determined by oil drop coalescence and breakup phenomena.

At low water flow rate and low oil phase fraction, the interactions between two oil drops and between oil drop and turbulent eddy could be neglected, and the area-averaged interfacial area concentration seldom changed along the flow direction. When the flow pattern transiting from dispersion oil in water to oil in water churn flow, the increase in oil phase fraction raised the contact time and collision frequency among oil drops, and further promoted the occurrence of oil drops coalescence phenomena. The interfacial area concentration decreased along the test section. However, at high water flow rate, the large turbulence fluctuation may decrease the contact time between oil drops, resulting in the decrease of the oil drop coalescence rate. Simultaneously, the enhanced turbulence promoted the breakup phenomena of oil drops, resulting in an increase of the area-averaged interfacial area concentration along the flow direction of the test section, as shown in Fig. 5 (d).

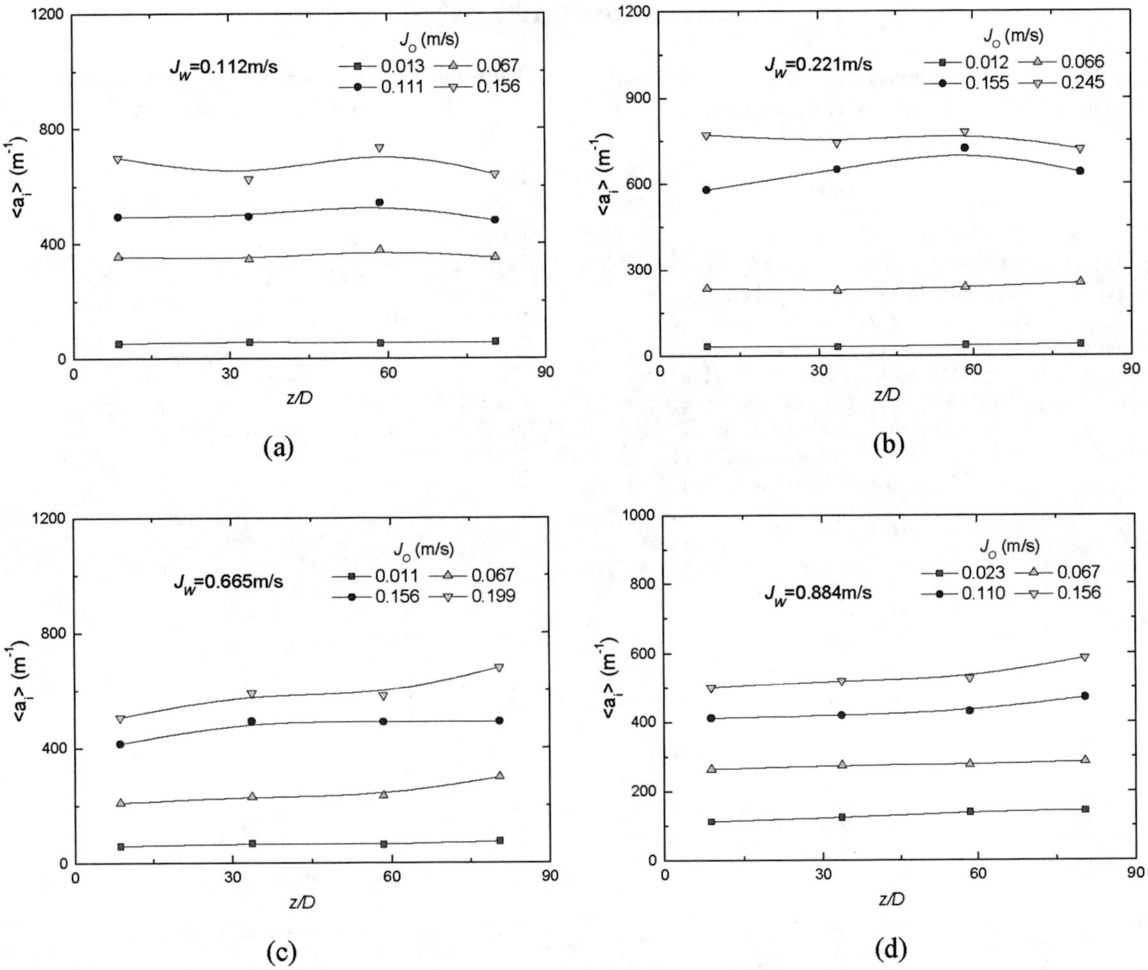

FIGURE 5. Development of area-averaged interfacial area concentration along the flow direction of the test section at different flow conditions.

CONCLUSIONS

Using double-sensor conductivity probe method, the distribution and transport characteristics of interfacial area concentration of oil-water dispersed flow in a vertical upward pipe were investigated experimentally. The results showed that, the profiles of interfacial area concentration, oil phase fraction and interfacial velocity at the first measuring location were more flat than those at other three locations. This is because the initial size of oil drops were mainly determined by the mixing chamber and the flow of oil-water dispersion were far from fully-development. At low water flow rate and low oil phase fraction, the interactions between oil drops and between oil drop and turbulent eddy could be neglected, and the area-averaged interfacial area concentration seldom changed along the flow direction. With an increase of oil phase fraction, the coalescence rates of oil drops increased gradually. However, at high water flow rate, the large turbulence fluctuation may decrease the contact time between oil drops, resulting in the decrease of the oil drop coalescence rate. Simultaneously, the enhanced turbulence promoted the breakup phenomena of oil drops, resulting in an increase of the area-averaged interfacial area concentration along the flow direction of the test section.

ACKNOWLEDGEMENTS

The author wishes to acknowledge the financial support of National Science Fund Project for Excellent State Key Laboratory (No.50323001), Key Project (No.50536020) and Creative Research Groups (No.50521604).

REFERENCES

1. X. J. Chen and L. J. Guo, *Int. J. Multiphase Flow*, 25, 1053-1072(1999).
2. W. H. Liu et al., *Chinese J. Chem. Eng.*, 11 (5), 491-496(2003).
3. T. Hibiki et al., *Nucl. Eng. Des.*, 184, 287-304(1998).
4. T. J. Liu and S. G. Bankoff, *Int. J. Heat and Mass Transfer*, 36 (4), 1061-1072(1993).
5. B. Farrar and H. H. Bruun, *Int. J. Multiphase Flow*, 22 (4), 733-751(1996).
6. J. Lovick and P. Angeli, *Chem. Eng. Sci.*, 59, 3105-3115(2004).
7. G. Elseth et al., *3rd Int. Symp. Two-Phase Flow Modeling and Exp.*, Pisa, 2004, pp22-24.
8. J. G. Flores, "Oil-water flow in vertical and deviated wells". Ph. D. Thesis, The University of Tulsa, 1997.
9. D. J. Zhao et al., *Int. J. Heat and Mass transfer*, 48 (10), 1926-1935(2005).
10. D. J. Zhao et al., *Int. J. Multiphase Flow*, 32, 1254-1268(2006).

Measurement of Refractive Index Profile Of GI-POF By Using Least Square Method

Huifen Jiang[1], Xiang'e Han and Renxian Li

School of Sciences, Xidian University, 710071, Xi'an, China

Abstract. This paper is devoted to the study on measurement of refractive index profile of graded-index polymer optical fiber (GI-POF) by light scattering. Using Generalized Airy theory and Debye series of an inhomogeneous cylinder, the scattering intensity distributions are obtained of Airy structure of rainbows for different refractive index profile. The results show that positions of Airy like peaks depend closely on refractive index profile of GI-POF. Since each order of rainbow penetrates it to different depths, such methods could be used to provide information the refractive index profile of GI-POF. For GI-POF with given diameter, positions of Airy like peaks of rainbows are simulated as a function of refractive index profile, which can be used to inverse unknown parameters of refractive index profile. The least square method is used in inversion of refractive index profile. The results obtained are agreement with theoretical values with high precision. The method has the advantages of non-instructive and on-line measurement, and can be used for the measurement of other inhomogeneous droplets.

Keywords: Light scattering; Graded-index polymer optical fiber (GI-POF); measurement of refractive index profile

INTRODUCTION

Many forms of communication systems have appeared over years with increasing demand of high-speed access to Internet. High-bandwidth graded-index polymer optical fiber is currently one of the promising candidates for such high-speed data transmission medium. It is well known that the bandwidth characteristic of GI-POF is dominated by the modal dispersion which strongly depends on the fiber structure, i.e. refractive index profile. For such a bandwidth analysis of GI-POF, it is necessary to quantitatively approximate the refractive index profile of GI-POF. Some main methods for measuring refractive index profile of GI-POF are applied. These include refracted near-field technique[1], transverse interferometric[2] method, direction deflection[3]and phase –shifting interferometry[4], etc. The above methods are only used for measuring GI-POF perform in the laboratory using complicated instruments, and have the limitations in sizing parameters.

GI-POF can be considered as an infinite inhomogeneous cylinder. Based on the improved algorithm of electromagnetic scattering of normally plane waves by multilayered cylinders [5], this paper is devoted to the study on measurement of refractive index profile of graded-index polymer optical fiber (GI-POF) by light scattering. The method has the advantages of non-instructive and on-line measurement, and inversion results with high precision.

This paper is organized as follows. Section 2 presents refractive index profile of GI-POF. Section 3 is devoted to the intensity distributions by light scattering as function of refractive index profile. Section 4 describes the applications of least square method in inversion, and discuss the inversion precision . The last section is the conclusion.

REFRACTIVE INDEX PROFILE OF GI-POF

It was already shown that [6-7] the optimum refractive index profile for a GI-POF that exhibits the largest bandwidth should be described by the well known power-law from shown in Eq. (1)

[1] Supported by the Key Project of Chinese Ministry of Education (No. 106149)

CP914, *Multiphase Flow: The Ultimate Measurement Challenge,* edited by X. Cai, Y. Wu, Z. Huang, S. Wang, and M. Wang

$$m(r)=\begin{cases} m_1\left[1-2\Delta(\frac{r}{a})^g\right]^{1/2}, & 0\le r<a \\ m_2 & r\ge a \end{cases} \tag{1}$$

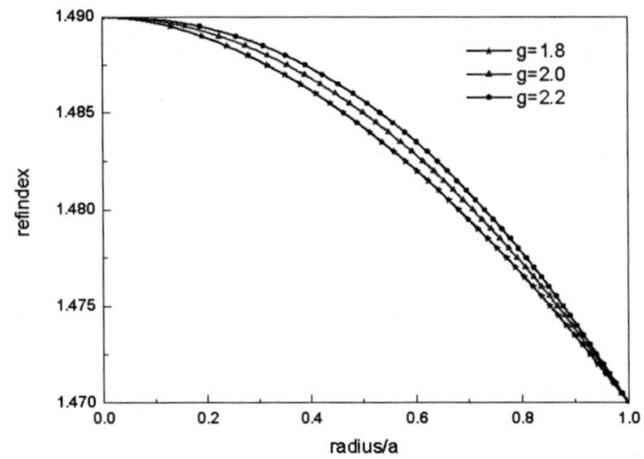

FIGURE 1. Refractive index profile

Where, m_1 and m_2 are the refractive indices of the core center and cladding, respectively, a is the core radius, and Δ is the relative index difference defined as

$$\Delta=\frac{m_1^2-m_2^2}{2m_2^2} \tag{2}$$

The parameter g, called the index exponent, determines the refractive index profile, as shown in Figure1, where m_1=1.49, m_2=1.47, $g=1.8, 2.0, 2.2$ respectively.

CHARACTERISTICS OF DISTRIBUTION OF SCATTERING INTENSITY OF GI-POF

GI-POF can be considered as an infinite long inhomogeneous cylinder. The spatial scattering intensity obtained by Generalized Lorenz-Mie theory [8] includes the information of the diameter and refractive index profile, so it can be used to measure the diameter and profile of the refractive index simultaneously. Fig.2 shows the distribution of scattering intensity of GI-POF simulated by Generalized Lorenz-Mie theory between the angle range of 0°～180°, with radius 500 μm and profile of refractive index m_1=1.49, m_2=1.47 and g=2.0. As we can see, the amplitude of intensity at different angles is obviously different, and that of some certain angle is relatively larger, such as the angles between 156° to 160°, what is we called the first rainbow, and the angles between 84°～ 95°, called the second rainbow. Since the rays of different orders of rainbow pass through different paths and provide information of different parts in the cylinder, it can be used to measure the diameter, (gradient of) refractive index, and etc.

Fig.3 gives the angular distribution of scattering intensity and its Airy structure of the first rainbow pattern of GI-POF, with radius $a=500\mu m$, refractive index of the core $m_1=1.49$, that of the cladding $m_2=1.47$, $g=2.0$ and $N_l=150$. Emphasis is put on that layers larger than 150 is enough to be considered as continue profile of refractive index. According to rainbow theory, the distribution of scattering intensity (Airy structure and ripple structure) depends closely on the diameter and the profile of refractive index of GI-POF. Position of each Airy peaks depends on refractive index profile and the distrance of different Airy peaks depends close on the diameter of cylinders.

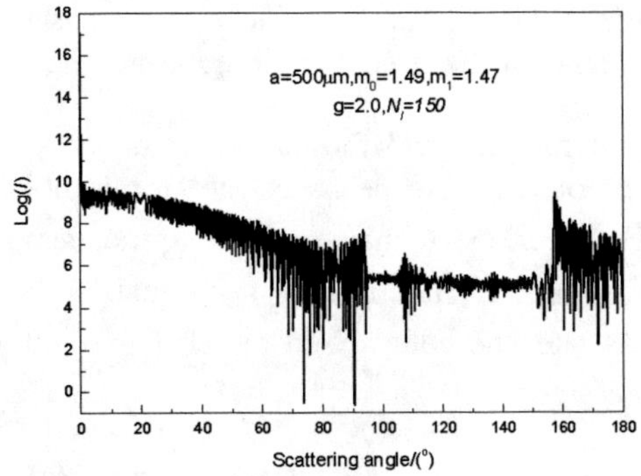

FIGURE 2. Distribution of scattering intensity of GI-POF by Lorenz-Mie theory

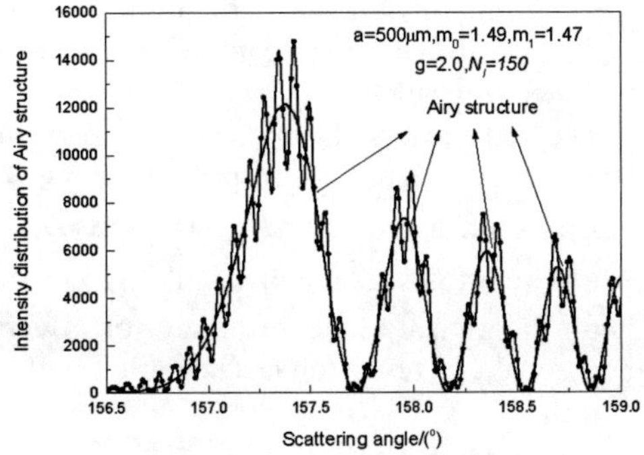

FIGURE 3. The first rainbow pattern of simulated GI-POF

The intensity distribution of Airy structure of rainbows can be obtained by Debye series[9] or Generalized Airy theory [10]. Fig.4 shows intensity distributions of Airy structure of the first rainbow as function of the index exponent parameter g for constant diameter and refractive index of the core and the cladding. As we can see, for GI-POF with given diameter, m_1 and m_2, the intensity distributions of Airy structures of the first rainbow depend on the index exponent parameter g. With g increasing, positions of each Airy-like peaks of Airy increase toward the larger degrees. Fig.5 shows the intensity distributions of Airy structure of the first rainbow as function of refractive index of the cladding, with constant diameter, m_1 and the index exponent parameter g. It is obvious that positions of Airy peaks depend closely refractive index of the cladding. Similarly, scattering intensity distributions of Airy structure of high order rainbows depend closely with refractive index profile.

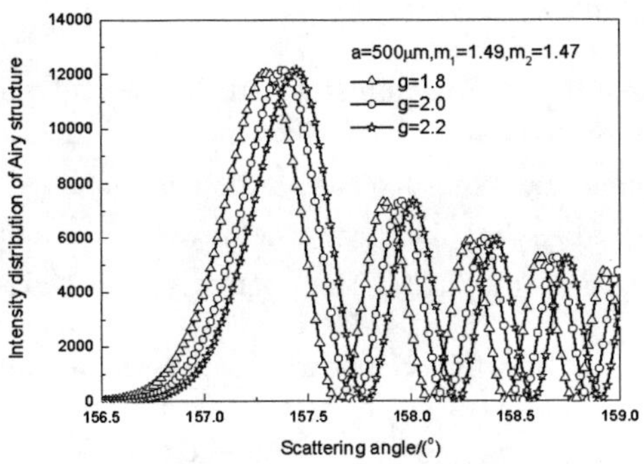

FIGURE 4. Intensity distributions of Airy structure with different exponent g for constant diameter and refractive index of the core and the cladding

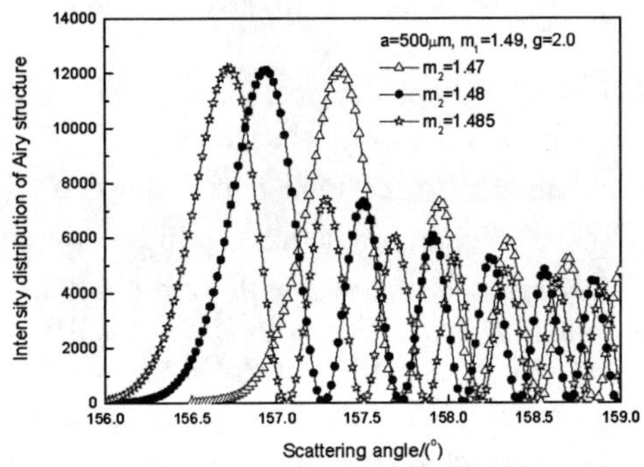

FIGURE 5. Intensity distributions of Airy structure with different m_2 for constant diameter and refractive index of the core and exponent g

INVERSION OF RESULTS BY LEAST SQUARE METHOD

The least square method is used in inversion of refractive index profile. The two-dimensional target function for optimization is expressed as follows:

$$Min \quad epsi0(b) = \sum_{j} \left\{ \left[a\max 1(j) - b\max 1(j) \right]^2 + \left[a\max 2(j) - b\max 2(j) \right]^2 \right\} \tag{3}$$

Where, $amx1$ and $amx2$ denote positions of Airy peaks of the first and second rainbows with different parameter b (index exponent g or surface refractive index m_2), and $b\max 1$ and $b\max 2$ are positions of Airy peaks of the first and second rainbows for a const b to be measured. When the function reaches the smallest values, the corresponding b can be obtained. For more parameters, the same method can be applied. The method has the advantages of non-instructive and can be used in on-line measurement, and can be used for the measurement of other inhomogeneous droplets.

Fig.6 (a) and (b) present the positions of Airy-like peaks of the first rainbow and the second rainbow as a function of profile parameter g respectively. As we can see, positions of Airy-like peaks of the first rainbow increase with parameter g increasing, and those of the second rainbow decrease with parameter g increasing. Using least square method in datum in Fig.6, parameter g can be obtained. The inversion g equals 1.996. Compared with theoretical value 2.0, the error is very small, and only 0.2%. The comparison of refractive index profile of theoretical and reversion is shown in Fig.10.

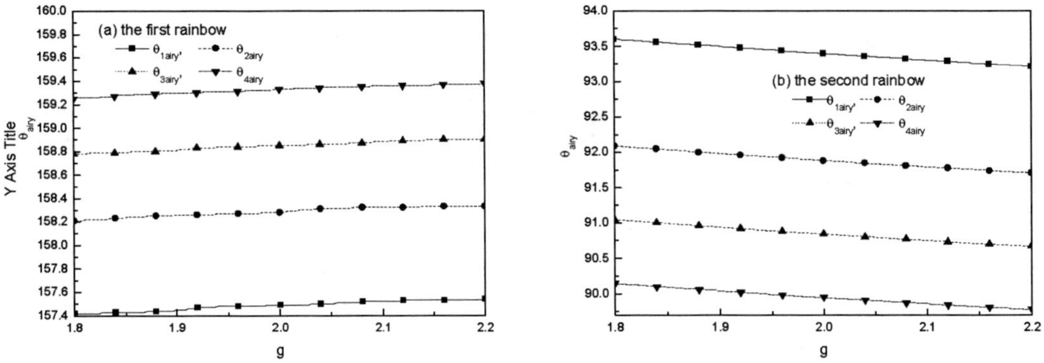

FIGURE 6. Positions of the first four Airy-like peaks as function of profile parameter g
(a) the first rainbow (b) the second rainbow ($a=300\mu m$, m_1=1.49, m_2=1.47, g=2.0, N_l=150)

Fig.6 (a) and (b) present the positions of Airy-like peaks of the first rainbow and the second rainbow as a function of profile parameter g respectively. As we can see, positions of Airy-like peaks of the first rainbow increase with parameter g increasing, and those of the second rainbow decrease with parameter g increasing. Using least square method in datum in Fig.6, parameter g can be obtained. The inversion g equals 1.996. Compared with theoretical value 2.0, the error is very small, and only 0.2%. The comparison of refractive index profile of theoretical and reversion is shown in Fig.10.

Fig.7 (a) and (b) present the positions of Airy-like peaks of the first rainbow and the second rainbow as a function of refractive index at the surface respectively. As we can see, positions of Airy-like peaks of the first rainbow decrease with m_2 increasing, and those of the second rainbow increase with parameter m_2 increasing. Using least square method in datum in Fig.7, parameter m_2 can be obtained. The inversion g equals 1.475, and agrees with theoretical value 1.475 without error.

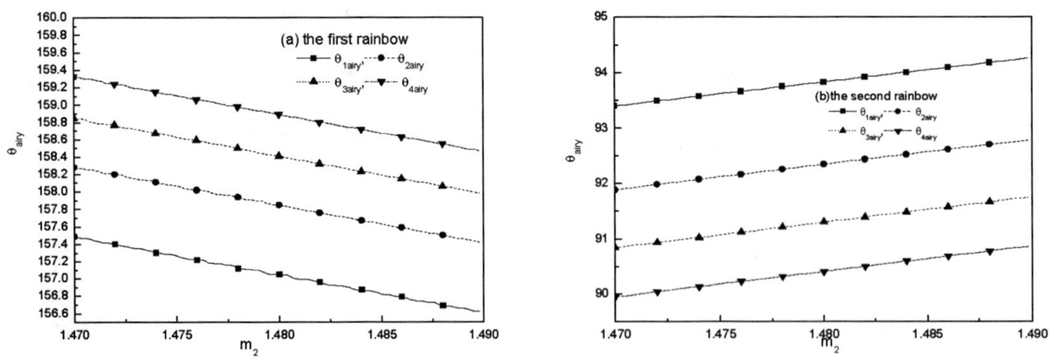

FIGURE 7. Positions of the first four Airy-like peaks as function of refractive index at the surface
(a) the first rainbow (b) the second rainbow ($a=300\mu m$, m_1=1.49, g=2.0, N_l=150)

Figs. 8 and 9 show the positions of the first Airy-like peaks of the first rainbow and the second rainbow as a function of the parameter g and the surface refractive index respectively. Similarly, positions of other Airy-like peaks can be obtained as function of the parameter g and the surface refractive index, which is omitted here. Using least square method in datum in Figs. 8-9, parameters g and m_2 can be obtained. The inversion g equals 1.996 and m_2 equals 1.47. Compared with theoretical value 2.0 and 1.47, the error is very small only in g (0.2%) . The comparison of refractive index profile of theoretical and reversion is shown in Fig.11.

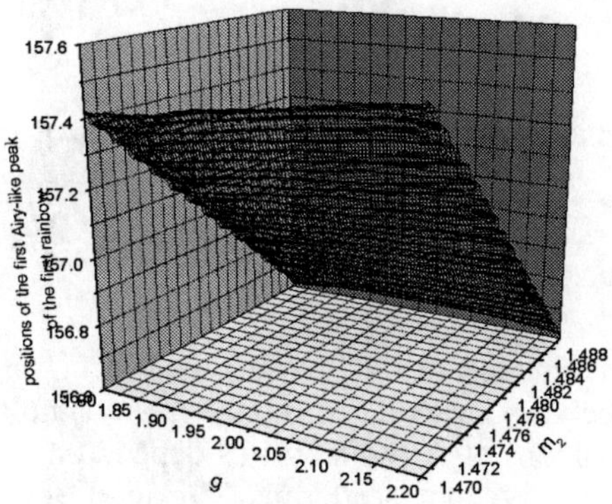

FIGURE 8. Positions of the first Airy-like peaks of the first rainbow as function of refractive index profile ($a=300\mu m$, m_1=1.49， m_2=1.47 , g=2.0， N_l=150)

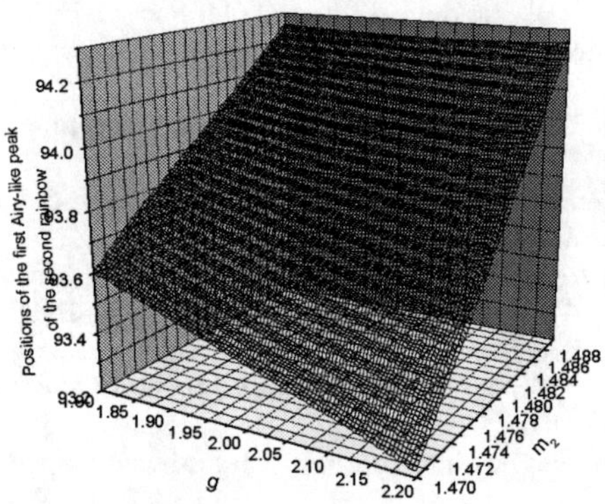

FIGURE 9. Positions of the first Airy-like peaks of the second rainbow as function of refractive index profile ($a=300\mu m$, m_1=1.49， m_2=1.47 , g=2.0， N_l=150)

We also simulated the variation of positions of Airy-like peaks of the first and second rainbows as functions of refractive index with different diameters, such as diameter ($d=1000\mu m$). The similar results can be obtained , which is omitted here. We will study in details.

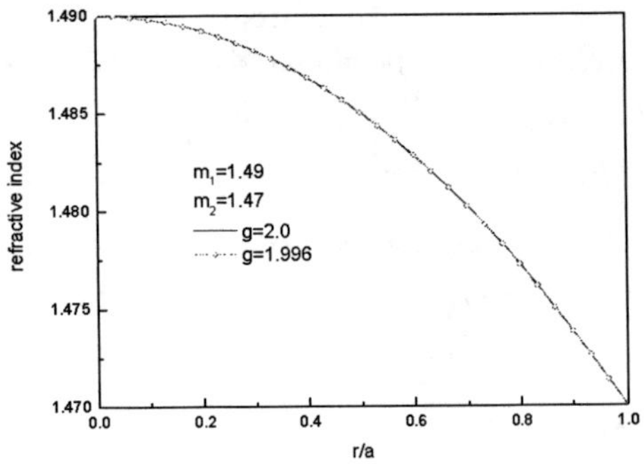

FIGURE 10. Refractive index profiles in theory and inversion and inversion

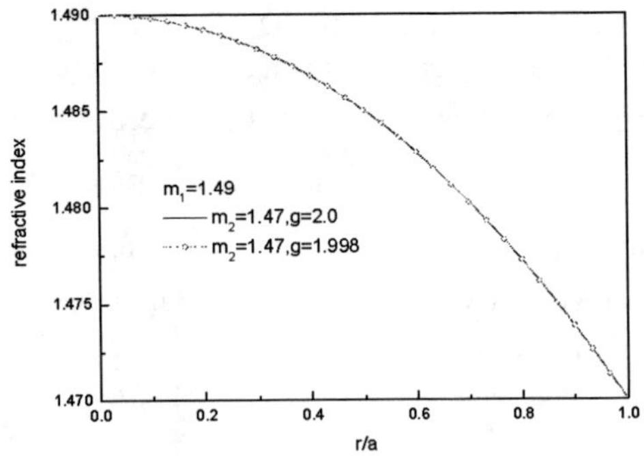

FIGURE 11. Refractive index profiles in theory and inversion

CONCLUSION

The Characteristics of intensity distribution of GI-POF are studied. Using Debye series and generalized Airy theory of inhomogeneous particles, the Airy structure of rainbows are simulated, and positions of Airy-like peaks can be obtained. Results show that positions of Airy-like peaks depend on refractive index profile. On the reverse, refractive index profile can be reversed by light scattering. The least square method is used in inversion, and refractive index profile is accessed with high precision. The method has the advantages of non-instructive and can be used in on-line measurement, and can be used for the measurement of other inhomogeneous droplets.

REFERENCES

1. KOPPEN C, SHI R F, et al. Properties of plastic optical fibers. J Opt Soc An B, Vol.15, No.2, 1998.
2. SUN Hui-gang, CHU Jiu-rong et al. Measurement of refractive index of plastic optical fibers. Optical Fiber & Electric Cable, No.4,2001.
3. KUZYK M G, GARVEY D W, et al. Characterization of single-mode polymer optical fiber and electrooptic fiber devices. Chemical Physics, 1999.
4. XU Long-yun, CHEN Jia-bi, et al. Method for measuring graded refractive index of polymer optical fiber and analysis of its error. Journal of China Jiliang University, Vol.15, No.1, 2004.
5. JIANG Huifen, HAN Xiang'e and LI Renxian. Improved algorithm of electromagnetic scattering by a multilayered cylinder and its application. 2007, Vol27. No.2 (To be published in ACTA OPTICA SINICA (in Chinese)).
6. R. Olshansky and D. B.Keck. Pulse broadening in graded-index optical fibers. Appl. Opt., vol. 15, no.2, pp.483-491,1976.
7. T. Ishigure, E.Nihei, And Y. Koike. Optimum refractive index profile of the graded-index polymer optical fiber, toward gigabit data links. Appl. Opt.,No.35, pp.2048-2053, 1996.
8. G. F. Bohren and D.R. Huffman. Absorption and scattering of light by small particles. Wiley New York, 1983.
9. LI Renxian, HAN Xiang'e et al. Debye Series of Normally Incident Plane Wave Scattering by an Infinite Multi-layered Cylinder. Appl. Opt., Vol45, No.24, 2006, pp: 6255-6262.
10. Maria Rosaria Vetrano, Jeronimus Petrus Antonius Johannes van Beeck, and Michel Léon Riethmuller. Generalization of the rainbow Airy theory to nonuniform spheres. OPTICS LETTERS, Vol. 30, No. 6, pp.658-660, 2005.

Particle Size Analysis By Transmission Fluctuation Spectrometry With Temporal Correlation: Theory And Simulations On Particle Overlapping

Bin Yu and Jianqi Shen

Institute of Science, University of Shanghai for Science and Technology, Shanghai 200093, China.

Abstract. Transmission fluctuation spectrometry with temporal or spatial correlation is being developed as a new method of particle size analysis, in which the particle size distribution and particle concentration can be measured simultaneously. Based on the assumption of low particle concentrations, an analytical expression of the transmission fluctuation correlation spectrum can be obtained with a layer model. It is found that the measurement matches the theory very well as long as the particle concentration is low enough. However, as the particle concentration increases, the measured transmission fluctuation spectrum deviates gradually from the theory, which is a result from the high concentration effects. The high concentration effects include particle overlapping and the monolayer structure. In this work, the theory and simulations on the particle overlapping are studied. As a result, an empirical expression is found.

Keywords: transmission fluctuation spectrometry; temporal correlation; particle overlapping
PACS: 42.62.-b, 42.68.-w

INTRODUCTION

Transmission signals measured on the flowing particle dispersion with a high spatial and temporal resolution show significant fluctuations, which include the complete information on particle size distribution and particle concentration. During the last years, the theory of transmission fluctuation spectrometry (TFS) has been developed as a method of particle size analysis [1-5]. After the transmission signal of a single beam has been subjected to a procedure of variable spatial and temporal averaging (or signal filtering), the transmission fluctuations are expressed in terms of the expectancy of the transmission square (ETS). By changing the averaging parameter, the ETS is obtained as a spectrum, which is related to the physical properties of the suspension and the process of signal averaging. Different techniques of the TFS have been developed, according to the method of signal averaging (i.e. spatial averaging, temporal averaging/signal filtering and their combination). Both the particle size distribution and the particle concentration can be measured simultaneously.

The transmission fluctuation spectrum may also be obtained with correlation techniques. One of these is the spatial correlation (transmission fluctuation spectrometry with spatial correlation, TFS-SC), in which two parallel narrow beams with a variable separation between each other are employed and the expectancy of the transmission product (ETP) is measured as a function of the beam separation [6]. The advantage of this version of TFS-SC is that the additional measurement of the flowing velocity is unnecessary and hence it is not strongly influenced by the velocity profile and velocity fluctuations in the measuring zone. However, the mechanical operation on the beam separation becomes challenging, especially for the measurements on small particles. The beams and the detectors should be able to move synchronously and fast enough in a micrometer scale. Furthermore, the beams should be consistent with each other both in the beam diameters and in propagating directions, which is very difficult to realize. So in real measurements, the temporal correlation is used with only one narrow beam [7-8]. The method is called as transmission fluctuation spectrometry with temporal correlation or auto-correlation (shortly TFS-TC or TFS-AC).

All these versions of the TFS theory are developed with a layer model and on the assumptions of geometric ray propagation, completely absorbent particles and low particle concentrations. Experimental studies show that the theoretical results match the measurements satisfactorily as long as the particle concentration is low enough.

CP914, *Multiphase Flow: The Ultimate Measurement Challenge,* edited by X. Cai, Y. Wu, Z. Huang, S. Wang, and M. Wang

However, with increasing particle concentration, the measured transmission fluctuation spectrum deviates gradually from the theory, which is caused by the high concentration effects [4, 5, 7, 8]. Similarly to what we have done before, effects such as multiple scattering and dependent scattering can be excluded so that only the steric interactions between particles are accounted for [9-11]. Steric interactions between particles are classified into those between particles within the same monolayer (i.e. the monolayer structure) and those between particles from different monolayers (known as particle overlapping), which cause different effects on the transmission fluctuation correlation spectrum. Theoretical and experimental studies show that, if the beam diameter is infinitely small compared to the particle size, the effect from particle overlapping can be excluded so that the transmission fluctuation spectrum is affected by the monolayer structure only [6,7]. However, while the beam diameter is not very small compared to the particle size, effects from particle overlapping are much stronger than those from the monolayer structure [8]. In most cases, the beam diameter is not so small, compared to the particle size. Therefore, in this work, we concentrate on the effects from particle overlapping for the case of an incident beam with a finite diameter (i.e. the beam diameter is comparable to or even a little larger than the particle size).

Due to the mathematical difficulties, particle overlapping is studied with simulation. As a result of the work, an empirical expression is found which can be used to extract the information on particle size distribution and particle concentration from the transmission fluctuation spectrum with temporal correlation.

PRINCIPLE OF THE MEASUREMENTS [7]

In the TFS-TC, the fluctuating transmission signal $T(t)$ of a narrow beam is recorded in a very long period of sampling time t_s. The expectancy of transmission product (ETP) is then calculated with a variable time difference (or the correlation time) τ :

$$\mathrm{ETP} = \lim_{t_s \to \infty} \frac{1}{t_s} \int_0^{t_s} T(t) T(t+\tau) dt \tag{1}$$

By varying the correlation time, the ETP is obtained as a spectrum, which can be further solved to extract the information on particle size distribution and particle concentration.

When the particle concentration is not high, the logarithm of the ETP of a 3-dimensional mono-dispersed particle system can be expressed as:

$$\ln \mathrm{ETP} = -\frac{1.5}{x} \Delta Z \cdot C_V \left[2 - \chi(\Delta, \Lambda) \right] \tag{2}$$

Here, x is the particle diameter, ΔZ is the beam path in the measuring zone, C_V is the particle volume concentration, $\Delta = v\tau/x$ (v is the flow velocity) is called as the dimensionless correlation time and $\Lambda = D/x$ (D is the beam diameter) is the dimensionless beam diameter (or beam-to-particle diameter ratio). $\chi(\Delta, \Lambda)$ is the transition function which is obtained as:

$$\chi(\Delta, \Lambda) = \int_0^{+\infty} J_0(2u\Delta) \cdot \left[\frac{2J_1(\Lambda \cdot u)}{\Lambda \cdot u} \right]^2 \cdot \frac{2J_1^2(u)}{u} du \tag{3}$$

The factor $\left[2J_1(\Lambda u)/(\Lambda u) \right]^2$ originates from spatial averaging over the cross section of the circular uniform beam, $2J_1^2(u)/u$ describes the property of the spherical particle and $J_0(2u\Delta)$ is the effect of temporal correlation.

Eq.(2) can be extended to a poly-dispersed particle system, expressed as:

$$\ln \mathrm{ETP}(\tau_j) = -\sum_i \frac{1.5}{\overline{x}_i} C_V(\overline{x}_i) \left\{ 2 - \chi\left[\Delta_{i,j}, \Lambda_i(Z, \overline{x}_i) \right] \right\} \Delta Z \tag{4}$$

τ_j ($j = 1, 2, \cdots, N$) is the variable time difference for signal correlation. $\overline{x}_i$ is the mean particle diameter in the ith fraction of particle size ($i = 1, 2, \cdots, M$) and $C_V(\overline{x}_i)$ is the corresponding volume concentration. $\Delta_{i,j} = v\tau_j / x_i$ is the dimensionless correlation time.

Eq.(4) shows that the logarithm of ETP is linear with respect to the particle concentration and the effects from the different particle size fractions superimpose linearly to form the logarithm of ETP. Therefore, particle size distribution and particle concentration can be retrieved by deconvoluting from the measurement of transmission fluctuation correlation spectrum.

EFFECTS FROM PARTICLE OVERLAPPING

1. A Brief Discussion On Particle Overlapping

Shielding effects can be traced back to near-field phenomena. Considering the near field diffraction, the radiation incident upon the second monolayer is strongly dependent on particle arrangement in the first monolayer. Even at a very low concentration, when the particle arrangement in each monolayer is random, there are statistically a certain part of particles in the second layer located in the shadows produced by those particles in the first layer. This interaction might be called as particle intersection or particle overlapping. As show in Fig.1a, the overlapping particles mimic an irregular particle with an equivalent size larger than the size of a single particle, thus widen the blockage of the beam (see Fig.1b) and modulate the composition of the transmission spectrum.

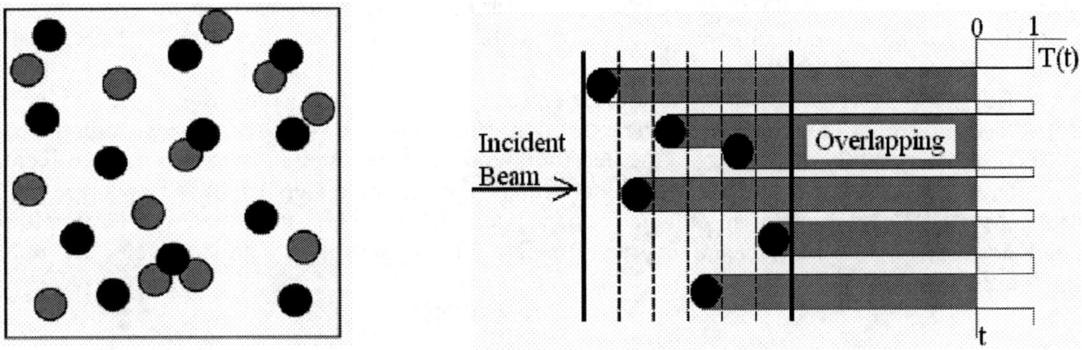

(a) (b)

FIGURE 1. Effects of particle overlapping.
(a) overlapping between particles in different layers; (b) the effect of particle overlapping: widened blockage of the beam

The high concentration effects can be quantified by the extinction of the irradiation *E*, which is approximately a product of the monolayer density β and the number of layers N.

$$E = -\ln T = -\ln(1-\beta)^N \approx \beta \cdot N \tag{5}$$

From this point of view, an increase of particle concentration is approximately equivalent to an increase of the layer number. However, the increase of particle concentration additionally leads to increasing effects from the monolayer structure. Therefore, in order to avoid any disturbance from the monolayer structure in the simulations, the particle concentration is assumed to be very low. Alternatively, this may be simulated either by assuming a large number of layers, each layer containing very few particles only, or by assuming a single, dense monolayer containing non-interacting particles, whereby particles are allowed to interpenetrate with each other. The dense monolayer containing N non-interacting particles is equivalent to a 3-dimensional particle system containing N layers, each layer containing one particle only and hence the monolayer density is

$$\beta = \frac{\pi x^2}{4 \cdot l_1 l_2} \tag{6}$$

As long as the dimensions of the monolayer (i.e. l_1 and l_2) are large enough, the monolayer density tends to zero so that the average transmission is expected to be:

$$e(T) = (1-\beta)^N \to e^{-\beta \cdot N} \tag{7}$$

2. Simulations With Beams Of Finite Diameters

In the simulations, another transition function of ETP is used, which is defined as $\Xi = \ln\{\mathrm{ETP}\}/\ln\{\mathrm{ET}\}$, whereby $\mathrm{ET} = 1-\beta$ is the average transmission [6]. Accordingly, the step height of the transition function is defined as:

$$\Xi_{SH} = \Xi(\Delta \to \infty) - \Xi(\Delta = 0) = 2 - \Xi(\Delta = 0) \tag{8}$$

The relationship between Ξ and $\chi(\Delta, \Lambda)$ is found to be $\lim_{\beta \to 0} \Xi = 2 - \chi$ in the limit of low particle concentration.

For beams with finite diameters ($\Lambda \neq 0$), particle overlapping has two effects on the transmission fluctuation correlation spectrum: the shift of the spectrum and the variation of the step height.

At the first step, we simulate a single, dense monolayer containing non-interacting particles, whereby particles are allowed to interpenetrate with each other. The simulated results are shown in Fig.2. The theoretical results are also plotted to make a comparison. It can be found that, as the extinct increases, the transmission fluctuation spectrum deviates gradually from the theory and all the simulated curves are below the theoretical one. That is to say, the step height of the transition function increases along with the increase of the extinction. Meanwhile the transition function gradually shifts to the range of larger dimensionless correlation time. Comparing Fig.2a with Fig.2b, we may find that the variation of the spectrum via the extinction depends on the value of the beam-to-particle diameter ratio Λ. This is because that the step height of the transition function depends on the beam-to-particle diameter ratio very closely. The smaller the beam-to-particle diameter ratio is, the larger the step height becomes (see Fig.2). Thus the overlapping particles play a role of an irregularly-shaped particle with a bigger particle size, leading to a smaller beam-to-particle diameter ratio and hence a bigger step height. The higher the extinction value is, the more serious the effect of particle overlapping is.

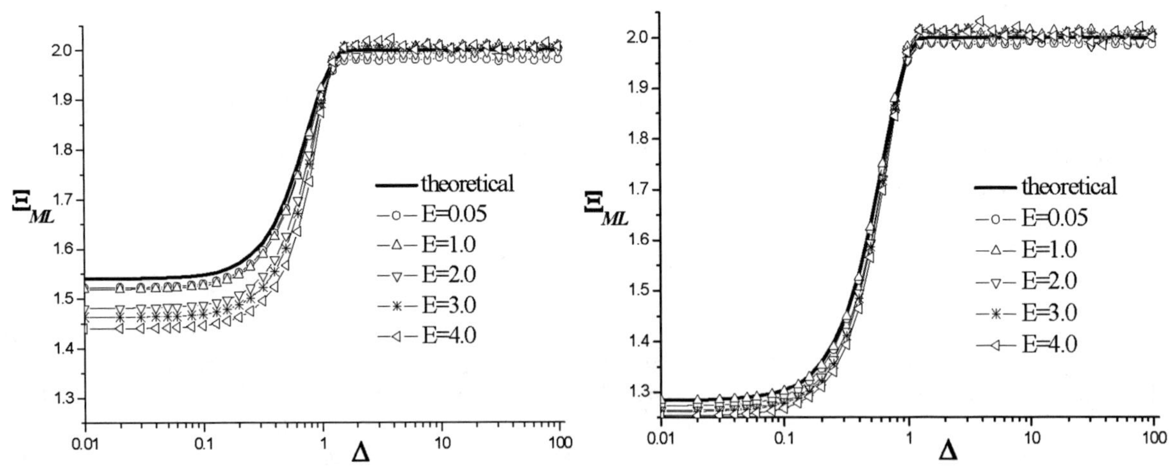

(*a*) $\Lambda = 1.0$ (*b*) $\Lambda = 0.5$

FIGURE 2. The simulated results on particle overlapping.

In Fig.3, the monolayer density is assumed to be $\beta = 0.0925$ and the beam-to-particle diameter ratios are respectively $\Lambda = 1.0$ and $\Lambda = 0.5$. So the increase of the extinction is determined by increasing the number of layers. The simulated transmission fluctuation spectrum deviates gradually from the theory as the extinction increases: the step height of the transition function tends to increase and the transition function shifts to the range of larger dimensionless correlation time gradually. The simulated results shown in Fig.3 are very similar to those in Fig.2 due to the fact that the monolayer density here is so low that the effects from the monolayer structure are not strong. Hence only the effects from particle overlapping are visible.

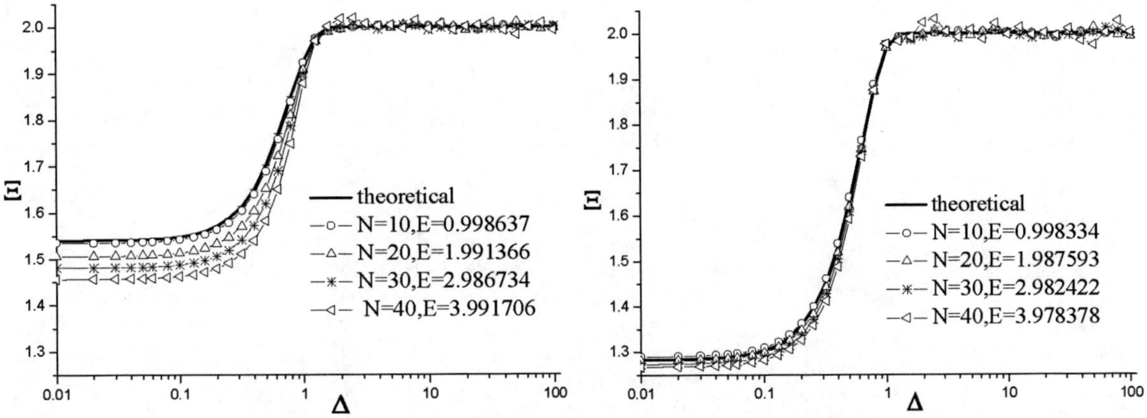

(a) $\Lambda = 1.0$ (b) $\Lambda = 0.5$

FIGURE 3. Transition functions of the ETP ($\beta = 0.0925$).

In order to see the shift of the transmission fluctuation correlation spectrum, the transition function is introduced into a normalized transition function, separating out the change of the step height. The normalized transition function $\Xi_N(\Delta)$ is defined as:

$$\Xi_N(\Delta) = \frac{\Xi(\Delta) - \Xi(\Delta \to 0)}{\Xi(\Delta \to \infty) - \Xi(\Delta \to 0)} + 1 \tag{9}$$

When $0 < \Lambda \leq 4$, effects from particle overlapping on the normalized transition function can be empirically expressed as the deviation from theoretical normalized transition function in low concentration limit, defined as:

$$\Delta\Xi_N(\Delta) = \left[\Xi_N(\Delta) - \Xi_N(\Delta, C_V \to 0) \cdot e^{-(\Lambda+1)\cdot\Delta}\right] \cdot e^{-\left[(4-\Lambda)+\frac{1}{\Lambda+1}\right]\cdot\Delta} \tag{10}$$

Here, $\Xi_N(\Delta, C_V \to 0)$ is the theoretical normalized transition function in low concentration limit. Fig.4 gives an example of the simulations.

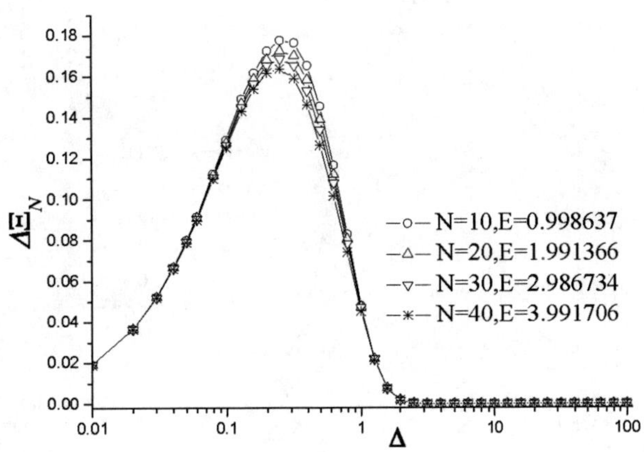

FIGURE 4. Deviation of the transition function from theory, corresponding to the simulated results shown in Fig.3a.

In order to take a deeper look, the deviation of the normalized transition function is factorized into a product of the Module M and the normalized deviation $\zeta(\Delta)$. M is the average of the maximum of the deviation of the normalized transition function $\Delta\Xi_N(\Delta)$:

$$\Delta\Xi_N(\Delta) = M \cdot \zeta(\Delta) \tag{11}$$

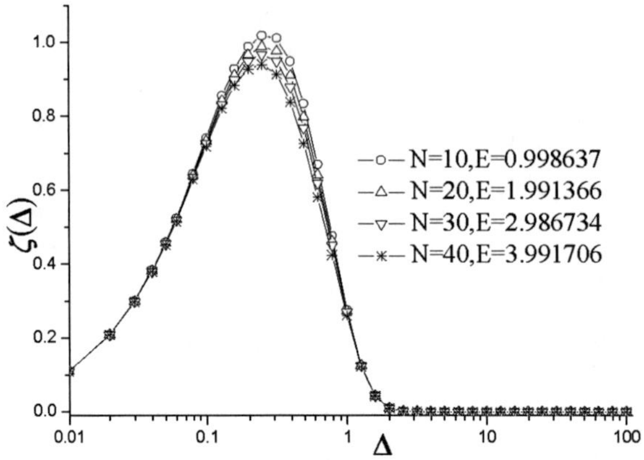

FIGURE 5. The normalized deviation, corresponding to the simulated results shown in Fig.4.

The normalized deviation $\zeta(\Delta)$ is given in Fig.5 with a peak in the range of Δ = 0. 2 to 0.4. Further more, it is found that the normalized deviation $\zeta(\Delta)$ does depend on the dimensionless beam diameter, which can be described with a translation of the dimensionless correlation time of:

$$\Delta' = \Delta \cdot \left[\frac{1 + 0.285 \cdot (\Lambda + 0.45) \cdot e^{-0.51 \cdot \Lambda \cdot \Delta}}{1 + 0.546 \cdot (\Lambda + 0.1)} \right] \tag{12}$$

and the distribution of the normalized deviation might be described with an empirical expression:

$$\zeta(\Delta') = \left(\frac{e}{a}\right)^a \cdot \xi \cdot e^{-\xi}, \qquad \text{here } \xi = \left(\frac{\Delta'}{b}\right)^c \tag{13}$$

The parameters are $a = 1.31843$, $b = 0.162711$ and $c = 0.860572$. Simulated results are shown in Fig.6 (in dots) to compare with the empirical expression (in line).

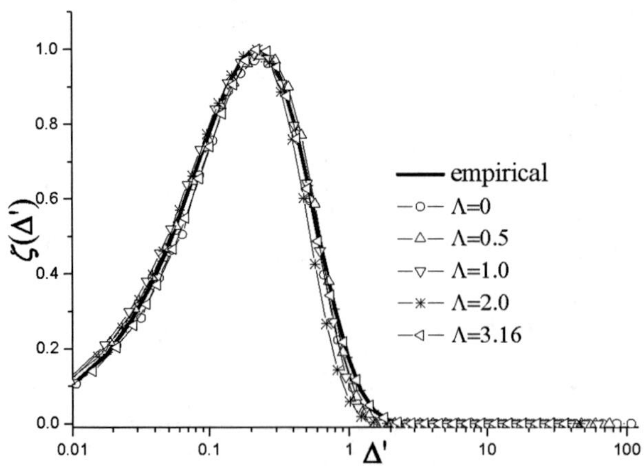

FIGURE 6. The normalized deviations. The abscissa is Δ' ($N = 20, E = 2.0$)

Fig.7 shows the dependence of M on Λ, with different values of E and N. As shown in Fig.7a, when N keeps unchanged (e.g. $N = 40$) and E is 0.05, 2.0 and 4.0 respectively, M increases along with the increase of Λ. These three curves are approximately consistent with each other. Similarly, when E is unchanged (e.g. $E = 0.05$) and N is 5, 10, 40 respectively, M increases along with the increase of Λ in Fig.7b. Therefore, M is related with the beam-to-particle diameter ratio Λ only and almost independent of E and N. This is quite different from what we obtained before in TFS with signal filtering [12]. The approximate expression is obtained

$$M = M(\Lambda) = k_1 \cdot \Lambda^2 + k_2 \cdot \Lambda + k_3 \tag{14}$$

here, $k_1 = 0.028, k_2 = 0.061, k_3 = 0.087$.

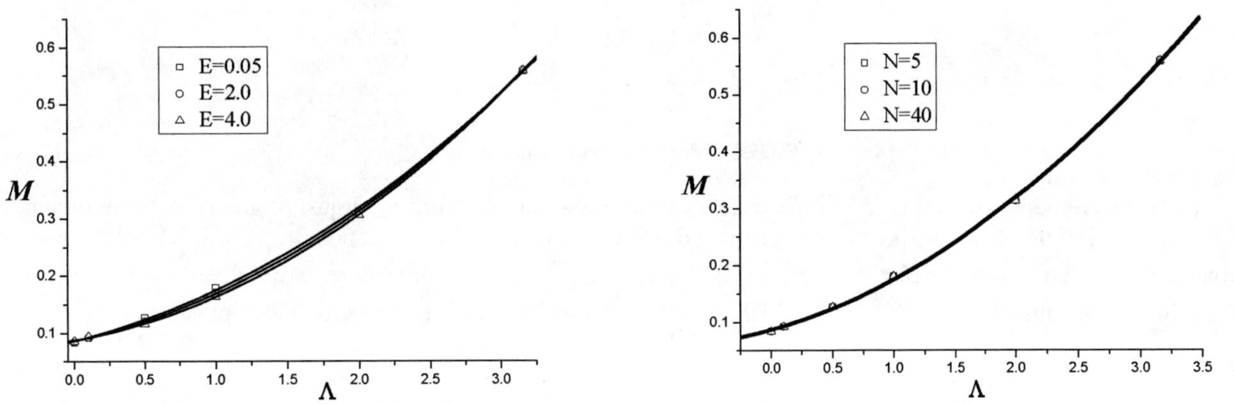

(a) $N = 40$ (b) $E = 0.05$

The curves correspond to $E = 0.05, 2.0, 4.0$ The curves correspond to N =5, 10, 40

FIGURE 7. The module of the normalized deviation

From the simulation above, we may conclude that effects from particle overlapping are closely related to the extinction E, the layer number N and the beam-to-particle diameter ratio Λ.

3. Experimental Validation Of The Empirical Expression

As a check of the empirical expression given in Eq.13, measurements are performed on the glass beads whose mean diameters are respectively $265\mu m$ and $375\mu m$. A focused Gaussian beam is used whose beam waist lies in the middle of the measuring zone. The average beam diameter is 50 μm and the pathlength of the beam is about $5.5mm$ [13]. In order to ensure constant measurement conditions, the particle suspension (i.e. the glass beads dispersed in water) is cycled in the circuit driven by a hose pump, which can be used to control the flow velocity in the measuring zone. The average particle flow velocity v is about 1.98m/s, measured with a Laser Doppler Anemometer, Polytec LDV380. The transmitted beam intensity or transmission is focused by a convex lens onto a photodiode (PBX65) with an effective area of $1mm \times 1mm$. The distance between the convex lens and the photodiode is about $700mm$. In order to obtain the spectrum of transmission fluctuation with temporal correlation, a high resolution digital oscilloscope (DL708 produced by YAKOGAWA) is applied to record the real-time transmission fluctuations with a high temporal resolution of $1e-6s$ or $1e-7s$, which is determined by both the size of the particles to be measured and the flow velocity. The total acquisition time of each group of experimental data ranges from $0.2-4s$ to prevent statistical deviation. The experimental setup is shown in Fig.8.

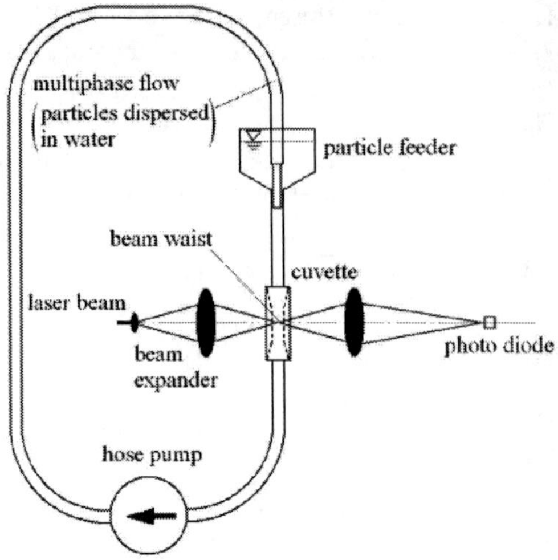

FIGURE 8. The experimental setup.

The measurements are executed at different particle concentrations. After the transmission signals are obtained, the transmission fluctuation spectra are calculated with Eq.(1) and the corresponding transition functions are obtained with the definition $\Xi = \ln\{ETP\}/\ln\{ET\}$. The following data procedure is described in Eqs. (10-13). Experimental results are plotted in Fig.9, which shows a complete match with the empirical expression.

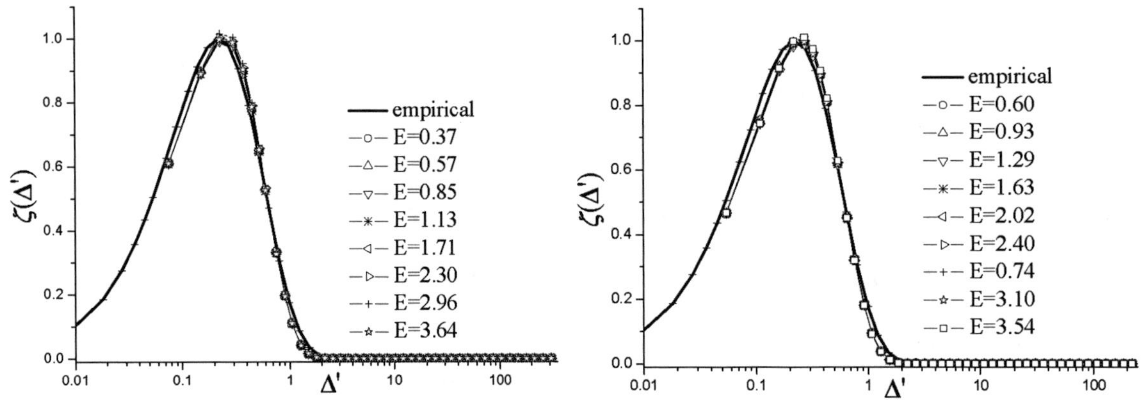

(a) glass 265μm (b) glass 375μm

FIGURE 9. Normalized deviations for the suspension of glass.

CONCLUSIONS

In this work, only the effects from particle overlapping are studied, i.e. the monolayer density is assumed to be very low so that effects from the monolayer structure can be excluded. For beams with finite diameters ($\Lambda \neq 0$), particle overlapping has two effects on the transmission fluctuation correlation spectrum: as the extinction increases, the step height of the transition function increases and the transition function shifts to the range of larger dimensionless correlation time gradually. It is found that these effects are closely related to the extinction E, the layer number N and the beam-to-particle diameter ratio Λ. An empirical expression for the normalized deviation is achieved from simulated data and it was proved by the measurements as well. Therefore, the empirical expression can be hopefully used to extract the information on particle size distribution and particle concentration from the transmission fluctuation correlation spectrum, whereby the particle concentration is moderate or even higher.

In usual cases, particle overlapping and the monolayer structure exist simultaneously, which causes the case more complicated. This will be later discussed in detail.

ACKNOWLEDGEMENTS

The idea of TFS-TC method was initiated during work on a project funded by DFG grant Ri 533/7-2 and is currently being investigated under NSFC 50376041 and the ShuGuang project 04SG49, which is gratefully acknowledged.

REFERENCES

1. Breitenstein, M.; Kräuter, U.; Riebel, U.: The Fundamentals of Particle Size Analysis by Transmission Fluctuation Spectrometry. Part 1: A Theory on Temporal Transmission Fluctuations in Dilute Suspensions. *Part. Part. Sys. Charact* **16,** 1999, pp. 249-256.
2. Breitenstein, M.; Riebel, U.; Shen, J.: The Fundamentals of Particle Size Analysis by Transmission Fluctuation Spectrometry. Part 2: A Theory on Transmission Fluctuations with Combined Spatial and Temporal Averaging. *Part. Part. Sys. Charact* **18,** 2001, pp. 134-141.
3. Shen, J.; Riebel, U.: The Fundamentals of Particle Size Analysis by Transmission Fluctuation Spectrometry. Part 3: A Theory on Transmission Fluctuations in a Gaussian Beam and with Signal Filtering. *Part. Part. Syst. Charact* **20**, 2003, pp. 94-103.
4. Shen, J.; Riebel, U.: Particle Size Analysis by Transmission Fluctuation Spectrometry: Experimental Results Obtained with a Gaussian Beam and Analog Signal Processing. *Part. Part. Syst. Charact* **20**, 2003, pp. 250-258.
5. Shen, J.; Riebel, U.; Breitenstein, M. and Kräuter, U.: Fundamentals of Transmission Fluctuation Spectrometry with Variable Spatial Averaging. *China Particuology* **1**, 2003, pp. 242-246.
6. Shen, J.; Riebel, U.; Guo, X.: Transmission Fluctuation Spectrometry with Spatial Correlation. *Part. Part. Syst. Charact* **22,** 2005, pp. 24-37.
7. Shen, J.; Riebel, U.; Guo, X.: Measurements on Particle Size Distribution and Concentration by Transmission Fluctuation Spectrometry with Temporal Correlation. *Opt. Lett* **30,** 2005, pp. 2098-2100.
8. Guo, X.; Riebel, U.: Experimental Study on Particle Size Distribution and Concentration Using Transmission Fluctuation Spectrometry with the Autocorrelation Technique. *Part. Part. Syst. Charact* **22**, 2005, pp. 161-171.
9. Shen, J.; Riebel, U.: Transmission Fluctuation Spectrometry in Concentrated Suspensions. Part 1: Effects of the Monolayer Structure. *Part. Part. Syst. Charact* **21,** 2004, pp. 429-439.
10. Riebel, U.; Shen, J.: Transmission Fluctuation Spectrometry in Concentrated Suspensions. Part 2: Effects of Particle Overlapping. *Part. Part. Syst. Charact.* **21,** 2004, pp. 440-454.
11. Shen, J.; Riebel, U.: Transmission Fluctuation Spectrometry in Concentrated Suspensions. Part 3: Measurements. *Part. Part. Syst. Charact* **22,** 2005, pp. 14-23.
12. XU Ya-min; SHEN Jian-qi, Simulation on High Concentration Effects of Transmission Fluctuation Spectrometry: *Chinese Journal of Lasers* **33(2)**, 2006, pp. 253-260.
13. Shen, J.; Cai, Xiao-shu.; Yu Bin, Particle Sizing by Transmission Fluctuation Autocorrelation Spectrometry. *Journal of Engineering Thermophysics* **27(5)**,2006, pp. 795-798.

Multi-scale Analysis of Flow Velocity of drag-reducing Surfactant Solutions[1]

Weiguo Gu, Dezhong Wang, Hanping Chen, Yanping Wang

School of Mechanical Engineering, Shanghai Jiao Tong University
800 Dong Chuan Road, Shanghai, 200240, China

Abstract. The fluid velocity in the straight channel was measured by Phase Doppler Animometry (PDA), and the characteristics of velocity fluctuations were studied in this paper. Because the cause of drag reduction was considered that the micelles damped the velocity fluctuations, the difference of fluctuating behavior between water and surfactant solution flow were analysed in different time scales. The results showed that the micelles could damp the lower-scale fluctuations effectively. The key zone where the drag was reduced was between the boundary layer and the core zone of the flow. The energy spectrum showed the kinetic energy distribution and the obvious difference existed in larger-scale fluctuations because the energy transportation between scales was changed.

INTRODUCTION

If adding a little surfactant (CTAC) to water in the straight pipe, more than 70% of friction drag was reduced when the fluid was in the turbulent state. [1] This experimental results indicated that less power was needed for driving the flow which meant great value in energy saving. The reason why drag reduction was caused attracted many researchers to study.

In turbulent water flow, the viscosity caused the loss of kinetic energy. If the loss was reduced, the power to offset it was therefore reduced. The kinetic energy dissipation was not only related with the viscosity, but also related with the changes characteristic of turbulent flow structures. As a result, study on the flow structures of the drag reduction fluid was very important. Nowadays, many consensus understandings were obtained: the fluctuations of velocity normal to the wall surface were damped and the streamwise velocity was enhanced; Reynolds shear stress was decreased and the stress deficit was found; the capability of heat exchange between the fluid and the wall was decreased, and so on.[2-4] These results helped to understand the cause of drag reduction.

The reason why the velocity fluctuations were damped was considered that surfactant additives formed rod-like micelles with network structure to damp the velocity fluctuations in the shear-induced structure (SIS). [5] The image of SIS formed in one surfactant solution was obtained by microscopy. [6] Consequently, the rules how the micelles damped the velocity fluctuations became the important and key topic to explain the mechanism of drag reduction.

It was known that the eddy in the turbulent flow was composed by eddies with different scales. The capability of micelles to weaken the velocity fluctuations was related with the scale of eddies. Some

[1] Contract grant sponsors: Specialized Research Funding for the Doctoral Program of High Education (20030248032) and the National Natural Science Foundation of China (No. 50476016).

CP914, *Multiphase Flow: The Ultimate Measurement Challenge,* edited by X. Cai, Y. Wu, Z. Huang, S. Wang, and M. Wang

researchers studied the scale characteristic of the surfactant solution flow by wavelet transform, but the scale range and detailed rules how micelles affected the flow structure were not studied in detail.[7]

This paper mainly studied on the characteristic difference of velocity fluctuations between water and surfactant solution in different time scales.

EXPERIMENTAL APPARATUS

Figure 1 showed a schematic view of the experimental apparatus for measuring the velocity of surfactant solution. The width (*W*) and height (*H*) size of the straight channel were 20 and 250 mm. The testing location was within the channel 200 times of W away from the inlet where the flow was considered as full development. The distance for full development was introduced in some papers was at least 130 times of hydraulic diameter.[8] The velocity testing point was along the direction normal to the wall surface with the relative height 0.5H, and 0.2 mm as spacing distance for first 10 points near the wall and 1 mm for next outer 7 points. The data collecting rate of PDA in this experiment was 1~1.5 kHz, and the time for continuous collecting in one location was 2 minutes. The tracing particle added to the fluid for velocity measurement was alumina with the size about 100 um.

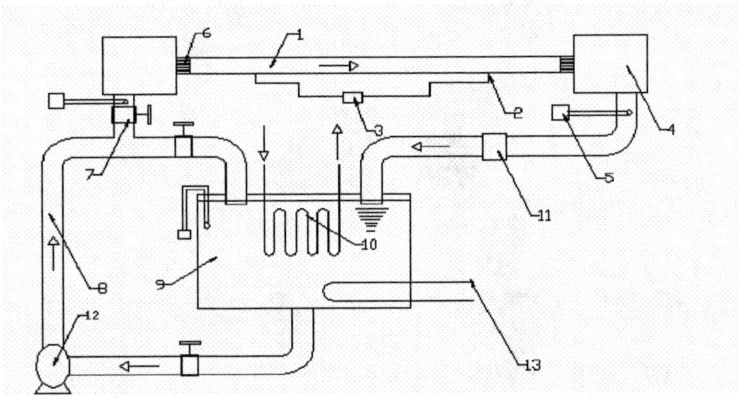

FIGURE 1. Experimental System and Main Apparatus

1.Flow Channel 2.Preasure Test Location 3.Manometer 4.Water Delay Tank 5.Thermometer 6. Commutating Grid 7.Valve 8.Water Pipe 9.Water Tank 10.Cooling Water Pipe 11.Flowmeter 12.Pump 13.Water Heater

The materials added to the fluid were Cetyltrimethylammonium Chloride (CTAC) and Sodium Salicylate with same Mol concentration. The mass concentration of CTAC was 90 ppm at which mass concentration the drag reducing performance was obvious. In the study by Kawaguchi, drag reduction was appeared obviously when the mass concentration reached 30 ppm. This difference was caused by different channel size and the working temperature. In this experiment, the environment temperature was about 30℃.

RESULTS AND DISCUSSION

The velocity was measured by PDA in water and surfactant solution flow with the same flux. The Reynolds Number of the flow is 1.4×10^4.

Mean Velocity

Figure 2 showed the mean streamwise velocity profile of the water and surfactant solution flow in the cross section of the channel. This profile showed change trend of the velocity in the flow, and the difference between water and surfactant solution was typical as same as what another researchers had found. The main difference was found near the wall.

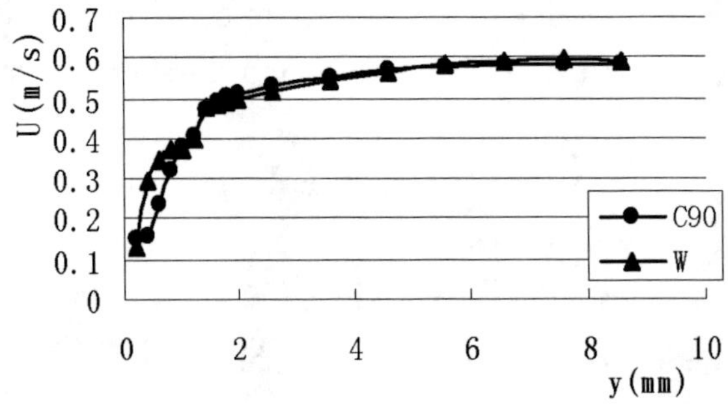

FIGURE 2. Mean Streamwise Velocity Profile

Wavelet Analysis

Wavelet transform was an important means to analyse the random signals in different scales of the independent variables. In this paper, the velocity fluctuations were transformed by wavelet function.

Velocity fluctuations:

$$u'(t) = U(t) - \overline{U} \tag{1}$$

U(*t*) was the steamwise velocity of the flow measured by PDA within 2 minutes. $\overline{U}$ was the mean velocity.

The wavelet transform coefficient *W*(*j*,*b*):

$$W(j,b) = \frac{1}{\sqrt{2^j}} \int u'(t) \cdot \varphi(\frac{t-b}{2^j}) dt \tag{2}$$

The relative wavelet coefficient:

$$W'(b) = W(j,b)/W_{max} \tag{3}$$

W_{max} was the maximal wavelet coefficient.

$\varphi(t)$ used in this computation was Marr wavelet function.

$$\varphi(t) = (1-t^2) \cdot e^{-t^2/2} \tag{4}$$

Figure 3 showed the relative wavelet coefficients *W'(b)* in some time scale (*j*). The location was 2 mm from the wall. When the time scale was smaller (e.g. *j*:-10 and -8), most value of *W'(b)* of surfactant solution was relatively nearer the zero line (*y*=0) than water. It was indicated that its velocity fluctuations in this scale were damped effectively. With the value of j increasing, this difference became less clear, because the micelles' capability was decreased to damp the larger-scale fluctuations, and later the reverse characteristics were showed. The reason why the fluctuations of surfactant flow in lager time scale were enhanced was relative with the kinetic energy dissipation.

The results in Figure 3 proved that the micelles had the probability to change part structures of the flow. The flow structures in boundary layer differed with that in the core turbulent area, so different characteristic existed by the change of the normal distance (*y*). Figure 4 showed the probability distribution of Wavelet coefficients with the different time scales and locations. Firstly, when the time scale was larger, the distribution had no obviously difference between both flows. Moreover, the fluctuating range was relatively lager in this scale which meant more kinetic energies existing in the lager-scale fluctuations. Secondly, surfactant solutions fluctuated more intensely than water with lower-scale near the wall. One reason was that the SIS structure in drag reduction solution was not formed successfully near the wall as the velocity was lower. Moreover, the fluctuations were much intense within the boundary layer because the wall disturbed the fluid. Lastly, the probability distribution showed no obviously difference in the core zone of the flow where the fluctuations with all scales were weak comparatively.

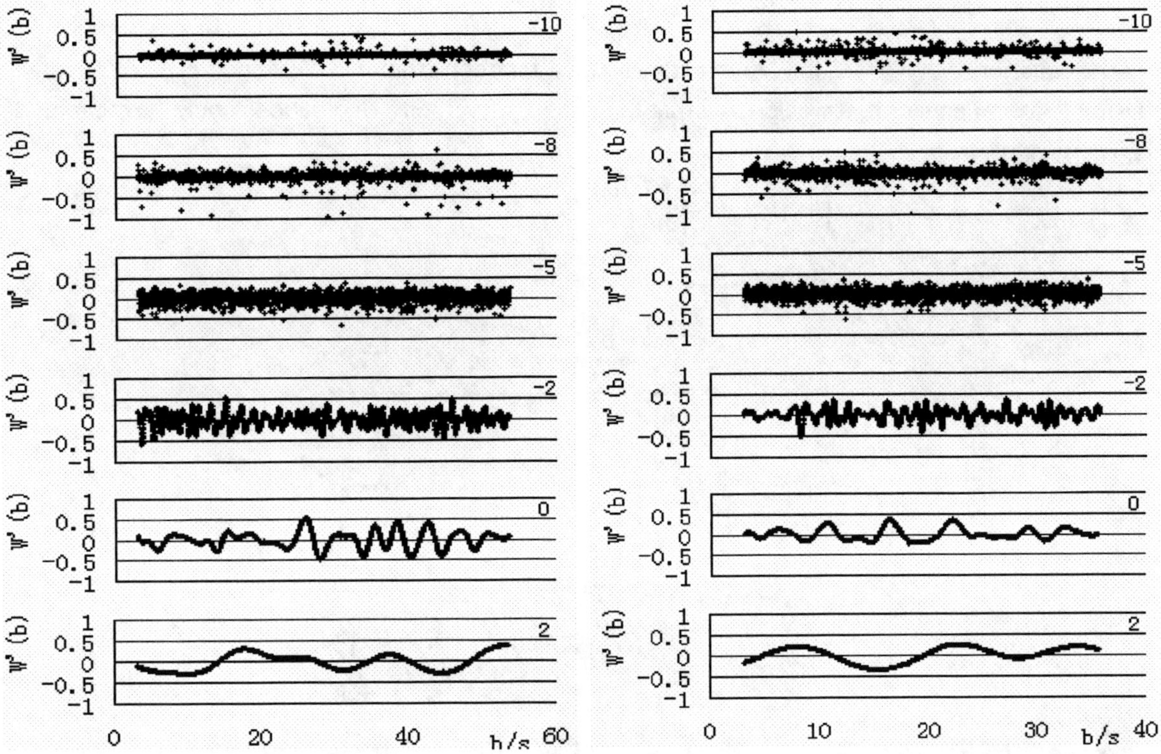

(a) CTAC Solution (b) Water

FIGURE 3. Wavelet Coefficient (y=2mm)

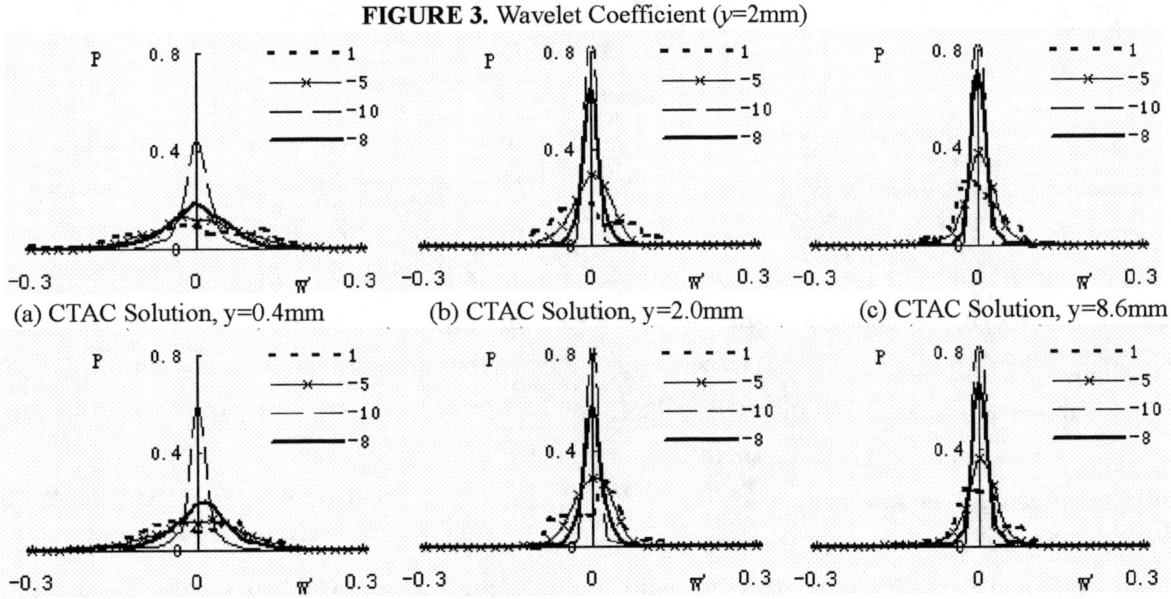

(a) CTAC Solution, y=0.4mm (b) CTAC Solution, y=2.0mm (c) CTAC Solution, y=8.6mm

(e) Water, y=0.4mm (f) Water, y=2.0mm (g) Water, y=8.6mm

FIGURE 4. Probability Distribution of Wavelet Coefficient

Wavelet Energy Spectrum Analysis

If the micelles damped the velocity fluctuations effectively, the kinetic energy dissipation was weak, which would cause different energy distribution characteristics in some scales. The wavelet energy in one scale was defined:

$$E(j)=\frac{1}{C_{\varphi}}\int\left|W(j,b)\right|^{2}db \tag{5}$$

$$C_{\varphi}=\int\frac{\hat{\varphi}(w)^{2}}{|w|}dw \tag{6}$$

In the equation, $\hat{\varphi}(w)$ was the Fourier transformed function from wavelet function $\varphi(t)$.

Using Parseval equation, it was proved that the energy of velocity fluctuations could be computed by wavelet coefficients.

$$\int\left|u'(t)\right|^{2}dt=\int\frac{E(a)}{a^{2}}da\,,\quad a=2^{j} \tag{7}$$

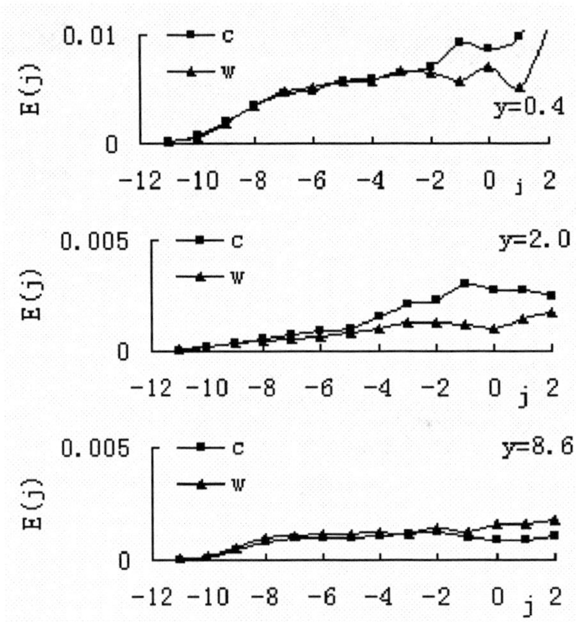

FIGURE 5. Wavelet Energy Spectrum

Figure 5 showed the wavelet energy spectrum of velocity fluctuations with different time scales. Near the wall, the value of wavelet energy was larger universally because of the intense fluctuations disturbed by the wall. In the core zone of the flow, the lower value of wavelet energy indicted that the velocity fluctuations were weak and the micelles almost did not change the flow structures. The obvious difference was appeared at the location y=2 mm where the area was called boundary layer transition. When *j* was greater than -5, the energy value of the surfactant solution was higher than the value of the water. It was well known that the turbulent kinetic energies were transported from lager scales to smaller scales. When the smaller-scale velocity fluctuations were damped by micelles, the energy dissipation was decreased which led the less kinetic energy transportation from the lager scales. So the higher value of the wavelet energy was proved the fact that the micelles damped the velocity fluctuations actively in this zone. As a result, it was considered as the key area to cause drag reduction. And -5 could be seen as the maximal scale number (*j*) under which the micelles could damped the fluctuations effectively, because when j was greater than -5, the energy was higher than water which indicated that the fluctuations were enhanced and not damped. In the inner of the flow, the new balance was formed between dissipation and energy transition. The reason why the fluctuations of surfactant solution were enhanced was the weakness of kinetic energy dissipation.

CONCLUSIONS

This paper studied the characteristics of the velocity fluctuations by the means of wavelet transform. The water and surfactant solution with the mass concentration 90 ppm were studied.

It was proved that the smaller-scale velocity fluctuations were damped effectively in the surfactant solution flow. And there was no obvious difference appeared form the lager-scales fluctuations. The weakness of the kinetic energy dissipation in smaller scale caused the higher kinetic energy appeared in lager scale, because the energy transportation was decreased between scales. The maximal scale number was estimated at -5.

In the boundary layer, the fluctuations were more intense and SIS structure was not formed normally because of the low velocity, as a result less weakness upon fluctuations was appeared compared with water. And in the core zone of the flow, the fluctuations were not intense enough, which cause no obviously characteristic difference compared with water. But the area between the tow zones, where the kinetic energy dissipation was decreased greatly, was the main zone for drag reduction being caused.

ACKNOWLEDGEMENT

The research was supported by Specialized Research Funding for the Doctoral Program of High Education (20030248032) and the National Natural Science Foundation of China (No. 50476016). The work was carried out at the key Laboratory of Power Machinery and Engineering of Education Ministry, P.R.C.

REFERENCE

1. Chara Z, Zakin JL, Severa M, Myska J, Turbulence measurements of drag reducing surfactant systems, Exp. Fluids, 16, 1993, 36-41;
2. F. -C, Y. Kawaguchi, Investigation on the characteristics of turbulence transport for momentum and heat in a drag reducing surfactant flow, Phys. Fluids, Vol.16, No.9, Sep.2004, 3281-3294;
3. Y. Kawaguchi, Y. Tawaraya, A. Yabe, K. Hishida, and M.Maeda, Turbulent transport mechanism in a drag reducing flow with surfactant additive investigated by two components LDV, Proc. 8th Int. Symp. Appl. Laser Tech. Fluid Mech., Lisbon, Portugal, 1996, 29-41;
4. G. Hetsroni, J. L. Zakin, and A. Mosyak, Low-speed streaks in drag-reduced turbulent flow, Phys. Fluids 9, 2397(1997);
5. J. L. Zakin, J. Myska, and Z. Chara, New limiting drag reduction and velocity profile asymptotes for nonpolymeric additives systems, AICHE J. 42, 3554(1996);
6. T.M. Clausen, P.K. Vinson, J. T. Minter, H. T. Davis, Y. Talmon, and W.G. Miller, Viscoelastic micellar solutions: microscopy and rheology, J. Phys. Chem. 96, 474(1992);
7. F. –C. Li , Y. Kawaguchi, and T. Segawa, Reynolds-number dependence of turbulence structures in a drag-reducing surfactant solution channel flow investigated by particle image velocimetry, Phys. Fluids 17, 075104(2005);
8. K. Gasljevic, E. F. Matthys. Experimental investigation of thermal and hydrodynamic development regions for drag reducing surfactant solution[J]. Trans. Amer. Soc. Mech. Eng., J Heat Transfer, 1997, 119: 80-88.

A New Method for Monitoring Branch Pipes of PCI into BF

Wang Yutao* Lu Zengxi* Yang Gang* Wang Shi* Ge Xueping+

*Key Laboratory of Integrated Automation of Process Industry (Northeastern University), Ministry Education, 110004, China;
+Ironmaking Plant, Baoshan Iron & Steel Co., Ltd, Shanghai, 200192, China)

ABSTRACT. On leaving the pulversing mill of blast furnace, the solid-air mixture is split into several different pipes each feeding an individual blast tuyere. Based on detected concentration with ECT and velocity of pulverized coal with correlation method, a new method of monitoring the states of the branch pipes of PCI into BF is introduced here. All the abnormal transporting states of coal powder, such as clogging, stopping and unimpeded transporting can be deduced and alarmed on line. The flowrate distributed to each branch pipe from general pipe, which is given as an extern signal, can be computed continuously to watch on the uniformity of pulverized coal. The method was run for test on a blast furnace of baosteel in December, 2000, and run in August 2000. Now it has been used on three of four blast furnaces in baosteel.

INTRODUCTION

Pulverized coal injection has been proved to be a useful method to decrease coak to metal ratio and control the temperature of blast furnace [1]. On leaving the pulversing mill pulverized coal is distributed into several different pipes and each pipe feeds an individual blast tuyere. Pulverized coal conveying in the branch pipes is a kind of typical gas/solid two-phase flow, having the characteristics of diverse flow regime and unstable combination of powder of granular materials to gas ratio, especially for dense phase conveying of PCI system. To reduce the cost of ironmaking and achieve the high combustion efficiency of blast furnace, it is essential to monitor the transporting state of pulverized coal in each pipe.

Now temperature difference and optics methods are usually used to identify the blockage of pulverized coal in the branch pipe of blast furnace and alarm on line. With temperature difference method [2] the temperature of pulverized coal is measured continuously by the temperature measuring units installed on the pipes with heat insulator and the temperature of branch pipe is measured by the detector without heat insulator. The temperature of pulverized coal is about 70°C, which is higher than the external environment. The temperature of branch pipe can drop when the blockage of pulverized coal happening. It may be applicable if all the branch pipes are installed in a close room. When the branch pipes are installed in an open place, the temperature of each branch pipe will vary with seasons, wind direction, and environment etc., it will also vary with different installed position. Therefore, the variation of temperature difference is difficult to be identified. This method has the fault of responding slowly for it can work only after detecting adequate falling values of temperature.

In some places optical method can also be adopted [3]. Optical detector is installed in the conical space in front of the peephole of the tuyere, which is not convenience to be taken down during the period of repairing device. This method is not appropriate under dense conveying for opaque solids-air mixture. Moreover, highly disordered flow regimes often occurring, such as pulsatile flowing, also help to limit this method.

The proposed method here can deduce all the abnormal flowing in the branch pipes, such as clogging, stopping and unimpeded transporting, and alarm on line under abnormal condition based on detected solid concentration and velocity. It has the advantage of judging the transporting state of branch pipe rapidly, in most cases it can work within 30 seconds. Another advantage is able to compute the flowrate distributed to each branch pipes from general pipe, which is given as an external signal, to monitor the uniformity of pulverized coal.

Industrial application proves that this method can alarm quickly whenever what type of abnormal flowing happens according to the applications on baosteel since August, 2000. Now it has been used on three of four blast furnaces of baosteel. It shows that this method is stable, the transporting state of branch pipes of BF can be identified accurately, and the distributed flowrate of branch pipes is worthy of belief.

CP914, *Multiphase Flow: The Ultimate Measurement Challenge,* edited by X. Cai, Y. Wu, Z. Huang, S. Wang, and M. Wang

MONITOR SYSTEM PRINCIPLE

Concentration Measurement Principle

Capacitance transducer is adopted here to measure the tiny change of phase concentration in monitor system. Here a capacitance sensor with symmetric side-guards is adopted. In order to overcome the drawbacks of capacitance phase concentration sensors, i.e., soft-field effect, the finite element analysis method of static electric field and the orthogonal design method are combined to optimize the configuration of capacitance concentration sensor. Homogeneous error (SVP) and relative sensibility are chosen as optimized parameters by comparison analysis. The influence of various design parameters on the sensitivity uniformity are also analyzed with capacitance phase concentration sensors different in configuration and physical parameters. The surface electrodes are helical so that the sensing field can be twisted 180° from one end of the sensor to the others, which helps to improve the sensitivity uniformity of sensor from 45.10% to 4.67% effectively [4]. The configuration of phase concentration transducer is shown in Fig. 1.

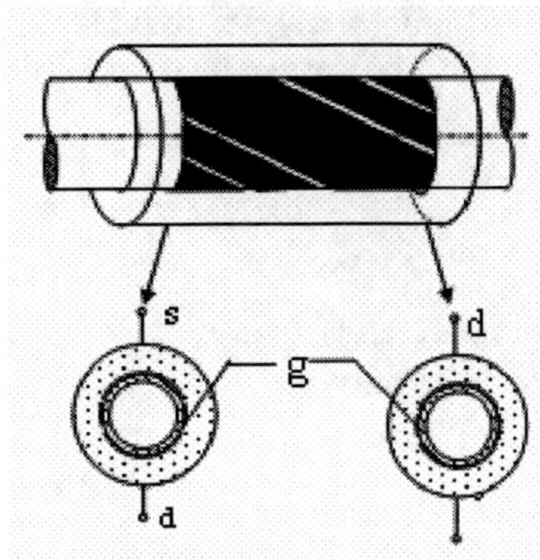

FIGURE 1. Configuration of concentration sensor

In Fig. 1, s notes source plate, d notes sensor plate, g notes side-guide plate.

Velocity Measurement Principle

Cross correlation technique is proved to be an appropriate method in velocity measurement of powder of granular materials for pneumatic conveying system [5]. Two optimized parallel surface electrodes capacitance sensor with eight pairs of plates are adopted here, the configuration of velocity transducer is shown in Fig. 2.

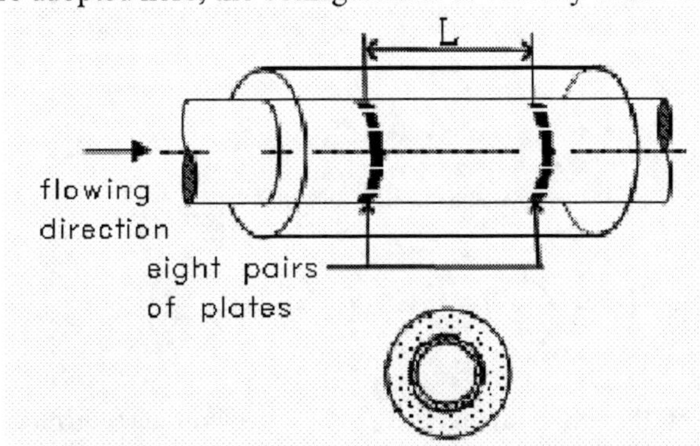

FIGURE 2. Configuration of velocity sensor

DIAGNOSIS OF MONITORING SYSTEM

Identification of Transporting State

The transporting state of pulverized coal can be deduced based on the detected concentration and velocity and some other auxiliary signals, such as temperature of plates. During diagnosis process velocity signal is chosen as primary parameter combined with concentration signal, and sometimes temperature signal is used as an amendatory parameter to improve the accuracy.

The diagnostic results can be completed during 2~3 data process period to do comprehensive analysis between accuracy and time. Now all abnormal state, such as clogging, stopping and unimpeded transporting, can be given within 30 seconds.

Self diagnosis is also available when there are wrong with concentration and velocity sensors and other hardware.

Distribution of the Flowrate

The uniformity of the flowrate of each branch pipe has direct impact on the combustion efficiency of coal. It has no available metering device for measuring the flowrate of branch pipes. It is also impossible to calibrate the flowrate of branch pipes on line in PCI system. However, the flowrate of each branch pipe can be computed if given the flowrate of general pipe and relative parameters of branch pipe flowrate, such as concentration and velocity. The flowrate of general pipe is given as an external signal from electronic scale.

The equivalent dielectric constant of solid-air mixture can be expressed using the following equation [6]:

$$\varepsilon = \frac{V_s}{V_0}\varepsilon_s + \frac{V_g}{V_0}\varepsilon_g \tag{1}$$

Where V_0 is the total volume of mixture, V_s and V_g are the volume of pulverized coal and compressed air respectively, ε_s and ε_g are the dielectric constant of pulverized coal and compressed air respectively.

When pulverized coal flows in the branch pipes, it can be considered that $V_s \ll V_g \cong V_0$, therefore, the above equation can be simplified as the following:

$$\varepsilon - \varepsilon_g = \frac{V_s}{V_0}\varepsilon_s \tag{2}$$

Dielectric constant ε is proportional to the capacity, i.e., $C=K\varepsilon$, K is the proportion factor between C and ε , the capacity difference between measured capacity of pulverized coal and compressed air can be expressed as $\Delta C=K\cdot\varepsilon_s\cdot V_s/V_0$, the volumetric concentration of coal is $\beta=(V_s/V_0)\cdot\rho_s$ (unit kg/m^3), ρ_s is actual density of coal, hence the following equation can be given:

$$\beta = \frac{\Delta C}{K\varepsilon_s}\rho_s \tag{3}$$

The flowrate of the *ith* branch pipe in PCI system can be expressed as:

$$Q_i(t) = A\upsilon_i \cdot \beta_i = \frac{A}{K\varepsilon_s}\rho_s \cdot \Delta C \cdot \upsilon_i \tag{4}$$

Where υ_i is the velocity of the ith branch pipe, ΔC_i is capacitance difference of the ith branch pipe, β_i is concentration of the ith branch pipe, A is cross-sectional area of branch pipe.

The flowrate of general pipe can be denoted as:

$$Q(t) = \sum_{i=1}^{N} Q_i(t) = \frac{A}{K\varepsilon_s}\rho_s \cdot \sum_{i=1}^{N}\Delta C_i \cdot \upsilon_i = \frac{\Delta W}{\Delta t} \tag{5}$$

Where Q(t) is the flowrate of general pipe, N is the number of branch pipes, $\Delta W/\Delta t$ is the variation of the weight of coal sampled from electronic scale per unit time.

At the same time the parameter of A and ρ_s and K and ε_s can be considered as constant for different branch pipe, thus the distribution flowrate of the ith branch pipe can be deduced as the following:

$$Q_i(t) = \frac{\Delta W}{\Delta t} \cdot \frac{\Delta C_i \cdot v_i}{\sum_{i=1}^{N} \Delta C_i \cdot v_i} \qquad (6)$$

CONFIGURATION OF MONITORING SYSTEM

The configuration of monitoring system can be divided into several parts, including branch pipe detector (including concentration and velocity transducer), data sampling and processing unit, data transmitting unit and IPC. Branch pipe detector is installed on the seven platform of BF, IPC is installed in the center control room of BF. The signals of velocity (including upstream and downstream signals), concentration and temperature of each branch pipes are transmitted with RS485. Other signals, such as the flowrate of general pipe, are also sampled and transmitted by RS485 to IPC. IPC can be communicated with other IPC using OPC protocol.

RESULTS

Main monitoring interface is shown as Fig. 3. and Fig. 4. It can indicate that different transporting state of different branch pipe with different color. The flowrate of different branch pipes is shown in Fig. 3, and the flowrate of general pipe is shown in the bottom.

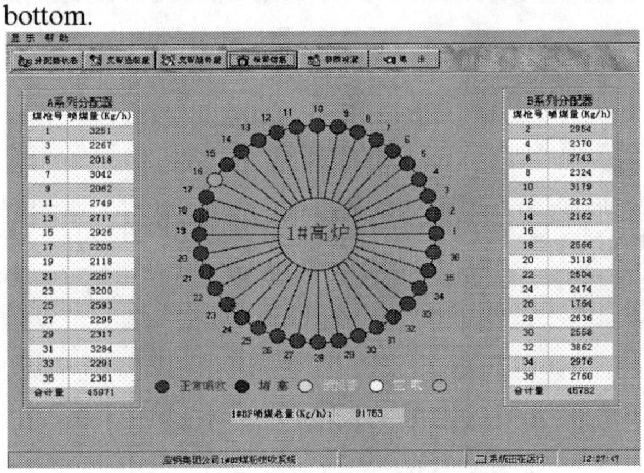

FIGURE 3. Main interface of monitoring system

The flowrate of general pipe is 91753kg/h, the flowrate of serie A is 4597kg/h, the flowrate of serie B is 45782kg/h, the flowrate of different series is about the same according to Fig. 3.

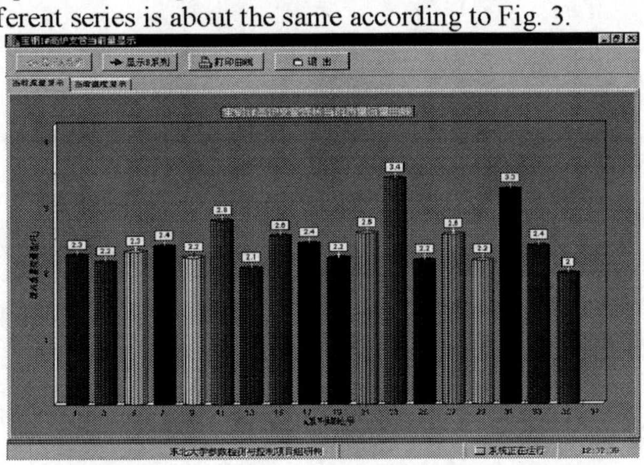

FIGURE 4. Bar graph of distributed flowrate of series A

Figure 4 shows the different flowrate of branch pipes for different pulverized series. The flowrate of the 23th and the 31th branch pipe is larger than other pipes in Fig. 4. It showed that disordered flow regimes occurred at this time.

At this time blockage was apt to occur, and these two branch pipes should be noticed.

Figure 5 shows the response curve of flowrate vs time comparing the proposed method with temperature difference method during occurred and removed blockage of 19th branch pipe. When blockage occurred the flowrate of 19th branch pipe decreases to zero at time 14:06:00 in Fig. 5(a), while it is about 14:13:00 in Fig. 5(b).

It is obvious that the proposed method can respond more rapidly comparing with time difference method. If the values of curve change slightly it indicates stable injection and accurate operation, otherwise, potential occurrence of abnormal state. The width of the sudden change denotes the duration of trouble. The history state of different branch pipe can also be found according to the charts of flowrate vs time.

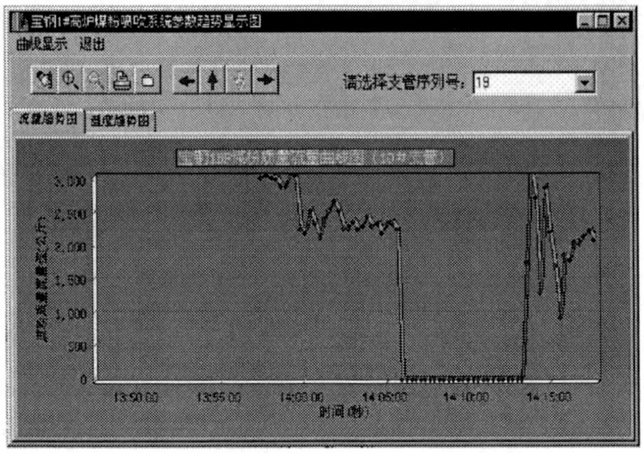

FIGURE 5(a). Response curve of proposed method occurring blockage

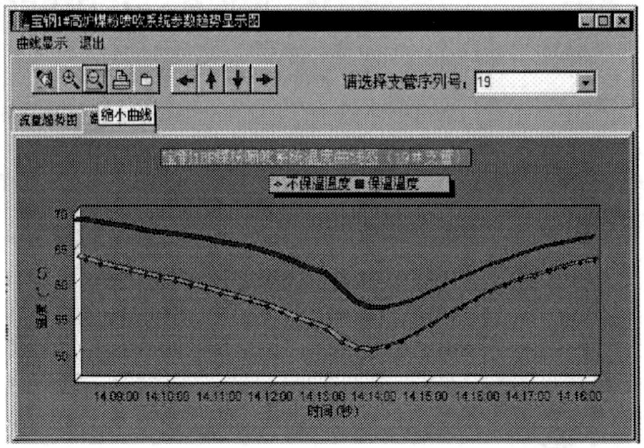

FIGURE 5(b). Response curve of temperature difference method occurring blockage

CONCLUSIONS

Based on detected velocity and concentration online monitoring system can predict the abnormal state of PCI system accurately, and find out the operation by the charts of flowrate vs. time of each branch pipe. It can also indicate the uniformity of the distributor on the tuyere at the same time. The monitor system for PCI pipes with concentration and velocity measuring units have worked well on baosteel for several years. Now it has been used on three of four blast furnaces of baosteel.

ACKNOWLEDGMENTS

The authors would like to thank the Liaoning Liaoning Natural Science Foundation Committee for the support on this project (grant 20062021). The authors would also like to thank Ironmaking Plant, Baoshan Iron & Steel Co., Ltd for their support.

REFERENCES

1. Ching-Ken Ho, Chi-Seng Chou, Sahn-Wen Du, China Steel Technology Report **10**, 81-86 (1996).
2. CHENG Zhengdong, YAN Dingliu, WANG Yuzhu, Iron and Steel **3**, 6-8 (2000).
3. XU Guanzhong, Steel Times **7**, 15-16 (1995).
4. Jin Feng, Lu Zengxi, Zhang Hongxi, Wang Shi, "Study of capacitance sensor for concentration measurement of gas/solid two-phase flow", MCGM'97 Conference Proceedings, Shenyang, 1997, pp. 129-132.
5. Yan Y. et al,. Measurement Science Technology **7**, 1687-1706 (1996).
6. Li Haiqing, Huang Zhirao, *Special Detection Technology*, Zhejiang University Press, Hangzhou, 2000, pp. 12-25.

Time-Serial analysis on Gas-Liquid Two-Phase Bubbly Flow using POD

Masaaki ISHIKAWA*, Kunio IRABU* and Isao TERUYA*

** Dept. of Mech. Sys., Univ. of the Ryukyus, 1 Senbaru, Nishihara, Okinawa, 903-0213, Japan*

Abstract. Bubbly flows are complicated flows because bubbles are deformable and the bubble-bubble interactions work in local area including a lot of bubbles. Such flow occurred turbulence in liquid phase by each bubble is called "pseudo turbulence". A PIV is most effective to measure complicated bubbly flow because of multi-dimensional measurement. Recently Dynamic PIV system which is consisted of high-speed camera and high-repetition laser is developed, and it is possible to capture time-serial images with higher resolution. In this paper POD (Proper Orthogonal Decomposition) is described as post-processing of bubbly flow. We confirmed it has the effect of temporal-spatio smooth filter.

Keywords: Bubbly flow, Infrared shadow technique, Dynamic PIV, POD, Time-serial analysis.

INTRODUCTION

Bubbly flow is seen in industrial fields such as the power plant, chemical plant, bio-reactor and so on. This bubbly flow is a very complicated flow because a bubble is deformable and the bubble-bubble interactions work in local area including a lot of bubbles. Such flow occurred turbulence in liquid phase by each bubble is called "pseudo turbulence". Recently there are many kinds of experimental [1], [2] and numerical studies [3], [4] for bubbly flow. Especially a PIV is most effective tool to measure bubbly flow because of multi-dimensional measurement. Recently Dynamic PIV system which is consisted of high-speed camera and high-repetition laser is developed, and it is possible to capture images with high resolution and time series. In our past research, we explained the combined techniques of IST (Infrared Shadow Technique) and Dynamic PIV and indicated to obtain instantaneous velocity vectors for liquid phase and bubbles with high accuracy [5]. However we haven't obtained time-serial flow structures in this bubbly flow. In this study we describe the POD (Proper Orthogonal Decomposition) known as structure analysis method and some results applied to bubbly jet flow are shown.

EXPERIMENTAL SETUP

Figure 1 shows the schematic diagram of experimental apparatus. Tank size is 600mm × 300mm × 30mm, inlet size is 30mm × 10mm. Tap water and Nitrogen gas are used. Bubbles are generated by 25 needles (outer and inner diameters are 1mm and 0.5mm, respectively) on the bottom. Water flow rate are 18L/min and gas flow rate are 0.45L/min and then void fraction are about 2.5%. A high-speed camera (Photron: FastCam APX120k, 1024×1024pixels, 2000fps), double-pulse laser (New-wave Research Inc.: Pegasus-PIV, Nd:YLF, 527nm, Dual heads, 1000Hz@sigle head), and infrared diode laser (Oxford inc.: 808nm, 500Hz, 4 pulses in burst mode) are used. Liquid phase are illuminated by the laser sheet from the side, and bubble images are illuminated by infrared shadow technique (IST) through the diffusion sheet. Tracer particle is high-porous polymer particle (Diaion HP20SS, Diameter: 60μm, Specific gravity: 1.02). The particle is courted by Rhodamine-B or -6G fluorescence dye. Bubbles and particles are captured by single high-speed camera.

CP914, *Multiphase Flow: The Ultimate Measurement Challenge,* edited by X. Cai, Y. Wu, Z. Huang, S. Wang, and M. Wang

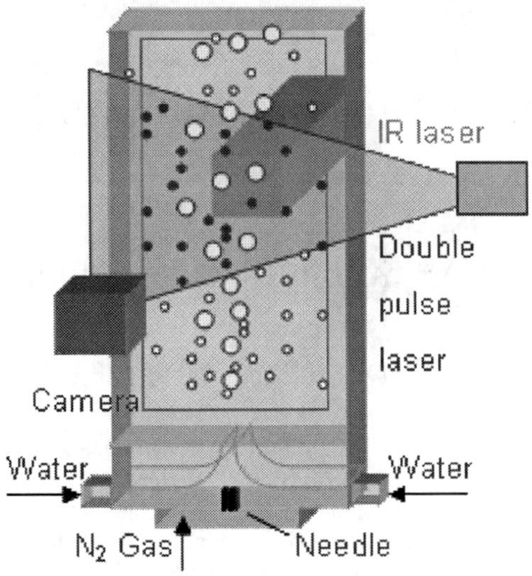

Figure 1. Schematic diagram of experiment apparatus

Figure 2 shows the signal timing chart for camera and lasers. Pulse period which is generated by pulse generator is 2kHz. The pulse transmits to a camera and two flip-flop devices which skip 1 pulse and 3 pulses, respectively. Signals of Each flip-flop device transmit to delay device and Infrared laser. Delay device controls the double-pulse fire signal. Delay time of the infrared laser is 50μs for the positive edge of the base pulse. Pulse interval of infrared laser is 0.5ms sequentially. Double-pulse laser is emitted at 150μs before and after the infrared laser pulse. By this setting, the velocity of liquid phase has second order accuracy for illuminated time of the infrared laser. And also bubble velocity has second order accuracy by using before and after images of the same illuminated time.

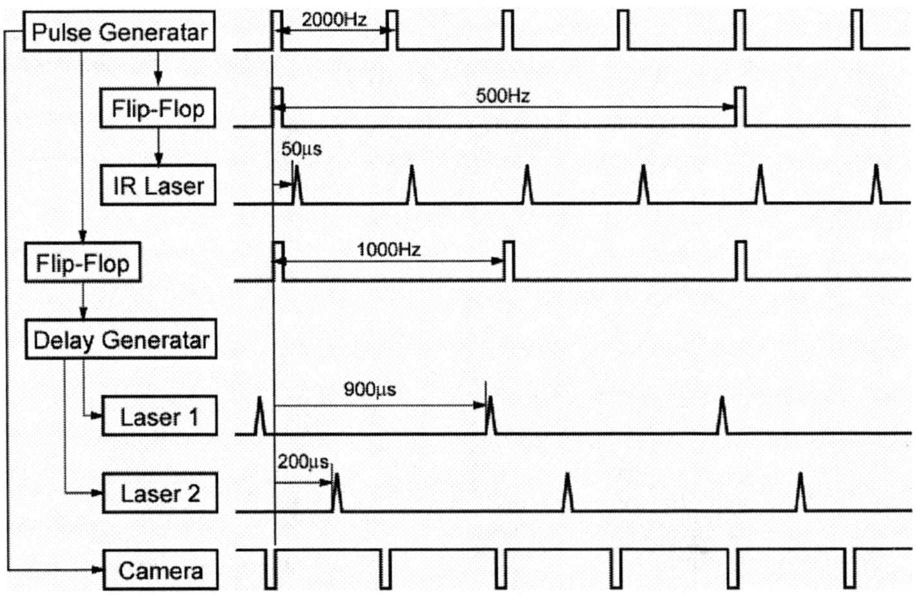

Figure 2. Signal timing chart

POD METHOD

Time-serial velocity vectors obtained by the Dynamic PIV have high temporal and spatio resolution. We hope that not only the spatio data but also temporal data were used in post-processing. In this paper we use POD (Proper

Orthogonal Decomposition) method which are evaluated by the characteristic of flow structure objectively and mathematically [6].

We explain about POD. Fluctuation vector $\mathbf{u}'(t,x,y)$ on a point (x, y) obtained by PIV is transformed to $\mathbf{u}'(f,x,y)$ by Fast Fourier Transformation (FFT). And a cross-correlation matrix $\mathbf{R}_{ij}(f)$ of fluctuation vectors $\mathbf{u}'(f,x,y)$ on the M points is given by the following equation.

$$\mathbf{R}_{ij}(f) = <\mathbf{u}'_i(f,x,y)\mathbf{u}'_j(f,x,y)> \quad (1)$$

where matrix size are $2M \times 2M$. If we thought eigenvalue issues of the matrix $\mathbf{R}_{ij}(f)$, we could give the following equation.

$$\mathbf{R}_{ij}(f)\Phi_m(f) = \lambda_m(f)\Phi_m(f) \quad (2)$$

where $\lambda_m(f)$ is eigenvalue, $\Phi_m(f)$ is eigenvector, and m stands for the mode ($m = 1, \cdots, 2M$).

If we used eigenvector $\Phi_m(f)$ which 1st - n th mode and 1st - f_n frequency, correct fluctuation vector $\tilde{\mathbf{u}}'(f,x,y)$ is obtained by the following equation.

$$\tilde{\mathbf{u}}'(f,x,y) = \sum_{f=1}^{f_n}\sum_{m=1}^{n}\left(\left(\mathbf{u}'(f,x,y)\Phi_m(f)\right)\Phi_m(f)\right) \quad (3)$$

The fluctuation vector $\tilde{\mathbf{u}}'(f,x,y)$ is transformed to $\tilde{\mathbf{u}}'(t,x,y)$ by inverse FFT. If we could select some frequency ranges and some mode ranges, we will be able to get many flow structures with some frequency ranges and some modes.

RESULTS AND DISCUSSIONS

Figure 3 shows the original images at the 18L/min and 0.45L/min in the in water and gas flow rate, respectively. This image is separated to a particle image and a bubble image, and velocities of liquid phase and bubble are calculated by PIV and PTV, respectively. Figure 4 shows velocity map of liquid phase velocity and bubble velocity. Reasonable velocity vectors of liquid phase and bubble are obtained.

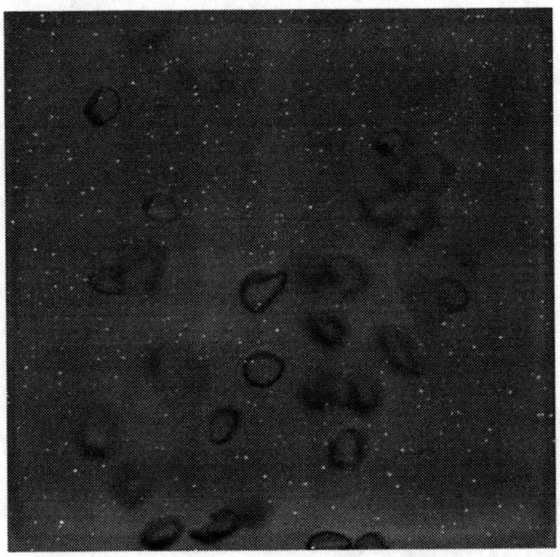

Figure 3. Original Image

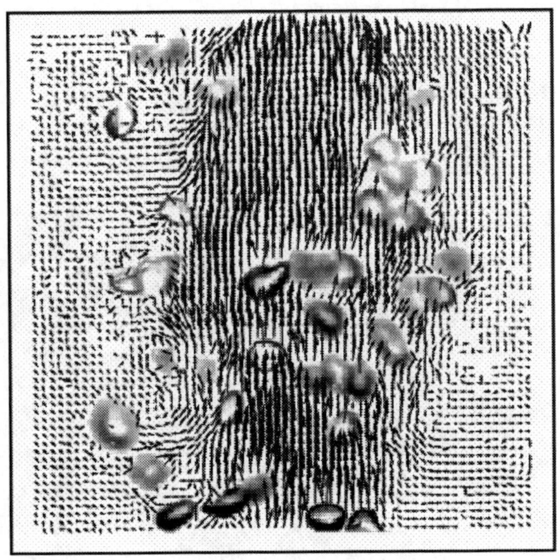

Figure 4. Velocity vector map for liquid phase and bubbles

Figures 5 show the time-serial velocity vector maps which is selected in the left side area from jet center in fig.4. The numbers of grid point are 15 × 31 girds and the numbers of time-serial vector maps are 1024. These time-serial vector maps are separated to 8 blocks including 128 vector maps. Then maximum mode is 8th mode and maximum frequency is 64.

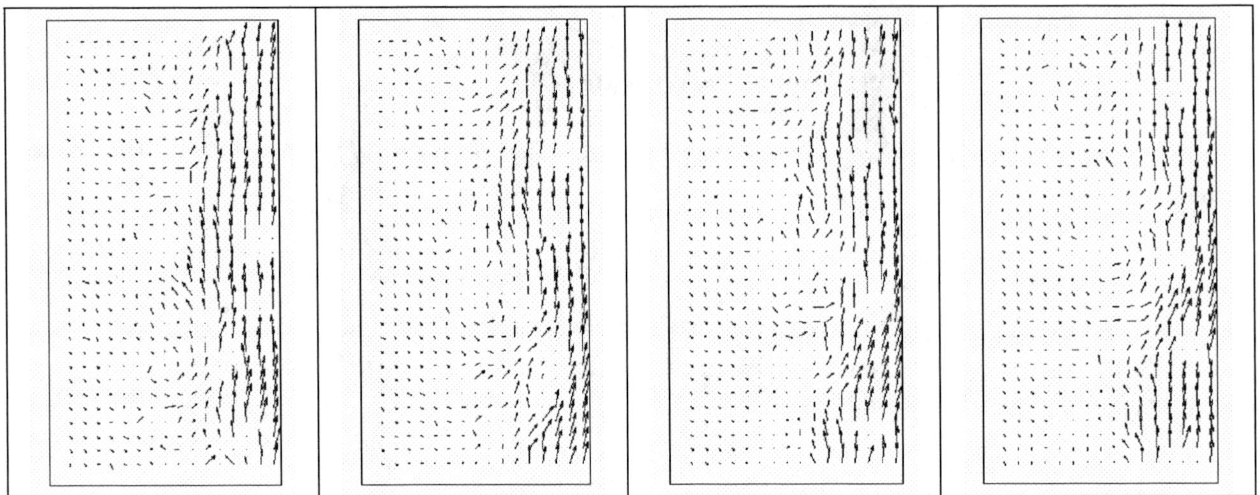

(a) $t = 0$ ms	(b) $t = 10$ ms	(c) $t = 20$ms	(d) $t = 30$ms

Figure 5. Time-serial velocity vector before POD

Figure 6 shows the relation among the eigenvalue, mode, and frequency. From this figure, eigenvalue at 1st mode has maximum. As the eigenvalue get larger mode, eigenvalue decreases monotonically. Here we defined the turbulent energy accumulate ratio η_n as follows:

$$\eta_n = \sum_{i=1}^{n} \lambda_i \Big/ \sum_{i=1}^{2M} \lambda_i \qquad (4)$$

The ratio at 1st - 4th mode occupies about 74% and also the ratio at 1st - 6th mode occupies more than 90%.

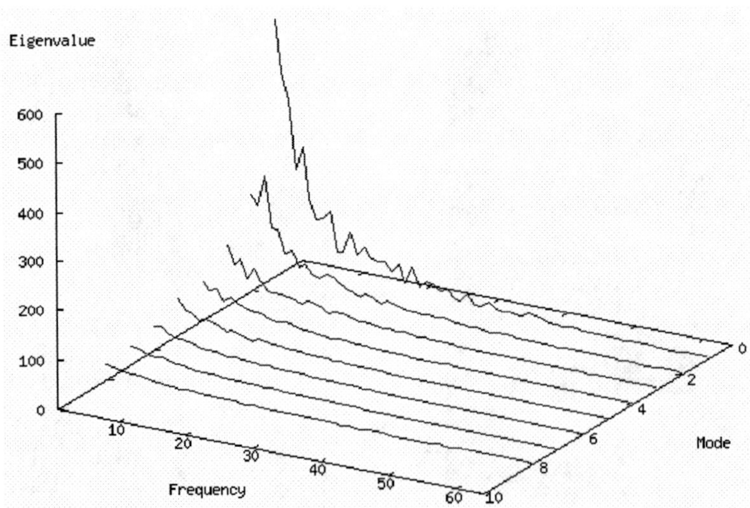

Figure 6. Eigenvalue for each frequency and mode

Figures 7 (a)-(d) show the velocity vector maps after POD processing with 1st - 4th modes and all frequency range. From these results, smooth velocity vectors are obtained. Needless to say, when low frequency was selected, the results were equivalent to low-pass filter. Figures 8 (a)-(d) show the velocity vector maps to change the range of mode. You can see that the smoothest velocity vectors is obtained by the low mode in fig.8 (b) and the vector map close to the original vector map is obtained by the high mode in fig.8 (d).The vector map in fig.8 (d). As the range of mode decrease, velocity vector get stronger spatio smooth filter. Thereby we can find that this POD gives a temporal-spatio smooth filter effect by selecting the range of mode and frequency and also the erroneous vectors reduced.

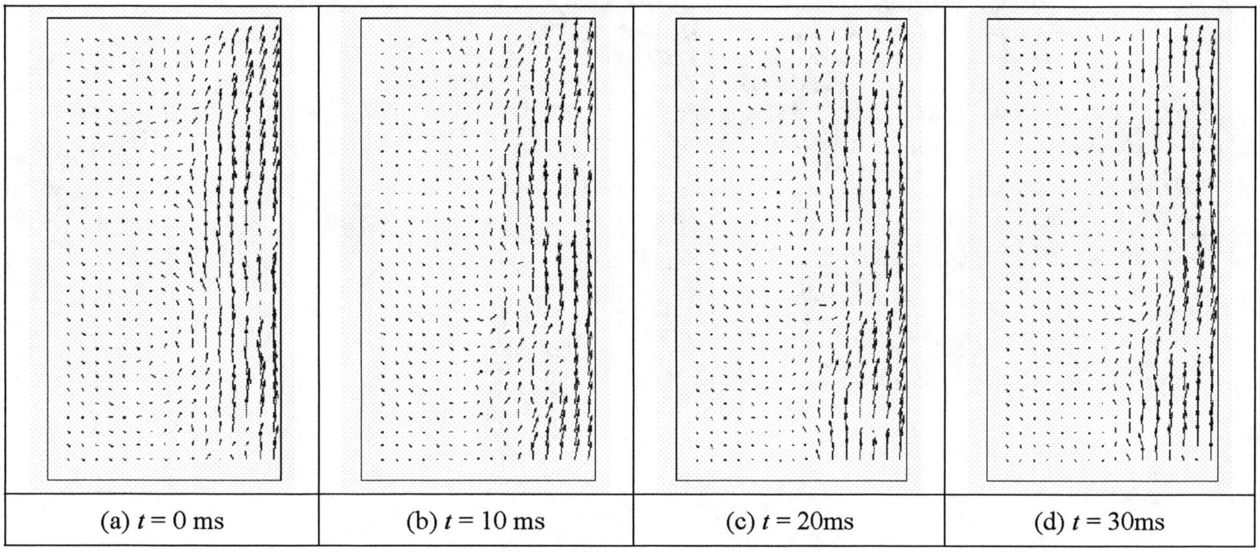

Figure 7. Time-serial velocity vector map after POD (1st - 4th mode, $f_n = 64$)

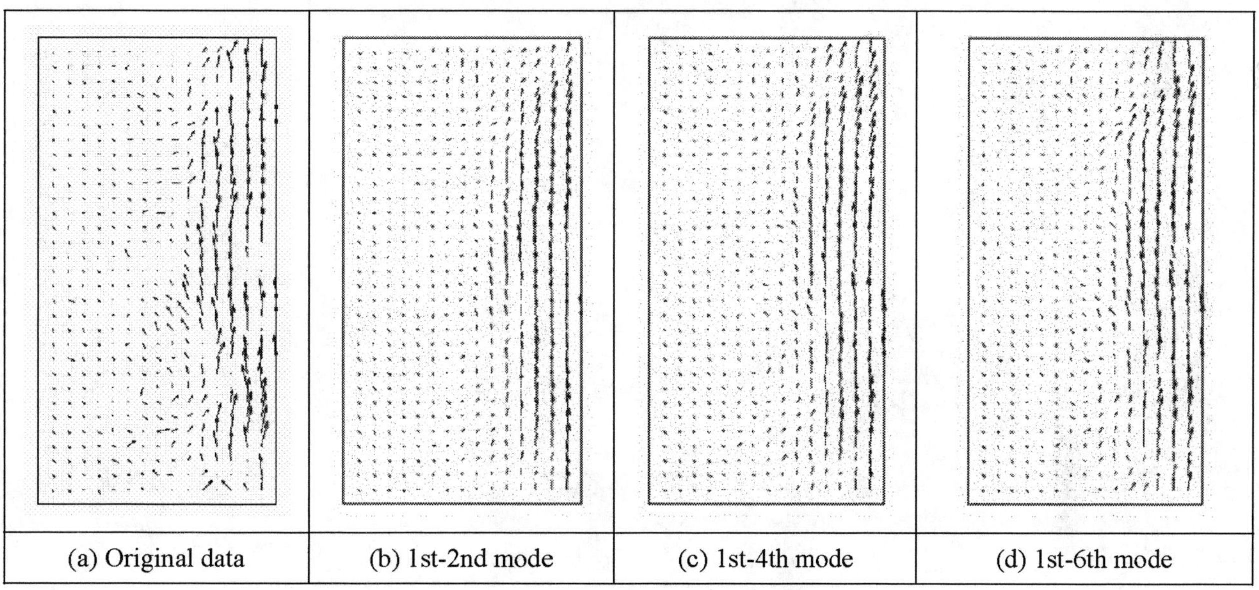

Figure 8. Velocity vector map at each range of mode ($f_n = 64$)

CONCLUSION

In this study we described the POD method and it was applied to time-serial PIV data in the bubbly jet flow. We showed that this POD had the effect of temporal-spatio smooth filter and erroneous vectors reduced. In future work, we will target to the flow structure classified by frequency and mode in bubbly jet flow.

REFERENCES

1. A. Fujiwara, Y. Danmoto, K. Hishida, and M. Maeda, *Experiments in Fluids*, **36**, 157-165, (2004).
2. R. Lindken, and W. Merzkirch, *Experiments in Fluids*, **33**, 814-825, (2003).
3. B. Bunner, and T. Tryggvason, *J. Fluid Mech.*, **495**, 77-118, (2003).
4. Y. Murai, and Y. Matsumoto, *Trans. JSME*, **62-593(B)**, 85-92, (1996). (in Japanese)
5. M. Ishikawa, K. Okamoto, H. Madarame, *Proceeding of VSJ-SPIE'04*, Abstract book: 26-27, (2004).
6. W. Bi, Y. Sugii, K. Okamoto, H. Madarame, *Measurement and Science Technology*, **14**, L1–L5, (2003).

Experimental Investigations on Performance of Collision-block-type Concentrator Using Fiber Optic Probe

Kefa CEN, Hua Yang, Kunzan QIU*, Zhanghua ZHOU, Ligang Zheng, Hao ZHOU

Zhejiang University, Institute for Thermal Power Engineering, State Key Laboratory of Clean Energy Utilization, Hangzhou, 310027, P. R. China

Abstract. The concentrating performance of a collision-block-type fuel rich/lean burner was experimentally investigated using a fiber optic measurement system in a two-phase-flow test facility of internal diameter 150 mm. The local solid concentration and particle size distributions were measured to investigate the separation performance of the fuel-rich/lean separator, also the mixing performance of the parallel fuel-rich and fuel-lean streams. The solid concentration of 0.42 kg/kg was employed in the experiments to investigate the influences of solid concentration on the concentrating performance of the separator. The measurements indicated that the collision-block-type separator always gave good concentrating performance under various inflow solid concentrations. The fuel-rich and fuel-lean streams do not mix together soon after leaving the burner, this leads to better ignition and combustion characteristics of the pulverized coal compared with the ordinary burner.

Keywords: Gas-solid flow, Concentration measurement, Fiber optic sensor, Fuel rich/lean burner, inter-particle collision

INTRODUCTION

Coal remains the major source of energy resource in the world. In recent years, a novel pulverized-coal combustion technology called horizontal fuel-rich/lean burner has been employed widely in power station boilers in China, with the benefits of burning low volatile coal more efficiently and decreasing the NO*x* emissions [1, 2].

In the conventional primary air of a pulverized coal burner, the mass ratio of the fuel/air mixture in the pneumatic pipe is normally about 0.3-0.6 kg (coal)/kg (air), but the optimum ratio for low volatile coals may be 0.6-1.4 kg (coal)/kg (air) as a result of the need to obtain stable flame [3]. When the horizontal fuel rich/lean burner is used in a tangentially fired boiler, a fuel- rich/lean separator needs to be employed in the fuel conveying line to divide the fuel/air mixture into two streams with various solid concentrations. Facing the high-temperature flame, the fuel-rich stream injects through the fuel-rich nozzle, this stream acts as a powerful stabilizer for its high fuel concentration. The fuel-lean stream injects through the fuel-lean nozzle locating between the fuel-rich nozzle and the furnace waterwall, oxidizing zones can be achieved near the waterwall, resulting in the resistance of the furnace against slagging and fouling by reason that oxidizing atmosphere results in a high ash fusion temperature. Seeing that the nonstoichiometry combustion occurs at the cross section of the furnace under fuel rich/lean combustion conditions, the NO*x* emissions concentrations can also be reduced, as verified by field data [3, 4].

Various means have been adopted to separate the fuel/air mixture, such as louver-type [4], collision-block-type [2] and bend-type [3]. In the work described here, the investigation focuses on a collision-block-type fuel-rich/lean burner. As shown in Figure 1, the collision-block diverts the primary –air flow and makes the pulverized coal particles rebound, thus changing their directions. The coal particles are then carried by the flow to be concentrated in the fuel-rich side over the central partition plate. As a result, the fuel-flow is divided into a fuel-rich flow in the left zone beside the partition plate and a fuel-lean flow in the fight zone. The fuel concentration ratio between the fuel-rich side and the fuel-lean side can be regulated continuously, by adjusting the height of the collision block.

* Corresponding author. Tel: +86-571-87952066; Fax: +86-571-87951616 *E-mail address:* qiukz@zju.edu.cn (QIU K).

CP914, *Multiphase Flow: The Ultimate Measurement Challenge,* edited by X. Cai, Y. Wu, Z. Huang, S. Wang, and M. Wang

The gas-solid flow characteristics in the fuel/rich lean burner and downstream of the burner exit are major criteria to evaluate the performance of the burner and the combusting system, access to detailed information on the two-phase flow structure is helpful to understand the separation performance of the separator and the interactions between the fuel-rich and the fuel-lean streams downstream of the burner exit. The availability of local particle concentration and particle size distributions can be employed to evaluate the different design possibilities and to test the theoretical modeling approaches.

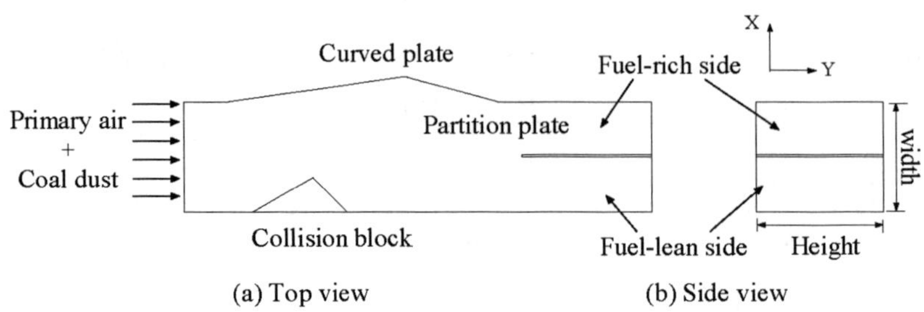

FIGURE 1. Sketch of collision block type fuel rich/lean burner.

Measurements of turbulent gas-solid two-phase flow have been conducted in several investigations. Optical techniques and tools for obtaining the measurements of local particle information were developed; These include Laser Doppler Velocimetry (LDV) [5], Phase Doppler Anemometry (PDA) [4], and fiber optic sensors[6].

The selection of the optical techniques and tools is affected by factors such as solids loading concentration and flow field geometry. In the gas-solid flow with a solid volume fraction greater than a few percent, the optical depth is limited to several particle diameters, which means that measurements using LDV and PDA are impractical in relatively dense suspension. As reported by Woodhead [7], the upper limit of particle volume fraction for light phase pneumatic conveying of pulverized coal measurements with LDV is only 0.03%, due to the light attenuation by the particles. While the typical volume fraction ratios (solid to air) are between 0.044 and 0.026% in the industrial pulverized coal conveying pipe [8]. In view of the local concentrating phenomenon in the fuel-rich/lean combustion technology, the volume fraction ratios (solid to air) at many locations will exceed the value of 0.03%. On the other hand, the PDA and LDV methods always require the restriction of the scale ratio of the model burner and clean environments, which results in their inconvenience for measuring industrial scale burner jet in dirty environment.

Fiber optic sensors have offered an alternative measuring method in circulating fluidized beds (CFBs) and pneumatic conveying systems, although they are intrusive. Fiber optic probes have been employed to measure the velocity and size of particle cluster and voidage in a CFB, to identify the flow regimes. Cai *et al* used a fiber optical instrument to conduct an in-line measurement of the coal concentration in the pulverized coal pipe using a light transmission fluctuation method [9]. The characteristics of a particle rope in the coal conveying pipe following the elbow were measured by fiber optic probes [6, 10-12]. Their Laboratory experiments were performed with pulverized coal and air in a 0.154 m diameter pipe, the particle concentration was determined by the obscuration of light over small distance by comparison with calibration data. The measurements obtained showed the overall pattern of the rope.

The present work describes laboratory investigations of fuel-rich/lean separation performance of the collision-block-type separator. A fiber optic measurement system was employed to obtain the solid concentration distributions both in the separator and downstream of the burner exit. Local solid concentration and particle size distributions were obtained. The experimental results showed the mixing mechanism of two parallel gas-solid two-phase flows with large fuel concentration difference.

FIBER OPTIC MEASUREMENT SYSTEM

Measurements of local solids concentrations and mean particle sizes were obtained using a transmissive fiber optic probe. The probe contained four glass fibers, which were placed in a 25-mm external-diameter ceramic tube. Two of the fibers were used to transmit light into the region of particle gas flow (See Figure 2). They were parallel

to the axis of the probe and located in holes on the vertical face of an opening notch 12 mm wide. The distance between the centers of fibers 1 and 2 was 7 mm. On the opposite face of the notch, two similar holes were opened to place the other two fibers. Fiber 3 transferred the transmitted light from fiber 1; and fiber 4 transferred the light from fiber 2 to the photodetectors. To avoid having the probe disturb the flow, the probe diameter must be sufficiently small. In the present work, the probe occupied ~4% of the cross sectional area of the measuring section.
Local particle velocities were measured with a cross-correlation technique using a standard fast Fourier transform (FFT) algorithm. The flight time of particles form the upstream beam to the downstream beam was estimated as follows:

$$C(\tau) = \frac{1}{T}\int_0^T u_1(t)u_2(t+\tau)dt \tag{1}$$

where the terms $u_1(t)$ and $u_2(t)$ are particle flow signal waveforms obtained from the two fiber-optic beams 1 and 2. The flight time of particles (τ_{max}) is the time lag (τ) where the cross-correlation function becomes maximum. Then, local particle velocity can be readily computed from:

$$U_p = \frac{L}{\tau_{max}} \tag{2}$$

where L is the optical distance between the laser beams 1 and 2.

Information on local particle concentration is obtained from the mean values of the signals with and without particles present.

$$I_p = I_m - I_0 \tag{3}$$

where I, the transmissive light signal intensity, is computed from the analog signal $u(t)$ for a time interval T.

$$I = \frac{1}{T}\int_0^T u(t)dt \tag{4}$$

Earlier investigations with fiber optic probes showed that the amount of light transmitted through a gas–solid flow was linearly dependent upon the volume concentration of particles [9]. The fiber optic probe was calibrated for particle concentration measurements conducted at an axial position *z/D* = 18 downstream of the elbow exit in the vertical upflow region of a gas–solid two-phase flow loop. A reasonably uniform particle distribution can be expected over the pipe cross-section at this location [12]. The probe was traversed over the pipe cross-section to obtain full information about flow nonuniformities and the average particle concentration. Measurements with the fiber optic probe were compared with those obtained with an isokinetic sampling device. This sampling device extracted a representative sample of particles to determine the mass flux; the suction velocity at the probe tip could be adjusted so as to be identical to the local gas stream velocity, so that the "isokinetic" condition was achieved. The detailed working principle of the isokinetic sampling probe can be found in [13]. The comparisons showed that the local particle mass fluxes obtained by the two measurement techniques were in agreement to within 10%.

Information on particle size was obtained from analysis of the fluctuation characteristics of the optical signals.

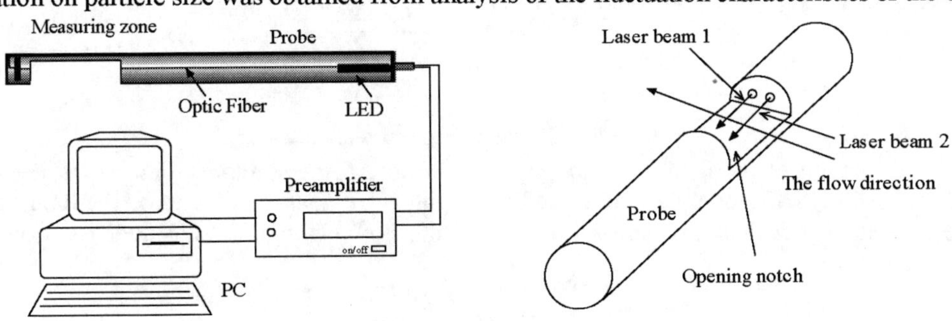

FIGURE 2. Fiber-optic measurement system and detail of fiber-optic probe design

A more detailed description of the principle of measurement of the local particle size can be found in [9]. Calibrations of the particle size measurement were made using two kinds of standard-size particles, one 74 μm in diameter and the other 9 μm in diameter. For each particle size, six tests with different particle concentrations were carried out. The results showed that the particle concentration and particle size measurements were nearly independent [14].

EXPERIMENTAL DETAILS

The pneumatic conveying test facility consists of a 6-m-long horizontal pipe and a 3.5-m-long vertical pipe (see Figure 3). The pipes are made of carbon steel pipes (0.15 m I.D.). A forced-draft fan provided a maximum air velocity of 40 m/s in the pipes. Particles stored in the hopper were discharged into the inlet air stream by means of a screw feeder calibrated for the test solid materials. The metering of the solids flow rate was accomplished by using a variable speed control. A cyclone is used to separate the solid particles from the air. The collected particles are fed to the hopper by an airlock. The remaining fine particles were removed by a fabric filter, and then fed back to the feeder hopper so that continuous operation and measurement were possible. Owing to the continuous circulation of particles in the loop, it is necessary to evaluate the attrition rate of the particles. In a series of experiments, the particles were extracted from the loop every 10 h of operation. Sieve analysis showed no significant attrition for 100 h period covered. After 100 h run of the loop, the particles were renewed.

The instrument ports on the wall of the measuring chamber allowed the fiber optic probe to traverse the chamber cross-section in both the lateral and the axial direction to measure the gas–solid jet flow downstream of the burner exit. For one cross-sectional measurement, more than 120 points were measured using the fiber optic probe to obtain detailed information about the gas-solid structure of the jet. As mentioned above, the use of fiber optic sensors is an intrusive measurement technique. The percentage of the cross-sectional area blocked by the probe in this work was ~4% to avoid having the probe disturb the flow.

The full industrial-scale fuel-rich/lean burner studied in the experiments was designed for a 300-MWe capacity utility boiler. It consisted of a collision-block-type concentrator and a coal nozzle in combination. The operational parameters of the industrial-scale burner are illustrated in Table 1. Due to the geometrical restrictions of the test rig, the burner must be scaled so that it can be installed in the test facility. A scale ratio of 1/2.708 was employed to make a model burner with a nozzle width of 160 mm and a height of 120 mm. A 3-mm thick partition plate was located at the center of the nozzle to separate the fuel-rich and fuel-lean streams.

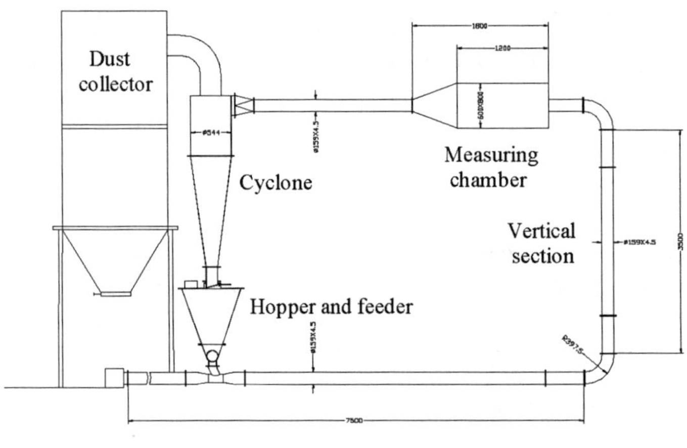

FIGURE 3. Sketch of laboratory facility

The aerodynamic scaling of the industrial burner should follow some similarity criteria, as listed in [15]: (1) geometric similarity and (2) self-modeling flows. The Euler number *Eu* does not depend on the Reynolds number *Re* when *Re* of the flow is greater than a certain value; that is, the flow enters the "self-modeling flow" region. The Reynolds number is defined as

$$Re = \frac{\rho U D}{\mu} \quad (5)$$

where ρ is the gas density, D is a representative length, U is the gas velocity, and μ is the viscosity of the gas. *Re* represents the ratio of the inertial force to the viscous force. *Eu* represents the ratio of the pressure difference to the inertial force, and is defined as

$$Eu = \frac{\Delta p}{\rho U} \tag{6}$$

where Δp is the pressure difference in the gas between the inlet and outlet of the flow.

When the flow enters the self-modeling flow region, the same flow phenomenon can be observed in both burners even though *Re* of the laboratory-scale burner flows are smaller than that of the industrial-scale ones. It has been reported that the flow enters the self-modeling flow region if the *Re* is greater than 10^4–10^5 [16]. *Re* of the model burner flow in our work was equal to 1.518×10^5, which means that the flow pattern of the turbulent field in the modeling burner was similar to that in the flow field of an industrial burner with a similar geometrical configuration.

Additional similarity criteria include (3) boundary condition similarity, (4) the condition that the Froude number remains constant with scale reduction, and (5) the condition that the Stokes number remains constant with scale reduction. The Froude number represents the ratio of inertial forces to gravitational forces on the airflow, and is defined as

$$Fr = \frac{U}{\sqrt{Dg}} \tag{7}$$

where *g* is the gravitational acceleration. The Stokes number *St* is defined as

$$St = \rho_p d_p^2 U/(18D\mu) \tag{8}$$

where ρ_p is the particle density, and d_p is the particle diameter.

St determined the density and mean size of the test particles in the laboratory test facility. As a scaling ratio of 1/2.708 was chosen in this work, talcum powder particles with a mean diameter of 14.3 μm and a true particle density of 2700 kg/m^3 were used as the material conveyed to satisfy the similarity requirement of *St*. The velocities at nozzle exit and in the conveying pipe were determined according to the scaling criteria. The experimental parameters are listed in Table 1, and the detailed comparison of the dimensionless numbers for the industrial and experimental scales are presented in Table 2. The collision block height was 60 mm which can lead a good separation performance. The influence of collision block height on the separation performance can be found in Ref. [17].

TABLE 1. The operational parameters of the industrial and laboratory burner

	Gas velocity at nozzle exit (m/s)	Exit area (mm×mm)	Air temperature (℃)	Atmospheric pressure (bar)	Particle mean diameter (μm)
Industrial	23	433×325	268	0.848	40
laboratory	14	160×120	20	1.0	14.3

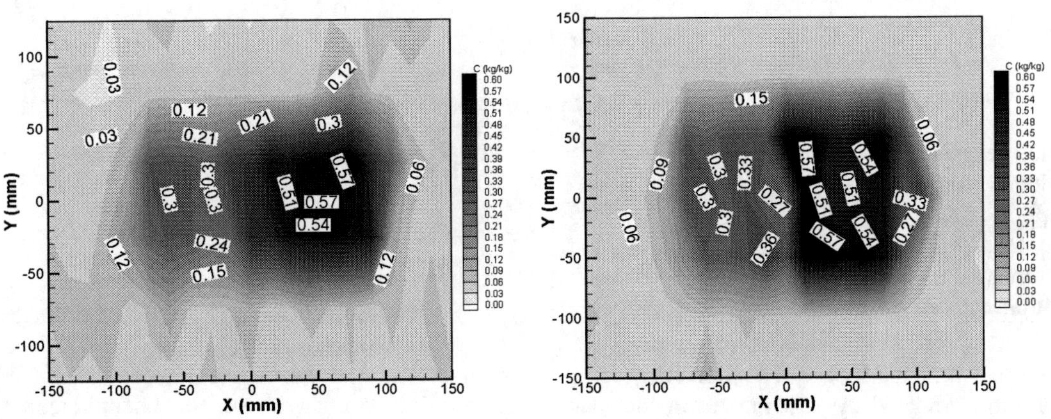

(a) 50 mm

(b) 100 mm

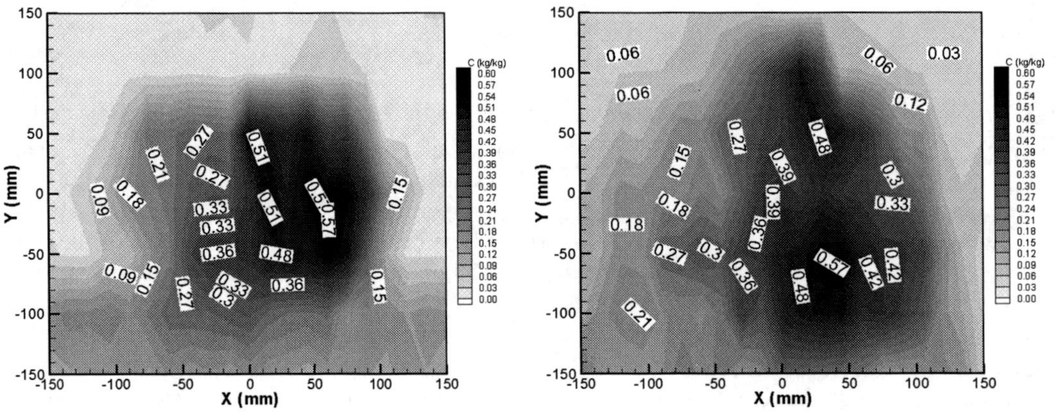

(c) 200 mm (d) 400 mm

FIGURE 4. Measured particle concentration distributions.

RESULTS AND DISCUSSION

Fig. 4 (a) shows a contour plot of the particle concentration measured at 50 mm distance downstream of the nozzle exit with the fiber optic probe. These measurements, obtained when the particle concentration was 0.42 kg/kg, showed that a good fuel-rich/lean separation was achieved. At the location of 50 mm downstream of the burner exit, the peak particle concentration in the fuel-rich stream was 0.607 kg/kg, while the peak concentration in the fuel-lean stream was 0.365 kg/kg, which resulted in a fuel-rich/lean ratio of 1.66 at this cross-section. The fuel-rich/lean ratios were 1.765 and 1.7 at the cross-sections 100 and 200 mm, respectively. The fuel-rich and fuel-lean streams merged at the cross-section at 400 mm, but the location of the peak particle concentration deviated by ~50 mm from the axis of the nozzle.

With the scaling ratio of 1/2.708 used in the experiments, the location 100 mm downstream of the nozzle exit in the model represents a location 270.8 mm downstream of the exit in the industrial burner. In a real combustion furnace, the coal particles begin thermal pyrolysis and then ignite at this location; such a fuel concentrating of the primary air flow will lead to enhancement of the ignition and further combustion of the coal particles, as well as a decrease in the NO_x emission.

TABLE 2. Comparison of the dimensionless numbers for industrial and laboratory scales

		Industrial scale	Laboratory scale
Reynolds	$Re = \dfrac{\rho U D}{\mu}$	1.687×10^5	1.518×10^5
Froude	$Fr = \dfrac{U}{\sqrt{Dg}}$	12.53	12.53
Stokes	$St = \rho_p d_p^2 U/(18 D \mu)$	0.197	0.172
Temperature (°C)	T	268	20
Gas velocity in the pipe (m/s)	U	25	15.2
Atmospheric pressure (bar)	P	0.848	1.0
Particle density (kg/m^3)	ρ_p	1200	2700
Particle mean diameter (μm)	d_p	40	14.3
Internal pipe diameter (mm)	D	406	150

The fuel-rich and fuel-lean streams merged at the cross-section at 400 mm, but the location of the peak particle concentration deviated by ~50 mm from the axis of the nozzle. Compared with the phenomenon of a centrally located particle concentration peak in an ordinary burner jet, such a deviation of the high-concentration zone is favorable for flame stability, and will result in strong heat transfer between the particles and the upstream hot gas in the chamber in a real combustion case, which can enhance the burnout of the coal particles.

Contours of the particle size distribution were also obtained at various cross sections downstream of the burner exit. Figure 5 presents the mean size distribution in the jet for various cases. The particles on the fuel-rich side always had larger mean diameters than those on the fuel-lean side.

It shows that the separator brought more quantity of the larger particles to the fuel-rich side than fuel-lean side; the larger particles can receive more intensive heating by the upstream flame as a result of facing the high temperature flame, which can enhance the burnout of the larger coal particles. The measurements also showed that the peak particle size decreases and the particle size distributions become more homogenous with the development of the burner jet. The two particle size peaks on the fuel-rich and fuel-lean sides merge into one peak at the cross-section 400 mm downstream, which behaves in the same way as the particle concentration contour plots. Comparing with the particle concentration distribution, it can also be found that the size uniformities between the fuel-rich and fuel-lean side were less serious than the concentration uniformities.

CONCLUSIONS

A fiber optic measurement system is an effective method to obtain the particle concentration and particle size distribution characteristics of the gas-solid two-phase jet flow. Because the fiber optic measurement system can be used for high particle concentration jet flow measurements, it can directly be employed for field measurements or the large-scale laboratory experiments, which model the practicable burner flow much better than the small-scale ones.

The separation performance of the fuel-rich/lean separator and mixing characteristics of the gas-solid two-phase burner jet were obtained in this work. Local solid concentration and particle size distributions were obtained. The measurements indicated that the collision-block-type separator always gave good concentrating performance under various inflow solid concentrations. The fuel-rich stream and the fuel-lean stream can maintain their continuous structure in long distance, ensuring the benefits of the horizontal fuel rich/lean combustion can be expected, which results in the flame stability and low NOx emissions. The solid size and concentration distribution have the same tendency. The particles with large diameter always exist in the zone where the particle concentration is high.

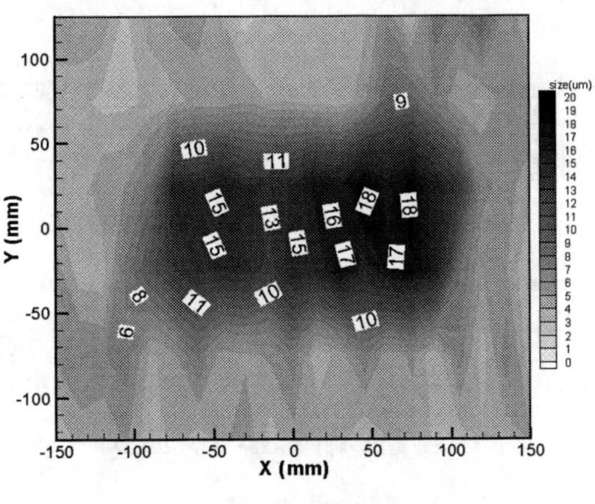

(a) 50 mm

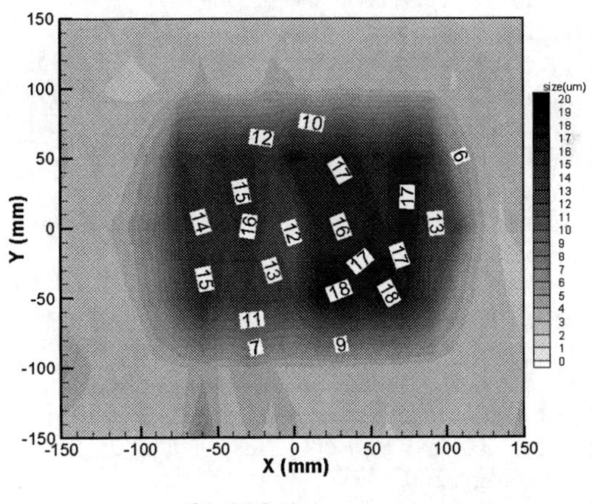

(b) 100mm

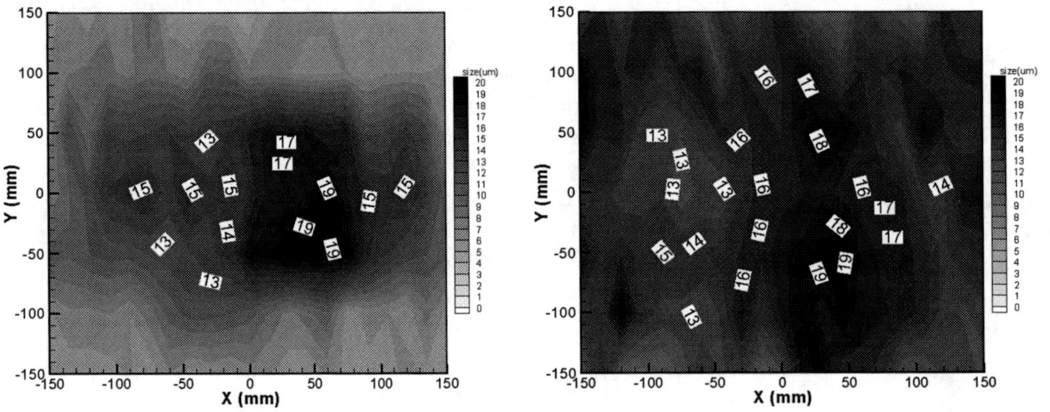

(c) 200 mm (d) 400 mm

FIGURE 5. Measured particle size distributions

ACKNOWLEDGEMENTS

Supported by National Natural Science Foundation of China (60534030, 50576081), program for Changjiang Scholars and Innovative Research Team in University.

REFERENCE

1. Zhou, H., K. Cen, and J. Fan, Experimental investigation on flow structures and mixing mechanisms of a gas-solid burner jet. Fuel, 2005. **84**: p. 1622-1634.
2. Zhou, H., K. Cen, and J. Fan, Detached eddy simulation of particle dispersion in a gas-solid two-phase fuel rich/lean burner flow. Fuel, 2005. **84**: p. 723-731.
3. Wei, X., T. Xu, and S. Hui, Burning low volatile fuel in tagentially fired furnaces with fuel rich/lean burners. Energy Conversion & Management, 2004. **45**: p. 725-735.
4. Xu, M., C. Sheng, and J. Yuan, Two-phase flow measurements and combustion tests of burner with continuously variable concentration of coal dust. Energy & Fuels, 2000. **14**(3): p. 533-538.
5. Fan, J., H. Zhao, and J. Jin, Two-phase velocity measurements in particle-laden coaxial jets. Chemical Engineering Journal, 1996. **63**(1): p. 11-17.
6. Schallert, R. and E. Levy, Effect of a combination of two elbows on particle roping in pneumatic conveying. Powder Technology, 2000. **107**: p. 226-233.
7. Woodhead, S., The measurement of particle velocity and suspension density in pneumatic coal injection system. 1992, University of Greenwich.
8. Giddings, D., et al., A 1/4 scale test facility for PF transport in power station pipelines. Fuel, 2004. **83**: p. 2195-2204.
9. Cai, X., et al., In-line measurement of pneumatcially conveyed particles by a light transmission fluctuation method. Flow Measurment and Instrumentation, 2005. **16**: p. 315-320.
10. Bilrgen, H. and E. Levy, Mixing and dispersion of particle ropes in lean phase pneumatic conveying. Powder Technology, 2001. **119**(2-3): p. 134-152.
11. Yilmaz, A. and E. Levy, Formation and dispersion of ropes in pneumatic conveying. Powder Technology, 2001. **114**(1-3): p. 168-185.
12. Yilmaz, A. and E.K. Levy, Roping phenomena in pulverized coal conveying lines. Powder Technology, 1998. **95**: p. 43-48.
13. Nguyen, T., A. Nguyen, and S. Nieh, An improved isokinetic sampling probe for measuring local gas velocity and particle mass flux of gas-solid suspension flows. Powder Technology, 1989. **59**: p. 183-189.
14. Cai, X., et al., A study of on-line measurement technology for size, concentration and velocity of pulverized coal. Power Engineering (in Chinese), 1999. **19**(6): p. 466-470.
15. Zhou, H., K. Cen, and J. Fan, Two-phase flow measurements of a gas-solid jet downstream of fuel rich/lean burner. Energy & Fuels, 2005. **19**(1-2): p. 64-72.
16. Cen, K., The study method and measurement techniques of boiler combustion experiments. 1987, Beijing: Water Conservancy and Electric Power Press (in Chinese).

17. Fan, J., et al., Numercial investigaion on two-phase flow in rich/lean pulverized coal nozzles. Fuel, 2000. **79**(14): p. 1853-1860.

On-Line Measurement of The Shape and Growth Rates of Organic Crystals in Reactors: Advances and Perspectives for Control

Xue Z Wang

Institute of Particle Science and Engineering, School of Process, Environmental and Materials Engineering, The University of Leeds, Leeds LS2 9JT, UK, Email: x.z.wang@leeds.ac.uk

Abstract. Organic crystals grown from solution are known to exhibit multiple morphology and habits which are of great importance to the end use property of the product such as the bioavailability and down stream processing such as in filtration and drying. The crystal morphology can also dictate other quality measures such as the size. This paper reviews recent developments in on-line crystal morphology measurement and control using various techniques including on-line imaging and image analysis, and XRD. On-line imaging was found to be the most promising technique in capturing with high fidelity crystal shape and polymorphic transitions in real-time. Process in imaging and image analysis techniques will also be reviewed for extracting the crystals from the image background especially for images with uneven background intensity, of needle-like crystals and of high concentration slurries. Discussion will be given to shape descriptors and automatic shape recognition. Preliminary results and challenges for estimating crystal growth rates and kinetics parameters for different crystal facets will also be discussed. Future research directions for obtaining 3D instead of 2D dimensional information using imaging and image analysis techniques will be explored. Finally a framework integrating morphology modeling, multi-dimensional population balance and computational fluid dynamics, with 3D imaging and image analysis will be presented which provides the basis for model predictive automatic control of the morphology of growing crystals.

Keywords: crystal morphology and shape measurement and control, model predictive control, morphology modeling, imaging, image analysis, population balance

INTRODUCTION

High value-added speciality chemicals such as pharmaceuticals are often manufactured in batch crystallization processes. The shape, size and polymorphic form are properties of great importance to crystalline drug products. It is known that certain crystal morphological forms and habits have been related to difficulties in dissolution rate, process hydrodynamics, solid-liquid separations, drug tableting, storage and handling, or in milling and grinding. Although there has been a large amount of research work on on-line measurement of other quality measures such as the size and concentration using various spectroscopy techniques including ultrasound, infrared, near infrared, and Uv spectroscopy, X-ray diffraction and Raman spectrometer1., the literature on monitoring crystal morphology is scarce, and so does that on crystal morphology control. This paper presents recent advances towards developing an enabling technique for real-time measurement and manipulation of the morphology of growing crystals through integrating on-line imaging, image analysis and morphology modeling, and proposes a framework and highlights challenges and future research needs for model predictive control of crystals morphology.

MORPHOLOGY MEASUREMENT: IMAGING AND IMAGE ANALYSIS

Laser diffraction techniques were investigated previously for the recognition of non-spherical particles with only limited success mainly due to the difficulty in obtaining a single-particle pattern in mixtures2,3.. Use of attenuation acoustic spectroscopy, Raman spectrometer, NIR and X-ray diffraction techniques though can detect polymorphs, but they cannot give detailed quantitative shape information. Several recent studies have demonstrated the effectiveness of using on-line imaging as an instrument for monitoring the shape of the crystals 4-14.. In our work, we investigated the use of an on-line imaging instrument developed by researchers of GlaxoSmithKline to monitor the onset as well as polymorph transitions during cooling crystallization of L-glutamic acid 5., developed an effective multi-scale image analysis technique for segmentation of the crystals from the complex background of image frames 6., and subsequently derived shape descriptors, classification techniques and novel process monitoring charts 4.. Figure 1 shows the on-line imaging system mounted on the outside wall of a 5 liter batch reactor which is able to take maximum 30 images per second of the pixel resolution of 480 × 640. Figure 2 shows polymorphic transition was captured in real-time during the cooling crystallization of L-glutamic acid.

CP914, *Multiphase Flow: The Ultimate Measurement Challenge,* edited by X. Cai, Y. Wu, Z. Huang, S. Wang, and M. Wang

On-line images of slurries with particles suspended in a solution pose much greater challenges to image analysis methods than images of particles obtained with off-line equipment. The major challenges lie in the fact that the slurries in a stirred reactor are in continuous motion, and that the variation of distances from the camera lens of particles captured in a snapshot makes some particles rather vague compared to others. In addition, the light effect and temporal changes of hydrodynamics within the reactor may lead to varied intensity in the image background. As a result a multi-step multiscale approach was developed which proved to be effective in extracting objects from the image background for images obtained by the GSK on-line microscopy system, as well as for the Lasentec's PVM probe 15., as demonstrated by Figures 3 and 4.

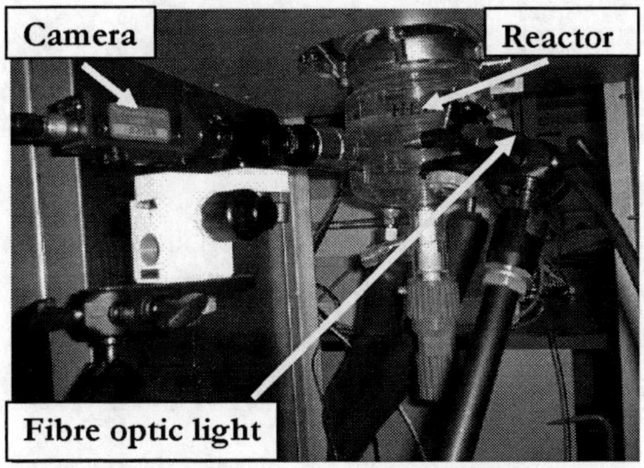

FIGURE 1. The on-line imaging system mounted on a 5 litre batch reactor

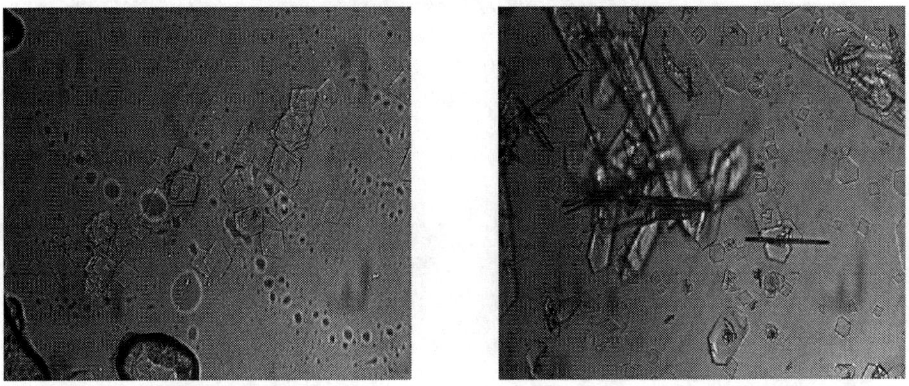

FIGURE 2. Polymorph transition for L-glutamic acid captured in real-time, from α form (left) to β form. The right figure shows mixed α and β indicating transition being taking place

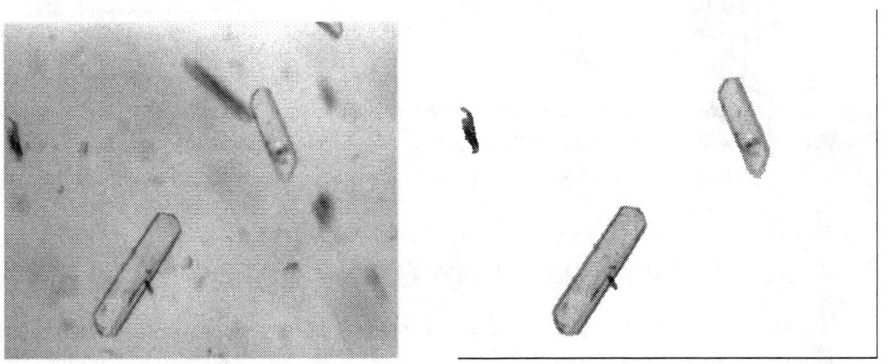

FIGURE 3. crystal images were effectively extracted from the background of an image frame obtained by the GSK imaging system

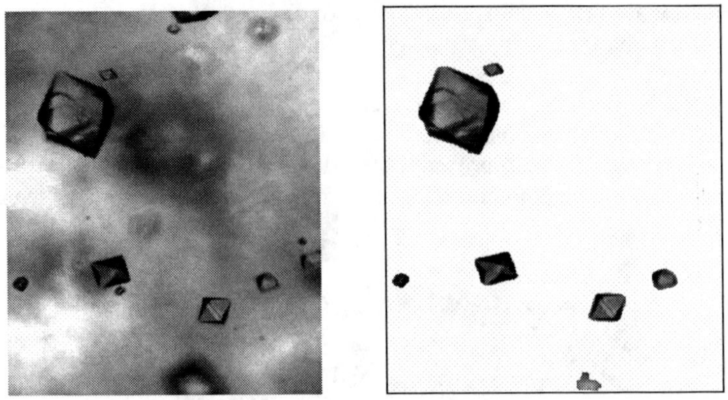

FIGURE 4. crystal images were effectively extracted from the background of an image frame obtained by the Lasentec PVM imaging system.

SHAPE DESCRIPTORS, SHAPE CLASSIFICATION AND MONITORING CHARTS

Following image segmentation, the next step is to classify the shapes. Shape descriptors should be invariant under rotation, translation and size scaling. It is also important to exclude descriptors that do not represent shape. After comparing various descriptors, we found that Fourier descriptors based on spectral transforms provide robust performance, accuracy and compact features and low computational complexity, therefore were used in our study 4.. The descriptors were then used by an unsupervised neural network algorithm, ART2 (adaptive resonance theory) to classify the crystals. Based on the results, advanced new monitoring charts were developed, as shown in Figures 5 (a) and (b).

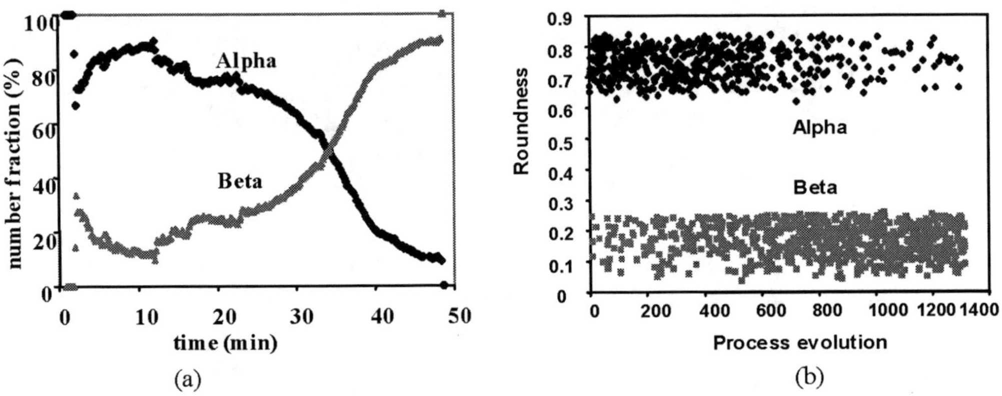

FIGURE 5. (a) crystal polymorphic transition charts in number percentage for the polymorphs α and β. Each point represents the previous ten minutes; (b) evolution of roundness values for shapes detected as α and β forms indicating transition of α to β for L-glutamic acid crystallization

FACETED CRYSTAL GROWTH RATES AND KINETICS

Given that the fundamental process of crystal growth and its associated kinetic control is surface controlled, the use of a single scalar parameter, particle size, usually defined as a volume equivalent diameter, i.e. based on a spherical assumption of particle shape can be misleading for a number of practical crystallization systems, notably pharmaceutical products, where facetted particles defined by non-unity aspect ratios 16.. Hence, measurement of the growth rate for each individual crystal surface in real-time and within processing reactors could open the way for the development of more effective processes and concomitant product quality control. Using on-line imaging and image analysis, a preliminary study was conducted on the estimation of the growth rates of needle-shaped crystals in two dimensions for β-form L-glutamic acid in cooling crystallization under a cooling rate of 0.100C/min 16.. The length and width of each needle-shaped crystal were measured every 60 seconds, ranging from 100 to nearly 200 μm in length and from 30 to 45 μm in width, and the values were used to estimate growth rates on both directions (Fig. 6). The growth rate in length was found to be 4 to 6 times greater than for the width. The (101) plane was found to be the fastest growing surface of the morphology studied and an attempt was also made to estimate its growth-kinetics parameters from measurements of length, whilst it was harder to estimate kinetics from measurements of width for other crystal faces. In the temperature range between 68.340C to 67.510C, the length growth rate is estimated as between 2.440×10-8 ~ 2.995×10-8 m/s, while the growth rate for the width is between 0.558×10-8 ~ 0.502×10-8 m/s. The capability to measure crystal growth rates in different directions could be used to estimate the parameters associated with growth kinetics in multi-dimensional directions. If a semi-empirical kinetic model is used, $R = k\sigma^{n}$, k ≈ 1.761×10-7 m/s, an d n ≈2.61. It was assumed for β L-glutamic acid, the growth rate in length is very close to the growth rate of the faces {101}.

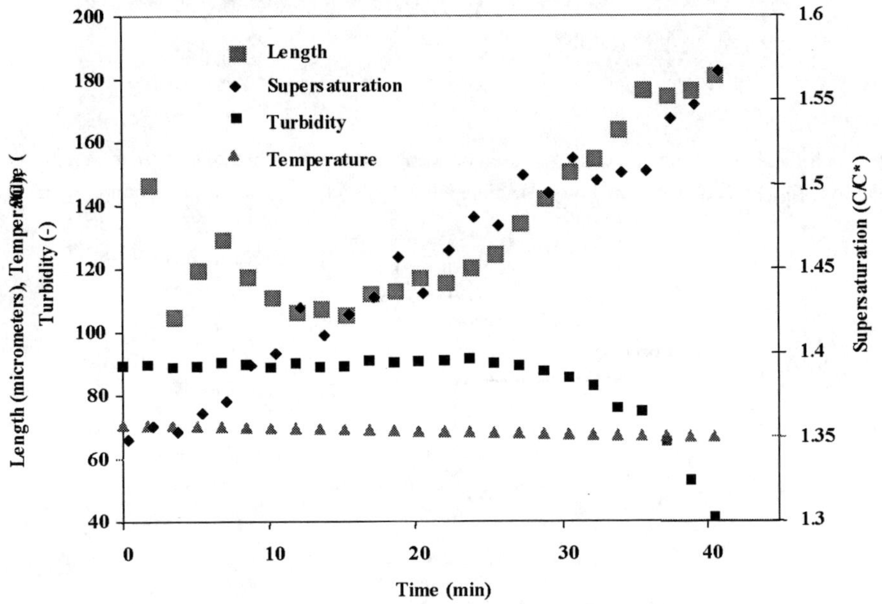

FIGURE 6. Crystal length evolution, plotted against supersaturation, temperature and turbidity. Each point represents the average of previous 60 seconds containing 300 images

PERSPECTIVES FOR MPDEL PREDICTIVE CONTROL OF CRYSTAL MORPHOLOGY

Model Predicted Crystal Morphology Control

The recently developed imaging and image analysis technique for real-time measurement of crystal morphology makes it possible to carry out automatic control of the morphology of crystals grown from solution. Since once an unfavorable polymorph or morphology is formed, to reverse the process e.g. through heating to dissolve and re-growing the crystals may not be a preferred option in industrial operation, advanced control based on shape predictive models is therefore attractive. Since morphology prediction is formulated for a single crystal, in order to consider the whole population of crystals in a reactor to give a statistical measure of the shape distribution, morphology modeling should be integrated with population balance (PB) modeling (as indicated in the next sub-section, this gap has yet to be filled), and very importantly, PB has to be conducted based on multi-dimensions, instead of being based on the assumption of an equivalent spherical shape with a single dimension of the diameter to represent the size distribution (Fig. 7). Multi-dimensional PB modeling depends on the number of facets of a crystal, each with its own growth rates. Some researchers 17,18. have experimented the idea of two dimensional PB modeling for needle shaped crystals. In early part of this paper we showed results of measuring the growth rates in two dimensions for needle-shape crystals. Not only being restricted to two dimensions, work so far has also assumed that there is only one polymorph in a reactor. If more than one polymorphs exist at the same time in the reactor, the problem clearly becomes more complicated because the transition between two polymorphs also needs to be taken into account.

Fig. 8 shows a conceptual framework highlighting the necessary components for developing a model predictive control strategy of crystal morphology. Key future research needs are highlighted below.

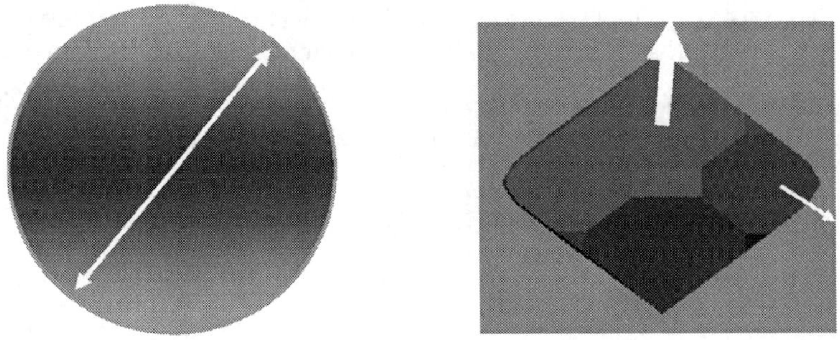

FIGURE 7. (a) traditional modelling and control: based on a spherical assumption, e.g. growth rate in m/s of diameter; (b) new modelling and control: based on multidimensional model, e.g. growth rates, m/s for each surface.

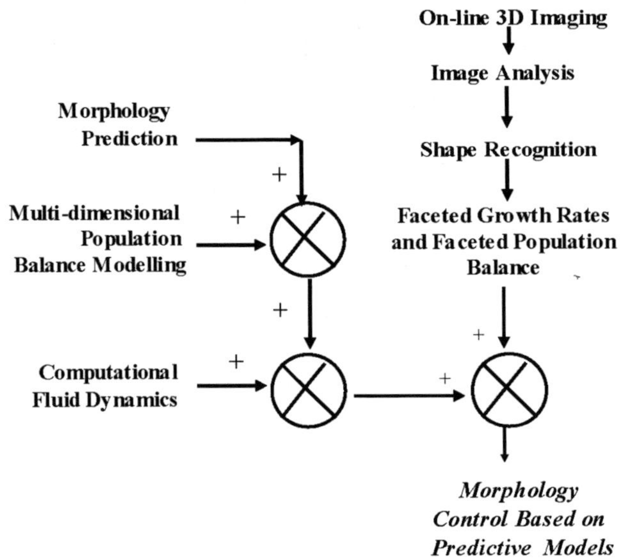

FIGURE 8. The necessary components for developing model predictive control strategy for crystal morphology

Shape Measurement

Although proved to be effective, so far the measurement of shape using on-line imaging has been restricted to giving 2-D images. Li et al 19. proposed to use a camera model to construct the 3-D shape from 2-D images, there are still a few obstacles to be overcome before it can become practical. We are experimenting the use of two video cameras to simultaneously image the same object so that the 3-D shape can be fully constructed.

Morphology Modeling

Traditionally, morphology prediction of crystals had almost assumed that the crystals were growing in vacuum without adequately considering process operational conditions such as cooling rates, supersaturation, solvents and impurity or additives etc though these factors are known to affect the polymorph and morphology. Progress are being made in improving the morphology models to take into account these factors and the work has been reviewed 20-24.. Success has been achieved on predicting the effect of solvent on the shape of several organic crystal systems. Despite the progress, there is still some way to go before robust models available that can fully take into account process conditions to give confident prediction of crystals shape. This does not mean however, this represents the weakest link that will jeopardize the whole framework for model predictive morphology control research. In fact, on-line shape measurement techniques could contribute to the development of morphology prediction models since they can be used for model validation.

Integration with CFD

The effect of mixing on crystal growth has long been recognized and computational fluid dynamics (CFD) has been integrated into PB modeling (though only for one-dimension based on the spherical assumption) 25. so that PB modeling can be conducted in different zones with varied mixing characteristics. A major challenge is computational speed. Research on either model reduction or data compression is needed in order to be able to effectively use the CFD data into crystal morphology model predictive control.

FINAL REMARKS

The recent developments in on-line shape measurement as well as in crystal morphology prediction and multi-dimensional population balance modeling opens the way for developing model-predictive control of the morphology as well as size of crystals grown form solution. This will need integration of on-line real-time 3-D shape measurement, multi-scale modeling of morphology, multi-dimensional PB modeling and CFD. Future research needs towards this goal are highlighted.

ACKNOWLEDGEMENTS

Financial supports from EPSRC for the Shape project (EP/C009541) and for the Chemicals Behaving Badly project (GR/R43860), and from Malvern Instruments Ltd for the Vision project and the IntelliSense partnership (www.intellisense.org.uk) are greatly appreciated.

REFERENCES

1 R.D. Braatz, M. Fujiwara, D.L. Ma, T. Togkalidou, D.K. Tafti, Simulation and new sensor technologies for industrial crystallization: A review, International Journal of Modern Physics B, Vol., 16, pp.346-353, 2002.

2 Z.H. Ma, H.G. Merkus, B. Scarlett, Extending laser diffraction for particle shape characterization: technical aspects and application, Powder Technology, Vol., 118, pp.180-187, 2001.

3 H. Yamamoto, T. Matsuyama, M. Wada, Shape distinction of particulate materials by laser diffraction pattern analysis, Powder Technology, Vol., 122, pp.205-211, 2002.

4 J. Calderon De Anda, X.Z. Wang, X. Lai, K.J. Roberts, Classifying organic crystals via in-process image analysis and the use of monitoring charts to follow polymorphic and morphological changes, Journal of Process Control, Vol., 15, pp.785-797, 2005.

5 J. Calderon De Anda, X.Z. Wang, X. Lai, K.J. Roberts, K.H. Jennings, M.J. Wilkinson, D. Watson, D. Roberts, Real-time product morphology monitoring in crystallization using imaging technique, AIChE Journal, Vol., 51, pp.1406-1414, 2005.

6 J. Calderon De Anda, X.Z. Wang, K.J. Roberts, Multi-scale segmentation image analysis for the in-process monitoring of particle shape with batch crystallisers, Chemical Engineering Science, Vol., 60, pp.1053-1065, 2005.

7 D.B. Patience, J.B. Rawlings, Particle-shape monitoring and control in crystallization processes, AIChE Journal, Vol., 47, pp.2125-2130, 2001.

8 M.J. Wilkinson, Jennings, K. H., Hardy, M., Non-invasive video imaging for interrogating pharmaceutical crystallization processes, Microscopy and Microanalysis, Vol., 6, pp.996-997, 2000.

9 P.A. Larsen, J.B. Rawlings, N.J. Ferrier, An algorithm for analyzing noisy, in situ images of high-aspect-ratio crystals to monitor particle size distribution, Chemical Engineering Science, Vol., 61, pp.5236-5248, 2006.

10 H.Y. Qu, M. Louhi-Kultanen, J. Kallas, In-line image analysis on the effects of additives in batch cooling crystallization, Journal of Crystal Growth, Vol., 289, pp.286-294, 2006.

11 K. Pollanen, A.W. Hakkinen, S.P. Reinikainen, A. Louhi-Kultanen, L. Nystrom, A study on batch cooling crystallization of sulphathiazole - Process monitoring using ATR-FTIR and product characterization by automated image analysis, Chemical Engineering Research & Design, Vol., 84, pp.47-59, 2006.

12 Y. Zhou, X.T. Doan, R. Srinivasan, Real-time imaging and product quality characterization for control of particulate processes, in: W. Marquardt, C. Pantelides (Eds.), Joint 16th ESCAPE and 9th PSE: Computer -Aided Chemical Engineering 21A, Elsevier, Garmisch-Partenkirchen, Germany, 2006, pp. 775-780.

13 X.Z. Wang, J. Calderon De Anda, K. Roberts, Advances in on-line monitoring and control of the morphological and polymorphic forms of organic crystals grown from solution, KONA (Powder and Particle), Vol., 23, pp.69-85, 2005.

14 X.Z. Wang, K.J. Roberts, J. Calderon De Anda, Advances and future directions in morphology monitoring and control of organic crystals grown from solution, in: W. Marquardt, C. Pantelides (Eds.), Joint 16th ESCAPE and 9th PSE: Computer -Aided Chemical Engineering 21B, Elsevier, Garmisch-Partenkirchen, Germany, 2006, pp. 1611-1616.

15 P. Barrett, B. Glennon, Characterizing the metastable zone width and solubility curve using lasentec FBRM and PVM, Chemical Engineering Research & Design, Vol., 80, pp.799-805, 2002.

16 J. Calderon De Anda, X.Z. Wang, K.J. Roberts, Real-time measurement of the growth rates of individual crystal faces using imaging and image analysis: a feasibility study on needle-shaped crystals of L-glutamic acid, Journal of Pharmacutical Sciences, submitted, Vol., 2005.

17 H. Briesen, Simulation of crystal size and shape by means of a reduced two-dimensional population balance model, Chemical Engineering Science, Vol., 61, pp.104-112, 2006.

18 F. Puel, G. Fevotte, J.P. Klein, Simulation and analysis of industrial crystallization processes through multidimensional population balance equations. Part 1: a resolution algorithm based on the method of classes, Chemical Engineering Science, Vol., 58, pp.3715-3727, 2003.

19 R.F. Li, G.B. Thomson, G. White, J. Calderon De Anda, X.Z. Wang, K.J. Roberts, Combining morphological modelling and online imaging technique for on-line crystal shape recognition in batch crystallisation processes, in: Chemical Engineering Congress 2005, Glasgow, Scotland, 2005, pp.
20 D. Winn, M.F. Doherty, Modeling crystal shapes of organic materials grown from solution, AIChE Journal, Vol., 46, pp.1348-1367, 2000.
21 G. Clydesdale, K.J. Roberts, R. Docherty, HABIT95 - A program for predicting the morphology of molecular crystals as a function of the growth environment, Journal of Crystal Growth, Vol., 166, pp.78-83, 1996.
22 X.Y. Liu, P. Bennema, Prediction of the growth morphology of crystals, Journal of Crystal Growth, Vol., 166, pp.117-123, 1996.
23 X.Y. Liu, E.S. Boek, W.J. Briels, P. Bennema, Prediction of Crystal-Growth Morphology Based on Structural- Analysis of the Solid-Fluid Interface, Nature, Vol., 374, pp.342-345, 1995.
24 S.B. Gadewar, M.F. Doherty, A dynamic model for evolution of crystal shape, Journal of Crystal Growth, Vol., 267, pp.239-250, 2004.
25 V. Kulikov, H. Briesen, W. Marquardt, Scale integration for the coupled simulation of crystallization and fluid dynamics, Chemical Engineering Research & Design, Vol., 83, pp.706-717, 2005.

Volterra Filter Forecasting Method Based on Chaotic Time Serried and its Application to Fire Smoke Flow

Zhang xiaoying*， Ma xiaoqian*， Wang hairong*

*Electrical Engineering College, South China University of Technology,
Wu Shan Road, Tian He District, Guang Zhou, 510640 China

Abstract. smoke movement was the main factor of the development and expanding of fire hazard and personnel casualty. Uncertain character of exotic environment and interior condition that effected smoke flow resulted that It was difficult to establish suitable model to simulate smoke flow. Firstly, this paper has analysis the characteristic of smoke movement. And maximal Lyapunov exponent of smoke movement basic variable (temperature, velocity and differential pressure)has been obtained. Positive maximal Lyapunov exponent of temperature has proved that smoke movement was chaotic. Negative maximal Lyapunov exponent of velocity was probably because that time interval of data was too big and part information has been loss. Then, the smoke flow character was recovered by real monitor data on the basis of Takens theorem and topological transformation. In order to do that, the delay time and embedding dimension of reconstructed phase has been got by the method of auto-correlation function and false neighbors respectively. Based on the mentioned above, reconstructed phase was obtained in delay coordinate. Points in reconstructed phase presented system evolution tendency and Volterra filter was established in reconstructed phase points. Finally, smoke flow parameter was forecast by a third-order Volterra filter. And the result has been caparisoned with experimental data. According the data from Multi-story building fire experiment, the delay time of temperature, differential pressure and velocity was respectively 8, 9 and 9, the embedding dimension of the three variable parameter were 3. The prediction has showed that third-order Volterra filter had intensive nonlinear approximate capacity, three variables relative errors of prediction were all small than 0.05

Keywords: fire hazard, smoke, time series, Volterra filter, forecast
PACS: 44.05.+e

INTRODUCTION

Fire hazard of high-rise building frequently happened with the increasing of the tall building's quantity. Smoke was the reaction production generating in the combustion process, which was the mixture of high heat solid and gaseous material. Smoke was mainly consisting of great smaller charcoal particle, incomplete combustion ash and other burning resolve results of combustible matter. The smoke would induced the human body to be poisoned, burned and suffocated, even lose the ability of escaping. So that 70% - 75% of the death in fire hazard was because of the smoke【1】. If incomplete combustion matter in the smoke mixing with air, continuing combustion or explosion would be caused. High heat of smoke would also induce new fire by the convection or radiation, cause inflammable matter in fire field rapidly catch fire and form flashover. Therefore, smoke movement was the main factor of the development and expanding of fire hazard. The prediction of smoke temperature, pressure or velocity was significant for the fire control.

At present, the research about fire smoke was focused on two aspects: smoke movement principle simulation and smoke component, concentration and toxic analysis depended on FED (fractional elective dose), TGAS (toxic gas assessment software) model and et. al The first one commonly divided the fire space into several compartments according to the structure of the building. Then, each compartment was divided into up and down two layers. And the smoke flow was simulated by zone, net work or field model. In net work model every compartment was regards as one control bulk, this method was applicable for smoke flow analysis in the compartments that far from fire field. Though zone model could analysis the dispersion of smoke, the researched space would be limited. The data got by field model was detailed, but the computed time was long and high-powered computation was needed. Both zone, net work and field model were all based on definite fire model, and simulated smoke flow on the PC【2-4】. Smoke flow was very complicated which involved in various fields, including aerodynamic, multiphase flows, mixture and transfer of turbulent, combustion and et. al. It was difficult to establish suitable model to simulate smoke flow. Because that the smoke parameter vary appeared nonlinear

CP914, *Multiphase Flow: The Ultimate Measurement Challenge,* edited by X. Cai, Y. Wu, Z. Huang, S. Wang, and M. Wang

character, this paper has recovered the smoke character by real monitor data on the basis of Takens theorem and topological transformation, and applied Volterra filter to predict the related parameter of smoke flow in order to offer data for rescue and dispersal of personnel.

CHOOSE OF FILTER MODEL

In the fire hazard of high-rise building, smoke flow was affect by expanding of smoke, funnel effect, blast pressure, air intake and circulating of ventilating system. These factor resulted that smoke temperature, pressure and velocity was acted by convection and turbulent. In practice, many similar nonlinear system variable was described by Volterra progression. Volterra model was established on chaotic time series, and could predict evolution tendency of complicated system. According to Takens theorem, one dimension time series included topological structure and system dynamics state character. In delay coordinate, one dimension time series could recovered variable dynamic state character by reconstructed phase 【5-7】. The reconstructed phase was showed as follows:

$$X(T) = [x(t), x(t+\tau), x(t+2\tau), \cdots x(t+(m-1)\tau)] \quad (1)$$

In the equation, $x(t)$ was system dynamics state variable at t time; τ was delay time; s; m was embedding dimension; for N digital points $x(t)$, in m dimension reconstructed phase M=N-(m-1)× τ dynamics state points was obtained. M dynamics state points presented evolution tendency $F(\cdot)$. The system could be predicted by evolution tendency, that is:

$$X(T+t') = F[x(t), x(t+\tau)x(t+2\tau), \cdots, x(t+(m-1)\tau)] \quad (2)$$

In the equation, time $T = t + (m-1)\cdot\tau$, t' was forward time step. Following, delay time and embedding dimension would be choose.

DELAY TIME

The choose of delay time τ must assure that dynamics state points in the reconstructed phase was depended each other. If the choose τ was small, reconstructed phase would be compressed and redundancy. In the reverse, if the choose τ was big, reconstructed phase would lose system original information and necessary connection. Auto-correlation function $C(\tau)$ was selected to determine the delay time【7-10】:

$$C(\tau) = \frac{1}{n-\tau} \cdot \frac{\sum_{i=1}^{n=\tau} (x(i)-\mu)(x(i+\tau)-\mu)}{\sigma^2} \quad (3)$$

In the equation, μ was the equalizing value of time series, σ was the standard deviation. equation (3) determined that auto-correlation function $C(\tau)$ was function about τ. When $C(\tau)$ was equal to 0 or ($1-1/e$) times of initial value, the corresponding τ was the delay time.

MIN EMBEDDING DIMENSION

If d was the fractal dimension, embedding dimension $m \geq 2d+1$ was a min embedding dimension of fractal[12-13]. But the calculation work capacity, noise of real monitor data of fractal dimension was great. false neighbors method was not need to determine delay time first and had good independence. Supposed that x_i' was neighbor point of x_i in m dimension phase, when dimension increased from m to m+1, if:

$$\frac{\left\| x_i' - x_i \right\|^{(m+1)}}{\left\| x_i' - x_i \right\|^{(m)}} > r \quad (4)$$

In the equation, denominator and numerator was respectively distance between x_i' and x_i in m dimension phase and m+1 dimension phase. The value of r was from 10 to 50. Considering the noise of time serried, the following equation was added:

$$\frac{\left\| x_i' - x_i \right\|^{(m+1)}}{\sqrt{\frac{1}{N}\sum_{n=1}^{N}(x_n - \overline{x})}} > 2 \qquad (5)$$

In the equation, $\overline{x} = \frac{1}{N}\sum_{n=1}^{N} x_n$ was equalizing value. The points meted the two equation above was false neighbors. The calculation began from m=2, until the rate of false neighbors was small than threshold value, the corresponding m was the min embedding dimension.

IMPLEMENTATION OF FILTER

According to the reconstructed dynamic model $F(\cdot)$, input was $X(n)$, output was $y(n) = X(n+1)$, so third-order Volterra filter model was[13-16]:

$$x(n+1) = F(x(n)) = h_0 + \sum_{m=0}^{+\infty} h_1(m_1)x(n-m) + \sum_{0}^{\infty}\sum_{0}^{\infty} h_2(m_1,m_2)x(n-m_1)x(n-m_2) + \sum_{0}^{\infty}\sum_{0}^{\infty}\sum_{0}^{\infty} h_3(m_1,m_2,m_3)x(n-m_1)x(n-m_2)x(n-m_3) \qquad (6)$$

In the equation, h_0、$h_1(m_1)$、$h_2(m_1,m_2)$、$h_3(m_1,m_2,m_3)$ was the nucleus of Volterra progression. Third-order Volterra filter had intensive nonlinear approximate capacity. If filter was directly implemented according to equation (6), the total coefficient of filter was 1+m+m2+m3. with the increasing of m, the quantities of filter coefficient increased with exponent. Therefore, product form with constant term was used in order to reduce the quantities of filter. Resolving and defining filter as:

$$y(n) = \left[\left[h_{1,0} + \sum_{i=1}^{m} h_{1,i}X(n,i)\right] \times \left[h_{2,0} + \sum_{m=1}^{m} h_{2,j}X(n,j)\right]\right] \times \left[h_{3,0} + \sum_{m=1}^{m} h_{2,k}X(n,k)\right] \qquad (7)$$

After being resolved, the quantity of filter was turned into 3m+3, and expansion equation was similar to equation (6). Otherwise, nonlinear normalization LMS algorithm was used to assure rapid convergence. The construction of the filter was showed as fig.1:

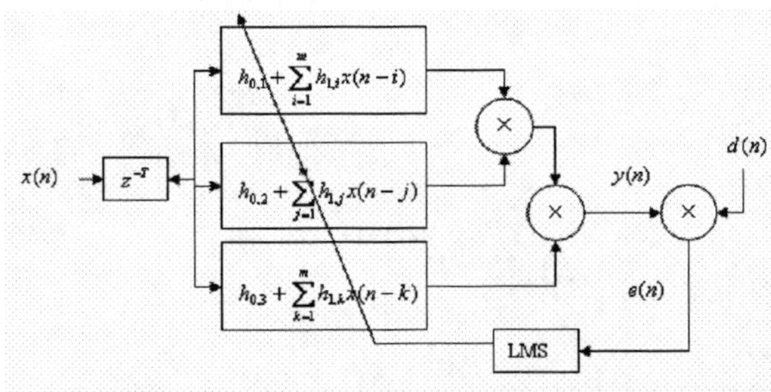

FIGURE1. The design of third-order Volterra filter

CHAOTIC CHARACTER OF SMOKE FLOW

The data from Multi-story building fire experiment of China Science & Technology University was used to analysis the smoke flow character. Experiment equipment【11】 was 5 stories building. The temperature was measured by 28 pieces of thermocouples at the diameter of 2mm, each 4 thermocouples make up a thermocouple tree, to measure the temperature in different height of room on fire. Differential pressure was measured by sensor. The temperature, differential pressure and velocity of measurement point C was choose as time series. The time series of temperature was 124 points, duration time was 1400 s; the time series of differential pressure was 94 points, duration time was 250 s; the time series of velocity was 86 points, duration time was 20 min. Max Lyapunov exponent and power spectrum of temperature was:

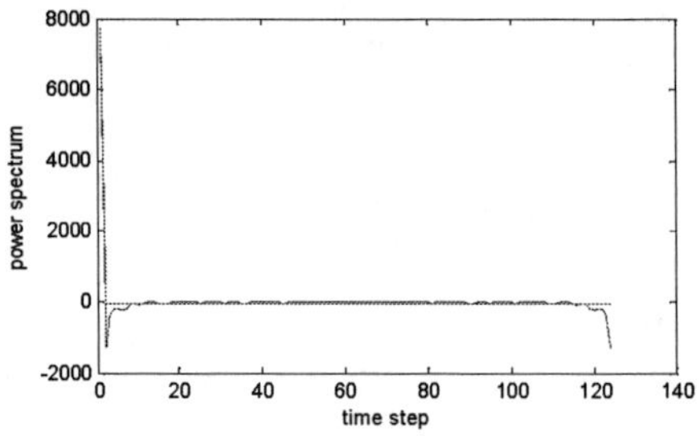

FIGURE2. Power spectrum

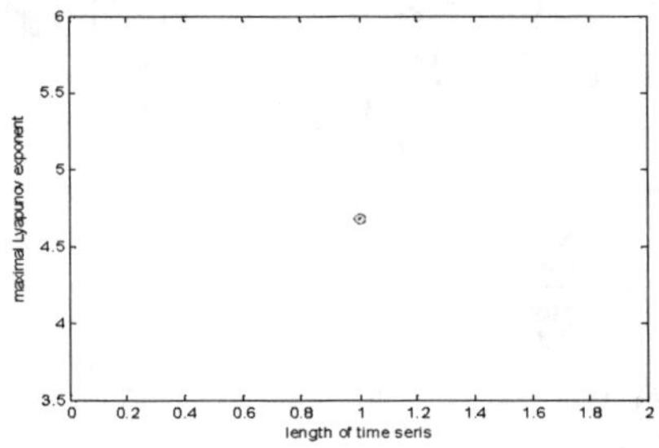

FIGURE3. Maximal Lyapunov exponent

Maximal Lyapunov exponent was calculated by Rosenstein method of small data size. In the calculation, average period was obtained by FFT changes. According to the power spectrum of temperature (fig.2), average period P=1. Positive maximal Lyapunov exponent of temperature has proved that smoke flow was chaotic. Negative maximal Lyapunov exponent of velocity was probably because that time interval of data was too big and part information has been loss. The maximal Lyapunov exponent of velocity and differential pressure showed as table 1:

TABLE1. The delay time, embedding dimension and maximal Lyapunov exponent

Time series	Delay time	Min embedding dimension	Max Lyapunov exponent
temperature	8	3	4.6797
Differential pressure	9	3	-1.8644
Velocity	9	3	-3.6505

VOLTERRA FILTER FREDICTION OF SMOKE FLOW

1/4 of time serried data was adopted as training data, 3/4 of time serried data was adopted as checking data. Delay time and min embedding dimension in filter model was choose as mentioned above, the results was showed as table 1. Then, smoke flow parameter was predicted by third-order Volterra filter. The prediction result and its error was showed as fig. 4-6:

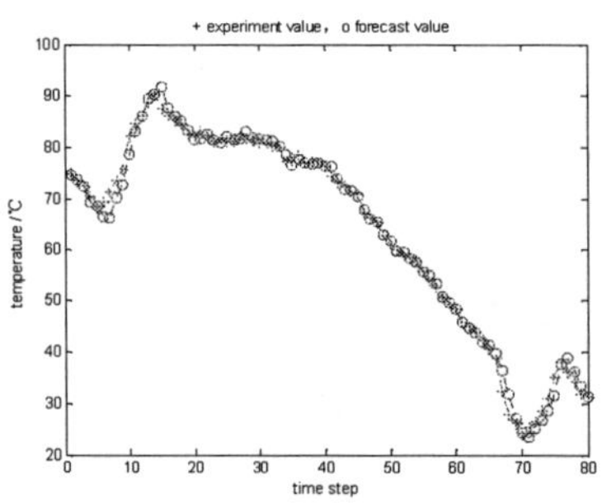

(a) Temperature forecast

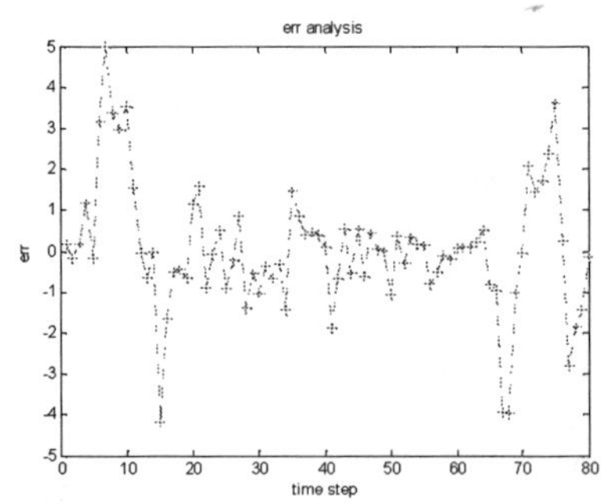

(b) Error analysis

FIGURE4. Forecast and error analysis of temperature

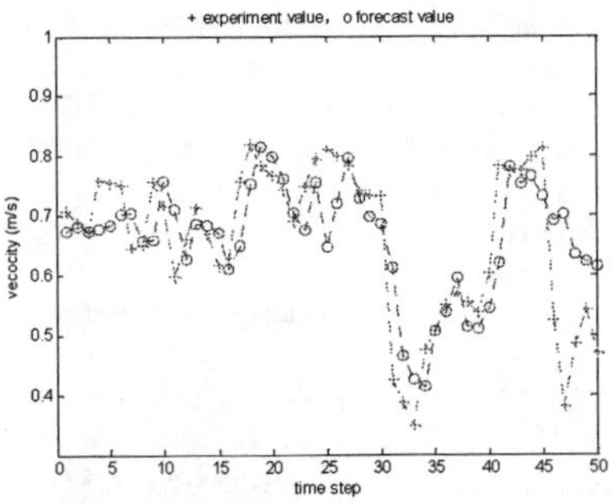

(a) Velocity forecast

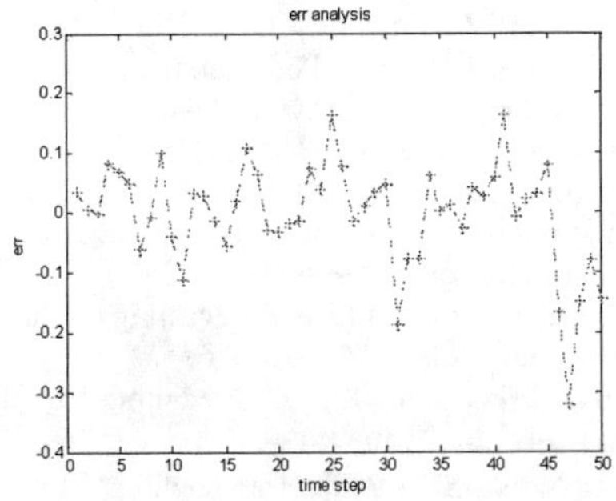

(b) Error analysis

FIGURE5. Forecast and error analysis of velocity

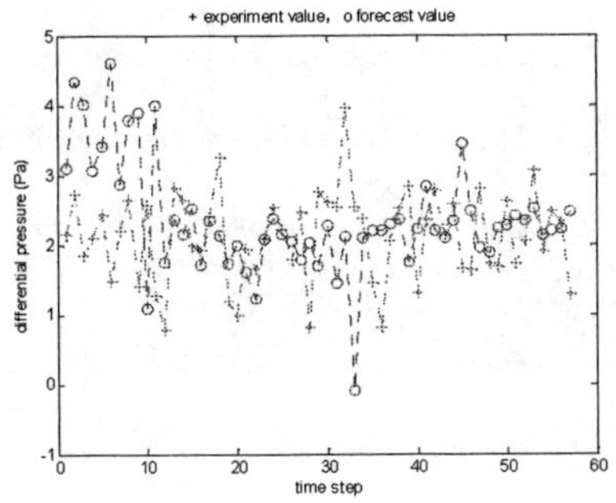

(a) Differential forecast

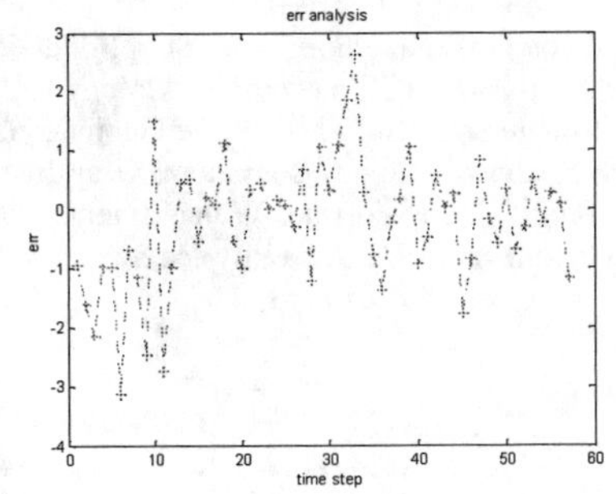

(b) Error analysis

FIGURE6. Forecast and error analysis of differential pressure

CONCLUSION

Phase reconstruct could recover system dynamic state character and evolution tendency by few training data, even complicated nonlinear phenomena that involved chaotic information. It has been founded that the prediction error of smoke temperature was the minimum. The positive max Lyapunov exponent has showed that heat convection caused temperature vary had strong non-characteristic. Therefore, third-order Volterra filter had intensive nonlinear approximate capacity. Meanwhile, three variables relative errors of prediction were all small than 0.05.

For time series, different forecast model could be established, for example Neural network. But Volterra filter was used for short-term forecast based on early monitor data. For the complicate system which was affected by uncertain exotic environment and interior condition, Volterra filter could reality recovered the system character. And the the structure and calculation method of Volterra filter was simple. The following work may be implement Long-term forecast by Volterra filter.

ACKNOWLEDGMENTS

This paper is sponsored by Guangdong Province Scientific Project 2005B324010001.

REFERENCE

1. Lu Zian, Lian Chenzhou, JI Chunsheng, and etc, Toxicity analysis of smoke from materials burnt in fires[J]. J Tsinghua Univ (Sci&Tech), 2004, 44(2): 278-281;
2. Qiu Xudong, Gao Fusheng, Wang Yanling, Review and evaluation on numerical simulation method of smoke movement under building fire[J]. Journal of Natural Disasters, 2005, 14(1): 132-138;
3. Zhou Xinquan, Ren Aizhu, Deng Yunfeng, Networks simulation of high-rise building fire[J]. China CivilEngineering Journal, 2005, 38(9): 68-73;
4. Zhou Ci, Yang Zhenhong, Guo Jinping, Numerical simulation of fire smoke flow properties in high-rise building[J]. Journal of Liaoning Technical University, 2003, 22(4): 524-526;

5. Qian Jinglin，Li Fugiang，Chen Bin and etc, Study on Volterra filter based on group method of datahandling and its application to flood forecast[J]. Journal of Zhejiang University, 2005, 39(1): 160-164;
6. Yang Shaoqing, Jia Chuanying, Two practical methods of phase space reconstruction[J]. Acta Physica Sinca, 2002, 51(11): 2452-2458;
7. Wang Yan, Xu Wei, Qu Jisheng, The algorithm and chick of phase space reconstruction based on the time series (1)[J]. Journal of Shangdong University, 2005, 35(4): 109-114;
8. D.Kugiumtzis, State space reconstruction parameters in the analysis of chaotic time series-the role of the time window length[J].Physica D, 1996, 95: 13-28;
9. Chuanwen Jiang , Tao Li b, Forecasting method study on chaotic load serieswith high embedded dimension[J]. Energy Conversion and Management, 2005, 46, 667–676;
10. Zhang Jiashu, Xiao Xianci, Predicting low dimension chaotic time series using Volterra adaptive filters[J]. Acta Physica Sinaca , 2000，49（3）：403-408;
11. Zhong Maohua, Li Peide, Liu Tieming, The experiment of smoke flow simulation of mutiny stories compartments on fire[J]. Science in China , Ser. E , Engineering & Materials Science, 2005, 35(5): 490~502;
12. Guo Shuangbing, Xiao Xianci, Determining rank of Volterra adaptive filters of chaotic time series[J]. Journal of Electronics and Information Technology, 2002, 24(10): 1334-1340;
13. Zhang Huajun, Han Chongzhao, Simplified Volterra series indentification method based on variable step size block least mean square filter[J]. Journal of Xi'an Jiaotong Uniersity, 2004 38(6): 583-586;
14. Ma Hongguangl, Li Xihai, Wang Guohua, Han Chongzhao and etc, Selection of Embedding Dimension and Delay Time in Phase Space Reconstruction[J]. Journal of Xi'an Jiaotong Uniersity, 38(4): 335-338;
15. Wei Baolin, Luo Xiaoshu, Wang Binghong and etc, A method based on the third order Volterra foilter for adaptive prediction of chaotic time series[J]. Acta Physica Sinca, 2002, 51(10): 2205-2210;

Determination of Minimum Agitation Speed for Complete Solid Suspension Using Four Electrode Conductivity Method

Vijay Shirhatti+, Mi Wang+, Richard Williams+ and Javier Romero Ortega*

+Institute of Particle Science and Engineering, University of Leeds, Leeds. LS2 9JT.
**Danone Vitapole, RD 128, 91767 Palaiseau Cedex. France.*

Abstract. In many solid liquid operations, it is required to keep all the solids in suspension to avoid localised reactions, prevent the solid build-up at a particular location on the base of the vessel or just avoid clogging the outlet port. In operations like dissolution of a solid in a liquid, the employment of agitation speed over the just suspension speed (Njs) does not necessarily increase the effective surface area of the solids and hence increase mixing efficiency. Many workers have proposed the correlations for the Njs but it is widely accepted that the phenomenon is best represented by Zweitering correlation. The discrepancies in the correlations are probably due to lack of accurate Njs measurement instruments. In this paper, a sensor based on the four electrode conductivity measurement has been proposed to directly measure the Njs. This sensor relies on the change in the electrical resistivity as the solids settle over the electrodes. A relationship to determine the Zweitering constant has been investigated for the Rushton impeller using this technique. A 5-10% discrepancy in Njs values is normally found between the sensor prediction and the visual determination.

Keywords: Minimum agitation speed, Njs, Solid liquid mixing, Conductivity

INTRODUCTION

The idea of just suspension speed (Njs) was first introduced by Zweitering (1958) to measure the minimum agitation speed required to keep all the solid particles in suspension in a stirred vessel. At Njs, all the solids are completely suspended and no particle rests on the bottom of the vessel for more than 1-2 s (Zweitering, 1958). Many workers (Baldi (1978), Nienow (1968), Wichterle (1988)) have proposed the correlations for the Njs but it is widely accepted that the phenomenon is best represented by Zweitering correlation.

The discrepancies in the correlations are probably due to lack of accurate Njs measurement instruments. Zweitering (1958) used visual methods which were simple but subjective. The actual value of the Njs varies among researchers who studied the same particle system. Zweitering (1958) gave a value of 10% error for his experiments relying on visual method. The Njs measurements become complicated for the systems whose suspensions are usually opaque even at lower concentrations and make it harder to visually measure the Njs.

Mak (1992) used Ultrasonic Doppler Flow meter (UDF) placed on the bottom of the vessel to directly measure the Njs. The probe measures the frequency shift of an ultrasonic signal reflected from the particle discontinuities in the flowing fluid. The maximum measurement error in Njs as recorded by Mak for UDF was about 10%.

Measuring the local solid concentration is also one of the popular methods to determine the Njs. Recently, Ricard et al (2005) used Electrical Resistance Tomography (ERT) with a linear sensor to measure the average resistance in the bulk of the mixing vessel to predict the Njs. They noted down the impeller speed at which the measured bulk solid concentration becomes constant. It was assumed that when the solid concentration in the bulk remains constant, all the solids have been suspended and no more solids are remaining on the bottom of the vessel. Ricard reported the maximum measurement error to be 10%. The use of a linear ERT probe may not give the exact impeller speed at which the particles have 'just' started to suspend (as per Zweitering's definition) but gives a fairly good idea of the Njs. Similar Njs experiments of measuring the local solid concentration by photometric method (Fajner et al, 1986) were also reported in the literature.

Zweitering (1958) used dimensional analysis to determine the correlation for the Njs and used his experimental data to determine the exponents of various suspension parameters (Eqn 1).

CP914, *Multiphase Flow: The Ultimate Measurement Challenge,* edited by X. Cai, Y. Wu, Z. Huang, S. Wang, and M. Wang

$$N_{js} = s\frac{\left(\frac{\mu_f}{\rho_f}\right)^{0.1} d^{0.2} X^{0.13}}{D^{0.85}}\left(g\frac{\rho_p - \rho_f}{\rho_f}\right)^{0.45} \tag{1}$$

Zweitering presented the plots of correlation parameter '*s*' against T/D at various impeller clearances. '*s*' is a constant (for particular impeller and T/D) in the correlation and is supposed to take into account the impeller clearance and the characteristic flow profile generated by the impeller (Eqn 2).

$$\text{i.e. } s = f\left(\frac{T}{D}, \frac{T}{C}\right) \tag{2}$$

For all impellers, but for 6-blade fan turbine he studied, the constant *s* decreases with decreasing clearance ratios. But for the 6-blade fan turbine impeller, the *s* parameter actually remains constant through a range of impeller clearances. This observation of Zweitering is rather surprising as many researchers like Armenante et al (1998) and Nienow (1968) (including present authors) have observed that the particles are suspended at lower impeller speed at lower clearances even for the 6-blade turbine than opined by Zweitering.

So, Armenante et al (1998) proposed the modified Zweitering correlation, in effect regressed their data in the Zweitering correlation using s as a function of T/D and T/C (Eqn 3)

$$s = a\left(\frac{T}{D}\right)^b \exp\left(c\frac{C_b}{T}\right) \tag{3}$$

C_b in eqn 3, is the clearance of the impeller from bottom of the vessel to the impeller bottom. The modified Zweitering as given by Armenante et al is given in Eqn 4 and 5.

$$N_{js} = \left[2.10\left(\frac{T}{D}\right)^{1.18} \exp\left(0.24\frac{C_b}{T}\right)\right]\frac{\left(\frac{\mu_f}{\rho_f}\right)^{0.1} d^{0.2} X^{0.13}}{D^{0.85}}\left(g\frac{\rho_p - \rho_f}{\rho_f}\right)^{0.45} \quad \text{for C/T > 1/5} \tag{4}$$

These empirical models present possible ways to estimate N_{js} based many known process conditions that may be difficult in practice. In this paper, we propose the use of a 4-electrode conductivity sensor to directly measure the N_{js} of the solid particles.

METHODS AND MATERIALS

We propose to use a 4-electrode conductivity sensor to measure the Njs. The 4-electrode sensor was essentially a conductivity measuring instrument employing one pair of electrodes to inject the current (source) and the other electrode pair to measure voltage (signal). The sensor was placed at the base of the vessel. The agitation speed was reduced gradually in steps and the change in voltage due to the solid settling was measured. The Njs was calculated based on the voltage profile.

The electrode construction can be referred from other articles (Wang 1995, 2005). Each electrode used in the experiment was 2 x 2 cm2 in dimension and was 1 cm apart from the adjacent electrode. An additional reference ground electrode was also provided and the dimensions were the same as that of the measuring electrodes. The two sets of 4-electrode sensors with the ground electrode were mounted on a flexible non-conductive rubber base of 5 mm thickness. The flexible rubber base was to ensure that the electrodes were placed firmly on the curvature of the base of the vessel. The whole arrangement was fixed to the base of the vessel (See figure 1).

The data collection from the sensor was automated by a data acquisition system (ITS P2000) and the AC (9.6 kHz) current injection and voltage measurement were controlled by the DAS. The optimum current injection of 5 mA was selected with the insurance of no signal saturation at any processing condition after carrying out various optimising studies on current injection quantities. In the experiments reported, particles used were white, spherical and had a diameter of 3 mm and a density of 1051 kg/m3 which were custom made for the settling experiments. The mixing vessel was made of Perspex material with a dished bottom and with 4 standard baffles (width=T/10). The vessel contents were agitated with an IKA power visc® mixer which was capable of communicating with a computer via RS232 port. The mixer was programmed to ramp down the agitation speed by 5 rpm. The contents of the vessel were stirred with an impeller (6-blade Rushton and A315 impellers) arranged in the centre of the vessel. The diameter of Rushton impeller used was 0.5T and that of A315 was 0.42 T. The experimental arrangement is shown in the figure 2.

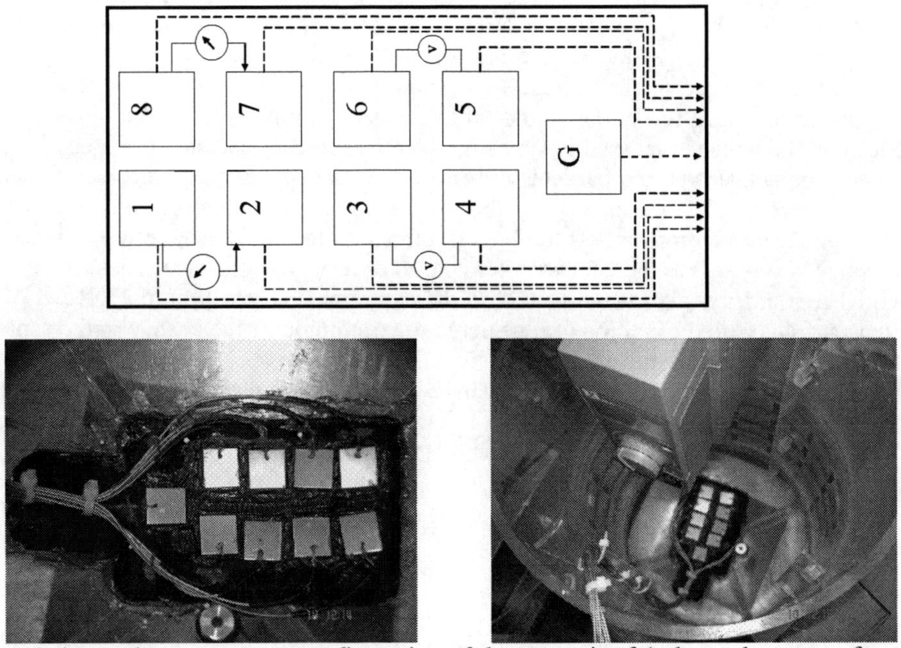

FIGURE 1. The construction and measurement configuration of the two pair of 4-electrode sensors for measurement of the N_{js}, (a) shows the schematic diagram of the sensors, (b) an enlarged photograph of the sensors (c) the sensor is placed on the bottom of the vessel where the voltage measurement electrodes are in the centre of the base of the vessel.

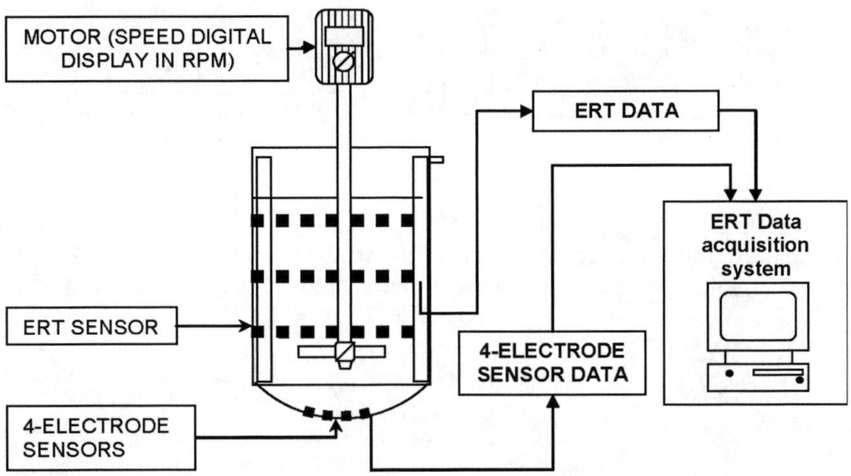

FIGURE 2. The experimental set-up showing the arrangement of Njs sensor on the base of the vessel.

The determination of the N_{js} was made from the 4-electrode sensor measurement as well as visual observation by two other researchers by increasing and decreasing the stirrer velocity by 5 rpm at regular intervals.

RESULTS AND DISCUSSIONS

When the impeller speed is close to the Njs, the location of the earliest particle settling depends on the characteristic flow profile established at the base of the vessel and also depends on the particle properties (Mak, 1992). The Rushton impeller, being a radial type, tends to make a particle heap at the centre of the base of the vessel; hence the four electrodes are arranged in the centre of the vessel in the present study.

Figure 3 shows the evolution of the voltage as the agitation speed is increased. At very low impeller speeds (relative to Njs), the solids are at rest at the bottom of the vessel. At this stage, the particles are very densely packed on the electrodes and the voltage recorded is very high. As the impeller speed is gradually increased and the speed begins to approach the actual Njs, the particles are in partial suspension and there is exponential decrease in the voltage as the resistance of the fluid nearest to the sensor decreases. When the impeller speed reaches Njs, the particles are in complete suspension and off the bottom offering little resistance between the electrodes (the voltage recorded at this stage is very less as there are very less particles resting on the electrodes).

An exponential first order decay curve is fitted to the experimental data (figure 3). A cut-off point of 5% of the maximum voltage at fully settled condition is considered to accommodate the change in the sensor voltage due to the presence of the particles closest to them during the impeller speeds just before reaching Njs. After the cut-off point, the particles are considered completely suspended. This 5% cut-off line, though arbitrary, is necessary to compensate the resistance offered by the particles closest to the electrodes. The Njs measurements made by conventional visual observations were also found to be in agreement with this experimental data analysis.

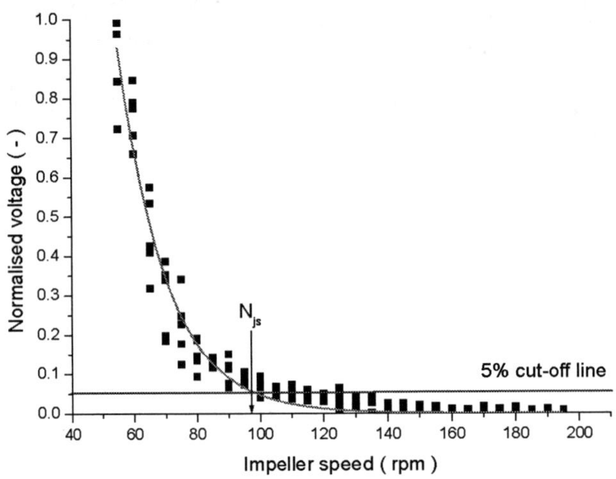

FIGURE 3. The experimental data is fitted with an exponential decay curve and the N_{js} is at the intersection of the 5% cut-off line and the decay curve. The recording of the voltage (at impeller speeds >Njs) shows a linear relationship with the bulk solid concentration. The voltage curve shows exponential trend when the N_{js} is being reached.

The particle settling curves are obtained as the normalised voltage vs. impeller speed and can be described by the correlation

$$y = A.e^{-Njs/B} \tag{5}$$

or can be rearranged as

$$N_{js} = -B\ln\left(\frac{y}{A}\right) \tag{6}$$

The y value is the chosen cut-off value of 0.05 (i.e. 5%) depending on the sensitivity of the sensor to particle concentration. The constants A and B are regressed with the experimental data (eqn 7 and 8) and are described as a function of impeller clearance and concentration.

$$A = \left(12.81\frac{T}{C} - 16.95\right)X^{14.08e^{-1.1\frac{T}{C}}} \tag{7}$$

$$B = 46.02e^{-0.39\frac{T}{C}} X^{0.24\frac{T}{C}-0.70} \quad (8)$$

Hence, we are able to propose a new correlation to present the relationship between Njs and process conditions based on 4-electrode measurement, which is given by eqn 9. For the convenience of the discussion, we called it as the conductivity correlation (CC)

$$N_{js} = -46.02e^{-0.39\frac{T}{C}} X^{0.24\frac{T}{C}-0.70} .\ln \frac{y}{\left(12.81\frac{T}{C}-16.95\right) X^{14.08e^{-1.1\frac{T}{C}}}} \quad (9)$$

Determination of s for Rushton turbine

Zweitering constant, s (eqn 1), is considered a constant which is a function of impeller clearance and impeller diameter during the calculation of Njs. The plots of s vs. T/D and s vs. C/T show the relationship between the two functions (figure 4).

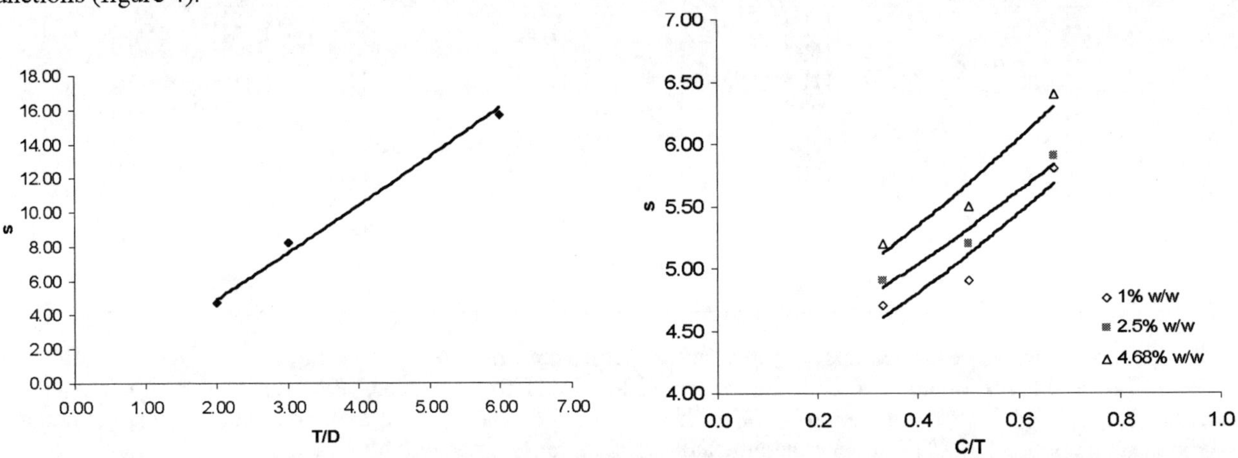

FIGURE 4 . (a) relationship between N_{js} and T/D at solid concentration of 1% w/w and impeller clearance of T/3 (b)The relationship between N_{js} and C/T at various solid concentrations

The regression of *s* vs. T/D and s vs. C/T data to determine the dependency is obtained from Armenante (eqn 3) and is given in eqn 11.

$$s = 1.74\left(\frac{T}{D}\right)^{1.09} \exp\left(0.41\frac{C}{T}\right) \quad (10)$$

The Njs correlation was drawn based on the Zweitering relationship as Zweitering performed over a thousand experiments to derive the relationships between the N_{js} and other process conditions. So the final relationship when the s is incorporated into the Zweitering relation becomes

$$N_{js} = \left[1.74\left(\frac{T}{D}\right)^{1.09} \exp\left(0.41\frac{C}{T}\right)\right] \frac{\left(\frac{\mu_f}{\rho_f}\right)^{0.1} d^{0.2} X^{0.13}}{D^{0.85}} \left(g\frac{\rho_p-\rho_f}{\rho_f}\right)^{0.45} \quad (11)$$

Various authors have contributed to the modification of the Zweitering relationship and the correlation obtained in the eqn 12 can be compared with other researchers in table 1.

TABLE 1: Comparison of constants in the function of s on T/D and C/T by various authors

	a	b	c
Armenante (1998)	2.10	1.18	0.24
This work	1.74	1.09	0.41

Nienow (1968)		1.28	
Chapman (1983)	-	-	0.71
Zweitering (1958)	-	1.50	Independent of C/T

Comparison between different correlation methods

The comparison was made between Njs predicted using the new conductivity correlation (CC) derived with the 4-electrode sensor method (Eqn 11) and the Njs estimated by Armenante correlation (AC) (modified Zweitering's correlation, Eqn 4) in table 2. Table 2 shows the results of CC match the trend observed by Armenante. The maximum error of 18.6% in the measurements is reflected at concentration 1% w/w and at clearance T/2. The average deviation of all the measurements is found to be in an acceptable range of below 6%. As can be observed in the table, the Njs speed is indeed less than those predicted by Zweitering at small impeller clearances.

TABLE 2: Rushton 6 blade impeller, diameter 0.5 T

Concentration (%w/w)	Clearance	N_{js} (RPM), CC	N_{js} (RPM), AC	Deviation %
1.00	T/1.5	98	95	-3.1
2.50	T/1.5	118	107.5	-9.7
4.68	T/1.5	121	116.6	-3.7
1.00	T/2	108	91	-18.6
2.50	T/2	112	103.5	-8.2
4.68	T/2	113	112	-0.0
1.00	T/3	87	88	0.0
2.50	T/3	95	99	4.0
4.68	T/3	101	107.7	6.2
Average deviation				5.94

Similar procedure as described above was adopted to measure the Njs of the same particle system with A315 impeller. The results were compared with the data published by Harrop et al (HC) (Harrop et al 1997). The table 3 shows that there is about 10% deviation between the HC and CC. Harrop et al addressed about 7.9% deviation in their measurement of s for A315 impeller and the authors' data fairly agree with their data.

TABLE 3: Njs for A315 impeller, D/T = 0.42, impeller clearance at C=T/4

concentration (%w/w)	Njs (RPM), CC	Njs (RPM), HC	Deviation (%)
1	100	110	9.0
2.5	135	124	-8.0
4.68	145	135	-7.4
Average deviation			8.13

Zweitering predicts a constant Njs over a range of impeller clearances, which is not what is observed by this and other authors. The reason for Zweitering to predict so could be because the difference between Njs at different clearances is too close to be observed by visual means.

CONCLUSION

A flexible 4-electrode conductivity sensor has been developed to directly measure the Njs in the solid-liquid mixing systems. The experimental results have shown that the sensor measurements and the visual measurements of Njs differ by 5-10%.

As the Njs is highly dependent on the size and shape of the particles the direct measurement sensor can be used to measure the 'characteristic' Njs with all the internals inside the mixing vessel and in a non-standardised mixing vessel. It can also be used in optically opaque mixtures and as a feedback control to monitor the varying requirements of Njs.

The relationship between the Zweitering constant s and other variables like T/D and T/C give an insight into the dependency of Njs on impeller clearance and diameter. A very low impeller clearance is to be avoided if impeller is bare metal and is not coated with non-conductive material (like powder coating) as it might affect the measurements (even though the impeller effect is subtracted during the reference measurements of the sensor).

ACKNOWLEDGEMENTS

The authors would like to thank the members of Danone Vitapole, France for their technical input and financial support throughout this research.

REFERENCES

ARMENANTE et al (1992), Determination of correlations to predict the minimum agitation speed for complete solid suspension in agitated vessels, *The Canadian journal of chemical engineering*, v76, pp 413-419.

BALDI et al (1978), Complete suspension of particles in mechanically agitated vessels, *Chemical engineering science*, V 33, pp 21-25

CHAPMAN et aL (1983), Particle-Gas-Liquid mixing in stirred vessels. Part I: Particle-Liquid mixing, Transactions institution of chemical engineers, V 61, pp 71-81.

FAJNER, MAGELLI, NOCENTINI and PASQUALI (1985), "Solids Concentration Profiles in Mechanically Stirred and Staged Column Slurry Reactor", *Chemical engineering research & design*, 63, 235–240.

HARROP et al (1997), Impact of suspended solids on the homogenisation of the liquid phase under turbulent conditions in a stirred vessel, Recents progress en genie des procedes, v11, pp 41-48.

ITS P2000 (2005), Industrial Tomography Systems Ltd, Manchester, UK

MAK (1992), Solid-liquid mixing in mechanically agitated vessels, Thesis, Department of chemical engineering, University college London, London.

MAXWELL J.C (1873), A Treatise on Electricity and Magnetism vol. 1, Clarendon Press, Oxford.

NIENOW (1968), Suspension of solid particles in turbine agitated baffled vessels, *Chemical engineering science*, 1968, V 23, pp 1453-1459.

RICARD et al (2005), Monitoring of multiphase pharmaceutical process using electrical resistance tomography, Chemical engineering research and design, 83 (A7), pp 794-805.

WANG (2005), Electrode models in electrical impedance tomography, Journal of Zhejiang University Science, pp 1386-1393

WANG et al (1995), The grouped node technique as a means of handling large electrode surfaces in electrical impedance tomography, *Physiological Measurements*, v16-A, pp 219-226.

WICHTERLE (1988), Conditions for suspension of solids in agitated vessels, *Chemical Engineering Science*. V 43, pp 467-471.

ZWEITERING (1958), Suspending of solid particles in liquid by agitators, *Chemical engineering science*, v8, pp 244-253.

NOMENCLATURE

a, A : Constant

b, B : Constant

s : Zweitering Constant

μ_f : Viscosity of the liquid (Ns/m^2)

N_{js} : Minimum agitation speed for complete suspension (rps)

C : Impeller clearance from the centreline of the impeller to base of the vessel (m)

C_b : Impeller clearance from the bottom of the impeller to base of the vessel (m)

H : Height of the liquid level in the vessel from the bottom of the vessel (m)

T : Diameter of the vessel (= H)

d : Diameter of the particle (m)

D : Diameter of the impeller (m)

ρ_p : Density of the particles (Kg/m^3)

ρ_f : Density of the liquid (Kg/m^3)

y : Cut-off point for the normalised voltage in settling plots on y-axis

X : Concentration of solids (%w/w)

<u>SESSION D</u>

OPTICAL, NON-INTRUSIVE TECHNIQUES

Application Of Phase Doppler Particle Anemometer For Studying Turbulent Gas-Particle Flows At Tsinghua University

Lixing Zhou

Department of Engineering Mechanics, Tsinghua University, Beijing, China
zhoulx@mail.tsinghua.edu.cn

Abstract. The phase Doppler particle anemometer (PDPA) made by both Aerometrcis Inc. and Dantec Inc. are used to measure the time-averaged gas and particle velocities, RMS values of gas and particle fluctuation velocities, gas-particle velocity correlations, gas turbulence modification and particle concentration for turbulent swirling and non-swirling gas-particle flows and also gas-particle flows with combustion in the Laboratory of Two-phase Flows and Combustion, Department of Engineering Mechanics, Tsinghua University. For strongly swirling gas-particle flows with swirl numbers greater than unity (s=1.5 and s=2.1), it is found that their gas-particle flow behavior is different among each other and is different from that for weakly swirling flows. For example, there are no central reverse flows and there are enlarged solid-body rotation zones for swirling flows with s=1.5. The behavior of swirling gas-particle flows depends on the swirl number and the flow configuration For sudden-expansion gas-particle flows it is found that the gas-particle velocity correlation has the distribution similar to that of gas and particle RMS fluctuation velocities and its magnitude is smaller than that of both gas and particle RMS fluctuation velocities. For turbulence modification in sudden-expansion flows it is found that the separating flows increase the gas-turbulence enhancement tendency in comparison with the pipe flows. For combusting gas-particle flows it is found that combustion reduces the velocity slip between the gas and particle phases.

Keywords: gas-particle flows, PDPA measurement, swirling flows, turbulence modification

INTRODUCTION

One of the important problems in many engineering facilities is swirling gas-particle flows. Swirling gas-particle flows are frequently encountered in cyclone separators, hydrocyclones, swirl combustors/burners and cyclone combustors. The gas and particle velocity, turbulence intensity and particle concentration in the two-phase flow field will remarkably affect the pressure drop, collection efficiency, flame stabilization, combustion efficiency and even pollutant emission. Most of studies are limited to single gas-phase flow field [1-3]. Yoshida et al. [1] studied the gas flow field in a cyclone separator using probe and hot-wire systems. The probe measurements were also made by Silva, Nebra et al. [2]. The studies by Ogawa [3] demonstrated that there was no distinct difference between the dust-laden gas flow and the pure air flow in a cyclone chamber. Obviously, probe measurements cannot give gas-particle two-phase flow field. Studies using conventional and modified laser Doppler anemometer (LDA) [4] show similar behavior of gas and particle flow fields (for example, Rankine vortex structures of tangential velocity profiles) and slip velocity between gas and particle phases in swirling flows. However, even the modified LDA measurements cannot give reliable results, because it is unable to identify particle sizes and the effect of particle size is included in the particle "turbulent" fluctuation. The advantage of measurements by phase Doppler particle anemometer (PDPA) is able to simultaneously give the gas and particle velocities, turbulence intensities, particle sizes and particle concentration. The PDPA was first used to study swirling gas-particle flow with a swirl number s= 0.47 by Sommerfeld and Qiu [5]. Their results demonstrate the different behavior of particles of different sizes in swirling flows. However, no PDPA measurements of strongly swirling gas-particle flows with swirl numbers greater than unity are reported. Within last ten years measurements of swirling gas-particle flows using PDPA, made by both Aerometrics Inc.& Dantec Inc., were conducted in the Laboratory of Two-phase Flows and Combustion, Department of Engineering Mechanics, Tsinghua University The PDPA measurements allow obtaining the detailed information of gas and particle time-averaged velocities and RMS values of fluctuation velocities, particle sizes and concentration.

The next problem is the gas-particle velocity correlation. The two-phase fluctuation velocity correlation <vivpj> is an important term in the two-phase turbulence modeling. Its role likes that of the pressure-strain term in the single-phase Reynolds stress equation. The relationship of the two-phase fluctuation velocity correlation

CP914, *Multiphase Flow: The Ultimate Measurement Challenge,* edited by X. Cai, Y. Wu, Z. Huang, S. Wang, and M. Wang

<vivpj> with the fluid Reynolds stress <vivj>, particle Reynolds stress < vpivpj >, and the particle size or relaxation time needs to be studied both theoretically and experimentally. The phase Doppler particle anemometer (PDPA) can measure not only the fluid and particle size and velocity, fluid and particle Reynolds stresses, but also the two-phase fluctuation velocity correlation. The measurements of <vivpj> using PDPA were done in a gas-droplet mixing layer [6] and the far field of an axi-symmetrical jet [7]. However, it is not clear about the behavior of two-phase fluctuation velocity correlation in separating or recirculating flows such as sudden-expansion flows. Hence, the PDPA measurements are used to study <vivpj> in sudden-expansion flows in our laboratory.

Furthermore, the problem of turbulence modification attracts significant attention of many investigators. A number of experiments were performed in simple flows, such as jet, pipe and channel flows, for example, by Tsuji et al [8] and Sommerfeld et al. [9]. It is not clear about the turbulence modulation effect in complex flows such as separating flows, recirculating flows and swirling flows. The turbulence modulation by particles in different regions of the flow field for sudden-expansion gas-particle flows was studied in our laboratory.

Finally, most of studies are made for isothermal gas-particle flows. A few works, such as the LDA has been used to measure the flow field in swirling combustion [10], but less work are done for using PDPA for measuring combusting gas-particle flows. Therefore, recently PDPA was used to measure swirling gas-particle flows with gas combustion in our laboratory.

EXPERIMENTAL SET-UP AND MEASUREMENT TECHNIQUE

The experimental set-up used for studying turbulent gas-particle flows is shown in Fig.1.

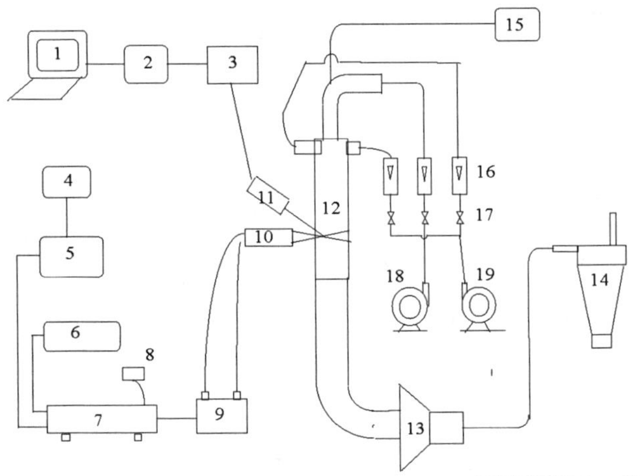

1. Data processor; 2. Signal processor; 3. Converter 4. Transformer; 5. Power supply; 6: water purifier; 7. Laser; 8. Laser controller; 9. Splitting group; 10. Emitter unit;11. Receiver; 12: Test section; 13. Flow stabilizer; 14. Cyclone separator; 15. Powder feeder; 16. Flow meter; 17.Valve; 18,19. Compressor

FIGURE 1. Experimental Set-up

The test section for measuring swirling gas-particle flows and two-phase velocity correlation in sudden-expansion flows is shown in Fig.2

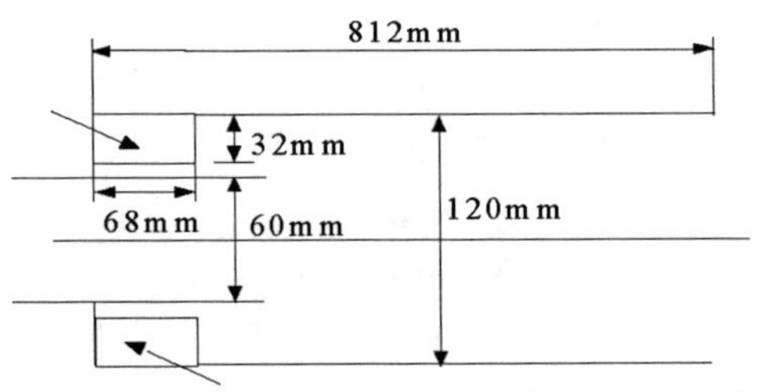

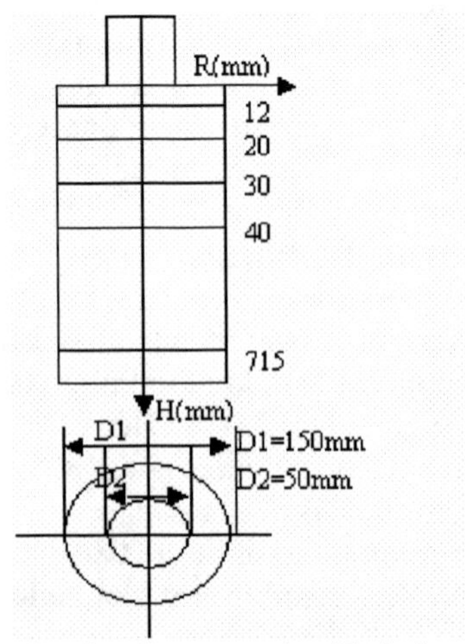

FIGURE 2. Test Section for Swirling Gas-Particle Flows **FIGURE 3.** Test Section for measuring Turbulence Modification

The test section for measuring turbulence modification in sudden-expansion flows with its geometrical sizes is shown in Fig. 3. The test section for measuring combusting gas-particle flows is given in Fig.4. The diameters of the fuel inlet tube, the primary-air tube, the secondary-air tube and the combustor are 8mm, 40mm, 58mm and 160mm respectively. The combustor length is 1000mm.

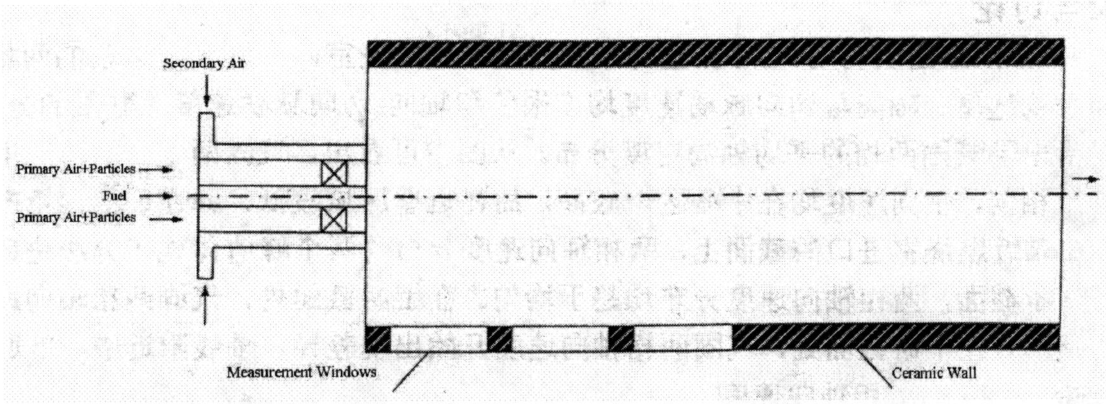

FIGURE 4. Test Section for Measuring Combusting Gas-Partice Flows

Both a 3-D PDPA made by Dantec Inc and also a 2-D PDPA made by Aerometrics Inc. were used to measure gas-particle flows. For the 3-D phase-Doppler particle anemometer (PDPA), a backward scattering arrangement is shown in Fig.5. It is a well-known and widely used technique. The particle velocity is measured based on the Doppler frequency shift. The particle size and concentration are measured based on the phase difference caused by Mie scattering and the number of particles passing the measuring volume during a certain time interval. The parameters of the optical system are shown in Tables 1 and 2

The inlet flow parameters for swirling gas-particle flows are given in Tables 3 and 4. The swirl number is defined as:

$$S = \frac{2\int_0^{D_3/2} \rho w u r^2 dr}{D_4 \int_0^{D_3/2} \rho u^2 r dr}$$

, where D_3 is the inlet diameter and D_4 is the chamber diameter.

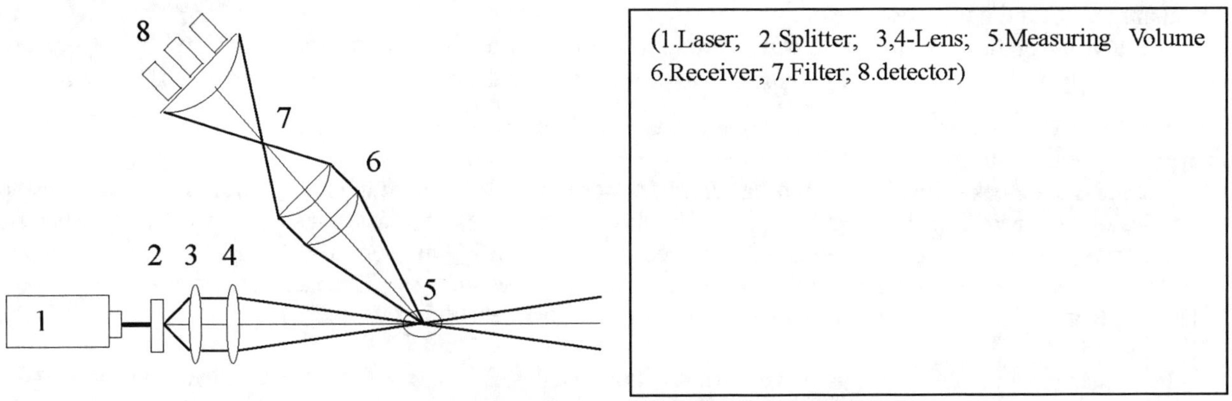

FIGURE 3. 3-D PDPA Instrumentation

TABLE 1.

Transmitting optics	Ux	Uy	Uz
Wave length of the laser(μm)	514	488	476
Frequency shift(MHz)	40	40	40
Beam separation(mm)	74	74	74
Diameter of measuring Volume(mm)	1.35	1.35	1.35
Focal length of the laser(mm)	500	500	500
Fringe separation(μm)	3.5	3.31	3.23
Fringe number	36	36	36

TABLE 2.

Receiving optics	
The maximum detected diameter	260.24 μm
The maximum detected concentration	$2 \times 10^5/cm^3$
Off-axial angle	147deg
Focal length of receiving length	600mm
The valuable sign level	-6db
The Maximum phase error	25deg
The maximum spherical error	15%

TABLE 3. Inlet Flow Parameters (Case 1, s=2.1)

Axial inlet velocity	0 m/s
Tangential inlet velocity:	10 m/s
Swirl number:	2.1
Particle mean diameter:	76.3 μm
Particle material density:	2400 kg/m3
Particle mass loading:	0.01 kg-solid/kg-gas

TABLE 4. Inlet Flow Parameters (Case 2, s=1.5)

Axial inlet velocity	5 m/s
Tangential inlet velocity	10 m/s
Swirl number	1.5
Particle mass loading	0.01 kg-solid/kg-gas

During the experiments the tangential inlet flow velocity was kept constant, and the swirl number is changed by changing the axial inlet flow velocity

For measuring the gas-particle velocity correlation in non-swirling sudden-expansion flows, three size groups (30, 50, 95 μm) of 10 μm width are chosen to express the particle phase, while particles with the size range of 0 to 10 μm are taken to represent the gas phase. The inlet volumetric flow rate is 212.4 m^3/h. The particle-loading ratio is about 0.005%.

For measuring gas turbulence modification, glass beads with a mean diameter of 150μm were used as heavy particles. Glass beads with sizes less than 10μm were taken as the gas-phase tracer in two-phase flows. Evaporating ethylene glycol droplets with a mean diameter of 5μm were used as the gas-phase tracer for single-phase gas flows. The inlet-flow parameters are: air flow rate 214.39m3/h; particle size150μm; particle material density 2400kg/m3; particle mass loading 0.24; ethylene glycol droplet size 5μm and material density 1.115kg/m3

For measuring combusting gas-particle flows, the gaseous fuel is methane with a flow rate of 1m3/h, the swirling primary air is with a flow rate of 8.67m3/h and a swirl number of 0.265, and the non-swirling secondary air is with a flow rate of 3.71m3/h.

MEASUREMENT RESULTS FOR SWIRLING AND SUDDEN-EXPANSION GAS-PARTICLE FLOWS

Figures 4 to 8 show the measurement results for cyclonic gas-particle flows with s=2.1 (case 1). Figures 4 and 5 show that the general flow behavior is w-shaped axial velocity profiles with an annular reverse flow zone and Rankine-vortex tangential velocity profiles of two phases. The particle tangential velocity lags behind the gas tangential velocity almost everywhere with the larger particles having lower velocities due to their higher inertia. The particle axial velocity lags behind the gas axial velocity only in the near-wall region. The measured fluctuation velocities (Figs.6, 7) show the central peak and the increased value near the wall, because of the large velocity gradient in the solid-body rotation region and the region near the wall. The comparison between these two figures shows that the axial fluctuation velocities are much smaller than the tangential ones, demonstrating the high anisotropy of turbulent fluctuations. The particle turbulent fluctuation is lower than the gas fluctuation in most regions both in axial and tangential directions, and the larger the particle size, the smaller the particle fluctuation. For particle concentration, a large amount of particles almost concentrates in the near-wall region (Fig.8) due to the strong centrifugal force.

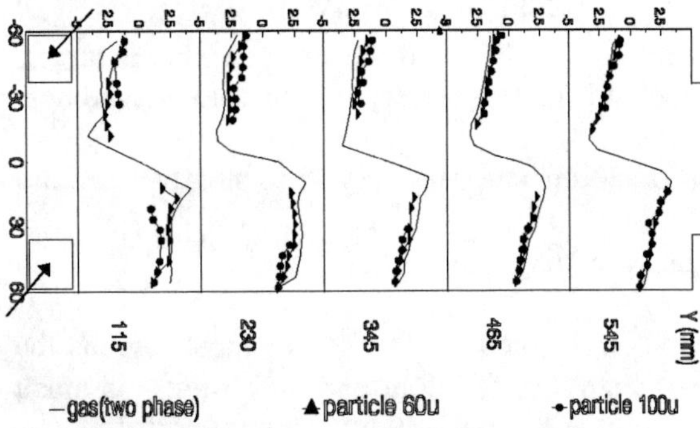

FIGURE 4. Two-phase tangential time-averaged velocities (m/s) (case 1)

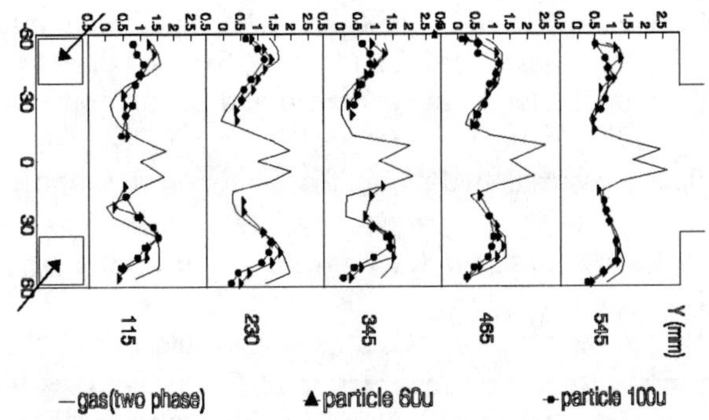

FIGURE 5. Two-phase axial time-averaged velocities (m/s) (case 1)

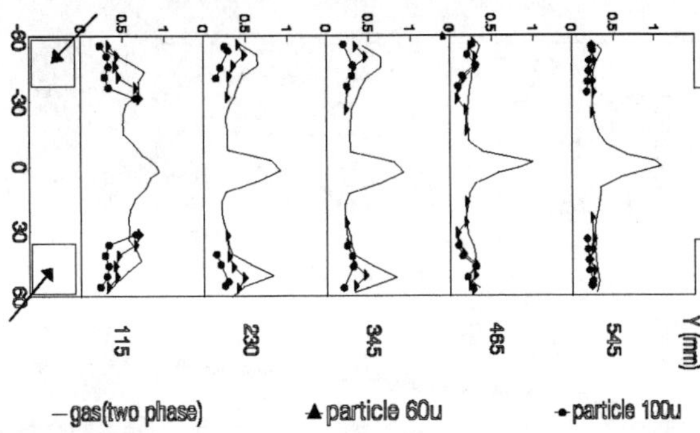

FIGURE 6. Two-phase tangential fluctuation velocities (m/s) (case 1)

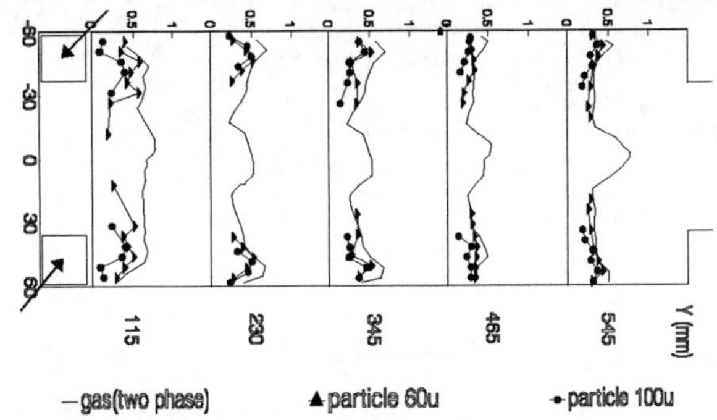

FIGURE 7. Two-phase axial fluctuation velocities (m/s) (case 1)

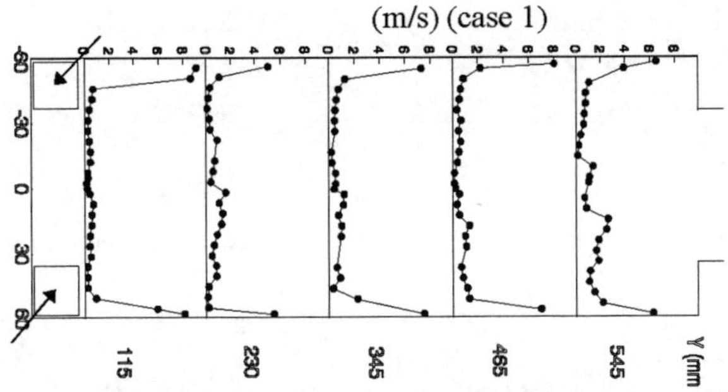
FIGURE 8. Particle number density (E+2/cc) (case 1)

Figures 9 to 13 show the measurement results for swirling gas-particle flows with s=1.5 (case 2). As seen from Fig. 9, in this case, the axial velocities of two phases have no w-shaped distributions as those for the cases of s=2.1, and are higher near the axis and lower with a small peak near the wall. The particle axial velocity lags behind the gas one in most regions, and the velocity slip between two phases increases with the increase in particle size, but the relative slip is not very large. In the near-wall region the particle axial velocity is slightly larger than the gas velocity. The particle tangential velocity lags behind the gas tangential velocity (Fig. 10), and the larger the particle size, the more the lag. In the first and second sections, the structure of rigid body rotation in tangential velocity profiles is not obvious because of injection of axial non-swirling flows. In the subsequent sections, due to the increasing effect of centrifugal force the tangential velocity distribution of two phases shows an obvious Rankine-vortex structure with the gradually vanishing effect of inlet conditions. Figure 11 gives the particle and gas axial fluctuation velocities with higher values near the wall and lower values near the axis. Unlike that in case 1, there is no peak value near the axis in this case. The particle axial fluctuation velocity is smaller than that of the gas phase, and decreases with the increase of particle size. These results are similar to those for case 1. Figure 12 indicates that the particle tangential fluctuation velocity is smaller than that of gas phase in most regions. This is similar to that for case 1. Comparison between Fig.11 and Fig.12 indicates that both gas and particle turbulent fluctuations are anisotropic, but the axial fluctuation exceeds the tangential fluctuation due to the fact that the axial velocity gradient is still larger than the tangential velocity gradient. So, this is qualitatively similar to that for swirling flows with s=0.47 reported in [7] but is opposite to the results for the case of s=2.1. Figure 13 gives the particle concentration distribution. Unlike the case of cyclonic flows, and similar to that for swirling flows with s=0.47, in this case particles at first concentrate in the near-axis zone due to the effect of axial inlet velocity, then gradually move to the wall under the effect of centrifugal force and turbulent diffusion, and finally concentrate in a thin layer adjacent to the wall. The phenomena that the particle axial time-averaged velocity exceeds the gas one, and the particle axial fluctuation velocity exceeds the gas one in reverse flow zone is owing to higher inlet axial velocity of particles for the case of low swirl number s=0.47 [7]. For the cases of s=2.1 and s=1.5, the inlet particle axial velocity is smaller, and the swirl number is larger, so the particle axial time-averaged and fluctuation velocities lag behind the gas corresponding velocities almost everywhere.

The measured two-phase fluctuation correlation of particles of 30μm and 50μm in sudden-expansion are given in Figs. 14 and 15. Obviously, the profile of $\langle u_i u_{pi} \rangle$ has the shape similar to those of $\langle u_i u_i \rangle$ and $\langle u_{pi} u_{pi} \rangle$. $\langle u_i u_{pi} \rangle$ takes its peak value in the shear layer as $\langle u_i u_i \rangle$ and $\langle u_{pi} u_{pi} \rangle$. In most regions the two-phase fluctuation correlation of 50μm particles is slightly less than that of 30μm particles. <uup> is much smaller than <uu> and <upup> <vvp> is also smaller than <vv> and <vpvp> but with much smaller difference. Similar to gas and particle turbulent Reynolds stress, the two-phase fluctuation velocity correlation is also anisotropic, <vvp> is smaller than <uup>..

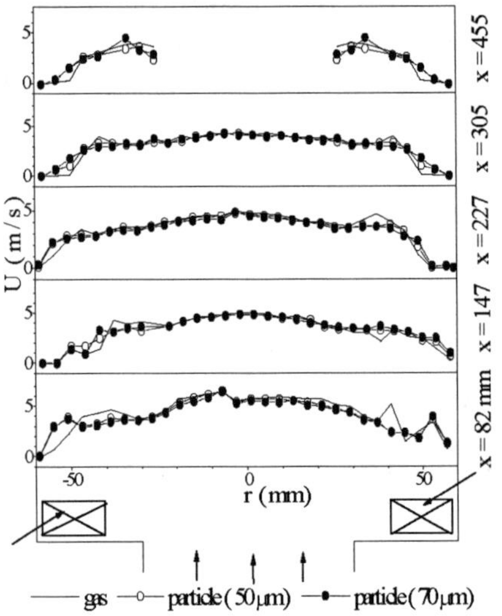

FIGURE 9. Axial Time-averaged Velocity (case 2)

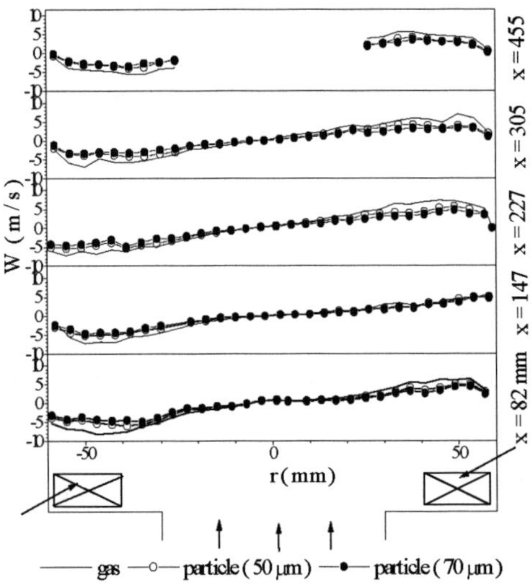

FIGURE 10. Tangential Time-averaged Velocity (case 2)

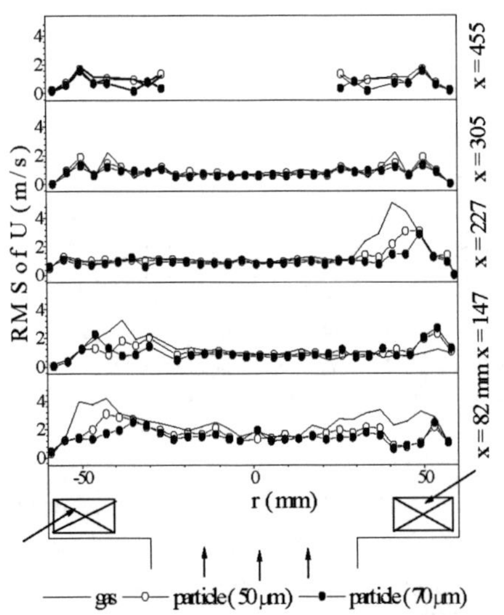

FIGURE 11. Axial Fluctuation Velocity (case 2)

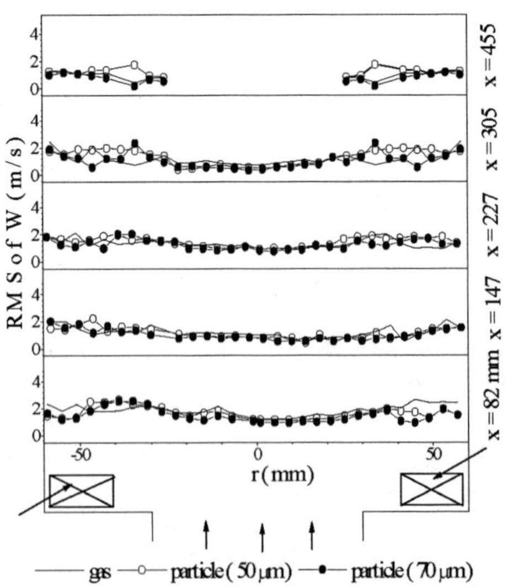

FIGURE 12. Tangential Fluctuation Velocity (case 2)

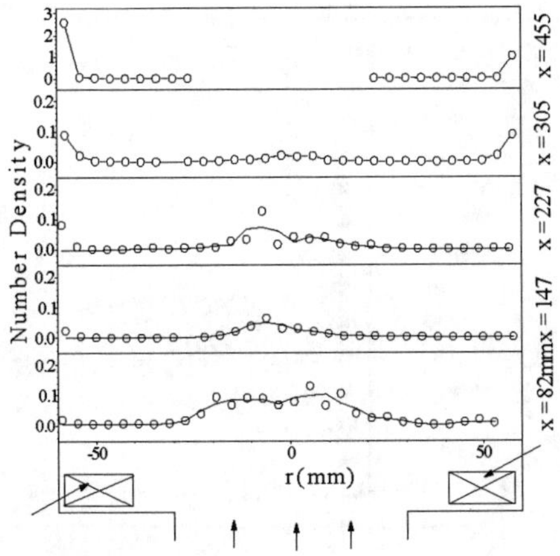

FIGURE 13. Particle Number Density ($10^{10}/m^3$) (case 2)

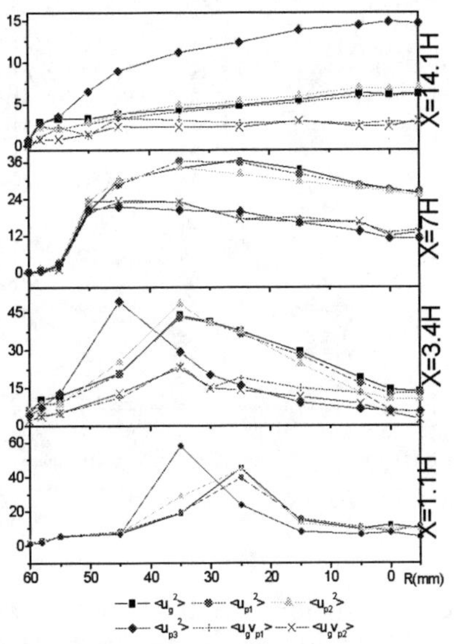

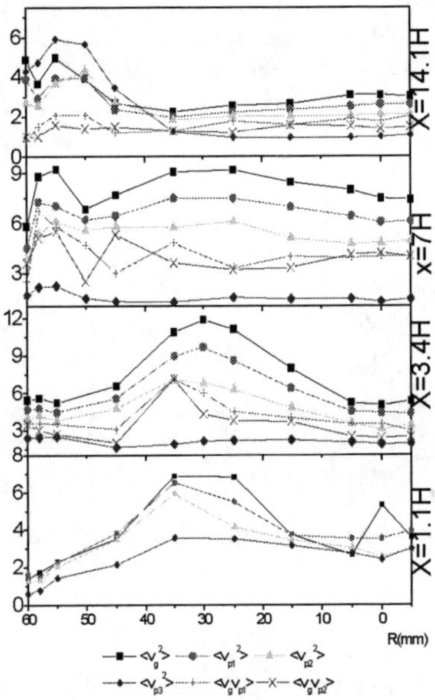

FIGURE 14. Gas and Particle Axial Reynolds Stresses and Two-phase Velocity Correlation (m2/s2)

FIGURE 15. Gas and Particle Radial Reynolds Stresses and Two-phase Velocity Correlation (m^2/s^2)

MEASUREMENT RESULTS FOR TURBULENCE MODIFICATION

The measured single-phase and gas-phase fluctuation velocities in sudden-expansion gas-particle flows are given in Fig. 16. It is seen that the turbulence modulation due to particles is different in different regions of the flow field. In the recirculation zone (Sections A and B), particles reduce gas turbulence in the near-wall reverse-flow region, while enhance gas turbulence in the near-axis forward-flow region. In the downstream and fully developed flow zones particles only attenuate gas turbulence. Figure 17 gives the axial fluctuation velocities for gas and particles of three size groups (60, 120 and 150μm) in two-phase flows. Obviously, in the recirculation zone, the particle turbulence is stronger than the gas turbulence in the reverse-flow region and is weaker than the gas turbulence in the forward-flow region, while in the downstream zone the particle turbulence is smaller than the gas turbulence everywhere.

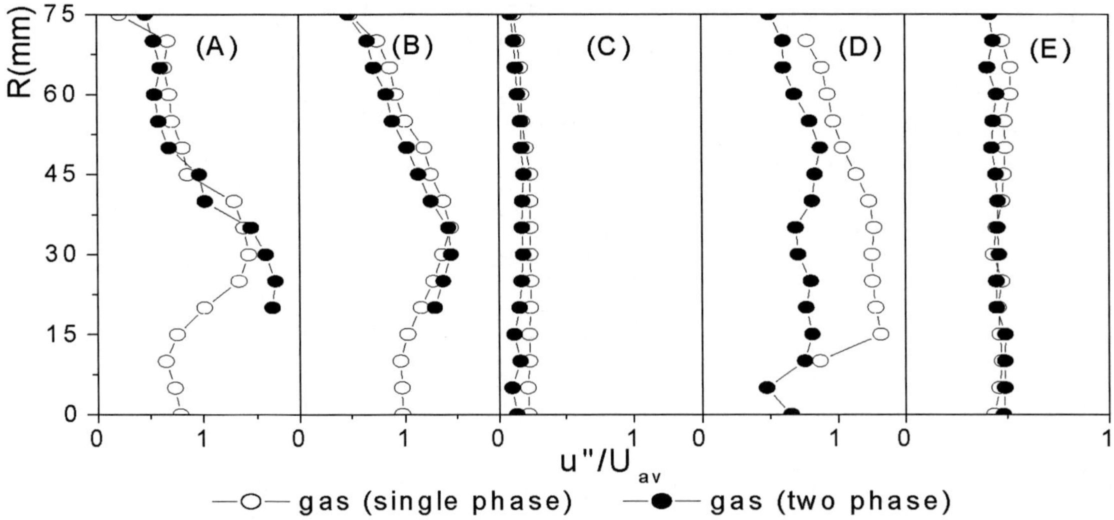

FIGURE 16. Gas Axial Fluctuation Velocity (m/s, A ,H=120mm, B,H=200mm, C,H=300mm, D,H=400mm, E,H=715mm)

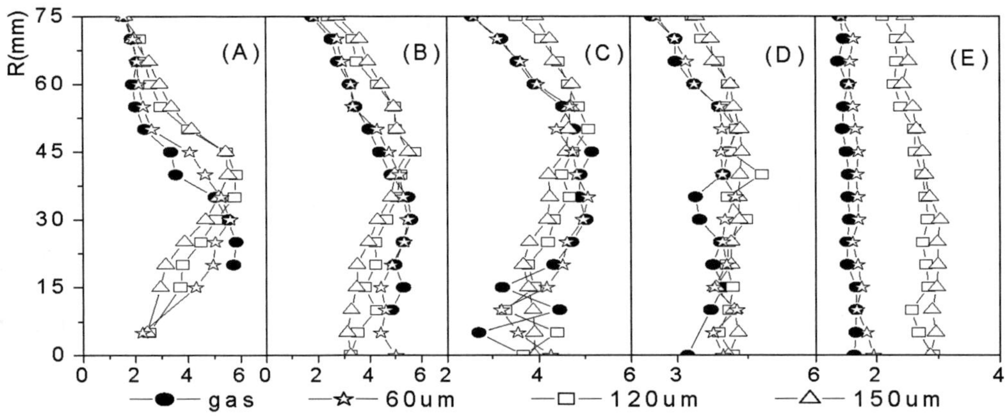

FIGURE 17. Axial Fluctuation Velocity (m/s, A ,H=120mm, B,H=200mm, C,H=300mm, D,H=400mm, E,H=715mm)

For the case studied, particles of 150μm mean diameter enhance gas turbulence in the forward-flow region of the recirculation zone and reduce gas turbulence in other regions. This is different from that observed in pipe flows where the 1mm and 3mm particles enhance gas turbulence, 200μm particles damp gas turbulence and 500μm particles enhance or reduce gas turbulence in different regions. Therefore, the flow separation increases the tendency of particle enhancing gas turbulence.

MEASUREMENT RESULTS FOR COMBUSTING GAS-PARTICLE FLOWS

Figures 18 and 191 give the measurement results of combusting swirling gas-particle flows in a swirl chamber. For axial two-phase velocities (Fig.18a), there are only weakly reverse flows in the near wall region. The velocity slip between two phases can be seen only at the cross section near the exit. The highest tangential velocities are located near the axis (Fig.18b), that is, the solid-body rotation zone is rather small. Also, there is only slight velocity slip between two phases. For the axial two-phase fluctuation velocities (Fig.19a), higher values are located near the axis, and the particle axial fluctuation velocity is higher than the gas one at the 3rd and 4th cross sections. The tangential two-phase fluctuation velocities (Fig.19b) also are higher in the near-axis region, and the particle fluctuation velocity is only slightly higher than the gas one at two cross sections. It is seen that in comparison with the isothermal gas-particle flows, for combusting gas-particle flows combustion reduces the velocity slip between two phases and the difference in turbulence intensities between two-phases. These results imply that combustion increases the particle drag force.

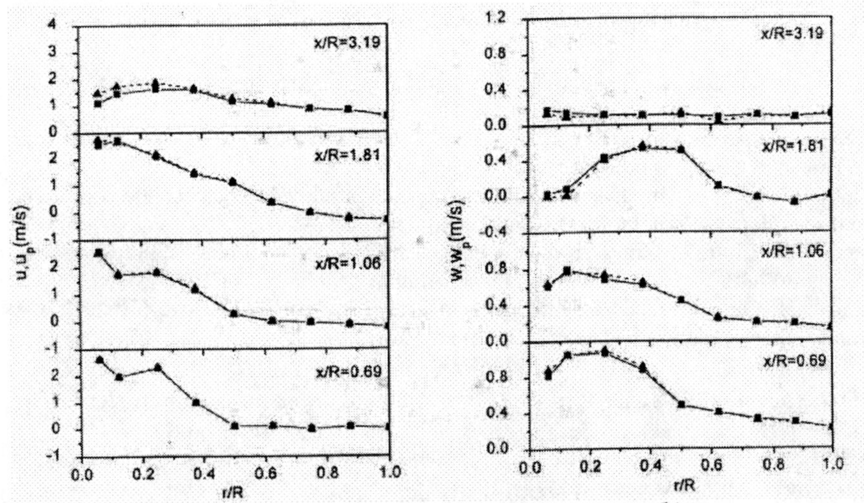

(a) (b)

FIGURE 18. Two-Phase Time-Averaged Velocities (a-Axial , b-Tangential, ▲ -Gas, ■ -Particle)

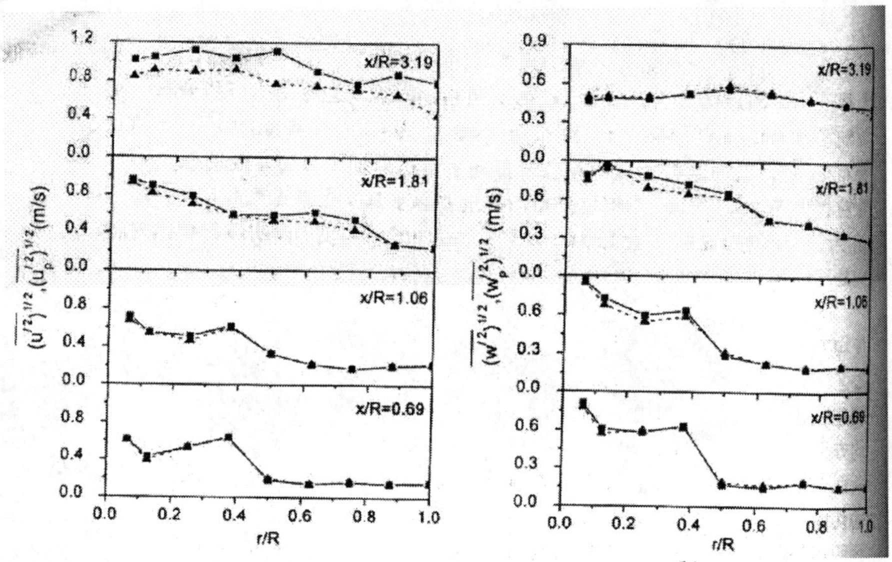

(a) (b)

FIGURE 19. Two-Phase Fluctuation Velocities (a-Axial , b-Tangential, ▲ -Gas, ■ -Particle)

CONCLUSIONS

The behavior of strongly swirling gas-particle flows is different from that of weakly swirling gas-particle flows.

The gas-particle velocity correlation has the distribution similar to that of gas and particle RMS fluctuation velocities and its magnitude is smaller than that of both gas and particle RMS fluctuation velocities.

The separating gas-particle flows increase the gas-turbulence enhancement tendency in comparison with the pipe flows.

Combustion reduces the velocity slip between two phases and the difference in turbulence intensities between two-phases

ACKNOWLEDGEMENT

This study was supported by the National Key Project of Fundamental Research, sponsored by the State Science and Technology Commission, PRC under the Grant G1999-0222-08 and the Project, sponsored by the National Natural Science Foundation of China under the Grant 19572043.

REFERENCES

[1] S. Yuu, T. Jotaki, Y. Tomita, and K. Yoshida, The reduction of pressure drop due to dust loading in a conventional cyclone, Chem. Eng. Science, 1.33:1573-1580, 1978.

[2] M. A. Silva, S.A.Nebra, Experimental study of the variation of the velocity fluid field in the cyclone with the solid concentration, 10th Brazilian Congress of Mechanical Engineering, 1989

[3] A. Ogawa, O. Seito, H. Nakabayashi, Distributions of the tangential velocities on the dust -laden gas flow in the cylindrical cyclone dust collector, Particle Science and Technology, p.17-28, 1988;

[4] B. Zhou, X.L.Wang, R.X.Li, L.X. Zhou, Swirling gas-particle flows and coal combustion in a spouting-cyclone combustor, Proceedings ASME, Heat Transfer Conference, Atlanta, 1993

[5] M. Sommerfeld and H.H. Qiu, Detailed measurements in a swirling particulate two-phase flow by a phase-Doppler anemometer, International J. Heat and Fluid Flow, 12: 15-32, 1991.

[6] M.R. Wang, and D.Y. Huang, Measurements of $\overline{u'_{gl}u'_{pl}}$ in mixing-layer flow with droplet loading, Atomization and Sprays, 5: 305-328, 1995

[7] F. Prevost, J. Boree, H.J. Nuglisch and G. Charnay, Measurements of fluid/particle correlated motion in the ar field of an axi-symmetric jet, Int. J. Multiphase Flow, 22(4), 1996

[8] Y. Tsuji and Y. Morikawa, LDV measurements of an air-solid two-phase flow in a vertical pipe, J. Fluid Mech., 139:417-434, 1984.

[9] J. Kussin, and M. Sommerfeld, Experimental studies on particle behavior and turbulence modification in horizontal channel flow with different wall roughness, Experiments in Fluids, 33:143-159, 2002.

[10] L.M. Pickett, R.E. Jackson, D.R. Tree, LDA measurements in a pulverized coal flame at three swirl ratios, Combustion Science and Technology, 143:79-107, 1999.

Volumetric Fraction Dynamic Measurement in Oil-Water-Gas Multiphase Horizontal Pipe Flow with Dual Energy Gamma-Ray

Donghui Li, Yingxiang Wu, Keren Wang and Xingfu Zhong

Institute of Mechanics, Chinese Academy of Sciences

Abstract. The problems of how to measuring the volumetric fractions of oil-water-gas multiphase flow are still a problem remaining to be solved in oil industry. With the technological development of nuclear radioactive inspection, dual-energy γ-ray techniques make it possible to investigate the concentration of the different components on the cross-section of oil-water-gas multiphase pipe-flow. The dual-energy Gamma-ray technique is based on the materials attenuation coefficients measurement. It is comprised of two radioactive isotopes of ^{241}Am and ^{137}Cs with emission energies of 59.5keV and 662keV. In order to measuring the material's attenuation dose rate, some nuclear instruments and data acquisition system were designed; a number of static and dynamic tests were carried out in the Multiphase Laboratory, Institute of Mechanics, Chinese Academy of Sciences. The oil-water-gas three phases of medium have been investigated to simulate different media volumetric fraction distributions on the experimental flow loop. The measurement results of attenuation intensities were obtained; the linear attenuation coefficients and the volumetric fractions were studied and measurement error was discussed in this paper as well.

Keywords: Volumetric fraction, Multiphase flow, Dual-energy γ-ray

INTRODUCTION

Oil-water-gas multiphase mixed transportation in oil pipes is extensively used in ocean oil industries. How to metering the volumetric fractions of oil-water-gas multiphase flow is important in oil industry. Related research work started since the 1980s. The measurement of component ratios in multiphase flows using γ ray attenuation was first suggested by Abouelwafa and Kendall (1980) [1], and the technique has been used in many current commercial multiphase metering systems. Single energy γ ray technique working as a densimeter is satisfactory in two components measurement of gas/liquid or oil/water pipe-flow if not considering the concentration distribution of the different components in the cross-section. But in oil-water-gas three phase flow, the situation is different. Dual energy γ ray techniques have been developed rapidly in the last decade, the different attenuation properties of the three phase's media in the oil pipe are used to obtain the components fraction information. Dual energy γ ray technique has been applied to a case with three basis materials having distribution of linear attenuation coefficients with an obvious variance. It is proved to be a very promising technique for the purpose of simple and fast estimating the volumetric fractions of oil-water-gas multiphase flow, and it becomes the constituent part of radioactive based multiphase meter or tomography [2].

CP914, *Multiphase Flow: The Ultimate Measurement Challenge,* edited by X. Cai, Y. Wu, Z. Huang, S. Wang, and M. Wang

THEORIES

Dual Energies Theories

The following absorption law describes the mathematical relation between the γ ray intensity I_0 radiated by the γ ray source and the remaining intensity I after passing through a massive object with a given length L and a density ρ (Minder and Liechti 1955[3]; Morneburg 1995[4]):

$$I = I_0 e^{-\eta(ZE)\rho L}$$

The absorption coefficient η is a function of the γ ray energy E and the atom number Z.
Under the γ ray energy E, then:

$$I(E) = I_0(E)e^{-\mu(E)L}$$

For an oil-water-gas three-phase flow, the attenuation coefficient of the mixture μ(E) is represented by

$$\mu(E) = \alpha\mu(E)_o + \beta\mu(E)_w + \gamma\mu(E)_g$$

where $\mu(E)_o$, $\mu(E)_w$ and $\mu(E)_g$ are the linear attenuation coefficients of the oil, water and gas and α, β, γ are the respective volumetric fractions. The transmitted intensity I through a thickness L of an oil/water/gas mixture is therefore

$$I_0 e^{-(\alpha\mu(E)_o + \beta\mu(E)_w + \gamma\mu(E)_g)L}$$

The build-up factor can be eliminated through a good collimation at the source and detector. If the transmitted fluxes I_1 and I_2 at two energies E_1 and E_2 are measured, the volume fractions can be calculated from the linear attenuation coefficients of the flow components at E_1 and E_2

$$\ln(\frac{I}{I_0})_1 = -[\alpha\mu(E_1)_o + \beta\mu(E_1)_w + \gamma\mu(E_1)_g]L$$

$$\ln(\frac{I}{I_0})_2 = -[\alpha\mu(E_2)_o + \beta\mu(E_2)_w + \gamma\mu(E_2)_g]L$$

$$\alpha + \beta + \gamma = 1$$

There are three equations with three unknown volumetric fractions α, β and γ. In practice, $\mu(E_1)_g$ and $\mu(E_2)_g$ can be taken as zero without appreciable error in the volume fraction measurements because of the low density of the gas phase.

The linear attenuation coefficients of water and mineral oil (kerosene) over the same energy range reveal differences in the photon absorption, which may be used to distinguish the two materials. Fig.1 shows that the photon attenuation is greater in water than in oil. This is because the oxygen has a higher atomic number than the carbon, and also because water (ρ=1.00 gcm^{-3}) has a higher density than most mineral oils (typically ρ= 0.80~0.90 gcm^{-3}).

Since the photoelectric interaction is a stronger function of photon energy and atomic number than the Compton interaction, one sees a great contrast in the linear attenuation coefficients of oil and water in the region where the photoelectric effect dominates the interaction cross-section of both materials [5].This is illustrated in Fig.2, where the relative difference in the linear attenuation coefficient is plotted as a function of photon energy for the same energy interval. It indicates that the γ ray measurement system relying on the photon attenuation in oil and water to distinguish the two materials can obtain best results in the energy region below 40 keV.

In this paper, the γ ray system is comprised of two radioactive isotopes of ^{241}Am and ^{137}Cs, which have emission energies of 59.5keV and 662keV. The lower γ ray energy is higher than 40 keV, so that the linear attenuation coefficients of two energies in oil and water are somewhat close than expected. The necessary condition for above equations to be linearly independent is:

$$\frac{\mu(E_1)_w}{\mu(E_1)_o} \neq \frac{\mu(E_2)_w}{\mu(E_2)_o}$$

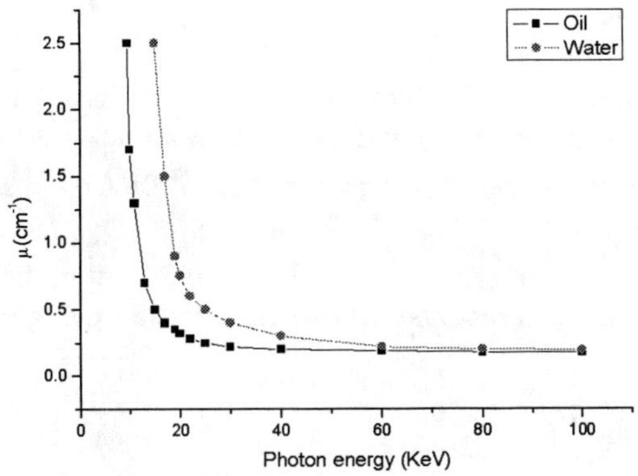

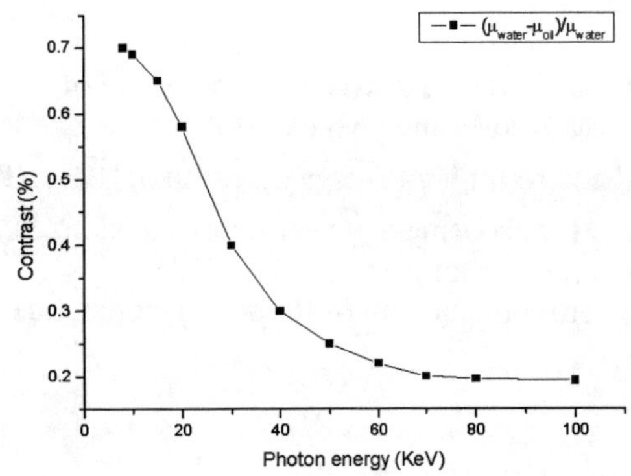

FIGURE 1. Water and oil attenuation coefficients **FIGURE 2.** Radiographic contrast between oil and water

But the too close linear attenuation coefficients of two energies may propagate small errors in the measurements of the intensities into large errors in the estimation of the thickness; this situation is avoided only by a proper selection of the energies.

Geometry Correction

A single beam dual energy γ ray was used as a densimeter in this measurement. The oil-water-gas three phase media have different densities, influenced by gravity, they stay in different layers in the horizontal pipe, and also have different dynamic distributions in the pipe. In other hand, because there are very complicated flow regimes in the oil-water-gas multiphase horizontal pipe flow, and the single beam γ ray could not cover the whole pipe cross section, it only pass through the centre of the pipe, so the measured results must be flow regime dependent, and need a geometry correction.

The correction is based on a layered distribution of three phases in horizontal pipe, with water in the bottom layer, oil in the middle layer and gas in the top layer. Let α, β and γ be the linear fractions of oil, water and gas, respectively, as shown in Fig.3. If R is the radius of the pipe, it can be shown that the half angle θ covered by the portion occupied by water will be:

$$\theta = \arccos(1 - \beta)$$

Therefore, the total volume occupied by water is:

$$V_{water} = \theta R^2 - R^2 \cdot \cos\theta \cdot \sin\theta = R^2(\theta - \cos\theta \cdot \sin\theta)$$

So the corrected volumetric fraction of water becomes:

$$\beta' = \frac{\theta - \cos\theta \cdot \sin\theta}{\pi}$$

Similarly, it is easy to get the corrected volumetric fractions of oil and gas, α' and γ'. In the following discussions, the superscript prime for the corrected volumetric fractions will be dropped.

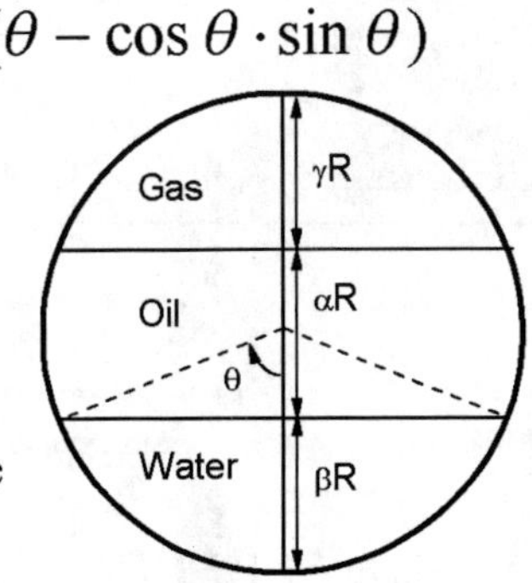

FIGURE 3. Oil-water-gas distributions.

EXPERIMENTAL SETUP

γ Ray Source

In this paper, the γ ray system is comprised of two radioactive isotopes of ^{241}Am and ^{137}Cs, which have emission energies of 59.5keV and 662keV. Both of radioactive isotopes were assembled and shielded in a thick lead pot to prevent the harmful high energy emission of ^{137}Cs. The radiation intensities of both isotopes are 100mCi and 20mCi. The reason for choosing a greater radiation intensity for ^{241}Am than ^{137}Cs is that ^{241}Am has a lower photon energy therefore weakens in the penetration of the measurement pipe. A collimated single γ ray beam of 20mm in diameter comes out from the bottom of the source pot and can be turn on/off by a mechanical switch to ensure the operation safety.

Scintillator Detector

An important step in the dual energy γ ray design is the selection of scintillation detectors. Two parameters are important while selecting the scintillation detectors: detection efficiency and decay constant of the scintillator. High detection efficiency is required to reduce the source strength and short decay constant is required for high count rate to avoid pulse pile up or saturation. NaI (Tl) crystal is most commonly used scintillator for its highest detection efficiency; it is very important for ^{137}Cs due to its high emission energy (662keV) and strong penetration ability.

A column crystal of Scintillator was made in size 40mm (H) ×40mm (D) and assembled with a photo multiplier tube (PMT). The total diameter of the detector is ϕ55mm and the length is 220mm. In addition, a collimation hole was designed of size of 30mm(W)×50mm(L)×150mm(H), which was mounted on the top of detector.

Nuclear Instrument

The nuclear instrument presented in this project was designed as a multi-channels instrument. The system operated in counter mode. It is comprised of high voltage source, amplifier, shaping amplifier and programmable counter data acquisition system. The diagram is shown in Fig.4.

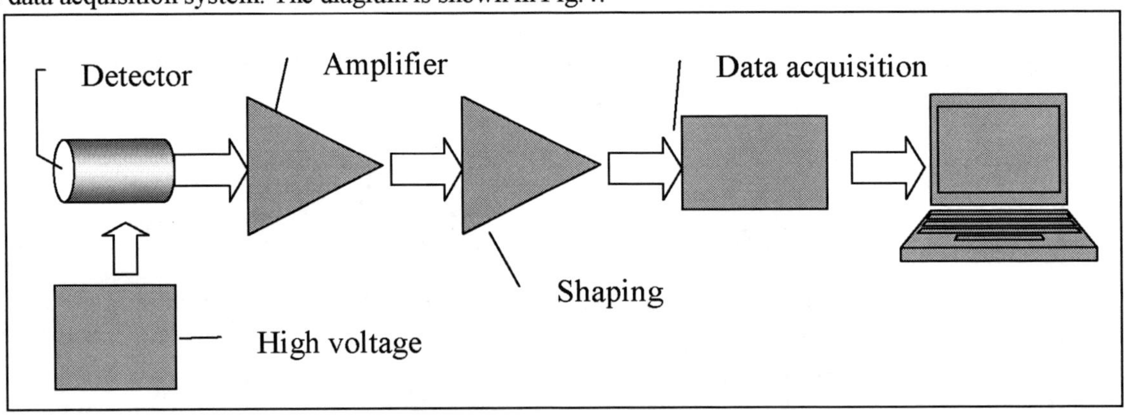

FIGURE 4. Nuclear instrument

Flow Loop

The flow loop is located in Multiphase Flow Laboratory, Institute of Mechanics. It is used to simulate oil-water-gas three phase flow. The flow loop is shown in Fig.5. The experiment section is comprised of a 20 metre long horizontal pipe, including 3m upstream pipe and 3m downstream vertical pipe; the pipe diameter is 50mm and made of transparent Plexiglass in order to observe the flow regime conveniently.

Water from the water tank is pumped to the experiment pipe with a water pump, and passes through an electrical-magnetic flow meter for measuring the flow rate; oil in the oil tank is pumped and metered by an oil pump and oil flow rate meter; gas comes from air compressor and is metered by a gas mass flow meter. Three kinds of media are mixed in the 'Y' type entrance of the experiment pipe. In the end of the experiment pipe, the fluid passes through a 'T' junction pipe in order to pre-separate part of gas and oil, then the remaining medium passes in a two class gravity separation tank. After a period of stablizing time, the oil and water are well separated and pumped to the water tank and the oil tank ready for recycling.

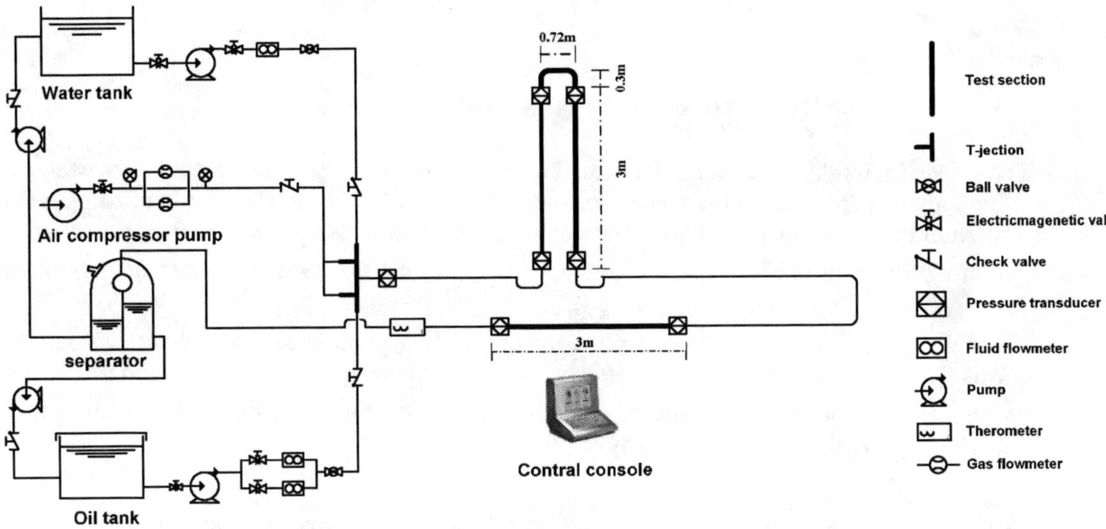

FIGURE 5. Multiphase flow loop

Static Calibration Rack and Vessels

Before the dynamic test, the dual energies γ ray system has to be statically calibrated. An experimental rack and 4 square vessels were designed to be a static test rack, which stood separately from the multiphase flow loop; it also can work as a static calibration rack. The rack is one meter high and supports the source pot on its top. Plexiglass vessels were designed to contain oil-water-gas three phase media, in size of 100mm(L)× 100mm(W)×600mm(H) and with wall thickness of 5mm. The vessel works in horizontal direction with total thickness of 110mm and a valid space of 100mm. Its position (height) between γ ray source and detector could be adjusted by two jacks depending on the experiment requirements. The rack is shown in Fig.6.

EXPERIMENTAL ARRANGEMENT

The oil-water-gas three phase flow is very complicated and it is impossible to test all the flow rate combinations for each phase. In order to study the three phase materials attenuation character, an experimental arrangement was designed as shown in Table 1, including total 6 groups of measurements in the experiment. All the flow rate data are the superficial flow rate obtained from the flow meter of each phase.

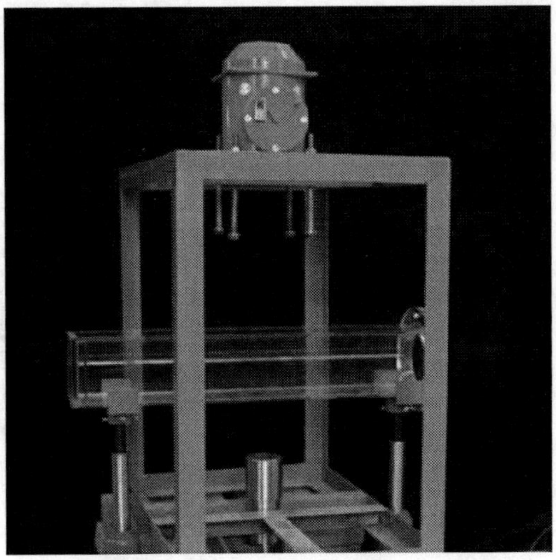

FIGURE 6. The static experimental rack

TABLE.1 (Gas1=20SL/min, Gas2=100SL/min)

Oil	Water							
$0.75m^3/h$	$2.5m^3/h$	$3.75m^3/h$	$5m^3/h$	$6.25m^3/h$	$7.5m^3/h$	$8.75m^3/h$	$10m^3/h$	$12.5m^3/h$
$1.5m^3/h$	$2.5m^3/h$	$3.75m^3/h$	$5m^3/h$	$6.25m^3/h$	$7.5m^3/h$	$8.75m^3/h$	$10m^3/h$	$12.5m^3/h$
$3.6m^3/h$	$2.5m^3/h$	$3.75m^3/h$	$5m^3/h$	$6.25m^3/h$	$7.5m^3/h$	$8.75m^3/h$	$10m^3/h$	$12.5m^3/h$

RESULTS AND DISCUSSIONS

Discussions on Measurement Accuracy

With a well collimated scintillation detector and a dual energy γ ray source, a series experiments were carried out on multiphase flow loop in IOM. The test curves on oil-water-gas three phase flow were obtained. A test curve of slug flow was shown in Fig.7. The dispersed areas are bubbles, the continuous area is the liquid phase, and the flow regime is shown on top diagram in Fig.7. The left diagram is the signal of ^{241}Am ; the right diagram comes from ^{137}Cs channel.

When the bubbles pass through the test section, the materials attenuation is reduced; then the count rate rises, when the liquid fills the whole test section of the pipe, the materials attenuation reaches it maximum, then the count rate drops down. In ^{137}Cs channel, the signal character is the same as ^{241}Am channel but its total attenuation is less because ^{137}Cs has higher emission energy.

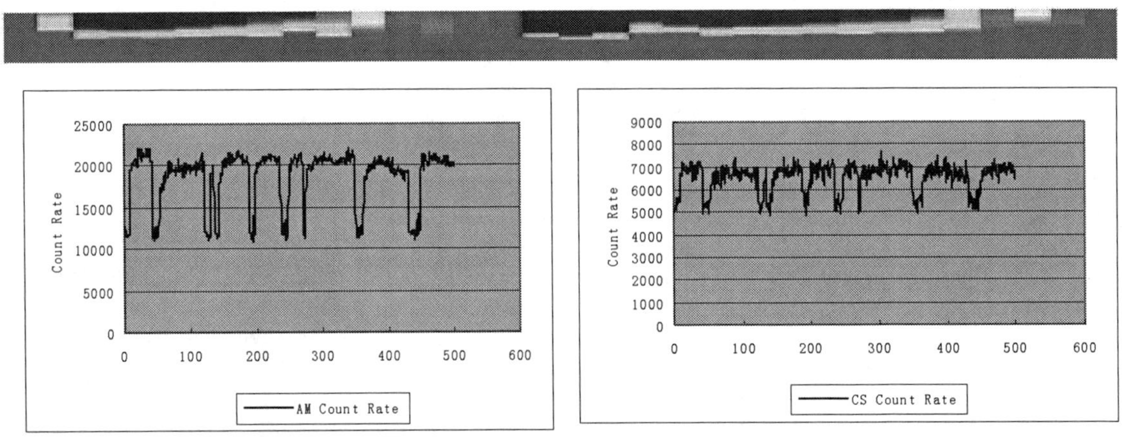

FIGURE 7. Measurements curves

There is statistical fluctuation in γ ray source emission intensity. This requires a long integral time in the measurement to reduce the error. The accuracy of both measurement channels is better than 3‰ in a single energy measurement when the integral time is longer than 200 seconds. But it becomes worse in a short integral time, for example, the accuracy drops to 1.5% in 10s and even to 5% in 0.1s integrate time. This means that there should be a significant error in dynamic measurement because of the short acquisition time. In fact, during dynamic measurement, due to the moving fluid, in order to obtain the short time flow detail, we must let the γ ray system to have high accuracy in short time measurement. The circumstances become even worse in dual energy three phase measurements. In the three phase volumetric fraction measurement, as mentioned in part 2 of this paper, the γ ray measurement system relying on photon attenuation in oil and water to distinguish the two materials would obtain a maximum discrimination in the energy region below 40 keV, but the lower energy isotopes of ^{241}Am used in this project has the emission energy of 59.5keV, therefore, the linear attenuation coefficients of two energies in oil and water is closer than expected. The too close linear attenuation coefficients may propagate small errors in the measurements of the intensities into large errors according to equations 1/2/3[6], thus a small error in a single energy measurement may be magnified into large error in the dual energy measurement, even up to 100%. The only way to solve this problem is to reduce the single energy primary measurement error, using a long integral time (more than 200s in this project). A well treated dual energy γ ray system has the potential to get the accuracy better than 90% in average in the oil-water-gas three phase system.

Results for Three Phase Flow

Six groups of oil-water-gas three phase flow were measured by the dual energy γ ray system. The results of three phase volumetric fraction of the oil-water-gas multiphase flow are shown in Fig.8.

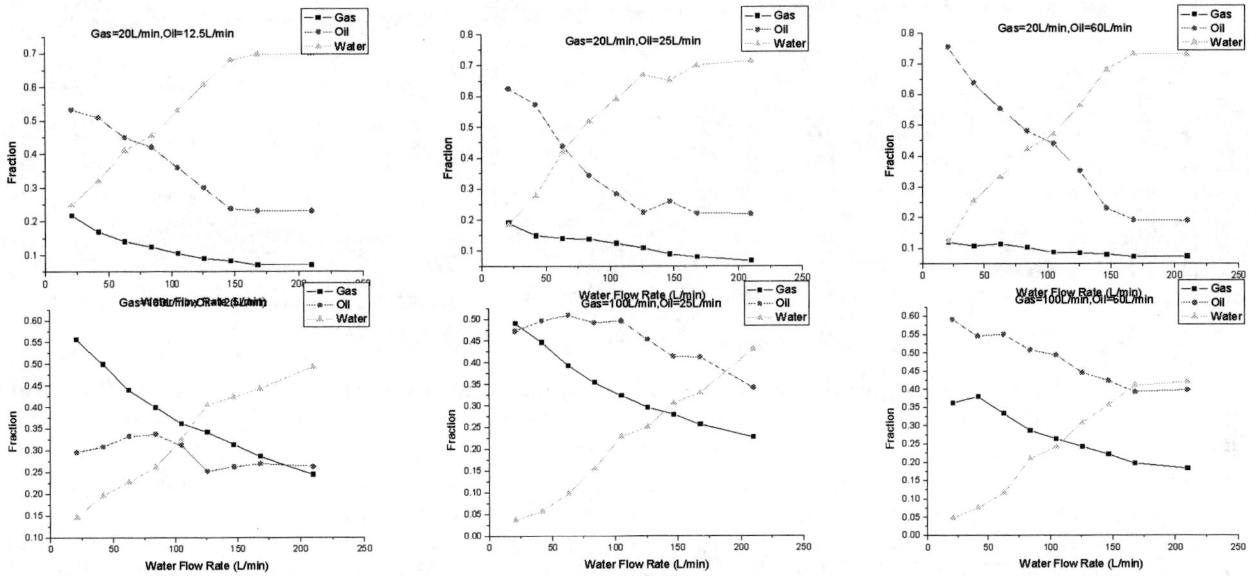

FIGURE 8. Volumetric fractions of oil-water-gas multiphase flow

In the top row of curves in Fig.8, the gas flow rate is fixed in 20L/min; in bottom row, the gas flow rate is increased to 100L/min. The oil flow rate is changed from 12.5L/min, 25L/min to 60L/min respectively in the diagrams of both rows from left to right in Fig.8. The water flow rate is increased from 21L/min to 210L/min in 9steps as shown on the X axis in each diagram. The sum of gas, oil and water volumetric fraction is equal to 1, and the measurement integral time is 200s in this experiment.

It is easy to find that with the increasing of the oil flow rate as shown in Fig.8 from left to right curves in both rows; the total liquid flow rate increased, and then the gas volumetric fraction decreased. This is because gas is a compressible fluid, when the total liquid fraction increased; the pressure inside the pipe goes up, thus the volume fraction of gas decreased.

The oil volumetric fraction also increased as expected as shown in Fig.8 from left to right curves in both rows. The oil flow rate is changed from 12.5L/min, 25L/min to 60L/min, and the dual energy γ ray system gives out a correct response to the fraction changes. This shows a reasonable measurement accuracy. But there are also some significant error points on these curves; some curves do not look smoothly as they should be.

The following diagrams show the oil to water ratio character in experimental measurement related to the oil to water superficial flow rate ratio. The results of oil to water ratio values of oil-water-gas multiphase flow are shown in Fig.9.

In oil-water-gas three phase measurements, it is easier to get better accuracy in gas/liquid measurement. But it is difficult in oil/water discrimination in oil-water-gas three phase. In this group of curves, the black lines of curves are the actual oil to water ratio values, measured by oil and water flow meter, and they are called oil and water superficial flow rate. The red lines of curves are experimental values of oil to water ratio; it is measured by γ ray system and obtained by equations 1/2/3.

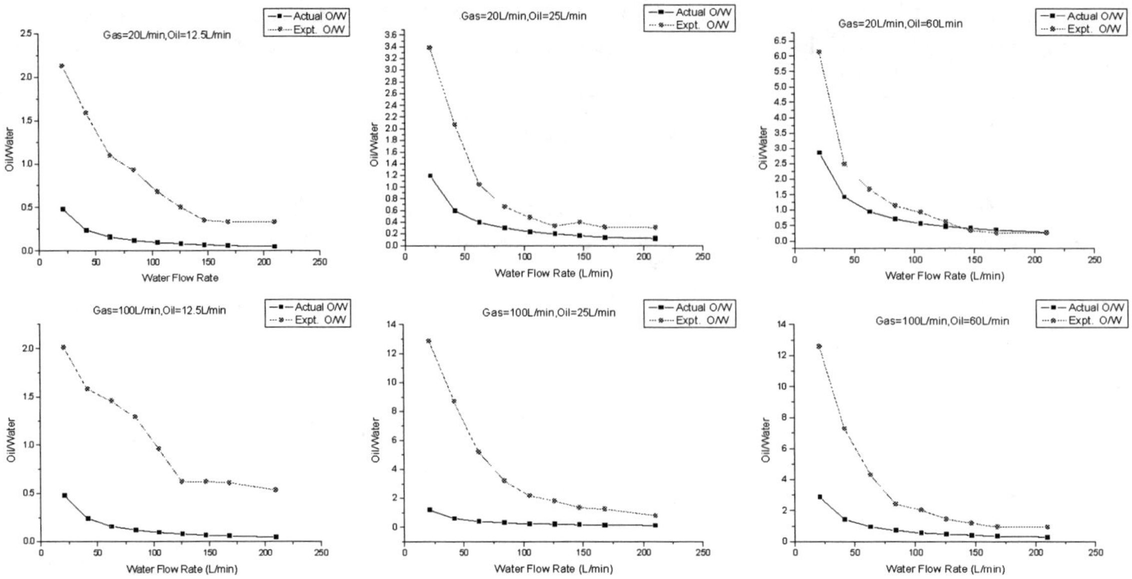

FIGURE 9. Oil to water volumetric fraction of oil-water-gas multiphase flow

From the column 1 in Fig.9, the oil flow rate is only 12.5L/min. this is a kind of water dominated flow, and the flow regimes are the stratified flow and long intermittent type flow. It is obvious that the oil to water ratio is much higher than the actual value. This is not a measurement error but it just reflects a phenomenon that the oil speed may be much smaller than water speed in the pipe because oil is more viscous than water, there must be more resistance on the pipe wall to oil especially when oil is less and flow as a dispersed phase in the pipe.

When the oil flow rate is increased to 25L/min even to 60L/min, the circumstances are changed. The water's domination becomes less and oil will not flow as dispersed phase. The oil flow resistance begins to influence the water flow speed; this may slow down the water speed, in other hand, the fast water will bring the oil to a faster speed until the oil and water get nearly the same velocity. Then we can find the red curves and the black curves going together in column 2 and column3 in Fig.9. Meanwhile, the oil and water volumetric fraction get the same with their superficial flow rate ratio.

The other circumstance is the oil dominated flow. This happens in the initial part of the curves in where the water flow rate is only 21L/min or 42L/min. In this flow rate of the water, there is no enough energy and enough touch area to push the oil, water only keeps the fast speed itself and leaves the oil behind. This is the reason why there are large differences between the initial parts of two curves in all the diagrams in Fig.9.

The next group of diagrams shows gas to liquid ratio character in the experimental measurement related to the gas and liquid superficial flow rate ratio. The results of gas to liquid ratio values of oil-water-gas multiphase flow are shown in Fig.10.

In this group of curves, the black lines of curves are the actual gas to liquid ratio values; the liquid flow rate is the total flow rate of water and oil, measured by the oil and water flow meter. The red lines of curves are experimental values of gas to liquid ratio; measured by the γ ray system, like the oil to water ratio measurement, the values are obtained by equations 1/2/3.

It is easy to find that the red line curves (experimental G/L) are lower than black line curves (actual G/L) in Fig.10. This is because the gas is a compressible fluid; the gas flow rate values were measured from the gas mass flow meter, this flow rate is under the standard atmosphere pressure. When the gas enters the pipe, there is a larger pressure inside the pipe, then the gas is compressed in volume, and the gas volumetric fraction at the pipe cross section is reduced as well. So the experimental measured gas to liquid ratio is always lower than the values computed from the superficial flow rate.

In the first row of curves in Fig.10, the gas flow rate is quite low, only 20L/min. The velocity differences between gas and liquid in the pipe are small. Then the red curves and black curves are quite close. But in the second row of curves, the gas flow rate increased to 100L/min.; then the gas transport speed increased significantly. The differences between the red curves to black curves become large.

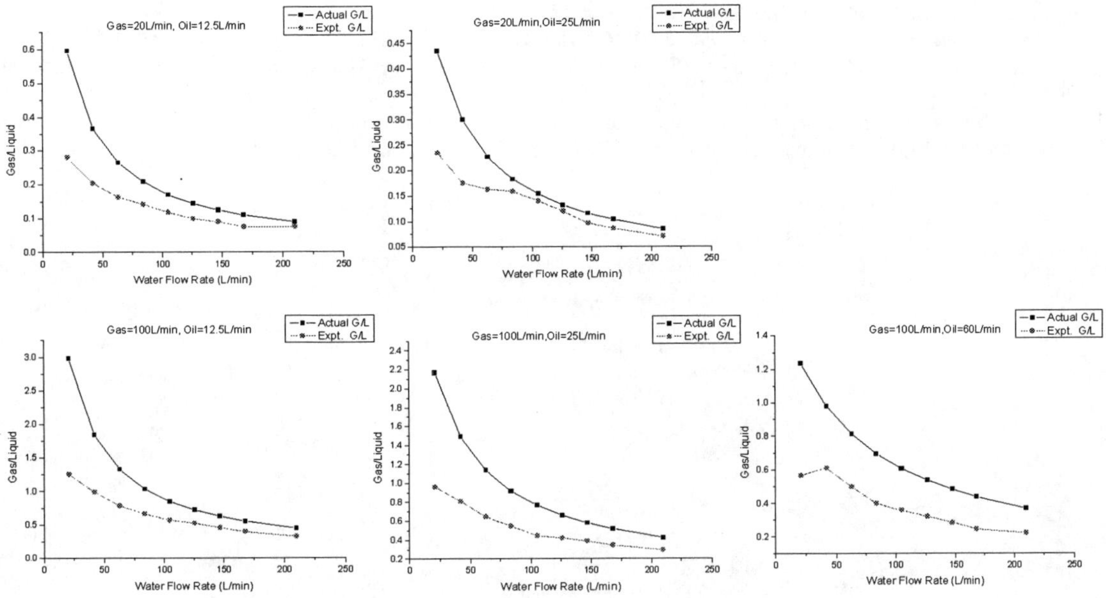

FIGURE 10. Gas to liquid volumetric fraction of oil-water-gas multiphase flow

CONCLUSIONS

Through the studies on the oil-water-gas three phase horizontal pipe flow with the dual energy γ ray system, it is clear that the oil-water-gas three phase flow is very complicated. Not only there are speed shifts and velocity differences between every phase media but also there are phase volumetric fraction changes in the cross section of the pipe. Until now there are no other more efficient and convenient way that can be used to measure the oil-water-gas multiphase flow. The dual energy γ ray is a promising technique for the purpose of simple and fast estimating the phase fractions of the oil-water-gas multiphase flow.

There is statistical fluctuation in γ ray source emission intensity. This requires a long integral time in measurement to reduce the error, and it is difficult to get short time measurement information, only the long time average results are reliable unless uses large activity γ ray source, but this will bring the new problems related to system cost and radioactive safety protections.

Dual energy γ ray techniques are based on materials attenuation coefficients measurement. In oil-water-gas three phase system, if the lower γ ray energy is above 40 keV (59KeV ^{241}Am in this project), the linear attenuation coefficients of two energies in oil and water is somewhat close to each other. It may propagate small errors in the measurements of the intensities into large errors in the estimation of the thickness through the equation 1/2/3. This is another limitation to the measurement accuracy.

Generally, there are still some difficulties in oil-water-gas three phases discrimination using dual energy γ ray system, the dynamic results of experimental measurement in this project could not be checked real timely due to the fact that there are no fast switch valves on the flow loop. But the dynamic measurement results and curves are reasonable and repeatable during the experiment..

Finally, a well treated dual energy γ ray system has the potential to get the accuracy better than 90% on averagely in oil-water-gas three phase system.

REFERENCES

1. M.S.A.Abouelwafa and E.J.M.Kendall, "The Measurement of Component Ratios in Multiphase System Using γ-ray Attenuation", *J. Phy. E: Sci. Instrum.* **13**, 341(1980)
2. T. Grassler and K.E. Wirth, "Dual-Energy X-Ray Tomography in Process Engineering–A Non-Intrusive Technique to Characterize Vertical Multiphase Flows" 2nd World Congress on Industrial Process Tomography, Hannover, Germany 2001.

3. W. Minder and A. Liechti, *Röntgenphysik*, Springer Verlag, Wien (1955).
4. H. Morneburg (Ed.), *Bildgebende Systeme für die medizinische Diagnostik*, Publicis MCD Verlag, München(1995)
5. M. J. Key, *Gas Microstructure X-Ray Detectors and Tomographic Multiphase Flow Measurement*. Ph.D Thesis, University of Surrey(1999).
6. Li Dong-hui, Wu Ying-xiang, Li Zhi-biao, Zhong Xing-fu, "Volumetric Fraction Measurement in Oil-Water-Gas Multiphase Flow with Dual Energy Gamma-Ray System", J Zhejiang Univ SCI 2005 6A(12):1405-1411.

Development of a Precise and in Situ Turbidity Measurement System

Kuanfang REN*, Feng XU*†, Jean-Marc Dorey‡ and Xiaoshu CAI†

* *UMR 6614/CORIA, CNRS - Université et INSA de Rouen, av. de l'Université, BP 12 76801 Saint Etienne du Rouvray, email : fang.ren@coria.fr*
† *Institute of Particle and Two-phase Flow Measurement Technology, University of Shanghai for Science and Technology, 516 Jungong Rd, Shanghai 200093, China*
‡ *EDF-R&D Department of Fluid Dynamics, Power generation and Environment, 6 Quai Watier, 78401Chatou*

Abstract. The turbidimetry is a technique based on the transmittance spectra of the light passing through the media containing of small particles. It permits to measure the size distribution of particles for size in the range of sub-micrometer or micrometer. But the inversion problem is one of the most important obstacle for its applications. Based on the Non-negative Least Square method, we have developed stable and rapid algorithm and a measurement system permitting to the temporal acquisition (in ms) and to realize in-line measurement. To ensure its performance, the sensibility and the stability of the system have been examined in different stages: the light source, the spectrometer and the variation of the media concentration according to the optics configuration. By the measurements in the laboratory and that of the wet steam in a turbine we show that such system permits to measure very precisely the variation of the volume fraction of the particle or the wetness of wet steam.

Keywords: Spectral extinction, in-situ measurement, wet steam, turbine.
PACS: 78.20.Ci, 07.07.Vx

INTRODUCTION

In many industrial processes, it is necessary to know the size of micrometric or sub-micrometric particles, such as the wet steam in turbine, the sprays in the motor combustion, the soot emitted by the automobile, or in general the aerosols, in order to optimize the process and to control the pollution.

Amongst the numerous techniques developed to this end: sedimentation, Scanning Mobility Particle Sizer (SMPS), Electrical Low Pressure Impactor (ELPI), etc. the optical metrology is particularly interesting since it is fast, non-intrusive and easy to realize an in situ measurement without mechanical perturbation to the process. Especially, the technique based on the extinction spectra permits to measure the wetness of wet steam in a turbine and to determine the size distribution of the particles without dilution of soot emitted from an automobile motor, with an acquisition time at the order of milliseconds and the processing time in seconds.

Many authors[1-4] have contributed to its development. E. E. Uthe[3] uses two or three wavelengths to measure the concentration and the Sauter mean diameter D_{32}. Such a method is based on the monotonic relationship between the extinction ratio and particle size and it can only be used for a narrow size distribution. When the size distribution is large or there are two or more populations more wavelengths are needed. Thanks to the development of linear CCD based spectrometer and the performance of the computer it is easy to realize a measurement system by using the spectra of several hundreds of wavelengths. The compact CCD based spectrometer with 2048 wavelength is now a conventional one.

One of the most important advantage of the turbidimetry is that in most cases it does not need dilution. However, in certain circumstances the concentration of the particles can be so high that the multiple diffusion effect becomes non negligible. The validity of the Beer-Lambert law should be examined carefully. In fact, the validity of this method depends not only on the concentration or the optical thickness, but also on the size of the particle and optical configuration. Swanson *et al*[5] have examined carefully these limits by theoretical prediction and by experiments.

CP914, *Multiphase Flow: The Ultimate Measurement Challenge,* edited by X. Cai, Y. Wu, Z. Huang, S. Wang, and M. Wang

They have shown that the optical thickness can be as large as 10 (the transmission is at the order of 10^{-5}) if the optical configuration is well designed.

In the turbidimetry, the refractive index of the particles are often assumed to be known, but it is not always the case. The refractive index of water, for example, depends on the temperature, the pressure and the wavelength. When we want to extend the measurement range on the size, the wavelength range of spectrum should be very large since the measurable range of particle size is about 5 λ_{min} ~ 5 λ_{max}. The variation of the refractive index of water for the wavelength from 0.25 μm to 1.0 μm can be as large as 0.05 for given temperature and pressure[6]. The influence of inaccurate estimation of refractive index on inversion results of particle size distribution (PSD) has been systematically studied by Wang *et al* [7] and they have shown that small inaccuracy ($\Delta m = 0.05$) in particle refractive index might cause a error up to 5~6 % to volume mean diameter D_{v50}.

The inversion problem have been a major obstacle in the turbidimimetry. Many authors have contributed to the improvement of algorithms. Twomey's method is one of the most widely applied algorithms[8]. To ensure that the distribution function is non-negative, Phillips-Twomey method is often combined with the non-negative least square (NNLS) method[9-11]. However, a difficulty in Phillips-Twomey-NNLS algorithm is to find an optimized Lagrange multiplier which serves as an equilibration point ensuring both the smoothness of PSD and the fitness of the measured and reconstructed extinction spectra. To achieve such an aim, the Generalized Cross-Validation (GVC) criterion and L-curve criterion are introduced for determination of the Lagrange multiplier. However, a drawback is that the GCV function might have a flat minimum so that it is difficult to find the best solution. As another efficient independent model inversion method is the optimized regularization technique (ORT) developed by Schnablegger and Glatter[12]. In their method, a set of cubic spline is introduced and each spline element represents a size distribution function within a certain size window. Then inversion becomes to determine the weight coefficients multiplied with these splines. Two algorithms (ORT and Twomey combined with NNLS) are integrated in our software. Furthermore, a justification on the extinction matrix according to the particle size has also been calculated to decrease the condition number of inversion matrix.

A system based on the extinction spectra has been developed in our laboratory by using a compact spectrometry (Ocean Optics). With a fast and stable inversion algorithm and a friendly interface this system permits a temporal acquisition, an in situ measurement and a simulation for a given size distribution.

In order to determine the performance of the system, the sensibility and the precision is to be studied. After a description of the measurement system in section 2, we present the experimental results in laboratory for the light source stability (section 3) and the attenuation of light by diluted milk liquid and the concentration of standard particle (section 4), and the wetness of wet steam in a turbine platform in EDF (section 5).

DESCRIPTION OF THE SYSTEM

The schema of a turbidity measurement system is shown in Fig. 1. The light from a source with wide range of wavelength is conducted by a bifurcation fiber : one passing by the medium to be measured and the other going directly to a two channel spectrometer. The transmittance spectra measured by the spectrometer are recorded thanks to the software TURACE, which permits the temporal acquisition with a frequency of 1000 Hz and the in-line inversion directly or a post processing. Since the measurable size of the particles is in the range of fifths of the wavelength to five times of the wavelength, it is worth using a source with a very large range of the wavelength to increase the measurable size range. In our system a light source of the combination of Halogen and deuterium lamps (DH 2000 BAL from Ocean Optics) is used. The relative flat spectrum from 230 nm to 1100 nm is then obtained. To minimized the instability of the light source, two channel system are used.

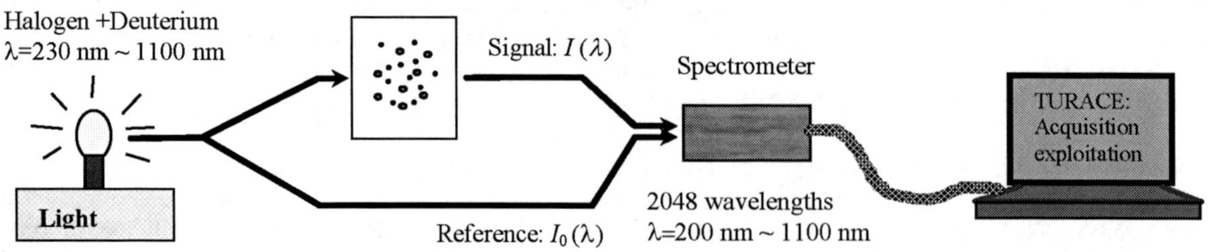

FIGURE 1. Schema of a turbidity system.

The measurement precision of such system depends on the stability of the light source, the noise of the surroundings, of the spectrometer and the stability of the media to be measured. We have examined the influences of each source of eventual influences.

STABILITY OF LIGHT SOURCE

First of all the background noise should be eliminated or minimized. By the measurement in different conditions, we find that after the correction of the electronic noise, the background noise is very small, less than 0.1 %. But it should be pointed out that to do the measurement with the Ocean Optics spectrometer (SD 2000 and H2000+) several acquisitions must be done before the real measurement (the background noise is about 2 % for the first acquisition).

The stability of the measured intensity depends on several factors: the stability of the light source, the media to be measured and the acquisition parameters of the spectrometer. Table 1 shows the standard deviation of intensity for different combination of parameters of one channel spectrometer (H2000+). We find that the standard deviation is in general less than 0.03 % and the deviation decreases with the sample number, the integration time and the interval time.

To minimized the instability of the light source, two channel system should be used. But in this case the sensibility of the two channels and the relation between the pixel number and the wavelength are not the same, the correction factors should be determined before the real measurement. Fig. 2 shows the dependence of the correction factors as function of wavelength which vary from 0.6 to 1.3. A temporal series of spectra measured after the correction is shown in Fig. 3. we find that The difference of measured intensities in the two channels is less than 0.001 %. It is important to point out that correction factors depend not only on the spectrometer and the parameters used by the spectrometer, but also on the optical configuration, so they must be measured for each configuration.

TABLE 1. Stability de la source (DH2000 BAL) avec le spectrometer HR 2000+.

Item	Condition 1	Condition 2	Condition 3
Integration time (ms)	5	5	5
Spectrum sample number	2	2	20
Boxcar on wavelength	2	2	2
Wavelength range	0.5 µm – 1 µm	0.5 ~1 µm	0.5 ~1 µm
Wavelength number	100	100	100
Interval of acquisition (ms)	100 ms	1000 ms	100 ms
Number of spectra	100	100	100
Mean intensity	1035.6	1037.6	1035.7
Standard deviation of intensity	0.35632	0.55912	0.17521
Relative Standard deviation of mean intensity	0.00035 %	0.00055 %	0.00017 %
Relative Standard deviation intensity for each	0 .023 %	0.024 %	0.008 %

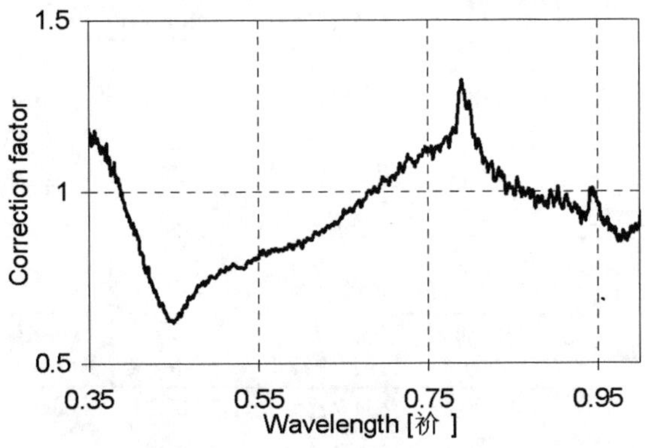

FIGURE. 2 Correction factors as function of wavelength

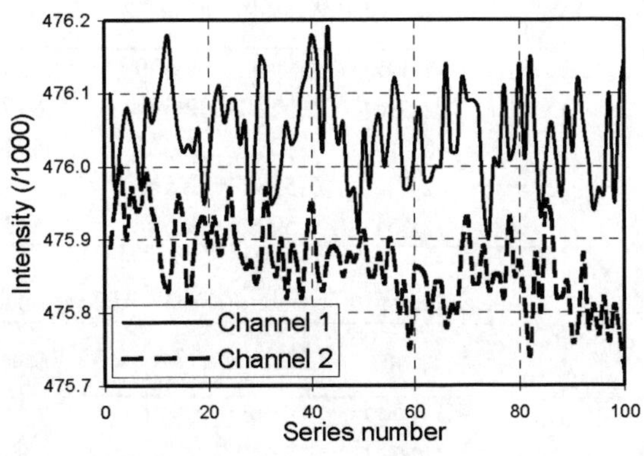

FIGURE. 3 Typical mean intensity evolution on the two channels.

SENSIBILITY OF THE SYSTEM TO THE VOLUME FRACTION

In order to study the sensibility of the system to a very small variation of concentration, we have used a diluted milk liquid and polystyrene spherical particles in water for which the variation of the concentration can be controlled very precisely (better than 0.01 %).

A tank of glass contained of water (300 ml) is located in the measurement channel and a quantity of milk or polystyrene spherical particles is added so that the transmittance is at about 50 %. To change the variation of the concentration, a little volume of water is added in the tank.

For each concentration of milk 3 series of 100 spectra are acquired and analyzed for statistical characteristics of the temporal evolution of the transmittance. We found that the stability of the intensities is excellent and the profile of the spectra is the same for different concentration.

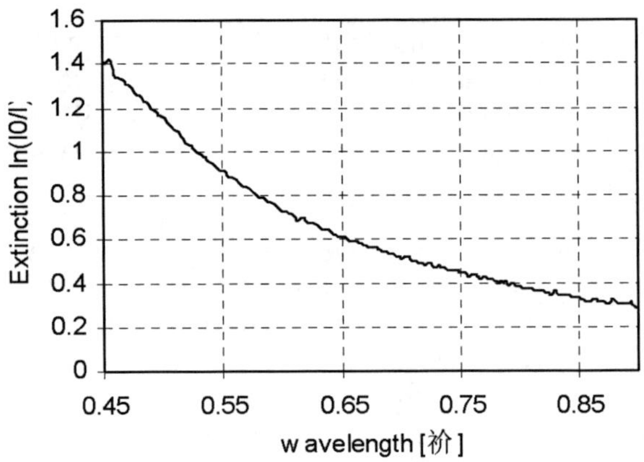

FIGURE 4. Extinction spectrum of diluted milk liquid of added water.

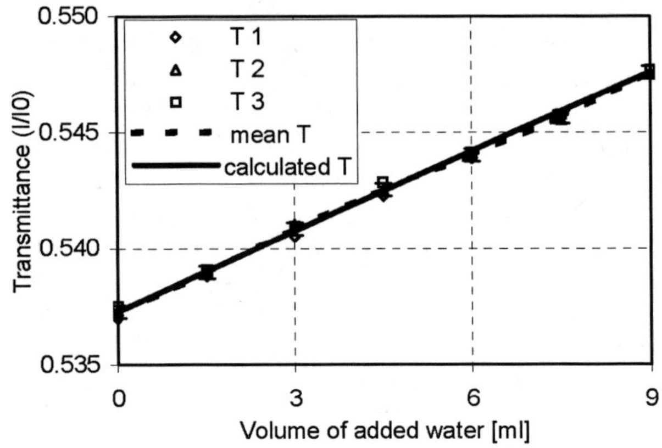

FIGURE 5. Measured transmittance as function of the volume

Fig. 4 present an example of the extinction spectra of diluted milk while the Fig. 5 shows the evaluation of the transmittance of diluted milk as function of the quantity of added water. The minimum graduation of the syringe is 0.01 ml and we can estimated with a precision of 0.003 ml. The relative incertitude due to the added water is 0.001 %. When we compare the measured transmittance to that predicted theoretically according to the Beer-Lambert law, we find that the deviation of less than 0.05 %.

Similar experiments have also been done for the standard polystyrene particle of two different sizes : d=0.5 µm and d=1.0 µm. For each particle size and each concentration 5 series of 100 spectra are acquired. Tables 2 and 3 present the measured volume fractions and the theoretical volume fractions predicted according to the Beer-Lambert law. The relative error on the volume fraction is found less than 0.5 %.

TABLE 2. Measured volume fraction (V/V0 10^{-5}) for particles of d=0.5 µm

Vol. Of added water (ml)	Meas. 1	Meas. 2	Meas. 3	Meas. 4	Meas. 5	Meas. mean f_v	Meas. fv / ref. (%)	f_v calculated	relative Error (%)
0	2.697	2.691	2.699	2.698	2.658	2.68650	0 (Ref)	2.6865	Ref.
0.15	2.680	2.652	2.696	2.635	2.658	2.66025	-0,98	2.6533	0.26
0.30	2.610	2.608	2.612	2.607	2.632	2.61475	-2,67	2.621	-0.24
0.45	2.605	2.578	2.583	2.574	2.596	2.58275	-3,86	2.5894	-0.26
0.6	2.551	2.573	2.537	2.543	2.542	2.54875	-5,13	2.5586	-0.39

TABLE 3. Measured volume fraction (V/V0 10^{-6}) for particles of d = 1.0 µm

Vol. of added water (ml)	Meas. 1	Meas. 2	Meas. 3	Meas. 4	Meas. 5	Meas. mean f_v	Meas. Fv / ref. (%)	f_v calculated	relative Error (%)
0.00	6.987	6.996	7.011	7.012	7.040	7.0092	0 (Ref)	7.0092	Ref.
0.15	6.904	6.912	6.962	6.947	6.947	6.9344	-1,07	6.9149	0.28
0.30	6.815	6.816	6.821	6.844	6.832	6.8256	-2,62	6.8231	0.04
0.45	6.758	6.756	6.779	6.772	6.761	6.7652	-3,48	6.7337	0.47
0.60	6.672	6.645	6.656	6.673	6.655	6.6602	-4,98	6.6466	0.20

MEASUREMENT OF WET STEAM

This system has been applied in the measurement of the wet steam generated by an installation plate-form in EDF. The schema of the facility is shown in Fig. 6. The overheated steam is cooled by water spraying and then conducted to a turbine. The wet steam generated by such facility is measured by the turbidity system. The plate-form is designed to produce wet steam with a very good stability and a very fine adjustment of turbine inlet temperature and pressure so that the concentration of droplets corresponds to the wetness that prevails at the exit of low pressure cylinder of large power plant steam turbines. The theoretical wetness at the measurement section can be inferred from the inlet conditions and the efficiency of the turbine (previously measured without cooling).

The measurement of the wetness and the size distribution have been done in several conditions, of inlet temperature with a step of about 0.5°C. We show hereafter an example of such a measurement. The upstream and downstream pressures are respectively 1.034 bar and 0.552 bar. The turbine inlet temperature at downstream is varied from 127.5°C to 130 °C. The rotational speed of the turbine is 38 750 tr/min.

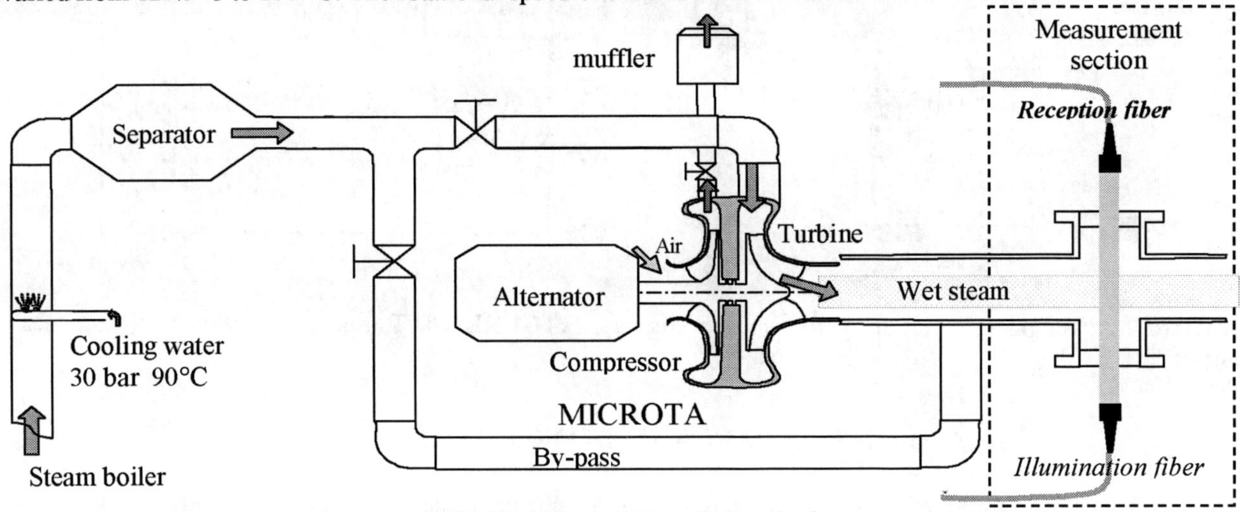

FIGURE 6. Schema of installation plat-form.

The parameters of the optical system are as followings:

- Light source: DH200 BAL (Deutérium + halogène) , λ = 0.2 µm ~1.1 µm
- Optical fiber: Ocean Optics BIF600 VIS/NIR
- Spectrometer : Ocean Optics SD2000, two channels, wavelength from 0.35 µm to 1.0 µm with 2048 pixels.
- Software: TURACE 5 for acquisition, processing and simulation. ORT algorithm has been used in the following data processing.

A typical measured size distribution (volume fraction) of the droplets is shown in Fig. 7. Each wetness corresponds to a mean value computed from 100 spectra acquired within 10 seconds, with 10 ms integration time for each. We found that the size distribution is quasi mono-modal in most cases and the Sauter mean diameter varies from 0.6µm to 0.8 µm according to the conditions. The measured wetness are compared in Fig. 8 to the theoretical wetness. The measured wetness shown here are a mean values obtained from 100 measured spectra and the theoretical wetness are predicted according to the pressure and the temperature at upstream and at downstream and the gain of the turbine. We find that the turbidimeter permits to measured the wetness with a precision of about 0.02 % for values in the range of 0.62% to 0.8%.

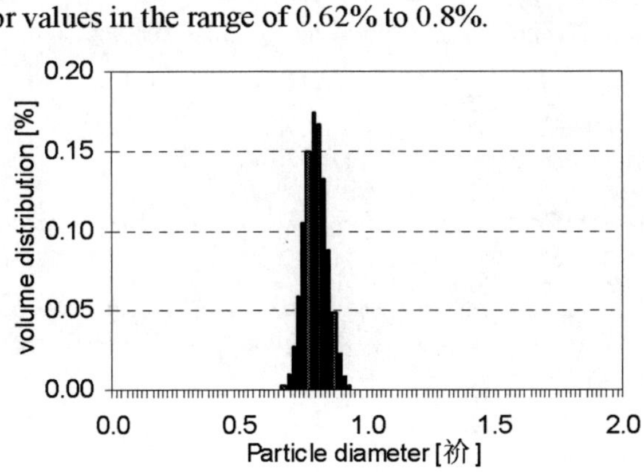

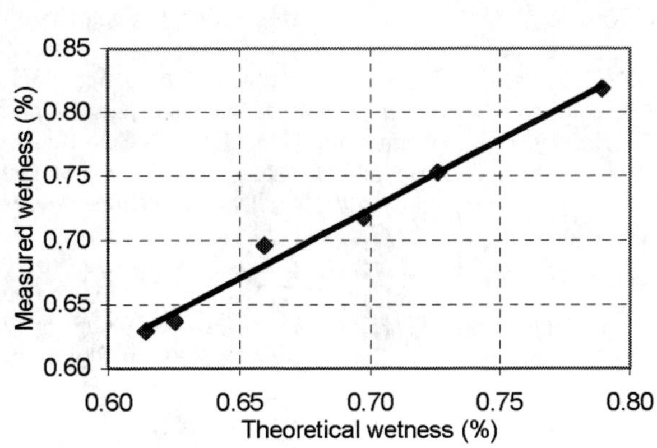

FIGURE 7. Particle size distribution of wet steam.

FIGURE 8. Comparison of the theoretical and measured wetness of wet steam.

Our measurement system permits also to obtain rapid temporal evolution of a particle system. The temporal evolutions of the wetness and the Sauter mean diameter of the droplets are shown in Figs. 9 and 10. The integration time in the acquisition is 100 ms and the useful wavelength range in the spectrum extends from 0.45 μm to 0.90 μm. The discrepancy between the measured and the theoretical wetness is about 0.02 % and that of the Sauter mean diameter is about 0.5 μm with a jump of 0.5 μm from 0.20 s to 0.25 s. The reasons of such evolution are to be examined.

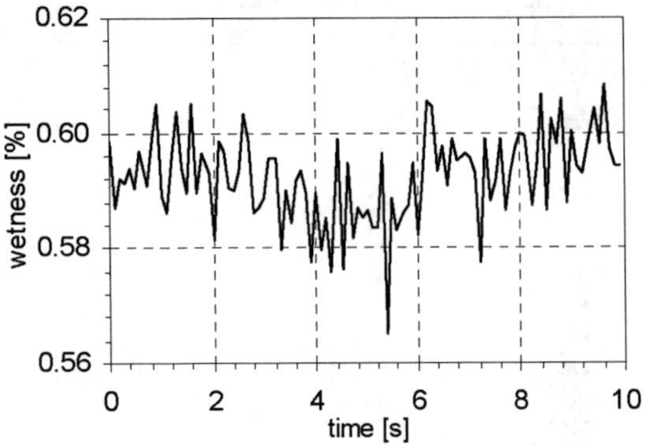

FIGURE 9. Temporal evolution of the steam wetness.

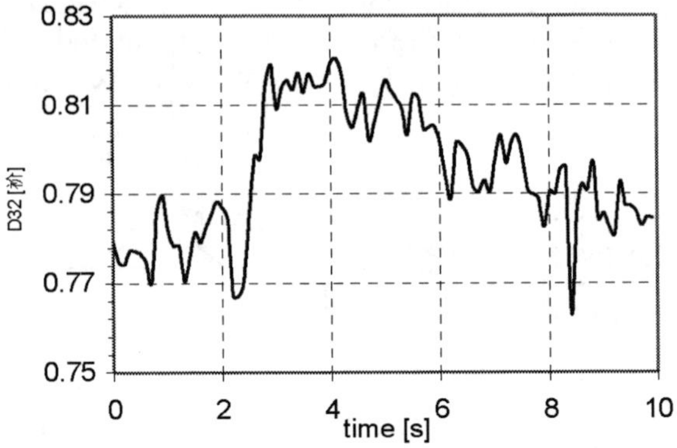

FIGURE 10. Temporal evolution of Sauter mean diameter.

DISCUSSION AND CONCLUSIONS

A optical turbidity measurement permitting an in line measurement system is presented. The performance of the system is carefully examined by the diluted milk and standard polystyrene spherical particle in the laboratory and by the wet steam generated from an installation plat-form of turbine. We have shown that the stability of the light source (HL 2000 and DH 2000 BAL) is very good: the standard deviation is better than 0.05%. A variation of 0.01 % on the extinction can be well measured with the system. The measurement of wet steam in a facility in EDF has shown that the precision the wetness can reach 0.02 % for values in the range of 0.62% to 0.8%.

ACKNOWLEDGMENTS

The authors acknowledge supports from the French Embassy in China for the joint thesis of Feng XU (project 4B2-007) and the Natural Science Foundation of China (50336050).

REFERENCES

1. P. T. Walters, *Appl. Opt.* 19, pp2353-2365 (1980).
2. N. Wang et al, *Optical particle sizing techniques and its applications* (in chinese). Atomic energy Press, Beijing, 2000.
3. E. E. Uthe, *Appl. Opt.* 21, pp454-459, (1982).
4. X. Cai, N. Wang, J. Wei, and G. Zheng. *J. Aerosol Sci.*, 23(7), pp749-757 (1992).
5. N. L. Swanson, B. D. Billard, and T. L. Gennaro, *Appl. Opt.* 38, pp5887-5893 (1999).
6. A. H. Harvey, J. A. Gallagher and J. M. H. Levelt Sengers , *J. Phys. Chem. Ref. Data,* 27(4) pp761-774 (1998).
7. N. Wang, J. Wei, C. Hong, and H. Zhang, *Part. Part. Syst. Charact.* 13, pp238-244 (1996).
8. S. Twomey. *Introduction to the Mathematics of Inversion in Remote Sensing and Indirect Measurement.* Elsevier Scientific Publishing Company, 1977.
9. C. L. Lawson and R. J. Hanson. *Solving Least Square Problem*. Prentice-Hall Inc., 1974.
10. W. H. Press, S. A. Teukolsky, W. T. Vettering, and B. P. Flannery. *Numerical Recipes,* Cambridge University Press, 1992.
11. X. Cai and N. Wang. *Proceed. 3th Int. Aerosol Confer.*, pp126-129, Kyoto, Japan, September 1990.
12. H. Schnablegger and O. Glatter, *Appl. Opt.* 30, pp4889-4896 (1991).

Research of the PIV Technology Used to Measure the Liquid-solid Two-phase Flow in the Pump

Yang Min-guan, Liu Dong, Yang Bo,and Kang Can

School of Energy and Power Engineering, Jiangsu University, Zhenjiang, 212013, China

Abstract. Based on single-phase PIV technique, combined with the image processing techniques, the PIV is used to measure the liquid-solid two-phase flow with salt out inside the centrifugal pump: using the image grabber card to collect the flow images and then sent to computer simultaneously via data line, we develop the liquid-solid image processing and analysis software to separate the particle from the background, and then use some arithmetic to survey and recognize the particle, the solid particle and the hollow glass particle that tracking the liquid was distinguished from its diameter. The experimental results show that the new method has advantages of lower cost and less dependence on hardware. Therefore, it lays the foundation for researching the liquid-solid two-phase flow with salt coagulation in the pump.

Key words: Centrifugal pump, Liquid-solid two-phase flow, Image processing, PIV

INTRODUCTION

Salt out is a process that solid particle separates from a single solution or mixing solution composing of natrium and kalium ion mainly. The feature of this phenomenon is that there are lots of little crystal particles in the solution whose average diameter is only 50 micrometers. In the view of hydrokinetics, this solution could be considered as liquid-solid two-phase flow. To study two-phase flow used to measure and analyse its whole flow field. As an advanced whole flow field, instantaneous state and no touch measurement technology, particle image Velocimetry technology has been widely used in all sorts of steady and unsteady single phase flow measuring. But in these years, native and foreign researchers began to measure liquid-solid two-phase flow, especially synchronous measure the two-phase's velocity. So the PIV measurement technology of liquid-solid two-phase flow is still on developing[1]. Usually, two methods are applied in two-phase flow measurement[2]: fluorescence particle method and particle diameter differentiation method. The principle of the first method is wavelength discrimination between solid particle and fluorescence particle, the measurement of the liquid velocities was performed by adding fluorescing particles with a mean diameter of 50 mm. This method depended on hardware more serious and used to measure tiny flow field. The principle of the second method is the differences of solid and liquid phases to laser scatter intensity. This method uses image processing method to separate the particles of solid and liquid phases, and then uses cross-correlation algorithm to calculate the velocity field of solid and liquid phases.A new kind of two-phase particle image processing and analysis method will be introduced, we writing WINDOWS application to manage the liquid-solid flow field in centrifugal pump.

EXPERIMENTAL SYSTEM AND PRINCIPLE

Experimental System

Fig.1 shows the experimental system. In order to meet the requirement of PIV system, we take the optimal design of centrifugal pump. The design flux of pump is $25\,m^3/h$, design head is 20m, and the rotational speed is 2890r/min. Both impeller and spiral casing back cover are made of glass, so the CCD camera could shoot the whole field in the impeller. To let the laser easy to enter the inside of impeller, a rectangle window was opened at the flank of spiral casing. The PIV system , purchased from TSI inc., consists of a two-pulsed Nd-Yag laser, a CCD camera, a synchronizer and a data analysis system .The Laser which wavelength is 532nm has a frequency of 15 Hz. The output power of the laser is 120mJ/puls, which

CP914, *Multiphase Flow: The Ultimate Measurement Challenge,* edited by X. Cai, Y. Wu, Z. Huang, S. Wang, and M. Wang

is sufficiently high for the experiment. The CCD camera was used to obtain digital images with 1248×1024 pixels, and can capture 3.75 pictures in one second. The particles used in these experiments are hollow glass particles with diameters of 8-10 micron.

The experiment is taken on rotating flow field, we have to use the shaft encoder and frequency divider. Shaft encode fix up on axes of motor. With every rotating of encoder, shaft encoder transmits 500 pulse signals which were divided by frequency divider to meet the speed collection of CCD.

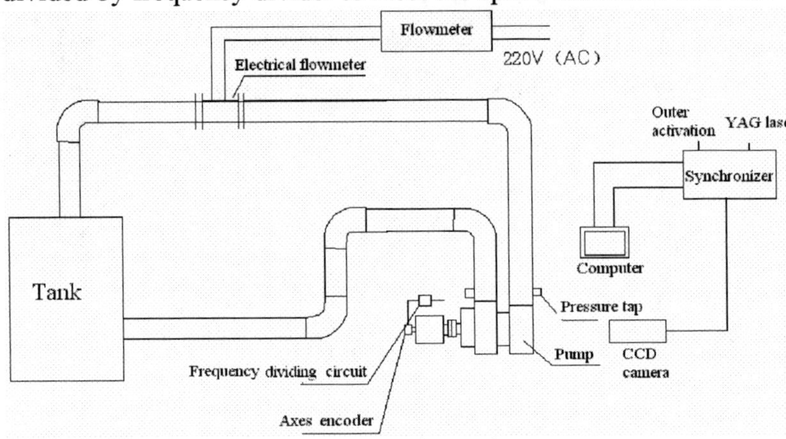

FIGURE 1. Experimental system

Basic Principle of PIV

The basic principle of PIV is to measure the displacement of particles within Δt, which include two directions: ΔX and ΔY, and then obtain the velocity of the particle as follow: U=ΔX/Δt, V=ΔY/Δt. The principle is showing in fig 2. Due to the features of the tracer particles, particle's velocity is the same as the velocity of flow motion. The displacement of particles should be as small as possible to let **Δx**/ Δt close to particle's velocity. That's to say, within the time of Δt, the tracer of particles are almost lines and particle's velocity in the tracer is almost invariableness, which should be satisfied with a proper time of Δt.

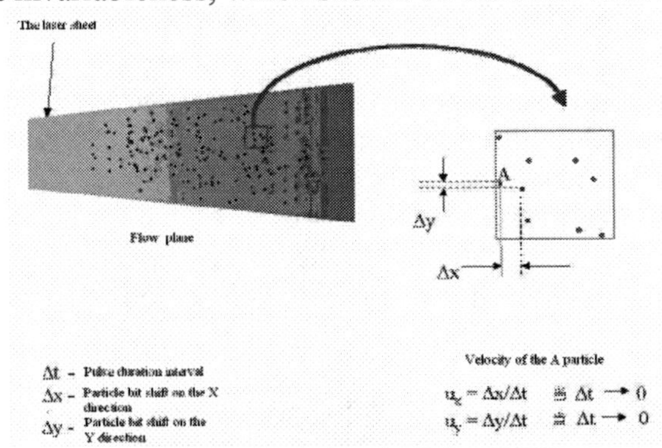

FIGURE 2. Basic principle of PIV

Principle of particle diameter differentiation

The crystal particles mixed with liquid in two- phase flow, it is hard to distinguish them and estimate the particle's velocity from the images which gained by PIV. However, the crystal particles are larger than the tracer particles in the liquid, so its scattering light is quite stronger. With the technology of images processing, we can get distributing graph of crystal particles and tracer particles respectively, then two–phase velocity field can be get by software of insight5.0.

In the course of image processing, the first step is to divide up the images, which divides the image into the prospect images element (particles) and the background images element (liquid) according to the difference between their gray scales. The basic method is to assure the threshold, when the gray is bigger than the assured value, we consider their as prospect image element; when the gray is smaller than the value, we consider their as the background image element. The selection of threshold is very important in image processing. The average diameter of the crystal particles is only 50 micron, so the contrast gradient of most images obtained in the experiment is not very large. Therefore we choose OTSU to select the image threshold, this method divides the image into two groups with a certain threshold, when the two groups have the larger variances, this gray scale is the threshold for segmenting the image into binary one. This method according to the probability statistics and least squares method. Based on the entire statistical property of the image, it is suitable for the segmenting of the image. The detail of the algorithm is as follows:

Supposed the mage can be divided into m-1grades, when the gray scale is i, the pels must be n, then the overall pels $N=\sum_{i=1}^{ni} n_i$, the probability of all pels value is $Pi = n_i/N$; then we can divide it into two groups with a integer K: $C_0=\{1,2,...,K\}$, $C_1=\{K+1,K+2,...,m\}$;The probability of C_0 is $w_0=\sum_{i=1}^{k} P_i = w(k)$, average value of C_0 is $\mu_0=\sum_{i=1}^{k} iP_i/w_0 = \mu(k)/w(k)$, The probability of C_1 is $\omega_1=\sum_{i=k-1}^{m} p_i = 1-\omega(k)$, average value of C_1 is $\mu_1=\sum_{i=k-1}^{m} ip_i/\omega_1 = [\mu-\mu(k)]/[1-\omega(k)]$ .The statistics average value of the gray scale is $\mu=\sum_{i=1}^{m} ip_i$, then $\mu=\omega_0\mu_0+\omega_1\mu_1$ .Then $\sigma^2=\omega_0(\mu_0-\mu)^2+\omega_1(\mu_1-\mu)^2=\omega_0\omega_1(\mu_1-\mu_0)^2$.Altering K from 1 to m, we can get the value K when the Variance is maximum, and the best threshold is $\max\sigma^2(k)$.

After obtains the threshold, we segmenting the image into binary one, and regard the prospect region as the isolated granule. And check all the particles in the image, and then mark all regions. That is to say, each connected region different marked differently, the methods of four connected region or eight connected region are common in use. Because the picture is very big, considering the processing speed of the algorithm and the synthesis processing effect, this paper selects four connected method to deal with the connected regions. Finds a basic spot in the image, search its region area with the four connected region method. In searching, marking the same particle as a certain sign. Go through the whole image, we can find all particle regions.

After marking the particle regions, we must identify all particles in the region. In this paper, we design an new algorithm for distinguishing particle diameter. The detail of the algorithm is as follows:

(1) Figure out the center of the particle by calculating all the particles regions;
(2) Find the edges of the particles from the center, from left to right, we find P1 and P2,from up to down, we find P3 and P4;
(3) Then we gain the diameter of the particle: R1=P2-P1; R2=P4-P3;
(4) Judge whether R1 is equal to R2 or not;

According to these steps, If R2 equals to R1, R1 is the diameter of particle; or it means that there is superposition between particles, which would be dealt with other ways, we are not discussing in this paper. After getting the diameters of all the particles, we distinguished the crystals particles and the tracer particles, supposed an intermediate value according to the different diameter between the solid and fluid particle, and then consider these bigger than the value as crystal particles, and the smaller ones as the tracer particle of the liquid phase. Save the different particles into different images, in this way we completed the separating of the two phase particles. Fig.3 is the interface of software we developed.

FIGURE 3. Interface of software to separate fluid and solid particles

SOFTWARE APPLICATION

Fig.4a is the original picture took by CCD which includes plenty of tracer particles and crystal particles. The mean diameter of tracer particles is 8 micron while the mean diameter of crystal particles is 50 micron. With the developed software, we assume the middle diameter of particle is 10 micron and the particles with diameter lower than 10 micron are tracer particles while the particles with diameter bigger than 10 micron are crystal particles. Fig.4b and Fig.4c are the liquid-solid granule fields which have been disposed respectively.

The liquid-solid two-phase absolute velocity fields processed by cross-correlation can be decomposed with the impeller's inner velocity triangle. Its relative velocity field is shown in Fig 5.

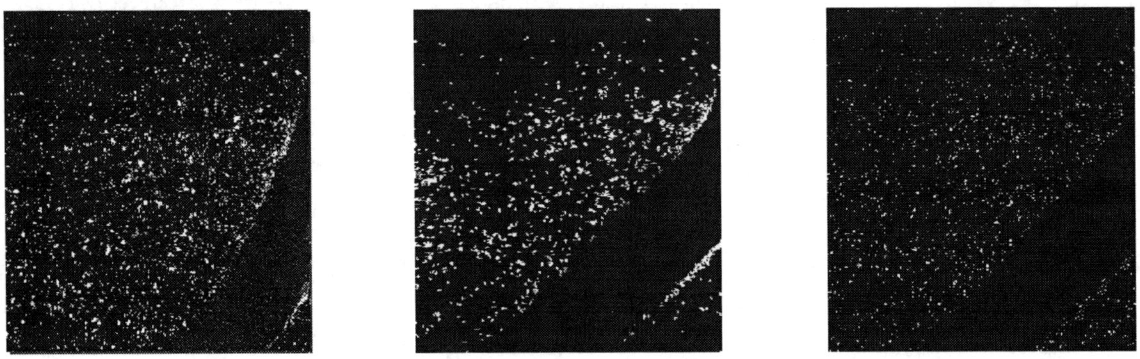

a b c

FIGURE 4. Experimental picture

During image processing, because of the big non-uniformity of solid particles' distribution, a great many small granules gather to generate bigger elements being mistaken for big granules as well as the blade reflection place. These have caused solid phase pellet velocity field distribute unevenly in the cross-correlation processing and inquiry area granules become less, some massively piles up in the certain place. We may use the method of Ensemble Average Correlation Processor for processing the data, which can make more satisfying results.

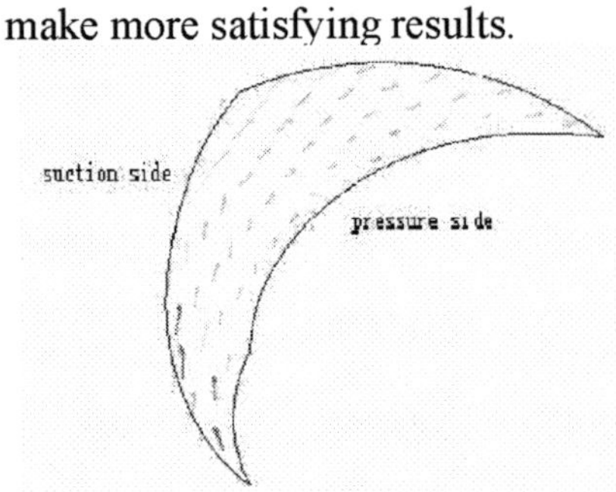

liquid relative velocity field

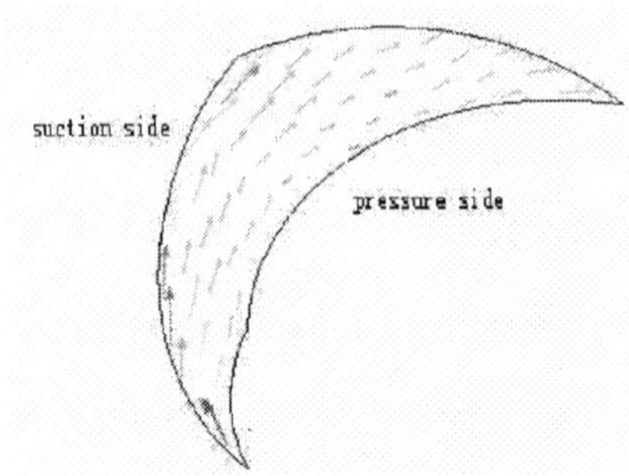

solid relative velocity field

(a) (b)

FIGURE 5. Relative velocity field

CONCLUSION

(1) In this paper, we developed a method of image processing to use the PIV technology to carry out the fluid-solid two phase flow study, obtained the threshold of the picture by the law of OTSU to discriminate the background and the granule. Recognition and separation of the fluid-solid granule was realized. The WINDOWS application procedure was developed, which had the friendly user surface and could achieve the multi-parameter adjustment;

(2) By using PIV and two-phase flow picture processing software, it is convenience to research inner fluid-solid two-phase flow of centrifugal pump with salting-out phenomenon. It has provided a helpful experimental technique to the investigation of analogous problems;

(3)In the process of image processing, the recognition question of granule sheltering from particles still requires the further research work.

ACKNOWLEDGMENT

The authors wish to acknowledge the financial support of the "National Natural Science Foundation of China", grant no. (50476068).

REFERENCES

1. Wang xilin,et al. Preliminary investigation of particle image velocimetry(PTV-PIV) technique in two-phase flow [J].Acta Mechanica Sinica.,1998(1):121～125.
2. Shi hui-xian, Wang qin-hui et al. The Application of PIV in Gas-Solid Multiphase Flow. [J].Power Engineering, 2002(1): 1589 — 1593.
3. YANG Min-guan, WANG Chun-lin,et al. On the Design of the Bittern Transfer Pump and Measures Of Preventing Crystal in use [J]. Journal of Jiangsu University, 2003.24(1):19-22 (in Chinese).
4. Kelder J D H, Dijkers R JH., van Esch B P M, et al.Experimental and theoretical study of the flow in the volute of a low specific-speed pump[J]. Fluid Dynamics Research, 2001(4): 267-280
5. Mehta M., Kadambi J R., Sastry S ,et al. Study of particulate flow in the impeller of a slurry pump using PIV[C]. Proceedings of the ASME Heat Transfer/Fluids Engineering Summer Conference 2004, 2004(1): 489-499
6. Sankovic, John M. Kadambi,.et al. PIV investigations of the flow field in the volute of a rotary blood pump[J]. Journal of Fluids Engineering, 2004(5): 730-734

on The Measurements of Particles Smaller Than 20 μM by Global Rainbow Refractometry

S. Saengkaew[(1,2], D. Bonin[(2] and G. Gréhan[(2]

1) *LEMTA, (CNRS, UMR 7563), 2, avenue de la Forêt de Haye, BP 160, F-54504 Vandœuvre-lès-Nancy cedex, France*
2) *LESP, UMR 6614/CORIA, CNRS/Université et INSA de Rouen, 76800 Saint Etienne du Rouvray, France*

Abstract. The measurement of the thermo-chemical characteristics of particles under evaporation or cooling is a challenge. Among others techniques, Global Rainbow Refractometry (GRR) is potentially applicable to a large variety of realistic media. This paper is focused on refractive index measurements of particles smaller than 20 μm which are especially important to extract droplet temperature in spray combustion.

Keyword : Spray temperature measurements, Rainbow refractometry

INTRODUCTION

During the last decades a large effort has been devoted, in the combustion field, to the measurement of the velocity and size of droplets in sprays. Several techniques have been developed to reach this aim as: LDV, PDA, PIV, diffractometry, ILIDS, digital holography, etc ... . However, in order to have a better understanding on droplet evaporation, the interest is focused on the measurement of the droplets temperature and composition. Nowadays, different teams are involved in this effort, by developing several techniques based on different physical properties as fluorescence, infra-red emission or elastic scattering to name few of them. Among these techniques, the global rainbow refractometry, introduced by van Beeck et al [1], is long of a large potential of application in realistic sprays. This technique is based on the analysis of the collective rainbow created by the summation of a large number of individual rainbows. The position of this collective rainbow is essentially dependant on the value of the refractive index of the particle, therefore on the temperature and composition of the studied droplets.

This paper is devoted to the numerical and experimental studies of the applicability of GRR to droplets smaller than 20 μm, which are of special interest in combustion systems. The paper is organized as follow. The section 2 recalls the basic of rainbow physics for one particle and discuss the implication of these behaviours for GRR. Section 3 is devoted to the numerical analysis of the GRR accuracy for small particles. Section 4 compiles some experimental results obtained from the sprays with a mean diameter smaller than 20 μm. Section 5 is a conclusion.

THEORETICAL BACKGROUND

Individual Rainbow

The rainbow, one of the most beautiful phenomena of the nature, has been at the origin of active researches. The rainbow position has first been explained by Descartes as created by rays experimenting one internal reflection in a droplet. In this framework, the rainbow location depends only on the droplet refractive index value. Later, Airy [2] improves the accuracy of the description by taking into account the fact that the particle possesses a finite size. A rigorous description of the scattering of a plane wave by a spherical particle, including rainbows, is given by Mie [3] in the framework of the Maxwell's equations. The Mie's theory is exact but all the contributions are mixed. Then, Debye's theory [4] has been introduced. This theory can be seen as a post-processing of the Lorenz-Mie's theory where the different contributions can be separated. Latest, Nussenzveig [5] develops the Complex Angular Momentum (CAM) theory which is claimed to be as accurate as Mie's theory but as fast as Airy's theory.

CP914, *Multiphase Flow: The Ultimate Measurement Challenge,* edited by X. Cai, Y. Wu, Z. Huang, S. Wang, and M. Wang

Figure 1 plots the intensity of the scattered light versus the scattering angle for one droplet of alcohol with a refractive index equal to 1.36 – 0.0 i and a droplet diameter equal to 60 µm. Three regions can easily be identified, corresponding to the first rainbow (created by the 'rays' experimenting one internal reflection), the Alexander's dark band (created by externally reflected light) and the second rainbow (created by the 'rays' experimenting two internal reflections). In the first/second rainbow regions the low frequency structure is created by 'rays' experimenting one/two internal reflection(s) while the high frequency fringes (called ripple) superposing on the rainbow pattern are due to interferences between these 'rays' and the externally reflected light.

When the refractive index value of the particle increases, the first rainbow moves toward a larger scattering angle while the second one moves down to lower scattering angle. It is the opposite when the refractive index decreases.

To extract the refractive index value and the size from rainbow signals, a classical solution is to remove the ripples by filtering the signal. However, this approach is not applicable to small particles. As the particle size decreases, the following modifications in the light scattering diagram region corresponding to the first rainbow are induced:

- The angular distance between the geometrical optics rainbow location and the maximum of intensity increases.
- The frequency of the ripple structure and the Airy frequency of the supernumerary peaks are closer.
- The Alexander's dark band dimension decreases, and disappears.
- The description of the rainbow by Airy's theory is less and less accurate.

These facts (ripple structure, Airy's theory limitation, sensitivity to shape modification, ...) induce that the processing of individual rainbow signal is especially difficult for small particles.

Other possibilities are to include the ripple structure in the processing strategy [6] or summing the rainbows created by a large number of droplets with different size [1], corresponding to the technique called Global Rainbow (GR).

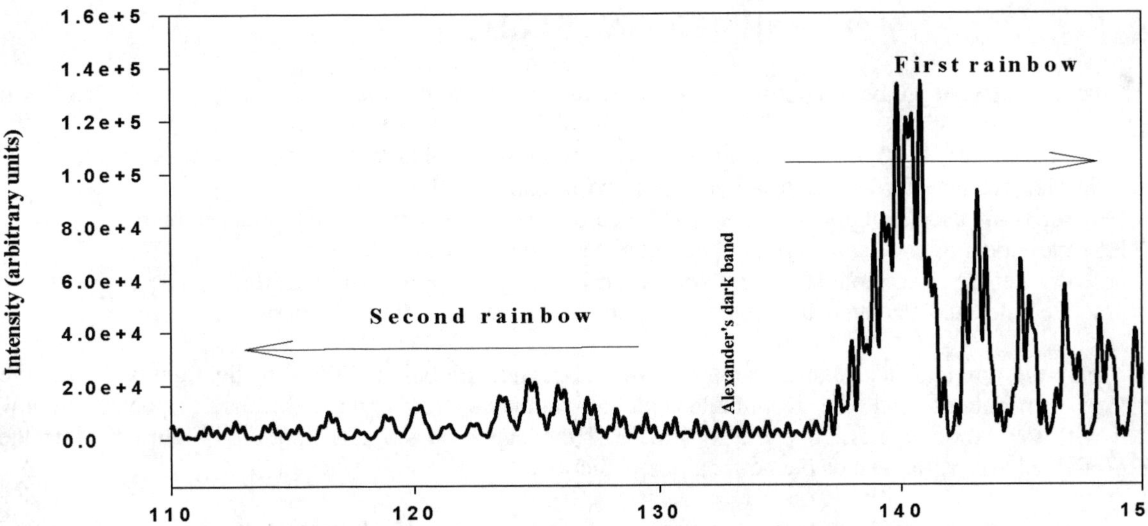

FIGURE 1. Rainbow created by one droplet

Global Rainbow

The global rainbow is created by the summation of individual rainbows created by the particles located in a section of a cloud. Figure 2 displays a typical global rainbow created by a spray of water droplets (m=1.333-0.0i) with a Gaussian distribution size distribution. The mean size is equal to 60 µm and the standard deviation is equal to 40 µm.

To extract the refractive index value and the size distribution a software has been written [6], based on the CAM theory developed by Nussenzveig [5,7]. The algorithm scheme is as follow:

1. A first value of the refractive index is estimated from the global rainbow position by using geometrical optics law.
2. With this refractive index value, a matrix is computed by using Nussenzveig's theory (assuming that only the p=0 and p=2 rays contribute to the signal) where each line corresponds to the scattering intensity in a given direction and each column is associated with the scattered intensity for a given diameter. Typically, the matrix is 80 lines by 40 columns.
3. By using the Non-Negative Least Square method, as described by Lawson and Hanson [8], a size distribution is computed.
4. Steps 2 and 3 are repeated to minimize the distance between the measured global rainbow and the recomputed global rainbow in order to obtain the best values for the refractive index and the size distribution.

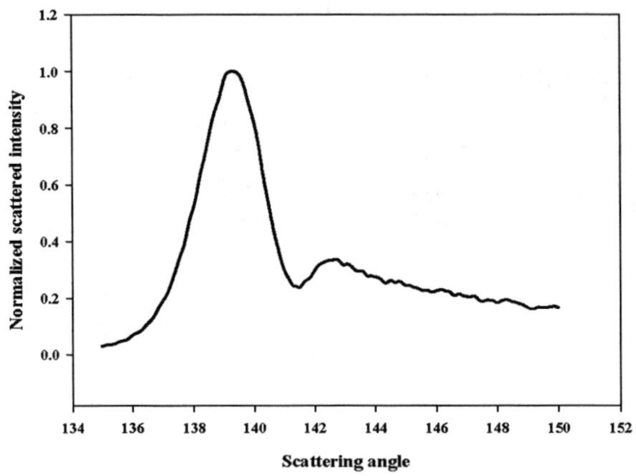

FIGURE 2 . a typical global rainbow for relatively large particles (d = 60 μm)

NUMERICAL STUDY

To study the accuracy of global rainbow technique with tiny particles, the following procedure has been carried out.

a) Series of diameter (1000 particles in this study) are constructed in agreement with a Gaussian distribution characterized by its mean diameter and its standard deviation.
b) These series are used as inputs in a Lorenz-Mie code to compute series of individual rainbows.
c) These individual rainbows are summed up to create synthetic global rainbows.
d) These synthetic global rainbows are inversed by using the code previously introduced. The extracted refractive index and size distribution are compared to the Lorenz-Mie code inputs.

Figures 3 plot the nominal size distributions and the associated global rainbows while figures 4 plot the extracted size distributions and the recomputed global rainbows (recomputed means a global rainbow recomputed with the extracted refractive index value and the extracted size distribution). Figure 5 plots the extracted refractive index value versus the nominal mean diameter.

Figure 3-b shows that when the mean diameter decreases, the location of the first rainbow maximum is shifted toward larger scattering angle. Furthermore, it is observed that the beginning of the global rainbow curves are less smooth when the diameter decreases. This behaviour has an effect on the accuracy of the refractive index measurements.

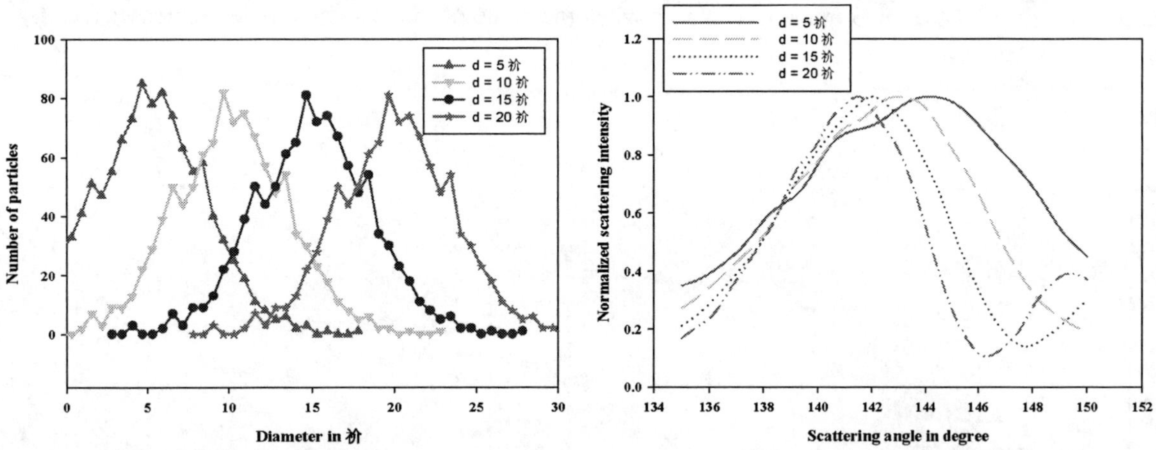

a) Initial size distribution b) Synthetic global rainbow

FIGURE 3. Nominal size distributions and associated synthetic global rainbows for a refractive index equal to 1.333.

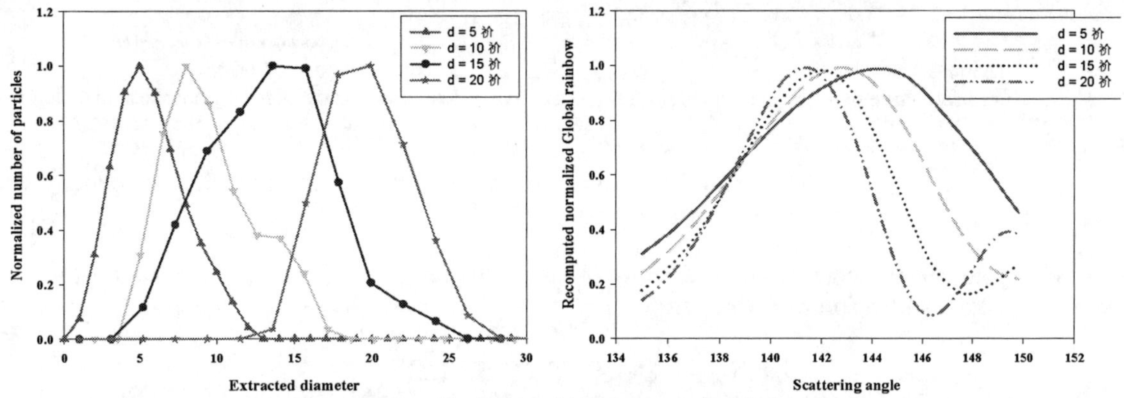

a) Extracted size distribution b) Recomputed global rainbow

FIGURE 4 . Extracted size distributions and associated recomputed global rainbows.

Figure 5 plots the deviation from the nominal refractive index value and the measured refractive index value versus the droplets size. For the case under study (water droplets), this divergence can be attributed to the perturbing effects coming from higher order interactions between the droplet and the beam. Then the refractive index measurements carried out on water spray of small droplets (d < 20 μm) must be corrected from the shift due to these effects which is directly dependant on the particle size.

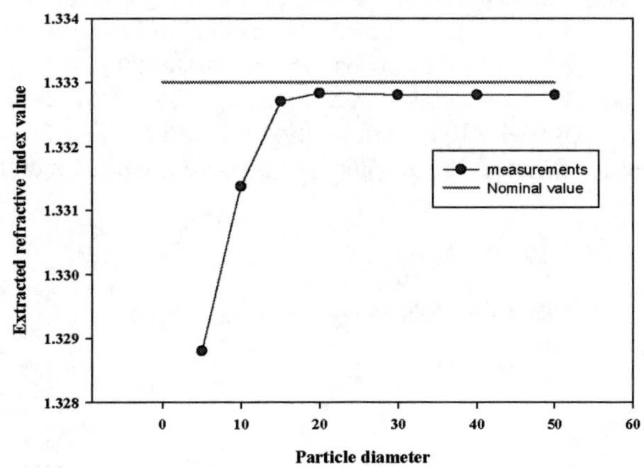

FIGURE 5 . Measured refractive index value versus nominal particle diameter.

For small particles (mean diameter equal to 5 μm), global rainbows have been computed and then inversed for nominal refractive index values running from 1.333 to 1.420. It must be noticed that the extracted size distributions are very close. Figure 6-a shows that the measurement of the refractive index is essentially good, however the accuracy of the measurement increases when the value of the refractive index increases (see figure 6-b).

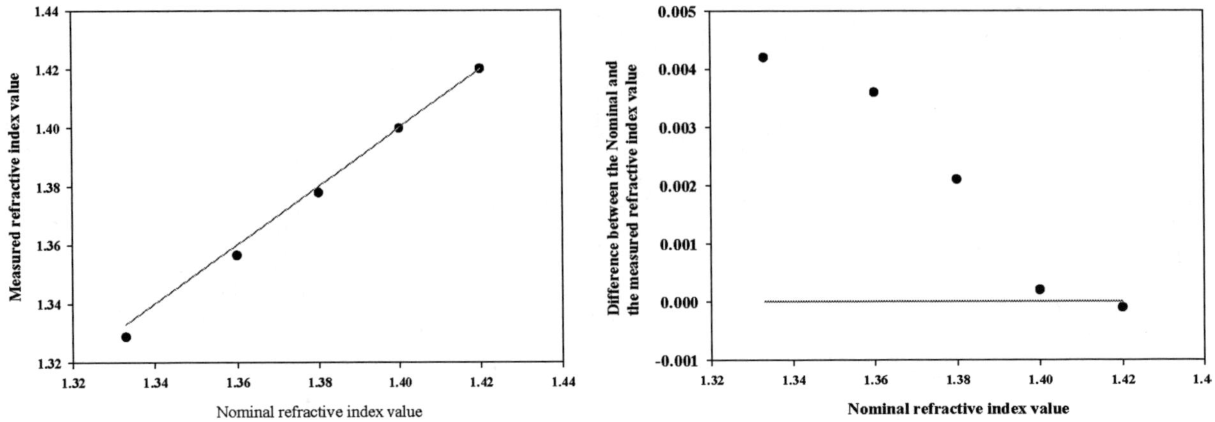

a) measured refractive b) measurement error

FIGURE 6 . Measured refractive index value versus nominal refractive index value for particles with a mean size equal to 5 μm.

Series of figures 4 to 6 prove that useful information on the refractive index and size distribution can be extracted from global rainbow created by particle as small as 5 μm.

The next step is the experimental study of a spray of small droplets. The next section describes preliminary experiments carried out with a spray of water droplets.

EXPERIMENTAL STUDY

A global rainbow set up has been realized in agreement with the one described by van Beeck et al [1]. The details of the set up are given in [6]. The figure 7 displays this setup. The light source is an Argon Ion laser (Spectra Physics 2016) which emits a beam with a diameter equal to about 1.5 mm at a wavelength equal to 0.5145 μm. The light scattered by droplets, created in this paper by a TSI 3076 generator, is collected by a first lens with a focal length equal to 50 mm. This lens images the spray on a diaphragm which selects a small region of the spray. A second lens with a focal length equal to 100 mm images the focal plane of the first lens on a screen. Then, only the global rainbow created by the particles selected by the diaphragm is observed.

Figure 8-a plots a recorded global rainbow image while figure 8-b plots its associated intensity versus the scattering angle. The particle size distributions measured by global rainbow refractometry and by PDA have been compared. The agreement between the two approaches is satisfactory, the mean diameter of the spray is equal to 5 μm. Figure 9-a plots the refractive index values extracted from 19 consecutive images. The measured refractive index values run from 1.324 to 1.3307, with an average value equal to 1.3273. This average value is close of 1.329, the value expected for 5 μm water droplet (see figure 5), which is materialized by a dashed line.

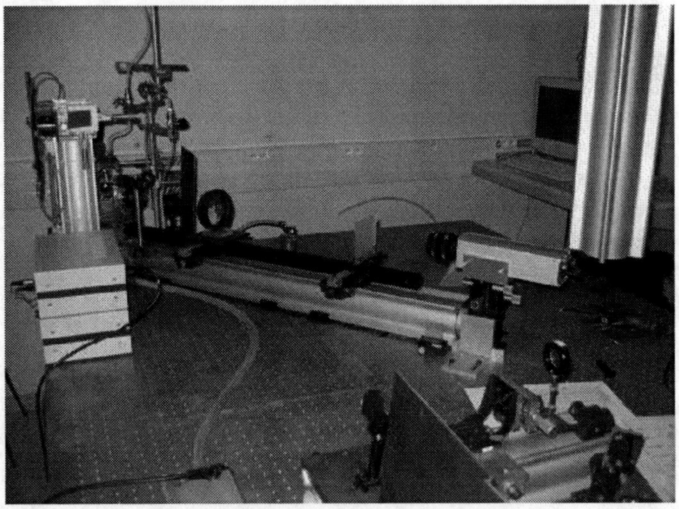

FIGURE 7 . The experimental setup

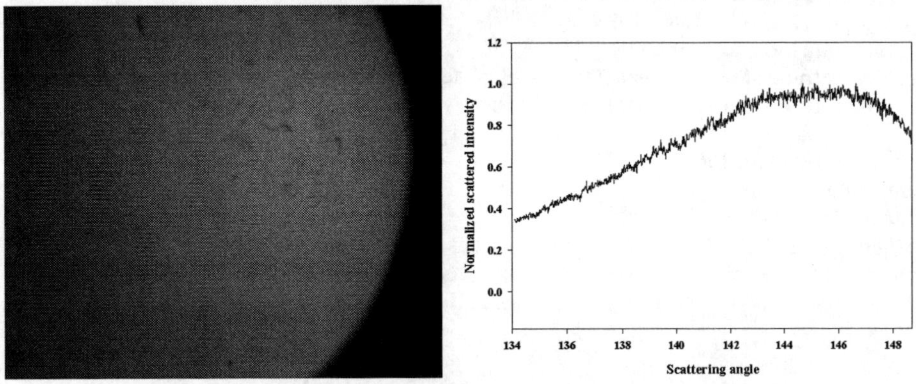

(a) (b)

FIGURE 8. Image of a recorded global rainbow, angular intensity behaviour

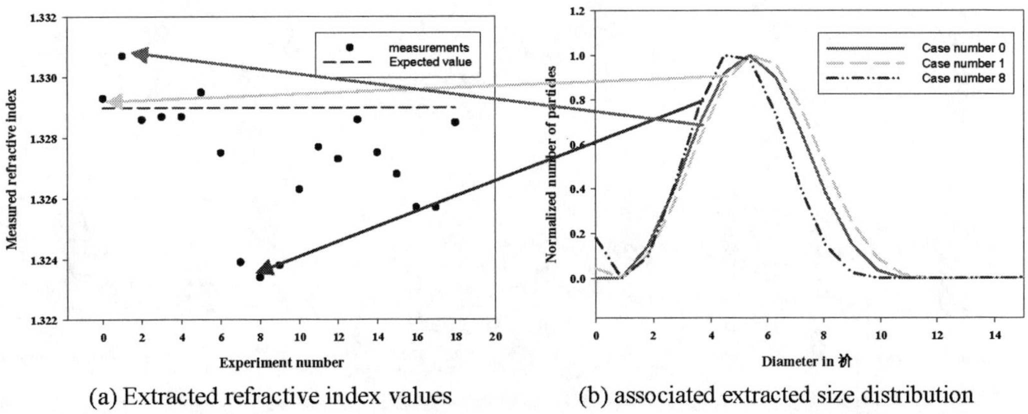

(a) Extracted refractive index values (b) associated extracted size distribution

FIGURE 9. Comparison between the global rainbow extracted size distribution and a size distribution simultaneously measured by PDA

Moreover, when the extracted size distributions are compared, it appears that the higher extracted refractive index values are associated with the larger particles, and the smaller extracted refractive index values are associated with smaller particles. This fact is illustrated by figure 9 where, for clarity reason, only case 0, 1 and 8 have been plotted.

This behaviour is in agreement with the dependence between the measured refractive index and the size as illustrated in figure 5.

CONCLUSION

The preliminaries results (numerical and experimental) presented in this paper prove that the simultaneous measurement of the particle size distribution and refractive index by global rainbow refractometry is possible for small particles (as small as 5 μm), when the inversion is carried out under the assumption that only the rays p=0 and p=2 contribute to the signal. More accurately, when the refractive index is larger than about 1.38, the extracted refractive index is directly the measurement while for refractive index smaller than 1.38 the extracted refractive index must be corrected of the higher order contributions. This is particularly important for refractive index smaller than 1.34. The next step will be to carry out measurements for particle size distribution and refractive index with several products covering a large range of applications, in a well controlled environment.

ACKNOWLEDGEMENTS

This work was made possible by the ASTRA Program of CNRS and ONERA, by a post-doctoral grant from CNRS, and by a PhD grant from region Haute-Normandie.

REFERENCES

1. J.P.A.J. van Beeck, D. Giannoulis, L. Zimmer and M.L. Riethmuller, Global rainbow refractometry for droplet temperature measurement, Optics Letters, 24, 1696-1698, 1999
2. G.B. Airy, Trans. Camb. Phil. Soc. 6 (1838) 379.
3. G. Mie, Beiträge zur Optik trüben Medien speziell kolloidaler Metalösungen, Ann. der Phys., 25, 377-452, 1908
4. P. Debye, Der Lichtdruck auf Kugeln von beliebigem Material, Ann. der Phys., 4, 57-136, 1909
5. Nussenzveig H.M., High-frequency scattering by a transparent sphere. II Theory of the rainbow and the Glory, Journal of Mathematical Physics, 10, 125-176, 1969
6. S. Saengkaew, Température de spray: sur le développement d'une technique de mesure par réfractométrie d'arc-en-ciel, Thèse de l'université de Rouen, 12 Octobre 2005
7. S. Saengkaew, T. Charinpanitkul, H. Vanisri, W. Tanthapanichakoon, L. Mées, G. Gouesbet and G. Grehan, Rainbow refractometry: on the validity domain of Airy's and Nussenzveig theories, Optics Communications, 7-13, 2006
8. C.L. Lawson and R.J. Hanson, Solving least squares problems, Prentice-Hall, Inc, 1974

The Impact of the PDA Measurement Method in Forward Scatter on The Concentration of Gas-Particle Two Phase Flow

Chen Lizhe[1], Chen Zhichao, Sun Shaozeng, Li zhengqi

School of Energy Science and Engineering, Harbin Institute of Technology, 92, West Dazhi Street, Harbin, 150001, P. R. China

Abstract. Use a two-phase test facility to calibrate the measured diameter and concentration of the PDA system in forward scatter. In the experiment, the standard pitot was used to measure velocity and the pressure sensor was used to gives the on-line weight of glass beads. The speed mean value and the concentration mean value of measurement section can be calculated. Surveying the concentration with different dimension PDA in forward scatter, by comparison with the demarcation result, the uncertainty basically in 30% scope, and the error tendency of 1D PDA, 2D PDA and the 3D PDA is increasing. Using the same dimension PDA to measure the concentration, with the concentration increasing, the experiment data is increasing. So the PDA system can be used to measure relative concentration. When the number of samples is bigger than 5000 at one point, the measured value tends to be steady and close to the true value.

Keywords: Two-phase flow, Phase-Doppler anemometry
PACS: 06.30.Dr

INTRODUCTION

With Laser-Doppler anemometry (LDA) and Phase-Doppler anemometry (PDA) developing, the characters of two-phase flow can be measured in detail [1-2]. Because the measurement about the particle concentration is complex, the uncertainty of flux is more than that of velocity and diameter.

In this paper a two-phase test facility was set up to calibrate the measured velocity, diameter and concentration of the PDA system. Some experts have already study the two-phase flow in the similar test facility [3-4]. So we only discuss the diameter, flux and uncertainty in this paper. This work gives some advice for some scientific research experiment with the PDA system.

EXPERIMENTAL

A three-dimensional PDA made by Dantec was used in this study. The instrument includes an argon ion laser, transmitter, receiving optics, signal processors, traversing system, computer system and a three-dimensional auto-coordinated rack. The PDA uses the proven phase Doppler principle for simultaneous non-intrusive and real-time measurements of three velocity components and turbulence characteristics, and makes use of new methods for phase differences between Doppler signals received by two detectors located at different positions. Several optical configurations from 0-500mm are radially available. All instrument settings, such as bandwidth and high voltage, are computer controlled. An analog-digital converter allows the computer to read the anode current of the photomultipliers. The combination of photomultiplier and particle velocity correlation bias can contribute to uncertainty. But the error is likely to be small. The overall uncertainties in measured values of the particle diameter and the particle concentration are 4 and 30% respectively, and the ranges of measurable sizes, velocities and particle concentrations are 0.5-1000μm, -500 to 500 m/s and 10-1012 particles/m3.

[1] Corresponding Author: clz@hit.edu.cn

CP914, *Multiphase Flow: The Ultimate Measurement Challenge,* edited by X. Cai, Y. Wu, Z. Huang, S. Wang, and M. Wang

The test facility is illustrated in Fig. 1. It mainly consisted of a sucker, a feeder, a cyclone separator and a pressure sensor. Glass beads were fed via the feeder into the pipe. The particle material density is 2 500 kg/m3. With the particle put in the water, the particle-size distribution was measured by laser in the physics of condensed matters center of Harbin Institute of technology. The particle-size distribution and the particle mean number diameter is 42μm. The feeder is on a pressure sensor which gives the on-line weight of the feeder. The particle mass flow rate can be controlled by the pressure sensor and the velocity in the pipe can be get by a standard pitot. Then the speed mean value and the particle concentration mean value of the test section can be calculated. The velocity of the test section is 5.18 m/s and the particle mass flow rate is 7.82 kg/h (the particle volume flow rate is 218 g/m3).

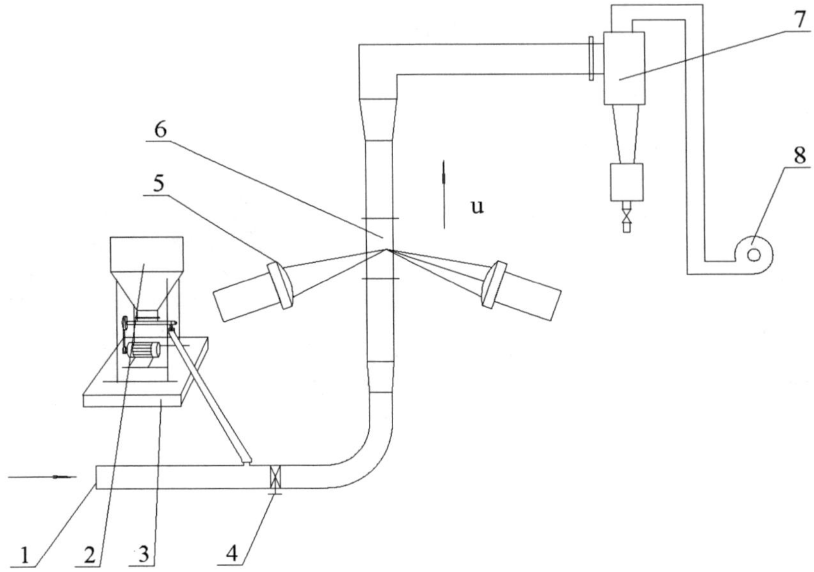

FIGURE 1. Schematic drawing of test facility
1. Inlet of the test facility 2. Feeder 3. Pressure sensor 4. Adjusting valve 5. Probes of the PDA 6.Test section 7. Cyclone separator 8. Sucker

The data criterion of PDA is 10 000 samples or 120 s per position. The particle size less than 5μm are sufficient for reliable statistic. The particle characteristics of the size distribution in the range from about 0 to 5μm are analyzed to trace the air flow. The particle number concentration is the percent of the particle number to the carrier phase volume and the particle volume flux is the volume of dispersed phase crossing the measurement volume per unit time per unit area.

Distributions of Particle Volume Fluxed and Concentrations

The particle mass flow and error is shown in Fig. 2 which was calculated by the particle volume flux and the particle number concentration measured with PDA. Accord to the particle volume flux measured by 1, 2, 3D PDA,

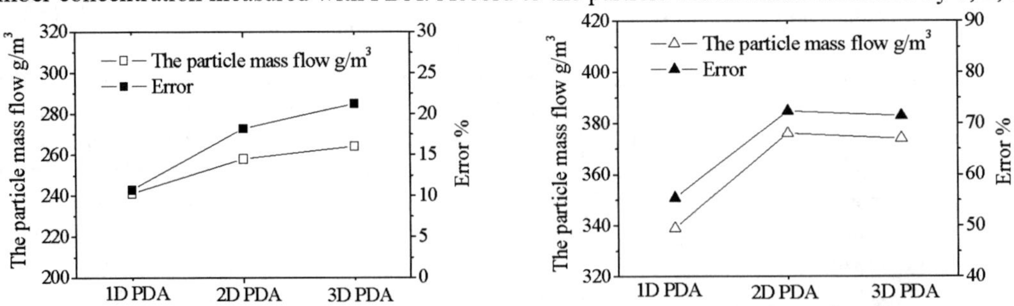

FIGURE 2. The particle mass flow and error measured by 1, 2, 3D PDA
a. Calculated by the particle volume flux b Calculated by the particle number concentration

the particle mass flow was respective 241g/m3, 258 g/m3 and 264 g/m3, and error of the results was 10.7 %, 18.2 % and 21.2 %. Accord to the particle number concentration measured by 1, 2, 3D PDA, the particle mass flow was respective 339 g/m3, 376 g/m3 and 374 g/m3, and the error of the results was 55.4 %, 72.4 % and 71.5 %. The uncertainty is in 30% scope in accordance with that given in the Dentec guide, and the error tendency of 1D PDA, 2D PDA and the 3D PDA is increasing.

Declines of the calculated particle mass flow with different particle mass flow is shown in Fig. 3. The measured particle mass flow is calculated by the particle volume flux. With the same dimensional PDA, the calculated particle mass flow increased when the particle mass flow increased. We can conclude that the PDA system can be used to measure relative concentration.

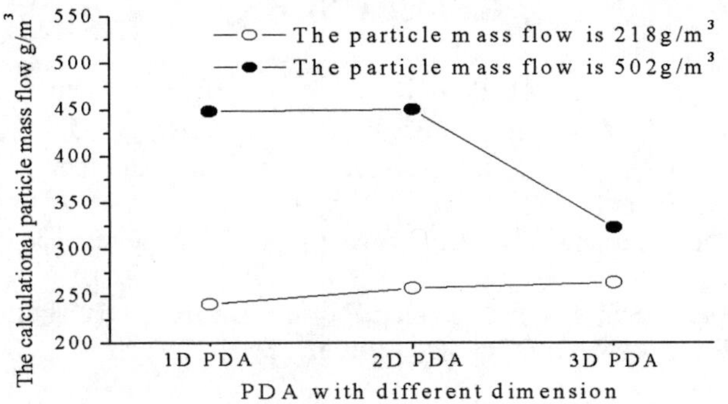

FIGURE 3. Declines of the calculated particle mass flow with different particle mass flow

Distributions of Diameter and The Particle Mass Flow With Different Samples at The Same Position

The sample of per position is an important factor to get reliable statistics. With the fluctuation of feeder and other factors, the velocity fluctuates around the flow velocity. The more samples per position is collected, the more the velocity is close to the flow velocity. It is impossible to get infinite samples for experiment time, cost and others. The distribution of diameter and the particle mass flow with different samples at the same position is given in Fig. 4. The number of samples collected at the same position is respective 200, 500, 1000, 2000, 3000, 5000, 10 000, 50 000 and 100 000.

When the number of samples is less than 3000 at one point, the measured value of the particle diameter and mass flux deviate from the true value. Not until the number of samples is bigger than 5000 at one point, the measured value does not tend to be steady and close to the true value.

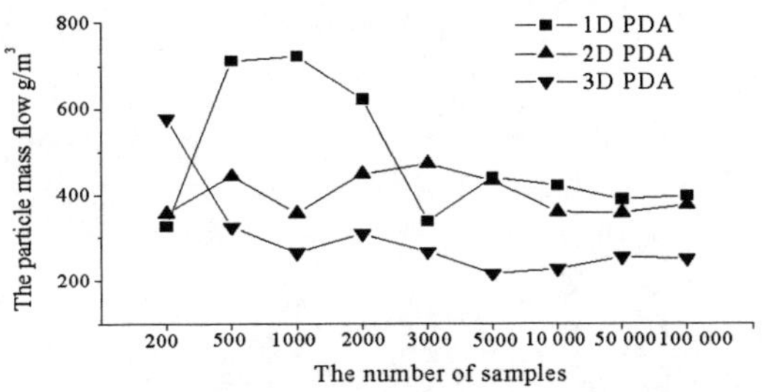

a. Distribution of the particle mass flow

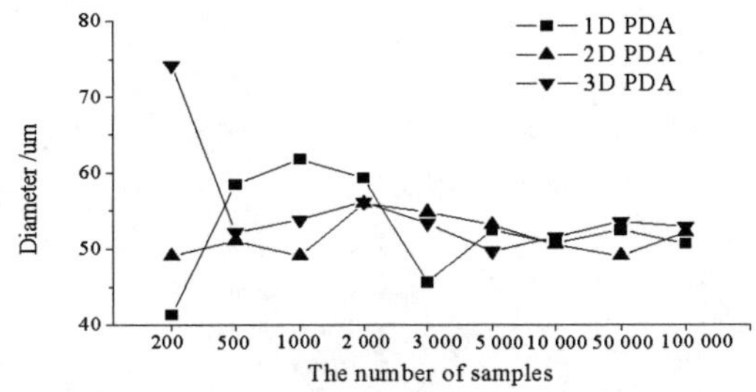

b. Distribution of diameter

FIGURE 4. Distribution of diameter and the particle mass flow with different samples

CONCLUSION

Surveying the concentration with different dimension PDA in forward scatter, by comparison with the demarcation result, the uncertainty basically in 30% scope, and the error tendency of 1D PDA, 2D PDA and the 3D PDA is increasing. Using the same dimension PDA to measure the concentration, with the concentration increasing, the experiment data is increasing. So the PDA system can be used to measure relative concentration.

When the number of samples is less than 3000 at one point, the measured value of the particle diameter and mass flux deviate from the true value. Not until the number of samples is bigger than 5000 at one point, the measured value does not tend to be steady and close to the true value.

ACKNOWLEDGMENTS

The present study was supported financially by the National Basic Research Program of China.

REFERENCES

1. M. Sommerfeld et al. Characterization of Particle－Laden, Confined Swirling Flow By Phase－Doppler Anemometry and Numerical Calculation. Int. J. Mult. Flow. 1993. 19(6)：1093-1127
2. I. Gillandt, U. Fristching, K. Bauckhage, (2001) Measurement of phase interaction in dispersed gas/particle two-phase flow. Multiphase Flow, 27: 1313-1312
3. J. Bao, S. L. Soo, (1995) Measurement of particle flow properties in a suspension by a laser system. Powder Technology, 85:261-268
4. LIU Chunrong, GUO Yincheng. Stability Analysis on Particle Uniform Distribution in Cocurrent-up Gas-Solid Flow. Journal of Jishou University (Natural Science Edition). 2006, 27(1): 63~7

Experimental Study on In-situ Concentration Monitoring of Flue Gas from the Fixed Pollution Source Based on DOAS

Tang Guanghua, Xu Chuanlong, Wang Shimin

School of Energy and Environment, Southeast University, Nanjing 210096, CHINA

Abstract. The air pollution is mainly derived from the flue gas from fixed pollution source and there has been no appropriate method to measure the concentration of flue gas in bad conditions for a long time. Based on the differential optical absorption spectroscopy (DAOS), an improved inversion algorithm, which is applied to the measurement of flue gas concentration at normal temperature and pressure, is put forward according to the characteristics of the flue gas from fixed pollution source and the experimental study of gas concentration measurement is performed combining the continuous transmittance spectrum from ultraviolet to visible region presented by avantes-2048 spectrometer with the absorption cross sections of sulfur dioxide and nitrogen dioxide derived form the experiment. Research results are shown the algorithm can not only be utilized to measure the concentration of sulfur dioxide with the interference of nitrogen dioxide but also be applied to simultaneously measure the concentration of sulfur dioxide and nitrogen dioxide in flue gas containing dust particles of high concentration.

Keyword: DOAS; fixed pollution source; in-situ measurement of flue gas concentration; experimental study

INTRODUCTION

Exhaust gas of automobiles, straw combustion and flue gas emission of many kinds of boilers and kiln make a great contribution to the air pollution in China at present and the air pollution is to a great degree derived from the fixed pollution sources such as the boilers of power plants and industry and kilns of cement factories etc[1-2]. Being the dust particles and frequent temperature change[3-5], there has been no appropriate method to monitor the concentration of flue gas for a long time. With the improvement of people's life and consciousness of environmental protection, preventing from the atmosphere being polluted is more and more becoming the subject of concerns. So it is necessary and urgent to develop an effective method to control air pollution and perform the in-situ and continuous measurement of flue gas concentration. DOAS was put forward by U.Platt et al[6] of institute of environmental physics of Heidelberg University in Germany until 1970's and then was widely used in air pollution monitoring. Being the advantage of simple theory and configuration, short response time, high measurement precision, non-invasive and in-situ measurement, some scholars in fields such as atmospheric physics and chemistry, geophysics and environmental protection show great interest and concern in DOAS. DOAS was applied to the measurement of flue gas concentration by Fu Jingye, Zheng Chaohui and Xie Pinhua et al[7-9], but dust particles was not considered and low concentration measurement of flue gas was not carried out in their experiment so there are many differences from the actual situation. An improved method of in-situ and non-invasive flue gas concentration monitoring was put forward according to the characteristics of fixed pollution source and measurement of gas concentration applying the algorithm was done on the basis of experiment setup at normal temperature and pressure.

BASIC PRINCIPLE OF DOAS

DOAS has been showing huge pushing effect on monitoring of atmospheric environment, especially on measurement of trace gas concentration, since it is put forward by U.Platt et al. Flue gas concentration can be precisely measured through DAOS which utilizes the configuration of differential absorption in the selected wavelength region to eliminate the interference of other gases and dust particles. So the characteristics of absorption spectroscopy of gases must be used as the transcendent knowledge of DOAS. The basic principle of DOAS is based

CP914, *Multiphase Flow: The Ultimate Measurement Challenge,* edited by X. Cai, Y. Wu, Z. Huang, S. Wang, and M. Wang

on the Lambert-Beer law, shown as figure1, and the model can be expressed as follows:

$$I(\lambda) = I_0(\lambda)\exp[-LC\sigma(\lambda)] \quad (1)$$

$I(\lambda)$—original light intensity, $I_0(\lambda)$—received light intensity, $\sigma(\lambda)$—absorption cross section

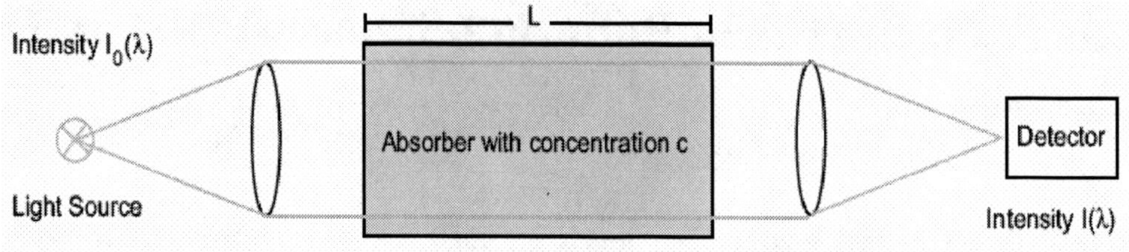

FIGURE 1. schematic diagram of Lambert-Beer law

The above model is based on the presumption described as below:

a. The incidence light is homochromous and parallel;
b. The concentrations of gases to be measured are sparse i.e. the interaction of molecules can be neglected;
c. The concentrations of gases to be measured are uniform;

CALCULATION MODEL

The gases to be measured usually coexist with the other gases and dust particles. There exist absorption of water vapor, response of CCD, Mie scattering and absorption of dust particles, Raileigh scattering reduced by themselves and so on. So the Lambert-Beer law can not be directly used and equation (1) must be modified applying the addition of the above law.

$$I(\lambda,l) = I_0(\lambda)\exp[\int_{l=0}^{l=L}(\sum_{i=1}^{m} c_i\sigma_i(\lambda,T,P) + \varepsilon_R(\lambda,l))dl] \quad (2)$$

Though the Mie scattering reduced by dust particles and Raileigh scattering reduced by gases don't belong to absorption process in equation (2), diffused light derived from them can not arrive at the detector. So the Mie and Raileigh scattering can be equally viewed as the absorption process[10]. $\sigma(\lambda)$ and c_i denote the absorption cross section depending on temperature and pressure at wavelength λ and the concentration at light path l of the ith gas respectively. $\varepsilon_R(\lambda,l)$ and $\varepsilon_M(\lambda,l)$ [10-11] denote the Raileigh and Mie scattering at wavelength λ and light path l respectively. The absorption cross section is split in two parts[10-13] which include the narrow one $\sigma_i'(\lambda)$ varying rapidly with wavelength and the broad one σ_{ib} varying slowly with wavelength and then can be expressed as:

$$\sigma_i = \sigma_i' + \sigma_{ib} \quad (3)$$

If equation (3) is put in equation (2), then calculation model of gas concentration can be deduced as equation (4) with the interference of the other gases and dust particles.

$$I(\lambda,l) = I_0(\lambda)\exp[-\int_{l=0}^{l=L}(\sum_i c_i\cdot\sigma_{ib}(\lambda,T,P)+\varepsilon_R(\lambda,l) + \varepsilon_M(\lambda,l))\cdot dl]\exp[-\int_{l=0}^{l=L}(\sum_i c_i\cdot\sigma_i'(\lambda,T,P))dl] \quad (4)$$

EXPERIMENT SETUP

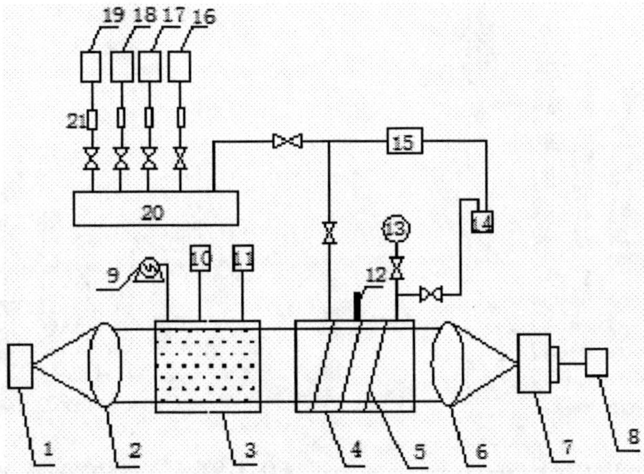

FIGURE 2. schematic diagram of gas concentration measurement based on DOAS

1 xenon lamp; 2、6 lens; 3 dust cell; 4 measurement cell; 5 belt for heating; 7 spectrometer; 8 pc; 9 compressor; 10 dust generator; 11 dust collector; 12 thermocouple; 13 manometer、14 bottle for absorbing flue gas; 15 flue gas analyzer; 16~18 NO、NO_2、SO_2 standard gases; 19 pure nitrogen gas; 20 pipe for mixing gases; 21 flowmeter;

The experiment setup is shown as diagram 2. The impulse and short arc xenon lamp is selected as light source which can supply the continuous spectrum between 200 and 1100 nm and the fiber spectrometer presented by Avantes company in Netherlands is applied to record the spectrum with resolution of 0.4 nm. The upper part of diagram 2 is used for obtaining the mixture of different component concentration and the lower part is used for measurement. In order to obtain the mixture of different component concentration and make gases completely mixed before entering the measurement cell, the standard gases and pure nitrogen gas flow into flowmeters in proportion and then are mixed in a 0.5-meter-length round pipe. The mixture is split in two: one passes through the flue gas analyzer and the other passes through measurement cell, finally both enters the bottle for absorbing gases. The ultimate concentration of gases is based on the data presented by analytical instrument for flue gas. The work flow of measurement part is described as below: Firstly, the light emitted by xenon lamp enters the 0.5-meter-length dust cell through optical fiber after it is collimated by lens 2, then the light enters the same length measurement cell, finally the light is received by the optical fiber spectrometer. Simultaneously, it is transformed into electric signal from light signal and processed by computer. It can be concluded from equation (2) that directly adding dust particles into measurement cell is the same as letting light passes through the same length dust cell before entering measurement cell. The system of adding dust particles is made up of dust generator, dust collector and small compressor. The compressor is used for generating turbulence to prevent dust particles of large diameter from depositing.

EXPERIMENTAL RESULTS AND DISCUSSIONS

Based on the experiment setup and algorithm in the article, the measurement of gas concentration is carried out according to real spectrum data which is mean value of many times measurement and acquired at normal temperature and pressure. The measurement results are shown from diagram 3 to 6 and in table 1 and 2. The measurement of sulfur dioxide concentration is performed from 290 to 310 nm wavelength region. If the concentration measurement of sulfur dioxide and nitrogen dioxide is carried out simultaneously, the wavelength region is from 290 to 500 nm.

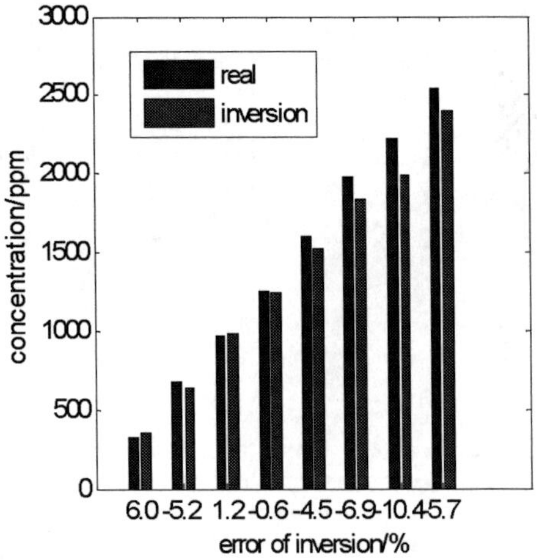

FIGURE 3. Measurement of sulfur dioxide concentration without other gases and dust particles

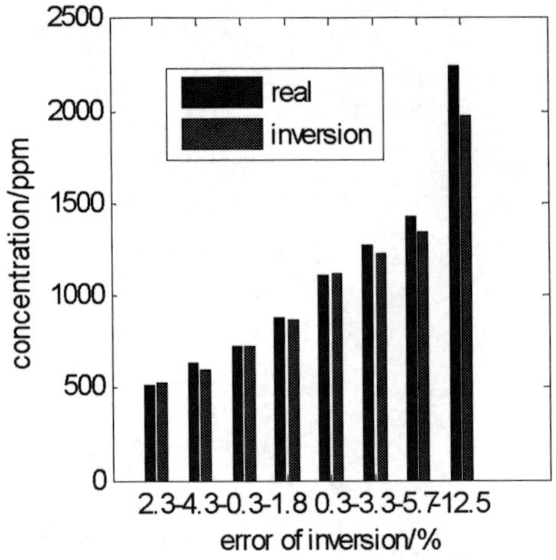

FIGURE 4. Measurement of sulfur dioxide concentration with interference of nitric oxide

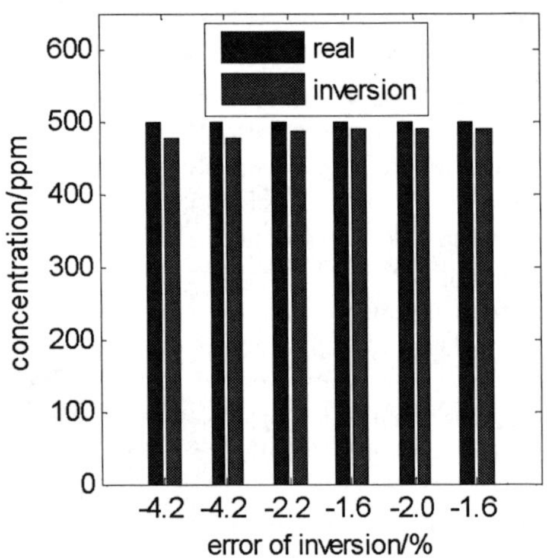

FIGURE 5. Repeated measurement of sulfur dioxide concentration With interference of nitrogen

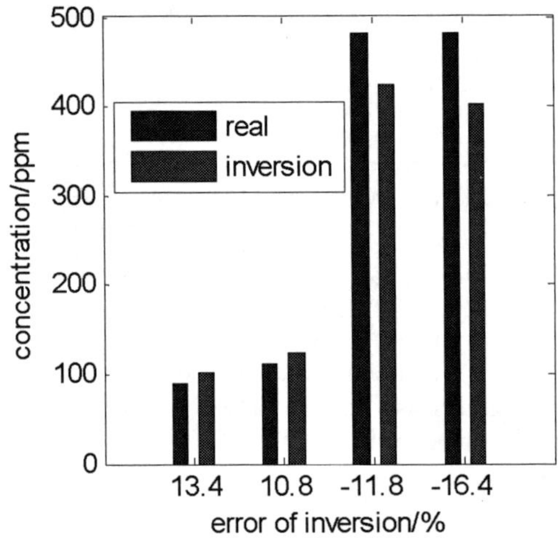

FIGURE 6. Measurement of sulfur dioxide concentration With interference of dust particles

As figure3 shows, the error of measurement result is small when the range of sulfur dioxide concentration is from 330 to 2550 ppm and generally will not exceed 10 percent. The response time of measurement is within 2 seconds. It is found that the error of measurement result will increase rapidly to 25 percent when the concentration of sulfur dioxide exceeds 4000 ppm. The reason is that the foundation of DOAS will not be applicable when the concentration is too high and the Lambert-Beer law is based on the premise that the concentration is neither too high nor too low i.e. when the absorbency is between 0.155 and 1[14], the relative error of concentration measurement is small and varies slowly. Figure4 is the measurement results of sulfur dioxide concentration between 520 and 2250 ppm with interference of nitric oxide. Conclusions can be drawn that interference of nitric oxide can be eliminated with the algorithm based on DOAS. In figure4, when the concentration of sulfur dioxide is 2250 ppm, the measurement error has climbed to –12.5 percent. After a host of experiments, it is found that huge error will be introduced by error of light path length when the measurement cell is not full of sulfur dioxide. Figure5 shows the results of repeated experiment with interference of nitrogen dioxide when concentration of sulfur dioxide is 500 ppm. It is demonstrated that the algorithm can realize fast and precise measurement and has good performance of anti-jamming. The measurement error gradually decrease to –1.6 percent with sulfur dioxide full of measurement cell slowly. The concentration measurement of sulfur dioxide is shown with interference of dust particles of high concentration in figure6. Measurement errors shown in figure6 are larger than the ones in figure3, figure4 and figure5 and the largest error of absolute value goes up to 16.4 percent. There are two aspects that lead to these large errors. One is there is no enough time to make the measurement cell full of gases before measurement so the real value of light path length is different from that used in calculation and the other is the concentration of dust particles is so high that the received intensity decreased rapidly compared with original light intensity. Figure7 illustrates received light intensity decreases to 20 percent of original one when concentration of sulfur dioxide is 111 ppm with interference of dust particles.

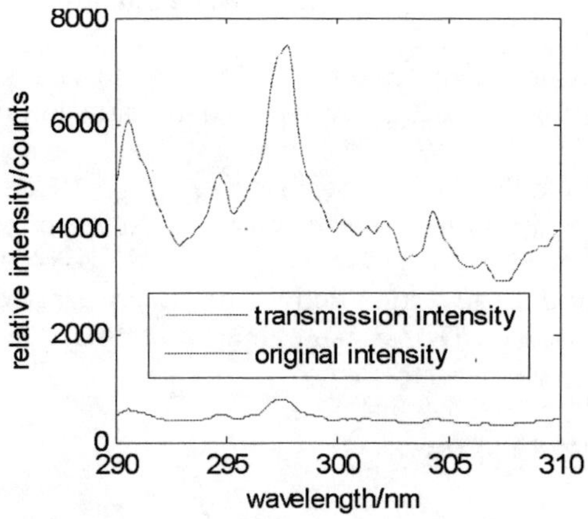

FIGURE 7. comparison between original and received light intensity

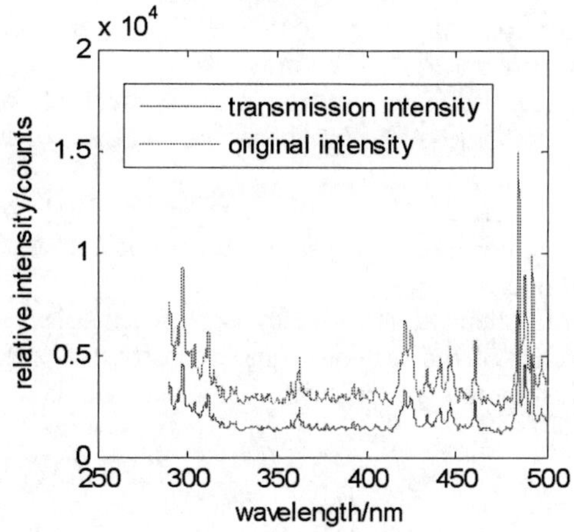

FIGURE 8. comparison between original and received light intensity

Concentration measurement results of sulfur dioxide are shown from figure3 to figure6. According to these results, conclusions can be drawn that interference introduced by other gases and dust particles can be eliminated with the algorithm based on DOAS. The results of simultaneous concentration measurement of sulfur dioxide and nitrogen dioxide are illustrated in table1 and table2. Table1 is carried out without dust particles. The algorithm can simultaneously measure concentrations of many gases precisely and the results can be achieved repeatedly. The largest measurement error of nitrogen dioxide is -14.4 percent and the reason is that measurement cell is not full of gases before measurement. Compared with table1, table2 considers effects of dust particles. Even though the concentration of dust particles is very high and concentrations of gases are very low, the measurement results are still precise. Being low concentration of gases, short light path length and high concentration of dust particles, received light intensity decreases to 50 percent of original one shown in figure8 when concentrations of sulfur dioxide and nitrogen dioxide are 90 and 183 ppm, 26 and 52 ppm respectively. So there is large errors in the measurement results. The concentration measurement of sulfur dioxide is more precise than that of nitrogen dioxide for the reason that the latter will decompose[15] after illumination of xenon lamp.

TABLE 1. Simultaneous measurement of sulfur dioxide and nitrogen dioxide without dust particles(ppm)

SO_2			NO_2		
real	inversion	error(%)	real	inversion	error(%)
500	492	-1.6	487	466	-4.3
	495	-1		417	-14.4
	490	-2		494	1.4
	515	3		492	1

TABLE 2. Simultaneous measurement of sulfur dioxide and nitrogen dioxide with interference of dust particles(ppm)

SO_2			NO_2		
real	inversion	error(%)	real	inversion	error(%)
500	502	0.4	487	456	-6.4
241	239	-0.8	487	457	-6.4
90	84	-6.7	183	142	-22.4
56	57	1.8	112	113	0.9
26	25	-3.8	52	70	34.6

CONCLUSIONS

Conclusions can be drawn according to the analysis and discussions of above measurement results:

1) Errors of measurement results are small and vary slowly when the concentration of sulfur dioxide is between 20 and 3000 ppm. Generally speaking, they will not exceed 10 percent.
2) Concentrations of sulfur dioxide and nitrogen dioxide can be measured simultaneously at high precision and fast response speed of 4 seconds.
3) Compared with concentration measurement of sulfur dioxide, larger errors will be introduced to nitrogen dioxide with the algorithm based on analysis of spectrum for the decomposition after illumination of xenon lamp.
4) Distinct errors will be introduced by error of light path length to the measurement results.
5) Satisfying measurement results can still be achieved when received light intensity decreases to 50 percent of the original one.

In a word, when the absorbency is between 0.155 and 1 and the attenuation derived from dust particles is less than 50 percent of the original light intensity, the algorithm based on DOAS can eliminate interference of other gases and dust particles to realize in-situ, real time and high precision measurement.

ACKNOWLEDGEMENTS

The authors would like to acknowledge the financial support of National Natural Science Foundation of China (Project No. 50576010).

REFERENCES

1 X.Xiangde, Z.Xiuji, S.Xiaohui, 35(supplement 1), pp. 1-19, 2005.

2 State Environmental Protection Administration of China, General Administration of Quality Supervision, Inspection and Quarantine of the People's Republic of China, No.2, pp.7-11, 2005.

3 M.JOHAN, R.ARNE, Journal of Quantitative Spectroscopy and Radiative Transfer, Vol.56, pp.187-208, 1996.

4 M.JOHAN, R.ARNE, Journal of Quantitative Spectroscopy and Radiative Transfer, Vol.56, No.2, pp. 209-224, 1996.

5 M.JOHAN et al., Journal of Quantitative Spectroscopy and Radiative Transfer, Vol.56, No.2, pp. 225-240 , 1996.

6 U.Platt et al., Journal of Geophysical Research, Vol.84, No.10, pp. 6329~6335, 1979.

7 Fu Jingye, Cai Xiaoshu, Zhang Yurong et al. System for Emissions Pollutant (SO_2) On-line Measurement and Monitoring. ACTA METROLOGICA SINICA, No.2, 2003.

8 Z.Zhaohui et al. Measurement of Flue Gas with Differential Optical Absorption Spectroscopy. Chinese Journal of Quantum Electronics, Vol.18, Supplemet, pp.65-69, 2001.

9 X.Pinhua et al, ACTA PHOTONICA SINICA, Vol.29, pp.271-276, 2000.

10 U.Platt, Air mornitoring by spectroscopic techniques. Switzerland: JOHN WILEY & SONS, 1994.

11 M.C.Plane and S.Nicola, Spectroscopy in environmental science, Switzerland: JOHN AND SONS, 1995.

12 Z.Yujun et al., Proceedings of SPIE, pp.4897, 2003.

13 J.Stutz and U.Platt, APPLIED APTICS, Vol.35, pp.6041-6035, 1996.

14 L.Quanchen and J.Yuejuan, Theory of spectroscopic instrument. BEIJING: BEIJING INSTITUTE OF TECHNOLOGY PRESS, 1999, pp.13.

15 L.Zafonte et al., *Environmental Science and Technology*, Vol.11, pp.483-487, 1977.

PIV Experimental Study On Electrostatic Spray Of Pressure-swirl Atomizer With Annular Electrode

Junfeng Wang [*] [+,1], Wook Ryol Hwang [*] and Tiqian Luo [+]

[*]*Research Center for Aerospace Parts Technology, School of Mechanical and Aerospace Engineering, Gyeongsang National University, 660-701, Jinju, South Korea*
[+]*School of Energy and Power Engineering, Jiangsu University, 212-013, Zhenjiang, China*

Abstract. The middle-size electrostatic spray generated by pressure-swirl atomizer and annular electrode, which could be applied in inside-room disinfections, was investigated through PIV experiment. PIV was applied to measure the flow of charged atomized particles. The flow field of uncharged and charged spray was compared and the change of spray flow filed was analyzed under different charging voltage. The experiment results indicate that the spray angular of charged spray increased and the particles disperse apparently compared with conventional spray. The velocity of charged particles in spray core become higher and the motion of particles in the domain around the electrode appears almost different change law when the spray was charged. The flow description of the charged particles spray under different charging voltage was presented in this paper. The measurement and analyses presented a research base for the application of electrostatic spray of the pressuer-swirl atomizer with an annular electrode.

Keywords: electrostatic atomization, spray, PIV, charged multi-phase flow
PACS:

INTRODUCTION

Due to many advantages, electrostatic atomization has been already applied in many spray fields such as agricultural spray, which can obtain more fine droplets and improve the deposition of pesticide on the plants so as to avoid more pollutant deposition in soil [1-2]. In industrial field, this technology was also applied in spray coating and combustion [3], dust control and etc. When the droplets were charged, the decreases of surface tension make the liquid break into more fine particles and the spray disperse more uniformly under the action of the electrical force. The electrostatic force make the motion of charged particles more complex for the couple of electricity field and flow filed especially in the case of turbulent multi-phase flow, which is so called charged multi-phase flow. Since the electrostatic atomization was found for more than one hundred years, many researches on its application and basic principle have been reported [4]. Normally the electrostatic atomization was generated after the liquid was atomized by conventional atomization method for example pressure atomization or gas-liquid twin-fluid atomizer. Hence electrostatic atomization actually happened in most industrial application is a kind of second time atomization. Of course some cases of micro spray application or basic theoretical research was also based on capillary [4]. The characteristics of atomization is almost different due to the atomizer's difference for electrostatic spray. At the same time, the differences of the application field make the properties of charged multi-phase flow also different completely. In industrial application, the electrostatic spray is usually in big size or middle size for example agriculture spray and combustion. Some basic research presented in literature mostly are based on the capillary's micro spray and it is still not quite clear for people to understand exactly the motion of charged particle with different mechanical atomizer in a big spray size due to the complex couple action of electricity field and flow field. Another factor affecting the motion of droplets is the mode of charging. Different charging method and the shape of electrode for electrostatic spray make the charge to mass ratio of droplets and the electrical field different, which

[1]Corresponding author: wangjunfeng@ujs.edu.cn

CP914, *Multiphase Flow: The Ultimate Measurement Challenge,* edited by X. Cai, Y. Wu, Z. Huang, S. Wang, and M. Wang

directly decided the motion of charged droplets. With the application spread of the electrostatic spray, the requirement of more deeply understanding of its flow field has been put forward to the researchers.

During the period of SARS (Severe Acute Respiratory Syndrome) in 2002, the electrostatic spay was begin to applied successfully in disinfections in China. In this application of charged spray, pressure swirl atomizer and annular electrode were used and the middle-size charged multi-phase flow is generated when fanner transported the charged droplets. In this paper the middle-size electrostatic spray generated by swirl atomizer and annular electrode, which could be applied in inside-room disinfections, was investigated through PIV experiments. An electrostatic spray experimental equipment with 0~13Kv charging system was built. Particle Image Velocimetry was applied to measure the flow of charged particles.

EXPERIMENTAL SYSTEM AND TECHNIQUES

The electrostatic sprayer (see Fig.1) we used in the experiment is a kind of annular (or corona) electrode inducting charging sprayer developed for disinfection, which is applied at the first time in the spraying equipment during the period of SARS (Severe Acute Respiratory Syndrome) in 2002, in China. The atomizer is a kind of pressure-swirl nozzle. By changing the distance between nozzle and electrode, a proper distance could be obtained to make the spray be charged effectively. To guarantee the effective charging of spray in the experiment, the charge to mass ration was measured through the equipment as Fig.2 shows and the value of charge to mass ration is 0.8mC/kg under 13kV charging voltage. A tap water used as a test liquid has the following electrical and physical properties as shown in table1. At a certain operating pressure of 0.3MPa and the specific flow rate of 2L/min, various charging voltages range from 0 to 13Kv by an AC high voltage electrostatic generator are used to obtain uncharged and charged spray. The spray and PIV measurement system is show in Fig.3. To protect the measuring and controlling equipments, all the apparatus such as computer, laser generator and synchronization et al are connected with ground to avoid the breakage caused by electrostatic induction. Two double-pulsed ND: Yag laser were used to emit a frequency-double beam, with a wavelength of 532nm, 120mJ energy per pulse. The CCD camera has a total of 1280×1048 pixels and a Nikon 60mmf/2.8D Micro Nikker lens was used to accompany with the camera.

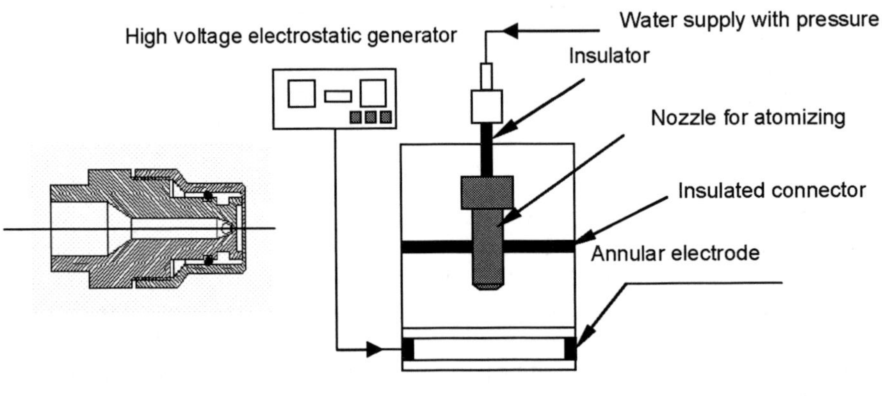

(a) (b)

FIGURE 1. Schematic of (a) pressure swirl nozzle; (b) annular electrode charging sprayer.

TABLE 1. Here Properties of fluid used in the experiments.

Surface tension (N/m)	Electrical conductivity (S/m)	Electrical permittivity (C^2/N·M)	Dynamic viscosity (kg/m.s)	Density (kg/m^3)
7.4×10^{-2}	4.5×10^{-5}	7.10×10^{-10}	9.0×10^{-2}	998

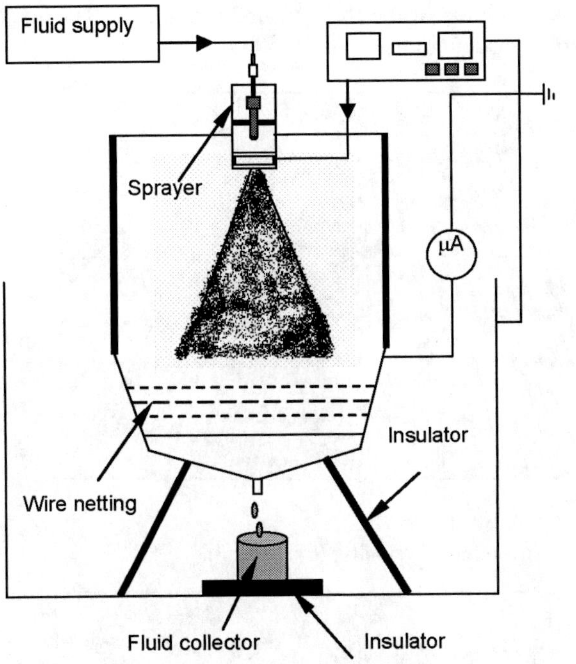

FIGURE 2. Equipment of charge to mass ratio.

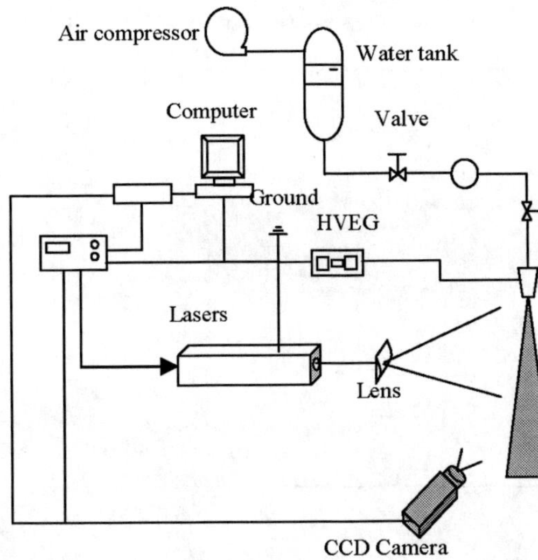

FIGURE 3. Sketch of charged spray and PIV measurement system.

The insight 6.0 software of TSI are applied in the PIV system, in which a direct cross-correlation interrogation method was used to analyses the spray image [5-6]. The interrogation window size is 36×36 pixels and the real size of the spray caught by the CCD camera is approximately 150×180mm2. We obtain 120 photos which supports 60 flow fields for every spray case.

RESULTS AND DISCUSSION

Charged And Uncharged Spray Flow Structure

In the present study, in order to observe the drop flow at the edge of the spray, the area with size of 150×180 mm2 around the nozzle was chosen as an observation domain which is deflected to one side of the central axle. Fig.4 shows the velocity vector distribution and streamline of uncharged spay. It can be seen that the highest velocity of the swirl spray is not in the spray center but near to the edge, which is different with the conventional jet flow. The advancing speed of the spray decreases with the spray time went on especially in the latter of the spray. The fluid around the nozzle was absorbed apparently by the spray due to the shear action of the high speed jet. From Fig.4, it can be found clearly that the streamline at the left upper corner shows the flow of the fluid around the nozzle.

Fig.5. shows the velocity vector distribution and streamline under 5kV charging voltage. Apparent change of the spray flow can be found when the spray was charged. The flow of the fluid around the nozzle changed completely under the electrical force. The strong electrical field exists around the nozzle and the charge droplets with low speed move along the electrical force direction (see Fig.4.b). So the movement of the charged droplet will affect the flow of the fluid around the nozzle as well as the absorption of the gas around the spray. At the same time, the droplets were accelerated by the electrostatic force in the center of the spray, which could be found easily through comparing the vector distribution of Fig.4 and Fig.5. It is interesting that the streamline of the spray flow around the nozzle under 5kV shows an opposite flow direction of uncharged spray. But under the acceleration of jet flow caused by electrostatic force, whether the air absorption of the spray weakened or not need to be studied more deeply because the air absorption of the spray affect the quality of spray combustion directly in a big degree. But it could be made sure that the spray angle increased due to the change of the flow direction of the fluid around the nozzle. The above result was also caused in some degree by the exclusion of charged particle in the spray center.

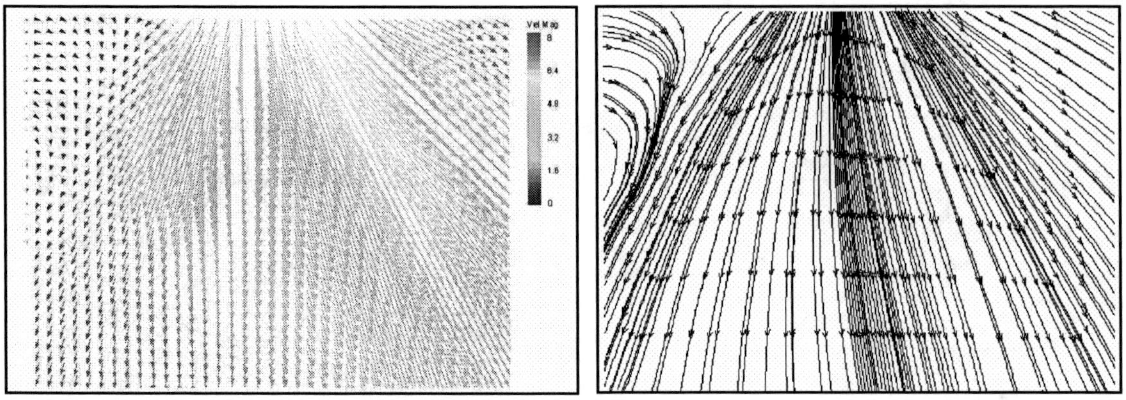

(a) (b)

FIGURE 4. Uncharged spray field: (a) vector distribution of velocity; (b) streamline.

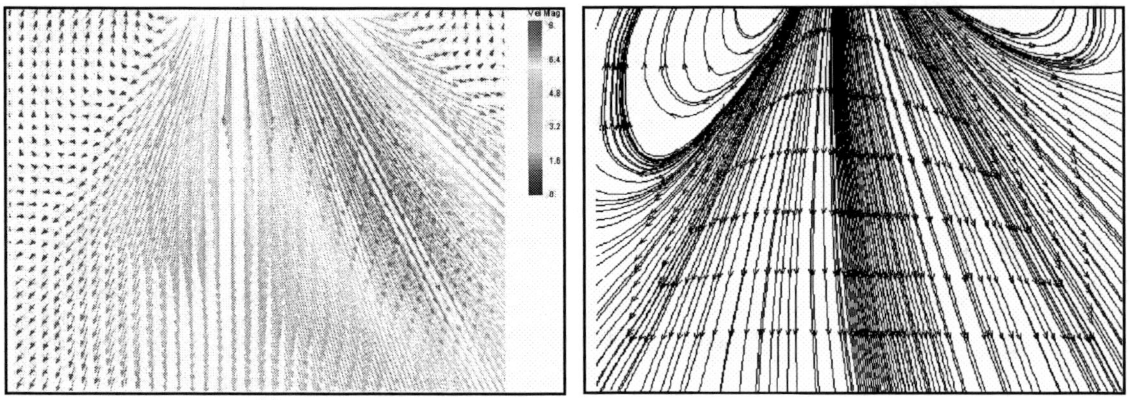

(a) (b)

FIGURE 5. Charged spray under 5kV: (a) vector distribution of velocity; (b) streamline

Charged Spray With Different Voltage

To investigate the charged spray under different charging voltage, we increase the charging voltage and get the flow information in the case of 5kV, 8kV, 10kV and 13kV charging voltage. When the charging voltage is beyond 13kV, the charging system begins to breakdown due to the discharge of electrode in the air. Fig.6 shows the velocity contour and streamline of charged spray under the voltage of 5kV, 8kV, 10kV and 13kV. It is interesting that the maximum velocity of the spray fluctuates with the increase of the charging voltage. It is easy to be found that the spray seemed to be accelerated when it was charged. But with the continuous increase of the charging voltage, the maximum velocity decrease oppositely under the voltage of 8kV. But the flow velocity of the fluid around the nozzle increase and the flow direction keep the same. In the case of 10kV, the maximum velocity of the charged spray become larger than the case of 8kV and seemed to be similar to the case of 5kV. In the center of the spray, the velocity in the case of 5kV and 10 kV appear to a similar distribution but the streamline shows a great difference. This change happened when the charging voltage reach 10 kV and keep the same situation to the voltage 13kv. From the distribution of the velocity contour of the fluid around the nozzle, it could be found that, the velocity of this area increase continuously with the increase of charging voltage.

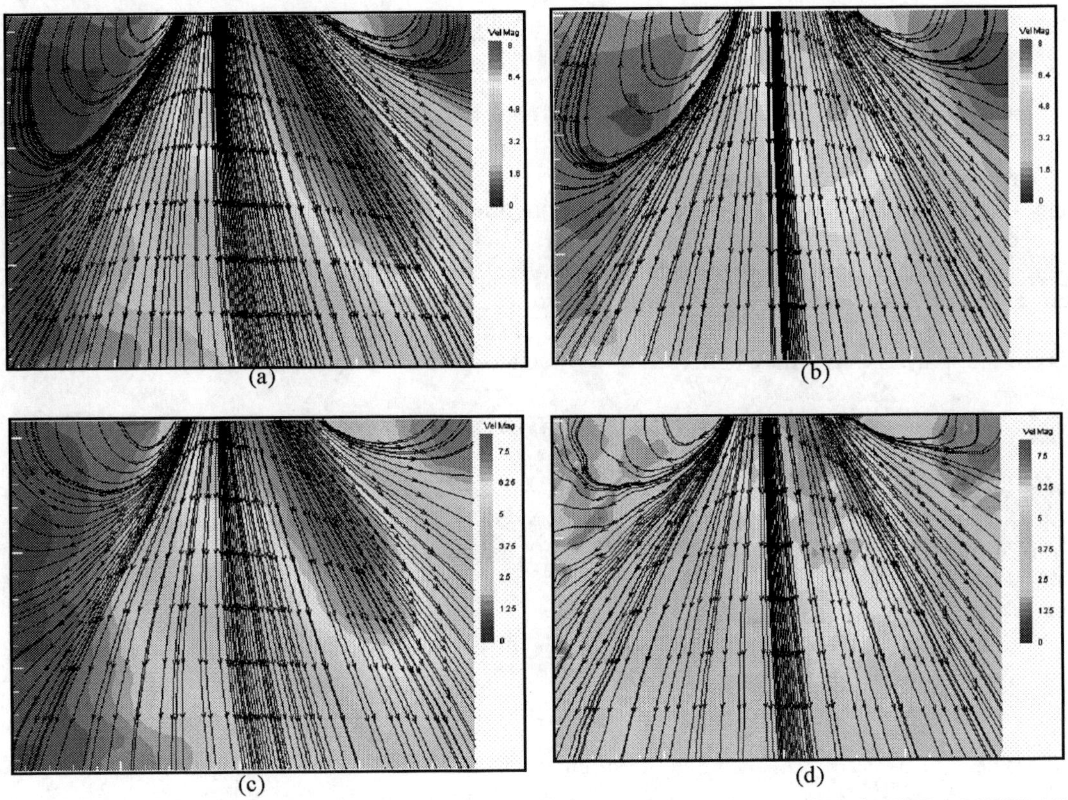

FIGURE 6. Velocity contour of charged spray under different voltage. (a) 5kV; (b) 8kV; (c) 10kV; (d) 13Kv

Therefore, with the charging voltage increase, the change of spray velocity in the area of the spray center and the area around the nozzle appear different law. The former shows a fluctuating change and the latter increase linearly. That means when the charge to mass ratio reach a certain value, the moving direction of the droplets around the nozzle could be change by the electrical force. This change is different with the situation of changing from uncharged spray to charged spray. The streamline of uncharged spray shows that the jet spray absorbs the fluid around the spray and the kinetic energy of the center spray are cut down by the absorbed fluid. When it was charged such kind of absorption seemed to disappear or weakened, which make the spray save more energy for transportation. Except for the action of electrical force, it is another reason why the spray was accelerated when it was charged. Normally, the kinetic energy of the particles around the nozzle is relative smaller than the one of center spray and their moving direction are easy to be change by the electrical force. With the increase of charging voltage, their moving direction is changed since electrical force on the particles is strengthened. From the comparison of streamline in the area (left upper corner of the figure) around the nozzle between the case of 5kV, 8kV and 10 kV, 13kV, an interesting phenomenon we could found is that the curvatures of these streamlines decrease when the charging voltage increases. It is a contravention that the strengthened electrical fore would make the droplets move along the electrical force direction when the charging voltage increase. One reason is that the droplets diameter distribution around the nozzle is also changed when the charge to mass ration increased. Based on the former research, the change of charge to mass ration is a main factor which mainly affects the Coulomb-break of the droplet. But till now people don't know clearly the mechanism of the affection of charge to mass ration on the droplets diameter distribution. Another reason is that the change is a results affected by the air movement around the spray. It need more deeply research in the future combined with the diameter measurement of charged spray.

According to some exiting research, the spray angle enlarged when the spray was charged [1,7]. It could be found clearly in the present experiment that the spray expanded when it was charged due to the repulsion of charged particles. Actually, when the spray is charged, it is very difficult to define the boundary of the spray especially under the annular inducting electrode since the charged droplets have high repulsion action with each other. In the center of the spray, the density of the particles is less than the uncharged one and the particle seem to distribute more uniformly. But in the area near the nozzle, the particle density increase and seem to distribute in the whole area outside the spray zone near the nozzle. So how to define the spray angle under charged cases of spray is also need to be more discussed.

CONCLUSIONS

A kind of electrostatic sprayer with annular electrode which could be applied in disinfection is developed and the flow of the spray is measured through PIV system. By changing the charging voltage, the flow properties of spray affected by voltage are analyzed and the comparison of flow field of the charged and uncharged spray is presented. The research indicates that the electrical filed could apparently change the flow field of spray especially the air absorption around the spray. The flow in the area of spray center and around the nozzle appears different change under the action of charging voltage. Through comparing with the conventional spray, the electrical field makes the charged droplets of the spray disperse more uniformly and cause an apparent spray expansion.

ACKNOWLEDGEMENTS

Authors here would like to acknowledge the Korea Research Foundation Grant funded by the Korean Government which supported this work (KRF-2005-005-J09902). The research was also funded partially by the National Natural Science Foundation of China (No.59776028) and Jiangsu Province Natural Science Foundation, China (No.03KJD4300064).

REFERENCES

1. G.N. Laryea, S.Y. No. Development of electrostatic pressure-swirl nozzle for agriculture applications. J Electrostatics 2003; 57, pp. 129-142.
2. J.F. Gaunt, J.F. Hughes, N. M. Harrison. Electrostatic deposition of charged insecticide sprays on electrically isolated insects. J Electrostatics 2003; 57, pp. 35-447.
3. J.F Wang. A fundamental investigation of electrostatic spray combustion. Ph.D thesis 2002, Jiangsu University,China.
4. R.P.A.Hartman, J.P. Borra, D.J. Brunner et al, The evolution of electro-hydrodynamic sprays produced in the cone-jet mode, a physical model. J Electrostatics 1999; 47, pp. 143-170.
5. Insight 6.0 software and user guide, TSI.corp. 2004.
6. C.E.Willert, M.Gharib. Digital particle image velocimetry. Exp Fluids 1991, 10, pp. 181-193.
7. G.N.Laryea, S.Y. No. Spray angle and breakup length of charged-injected electrostatic pressure-swirl nozzle. J Electrostatics 2004, 60, pp. 37-47.

A Piv Measurement and Numerical Simulation of Gas-Liquid Two-Phrase Flow in An Aeration Tank

Wen CHENG[1,2], Chun-di YANG [2], Yuichi MURAI [3], Fujio YAMAMOTO[4]

[1] *State Key Laboratory of Multiphase Flow in Power Engineerig; Xi'ann Jiaotong University, Xi'an 710049, China, Email wencheng@xaut.edu.cn*
[2]*Environment Research Institute, Xian University of Technology, Xian 710048, China*
[3]*Div. of Mec. Science, Graduate School of Engineering, Hokkaido University, Japan, murai@eng.hokudai.ac.jp*
[4]*Graduate School of Engineering, University of Fukui, Bunkyo 3-9-1, Fukui 910-8507, Japan, yamamoto@fv.mech.fukui-u.ac.jp*

Abstract. Aeration plays an important role in the treatment of activated sludge. The aeration performance is deeply affected by the movement law of bubble liquid flow in aeration tank. In order to discuss the influence of gas-phase, liquid-phase motions in an aeration tank on the sewage disposal, three kinds of boundaries and initial conditions were applied in the experiment to study the law of two-phase flow. A developed two-equation turbulence model was employed in this paper. Results obtained by experiment are compared with those from the mathematical model. It is found that the results of measurement and simulation agree with each other.

Keywords: aeration tank experiment numerical simulation

INTRODUCTION

Aeration plays an important role in the treatment of activated sludge due to the interactions among bubbles and sewage in an aeration tank. The aeration performance is concerned directly with the efficiency of sewage disposal. Both physical experiments and numerical simulations have been done to improve the efficiency, some new methods were offered by experts and scholars. Since the irregularity motion in two-phase flow [1] is higher than that in single-phase turbulent flow, there remain a lot of unclear phenomena at present about the exchanges of mass, momentum and energy between phase as well as the interaction among turbulent components. However, the two-equation model [2] has better predictability in practice in comparison with other models from engineering point of view. The oxygen transfer efficiency in the aeration tank, which is the key reactor of the activated sludge treatment system, has immediate relations with the cost and efficiency of the wastewater treatment. Based on the achievements of other scholars and the observation results of the two-phase flow experiment, this paper aims at establishing the two-equation turbulence model. Thus it provided theoretical basis for the rational designing, accessing and controlling of the aeration tank operation.

PIV EXPERIMENTAL SET-UP AND NUMERICAL SIMULATION

Experiment Set-up

Fig.1 shows a schematic diagram of the present experiment set-up. The plane bubble plume is generated in a rectangular test tank 300 mm in width, 800 mm in height, and 40 mm in horizontal depth. The tank is filled with tap water at room temperature, and hence the bubble interface should be considerably contaminated during the experiment. The air bubbles are injected into the bottom of the tank through three rows of capillary tubes. Each row has 17 tubes with external diameters of 0.19 mm and internal diameters of 0.13 mm. The tubes produce bubbles of around 1.0 to 2.5 mm in diameter dependent on the gas flow rate. The width of the bubble injection area is 80 mm.

A CCD camera is fixed in front of the test tank at a distance of 4 m. The shooting area covers the whole tank.

CP914, *Multiphase Flow: The Ultimate Measurement Challenge,* edited by X. Cai, Y. Wu, Z. Huang, S. Wang, and M. Wang

Two halogen lights are fixed on the back of the tank. The positions of the lights are determined by the illumination angle of the spherical bubbles - that is, around 115^0 between the camera's viewing axis and the lighting direction. At this angle all bubbles in the tank scatter the light well in the direction of the camera regardless of the bubble size. If instead a backlighting method were to be applied, the bubble centroids and the shapes would be better captured. However, this is not a practical way for capturing the whole region of the tank because a major part of the bubble shadow image does not show up in the CCD image due to defocusing and overlap of the shadows. In the case of high void fraction, we confirmed that the direct lighting method at the chosen illumination angle led to better recording of bubble motions.

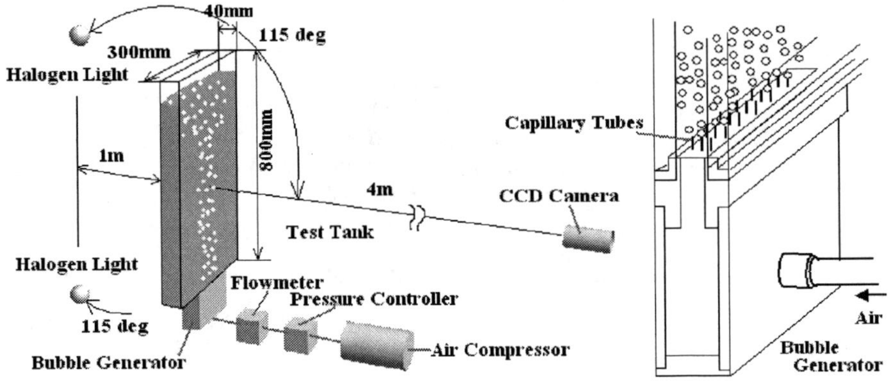

FIGURE 1. Experimental set-up

TABLE 1: Experimental conditions

Case	Q/ $(10^{-5}m^3/s)$	H/m
Liquid: water; Gas: air (ρ:1.28 kg/m3, T: 13-15℃)		
Coefficient of kinematics viscosity （υ）： 10^{-6} m2/s,		
Case 1	0.99	0.451
Case 2	7.04	0.617
Case 3	8.88	0.31

Changing the gas flow rate and the water lever in the tank, various structures and unsteady flow patterns appear. The gas buddle behaviors are recorded via the CCD camera at a frame rate of 30fps. 1200 consecutive images are processed to calculate the time related statistic quantities of the bubble velocity fields [3].

Numerical Simulation

Mathematical model

With the development of computer and experiment measurements, making use of numerical simulation is more and more important in studying two-phase flow. At present the most popular model applied solves phases cordially. In this paper the Inter Phase Slip Algorithm [4] is applied.

The scalar diffusion equation, continuous equation, momentum equation and energy equation were introduced by Cheng et al [5]. While bringing in these equations, many other governing conditions were used to complete the building of model.

The state and basic equation can be written as following:

$$\rho_\alpha = \rho_\alpha(T_\alpha, p_\alpha) \tag{1}$$

$$h_\alpha = h_\alpha(T_\alpha, p_\alpha) \tag{2}$$

Governing equation of the two phases' volume fraction is

$$r_\alpha + r_\beta = 1 \tag{3}$$

Where ρ is density; U is liquid velocity; P is intensity of pressure; B is volume force; T represents temperature; t is time; μ is dynamic viscosity; Γ is drag coefficient; λ is heat exchange coefficient. α 、β represents gas and liquid phase respectively. Changing α for β , liquid phase equation is obtained. N_P is the total number of phases. γ_α is volume fraction of gas phase and γ_β is the liquid phase's. H_α is the static enthalpy, $H_\alpha = h_\alpha(T_\alpha)$.

Pressure governing equation is

$$p_\alpha = p_1 = p \tag{4}$$

In the equation, $C^d_{\alpha\beta}$ is the drag parameter between α and β .

$$C^d_{\alpha\beta} = \frac{3}{4}\frac{C_D}{d}\gamma_\beta \rho_\alpha \left|U_\beta - U_\alpha\right| \tag{5}$$

C_D is a drag coefficient depending on Reynolds number, $C_D = C_D$ (Re)

$$\mathrm{Re} = \frac{\rho_\alpha U d}{u_\alpha} \tag{6}$$

Where u_α is the velocity of continuous phase α .

In laminar flow region, 0≤Re≤0.2, C_D is determined by storks formula:

$$C_D = \frac{24}{\mathrm{Re}}.$$

In transition region，0.2≤Re≤500-1000，the relation between C_D and Re was determined by different empirical formulas as following:

$$C_D = \frac{24}{\mathrm{Re}}\left(1 + 0.15\mathrm{Re}^{0.687}\right)$$

$$C_D = \frac{24}{\mathrm{Re}}\left(1 + 0.15\mathrm{Re}^{0.75}\right)$$

$$C_D = \frac{24}{\mathrm{Re}} + 5.48\mathrm{Re}^{0.573} + 0.36$$

In the turbulent region, 500-1000≤Re≤1-2×10^5, C_D =constant=0.44.

Grid

In order to fit the boundary shape of the aeration tank, the Delaunay triangulation method [6] was adopted to divide the calculation region into triangle meshes. An aeration tank model which aspect ratio is two is discredited into 247910cells, 123957grid nodes. With the bottom of tank is entrance, the upper device is exit, and liquid in tank is static. Keeping a certain gas flow, simulate the gas-liquid two-phase flow in an aeration tank.

RESULTS AND DISCUSSION

The proposed method is applied to the measurement of bubble velocity vector fields of bubble plumes. The applicability of typical PIV algorithm [7], [8] is examined. The target image is a bubble plume in a rectangular tank in strong turbulent flow conditions. Results obtained by different methods are shown in Fig.2 and Fig.3.

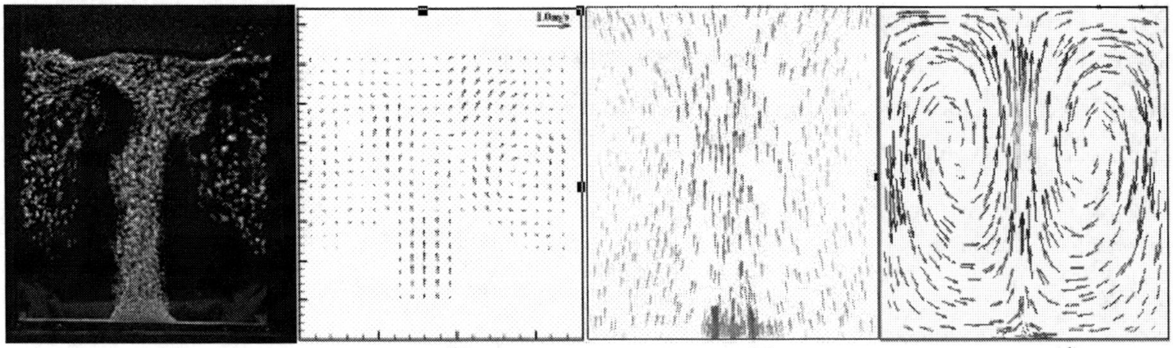

(a) (b) (c) (d)

FIGURE 2. Instantaneous velocity field of case

(a) Case 1

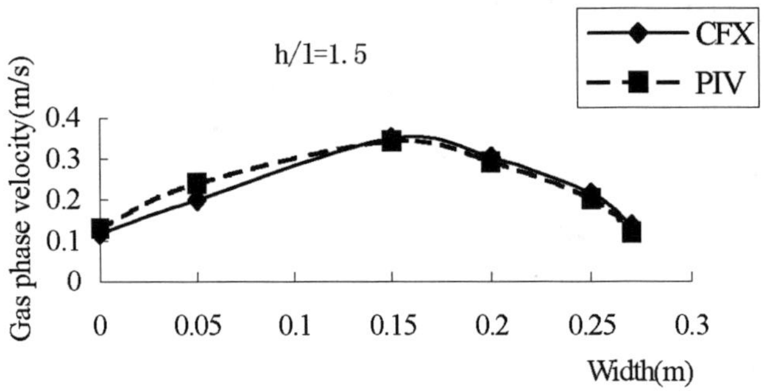

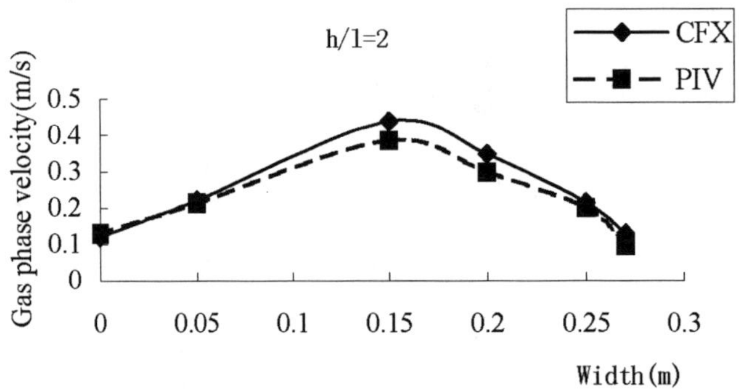

(b) Case 2

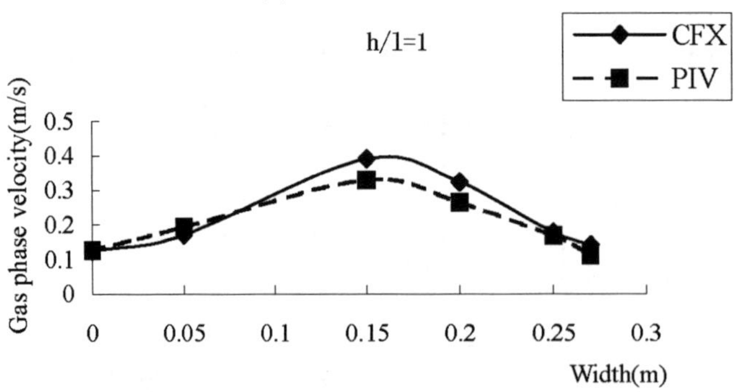

(c) Case 3

FIGURE 3. Velocity compare of the results of numerical simulation and PIV

Fig.2 (a) shows a time-dependent bubble distribution of case 3 taken by a CCD camera at t =25s. Bubbles rise along installment middle line. There are two circulation circles with "the cooling tower" vortex fluxion structure on both sides of the bubble plume. A small part of air bubble overflow installment, the majority of air bubbles enter stable vortex structure. Fig.2 (b) stands for the instantaneous velocity field gained by calculation using IFFFT (Inverted Fast Fourier Transform) PIV algorithm. Two circulation circles in velocity field were shown clearly, this inverse analytical method was applied properly. The instantaneous velocities of gas phase and liquid phase obtained by simulation are shown in Fig.2 (c) and Fig.2 (d), respectively.

Velocity compare of the results of numerical simulation and PIV was shown in Fig. 3. According to the comparison of velocity, the differences between the results of simulation and PIV are inconspicuous; bringing digital image processing technology into the aeration tank model is feasible. Reasons of errors existing are extensive; they mainly include the set of boundary conditions of numerical simulation, the selection of diagnosis window of PIV, errors in processing and installing experimental device, et al.

From the results we can also find that when diameters of capillary tubes keep invariable, the greater the simulated gas aeration tank vapor pressure, the two-phase gas-liquid mixing more strong, turbulent intensity getting bigger, more fully makes gas-liquid mass transfer, the higher the efficiency of sewage disposal we can get. When flows are in case 3, the bubbles in the equipment will have longest rest time and stronger turbulent intensity, so oxygen transferring speed and efficiency will be advanced greatly. When we design aeration tank, we must make gas-liquid two-phase flows in case 3.

Both of the results show the velocity field successfully, and they agree well with each other. This shows that the creation of gas-liquid two-phase flow model is a success.

CONCLUSION

A three-dimensional two-phase turbulent flow model has been established herein by generalizing the two equations model. The following points have been clarified from this study.

1) Different flow patterns of carrier phase could be generated depending on the boundary conditions at the bottom and top of the aeration. CFX software is used to simulate gas-liquid two-phase flow fields of aeration tank device. The results are found and offered as visualization form. The void fraction distribution in the aeration tank shows good agreement between the experimental data and results obtained from the present numerical model.
2) Bubble flux is the important factor of affecting the bubble size and velocity when water height and aperture are homology. When bubble flux or velocity increases, bubble volume becomes augmented, and the bubble velocity increases obviously. This conclusion laid a foundation for the theoretical research of the aeration tank operation and provided methods for the operation of different aeration systems under various operation conditions.

REFERENCES

1. Qiao Wei-yang, Wang Zhan-xue, Cai Yuan-hu, Efficient Cell Centered Multigrid Scheme for the Three-Dimensional Navier-Stokes Equations, Chinese Journal of Aeronautics, Vol.15, pp.193 –199, 2002.
2. W. Cheng, X. Zhou, C. Song, Experimental and Numerical Simulation of Gas-liquid Two-phase Flow in an Aeration Tank, J. Hydrondyn, Vol.12, pp.32-35, 2001.
3. Sumit R. Maloo, A Rule-Based Approach To Quadrilateral Mesh Generation, The University of Toledo (The Master of Science degree in Industrial Engineering), December.2001.
4. Kiger K T, Pan C, PIV Technique for the Simultaneous Measurement of Dilute Two-Phase Flows, Journal of Fluids Engineering, Vol.122, pp.811-818, 2000.
5. R J Adrian, Particle-Imaging Techniques for Experimental Fluid Mechanics, Annual Review Fluid Mechanics, Vol.23, pp.261-304, 1991.
6. Xiang Qun Song, The Survey of PIV and Its Applications in Multiphase Flow, Doctor paper in Fukui University, 1998.
7. Murai Y., Matsumoto Y., Song X. Q., Yamamoto F., Numerical Analysis of Turbulence Induced by bubble Buoyancy, JSME. Int. J. Ser. B, 43(2): 180-187, 2000.
8. Wen Cheng, Xiao-de Zhou, Ce Song, Tao Min, Experiment and Numerical Simulation of Three-phase Flow in an Aeration Tank, Journal of Hydrodynamics, Ser.B, Vol.4, pp.118-123, 2003.
9. Wallis G B, One-Dimensional Two-phase Flow, McGraw-Hill, New York, 1969.
10. Rudinger G, Fundamentals of Gas-particle Flow, Elsevier Scientific Publishing Co. Amsterdam, 1980.

Glare Spot Phase Doppler Anemometry

Camille Hespel*, Kuanfang Ren*, Gérard Gréhan* and Fabrice Onofri†

** UMR 6614/CORIA, CNRS - Université et INSA de Rouen, av. de l'Université, BP 12*
76801 Saint Etienne du Rouvray, email : fang.ren@coria.fr
† UMR 6595/IUSTI, CNRS-Université de Provence, Polytech' Marseille-DME,
Technopôle Château-Gombert, 5 rue E. Fermi, 13453 Marseille

Abstract. The Phase Doppler anemometry has been developed to measure simultaneously the velocity and the size of droplets. The measurement of the refractive index would be also interesting since it depends on the temperature and the composition of the particle and its measurement permits both to increase the quality of the diameter measurement and to obtain information on the temperature and/or the composition of the droplets. In this paper, we introduce a Glare Spot Phase Doppler Anemometry which uses two large beams. In this case, the images of the particle formed by the reflected and refracted light, known as glare spots, are separated in space. When a particle passes in the probe volume, the two parts in a signal obtained by a detector in forward direction are then separated in time. If two detectors are used the phase differences between two signals, the distance and the intensity ratio of reflected and refracted parts can be obtained and they provide rich information about the particle diameter and its refractive index, as well as its velocity. This paper is devoted to the numerical study of such a configuration with two theoretical models: geometrical optics and rigorous electromagnetism solution.

Keywords: Phase Doppler anemometry, glare spot, particle imaging, refractive index, dual burst PDA.
PACS: 42.25.Fx, 78.20.Ci, 40.80.Fg

INTRODUCTION

In the research of two phase flow, the measurement of the velocity and the size of the droplets is essential. The Phase Doppler Anemometry (PDA) is one of the most important and the most used techniques to reach this end because it gives the correlation between the size and the velocity. But in certain circumstances, as in spray combustion or mixing, we need also to have access to the particle temperature/refractive index or mixing fraction.

With an elastic light scattering based technique such as PDA or rainbow refractometry, the only way to get information about the temperature or the material properties of a droplet or a particle is to measure its optical refractive index. The rainbow refractometry can measure precisely the refractive index (until 2.0 10^{-3} for nearly mono-dispersed droplets[1]) and the mean refractive index for poly-dispersed clouds[2]). But it does not give information about the velocity of the particle.

There exits also non elastic light scattering based techniques such as the Laser Induced Fluorescence (LIF) or Raman spectroscopy, to measure the temperature and particle composition. But theses diagnostics are quite difficult to implement. LIF, for example, based on the dependence of fluorescence spectra on the temperature but dye must be added in the droplets. The properties of fluorescence also depends strongly on the pressure and composition of the particles. Furthermore, those techniques require complementary methods to access to the particle size and diameter (such as PDA). Then the *in situ* estimation of the refractive index of the scattering particles with the PDA measurement is a challenge to both increase the PDA size measurement accuracy and extract correlated information on the particle velocity, size, temperature and/or composition.

Different techniques have been developed[3,4], extended PDA[3,5], Dual Mode PDA[6], Dual Burst[7,8], coupling of PDA and rainbow refractometry[9], coupling of PDA and LIF[10] which give velocity, size and refractive index of droplets. Among them, the Dual Burst technique is one of the more promising one. It is based on the temporal separation between the reflected and the refracted light of an individual particle thanks to a small probe volume formed by two strongly focused laser beams. So it imposes that the diameter of the particle must be of the same

CP914, *Multiphase Flow: The Ultimate Measurement Challenge,* edited by X. Cai, Y. Wu, Z. Huang, S. Wang, and M. Wang

order or greater than the beam waist diameter and the measurement quality depends on the longitudinal location of the particles in the probe volume because of the strong curvature of the wave front.

To avoid the sensitivity to the wave front, we propose in this paper an innovative solution which is based on the separation of the light reflected and refracted by the particle in space[11,12]: two large beams are used to create the probe volume and the detectors are located in the image plane of the probe volume, in the forward direction at an angle where the intensities of reflection and of refraction of the first order are about at the same order. In fact, as the detector is located at the image plane, the signal is composed of two parts, one due to the reflection and the other due to the refraction. This configuration, called Glare Spot phase Doppler anemometry (GS-PDA) will be studied by numerical simulation with geometrical optics model (GOM), which permits to understand the principle and treat the problem in a simple way, and with a full electromagnetic model (EIM) thanks to a simulation code of imaging of small particle. The sensitivity of GS-PDA to the refractive index and diameter of the particle will be examined in detail for different configurations.

The paper is organized as follows. A brief presentation of the glare spots and the GS-PDA configurations will be given in the section 2. The section 3 is devoted the geometrical optics model of GS-PDA. The full electromagnetic imaging model will be introduced and applied in the section 4 to investigate the possibility to measure the droplet refractive index and diameter. The conclusions are given in the last section.

GLARE SPOTS AND GS-PDA

When a particle is illuminated by a beam of light, it scatters the light in all space. If the particle is much larger than the wavelength of the incident beam we can observe distinctively at certain angle two bright points on the particle, which are called glare spots[13]. The formation of the glare spots can be explained in a geometrical optics point of view as due to the separation of reflected and refracted light.

This principle can serve to separate the reflected and refracted light in particle measurements. We consider a particle of diameter d and refractive index m illuminated by a laser beam. The scattered light is collected by a lens of diameter ϕ and focal distance f located in the direction θ relative to the incident beam (Fig. 1). Two bright points are then formed on the image plan. One (A' in Fig. 1) is created by the refracted light and the other (R') by the reflected light. When a particle is illuminated by two incident beams as in the classical Phase Doppler Anemometry (Fig. 2), the light due to the reflection and the refraction from the two beams interfere and form two spots on the image plane. When a particle passes in the probe volume, the detector(s) located on the image plane will produce a signal composed of two parts: one due to the reflection and the other due to the refraction.

The detailed analysis of such a configuration will be given in this paper by using the geometrical optics model (GOM) and the full electromagnetic imagine model (EIM) in the following sections.

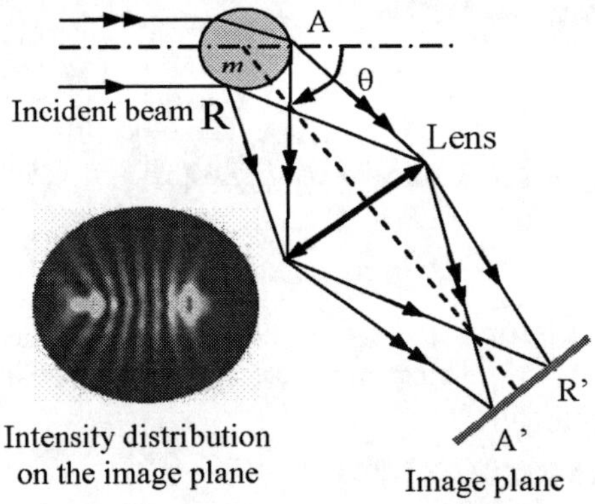

FIGURE 1. Off-axis imaging and intensity distribution on the image plan.

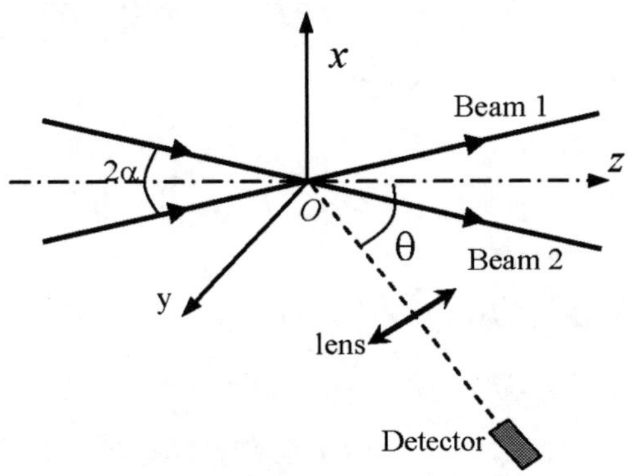

FIGURE 2. Schema of a GS- PDA.

GEOMETRICAL OPTICS MODEL

The geometrical optics is the most simple and instructive approach to understand the behavior of GS-PDA. At the first stage, we do not take into consideration the lens and the polarization of the incident wave.

When a particle of refractive index m and diameter d is illuminated by a plane wave of wavelength λ, the glare spots' positions can be determined by the refracted and reflected rays. The positions of the glare spots depend then on the refractive index, the diameter of the particle, and the detection angle θ. The detector being far from the particle, we suppose that the reflected and refracted rays are parallel for the simplicity.

For the convenience, we define the reference ray as the line coming from the source at infinity, passing by the centre of the particle and goes out in the direction of immerging rays (Fig. 3). The distances between the reference ray and the reflected or refracted ray is noted respectively by P_r, P_a (the subscript r and a represent respectively reflection and refraction). P_r and P_a are then given by:

$$P_r = \frac{d}{2}\cos\frac{\theta}{2} \qquad P_a = \frac{dm}{2}\sin\left(\frac{\theta}{2}\right) \Big/ \sqrt{1+m?-\ m\cos\left(\frac{\theta}{2}\right)}$$

It is clear that the distance P_r is independent of the refractive index but P_a depends on the refractive index of the particle[14,16]. The distance between the two spots $D_s = P_r + P_a$ is proportional to the diameter and depends also on the refractive index and the collection angle. The intensities of the reflected and refracted light can be also calculated according to the geometrical optics[14].

When the particle is illuminated by two beams with an angle 2α between them, the reflected and refracted light from each beam interfere. The optical path differences for the reflected and refracted rays are proportional to the size of the particle, the intensities on the image plane for the reflection and the refraction are given by:

$$I_r = T_r d^2 \cos^2\left(\frac{A_r d}{2}\right) \qquad I_a = T_a d^2 \cos^2\left(\frac{A_a d}{2}\right)$$

where T_r and T_a are two coefficients depending on the reflection and refraction Fresnel coefficients and the optical configuration. The figure 4 shows the behaviour of the reflected and refracted light intensities from a water droplet ($m = 1.33$) illuminated by two planes waves with a half-angle $\alpha = 5°$, the collection angle is $\theta = 66.5°$. The two factors A_r and A_a are respectively equal to $A_r = 0.299\pi$ and $A_a = 0.3533\pi$. The intensity of the spots varies periodically with the particle diameter. The corresponding periods are then 6.689 μm and 5.661 μm for reflection and refraction respectively.

This simple model of geometrical optics predicts the skeleton of the behaviour of the glare spots in the GS-PDA. The significant parameters as the lens dimension, the focal length, and the high order refraction effects have not been taken into account. These questions will be examined in the next section with a full electromagnetic model.

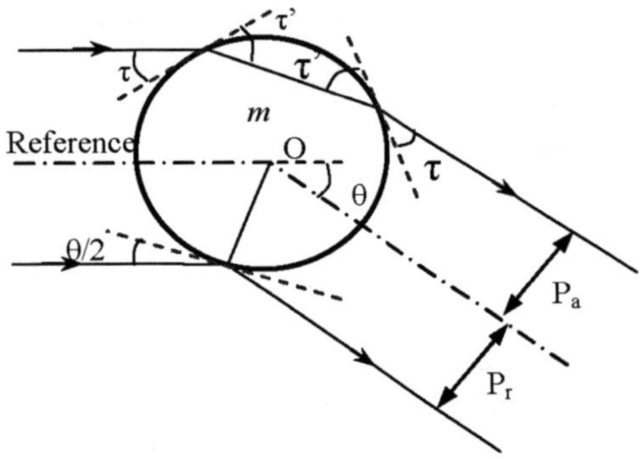

FIGURE 3. Optical paths.

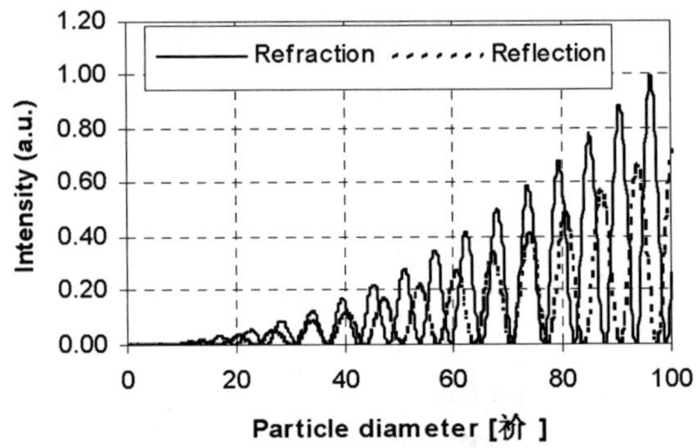

FIGURE 4.. Intensities as function of particle diameter predicted by geometrical optics.

ELECTROMAGNETIC IMAGING MODEL

In this section an electromagnetic imaging model based on full solution of Maxell's equations and Huygens-Fresnel diffraction principle will be used. The model will be presented in the first subsection and it will then be applied to one or two beam(s) configurations of a GS-PDA system.

The electromagnetic imaging model (EIM) is based on the work of Ren et al.[17,18]. A particle of diameter d and refractive index m located arbitrarily in a laser beam of wavelength λ is considered. The calculation of the intensity on the image plane is composed of three steps. The scattered field on the surface of the lens is firstly calculated by the Lorenz-Mie theory or generalised Lorenz-Mie theory[19]. The role of the lens is then taken into account by the thin lens transformation[20]. The intensity on the detector is finally calculated by Huygens-Fresnel integration[21,22,17] of the electromagnetic field on the exit surface the lens.

Glare spots of a particle illuminated by one beam

When a particle is illuminated by one beam, two glare spots are formed on the image plane. Their intensities and the distance between them depend on properties of droplets (size and refractive index) and the optical configuration of the system.We consider a water droplet illuminated by a plane wave of wavelength 0.488 μm. The detector is located at $\theta = 66.875°$, on the image plane, 400 mm from the particle. The diameter and the focal length of the collection lens are respectively 15 mm and 200 mm. The magnification of this configuration is then equal to 1.

The evolutions of the maximum intensities and the distance between the spots due to reflection and refraction can be simulated by the full EIM. Fig. 5 shows the distance between the spots for a particle of refractive index 1.33. We find that when the diameter is greater than a critic value, about 20 μm in the present case,, the distance is proportional to the diameter of the particle and the agreement between EIM and GOM is rather good. So the distance between the spots can be used to determine the diameter of the particle. Figs. 6 displays the distance between the glare spots formed by reflection and refraction as function of the refractive index. The distance predicted by the EIM is in good agreement with GOM. The error is less than 1.0% for the particle of 100 μm and less than 5% for the particle of 50 μm.

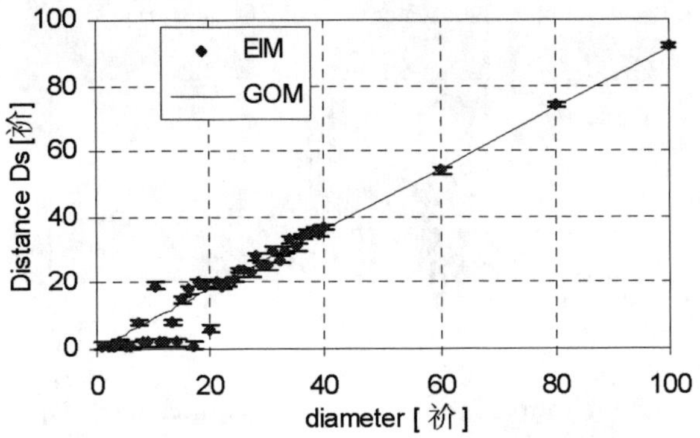

FIGURE 5. Evolution of the spots' distance as function of the particle diameter.

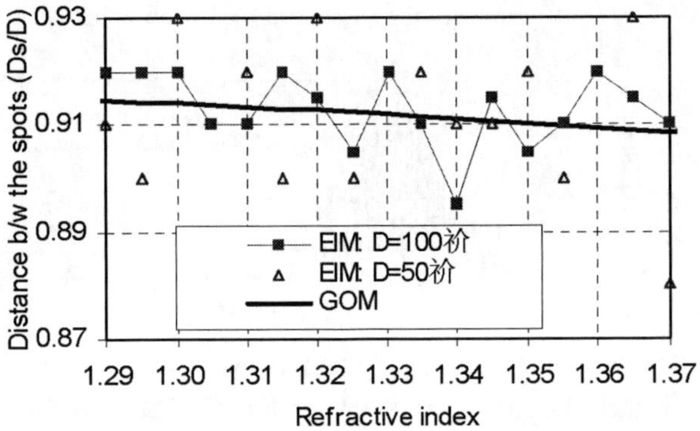

FIGURE 6. Evolution of the spots' distance as function of the refractive index.

Glare Spots Phase Doppler Anemometry

When the particle is illuminated by two beams, the total intensity on the detector is the sum of the electric fields scattered by the particle from the two beams. The two scattered fields then interfere on the detector. When a particle passes through the probe volume, the two spots' images pass successively the detector and the corresponding signals will be recorded. A typical GS-PDA signal simulated by EIM is presented in Fig. 7. A particle of 80 μm passes along x axis through the probe volume formed by two plane waves of wavelength 0.488 μm with a half angle of 5° between them (Fig. 2). The spacing in this configuration is equal to 2.8μm. The velocity of the particle is assumed to be 1.0 m/s and the period of the signal is then 2.8μs. The two parts due to the reflected and refracted light are well distinct in the signal. By the Fourier transformation we obtained the main frequency of the signal f=0.357 MHz. We can also obtained the pedestals of the signal by filtering the high frequency and extract the distance between the two spots and the intensity ratio. It is important to note that this signal is similar to the signals obtained by a dual burst PDA in which two tightly focalized beams are used[7,8].

The distance between the two spots and their intensity ratio as function of the particle size and its refractive index can then be extracted from this signal (Figs. 8 and 9). The detector in these simulations is located at θ = 67.574° so that the reflected and refracted light from a sphere of water (m=1.33) is of the same order of intensity.

When the particle moves in x direction, perpendicularly to the bisect of the two incident beams, the time delay between the two glare spots T_s is related to the displacement of the particle D_x and the distance between the spots D_s:

$$D_s = D_x \cos\theta = V_x T_s \cos\theta$$

By taking into account this effect, the particle diameter can then be obtained from the time delay of the two spots provided that its refractive index and the direction of particle movement are known.

For the case shown in Fig. 7, the time delay between the two maxima in the pedestal is T_s=192 μs which corresponds to a displacement of D_x = 192 μm. Considering the angle position of the detector, the distance between the two spots D_s = 72.6 μm can then be deduced and it is in agreement with that predicted in Fig. 5.

In Fig. 8, the distances between the two spots D_s predicted by GOM and EIM are compared for a water droplet. The agreement is found very good. Therefore the distance D_s permits to determine precisely the diameter of the particle provided that the refractive index is known.

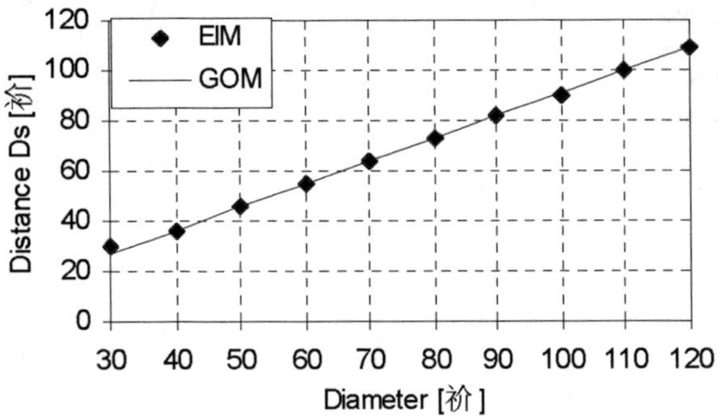

FIGURE 8. Distance between the spots as function of diameter for a water droplet (m=1.33).

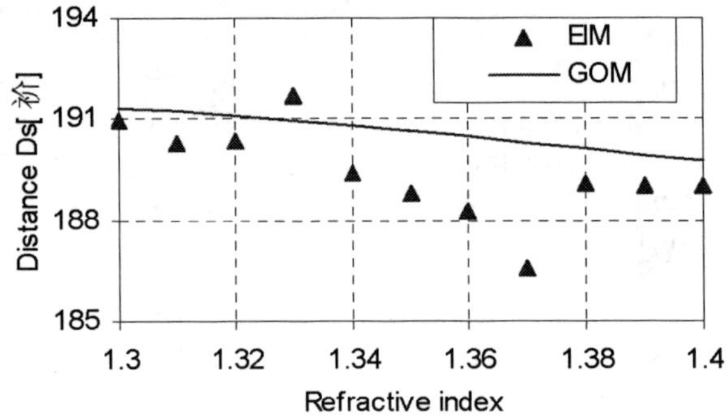

FIGURE 9. Distance between the two spots as function of the refractive index for a water droplet (d= 80 μm) μm).

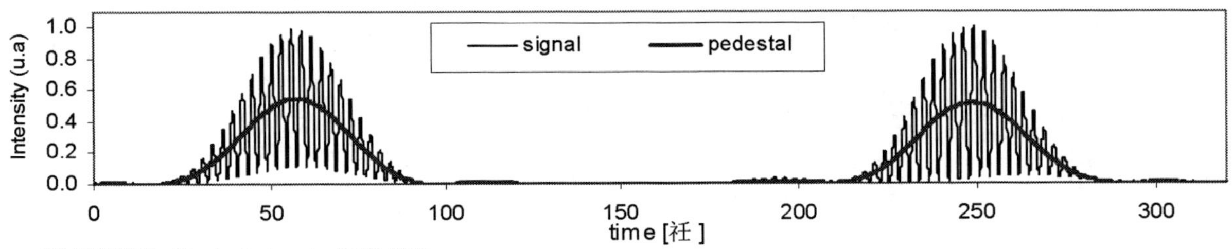

FIGURE 7. Typical signal of GS-PDA.

On the other hand, since the position of the refraction spot and the ratio of the two maximum intensities depend both also on the refractive index, we could expect to obtain information about refractive index from these values, if the diameter is known. Fig. 9 show the distance D_s as function of the refractive index for a water droplet of d = 80 μm. We find that the distances D_s depends little on the refractive index and the distance obtained from EIM signals is in general smaller than that obtained by GOM, but the difference is less than 2 %. Nevertheless, the fluctuation of the intensity ratio as function of the refractive index is too strong to be used to determine the refractive index by a linear relation.

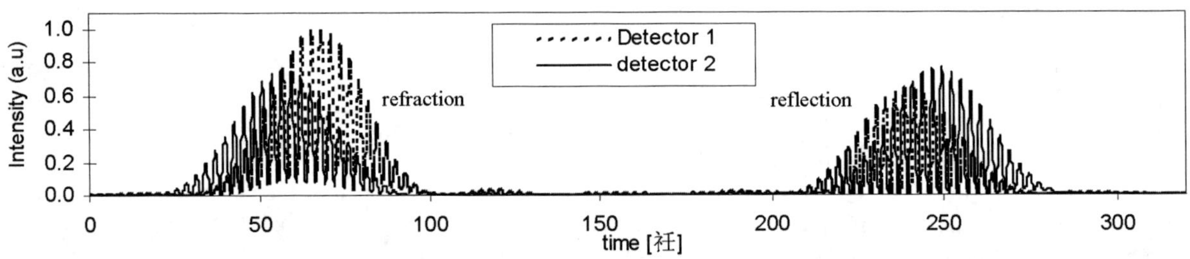

FIGURE 10. GS-PDA signals obtained by two detectors located respectively at θ = 65.426° and 67.574° with a half angle of 5° for a water droplet (m=1.333, d=80 μm) illuminated by two plane wave of wavelength 0.488 μm.

When two detectors are used in GS-PDA, two signals will be obtained as in the dual burst PDA [7,8]. Besides the distances D_s and intensity ratios as described above, we can also extract the two phase differences between the two signals separately for reflection and refraction parts.

Fig. 10 shows two signals simulated by EIM. We remark that the distances between the two intensity maxima are not the same because the two detectors are not symmetrical to the particle trajectory. The phase differences extracted from the signals are shown in Fig. 11. We find that the phase differences are relatively stable around the maximum intensity and the average phase differences for reflection and refraction are respectively 43° and 24°.

It should be noted that the relation between the particle diameter and the phase differences is independent of the particle trajectory. It is therefore possible to determine the particle size and the two components of velocity by the distances between the spots and the phase differences between the two signals.

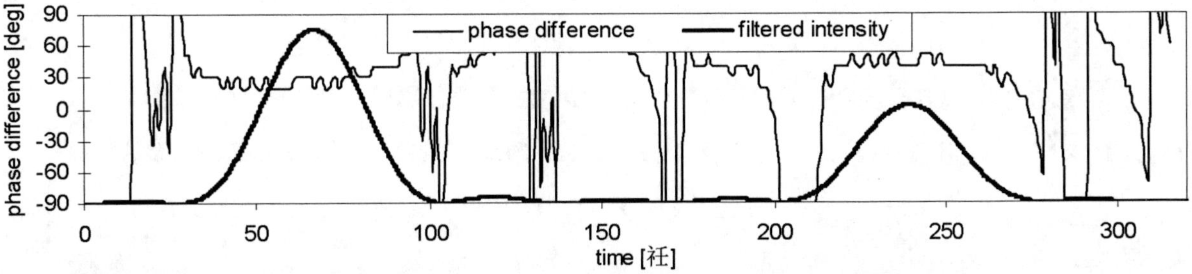

FIGURE 11. Pedestal and phase difference between the two signals extracted from the signals in Fig. 11.

CONCLUSIONS

The principle of the Glare Spots Phase Doppler Anemometry is introduced and studied with a geometrical and a full electromagnetic imaging model. The GS-PDA uses two large beams and two detectors which are located on the image plane of the collection lens. The signals obtained with such system are composed of two parts in each of them: one due to reflection and the other to refraction. This new configuration provides richer information about the particle to be measured: two distances between the spots, two intensity ratios of the spots, two phase differences as well as the frequency of the signals can be extracted for the determination of the particle size and the refractive index as well as the two components of velocity. The particle diameter and two components of velocity could be determined by the spots' distances, the phase differences and the frequency of the signals if the refractive index is known. However the relation between the refractive index and the measured values (phase differences, spots distances and intensity ratios) are not linear in the planar configuration. The experimental validation is to be done and other configurations should be investigated for a real measurement of the diameter and the refractive index.

ACKNOWLEDGMENTS

This work partially is supported by the inciting joint action, ACI "pollution, Non-pollution", VITAMA. project 47. This ACI is financed by the FNS, CNRS and the DGA.

REFERENCES

1. J. P. A. J. van Beeck and ML Riethmuller, *Appl. Opt.*, 34(10) (1995).
2. J. P. A. van Beeck , T. Grosges and M.G. De Giorgi , *App. Opt.* 42(19), pp4016-4022, (2003).
3. H. E. Albrecht, N. Damaschke, M. Borys and C. Tropea, *Laser Doppler and Phase Doppler Measurement Techniques,* Berlin: Springer (2003).
4. G. Brenn and F. Durst, *Atomization and Sprays*, 5, pp545-567 (1995).
5. A. Naqwi, F.Durst, and X. Liu, *Part. Part. Syst. Charact.,* 8(33), pp16-22, 1991.
6. C. Tropea, T-H. Xu, F. Onofri, G. Gréhan, P. Haugen, *PPSC*, 13, pp165-170 (1996).
7. F. Onofri, T. Girasole, G. Gréhan, G. Gouesbet, G. Brenn, J. Domnick, T.H. Xu, C.Tropea, *PPSC*, 13, pp112-124, (1996).
8. F. Onofri, L. Bergounoux, J-L. Firpo, J. Mesguish-Ripault, *Appl. Opt.*, 38:4681-4690, (1999).
9. S. V. Sankar, K. M. Ibbrahim, D. H. Buermann, M. J. Fidreich, W. Bachalo, *Proc. 3rd OPC*, Yokohoma, 23-26 Aug. 1993.
10. A. Delconte, D. Blondel, F. Lemoine, *ICLASS*-2006, Kyoto, Japan, Aug.27-Sept.1, 20s06.
11. J. K Schaller, S. Wassenberg, D. K. Fiedler, C. G. Stajanoff, *Proc. ICLASS*, Rouen pp443-450,1994 .
12. J. K Schaller, S Wassenberg, D.K Fiedler, C.G Stajanoff, *Proc SPIE*, 2248, 318-324, 1994.
13. H. C. Van de Hulst and R. T. Wang , *Appl. Opt.*, 30(33), 4755-4763, (1991).
14. H. C. van de Hulst, *Light Scattering by Small Particles,* Dover Publications Inc., New York (1981).
15. F Durst, G Brenn and T. H. Xu, *Meas. Sci. technol.* 8, 1203-1221 (1997).
16. S. Dehaeck, J.P.A.J. van Beeck1and M. L. Riethmuller, 12th *Int. Sym. Appl. Laser Tech Fluid Mech.*, Lisbon, 12-14 Jul. 2004.
17. K. F. Ren, *Diffusion des faisceaux feuille laser par une particule sphérique,* Ph. D. thesis, University of Rouen, France, 1995
18. K. F. Ren, D. Lebrun, C. Ozkul, A. Kleitz, G. Gouesbet and G. Grehan, *PPSC*, 13, pp. 156-164, (1996).
19. G. Gouesbet, B. Maheu, G. Gréhan, *J. Opt. Soc. Am. A* 5(9), 1427-1443, (1988).
20. J.W. Goodman, *Introduction à l'optique de Fourier et à l'Holographie,* Masson and Cie, Paris, (1972).
21. H. H. Hopkins, *Proc. Phys. Soc. B,* 70:1002-1005 (1957)
22. J. J. Stamnes. *Wave in Focal Regions*. Adam Hilger, 1986.

Design and testing of a new miniature combined optical/pneumatic wedge probe for the measurement of steam wetness

M. Schatz, M. Casey

Institut für thermische Strömungsmaschinen und Maschinenlaboratorium (ITSM), Universität Stuttgart, Stuttgart, Germany, Schatz@ITSM.Uni-Stuttgart.de

Abstract. A new combined optical/pneumatic probe design has been developed in order to enhance steam wetness measurements on the basis of the light extinction method. The new design is a consequent further development of earlier probes. The probe diameter has been kept as small as 10mm , so that it can be used in a model steam turbine of the institute. Measuring steam wetness and droplet spectra contributes to better understanding of condensation phenomena in steam turbines and helps to validate numerical results. In addition, calculating the power output of the turbine allows an approximate validation of the wetness measurements, thus increasing the reliability of the data acquisition and analysis.

Keywords: multiphase flows, wet steam, light extinction measurement
PACS: 47.55.–t

INTRODUCTION

In steam power plants, the working fluid expands into the two-phase area of the h-s-Diagram. This means that condensation occurs at some stage of the expansion process. Due to the high expansion velocity found in turbine stages, condensation occurs not at the saturation line as expected for an equilibrium process, but a non-equilibrium metastable state is achieved, where the steam still has a single phase. The steam is subcooled and small agglomerations of water molecules occur randomly. These agglomerations are called nuclei, and when the subcooling is high enough, they do not fall apart again, but grow very rapidly due to steam condensing on their surface. Thus, a two-phase flow with fog droplet emerges, whereby typical droplet radii are 10^{-7} m. This process, in which the steam is almost instantaneously reverted to equilibrium comes with a diabatic heat addition to the flow when the latent heat is released. After equilibrium is restored, condensation continues to occur on the droplets formed as the flow is expanded further below the saturation line in the following turbine stages. If the expansion velocity becomes high there may not be enough droplets to absorb the condensing steam and a new subcooling of the flow resulting in secondary nucleation may occur. This is very often the case in the tip region of the final stages of steam turbines.

Condensation and the resulting two-phase flow entails thermodynamic and mechanical losses, which leads to a deterioration of the turbine efficiency. The losses can be divided in thermodynamic losses due to the release of latent heat to the flow, producing an entropy rise, and mechanical losses. These result from the fact that the large droplets do not follow the main flow. Thus, droplets are either deposited on the stator blade surfaces or hit the rotor blades, causing braking losses. The droplets that are deposited on the stator blades run to the trailing edges and are re-entrained into the flow in the form of so-called coarse water, i.e. droplets with a radius ranging from 10^{-5} to 10^{-3} m and when those droplets hit the rotor blades they cause severe blade erosion, limiting the turbine lifetime considerably.

In the tip regions of the last stages of steam turbines, where the circumferential velocities are very high, the flow becomes transonic. This is a state where small disturbances, such as the latent heat release of the condensing steam, may cause aerodynamic shocks, causing even higher losses and affecting operational reliability, as instationary phenomena such as moving shock systems may emerge and cause blade vibration.

CP914, *Multiphase Flow: The Ultimate Measurement Challenge,* edited by X. Cai, Y. Wu, Z. Huang, S. Wang, and M. Wang

Another problem caused by the two-phase flow found in LP-turbine exits is the difficult determination of the expansion end point, needed to calculate turbine efficiency. In a single-phase flow the static pressure and temperature suffice to calculate the relevant thermodynamic properties such as enthalpy and entropy, whereas in wet steam flow the knowledge of steam quality is indispensable, as long as the overall mechanical power output of the turbine cannot be measured. This problem may seem to be of purely academic nature, yet turbine efficiency is an important parameter for both turbine manufacturers and plant operators. In normal practice, turbine manufacturers need to guarantee a certain turbine efficiency and if the turbine efficiency does not comply with the efficiency promised by the manufacturer severe penalties will result. On the other hand, if the machine performs better than promised, the manufacturer may be entitled to demand a better remuneration.

As a consequence, both manufacturers and plant operators are interested in a reliable and accurate determination of the expansion end point in order to calculate turbine efficiency. This is mirrored in the fact, that special standards regulating the determination of turbine efficiency exist [1,2,3].

MEASUREMENT TECHNIQUES

In the German standard “VDI 2043” [4], a number of methods for the determination of steam wetness fraction, together with their relative measurement uncertainty, are listed. These methods include energy balance (relative uncertainty according to VDI 2043: 10-30%), tracer methods (0,1-2%), optical methods (3-15%) and calorimetric test methods (10-30%).

Other methods not listed in this standard are microwave methods, which are basically calorimetric methods, and the measurement of static charge [5].

Optical methods for wetness determination can be divided into scattering methods and extinction methods. The underlying physical principle is the same for both methods and shall be explained quickly with emphasis on light extinction. For further details about the light scattering method, the interested reader is referred to [6].

When light is passing a body with a refractive index different from that of the surrounding medium, it is absorbed and scattered by the body. Thus, the amount of light received by a detector is smaller than that emitted by the light source. This phenomenon is described by the equations of Maxwell. Mie was the first to solve these equations under the assumptions of parallel light and spherical shape of the body. In his theory, he showed that light scattering is only a function of the ratio of the real part of the refractive indices and the ratio of droplet diameter over wavelength [7]. This parameter is called the Mie-coefficient:

$$X_{Mie} = \frac{\pi D}{\lambda} \tag{1}$$

The imaginary part of the refractive index is a measure for absorption. The attenuation of light due the combination of scattering and absorption is also known as extinction.

Water droplets emerging and growing due to condensation in steam turbines are usually in the range of 0.1 to 5 microns [9]. Droplets of this size are assumed to be spherical and not deformed by shear forces. Moreover, if light in the wavelength region from 350 nm to 1200 nm is used, absorption is very low and can be neglected [8]. Thus, light extinction by water droplets in steam turbines is reduced to scattering and can be described by the Mie-theory.

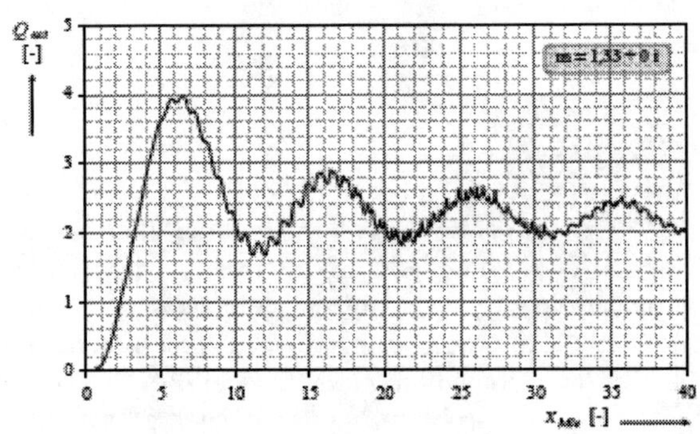

FIGURE 1: Extinction efficiency vs. Mie-coefficient

Therefore, the scattering cross section which is defined as the detector area that has to be covered to reduce the intensity of unscattered light by the same amount which a particle scatters is equal to the extinction cross section. The extinction efficiency is defined as the extinction cross-section divided by the geometrical cross-section of the scatterer.

$$Q_{ext} = \frac{4 \cdot C_{ext}}{\pi D^2} \qquad (2)$$

The extinction coefficient for water droplets in steam is plotted over the Mie-coefficient in fig. 1. Here it is important to see that for values of $X_{Mie} < 1$ the extinction efficiency is very low. On the other hand, for Mie-coefficients greater than 35, the extinction efficiency oscillates around a value of 2. This is important for data analysis.

The intensity of light at a specified wavelength λ scattered by a number N of droplets of diameter D in an optical path is, according to the theory of Mie

$$I_{sca} = I_0 \cdot Q_{ext} \cdot \frac{\pi D^2}{4} \cdot N \qquad (3)$$

If the distribution of the particles is not monomodal, but polymodal, we get

$$I_{sca} = I_0 \cdot \int_{D\min}^{D\max} Q_{ext} \cdot \frac{\pi D^2}{4} \cdot N(D)dD \qquad (4)$$

This form of integral is called a Fredholm integral of the first kind. On the other hand, from the law of Bouger used in spectroscopy, we know that the intensity of light I of a light beam with unscattered intensity I_0 going through a suspension of the length l can be described by the following function:

$$I = I_0 \cdot e^{-k \cdot c \cdot l} \qquad (5)$$

Where k is a specific attenuation constant and c is the concentration of particles, i.e. the number of particles per unit volume.

Combining eq. 4 and eq. 5 we get

$$\ln\left(\frac{I_0}{I}\right) = l \cdot \int_{D\min}^{D\max} Q_{ext} \cdot \frac{\pi D^2}{4} \cdot N_c(D)dD \qquad (6)$$

If the distribution of droplets is polymodal, it is not sufficient to measure the extinction of light of one wavelength only. A range of wavelengths must be measured to obtain enough information about the droplet spectrum. The resulting set of integrals for m wavelengths can be approximated numerically for n droplet diameter classes and written in matrix form:

$$\vec{g} = \underline{A} \cdot \vec{f} \qquad (7)$$

Here, $\underline{A}$ is a m x n-Matrix containing all known parameters such as Q_{ext} and the given diameters, g is a m-vector consisting of the measured extinction for each wavelength and f is the solution vector with the number distribution.

Solving eq. 7 directly is theoretically possible by inverting the matrix A, yet Fredholm integrals of the first kind are extremely sensitive against noise, i.e. a direct solution from measurement data is not possible. Hence, the set of equations is solved using the method of Phillips [10] and Twomey [11]. This method, however, is just an analytical approach, thus the solution may also incorporate negative numbers of droplets, which is unphysical. For this reason, a standard non-negative least squares algorithm [12] is used which solves the problem in eq. 7 in such a form, that $f_i \geq 0$.

With all these methods and algorithms at hand it may seem very simple to do droplet sizing using the light extinction method, but measurements in wet steam, particularly in steam turbines, are a very complex task.

First of all because of the limited access to the flow within a steam turbine, secondly because of the partly transonic flow, and thirdly because of 3D-effects such as varying yaw and pitch angles over span. Since the presence

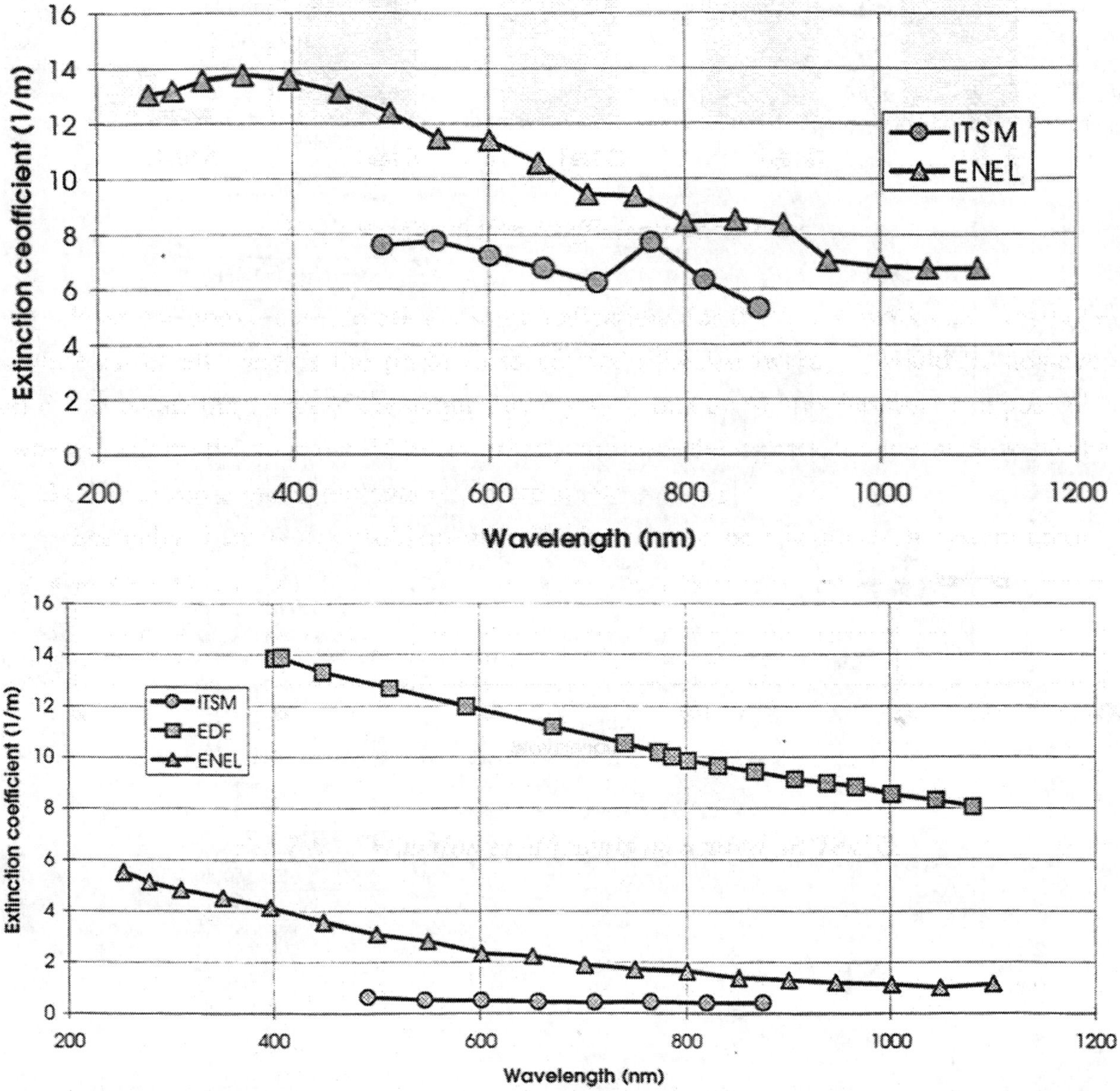

FIGURE 2: Measured extinction vectors for two operating points of the installation in [13]

of a wetness probe affects the flow field around it, the droplets suspended in the flow are also affected, and a badly designed probe will yield erroneous measurements.

This problem can be very well explained using the measurement data obtained during the STEAMQUAL EU-Project [13]. In this project, a number of European research groups from universities and industry met to do comparative tests in order to improve the quality of wet steam measurements. Within this project, wet steam measurements under laboratory conditions, i.e. in a steam wind tunnel equipped with a supersonic nozzle to incept condensation were done. This is a much simpler set-up than measurements in a steam turbine, yet some of the data acquired by the various researchers were completely different, as can be seen in fig. 2, although they had all measured under the same conditions. The only conclusion of this observation is, that, apart from possible errors in the optical set-up, there must have been a negative influence of the probes on the two-phase flow passing it.

Another interesting outcome of the STEAMQUAL project was, that the different methods and software codes for the evaluation of the data acquired gave quite different results for the same data input, shown in fig. 3. This clearly shows the difficulties of wet steam measurements in steam wind tunnels or even turbines: first of all, data acquisition must occur in the most perfect way, causing the least disturbance of the flow, thus avoiding any source of error. Second, reliable approaches to analyse the measurement data must be developed and validated.

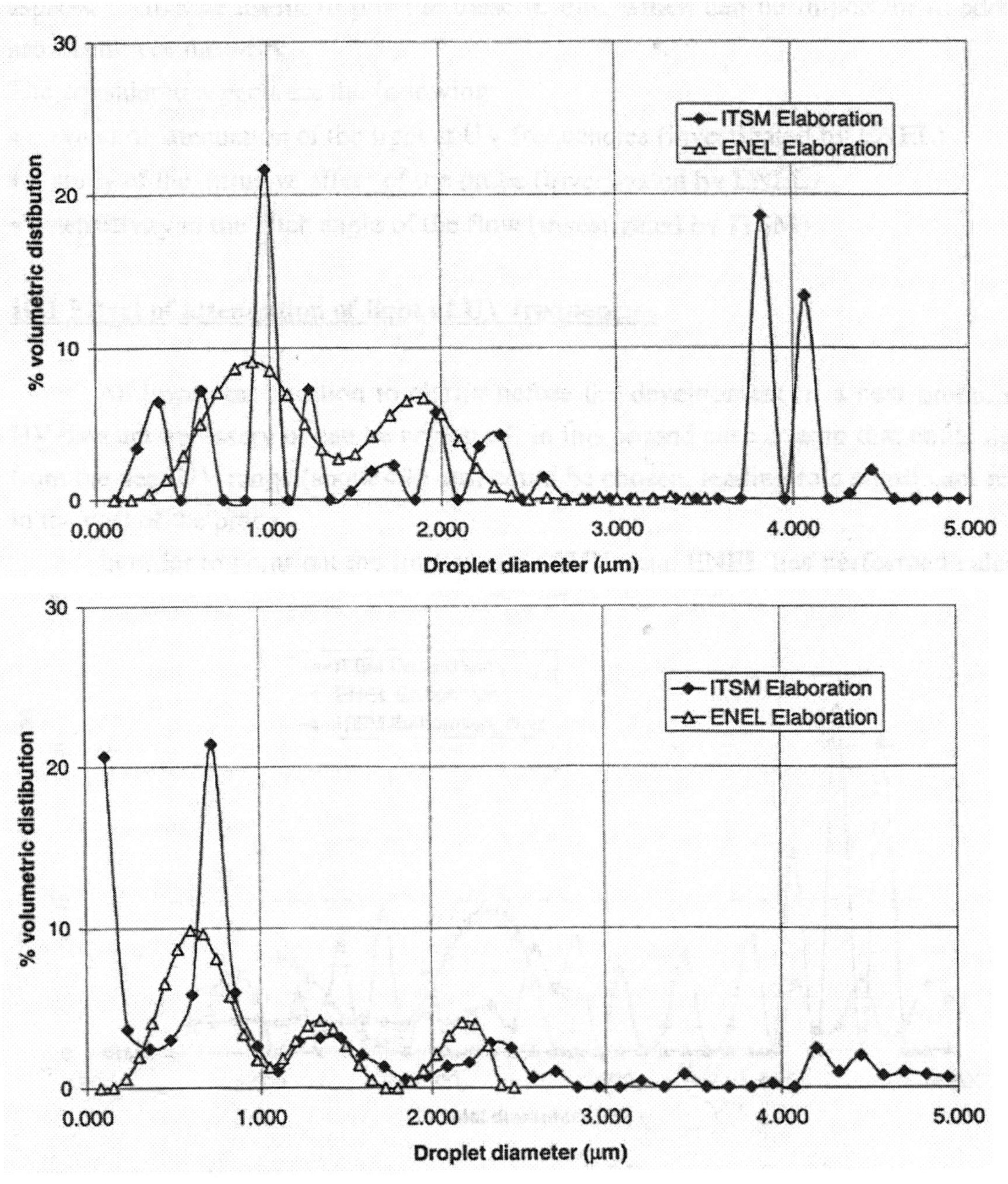

FIGURE 3: Droplet spectra for measured extinction, calculated with different software codes

THE FINAL STAGE TEST STAND AT ITSM

The Institute of Thermal Turbomachinery and Machinery Laboratory of the University of Stuttgart operates a test stand for the evaluation and testing of LP final stage bladings. It is a three stage turbine, where the last stage is a scaled version (between 20% and 25% of full size) of an original blading and the two preceding stages only serve to generate the same flow field (in terms of flow angles and Mach-number) for the last stage as found in a full-scale turbine [14]. A meridional section of the test stand together with the measurement locations of the last stage is shown in fig. 4.

Apart from the measurement locations behind every blade row, the model turbine also has dynamometers for torque measurements. In addition, the losses of the bearings of the turbine can be determined. Thus, the power output of the turbine can be calculated. The relation between power output and enthalpy drop of the working fluid in a turbine can be expressed by the formula

$$P = \dot{m}\,\Delta h \tag{8}$$

When power output and steam mass flow are known, the exit enthalpy (i.e. the expansion end point) of the steam can be calculated. This, together with the static pressure and temperature at the exit allows the determination of steam quality from the well-known Mollier-chart. This means that the model turbine at ITSM allows the determination of steam wetness independently from any wetness measurement, hence giving the opportunity to

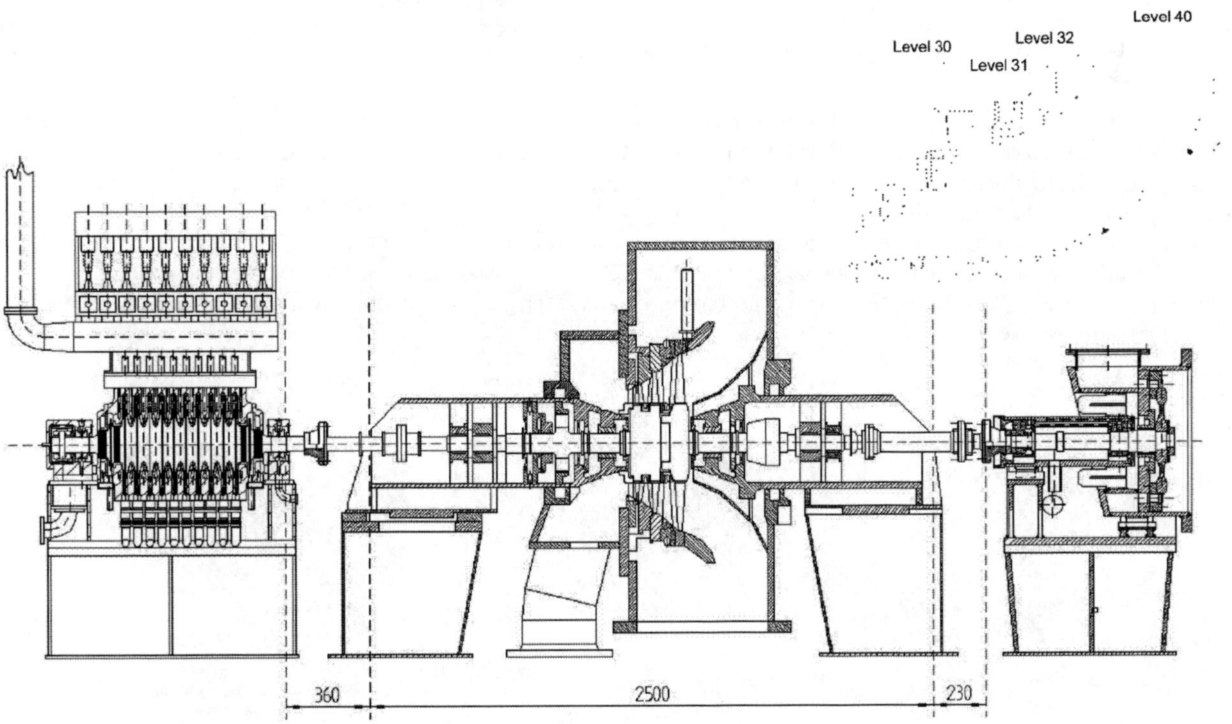

FIGURE 4: Meridional section of the test stand at ITSM and measurement locations

validate wetness measurement equipment and methods. The only drawback of the installation is the fact, that due to the relatively small size of the model turbine the maximum shaft diameter of any flow or wetness probe is limited to 10 millimetres to reduce probe blockage effects.

Measuring wetness in the model steam turbine is not only done for validation purposes, but also to determine droplet spectra in order to enhance the understanding of the influence of condensation on the flow through the turbine and contribute to the improvement of blading geometry and calculation methods, meaning that there is a mutual benefit for both. The development of wetness probes at ITSM was driven by this task.

DEVELOPMENT OF WETNESS PROBES AT ITSM

The beginning of the development of wetness probes at ITSM dates back to the 1980's and the first probes consisted only of a shaft with a broad slit which formed the measurement volume [15].

In the beginning, the probes were made solely for the purpose of wetness measurement, meaning the flow field had to be measured separately to determine flow angles before measuring wetness, until Renner [16] started to combine a standard wedge probe with an optical set-up, combining pneumatic and optical measurements in one probe (fig. 6). This probe allowed the simultaneous measurement of total and static pressure, temperature, yaw and pitch angle and finally droplet spectra, which, apart from saving time and money, increased accuracy due to an improved positioning of the probe. However, this combined probe still had a number of disadvantages, the greatest of all being the fact that the optical measurement volume was still within a slit in the probe shaft. With this design, the accuracy of the measurement can be deteriorated by flow separation bubbles occurring at the edges of the slit because of misalignment. While the yaw angle can be corrected by turning the probe in the flow, there is no chance of correcting for the pitch angle. For the measurement in LP-turbines, where a high density gradient causes pitch angles of up to 40 degrees, this means a severe restriction of applicability.

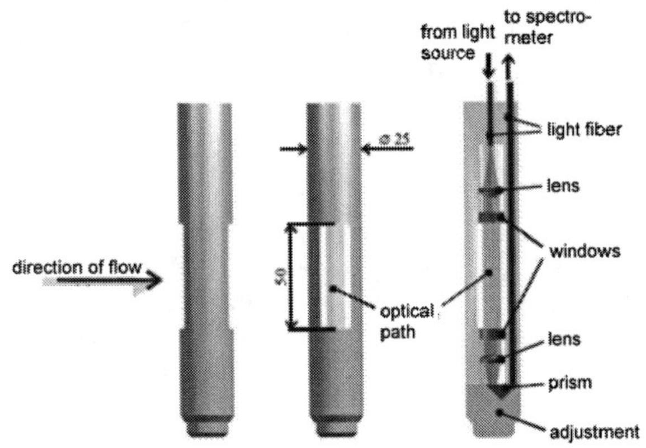

FIGURE 5: First design of wetness probes at ITSM

In his dissertation [16], Renner has studied the influence of misalignment on the flow through the slit using water channel studies as well as flow visualisations with Laser-2-Focus measurements. These experiments clearly show a separation bubble within the optical path, shortening the path length l to an unknown value and deterioration the optical signal.

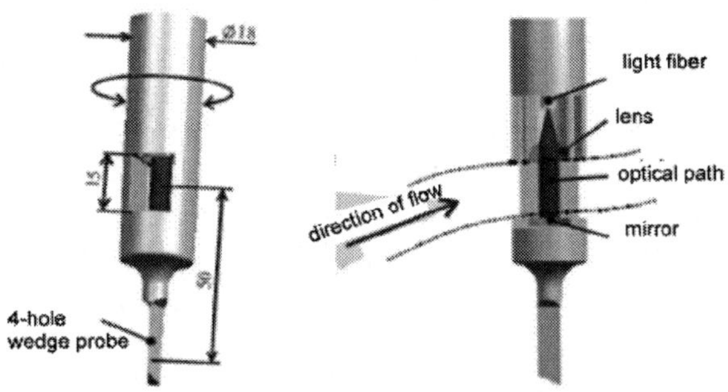

FIGURE 6: First combined optical/pneumatic probe [16]

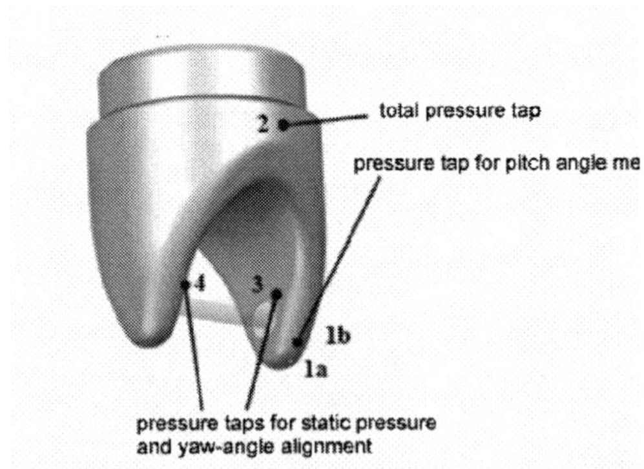

FIGURE 7: Second generation of combined probes, described in [17]

Seibold [17] attempted to develop the concept of combined probes further by switching from the wedge design to two elliptical appendices of a cylindrical probe (fig. 7). This design reduced the interaction between the two-phase flow and the probe greatly as it was very tolerant against yaw- angle misalignment and did not restrict pitch angle at all. The orientation of the optical measurement volume changed from vertical to horizontal. The idea behind this was to improve the radial resolution of the measurement. Moreover, pressure and optical measurement where now at exactly the same radius, giving a truly simultaneous reading. Unfortunately, the elliptical geometry proved to be unsuitable to measure pressure in transonic flows accurately, as the flow field around it depends on the Mach-number and is also susceptible to variations of Reynolds-number.

Moreover, the optical layout was not really suitable for the rugged conditions in steam turbines and too complex to be miniaturized, so that there was only one prototype with a shaft diameter of 25 millimetres to be used in full scale turbines.

INTRODUCTION OF THE THIRD GENERATION COMBINED PROBE

The third generation of combined probes now aims to get the best of both worlds, uniting the superior characteristics of wedge probes for pressure measurement in transonic flows and an arrangement of the optical measurement volume such that flow separation due to yaw angle and especially pitch angle misalignment is reduced to a minimum. In addition, the goal of miniaturisation has been kept in mind during the design process and the selection of all components. The probe geometry is shown in fig. 8.

The wedge geometry has been chosen because the flow field around it is independent of Reynolds number, i.e. flow separation only occurs at the corners of the geometry. Since a comparatively small flow angle has been chosen, the acceleration of the flow near the probe due to blockage effects is reduced to a minimum. Furthermore, a wedge geometry is very well suited for measurements in transonic flows, as the behaviour of detached and bounded shocks is well defined. In terms of gasdynamics, a wedge is a very simple geometry

The wedge geometry has been shifted out of the center axis to obtain more space for the installation of the optical components. This way, the measurement volume is now only confined vertically on one side, and the level of both pressure and optical measurement is in the same radius, giving a truly simultaneous reading. For the optical axis, a vertical orientation has been chosen, as this helps to keep the shaft diameter as small as possible. The consequence of this design is that the optical path is again situated between two walls, having a negative influence on the pitch angle tolerance, yet both ends of the optical path have been chamfered with an angle of 15 and 22 degrees respectively, so that the range of pitch angles encountered in the flow through the last stages of LP-turbines is covered.

To confirm the advantageous features of the probe it has been examined by CFD. The flow field around the wedge geometry has been computed with the standard CFD-code CFX 10 from ANSYS in 2D-simulations in order to check the appropriateness of the design for measurements in transonic two-phase flow. The main task was to identify if the interaction of the probe with the flow has any influence on the quality and accuracy of the optical measurement. This could happen due to two different mechanisms: first of all a misalignment of the probe, causing flow separation at the front edge of the probe. The question here is, whether the separation bubble reaches into the optical path or influences the flow going through the measurement volume.

The second mechanism possibly causing a deterioration of the optical measurement is the occurrence of aerodynamic shocks in transonic flow, causing a strong increase of pressure. If the pressure behind the shock is higher than saturation pressure, small droplets could be evaporated again as their thermal inertia is very small.

Steam at a pressure of 100 mbar and a temperature of 55°C, with real gas properties calculated from the IAPWS-F97 tables has been chosen as working fluid for the simulation. As inlet boundary condition, flow velocity and angle were used, covering a Mach-number range from 0.6 to 1.4 and a yaw angle range of 10 degrees. Condensation of steam was modelled using an equilibrium condensation model.

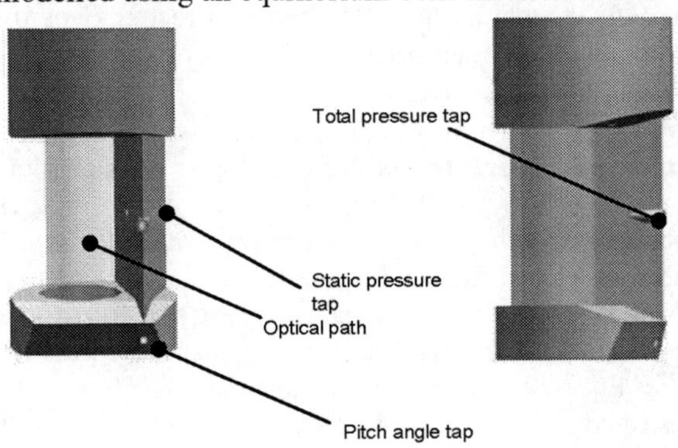

FIGURE 8: New combined probe design

First the results for a free-stream Mach number of Ma=0.6 shall be presented.

For a flow angle of 0 degrees, shown in fig.9 the periphery of the optical path, illustrated by the small circle next to the wedge geometry, is influenced by the vortex street formed by the flow separation at the corner. This can be seen in the density gradient plot and in the velocity field plot. The pressure drop due to the vortices formed in the separation

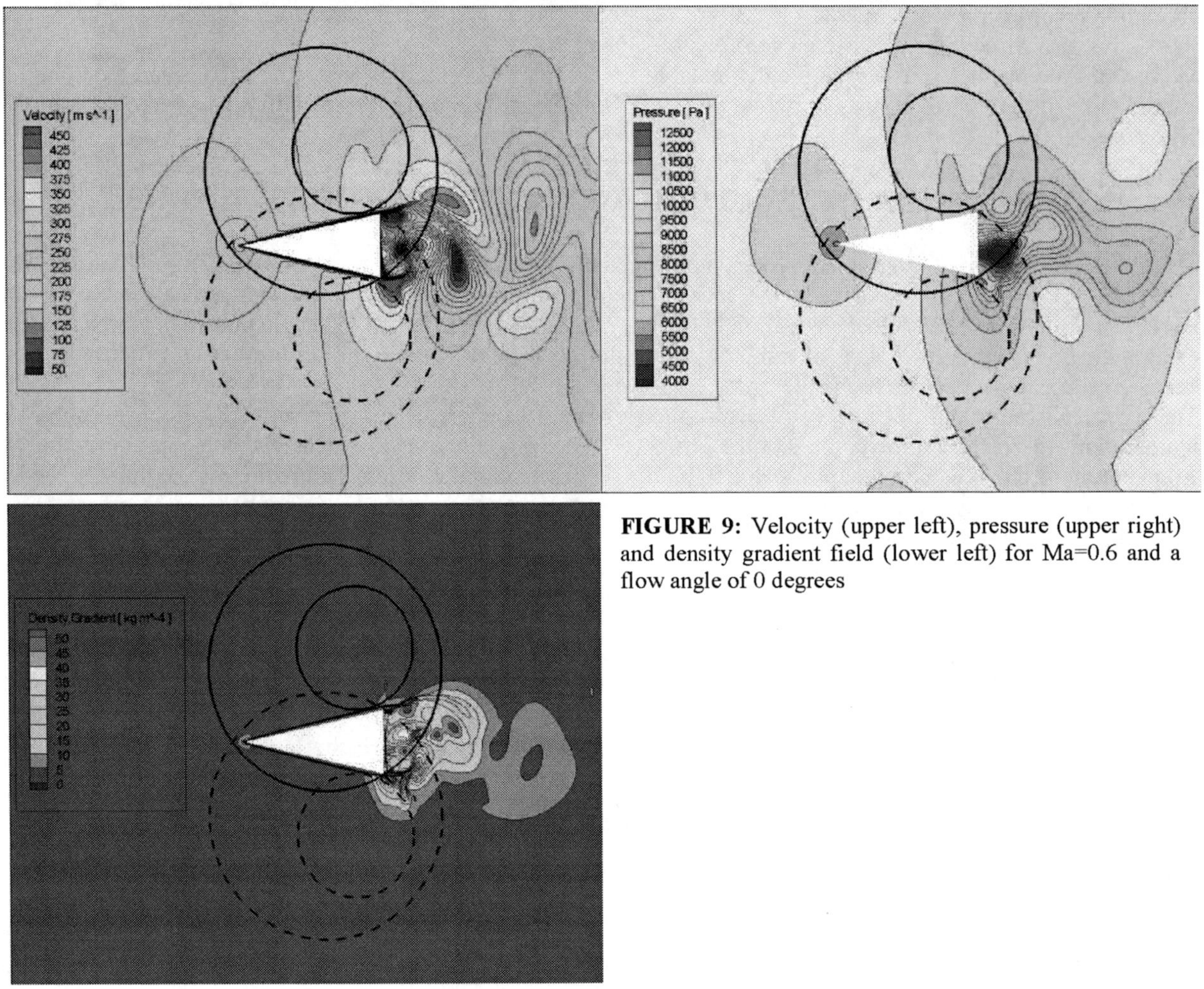

FIGURE 9: Velocity (upper left), pressure (upper right) and density gradient field (lower left) for Ma=0.6 and a flow angle of 0 degrees

area does not reach into the optical path, hence there is no danger of secondary nucleation during measurement. The location of the optical path is far enough away from the wedge so that the temperature increase due to viscous effects in the boundary layer has no consequence for the measurement either.

For a flow angle of +5 degrees, illustrated by the solid line circle in fig. 10, the flow through the optical path is less affected by the vortex street forming behind the wedge. The flow velocity is uniform. For an angle of –5 degrees, described by the dashed line circle in fig. 10, the optical path is also hardly affected. This means that a misalignment of up to ±5 degrees even improves the quality of the measurement. This surprising result is related to the changed structure of the vortex street from the wedge at low incidence.

For a flow angle of ±10 degrees, the perturbation of the flow due to the misalignment will also affect the optical measurement, as can be seen in fig 11. For +10 degrees, the separation bubble that forms at the tip of the wedge reaches into the optical path (solid circle). This can be seen in the density gradient plot in fig. 11. The flow around the separation bubble is strongly accelerated and becomes transonic. The pressure drop connected with this strong acceleration could possibly lead to a small area of condensing flow, thus distorting the optical measurement.

For a free stream Mach number of Ma=1.1, shown in fig. 12 , the aerodynamic shock that forms in front of the probe can be clearly identified in the velocity plot as well as in the Mach number plot. At the corner of the wedge geometry, the flow is accelerated to supersonic velocities due to a Prandtl-Meyer expansion.

Of course, the pressure rise due to the shock, illustrated in fig. 12, could partly lead to an evaporation of very small droplets which possess a low thermal inertia. Yet the shock is located relatively close to the probe, thus limiting this effect to a certain extent. Keeping the high velocity of the flow in mind, the density gradient within the optical path is relatively low and will very probably not affect the measurement. The high turning at the edge of the geometry again incepts acceleration to a supersonic state.

Much care has been taken to improve the miniature optical lay-out of the probe, which has been optimised using OSLO, a standard optical software. With this improved lay-out, the measurement works accurately over a wide range of wavelengths.

Due to the thorough aerodynamic and optical layout, this combined pneumatic/optical probe has a shaft diameter of only 10mm while having full functionality. Therefore, it can be used throughout the whole test rig at ITSM including the intrablade row traverse planes in the last stage, whereas measurements in full scale turbines can be done by simply scaling the probe.

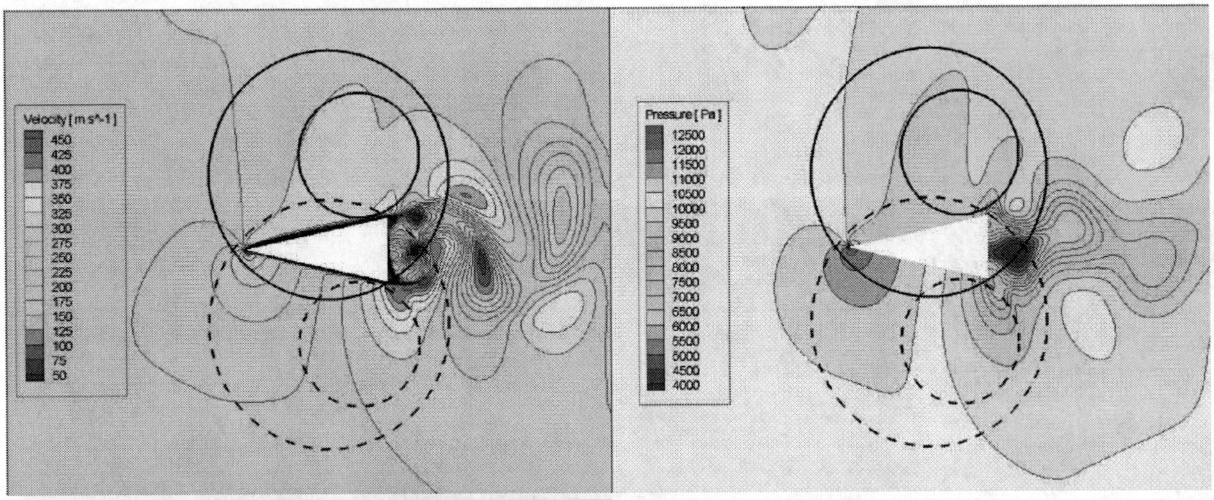

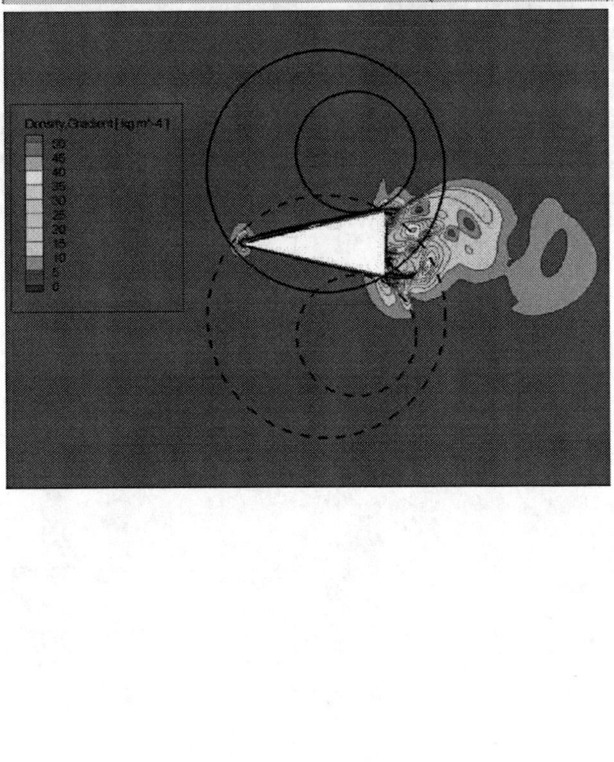

FIGURE 10: Velocity (upper left), pressure (upper right) and density gradient field (lower left) for Ma=0.6 and a flow angle of ±5 degrees

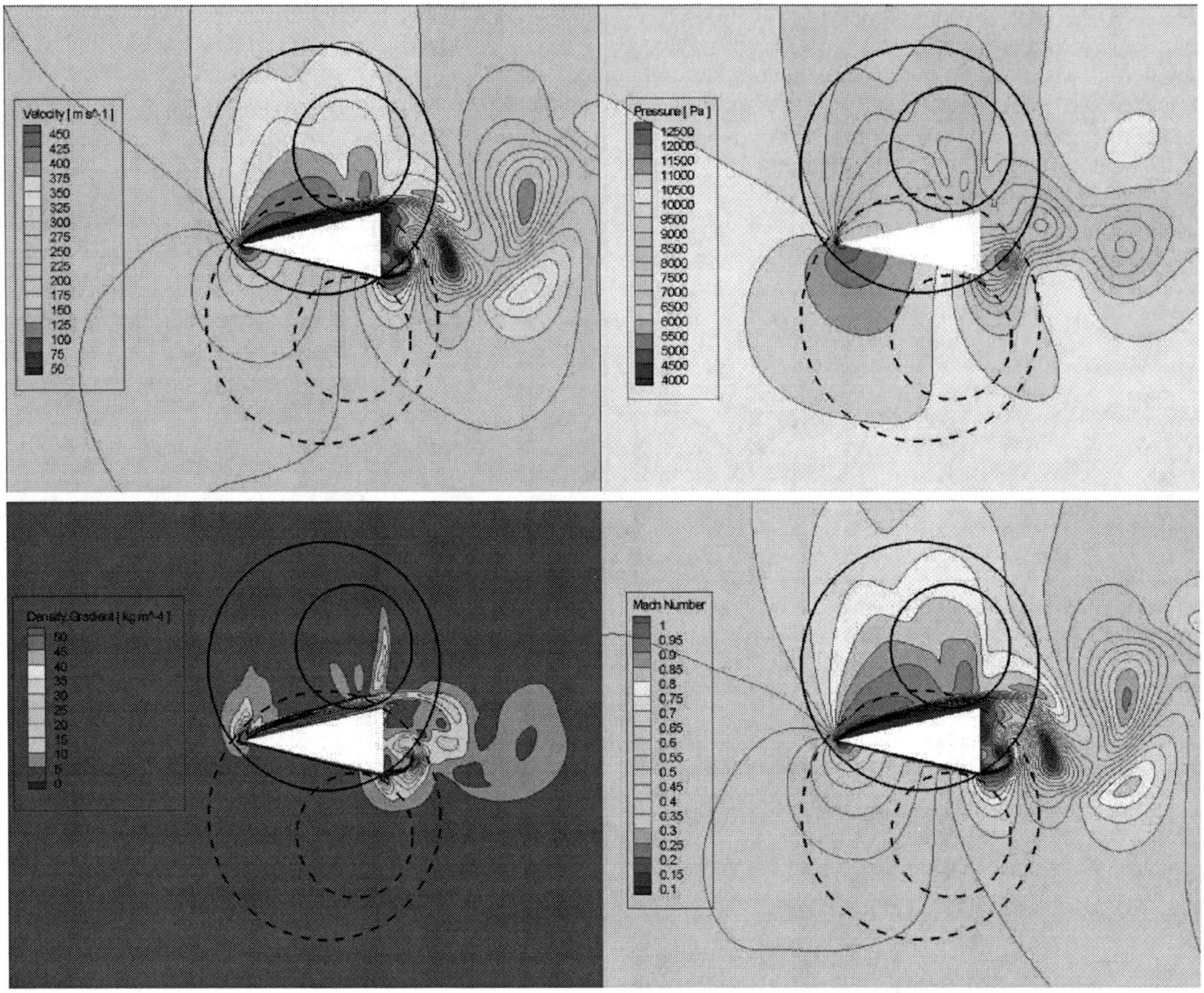

FIGURE 11: Velocity (upper left), pressure (upper right), density gradient field (lower left) and Mach number field (lower right) for Ma=0.6 and a flow angle of ±10 degrees

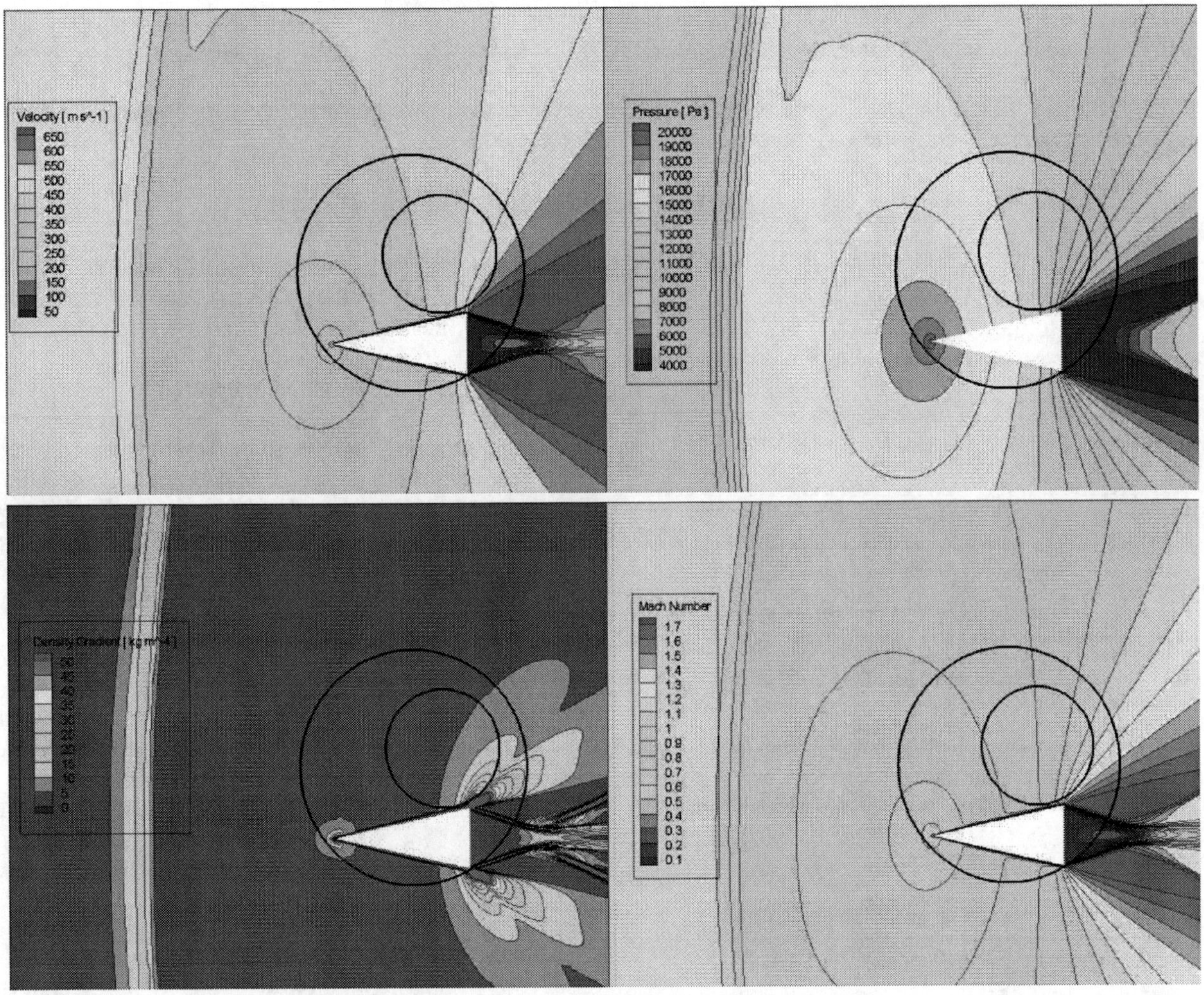

FIGURE 12: Velocity (upper left), pressure (upper right), density gradient field (lower left) and Mach number field (lower right) for Ma=1.1 and a flow angle of 0 degrees

DATA ACQUISITION AND ANALYSIS

For data acquisition, a two-channel SD2000 CCD-spectrometer from OceanOptics is used. The minimum integration time of the spectrometer is 3ms, corresponding to a sampling frequency of 333 Hz. The non-linearity of the CCD-arrays is corrected using a software tool from the same company. The spectrometer is connected to a standard PC by USB. For spectrometer control and data storage and visualisation, an in-house code has been programmed using Visual Basic.

Since the flow in turbines is inherently instationary and the measurement of fog droplets may also be impaired by big droplets separating from turbine blades or other surfaces, it is not useful to make single scans for data acquisition, but to use a number of scans and average them. In addition, making multiple scans allows a statistical treatment of the measurement.

When the optical data is pre-processed, an adaptation of the Shapiro-Wilks W-test [18,19] is used to check all samples for normality. If the W-test indicates that the samples contain outliers, these are identified and removed using the well-known Nalimov test [20] and the set f remaining samples is again checked for normality.

Another criteria for the quality of the data is the relative standard deviation of the measurement. This parameter is used to identify the upper and lower limit of the wavelength spectrum and together with the W-test, it helps to determine whether instationary phenomena are to be expected at the measuring point in question. Should the W-test indicate a normal distribution, whereas the relative deviation is high for the whole wavelength range, one can assume that some form of oscillation is present and should be investigated further by instationary measurement with higher sampling frequencies.

Finally, the fact that the refractive index of both steam and especially water varies with wavelength has been compensated by approximating the variation with a fourth order polynomial.

FIRST RESULTS

As a first demonstration of the functionality of the probe, wetness measurements made in the diffuser exit of the model steam turbine at ITSM shall be presented. The probe traversing unit was located outside of the turbine to facilitate access to the probe during the first tests. Further tests in other measurement locations are planned, where the probe will be situated within the turbine casing (see fig. 4).

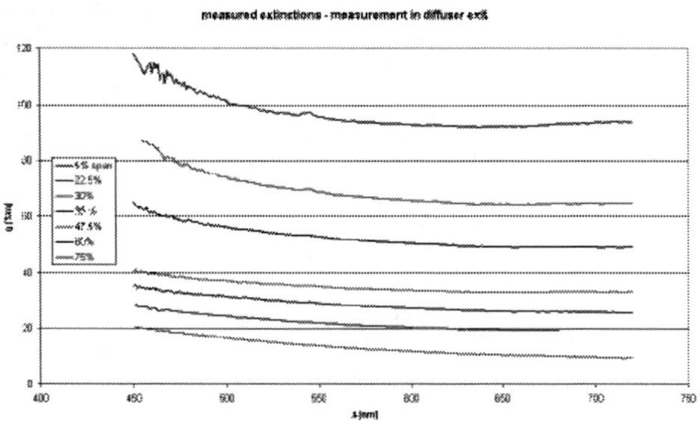

FIGURE 13: Measured extinction at selected locations

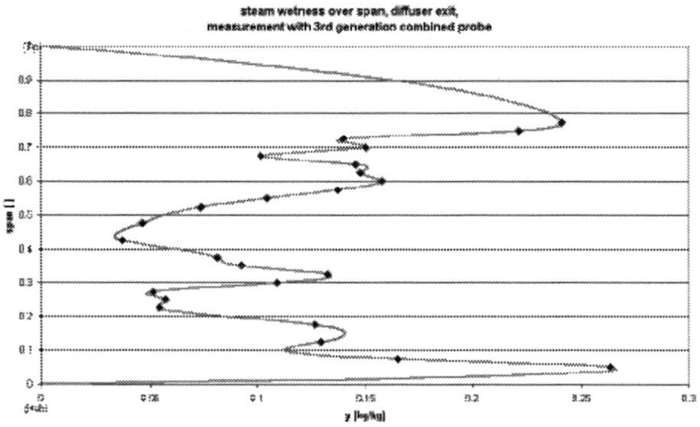

FIGURE 14: Distribution of wetness over span

Fig. 13 shows the measured extinctions at selected span levels. Apart from the measurement at 5% span, where signal shows oscillations in the range of 450 nm up to 480 nm, the signal is very smooth. Towards the diffuser walls (5% and 75 % span), extinction becomes very high. Generally, the run of all extinction curves is similar, suggesting a relatively homogeneous droplet distribution over span. The steam wetness distribution over span determined from these extinction measurements is presented in fig. 14. The diamonds represent the results from the measurements, the line is a cubic-spline interpolation. It can be seen that the wetness distribution is not homogeneous. A possible explanation of the wetness curve as measured could give too much information about the commercially sensitive blading geometry and is therefore omitted as well as the value of the overall wetness, which deviates from the wetness determined with eq 8 by about 13 %. Of course, this result is not completely satisfying, but it is a good starting point. Fig 15 shows the calculated droplet spectra for different measurement locations. It can be seen, that small droplets with a diameter of about 0.5 microns dominate, but there are also bigger droplets with a diameter of about 2 microns. It is assumed, that the bigger droplets were formed in the preceding stages, whereas the smaller droplets originate from secondary nucleation in the last stage. Towards the walls, a population of even bigger droplet emerges.

CONCLUSIONS

The accurate determination of steam quality is a very important task, as it is indispensable for the calculation of turbine efficiency. Therefore, the existing probe designs at ITSM have been evaluated and improved further. A new combined optical and pneumatic probe has been designed and tested, allowing for the simultaneous measurement of wetness and pressures, thus making it possible to get the thermodynamic state of the working fluid with one measurement only.

For the application in the model steam turbine of the Institute of Thermal Turbomachinery and Machinery Laboratory at the University of Stuttgart, the probe design has been miniaturised to a diameter of only 10 mm.

In the course of the design, the appropriateness of the probe geometry for wetness measurements has been studied. It has been shown, that due to the small probe geometry, the interaction between flow and probe is kept to a minimum, and that the optical path has been located in such a way, that the measurement of wetness is not distorted by the interaction between flow and probe. In addition, the applicability of the probe for measurements in subsonic and supersonic flows has been demonstrated.

The data acquisition and analysis has been improved, introducing statistical methods to identify and remove outliers.

First measurements in the model steam turbine of the institute have been done and yield satisfying results. Further measurements at different positions within the turbine are planned.

Moreover, a probe for measurements in full-scale turbines has already been designed and will be tested in the near future.

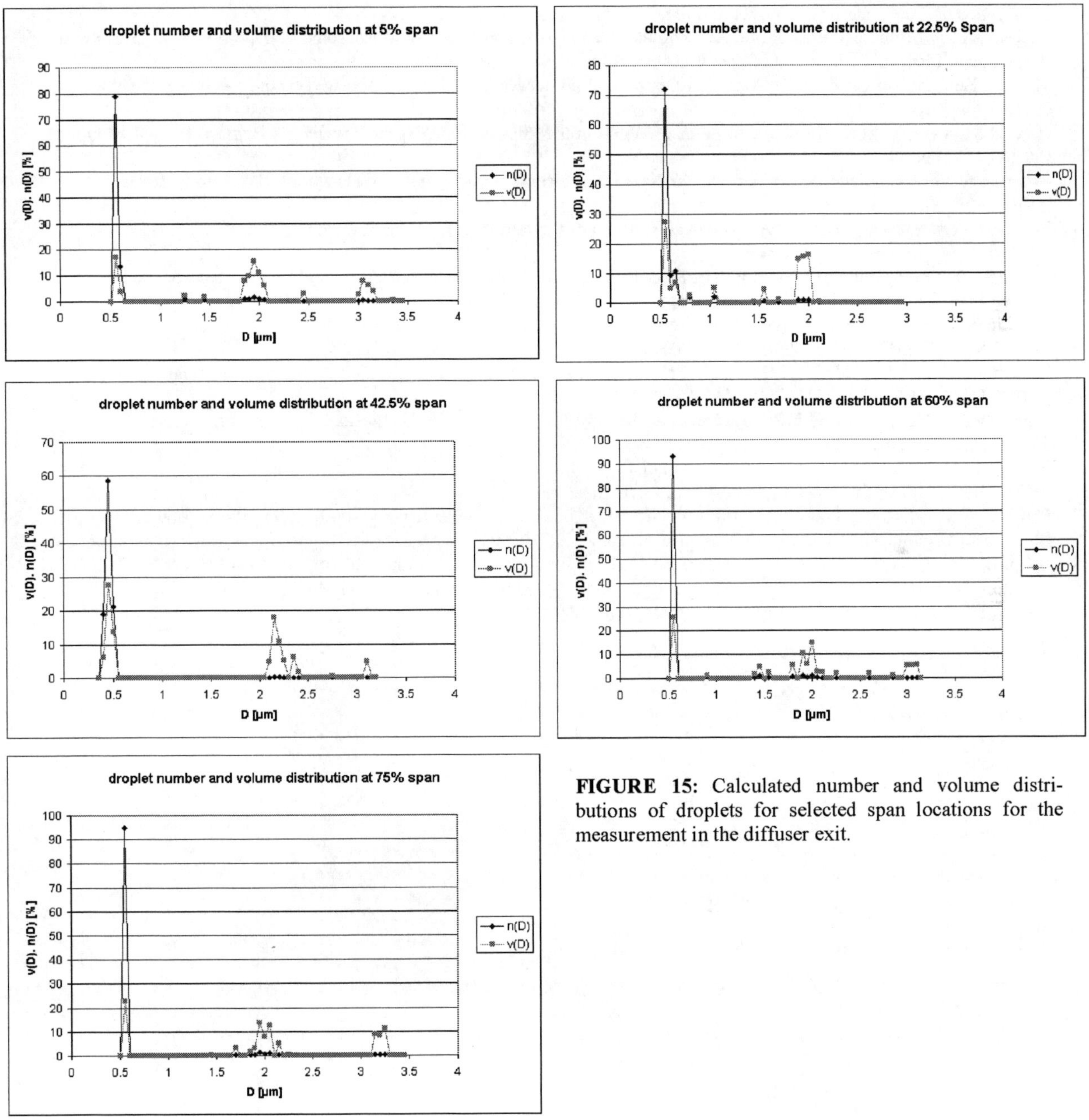

FIGURE 15: Calculated number and volume distributions of droplets for selected span locations for the measurement in the diffuser exit.

ACKNOWLEDGEMENTS

The authors wish to thank Dipl.-Ing. (FH) Christoph Heinz for the provision of CFD-results.

REFERENCES

1. DIN1943: Wärmetechnische Abnahmeversuche an Dampfturbinen, Deutsches Institut für Normung, 1975
2. IEC 953: Rules for Steam Turbine Thermal Acceptance Tests, 1990
3. ASME PTC 6 – Steam Turbines, 2004
4. VDI 2043 : Messung der Dampfnässe – Measurement of Steam Wetness Fraction, Verein dt. Ingenieure, 1999
5. Petr, V.; Kolovratnik, M.: Electrostatic Charge of Fine and Coarse Droplets in LP Steam Turbines, Proc. of the14th Int. Conference on the Properties of Water and Steam, 2004
6. Ederhof, A.: A Light Scattering Probe for Droplet Size and Wetness Fraction Measurement in Two-Phase Flow, Two-Phase Steam Flow in Turbines and Separator, Hemisphere, New York, 1976
7. Mie, G.: Beiträge zur Optik trüber Medien, speziell kolloidaler Metalllösungen, Ann. d. Physik, IV, No. 25, 1908, in German language
8. Hale, G.M., Querry, M.R.: Optical Constants of Water in the 200-nm to 200-μm Wavelength Region. J. Applied Optics Bd. 12 (1974) Nr. 8
9. Gyarmathy, G.: Grundlagen einer Theorie der Nassdampfturbine, Ph.D.-Thesis, ETH Zürich, 1962, in German language
10. Phillips, D.L.: A Technique for the Numerical Solution of Certain Integral Equations of the First Kind, J. ACM No.9, 1962
11. Twomey, S.: On the Numerical Solution of Fredholm Integral Equations of the First Kind by Inversion of the Linear System Produced by Quadrature, J. Ass Comput. Mach. No 10, 1963
12. Lawson, C.L.; Hanson, R.J.: Solving Least Squares Problems, Philadelphia : SIAM, 1995
13. STEAMQUAL SMT4 CT-97 2203, WP1: Benchmark Tests at Laboratory Level, 1998
14. Völker, L.; Casey, M.; Neef, M.; Stüer, H.: The flow field and performance of a model low pressure steam turbine, ETC 6, Lille, 2006
15. Wang, N.N.: „Nässemessung in Niederdruckturbinen mit Lichtsonden“, Ph.D.-thesis, University of Stuttgart, Germany, 1982, in German language
16. Renner, M.: Entwicklung eines optisch-pneumatischen Messsystems für Zwei-Phasen-Wasserdampfströmungen, Ph.D.-Thesis, University of Stuttgart, 2002, in German language
17. Seibold, A.:Entwicklung einer hoch auflösenden Sonde zur simultanen Erfassung von Kondensations- und Strömungsvorgängen in Niederdruck-Dampfturbinen, Ph.D.-Thesis, University of Stuttgart, 2004, in German language
18. Shapiro, S.S., Wilk, M.B.: An Analysis of Variance Test for Normality, Biometrica Vol. 52 No. 3 / 4, 1965
19. Royston, J.P.: Algorithm AS181: The W-Test for Normality, Appl. Statistics Vol. 31 No. 2, 1982
20. Kaiser, R.: Errors in Chromatography, Chromatographia Vol. 4 No. 3, 1971

NOMENCLATURE

$\underline{A}$	m x n-matrix
c	concentration, $1/m^3$
C_{ext}	extinction cross section, m^2
D	droplet diameter, m
$\vec{f}$	n-vector with droplet numbers
$\vec{g}$	m-vector with measured extinctions
Δh	enthalpy difference, kJ/kg
I	measured intensity, W/m^2
I_0	intensity of undisturbed light beam, reference intensity, W/m^2
I_{sca}	intensity of scattered light, W/m^2
k	specific attenuation coefficient, m^2
l	length of optical path, m
λ	wavelength, m
n	diameter class index
N	number of droplets
N_c	number of droplets per unit volume, $1/m^3$
m	wavelength index
$\dot{m}$	Mass flow, kg/s
P	Power, W
Q_{ext}	Extinction efficiency
X_{Mie}	Mie coefficient

Exploration on the Testing Method of Flow Field in Near-Wall Area with PIV

Shunsen WANG, Jingru MAO, Guanwei LIU

Institute of Turbomachinery Xi'an Jiaotong University, 710049 Xi'an, China

Abstract. High velocity gradient, non-uniform distribution of tracer particles and strong wall reflection of light are three major factors to limit the application of PIV in the near-wall area. Systematically investigating on PIV measurement error caused by cross-correlation algorithm and velocity gradient of shear flow have been done, and the results indicate that the relative error of PIV is in proportion to the velocity gradient and the width of interrogation window along the normal direction of vector, but it is in inverse proportion to the local velocity. Further analysis on the error of synthetic cross-correlation images suggested that the relative error of PIV is weakly depended of the particles distribution when the image reconstruction and interrogation window deformation are used, and a new criterion for judging tolerable measurement range of PIV on high velocity gradient flow field is presented based on above conclusion and correction coefficient for relative error. In addition, Weakened laser intensity, thinned laser light sheet and reduced interrogation size was used to better the vector field of particles concentration area, fluorescence technique, light filter technique and particular image processing method were used to resolve wall reflection of light and to improve the vector of field insufficient particles area, and these measures were proved to be valid by experiments. Above investigations widen the measurement range of PIV on the flow field of near-wall area.

Keywords: Near-wall area; Flow visualization; Particle image velocimetry (PIV); Measurement errors; Cross-correlation; Velocity gradient; interrogation window.
PACS: 47.80.-v

INTRUDUCTION

PIV velocity measurement techniques break through the limitation of one-point measurement by combined flow visualization and cross-correlation algorithm. The major advantages of PIV over the traditional techniques such as hot-wire anemometry (HWA) and laser-Doppler velocimetry (LDV) is that it can provide instantaneous velocity fields [1]. The near-wall area is a challenge for all kinds of optical fluid-flow measurement because of high velocity gradients and flow visualization defects; PIV is not an exception. In recent years, many investigations on this problem have been carried out and we can clarify them into the following two major aspects

(1) Measurement error caused by high velocity gradient in the near-wall area.

The cross correlation methods are found to be very accurate for nearly uniform flows. However, the accuracy is drastically reduced in the presence of high velocity gradients [2]. A solution to this problem was proposed by Huang et al. [3, 4]: deformation of the correlation windows according to the velocity gradients, in an iterative process. Error tests have been carried out by Fincham and Delerce [5] and Scarano and Riethmuller [6]. In all these approaches, the calculation of the first correlation poses a major difficulty, since the velocity gradients are still unknown. Lin and Perlin [7] proposed a method of achieving this first step, through a complex algorithm that reduces the size of the windows. In addition, iterative methods with window deformation [8, 4] and image deformation [9, 10] can also reach a slightly higher spatial resolution than the base correlation [11], e.g. with higher-order deformations, which better fit the velocity gradients [12]. Florio and co-workers present a procedure for analyzing images obtained in PIV. This consists of weighting, re-shaping and re-orientating the interrogation windows to improve the measurements in flow regions exhibiting velocity differences in space as in shear flows [13]. However, not all the flow with high velocity gradient could be accurately measured by PIV even if above various techniques were used. Accurate results depend on the proper PIV parameter. Therefore, it is very imperative for PIV user about how to

CP914, *Multiphase Flow: The Ultimate Measurement Challenge,* edited by X. Cai, Y. Wu, Z. Huang, S. Wang, and M. Wang

judge the tolerable range of PIV and how to select proper PIV parameter on high velocity gradient flow. These problems will be discussed in the section 2 and 3 of this paper.

(2) Flow visualization defects in the near-wall area.

The main phenomenon is non-uniform distribution of tracer particles and strong wall reflection of light, and it is particularly severe for the flow which the centrifugal force can not be neglected, such as the flow in bend passage. Some tracer particles concentrate on or adhere to the pressure surface of passage in the effect of inertial centrifugal force and generate continuous dispersion speckle when the pulsed laser is lightened, and results in image recognition barrier. The insufficient particles caused by flow separation near the suction wall can't meet the demand of PIV for the number of particles within an interrogation window (IW). In addition, wall reflection of light can reduce the signal noise ratio. These problems cause large PIV measurement error. A series of measures for resolving or bettering these defects are presented in the section 4 of this paper, and their feasibility has been verified by means of systematic experimental research. These investigations widen the PIV measurement on the flow field of near-wall area, and improve the performance of PIV.

ANALYSIS ON MEASUREMENT ERROR OF PIV

The Particle-tracking techniques of PIV rely on the two assumptions, which the particles velocity match fluid velocity and all particles within an IW have moved homogeneously in the same direction and the same distance. However, improper tracer particles and non-uniform flow fields may invalidate this assumption and result in additional error. Therefore, the relative measurement error of PIV consist of three parts and the equation can be written as

$$\gamma = 1-(1-\gamma_a)(1-\gamma_s)(1-\gamma_n) \tag{1}$$

Where γ_a is the cross correlation algorithm error, and it is only source of error when the flow condition meet above two assumptions; γ_n is the error caused by inhomogeneous particles motion within an IW, and the section 3 of this paper focuses on this problem; γ_s is the error from the slip of tracer particles and fluid and it will be solely discussed in another our paper.

Almost all the flow fields measured by PIV are non-uniform, and the velocity gradient is the best embodiment on its non-uniformity. There are four components: $\partial u/\partial x$, $\partial u/\partial y$, $\partial v/\partial x$, $\partial v/\partial y$ for the velocity gradient of two dimension flow field, and all the components are acting on γ_n, however, only $\partial u/\partial x$ and $\partial v/\partial y$ are relevant to γ_s. In the following investigations, no slip between the particles and the fluid, namely, γ_s=0 is supposed.

The error γ_a in the PIV estimates stems from the timing of the laser light sheet pulses Δt and the estimate of the particle displacement. The laser timing is very accurate with the development of electronic control technology and does not significantly contribute to the error in the velocity estimates. The accuracy of PIV displacement estimates σ_D is primarily determined by the ratio of the correlation peak width d_r to the particle displacement (unit: pixel) across the IW D [16]:

$$\gamma_a \approx \sigma_D$$

$$\sigma_D = \frac{d_r}{D} \tag{2}$$

$$D = \Delta t \times \vec{U} \tag{3}$$

$\vec{U}$ is the particles velocity within an IW. The full-scale measurement error, which will be the lowest error, is limited by the maximum displacement permitted on the IW. By using interpolation techniques, the correlation peak position can be estimated to sub-pixel accuracy [17]. Nominal values for the peak position error are on the order of 0.1 pixels for 3-point Gaussian peak estimators and particle images spanning 1-2 pixels. To avoid aliasing in the correlation plane, the displacement search region must be restricted to 1/4 of the IW size [18]. According to the equation (2), we could get the conclusion that the value of D should try to approach 1/4 of the IW size in order to improve the measurement precision. Assuming a PIV IW size of 64^2 pixels with 16 pixel particle displacements, the relative error in the displacement measurement is approximately 0.1/16 =0.6% of full scale, and if D=8 pixels, the relative error is approximately 0.1/8 =1.25% of full scale. It should be noticed that reducing IW can not improve the accuracy of displacement estimates.

Large displacements between exposures are desirable in order to achieve high-accuracy velocity estimates. By incorporating IW shifting in the data processing procedure, high-accuracy velocity estimates can still be obtained. In cross-correlation processing, the second correlation IW can be spatially shifted with respect to the first IW by an amount equal to the mean flow displacement between exposures, which keep the correlation peak at the center of the correlation plane and minimize any distortion effects caused by the windowing of the FFT-based correlation operation. IW image shifting ensures that most of the particles in the first exposure IW are also in the second exposure IW, which results in an improved signal to noise ratio. Employing IW shifting also enables the use of smaller interrogation regions, provided the smaller IW still contain a sufficient number of particles to provide a correlation result. Therefore, Employing IW shifting enables the measurement of displacements larger than prescribed by the 1/4 rule [16], and the accuracy can be markedly improved.

However, it is very difficult to keep all the particle displacement D approaching 1/4 of the IW size by selecting proper Δt because the majority of flow field measured by PIV is non-uniform and different area demands different Δt, and how to resolve this problem will be discussed in the section 4 of this paper.

MEASUREMENT ERROR CAUSED BY VELOCITY GRADIENT

When dealing with the highly sheared flow, the deformation of the particle image pattern between the two exposures becomes critical: the correlation peak becomes broader and multiple peaks can occur in the direction where the velocity exhibits the larger variation along the interrogation window. This behavior is undesirable since it leads to a higher measurement noise for those regions where interest is often focused, such as boundary layers, shear layers or vortices [8].

The flow in near-wall area is typical shear flow, although the velocity gradient is not the constant, reducing its flow model to the uniform shear flow is conservative and dependable for the measurement error estimates. In fact, the flow is close to the uniform shear flow within an IW with PIV. Fig.1 shows the particles motion within an IW for the 2-D uniform shear flow. The mathematical expression for its flow characteristic is

$$\frac{du}{dy} = k \quad (k \text{ is constant})$$

The deformation rate of particles image (pixel/ pixel) can be defined as

$$\varepsilon = \frac{x_1 - x_2}{\Delta y} \tag{4}$$

It quantifies the inhomogeneous particles motion and has direct relation to the γ_n. The relative error increases with the increase of ε. The relation of ε and velocity gradient can be acquired by analyzing on uniform shear flow

$$\frac{du}{dy} = \frac{u_1 - u_2}{\Delta y} = \frac{x_1 - x_2}{\Delta y \Delta t} = \frac{\varepsilon}{\Delta t} \tag{5}$$

Obviously, deformation rate is the function of Δt, and we can widen the measurement range of PIV on velocity gradient by selecting small Δt. However, the increase of cross-correlation algorithm error caused by small Δt must be considered. According to the original definition, the relative error on velocity within an IW can be written as

$$\gamma = \frac{U_t - U}{U} \times 100\% \tag{6}$$

Where U_t is the processing result of PIV, U is the true value of velocity in the center position of IW. The value of U_t is depend on the localization of the available information due to the sampling by tracer particles weights the velocity estimates towards bright particles and densely seeded parts of the flow [8] (fig. 2). It is very difficult to accurately estimate the value of γ, however, we know that $\frac{\Delta Y}{2}\frac{du}{dy}$ is the velocity difference of edge and center position of IW, and this is also the maximum difference of U_t and U for the shear flow. Therefore, the following equation can be obtained

$$\gamma < \left|\frac{\Delta Y}{2U}\frac{du}{dy}\right| \quad \text{or} \quad \gamma = c\left|\frac{\Delta Y}{2U}\frac{du}{dy}\right| \quad (c<1.0) \tag{7}$$

Where ΔY is the width of IW along the normal direction of velocity, c is correction coefficient for relative error. It is available to consider above equation as the criterion for judging the tolerable measurement range of PIV on velocity gradient. It can be seen that the measurement range is different for different value of U. In addition, the range can be enlarged by selecting small ΔY. A rectangular IW, namely, $\Delta X : \Delta Y = m$, $m>1$ is best solution for this problem in respect that tracer particles must be sufficient within an IW when ΔY is decreased. The measurement range of PIV on the velocity gradient can be obviously widened by selecting proper m. However, the most imperative thing is to obtain the value of c at present.

For verifying the above relation of velocity gradient, $\boldsymbol{\Delta Y}$, $\boldsymbol{c}$ and relative error of PIV, five groups cross-correlation images have been constructed, they have following characteristics

$$U = \frac{\partial u}{\partial y}Y \quad \text{or} \quad \frac{\frac{\partial u}{\partial y}}{U} = \frac{1}{Y} \tag{8}$$

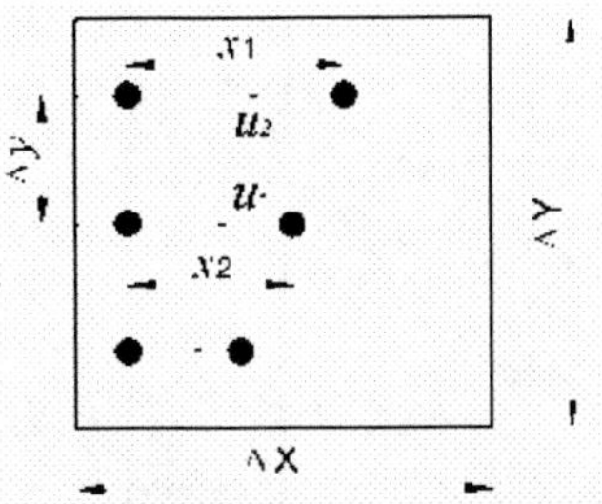

FIGURE 1. Scheme for particles motion within an IW for the shear flow

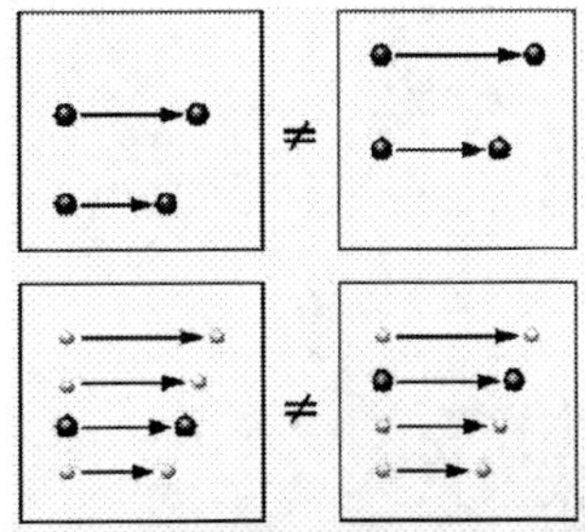

FIGURE 2. Illustration of the influence of locally available information in gradient flows: localization of the velocity information due to the random spatial sampling (top) and the weighting towards bright particles (bottom) [8]

There are 16 columns and 21 rows dot matrix in the region of 1.2mm×1.0mm for every image. All particles move along with the positive X direction. The scale of image is 1pixel=0.00286mm, and single particle possesses about 10 pixels. In order to show clearly particles motion of different position, two frames cross correlation image are combined into one group, as shown in fig.3. The velocity gradient for five groups images cover 20m/s·mm to 60m/s·mm.

The initial displacement estimation is obtained by a conventional cross-correlation analysis. With an existing velocity field the imaginary deformation can be calculated for each pixel and determines the intensity at the new position by bilinear interpolation. Multi pass iteration and decreasing IW size techniques are used.

The initial reference vector come from the IW of 128×128 pixel^2 and 87.5% overlap, there are three kinds different setting for the final vector with 128×128 pixel^2 IW and 87.5% overlap, 64×64pixel^2 IW and 75% overlap, 32×32pixel^2 IW and 50% overlap, respectively. Fig.4 shows the variation of velocity along the Y direction of fig.1 and provides the comparison of the cross-correlation processing results on synthetic images with 64×64pixel^2 IW and given value of velocity, and they have good coincidence. Fig.5, fig.6 and fig.7 show the relative error of cross-correlation results corresponding with above three kind different setting for IW size, respectively.

There are three 64×64pixel^2 interrogation windows marked 1, 2 and 3 is showed in fig.2 and red mark is the IW center. It can be seen that particles distribution along the direction of velocity gradient is always nonsymmetrical within an IW although it is uniform for whole flow field. In this paper, we select one column dots to analyze the particles distribution along the direction of velocity gradient, and the results was showed in Fig.8 (a), (b) and (c) for three kinds ΔY, respectively. All IW center was moved to 0 pixel position of vertical coordinate; and the horizontal coordinate is the same as that of fig.3. Some conclusions can be drawn from above figures.

(1) For the IW of 64×64pixel^2 and 32×32pixel^2, the relative error is less than ±1% for the majority of region and this is within the error range of γ_a. Large error occurs when U is small, or particles distribution is obviously nonsymmetrical. However, the IW of 128×128 pixel^2 has resulted in large uncertainty.

(2) According to the equation (8), the value of $U/(du/dy)$ is same for five kinds velocity gradient in the same Y position, so the relative error should have similar variation trend for the flow of different velocity gradient, and it has been proved by fig.5~fig.7.

(3) The value of c is closely related to the cross-correlation algorithm, and its range is 0.10~0.05 for above processing results.

Comprehensive analysis on above results, it can be seen that the relative error is weakly depended of the particles distribution and its range is still acceptable for engineering application only if proper ΔY is selected. We should owe this to the image reconstruction and interrogation window deformation. These methods have weighted the particles intensity in the interrogation window center for shear flow. Therefore, above results is still important reference on real CCD images, although they are based on the synthetic images.

FLOW VISUALIZATION DEFECTS IN NEAR-WALL AREA

The particles distribution and velocity vector in near-wall area are different from those in main flow area, as discussed in the section 1. It is improper to apply same processing method and parameter setting for various areas and will results in large error in some areas. So it is advisable for PIV to separate different area from the whole flow field and to take different processing measures.

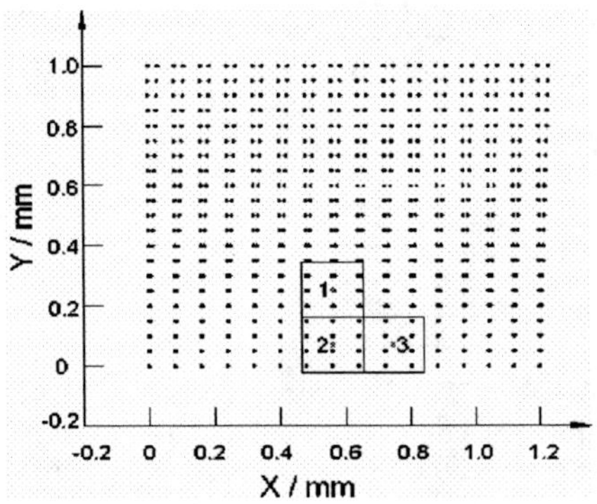

FIGURE 3. Schematic of dot matrix images for shear flow

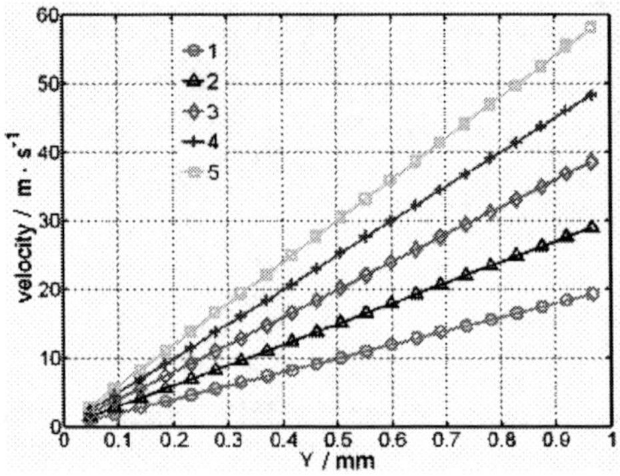

FIGURE 4. Comparison between results of cross-correlation and exact velocity, real line presents the exact velocity, and mark represents the results of cross-correlation

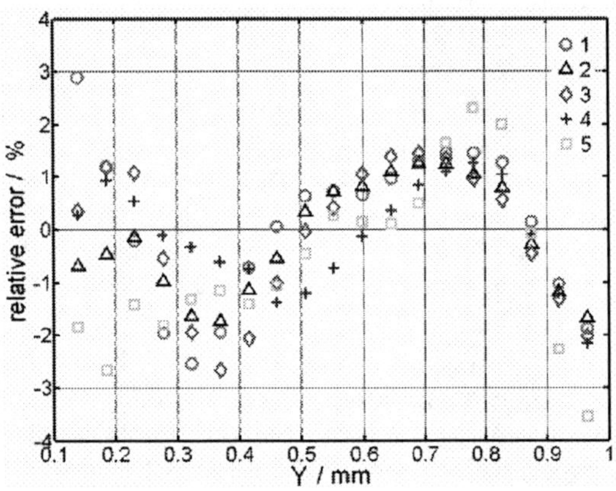

FIGURE 5. Relative error of different velocity gradient flow with the IW of 128×128 pixel^2

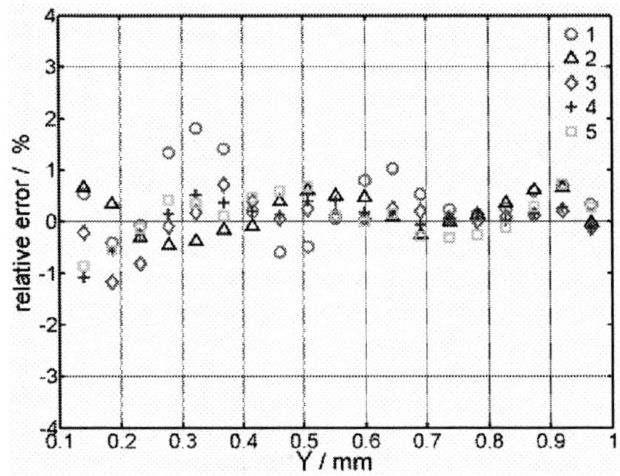

FIGURE 6. Relative error of different velocity gradient flow with the IW of 64×64 pixel^2

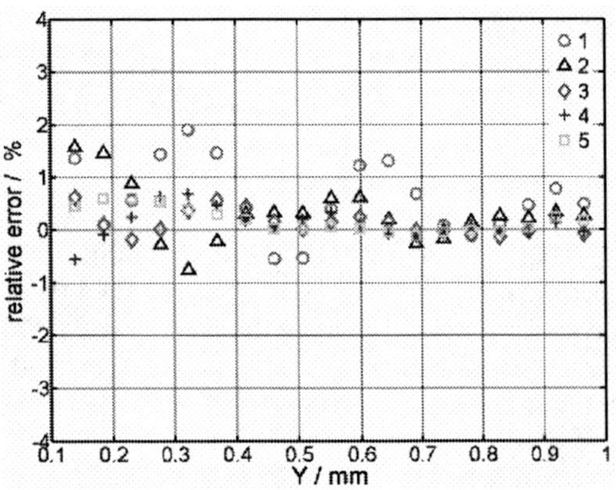

FIGURE 7. Relative error of different velocity gradient flow with the IW of 32×32 pixel^2

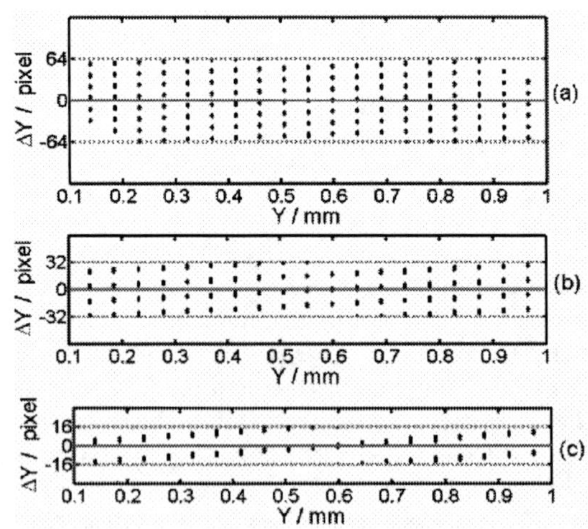

FIGURE 8. Particles distribution along the Y direction for every IW, red real line represents the IW center, and broken line represents the boundary of IW. (a) for the IW of 128×128 pixel^2, (b) for the IW of 64×64 pixel^2, and (c) for the IW of 32×32 pixel^2.

Fig.9 shows an example of a typical PIV experiment including near-wall area flow, which it is the flow within the bend passage. The whole flow field can be divided into three areas: A is particles concentration area, B is insufficient particles area and C is main flow area as shown in figure 9. Fig.10 (a) shows the velocity field of fig.9 obtained by combining standard cross-correlation algorithm with IW shifting and multi pass iteration technique, and the IW size is 32^2 pixels. Clear vector defects occur in the A and B region. The remainder of this paper focus on the how to eliminate above defects. Some particular methods will be used based on the investigation on error in section 2 and 3.

For the area A, continuous dispersion speckle come from the particles overlap and mutual dispersion between particles, and weakened laser intensity and thinned laser light sheet are available for improving the signal noise ratio of particles image. In view of sufficient particles number and small velocity in this area, reduced IW size and large Δt are used to better the match of IW on bend boundary and to improve the accuracy of PIV. However, these measures are improper for the main flow area C. Therefore, the final velocity field comes from the combination of the vector field in main flow area which is obtained under the condition of normal laser parameter setting and in near-wall area according to peculiar parameter setting. In the meantime, mask technique is used to separate different area. This paper has ever attempted to obtain the whole velocity field by means of combining the image of main flow area with that of near-wall area firstly and then to carry out cross-correlation analysis, but the results were not well. In addition, improving spatial resolution by raising the magnification of CCD lens can also reduce the tracer particles dispersion.

For area B, we add many cross-correlation images which background has been subtracted to one frame original images in order to gain enough particles number firstly, then, carry out the cross-correlation processing on composite image and acquire the vector field. It is very important that the Δt of added image must be the same as the original image.

Wall reflection of light is the familiar problem on PIV velocity measurement in the near-wall area. Fluorescent material can make the wavelength of reflection and dispersion light different from the wavelength of incidence light. In this paper, we paint the fluorescent material on the wall and then applying the band pass filter to prevent the light which is different from the wavelength of incidence light into CCD. A lot of experiments have indicated that it is valid for clearing up wall reflection of light under the normal temperature condition. However, for the high temperature flow field, fluorescent material loses its function, and some other method such as background subtraction, normalized cross correlation function, Whittaker image reconstruction and etc. are used to weaken wall reflection of light and to improve the signal noise ratio.

In addition, the image reconstruction and interrogation window deformation are used in the cross-correlation process on the near-wall area. The fig.10 (b) is the experimental results by using all kinds of above-mentioned measures on the flow shown in fig.9. The contrast results of fig.10 (b) and fig.10 (a) indicate that above-mentioned measures are valid for improving the flow field measurement in the near-wall area. The defects of vector field in area B are eliminated and the accuracy and reliability in area B is also improved.

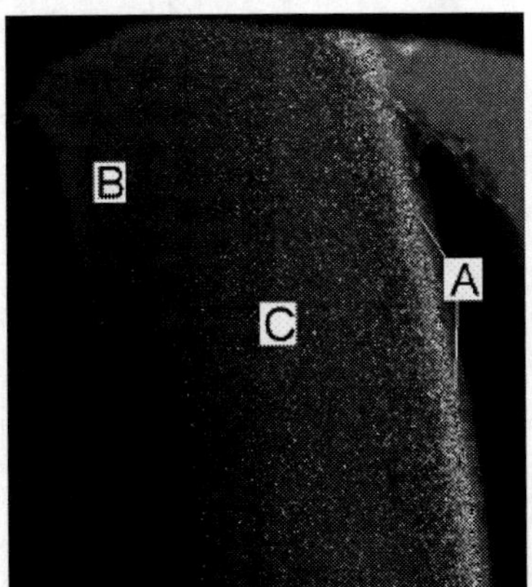

FIGURE 9. The CCD particles image including particles concentration area, insufficient particles area and main flow area

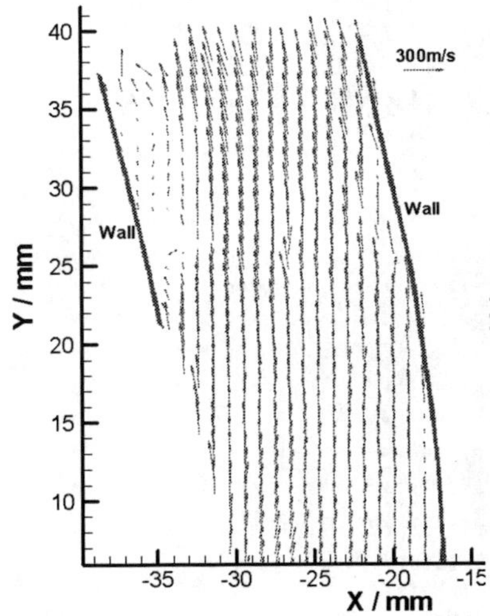

(a)

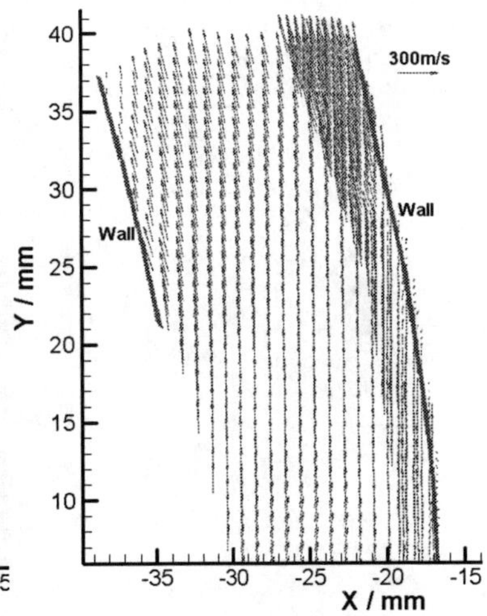

(b)

FIGURE 10. The velocity field obtained based on different processing method for the flow shown in fig.9, (a) normal cross-correlation with IW shifting and multi pass iteration technique; (b) special process for near-wall area in the basis of (a)

CONCLUSIONS

(1) For the flow in near-wall area, the relative error of PIV is in proportion to the velocity gradient and the width of interrogation window along the normal direction of velocity components, but it is in inverse proportion to the local velocity U. The correction factor for relative error is obtained and its range is 0.10~0.05. These conclusions can be used to select proper PIV parameters for the flow with strong velocity gradient.

(2) The relative error of PIV is weakly depended on the particles distribution when the image reconstruction and interrogation window deformation are used. Selecting proper interrogation window size and Δt, particularly, rectangular interrogation window can widen the measurement range of PIV on the velocity gradient.

(3) Taking different processing method for different area is available for resolving the measurement difficulty of PIV in near-wall area. Weakened laser intensity, thinned laser light sheet and reduced IW size can improved the vector field of particles concentration area. The method of adding many images which background has been subtracted to original image is used to eliminate the vector field defects of insufficient particles area.

ACKNOWLEDGMENTS

This project was supported by National Natural Science Foundation of China (NSFC) (No. 50476051; 50323001).

REFERENCES

1. H.Huang, D.Dabiri and M.Gharib, *Measurement Science & Technology* 8, 1427-1440(1997)
2. P.Meunier and T.Leweke, *Exp. Fluids* 35, 408-421(2003).
3. H.T.Huang, H.E.Fiedler and J.J.Wang, *Exp. Fluids* 15, 168-174(1993).
4. H.T.Huang, H.E.Fiedler and J.J.Wang, *Exp. Fluids* 15, 263–273(1993).
5. A.Fincham and G.Delerce, *Exp. Fluids [Suppl]* 29, S13–S22(2000).
6. F.Scarano and M.L.Riethmuller, *Exp. Fluids* [Suppl] 29, S51–S60(2000).
7. H.J.Lin and M.Perlin, *Exp Fluids* 25, 431-444(1998).
8. F.Scarano, *Meas. Sci. Technol.* 13, R1-R19(2002).
9. O.T.Tokumaru and P.E.Dimotakis, *Exp. Fluids* 19, 1-15(1995).
10. K. Jambunathan, X.Y.Ju, B.N.Dobbins and S.Ashforth-Frost, *Meas. Sci. Technol.* 6, 507-514(1995).
11. H.Nobach and C.Tropea, *Exp. Fluids* 39, 614-622(2005).
12. F.Scarano F, *Meas. Sci. Technol.* 15, 475-486(2004).
13. D.Di.Florio, F.Di. Felice and G.P.Romano, *Meas. Sci. Technol.* 13, 953-962(2002).
14. J.M.Bergthorson and P.E.Dimotakis, *Exp. Fluids* 41, 255-263(2006).
15. A.Melling, *Meas. Sci. Technol.* 15, 1406–1418(1997).
16. M.P.Wernet, *Optics & Laser Technology* 32, 497-525(2000).
17. M.P.Wernet, Fourth International Conference on Laser Anemometry, pp. 327-35, 1991.
18. R.J.Adrian, *Int, J, Heat Fluid Flow* 7, 127-45(1986).

Laser-induced incandescence of soot particles in the flue gas evolved by coal pyrolysis

Linghong CHEN*[1], Annie GARO[+], Kefa CEN*, Gerard GREHAN[+]

*State Key Laboratory of Clean Energy Utilization, Zhejiang University, No.38 Zheda road, Hangzhou, 310027, China, *Tel: 86-571-87952647, Email: chenlh@cmee.zju.edu.cn

[+]Laboratoire d'Electromagnétisme et Systèmes Particulaires, CORIA/CNRS UMR 6614, C.N.R.S., Universite & INSA Rouen, 76801,Rouen, France

Abstract. For the measurement of solid fine particles in pyrolysis systems, laser-induced incandescence (LII) has been proved to be a powerful technique for obtaining volume fraction and primary particles sizes of soot particles. In this paper, a Monte Carlo simulation is developed to analyze the influence on the detected LII flux of soot particles with the presence of coal pyrolysis products between a detector and a laser-induced incandescence control zone. The analysis corresponds to a first approach of the measurement technique, when two distinct aerosols are present around the LII control volume. The evolved flue gas of coal pyrolysis is assumed to be composed of N_2, CO_2, CO, CH_4 and fine carbon particles (soot and fly-ash particles), exhibiting spectrally dependent absorption, and anisotropically light scattering. The developed Monte Carlo simulation is applied to analyze successive absorption and scattering, especially multiple scattering events in any sequence during the propagation of LII emission. The results show that the influence of the flue gas on the detected fluxes is not uniform along the whole wavelengths spectrum. The measurement bias of LII technique for soot particles is presented when modifying coal pyrolysis products composition.

INTRODUCTION

There is an urgent need to control nanoscaled particulates from the pyrolysis sources such as power plants, diesels trucks and buses, cars, etc. This requirement is driven by a fundamental desire to preserve all the life's health and the environment in the global world and it is being forced by recent environmental standards and regulations. Besides these concerns, this increased need is also because of an adverse influence of these nanoscaled particles on the performance of the power generation systems.

Laser diagnostic techniques have been proved to be powerful tools for extracting information of nano-particles in a non-intrusive way with an extremely high time resolution. The technique of Laser-induced incandescence (LII) [1-6] generally applied soot particles has received increased attention during the last decade. This theoretical approach is confirmed by various measurements in the premixed and diffusion flames, in the diesel exhaust as well. Produced by hydrocarbons pyrolysiss,

CP914, *Multiphase Flow: The Ultimate Measurement Challenge,* edited by X. Cai, Y. Wu, Z. Huang, S. Wang, and M. Wang

soot particles consist of aggregates of spherical primary particles of about 30-100 nanometers, mainly formed of amorphous carbon. With the LII method, the soot particles within the laser beam path are rapidly heated to their vaporisation temperature (approximately 4000K to 4500K), which is achieved by means of a highly energetic laser pulse. The enhanced thermal radiation generated, i.e. incandescence is then detected. For the simple case of emission due to thermal radiation only, the homogeneous isotropic thermal radiation in any point of this volume can be described by the Planck law [7], assuming that an elementary volume is large compared to the size of particles. The LII signal depends on the excitation and detection wavelengths, the excitation intensity, detection timing duration, and the soot volume fraction.

During the LII measurement procedure, the common assumption is usually that no scattering or absorption by ambient particles and gases affects the soot particles. However, detected signals may be interfered by scattering and absorption in the media when the ambient soot concentration is high or there exists high scattering particles such as fly-ash, coal etc. Recently a study of the bias generated by light scattering on pyrometry measurements was published for a one-dimensional on-line measurement through a temperature gradient [8]. To our best knowledge, no research has been reported on the prediction of LII signal propagation in the media simultaneously exhibiting absorption and anisotropically scattering.

In the present study, a 3-dimensional (3D) Monte Carlo (MC) simulation is developed to analyze absorption and scattering effects on the LII signal propagation within a cube of the coal pyrolysis products (i.e. a mixture of gases, soot and fly-ashes particles). The soot particles show strong absorption but weak scattering; the fly-ash particles show weak absorption and strong scattering. In the following paragraphs, a test case of LII within the coal pyrolysis products is described, and the optical properties of soot and fly-ash particles from the coal pyrolysis are discussed. Then a 3D MC algorithm is described including the description of emissive source, interface crossing, absorption, scattering, and heat flux. The different levels of scattering and absorption effects on the heat flux are then discussed. The detected flux showing the influence of particles and the assumptions related to their absorption and scattering behavior is presented for different particles sizes and concentrations. The conclusions are drawn in the last paragraphs.

NUMERICAL METHOD

For the absorption, emitting and scattering mixture of coal pyrolysis products: the mixture of gases and particles in local thermodynamic equilibrium is enclosed in a three-dimensional enclosure, the spectral RTE [2] is written by:

$$\frac{\partial I_\lambda(\vec{x},\vec{\Omega})}{\partial x} = k_{abs,\lambda} I_\lambda^0[T(\vec{x})] - k_{ext,\lambda} I_\lambda(\vec{x},\vec{\Omega}) + k_{sca,\lambda}\int_{\Omega'=4\pi} I_\lambda(\vec{x},\vec{\Omega}') f_\lambda(\vec{x},\vec{\Omega}'\rightarrow\vec{\Omega})d\Omega' \quad (1)$$

where, $I_\lambda^0[T(\vec{x})]$ is the spectral intensity blackbody intensity at temperature T. $I_\lambda(\vec{x},\vec{\Omega})$ is the spectral intensity at the location $\vec{x}$ along the direction $\vec{\Omega}$. $k_{abs,\lambda}$, $k_{sca,\lambda}$ and $k_{ext,\lambda}$ are the spectral coefficients of absorption, scattering and extinction, respectively. $f_\lambda\,(\vec{\Omega}'\rightarrow\vec{\Omega})$ the phase function of the medium.

To solve Eq. (1) in the gas flue from coal pyrolysis, the field distributions of the absorption and scattering coefficients and the phase functions of the medium are needed. The radiative characteristics of the gas phase are simply introduced by a prescribed spectral absorption coefficient. The particulate phase may consist of soot and fly-ash particles in varied proportions.

In the flames, solid soot and fly-ash particles are localized in rich hydrocarbon zones in presence of PAHs (polynuclear aromatic hydrocarbons), which may be absorbed on the soot solid surface. When using optical diagnostics to characterize soot particles, the selection of collected wavelengths is generally performed avoiding the PAHs fluorescence and the C_2 emission. The value of 405 nm (+/- 5nm) is often selected as the wavelength bandwidth to collect LII signal. The way to introduce the absorption of gas components is shown in our numerical tool [17].

The complex refractive index of fly-ash particle [9] is assumed as: m=1.5-0.005i with monodisperse spheres $d_{fly\text{-}ash}$=1~4.3 μm. The complex refractive index of soot particles is the object of a large number of publications [10-12]. We chose the value: m = 1.57-0.56i. The soot particles are considered as monodisperse spheres with diameter d_{soot} between 0.001~0.1 μm. Table 1 lists the optical properties of soot and fly-ash particles computed with the Mie formulation [17]. At the wavelength of 405 nm, the soot particles show strong absorption while fly-ash particles a high scattering. Figure 1 shows the scattering phase functions of particles calculated by Mie Theory.

TABLE 1. Optical properties of soot and fly-ash particles with the incident wavelength λ=405 nm

	Soot particles			Fly-ash particles		
d_p (μm)	0.001	0.03	0.1	1	2	4.3
C_{ext} (m^2)	6.4×10^{-21}	1.82×10^{-16}	8.94×10^{-15}	1.39×10^{-12}	7.15×10^{-12}	3.03×10^{-11}
C_{sca} (m^2)	1.7×10^{-27}	1.22×10^{-18}	1.45×10^{-15}	1.21×10^{-12}	6.25×10^{-12}	2.30×10^{-11}
C_{sca} / C_{ext}	2.6×10^{-7}	0.00668	0.162	0.871	0.874126	0. 76 1.
$\alpha = \frac{\pi d}{\lambda}$	0.008	0.2327	0.775	7.75	15.50617	33.3

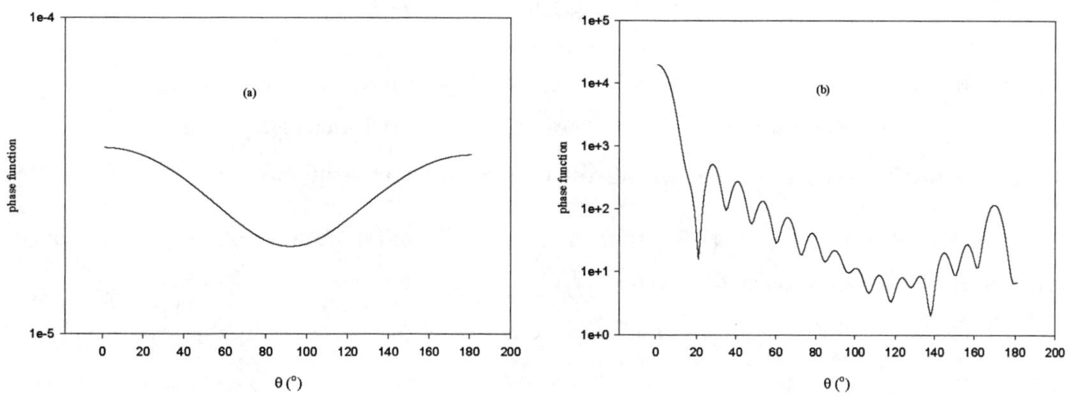

FIGURE 1. The scattering phase functions of particles using Mie theory at the wavelength of 405 nm. (a) the diameter of the soot particles: d_{soot}=0.03 μm; (b) the diameter of fly-ash particles: $d_{fly\text{-}ash}$=2 μm.

The main flow chart of Monte Carlo method to solve equation (1) is described in Fig. 2. To test the concept, a linear Cartesian grid is established for a cubic domain, which is divided into several sub-volumes, corresponding respectively to the coordinates x, y, k. A local referential (x', y', z') is defined in each sub-volume and the origin of the referential is assumed to be in the centre of this sub-volume. Then the initial emissive location in the local referential (x', y', z') is:

$$\vec{p} = \begin{bmatrix} 0.5\times(2R_1-1)\Delta L_{x'} \\ 0.5\times(2R_2-1)\Delta L_{y'} \\ 0.5\times(2R_3-1)\Delta L_{z'} \end{bmatrix}_{(x'y'z')} \tag{2}$$

where R_1, R_2, R_3 are equidistributed random number between 0 and 1. The values of $\Delta L_{x'}$ $\Delta L_{y'}$ and $\Delta L_{z'}$ are the side lengths of the sub-volume along x'-axis, y'-axis and z'-axis, respectively.

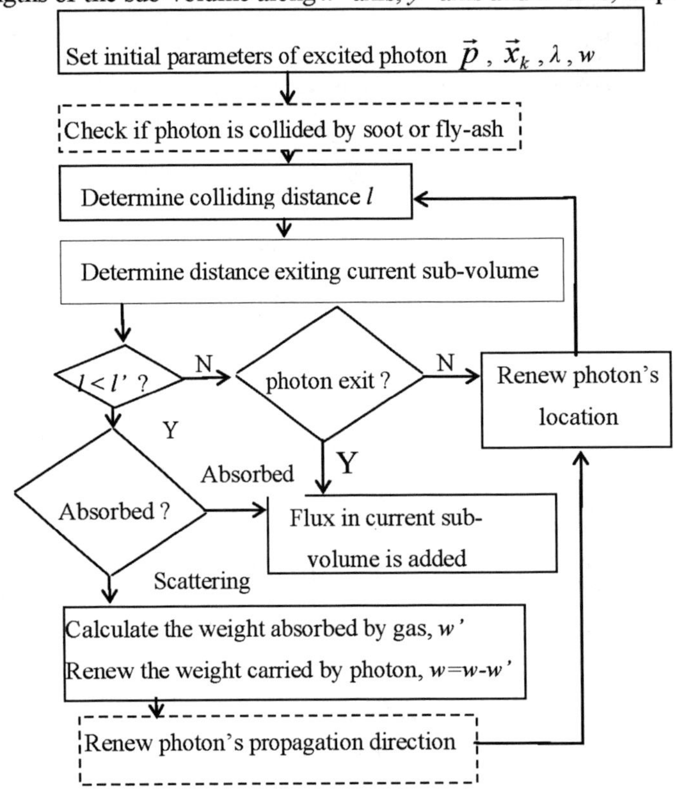

FIGURE 2. The flowchart of the MC simulation for describing the incandescence emission of a photon excited by the laser beam through a three-dimensional absorbing and scattering medium.

The initial emissive direction $\vec{x}_k$ is described with a polar angle θ and an azimuthal angle φ, which come from arbitrary reference axes [13]. Since the angles of θ and φ are independent, and the integral is separable, then the angles can be determined by:

$$\theta = \cos^{-1}(1-2R_4) \quad 0 \le \theta \le \pi \tag{3}$$

$$\varphi = 2\pi R_5 \qquad 0 \le \varphi \le 2\pi \tag{4}$$

So, the x'-axis, y'-axis, z'-axis components of unit direction vector in the in the local referential are:

$$\vec{x}_k = \begin{bmatrix} \sin\theta\cos\varphi \\ \sin\theta\sin\varphi \\ \cos\theta \end{bmatrix}_{(x'y'z')} \tag{5}$$

The initial statistical weight w, carried by each photon in the interval $[\lambda, \lambda+\Delta\lambda]$ is by:

$$w = 4\varepsilon_\lambda V_{ijk} F_\lambda^{\lambda+\Delta\lambda} \sigma T_{ijk}^4 / N_\lambda \tag{6}$$

where ε_λ represents the spectral emissivity. V_{ijk} is the volume of the sub-volume (i, j, k). $F_\lambda^{\lambda+\Delta\lambda}$ is the fraction of blackbody radiative power in the interval $[\lambda, \lambda+\Delta\lambda]$ [13]. σ is the Stefan-Boltzmann constant. T is the temperature of soot particles in the sub-volume (i, j, k). N_λ is the number of photons in the interval $[\lambda, \lambda+\Delta\lambda]$of the sub-volume (i, j, k). One million photons are assumed to be launched from each wavelength interval in each sub-volume of the laser-heated zone.

During the process of tracking a photon through the medium, the calculation of intersection mainly includes determination of the intersection location on the interface of the current sub-volume, a new sub-volume for photon to enter, and its launching coordinates and direction related to the local referential (x', y', z') in the new sub-volume. The distance l' is needed to be determined, which is the distance from the photon's launched location to the interfaces around the current sub-volume along its propagation. The intersection location is deduced from the planar nature of the interfaces after several iterations [14].

After being emitted with a probabilistic emission law, photon may have be absorbed or scattered several times inside the cubic domain. The travelling distance for a photon before being collided within the medium is decided by,

$$l = -\frac{\ln R_6}{k_{ext}} \tag{7}$$

where R_6 is equidistributed random number between 0 and 1, k_{ext} extinction coefficient of particles.

A random number R_7 is used to be compared to the scattering albedo a to decide whether photon is absorbed or scattered. If R_7 is greater than the value of albedo, then the photon is absorbed by the particle, otherwise the photon is scattered. The direction after scattering is according to the scattering phase function. Whenever a photon is emitted, a relative coordinate system is adapted on it. The z''-axis of the system is identical to the propagation direction of the photon. In this system, the phase function is symmetric with respect to the azimuth angle. Under this arrangement and when scattering occurs, $f(\Omega, \Omega')$ for spherical particles has been reduced to $f(\theta)$. The azimuthal angle φ is calculated with eq. (4), and the polar angle θ is sampled with random number R_8:

$$R_8 = p(\theta) = \frac{\int_0^{\theta} f(\theta)\cdot \sin\theta \cdot d\theta}{\int_0^{\pi} f(\theta)\cdot \sin\theta \cdot d\theta} \tag{8}$$

Because the above equation has no analytic expression for the general expression of phase function, the value $p(\theta_i)$ is usually stored in a table for discrete value $\theta_i (i=1,180)$. For a random number R_8, the value θ_i is searched in this table, and $p(\theta_i)<R< p(\theta_{i+1})$, then θ is deduced by linear interposition between θ_i and θ_{i+1}.

RESULTS

Absorbing and emitting medium

A basic test case is taken to represent globally the domain of pyrolysis products downstream the flame front of coal pyrolysis with dimensions 50 mm × 50 mm × 50 mm, in Fig. 3. The incident laser beam enters in the middle of the vertical side. The zone around the y-axis corresponds to the laser beam diameter of about 2 mm. We assume that the heated particles are at an identical temperature, all along the laser beam path, neglecting the possibility of on-line extinction of the laser excitation. The temperature of laser-heated soot particles T_{lii} is 4000K. Outside of the cube, the walls are chosen like a black and cold envelop. There is no influence of the external medium.

Typical chosen values are adopted for the cubic domain at atmospheric pressure: (1) At the laser-heated zone: the value of the gases absorption coefficients $k_{abs,gas}$ is assumed to be 0.001m^{-1}, the diameter of soot particles is d_{soot}=30 nm with 0.14 ppm m^3/m^3, no fly-ash particles exist; (2) outside the laser-heated zone: the value of the gases absorption coefficients $k_{abs,gas}$ is 0.001m^{-1}, the diameter of soot particles is d_{soot}=30 nm with 1.41 ppm m^3/m^3, the diameter of fly-ash particles is $d_{fly\text{-}ash}$=2 μm with 8.37 ppm m^3/m^3.

The LII signal is observed in a plane at 90° (plane X_{max},) after a mean direct path δ(= 25 mm) in the medium, which corresponds to the distance between the "lii" zone and the observation point.

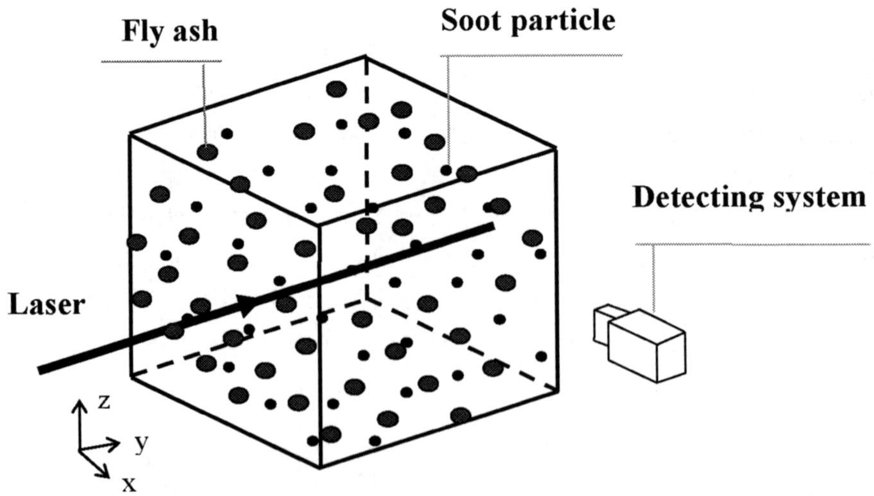

FIGURE 3. The configuration of the MC numerical simulation.

The influence of coal pyrolysis products

In this paper we assume the soot and fly-ash particles as main contributors to radiative heat transfer of LII during the pyrolysis of coal. As a base case, a cube contains only flue gases, then soot and fly-ash particles are progressively introduced. The cube is shown in Fig. 3. The radiative properties of coal pyrolysis products are calculated by Mie theory, shown in Table 1. The effect of soot and fly-ash on the surface flux of LII, as calculated with MC method, is shown in Fig. 4. A typical decrease of the wall flux then soot particles is added to the flue gas is observed. In contrast, when fly-ash, which has extinction in the same order of soot particles, is added in the mixture, the result is as apparent increase of the wall flux. This is due to the low scattering albedo of soot particles which allows an absorption coefficient almost an order of magnitude greater than that of fly-ash particles.

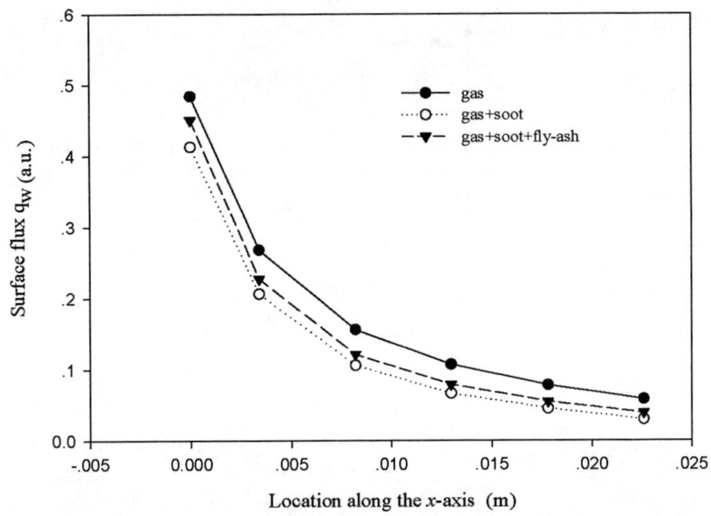

FIGURE 4. Effect of coal pyrolysis products on the distribution of the radiative wall flux is predicted by the MC method. The coefficient of gas, $k_{abs,gas}$=0.001 m^{-1}. The diameter of the soot particles d_{soot}=30 nm with 0.14 ppm m^3/m^3 in the laser-heated zone, and 1.41 ppm m^3/m^3 in the other zones. The mean diameter of the fly-ash particles $d_{fly\text{-}ash}$=2 μm with 8.37 ppm m^3/m^3 outside the laser-heated zone.

The effect of particles sizes

In order to investigate the particle size effect on the LII signal propagation , the MC simulation is made when the volume outside the laser-heated zone exists soot particles and fly-ash particles, respectively. The particles at the laser-heated zone are assumed as the soot particles with the diameter of 30 nm and the volume fraction of 0.413 ppm m^2/m^2 at the following MC simulations.

Figure 5 shows the calculated surface flux at the positive *x*-axis at several diameters of soot particles outside the laser-heated zone are investigated: 1 nm, 30 nm, 60nm, and 100 nm respectively. We can find the obtained surface fluxes decrease exponentially with the increase of the concentration of soot particles. The similar tendency is shown for the fluxes at these sizes of the soot particles.

Figure 6 is the calculated surface flux at the positive *x*-axis at several diameters of fly-ash particles outside the laser-heated zone: 1μm, 2 μm, and 4.3 μm, respectively. With the increase of the fly-ash concentration, the surface LII flux decrease at the several sizes of particles. At the same concentration of particles, the results show the value of the surface flux decreases with the increase of the size of the fly-ash particles.

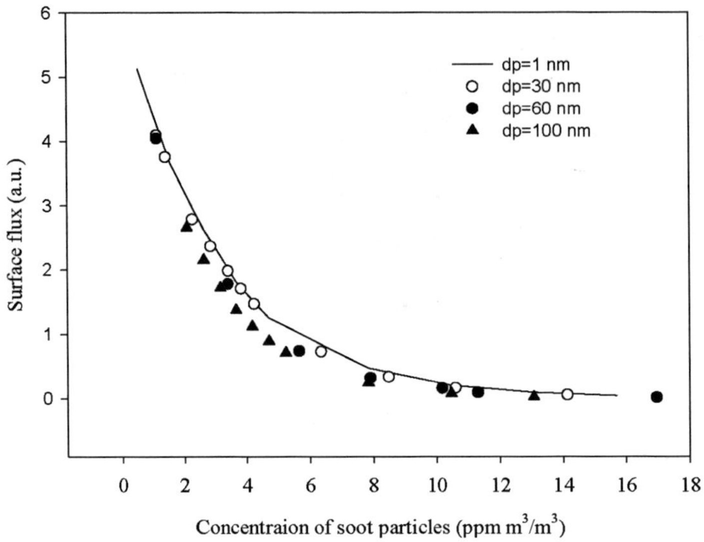

FIGURE 5. The calculated LII flux arriving on the surface of medium at +*x*-axis versus the volume concentration of soot particles outside the laser-heated zone.

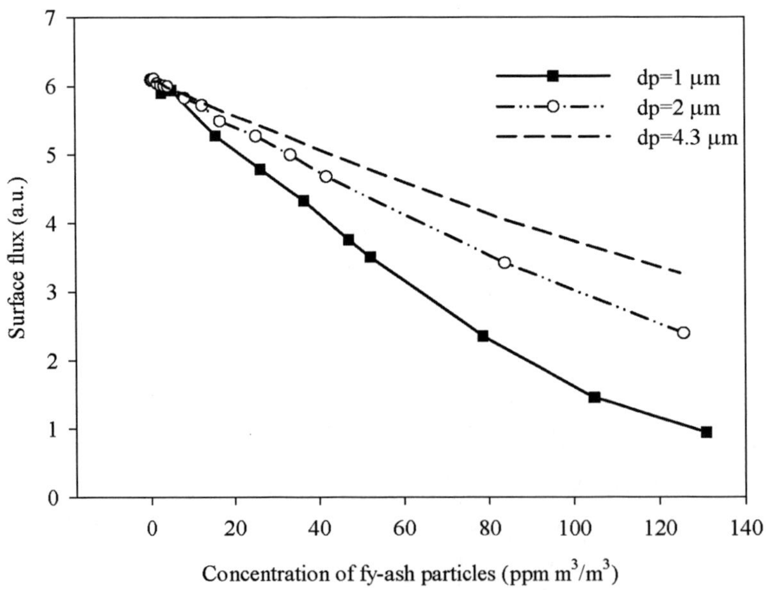

FIGURE 6. The calculated LII flux arriving on the surface of medium at +*x*-axis versus the volume concentration of fly-ash particles outside the laser-heated zone.

The influence of scattering

For each set of volume fractions, the calculations are repeated considering isotropic, anisotropic as well as neglecting of scattering. The wall fluxes at the positive *x*-axis are calculated with the MC method. The obtained esults are presented in figures 7 and 8 for the soot and fly-ash particles respectively.

On figure 7, the surface flux decreases exponentially with the increase of the concentration of soot particles. The scattering effect on the fluxes shows the similar tendency. This phenomenon is explained by the strong absorption characteristics of soot particles, show in Table 1 in the previous paragraph. The relation of flux and volume fraction can be calculated according to Beer-lambert law [4]. From the figure, the scattering effect can be neglected.

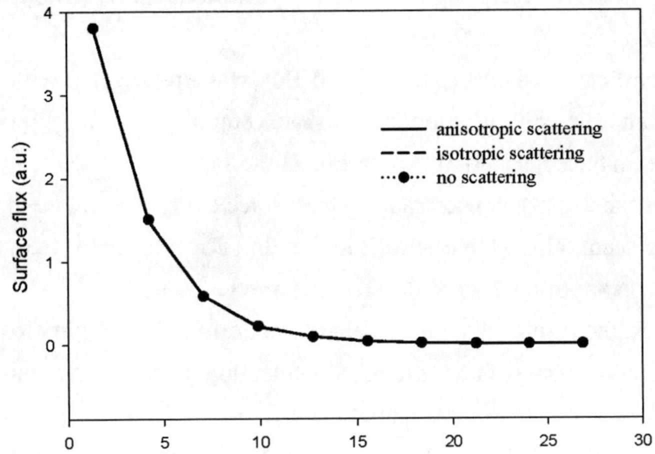

FIGURE 7. Effect of the assumption concerning the scattering behavior of soot particles on the LII flux arriving on the surface of medium at +*x*-axis. The particles existing outside the laser-heated zone are soot particles: d_{soot}=30 nm.

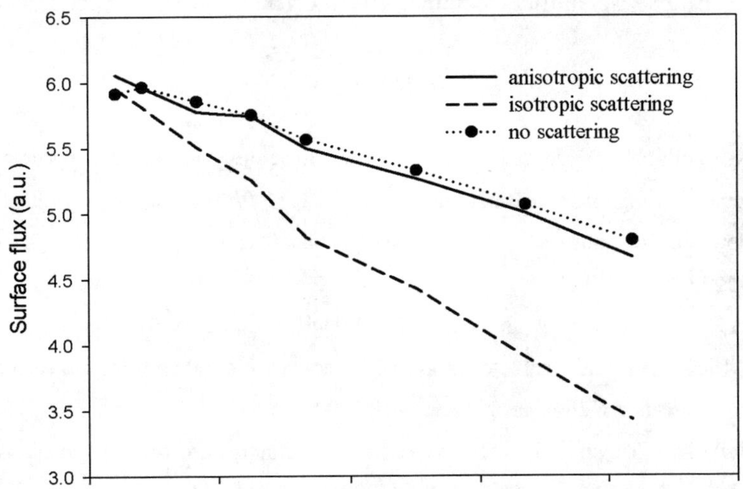

FIGURE 8. Effect of the assumption concerning the scattering behavior of fly-ash particles on the LII flux arriving on the surface of medium at +*x*-axis. The particles existing outside the laser-heated zone are fly-ash particles: $d_{fly\text{-}ash}$=2 μm.

Compared with soot particles, the larger discrepancies among the curves corresponding to anisotropic, isotropic and neglecting of scattering occurred for the fly-ashes particles, shown in Fig.8. In addition, the curve of non-scattering calculations is closer to those of anisotropic scattering, compared to the curves of isotropic scattering. This is explained by the highly forward scattering behavior of the fly-ash particles, as mentioned in the previous paragraph.

CONCLUSION

The effect of absorption and scattering of coal pyrolysis products on LII signal of soot particles are investigated with 3D Monte Carlo method.

The effect of soot particle sizes on the surface LII flux was investigated with the diameter in the range of 1 and 100 nm, and the concentration is less than 18ppm m^3/m^3. We found the size effect of soot particle at this range on the detected LII flux is almost the same.

With the increase of the fly-ash concentration, the surface LII flux decrease at the several sizes of particles. At the same concentration of particles, the simulated results show the value of the surface flux decreases with the increase of the size of the fly-ash particles.

For the soot particles, the results show that realistic scattering of soot particles can be assumed as isotropic or neglecting the scattering. Therefore, both neglecting of scattering and isotropic scattering for the soot particle are reasonable approximate approaches.

For the fly-ash particles, neglecting scattering gives results closer to anisotropic scattering than isotropic scattering.

ACKNOWLEDGMENTS

This study is supported by National Natural Science Foundation of China (No.60534030), and financed by the scholarship of French Embassy in China and by a Doctoral Grant from French Embassy in China ("Bourse Scientific Doctorale en Alternance" n° de l'Ambassade de France en Chine).

REFERENCES

1. B. Axelsson, R. Collin, P. E. Bengtsson, Laser-induced incandescence for soot particle size and volume fraction measurements using on-line extinction calibration, *Appl. Phys. B* **72**:367-372 (2001).
2. C. J. Dasch, New soot diagnostics if flames based on laser Vaporization of soot, *20th Symp. (Int.) on Pyrolysis*, 1231-1237 (1984).
3. D.R. Snelling, G.J. Smallwood, R. A. Sawchuck, S. W. Neill Gareau, W. L. Chippior, F.Liu, Ö. L. Gülder, W. D. Bachalo, In-situ real-time characterization of particulate emissions from a diesel engine exhaust by laser-induced incandescence, *SAE paper* No. 2000-01-1994.
4. R.L. Vander Wall, T.M. Ticich, A.B. Stephens, Can soot primary size be determined using laser-induced incandescence, *Pyrolysis and Flame* **116**: 291-296 (1999).
5. R.L. Vander Wall, K.J. Weiland, Laser-Induced Incandescence: Development and Characterization Towards a Measurement of Soot-Volume Fraction, & quot; *Appl. Phys. B*, **59**: 445-452 (1994).

6. L.A. Melton, Soot diagnostics based on laser heating, *Appl. Opt.*, **23**: 2201-2208 (1984).
7. R. Siegel, J.R. Howell, Thermal Radiative Heat Transfer, Hemisphere Publishing Corporation, Tokyo: Mc Graw-Hill, 1972.
8. J.J. Murphy, C.R. Shaddix, Influence of scattering and probe-volume heterogeneity on soot measurements using optical pyrometry, *Pyrolysis and Flame* **143**: 1-10 (2005).
9. F. Liu, J. Swithenbank, The effects of particles size distribution and refractive index on fly-ash radiative properties using a simplified approach, *Int. J. Heat Mass Transfer* **36**: 1905-1912 (1993).
10. C. Janzen, H. Kleinwechter, J. Knipping, H. Wiggers, P. Roth, Size analysis in low-pressure nanoparticles reactors: comparison of particle mass spectrometry with in situ probing transmission electron microscopy, *J. Aerosol Science* **33**:833-841 (2002).
11. J. Keskinen, K. Pietarinen,M. Lehtimäki, Electrical low pressure impactor, *J. Aerosol Science* **23**:353-360 (1992).
12. E.O. Knutson, K.T. Whitby, Aerosol classification by electric mobility: Apparatus, theory, and application, *J. Aerosol Science* **6**:443-451 (1975).
13. D.R. Snelling, G.J. Smallwood, I.G. Campbell, J.E. Medlock, Ö.L. Gülder, Development and application of laser-induced incandescence (LII) as a diagnostic for soot particulate measurements, AGARD Conference Proceedings 598, Advanced Non-Intrusive Instrumentation for Propulsion Engines, Brussels, Belgium, 23-1 to 23-9 (1997).
14. P. Roth, A. V. Filippov, In situ ultrafine particle sizing by a combination of pulsed laser heat up and particle thermal emission, *J. Aerosol Science* **27**: 95-104 (1996).
15. D.R. Snelling, G.J. Smallwood, R.A. Sawchuck, S.W. Neill, D. Gareau, W. L. Chippior, F. Liu, Ö.L. Gülder, W.D. Bachalo, Particulate matter measurement in a diesel engine exhaust by laser-induced incandescence and the standard gravimetric procedure, *SAE paper* No. 1999-01-3653
16. T. Ni, J.A. Pinson, S. Gupta, R.J. Santoro, Two-dimensional Imaging of Soot Volume Fraction by the Use of Laser-Induced Incandescence, & quot; *Appl. Opt.* **34**: 7083-7091 (1995).
17. L.H. Chen, Study by numerical simulation of impact of multiple scattering on participating media, Ph.D. thesis at Rouen University in France, (Oct. 2005).

NOMENCLATURE

a	ratio of scattering to extinction $=k_{sca}/k_{ext}$
$C_{ext}\ C_{sca}$	cross sections of extinction and scattering, respectively [m^2]
d_{soot}	diameter of soot particles [nm]
$d_{fly\text{-}ash}$	diameter of fly-ash particles [μm]
F_λ	fraction of blackbody radiative power in the interval [λ, $\lambda+\Delta\lambda$]
f_v	volume fraction of soot particles [ppm]
$f()$	scattering phase function
i, j, k	node numbers of sub-volume along x-axis, y-axis and z-axis
$k_{abs}, k_{ext}, k_{sca}$	coefficients of absorption, extinction, scattering, respectively [m^{-1}]
l	distance between emitting and colliding location for photon [m]
l'	distance between emitting and interface of current sub-volume [m]
$\Delta L_{x'}\ \Delta L_{y'}\ \Delta L_{z'}$	side length of sub-volume along x'-axis, y'-axis, z'-axis respectively [m]
m	complex refractive index
N_λ	number of photons in the interval [λ, $\lambda+\Delta\lambda$]of the sub-volume (i, j, k)
N_p	concentration of soot particles [$/m^3$]
$\vec{p}$	location of photon
q	radiative heat flux per unit area [kW/m^2]
∇q	divergence of radiative heat flux per unit volume [kW/m^3]
$Qnet$	power crossing the surface interface [kW]
$R_i(i=1,8)$	equidistributed random number between 0 and 1

T_{lii}	temperature of laser-heated particles [K]
$\vec{x}_k$	propagating direction of photon
V_{ijk}	volume of the sub-volume (i, j, k)
w	Energy weight carried by photon
α	dimensionless optical size parameter $=\pi d_p/\lambda$
ε	emissivity
θ	polar angle [rad]
σ	Stefan-Boltzmann constant
Ω	solid angle [sr]
λ	wavelength [nm]
φ	azimuthal angle [sr]
δ	distance between the "lii" zone and the observation point [cm]

Measurements on CO Concentration and gas temperature at 1.58 um with Tunable Diode Laser Absorption Spectroscopy

Fei WANG *, Ning LI, Qunxing HUANG, Jianhua YAN, Kefa CEN

State Key Laboratory of Clean Energy Utilization, Zhejiang University, Hangzhou, 310027, China

Abstract. The technique based on tunable diode laser absorption spectroscopy (TDLAS) has been developed to realize the real-time and dynamic measurement of the combustion-generated pollutants. In this paper the fundamental of concentration and temperature measurement with TDLAS is presented. To validate the method, an experimental system of TDLAS is set up. A simple VCSEL diode without temperature controller is used for direct absorption and WMS setup. Restricted by modulation voltage and noise from diode laser and detector, concentration of 0.2% for 1 m path length can be detected. Temperature measurement in tube furnace is also carried out by CO2 absorption lines near 1.58um, and below 800K the difference between calculation by TDLAS and temperature signal by K-thermocouple is less than 60K. The method to eliminate the effects of the particles is also discussed here. By using three-order polynomial program, the measurement can be not affected by particles and power fluctuation.

Keywords: Three-dimensional temperature distribution; Radiativ e energy image; CCD camera; Ill-posed equations; LSQR.

INTRUDUCTION

The technique based on tunable diode laser absorption spectroscopy (TDLAS) is an effective way to measure the gas concentration and temperature. A lot of research has been carried out in MIR range, but more and more attention has been paid to the measurement in NIR range, especially near 1.5um, corresponding to the communication band. Most optical instruments in this range are cheap, simple, and can operate within room temperature. That provides the possibility to popularize TDLAS in various areas, which need parameters of gas concentration or temperature on-line.

J.Wang[1] has achieved the measurement of CO concentration in vehicle exhausts and combustion pollutants at 2.3 um. Compared with the equipment cost, it is very attractive to carry out the measurement on the second-overtone band of CO near 1.5 um, although the linestrength in this range is much weaker. Because VCSEL diode can scan a wider wavelength range than DFB laser, it provides more choices to select the suitable absorption lines for gas sensing. The VCSEL diode at 1580 nm without temperature controller is used here for CO concentration measurement. For the purpose of multi-parameter measurement by using single diode laser, the gas concentration measurement is also carried out by the same laser with the help of CO_2 absorption lines near 1.58um.

THEORY

The fundamental theory for TDLAS is embodied in the Beer–Lambert law, which is described as equation (1)[2]:

$$\frac{I_t}{I_0} = \exp(-PS(T)\phi XL) \qquad (1)$$

where I_t and I_0 are the transmitted and initial power; $S(T)$ is the linestrength; ϕ is the lineshape function, whose integrated value over frequency is equal to unity, describing the shape of the measured spectral line;

CP914, *Multiphase Flow: The Ultimate Measurement Challenge,* edited by X. Cai, Y. Wu, Z. Huang, S. Wang, and M. Wang

P is the total pressure of gas and L is the path length; X is the mole fraction of the absorbing species. For direct absorption measurement, the concentration X can be obtained by integrating the absorbance over frequency:

$$X = \frac{\int_{-\infty}^{+\infty} -\ln(\frac{I_t}{I_0})}{PS(T)L} \qquad (2)$$

In the low concentration and much noise measurement, wavelength modulation spectroscopy (WMS) is more suitable than direct absorption measurement. By adding a modulation signal to the laser module, the gas concentration measurement can shift detection to higher frequency. With phase-sensitive detection technique, the detection bandwidth is reduced and the noise in the measurement will be much smaller. Moreover, excess laser noise and thermal noise are smaller at higher frequencies.

By combining the sinusoidal modulation and ramp scanning signals together and sending them to the laser module, the instantaneous laser frequency will be represented by:

$$v = \bar{v} + a\cos(wt) \qquad (3)$$

where a is amplitudes of modulation around $\bar{v}$. When the modulated laser passes the gas media and then impinges onto the detector, nth harmonic signal can be obtained from a lock-in amplifier. Fig.1 shows direct absorption and nth even harmonic signals:

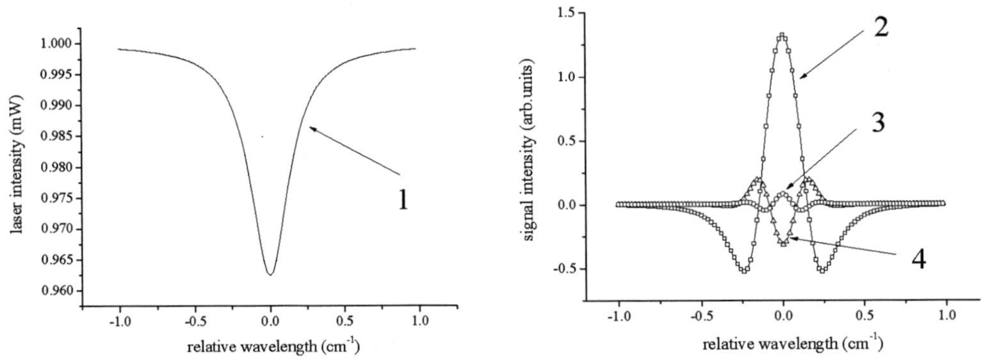

FIGURE 1. Simulation of direct absorption signal and WMS harmonic signals. (1. Direct absorption signal 2. 2f signal 3. 4f signal 4. 6f signal.)

Usually the even harmonic signals are used for gas concentration measurement, and odd harmonic signals for locating absorption line position. Because the peak value of 2f signal is much higher than others, it is often used in the actual measurement. For the transition with a Lorentzian lineshape, we can use the 2f peak value to infer the gas concentration according to the following equation[1]:

$$XL \propto \frac{\Delta v \times P_{2f}}{I_0 \times S}\left\{\frac{2}{m^2}[\frac{2+m^2}{(1+m^2)^{1/2}} - 2]\right\}^{-1} \qquad (4)$$

where m is the ratio of modulation, and defined as the ratio of the modulation amplitude to the linewidth. During the measurement, laser power will also be modulated, which often referred to residual amplitude modulation (RAM). Modulated laser power introduces asymmetry of 2f signal, but have no effect on the magnitude of the 2f signal at resonance[3], which can be seen from Fig.2.

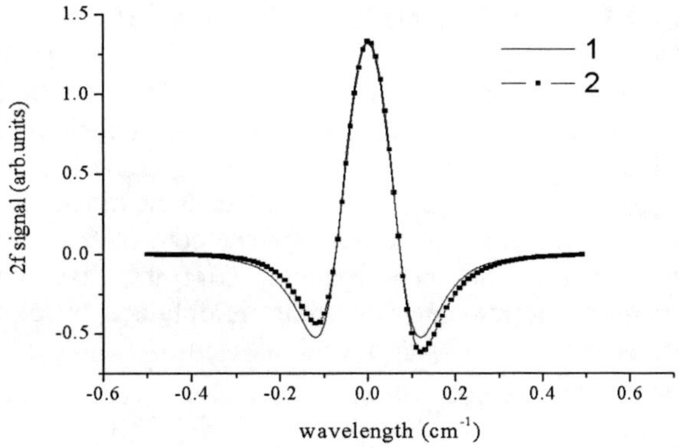

FIGURE 2. Simulation of RAM effect on 2f signal (1. 2f signal without RAM effect 2. 2f signal with RAM effect)

If other parameters are fixed, the 2f peak value will be linear with the gas concentration. After calibration, the exact number can be obtained. Compared with direct absorption measurement, WMS is inherently much less vulnerable to hostile environments and don't need to know the exact value of baseline. And if we use some ways to remove I_0 in equate (4), WMS can be immune to the particle effect or laser power fluctuation in hostile environment, so that the system robustness is improved.

LINE SELECTION

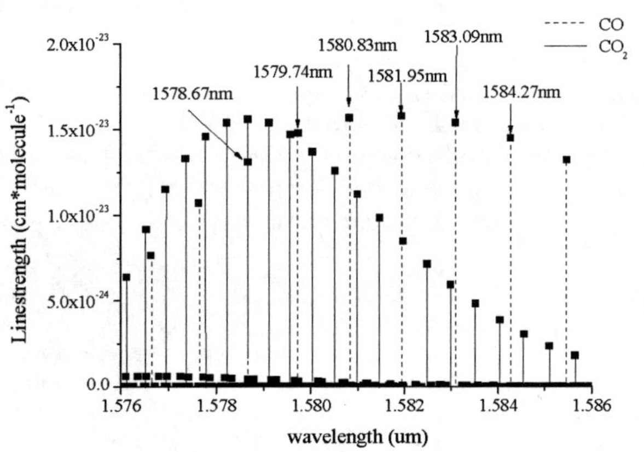

FIGURE 3. CO and CO_2 absorption lines between 1576 nm and 1586 nm based on HITRAN database[4]

The TDLAS measurement is based on one absorption line, so it is important to select a suitable one with enough strong linestrength and no interference of other absorption lines. For CO, the fundamental vibrational band near 4.6 um, and the first-overtone band near 2.33 um are strong enough for measurement, but the equipment in these regions are more expensive. The second-overtone band near 1.55 um, which is often used in communication industry, provides the possibility to lower the price of equipment, simplify the operation. Moreover linestrength of H_2O in this region is very weak. Although the lines at 1.55 um are weaker than those in MIR range, they are strong enough for normal application. As Fig.3 shows, the CO absorption lines at 1579.74nm, 1580.83nm, 1583.09nm and 1584.27nm can avoid the inference of CO_2 at normal pressure, which often exists with CO, and these lines can be easily got using VCSEL diode at 1580 by changing the input current. Especially lines at 1579.74nm, 1580.83nm, not only are suitable for CO measurement, but also can be used for CO_2 measurement simultaneously because of the strong absorption lines nearby at normal temperature measurement. The line at 1579.74 nm is used here for CO measurement, because of the better linear relationship of VCSEL between laser power and input current.

CONCENTRATION MEASUREMENT BY DIRECT ABSORPTION

For direct absorption measurement, a ramp signal (20Hz) from the function generator (Tektronix-AFG3102) is added to the laser diode controller, and then the laser wavelength from VCSEL diode (Vertilas-VCL1580) is changed by input current. The wavelength range of VCSEL diode is wide, more than 5 nm at one single mode, and the price is lower than other lasers. But there are also some disadvantages. Because of low price, there is no temperature controller in this diode, the wavelength and power may fluctuate a little during the measurement. After the laser passes through the long-path absorption cell about 10 m long, the transmission signal is obtained by detector (NewFocus-2033), and analog signal from detector is sent to NI-DAQ device. Two mass flow controllers are used to obtain the desired concentration of CO mixed in air.

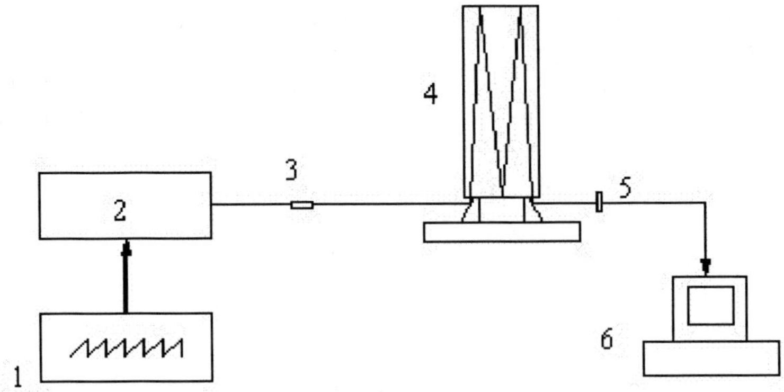

FIGURE 4. Direct absorption system (1. Function generator 2. VCSEL diode 3. Collimating lens 4. 10 m path length cell 5. Detector 6. DAQ system)

The scanning frequency from function generator in the experiment is 40Hz, and 20-sweep average is used for minimizing the random error. Because the laser power changes with the input current, a three-order polynomial fit program is used to obtain the baseline of absorption signal. Fig.5 shows the raw data and fit of baseline:

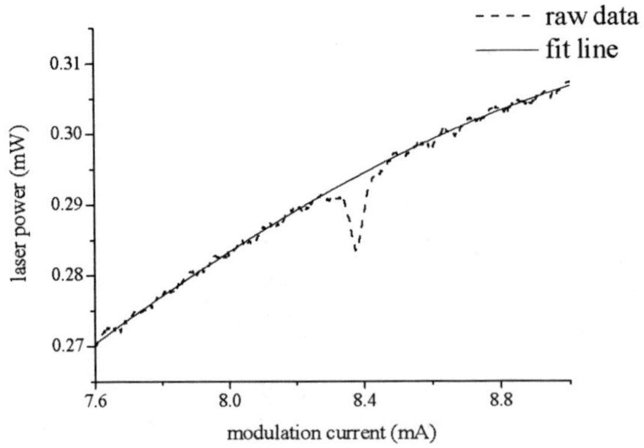

FIGURE 5. Baseline fit of absorption signal

With the help of Voigt fit function and parameters from HITRAN database, the concentration can be calculated. Fig.6 shows the result of CO at 1579.74 nm with concentration of 2.3% in air. The error between the calculation value and fixed value is 0.11%, which means relative error of about 5%. Because there is no temperature controller on this VCSEL diode, power instability and noise would contribute to the error a lot. As the concentration gets lower, the absorption signal is weaker. Especially in hostile environment, it is hard to distinguish the absorption signal from noise at low concentration.

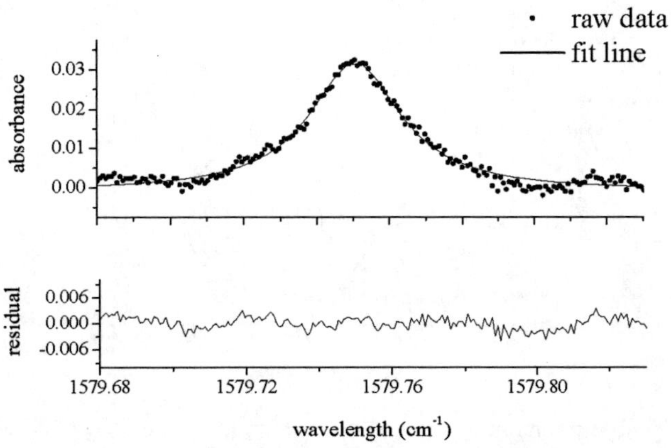

FIGURE 6. Measured CO lineshape at 1579.74 nm recorded in mixed gas of air and CO corresponding to 2.3% concentration (T=296 K, P = 1 atm, L =1000 cm)

CONCENTRATION MEASUREMENT BY WMS

Though direct absorption measurement is simple to implement, and can obtain the absolute concentration without calibration, it need the exact value of baseline, and is prone to be disturbed by noise. Especially at low concentration measurement, a small change of baseline would cause a very huge error. Compared with direct absorption, WMS is more suitable for in situ measurement. Because it is sensitive to curvature, we don't need to know the location of baseline. Moreover, WMS measurement can be carried out with high SNR, and can be immune to the particle disturbance or laser power fluctuation effect.

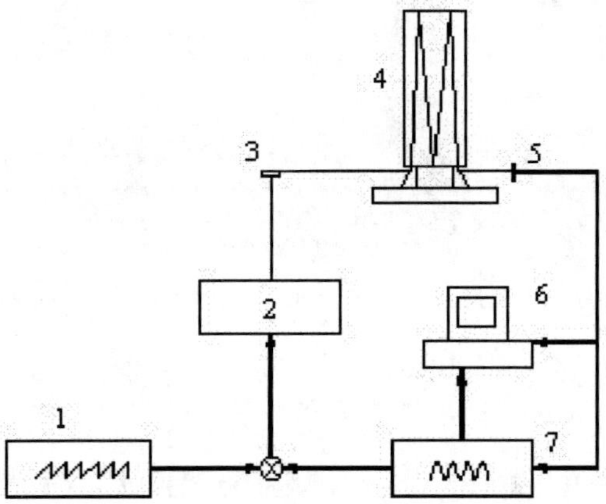

FIGURE 7. WMS system (1. Function generator 2. VCSEL diode 3. Collimating lens 4. White cell 5. Detector 6. DAQ system 7. Lock-in amplifier)

For WMS measurement, the ramp signal (20Hz) and the sinusoidal modulation signal (10KHz) from lock-in amplifier (Stanford-SR830) are mixed and sent to the laser diode controller. After the transmission signal passes through the absorption cell and impinges on the detector, a power splitter (Minicircuits-ZFRSC2050) is used for dividing the analog signal into two parts: one part is sent to the lock-in amplifier to obtain the 2f signal; the other one is directly sent to DAQ device to monitor the change of laser power.

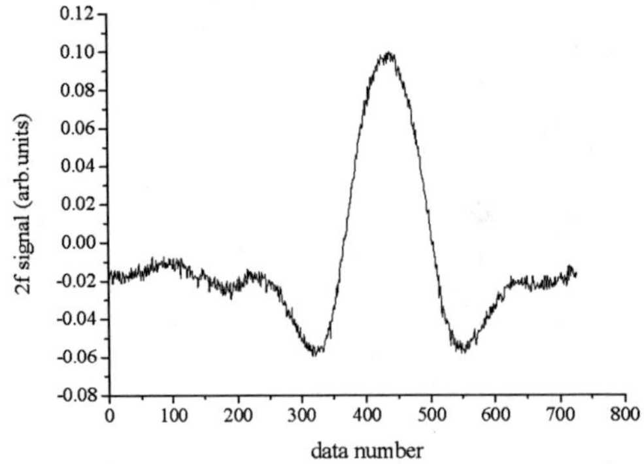

FIGURE 8. 2f signal of CO absorption line at 1579.74 recorded in mixed gas of air and CO corresponding to 0.27% concentration (T=296 K, P = 1 atm, L =1000 cm)

Fig.8 shows the 2f signal of 0.27% concentration with SNR about 10, and there is no obvious asymmetry of this signal. Because the concentration is linear with 2f peak value, the 2f signal can represent the concentration value after calibration. The minimum detectable concentration of CO in this system is nearly 200ppm in the experiment, which means 0.2% concentration for 1 m path length. There are some restrictions for the lower detectable concentration in this system. Besides the noise caused by laser diode and detector without temperature controller, the modulation index is too large to get the best 2f signal. The modulation voltage in the experiment is 4mV, which is also the minimum modulation amplitude of SR830, but the corresponding modulation index is still large than 2.2, which corresponds the maximum of 2f peak value. Fig.9 shows how 2f peak value changes in simulation and experiment.

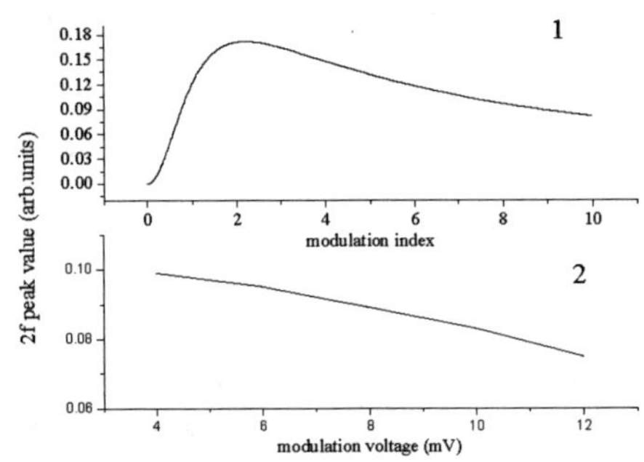

FIGURE 9. 2f peak value VS modulation index (voltage) (1. Simulation of 2f peak value as a function of modulation index 2. 2f peak value as a function of modulation voltage (concentration=0.27%))

When the concentration gets smaller, the modulation index would be larger, and the 2f peak value is far less than optimal one. So the detectable concentration would be improved if the modulation index can be kept about 2.2.

TEMPERATURE MEASREMENT

The measurement of gas temperature can be realized based on the variety of linestrength at different temperature. The relationship between linestrength and temperature can be expressed as:

$$S(T) = S(T_0)\frac{Q(T_0)}{Q(T)}\exp[-\frac{hcE"}{k}(\frac{1}{T}-\frac{1}{T_0})]\times[\frac{1-\exp(-hc\nu_{0,i}/kT)}{1-\exp(-hc\nu_{0,i}/kT_0)}] \quad (5)$$

The last term accounts for stimulated emission and is negligible at wavelengths below 2.5 um and temperatures below 2500 K. The variation trends of linestrength are different for absorption lines over the selected temperature range because of lower energy $E"$.

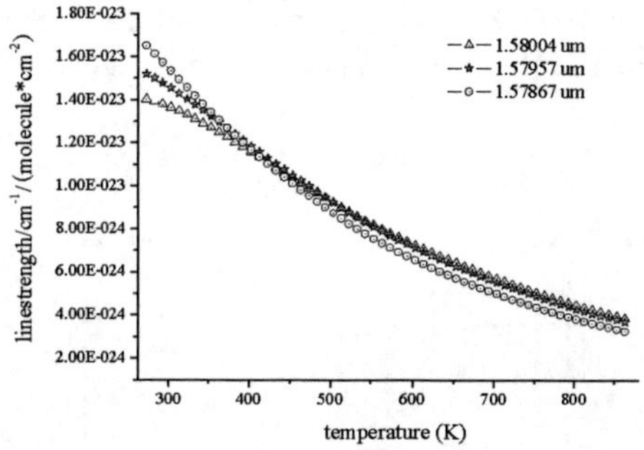

FIGURE 10. linestrength variation at different wavelength and temperature

Based on the different variation trends of two absorption lines, the temperature measurement can be carried out. The ratio of linestrength of two absorption lines can be reduced to a function only of the temperature T:

$$R = (\frac{S_1}{S_2})_T = (\frac{S_1}{S_2})_{T0} \times \exp[-\frac{hc\Delta E}{k}(\frac{1}{T} - \frac{1}{T_0})] \tag{6}$$

Because WMS is very useful to depress the noise and improve SNR, it is also introduced in temperature measurement. According to equation (2-4), the 2f signal is proportional to the linestrength, so it implies 2f signal can be treated as linestrength when other parameters are fixed:

$$R = (\frac{2f_1}{2f_2})_T = (\frac{S_1}{S_2})_T \tag{7}$$

As can be seen from equation (4), several factors affect temperature measurement: laser power, line width and modulation index. Laser power I_0 can be obtained by polynomial fit, which is detailed at chapter 3. Line width depends on the absorption line, temperature and pressure. For a lot of measurement the pressure is relatively stable, and it is fixed to be normal pressure in our experiment. Effect of temperature change on linestrength can be reduced by selecting appropriate absorption lines, whose temperature coefficients are nearly the same. According to the parameters from HITRAN database, the effect of temperature on the ratio of 2f signal in the range of 300-1000 K is less than 10% for CO_2 absorption lines near 1.58um. The effect of modulation index m is described in Fig.9, and it is fixed at 2.2 for both absorption lines in the experiment.

CO_2 is found abundantly after combustion, and near 1.58um, there are several absorption lines are suitable for TDLAS measurement, as shown in Fig.3. For the purpose of multi-parameter measurement with single laser, 1.58 um VCSEL diode laser is also used here to complete the temperature measurement with the help of CO_2 absorption lines at 1577.36 nm and 1580.04 nm. Fig.13 shows the experiment setup of temperature measurement in tube furnace. Laser beam from collimator lens is sent into quartz tube, which contains mixture of CO_2 and air. During the measurement, the quartz tube is heated by furnace, and the temperature and duration are controlled by computer. A K-thermocouple is used to measure furnace temperature for the comparison with TDLAS calculation result. Laser beam passes through the heated gas, and then collected by detector. After scanning the two absorption lines, the ratio of 2f signal is got and the temperature is calculated.

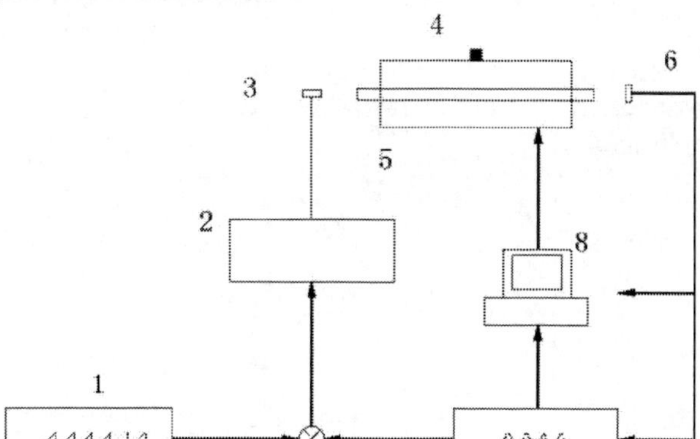

FIGURE 11. temperature measurement setup (1. Function generator 2.Laser diode 3. Collimating lens 4. Thermocouple 5. Tube furnace 6. Detector 7. Lock-in 8. PC)

In the experiment, CO_2 gas temperature is measured during the range of 300 to 800K. The calculated temperature based on TDLAS measurement is show in Fig.12 and compared with signals from K- thermocouple:

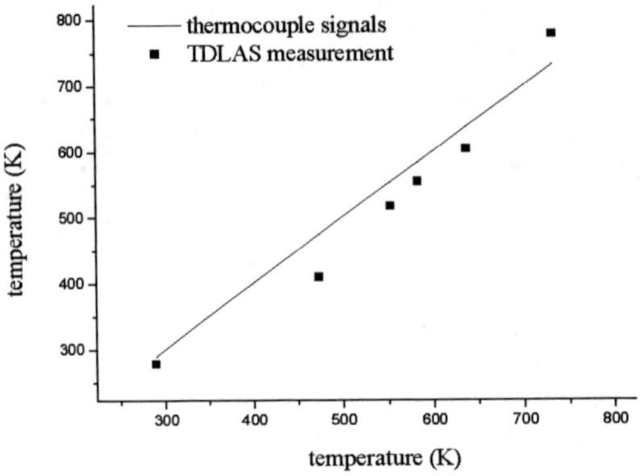

FIGURE 12. comparison of temperature signal between TDLAS and thermocouple

Below 800K, the temperature difference between calculation by TDLAS and measurement by K- thermocouple is less than 60K. And TDLAS results are a little lower than temperature from K- thermocouple. Because the position of thermocouple is in the middle of the furnace, where the temperature is higher than the average, the difference between TDLAS results and average would be smaller. The measurement difference is greater at higher temperature because of the interference of other absorption lines, whose linestrength can't be neglected, as shown in Fig. 13.

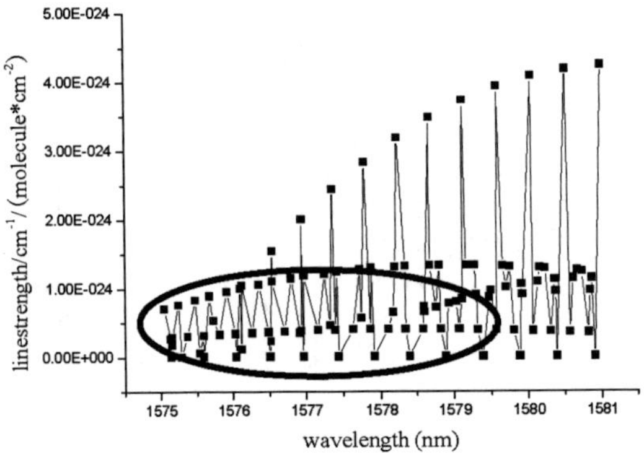

FIGURE 13. CO_2 absorption lines near 1.58um at 830 K

Particle Effect On Measurement

For in-situ gas concentration measurement, there may be some particles in the target gas. It may add much difficult to direct absorption measurement, but affect the WMS technique a little. Fig.14 shows the experiment setup of particle effect on gas concentration.

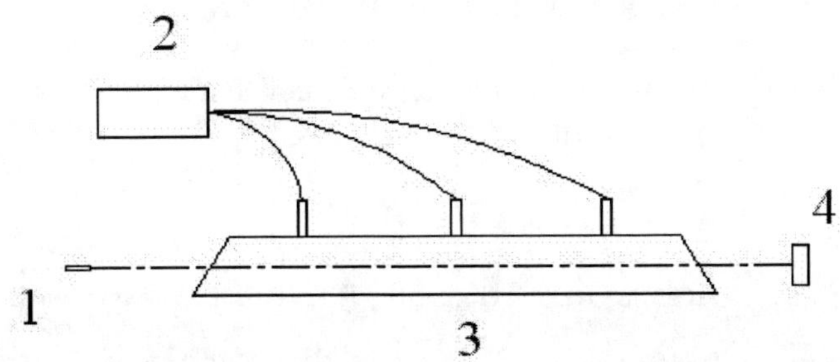

FIGURE 14. Particle effect experiment (1. Laser 2. Coal box 3. 1 m absorption cell 4. Detector)

Coal (20um) is used here to simulate the effect of particle on WMS measurement in 1 m absorption cell. Before the gas with fixed concentration gets into absorption cell, it passes through a box which contains some coal first, and then the gas would take the coal into the cell together. Because of the gravity, the coal would deposit gradually. Curve 1 and 2 of Fig.15 shows the variety of 2f signals from lock-in amplifier and laser power obtained by detector during the 3 minutes since the coal get into the cell. In the experiment, the 2f signal from lock-in and laser power signal from power splitter are obtained by DAQ system simultaneously. The value of laser power can be got by a three-order polynomial program. With the help of the program in LabVIEW, the concentration value can be obtained on-line.

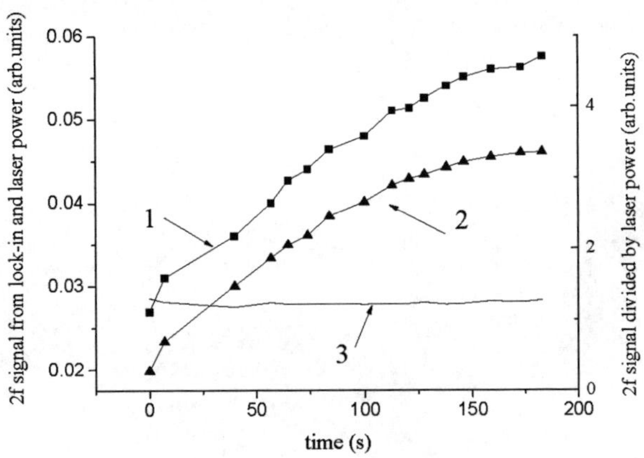

FIGURE 15. Particle effect on signal (1. 2f signal from Lock-in 2. Laser power 3. 2f signal divided by laser power)

In Fig.15, it is clear that curves 1 and 2 have the same trend during the coal powder deposits in the cell, and their linear fitting coefficient is 0.995 by using least-squares procedure, demonstrating better linear relationship between 2f signal from lock-in and laser power. According to equation (2-3), if we use the 2f signal divided by laser power I_0, the "calibrated" 2f signal varies only 5%, while 2f signal from lock-in varies from 0.025 to 0.058. It means that the 2f signal divided by laser power can represent the real concentration of gas, and particles have no obvious effect on the WMS measurement, which can be seen from curve 3 of Fig.15.

CONCLUSION

Gas concentration measurement using TDLAS in NIR range has been paid more and more attention, because of the advantage of low equipment cost and simple operation. CO concentration and gas temperature measurement are realized with a VCSEL diode at 1.58um. The minimum detectable concentration can be 0.2% at 1 m path length in this experiment system using WMS technique. The concentration can be improved if the noise from optical equipment can be depressed and the modulation voltage can be smaller to keep the modulation index optimal. By scanning CO_2 absorption lines at 1577.36 nm and 1580.04 nm, temperature measurement can be carried out below 800K, and the difference between TDLAS calculation and signal from thermocouple is less than 60 K. Moreover, the measurement with WMS can be less affected by particle in target gas or laser power fluctuation, which makes it suitable for in-situ measurement.

ACKNOWLEDGEMENTS

This work has been supported by National Natural Science Foundation of China (60534030) and Program for Changjiang Scholars and Innovative Research Team in University (IRT0434).

REFERENCES

1. J.Wang, M.Maiorov, J.B.Jeffries, D.Z.Garbuzov, J.C.Connolly, R.K.Hanson, *Measurement Science & Technology* **11**, 1576-1584 (2000).
2. N.Li, F.Wang, J.H.Yan, Z.Y.Ma, K.F.Cen, *Proceedings of the CSEE* **25**, 121-126 (2005).
3. P.Kluczynski, O.Axner, *Applied optics* **38**, 5803-5815 (1999).
4. L.S.Rothman, C.P.Rinsland, A.Goldman, S.T.Massie, D.P.Edwards, J.M.Flaud, A.Perrin, C.Camy-Peyret, V.Dana, J.Y.Mandin, J.Schroeder, A.Mccann, R.R.Gamache, R.B.Wattson, K.Yoshino, K.V.Chance, K.W.Jucks, L.R.Brown, V.Nemtchinov, P.Pvaranasi, *Journal of Quantitative Spectroscopy & Radiative Transfer* **60**, 665-710 (1998).

Resonance Shift and Q-Factors of Distortional Liquid Droplet

Yiping Han*, Jianyong Liu*, Loic Mées†, Gerard Gouesbet† and Gerard Gréhan†

*School of Science, Xidian University, Xi'an, China. Email : yphan@xidian.edu.cn
†LESP, UMR CNRS 6614/CORIA, CNRS/Université et INSA de Rouen, FRANCE

Abstract. The resonance spectrum for the distortional liquid droplet illuminated by plane waves is rigorously computed. The behaviors of the resonance shift due to droplets depart from sphericity are discussed. The expressions for Q-factors of resonances shift are presented by means of curve fitting of the numerical results.

INTRODUCTION

To measure the size of scatterers with a very high accuracy is interested in many fields of fundamental physics as well as in applied sciences. When the particles are spherical, one can use techniques based on the analysis of the scattering spectra at one wavelength as the scatterer size is large enough. Nevertheless, when the scatterer size decreases, the sensitivity of this approach decreases too. For spheres in the range of a few microns to 100 microns, the scattering exhibits a complicated angular structure owing to the existence of many partial waves, and also a rich frequency dependence with many sharp resonances, commonly called morphology dependent resonances. The approaches based on Morphological Dependant Resonance are the most powerful to capture the main features of processes if one knows the locations, Q-factors of these resonances[1-4]. In many cases, however, the particles are not perfect spheres. For example, a falling droplet will be slightly imperfect, may not be exactly spherical but may be described by spheroid.

In this paper, we discuss the resonance behaviour when the scatterer departs from perfect sphericity. A previously developed theoretical method for determining the light scattering is used to investigate the resonances for deformed droplets. Some features of resonance spectra for both spherical and deformed droplets are given.

THEORETICAL BACKGROUND

The incident field is assumed to propagating along the x axis. The scatterer is located at the origin of the coordinate system. The Maxwell's equations are rigorously solved for some symmetric scatterer in the framework of a variable separation technique, such as sphere, spheroid, etc.

For the spherical droplet, the normal modes of a dielectric sphere are described by a set of vector spherical harmonics, the coefficients a_n and b_n are the amplitudes of the TM and TE modes, respectively. They are functions of the relative index of refraction m of the sphere and the size parameter x according to the following formulas[7,8]

$$a_n = \frac{m^2 j_n(mx)[xj_n(x)]' - \mu_1 j_n(x)[mxj_n(mx)]'}{m^2 j_n(mx)[xh_n^{(1)}(x)]' - \mu_1 h_n^{(1)}(x)[mxj_n(mx)]'} \tag{1}$$

$$b_n = \frac{\mu_1 j_n(mx)[xj_n(x)]' - j_n(x)[mxj_n(mx)]'}{\mu_1 j_n(mx)[xh_n^{(1)}(x)]' - h_n^{(1)}(x)[mxj_n(mx)]'} \tag{2}$$

CP914, *Multiphase Flow: The Ultimate Measurement Challenge*, edited by X. Cai, Y. Wu, Z. Huang, S. Wang, and M. Wang where m is the refractive index and $j_n, h_n^{(1)}$ are the spherical Bessel functions and Hankel functions of the first kind, respectively. The peaks are due to resonances in the scattering coefficients in the case of the approaches based on variable separation.

As a droplet is falling in air, the deformed droplets can be described as prolate spheroid. The coefficients ($a_{nl}, b_{nl}, c_{nl}, d_{nl}$) in the prolate spheroidal system can be obtained, for each value of l, in the form[5,6]:

$$\sum_{n=l}^{\infty} i^n \begin{bmatrix} V_{nl}^{(3),t} & U_{nl}^{(3),t} & -V_{nl}^{(1),t} & -U_{nl}^{(1),t} \\ Y_{nl}^{(3),t} & X_{nl}^{(3),t} & -Y_{nl}^{(1),t} & -X_{nl}^{(1),t} \\ U_{nl}^{(3),t} & V_{nl}^{(3),t} & -mU_{nl}^{(1),t} & -mV_{nl}^{(1),t} \\ X_{nl}^{(3),t} & Y_{nl}^{(3),t} & -mX_{nl}^{(1),t} & -mY_{nl}^{(1),t} \end{bmatrix} \begin{bmatrix} a_{nl} \\ b_{nl} \\ c_{nl} \\ d_{nl} \end{bmatrix} = -\sum_{n=l}^{\infty} i^n \begin{bmatrix} G_{nl,TE} U_{nl}^{(1),t} + G_{nl,TM} V_{nl}^{(1),t} \\ G_{nl,TE} X_{nl}^{(1),t} + G_{nl,TM} Y_{nl}^{(1),t} \\ G_{nl,TM} U_{nl}^{(1),t} + G_{nl,TE} V_{nl}^{(1),t} \\ G_{nl,TM} X_{nl}^{(1),t} + G_{nl,TE} Y_{nl}^{(1),t} \end{bmatrix}$$

$$(l = 0, 1, 2, \ldots; \quad t = 0, 1, 2, \ldots) \qquad (3)$$

The parameters, $U_{nl}^{(j),t}$, $V_{nl}^{(j),t}$, $X_{nl}^{(j),t}$, $Y_{nl}^{(j),t}$ (j=1 or 3, according to the radial functions $R_{nl}^{(j)}(c,\zeta)$ of the first or third kinds) are given in [6] The coefficients $a_{nl}, b_{nl}, c_{nl}, d_{nl}$ in Eq. (3) are exact solutions from scattering theory. The numerical values of the coefficients under plane wave for spherical and distortional droplet are thereafter evaluated.

RESONANCES IN A SPHERICAL DROPLET

The normal modes of a spherical liquid droplet are described by a set of vector spherical harmonics. Figure 1 displays scattering spectra when a perfectly spherical droplet with a diameter equal to 6 μm and a refractive index equal to 1.33 is illuminated by a spectrum of plane waves. The collecting angle is located at 44°.

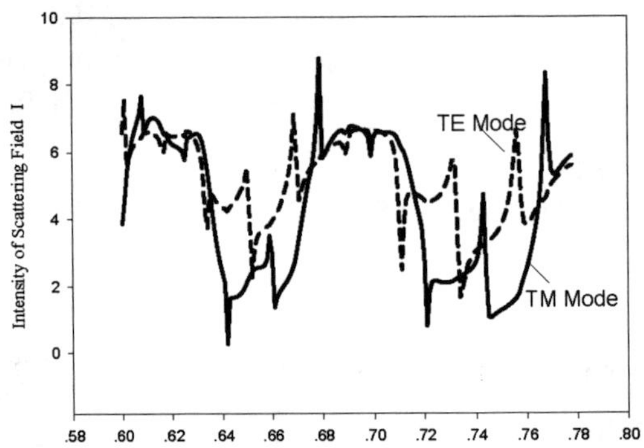

FIGURE 1. The TE and TM spectra for a water droplet (m = 1.33) with a diameter d equal to 6 μm illuminated by a spectrum of plane waves.

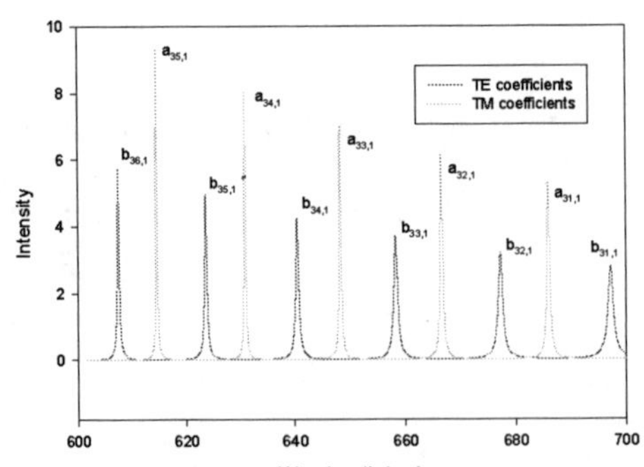

FIGURE 2. Individual scattering coefficients computed for d = 6 μm and m = 1.33

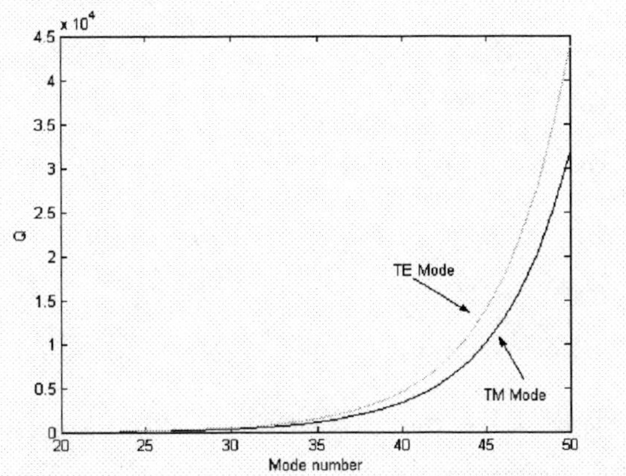

FIGURE 3. Quality factor versus mode numbering the case of a water droplet (m = 1.33) with a diameter equal to 6 µm.

Figure 2 displays the amplitude and location of the electric and magnetic coefficients for a water droplet with a diameter equal to 6 µm, for illuminating wavelengths between 0.6 and 0.7 µm.

The quality factor Q can be computed for these peaks by computing the individual behaviour of scattering coefficients for a mode number n and an order number l, and using

$$Q = \frac{\lambda_{n,l}}{\Delta\lambda_{n,l}} \tag{4}$$

where $\Delta\lambda_{n,l}$ is the full width at half maximum (FWHM) of the resonance peak with center at $\lambda_{n,l}$. The numerical computations of the locations as displayed in Fig.3.

THE RESONANCES FOR THE DISTORTIONAL DROPLETS

When falling droplet slightly depart from sphericity, shape distortion destroys the original spherical symmetry, and lifts the original degeneracy of $(2n+1)$-multiplets. And the resonance wavelength will be shift with respect to the distortion, as displays in Fig. 4. When ellipticity a/b is increased, the location of the resonance shifts to larger wavelengths, the strength of peaks decreases, which means the strength is sensitive to shape distortion. The locations of resonance peaks, for some $b_{n,l}$ as a function of ellipticity, are displayed in figure 5.

The figure 6 exemplifies that Q-values decrease rapidly when the ellipticity increases. In order to capture the main features of processes dominated by MDR's rapidly, we used numerical results from our theory to give relatively simple empirical expressions for Q-factor of resonance by curve fitting.

$$Q_n = \frac{\lambda_{n,1}}{\Delta\lambda_{n,1}} = Q_0 \frac{(1-e/6)}{\{\, 1+(0.02n-0.37)\cdot e+[0.113n^2(m^2-1)-3.5n+45.6]\cdot e^2 \,\}} \tag{5}$$

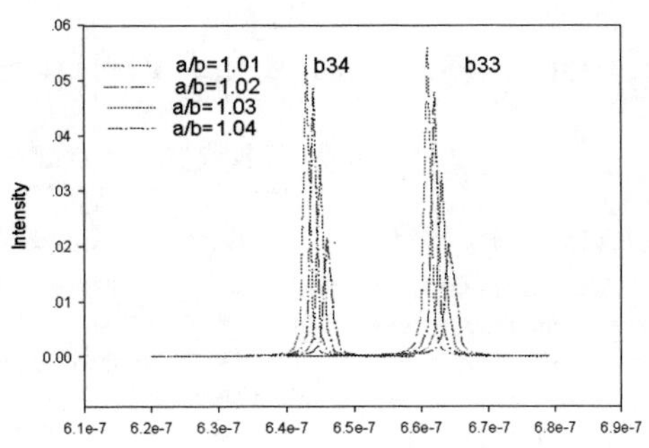

FIGURE 4. The scattering coefficient versus wavelength for several ellipticities running from 1.01 to 1.04. The wavelength step is equal to 0.0001×10^{-7} m, and m = 1.33 , $r_v = 3\mu m$

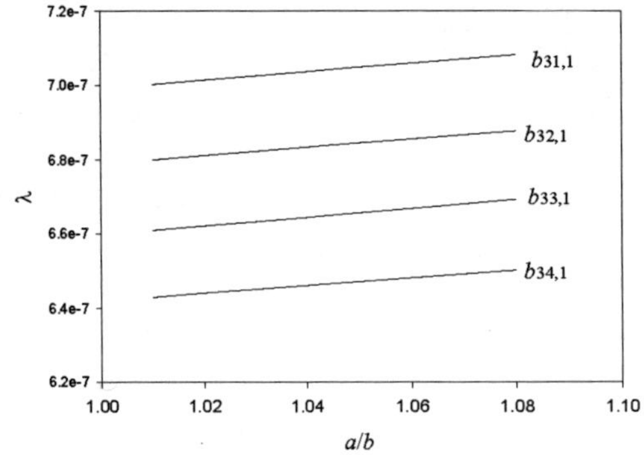

FIGURE 5. Locations of the resonance of scattering coefficients $b_{n,1}$ (n = 31 .. 34) obtained numerically for $m = 1.33$, $r_v = 3\mu m$.

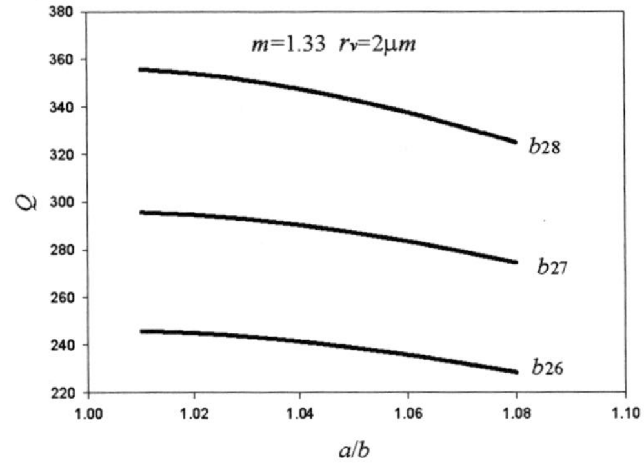

FIGURE 6. The values of Q versus a/b for $m = 1.33$ $r_v = 2\mu m$.

In which $Q_0 = \frac{\lambda^0_{n,1}}{\Delta\lambda^0_{n,1}}$ is the quality factor for perfect sphere. The accuracy of Q values predicted by this formula was found to be better than 97%.

CONCLUSION

We have considered the resonance behavior for a distortion of droplet. The locations, quality factors, of resonances are discussed. The results shows that location of the resonance shifts to larger wavelengths as distortion of droplet is increased. Also the strength of resonance is sensitive to shape distortion. These results will be useful for interpreting experimental data without using the complication of the full scattering theory.

ACKNOWLEDGEMENTS

This work is supported by Nation Nature Science Foundation of China (No: 60371019) and NECT-04-0949.

REFERENCES

1. P R. Conwell, P W. barber, C K. Rushforth, *J. Opt. Soc. Am. A.* **1**, 62-67 (1984).
2. J A Lock, *Opt. Lett.* **24**, 427-429 (1999)
3. H.M Lai, P. T. Leung, and K. Young, *Physical Review A* **41**, 5187 (1990).
4. H. Lai, C. Lam, P. Leung, and K. Young, *J. Opt. Soc. Am. B* **8**, 1962-1970 (1991)
5. Y. Han, G. Gréhan and G. Gouesbet, *Appl. Opt.* **42**, 6621-6629 (2003)
6. Y. Han, Z.S. Wu , *Appl. Opt.* **40**, 2501-2509 (2001).
7. C. F. Bohren, D.R. Huffman, *Absorption and Scattering of Light by Small Particles* (Wiley, New York, 1998).

8. J. R. Probert-Jones, *J. Opt. Soc. Am. A* **1**, 822 (1984).

Developments of an Adaptive Stereoscopic-Piv System For The Analyses Of The Flow-Structure Interactions(Fsi) Of Air-Lifted Bodies

Doh, D.H.[†], Hwang, T.G.[*], Jo, H.J.[**], Takei, M.[***], Tanaka, K.[****], Uemura, T.[*****]

† *Div. of Mech. & Info. Eng./Korea Maritime Univ., Yeongdo-ku/Busan/Korea, doh@mail.hhu.ac.kr*
* *Dept. of Mech. Eng./Georgia Institute of Tech., Georgia/USA, andrew@bada.hhu.ac.kr*
** *Div. of Ocean Systems Eng./Korea Maritime Univ., Yeongdo-ku/Busan/Korea, hjjo@mail.hhu.ac.kr*
*** *Okamoto Industries Inc./Bunkyo-ku, Tokyo/Japan, cpt.tanaka@imx.okamoto-inc.co.jp*
**** *Department of Mechanical Engineering, Nihon Univ./Tokyo/Japan, takei@mech.cst.nihon-u.ac.jp*
***** *Faculty of Engineering/Kansai Univ./Osaka/Japan, uemura@kansai-u.ac.jp*

Abstract. An adaptive stereoscopic-PIV system has been constructed for flow analyses around a cylinder body (length=60mmm, diameter =10mm, polystyrene). The body is floated by the swirling flows in a vertical pipe with a length of 600mm and an inner diameter of 100mm. Four optical sensors(LEDs) were installed onto the pipe wall to detect the posture of the cylinder body. The constructed stereoscopic-PIV system is activated by the signals of the optical sensors when the cylinder is in upright posture. The measurement system consists of two cameras, a Nd-Yag laser and a host computer. Flow characteristics around the floating cylinder body and the motional characteristics of the body itself are investigated when the body becomes upright posture. It has been verified that the motion of the floating cylinder becomes stable when the azimuthal velocity component of the swirl flow is maintained at stable states.

Keywords: Adaptive Stereoscopic-PIV, Floating Cylinder Body, Swirl Flow

INTRODUCTION

Pneumatic conveyance systems have been widely adopted for transportations of industrial materials [1] and are more important in ocean application such as in risers. However they have problems of deterioration of the material quality due to collisions of the materials to the pipe wall, further low transportation efficiency. Studies on pneumatic conveyance systems with swirl flows have been performed with goals of avoiding the deterioration of the material quality. Li and Tomida [2] investigated the flow characteristics of the polyethylene particles and polyvinyl ones to report the pressure drops in a horizontal pipe. However, their results were restricted to spherical materials. In some practical applications, the material that should be conveyed are cylindrical or complex shapes such as ocean minerals. The cylindrical materials moving in the swirling flow has a characteristic that its axis is orientated to the main flow direction in a certain flow condition. It was shown that the stable orientation of the cylindrical body to the main flow could be attained by using a swirling flow [3].

Horii et al. [4] used a special swirling flow which had a steep axial velocity profile with swirling motion in order to transport particles without touching the inner wall, which resulted in a high performance transportation of the cylindrical materials. These studies are mainly focusing on the orientation efficiency. And there are few studies that report the flow characteristics around the cylindrical material body. Yin et al. [5] reported a theoretical study on the force balances acting on a cylinder material in a swirl flow. Unfortunately, experimental investigations have not been reported on the flow characteristics around the rotating cylindrical material in a swirling flow.

The purpose of the present study is to provide experimental investigation results on the flow characteristics around a cylindrical body floated in a vertical swirl flow and on the motional information of the body itself through constructions of a new 'Adaptive Stereoscopic-PIV system'.

CP914, *Multiphase Flow: The Ultimate Measurement Challenge,* edited by X. Cai, Y. Wu, Z. Huang, S. Wang, and M. Wang

ADAPTIVE STEREOSCOPIC-PIV SYSTEM

Hardware System

The adaptive stereoscopic-PIV system has been constructed to measure the flow characteristics around the cylindrical body which is floated freely by the swirl flow in a vertical pipe. The polystyrene cylinder (l_c = 60 mm, d_c = 10 mm, density 220 kg/m^3) has been used as the cylinder body. Fig. 1 (a) shows the concept of the adaptive stereoscopic-PIV system. Four-LED-sensor unit detects the light reflected from the surface of the floating body, and it activates the conventional stereoscopic-PIV system when the cylinder body passes through the sensing zone in nearly vertical posture. The sensor unit is composed of four reflection type fiber sensors. The sensor unit was embedded in the pipe wall as shown in Fig. 1(b). The laser light sheet used for flow visualization around the body passes through the same vertical line of the sensor array. The red LED frequency is 680 nm.

Fig. 1(b) shows the condition at which all four fiber sensors(OMRON, E3X-NA11F) detect the reflected lights from the cylinder body, which indicates the cylinder body floated and rotated by the swirl flows in the pipe is in upright condition (posture). The length of the sensor unit l_s = 57 mm almost corresponds to the length of the rotating cylinder, l_c = 60 mm. Fig. 1(c) shows the condition at which only two fiber sensors detect the cylinder body, which means the cylinder body rotates inside of the pipe with non-oriented posture (unstable). In the non-orientated case, the sensor unit does not activate the stereoscopic-PIV system. Since the response time of the fiber sensor was 20μs which is small enough than the body's collapsing time, the PIV system could be always activated adaptively in the same posture of the body without a fail. The detection length of the sensor unit has been adjusted to w_s = 10 mm from the pipe inner wall to attain maximum light signals reflected.

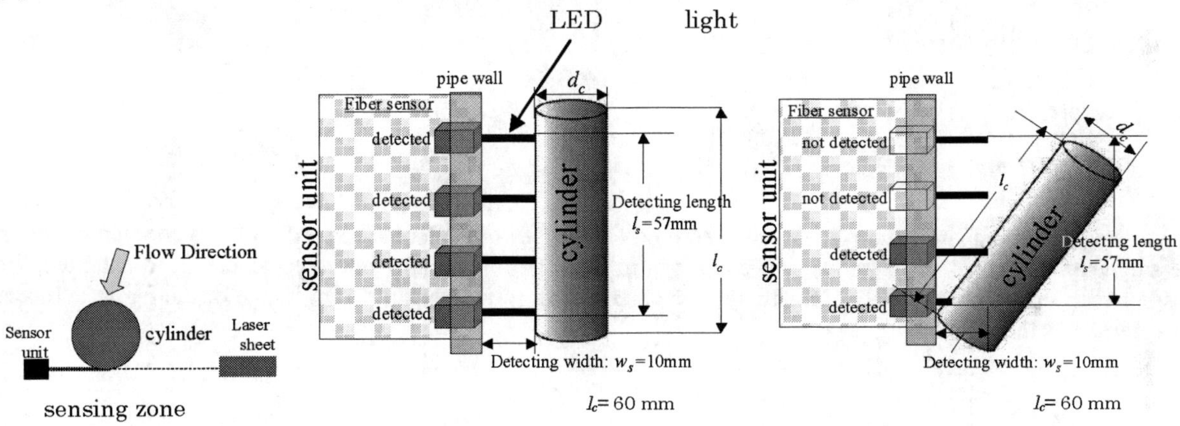

(a) Location of sensor detection (b) Orientated condition (c) Non-orientated condition

FIGURE 1. Adaptive sensing system

Software System

In this study, a hybrid stereoscopic-PIV is used. The hybrid stereoscopic-PIV introduces, a geometric transformation to consider the refraction and the aberration of the lens and the camera calibration method used for the 3D-PTV measurement developed by Doh et al. [6]. In order to attain three-dimensional measurement with two cameras, it is necessary to know their camera parameters. The ten-parameter method [6] is used. Since the arrangement of the cameras has a large view angle, the obtained images are strongly distorted, which makes the camera calibration difficult. To overcome this matter, a two-dimensional image transformation on each camera image: warping, is carried out before using the original images for the camera calibration as shown in Eq. (2.1).

$$x = \frac{a_{11}x_s + a_{12}y_s + a_{13}}{a_{31}x_s + a_{32}y_s + 1},\ y = \frac{a_{21}x_s + a_{22}y_s + a_{23}}{a_{31}x_s + a_{32}y_s + 1} \tag{2.1}$$

x, y : coordinates before transformation
x_s, y_s: coordinates after transformation

The calibrator image is shown in Fig. 2 and Fig. 3 before and after transformation. In this study, the image at the center physical plane ($Z = 0$ mm) was used to calculate the elements of the a_{ij} matrix.

In the ten-parameter method, 10 parameters (6 exterior parameters: l, α, β, γ, m_x, m_y, four interior parameters: c_x, c_y, k_1, k_2) are obtained. (α, β, γ) represents the tilting angles of the photographic coordinates axes against the absolute axes. Fig. 4 shows a coordinate relation when the photographical axes is set parallel to the absolute coordinate by the tilting angles (α, β, γ). (X, Y, Z) represents the absolute coordinate, and (x, y, z) represents the photographical coordinate of the image centroid of the calibration targets. The l means the distance between the origin O (0, 0, 0) and the principal point (X_0, Y_0, Z_0) of the camera.

The coordinate (X_m, Y_m, Z_m) represents the point P position of the calibrator when the camera coordinate is rotated with the tilting angles to make the collinear set in one line as shown in Fig. 4. The m_x and m_y means the point at which the normal vector from the origin O (X_0, Y_0, Z_0) of the camera coordinate meets with the X-Y plane. The collinear equation for every point between the two coordinates is expressed as Eq. (2.2). The c_x and c_y are the focal distances for x and y components of the coordinate. Δx and Δy are the lens distortions as expressed as Eq. (2.3). The Eq. (2.2) can be converted to the following Eq. (2.4).

$$x = c_x \frac{X_m - m_x}{\sqrt{l^2 - m_x^2 - m_y^2} - Z_m} + \Delta x \ , \ y = c_y \frac{Y_m - m_y}{\sqrt{l^2 - m_x^2 - m_y^2} - Z_m} + \Delta y \tag{2.2}$$

$$\Delta x = (x/r) \times \left(k_1 r^2 + k_2 r^4\right) , \ \ \Delta y = (x/r) \times \left(k_1 r^2 + k_2 r^4\right) , \ r = \sqrt{x^2 + y^2} \tag{2.3}$$

$$F = c_x \frac{X_m - m_x}{\sqrt{l^2 - m_x^2 - m_y^2} - Z_m} - (x - \Delta x) = 0 \qquad G = c_y \frac{Y_m - m_y}{\sqrt{l^2 - m_x^2 - m_y^2} - Z_m} - (y - \Delta y) = 0 \tag{2.4}$$

Since this equation is a strong non-linear, an improved Gauss-Newton calculation method [7] is adopted to obtain all necessary parameters using the above two equations in Eq. (2.4). Once all camera parameters are obtained, the relations between the photographic coordinate and the absolute coordinate of the target image or the particle image can be expressed as the following Eq. (2.5).

$$\begin{bmatrix} X \\ Y \\ Z \end{bmatrix} = M_M^{-1} \begin{bmatrix} X_m \\ Y_m \\ Z_m \end{bmatrix} \tag{2.5}$$

Where, M_M is the matrix for the rotational transformation is expressed in the form of Eq. (2.6).

$$X_m = \frac{x - \Delta x}{c_x} t + m_x \, , \ Y_m = \frac{y - \Delta y}{c_y} t + m_y \, , \ Z_m = d - t \ , \ d = \sqrt{l^2 - m_X^2 - m_y^2} \tag{2.6}$$

If the camera center is set to a vector (X_0, Y_0, Z_0), the collinear equation for one target (or particle) can be expressed as P(X, Y, Z) = ($a_1 t + X_0$, $a_1 t + Y_0$, $a_3 t + Z_0$). The cross-sectional points constructed from the following two collinear equations for the two cameras are defined as the three-dimensional positions in the absolute coordinate.

$$\begin{aligned} \mathrm{A}(X, Y, Z) &= \mathrm{A}(a'_{11} t + b'_{11} \ , \ a'_{12} t + b'_{12} \ , \ a'_{13} t + b'_{13}) \\ \mathrm{B}(X, Y, Z) &= \mathrm{B}(a''_{21} s + b''_{21} \ , \ a''_{22} s + b''_{22} \ , \ a''_{23} s + b''_{23}) \end{aligned} \tag{2.7}$$

Where, t and s are obtained by the least square method (LSM). Since the cross-sectional points do not always intersect on one point, the below Eq. (2.8) is used for the definition of the last three-dimensional position of the targets (or the particles), which implies the center of the shortest distance between the two collinear equations.

$$\begin{bmatrix} X_P \\ Y_P \\ Z_P \end{bmatrix} = \frac{1}{2}\left\{ \begin{bmatrix} X_A \\ Y_A \\ Z_A \end{bmatrix} + \begin{bmatrix} X_B \\ Y_B \\ Z_B \end{bmatrix} \right\} \tag{2.8}$$

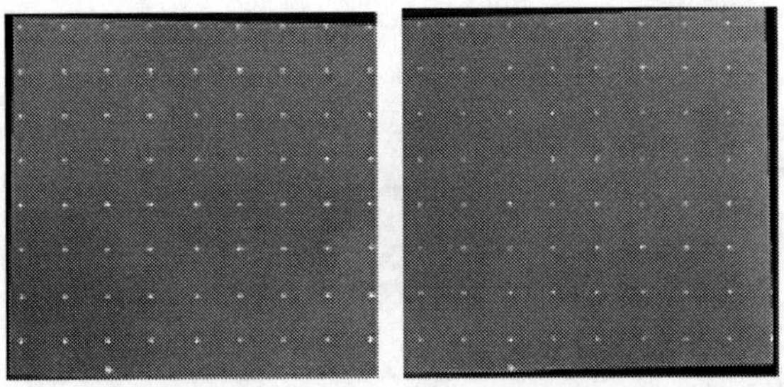

(a) Camera 1 (b) Camera 2

FIGURE 2. Calibrator image before transformation

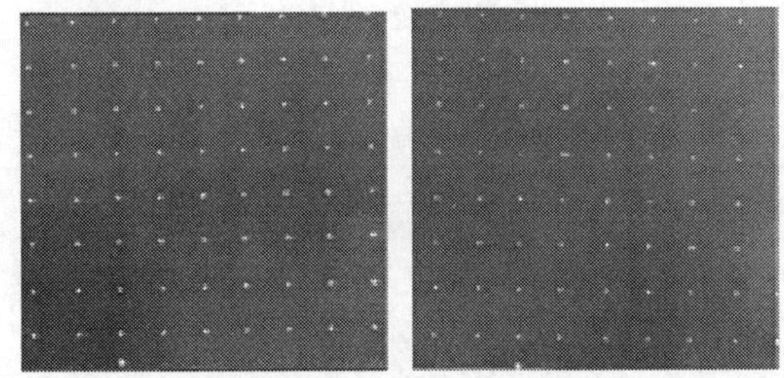

(a) Camera 1 (b) Camera 2

FIGURE 3. Calibrator image after transformation

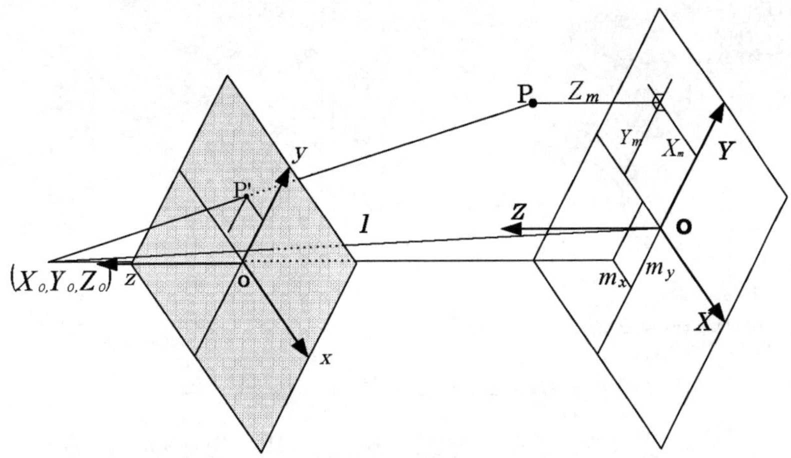

FIGURE 4. The camera parameters on the absolute coordinate and the photographical coordinate

Where, X_A, Y_A and Z_A represent the absolute coordinates for camera A defined by Eq. (2.7). X_B, Y_B and Z_B represent the absolute coordinates for camera B. After obtaining the positions of the vector grid points (vector start points), the three-dimensional vector terminal points are calculated by combining the two-dimensional vector terminals of each camera's image. The two-dimensional vectors are obtained by the conventional gray level cross-correlation method [8].

MEASUREMENT SYSTEM AND PROCEDURES

Fig. 5 shows the overall measurement system. Two digital cameras (Kodak ES1.0, 1k x 1k) were used for the stereoscopic measurements and arranged as seen in the figure. Smoke (Glycol , d=0.5~ 1μm)was used as seeding particles and it was let into the main pipe (D=100mm, L=1100mm). Nd-Yag laser (120mJ, 15Hz) was used for flow field visualizations. The swirl flow is generated by suctioning the main pipe air at top with a blower. The air inlet is made at the bottom of the main pipe through four small horizontal pipes (d=10mm) as seen in the Fig. 5. As mentioned previously the floating body was put into the swirl flow field in order to investigate the flow properties at which its posture becomes upright rotating at a certain height from the bottom of the main pipe with stability. Its posture was sensed by the four optical sensors (LED, reflection type) attached on the pipe wall. The height of the cylinder's rotating position of the cylinder was controlled by adjusting the blower's speed control system. Four cases were tested as seen in the Table 1. Blowing condition implies that the flow was generated by the blower installed at the inlet of the main pipe. On the contrary, suction condition implies the blower was installed at the outlet of the main pipe. The flow rate was controlled by the blower's speed control using the motor inverter of the blower.

Table 1. Tested four experimental conditions

Blowing condition	**Suction condition**
- Floating cylinder in	- Floating cylinder in
- Floating cylinder out	- Floating cylinder out

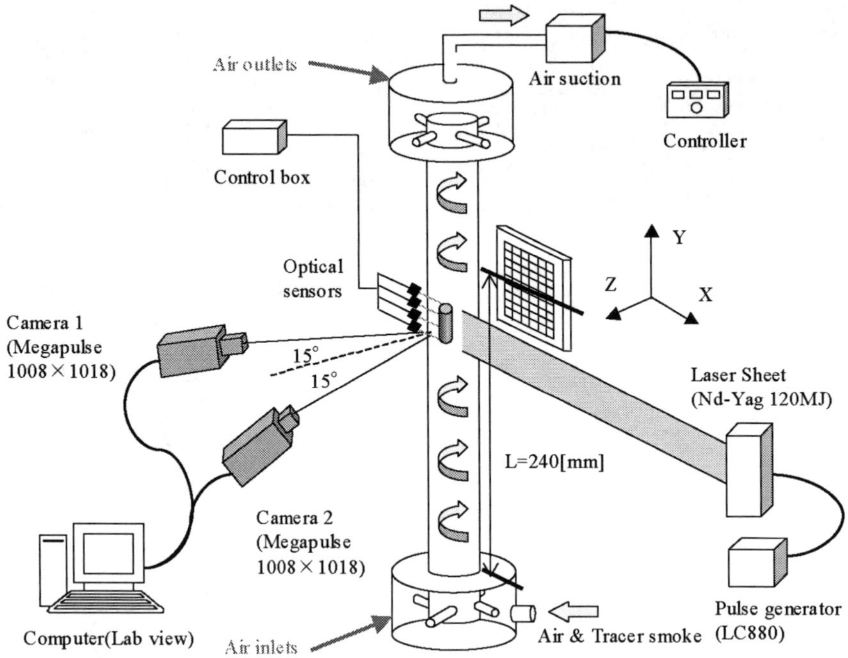

FIGURE 5. Overall experimental system

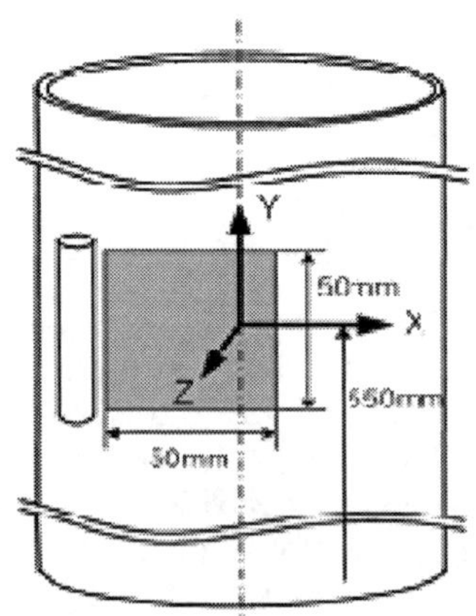

FIGURE 6. Measurement region

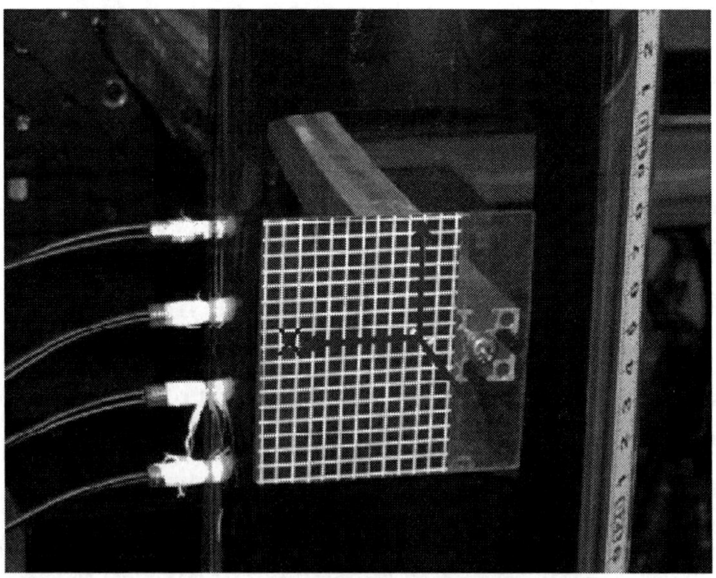

FIGURE 7. Camera calibrator

The flow rate was maintained at $1.7\text{x}10^{-3}$ m^3/sec. The flow rate was measured with a flow meter installed at the outlet of the main pipe. The Reynolds number with the diameter of the main pipe is about 4240. Fig. 6 shows the measurement region. X is the horizontal direction of the vertical pipe. Y is the vertical direction of the vertical pipe.

And Z is the azimuthal direction of the vertical pipe. The measured velocity components U, V and W were defined as the horizontal, vertical and azimuthal components, respectively. The measured velocities were normalized with the velocity obtained from the flow rate, 0.42m/sec. Before commencing the main experiments the camera calibration was carried out. A flat plate (56mm x 48mm) was attached onto a traverse and was moved to 9 positions to Z-axis direction (-12mm, -9mm, -6mm, -3mm, 0mm, 3mm, 6mm, 9mm, 12mm) with 3mm distance at which the images of the plate were captured for camera calibration. Fig. 7 shows the camera calibrator.

RESULTS AND DISCUSSIONS

Fig. 8 shows a picture taken at an instance when the floating cylinder was maintained at a stable but not passing through the measurement zone. Fig. 9 shows instantaneous raw images taken by the two cameras in Fig. 5. Fig. 10 shows mean three-dimensional vector field. The number of instantaneous vector fields for averaging is 96. Every instantaneous vector field was obtained at the time when the bar came into the same posture regions. That is, an adaptive sampling for vector field was carried out and averaging was made at the same condition where the rotating bar comes into a certain area in the swirl flow. The two spots seen at the lower part in Fig. 10 are error vectors due to reflections. Fig. 11 shows the horizontal velocity profiles when the flow is blowing condition with the floating cylinder in and out conditions. There is a slight discrepancy between the two profiles even if the flow rate were maintained at the same. This is due to the existence of the floating cylinder. When the floating cylinder is rotating inside of the main pipe, W component increased along the whole radial direction. V component that is a vertical component direction becomes decreased from x/R=0.3 to the floating cylinder's location while U component is still maintained in this region except the region over x/R=0.75. Fig. 12 shows the vertical velocity profile at x/R=0.8, which means the location of nearest vertical line along the floating cylinder. U, V and W components decreased along the whole zone except y/R=-0.1~0.05 for W component.

Fig. 12 shows the horizontal velocity profiles when the flow is suction condition with the floating cylinder in and out conditions. W components show similar profiles for both cylinder in and out conditions. U and V components shows decreased profiles over the whole x/R locations. Fig. 13 shows the vertical velocity profile at x/R=0.8 when suction condition. W component shows uniform profile showing different shape from that in the case of blowing. U, V and W components show homogeneous profiles along the whole y/R locations showing different profiles from those in the case of blowing, which implies that suction condition is better for maintaining the floating cylinder at stable posture during operation rather than the case of blowing.

FIGURE 8. Experimental picture taken at an instance

(a) viewed by camera 1 (b) viewed by camera 2

FIGURE 9. Instantaneous raw image

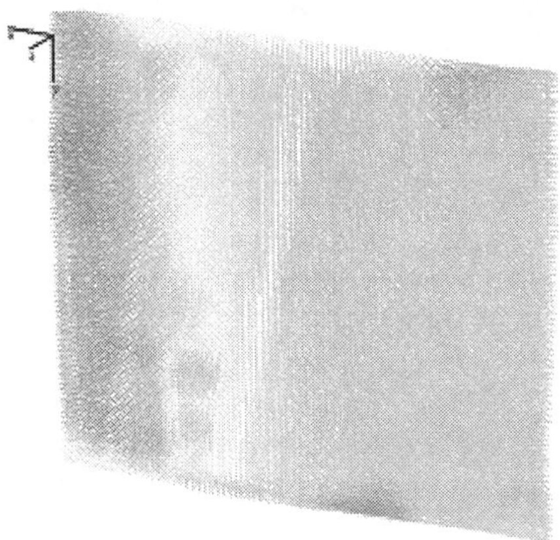

FIGURE 10. Mean velocity field(Blowing with no floating cylinder)

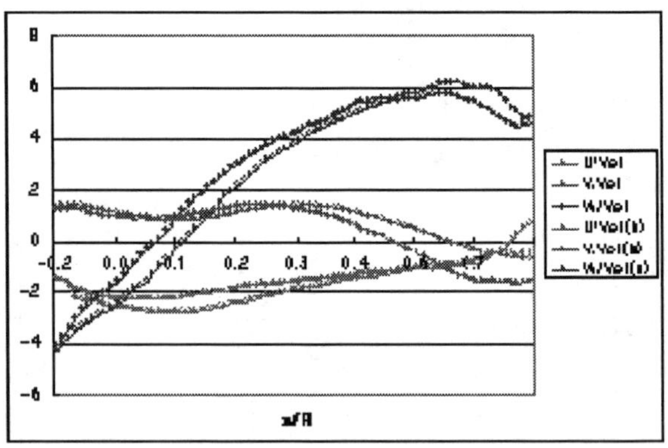

FIGURE 11. Horizontal velocity profile at y/R=0.0(blowing condition)

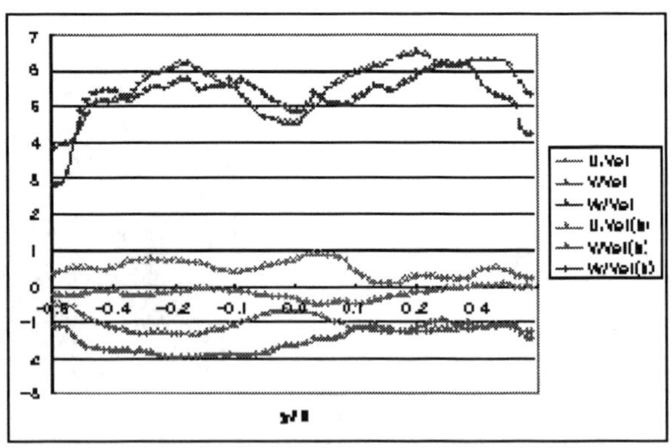

FIGURE 12. Vertical velocity profile x/R=0.8 (blowing condition)

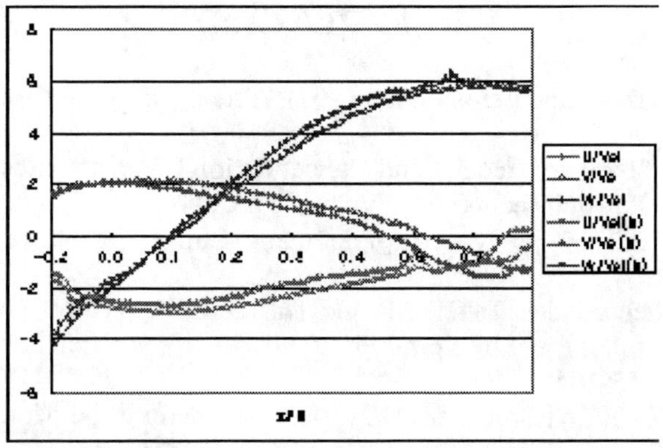

FIGURE 13. Vertical velocity profile x/R=0.8 (blowing condition) Horizontal velocity profile at y/R=0.0(suction condition)

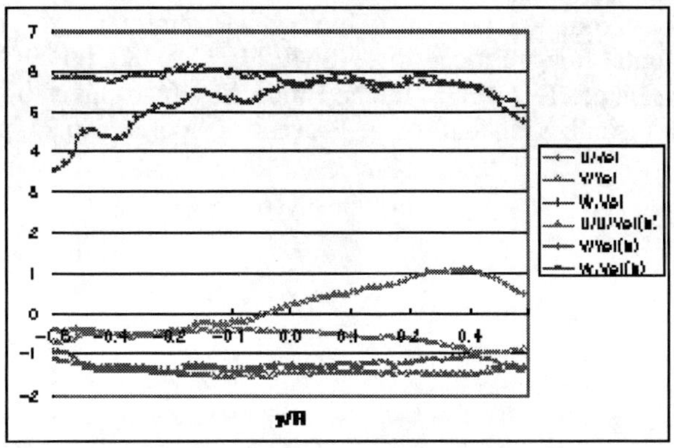

FIGURE 14. Vertical velocity profile x/R=0.8(suction condition)

CONCLUSION

An adaptive stereoscopic-PIV system has been developed for measuring the flow characteristics around the floating cylinder in a swirl flow.

All sensors detected the rotating cylinder's posture and its location well. When any one of the four sensors does not detect a light signal reflected from the cylinder, the posture of the floating body has been considered as non-orientated. Three velocity components profiles have been successfully measured at the same cylinder's posture with the constructed system. It was seen from the experiments that the height of the floating cylinder body was maintained at the same level regardless of blowing and suction conditions. For the case of suction, all velocity profiles of U, V, W are similar regardless of body in and out cases.

W component was dominant than other components and played crucial roles in stabilizing the posture of the floating body.

It was seen that the turbulent properties were smaller in the case of suction than in the blowing case, which implies that suction condition is better for controlling the floating cylinder in stable parallelization of the floating body.

The peripheral self-rotating speed of the cylinder body was calculated from the image shifts of the body seen in the experimental images, and it was about 0.42 m/sec. For complete analyses on the force balances acting on the floating body, the equation $F_C = F_M + F_S + \alpha(static\ \ pressure\ \ force)$ should be satisfied. Here F_c is centrifugal force, F_M is Magnus force and F_s is Suffmann force. The static pressure force can be obtained from the velocity vector field obtained from the stereoscopic PIV measurements. In this study, the flow characteristics only have been reported, and detailed calculation results on the force balances are to be appeared with another subject.

ACKNOWLEDGEMENTS

This work was supported by Korea Research Foundation (KRF-2006-521-D00078).

REFERENCES

1. Klinzing, E. G, Marcus D R, Rizk F, and Leung S. L, (1997), "Pneumatic Conveying of Solids", Chanpman & Hall, second edition.
2. Li, H. and Tomita, Y, (2000), "Particle Velocity and Concentration Characteristics in a Horizontal Dilute Swirling Flow Pneumatic Conveying", Powder Technology, Vol.107, pp.144-152.
3. Gould, I V, and Aust, M I E., (1983) "Vaccum-Pneumatic Conveying of Raw Wool", Mech., Eng. Trans., Vol.8, pp.77-83.
4. Horii, K., Matsumae, Y., Cheng, X.M, Takei, M. and Hashimo, B., (1989) "A Study of Spiral Flow (Part3) Opening and Orientation Control of Fiber by Spiral Flow", Trans. of the Japan Society for Aeronautical and Space Sciences, Vol.32, No.98, pp. 155-164.
5. Yin, C, Rosendahl, L, Kar, K. S. and Sorensen, H., (2003) "Modelling the Motion of Cylindrical Particles in a Nonuniform Flow", Chemical Engineering Science, Vol.58, pp.3489-3498.
6. Doh, D.H, Hwang, T.G. and Saga, T., (2004), "3D-PTV measurements of the wake of a sphere", Measurement Science and Technology, Vol.15, pp.1059-1066.
7. Kobayashi, T., Saga, T., Sekimoto, K., (1989), "Velocity Measurement of Three-dimensional Flow around Rotating Parallel Disks by Digital Image Processing, ASME FED Vol. 85, pp.29-36.
8. Kimura, I., Takamori, T. and Inoue, T., (1986) "Image Processing Instrumentation of Flow by using Correlation Technique", Journal of Flow Visualization and Image Processing, Vol. 6, No.22, pp.105-108.

SESSION E

INDUSTRIAL APPLICATION AND OTHERS

Conductance Sensor for Measurement of the Fluid Watercut and Flowrate in Production Wells

Liu-Xingbin*, Hu-Jinhai*, Huang-Chunhui*, Wang-Min†, Dai-Xuefei†

*Daqing Logging &Testing Services Company , Ranghulu District ,Daqing , Heilongjiang ,China
†Department of Electronic Science , Daqing Oil College , Heilongjiang ,China

Abstract. A new type of conductance sensor is designed to measure both the flowrate and the water cut simultaneously in oil/water two phase flow in oil production wells. With the conductance sensor, the watercut is obtained by the ratio of the conductivity of the oil/water mixture to that of the water, and the flowrate is gotten by cross-correlation flowrate measurement, which can figure out the transit time of the fluid flowing noises passing through the upstream and downstream sensors in the conductance sensor. In the paper, the principles of measuring flowrate and watercut are introduced. The results of dynamic experiment in multiphase flow loop and the application in field of the technology are introduced in it too.

Keywords: production well conductance sensor cross-correlation flowrate watercut multi-phase flow

INTRODUCTION

The field application indicates the effect of the conductance sensor [1] in measurement of the watercut is obviously better than those of other domestic tools used in oil wells with high watercut. And the conductance correlation flowmeter using the conductance sensor with the upstream and downstream sensors [2] [3] is a novel flowmeter without moveable and intrusive elements. So the production well logging tool used the same conductance sensor in measurement of flowrate and watercut of the fluid has obvious advantages and can overcome the shortcomings of spinner flowmeter of unstable instrument constant resulted from friction. At present, the flowrate and the watercut are measured with different sensors usually. For example, the spinner flowmeter is used to measure flowrate, and capacitance sensor or conductance sensor is used to measure watercut. So the technology is a supplement or replacement of the present measurement methods.

THE PRINCIPLES OF THE CONDUCTANCE SENSOR

MEASURING FLOWRATE AND WATERCUT

Principle Of The Flowrate Measurement[6] [7]

When the fluid flows through the sensors, the variation of the oil/water mixture conductivity causes the variation of output of the upstream sensor and the down stream sensor. The signals from the sensors are transmitted into the signal processing circuit, which can demodulate random flow noise signals x(t) and y(t). Cross-correlation function can be obtained through cross-correlation calculation, with the time delay τ as its variable. The cross-correlation function is:

$$R_{xy}(\tau) = \lim_{T\to\infty} \frac{1}{T} \int_0^T x(t)y(t+\tau)dt \tag{1}$$

The peak value of the cross-correlation function represents the greatest resemblance of the two flow noise signals. The corresponding time τ_0 of the peak value is called transit time. Given the distance between the two sensors L, the correlation velocity Vcc is derived by the formula (2)

CP914, *Multiphase Flow: The Ultimate Measurement Challenge,* edited by X. Cai, Y. Wu, Z. Huang, S. Wang, and M. Wang

$$v_{cc} = L / \tau_0 \qquad (2)$$

Principle Of Watercut Measurement

According to the electricity principle, the voltage amplitude between the two measuring electrodes is inversely proportional to the conductivity of the fluid flowing through the sensor. Given that the conductivity of mixture is σ_m, the conductivity of pure water is σ_w.

The ratio of σ_m to σ_w can be obtained through Maxwell formula

$$\sigma_m / \sigma_w = 2\beta/(3-\beta) \qquad (3)$$

where ß is the volume fraction of the continuous conductive phase in two-phase flow. In oil/water mixture, it is called water holdup.

DYNAMIC EXPERIMENTS IN MULTIPHASE FLOW LOOP

Analysis Of Watercut Measurement

Figure 1 gives out the relationship between the relative responses and flowrate at different watercuts. Figure 1 indicates that when flowrate is higher than 10m^3/d the meter has good water cut resolution and the slippage effect isn't obvious. When flowrate is lower than 10m^3/d, the resolving power decreases. That is because the slippage effect increases along with the decrease of flowrate. According to figure 1, the watercut can be figured out when the relative response and flowrate are known.

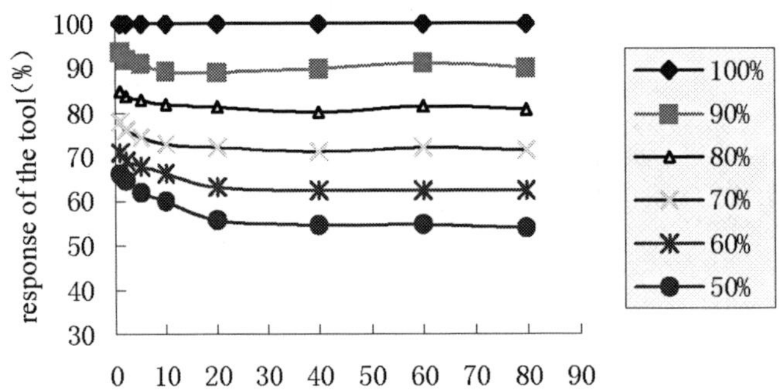

FIGURE 1. The relationship between the relative responses and flowrate at different water cuts

The Analysis Of Experiment Results Of The Flowrate Measurement

Figure2a and figure2b show the relationship between correlation velocity and standard flowrate when the flowrate is below 7m^3/d and above 7m^3/d. From the figures, we can see that the correlation velocity increases with the increase of flowrate from 1 m^3/d to 100 m^3/d.

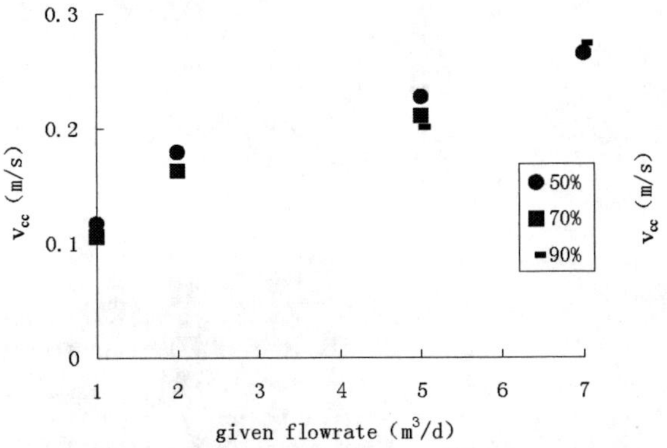

FIGURE. 2a Relationship between correlation velocity and standard flowrate below 7 m^3/d

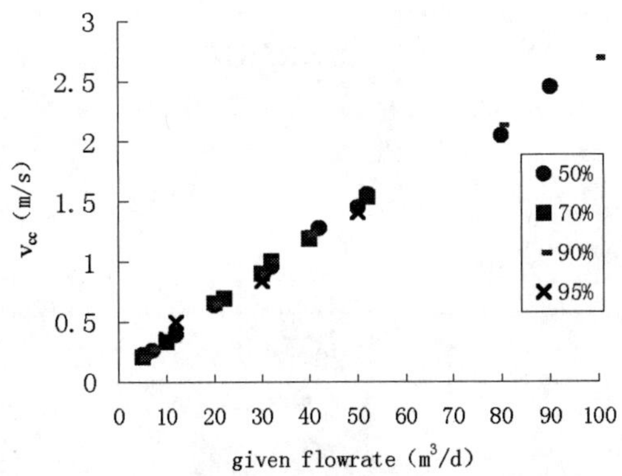

FIGURE. 2b Relationship between correlation velocity and standard flowrate above 7 m^3/d

The figure 2a indicates that the tool has a good resolution even the flowrate is very low. In the figure 2b, when flowrate increases from 10 m^3/d to 12 m^3/d and from 20 m^3/d to 22 m^3/d, the correlation velocities increase obviously. So the sensor can resolve the flowrate difference with 2m^3/d.

ANALYSIS OF FIELD APPLICATION

Repeatability And Consistency Of Field Logging

Figure 3a and 3b are separately curves of flowrate and water cut measured at depth of ××m of the oil well N1-J3-××; figure 6c and figure 6d are separately cures of flowrate and water cut measured at the same depth. By contrasting figure 3a with figure 3c, the repeated measurement error of flowrate is 0.8m^3/d; by contrasting figure 3b with figure 3d, the repeated measurement error of water cut is 2.7%. The contrast results demonstrate that the tool has a good repeatability.

Figure 3e and figure 3f are separately flowrate curve and water cut curve measured at depth of ××m with No.2# tool in oil well N3-10-××; figure 3g and figure 3h are corresponding curves measured at the same depth with No.4# tool. By contrasting figure 3e with figure 3g , the consistence error of flowrate is 1.7m^3/d, by contrasting figure 3f with figure 3h, the consistence error of water cut is 2.9%. The contrast results demonstrate that the tool has a good consistency.

Analysis Of Field Application

In the oil well N1-J3-××, the flowrate and the water cut measured in wellbore are 65.6m^3/d and 84.1% respectively; the flowrate and the water cut metered at the surface are 62 m^3/d and 82% respectively. The values measured in wellbore and those measured at the surface are consistent. In the oil well N3-10-××, the flowrate and the water cut measured in wellbore are 42.9 m^3/d and 95%, the flowrate and the water cut metered at the surface are 40 m^3/d and 93%. Also the values measured in wellbore and those measured at the surface are consistent. These indicate that the measuring results are right.

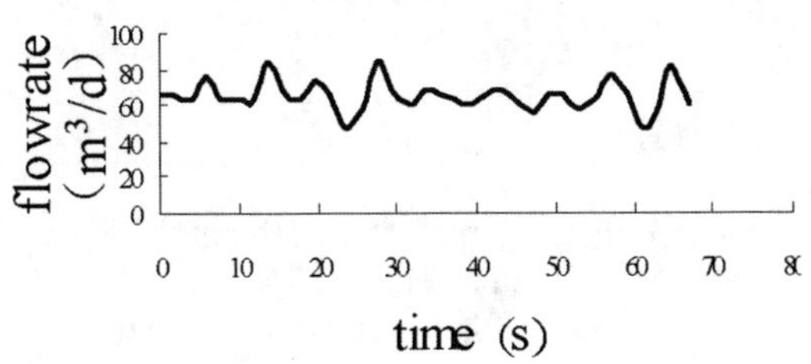

\\

\\

FIGURE. 3a flowrate curve

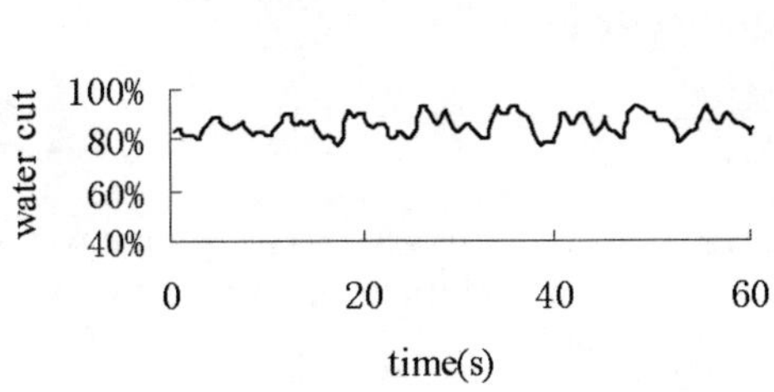

FIGURE. 3b water cut curve

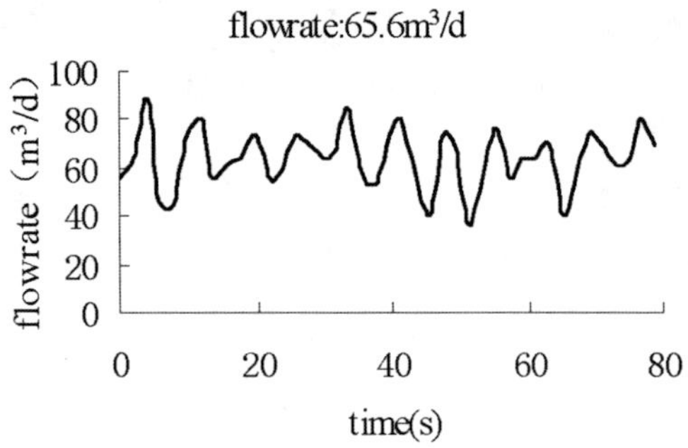

FIGURE. 3c flowrate repatition cure

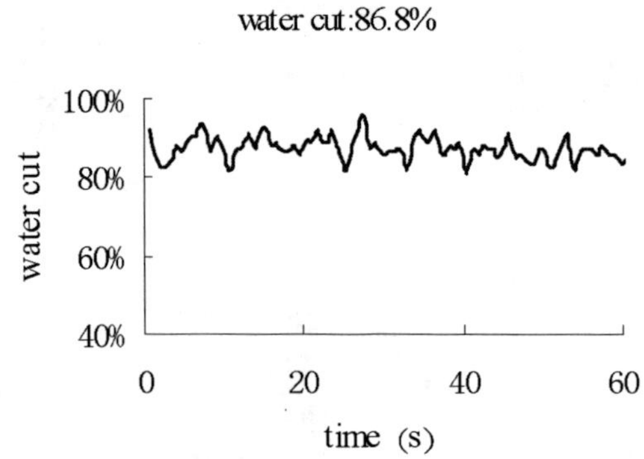

FIGURE. 3d water cut repeatition

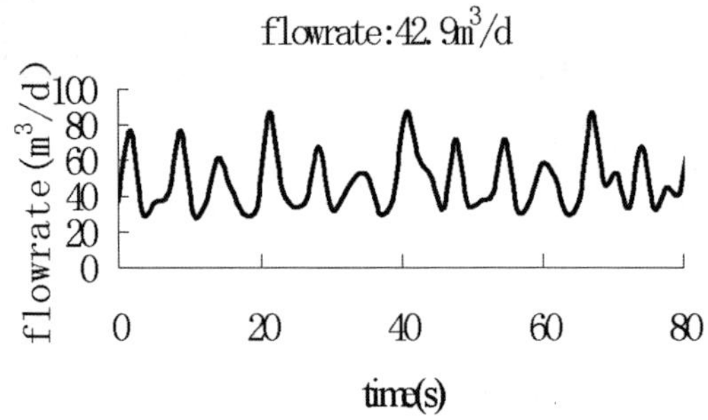

FIGURE. 3e 2 flowrate cure

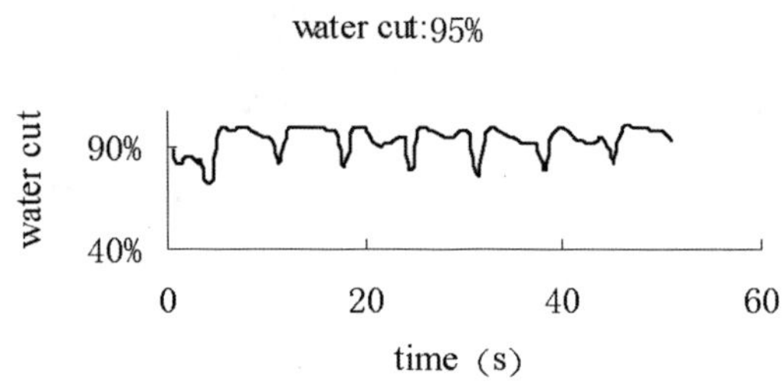

FIGURE. 3f 2#water cut

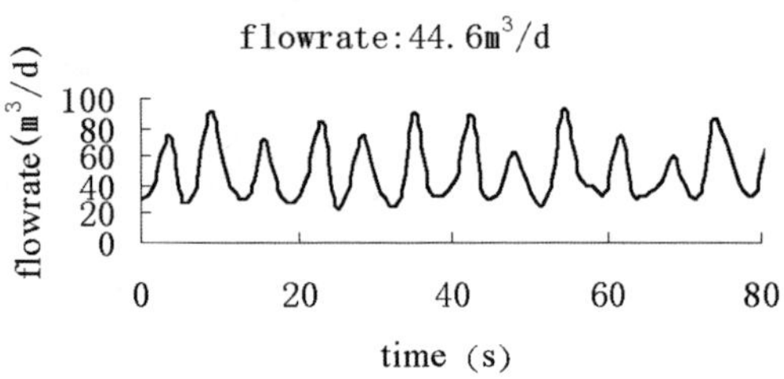

FIGURE. 3g 4#flowrate repeatition cure

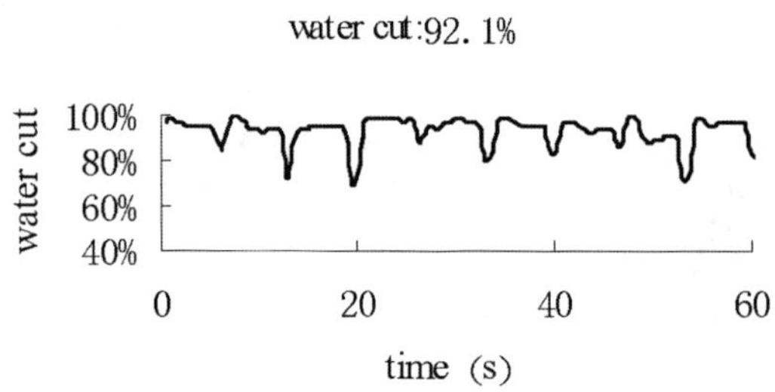

FIGURE. 3h 4#water cut repeatition cure

In the well S2-3-××, the total water cut and the produced liquid measured with the meter are 92% and 68 m^3/d respectively. The water cut and the flowrate measured at the depth of 943m are 93.3% and 47.8 m^3/d respectively, and at the depth of 948m are 88.5% and 29.3 m^3/d. The interpretation result shows that the productive formation between 948m and 943m is a formation producing water with producing water about 18 m^3/d.

The conductance correlation flowrate well logging tool is compared with a spinner flowmeter in well N3-4-××. During the measurement, the flowing sand frequently clogged the spinner flowmeter. At last it failed to work. On the other hand, the tool which has no movable and no intrusive elements isn't affected by grains of the sand, so the complete measurement results were obtained. It indicates the tool can compensate the disadvantage of tools with spinner flowmeter.

CONCLUSION

1.The flowrate measurement model and the water cut measurement model are gotten on the multiphase flow loop.
2.The field applications indicate that the tool has good repeatability and consistency.
3.This technology is not affected or less affected by sand during work. The application indicates the conductance correlation flowrate meter has very good application prospect and market prospect. It is an effective supplement of the present production profile well logging tools.

REFERENCES

1 Liu Xingbin, Downhole Measurement of Oil/Water Two-phase Flows, Ph.D. Dissertation of Harbin Institute of Technology, March 1996.
2 Hu Jinhai, Liu Xingbin, Zhang Yuhui, A New Flowmeter for Downhole Flowrate Measurement of Oil/Water Two-phase Flow, Well logging Technology, Vol. 23 N0.4, 1999.
3 Xu Ling-an, Cross Correlation Flow Measurement Techniques, Tianjing University Publishing, 1988.
4 Liu Xingbin, Hu Jinhai, et al, An Novel Impedance Sensor for Measuring Water Fraction In An Oil/Water Two-phase Flowset, FLOWMEKO'1998, Lund, Swenden.
5 Liu Xingbin, Impedance Water Holdup Meter for Downhole Measuring Water Fraction in Oil Wells, China Patent 98250463.2, 1999.
6 Liu Xingbin et al, Conductance Cross-correlation Flowmeter for Measurement in An Oil/water Two Phase Flow, FLOWMEKO'1996, Beijing, 1996.
7 Liu Xingbin, Conductance Cross-correlation Flowmeter Used for Measuring Oil/water Two-phase Flow in Wellbore, China Patent (pending)
8 Hu Jinhai, Liu Xingbin, Zhang Yuhui, A New Flowmeter for Downhole Flowrate Measurement of Oil/Water Two-phase Flow, Well logging Technology, Vol. 23 N0.4, 1999.
9 Liu Xingbin, Hu Jinhai, et al, Field tests of conductance cross-correlation flowmeter in measurement of oil/water two phase flow in high water production well ,FLOMEKO2000,Salvador,Brazil.

Spray Characteristics Determination of Water Mist for Suppressing Pool Fire

Li G. Zheng [*+], Ming G. Yu [*], Shui J. Yu [*], Zhi C. Liu [*]

[*] *School of Safety Science and Engineering, Henan Polytechnic University, Jiaozuo 454003 Henan, China*
[+] *Institute for Thermal Power Engineering, Zhejiang University, Hangzhou 310027 Zhejiang, China*

Abstract. To optimize the design of a water mist fire suppression system or to provide input data and validation data for field models, the characteristics of the water mist field are needed. This data typically consists of the distribution of droplet sizes and velocities. In this research a LS2000 Laser Sizer has been proposed to simultaneously determine spray characteristics of 408 kinds of self-developed nozzles for fire suppression. The results of LS2000 Laser Sizer measurements to characterize the water mist produced by nozzles for suppressing pool fire are presented. These data can provide a foundation to the design and optimization of additive-containing WMFSS. To explore the effect of the water pressure on the droplet size distribution and spray velocity, data from three typical pressures of 0.9MPa, 1.2MPa and 1.5MPa have been obtained. With the measured average size and distribution of droplets, a comparison between three nozzles was made in the pressure of 1.5MPa to investigate the influence of the nozzle design parameters on atomizing characteristics. The results show that representative size of droplets decreases slightly and the water flow rate increases with the water pressure to the nozzle head P_w, and the droplet velocity of water mist is approximately directly proportional to P_w. The trends that the nozzles with the larger D_{32} have a wider range of droplet diameter distribution and their volume fraction curves shift towards the large diameter direction are also found in measurements. This will be favorable to the extinguishment of kerosene pool fire. The water flux density distribution results demonstrated that the water flux density over the cross section was very non-uniform when the water mist nozzle is improperly designed.

Keywords: water mist; fire suppression; two-phase flow; measurement

INTRODUCTION

Water mist is currently receiving considerable attention as a potential replacement for Halon fire suppression agents that are required to be phased out due to their impact on the environment. Water mist refers to a water spray for which the diameter $D_{V0.99}$, for the flow-weighted cumulative volumetric distribution of water droplets, is less than 1000 μm. Water mist fire suppression systems (WMFSS) have been demonstrated to have a number of advantages, including low cost, efficacy in suppression of flammable liquid pool and spray fires, absence of toxicity or adverse environmental effects, and potential efficacy as inerting or explosion suppression systems [1].

The effectiveness of the WMFSS at controlling a fire is governed by the spray characteristics, e.g., spatial distributions of drop size, drop velocity, mass flux, water flux density, spray angle. Moreover, with the rapid increase in more sophisticated modeling of the dynamics of fires, it is now possible to include the effect of water mist on the fire dynamics to account for the complex interaction of this multiphase combustion process. To include the effect of water mist on the fire dynamics, it is necessary to provide spray characteristics of the water mist produced by the nozzle. The breakup of the water sheet exiting the nozzle, however, is not completely understood in the literature, and droplet size and velocity can not be predicted to a high level of certainty. As noted in Ref.[2], experimental measurement is the way with the best confidence to determine characteristics of water mist.

There are many methods currently available for measuring the size and/or velocity distributions of water mist, and most of them are non-intrusive optical technology. J. H. Liu and co-workers [3] had been studying the water mist characteristics using three dimensional Laser Doppler Velocimetry (LDV)/Adaptive Phase Doppler Velocimetry (APV). J. F. Widmann and D. T. Sheppard [4-5] presented Phase Doppler Interferometry (PDI) and Particle Image Velocimetry (PIV) measurements in water sprays produced by residential fire sprinklers. L. J. Yang and W. Wang[6] demonstrated that Malvern droplet size measurement device can be used to measure atomizing

CP914, *Multiphase Flow: The Ultimate Measurement Challenge,* edited by X. Cai, Y. Wu, Z. Huang, S. Wang, and M. Wang

characteristics of two-phase flow emulsification water mist nozzles. X. S. Wang and X. P. Wu [7] had carried out measurements of full field velocity in a water mist based on Digital Particle Image Velocimetry (DPIV). A. D. Putorti Jr [8,9] reported first the results of water spray from a standard orifice using optical array probe and later simultaneous measurements of drop size and velocity in large-scale sprinkler flows using Particle Tracking Velocimetry (PTV) and Laser-Induced Fluorescence (LIF) in his Ph.D thesis. Global rainbow thermometry, based on the elastic light scattering phenomenon, is recently used to characterize a non-isothermal water spray [10].

This paper presents an application of a commercially available, non-intrusive, droplet size and velocity measurement technology that has recently been shown to be effective in characterizing the water mist. Experimental data obtained from 408 kinds of pressure swirl type nozzles for suppressing pool fire are presented, and effects of several factors on atomizing characteristics of these nozzles are discussed.

EXPERIMENTAL SETUP

FIGURE1 shows a schematic diagram illustrating the measurement system. The experimental facility consists of water supply system, swirl type nozzles, *LS-2000* laser Sizer and data acquisition software. The water supply system makes of a water tank, a centrifugal pump and auxiliary pipeline. The pressure swirl nozzle to be studied is attached to the water pipeline, which provides the water to the nozzle. *LS-2000* lase Sizer measures atomizing characteristics, and data acquisition software records and saves measurement data.

The droplet size distribution (diameter and range), the spray angle and spray velocity can be measured by *LS-2000* sizer. The principle of the LS2000 laser sizer is shown in FIGURE2. The instrument, based on Mie's theory, can obtain the mean spray drop size and its distribution. Light beam is emitted by a laser and expanded by lens. When passing through spray droplets, light is scattered by spray droplets, and the scattered light is collected by a photo-detector. By solving a so-called inversion problem, droplet size distribution can be determined from theoretic Mie scattering matrix and collected luminous energy.

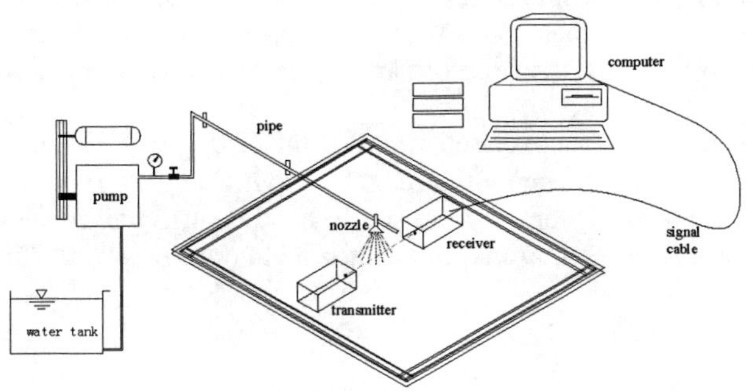

FIGURE 1. Schematic of the measurement system for droplet size

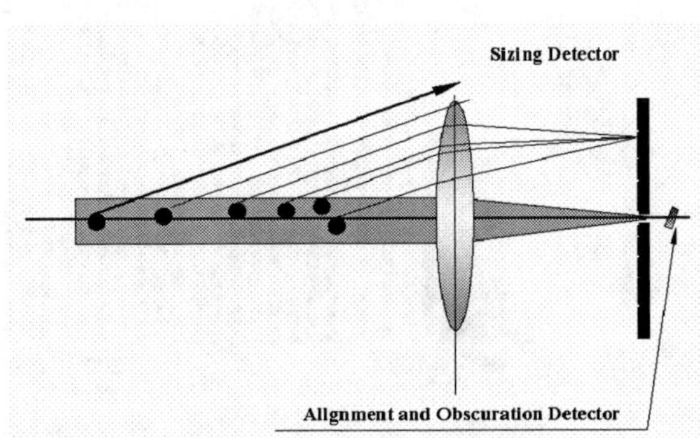

FIGURE 2. Principle of droplet size measurement

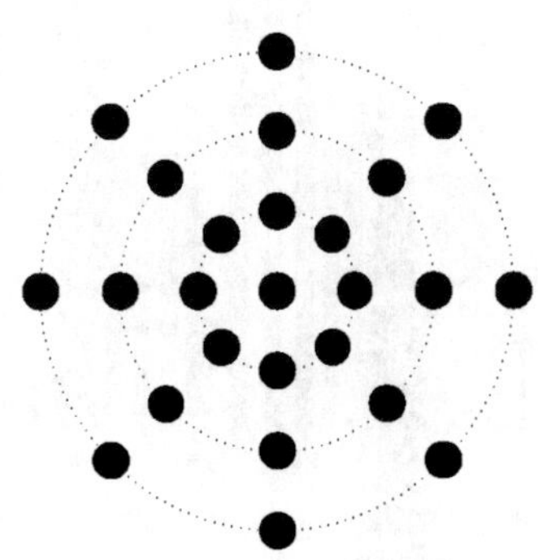

FIGURE 3. Configuration of cup network for water flux density

Average spray velocity at the fixed cross section away from the nozzle exit can be determined by correlation method. The spray angle can also be automatically analyzed and obtained by using the image processing method.

The water flux density was also measured in the present study. A simple way to characterize spray density is adopted, i.e., total flow rate per unit area [11]:

$$\text{Avg. Area Density, Da=Qtotal/Atotal [Lm-2min-1]} \tag{1}$$

Density expressed in this way (i.e. Equation (1)) can be measured by cup collector method, i.e., collecting spray on sample surface areas, over a known time interval. This method is applied in the present study because it can be relatively easily measured. The configuration of a circular measurement network for the water flux density was used throughout the experiments, as illustrated in FIGURE 3.

In this study, the spray characteristics of 408 nozzles with different flow channel diameter and parameters of screw have been carried out.

RESULTS AND DISCUSSION

Spray Characterization

The atomizing characteristics investigated in this research include the spray droplet size distribution (diameter and range), the spray angle, spray velocity and water flux density distribution.

As noted by A. Jones and P. F. Nolan [12], for a comprehensive picture of the spray, it is usual to quote an average diameter and to provide an idea of the distribution of the droplets in various size ranges, either in terms of a histogram or a cumulative distribution curve.

The volume fraction and cumulative volume fraction distributions of the representative size (D_{32}) of a certain nozzle under the operating pressure of 1.5MPa obtained with the *LS-2000* laser Sizer are shown in FIGURE 4 and FIGURE5. FIGURE 4 presents the representative size distribution. The sauter mean diameter is 117μm. The size distribution presented in FIGURE4 shows that majority of the droplets have diameters between 90 and 125μm. The cumulative distribution is given in FIGURE5, which illustrates that up to 91% of the droplets have diameters below 175μm.

The angle of the nozzles developed in this research varied from 17 degree to 84 degree. Most of the nozzles have a spray angle larger than 45 degree. Fire tests show that nozzles having a spray angle larger than 45 degree will have enough coverage to protect the fire source. Some of the single head centrifugal nozzles are of small spray angle. In order to have big protective area, the combinational heads nozzle should be used. This work will be carried out in future.

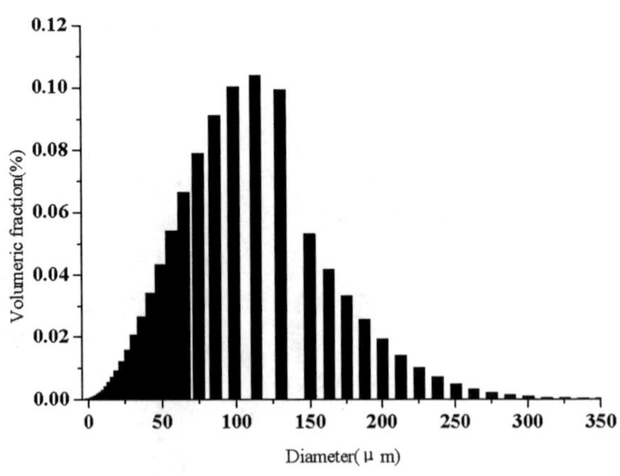

FIGURE 4. Representative size distribution

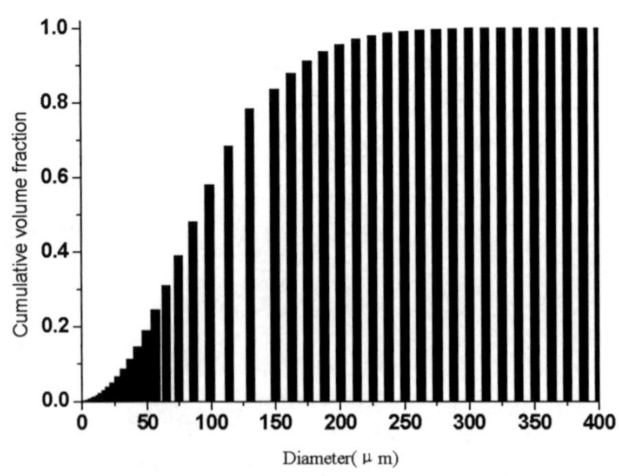

FIGURE 5. Cumulative volume fraction distributions

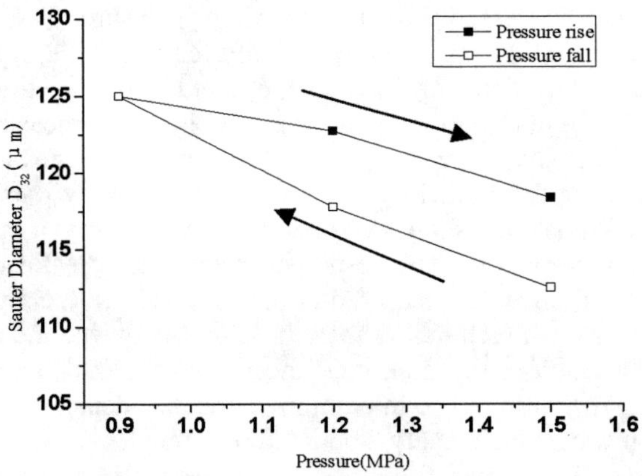

FIGURE 6. SMD .vs. water pressure

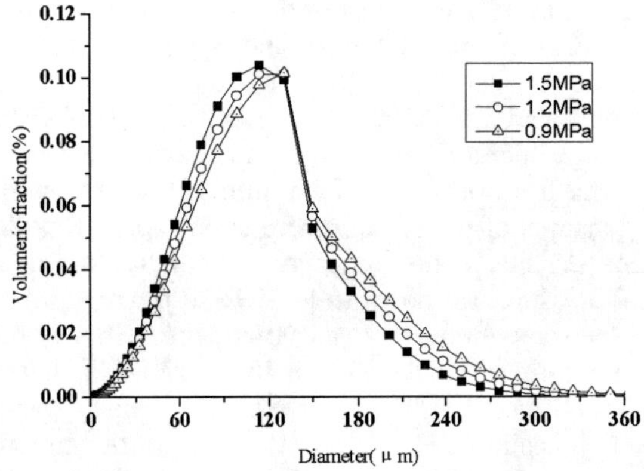

FIGURE 7. SMD distribution at various operating pressures

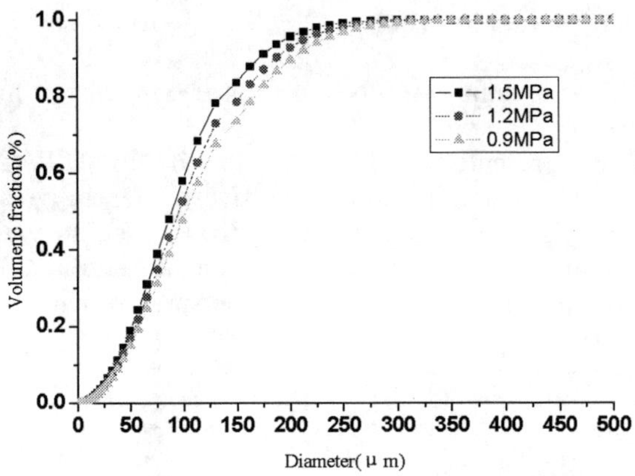

FIGURE 8. SMD cumulative distribution at various operating pressures

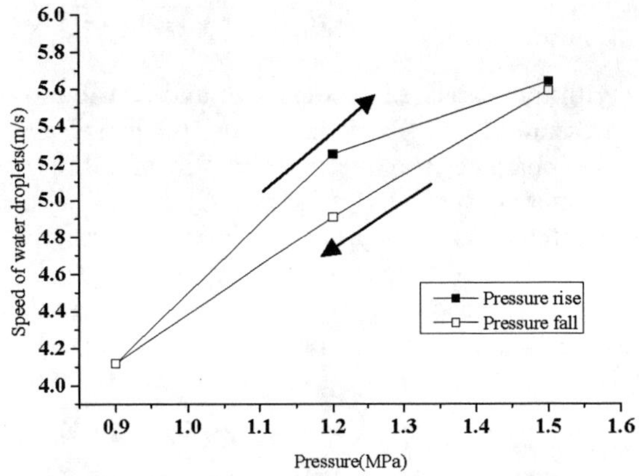

FIGURE 9. Variation of water mist speed with water pressure

Effect of Water Pressure

To explore the effect of the water pressure on the droplet size distribution and spray velocity, measurements were obtained in the spray produced by a certain nozzle under various operating pressures. The water pressure to the nozzle head was varied from 0.9MPa to 1.5MPa.

FIGURE6 presents the plot of SMD as a function of water pressure. The representative size decreases slightly with water pressure. It can be concluded that water pressure has little influence on SMD. Volume fraction and cumulative volume fraction of droplet size from three typical pressures of 0.9MPa, 1.2MPa and 1.5MPa have been obtained, as shown in FIGURE7 and FIGURE9. With the increase of water pressure, the distribution curve shifts towards the small diameter direction or negative direction of horizontal axial.

FIGURE9 plots water mist speed as a function of water pressure. The speed varied from 4.11m/s to 5.64m/s with respect to water pressure to the nozzle from 0.9MPa to 1.5MPa, varied form 5.59m/s~4.11m/s with respect to water pressure from 1.5MPa to 0.9MPa. Velocity data presented in FIGURE9 follow the law that the velocity of water mist is approximately directly proportional to water pressure.

The data in FIGURE6 and FIGURE9 indicates that a little difference in the SMD and velocity measurements obtained from pressure rise process (varied from low pressure to high pressure) and from pressure fall process (from high pressure to low pressure) appeared. Comparative analysis shows differences at the same pressure are within 6μm for D_{32} and 0.34m/s for speed, respectively. In other words, D_{32} and droplet velocity obtained with *LS-2000* can give relatively good repeatability, as illustrated in FIGURE6 and FIGURE9.

Contour plot of the water flux density distribution under various water pressures measured at 1m away from the nozzle exit is shown in FIGURE10. The total water flux was measured by collecting water with containers placed on the floor 1m away from the nozzle exit. The water flow rate increased with the operating pressure. The water volume flow rates are 2.09lmin^{-1} and 3.44lmin^{-1} under the operating pressures of 0.5MPa and 0.8MPa, and the corresponding mean water flux density considering the 0.9m diameter of the measured cross section are 3.28lm^{-2}min^{-1} and 5.41lm^{-2}min^{-1}, respectively. FIGURE10a is the distribution of the water flux density measured at 0.5MPa of the operating pressure of the nozzle, the water flux density at various points of the cross section was very non-uniform, from 0.71 to 8.47 lm^{-2}min^{-1}. It shows that because of the swirling effect of the screw body located in the flow channel of the nozzle, there is one water flux density peak at the below zone of the cross section, and the water flux density in the upper zone of the cross section is very low. The results measured at 0.8MPa of the operating pressure, as shown in FIGURE10b, are similar with the results measured at 0.5MPa. Due to the swirling effect of the screw body, the water flux density in the upper zone is much higher than those in the lower zone at this section downstream of the exit of the nozzle. At this cross section, the measured mean water flux density was 5.41lm^{-2}min^{-1}, whereas the peak water flux density was 14.17lm^{-2}min^{-1}. The maximum water flux density over the cross section is almost 11.61 times the value of the minimum water flux density, about 1.22lm^{-2}min^{-1}.

The data in FIGURE10 demonstrated that the water flux density over the cross section was very non-uniform when the water mist nozzle is improperly designed.

Effect of Nozzle Design Parameter

With the measured average size and distribution of droplets, a comparison between three nozzles was made in the pressure of 1.5MPa, as shown in FIGURE11 and FIGURE12.

The characteristic mean sizes for the three nozzles are presented in FIGURE 11 and FIGURE12. The representative size D_{32} of nozzle A, B and C are 78μm, 113μm and 171μm, respectively. Generally, another representative size $Dv_{0.99}$ in water mist is popularly used. $Dv_{0.99}$ is defined as the droplet diameter below which the size of 99% of the volume of the spray is. The $Dv_{0.99}$ of the above three nozzles are 162μm, 270μm and 370μm, respectively. According to the definition of less than 1000μm of water mist by NFPA, the spray produced by the above three nozzles can be classify into water mist.

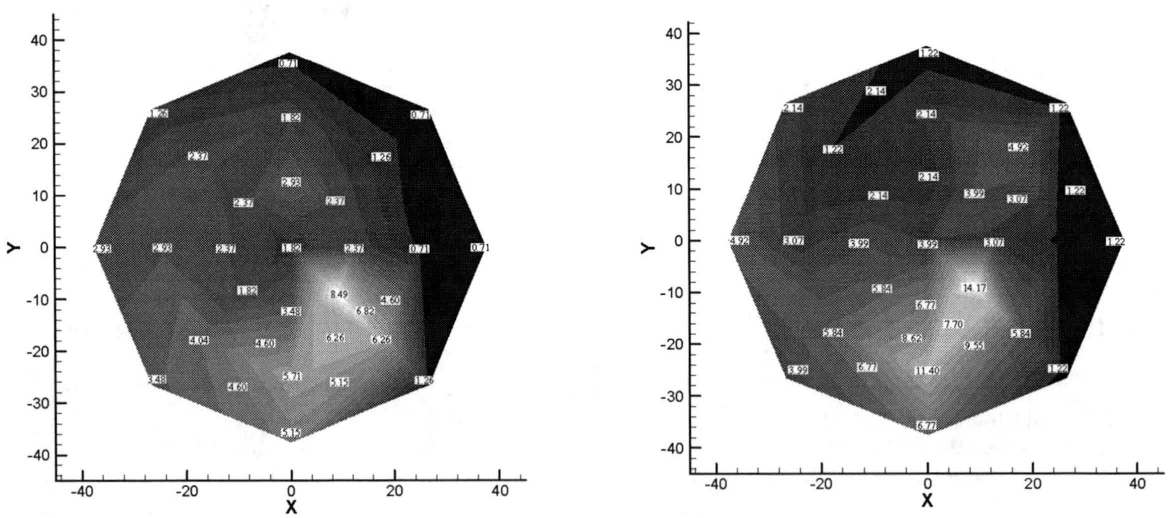

(a) p_w=0.5MPa (b) p_w=0.8MPa

FIGURE 10. Contour plot of the water flux density measured at 1m away from the nozzle exit

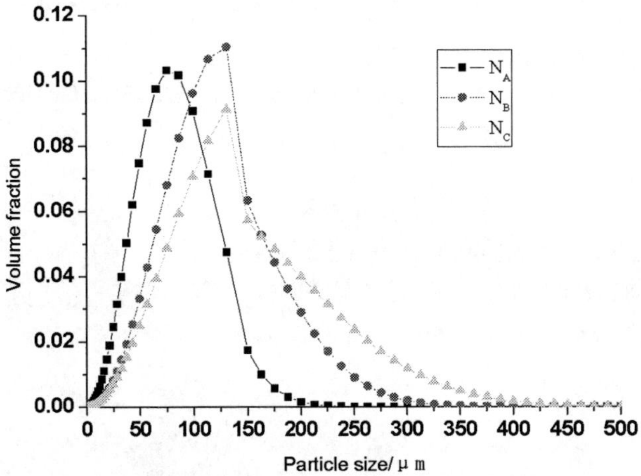

FIGURE 11. Size distribution of four nozzles

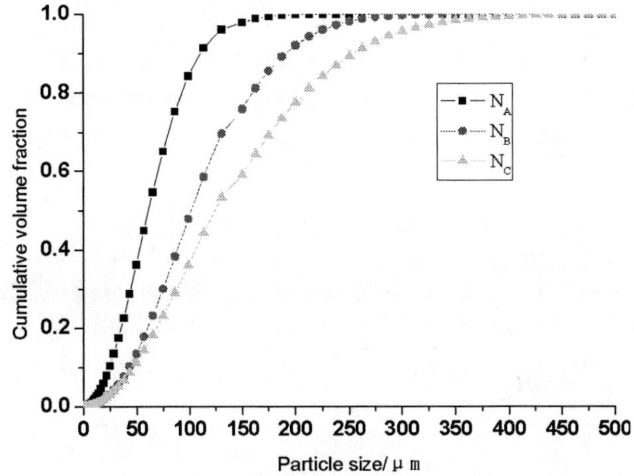

FIGURE 12. Cumulative distribution of four nozzles

The data in FIGURE 11 indicate that the trends that the nozzles with the larger D_{32} have a wider range of droplet diameter distribution and its volume fraction curve shifts towards the large diameter direction are also found in measurements.

CONCLUSION

With a *LS-2000* laser sizer and image processing method, 408 kinds of nozzles for suppressing fire have been experimentally characterized. The results indicate that LS2000 is a useful method of providing information to the optimal design of the nozzle.

The effect of the operating pressure on the mean drop size, velocity, water flow rate and water flux density distribution was investigated. The data indicate that the characteristic sizes decreases slightly with the operating pressure P_w, and the velocity of water mist is approximately directly proportional to P_w. The water flow rate increases with the water pressure. The trends that the nozzles with the larger D_{32} have a wider range of droplet diameter distribution and its volume fraction curve shifts towards the large diameter direction are also found in measurements. The water flux density distribution results demonstrated that the water flux density over the cross section was very non-uniform when the water mist nozzle is improperly designed.

ACKNOWLEDGEMENTS

The authors would like to acknowledge the support provided by NSFC (Nos.50274061, 50476033), the Outstanding Talents Innovative Foundation of Henan Province of China (No.0421000800) and NSF of HPU (No. 646102). The contributions of all graduate students involved in these projects are greatly appreciated for their invaluable work in conducting experiments and extracting the raw data from database. Many thanks to Prof. XS Cai, Prof. JQ Shen, Mr. JF Li and Mr. YS Tang of USST for the use of LS2000 Laser Sizer.

REFERENCES

1. G. Grant and J. Brenton, *Prog. Energ. Combust. Sci.* 26, 79-130(2000).
2. M. H. Fan and H. Zhou, *J. Zhejiang Univ.(Engineering Sci.)* 39, 1431-1434(2005).
3. J. H. Liu and G. X. Liao, *Laser J.* 21, 58-60(2000).
4. J. F. Widmann, *Fire Safety J.* 36, 545-567(2001).
5 J. E. Widmann and D.T.Sheppard, *Fire Technol.*37, 297-315(2001).
6. L. J. Yang and W. Wang, *J. Beijing Univ. Aeronaut. and Astronaut.* 28, 413-416(2002).

7. X. S. Wang and X. P. Wu, *J. Exp. Mechanics* 18, 6-11(2003).
8. A. D. Putorti Jr, *Ph.D thesis*, University of Michigan, Gaithersburg, MD, 2004.
9. A. D. Putorti and T. D. Belsinger, “Determination of Water Spray Drop Size and Speed from a Standard Orifice, Pendent Spray Sprinkler” in *National Instit. Stand. Technol.*, Rept.of Test, MD, May, 1999.
10. M. R. Vetrano and S. Gauthier, *Exp. Fluids* 40, 15-22(2006).
11. J. R. Mawhinney, “Engineering Criteria for Water Mist Fire Suppression System” in *Water Mist Fire Suppression Workshop*, Proc. of Water Mist Fire Suppression Workshop, 1993, pp.37-75.
12. A. Jones and P. F. Nolan, *J. Loss Prevention Proc.* 8, 17-22(1995).

Effect of Slug Flow on Pipe Vibration in Horizontal Pipeline

ZHONG Xingfu[1] WU Yingxiang[1] LI Donghui[1] LI Zhibiao[1] LI qiang[2]

1 Institute of Mechanics, Chinese Academy of Sciences No.15 Zhongguancun Road, Beijing 100080, China
2 Jinghe Industry ,Xi'an Shaanxi, Changqing Well Logging,China petroleum logging Co.LTD. 710201,China

Abstract. The vibration characteristics of the pipeline in gas liquid flow are investigated in this paper. The pipeline vibration which responds to pressure sensors and 3D laser measurement was analyzed. It shows that pipeline vibrations decrease in severe slug flow when gas flow rate increases and liquid flow rate keeps a constant. But when liquid flow rate increases and gas flow rate keeps a constant, at the first stage, the vibration increases and is up to a max amplitude, especially, in the slug flow. At the second stage, the amplitude of pipeline vibration decreases but the vibration frequency increases.

Key words: Slug flow Vibration Horizontal pipeline

INTRODUTION

The pipeline vibration can occur in petroleum, natural gas and chemical industry because of unsteady flow, the change of flow direction, pipe diameter, etc. This problem will affect safe running of pipeline severely if it is not handled well. Therefore, the pipe vibration arouses more and more people's attention. But nowadays, the study on it is mainly focused on the pipe vibration due to mechanical vibration source. However, there is a kind of pipelines in engineering with gas-liquid flow inside and whose both inlet and outlet are without pump and compressor, and they occur severe vibration. The reasons of this situation are too difficult to analyze.

Fluid vibration source in the pipeline has been analyzed by theoretical methods before[1]. Because of adopting some simplifications and assumptions, the results from these methods are only used to references. Generally, the vibration of the pipe system happens because of mechanical vibration, fluid vibration, vibration of themselves, earthquake and so on. Fluid vibration includes fluid pulsation, gas-liquid flow vibration, high-speed flow vibration and flow vibration outside the pipeline[2]. The first three kinds of fluid vibration are mainly analyzed in this paper.

The vibration of gas-liquid flow pipeline is studied by experiment in this paper. The main purpose of the study is to know the mechanism of this type of vibration, and to adopt reasonable steps to decrease the vibration. By means of changing gas-liquid flow-rate in experiment, the different flow patterns can be observed. After flowing steadily, the laser measurement is used to measure the displacement of the pipe. The influence of flow parameters on the pipeline vibration is analyzed, and the displacement and amplitude of the pipeline are computed. The experimental result indicates that the fluid flow plays an important role in pipe vibration response and must to be taken into consideration when designing the pipe. Besides, one of the main reasons to cause the pipeline vibrate is that flow pressure, velocity and density change periodically with time.

EXPERIMENTAL SYSTEM AND METHOD

The experiment is completed at the multiphase flow loop facility in IMECH. The pipeline of the facility is made of transparent organic glass. The flow patterns can be observed directly. The inner diameters of pipes are 50mm, and the length are 28m. The experimental mediums are air, water and oil. Gas and water are supplied by compressor and water tank separately. They go through respective flow meters and are mixed by T-type pipe in front of test pipe. At last, they are separated after flowing across the test section. Water returns to tank for re-use. Gas superficial velocity is 0~5 m/s, and liquid superficial velocity is adjustable within the scope from 0.2 to 1.2 m/s.

CP914, *Multiphase Flow: The Ultimate Measurement Challenge,* edited by X. Cai, Y. Wu, Z. Huang, S. Wang, and M. Wang

FIGURE 1 is the sketch of test section. Fig 1a is the photo of section AC and is enlarged in fig 1b. The distance from upper stream A to downstream B of horizontal section is 1m. The distance from section A to inlet is 22m and C to outlet is 5m. Three pressure sensors are installed on both section A and C, in order to obtain the pressure signals and differential pressure signals.

Point B is the measurement point of pipe vibration that is got from the middle point of section AC and jointed rigidly with AC. Pipeline's vibration in the X, Y and Z direction when two-phase fluid flows from A to C is obtained by measuring the vibration displacement of point B in these three directions. This displacement is measured by optoNCDT laser instrument, which resolution is 20μm and accuracy 0.01%FSO.

Different flow patterns are obtained in the pipeline by changing given flow rate of gas-liquid. They can be observed and recorded after flow is steady. At the same time, the signals of pressure, differential pressure and vibration distance of the pipeline are recorded, too. All the values about signals are got from DAQP acquisition system which sample frequency is 50Hz. The process of the experiment is that gas flow rate(10，50，100，200，300，400，unit: %25 l/min) increases gradually under the condition of constant liquid flow rate(4，6，10，20，40，60，70，unit: %25m^3/h), and then the data are gathered and recorded. The units of measurement in this paper are the ones of the meters used in the experiment. Without special explanation, the unit of liquid flow rate w is %25m^3/h, and the liquid flow rate is %25 l/min.

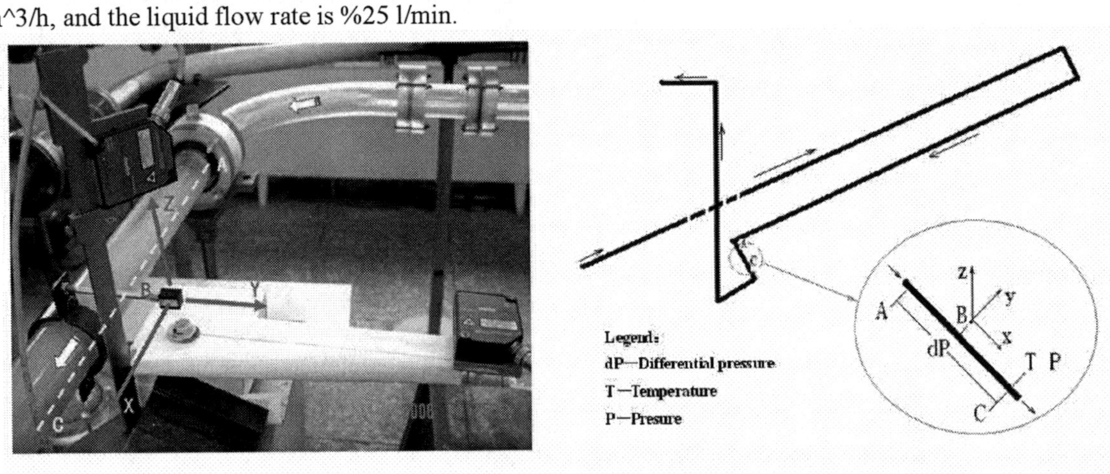

a、The photo of experimental pipeline and test section AC b、A sketch map of test section AC

FIGURE 1. The experimental flow pipeline

THE ANALYSIS OF FLOW PATTERNS IN THE PIPE

During the two-phase flow, flow patterns are changed with phase velocity，density, viscosity and surface tension of the individual phase. It not only affects flow characteristics, mass and heat transfer, and vibration characteristics, but also affects the system's dependability, safety and efficiency. It is well-known that fluid pulsation appears when pipe flow's parameters such as pressure, velocity and density change periodically with time. Slug flow pattern is a kind of typical fluid pulsation. It's one of the main reasons that arouse the pipe and the attached instruments vibrate.

In order to analyze the vibration caused by the fluid flow in pipeline, first of all, the flow patterns are observed and recorded carefully. FIGURE.2 gives out the flow patterns and the extension of experimental parameters. Its horizontal axis is liquid flow rate, and vertical axis is gas flow rate. The flow patterns which can be seen during the experiment are bubble flow, slug flow, stratified flow and wavy flow.

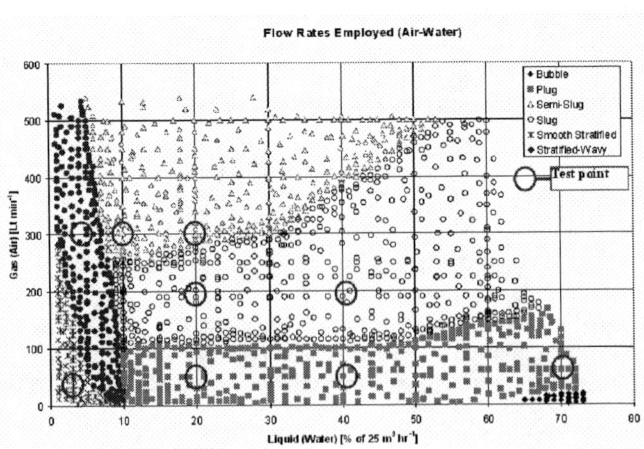

FIGURE 2. Flow pattern in horizontal pipeline(ID=50mm)

According to this figure, the needed flow patterns can be obtained manually. Due to gravity, gas is mainly at the top of the pipe, and liquid is at the bottom. Usually, gas-liquid two-phase horizontal flow is defined into following flow patterns: bubble flow, slug flow, stratified flow and wavy flow.

Flow patterns not only can be observed directly, but also can be analyzed from measured signals. The fluctuations of pressure drop are different under different flow patterns, so pressure changes reflects the flowing situation at certain degree. For example, , pressure fluctuation is large when slug flow happens. FIGURE.3 is fluctuation response of vibration displacement and differential pressure in Y, Z direction which are simultaneously measured when w=4, g=50. Flow patterns, pressure and pipe vibration are corresponding to each other. To this slug flow, the length of gas slug is about 9-13m and the tail is long, while the water slug is short. The frequency of the gas slugs' appearance is approximately 0.14 Hz, which agrees with the PSD analysis result very well. The length of water slug is 3-4 m, in which there are little gas bubbles. When there is no gas or liquid slug, and the water level is at the 2/5 position of the pipe, a part of water flows back after a large water slug passes test section. FIGURE.3b shows an integrated slug flow that the liquid slug section AC and the gas slug section. Here the liquid slug section AC is analyzed first, then the gas slug CE.

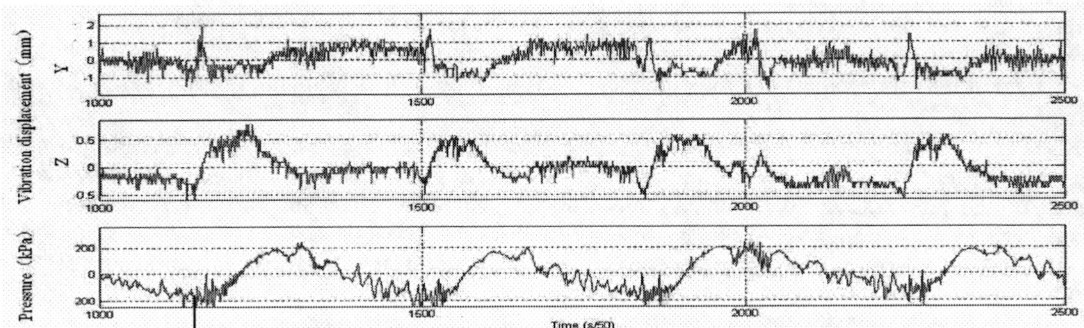

FIGURE 3a Vibration and differential pressure in Y，Z direction when w=4，g=50

X-axis：Time(1/50 s), Y-axis：Vibration displacement(mm)

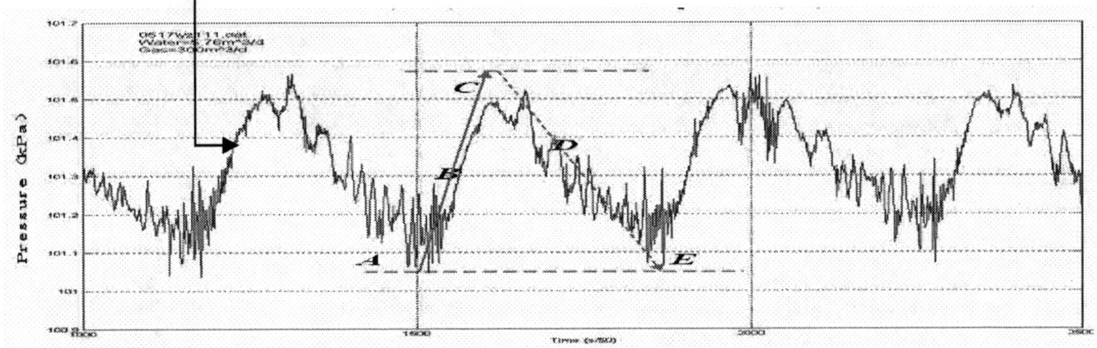

FIGURE 3b Typical pressure fluctuate(w=4，g=50)

X-axis：Time(1/50 s), Y-axis：Pressure(kPa)

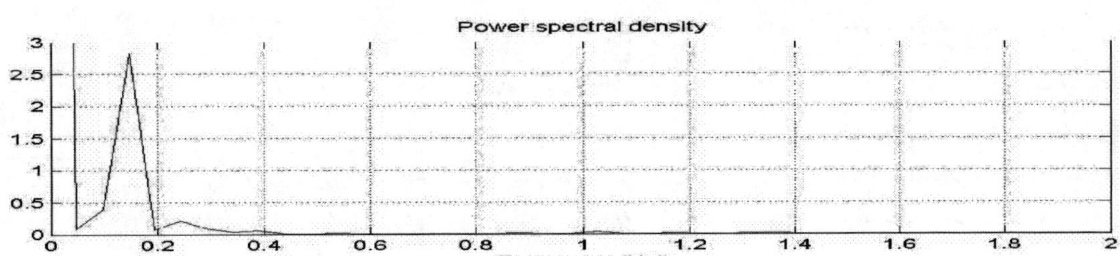

FIGURE 3c The PSD of pressure fluctuate (f=0.14Hz，w=4，g=50) X-axis：Frequency(Hz), Y-axis：Energy

FIGURE 3. Vibration displacement and differential pressure in Y，Z direction in slug flow (w=4，g=50)

The point A of section AC in FIGURE 3b is the head of a liquid slug and point C is tail. When the liquid slug reach the pressure sensor, pressure rapidly increases from point A (low value) to point C(high value). The section BC is the pure liquid slug, of which pressure fluctuation is smaller than the one of section AB. After the liquid slug passes away the test pipe, the pressure reduce and the flow come into stage of gas slug.

The point C of section CE in FIGURE 3b is the front of the gas slug and point E is tail. When the gas slug reaches the pressure sensor, pressure decreases from point C(high value) to point E(low value). liquid holdup is different at different place of the same gas slug, the pressure fluctuation in gas slug's is larger than the liquid slug's. During the process of gas slug flow CE in Fig 3b, the pressure fluctuation is always large, and descending of pressure dorp is slow relatively. Until the gas slug CE has passed the pipeline, the pressure begins to ascend, and the flow comes into the next stage.

ANALYSIS OF HORIZONTAL PIPE VIBRATION CAUSE BY TWO-PHASE FLOW

The vibration of the pipe system happens generally because of mechanical vibration, fluid vibration, vibration of themselves and earthquake and so on. Fluid vibration includes fluid pulsation, gas-liquid flow vibration, high-speed flow vibration and flow vibration outside the pipeline. The first three factors of them are principal in the analysis of this paper.

FIGURE3a shows the differential pressure and pipe's vibration in Y, Z direction when the slug flow occurs. It can be seen that when liquid slug got to the test point (liquid slug is section AC in FIGURE.3b), the vibration displacement of the pipe was up to the maximum. Gas slug has little effect on the pipe's vibration. Vibration becomes smaller and smaller after liquid slug passes.

FIGURE 4 gives vibration's displacement in X, Y, Z direction with gradually increasing gas flow rate (g=50,100,300) on condition that liquid flow rate is constant (w=4). It indicates that the pipe's vibration decreases when gas flow rate increases. Fig.5 is vibration's displacement in X, Y, Z direction when liquid flow rate increases (w=4,6,20,40,70) under condition of constant gas flow rate (g=50). The pipe's vibration becomes severe gradually as the liquid flow rate goes up, and there is the max amplitude. Then the amplitude of pipeline comes down, but the frequency increase.

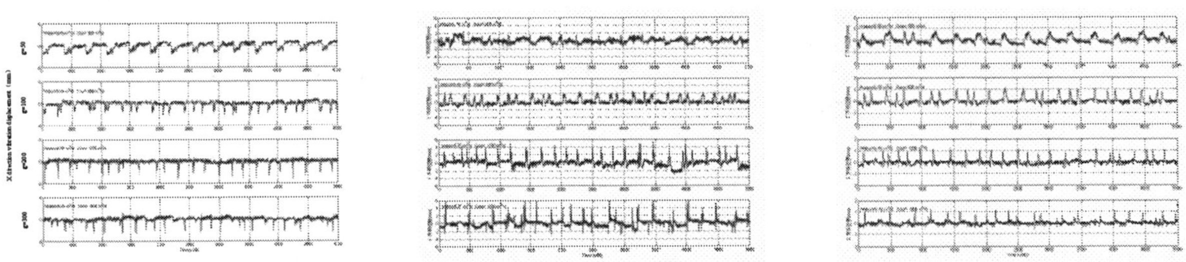

X direction vibration displacement(mm) Y direction vibration displacement(mm) Z direction vibration displacement(mm)

FIGURE 4. The 3D vibrations of pipeline in horizontal slugging flow when gas flow rate increase and liquid flow rate keep a constant(w=4, g=50，100，200，300), X-axis：Time(1/50 s), Y-axis：Vibration displacement(mm)

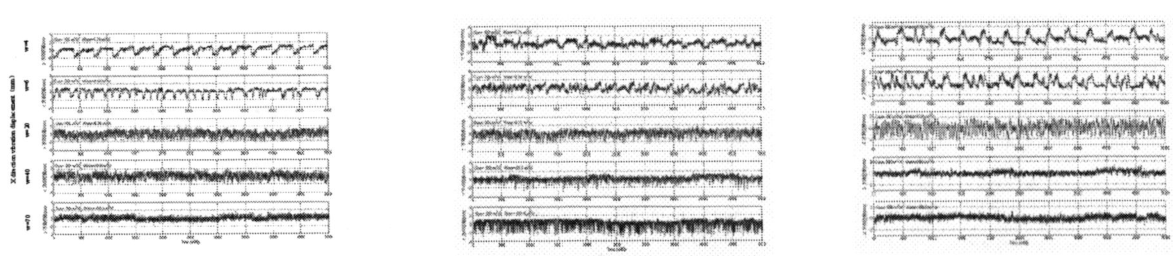

X direction vibration displacement(mm) Y direction vibration displacement(mm) Z direction vibration displacement(mm)

FIGURE 5. The 3D vibrations of pipeline in horizontal slugging flow when the liquid flow rate increase and gas flow rate keep a constant(g=50, w=4， 6， 20， 40， 70), X-axis： Time(1/50 s), Y-axis： Vibration displacement(mm)

CONCLUTION

Gas-liquid two-phase flow in pipeline is common in the petroleum chemical industry equipment. In the process of gas-liquid flowing, some pipes vibrate severely owing to the changes of flow parameters, pipe's construction and arranging ways. Feebleness pipe vibrations could not damage the equipment, but more intense vibration may cause pipes' construction and the equipments at upper stream and downstream harmful, even to make them stop running. Through experimental researches on gas-liquid two-phase flow in the horizontal pipe, some experimental results were obtained, but the mechanism is not further analyzed that the fluid flowing parameters affect pipe's vibration.

Initial experimental results indicate that the pipe vibration gets weaker and weaker when liquid flow rate keeps constant, and gas flow rate increases. Whereas, when gas flow rate is constant, the vibration becomes more violent gradually with liquid flow rate increasing, and then the vibration has the max amplitude. Subsequently the amplitude of pipe's vibration decreases, but vibration's frequency increases. Seen from the measuring results, flow parameters such as flow pattern and pressure respond to pipeline's vibration. The relationship between these parameters is await to be further analyzed.

ACKNOWLEDGEMENTS

The author is grateful to Simon Philip (School of Chemical Engineering and Analytical Science, The University of Manchester Manchester M60 1QD， United Kingdom) and Dr. Jy XU for their flow pattern experimental used in this paper.

REFERENCES

1. Tan ping, FU xingjun et al. Study on Vibration of Conveying Water Piping System in Power Plant, TURBINE TECHNOLGY, Vo1.46, No.2, 2004.
2. Zhang Duqing, Zhang Guangcheng et al. Treatments& Reasons for Vibration of the Pipe in Power Plants, SHANDONG DIANLI JISHU, Shan Dong Power Technology. Vo1.47, No.1, 2006.

Study on the Horizontal-well Injection Profile Logging Interpretation Technology

Li Qiang, Yao Xugang, He Xiaolu, shen linshu, Xu Qingying, Liu Dongming
Liu hongsheng

(Jinghe Industry Xi'an Shaanxi, China, Changqing Well Logging, China petroleum logging Co.LTD. 710201)

Abstract. In order to get higher injection capacity and raise the coefficient of water driving waves and accelerate the speed of oil extraction. Changqing field carries on the development of horizontal-well infusion exploitation in the XXQ sandstone layer oil pool district. In compare with the traditional vertical-well affusion, the effect of comprehensive result of horizontal-well infusion exploitation will enhance as five times more as the current value. Because the flow of horizontal-well varies more in compare with the level-well , many horizontal-well logging data is hard to explain from the normal regulations ,because the influence of the dynamic state of well hole and the size of it. Basing on the flow state of horizontal-well and the analysis of the layer of the low degree state and the annular flow and the turbulent flow ,in order to get the parameters. To make attempt research to the quantitative interpretation of the horizontal-well.

Key words: horizontal-well , production logging, flow state, logging interpretation

INTRODUCTION

Usually the injection profile logging is carried on in the level-well or the inclined-well. Through several decades of development, no matter the logging technics or the logging interpretation technology is perfect as days by. With the development of oil-gas field exploitation, the rapid development of the drilling technology , horizontal-well drilling technology made great progress. In addition to the completion of horizontal-well petroleum reservoir engineering, drilling, completion logging, perforation production increase methods, the number of drilling increases daily. So we are confronted with the problem of horizontal-well injection profile logging (but the study of horizontal-well injection profile has not been done).

Usually the interpretation technology of the level-well is not suitable for the horizontal-well. Many horizontal-well logging data is hard to explain from the normal meaning .Because the flow of horizontal-well varies more in compare with the level-well and the influence of the dynamic state of well hole and the size of it. So it's a trend to study the horizontal-well injection profile logging technics and interpretation technology.

BASIC PROCESS OF HORIZONTAL-WELL INJECTION PROFILE LOGGING

In the vertical-well or the inclined-well which the inclination angle.

is not very big, it's usually to carry on logging by the gravity of the instrument to send the instrument to the purpose layer of well bottom. In the level-well it can only send the instrument to the hole inclination of the 40-60 degree, in the segment of the horizontal ,it must depend tools to send the production logging instrument.

At present, it adopts continuous pipeline transmission and retractor transmission methods to carry on horizontal-well production logging construction among nations. Through experiment and contrast,zp1,zp2 well adopt retractor transmission to carry on horizontal-well profile injection logging.

Retractor is made up of revolve short stanza, the tension short stanza, thimble joint localizer, two ball bearing centralizer, electronics circuit ,timing stanza, drive motor, the main retractor, the auxiliary retractor .In the process of the production logging of the level-well and big inclined well, when the production instrument is sent to some inclined segment ,it can not be transferred to the measured sections by gravity ,then retractor tools will send the production logging instrument to the certain place.

CP914, *Multiphase Flow: The Ultimate Measurement Challenge,* edited by X. Cai, Y. Wu, Z. Huang, S. Wang, and M. Wang

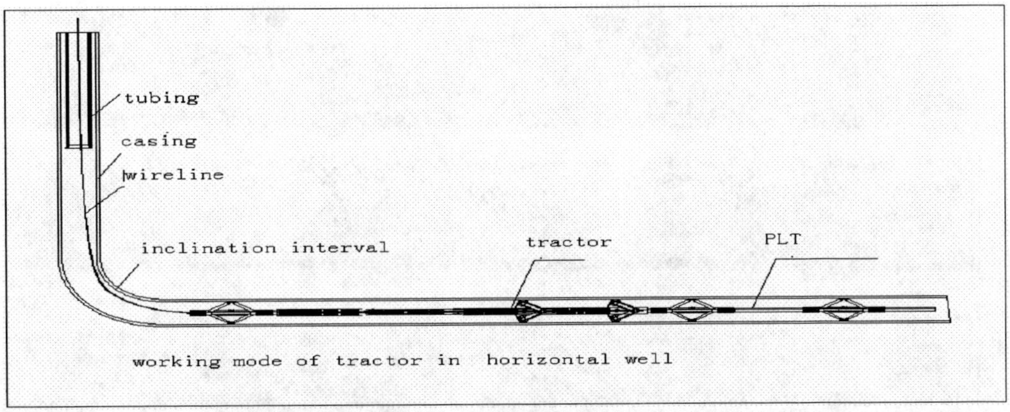

FIGURE 1. Working status in the horizontal well

LOGGING INTERPRETATION FOUNDATION FOR HORIZONTAL-WELL INJECTION PROFILE

Single phase flow has two different flow status due to different flow velocities, which are layer flow and other flow. In the case of layer flow the flow velocity is zero near to the pipe surface, the flow velocity is fastest in the pipe center, fluid member flow forward without disturbance. In the case of the other flow the flow velocity is also zero near to the pipe surface, there is very thin layer which belongs to the layer flow, fluid member flow forward irregularly.

Analysis of layer flow for single phase flow(horizontal).

LOGGING INTERPRETATION METHOD FOR HORIZONTAL-WELL INJECTION PROFILE

Based on the relationship between wellbore path and reservoir space, the method of interpretation and evaluation for horizontal well is discussed in this study.

1. Interpretation of relationship between wellbore path and reservoir space is studied.
2. How to use Interpretation result of relationship between wellbore path and reservoir space for development.
3. Based on the porosity, permeability and saturation curves of reservoir, the change of flow quantity, well temperature and pressure are systematically analyses. The satisfied result is achieved by combining the structure of well depth and the change of horizontal well.

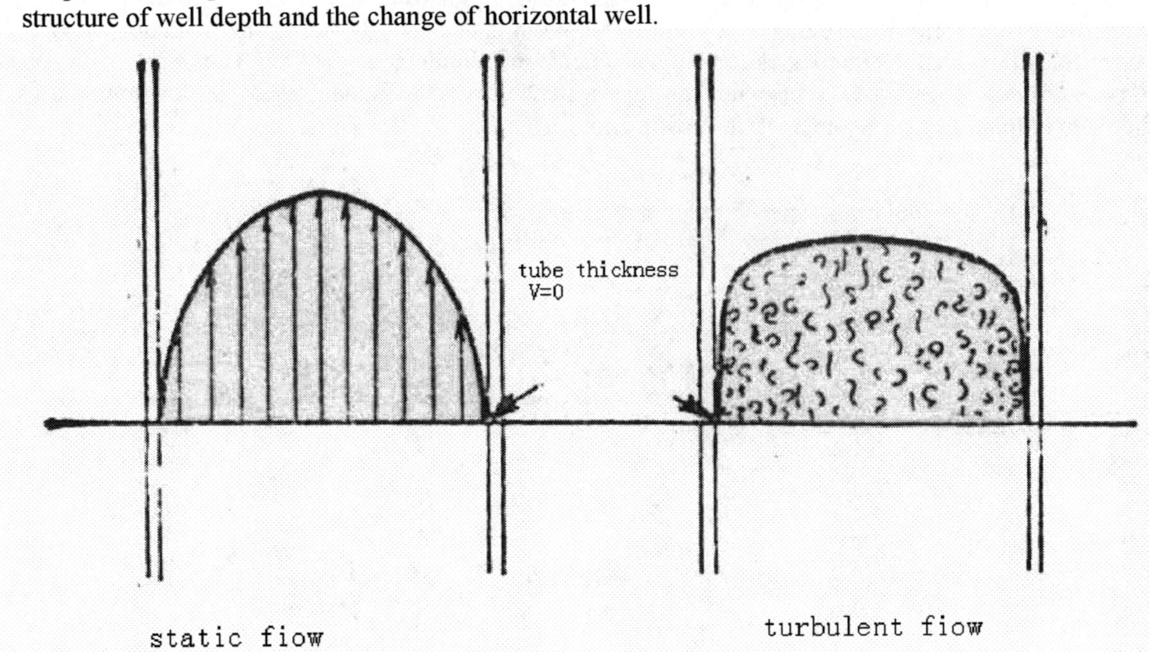

FIGURE 2. The flow speed distribution of Laminar and rapids in the vertical casing

FIGURE 3. Fluid horizontal flow with different flow quantity

Jet flow theory is applied in the data processing. Jet flow is a stream of fluid influx into accepting fluid and a miscible flow regime through the throttle with its dimension smaller than the space packed by accepting fluid. Assuming that a stream of fluid with density p_o jets from throttle plumb into a static and homogeneous environment with density p_a, and its imposed buoyancy is（p_a—p_o）*g , where g is the gravitational acceleration 。 if $p_a > p_o$,buoyancy directs up and is termed as positive buoyancy. Jet flow effected by positive buoyancy is positive buoyancy jet flow. If $p_a < p_o$, buoyancy directs down and is termed as negative buoyancy, Jet flow effected by negative buoyancy is negative buoyancy jet flow. if $p_a = p_o$， no buoyancy but initial momentum exists, it is pure jet flow. If no initial momentum exists, but effected by buoyancy, it is scrolling flow positive buoyancy jet flow agrees with the real flow types, so the numerical simulation for downhole flow types is established by positive buoyancy jet flow.

Interval with standing liquid in well bore approximates the computation region with gas and water flowing(figure 3). Natural gas enters into interval with standing liquid from the left inlet port, and overflows from free liquid after mixing with liquid in interval with standing liquid。Assuming well bore diameter D=0.16m, diameter for inlet port S=0.018m and length for interval with standing liquid L=1m. Since it is difficult to determine free liquid form from region with standing liquid, it is temporarily substituted with horizontal plane in this paper, Such flow can be regarded as two-dimensional axial symmetry in view of symmetry of flow behavior. If two phase fluid is regarded as a miscible flow, at constant flow state, Renald mean motion equation for axial symmetric two dimensional imcompressive flow in coordinates is:

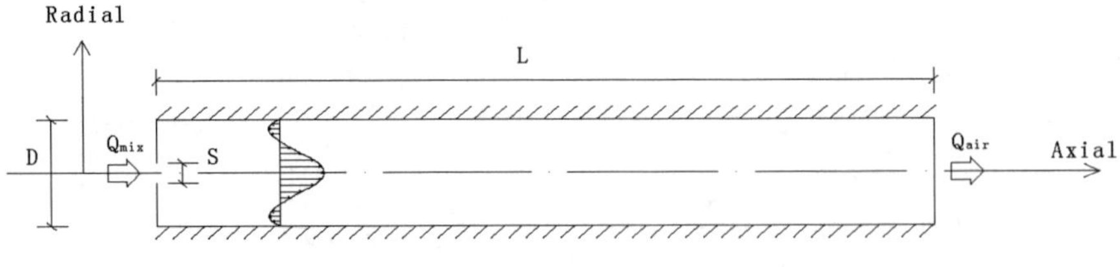

FIGURE 4. Physical flow region vs coordinate

continuous equation:

$$\frac{\partial \overline{u}}{\partial x} + \frac{\partial \overline{v}}{\partial r} + \frac{\overline{v}}{r} = 0 \tag{1}$$

motion equation:

$$\begin{aligned} \overline{u}\frac{\partial \overline{u}}{\partial x} + \overline{v}\frac{\partial \overline{u}}{\partial r} = & -\frac{1}{\rho_m}\frac{\partial \overline{p}}{\partial x} + \frac{\partial}{\partial x}(v_{eff}\frac{\partial \overline{u}}{\partial x}) + \frac{1}{r}\frac{\partial}{\partial r}(v_{eff} r \frac{\partial \overline{u}}{\partial r}) \\ & + \frac{\partial}{\partial x}(v_{eff}\frac{\partial \overline{u}}{\partial x}) + \frac{1}{r}\frac{\partial}{\partial r}(v_{eff} r \frac{\partial \overline{v}}{\partial x}) \end{aligned} \tag{2}$$

$$\bar{u}\frac{\partial \bar{v}}{\partial x}+\bar{v}\frac{\partial \bar{v}}{\partial r}=-\frac{1}{\rho_m}\frac{\partial \bar{p}}{\partial r}+\frac{\partial}{\partial x}(\nu_{eff}\frac{\partial \bar{v}}{\partial x})+\frac{1}{r}\frac{\partial}{\partial r}(\nu_{eff}r\frac{\partial \bar{v}}{\partial r})$$
$$+\frac{\partial}{\partial x}(\nu_{eff}\frac{\partial \bar{u}}{\partial r})+\frac{1}{r}\frac{\partial}{\partial r}(\nu_{eff}r\frac{\partial \bar{v}}{\partial r})-2\nu_{eff}\frac{\bar{v}}{r^2} \quad (3)$$

where $(\bar{u},\bar{v})$ are miscible fluid velocity component at directions (x, r) respectively;p is intensity of pressure of flow field. $\nu_{eff}=\nu_m+\nu_t$ is equivalent motion coefficient of viscosity for turbulent miscible fluid; ν_t is viscosity coefficient for turbulent eddy,which will be discussed later,motion viscosity coefficient of equivalent molecule for miscible fluid and can be computed as follows:

$$\nu_m=(1-Y_w)\nu_{gas}+Y_w\nu_{water}$$

where Y_w is water holdup(water concentration), computed in terms of mass conservation relationship:

$$\frac{\partial(Y_W\bar{u})}{\partial x}+\frac{\partial(Y_W\bar{v})}{\partial r}+\frac{Y_W\bar{v}}{r}=0 \quad (4)$$

similarly, miscible fluid density is computed with following equation:

$$\rho_m=(1-Y_w)\rho_{gas}+Y_w\rho_{water}$$

In the equations, it is neglected that gas and water slide with the same velocity, assuming that two phase fluid flow in the same speed .

TURBULENT MODEL

In standard k-epsilon double equation turbulent model, relationship between turbulent eddy viscosity coefficient ν_t with turbulent kinetic energy k and its dissipative rate ε :

$$\nu_t=C_\mu k^2/\varepsilon, \quad (5)$$

C_μ=0.9 is empirical constant。turbulent kinetic energy k and its dissipative rate ε are computed by the equations as follows:

$$\bar{u}\frac{\partial k}{\partial x}+\bar{v}\frac{\partial k}{\partial r}=\frac{\partial}{\partial x}[(\nu_m+\frac{\nu_t}{\sigma_k})\frac{\partial k}{\partial x}]+\frac{1}{r}\frac{\partial}{\partial r}[(\nu_m+\frac{\nu_t}{\sigma_k})r\frac{\partial k}{\partial r}]+P_k-\varepsilon \quad (6)$$

$$\bar{u}\frac{\partial \varepsilon}{\partial x}+\bar{v}\frac{\partial \varepsilon}{\partial r}=\frac{\partial}{\partial x}[(\nu_m+\frac{\nu_t}{\sigma_\varepsilon})\frac{\partial \varepsilon}{\partial x}]+\frac{1}{r}\frac{\partial}{\partial r}[(\nu_m+\frac{\nu_t}{\sigma_\varepsilon})r\frac{\partial \varepsilon}{\partial r}]+\frac{\varepsilon}{k}(C_{\varepsilon1}P_k-C_{\varepsilon2}\varepsilon) \quad (7)$$

$$P_k=\nu_t\{2[(\frac{\partial \bar{u}}{\partial x})^2+(\frac{\partial \bar{v}}{\partial r})^2]+2(\frac{\bar{v}}{r})^2+(\frac{\partial \bar{u}}{\partial r}+\frac{\partial \bar{v}}{\partial x})^2\}$$

where $(C_{\varepsilon1}, C_{\varepsilon2}, \sigma_k, \sigma_\varepsilon)$ corresponds with (1.44, 1.92, 1.0, 1.3) respectively.

APPLICATION

In order to get higher injection capacity and raise the coefficient of water driving waves and accelerate the speed of oil extraction. Changqing field carries on the development of horizontal-well infusion exploitation in the XXQ sandstone layer oil pool district. The results are shown in the following figures.

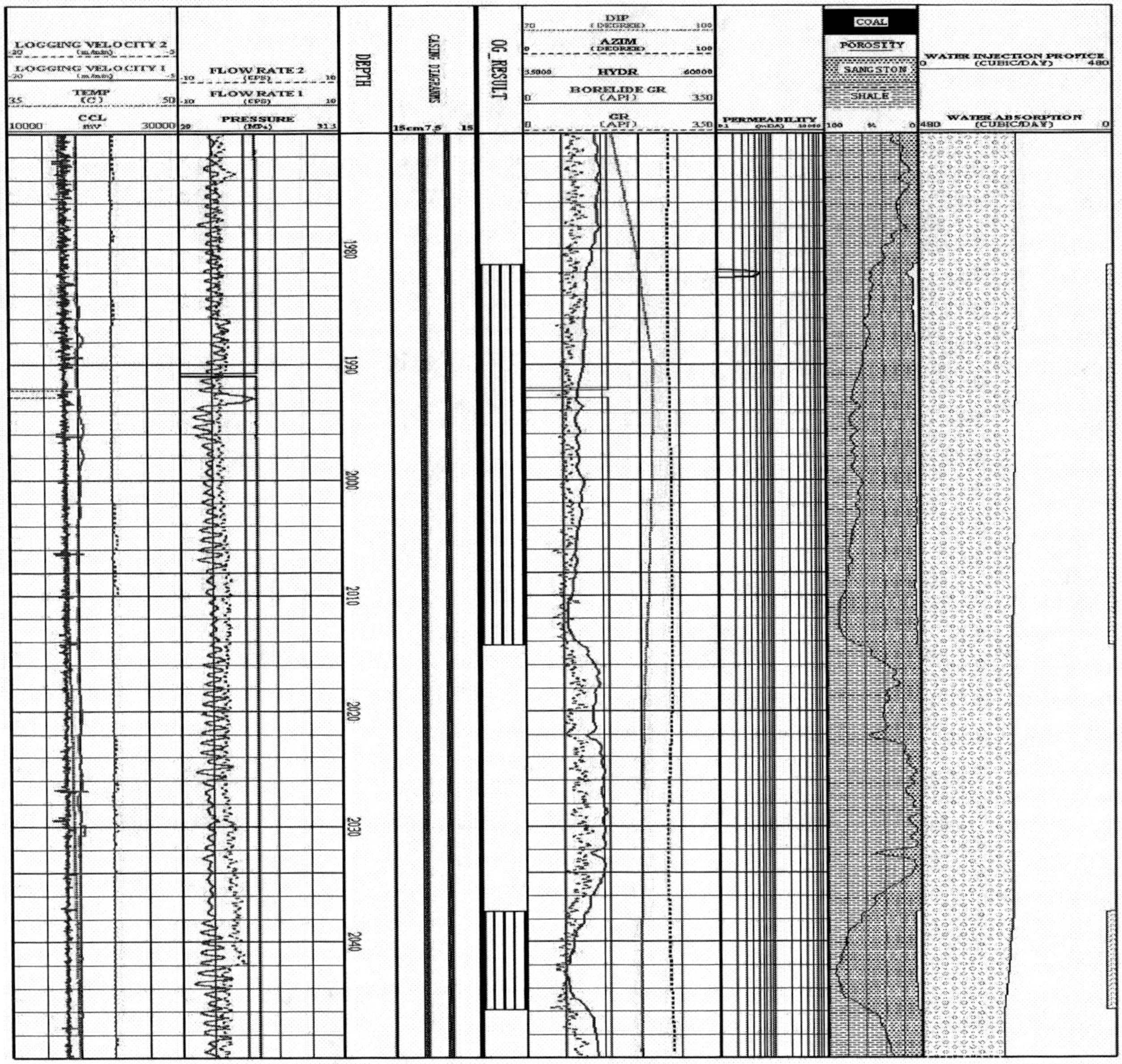

FIGURE 5.ZP-××well logging curve and interpretation (1)

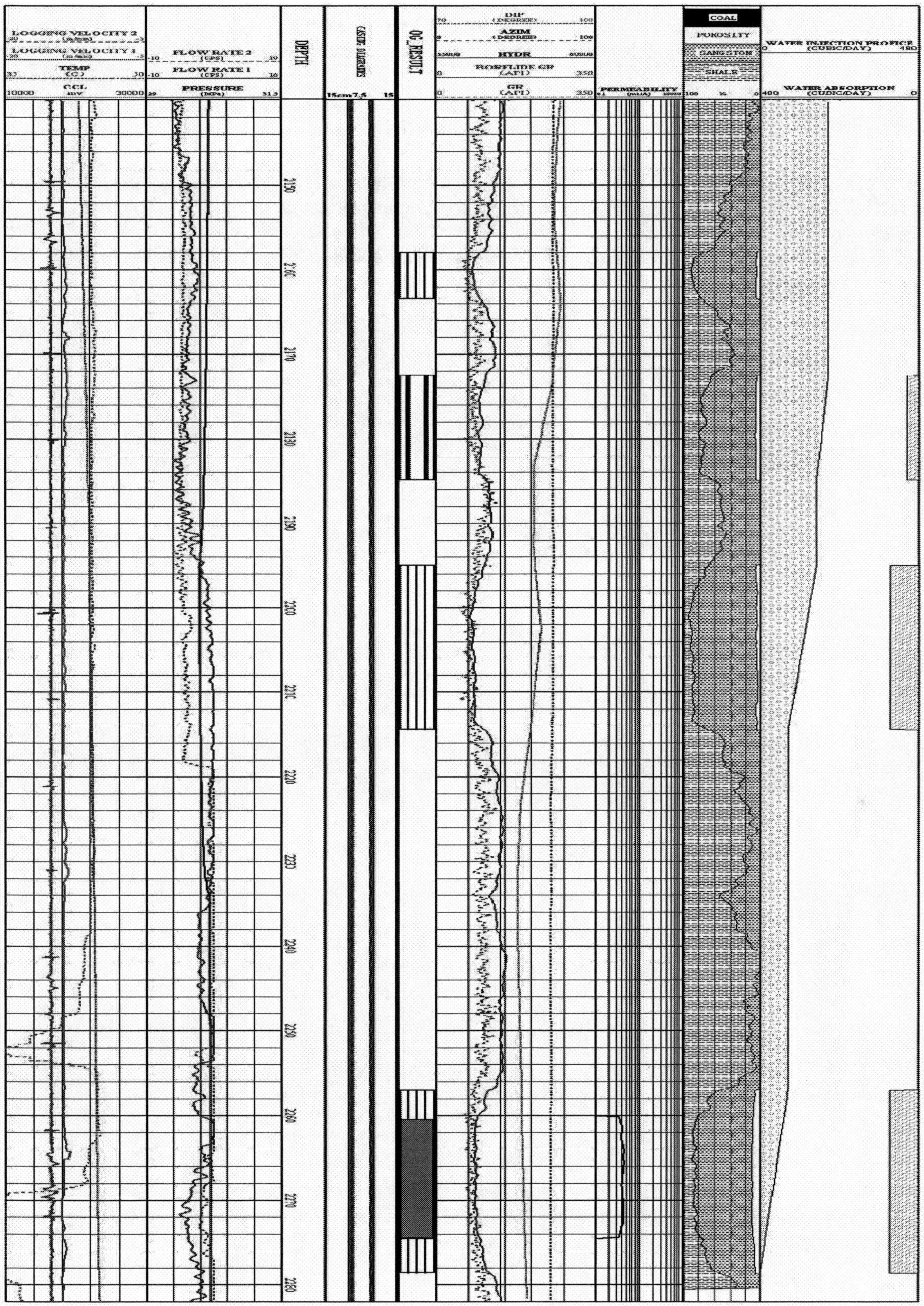

FIGURE 6. ZP-××well logging curve and interpretation (2)

CONCLUSION

Changqing oilfield carried on the development of horizontal-well infusion exploitation in the XXQ sandstone layer oil pool district. The satisfied results is achieved. However, the following problems should be solved.

1. Interpretation of relationship between wellbore path and reservoir space
2. At present many applied models is not adopted for the special situation.the interpretation method should be improved.

REFERENCES

1. Dai Yugang. the Application of Tugger in horizontal well production log. Petroleum Instrument.2005,19(4)
2. Hou Yueming. Study on Horizontal Well Production logging interpretation method. Welllogging Technology.2004,28(1)
3. Guo Haimin. Production Log Introductory Remarks [M].peking: Oil industry publishing house,2003.

A New Ice Slurry-System Combined with Liquid-Dehumidifying

Xiuwei Li*, Xiaosong Zhang*

*School of Energy and Environment, Southeast University, Nanjing, 210096, Jiangsu, PRC

Abstract. Ice is widely used in many fields of today's life. However, we have to pay a lot of energy. in order to produce the ice we need. Since that, we propose a new ice-slurry producing system in this paper, which produces ice using the evaporating freezing method combined with liquid- dehumidifying. The system consists of three major parts, known as the pre-cooling part, the liquid-dehumidifying part and the evaporating-freezing part. We did simulating calculation on evaporating freezing process, and the result exposes the factors that have impacts on this process and how they influence it. Also, we analyze the coefficient of the performance theoretically, from which it could be concluded that this system is an energy-saving compared to a traditional ice-producing system.

Keywords: ice-slurry, evaporating-freezing, liquid-dehumidifying, COP
PACS: 47.61.Jd Multiphase flows

INTRODUCTION

It has been a long time that the ice is widely used in many fields of human's life. Especially, the ice-storage method is meaningful of energy-saving in the field of refrigeration and air-conditions, as we are facing so many problems caused by the short of energy [1]. As to the ice-storage, we should first produce a large amount of ice particles or ice slurry. Generally speaking, the traditional method of producing ice could be divided into two kinds; one produces ice with direct contact of water and cooling mediums, and the other is an indirect contact style. The latter one can be further divided into two parts: the static ice-producing and the dynamic ice-producing [2]. However, all of these methods could be a large consumer of energy, mostly the electronic power. The reason is that the refrigeration cycles engaged to produce ice have wasted a lot of energy, because they just afford the cooling load while the heat of the condenser could only be rejected [3] Difference from that, Yunus Cerci proposed a new evaporating-freezing cycle to produce ice [4]. In that process, it evaporates the water of the temperature around 0℃ in the atmosphere which is of very low humidity (That means a very low vapor pressure). During evaporation, the sensible heat of the water turns into the latent heat, which cools the water itself and forms ice finally. The atmosphere whose humidity increases because of the evaporation of water should be dehumidified for reutilization, and he engaged a solid adsorption to work out. That ice-producing method improves the performance of the ice-producing system. H. T. Shin et al did a lot of work on the process of evaporating freezing, both the theoretical simulation calculation and the experimental analysis [5]. Considering all of these and the advantage of the liquid-dehumidifying method compared to the solid adsorption dehumidifying method [6], we propose a new ice-slurry producing system combined with liquid-dehumidifying. That makes a good use of rejected heat as a low grade heat source to drive the dehumidifying part, and it does enhance the COP.

SYSTEM DESCRIPTION

Principle of Evaporating-freezing

The water droplets are sprinkled and evaporate in the atmosphere as described in Fig. 1, and the vapor pressure of

CP914, *Multiphase Flow: The Ultimate Measurement Challenge,* edited by X. Cai, Y. Wu, Z. Huang, S. Wang, and M. Wang

which is much lower than the three-point saturation pressure of water. During this process, the sensible heat of the water droplet itself turns into the latent heat as the phase transition happens. The temperature of the water droplet decrease and the humidity increases. Finally, the water droplets which are overcooled enough form themselves to ice particles.

System Introduction

The flow chart is show as Fig.2 on the whole. The system is composed by the pre-cooling part (it includes two subparts as the evaporating part and the condensing part), the liquid-dehumidifying part (it includes two subparts as the dehumidifying part and the regenerating part) and the evaporating- freezing part. The water used for producing ice first flows through evaporating part, which is the subpart of the pre-cooling part, and the temperature of the water drops to a certain value. Then the water is sprinkled in the chamber of evaporating-freezing part in the form of droplets. Meanwhile, the air that would be sent to that chamber is primarily dehumidified as it passed the liquid-dehumidifying part. Through that, the moisture of the air is removed and the concentration of the liquid desiccant decreases. After that, the vapor pressure of the air comes to be lower than 609Pa and be sent to the evaporating freezing part. Finally, the pre-cooled water droplets spray in a chamber with the low vapor pressure atmosphere, and the ice-slurry formation process begins.

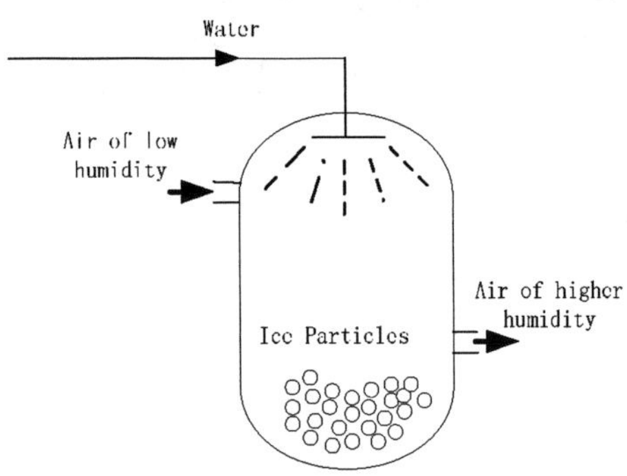

FIGURE 1. Freezing-evaporation process

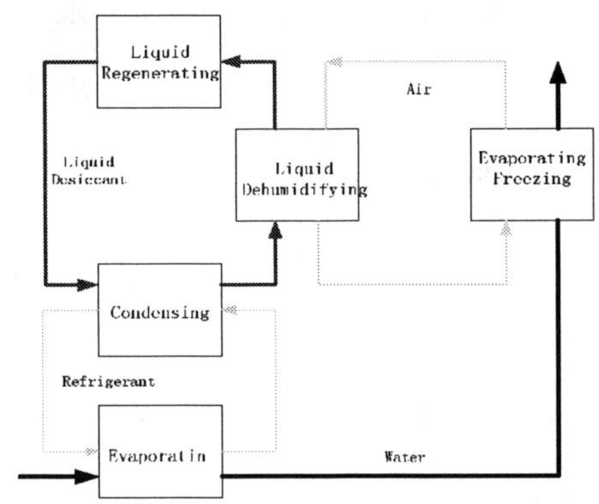

FIGURE 2. System flow chart

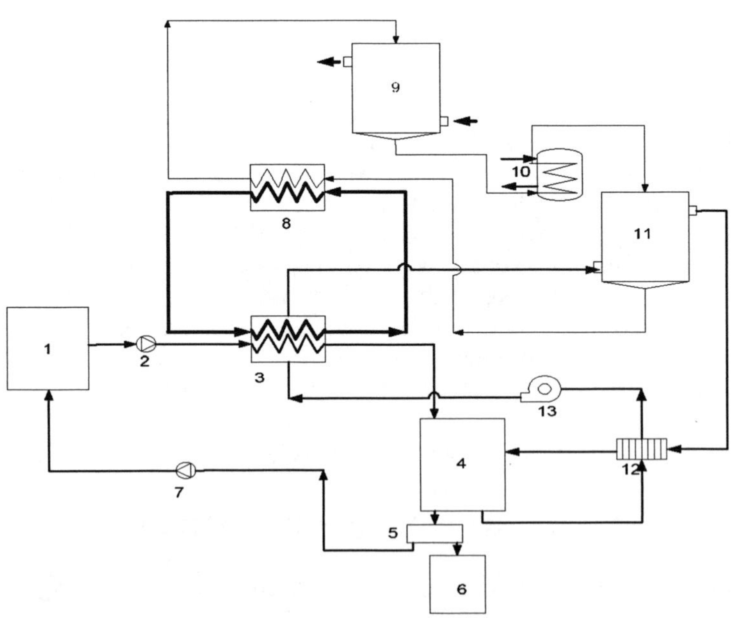

1. Water source 3. Evaporating subpart 4. Evaporating-freezing part 5. Ice-water separator 6. Ice-storage tank 8. Condensing sub-part 9. Regenerator

10. Heat-exchanger 11. Dehumidifier 12. Heat-exchanger

FIGURE 3. System flow chart (complete)

Since the concentration of the liquid desiccant decreases after the dehumidifying, it loses its ability to dehumidify again. To driving the cycle, we should enhance the concentration of the liquid desiccant by regenerating. Therefore, we reutilize the rejected heat from the condensing part, which is the subpart of the pre-cooling part, or this heat may be usually wasted. By that, the concentration rises and the liquid desiccant recollect its ability of dehumidifying. This system is described completely in Fig.3. The ice-particles are separated from the mixture of ice and water, and it would be stored in the ice storage tank. The water that doesn't turn into ice would be back to the cycle and reused. The air of enhanced humidity rejected from the chamber of evaporating-freezing part is reused. And it will pass the evaporating part to dehumidify for the first time before further dehumidifying in liquid-dehumidifying part.

THEORETICAL ANALYSES

Evaporating-freezing Model

Mass Equation

According to the diffusion-controlled evaporation model [5, 7], the mass equation of the water droplets evaporating is as follows:

$$\dot{m} = 4\pi a^2 D_v \left(\frac{dc}{dr}\right)_a \tag{1}$$

, where $\dot{m}$ represents the mass of the water evaporating per second, and the dc/dr is the radial gradient of the vapor concentration, from equation (1), we can get equation (2):

$$\dot{m} = -4\pi a D_v (c_a - c_\infty) \tag{2}$$

By substituting $c = \dfrac{MP}{\overline{R}T}$ into the equation above, we obtain the final mass equation as follows:

$$\frac{dm}{dt} = 4\pi a D_v \frac{M}{\overline{R}}\left(\frac{P_a}{T_a} - \frac{P_\infty}{T_\infty}\right) \tag{3}$$

Energy Equation

The temperature of a droplet in an infinite space would be determined by balancing between the latent heat loss due to its evaporation and the heat conduction decided by the temperature difference between the droplet surface and the surroundings.

The Q_e, which represents the heat released by evaporation is:

$$Q_e = h_{fg}\dot{m} = 4\pi a^2 D_v h_{fg}\left(\frac{dc}{dr}\right)_a \tag{4}$$

, while the Q_c, which represents the heat obtained from surroundings is:

$$Q_c = 4\pi a^2 k_g \left(\frac{dT}{dr}\right)_a \tag{5}$$

, and $\left(\dfrac{dT}{dr}\right)_a = \dfrac{T_\infty - T_a}{a}$.

Derived from the equations (3)、(4) and (5), the temperature variation of the droplet is obtained as follows:

$$\frac{dT}{dt} = \frac{3\left[\frac{D_v h_{fg} M}{R}\left(\frac{P_a}{T_a} - \frac{P_\infty}{T_\infty}\right) - k_g (T_\infty - T_a)\right]}{r^2 \rho C_l} \tag{6}$$

COP Analysis

Considering the mass and energy equations to derive the COP of this system, we obtain the equations as follows:

$$Q_0 = C_l m (t_1 - t_2) \tag{7}$$

$$m = m_d + m_i \tag{8}$$

$$C_l m_i (t_2 + \Delta t) = m_d h_g \tag{9}$$

, where t_1 is the temperature of the water at the entrance of the evaporating subpart, and t_2 is the temperature of the water out of this part, while Δt represents the overcooled degree for producing ice in this system. From these three equations, we can get the COP of this system.

$$COP_{ice} = \frac{m_i}{W} = \frac{m_i}{\frac{Q_0}{COP}} = \frac{COP}{C_l\left(1 + \frac{C_l(t_2 + \Delta t)}{h_g}\right)(t_1 - t_2)} \tag{10}$$

Meanwhile, for a traditional ice-producing system, we have the COP as:

$$COP_{ice}' = \frac{m_i}{W} = \frac{m_i}{\frac{Q_0}{COP}} = \frac{m_i COP}{m_i C_l (t_1 + \Delta t)} = \frac{COP}{C_l (t_1 + \Delta t)} \tag{11}$$

, and:

$$\frac{COP_{ice}}{COP_{ice}'} = \frac{t_1 + \Delta t}{\left(1 + \frac{C_l(t_2 + \Delta t)}{h_g}\right)(t_1 - t_2)} = \frac{t_1 + \Delta t}{(1 + 0.00168(t_2 + \Delta t))(t_1 - t_2)} \approx \frac{t_1 + \Delta t}{t_1 - t_2} > 1 \tag{12}$$

So it can be seen obviously the COP of this system is pretty higher than a traditional one, and that do make sense for energy-saving.

RESULTS AND DISCUSSION

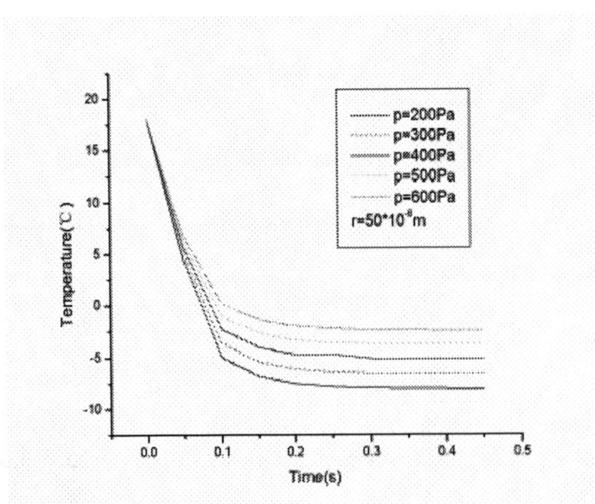

FIGURE 4. Variation of the temperature of the water droplet under different vapor pressure

The curve of temperature variation of the water droplet under different vapor pressure is exposed in Fig.4, and it agrees well with the theoretical and experimental results obtained from the reference [5]. The temperature of the water droplet falls sharply at the beginning, but with the time going, the slope of the curve becomes smooth and it is almost flat finally. That is due to the pressure drop of the water droplet, and the value of this pressure gets close to that surroundings. Therefore, the driving power for mass transfer reduces, and that leads to a slower drop of the temperature.

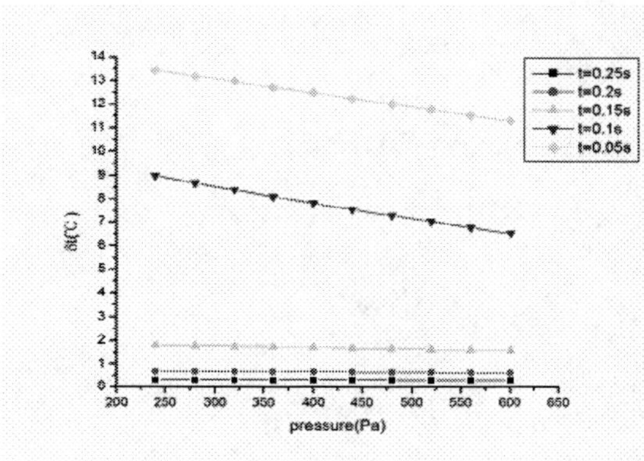

FIGURE 5. Influence of surrounding vapor pressure temperature drop of the droplet

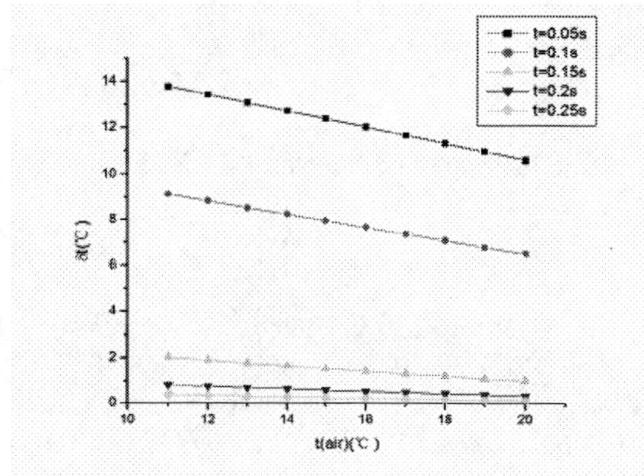

FIGURE 6. Influence of the atmosphere temperature on on temperature drop of the droplet

With 0.05s as the time space, it is showed from Fig.5 to Fig8 that the relationship between the temperature drop of the water droplet and the vapor pressure surroundings, the temperature of the air at the entrance of the evaporating-freezing chamber, the initial temperature of the water droplet and its initial radius. And we found that the whole process of evaporating-freezing is done in a few seconds, and the main drop of the temperature of the water droplets happens in the early times of this process. Compared to other parameters, the radius of the droplet plays a more important role. In the early time the temperature of the droplet of smaller radius drops more. However, the other parameters, like the vapor pressure of the boundary layer of the droplet come to be the main factor influencing the process with the time going. And we can see that the temperature drop of the smaller droplet is even less than that of the bigger one. Also, it presents that the droplet of higher initial temperature seems have a larger drop of the temperature. And the reason for this could be that: for one thing the droplet of high temperature owns higher vapor pressure of the boundary layer, which means the driving power for mass transfer is stronger; for another thing, the temperature difference between the droplet and the atmosphere surroundings reduces, which caused the reduction of the heat transferred from the surroundings to the water droplet.

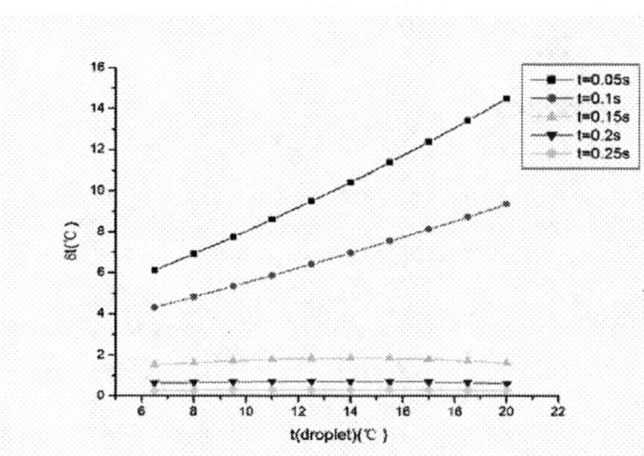

FIGURE 7. Influence of initial temperature of the droplet on temperature drop of the droplet

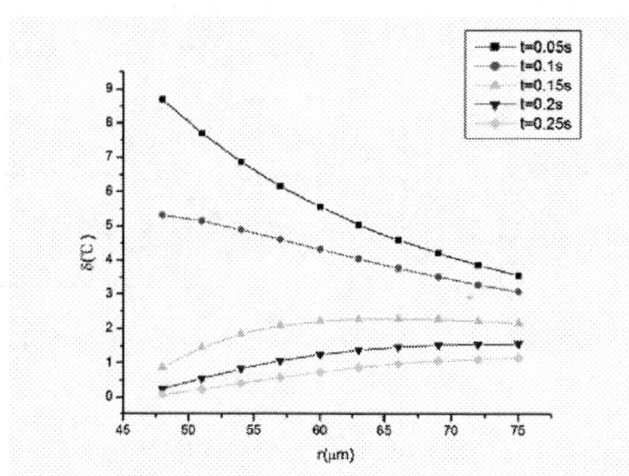

FIGURE 8. Influence of the initial radius of the droplet on temperature drop of the droplet

CONCLUSIONS

On one hand, this ice-slurry producing system appears to make a good use of the latent heat of the evaporation in order to cut down the load of sensible heat that the refrigeration should afford; on the other hand, it does improve the COP by reutilizing the rejected heat in conjunction with the liquid dehumidifying. We have the simulation calculation, and that would be useful to make further research on this system, and we recommend more study on the complicated process of the evaporating freezing and more work on the cooperation of this multi-component system through experiments and theoretical analysis.

ACKNOWLEDGEMENTS

This research is granted by the Natural Science Foundation of China (Contract no: 50376052)

REFERENCES

1. Yunliang Fan, Xianliang Zhao and Zhihu Xu. *Research and Information of Energy*, **13**, 54-58(1997).
2. Kaiyang Qu and Yi Jiang. *HV&AC*, **28**(3), 31-36(1998).
3. Shugang Wang, Ruzhu Wang, Jingyi Wu and Yixiong Xu. *Journal of Engineering Thermo physics*, **24**(4), 567-569(2003).
4. Yunus Cerci. *International Journal of Heat and Mass Transfer*, **46**, 2967-2974(2003).
5. H.T.Shin, Y.P.Lee and J.Jurng. *Applied Thermal Engineering*, **20**, 439-454(2000).
6. Xiaosong Zhang, Xiufeng Fei, Mingheng shi and Yiran Cao. *Journal of Southeast University*, **33**(1), 1-4(2003).
7. J.P.Hindmarsh, A.B.Russell and X.D.Chen. *International Journal of Heat and Mass Transfer*, **46**, 1199-1213(2003).

NOMENCLATURE

a	radius of a droplet	M	molecule weight, kg/mol
COP	coefficient of performance	P	pressure, Pa
C_l	specific heat, kJ/kg	$\bar{R}$	universal gas constant, N.m/(mol.K)
D_v	diffusion coefficient of vapor, m^2/s	r	radial length, m
k_g	thermal conductivity, kw/(m.K)	T	temperature, K
h_{fg}	latent heat of vaporization, kJ/kg	ρ	density, kg/m^3
m	mass of water, kg/s	**Subscripts**	
m_i	mass of ice produced, kg/s	a	value at the droplet surface($r = a$)
m_d	evaporating rate of the droplet, kg/s	∞	surrounding

Experimental Study on Motion of Inhable Particles in Acoustic Field Based on Microscopical Visualization

Zhao Bing, Leng Yin, Chen Houtao, Yao Gang and Shen Xianglin

Key Laboratory of Clean Coal Power Generation and Combustion Technology of Ministry of Education, Southeast University, 210096Nanjing,China

Abstract. The experimental device used to explore the dynamical behavior of inhalable particles, PM2.5, was set up by means of a visualization method based on a microscope and pulsed laser illumination. The aerosol generator based on the principle of fluidization bed generated inhalabel particles, emitted from a coal-fired power plant. Trajectories and shapes of the particles in the acoustic wave field were visualized and recorded by a high-rate CCD camera. The image recorded the details of particle motion, which showed that the trajectories of particles in an acoustic wave field swept much greater area because of the acoustic entrainment in comparison with the particulate trajectories under without acoustic excitation. From the images of particles illuminated by pulse laser, the single particles could be distinguished form the aggregates. Moreover three ways to determine the size of an observed object was introduced which made it possible to study the quantificational dynamic behaviors of an observed particle. It was found that to fine particle the way based on acoustic entrainment was the most reliable among these three ways.

Keywords: visualization; inhalable particle; acoustic wave field; entrainment.
PACS: 42. 30. -d

INTRODUCTION

Inhalable particles suspending in the atmosphere have become the most important pollutant. Fine particles can easily absorbed in human breathing system. They often carry poisonous matter such as chemicals, viruses and heavy metal elements et al, so they greatly threaten human health. In addition, as a source of light pollution, they can reduce air visibility. Up to now, suspending particles have done great harm to human health and environment [1,2,3]. The inhalable particles emitted from coal-fired power plants are the important source of the inhalable particles in atmosphere. Conventional ash removal devices cannot remove them efficiently because of their very small sizes. As a potential effective method, acoustic agglomeration excited wide interests of researchers. The acoustic agglomeration is used as a preprocessing means in practice. The inhalable particles are aggregated by acoustic wave, forming larger agglomerations, so as to be removed more efficiently by conventional devices.

For decades many studies have showed that strong acoustic wave could promote particle agglomeration [4,5,6,7,8,9,10]. In order to explore the acoustic agglomeration mechanisms of particles, Experiment researches on the particle dynamics in acoustic field by visualization technology began in the 60s last century. Buranov & Eknediosyants observed in their experiments Pseudo-collision of polydisperse particles, which were brought together but remained separated at a certain distance without contact. Hoffmann & Koopmann [11,12,13] tried to use particles with certified diameters $1.5\mu m$, $8.1\mu m$ and $21\mu m$ as experimental objects, respectively, but 1.5 *um* microspheres could not be successfully dispersed and dropped as large agglomerates during the feeding The $22.1\mu m$ microspheres did not encounter such problems, but stayed relatively stationary in the acoustic field due to their larger mass, thus the majority of their experiments was carried out using the 8.1μm microspheres. Hoffmann & Koopmann[11,12,13] observed the same phenomenon of Pseudo-collision, but also found several particle agglomerates formed along the acoustic axis which were regarded as the agglomeration mechanism of "acoustic wake effect". Gonzalez & Hoffmann [14] verified acoustic entrainment equation using $8.1\mu m$ particles, but ignored slippage influence between particle and fluid.

CP914, *Multiphase Flow: The Ultimate Measurement Challenge,* edited by X. Cai, Y. Wu, Z. Huang, S. Wang, and M. Wang

Up to now, acoustic agglomeration mechanisms have not been known clearly due to the complexity of acoustic agglomeration process and the differences in experiment conditions and parameters by different researchers. For example, the smallest particle used in experiment mentioned above is 8.1μm,so it is not sure whether regulations for large particles can apply to small particles, when there is the remarkable influence of slippage between inhalable particle PM2.5 and fluid. Furthermore, most theories about agglomeration mechanisms have not been verified through experiments. In this present work particle motion behavior was experimentally studied by means of a visualization method based on a microscope and pulsed laser illumination. The image recorded by a high-rate CCD camera showed the details of particle motion. In addition, a reasonable way to decide the diameter of particle observed in the acoustic field was put forward. All the work is the base to further quantification research on the dynamical behavior and on the acoustic agglomeration mechanism of inhalable particle in acoustic field.

EXPERIMENTAL SETUP

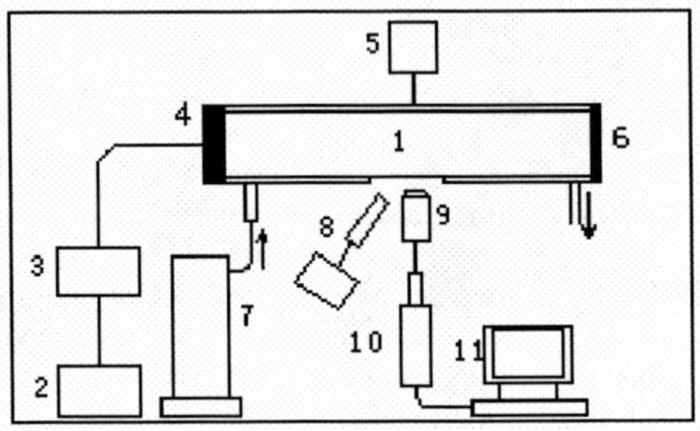

1 agglomeration chamber, 2 signal generator, 3 power amplifier, 4 loudspeaker, 5 transducer, 6 reflecting-plate, 7-aerosol generator, 8 YAG laser, 9 microscope, 10 high-speed CCD, 11 PC

FIGURE 1. Schematic diagram of experimental setup

The experiment setup shown in Figure 1 is composed of three main parts: acoustic agglomeration chamber, aerosol generator and particle visualization system. The acoustic agglomeration chamber is a 400mm long square organic glass pipe, whose cross section is 20x20mm^2. A 100mm wide window made of 0.5mm thick optical quartz glass is located on a sidewall of the chamber as the observation window of the microscope. A loudspeaker is set at one end of the chamber. A power amplifier to drive the loudspeaker amplifies a sine signal produced by a signal generator. A reflecting plate with the same cross section as the chamber is set at an appropriate location near the other end of the chamber where the phase of the acoustic wave is zero, so as to form a standing wave field in the chamber. The SPL(sound pressure level) in the chamber is measured by a transducer. In the experiment the acoustic frequency was always less than the cutoff frequency of the pipe to eliminate the high frequency harmonious waves, and ensure the mono-frequency character of the acoustic wave field.

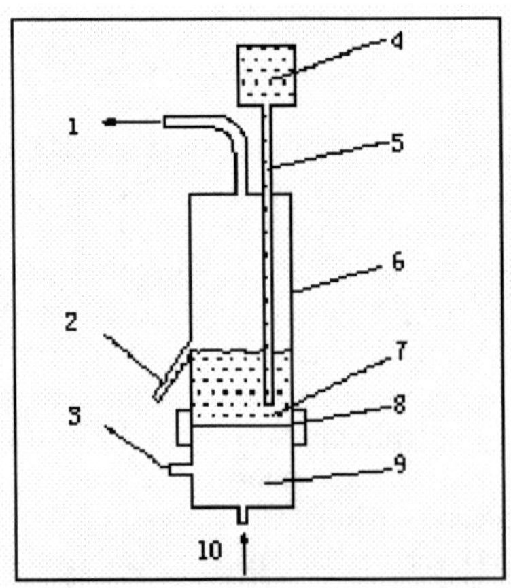

1 Aerosol outlet; 2 Overflow pipe; 3 Pressure measure orifice; 4 Particle reservoirs; 5 Particle inlet pipe; 6 Fluid bed; 7 Bed matter; 8 Wind distribution board; 9 Wind chamber; 10 Nitrogen inlet

FIGURE 2. Aerosol generator based on fluid bed

The aerosol generator that is shown in Figure 2 provides test particles. The bronze powder of $80\mu m$ in mean diameter is used as the bed material. Fly ash particles taken from a coal-fired power plant are mixed with bronze powder, and then fed into the fluid bed to generate the aerosol containing test particles. A steady bed height can be maintained through an overflow outlet. Nitrogen was used as fluidizing gas. Test particles provided by the generator, having the size range from $0.1\mu m$ to $10\mu m$ in diameter, are led in the agglomeration chamber to be investigated, and then vented.

The particle visualization system includes a 5W YAG laser, a microscope with high-rate CCD video camera and a PC. The illuminating Laser beam is pulsed in the frequency of 20 kHz that may divide the continuous particle trajectory into a series of discrete particle image, so that not only the trajectory but also the particle shape can be visualized. Particles and its trajectories are observed through the microscope, and recoded by the high-rate CCD video camera. The type of the microscope is Nikon ECLIPLE55i. The high-rate CCD camera is made by Photron Company, which frame rates are from 30fps (frames per second) to 10000 fps. Its the highest resolution is 512 x 480 pixels.

EXPERIMENTAL RESULTS AND DISCUSSIONS

Particle Gravity Settlement Without Sound Wave Excitation

Particle trajectories without the acting of acoustic wave were firstly observed for facility test. Figure 3(a) and 3(b) show free settlement trajectories of a particle and an aggregate of two particles, respectively. The resolution of the pictures is 512x480, so each pixel covers $0.375\mu m$. The figure indicated that the experiment setup could observe movements of objects with diameters around $1\mu m$.

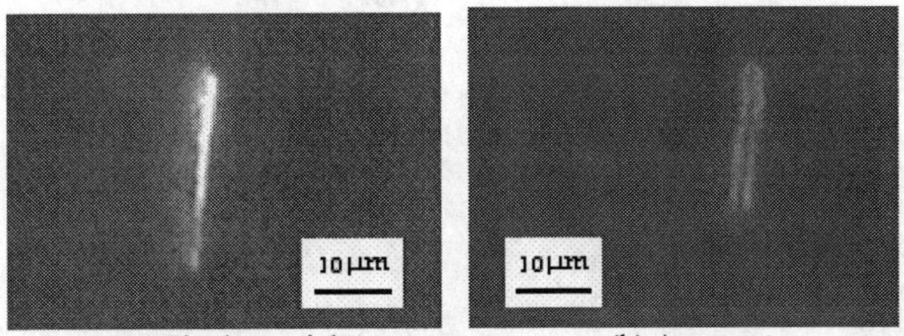

(a) Single particle (b) Aggregate

FIGURE 3. Gravity settlement trajectories of particle and aggregate (Frame rate: 30 fps; Shutter speed: 1/30s)

Particle Trajectories in a Horizontal Acoustic Wave Field

Particle trajectories of fly ash from a power plant in a horizontal acoustic field were recorded in the experiment setup. Two typical trajectories are shown in Figure 4. In each picture of figure 4 the bright spots marked locations of an observed particle at different time that described the trajectory of the particle. Figure 4 (a) and (b) were taken in an acoustic field of 3kHz frequencies under 139dB and 140dB of the sound pressure lever, respectively. In comparison with figure 3 that was without acoustic action, in addition to the gravity settlement, the particle also oscillated horizontally because of the acoustic entrainment. At the same time, the center of the horizontal oscillation drifted horizontally, so that the trajectory looked like a parallelogram. The oscillation greatly increased the area swept by the particle in the field that could be beneficial to the particle agglomeration.

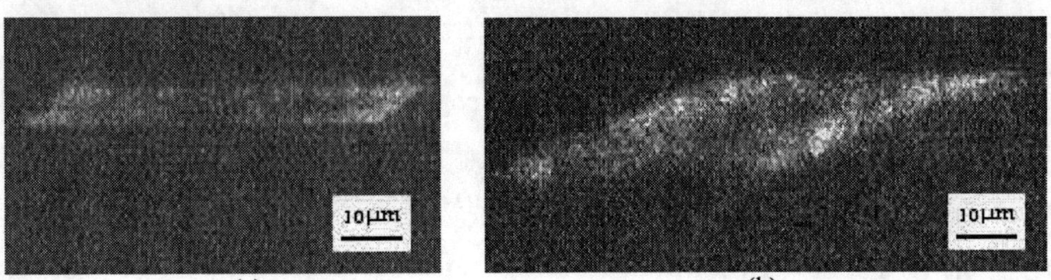

(a) (b)

FIGURE 4. Trajectories of particles in a horizontal acoustic field (Frame rate: 30 fps; Shutter rate: 1/30s; frequency: 3 kHz)

It also could be seen in figure 4(a) and 4(b) that the middle part of the parallelogram is less bright than its two sides which implied that the particle oscillation velocity changed regularly in the sine acoustic field. The oscillation velocity of the particle decreased gradually from the oscillation center to the two terminals at which the particle oscillation velocity were zero and changed its direction to the oscillation center. A lower oscillation velocity could result in a stronger exposure and caused a brighter image.

Being Distinguishable between a Single Particle and an Aggregate

Pictures (a) and (b) in Figure 5 present the trajectories of a particle and an aggregate in an acoustic field with the frequency of 3 kHz and the sound pressure level of 138dB, respectively. The trajectories of particles are the shape of "S", which is caused by the horizontal entrainment of the sound wave and gravity sedimentation. Because Pulse laser with the frequency being 20kHz is used to illuminate the particles, as can be seen from the figure 5 (a) and (b), the single particle and the particulate aggregates can be distinguished definitely. Hoffmann [12] adopted glass micro spheres with diameter of 8.1μm in his visualization experiment. The thick of trajectories of these particles was 22~33μm when only gravity sedimentation. As pointed by Hoffmann, particle diameter cannot be obtained from the thickness of gravity sedimentation trajectory in the picture by ruler measurement. One of the causes maybe was that the single particle was hard to distinguish from the aggregate. To distinguish between particle and aggregate can present useful information in order to further study the morphology of aggregate and to compare both dynamical behaviors.

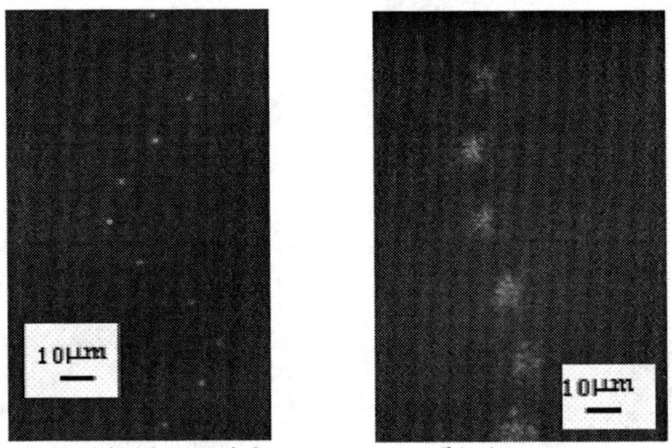

(a) Single particle (b) Aggregate

FIGURE 5. Trajectories of particle and aggregate in a horizontal acoustic wave field (Frame rate: 30 fps; Shutter rate: 1/30s; Frequency: 3 kHz)

DETERMINATION OF PARTICLE DIAMETERS

In visualization experiments it is almost impossible to know the size of the observed particle in advance since its randomicity. Except for using the size - known particle as the test particle, but it is very difficult and restricted. Thus, it is important to determine the size of an observed particle that is the basic parameter to study its dynamic behaviors. In general, there are three ways to determine the particle diameter according to its visualized image.

Three Methods to Determinate the Diameters of Particle

Optical method

In Figure 4 bright spots are images of fine particles that result from the laser beam pulse. A ruler image taken under the same condition can be used to measure the geography size of the bright spot. In addition, the pixels a spot covers can also be used to determine the particle diameter.

Based on gravity settlement

According to the balance between gravity and viscosity resistance of a sphere particle in the case that Reynolds number is less than 1, the particle terminal velocity is

$$v_t = g\tau \tag{1}$$

Where, g is gravity acceleration, τ is the relaxation time of particle, defined as

$$\tau = \frac{\rho_p d_p^2 C}{18\mu} \tag{2}$$

Here, ρ_p is particle density, d_p is particle diameter. C is Cunningham coefficient, and μ is fluid dynamic viscosity. On the other hand, based on the visualized trajectory of the particle, the settlement velocity of the particle is

$$v_t = \frac{s}{t} \tag{3}$$

Where, s is the settlement displacement that can be measured from the trajectory in a frame; t is the exposure time determined from the frame rate/shutter rate of video camera. Substituting equation (2) and (3) into equation (1), we have the particle diameter

$$d_p = \sqrt{\frac{18\mu s}{\rho_p g C t}} \tag{4}$$

Based on acoustic entrainment

Particle diameters can also be derived from their acoustic entrainment characteristics. The acoustic oscillation displacement amplitude of an acoustic field can be obtained by [14]

$$\varepsilon_a = \frac{v_a}{\omega} = \frac{\sqrt{2}p_e}{2\pi f \rho_0 c_0} \tag{5}$$

Where c_0 is sound velocity, p_{ref} is reference sound pressure, p_{ref}=0.00002Pa, v_a is acoustic velocity, ω is the round frequency of the sound source. In equation (5) the unknown p_e is the sound pressure that can be determined by

$$SPL = 20\log_{10}\frac{p_e}{p_{ref}} \tag{6}$$

Where SPL (sound pressure level) can be measure with an instrument in the experiment. The particle oscillation displacement amplitude ε_e of the entrained particle in the acoustic field can be obtained from its visualized trajectory image in the experiment, Substituting the ε_a and ε_e into the BFH equation [15] which gives the particle entrainment coefficient η (that is the ratio of ε_e to ε_a), we have

$$d_p = 4\sqrt{(\eta^{-2} - 1)\left(\frac{9\mu}{\pi f \rho_p}\right)^2 \cdot C^{-2}} \tag{7}$$

Evaluation and Comparison among the Three Methods

Each method mentioned above has its own advantage and shortcoming. It is most convenient to measure particle diameters using the ruler. However, as a pixel of the CCD camera covers $0.375\mu m$ solute error is about from 1 to 2 pixels, that is about $0.75\mu m$. This will bring great error to fine particles. On the other hand, as the observed particle diameter approximates the wavelength of laser, diffraction might blur the edge of the particle image. Thus, this method can only estimate fine particle diameters.

Particles evaluated based on the gravity settlement can get its more precise size sometimes. However the method is prone to be disturbed. The gravity settlements of fine particles may be affected if some vertical streams exist in the flow field. For example, according to the gravity settlement method, the diameter of the particle in Figure 3(a) is $3.3\mu m$,and particles in figure 4(a) and 4(b) are $1.61\mu m$ and $2.70\mu m$, respectively, which matched the results using the optical method. However, for the diameter of the particle in Figure 5(a), it is $1.5\mu m$ based on the optical method, and the settlement method gives $38\mu m$ that is an obviously unreasonable size as a macroscopic stream increases the dropping distance of the particle. In this case the gravity settlement method is invalid.

In an acoustic fields with higher SPL, an acoustic entrainment force exerted on a fine particle, unlike gravity, is less affected by the stream disturbance, and the larger horizontal displacement is almost unaffected by the border blur of the particle image. So, it is a suitable way to obtain particle diameters in the visualization experiment. For example, in this way, diameters of particles in figure 4(a) and (b) are 1.61μm and 2.69μm, respectively, which accord with the results using the gravity settlement method in undisturbed flow fields. Although the gravity settlement method gives a unreasonable size for the particle that the size magnetite was 58μm in figure 5(a) due to stream disturbances, the acoustic entrainment method determines that its diameter is 2.50μm that matches the result using the optical method.

In a word, above-mentioned three ways to determine the particle size could avoid knowing particle size in advance in the experiment of fine particle dynamics in an acoustic field, which would promote further and deeper investigations in the field. One of the three ways could validate with others, and as for the fine particles the most reliable way among them is the method based on the acoustic entrainment.

CONCLUSIONS

Trajectories of fly ash particles from a coal-fired power plant with the diameter about 2μm in acoustic wave fields were taken by the microscopical visualization method to investigate the motion behavior of inhalable particles, PM 2.5. The pulsed laser beam illumination divided the continue trajectory into a series of discrete images of observed object in recorded pictures. This made it possible to obtain both the trajectory and the shape of an observed object at the same time, as a result, the motion details of inhalable particle in an acoustic wave field could be obtained and the single particles could be distinguished form the aggregates.

The size determination of an observed particle is significant to particle dynamics investigation since it is a basic parameter. However, it is difficult to know the size of the observed particle in advance. Three ways to determine the observed particle diameter according to the trajectory taken in the microscopical visualization method were introduced. In comparison with the optical method and gravity settlement, the method based on acoustic entrainment could provide more accurate particle diameter.

ACKNOWLEDGMENTS

The authors would like to acknowledge the financial support of the present study under the Special Funds for Major State Basic Research Projects of China (Project No. 2002CB211600)

REFERENCES

1. T.Arthru Degaetano and M. Owen Doherty, *Atmospheric Environment*, **38**, 1547-1558 (2003).
2. A.D. Shendriker and W.K. Steinmetz, *Atmospheric Environment*, **37**, 1383-1392(2003).
3. Michele Gingliano, Giovanni Lonati et al, *Atmospheric Environment*, **39**, 2421-2431 (2005).
4. E. P. Mednikov, Acoustic coagulation and precipitation of aerosols. New York: Consultants Bureau, 1965,pp. 152-153.
5. G.Reethof, *Transactions of the ASME*, **110**,552-557(1988).
6. Song, L., Modeling of acoustic agglomeration of fine aerosol particles, Ph.D. thesis. The Pennsylvania State University, 1990.
7. Ph. Capéran, J. Somers and K. Richter, *Journal of Aerosol Science*, **26**, 275–276(1995).
8. J. A. Gallego, E. R. Sarabia, G. Rodriguez et al, *Environment Science and Technology*, **33**, 3843–3849(1999).
9. I. Gonzalez, T. L. Hoffman and J. A. Gallego, *Acta Acoustica*, **88**, 19–26(2002).
10. E. R.Sarabia, L. Elvira-Segura, I. Gonzalez-Gomez, et al, *Ultrasonics*, **41**, 277-281(2003).
11. T. L. Hoffmann, “Visualization of Particle Interaction and Agglomeration in an Acoustic Field”, Ph.D. Thesis, The Pennsylvania State University, 1993.
12. T.L. Hoffmann and G. H. Koopmann, A new technique for visualization of acoustic particle agglomeration, *Review of Scientific Instruments*, **65**, 1527-1536(1994).
13. T.L. Hoffmann and G. H. Koopmann, *Journal of acoustic society of American*, **99**, 2130-2141(1996).
14. Samuel Temkin, Elements of Acoustics, John Wiley & Songs, Inc., New York, 1981,pp253-260.
15. I. Gonzalez & T. L. Hoffmann, *Journal of Aerosol Science*, **21**, 1461-1468(2000).

Research on Resistance Properties for Dense Phase Pneumatic Conveying of Pulverized Coal in Vertical Pipe at Different Pressures

Wenhao Pu Changsui Zhao* Yuanquan Xiong Cai Liang Xiaoping Chen
Chunlei Fan Peng Lu

(Key Laboratory of Clean Coal Power Generation and Combustion Technology of Ministry of Education, Thermoenergy Engineering Research Institute, Southeast University, Nanjing 210096,Jiangsu China)

Abstract: Experiments on pressure drops in vertical pipe in dense-phase pulverized coal pneumatic conveying system were conducted under the pressure range from 2.1 MPa to 3.6 MPa. The nitrogen, as conveying gas, was introduced into the system via a header connected with sixteen nitrogen cylinders. Gas flow rate was controlled with a needle valve, and three flow meters were used to measure the volume flow rate of nitrogen. The dosing hopper was suspended on three load cells, which enable on-line weighing the solids collected in the hopper. Pressures and pressure drops were measured with pressure transducers and KELLER differential pressure transducers, respectively. All the signals were connected to a computer. Sampling time interval was from 5 milliseconds to 1 second.

The experimental results indicate that at the pressure range from 2.1 MPa to 3.6 MPa when the solids mass flow rate is kept constant and the superficial velocity increases, the pressure gradient first decreases, then starts to increase and when the solids mass flow rate and superficial velocity is constant, the pressure gradient increases as the transport pressure is increased. Based on Barth's pneumatic conveying theory, an empirical correlation of the additional pressure drop coefficient in vertical pipe with Froude number and the solid/gas density ratio was proposed under different pressure. Comparing the pressure drops predicted by the correlation with the experimental data under the pressure of 2.13MPa, it is found that the calculation results are in good agreement with the experimental data.

Keywords: Dense-phase; Pneumatic conveying; Additional pressure drop coefficient

INTRODUCTION

Pneumatic conveying is the use of gas to transport particles through a pipeline. It has been successfully used for many years in the chemical and process industries for the transport of materials such as flour, granular chemicals, lime, plastic chips and coal. In most industrial installations, it is generally known as lean phase conveying. Transport velocities are high, which results in high attrition rates and high power consumption [1]. Dense phase pneumatic conveying overcomes these disadvantages. When the less amount of nitrogen is used to feed coal into a coal gasification reactor, the synthetic gas is easier to produce methanol and synthesized ammonia. Nowadays, coal gasification technology plays an important role in clean coal technology while dense phase pneumatic conveying of pulverized coal is one of the most important technologies.

In a pneumatic conveying system the gas density decreases along the transport line due to the falling pressure of the conveying gas. The superficial velocity must increase along the pipe, which may result in a change in the flow pattern and hence the pressure gradient. If the pressure gradient could be predicted, it is of great significance to design and optimize the pneumatic conveying systems. Because of the complexity of the pneumatic conveying, it is difficult to obtain a unified correlation. Therefore, it is necessary to carry out the experiments to obtain resistance properties for different solid materials.

Seville et al. [2] state that "at a mechanistic level, the distinction between hydraulic and pneumatic conveying lies in the ratio of the density of the conveyed solids to that of the conveying fluid". In a hydraulic conveying system, the

* Corresponding author Tel.: +86-25-83793453; fax: +86-25-83793453. E-mail address: cszhao@seu.edu.cn

CP914, *Multiphase Flow: The Ultimate Measurement Challenge,* edited by X. Cai, Y. Wu, Z. Huang, S. Wang, and M. Wang

density of the fluid is normally similar to that of the solids and the ratio of densities is constant. In pneumatic conveying, the solids density is typically three orders of magnitude greater than that of the fluid and the ratio of densities changes along the pipe. As a result, the significance of particular flow phenomena is different in each type of system. The effect of the density ratio can be seen in the number of particle wall collisions, which are reduced in hydraulic systems due to the magnitude of the lift force on the particles near the wall [3]. In pneumatic conveying at high pressure, the gas density increases linearly with the increment of pressure. When the pressure of nitrogen reaches 3MPa, nitrogen density is 30 times greater than that at ambient pressure and the solids density is one order of magnitude greater than that of nitrogen. The solid/gas density ratio is 39 which lie between that of hydraulic conveying and pneumatic conveying at ambient pressure.

Geldart and Ling [4] carried out experiment research on the total pressure drops of high pressure dense phase conveying of fine coal. Xiong et al. [5, 6] studied pressure drops for vertical and horizontal pipe in dense phase pneumatically conveying pulverized coal at high pressure, but they did not take account of pressure changes.

Experiments on pressure drops in vertical pipe in dense-phase pulverized coal pneumatic conveying system were conducted under the pressure range from 2.1 MPa to 3.6 MPa. This paper presents an empirical correlation of the additional pressure drop coefficient in vertical pipe with Froude number and the solid/gas density ratio under different pressure.

EXPERIMENTAL SET-UP AND PROCEDURES

The experiment apparatus used in this study is illustrated in Fig. 1.The equipment is capable of conveying batches of up to 350kg of pulverized coal using nitrogen at pressure of up to 4.0MPa. Both hoppers are identical feeder or receiver vessels. The nitrogen, as conveying gas, was introduced into the system via a header connected with sixteen nitrogen cylinders. Gas flow rate was controlled with a needle valve, and three flow meters were used to measure the volume flow rate of nitrogen. A derivation of the nitrogen feeding system connected to the pipeline allows us to control the solid volume fraction. The dosing hopper was suspended on three load cells, which enable on-line weighing the solids collected in the hopper. The pressure of the reception hopper is controlled by electro valve. Pressures and pressure drops were measured with pressure transducers and KELLER differential pressure transducers, respectively. All the signals were connected to a computer. Sampling time interval was from 5 milliseconds to 1 second.

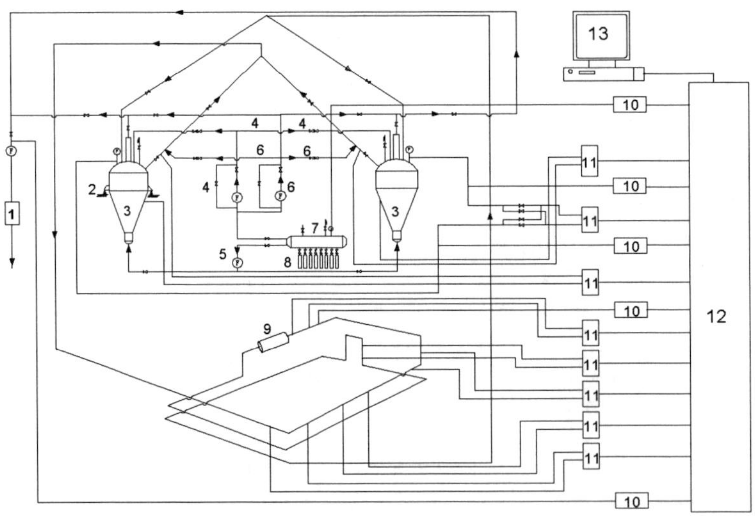

FIGURE 1. Illustration of the dense phase pneumatic conveying system of pulverized coal
1-electro valve 2-weigh cells 3-hoppers 4- pressurizing gas 5-fluidizing gas 6- supplement gas 7-header 8-nitrogen cylinder 9-visual section 10- pressure transducer 11- differential pressure transducer 12-A/D converter 13- computer

The inner diameter of the stainless steel pipe is 10mm and the total length is 53m. The density of pulverized coal is 1350kg/m^3 and mean diameter is 37μm.

EXPERIMENTAL RESULTS AND DISCUSSION

According to the Barth's pneumatic conveying theory, the pressure drop ΔP along the length ΔL of the pipe can be split into two parts where the ΔP_g is gas friction, and usually calculated as for single phase flow conditions and ΔP_s is the resistance of the particulate material.

$$\frac{\Delta P}{\Delta L} = \frac{\Delta P_g}{\Delta L} + \frac{\Delta P_s}{\Delta L} \tag{1}$$

The pressure gradient due to gas friction is given as

$$\frac{\Delta P_g}{\Delta L} = \lambda_g (\mathrm{Re}_g) \frac{\rho_g u_g^2}{2D} \tag{2}$$

The gas phase friction factor λ_g is assumed to be the same in the gas-solids flow as when only gas is flowing in the tube.

$$\lambda_g = 0.316 \mathrm{Re}^{-0.25} \tag{3}$$

The pressure drop due to the particle is given as

$$\frac{\Delta P_s}{\Delta L} = \lambda_s \mu \frac{\rho_g u_g^2}{2D} \tag{4}$$

Where λ_s is the additional pressure drop coefficient and μ is solid/gas mass flow ratio. Substitution of (2) and (4) into (1) gives the pressure drop

$$\frac{\Delta P}{\Delta L} = (\lambda_g + \lambda_s \mu) \frac{\rho_g u_g^2}{2D} \tag{5}$$

The pressure gradient curve under different solids mass flow rate and different pressure is illustrated in Fig. 2-4. The trends in these figures are rather clear. When the solids mass flow rate is kept constant and the superficial velocity increases, the pressure gradient first decreases, then goes through a minimum point, and subsequently starts to increase. The effect of the magnitude of the solids mass flow rate is also clear. In Fig. 2, when the solids mass flow rate and superficial velocity is constant, the pressure gradient increases as the transport pressure is increased. An increase in the transport pressure leads to an increase of gas density, which causes the loss increase due to gas friction, while the pressure drop due to the particle is almost constant. When the transport pressure is the same at any constant gas velocity, the pressure gradient increases as the solids mass flow rate is increased, as shown in Fig. 3-4. Comparing Fig.3 and Fig. 4, it can be seen that the effect of the solids mass flow rate on the pressure gradient is greater than that of the transport pressure.

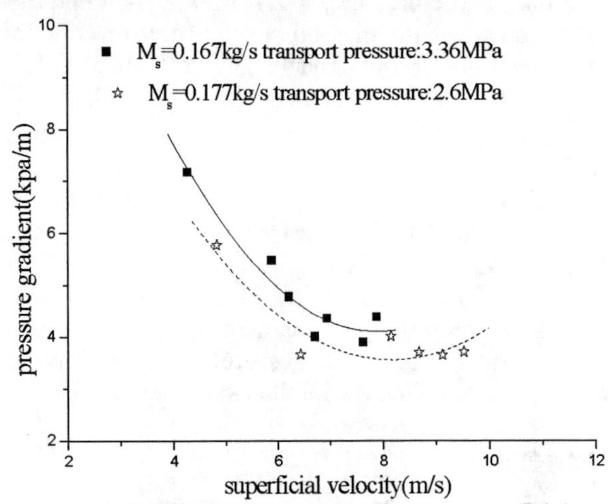

FIGURE 2. Pressure gradient vs. superficial

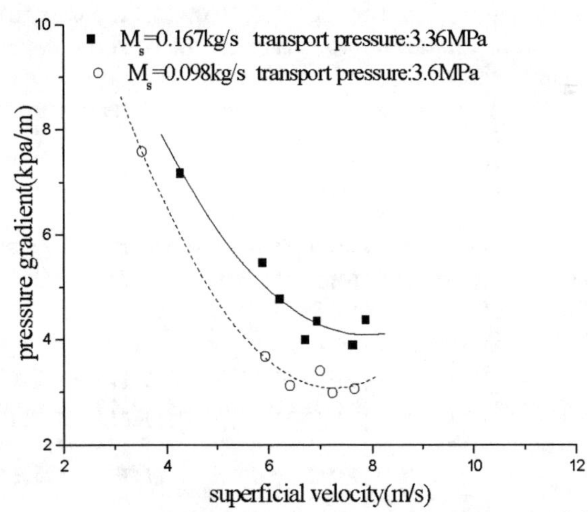

FIGURE 3. Pressure gradient vs. superficial

Usually the dimensional groups for pneumatic transport are used to obtain the additional pressure drop coefficient. However, there is no unified equation now due to the different systems and material properties. When the transport pressure changes from 2.1MPa to 3.6MPa, nitrogen density increases with increase of the transport pressure. The relative velocity between the gas and solid phase phases, the most significant feature, changes as the inter-phase force varies. Therefore, the pressure gradient is affected by the transport pressure. Based on the dense phase pneumatic conveying experimental results, Xiong et al. [5, 6] verified the correlation of the additional pressure gradient coefficient proposed by Zhou [7], who regarded that the additional pressure gradient coefficient related to

Froude Number. Molerus et al. [8, 9] regarded that the term $\sqrt{\frac{\rho_s}{\rho_g}}$ represented the influence of the fluid on the particle transport under fully suspended flow conditions. According to the experimental results, the additional pressure gradient coefficient in vertical pipe is obtained by equation (1)-(4). The experimental correlation of the additional pressure gradient coefficient is obtained by regression.

$$\lambda_s = 4.39\sqrt{\frac{\rho_s}{\rho_g}}Fr^{-2.87} \tag{6}$$

When the transport pressure is 2.13MPa and the solids mass flow rate 0.171 kg/s, experiments on dense-phase pulverized coal pneumatic conveying were conducted. Experimental results were shown in Fig. 5 and the trends are

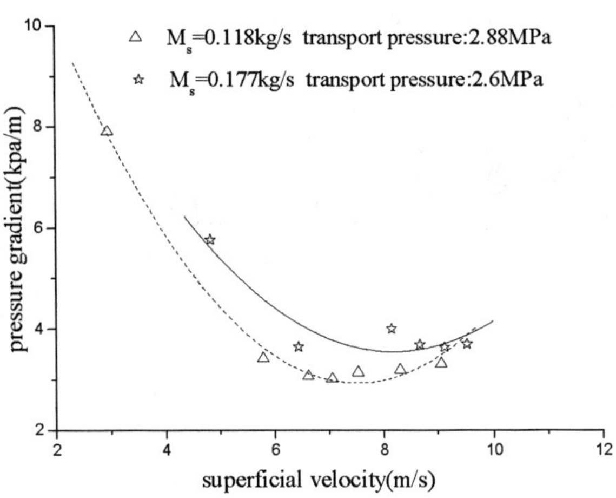

FIGURE 4. Pressure gradient vs. superficial velocity

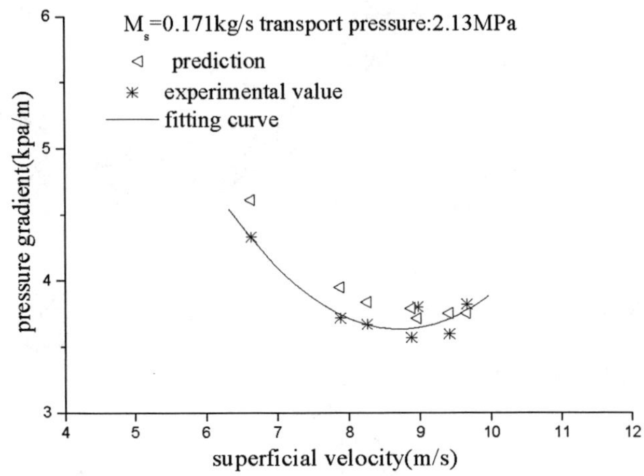

FIGURE 5. Pressure gradient vs. superficial velocity

the same. When the solids mass flow rate is kept constant and the superficial velocity is increased, the pressure gradient first decreases, then goes through a minimum point, and subsequently starts to increase. Comparing the pressure gradient predicted by equation (6) with the experimental data at the pressure of 2.13MPa, it is found that the calculation results are in good agreement with the experimental data and the maximum relative error is below 6.5%. It manifests that the correlation of the additional pressure gradient coefficient considers the influence of fluid mechanics on the pneumatic conveying and predicts the pressure gradient at different transport pressure well.

CONCLUSIONS

Extensive experiments have been done on dense phase pneumatic conveying of pulverized coal at high pressure in pipelines. It is concluded from experiments that when the solids mass flow rate is kept constant and the superficial velocity increases, the pressure gradient first decreases, then starts to increase and when the solids mass flow rate and superficial velocity is constant, the pressure gradient increases as the transport pressure increases.

Based on Barth's pneumatic conveying theory, a new correlation of the additional pressure drop coefficient in vertical pipe was developed for different pressure. Predictions are in good agreement with the experimental data.

ACKNOWLEDGEMENTS

The authors gratefully acknowledge the financial support from the National Key Program of Basic Research in China 2004CB217702.

REFERENCES

1. K. Konrad. Dense phase pneumatic conveying: A review [J].Powder Technology, 1986, 49:1-35
2. J.P.K. Seville, U. Tuzun, R. Clift, Processing of Particulate Solids, Chapman & Hall, London, 1997.
3. K.C. Wilson, G.R. Addie, A. Sellgren, R. Clift, Hydraulic Conveying of Solids with Centrifugal Pumps, 2nd edn., Chapman & Hall, London, 1996.

4. D. Geldart, S.J. Ling, Dense phase conveying of fine coal at high total pressure [J] Powder Technology, 1990, 62:243-252
5. XIONG Yuan-quan, ZHAO Bing, SHEN Xiang-lin, A Study on the Resistance Properties for Vertical Pipe in Dense Phase Pneumatic Conveying of Pulverized Coal Under High Pressures [J]. Proceedings of the CSEE, 2004, 24(9): 248-251.
6. XIONG Yuan-quan, ZHAO Bing, SHEN Xiang-lin, Resistance Properties for Horizontal Pipe in Dense Phase Pneumatic Conveying of Pulverized Coal Under High Pressures [J]. Journal of Combustion Science and Technology, 2004, 10(5): 428-432.
7. Zhou Jiangang, Zhang Shu, Experimental Study on the Pressure Drop of Pneumatic Conveying Pulverized Coal in a Pipeline (II) [J]. Journal of Iron and Steel Research, 1994, 6(4): 1- 7.
8. O. Molerus Overview: Pneumatic transport of solids [J] Powder Technology, 1996, 88:309-321
9. O. Molerus U. Heucke Pneumatic transport of coarse grained particles in horizontal pipes [J] Powder Technology, 1999, 102:135-150

Development Characteristics of Velocity Transports in An Isothermal Heated Drag-Reducing Surfactant Solution Flow

Hongxia Zhang*, Dezhong Wang+, Hanping Chen and Yanping Wang

*School of Mechanical Engineering, Shanghai Jiao Tong University, 800 Dongchuan Road, Shanghai 200240, P. R. China. *E-mail address: hxgmzhang@sjtu.edu.cn*

Abstract. The development characteristics, turbulence transports for stresses and kinetic energy of a cetyltrimethyl ammonium chloride (CTAC) surfactant solution for a two-dimensional channel flow have been experimentally investigated. Time mean velocity and fluctuating velocity are measured using a Phase Doppler Anemometry (PDA) at the Reynolds number 1.78×104 and isothermal heated temperature 31℃. Although mean velocity profiles at three cross sections show that the fluid is almost fully developed, the peak location of fluctuating intensity for the CTAC solution is slightly away from the wall downstream from the fluid and the peak location of fluctuating intensity is observed at far away from the wall than that of water. The location where the velocity gradient has its maximum, the fluctuating intensity does not get the high value. The elastic shear stress contribution to the total shear stress is 15 percents to 36 percents and it gets to the maximum near to the wall. The surfactant elastic shear stress is almost a liner function of the height of the channel, which means that the elastic stress contribution of the different cross locations is approximately the same. The fluctuating surfactant stress work is negative and the fluctuating elastic shear stresses produce rather than dissipate kinetic energy.

Keywords: surfactant solution, drag reduction, turbulent kinetic energy, channel flow
PACS: 47.50.-d

INTRODUCTION

It is well known that a drag reduction of up to 80% can be achieved by the addition of a small amount of a surfactant or polymer into turbulent flow[1,2]. Recently, surfactants have been widely accepted to the most practical drag reducing additives in distinct heating and cooling systems (DCH) for reducing pumping power because they are quite stable and no mechanical degradation as compared with polymers[3]. However, the mechanism of the occurrence of drag reduction with the use of surfactant additives has not been clarified.

Some points and consensus have been reported concerning the different characteristics in turbulent structures between drag reducing surfactant solution flow and those of Newtonian flow. Sadanandan and Surehkumar[4] suggested that the damping of the wall normal velocity fluctuations and enhancement of streamwise velocity fluctuations. Hetsroni et al. [5] investigated the space between low-speed streaks near the wall in turbulent drag reducing flow. B.Yu [6] also numerically reported this using Giesekus viscoelastic model. Kawaguchi [7] reported that the turbulent shear stress of a surfactant solution flow becomes almost zero and the viscous stress exists only in the near-wall region, while high-level velocity fluctuations can be observed. The decorrelation between the wall-normal and streamwise velocity components results in stress deficit have been reported by Warholic[8]. It can be seen that the mechanisms of the drag reducing phenomena are of great significance for controlling drag reduction in application are still hotly debated.

As investigated by Suzuki, the formation time of the micellar network structure is quite long, and the surfactant flow development becomes very slow, even more than 1000 times the hydraulic diameter. In this study, experiments

+ Corresponding author. E-mail address: dzwang@sjtu.edu.cn
Contract Grant Sponsor: Specialized Research Funding for the Doctoral Program of High Education (20030248032), the National Natural Science Foundation of China (50476016)

CP914, *Multiphase Flow: The Ultimate Measurement Challenge,* edited by X. Cai, Y. Wu, Z. Huang, S. Wang, and M. Wang

have been carried out for the development characteristics of the surfactant solution flow in an isothermal heated fluid using PDA. The shear stress deficit and the turbulent kinetic budget are studied in detail additionally.

EXPERIMENTAL PROCEDURE

Test Channel

A closed-circuit water channel is used in the present study. It consists of a water tank, with a volume of $1m^3$, a circulating pump, a two-dimensional channel, pipeline, and valves. The fluid temperature in the tank can be controlled within $\pm 0.5K$ by a heating and cooling system (using cooling water and heating coils). The test channel is made of transparent acrylic resin, with a height (H) of 0.02m, a width of 0.25m and a hydraulic diameter of 0.037m. A turbine flow meter (LWGY280A) with uncertainty of 0.5% is adopted for measuring the flow rate. A differential pressure meter with a precision of 0.25% is used to measure the pressure drop. The shear stress is calculated from the pressure difference between two fixed points with the distance of 3m in the streamwise direction. The velocity measurement station is in a central plane in the spanwise direction and is located at the three cross sections 208H, 230H and 250H down stream from the channel entrance, respectively. Schematic diagram of the test section is shown in figure.1.

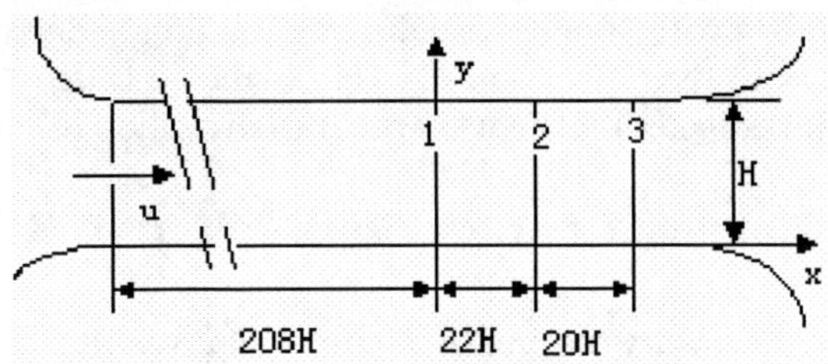

FIGURE 1. Diagram of 2D- channel

Working Fluid

In this paper, cetyltrimethyl ammonium chloride (CTAC) with the chemical formula of $C_{16}H_{33}N(CH_3)_3Cl$ is dissolved in the tap water. Sodium salicylate (NaSal) is added to the solution with the same molar concentration of the CTAC to provide the acidic icons. The solution with 120ppm (parts per million) weight concentration of the CTAC is tested, which has effective drag reduction in a relative low Reynolds number (1.78×10^4). Note that the Reynolds number of the surfactant solution is based on the solvent viscosity, bulk velocity, and hydraulic diameter. In the present study, the isothermal heated temperature of the fluid is 31℃. Details of the flow parameters are given in table 1.

TABLE 1. Test parameters

	Distance of the entrance	u_τ (m/s)	DR (%)
Water W1	208H	0.0316	0
W2	230H		
CTAC C1	208H	0.0125	56
C2	230H		
C3	250H		

PDA System

Using Phase Doppler Anemometry (PDA) measurement system (DANTEC Co Ltd., model 58N81), the fluctuating velocity distributions at three cross sections are measured. The streamwise, spanwise and wall-normal velocity components are measured by using blue (488nm), green (514nm), and violet (476.5nm) laser beams, respectively. In probe 1, two green beams are placed in horizontal plane to measure the streamwise velocity components, while two blue laser beams are placed in the vertical plane to measure the spanwise velocity components. The Doppler frequency shift signals from the green and blue beams are received by probe 2, which emits two violet beams and measures the wall-normal velocity components. The Doppler frequency signals of the violet beams are received by probe 1. The fringe mode with frequency shift of 40 MHz produced by Bragg cell is used to eliminate directional ambiguity in velocity measurement. The optics to focus the laser beam into the channel center includes focusing front lens with a focal length of 500mm. Backscatter mode is adopted. At each measurement location, about 5000 to 20000 data samples are measured. Our experimental results are compared with results obtained with water.

RESULTS AND DISCUSSION

Drag Reduction (DR) Characteristics

For this temperature, the characteristic of frictional drag $f = \tau_w/(1/2\rho U_b^2)$ and drag reduction rate $DR = (f_{water} - f_{CTAC})/f_{water}$ of the tested CTAC solution versus Reynolds number is shown in figure. 2. In figure. 2, the friction factor in a channel with hydraulic diameter $D = 44mm$ for the same surfactant drag reduction is also included (reported by F-C Li.[9]), which shows a dependence of flow characteristics on channel geometry. In our present study, surfactant solution with 30ppm has not produced drag reduction phenomena, while F-C Li. get a satisfying drag reduction effect with their test channel. This is the so-called "diameter effect" for drag-reducing fluids and K. Gasljevic et al.[10] investigates the diameter scaling correlation using polymer solution with pipe fluid.

Development Characteristics Of CTAC Solutions

The measured mean streamwise velocity (normalized with u_τ) for both water and the CTAC solution flows are presented in figure. 3. In this figure, dash line and solid line respectively indicate the log laws for Newtonian turbulent flow and for drag-reducing surfactant flow suggested by Zakin as follows.

$$U^+ = 2.5\ln y^+ + 5.5 \tag{1}$$

$$U^+ = 23.4\ln y^+ - 6.5 \tag{2}$$

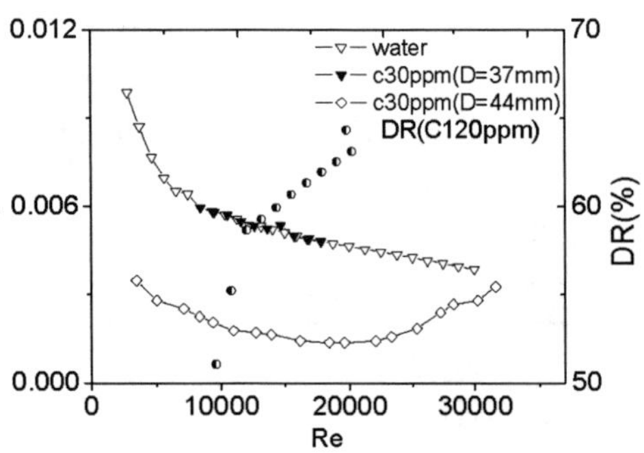

FIGURE 2. Friction factor and drag reduction profiles

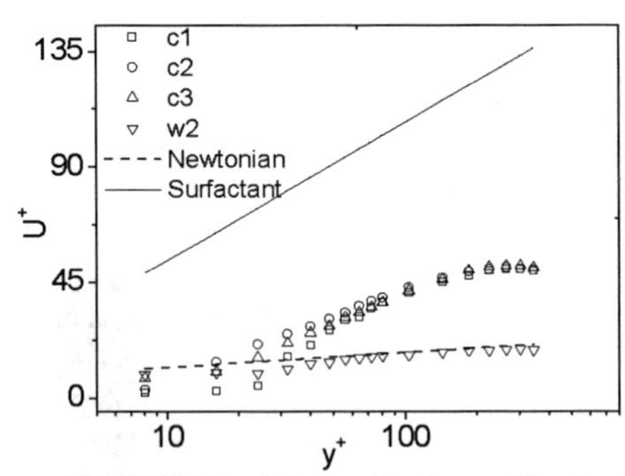

FIGURE 3. Mean velocity profile development

It is seen that the measured mean velocity curves for water flow is in close agreement with the line of the Newtonian turbulent flow. For the CTAC solution flow, the velocity profiles at different cross sections show that it has not approached the maximum drag reduction at this measured condition. However, the surfactant fluid flow grows in the core flow region of $y^+ > 50$ and it has found to be almost fully developed. While S.Hiroshi [11] investigates the surfactant solution flow has not fully developed even at their last cross section of 112D. It may be that the micellar network strucuture of CTAC solutions can be form at the relative short time. In odder to research the development characteristics in detail, other fluctuating intensity characteristics will be discussed at the three cross sections.

Figure.4 shows the distribution of the fluctuating intensity of the streamwise velocity component (normalized with mean streamwise velocity) at each measuring cross section. There are obvious differences at three cross sections compared with the mean velocity profiles. It is seen that the fluctuating intensity has a peak value at the measuring point closed to the wall for both CTAC solution and water flows and then decreased monotonically till to one steady value. It is also observed that the peak location of fluctuating intensity for the CTAC solution is slightly away from the wall downstream from the fluid. At the same cross section, the peak location of fluctuating intensity is observed at far away from the wall than that of water.

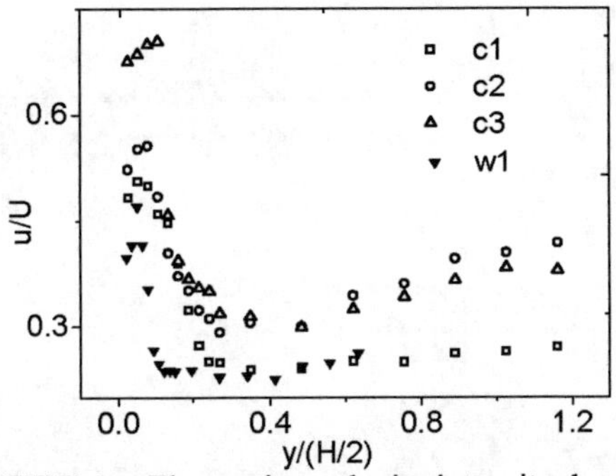

FIGURE 4. Fluctuating velocity intensity development

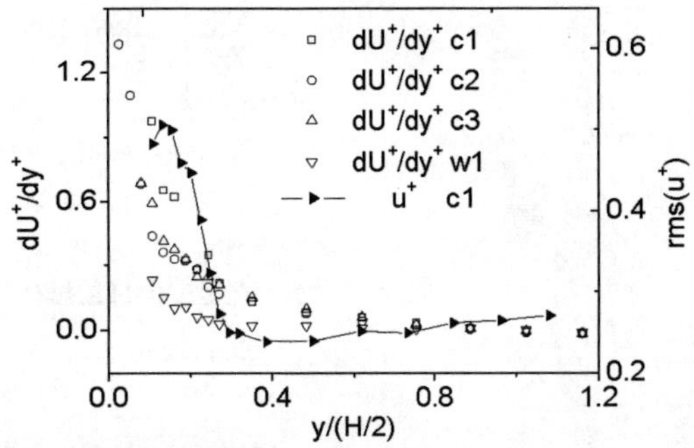

FIGURE 5. Mean velocity gradient development

In order to discuss the mean streamwise velocity profiles in detail, the distributions of the gradient of the mean velocity in the y direction will be analyzed. Clearly, the mean velocity gradient has a peak value close to the wall at each cross section similar to the streamwise velocity fluctuating intensity. Form the figure.5, it is seen that at the location where the velocity gradient has its maximum, the fluctuating intensity $rms(u^+)$ does not get the high value. The velocity gradient of the CTAC solution flow is lager than that of the water flow especially close to the wall. As Y.Kawaguchi [7] reported that the production term of the turbulent kinetic energy $-\overline{u^+v^+}(dU^+/dy^+)$ was reduced in the drag reducing surfactant flows, and at high DR level the reduction was quite significant. On the other words, it also indicates that the depression of the Reynolds shear stress in CTAC solutions flows.

The profiles of $rms(v^+)$ for surfactant solution flow, as show in figure.6, are all depressed and the depression becomes increasingly serious downstream from the fluid. The main conclusion of the velocity fluctuation measurements is that drag reduction leads to modification of turbulent structure.

Figure.7 shows that the addition of the CTAC results in a strong decrease of the Reynolds shear stress τ_t^+ and the turbulent momentum transfer could be obviously depressed. The reduction of the wall-normal turbulence intensity alone could not explain the reduction in the Reynolds stress τ_t^+ completely. This is attributed to the decorrelation between the two velocity components, u and v. The correlation can be described as $\rho = \overline{uv}/(\sqrt{\overline{u^2}}\sqrt{\overline{v^2}})$.

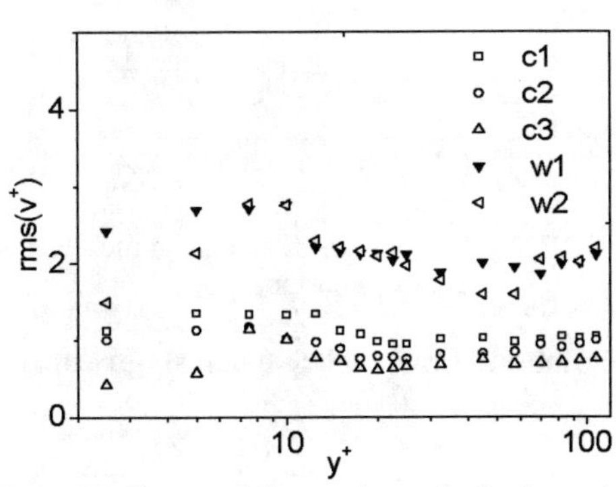

FIGURE 6. Wall-normal fluctuating velocity intensity development

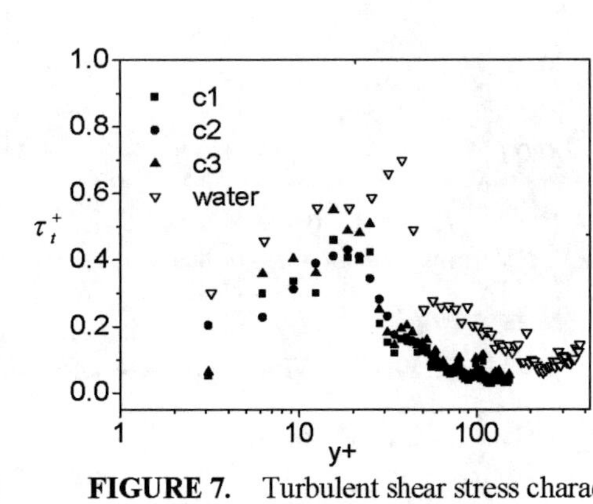

FIGURE 7. Turbulent shear stress characteristics

Shear Stress Contributions And Kinetic Energy Budget

We will discuss the various contributions to the mean shear stress. Mean shear stress can be written as the sum of the Reynolds stress, viscous shear stress and surfactant shear stress. Note that, the following researches on the shear stresses and kinetic energy characteristics are measured at the last cross section (fully developed equations for the analysis will be used). For the dimensionless values of these stresses (normalized with τ_w) then follows:

$$\text{Mean shear stress: } \tau_m^+ = \frac{2(H-y)}{H} \tag{3}$$

$$\text{Reynolds shear stress: } \tau_t^+ = \frac{1}{u_\tau^2}\overline{uv} \tag{4}$$

$$\text{viscous(solvent) shear stress: } \tau_v^+ = -\frac{1}{\rho u_\tau^2}\eta_v \frac{dU}{dy} \tag{5}$$

$$\text{elastic shear stress: } \tau_e^+ = \tau_m^+ - \tau_t^+ - \tau_v^+ \tag{6}$$

First we discuss the case of the water. As shown in figure.8, the theoretical sum of the shear stresses is a line across the channel height. Reynolds shear stress τ_t^+ and viscous shear stress τ_v^+ have been computed by the measured velocity values. The total of the two components $\tau_t^+ + \tau_v^+$ is shown and compared to the theoretical men shear stress τ_m^+, which agrees well.

From figure.9, we can see that the sum of the turbulent and the solvent stresses no longer equals to the mean stress in the drag reducing flow, which is called Reynolds stress deficit. The profiles of CTAC stress contributions show that the elastic shear stress takes the role from turbulence. The elastic shear stress contribution to the total shear stress is 15percents to 36percents and it gets to the maximum near to the wall. It is also shows that the surfactant stress is almost a liner function of the height of the channel, which means that the elastic stress contribution of the different cross location is approximately the same. Therefore, we can consider describing the drag reducing fluid using liner elastic stress model and we would gradually research this topic later.

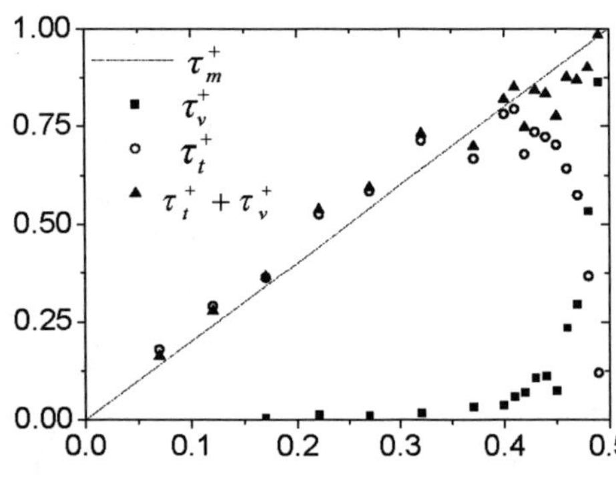

FIGURE 8. Various contributions to the shear stresses of water

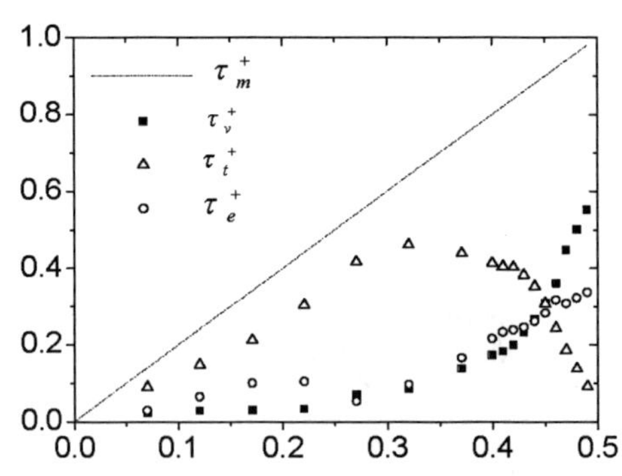

FIGURE 9. Various contributions to the shear stresses of CTAC solutions

Now we turn to the energy budget of the surfactant solution flow. The balance of mean kinetic energy $\frac{1}{2}\rho(U^2+V^2)$, gives

$$0=\underbrace{-U\frac{\partial P}{\partial x}}_{P_u}+\underbrace{\frac{\partial}{\partial y}(U((-\rho\overline{uv})+\eta_v\frac{\partial U}{\partial y}+\tau_e))}_{T_u}+\underbrace{\rho\overline{uv}\frac{\partial U}{\partial y}}_{D_u}\underbrace{-\tau_v\frac{\partial U}{\partial y}}_{V_u}\underbrace{-\tau_e\frac{\partial U}{\partial y}}_{E_u} \quad (7)$$

P_u is the production of mean flow energy by imposed pressure gradient. When averaged over the channel cross-section, as to the channel symmetry around the centerline, the contribution of T_u to the area averaged energy budget is zero. The three terms in T_u represent the transportation by Reynolds stress, mean viscous stresses and mean surfactant stresses, respectively. D_u is called deformation work. V_u is the viscous dissipation by the mean flow and E_u is the dissipation by the mean surfactant stresses.

A balance of the mean turbulent kinetic energy $\frac{1}{2}\rho\overline{(u^2+v^2)}$, gives

$$0=\underbrace{-\rho\overline{uv}\frac{\partial U}{\partial y}}_{P_k}+\underbrace{\frac{\partial}{\partial y}(-\frac{1}{2}\rho\overline{vu_iu_i}-\overline{pv}+\frac{1}{2}\eta_v\frac{\partial}{\partial y}\overline{u_iu_i}+u_i\tau_e)}_{T_k}\underbrace{-\overline{\tau_{ij}\frac{\partial u_i}{\partial x_i}}}_{\varepsilon_s}\underbrace{-\overline{\tau_e\frac{\partial u_i}{\partial x_i}}}_{\varepsilon_e} \quad (8)$$

The Reynolds decomposition is used for the velocity vector $u_i=(u,v)$ and u_i is velocity vector. P_k is the production of turbulent kinetic energy. T_k is the transportation due to velocity fluctuation, pressure fluctuation, viscous stresses fluctuating and surfactant stresses fluctuating, respectively. Similar to T_u, T_k do not contribute to the balance when integrating over the channel cross section. The last two items are the solvent viscous dissipation and the added surfactant stresses work.

In averaged budget, both V_u and ε_s are always positive. Equation (7) indicates that the energy input by the pressure gradient is directly dissipated by viscosity and becomes turbulent production in equation (8). In equation (7), D_u, $-E_u$ and $-V_u$ are negative. In equation (8), P_k is positive whereas ε_e could be negative or positive. A minus value results in a production of velocity fluctuations.

Last we discuss the budget of the turbulent kinetic energy. P_k can be measured directly. ε_s can be approximate by $\varepsilon_s\approx c_D\rho\frac{e^{3/2}}{D}$ [12], where $c_D=7.21$ is a dimensionless constant for Newtonian fluid and $e=\frac{1}{2}(\overline{u^2}+\overline{v^2})$. In this paper, we presume that c_D remains this constant for the turbulent surfactant flow. The results for water and surfactant flow are given in table. The integrated quantities are all normalized by $\eta_v u_\tau^2$.

TABLE 2. Various turbulent kinetic energy budgets for water and CTAC solution

	Water	**C3**
$(\int_S P_k)^+$	72810	43927
$(\int_S \varepsilon_v)^+$	72810	58949
$(\int_S \varepsilon_e)^+$	0	-15022

We find the fluctuating surfactant stress work is negative, which shows the fluctuating elastic shear stresses produce rather than dissipate kinetic energy. Warholic et al[13] research the processes using polymer solutions, in which experiments the production of kinetic energy is zero.

CONCLUSIONS

Development characteristics, turbulence transports for stresses and kinetic energy of a surfactant solution for a two-dimensional channel flow has been experimentally investigated, and the following results have obtained.

(1) Although mean velocity profiles at three cross sections show that the fluid is almost fully developed, the peak value of fluctuating intensity for the CTAC solution is slightly away from the wall downstream from the fluid. At the same cross section, the peak location of fluctuating intensity is observed at far away from the wall than that of water.

(2) The location where the velocity gradient has its maximum, the fluctuating intensity does not get the high value.

(3) The surfactant stress is almost a liner function of the height of the channel, which means that the elastic stress contribution of the different cross locations is approximately the same, so we can consider describing the drag reducing fluid using liner elastic stress model and we will gradually research this topic numerically next.

(4) The fluctuating surfactant stress work is negative, which shows the fluctuating elastic shear stresses produce rather than dissipate kinetic energy.

LIST OF SYMBOLES

$(\cdot)^+$ normalized by the inner wall variable

$\overline{(\cdot)}$ average value

$rms(\cdot)$ root-mean-square value

u_τ friction velocity, $u_\tau = \sqrt{\tau_w / \rho}$

τ_w wall shear stress $\tau_w = \dfrac{D}{4}\dfrac{\Delta P}{\Delta x}$

f friction factor

U_b bulk velocity

U mean streamwise velocity

P mean pressure

u streamwise velocity fluctuation

v wall-normal velocity flucuation

ρ solvent density

η_v solvent viscosity

H channel height

D hydraulic diameter

x streamwise direction

y wall-normal direction

z spanwise direction

ACKNOWLEDGEMENTS

We are grateful to the Key Laboratory of Power Machinery and Engineering of Education Ministry. P.R.C to provide the PDA for velocity measurements of the surfactant solutions.

REFERENCES

1. Gyr and H.W.Bewersdorff, Drag Reduction of Turbulent Flows by Additives, Kluwer, Dordrecht, The Netherlands, 1995.
2. J.L. Zakin, B.Lu and H.W.Bewersdorff, Surfactant Drag Reduction, Rev Chem Eng, vol.14, pp.253-320,1998.
3. B. Yu and Y. Kawaguchi, Direct Numerical Simulation of Viscoelastic Drag-reducing Flow: A Faithful Finite Difference Method, J. Non-Newtonian Fluid Mech,vol.116, pp.431,2004.
4. P.Sadanandan and R.Surehkumar, Viscosity Effects on the Stability of Wall-bounded Shear Flows, Phys.Fluids, vol.14, pp41, 2002.
5. G. Hetsroni, J.L.Zakin and A.Mosyak, Low-speed Streaks in Drag-reduced Turbulent Flow. Phys. Fluids, 9, pp.2397-2404, 1997.
6. B.Y and Y.Kawaguchi, DNS of Fully Developed Turbulent Heat Transfer of a Viscoelatic Drag-reducing Flow, International Journal of Heat and Mass Transfer, vol.48, pp.4569-4578, 2005.
7. Kawaguchi, Y.Tawaraya , A. Yabe,K.Hishida and M.Maeda, Turbulent Transport Mechanism in a Drag Reducing Flow with Surfactant Additive Investigated by Two Component LDV, In:Proceedings of the 8th International Symposium Application of Laser Techniques to Fluid Mechanics, Lisbon, 29.4.1–29.4.7,1996.
8. W.D.Warholic, G.M.Schmidt and T.J.Hanratty, The Influence of a Drag-reducing Surfatant on a Turbulent Velocity Flied, J, Fluid Mech, 388, pp.1, 1999.

9. F-C Li, D-Z Wang, Y. Kawaguchi, etc., Simultaneous Measurements of Velocity and Temperature Fluctuations in Thermal Boundary Layer in a Drag-reducing Surfactant Solution Flow, Experiments in Fluids, 36, pp131-140, 2004.
10. K. Gasljevic, G. Aguilar and E.F. Matthys, An Improved Diameter Scaling Correlation for Turbulent Flow of Drag-reducing Polymer Solutions, J. Non-Newtonian Fluid Mech, 84, pp 131-148,1999.
11. S.Hiroshi, Hong-Phuc Nguyen, N.Tomoe and U.Hiromoto, Development Characteristics of Fluctuating Velocity Field of Drag-reducing Surfactant Solution Flow in a Duct, Rheol Acta, 44, pp 457–464, 2005.
12. H.Tennekes, and J.L.Lumley, A First Course in Turbulence, MIT Press, Cambridge, MA, 1973.
13. M.D.Warholic, H.Massah and T.J.Hanratty, Influence of Drag-Reducing Polymers on a Turbulence: Effects of Reynolds Number, Concentration and Mixing, Exp. Fluids, 27, pp 461,1999.

Experimental Investigation On Flow Characteristics Of Pulverized Coal Dense-Phase Pneumatic Conveying At High Pressure

Cai Liang, Changsui Zhao, Xiaoping Chen, Wenhao Pu and Peng Lu

Key Laboratory of Clean Coal Power Generation and Combustion Technology of Ministry of Education, Southeast University, Nanjing 210096, China

Abstract. Experiments of pulverized coal dense-phase pneumatic conveying using nitrogen were carried out in an experimental setup at the pressure up to 3.7MPa and the solid-gas ratio up to 700kg/m^3. The influences of total conveying differential pressure, the pulverized coal moisture content, injected velocity and pressure on the characteristic parameters were investigated. The results indicate that pulverized coal mass flow rate increases with increase in total conveying differential pressure or conveying pressure. The solid-gas ratio rises with increase in total conveying differential pressure and increases at first and then declines as injected velocity increases. With increase in the pulverized coal moisture content, the pulverized coal mass flow rate and the differential pressure per unit length decrease under the same experimental condition.

Keywords: pneumatic conveying; differential pressure; mass flow rate; solid-gas ratio.
PACS: 47.61.Jd

INTRODUCTION

The pneumatic conveying has been successfully used in the chemical engineering, energy, metallurgy and other industrial processes with advantages, such as reliability, flexibility of layout, ease of automation, low maintenance, hygienic and environmental friendly. In recent years, numerous studies, for instance, experiment, signal analysis and numerical simulation, have been conducted on different pneumatic conveying systems to characterize the flow behavior of the gas-solids two-phase flow in the pipes with different sizes and for different pipe bends. Many valuable achievements on the characteristics of pneumatic conveying have been obtained. Gong [1] investigated the flow characteristic of high solid loading pneumatic conveying under low pressure and obtained some fitted empirical equation. Wypych[2] focused work emphasis on semi-empirical techniques to predict pressure drop with good accuracy; back-pressure unblocking techniques and so on. Li[3] used the photographic technology to measure the particle velocity and concentration profiles in the horizontal dilute-phase pneumatic conveying. Masona[4] applied the sub-models to describe transfer of mass, momentum and energy between the gas and solids in a layer. The prediction of pressure profile and the predicted depth of the dense layer show reasonable agreement with the experimental observations. Laouar's[5,6] aims were to characterize the differential pressure in dense-phase pneumatic conveying line at a very low velocity and a general differential pressure law was obtained which proves to be independent of both flow regimes and pipe diameter.

Generally, many of those researchers mainly work in the low pressure and dilute-phase pneumatic conveying [1-14]. The influence of the pulverized coal moisture content on the conveying characteristic is few reported. At the present time, the large-scale coal gasification technology is gaining attention and developed, and dense-phase pneumatic conveying of pulverized coal under high pressure is one of the key technologies in it [7, 8]. Because of low velocity and high solid concentration under the high pressure in transportation, the gas–solid two-phase flow

To whom correspondence should be addressed: Tel.: +86 25 83793453; Fax: +86 25 83793453.
E-mail:cszhao@seu.edu.cn

CP914, *Multiphase Flow: The Ultimate Measurement Challenge,* edited by X. Cai, Y. Wu, Z. Huang, S. Wang, and M. Wang

becomes very unsteady and complicated [1, 2], and unsteadiness of flows often cause blockage and pipe vibration. Pulverized coal moisture content is easy to result in fluctuation of mass flow rate and blockage of conveying pipeline. Very few references and experiences in this field are available. Theories and empirical formulas of dense-phase pneumatic conveying under high pressure still don't ideate and the flow characteristics of the high pressure conveying process in gas–solid system are not fully understood. Both experimentation and theory in this field still need the further research.

This paper is devoted to experimental investigation on the flow characteristics in dense-phase pneumatic conveying under high pressure. Experiments of pulverized coal dense-phase pneumatic conveying using nitrogen were carried out in an experimental setup at the pressure up to 3.7MPa and the solid-gas ratio up to 700kg/m3. The influences of total conveying differential pressure, the pulverized coal moisture content, injected velocity and pressure on the characteristic parameters were investigated, aiming at gaining the evolutional laws of pulverized coal mass flow rate and solid-gas ratio.

EXPERIMENTS

The pressurized experimental facility is shown schematically in Fig.1. High pressure nitrogen from the buffer tank is divided into pressurizing gas, fluidizing gas and supplement gas. The feeding hopper adopts the bottom-fluidization and top-discharge arrangement. Pulverized coal in the feeding hopper is fluidized by fluidizing gas and enters the conveying pipeline through the accelerating segment. Supplement gas is imported to enhance the conveying ability of gas at the outlet of the feeding hopper. In order to adjust pulverized coal moisture content, water through vacuum pump is injected into the pulverized coal in the conveying pipeline at the constant proportion. Pressurizing gas is used to regulate pressure in the feeding hopper in the conveying process. The pressure of the receiving hopper is controlled by the motor-drived control valve. Each of the feeding hopper and receiving hopper have a capacity of 0.648m3. The conveying pipeline (vertical section and horizontal section, as well as the bend) is made of a smooth stainless steel tube with an inside diameter of 10mm and a length of about 45m. The gas volume rates are measured by the metal tube variable-area flow meter, and the fluctuation of solid mass flow rate is gained by the weigh cells. Pressure and differential pressure are measured by the semiconductor pressure transducers with frequency response of 200Hz and precision of 0.3%. Signals of differential pressure, pressure, weight and gas volume flow rate are obtained by a multi-channel sampling system and then are sent to a computer through an A/D converter. The high-speed camera is employed to photograph the flow regimes through visiuallable test section. Pulverized coal with the mean diameter of 36μm and the density of 1350kg/m3 is used as test particles. Conveying gas is N2 with the maximum pressure of up to 4.8 Mpa.

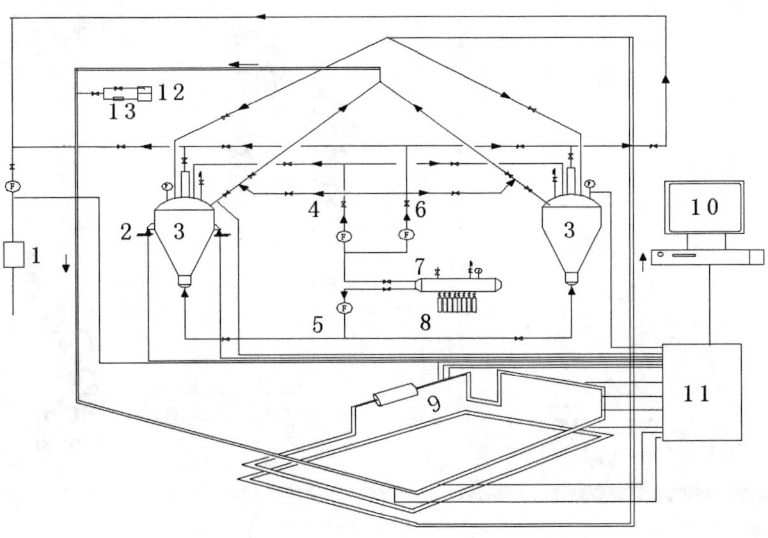

FIGURE 1. Schematic diagram of Dense-Phase Pneumatic Conveying of Pulverized Coal under High Pressure
1-Motor-drived control valve 2-Weigh cell 3-Hopper 4- Pressurizing gas 5-Fluidizing gas 7-buffer tank
8-Nitrogen cylinder 9- Visiuallable test section 10-Computer 11-Sensor and A/D converter 12-Water
13- Vacuum pump

RESULTS AND DISCUSSION

The experiments concerning the conveying characteristics in the dense-phase pneumatic conveying were performed. The effects of the total conveying pressure drop, gas flow rate and pulverized coal moisture content on the conveying characteristic were revealed. In order to eliminate influence of supplement gas, the solid-gas ratio was calculated before supplement gas was injected into conveying pipe.

The Effect Of The Total Conveying Pressure Drop And Fluidizing Gas Flow Rate On Mass Flow Rate Of Pulverized Coal

Relationships between the total conveying differential pressure and mass flow rate are plotted in the Fig.2. Mass flow rates under pressure levels rise with increase in the total conveying differential pressure. The total conveying differential pressure supplies power for pneumatic conveying and determines the magnitude of transferable energy in pneumatic conveying. In the experiment, the fluidizing gas volume rate and pressure in the feeding hopper keep constant. Total conveying differential pressure is adjusted via changing pressure in receiving hopper. When pressure in the receiving hopper decreases, the total conveying differential pressures increase and conveying capacities of unit volume gas enhance. Increasing total conveying differential pressure results in increment of the pulverized coal mass flow rate[1]. Figure.2 shows that the higher the conveying pressures are, the more the mass flow rates are under the same experimental condition. The pressure of the conveying gas determines the gas energy and carrying capacity under the constant differential pressure.

The varying tendency of pulverized coal mass flow rate with increase in the fluidizing gas volume rate is shown in Fig.3. It can be seen that the increase in the fluidizing gas volume rate augments the mass flow rate of pulverized coal. In the experiment, the fluidizing gas and pressurizing gas are injected into the feeding hopper. Fluidizing gas is applied to fluidize coal particles through a bowl-type gas distributor at the bottom of the feeding hopper. At the certain velocity, coal particles are suspended and appear properties of the fluid in order to be conveyed more easily. When fluidizing gas volume rate is very small, fluidizing effect of pulverized coal around discharge orifice is quite limited and coal particles need to consume more energy in the moving process. Consequently, mass flow rate of pulverized coal in pneumatic conveying is low for very small fluidizing gas volume rate is. Along with increase in the fluidizing gas volume rate, gas-solid two-phase flow appears like fluid and easy to be conveyed. In addition, rising outlet velocity leads to increase in the conveying capacity .All of those imply that mass flow rate increases with increase in the fluidizing gas is inevitable[1,2,15]. It can be found from the Fig.3 that the effect of the fluidizing gas on mass flow rate is greater when the fluidizing gas volume rate is less and slows down as the fluidizing gas volume rate becomes bigger.

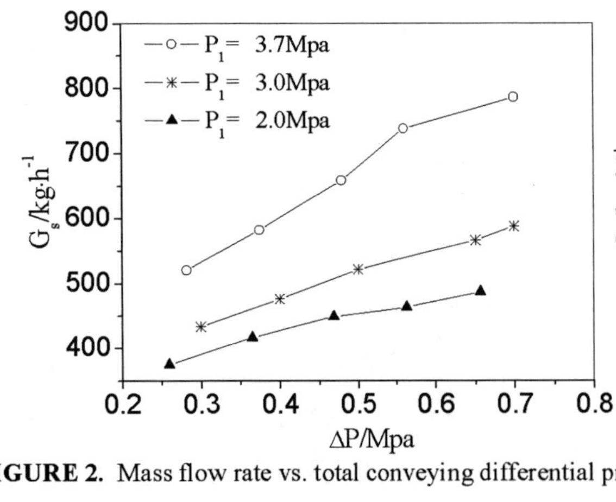

FIGURE 2. Mass flow rate vs. total conveying differential pressure

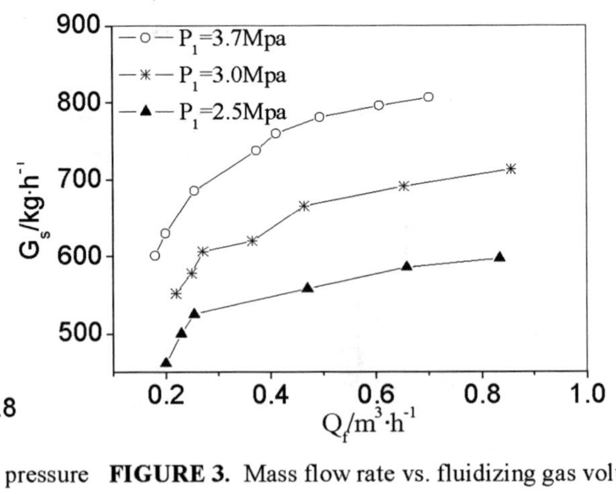

FIGURE 3. Mass flow rate vs. fluidizing gas volume rate

The Effect Of Injected Velocity And Total Conveying Differential Pressure On Solid-gas Ratio

Solid-gas ratio is defined as the ratio of solid flow rate and gas flow rate, and divided into mass flow ratio, volume flow ratio and mass/volume flow ratio. Mass/volume flow ratio is chosen to reflect the conveying capacity of unit volume gas in the paper. The influences of the injected velocity and total conveying differential pressure on solid-gas ratio are shown in Fig.4 and Fig.5.

Injected velocity is used to denote velocity of two-phase flow entering the accelerating pipeline from the feeding hopper. Figure.4 shows varying tendencies of solid-gas ratio for the various injected velocity. In the experiment, the total conveying differential pressure, pressures of the feed hopper and receive hopper keep constant. Solid-gas ratios increase at first and then decrease with increase in injected velocity. Regarding to the maximum solid-gas ratio the left of the Fig.4, when injected velocity is very small, coal particles need to consume more energy in the moving process and are more difficult to be conveyed, thus conveying capacity of the unit volume gas and solid-gas ratios are very low. As injected velocity increases, gas volume rate for fluidizing coal particles enhances and pulverized coal around discharge orifice appears fluidal characteristic. Increment of solid mass flow is larger than increment of gas volume rate and therefore solid-gas ratios increase. To the right of the maximum solid-gas ratio, the effect of injected velocity on solid mass flow rate slows down; increment of gas volume becomes greater and thus solid-gas ratios decrease with increasing injected velocity.

Relations between conveying differential pressure and solid-gas ratio are described in Fig.5. In the experiments, fluidizing gas volume rate, supplement gas volume rate and pressure in the feeding hopper keep the same. Total conveying differential pressure is adjusted through changing pressure in receiving hopper. Solid-gas ratios rise with the increase in the total conveying differential pressure. When pressure in receiving hopper decreases, the total conveying differential pressure increases and conveying capacity of unit volume gas enhances. Though conveying gas volume expands because of reducing pressure in receiving hopper, it is much less than the growing amplitude of mass flow rate of pulverized coal. In brief, solid-gas ratio always rises with increment of the total conveying differential pressure when keeping other operating parameters constant. It can be found from the Fig.6 that effect of differential pressure on solid-gas ratio is greater under the low differential pressure, and it goes down when the total conveying differential pressure is higher.

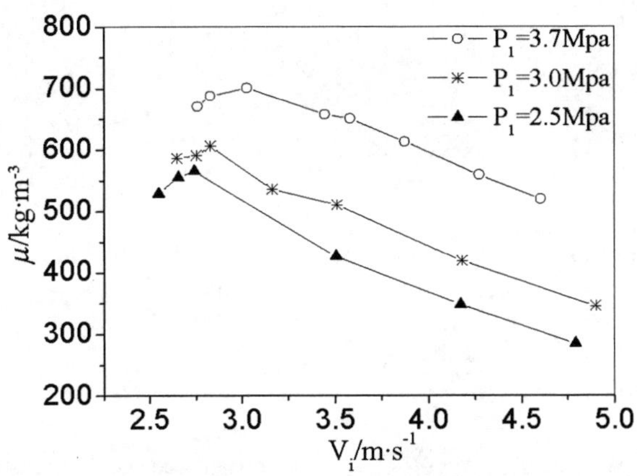

FIGURE 4. Solid-gas ratio vs. injected velocity

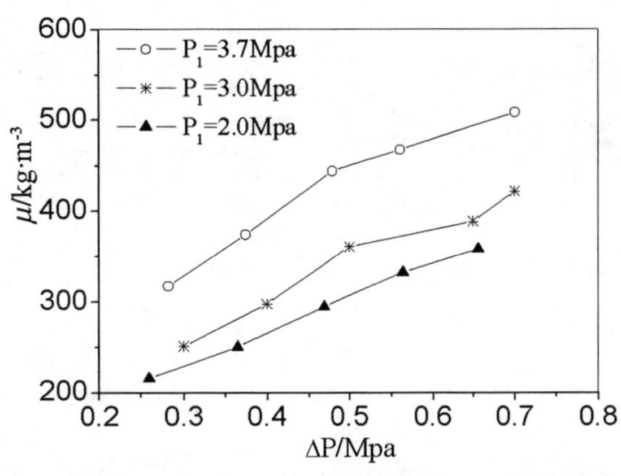

FIGURE 5. Solid-gas ratio vs. total conveying differential pressure

The Effect Of Coal Moisture Content On The Conveying Characteristic

Moisture content is important parameter of influenced powder flowability in pneumatic conveying and effect on friction properties, flowability, dispersibility and briquettability. As moisture content of pulverized coal is higher, free water among pulverized coal particles exist as felted water and sphenoid water. Coal particles appear adhesion traction and form Water Bridge because interfacial force of free water shrinks. Among the coal particles prilling phenomena and easily aggregated into the larger particles. All of these factors result in the increment of friction coefficient and viscosity. Atomized water is evenly injected into the pulverized coal to adjust the pulverized coal moisture content. All of the operating parameters except pulverized coal moisture content keep stabilization. The varying tendencies of characteristic parameters with the increase in moisture content in pulverized coal are shown in Fig.6 and Fig.7.

Figure.6 gives the related curves between pulverized coal moisture content and mass flow rate at two total conveying differential pressures. Mass flow rates decrease with the increase in the pulverized coal moisture content. When moisture content is lower, both viscosity of two-phase flow and friction coefficient are smaller. Conveying unit mass pulverized coal only needs to consume the less energy, therefore, conveying gas with the same energy can carry more pulverized coal than it with the greater moisture content. Along with moisture content increasing, the small size coal particles easily aggregated into the larger particles which result in spherical degree of coal particles becoming worse and rising in the viscosity of gas-solid two-phase flow. Pulverized coal needs more energy to be fluidized and conveyed and thus conveying capacity per unit gas decreases. Upon the reasons, it indicates that increment of pulverized coal moisture content leads to decrease in the mass flow rate. As the moisture content is greater than 6%, the conveying pipe usually causes the blockage.

The differential pressures per unit length of horizontal conveying pipe for various pulverized coal moisture content are depicted in Fig.7. All the experiments are performed under the same operating parameters except the total conveying differential pressure. It can be found that the differential pressure per unit length decreases with increase in the pulverized coal moisture content. As pulverized coal moisture content rises, the water bridge is formed among the particles and pelleted movement induces the fine coal particles to become the bigger particles. Because conveying gas volume rate retains unchanged, solid-gas ratio reduces. Although differential pressure tends to increase with the increment of coal particle size and two-phase flow viscosity, the influence of these factors on differential pressure is weaker than mass flow rate on it. It is very obvious that the differential pressures per unit length decrease with increase in the pulverized coal moisture content.

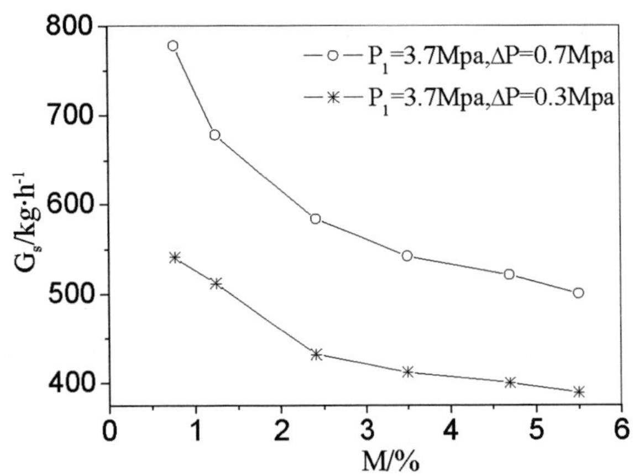

FIGURE 6. Mass flow rate vs. moisture content

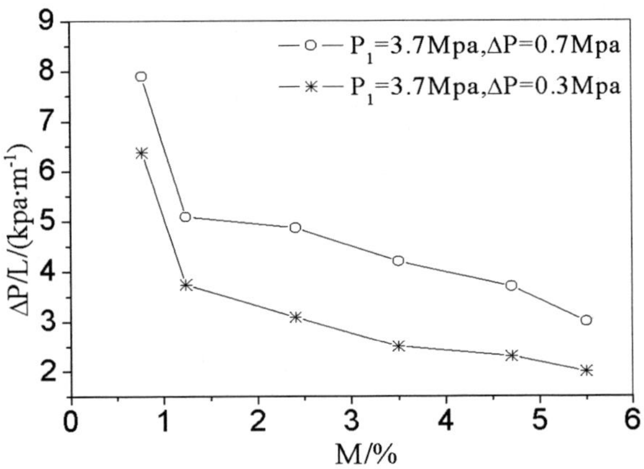

FIGURE 7. Differential pressures per unit length of horizontal conveying pipe vs. moisture content

CONCLUSIONS

The influences of different conveying differential pressure, injected velocity and pulverized coal moisture content on characteristic parameters were revealed. A number of valuable data and operating regularities were obtained, which are very useful to understand the complex characteristics of two-phase flow in pneumatic conveying. Based on the experiments and theory analysis, the main results can be summarized as follows.

(1). Mass flow rate rises with increase in the total conveying pressure drop. The higher the conveying pressure is, the more the mass flow rate is under the same experimental conditions. The increase in the fluidizing gas volume rate augments the mass flow rate of pneumatic conveying. The effect of the fluidizing gas on mass flow rate is great when the fluidizing gas volume rate is small and goes down as the fluidizing gas volume rate is greater.

(2). For the constant total conveying differential pressure, pressures in the feeding hopper and receiving hopper , solid-gas ratio increases at first and then decreases with increase in injected velocity; effect of differential pressures on solid-gas ratio is strong under the low differential pressure, and it goes down when the total conveying differential pressure is higher.

(3). When all the operated parameters except pulverized coal moisture content keep constant, mass flow rate and the differential pressure per unit length decrease with increase in the pulverized coal moisture constant. For the coal moisture of over 6%, the blockage usually happens in the conveying pipe.

ACKNOWLEDGMENTS

Financial support from the National Key Basic Research Program of China (No. 2002CB211600) and Excellent Ph. D. Thesis of Southeast University were sincerely acknowledged.

REFERENCES

1. X. Guo and et al, High solids loading pneumatic conveying of pulverized coal, *Journal of Chemical Industry and Engineering (China)*,Vol.**57**, pp. 640-644,2006.
2. P.W. Wypych and J. Yi, Minimum transport boundary for horizontal dense-phase pneumatic conveying of granular materials, *Powder technology*, Vol.**129**, pp. 111-121,2003.
3. H. Li and Y Tomita, Particle velocity and concentration characteristics in a horizontal dilute swirling flow pneumatic conveying, *Powder Technology*, Vol.**107**, pp. 144-152, 2000.
4. D J. Masona, A. Lev, A model for non-suspension gas-solids flow of fine powders in pipes, *International Journal of Multiphase Flow*, Vol.**27**, pp. 415-435, 2001.
5. S. Laouar, Y. Molodtsof, Experimental drop in dense phase pneumatic transport at very low velocity, *Powder Technology*, Vol.**95**, pp. 165-173, 1998.
6. S. Laouar, and Y. Molodtsof , Experimental characterization of the pressure drop in dense phase pneumatic transport at very low velocity, *Powder technology,* Vol.**95**, pp. 165-173,1998.
7. X. Shen and Y. Xiong, Experimental study on dense-phase pneumatic conveying of pulverized coal at high pressures, *Proceedings of the CSEE*, Vol.**25**, pp. 103-107 ,2005.
8. O. Molerus, Overview: pneumatic transport of solids, Powder technology, Vol.88, pp. 309-321, 1996.
9. J. Hong, Y. Shen and Y. Tomita, Phase diagram in dense phase pneumatic transport, *Powder Technology*, Vol.**84**, pp. 213-219, 1995.
10. H. Li, Application of wavelet multi-resolution analysis to pressure fluctuations of gas-solid two-phase flow in a horizontal pipe, *Powder Technology*, Vol.**125** pp. 61-73, 2002.
11. T. Singer, Dense-phase pneumatic conveying：Applications, system design，and troubleshooting, *Powder and Bulk Engineering*, Vol.**17**, pp. 27-33, 2003.
12. P E. Solt, Comparing dilute- and dense-phase conveying, *Powder and Bulk Engineering*, Vol.**17** pp. 64-68, 2003.
13. D. Geldart, Ling S. J. Dense phase conveying of fine coal at high total pressure, *Powder Technology*, Vol.**62**, pp. 243-252, 1990.
14. Y. Xiong, B. Zhao, X. Shen, Study on the resistance properties for vertical pipe in dense phase pneumatic conveying of pulverized coal under high pressures, *Proceedings of the CSEE*, Vol.**24**, pp. 248-251, 2004.
15. F J. Cabrejos and G E. Klinzing, Characterization of dilute gas-solid flows using the rescaled range analysis, *Powder Technology*, Vol.**84**, pp. 139-156, 1995.

NOMENCLATURE

$\Delta P/\Delta L$——differential pressure per unit length，[$kpa \cdot m^{-1}$]
ΔP——the total conveying differential pressure，[Mpa]
P_1——pressure of the feed hopper, [Mpa]
P_2——pressure of the receive hopper, [Mpa]
G_s——mass flow rate of pulverized coal，[$kg \cdot h^{-1}$]
V_i——injected velocity，[$m \cdot s^{-1}$]
μ——solid-gas ratio，[$kg \cdot m^{-3}$]
Q_f——fluidizing gas volume rate，[$m^3 \cdot h^{-1}$]
M——pulverized coal moisture, [%]

A Study on Turbine Unsteady Flow via PIV and Hotwire

LI Jun*, SU Ming

School of Mechanical and Power Engineering, Shanghai Jiaotong University, Shanghai, China

Abstract. Unsteady wake represents blade pneumatic load and affects turbine characteristic greatly, while secondary flow loss is the main part of turbine endwall loss. The appearance, attenuation of unsteady wake and their interaction with downstream secondary flow have significant effect on turbine design and characteristic.
Experiment measurement and numerical simulation are adopted to study secondary flow and wake. Upstream rotating blade can be replaced by cylinder with the same wake velocity and stagnant pressure loss. In experiment, two inlet velocities are taken to study the effect of inlet flux on blade characteristic. At the same time, upstream cylinders move quarter of circumferential pitch discretely and the influence of upstream cylinder wake to downstream blade is researched. Afterwards, two hollow boards are used to cover part of blade height to show tip and bottom secondary vortexes alternation under different relative blade height.

First, the system integrating method of hotwire measurement is developed with a rotating slanted hotwire in order to measure three-dimensions flow field at blade outlet. The procedure for solving the mean velocity vector involving the least-squares technique can be accomplished via the LSQNONLIN optimization function of Matlab. Then the whole velocity field of horizontal-flow and perpendicular plane has been measured with a Particle Image Velocimetry (PIV). Finally, numerical calculation is carried out. The result shows coincidence between experiment and calculation.

It is found that outlet secondary flow is obvious and severe at higher inlet velocity and smallest relative blade height. When the upstream rotor wake impinges upon the leading edge of the down stream stator, the corresponding stage efficiency will be higher; when the upstream wake is transferred into the mid passage of the downstream stator, the corresponding stage efficiency will be lower. As a result, the optimal blade circumferential layout is an effective way to increase turbine pneumatic efficiency. The numerical calculation shows that pressure pulse intensity on pressure side weakens rapidly along blade profile, while pressure pulse intensity on suction side weakens along profile first and then strengthen due to unsteady separation on trail edge.

Key words: unsteady; secondary flow; wake; three-dimension flow field; hotwire; PIV

INTRODUCTION

In the past, unsteady pneumatic design course is based on steady flow hypnosis. But there is much unsteady recourse in practical work, such as inlet guide vane disturbance, suddenly pressure pulse after combustion, instant temperature lift, and pressure fluctuation after weapon launch. Many basic experiments show unsteady inside and outside of turbomachinery has obvious effect on turbine blade pneumatic characteristics. There are several percent efficiency differences between unsteady and steady method, so a universe viewpoint in turbomachinery is that unsteady flow should stand on unsteady analysis system.

The turbomachinery periodical unsteady includes the following category: The first is inherent unsteady flow analysis. Relative movement between blade and vane causes a majority of unsteady phenomena. Although the inlet flow is steady, turbomachinery itself will cause period flow pulse, which can be broken into: Potential flow, Wake interference, Viscidity vortex and endwall transverse flow. The second kind is low frequency dynamic unsteady. Inlet unsteady pressure and temperature pulse will cause the next blade's characteristic change. The general example is the starting course, acceleration course and deceleration course of engine. The third is condition unsteady, which also can be called as flustration and relate with coupling.

Secondary flow loss is the main part of turbine endwall loss, which is formed by flow tunnel turn, fluid viscosity and traverse pressure gradient of adjacent blade concavo-convex side. The detailed flow process of blade secondary flow is important to develop theoretical computation method and improve turbine pneumatic characteristic. On the other side, unsteady wake represents blade pneumatic load and affects turbine characteristic greatly. The appearance, attenuation of unsteady wake and their interaction with downstream secondary flow have significant effect on turbine design and characteristic. This paper gives experiment and numerical simulation to study the secondary flow

CP914, *Multiphase Flow: The Ultimate Measurement Challenge,* edited by X. Cai, Y. Wu, Z. Huang, S. Wang, and M. Wang

and wake.

EXPERIMENT EQUIPMENT

Cylinder conversion system is adopted to simulate the inherent effect between vane and blade in turbomachinery. Pfeil&Eifler consider cylinder's wake structure is equal to that of blade by the same drag coefficiency [1]. The expression can be written by:

$$Y_P = \frac{c_d d}{s_{bar} \cos \beta}$$

In the expression, s_{bar} is distance among cylinders, β is blade inlet angle, c_d is cylinder drag coefficient. If cylinder's axial distance matches blade's, their wake width is almost the same. So cylinder diameter d can be decided by turbine blade profile loss coefficient Y_P. Among them, $c_d = 1$. As long as blade axial distance and blade pitch are true, wake velocity loss and wake distance are almost the same.

Air turbine vane profile is 25NSP, four vanes are used to give additional flow channels to ensure the parameters reliability of center channel. The wake of upstream cylinder simulates that of upstream blade. Fig.1 show the blade basic geometry, such as pitch, axial chord and cylinder. Fig.2 show the experiment equipment and vane photo.

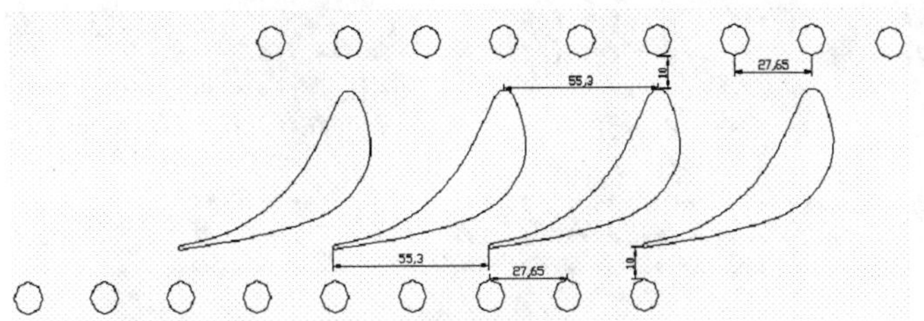

FIGURE 1. Vane profile geometry parameter [unit:mm]

Vane number	4
Pitch chord ratio	120/79.7 mm
Axial chord length	49.2 mm
Install angle (axial	51.1°
Airflow inlet angle	0°
Pitch/chord length	55.3/79.7 mm

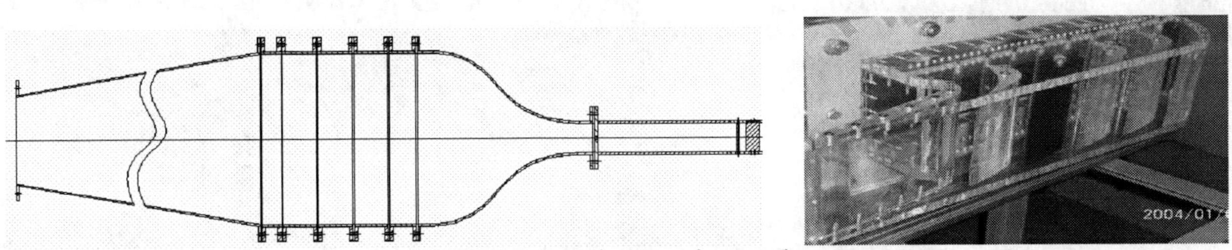

FIGURE 2. Experiment equipment photo

Blade experiment is tested at Shanghai Jiaotong University Turbomachinery Institute. Low-speed blow wind tunnel is used as inlet. Filter mesh is used to decrease the inlet flow turbulence. Danfoss VLT5000 digital transducer is used to control fan rotation speed and air flux. The transducer's frequency conversion range is 0~50Hz.

PIV and hotwire are used to measure turbine outlet three-dimension velocity flow field in this paper. Through PIV measurement, the whole outlet plane velocity is got at the same time, while hotwire can provide three-dimension velocity scatter point. Both experiment result can complement and testify each other.

PIV SYSTEM AND RESULT ANALYSIS

PIV (Particle Image Velocimetry) is a new non-contact velocity measurement technology developed in the past ten years, which is efficient to show instant velocity field and vortex vector. The system used in this experiment is DANTEC laser product, which is made up of four parts: slice lamp-house, photograph shoot system, signal disposal and Flow Map software. In the meanwhile, slice lamp-house is dual resonance Nd: YAG laser, the biggest work frequency is 10Hz, each pulse energy can arrive at 200mJ; pulse light adopts Qswitch trigger, width is 5.0~10.0ns, output laser beam wavelength is 532um(green light). The usage of two independent laser producers can easily choose time interval between two-laser pulses at microsecond grade, which enlarge the velocity measurement region.

Then laser beam passes through optical component and can be transformed to 1mm width slice light. At the same time, Kodak Mega plus ES1.0 cross correlation digital camera is used to identify the particle transducer with 256 degree grey mode. There is standard 60mm lens in the front of camera lens (Nikon Micro 60/2.8), which can ensure the accordance of CCD camera's focus and target plane.

Fig.3 show PIV three-step operation frames. The first step is dual-exposal to capture the particle in flow field. The second step is image analysis and velocity pickup. The third step is velocity vector display. Fig.4 gives the practical chart of PIV instrument.

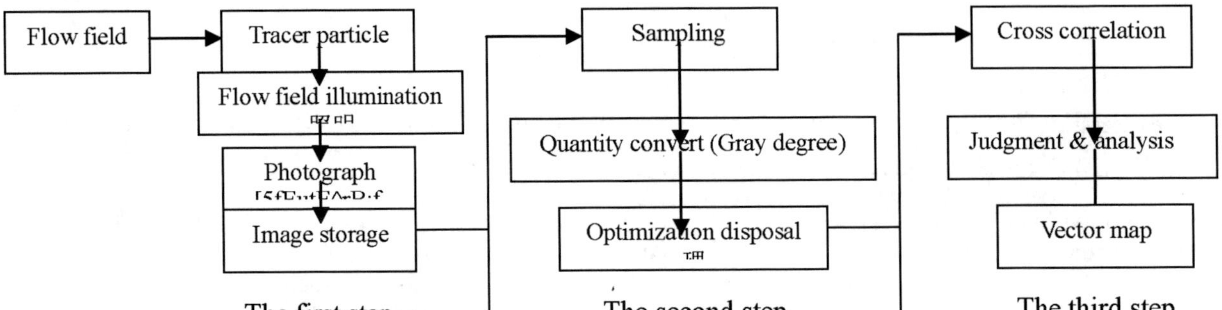

FIGURE 3. PIV three-step operation frames

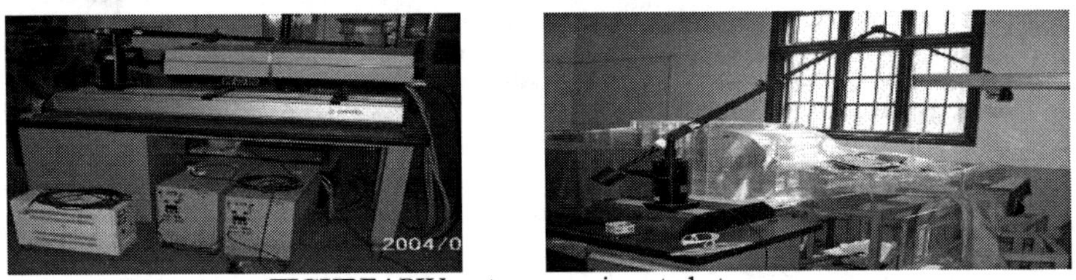

FIGURE4.PIV system experiment photo

To study the cylinder wake on downstream velocity field, cylinder row situate different circumferential position such as no cylinder, 0/4T, 1/4T, 2/4T, 3/4T (T is blade pitch), which move a quarter of pitch each time. Fig.5 shows PIV result.

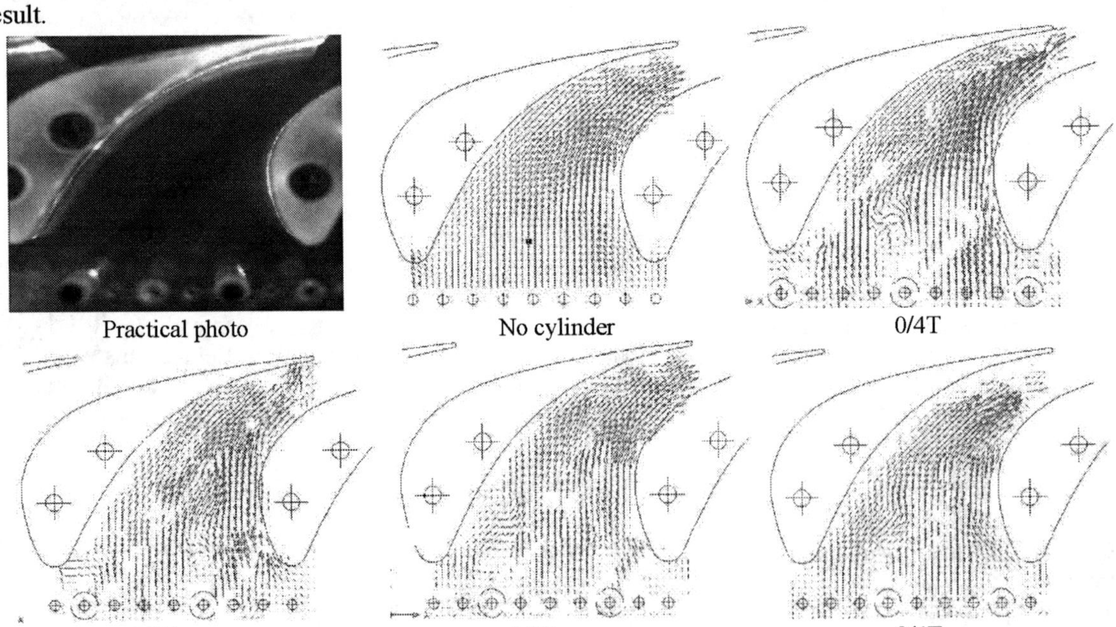

FIGURE 5. PIV result: unsteady wake

From the chart, when there is no cylinder upstream, airflow turns smooth. When there are cylinders upstream, cylinder wake is swallowed into blade flow channel and impugn on different channel location. The phenomena can be explained as follows: upstream cylinder wakes flow into channel and mix with mainstream to form large region low loss region. When upstream wake is transferred into the mid passage of the downstream stator, it will certainly increase the flow loss of next blade due to low energy mixing with mainstream. When the upstream rotor wake impinges upon the leading edge of the downstream stator, these low energy fluid are dissipated into blade boundary layer, which has little influence on mainstream, so the flow loss of next vane change little. As a result, when the upstream rotor wake impinges upon the leading edge of the down stream stator, the corresponding stage efficiency will be higher; when the upstream wake is transferred into the mid passage of the downstream stator, the corresponding stage efficiency will be lower.

Secondary flow loss is the main part of turbine endwall loss, which is formed by flow tunnel turn, fluid viscosity and traverse pressure gradient of adjacent blade concavo-convex side. The stagnant particle in endwall boundary suffers transverse pressure gradient and then flow from blade abdomen to blade backside. At the same time, reverse flow of mainstream come into being and finally make secondary flow form. According to literature, blade height is the most important factor to endwall[2][3]。So two hollow planes are used to change the blade length and gain different blade height in experiment. The distance between hollow planes are 30mm、60mm、90mm, Fig.7 shows blade outlet velocity flow vector respectively.

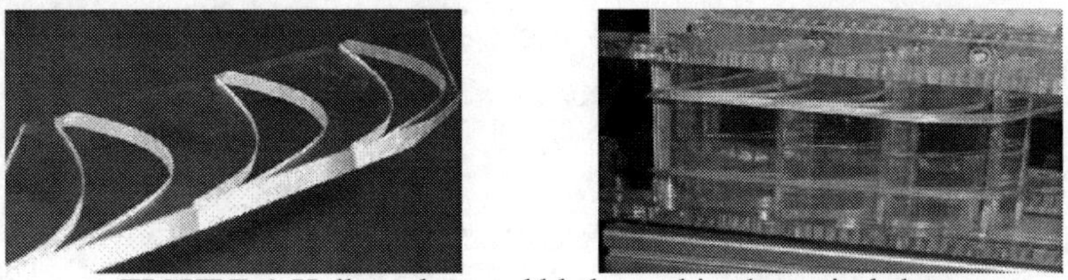

FIGURE 6. Hollow plane and blade combined practical chart

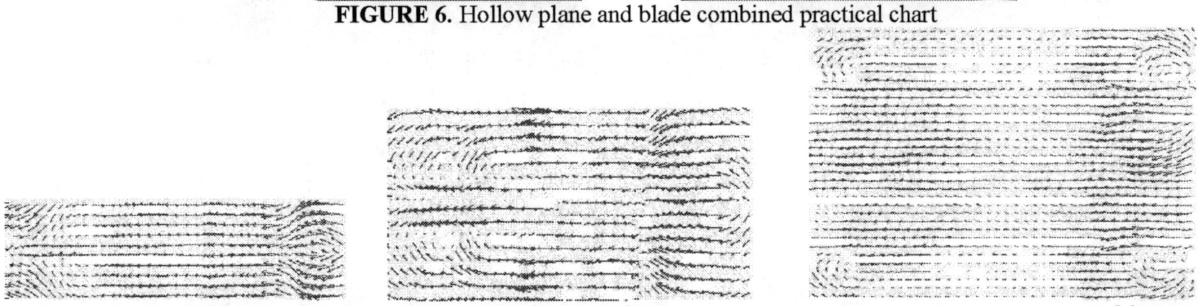

30mm 60mm 90mm

FIGURE 7. PIV experiment result of different blade height

The result shows that endwall secondary flow converges to biggest vortex region at smallest blade height (30mm), while secondary flow loss increase obviously and has large intensity. While at largest blade height (90mm), the secondary vortex of upper and bottom endwall separate, which posses less portion of endwall loss and has small intensity. So we can draw the conclusion that endwall loss has linearly inverse proportion to blade height.

HOTWIRE EXPERIMENT AND RESULT ANALYSIS

Single slanted hotwire measures three-dimension velocity field by its magnitude and direction sensitivity. When hotwire rotates towards its axial, the function of hotwire probe as following:

$$V_e(\gamma)=f(U,V,W,\gamma_i) \qquad i=1,2,3...n \quad (n\geq 3) \tag{1}$$

If hotwire probe rotate n degree, equation (1) constitute n non-linear equation group. Because rotation angle γ_i is known, equation group has three unknown parameters U ,V ,W . Hotwire probe only need rotate three positions for solving the equation group. In the experiment, hotwire probe rotate towards its axes by six degree position: 0° ,60° ,120° ,180° ,240° ,300° to calculate the three-dimension mean velocity value, the addition three rotation angle are used to increase measurement veracity.

Fig.8 show relationship between the single slanted probe coordinate and absolute coordinate. In single slanted probe coordinate, Jorgensen equation [4] should be obeyed:

$$V_e^w = U_N^2 + k_\alpha^2 U_T^2 + k_\beta^2 U_B^2 \tag{2}$$

U_T ,U_N ,U_B are velocity portion descripted in Fig.8; k_α is pitching coefficient; k_β is deflective coefficient.

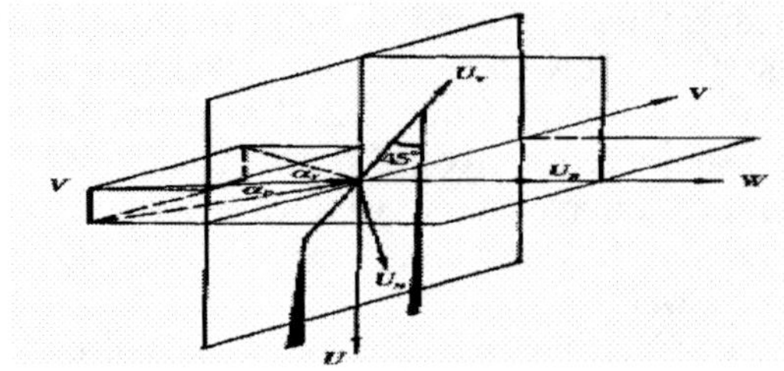

FIGURE 8. Single slanted probe coordinate(U_N, U_T, U_B) and absolute coordinate(U, V, W)

The velocity portions relationship between probe coordinate and absolute coordinate can be expressed by expression(3), in the meanwhile single slant angle is 45°, γ is single slanted probe rotate angle towards itself axes from initial position along anticlockwise, in this paper $\gamma = 60^\circ$.

$$\left.\begin{aligned} U_N &= U\sin 45^\circ + (V\cos\gamma - W\sin\gamma)\cos 45^\circ \\ U_T &= -U\cos 45^\circ + (V\cos\gamma - W\sin\gamma)\sin 45^\circ \\ U_B &= W\cos\gamma + V\sin\gamma \end{aligned}\right\} \quad (3)$$

(3) is taken into (2): $V_e^2 = A_1U^2 + A_2V^2 + A_3W^2 + A_4UV + A_5UW + A_6VW$. In the meanwhile, reference [1] proposes pitching coefficient $k_\alpha = 0.2$ and deflective coefficient $k_\beta = 1.2$.

$$\left.\begin{aligned} A_1 &= 0.5 + 0.5k_\alpha^2 \\ A_2 &= 0.5\cos^2\gamma(1 + k_\alpha^2) + k_\beta^2\sin^2\gamma \\ A_3 &= 0.5\sin^2\gamma(1 + k_\alpha^2) + k_\beta^2\cos^2\gamma \\ A_4 &= \cos\gamma(1 - k_\alpha^2) \\ A_5 &= \sin\gamma(k_\alpha^2 - 1) \\ A_6 &= \sin 2\gamma(-0.5 - 0.5k_\alpha^2 + k_\beta^2) \end{aligned}\right\} \quad (4)$$

Different output voltage can be achieved when hotwire rotate towards its axial, so effective cooling velocity V_e can be settled by (1). The procedure for solving the mean three-dimension velocity vector involving the least-squares technique can be accomplished via the LSQNONLIN optimization function of Matlab.

This paper uses DANTEC CTA (Constant Temperature Anemometer). Hotwire probe is 55P12 single slanted hotwire. Streamline unit, Lightweight displacement unit, UEI2WIN30/DSA/D board and PC constitute the high-speed data disposal system. Fig.9 give hotwire measurement system frame. In the meanwhile, transverse module unit make the stepping motor rotate probe towards its axes to measure the effective cooling velocity under different angles. Fig.10 gives hotwire system experiment practical photo.

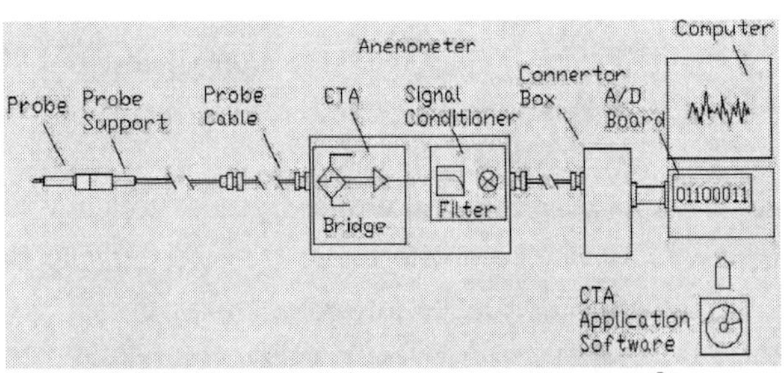

FIGURE 9. Hotwire measurement system frame

FIGURE 10. Hotwire system experiment photo

Two different inlet flux (transducer is 10Hz at low speed, 50Hz at high speed) and three different blade height (30mm, 60mm, 90mm) are experimented. Fig.11 and Fig.12 show circumferential velocity (v) and axial velocity (w). From Fig.11, wake region is sharper and obvious at faster velocity (25Hz), while it is smoother at slower velocity (10Hz). Because inlet flux (pneumatic load) is the key factor to blade traverse pressure gradient, which finally decide the endwall loss. On the other hand, the circumferential velocity is lowest at smallest blade height (30mm). Because circumferential velocity will increase along with blade height, then top and bottom secondary flow converge a bigger vortex region. The obvious increase of secondary flow destroys the initial flow structure, leading the increase of mainstream compensation to satisfy the continuum equation, whose direction is reverse to mainstream, so it decrease the traverse flow velocity of mean section plane.

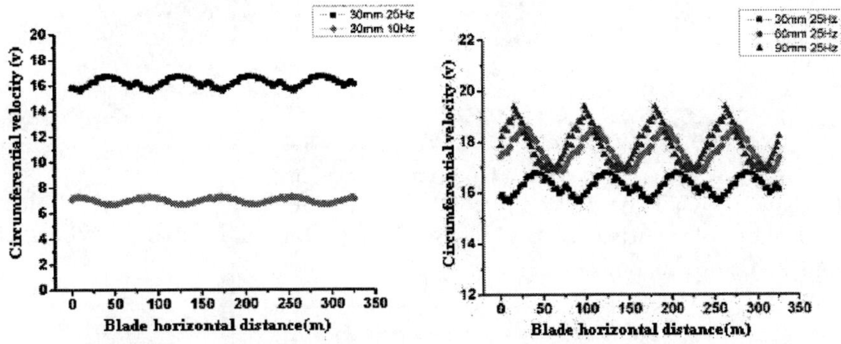

FIGURE 11. Circumferential velocity (v) in blade outlet plane

Fig.12 show the axial velocity (w). Wake region is also sharper and more intensified under larger inlet flux. Under the same inlet condition, smallest distance between hollow planes (30mm) get highest outlet velocity, because secondary flow is more severe at smallest blade height, the blade loss increase sharply and destroy the order flow structure to change the flow direction and increase the axial velocity perpendicularity to blade outlet plane.

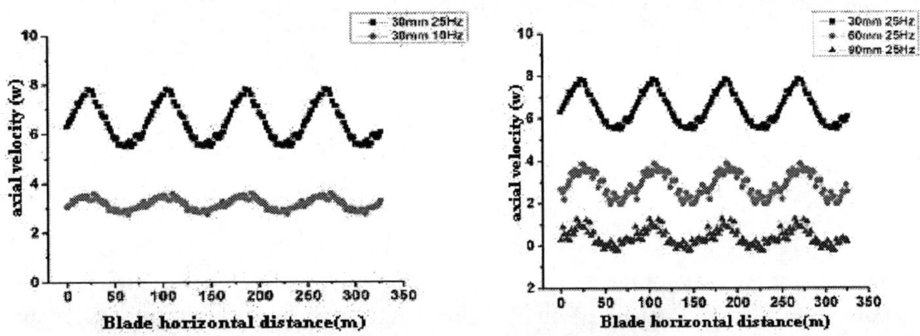

FIGURE 12. Axial velocity (w) perpendicular to blade outlet plane

Radial flow velocity is shown in Fig.13. From the point of view, radial flow velocity at of blade top and bottom outlet plane is bigger and secondary flow is stronger at high inlet flux and little blade height.

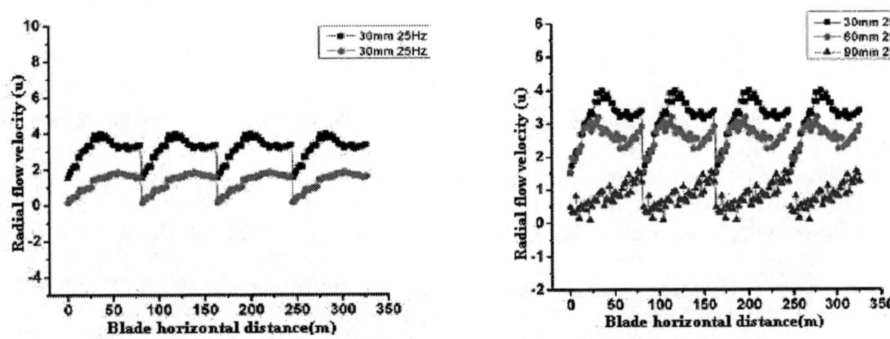

FIGURE 13. Radial flow velocity (u) in blade outlet plane

We can draw the conclusion that the decrease of blade height and the increase of pneumatic (inlet flux) will both enhance the blade secondary loss. Under this circumstance, blade traverse pressure gradient increase is the main reason, which deserves more attention on endwall loss under this circumstance.

CFD RESULT

Numerical simulation is more and more important in project design, optimization and exploration. This paper uses CFD software FLUENT to analyze the effect of unsteady phenomenon on blade boundary layer of 3-D blade model. Fig.14 shows 3-D entity model and grid.

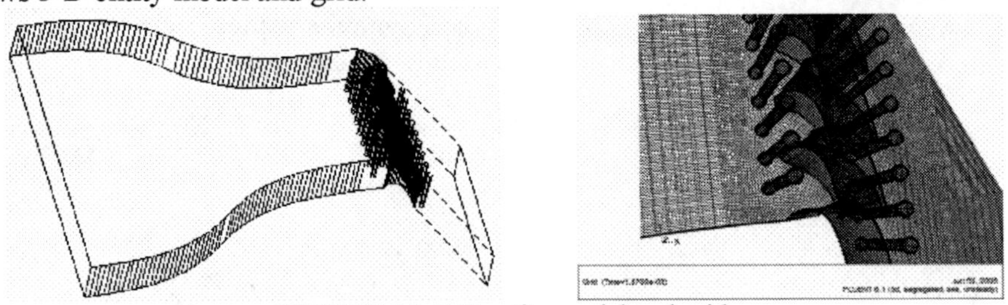

FIGURE 14. 3-D entity model and grid

Fig.15 and Fig.16 show the whole field velocity chart and pressure distribution curve of blade middle part at one cylinder position. Data show blade wake have obvious effect on pressure curve of cylinder behind. Pressure of pressure side is larger than that of suction side.

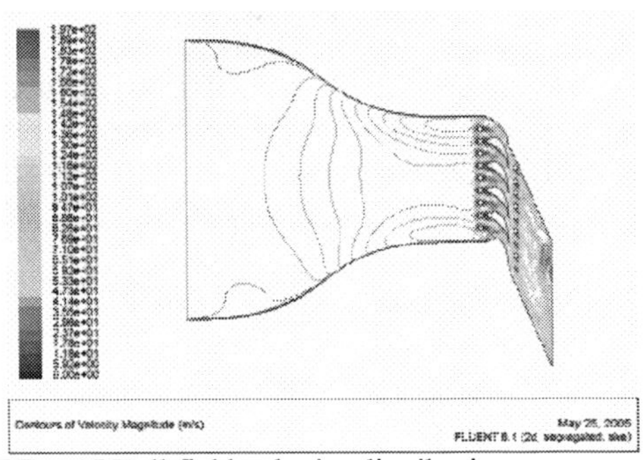

FIGURE 15. Full field velocity distribution

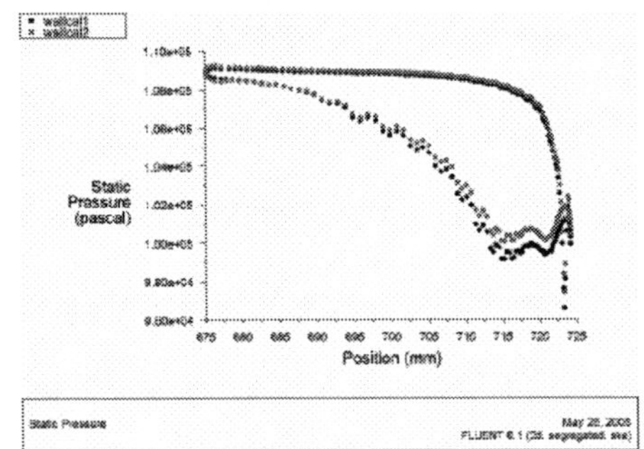

FIGURE 16. Blade surface static pressure distribution

CONCLUSIONS

Hotwire and PIV are used to measure the entire velocity field along blade height and outlet plane. It is found that outlet secondary flow is obvious and severe at small blade height and tip and bottom secondary vortexes changes under different blade height. In addition, when the upstream rotor wake impinges upon the leading edge of the downstream stator, the corresponding stage efficiency will be higher; when the upstream wake is transferred into the mid passage of the downstream stator, the corresponding stage efficiency will be lower. Experiment result gives effective reference to pneumatic performance of turbine blade.

REFERENCES

1.Reda R.Mankbadi, 1989, A study of unsteady rotor-stator interactions [J], ASME JOURNAL OF Turbomachinery, Vol. 111, pp. 394-400

2. T.Korakianitis, 1992, On the prediction of unsteady forces on gas turbine blades: part 1- Description of the approach [J] ASME JOURNAL OF Turbomachinery, Vol. 114, pp. 114-131

3. O.P. Sharma, G.F.Pickett, 1992, Assessment of unsteady flows in turbines [J], ASME JOURNAL OF Turbomachinery, Vol. 114, pp. 79-90

4. V.Schulte, H.P.Hodson, 1998, Unsteady wake induced boundary layer transition in high lift LP turbine [J], ASME JOURNAL OF Turbomachinery, Vol. 120, pp. 28-35

Experimental Study on Capture of PM_{10} Emitted from Coal Combustion with High Gradient Magnetic Field

Duanfeng Lu, Changsui Zhao†, Xin Wu, Yongwang Li, Song Han

Key Laboratory of Clean Coal Power Generation and Combustion Technology of Ministry of Education, Southeast University, Nanjing 210096, China, 05498119@163.com

Abstract. Experiments on capturing PM10 emitted from coal combustion with high gradient magnetic field were carried out for the first time. A new fluidized bed aerosol generator was developed for generating fly ash aerosol with diameter less than 10μm constantly. The variation in the particle number concentration caused by high gradient magnetic field was measured with the Electrical Low Pressure Impactor (ELPI). Fly ash particles from combustion of three kinds of coal were sampled. Particle saturation magnetic moment is 0.37emu/g, 1.25emu/g, 2.00emu/g respectively. The results show that for the particles in the size range of 0.1μm~10μm particle capture efficiency varies from 25% to 40%. The particle with either larger or smaller size has higher capture efficiency, and the particle with medium size (1μm~3μm) has lower capture efficiency. The particle capture efficiency rises with increase in the particle magnetization, the magnetic field gradient and the filling ratio of ferromagnetic medium, and it reduces with increase in aerosol velocity. The present study indicates that high gradient magnetic separation is an effective way to control fine particle emission from coal combustion.

Keywords: PM10, high gradient magnetic field, coal combustion, ELPI.
PACS: 42.68.Jg

INTRODUCTION

Particulate matter air pollution is hazardous to the environment, can affect atmosphere visibility, and cause acid rain and photochemical fog. The particulate matter easily adsorbs acid matter, organic compound and germina etc. It is harmful to people with cardiovascular morbidity, cancer and respiratory morbidity (Dingenen et al., 2004; Niemi et al., 2004; Long et al., 2003).

Fly ash particles emitted from coal combustion are the important particulate matter pollution sources. The particles contain many trace elements such as Hg, As, Se, Pb, Cu, Zn, Cd. The smaller the diameter of fly ash is, the more the relative enrichment of trace elements is (Constance et al., 2000; Huang et al., 2003). PM_{10}, which are hardly removed by standard dust separation device, are much harmful to human health, especially PM2.5. As a developing country, China is an important producer and consumer of energy source (Zhou and Zhu, 2005). And the structure of energy sources, given priority to coal, makes particulate matter pollution issue more serious. It is urgent to develop new method to control inhalable particle emission.

The basic principle of high gradient magnetic separation is as follows: an additional magnetic field occurs around the ferromagnetic medium existing in the uniform magnetic field, and a high gradient magnetic field comes into being by reciprocating between the additional magnetic and the external magnetic field, then magnetized particles are captured by the medium under the magnetic force when they pass through the high magnetic field (Luborsky, 1975; Svoboda, 2001). High gradient magnetic separation (HGMS) has the advantage of high recovery of weakly magnetic particles (Oberteuffer, 1973, 1974; Svoboda, 1987). Metal oxides, such as ferric oxides, are main components of fine particles emitted from coal combustion (Zhao et al., 2006; Shan and Lu, 2005). The content of ferric oxides in fly ash is normally about 4.1%~23.5% (Zhang et al., 2005; Chen et al., 2004). Ferric oxides in fly

† To whom correspondence should be addressed.
E-mail: cszhao@seu.edu.cn

CP914, *Multiphase Flow: The Ultimate Measurement Challenge,* edited by X. Cai, Y. Wu, Z. Huang, S. Wang, and M. Wang

ash exist in forms of α-Fe_2O_3, γ-Fe_2O_3 and Fe_3O_4 (Sun. et al., 2005). Both γ-Fe_2O_3 and Fe_3O_4 can be magnetized to possess high saturation magnetic moment. At this point, using magnetic separation technique to control emission of fine particles emitted from coal combustion is a valuable method. HGMS is used in industry and environment engineering field (Kenmoku and Sakata, 1984; Karapinar, 2003), but reports on capturing fly ash particles in high gradient magnetic field are rare.

The information on experiments of capturing fly ash particles in high gradient magnetic field is given in this paper. The effects on capture efficiency by particle magnetization, magnetic flux density, aerosol velocity and fill ratio of ferromagnetic media are studied. The results show that high gradient magnetic separation is an effective way to control fine fly ash particle emission from coal combustion.

EXPERIMENTAL

Apparatus

As shown in Fig.1, the experimental system consists of an aerosol generator system, a high gradient magnetic field and a measuring system. Aerosol generator is based on fluidized bed principle, which is developed by our laboratory, for the purpose of generating a constant output of inhalable particles. Bed material in the fluidized bed aerosol generator is stainless steel beads with diameter of 100μm ~120μm. Fly ash particle mixed with bed material is added into the fluidized bed. The aerodynamics characteristic of fly ash particles is strongly different from that of bed material because substantial differences on their size and density. Because of different terminal velocities for fly ash and bed material, fly ash particles can be carried out by fluidizing gas through controlling the gas velocity. In this experiment, fluidizing gas is pure nitrogen, and the diameters of generated aerosol particles are smaller than 10μm. The aerosol channel across the high gradient magnetic field is quadrate with dimension of 1200mm*50mm*50mm. Ferromagnetic media wires in the form of grid are fixed inside the channel. The distance between two adjacent wires is 1.4mm. The magnet used in the experiment is electromagnet with variable air gap and two yokes. Measuring apparatus is electrical low pressure impactor produced by Dekati Ltd. Finland, which detects particle concentration through charging, inertial classification, and electrical detection. ELPI primarily consists of a corona charger, low pressure cascade impactor and multi-channel electrometer. It can detect the variation in particle concentration and distribution in real-time (Keskinen et al., 1992).

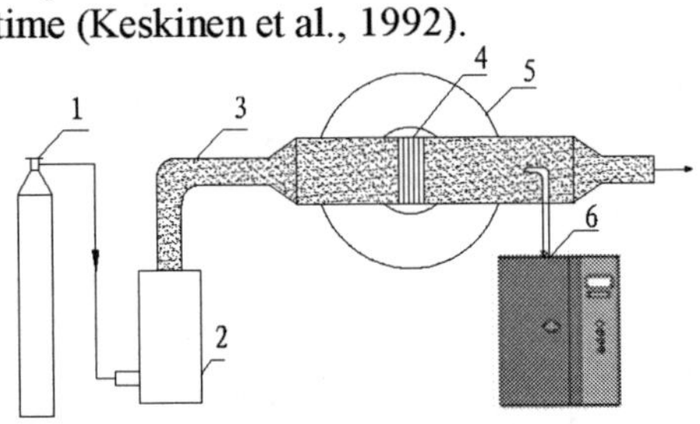

FIGURE 1. Schematic diagram of experimental system
1. N_2 cylinder 2. aerosol generator 3. experimental channel 4. Ferromagnetic medium 5. electromagnet 6. ELPI

Materials

Three kinds of fly ash particles emitted from combusting different bituminous coal, which are respectively from Dongsheng(S1), Datong(S2), and Xuzhou(S3), were sampled. Magnetic characters of the samples are shown in Fig.2. The samples are saturatedly magnetized while the external magnetic flux density is up to 500mT, and the saturation magnetic moments are 2.00emu/g, 1.25emu/g, and 0.37emu/g, respectively. After magnetization, all the samples have weak remanence and coercive force. Fig.3 shows the aerosol particle size distributions of three samples produced by aerosol generator. q in the figure is the ratio of the particle number in each size range to total particle number. Particles with diameter of 0.1μm~1μm are major part of sample S1, while particles with diameter 0.3μm~3μm are major part of sample S2 and S3. The mean diameter is 0.45μm, 0.53μm, and 0.82μm , respectively.

Method

Firstly, the samples are dried and mixed with bed material in the mass ratio of 10%. Then, the mixture is added into the aerosol generator. Subsequently, the valve of N_2 cylinder is turned on to control gas velocity. ELPI is turned on to detect the concentration of fly ash particles. Finally the electromagnet can be switched on to carry out the experiments of capturing fly ash particles after the fly ash aerosol concentration becomes steady. By varying correlative parameters, including magnetic flux density(B), aerosol velocity(v) and fill ratio of ferromagnet medium(R_f), the corresponding capture efficiency can be attained.

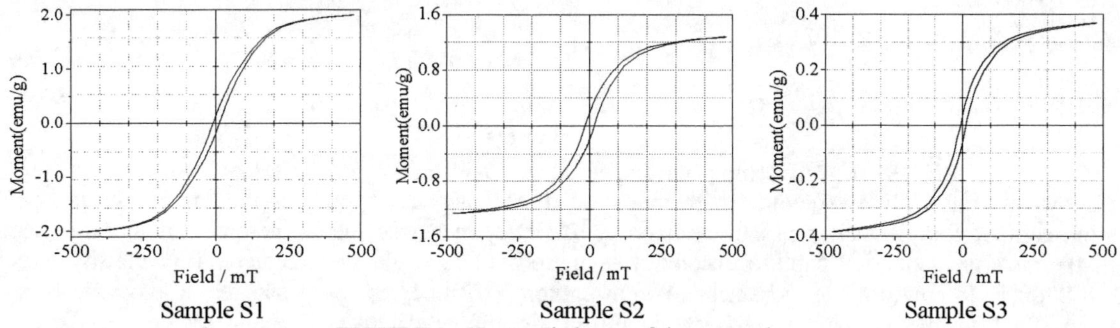

FIGURE 2. Hysteresis loops of three samples

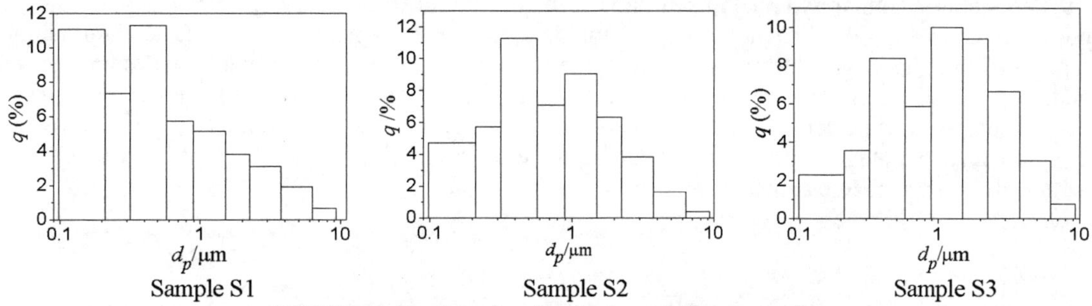

FIGURE 3. Particle size distribution of three samples

RESULTS AND DISCUSSIONS

Time series of particle capture efficiency for sample S1 are shown in Fig.4. Particle concentration decreases gradually with time after electromagnet is turned on. The concentration decreases slowly in the first 5 seconds and then decreases rapidly. Finally it becomes steady in 25 seconds. Particles with diameter between 1μm and 3μm have comparatively low capture efficiency while the other particles have high capture efficiency. The same results have been gained in the experiments using other two samples under the same condition. Fly ash particles are magnetized when they pass through the magnetic field. Magnetized particles are captured by magnetic media under the influence of magnetic force, drag force and Brown force. The magnetized particles tend to agglomerate especially among small particles and big particles. At the same time, the big particles are influenced greatly by magnetic force which dominates particle movement in the magnetic field. So the capture efficiency for both big particles and small particles is comparatively high. For example, the capture efficiency is 34% for particle with diameter of 8.08μm and 30% for particle with diameter of 0.20μm, but it is 25% for particle with diameter of 1.94μm.

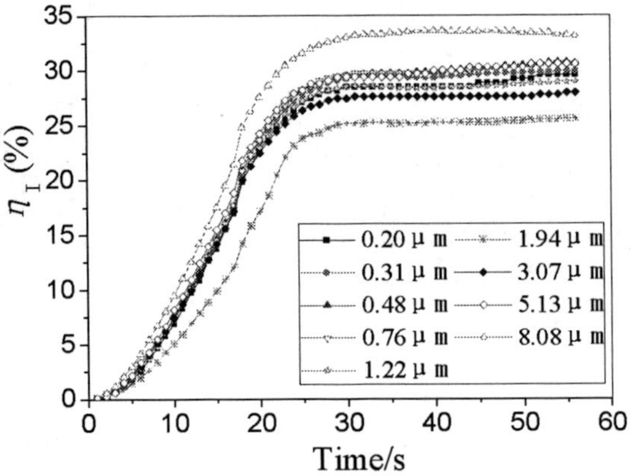

FIGURE 4. Time series of capture efficiency (B=600mT, v=0.068m/s, R_f=2.4%)

Variations in total capture efficiency with magnetic flux density for three samples are shown in Fig.5. With increase in magnetic flux density, total particle capture efficiency increases and its increment reduces gradually. For example, the increment of total capture efficiency is merely 1.5% while the magnetic flux density varies from 680mT to 800mT. In contrast, the increment of total capture efficiency is about 8% while magnetic flux density varies from 80mT to 200mT. Under the same condition, the higher the particle magnetization is, the higher the capture efficiency is. Capture efficiency of sample S1 is the highest. That of sample S2 is next. That of sample S3 is the lowest. This status is similar with that of saturate magnetization for three samples.

Variation curves of capture efficiency for different size particles with magnetic flux density for sample S1 are shown in Fig.6. While the magnetic flux density is in the range of 0mT~500mT, fly ash particles are not saturatedly magnetized. On this occasion, both the magnetic field gradient and magnetization of fly ash particles rise with increase in the magnetic flux density. It makes capture efficiency increase remarkably. While the magnetic flux density is more than 500mT, fly ash particles are saturatedly magnetized. And the capture efficiency increases slightly with further increase in magnetic flux density.

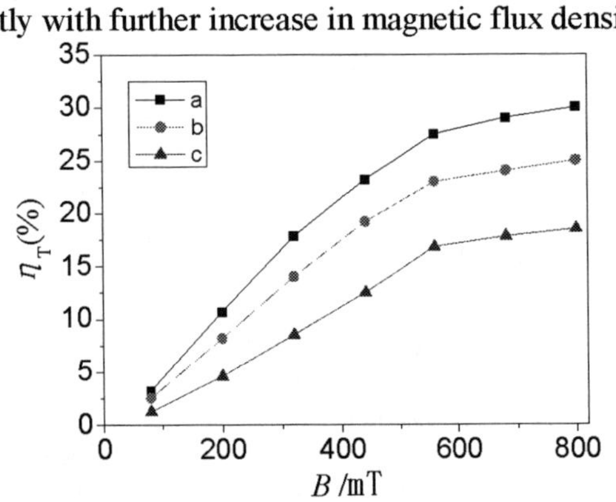

FIGURE 5. Effect of magnetic flux density on total capture efficiency(v=0.068m/s, R_f=2.4%)

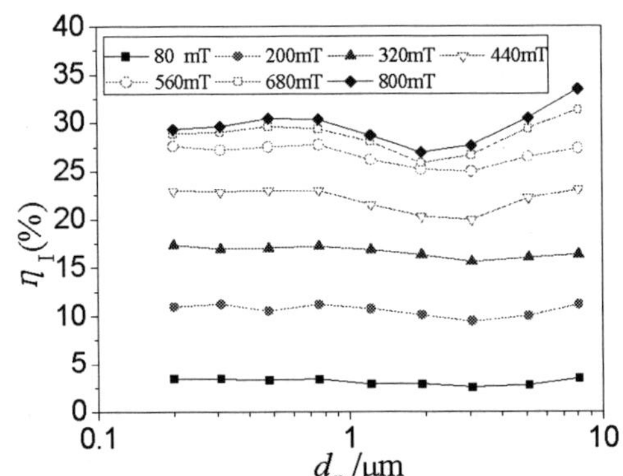

FIGURE 6. Effect of magnetic flux density on capture efficiency for different size particles(v=0.068m/s, R_f=2.4%)

Variation curves of total capture efficiency with aerosol velocity for three samples are shown in Fig.7. For the test section with a certain length slow aerosol velocity means long residence time for fly ash particles, which has an active effect on capture efficiency. While aerosol velocity goes up, the residence time reduces and the drag force acting on particles by transportation gas increases. It results in reducing particle capture efficiency remarkably. As shown in the figure, the capture efficiency decreases by 14%~16% while aerosol velocity varies from 0.045m/s to 0.1m/s.

Variation curves of capture efficiency for different size particles with aerosol velocity for sample S1 are shown in Fig.8. The particle diameter corresponding to the lowest capture efficiency tends to shift larger with aerosol velocity increasing. For example, the particle diameter with the lowest capture efficiency shifts from 1.22μm to 1.94μm while the aerosol velocity varies from 0.045m/s to 0.1m/s.

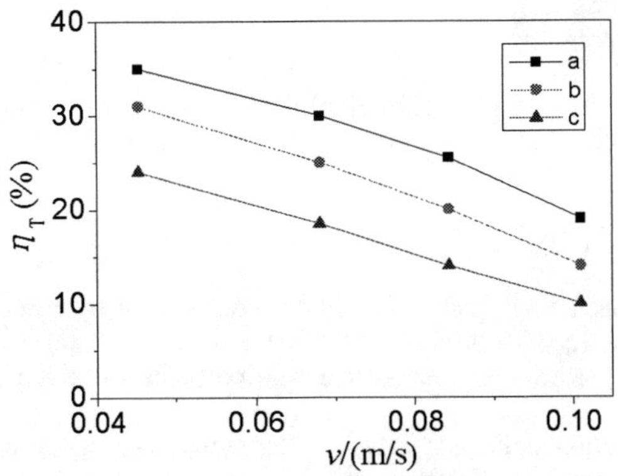

FIGURE 7. Effect of gas velocity on total capture Efficiency (B=800mT, R_f=2.4%)

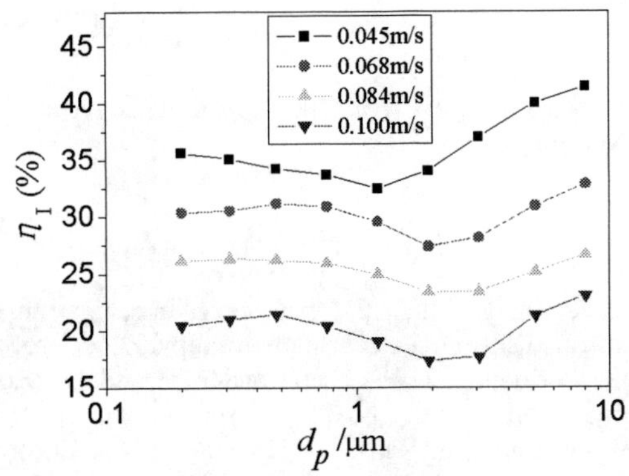

FIGURE 8. Effect of gas velocity on capture efficiency for different size particles(B=800mT, R_f=2.4%)

Variation curves of total capture efficiency with magnetic medium fill ratio R_f for three samples are shown in Fig.9. With R_f increasing, the space for capturing fly ash particles is enlarged to improve particle capture efficiency. The capture efficiency increases by 18%~21% while R_f varies from 0.56% to 2.24%. But the increment of capture efficiency gradually reduces with increase in R_f. Variation curves of capture efficiency for different size particles with R_f for sample S1 are shown in Fig.10. With R_f increasing, the increment of capture efficiency for particle with diameter between 1.0μm and 3.0μm is less than that for other particles. For example, capture efficiency increment for particle with diameter of 1.22μm is 18% while R_f varies from 0.56% to 2.24%, but it is 25% for particle with diameter of 8.08μm at the same condition.

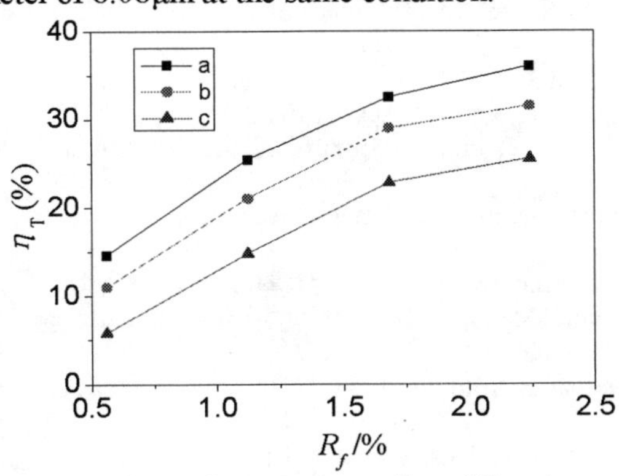

FIGURE 9. Effect of magnetic medium fill ratio on total capture efficiency (B=800mT, v=0.045m/s)

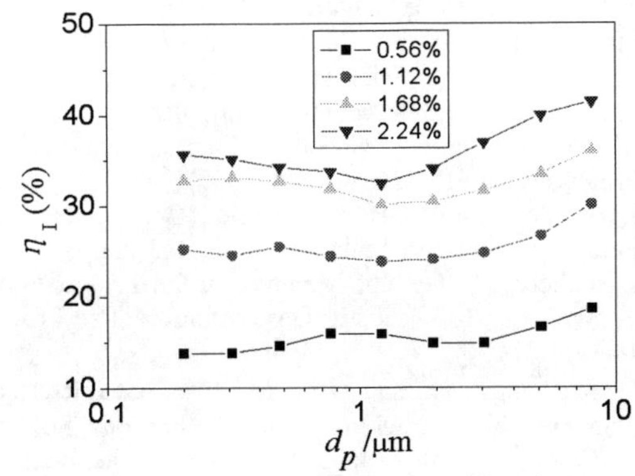

FIGURE 10. Effect of magnetic medium fill ratio on capture efficiency for different size particles (B=800mT, v=0.045m/s)

CONCLUSIONS

In high gradient magnetic field, fly ash particles are magnetized and captured by ferromagnetic media. After the electromagnet is turned on, the particle concentration decreases slowly in the first 5 seconds and decreases rapidly whereafter. Finally it becomes steady in 25 seconds. High magnetic flux density causes high magnetic field gradient, which has an active effect on particle capture efficiency. Under the same condition, the higher the particle moment is, the higher the particle capture efficiency is. Either low aerosol velocity or high fill ratio of ferromagnetic media has an active effect on particle capture efficiency. The capture efficiency for particles with diameter between 1.0μm and 3.0μm is lower, and it is higher for other particles with smaller or larger size. Particle capture efficiency rises with increase in the magnetic flux density, the particle magnetization and the fill ratio of ferromagnetic media, and reduces with increase in aerosol velocity. The results show that high gradient magnetic separation is an effective way to control fine ash particle emission from coal combustion. And further research is highly desirable for improving the capture efficiency.

ACKNOWLEDGMENTS

Financial support from the National Key Basic Research Program of China (No. 2002CB211600) was sincerely acknowledged.

REFERENCES

1. Y. Z. Chen, N. Shah, E. Frank etc., "Investigation of primary fine particulate matter from coal combustion by computer-controlled scanning electron microscopy," Fuel Processing Technology, 85, 743(2004)
2. L. S. Constance, J. H. Joseph and F. S. Adel., "Emissions of mercury, trace elements, and fine particles from stationary combustion sources," Fuel Processing Technology, 65, 263 (2000)
3. R.V. Dingenen, F. Raes, J. P. Putaud etc. "A European Aerosol Phennomenology-1: Physical Characteristics of Particulate Matter at Kerbside Urban, Rural and Background Sites in Europe," Atmospheric Environment, 38, 2561(2004)
4. Y. J. Huang, B. S. Jin, Z. P. Zhong etc., "Emission Features of Trace Elements in A Pulverized Coal Boiler," Proceedings of the CSEE, 23, 205(2003)
5. N. Karapinar, "Magnetic separation of ferrihydrite from wastewater by magnetic seeding and high-gradient magnetic separation," Int. J. Miner. Process., 71, 45(2003)
6. Y. Kenmoku and S. Sakata, "HGMS Technique Application to Environment Improvement Resulting from Material Recycling System(I)-Recovery and Recycling of Glass Grinding Sludge," IEEE Transaction on Magnetics, 20, 1198(1984)
7. J. Keskinen, K. Pietarinen And M. Lehtimaki, "Electrical Low Pressure Impactor," Aerosol Sci.,23, 353(1992)
8. R. W. Long, N. L. Eatough, N. F. Mangelson etc., "The measurement of PM2.5, Including Semi-volatile Components in the EMPACT Program: Results from The Salt Lake City Study," Atmospheric Environment, 37, 4407(2003)
9. F. E. Luborsky., "High Gradient Magnetic Separation: Theory Versus Experiment," IEEE Transactions on Magnetics, 11, 1696(1975)
10. J. V. Niemi, H. Tervahattu, H. Vehakamaki etc., "Characterization and Source Identification of A Fine Particle Episode in Finland," Atmospheric Environment, 38, 5003(2004)
11. J.A. Oberteuffer, , "High gradient magnetic separation," IEEE Trans., 3, 303(1973)
12. J.A. Oberteuffer, , "Magnetic separation: A review of principles, devices, and applications," IEEE Trans. 2, 223(1974)
13. H. D. Shan and S. G. Lu, "Mineral Magnetism of Power-plant Fly Ash and Its Environmental Implication," ACTA Mineralogica SINICA, 25, 141(2005)
14. J. M. Sun; Q. YAO, H. Y. Liu . etc., "Microstructure and mineral characteristics of ferruginous spheres in fly ash," Journal of Fuel Chemistry and Technology, 33, 263(2005)
15. J. Svoboda, , "Magnetic Methods for the Treatment of Minerals," Developments in Mineral Processing, 8(1987)
16. J. Svoboda, "A Realistic Description of The Process of High Gradient Magnetic Separation," Minerals Engineering, 14, 1493(2001)
17. C. F. Zhang， Q. Yao and J. M. Sun., "Characteristics of particulate matter from emissions of four typical coal-fired power plants in China," Fuel Processing Technology, 86, 757(2005)
18. Y. C. Zhao., J. Y. Zhang, Q. Gao etc., "Chemical Composition and Evolution Mechanism of Ferrospheres in Fly Ash from Coal Combustion," Proceedings of the CSEE, 26, 82(2006)
19. Q. F. Zhou, Y. H. Zhu., "Energy situation in China based on statistical review of world energy in 2005," Energy of China, 27, 40(2005)

Experimental investigation on agglomeration of coal-fired PM_{10} in uniform magnetic field

Yongwang Li, Changsui Zhao*, Xin Wu, Duanfeng Lu, Song Han

Key Laboratory of Clean Coal Power Generation and Combustion Technology of Ministry of Education, Southeast University, Nanjing 210096, China

Abstract. Agglomeration experiments on fly ash particles in the size range of 0.023−9.314 μm were conducted in a uniform magnetic field. The fly particles were produced from combustion of bituminous coal originated in Dongshen, China. A dedicated fluidized bed aerosol generator was developed to disperse particles to generate aerosol with constant rate. The aerosol particles from the generator underwent agglomeration during passing through the magnetic field. The variation in particles number concentration induced by agglomeration was measured in real time by Electrical Low Pressure Impactor (ELPI). The effects of the particle size, the magnetic flux density, the particle residence time in the magnetic field, the total particle mass concentration and the average gas velocity on particle agglomeration were examined. Experimental results indicate that the removal efficiencies for the midsized particles are higher than those for the smaller and the bigger ones. Single-sized and total particle removal efficiencies can be enhanced by rasing the magnetic flux density, the total particle mass concentration, the particle residence time in the magnetic field or reducing the average gas velocity. When particles are satuatedly magnetized, the magnetic flux density has no effect on their agglomeration.

Keywords: Fine particles; Coal combustion; Agglomeration; Magnetic field; Removal efficiency
PACS: 42.68.Jg

INTRODUCTION

Recently, hazards caused by suspended particulate matter (SPM), especially by particles smaller than 10μm (PM_{10}), are getting a matter of environmental concern. PM_{10} emission sources such as coal-fired power plants contribute significantly to the ambient aerosol loading in the atmosphere [1–3]. Because of their tiny size, light weight and enormous number, conventional dust removal facilities are little effective on the very fine fly ash particles from coal combustion. There are two ways to remove the fine fly ash particles from flue gas, one is to remove the particles directly with new-style facilities, another is to aggregate the particles, make the particles large enough first, and then remove them by conventional facilities. Particle agglomeration is the process of particle merging induced by their collision. Iron oxide is one of the main components constituting fly ash particle [4, 5], so the particle can be magnetized easily in magnetic field [6]. In a magnetic field, the magnetized particles come into close proximity and adhere together under the inter-particle magnetic dipole force and other primary forces such as drag, Brownian, gravity and van der Walls. Thus, agglomeration of the fine fly ash particles in a magnetic field is a promising method of removing the fine fly ash particles from flue gases. The investigation of particle agglomeration in magnetic field has become one of the National Basic Research Programs of China.

There are some studies done on particle collision and agglomeration in magnetic fields. Smoluchowski [7] investigated the agglomeration of airborne particles in static gas medium, and developed the Brownian agglomeration model for evaluating the agglomeration coefficient between two interacting particles. Based on the Brownian agglomeration model, Svoboda [8] considered the flocculation of weakly magnetic quasi-colloidal particles in suspensions, using an inter-particle force potential that includes electrical double layer, van der Walls and magnetic force. Tsouris and Scott [9] carried out flocculation experiments of paramagnetic particles under the

* Correspondence author. Tel.: +86 25 83793453; fax: +86 25 83793453. *E-mail address:* cszhao@seu.edu.cn (C.S. Zhao)

CP914, *Multiphase Flow: The Ultimate Measurement Challenge,* edited by X. Cai, Y. Wu, Z. Huang, S. Wang, and M. Wang

influence of a strong magnetic field. They found that flocculation has a strong dependence on particle size and on magnetic susceptibility. Chin *et al.* [10] studied the flocculation kinetics of mixtures of magnetite and polystyrene in a stirred tank under turbulent shear flow conditions. The results suggested that the flocculation rate was enhanced by increasing the agitation speed.

Although several studies have been carried out on the agglomeration of magnetic particles, most of the studies were dealing with magnetic particles with bigger sizes in liquid medium. To better understand the agglomeration properties of fine fly ash particles from coal combustion in uniform magnetic fields, and to provide a solid basis for the application of the magnetic field agglomeration technology, this paper presents agglomeration experiments on fine fly ash particles in the size range of 0.023−9.314 μm in a uniform magnetic field. The effects of particle size, magnetic flux density, particle residence time in the magnetic field, total particle mass concentration and average gas velocity were examined.

EXPERIMENTS

Lab-scale setup and procedure

Figure 1 is a schematic diagram of the experimental system for study particle agglomeration. It consists of two HEPA filters, a fluidized bed aerosol generator, an electromagnet and a measurement apparatus. One of the two HEPA filters filtered the nitrogen flowing into the fluidized bed aerosol generator and the dilutor, while the other was used to filter the aerosol coming from plexiglas tube placed between two magnetic heads.

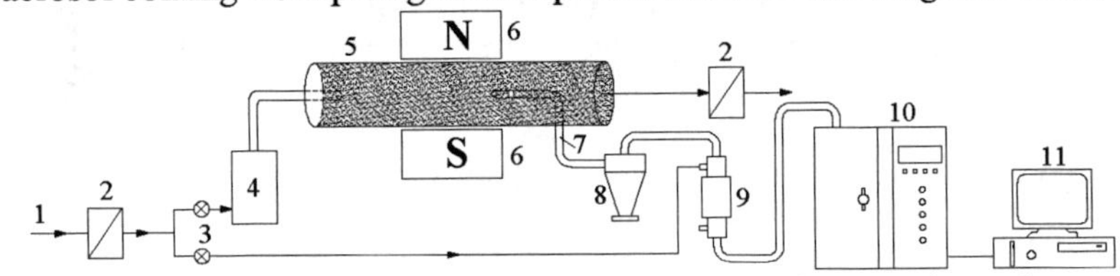

FIGURE 1. Schematic diagram of magnetic agglomeration experimental system.

1.Nitrogen 2.HEPA filter 3.Flowmeter 4.Fluidized bed aerosol generator 5.Polycarbonate tube 6.Magnetic heads 7.Sampling probe 8.Sampling cyclone 9.Dekati dilutor 10.Electrical Low Pressure Impactor (ELPI) 11.PC for data acquisition

The fluidized bed aerosol generator [11], shown in Figure 2, is a relatively simple apparatus for dispersing fly ash particles and generating aerosol. It consists of a dense bed, large particles (bronze beads) mixed with a solid powder (in this case, fly ash). Nitrogen gas passing through the bed from the nitrogen distributor causes the bed to fluidize. The bronze beads move and mix freely, grinding up the fly ash agglomerates. The smaller fly ash particles subsequently are entrained in the nitrogen flow, while the heavier bronze beads and the larger fly ash particles settle back to the bed. The fresh mixtures of bronze beads/fly ash are fed to the fluidized bed continuously through a vertical tube which is connected to the feed reservoir. In order to ensure that the bed maintains a relatively constant height, an overflow outlet is located at the top of the bed so that excess mixtures can be withdrawn continuously. The concentration of fly ash particles in the aerosol coming out from the generator can be adjusted by changing the mass ratio of bronze beads to fly ash, nitrogen flow rate through the bed, or the feed rate of the mixture.

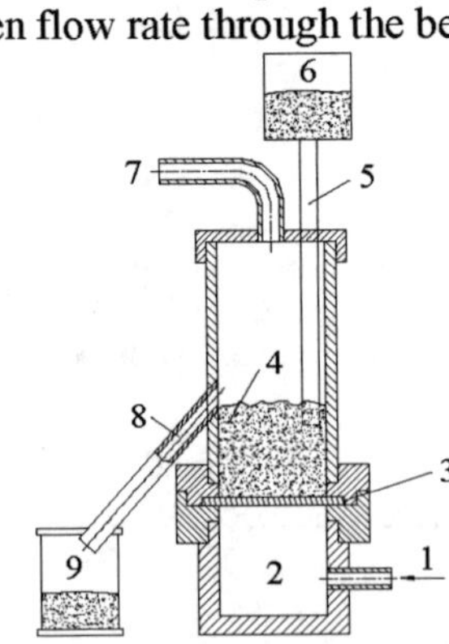

FIGURE 2. The fluidized bed aerosol generator.

1.Nitrogen 2.Nitrogen plenum chamber 3.Nitrogen distributor 4.Bed 5.Verticle tube 6.Fly ash/bronze reservoir 7.Aerosol outlet 8.Overflow outlet 9.Fly ash/ bronze excess

An electromagnet (SB-100, China) generates a uniform magnetic field between its two magnetic heads. The diameter of the two magnetic heads is the same 100 mm, and the maximum magnetic flux density is 1.4T. A plexiglas tube with an inside diameter of 50 mm was placed between the two magnetic heads as a passage for the aerosol particles. Agglomeration occurs inside this tube when the fly ash particles passing through the magnetic field.

An Electrical Low Pressure Impactor (ELPI, Dekati Ltd., Finland) was employed to monitor particle size distribution and concentration in real time with 13 stages (12 channels). Table 1 gives cut off diameter (D_{50}) of each stage and geometrical mean diameter (D_i) of each channel. The working principle of the ELPI is based on inertial classification of electrically charged particles and the determination of the number of particles according to the electrical charge carried by different sized aerosol particles. The sampling probe has a ferrite foil magnetic field shield so that the external magnetic field will have no effect on the particles inside the sampling probe. The sampling probe can move along the axial direction of the polycarbonate tube in order to change the distance from the boundary of magnetic field to the sampling point. To remove particles larger than 9.314 μm from the sample gas flow, a cyclone was connected between the sampler and the Dekati dilutor. The dilutor enables sampling from high particle concentration over a long period of time. The dilutor has a constant dilution ratio of 8.18.

TABLE 1. Impactor properties.

Stages	1	2	3	4	5	6	7	8	9	10	11	12	13
Aerodynamic D50% (μm)	0.023	0.030	0.050	0.098	0.211	0.317	0.576	0.891	1.505	2.243	3.758	6.285	9.314
Aerodynamic Di (μm)	0.026	0.039	0.070	0.144	0.259	0.427	0.716	1.158	1.837	2.903	4.860	7.651	

Fly ash particles aerosolized by the aerosol generator are transported into the plexiglas tube by gas flow. The agglomeration of the fly ash particles occur in the magnetic field. The number concentrations of the fly ash particles both before and after agglomeration process were measured by the ELPI on line

Particles

The fly ash particles used in this study were sampled from the hopper of the fourth electrical field of electrostatic precipitators (ESP) for tangential bituminous coal-fired utility boilers in Dalateqi Power Plant, China. The coal fired in the boiler is from Dongsheng, China. Fig. 3 shows SEM micrographs of the aggregates formed by the fly ash in the uniform magnetic field. It can be seen that the fly ash particles are basically spherical. Ash particles with different sizes aggregate together into large groups. Fig. 4 shows the original particle size distributions, whose number median diameters (d_{nm}) is 0.151μm. Fig.5 illustrates particles' hysteresis loops measured by the Vibrating Sample Magnetometer (Lake Shore, 7400). It indicates that particles exhibite soft ferromagnetic property with high magnetization and low coercivity. All experiments were carried out at room temperature.

FIGURE 3. SEM micrographs.

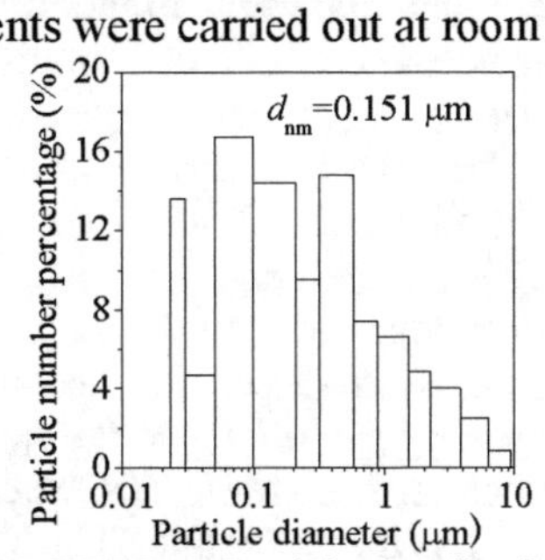

FIGURE 4. Particle size distribution.

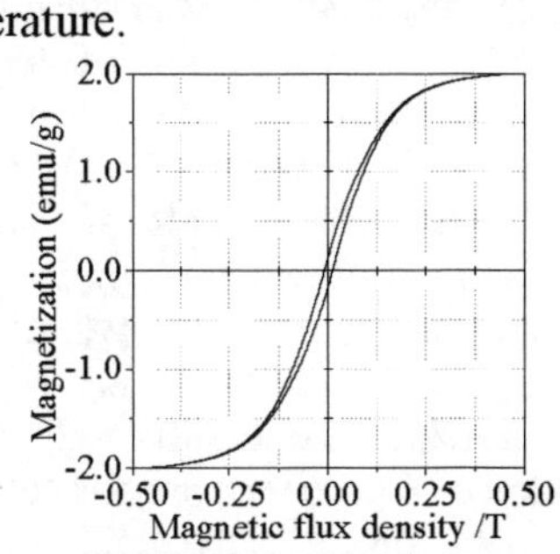

FIGURE 5. Particle hysteresis loops.

RESULTS AND DISCUSSIONS

Particle removal efficiency

Binary collision-agglomeration is the dominating mode of particle agglomeration, which results in variation of particle number concentration [12]. General Dynamic Equation (GDE) [13] can be used to predict the varying rate of particle number concentration:

$$\frac{\partial n(v,t)}{\partial t}=\underbrace{\frac{1}{2}\int_{v_{\min}}^{v}\beta(u,v-u)n(u,t)n(v-u,t)\mathrm{d}u}_{\text{aggregation gain}}-\underbrace{n(v,t)\int_{v_{\min}}^{v_{\max}}\beta(u,v)n(u,t)\mathrm{d}u}_{\text{aggregation loss}} \tag{1}$$

Where β (u, v) is the agglomeration coefficient (m^3/s) between two particles whose volumes are u and v. n (v, t) is the size distribution function. n $(v, t)\mathrm{d}v$ is the number concentration (particles/m^3) of particles in the volume range [v, v+dv] at time t. $v_{\min}$ is the volume of the smallest particles exists in the gas medium at time t, and $v_{\max}$ is the volume of the largest particles exists in the gas medium at the same time. The first term on the right-hand side of Eq. (1) represents the conversion rate of particles with volume u and volume v-u into aggregates of volume v. The second term represents the conversion rate of particles of volume v to particles with larger size through collisions with particles of all sizes in the system.

Removal efficiency (η_s) of single-sized particles is the percentage of decreasein its number concentration over a certain time interval. The total particle removal efficiency (η_t) is the percentage of decrease in total particle number concentration in the size range of 0.023−9.314 μm. Within a size range, the variation in particle number concentration is caused by the following four kinds of agglomerations:

(1) A particle in this size range aggregates with a smaller particle belonging to a smaller size range. The resulting new particle may remain in this size range or it may move to a size range for bigger particles.

(2) Two particles in the same size range aggregate. The resulting new particle may remain in this size range or it may move to a size range for bigger particles.

(3) A particle in this size range aggregates with a bigger particle from any larger size ranges. The resulting new particle moves to a size range for bigger particles.

(4) Two particles both from smaller particle size ranges aggregate. The resulting new particle adds to this size range.

Therefore, if agglomeration coefficient and particle number concentration do not change with particle size, the relative reduction in particle number concentration decreases with the increase in particle size. That is, the removal efficiencies of smaller particles are higher than that of the bigger ones.

The physical meaning of the agglomeration coefficient is an effective collision volume between two interacting particles per unit time [14]. For magnetized particles, the bigger the sizes are, the stronger the inter-particle magnetic dipole force between two particles is. Thus, the agglomeration coefficient increases as the particle size increases. Furthermore, the particle number concentrations of different sized particles are different (Fig.4). Fig.6(a) and Fig.6(b) show the effect of magnetic flux density on the single-sized and total particle removal efficiencies and particle number median diameter when particle residence time in the magnetic field (t), average gas velocity (u_f) and particle mass concentrations (C) are held constant. It is shown that the removal efficiencies of the midsized particles are higher than those of the smaller and the bigger particles (Fig.6 (a)), which was due to the combined effect of the agglomeration coefficients and the particle size distributions. Particle magnetization increases with an increase in magnetic flux density, until it reaches its saturation value. The increase in particle magnetization strengthens the inter-particle magnetic dipole force. Therefore, higher magnetic flux density would result in higher single-sized and total particle removal efficiencies, as shown in Fig.6 (a) and Fig.6 (b). When particle magnetization becomes saturated, further increase in magnetic flux density no long has effect on particle agglomeration and total particle removal efficiency, as shown in Fig.6 (b). Fig.6 (a) also shows that the particle size corresponding to the highest single-sized removal efficiency decreases slightly with the increase in particle magnetic flux density. That is also due to the combined effect of the agglomeration coefficients and the particle size distributions.

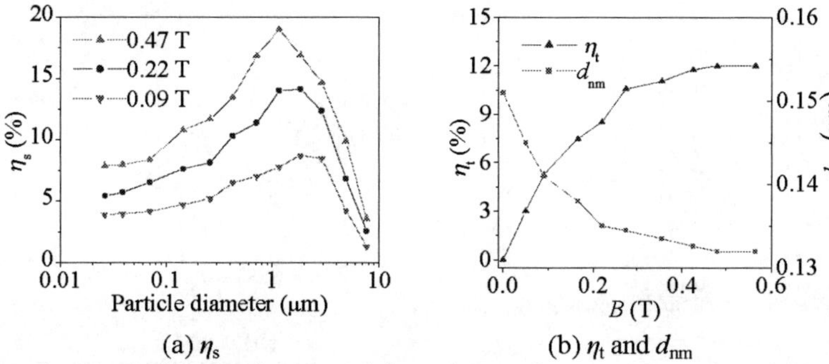

(a) η_s (b) η_t and d_{nm}

FIGURE 6. Effect of magnetic flux density on particle agglomeration at t=1.2 s and u_f= 0.08 m/s and C=0.96g/m^3.

Because the density of the fly ash particles is constant, higher total particle mass concentration corresponds to higher particle number concentration. Thus, increase in total particle mass concentration will increase particle collision frequencie. Therefore, particle removal efficiency will increase as the total particle mass concentration increases, as shown in Fig.7 (a) and Fig.7 (b).

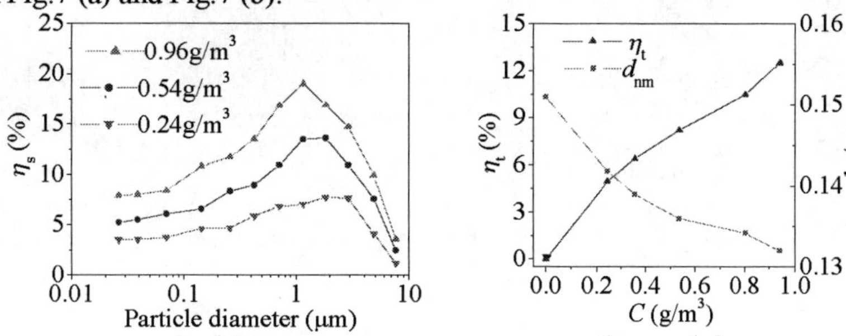

(a) η_s (b) η_t and d_{nm}

FIGURE 7. Effect of total particle mass concentration on particle agglomeration at B=0.47T, t=1.2 s and u_f=0.08 m/s.

Besides magnetic flux density and total particle mass concentration, particle residence time in the magnetic field and average gas velocity affect the agglomeration also. Keeping other factors in the agglomeration process constant, longer particle residence time in the magnetic field can lead to more collision, so that single-sized and total particle removal efficiencies will increase, as shown in Fig.8 (a) and Fig.8 (b). Increasing average gas velocity results in higher gas velocity gradient. Therefore, when particle residence time in magnetic field remains constant, shear-induced agglomeration is intensified as the average gas velocity increases [15]. As a result, particle removal efficiency increases, as shown in Fig.9 (a) and Fig.9 (b). However, the physical dimensions of the electromagnetic heads limit the maximum gas flow distance (equals $u_f \times t$) in the magnetic field. Thus, the maximum particle residence time in magnetic field decreases as the average gas velocity increases. Fig.10 (a) and Fig.10 (b) give the combined effect of average gas velocity and particle residence time in magnetic field on the single-sized and total particle removal efficiencies when the distance in which the aerosol traveled is a constant. It shows that the removal efficiency decreases with the increase in the average gas velocity. Moreover, Fig.8, Fig.9 and Fig.10 also indicate that the higher the total particle removal efficiency is, the smaller the particle size corresponding to the maximum particle removal efficiency is.

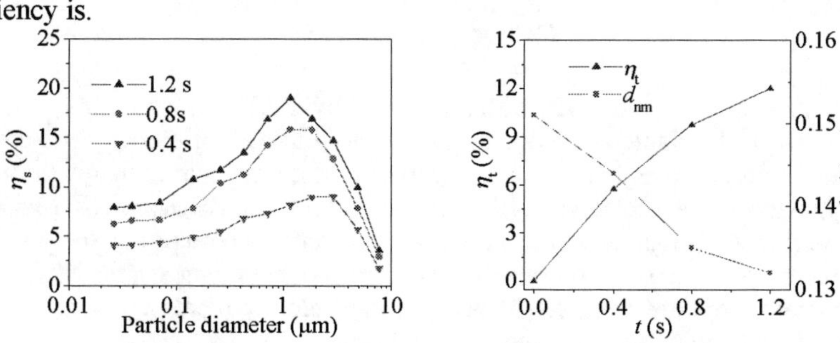

(a) η_s (b) η_t and d_{nm}

FIGURE 8. Effect of particle residence time in magnetic field on particle agglomeration at B=0.47 T, u_f=0.08 m/s and C=0.96g/m^3

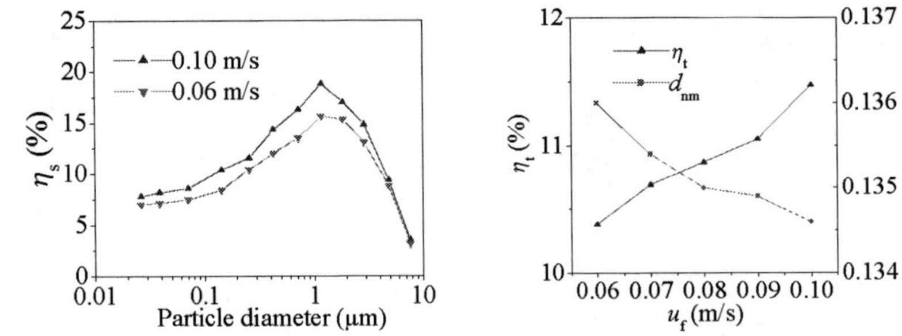

(a) η_s (b) η_t and d_{nm}

FIGURE 9. Effect of average gas velocity on particle agglomeration at B=0.47 T and t=1.2 s and C=0.96g/m^3

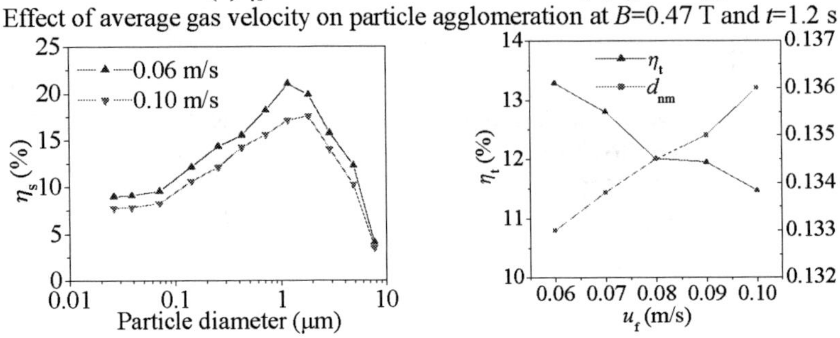

(a) η_s (b) η_t and d_{nm}

FIGURE 10. Combined effect of average gas velocity and particle residence time on particle agglomeration at B=0.47 T, $u_f \times t$=0.1 m and C=0.96g/m^3

Particle number median diameter

Because the removal efficiencies of different sized particles are different, the size distribution for the particles in the size range of 0.023-9.314 μm changes during agglomeration process. The number median diameter, d_{nm}, is a common way to describe particle size distribution, which means that 50% of the particles are smaller than the d_{nm} and 50% of the particles are bigger than the d_{nm}. Fig.6 (b) shows that particle number median diameters decreases with the increase in magnetic flux density, and it remaines constant when particle magnetization reachs saturation. Other effects, such as total particle mass concentration, particle residence time in magnetic field, average gas velocity, and the combined effect of the residence time and average gas velocity on particle number median diameter also support that conclusion, as shown in Fig.7 (b)-Fig.10 (b). The decrease in particle number median diameter induced by particle agglomeration indicates that bigger particles are easier to aggregate than the smaller ones. It can be concluded from these results that the number median diameter decreases with the increase in the total particle removal efficiency.

CONCLUSIONS

The experimental results show that the midsized particle removal efficiency is higher than those of the smaller and bigger particles. Moreover, the single-sized and total particle removal efficiencies increase with the increase in magnetic flux density, total particle mass concentration, particle residence time in magnetic field or the reduction of the average gas velocity. After particle magnetization reaches saturation, the magnetic flux density has no further effect on removal efficiency. Although higher average gas velocity can improve the single-sized and the total particle removal efficiencies at a constant particle residence time in the magnetic field, the maximum particle residence time in the magnetic field decreases as the average gas velocity increases, which results in the decrease in the maximum single-sized and total particle removal efficiencies. The particle size corresponded to the highest single-sized removal efficiency decreases with the increase in total particle removal efficiency. Particle number median diameter changes with particle agglomeration, and it decreases as the total particle removal efficiency increases. It implies that the bigger particles aggregate easier than the smaller ones.

Experimental results indicate that particle removal efficiency is lower than desired for removal with conventional facilities. Further studies are planed with the use of magnetic seed particles and lower average gas velocity and/or stronger magnetic field in order to improve particle agglomeration and increase particle removal efficiency.

ACKNOWLEDGMENTS

Financial support from the National Key Basic Research Program of China (No. 2002CB211600), Innovative Project for Graduate students in universities of Jiangsu province and Excellent Ph. D. Thesis of Southeast University were sincerely acknowledged. The authors also expressed sincere gratitude to Dr. Anthony, Leckner and Jia for presenting us valuable advices.

REFERENCES

1. C. Saskia and D. Z. Ven, *Atmospheric Environment* **32**, 3717-3729(1998).
2. Y. W. Li, G. L.Yang, F. G. Li, et al., *Opto-Electronic Engineering*, **32**, 54-58 (2005).
3. Y. L. Jin, G. L. He, F. Liu, et al., *Journal of Hygiene Research*, **31**, 342-348 (2002).
4. X. Guo, C. G. Zheng, T. Sun, *Journal of Combustion Science and Technology*, **11**, 192-195 (2005).
5. J. M. Sun, *Acta Mineralogical Sinica*, **21**, 14-18 (2001).
6. Y. C. Zhao, J. Y. Zhang, Q. Gao, et al., *Proceedings of the CSEE*, **26**, 82-86 (2006).
7. M. Smoluchowski, *Zeitschrift für Physikalische Chemie*, **92**, 129-188 (1917).
8. J. Svoboda, *IEEE Transactions on Magnetics*, **18**, 796-801 (1982).
9. C. Tsouris, T. C. Scott, *Journal of Colloid and Interface Science*, **171**, 319-330 (1995).
10. C. J. Chin, S. Yiacoumi, C. Tsouris, *Journal of Colloid and Interface Science*, **206**, 532-5454 (1998).
11. Y. W. Li, C. S. Zhao, X. Wu,et al., *Journal of Southeast University*, **35**, 742-745 (2005).
12. V. N. Piskunov, A. I. Golubev, *Journal of Aerosol Science*, **33**, 51-63 (2002).
13. V. N. Piskunov, A. I. Golubev, J. C. Barrett, et al., *Journal of Aerosol Science*, **33**, 65-75 (2002).
14. Y. Koizumi, M. Kawamura, F. Tochikubo, et al., *Journal of Electrostatics*, **48**, 93-101 (2000).
15. C. J. Chin, S. Yiacoumi, C. Tsouris, *Journal of Colloid and Interface Science*, **206**, 532-545 (1998).

Experimental Studies on the Measurement of Oil-water Two-phase Flow

Longbo Ma, Hongjian Zhang, Yuefang Hua, Hongliang Zhou

Department of Control Science and Engineering, Zhejiang University, Hangzhou 310027, Zhejiang, China

Abstract. Oil-water two-phase flow measurement was investigated with a Venturi meter and double-U Coriolis meter in this work. Based on the Venturi differential pressure and the quality of two-phase flow, a model for measuring oil-water mass flow rate was developed, in which fluid asymmetry of oil-water two-phase flow was considered. However, measuring the quality of two-phase flow on-line is rather difficult at present. Though double-U Coriolis meter can provide accurate measurement of two-phase flow, it can not provide desired respective mass flow rate. Therefore, a double-parameter measurement method with Venturi meter and double-U Coriolis meter is proposed. According to the flow rate requirement of Venturi, a new flow regime identification method based on Support Vector Machine (SVM) has been developed for the separated flow and the dispersed flow. With the Venturi model developed in this paper and mass flow rate of oil-water mixture measured with double-U Coriolis meter, mixture mass flow rate, oil mass flow rate and water mass flow rate could be obtained by the correlation. Experiments of flow rate measurement of oil-water two-phase flow were carried out in the horizontal tube with 25mm inner diameter. The water fraction range is from 5% to 95%. Experimental results showed that the flow regime could be identified well with SVM, and the relative error of the total mass flow rate and respective mass flow rate of oil-water two-phase flow was less than $\pm 1.5\%$ and $\pm 10\%$, respectively.

Keywords: Venturi meter; Water fraction; Water hold up; Oil-water two-phase flow; Double-U Coriolis meter
PACS: 07.90.+c

INTRODUCTION

Oil-water two-phase flow is common in the petroleum industry, and yet their measurement nearly always presents difficulties to the process engineer. The traditional method to measure the oil-water two-phase flow is to separate the mixture, and then to measure the individual phase flow rates through single-phase meters. However, the separation devices are very bulky, costly and can not monitor the oil well in real time. An alternative solution is the multiphase flow metering system, usually consisting of a combination of devices for phase fraction measurement and total flow rate measurement.

The relationship between differential pressure, quality and mixture flow rate should be known for measuring the flow rate by means of differential pressure. Usually, the differential pressure signal is obtained with throttle devices, such as orifice, Venturi, nozzle, and so on. In the last few decades, several equations based on differential pressure signals for gas-liquid mixtures have been developed and some typical equations have been proposed by Murdock [1], Chisholm [2], Lin [3] and Zhang[4, 5]. However, oil-water two-phase flow has not been explored to the same extent as the gas-liquid two-phase flow. Only limited research work on flow measurement has been done to such flow with throttle devices. Abdullah [6] carried out the flow measurement of two-phase flow oil-in-water using wedge meter and segmental orifice meter, and drew a conclusion that wedge meter and orifice meter were feasible flow-measuring devices for oil and water flow. Skea [7] investigated experimentally the effect of two-phase flow on the performance of Venturi meter. Therefore, it is necessary to do some research work about the application of throttle devices to flow measurement of oil-water two-phase flow. Compared with other kinds of throttle devices,

CP914, *Multiphase Flow: The Ultimate Measurement Challenge,* edited by X. Cai, Y. Wu, Z. Huang, S. Wang, and M. Wang

Venturi meter has little influence on flow regimes, the smallest pressure loss, and the shortest straight pipe upstream and downstream. As a result, Venturi meter was adopted to measure the oil-water flow in this paper.

Double-U Coriolis meter consists of a vibrating flowtube and an electronic transmitter to maintain vibration and to perform measurement calculations, which can make direct measurement of mass flow of fluid and give the density of fluid as an additional measurement. Fairly accurate flow measurement for single-phase can be provided with double-U Coriolis meter, usually to 0.2%[8]. However, only little (rather limited) research work has been carried out with double-U Coriolis meter. Liu [8] used a neural work to correct the mass flow measurement of two-phase flow, based entirely on internally observed parameters of double-U Coriolis meter, keeping errors to within 2%. Zhang [9] investigated the relative errors of mass flow rate of oil-water two-phase flow with double-U Coriolis meter, and developed a new calculation model of water fraction based on a dimensionless variable. In order to get better results with double-U Coriolis meter, it is essential to study the flow measurement of oil-water two-phase flow further.

Venturi meter has the characteristics of simple design, low cost and high reliability. However, the acquisition of some parameters of the Venturi model in real time is rather difficult. On the other hand, double-U Coriolis meter does well in the total mass flow rate measurement of oil-water mixture, but does badly in the respective mass flow rate measurement. In order to overcome the disadvantages of Venturi meter and double-U Coriolis meter, the aim of this investigation is to combine a Venturi meter with a double-U Coriolis to measure the total oil-water mass flow rate, and to develop a new measurement model from which the flow quality and respective mass flow of oil-water two-phase flow can be obtained simultaneously.

MEASUREMENT THEORY

Venturi Mode

In single-phase flow with thermal equilibrium, the mass flow rate and the pressure drop across the convergence section of the meter can be described with a homogeneous model

$$G_V = \frac{CA}{\sqrt{1-\beta}}\sqrt{2\Delta p \rho_V} \tag{1}$$

where G_V is the mass flow rate; C is the Venturi discharge coefficient; A is the area of the Venturi throat; β is the throat-to-pipe diameter ratio; Δp is the differential pressure between the upstream pressure and the throat; ρ_V is the density of the flowing fluid.

In two-phase flow, the two-phase mass flow rate and differential pressure can be expressed in the form of Eq.(1) if an appropriate two-phase fluid density is used in place of the single-phase fluid density. The homogeneous flow model treats the two-phase flow as if it were a single-phase flow. Using the homogeneous equilibrium model, which assumes that the oil and water have been mixed evenly and are in thermal equilibrium, the two-phase fluid density is given by

$$\frac{1}{\rho_V} = \frac{\chi_o}{\rho_o} + \frac{1-\chi_o}{\rho_w} \tag{2}$$

where ρ_V is the homogeneous density of oil-water mixture; χ_o is the quality of the two-phase flow, i.e. the ratio of the oil to total mass flow rate; ρ_o is the density of oil; ρ_w is the density of water.

In fact, the two-phase fluid is not homogeneous, so Eq.(2) can be amended as

$$\frac{1}{\rho_V} = \frac{\chi_o^n}{\rho_o} + \frac{1-\chi_o^n}{\rho_w} \tag{3}$$

where n is the modified coefficient of the average density.

By substituting Eq.(3) into Eq. (1), Eq. (1) can thus be written as

$$G_V = \frac{CA}{\sqrt{1-\beta^4}} K\sqrt{2\Delta p \rho_w} \tag{4}$$

where $K = \sqrt{\dfrac{1}{1+(\dfrac{\rho_w}{\rho_o}-1)\chi_o^n}}$ is coefficient; n is coefficient and dependent on the test condition.

Mixture mass flow rate can be calculated from Eq.(4) if χ_o and differential pressure Δp can be measured simultaneously. However, measuring the quality χ_o of the two-phase flow on-line is rather difficult at present and the applicability of Eq.(4) is affected by the flow regime in some extent. Alternative solutions have to be researched.

Double-U Coriolis meter

Double-U Coriolis meter can make direct measurement of oil-water two-phase mass flow, and the mass flow equation can be described as

$$G_C = \frac{k_s}{8r^2}\Delta t \tag{5}$$

where G_C is the mass flow rate; k_s is the angular spring constant of the flow tube; r is the radius of the U-bend of the tube; Δt is the corresponding time of phase deviation signal of double U-double-U Coriolis meter.

k_s and r are constants, so Eq.(5) can be written as

$$G_C = K_0 \Delta t \tag{6}$$

where K_0 is sensor-dependent calibration constant.

Double-U Coriolis meter can also measure the real density of oil-water two-phase flow, and the real density equation of liquid is given by

$$\rho_C = \frac{k_s T^2}{4\pi^2 V} - \frac{M_{\text{tube}}}{V} \tag{7}$$

where ρ_C is the real density of oil-water two-phase flow; T is the resonance period of the tube; M_{tube} is the mass of vibration tube; V is the volume of fluid in the tube. Moreover, k_s, M_{tube} and V are constants.

Therefore

$$\rho_C = K_1 T^2 - K_2 \tag{8}$$

where K_1 and K_2 are constants.

The real density can be converted to water hold up according to the following formula

$$Y = \frac{\rho_C - \rho_o}{\rho_w - \rho_o} \tag{9}$$

where Y is water hold up.

The water volume fraction is defined as

$$K_w = \frac{Q_w}{Q_w + Q_o} \tag{10}$$

where K_w is the water volume fraction; Q_w is the volume flow rate of water; Q_o is the volume flow rate of oil.

The water hold up Y is not equal to the water volume fraction K_w, because there exists the slip velocity between oil phase and water phase.

The total volume flow rate of oil-water two-phase flow can be derived from Eq.(6) and Eq. (8), and then the oil and water flow rates can be calculated individually if the water hold up is substituted for the water volume fraction. However, such method would result in large errors due to the existence of the slip velocity between oil and water. Thus, the double-parameter measurement method is proposed.

3. Double-parameter measurement method

According to the continuity equations, the relationship of mass flow rate of oil-water mixture can be derived as

$$G_V = G_C = G \tag{11}$$

where G is the total mass flow rate of oil-water mixture.

By substituting Eq. (11) for Eq.(4), Eq.(4) can thus be written as

$$\chi_o = \left[\frac{\dfrac{c\Delta P \rho_w}{G^2} - 1}{\dfrac{\rho_w}{\rho_o} - 1}\right]^{\frac{1}{n}} \tag{12}$$

where $c = \frac{2C^2 A^2}{1-\beta^4}$ is constant.

The mass flow rates of oil and water can be respectively described as

$$G_o = G\chi_o \tag{13}$$

$$G_w = G - G_o \tag{14}$$

where G_o is the mass flow rates of oil; G_w is the mass flow rates of water.

If the coefficient n of Eq.(12) is known, all the other parameters can be obtained after the acquisition of Δp and G. Therefore, the coefficient n is the key parameter, which is associated with the flow regime and water holdup of oil-water two-phase flow.

EXPERIMENTAL FACILITIES AND CONDITIONS

Experiments were carried out in the oil-water two-phase flow apparatus, as shown in FIGURE1. Oil and water were pumped separately from two storage tanks. Before they were mixed, the pressure of oil and water was stabilized with an oil-tank and a water-tank, and then the flow rate was measured individually. Gear meter was used to measure the volume flow of oil phase. If the density of oil was known, the actual mass flow of oil could be calculated. Electromagnetic flow meter was used to measure the volume flow of water phase. If the density of water was known, the actual mass flow of water could be calculated. After being measured, oil and water were mixed in the two-phase flow mixer shown in Fig.1, and then the mixture of oil-water flowed into a long flow developing section. The flow developing section was determined to be sufficient to obtain fully developed flow for the flow regimes in the experiment. After sufficiently mixing in the flow development, the mixture entered the Venturi meter and double-U Coriolis meter. The Venturi meter was mounted on a horizontal pipe. The diameter of the Venturi meter was 25mm with β =0.677. The differential pressure between the inlet and the throat of the Venturi meter was measured by Keller differential pressure cell connected to the Venturi . The output signals of the differential pressure cell were collected by a data collection card and computer . The sample frequency was 200Hz. Double-U Coriolis meter was used to measure the mass flow rate and the real density of oil-water mixture flow.

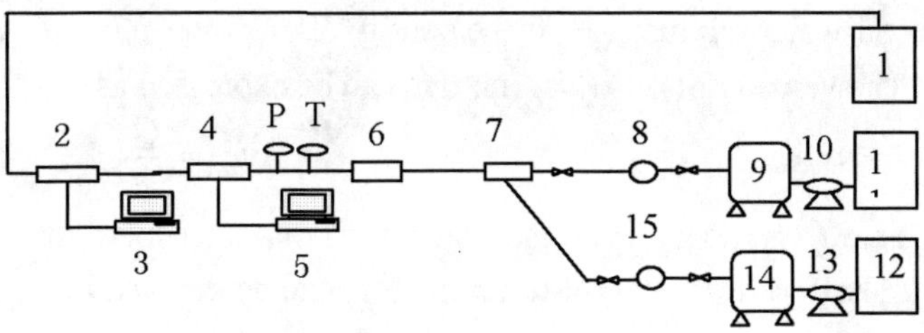

FIGURE 1. schematic diagram of experimental apparatus. 1.storage tank 2.coriolis meter 3.display device 4.venturi meter and differential pressure transmitter 5.computer 6.observation tube 7.mixing arrangement 8.electrimagnatic 9.water surge tank 10.water pump 11.water tank 12.oil tank 13.oil pump 14.oil surge tank 15.gear flowmeter

Experiments were performed at normal temperature. The test pressure of the oil-tank ranged from 0.3 to 0.4MPa and the test pressure of the water-tank was 0.4MPa. The oil flow rate was from 0.25 to 4.0 $m^3 \cdot h^{-1}$, while the water flow rate was from 0.2 to 6.51$m^3 \cdot h^{-1}$. The density of oil was 836kg·m^{-3} and the density of water was 998 kg·m^{-3}.

During each experiment, the oil flow rate was kept as a constant, and the water flow rate was gradually increased to change the total flow rate and water fraction. Data were acquired when the flow became stable.

In the experiments, data obtained are as follows: the actual volume flow rate of water, the actual volume flow rate of oil, real density of oil-water mixture, total mass flow rate of oil-water mixture and differential pressure signals of the Venturi. Because the density of oil and water was known, the actual mass flow rate of oil and water could also be calculated respectively.

FLOW REGIME IDENTIFICATION

Flow regime is an important parameter of oil-water two-phase flow, which not only reflects the hydrodynamic characteristics but also affects the coefficient C of Venturi model. Especially, the coefficient C can not be kept as a constant in some flow regimes. Therefore, it is important to identify the flow regime of oil-water two-phase flow.

Flow Regime Classification

The main flow regimes observed in horizontal oil-water two-phase flows can be classified as follows: stratified (S), stratified-mixed/oil layer (SMO), stratified-mixed/water layer (SMW), mixed-oil cont. (DO) and mixed-water cont. (DW) [10]. In order to measure the flow rate with Venturi tube better, the coefficient C should be kept as a constant. However, because of the low flow rate of oil-water mixture, the coefficient C is variable in the first three regimes, which is invariable in the last two ones. Therefore, the Venturi model fits DO and DW. According to the requirement of flow rate measurement, oil-water mixture was classified into two kinds of flow regimes: the separated flow (including S, SMO and SMW) and the dispersed flow (including DO, DW).

Relationship Between Flow Regime and Renolds

The Renolds of oil-water mixture can be described as

$$Re = \frac{dv_m \rho_m}{\eta_m} \tag{15}$$

where Re is the Renolds of oil-water mixture; d is the upstream internal diameter of Venturi tube; v_m is the velocity of oil-water mixture; ρ_m is the density of oil-water mixture; η_m is the viscosity of oil-water mixture.

The velocity of oil-water mixture can be expressed as

$$v_m = \frac{Q}{A} \tag{16}$$

where Q is the total volume flow rate of oil-water mixture; A is the internal diameter of the tube.

The density ρ_m of oil-water mixture can be denoted by the real density ρ_C measured with the double-U Coriolis meter . Therefore the total volume floe rate Q can be calculated from the following equation:

$$Q = \frac{G_C}{\rho_C} \tag{17}$$

Dukler et al. [11] for gas-liquid flows viscosity proposed averaging in terms of flow volume fractions of the phases. By applying the same principle in oil-water flows, the mixture viscosity η_m can be given by

$$\eta_m = (1 - K_w)\eta_o + K_w \eta_w \tag{18}$$

where η_o is the viscosity of oil; η_w is the viscosity of water.

If the real density ρ_C is measured by double-U Coriolis meter, Renolds of oil-water flow can be calculated from Eq.(15) to Eq.(18). Renolds distribution of separated flow and dispersed flow is listed in Table 1. It can be found that Renolds number is 0.23×10^4 to 1.01×10^4 for separated flow and 0.90×10^4 to 7.48×10^4 for the dispersed flow. From the above, there is an overlapped range of Renolds between separated flow and dispersed flow, and this indicates that the Renolds number can not distinguish separated flow from dispersed flow completely.

Table 1 Renolds Distribution of Oil-water Flow

Flow regime	Renolds number($\times10^4$)
Separated flow	0.23 ~ 1.01
Dispersed flow	0.90 ~ 7.48

Relationship Between Flow Regime and Differential Pressure Signals

The average differential pressure of the Venturi is

$$\Delta p = \frac{1}{n}\sum_{i=1}^{n} \Delta p_i \tag{19}$$

where Δp is the average differential pressure of the Venturi; n is the data length; Δp_i is the transient value of the differential pressure.

The root mean square (RMS) of the differential pressure signals can be described as

$$\delta p = \left[\frac{1}{n}\sum_{i=1}^{n} (\Delta p_i - \Delta p)^2\right]^{1/2} \tag{20}$$

where δp is defined as the pulse swing of the differential pressure, which reflects the fluctuating extent of the differential pressure signals.

R is defined as the normalized parameter of δp in the following

$$R = \frac{\delta p}{\Delta p} = \left[\frac{1}{n}\sum_{i=1}^{n}\left(\frac{\Delta p_i}{\Delta p} - 1\right)^2\right]^{1/2} \tag{21}$$

where R denotes the relative fluctuation of the differential signals.

FIGURE2. displays the relationship between R and Re. From FIGURE 2., it is noticed that with the variation of Re:

· At low Renolds of oil-water mixture ($Re < 0.90 \times 10^4$), R will rapidly decrease with an increase of Re because δp and Δp increase simultaneously, and the increment of δp is lower than that of Δp; At higher Renolds ($Re > 0.90 \times 10^4$), R is relatively unaffected by Re, because the increase of δp is nearly equal to that of Δp.

· The separated flow pattern prevails with R>0.04, while the dispersed flow pattern prevails as R<0.04.

· Near R=0.04, several data points of the separated flow pattern overlap that of the dispersed flow pattern, which complies with the Renolds number of Table 1.

As can be seen from the above, it is known that the parameter R can tell the separated flow from the dispersed flow but it can not exactly distinguish them. As it is with the Renolds number.

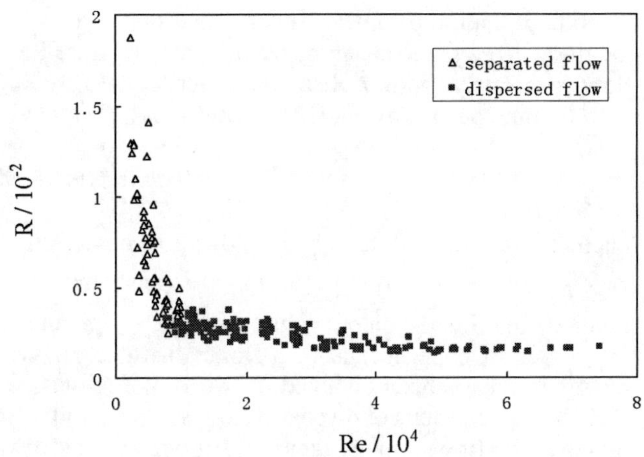

FIGURE 2 The variation trends of R vs. Reynold

Flow Regime Identification

According to the above analysis, neither Renolds nor the differential pressure pulse signals R can distinguish separated flow from dispersed flow completely. In order to identify the separated and the dispersed flow regime better, Support Vector Machine (SVM) has been used for the flow regime identification.

Support Vector Machine (SVM)

SVM has complete mathematical theory fundament and is an effective and general method for representing complex function in high dimensional space, which is suitable for flow pattern recognition *etal.* It has been used successfully in many traditional applications dominated by neural network.

Given a training set $\{x_i, y_i\}$, where i=1, 2, 3, …, N, $x_i \in R^n$ is the ith input vector with known binary target $y_i \in \{-1, +1\}$, the SVM classifier satisfies the following conditions

$$y_i[(\omega \bullet \varphi(x_i)) + b] - 1 \geq 0, \quad i=1,\ 2,\ 3,\ \ldots,\ N \tag{22}$$

where $\varphi : R^n \to R^m$ is the feature map mapping the input space to a usually high dimensional feature space where the data points become linearly separable by a hyperplane defined by the pair ($\omega \in R^m, b \in R$) [12]. The classification function is then given by

$$K(x_i, y_i) = \varphi(x_i)^T \varphi(x_i) \tag{23}$$

In order to allow for the violation of (22), slack variables ξ_i is introduced such that

$$y_k[\omega^T \varphi(x_k) + b] \geq 1 - \xi_k, \quad k = 1,2,3,\ldots, N, \ \xi_k \geq 0, \ k = 1,2,3,\ldots, N \tag{24}$$

and consider the following minimization problem

$$\mathrm{Min}\, J(\omega, \xi) = \frac{1}{2}\omega^T \omega + C\sum_{i=1}^{N} \xi_i \ , \tag{25}$$

subject to the constraints

$$\begin{cases} y_i[\omega^T \varphi(x_i) + b] \geq 1 - \xi_i, & i = 1,2,3,\ldots, N \\ \xi_i \geq 0, & i = 1,2,3,\ldots, N \end{cases} \tag{26}$$

where C is a positive constant parameter used to control the tradeoff between the training error and the margin; if $y_i = +1$, $y_i[\omega^T \varphi(x_i) + b] \geq 1$ comprises the given $\omega^T \varphi(x_i) + b] \geq +1$, if $y_i = -1$, $\omega^T \varphi(x_i) + b] \leq -1$. The binary classification problem can be posed by Eq. (25) and Eq. (26).

Feature Parameter Selection

Selection of feature parameter is very important when SVM is used for the flow regime identification. Though either the normalized parameter R or the oil-water mixture Renolds Re can not distinguish separated flow from dispersed flow completely, both of them can describe the flow regime well; In addition, the water holdup Y influences the flow regime to some extent [13]. Therefore, R, Re and Y are used as feature parameters.

Training and Identification Results

Schematic diagram of regime identification based on SVM is shown in FIGURE.3. As shown in Fig.3, Δp, G_C, ρ_C and d were used as the initial input values, R, Re, and Y as the input values of SVM after the character rization, and S as the output values of SVM. R, Re, and Y can be obtained through Eq.(21), (15) and (9) respectively.

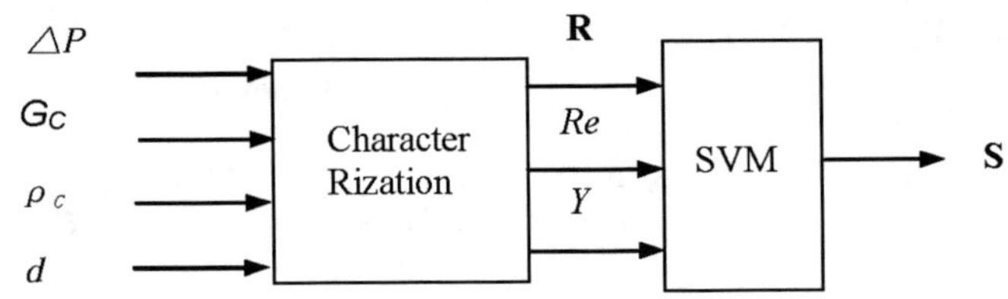

FIGURE 3. Schematic diagram of regime

239 groups of experimental data from oil-water two-phase flow have been obtained in two flow regimes, including 172 groups of dispersed flow and 67 groups of separated flow. Among them, 111 groups are used as the training data, the others as testing data. Through training and study, the flow regime identification results I with SVM are obtained, as shown in Table 2. From Table 2, it can be found that the identification fraction is 94.28 % for separated flow, 97.84 % for dispersed flow. In order to prove the universality of SVM, further training and study is done, in which the former testing data are regarded as the training data, the former training data as testing data. The flow regime identification results II with SVM are shown in table 3. As can be seen in Table 3, the identification fraction is 93.75 % for separated flow, 96.15 % for dispersed flow. The identification error between the two results came from the transitional area from the separated flow to the dispersed flow. During this area, the error occurred when the observed separated flow through the observation pipe was identified as the dispersed flow with SVM or when the observed dispersed flow was identified as the separated flow. These data resulting in the error have little influence on the flow rate measurement. Therefore, the flow regime identification results can be applied to flow rate measurement.

TABLE 2 Flow Regime Identification Results I

Flow regime	Identification fraction
Separated flow	94.28%
Dispersed flow	97.84%

TABLE 3 Flow Regime Identification Results II

Flow regime	Identification fraction
Separated flow	93.75%
Dispersed flow	96.15%

According to the above results, SVM really has better performance for identifying separated flow regime and dispersed flow regime. Feature parameters chosen have a good description of flow regimes information too.

FLOW RATE MEASUREMENT

The total mass flow rate of oil-water mixture measured by double-U Coriolis can be calculated from Eq.(6). The measurement relative error of the mixture mass flow rate can be obtained based on the measured value and the actual total mass flow rate which is calculated according to the standard flow meter. The distribution of the measurement relative errors vs. Renolds and water fraction are shown in FIGURE 4. and FIGURE 5. respectively. From the two figures, the relative error of the mass flow rate is less than $\pm 1.5\%$; as shown in FIGURE 4., the relative error deceases gradually with the increase of Renolds, which indicates that the Renolds influences the flow measurement of Coriolis meter to some extent; and from FIGURE 5., it can be seen that the water fraction has little influence on the relative error. According to the above analysis, the double-U Coriolis meter can provide accurate mass flow rate measurement of oil-water two-phase flow.

For the disperse flow identified by SVM, the total volume flow rate of oil-water two-phase flow measured by double-U Coriolis meter can be derived from Eq.(6) and Eq. (8), and then the oil and water mass flow rates can be calculated individually if the water holdup Y is substituted for the water volume fraction K_w.

The relative errors of oil mass flow rate G_o and that of water mass flow rate G_w are shown in FIGURE 6. respectively. As shown in FIGURE 6., with the increase of water fraction, the relative error decreases accordingly when water fraction is below 35%; and the value is less than ±8% with water fraction above 35%; on the contrary, the relative error of oil is less than ±8% when water faction is below 80%; moreover, with the increase of water fraction which is above 80%, the value increases. The above relative error is caused by the existence of slip velocity between oil and water.

In order to solve the above problem, double-parameter measurement method with the Venturi meter and the double-U Coriolis meter has been developed.

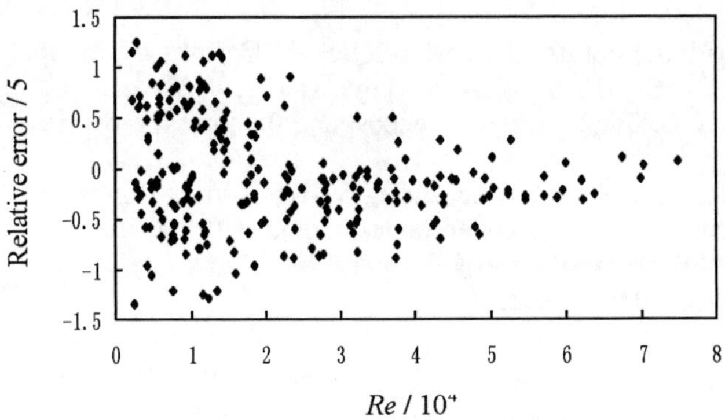

FIGURE 4. The relative error of total mass flow rate of Coriolis vs. Renolds

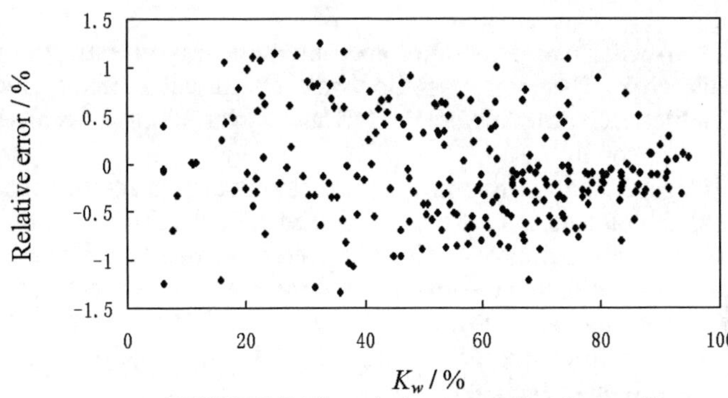

FIGURE 5. The relative error of total mass flow rate of Coriolis vs. water fraction

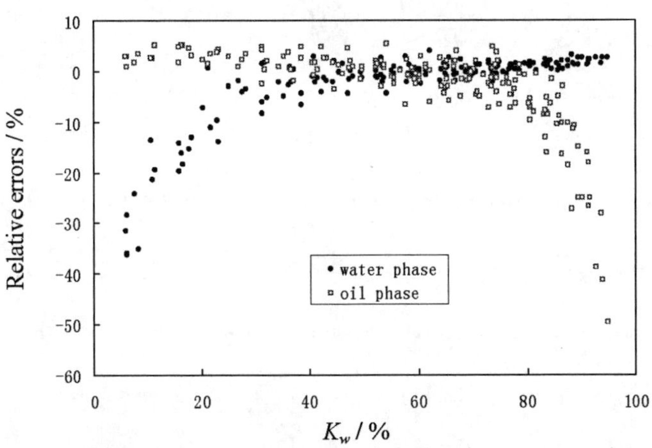

FIGURE 6. The relative error of oil and water respective flow rate calculated directly by water holdup vs. water fraction

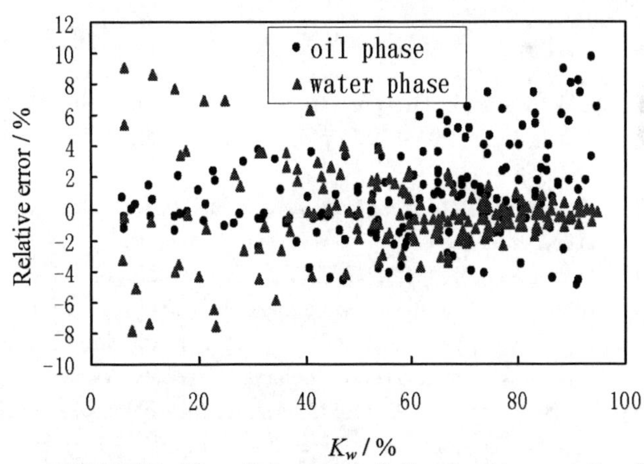

FIGURE 7. The relative error of oil and water respective flow rate calculated vs.water fraction

The actual value of K can be calculated from Eq.(4) according to the measured Δp and G, and the actual quality χ_o of the two-phase flow can be calculated from the actual value G_o and G_w of oil-water two-phase flow. Using the experimental data, the coefficient n of Eq.(12) can be found based on the actual quality χ_o and K, and n=0.95+0.1$\sqrt[3]{\chi_o}$. Using the interpolation method, the quality χ_o of the two-phase flow can be calculated from Eq.(6) and Eq.(12).

The mass flow rate G_o of oil and the mass flow rate G_w of water can be calculated with the dispersed data identified by SVM according to Eq.(13) and Eq.(14), respectively. The relative error of oil flow rate and water flow rate vs. water volume fraction are shown in FIGURE 7.. As shown in FIGURE 7., the relative errors of the calculated mass flow rate for both oil and water are all within $\pm 10\%$ except some measured values.

According to the above results, double-parameter measurement method with the Venturi meter and the double-U Coriolis meter has better performance for oil-water dispersed flow.

However, the double-parameter measurement method is not suitable for the measurement of separated flow, because the coefficient C of the Venturi is not a constant. Moreover, the flow rate of both oil and water is much lower, though the trend of relative error vs. water fraction is similar to that of the dispersed flow. Therefore, the water hold is substituted for the water volume fraction to calculate the respective mass flow rate of oil-water separated flow. As can be seen from the above analysis, the flow regime identification is very important in the measurement of oil-water two-phase flow.

CONCLUSIONS

Oil-water two-phase flow measurement was investigated with a Venturi tube and a double-U Coriolis meter in this work. The total mass flow rate of oil-water mixture was measured at relatively high accuracy by means of double-U Coriolis meter. Venturi meter was adopted because of its simple design, low cost and the lowest pressure loss especially.

Considering the two-phase fluid asymmetry, a revised model of Venturi for measuring oil-water mass flow rate was developed. Based on the revised model of Venturi, double-parameter measurement method with the Venturi meter and the double-U Coriolis meter has been developed. With the above method, the oil mass flow rate, water mass flow rate and the total mass flow rate of oil-water mixture could be calculated.

According to the flow rate requirement of Venturi, a new flow regime identification method based on SVM for the separated flow and the dispersed flow has been developed. With this method, the flow regime was identified with relatively high accuracy.

The above conclusions were derived from experiment results with oil-water two-phase flow. The two-phase flow rate can be measured on-line with the equations derived in this study. To prove their general validity, additional experiments with different pipe diameters, fluid densities and two-phase mixtures are required. This work may be instructive in further study of oil-water mixture measurement using Venturi and double-U Coriolis meter.

REFERENCES

1. J. W. Murdock, Two-phase flow measurements with orifices, Journal of Basic Engineering 84 (1962): 419-433.
2. D. Chisholm, Pressure drop during steam/water flows through orifices, Journal Mechanical Engineering Science 16 (1974): 353-355.
3. Z. H. Lin, Two-phase flow measurements with sharp-edged orifices, International Journal of Multi-phase Flow 8 (6) (1982): 683-693.
4. H. J. Zhang, S. J. Lu, G. Z. Yu, An investigation of two-phase flow measurement with orifices for low-quality mixtures, International Journal of Multiphase flow 18 (1) (1992): 149-155.
5. H. J. Zhang, W. T. Yue, Z. Y. Huang, Investigation of oil-air two-phase mass flow rate measurement using Venturi and void fraction sensor, Journal of Zhejiang University Science 6A (6) (2005): 601-606.
6. A. Buhidma, R. Pal, Flow measurement of two-phase oil-in-water emulsions using wedge meters and segmental orifice meters, The Chemical Engineering Journal 63 (1996): 59-64.
7. A. F. Skea, A. W. R. Hall, Effects of water in oil and oil in water on single-phase flowmeters, Flow Measurement and Instrumentation 10 (1999): 151-157.
8. R. P. Liu, M. J. Fuent, M. P. Henry, M. D. Duta, A neural network to correct mass flow errors caused by two-phase flow in a digital double-U Coriolis mass flowmeter, Flow Measurement and Instrumentation 12 (2001): 53-63.
9. X. G. Zhang, Q. M. Zhang, D. Wang, Z. H. Lin, Mass flowrate and water volume fraction measurements using double-U Coriolis flowmeter in oil-water two-phase flow, Journal of Engineering Thermophysics (in Chinese) 25 (6) (2004): 983-985.
10. P. Angeli, G. F. Hewitt, Flow structure in horizontal oil-water flow, International Journal of Multiphase Flow 26 (2000): 1117-1140.
11. A. E. Dukler, M. Wicks, R. G. Cleveland, Pressure drop and hold-up in two-phase flow, American Institute of Chemical

Engineers Journal 10 (1964): 38-51.

12. K. S. Chua, Efficient computations for large least square support vector machine classifiers, Pattern Recognition Letters 24 (2003): 75-80.
13. T. J. Wu, L. J. Guo, W. H. Liu, X. M. Zhang, An experimental study on flow patterns and their transitions for oil-water two-phase flow in horizontal pipes, Journal of Engineering Thermophysics (in Chinese) 23 (4) (2002): 491-494.

Estimation of Liquid Wall and Interfacial Shear Stress in Horizontal Stratified Gas-liquid Pipe Flow

Yiping Liu, Hua Zhang, and Jing Wang

Institute of Thermal Physics of Engineering, School of Mechanical Engineering, Shanghai Jiaotong University, Shanghai, China

Abstract. A modified two-phase shear stress calculation method for pipe flow problems is explored. A force balance has been set up on the control volume of liquid phase to determine the interfacial friction factor by employing both the measured pressure gradient and liquid height. The gradient of height of liquid layer has been taken into account, which is suitable for the case where the interface may be smooth, rippled or wavy. The correlation of model indicates that the careful estimation for liquid-wall shear stress is necessary, and the assumption of a stationary liquid element is not applicable for the case of higher gas flow rates. The interfacial friction factor evaluated indirectly from experimental liquid height and pressure loss measurements, which are obtained in 50mm ID pipeline for air and water in cocurrent stratified flow, is used to achieve its correlation with the combination of characteristic parameters. The evaluation of new correlation has been conducted by the comparison of the predicted pressure drop with the experimental data. The performance of correlation depends on the form of the gas-liquid interface.

Keywords: Gas-liquid pipe flow; Friction factors; Interfacial shear stress.
PACS: 47.55.-t

INTRODUCTION

Two-phase pipe flows occur in many industrial applications, such as condensers, and evaporators, chemical processing equipment, nuclear reactors, and oil pipelines. A variety of basic mechanistic flow models for predicting the pressure gradient and liquid loading characteristics of these types of flows to assist in design calculations have emerged over the past three decades, especially for the stratified and slug flow regimes. In these models, an understanding of the effects of interaction of a moving gas with a liquid surface presents an important problem to investigators, so most of basic assumptions and empirical closure equations relate to the evaluation of interfacial shear stresses.

Two assumptions, smooth pipe single-phase Blasius friction for the liquid and gas wall shear stresses, and the friction factor of the gas-liquid equivalent to the friction factor of the gas flowing against the wall, made in the pioneer work of Taitel and Dukler[1], is now commonplace in mechanistic models of stratified- and slug-type flows in horizontal pipes. Andritsos and Hanratty[2] demonstrated that this did not hold true at higher gas velocities and proposed a correction factor. This factor reduced the error in the momentum equation to within 10% of the experimental values, while the prediction from Taitel & Dukler was on the order of 50%.

For the equilibrium stratified flow, the momentum equations of two phases are written as

$$-A_{\mathrm{G}}(\mathrm{d}P/\mathrm{d}x)-\tau_{\mathrm{WG}}S_{\mathrm{G}}-\tau_{\mathrm{i}}S_{\mathrm{i}}=0 \tag{1}$$

$$-A_{\mathrm{L}}(\mathrm{d}P/\mathrm{d}x)-\rho_{\mathrm{L}}g(\mathrm{d}h_{\mathrm{L}}/\mathrm{d}x)-\tau_{\mathrm{WL}}S_{\mathrm{L}}+\tau_{\mathrm{i}}S_{\mathrm{i}}=0 \tag{2}$$

$$\tau_{\mathrm{WG}}=0.5f_{\mathrm{G}}\rho_{\mathrm{G}}u_{\mathrm{G}}^{2} \qquad \tau_{\mathrm{WL}}=0.5f_{\mathrm{L}}\rho_{\mathrm{L}}u_{\mathrm{L}}^{2} \qquad \tau_{\mathrm{i}}=0.5f_{\mathrm{i}}\rho_{\mathrm{G}}(u_{\mathrm{G}}-u_{\mathrm{L}})^{2} \tag{3}$$

CP914, *Multiphase Flow: The Ultimate Measurement Challenge,* edited by X. Cai, Y. Wu, Z. Huang, S. Wang, and M. Wang

It is the common practice that both the measured pressure gradient and liquid height are employed for the calculation of interfacial shear stress. In this case, the flow is commonly assumed to be fully developed, as a result, the gradient of height of liquid layer is omitted, and arithmetic average height on different locations is employed for a series of theoretical analysis. However, this assumption is improper as non-uniform flow occurs. Such non-uniform flow can affect liquid holdup measurement and flow pattern transitions, and can incur error because holdup varies along the pipeline [3].

In this study, with reference to the liquid control volume, a force balance in the horizontal direction is set up and the interfacial level gradient has been taken into account. By this approach, the interfacial shear stress evaluated indirectly from experimental liquid height and pressure loss measurements is provided for the estimation of the interfacial friction factor.

BASIC MODEL

Two-phase pipe flows occur in many industrial applications, such as condensers, and evaporators, chemical processing equipment, nuclear reactors, and oil pipelines. A variety of basic mechanistic flow models for predicting the pressure gradient and liquid loading characteristics of these types of flows to assist in design calculations have emerged over the past three decades, especially for the stratified and slug flow regimes. In these models, an understanding of the effects of interaction of a moving gas with a liquid surface presents an important problem to investigators, so most of basic assumptions and empirical closure equations relate to the evaluation of interfacial shear stresses.

Referring Fig. 1, including hydrostatic forces in the liquid phase, a force balance about the element of liquid phase with the length of δx in the axial direction yields

$$\begin{aligned}&\int_0^h [\rho_L g(h-y)]2b(y)\,dy + PA_L + S_i\delta s\tau_i\cos\theta + (P+\delta P/2)S_i\delta s\sin\theta + \rho_L A_L u_L^2 \\ &= \int_0^{h+\delta h}[\rho_L g(h-y)]2b(y)\,dy + (P+\delta P)(A_L+\delta A_L) + \rho_L A_L u_L(u_L-\delta u_L) + \tau_{WL}S_L\delta x\end{aligned} \tag{4}$$

and the one-dimensional equation of mass conservation for the liquid phase is

$$A_L u_L = (A_L + \delta A_L)(u_L - \delta u_L). \tag{5}$$

Since the interface half-width b varies with the liquid level, $b = b(y)$, and for small δh, $\delta A_L \approx S_i\,\delta h$. If both P and δP are assumed to be constant in the y-direction at a given location along axial direction, and the small order terms are neglected, Equation (4) can be expressed as follows:

$$S_i\delta x\tau_i + \delta PS_i\delta h/2 - \delta PA_L + \rho_L u_L^2\delta A_L - \tau_{WL}S_L\delta x = \int_h^{h+\delta h}[\rho_L g(h-y)]2b(y)\,dy = F_{net} \tag{6}$$

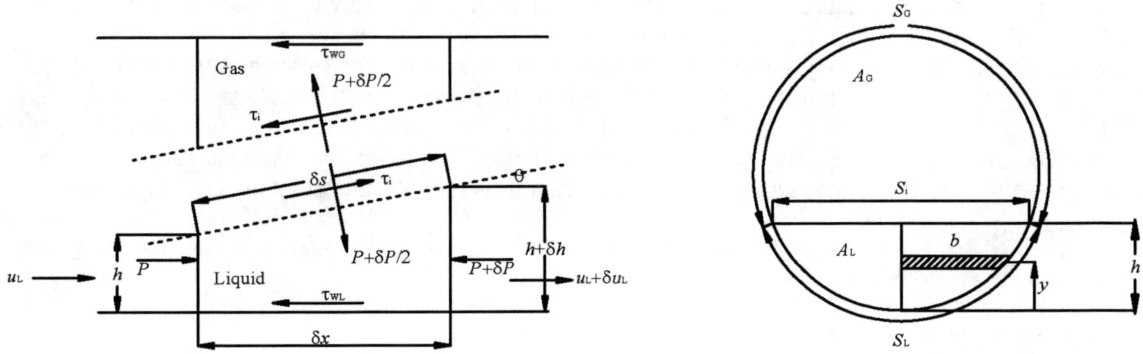

FIGURE 1. Control volume for force balance analysis.

So the interfacial shear stress can be calculated by above expression. Note that the term on the RHS of Equation (6) is the net hydrostatic force acting on the liquid element in the axial direction. The expression of b about the liquid level y is given by

$$b(y) = 0.5D\sqrt{1-(2y^*-1)^2} \tag{7}$$

where y^* is the dimensionless level (y/D), h^* is the dimensionless liquid height (h/D). Two auxiliary definitions, $\varphi = 2(h+\delta h)/D-1$, $\psi = 2h/D-1$, are introduced for the integration of Equation (6). Thus, F_{net} is given by

$$F_{\text{net}} = \rho_{\text{L}} g D^3 \left\{ \frac{1}{8}\left[\psi^2\sqrt{1-\psi^2} - \varphi^2\sqrt{1-\varphi^2} + \psi\sin^{-1}\psi - \varphi\sin^{-1}\varphi\right] + \frac{\pi}{16}(\psi-\varphi) + \frac{1}{12}\left[\left(1-\psi^2\right)^{1.5} - \left(1-\varphi^2\right)^{1.5}\right]\right\} \tag{8}$$

The cross-sectional area of liquid phase, A_{L}, and the perimeters of gas-liquid interface, S_{i}, is given by

$$A_{\text{L}} = 0.25D^2\left[\pi - \cos^{-1}(2h^*-1) + (2h^*-1)\sqrt{1-(2h^*-1)^2}\right] \tag{9}$$

$$S_{\text{i}} = D\sqrt{1-(2h^*-1)^2} \tag{10}$$

Substituting Equations (9) and (10) the following relations is obtained for the interfacial shear stress:

$$\tau_{\text{i}} = \frac{F_{\text{net}}}{S_{\text{i}}\delta x} + \frac{\delta P}{\delta x}\left[\frac{A_{\text{L}}}{S_{\text{i}}} - \frac{\delta h}{2}\right] - \frac{\rho_{\text{L}} u_{\text{L}}^2 \delta h}{\delta x} + \frac{\tau_{\text{WL}} S_{\text{L}}}{S_{\text{i}}} \tag{11}$$

thus, it can be determined for each set of values of $[u_{\text{SG}}, u_{\text{SL}}, \delta p/\delta x, \delta h/\delta x]$ from the experimental data.

EXPERIMENT

Flow Loop

Figure 1 is a schematic of the test section and flow loop used to obtain the measurements presented in this paper. A centrifugal pump was used to circulate water from the tank, through a pair Rousemount CMF300M355NU Coriolis mass flow meter installed in series, and to the test section liquid entry section. A frequency conversion compressor was used to circulate air, from a metal tube rotameter and vortex flow meter installed in series, and to the test section gas inlet piping. The manufacturer-stated accuracies of the water and gas flow meters were ±0.1 and ±1 %, respectively. The entry sections for the two phases were simple pipe tees with air introduced in the branch and water in the run. In this study the air-water mixture was developed in a 50 mm ID steel tube with a length of 8.5 m. The observation section was a clear Plexiglas tube, 2.2 m long and 50 mm ID. This allowed for a length to diameter (l/d) ratio of 214. Experiment data such as liquid height and pressure loss were acquired at an axial location of about l/d = 174, which ensured that the two-phase mixture was fully developed at the measuring point.

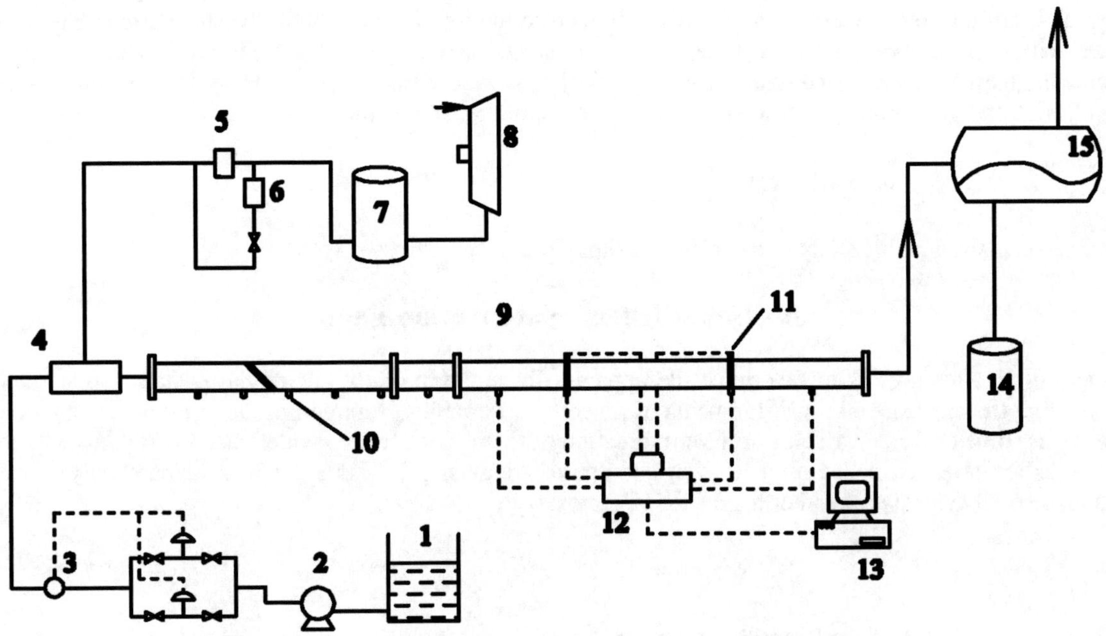

FIGURE 2. Experimental equipment of gas-water horizontal flow.

Instrument and Measurement

The base liquid height in the stratified flow was measured using parallel wire probes. The principle of this technique is that the conductance between two parallel wires increases linearly with the height of their immersion in a conductive solution. Four annularly curved parallel wires, divided into two groups which were used to obtain liquid level gradient, were expanded over the entire cross section of the pipe. They were sampled simultaneously in order to study the transient behavior of the liquid layer

A serial of KELLER PA23e pressure transducers, which have a frequency response of up to 5kHz, were connected to the bottom of the pipe to measure the pressure drop along the test section. Among them two transducers, 1m apart, were located just below the two groups of parallel wire probes. A NI Model PCI-6071 data acquisition system was used to record the flow rates and pressure signal with the sampling frequency of 2500Hz per channel.

RESULTS AND DISCUSSION

Liquid Wall Friction Factor

For stable stratified pipe flows with a moving liquid phase, if both u_L and δh are extremely small, these terms are expected to become insignificant when compared with the hydrostatic terms in Equation (11). Further, if the liquid flows so slowly that the gas stream can be assumed to run over a stationary liquid, Equation (11) reduces to

$$\tau_i = \frac{F_{net}}{S_i \delta x} + \frac{\delta P}{\delta x}\left[\frac{A_L}{S_i} - \frac{\delta h}{2}\right] \tag{12}$$

then the direct calculation of τ_i from aforementioned variables is available without the need for any empirical correlation. However, for the consideration of the various gas-liquid interface forms, it is necessary to obtain the value of the liquid-wall shear stress.

In two-phase flow computation, it is common practice to model the fluids wall friction factors using Blasius equations. Kowalski [4] used hot-film anemometry to measure the gas Reynolds stress profiles for air-water and

Freon-water stratified flows, which were then extrapolated to the interface to obtain the shear stress. The result was compared with typical method: from a momentum balance, using the wall shear stresses and void fraction measurements. Kowalski suggested that measurements of f_G were accurately predicted by Blasius equation, but the liquid-wall friction factor did not follow the Blasius relation, but a correlation of the type

$$f_L = 0.263[(1-\alpha)\mathrm{Re}_{SL}^{-0.5}] \qquad \mathrm{Re}_{SL} = u_{SL}D/\nu_L \tag{13}$$

This expression is used in this study to calculate the liquid-wall shear stress τ_{WL}.

Gas-liquid Interfacial Friction Factor

The aim of this research is to determine the apparent formulation of the interfacial friction factor f_i, which is estimated using the experimental measurements of a series of variables. An appropriate correlation may be derived to express f_i as some of these variables or the combination of them. Some of the widely used correlations available in the literatures are obtained based on a flowing gas phase. Agrawal [5] assumed a hydrodynamically smooth gas-liquid interface and evaluated the interfacial friction factor as

$$f_i = 1.293\,\mathrm{Re}_G^{-0.57} \tag{14}$$

Shoham and Taitel [6] studied stratified turbulent-turbulent gas-liquid flow in horizontal and inclined pipes and proposed a constant value (corresponding to a fully developed rough interface due to the presence of small waves):

$$f_i = 0.00142 \tag{15}$$

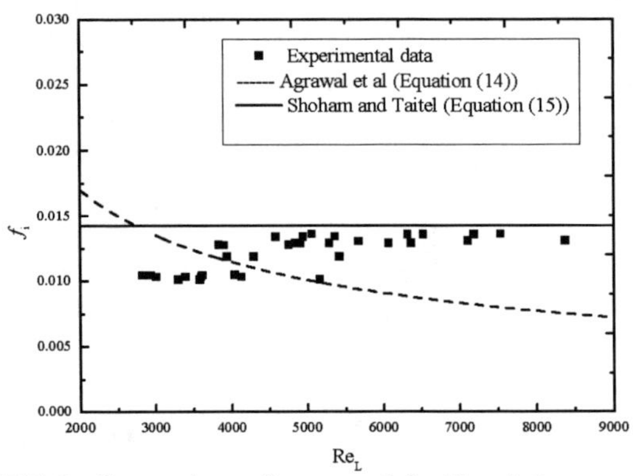

FIGURE 3. Comparison of measured f_i with existing correlations.

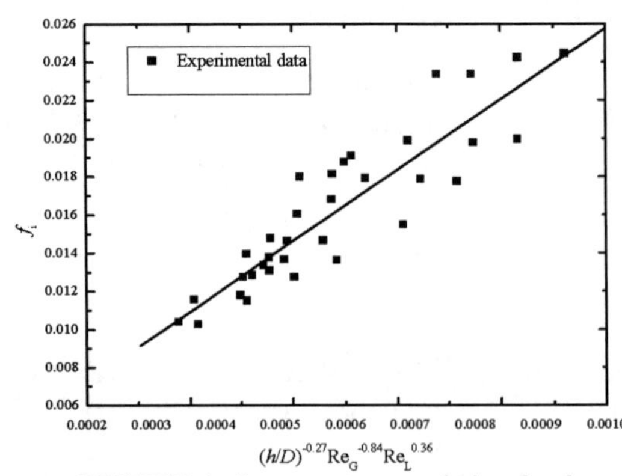

FIGURE 4. 3-parameter correlation for f_i.

Measurements of f_i are plotted against the gas phase Reynolds number in Fig. 3. The solid and dot lines represent two correlations listed above. Neither of these models predicts satisfactorily. Including more characteristic parameters is needed for the achievement of an improvement in the quality of prediction. According to the result of Pimsner and Toma [7], with the friction velocity, u_τ, the velocity profile may be expressed as the combination of the liquid height and gas phase and liquid phase Reynolds number.

$$u/u_\tau = F(D/h, \mathrm{Re}_\mathrm{G}, \mathrm{Re}_\mathrm{L}) \tag{16}$$

If an approximation is applied here

$$u_\tau/u \approx \sqrt{f_\mathrm{i}/2} \tag{17}$$

Equation (16) may be expressed as

$$f_\mathrm{i} = F[D/h, \mathrm{Re}_\mathrm{G}, \mathrm{Re}_\mathrm{L}] \tag{18}$$

so, a power relation about the interfacial friction factor is as follows

$$f_\mathrm{i} = a + b(D/h)^c \,\mathrm{Re}_\mathrm{G}^d \,\mathrm{Re}_\mathrm{L}^e \tag{19}$$

By the regression methods, the constant *a*, *b*, *c*, *d*, *e*, can be evaluated to fit the linear logarithmic form of the formulation (as shown in Fig. 4). For present experimental data, the interfacial friction factor is formulated by

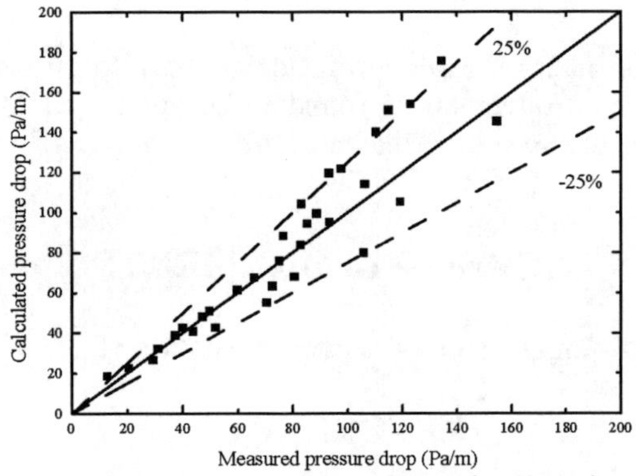

FIGURE 5. Comparison between calculated pressure drop using Equation11 and experimental values.

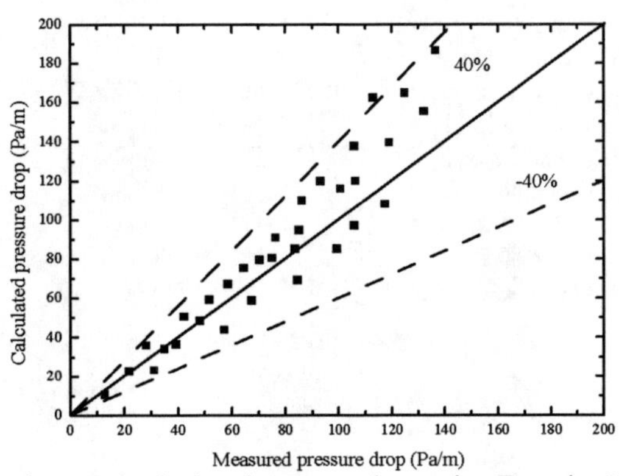

FIGURE 6. Comparison between calculated pressure drop using Equation11 and experimental values.

$$f_i = 2.56 \times 10^{-3} + 23.19(h/D)^{-0.27} \mathrm{Re}_G^{-0.84} \mathrm{Re}_L^{0.36} \quad (20)$$

The methodology used to calculate the liquid-wall and gas-liquid interfacial friction factors in this study requires evaluation against experimental data. Therefore, values of these two kinds of friction factor are substituted into one-dimensional momentum balance (Equation (1) or Equation (2)) for stratified flow to give pressure loss predictions. Comparison between calculated pressure loss using current procedure and experimental data are shown in Fig. 5 and Fig. 6, which correspond to normal (Equation (11)) and simplified (Equation (12)) correlations of interfacial shear stress respectively. For the former, scatters fall into an error band of approximately ±25%. The improvement over simplified correlation, which has an error band of approximately ±40%, can be attributed to the consideration of liquid level gradient and the proper estimation of liquid-wall shear stress. However, the data are overpredicted by current correlation. Especially, agreement is found to be poor for larger pressure drop, which is related to the case where larger amplitude waves are prevalent.

CONCLUSIONS

An investigation of the gas-liquid interfacial shear stress in horizontal pipe is conducted. Models are set up on a volume element of liquid phase to correlate target variable with liquid-wall shear stress, pressure drop and liquid height. Special aspect is that liquid level gradient has been accounted for, which can affect liquid hold-up measurement, and can incur error because holdup varies along the pipeline. The estimation of interfacial shear stress, along with its characteristic parameter, interfacial friction factor, is conducted on the experiment of air-water flow in a horizontal pipe. By regression analysis, the correlation about interfacial friction factor is obtained. As shown in results, combining dimensionless numbers, such as dimensionless liquid level and Reynolds number of each phase, in the form of exponent, it fits the measured interfacial friction factor well.

It is found that the liquid-wall shear stress is necessary in this model. Since the absence of direct measurements of these stresses, empirical correlation about liquid-wall friction factor are employed to support the estimation of interfacial shear stress.

The correlations about gas-liquid interfacial friction factor are used to provide the predicted pressure drop for the comparison with the measurement. Agreement is found to be good over the ranger of lower pressure drop. Overestimation for the case of larger pressure drop indicates the occurrence of larger amplitude waves which arise from higher gas flow rates.

ACKNOWLEDGEMENTS

This work was supported by the National Natural Science Foundation of China (No. 50476015).

REFERENCES

1. Y. Taitel and A. E. Dukler, *AIChE J.* **22**, 47-54 (1976).
2. Y. Taitel and A. E. Dukler, *AIChE J.* **22**, 47-54 (1976).
3. A. A. Bishop and S. D. Deshpande, *Int. J. Multiphase Flow* **12**, 957-975 (1986).
4. J. E. Kowalski, *AIChE J.* **33**, 274-281 (1987).
5. S. S. Agrawal, G. A. Gregory and G. W. Govier, *Can. J. Chem. Engng.* **51**, 280-286 (1973).
6. O. Shoham and Y. Taitel, *AIChE J.* **30**, 377-385 (1984).
7. V. Pimsner and P. Toma, *Int. J. Multiphase Flow* **3**, 273-284 (1977).

A Study of Ultrasonic Radiation Dissipation in Powder Processing System

Liu Chengwu, Wang Jingdai, Yang Yongrong

Department of chemical and biochemical engineering, Yuquan Campus, Zhejiang University, Hangzhou, 310027, Zhejiang province, China

Abstract. A model of ultrasonic radiation damping was proposed to describe the dissipation of ultrasonic vibration energy in solid processing system on basis of vibration damping theory. Four kinds of polyethylene particles with a diameter from 0.15mm to 0.85mm and bulk densities from 499.6 $kg{\cdot}m^{-3}$ to 252.5 $kg{\cdot}m^{-3}$ accordingly, were used as powder material and filled in a plexiglas storage bin. Ultrasonic energies absorbed by these kinds of particles were detected by the receiver at different levels. The results showed that the longer distance between receiver and source of ultrasonic vibration and higher bulk density of powder material would lead to higher energy absorption. It was also found that the dissipation of ultrasonic vibration energy along a bar of object as the internals was much higher than that along the wall of storage bin because of more radiation area from the object. The experimental result can be predicted by the radiation damping model proposed in fairly good accuracy. Furthermore a method of level gauging could be developed to meet the requirement of many applications, for instance, an online measurement of particle material and replace other level gauging methods such as γ-ray.

Key words. radiation damping; ultrasonic; level gauging; polyethylene

INTRODUCTION

Many containers or reactors contain usually inflammable and explosive liquid or solid particles in industrial production. When its expected level gauging system fails, its maintenance is generally complicated and time-consuming due to difficulties in the substitution of the trouble parts. This will be seriously harmful to the production efficiency and long-term operations, and may even cause the shut-down of plant for inspection of level gauging, especially in the case of the materials with a dangerous property. Available level gauging instrumentation include steel strip type, capacitance type, servo type, radar type, static pressure type, gamma rays and ultrasonic echoes level gauging system. In addition to gamma rays, many other level gauging systems need to interact directly with material tested. It is inconvenient for maintenance of instrument and sometimes dangerous. At present a non-intrusive level gauging method is often needed based on the ultrasonic vibration damping in commercial application.

A study was carried out about the ultrasonic energy dissipation in solid processing system. Many influence factors were discussed in this paper, for instance, particles' level in storage bin, the bulk density of particles and the ultrasonic transmission mediums. In this way, a new type of level gauging system can be designed based on the more accurate theory

EXPERIMENTAL

CP914, *Multiphase Flow: The Ultimate Measurement Challenge,* edited by X. Cai, Y. Wu, Z. Huang, S. Wang, and M. Wang

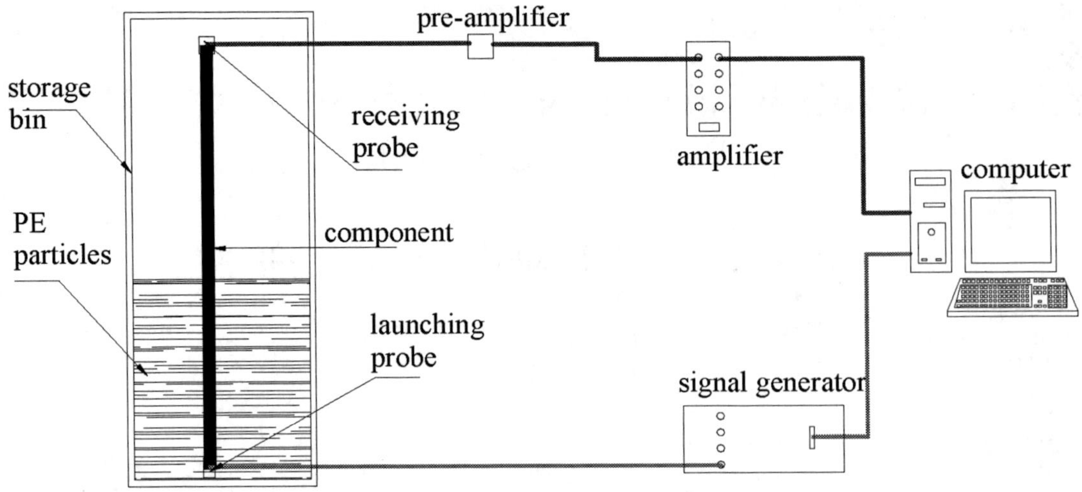

FIGURE 1. Schematic diagram of experimental apparatus

A cylindrical column made by plexiglass, which is 150 mm in diameter, 5 mm in wall thickness, was used as storage container in the experiment. A bar of plexiglass internals is 600 mm long, 20 mm wide and 5 mm thick as a track of ultrasonic conduction. An ultrasonic vibration source was placed in the bottom of internal bar and an ultrasonic vibration receiving probe installed in the top of the bar with a distance of 550 mm between the two probes' center as shown in Figure 1.

If ultrasonic vibration probes installed on the external surface of storage bin, it can significantly simplify installation and maintenance of the probe for commercial application, and can become a level gauging method with safety and non-intrusiveness. Therefore, the dissipation of ultrasonic vibration energy was investigated along the outer wall surface of storage bin as shown in Figure 2.

Experimentally ultrasonic vibration was produced by means of an excitation apparatus and received by a probe made of piezoelectric ceramic, which is 20 mm in diameter. the excitation probe is customized to have a resonance frequency of 49.05 kHz, and 73.9 Ω in equivalent impedance and 4.94 nf in electrostatic capacitance. The receiving probe is accordingly customized to have 49.05 kHz in resonance frequency, 40 dB frequency bandwidth of

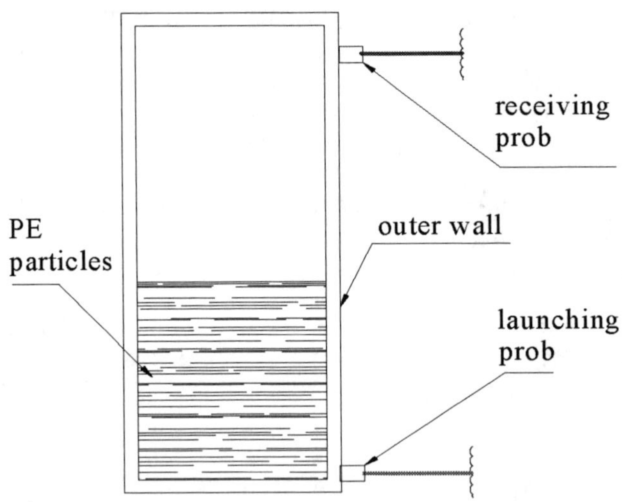

FIGURE 2. Schematic diagram of non-intrusive level gauging

TABLE 1. Property of HDPE powder

HDPE	Melt Index (g/10min)	Density ($kg{\cdot}m^{-3}$)	Porosity	Particle Size (μm)
DGDB 2480	9.6	944.0	0.40~0.45	511.6

TABLE 2. Bulk density of HDPE powder for each sieves sizing

No.	Particle Size (mm)	Bulk Density (kg·m^{-3})
1	0.85	252.5
2	0.60	320.8
3	0.45	382.0
4	0.15	499.6

30-140 kHz, with a sensitivity greater than 80 dB. Ultrasonic signal excitation/detection system (Ultrafinder v1.0) is made by UNILAB of Zhejiang University. The excitation part includes FPGA controlling module, signal conditioning modules, power modules and excitation control software. And the receiving part includes pre-amplifier, main amplifier, signal collection device and signal processing software. High density polyethylene powders are used as testing materials whose properties are shown in Table 1 and Table 2.

VIBRATION DAMPING THEORY

When the ultrasonic vibration transmission happens within an object, some of them will transfer into the gas-solid system around across the object interface. It means that the energy of vibration within the object will decrease significantly due to dissipation through gas-solid materials around in addition to minor loss by thermal movement in the object. It can be considered as the existence of a negative damping force, which names the radiation damping[1], with a material damping as shown in Figure 3.

The radiation damping from object to gas phase can be ignored because it is far less than the radiation damping from object to solid particles. When the vibration of the object surface is weaker than the limits of its maximum elasticity, the total resistance of vibration is expressed as

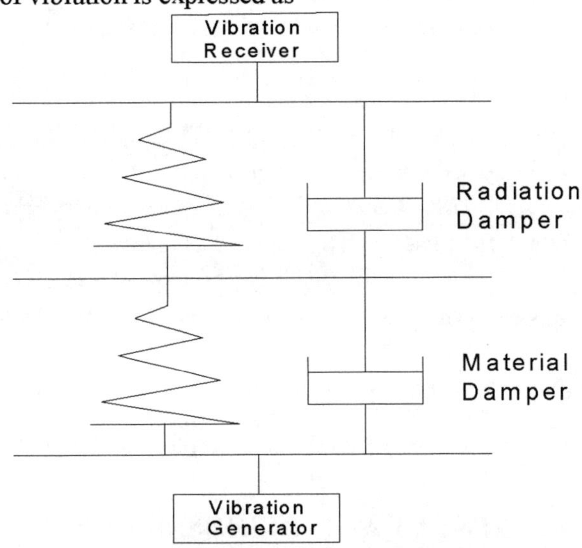

FIGURE 3. Model of energy dissipation

$$f = kxH + c_m vL + c_r vH \tag{1}$$

Where H is height of particles material, L is the length of object, v is vibration velocity, k is elasticity coefficient, cm is material damping coefficient, cr is radiation damping coefficient.

The expression of elasticity coefficient k is:

$$k = \pi G \tag{2}$$

Where G is shear modulus of particles.

Material damping coefficient cm can be calculated by using the modified model which was proposed by Hardin [2]

$$c_m = 2\pi r j P_r \xi \sqrt{\rho_m G} \tag{3}$$

Where r is equivalent length of object, Pr is signal power, j is factor of signal power, ρm is density of object material, ξ is damping ratio.

Radiation damping coefficient cr can be obtained from the modified model which was proposed by Randolph[3]:

$$c_r = 2\pi rjP_r\sqrt{\rho_B G} \tag{4}$$

Where ρm is bulk density of particles.

Surface vibration displacement x can be supposed in this paper as:

$$x = a\sin(\omega t) \tag{5}$$

Where a is vibration amplitude, ω is vibration frequency.

So the expression of vibration velocity v is written as:

$$v = a\omega\cos(\omega t) \tag{6}$$

A comprehensive relationship between total resistance and vibration amplitude and frequency, properties of object and particle material, the object length submersed in the solid particles and the total length of object itself can be obtained by substituting equation 2 to 6 into equation 1:

$$f = a\pi GH\sin(\omega t) + 2a\omega\pi rjP_r\xi L\sqrt{\rho_m G}\cos(\omega t) + 2jP_r a\omega\pi rjP_r H\sqrt{\rho_B G}\cos(\omega t) \tag{7}$$

Energy dissipated by damping in one cycle T is:

$$W' = \int_0^T fvdt \tag{8}$$

Substituting equation 6 and 7 into equation 8:

$$W' = 2a^2 jP_r\omega\pi^2 r\xi L\sqrt{\rho_m G} + 2a^2 jP_r\omega\pi^2 rH\sqrt{\rho_B G} \tag{9}$$

Since the distance between excitation probe and receiving probe is fixed, and the frequency of ultrasonic vibration produced is also constant. So the first term in equation 9 is constant. When the particles level rises, the reduction of signal energy received is only correlated with the second term of equation 9. So the differential equation of dissipated energy is:

$$dW = 2\omega\pi djP_r G^{1/2}\rho_B^{1/2}a^2 dH \tag{10}$$

The differential equation of vibration energy dissipated in particles phase is:

$$dW = -k_v ada \tag{11}$$

Combining equation 10 into equation 11:

$$a = A\exp(-2\omega\pi djP_r G^{1/2}k_v^{-1}\rho_B^{1/2}H) \tag{12}$$

So the relationship between the dissipated energy and level (height) of particle around the object is:

$$\Delta W = \omega\pi djP_r G^{1/2}\rho_B^{1/2}A^2 H\exp(-4\omega\pi djP_r G^{1/2}\rho_B^{1/2}k_v^{-1}H) \tag{13}$$

After normalization of amplitude A, the model can be deduced as below:

$$E = 1 - \omega\pi djP_r G^{1/2}\rho_B^{1/2}H\exp(-4\omega\pi djP_r G^{1/2}\rho_B^{1/2}k_v^{-1}H) \tag{14}$$

If the probes installed on the outside of the wall of the storage bin, equation 10 will be changed into:

$$dW = \omega\pi djP_r G^{1/2}\rho_B^{1/2}a^2 dH \tag{15}$$

So the final formulation of the model can be expressed as follow:

$$E = 1 - \frac{1}{2}\omega\pi djP_r G^{1/2}\rho_B^{1/2}H\exp(-2\omega\pi djP_r G^{1/2}\rho_B^{1/2}k_v^{-1}H) \tag{16}$$

RESULT AND DISCUSSION

Relationship between Gauging Level and Vibration Energy Dissipation

When excitation probe is installed on the surface of a vertical object 20mm above the bottom end, and its distance to the receiving probe on the top end of object is 820mm. The frequency of signal produced is 40 kHz. Sampling was performed by using super A/D modulus with a frequency of 200 kHz.

Relationship between gauging level of solid particles and the energy received on the top end is shown in Figure 4. The solid line shows the model prediction in which ωπdjPrG1/2ρB1/2 as coefficient A andωπdjPrG1/2ρB1/2kv-1 as coefficient B according to equation 14. It can be shown that agreement is very well with a very low SSE (sum of the squares of errors) of 0.001697 and higher R2(correlation coefficient) of 0.9933. Coefficient A is 0.001336 and B is 0.0006112.

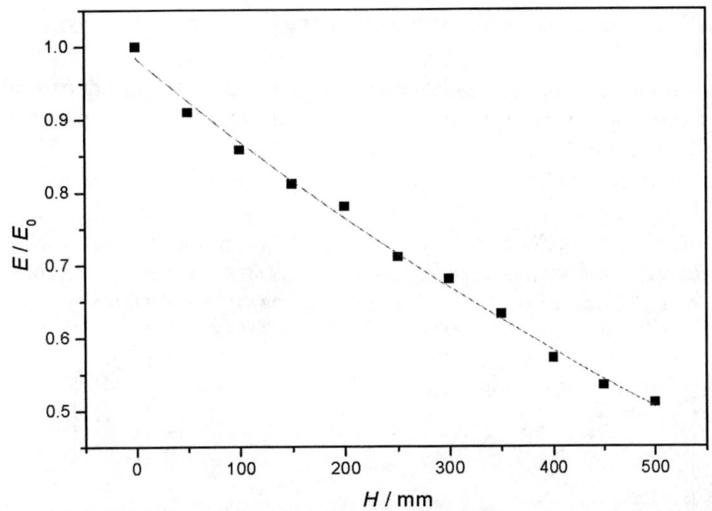

FIGURE 4. Relationship between gauging level and vibration energy dissipation

TABLE 3. Comparison of model parameters between object and wall

	A	*B*	SSE	R^2
Object	0.003167	0.001202	0.001149	0.9987
Wall	0.002404	0.0007038	0.003343	0.9954

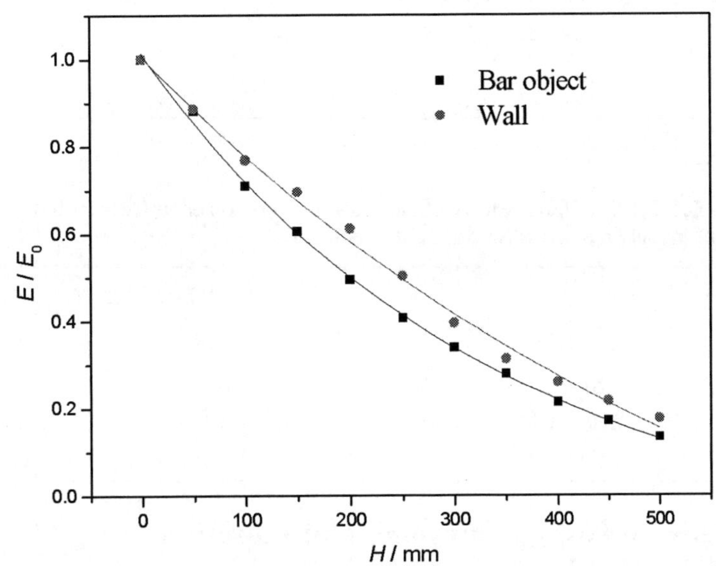

FIGURE 5. Comparison of vibration energy dissipation between object and wall

Effect of Installation on Vibration Energy Dissipation

If the excitation probe and receiving probe were installed on the outside wall of the bed or storage bin, it will efficiently facilitate the installation and maintenance of gauging level and improve its safety and non-intrusiveness.

The Comparison of energy attenuation of ultrasonic vibration through the single object and outer wall of storage bin is shown in Figure 5. The vibration decays more on the single object than on the wall because there are at least two sides of area contacting with solid material for single object while there is only inner area of wall facing solid particles. More area contacting with solid particles more energy losses there are. The coefficients of A and B from model are listed in Table 3. The variation of coefficients can also arise from the different path in single object and wall material.

Effect of Bulk Density of Powder on Vibration Energy Dissipation

Ultrasonic energy dissipation model proposed in Equation 16 has shown the effect of bulk density of particles on the radiation dissipation of ultrasonic vibration energy. Four different sizes of PE particles are chosen to study the impact of bulk density on the vibration energy in Figure 6.

As shown in Figure 6, four different lines discriminate the effects of bulk densities on the dissipation of ultrasonic vibration energy. Along with the reducing size of particles, bulk density increasing, the particles absorb more energy of ultrasonic vibration. So the decay of ultrasonic vibration along the wall increases for particles with small size. The Coefficients of A, A/ρB1/2, B and B/ρB1/2 are shown in Table 4. According to Equation 16, A/ρB1/2 and B/ρB1/2 should be independent on the bulk density of particles when other factors kept constant. The model can be validated by the small variations of A/ρB1/2 (less than 25%) and B/ρB1/2 (less than 3%).

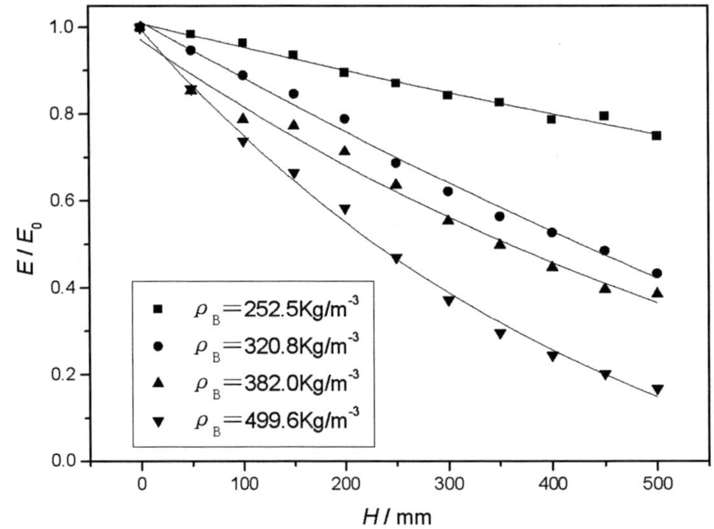

FIGURE 6. Effect of bulk density of particles on vibration energy dissipation

TABLE 4. Comparison of model parameters for different bulk density

ρ_B	A	$A/\rho_B^{1/2}$,	B	$B/\rho_B^{1/2}$	SSE	R^2
252.5	1.18×10^{-3}	0.74×10^{-4}	6.43×10^{-4}	4.05×10^{-5}	0.011	0.97
320.8	1.46×10^{-3}	0.81×10^{-4}	7.31×10^{-4}	4.08×10^{-5}	0.0031	0.99
382.0	1.87×10^{-3}	0.96×10^{-4}	8.04×10^{-4}	4.11×10^{-5}	0.0071	0.98
499.6	2.25×10^{-3}	1.00×10^{-4}	9.30×10^{-4}	4.16×10^{-5}	0.0028	0.99

Relationship between Position of Probes and Vibration Energy Dissipation

The results and discussion above are carried out under the condition that the point of measurement is on the same vertical line of excitation point on the wall of storage bin. However, It should be modified if the measurement point is not located in the same vertical line on the wall of storage bin due to the change of transmission path on the wall surface.

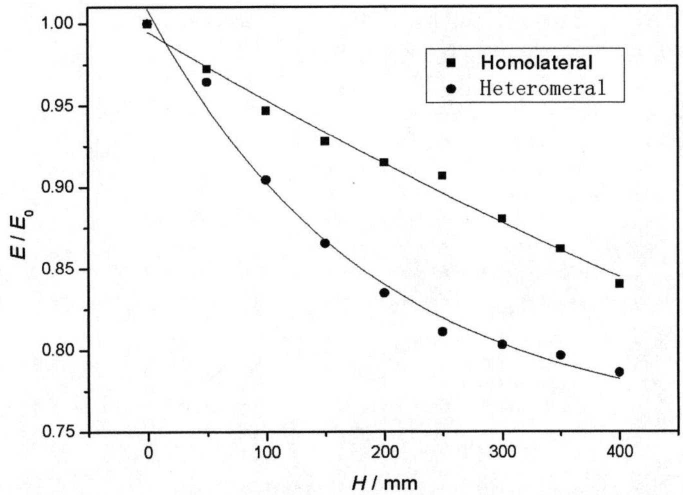

FIGURE 7. Vibration energy dissipation at heteromeral line on the wall

The ultrasonic energy dissipation along with the vertical line opposite to the excitation point on the wall surface of storage bin is shown in Figure 7. The ultrasonic energy received at homolateral line is significantly stronger than at heteromeral line. It is because the route length of transmission of the ultrasonic vibration is just straight distance between excitation and measurement when they are in same homolateral line, meanwhile route length of transmission on the wall surface will be quite a long especially when the probes are in heteromeral position. And this difference increases with the increase in diameter of storage bin. So the homolateral position of probes will be a good choice in practical use.The ultrasonic damping theory proposed here can be also used for the level gauging of gas-solid fluidized bed. The results related will be presented otherwhere.

CONCLUSION

A model of ultrasonic energy dissipation proposed in this paper can describe the relationship between the dissipation of ultrasonic vibration energy, particles' level in storage bin, the bulk density of particles and the ultrasonic transmission mediums based upon the theory of ultrasonic radiation damping.

The results showed that the higher material level and bulk density would lead to higher energy absorption. The dissipation of ultrasonic energy by probing on internal bar submersed in the solid particles of storage bin was more than that on the wall of storage bin because of the internal bar with more radiation area. The receiving probe receives more energy when the two probes are in homolateral position of the wall surface of storage bin. The proposed model can predict the experimental result in fairly good accuracy. Therefore the model of ultrasonic energy dissipation proposed can be used for level gauging in many important application of powder processing. And this novel kind of measurement could become an accurate online method to replace other level gauging methods such as γ-ray.

ACKNOWLEDGEMENTS

The authors are grateful to the National Natural Science Foundation of China (20490205) for financial support of this work.

REFERENCES

1. DING Wenjing, Theory of Shock Absorption, Tsinghua University Press, Beijing, 1988, pp.186-187.
2. Hardin B. O., Drnevich V. P., Shear Modulus and Damping in Soils: Design Equations and Curves, ASCE, Journal of Soil Mechanics and Foundation Engineering, Vol. 98, SM7, 1972, pp.667-692.

3. Randolph M. F., Simons H. A., An Improved Soil Model for One-dimensional Pile Driving Analysis, Proc. of the 3rd Intl. Conf. On Numerical Methods in Offshore Piling, Nantes, 1986, pp.3-17.

Multiphase Flow Measurement System of Oil Well

Zhiyao Huang*, Chaohong He +, and Qilin Liang¶

*National Laboratory of Industrial Control Technology Department of Control Science and Engineering, Zhejiang University, Hangzhou 310027, P.R.China, zyhuang@iipc.zju.edu.cn
+Department of Chemical and Biochemical Engineering, Zhejiang University, Hangzhou 310027, P.R.China,
¶Hangzhou Chuanglian Electronic Technology Ltd., Hangzhou 310013, P.R.China

ABSTRACT. A new multiphase flow measurement system of oil well was developed. This measurement system was based on the combination of a separator, two level meters and three commercial flowmeters. The separator separated the crude oil into three components: gas, water and oil-water mixture. By means of the automatic control of two interface levels (the oil-water interface level and the oil-gas interface level), three components were measured by the corresponding commercial flowmeters. The developed measurement system had been tested at Shengli Oilfield in China. The test results show that the developed measurement system is effective. It is suitable for the flowrate measurement of Chinese oil well with high water fraction and its accuracy is also satisfactory.

Keywords: Flowrate measurement, Multiphase flow, Separator, Oil well
PACS: 42.62.Cf, 47.61.Jd, 47.85.M-

INTRODUCTION

Multiphase flow has a wide range of applications in petroleum industry. Exact knowledge about how much oil/gas/water are supplied from the oil well is significant for production allocation, reservoir management and optimum use of chemicals. Ideally, accuracy of ±5% in cumulative flowrate of each phase and measurement independence of flow pattern are expected. Recently, many methods and multiphase meters have been developed for the measurement of multiphase flow. Unfortunately, few of them can meet the mentioned requirements [1-3].

The conventional separator-based measurement method is still a useful choice at oilfield. The crude oil is firstly separated and then the oil/gas/water components are measured by the commercial instruments. However, the conventional separators have some disadvantages such as the expensive cost, the huge volume, the low automation level and the considerable maintenance. Meanwhile, the Chinese oil well with high water fraction (typically, over 90%) causes additional difficulty of the multiphase flow measurement.

The aim of this work is based on the separator-based measurement system method to develop an automatic online multiphase flow measurement system with attractive cost and small volume.

DESIGN OF MEASUREMENT SYSTEM

The measurement system is schematically shown in Fig 1. The separator separates the crude oil into three components: gas, water and oil-water mixture. The flowrate measurement of the multiphase flow is implemented by three appropriate commercial flowmeters in three pipelines, including gas pipeline, oil pipeline and water pipeline. The gas flowrate in the gas pipeline is measured by a gas flowmeter. The water flowrate in the water pipeline is measured by an electromagnetic flowmeter. The oil-water mixture firstly overflows into the oil tank and then flows into the oil pipeline. The oil-water mixture density and the oil-water mixture mass flowrate are obtained by a Coriolis flowmeter. Therefore the oil flowrate, the gas flowrate and the water flowrate of the crude oil can be determined.

CP914, *Multiphase Flow: The Ultimate Measurement Challenge,* edited by X. Cai, Y. Wu, Z. Huang, S. Wang, and M. Wang

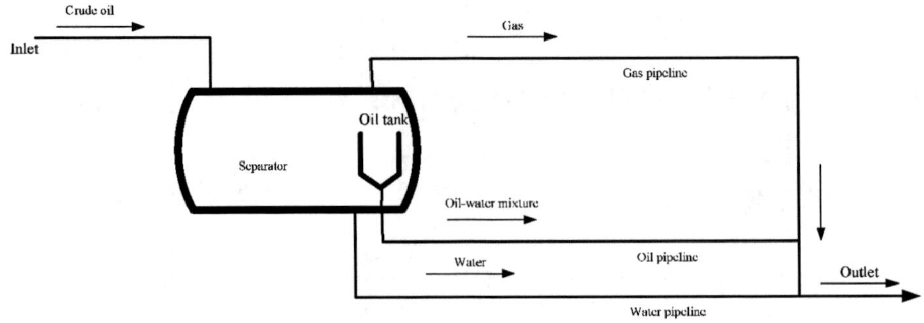

FIGURE 1. The multiphase flow measurement system of oil well

Obviously, the measurement of the oil-water mixture in the oil pipeline is most important. The Coriolis mass flowmeter can provide both the mass flowrate and the density of the oil-water mixture in the oil pipeline. The oil and water fraction in the oil pipeline are determined as follows:

$$x_o = \frac{\rho_w - \rho}{\rho_w - \rho_o} \quad (1)$$

$$x_w = \frac{\rho - \rho_o}{\rho_w - \rho_o}. \quad (2)$$

Where x_o and x_w are the oil fraction and water fraction of the oil-water mixture in the oil pipeline respectively, ρ , ρ_o and ρ_w are the densities of the oil-water mixture, oil and water respectively. Based on the fact that the quantity of oil in the water pipeline is little, the oil flowrate and the total water flowrate can be calculated by:

$$m_o = m \cdot x_o \quad (3)$$

$$m_w = m \cdot x_w + Q_w \cdot \rho_w \quad (4)$$

Where, m_o is the oil mass flowrate, m_w is the total water mass flowrate, m is the total mass flowrate of the oil-water mixture in oil pipeline obtained by the Coriolis mass flowmeter, Q_w is the volume water flowrate in the water pipeline obtained by the electromagnetic flowmeter. Meanwhile the pressure sensor and the temperature sensor are introduced to implement the compensations of the pressure and the temperature.

SEPARATOR

Obviously, the separator is an important unit of the measurement system. In order to improve the measurement accuracy, the design aim of this separator is that the oil fraction of the liquid in the water outlet should be low enough to be ignored and the oil fraction of the liquid in the oil outlet is relatively high (usually, ⩾30%) and stable. Also, the separator should have the advantages of small volume and high efficiency.

Figure 2 shows the structure of the separator. The crude oil firstly flows into the centrifugal degasser where the gas-liquid is separated based on the principal of centrifugal force. The energy dissipation plate can increase the retention time of the liquid and improve the flow stability by dissipating the liquid energy. The rectifying device is normally utilized to speed the oil-water separation and stratification process. At the gas outlet of the separator, a wire mesh and degassing baffles are installed to remove the liquid droplets.

An oil-water interface will be formed because of the gravity effect. The oil-water interface level can be detected by a level meter. The level will be controlled lower than the brim of the oil tank to ensure most of the oil flows into the oil pipeline and most of the water flows into the water pipeline. Because of the high efficiency of the separator, the oil fraction of the oil-water mixture which flows out of the outlet of the oil tank is more than 30%, and little oil flows into the water pipeline.

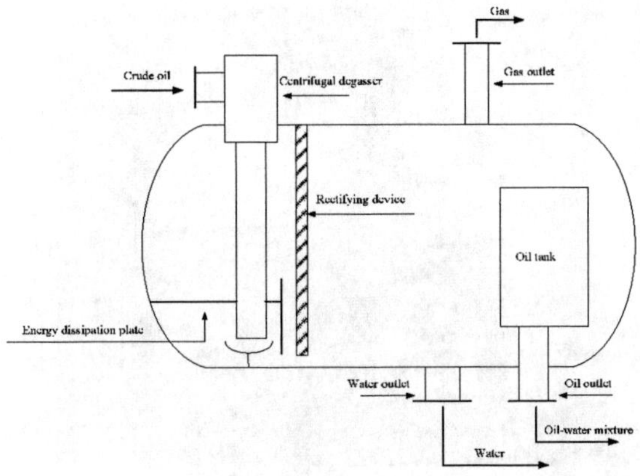

FIGURE 2. The structure of the separator

CONTROL SCHEME

To meet the requirement of the multiphase flow measurement, the control unit should ensure the stabilities of the pressure of the separator, the level of the oil-water interface and the level of the gas-liquid interface in the oil tank.

The pressure control is mainly implemented by adjusting the gas flowrate in the gas pipeline. The level control of the oil-water interface is mainly implemented by adjusting the water flowrate in the water pipeline. The level control of the gas-liquid interface in the oil tank is implemented by adjusting the oil-water mixture flowrate in the oil pipeline.

The conventional PID control is adopted as the main control scheme:

$$u(t) = K(e(t) + \frac{1}{T_1}\int_0^t e(t)dt + T_2\frac{de(t)}{dt}) \tag{5}$$

Where, $u(t)$ is the output of PID control, K is the proportional coefficient, T_1 is the integral coefficient, T_2 is the differential coefficient, and $e(t)$ is the deviation between the set value and measurement value.

Meanwhile, due to the complexity of the oil well production, the flowrate of the crude oil sometimes fluctuates dramatically. That will cause the pressure and two levels to greatly deviate the set values. So, the two state control scheme (including high-limit control and low-limit control) is also introduced to guarantee the safety and stability of our measurement system.

INDUSTRIAL TEST RESULTS

The developed measurement system had been tested for three months at Shengli Oilfield in China. Figure 3 is the practical measurement system. Four oil wells were tested and the water fractions of the wells were greater than 90%. The test results showed that the developed $0.18\,m^3$ (the volume of the separator) measurement system could process the crude oil with maximum liquid flowrate of $400\,m^3/day$ and maximum gas flowrate of $650\,Nm^3/day$. The measurement errors of the gas cumulative flowrate, the oil cumulative flowrate and the cumulative water flowrate were less than $\pm 5\%$, $\pm 3\%$, and $\pm 3\%$, respectively.

FIGURE 3. The practical measurement system

CONCLUSIONS

A multiphase flow measurement system for the oil well is developed. The test results show that the developed system is effective and has the following advantages:

1) The developed system is suitable for the measurement of oil well with high water fraction, especially for Chinese oil well (typically, over 90%).
2) Compared with conventional separation methods, this developed measurement system is small, low cost and compact. The developed measurement system with a small volume of 0.18 m^3 can process the crude oil with maximum liquid flowrate of 400 m^3/day and maximum gas flowrate of 650 Nm^3/day.
3) The flowrate measurement errors of the gas cumulative flowrate, the oil cumulative flowrate and the water cumulative flowrate are less than $\pm 5\%$, $\pm 3\%$, and $\pm 3\%$, respectively.

ACKNOWLEDGEMENT

The project is supported by National Natural Science Foundation of China under Grant No.50576084 and No.60532020.

REFERENCES

1. R. Thorn, G.A. Johansen, and E.A. Hammer, *Measurement Science & Technology* **8**, 691-701 (1997).
2. G. Oddie and J.R.Anthony Pearson, *Annual Review of Fluid Mechanics* **36**, 149-172 (2004).
3. A.W. Jamieson, *Measurement +control* **32**, 5-9 (1999).

Methods of Evaluating Bubble Boundary Definition Using Characteristic Parameters of Bubble Boundary

Shao JianBin, Chen Gang, Li Guodong

The Institute of Hydraulics, Institute of Water Resources and Hydro-electric Engineering, Xi'an University of Technology, P.R. China, 710048

Abstract. Taking image with high definition is the basis to study aerated water flow using image measurement method. It was found that gas-water flow have complicated optical properties because both bubbles and water are transparent, and bubbles are featured with continuous deformation. Even under the same illumination conditions, the characteristics of the captured bubble images are very different. By now, human intuition and analysis still play a central role in the choices of the best lighting schemes. Nevertheless, the decision made within a person's mind is not unreliable. In this paper, we presented a series of quantitative evaluation methods to identify the imaging quality of bubble by brightness feature analysis of water, bubbles and interface between them. As an example, an optimized lighting scheme was determined via selection from back-lighting, side-lighting, normal-lighting and other lighting schemes with different angles. It is shown that our method is effective to optimize photography, reduce the number of pictures and obtain high quality images.

Keywords: bubble, water-air flow, image measurement

INTRUDUCTION

Two-phase gas/liquid flow is of great importance in a wide range of technical applications, for example in the scientific research of high-speed hydraulics. Although lots of research works have been done in the past years, this kind of flow is still far from being understand for its deformable interface and for its compressible air phase. Restrained by the experimental methods, the design of gas/liquid flow systems have to use the empirical relations established by a great amount of observations and approximate measurements. Therefore, developing measuring technology is the most important things in the research of two phase flow.

Digital image measurement method was developed based on the extending of digital particle image velocimetry (DPIV), which represents the development trends of two-phase gas/liquid flow experiment techniques. As the development of image measurement technique, some problems that difficult to solve with traditional method are overcame, such as the investigation of bubble velocities , the numbers and the shape of air bubbles, as well as the air entrainment concentration distribution. Image measurement technique depends largely on the clear images picked up by the high-speed digital camera. Blurred images often make the extracted bubbles inaccurate, which will further leads to large errors in the calculation of the bubble velocities, the distribution of the concentration of air, etc.

In order to obtain the high visibility image of gas-water flow, one routinely need to adjust the brightness of light, the angle of the beam, and the parameters of camera system again and again, and record lots of photographs for the best technical option. However, it is very difficult to determine which scheme is the best, because this is concerned with problem of how to list the order of the bubble images according to their boundary definition. The evaluation methods based on the human intuition and analysis can not always select out the best image suitable for computer reorganization. In addition, the selected result is often uncertain. Therefore, it has a significant science value to study the quantitative methods for analyzing the imaging quality.

This work was supported by the National Natural Science Foundation of China (Grant No: 50579085), the Research Fund for the Doctoral Program of Higher Education (Grant No: 20060700001)

CP914, *Multiphase Flow: The Ultimate Measurement Challenge,* edited by X. Cai, Y. Wu, Z. Huang, S. Wang, and M. Wang

METHODS FOR BUBBLE IMAGE QUALITY EVALUATION

The term *definition* applies to an gray image means the distribution of the gray values of pixels tend to occupy the entire range of possible gray levels and, in addition, the gray values are distributed homogeneously in each sub region as well as have obvious distinction in different regions.[1] In two-phase gas/liquid flow experiments, too dim or too bright illumination may cause the gray values to be limited within a relatively narrow range, so that the gray value difference between water phase and gas phase become too small for computer to segment bubbles rightly. Consequently, the broader the gray level ranges of the image are, the more excellent the bubble image is.

In the study of iris image quality evaluation, the average height of pupil boundary is calculated for evaluating the definition of the iris image, [2] which can pick out high quality iris image effectively. We are aware that this method can be well suitable for the evaluation of bubble boundary definition. As will be seen, the preliminary studies reported here confirm our expectations. In experiments it is found that the more excellent the image is, the difference of gray lever between regions of bubble boundary and of water is bigger. This can serve as the second rule for evaluating bubble boundary definition.

Because the sharpness of bubble boundary is often different, using maximum gradient of bubble boundary to evaluate bubble boundary definition can get well effect. The preliminary studies show that the bigger the bubble boundary's gradient is, the easier for computer to identify the boundary of gas-water interface. This is the third rule of our methods.

FIGURE 1. Air bubble image

Let's take an example to explain the meaning of these three standards. Figure 1 shows a grasped bubble image. The distribution of the gray values is represented by a histogram (Fig.2), where the x-axis corresponds to the gray scale (usually 0~255) and the y-axis represents the occurrence frequency. The difference between the maximum value, Max, and the minimum value Min of image gray values divided by the maximum scope, Ran (usually 256) defines the relative gray scope number, that is

$$CF = \frac{Max - Min + 1}{Ran} \times 100\% \tag{1}$$

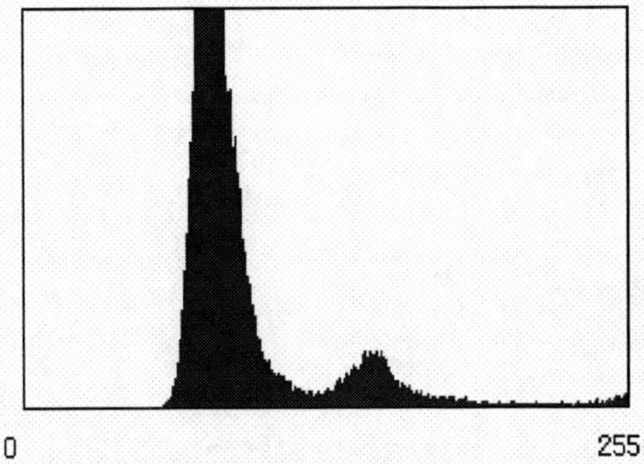

FIGURE 2. Corresponding histogram of Fig .1

The bigger the CF value is, the wider the bubble image scope is, so that the details formerly hard to take become clear in the bubble image. Therefore, in the experiments, one may modulate the illumination strength as well as the camera shutter and so on according to CF value, to ensure the image to have better gray scale distribution range.

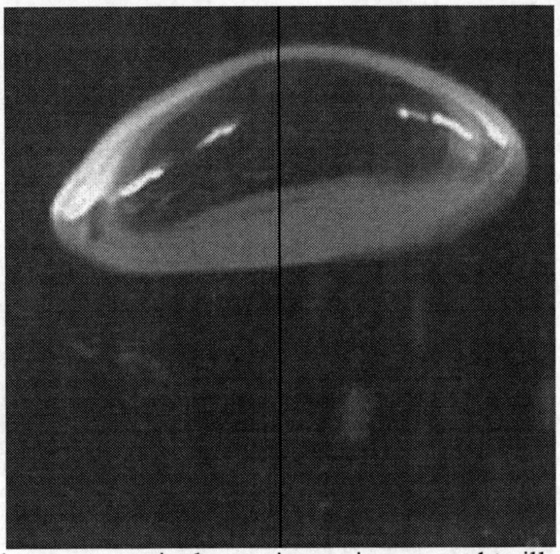

FIGURE 3. A scan line from bottom to top in the continuous image, used to illustrate the concept of sampling

As shown in Fig.3, taking the gray intensities from bottom to top along a line which pass through the center of the bubble, the result is plotted in Fig.4. We can see that the pixels from 0 to 131 and from 241 to 255 represent water phase (background of image), the pixel range between 136 and 234 represent the bubble, while the pixels 132～135,234～240 belonging to the bubble boundary (foreground of image). Let W be the gray level value of the pixels belong to the background of image, which always have small values in the image, but do not include the pixels belonging to the inner of bubble that also have little gray level value in the image. Let G represents the gray level values of the pixels belong to the foreground of image (bubble boundary). After calculating the average gray values G_{pj} of the two boundaries and background average gray value W_{pj} we can define

$$CH = G_{pj} - W_{pj} \tag{2}$$

to be their difference.

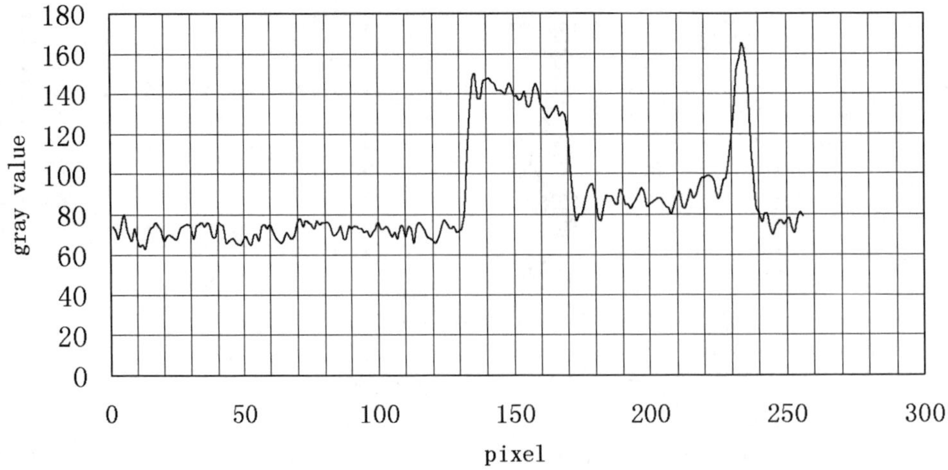

FIGURE 4. Gray values along the scan line in Fig 3

In fact, the bubble boundaries are generally rather irregular. Even for a same bubble, some parts of boundary are clear that can be easily identified, and have larger CH , while other parts of boundary are mistiness, even merging into with the background of the image, in these areas the CH value is very small and even close to zero. Therefore, following a pre-defined order we take samples along the several different lines at equally spaced angles, each of which through the center of bubble. The schematic paths are given in Fig.5. Calculating each CH value of all simples, then the average value $\overline{CH}$ can be obtained. Larger $\overline{CH}$ implies that the difference between the bubble boundary and image background is significant, i.e. that is the bubble boundary has higher definition.

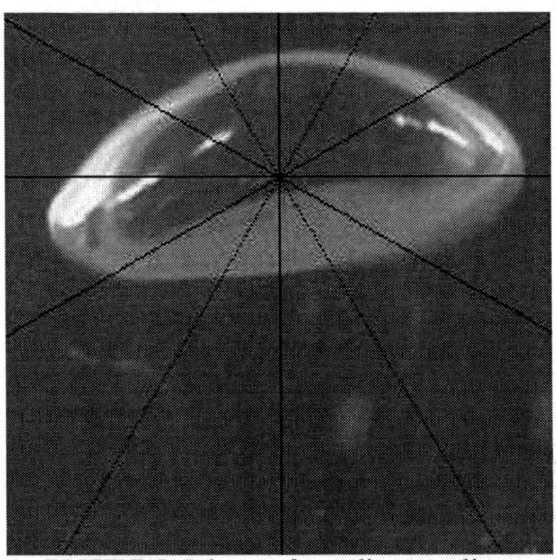
FIGURE 5. Scheme of sampling scan lines

The sharpness of bubble boundary, TD, is defined as the radio of ΔG to the width of bubble boundary, D,

$$\mathrm{TD}=\frac{\Delta G}{D}=\frac{\left|G_{\max}-W_{\mathrm{pj}}\right|}{D} \tag{3}$$

where ΔG is difference of the maximum average gray level value for two bubble boundary, $G_{\max}$, and the average gray level value of image background, W_{pj}.

The same as previous section, due to the irregular boundaries of bubble image, we also take the average of

TD around the bubble boundary, $\overline{TD}$, as the segmentation quality criteria. Larger $\overline{TD}$ means that the gray value transition from water to bubble boundary in the image increases abruptly over a comparatively short distance. In the image measurement, different processing method may leads to different result. For example, measuring the volume of a small bubble with different threshold value may results in the significant different volume values. For the bubble image that has bigger $\overline{TD}$ value, the computer recognition is independent of binary threshold. Therefore, in the experiment especially for the measurement of air entrainment concentration distribution, the bigger $\overline{TD}$ is, the higher the definition of the bubble boundary is.

EXPERIMENT

Experiment Equipment

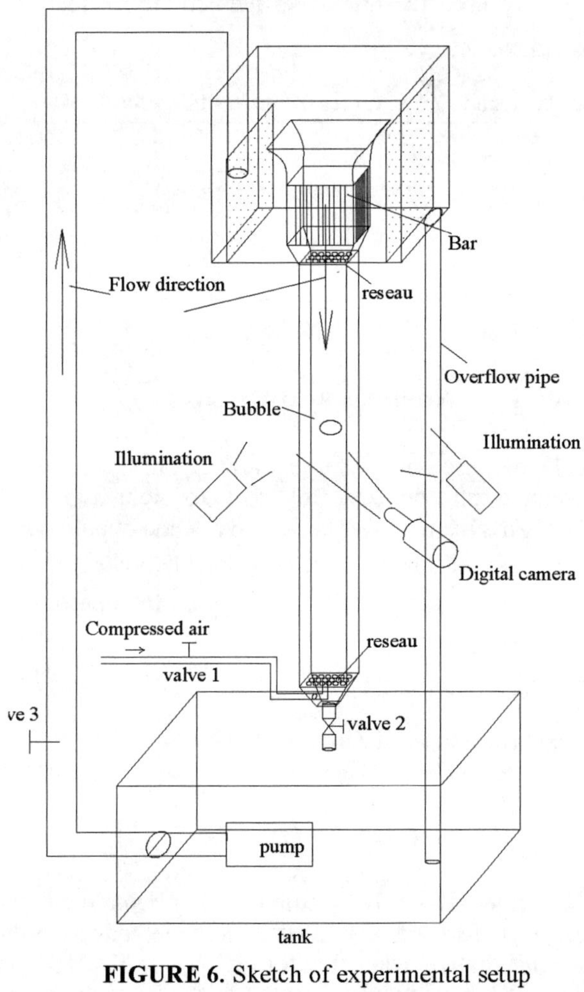

FIGURE 6. Sketch of experimental setup

The general layout of the test facility is shown in Fig.6, [3] which was made of PMMA. Valve 1 is used to convey air bubble into the pipe. Using valve 2 and valve 3 jointly can control the velocity of downward flow in the pipe. As the water velocity in the pipe reaches a certain value, the hydrodynamic forces and the buoyancy forces acting on the bubbles achieve balance, such that the bubbles stay in a relatively fixed region. This make repeated taking pictures on an identical bubble under different lighting schemes possible. The NIKON D100 (SLR) digital camera with a max resolution of 3008×2000 was used in our experiments.

Experiment Results

In order to verify the validity of the developed methods, the impact of different illuminations on the boundary definition of an identical bubble was conducted with a series of pictures taken from different lighting schemes such as back-lighting, side-lighting, normal- lighting and other lighting with different angle.

Lighting scheme 1: Back-lighting. A 240 W diffusion light cover the bubble with uniform intensity.

Lighting scheme 2-6: Two 300W Halogen Lamps with inclinations of 0^0,30^0,45^0,60^0, 90^0 serves as Dual Light Source, which located with photography area in the same horizontal plant.

Lighting scheme 7: Using the on-camera flash to take pictures from normal side.

In each scheme, the first step is to select best distance between the front plane of lens and specimen, and the shutter speed of camera, the ones with max CF values will be consider as the best options. In all these lighting schemes described above, the scheme 6 and 7 were firstly excluded because their CF values are much smaller than other schemes, and the grasped pictures are of illegibility.

However, the qualities of images obtained by scheme 1-5 are all well, and difficult to distinguish just according to the human's intuition. Therefore, we used the other two methods to evaluate these schemes. The calculated $\overline{CH}$ and $\overline{TD}$ value was listed below.

TABLE 1. The calculated $\overline{CH}$ and $\overline{TD}$ values of images in each scheme.

	Scheme 1	Scheme 2	Scheme 3	Scheme 4	Scheme 5
$\overline{CH}$	46.25	63.50	77.93	56.5	43.7
$\overline{TD}$	5.03	3.44	6.1	5.4	5.16

We can see that scheme 3 is the best scheme being with both maximum $\overline{CH}$ and $\overline{TD}$ values.

Analysis and Discuss

1) For back-lighting(scheme 1), because both water and air are transparent, in order to avoid lights directly entering into lens, diffusion light sources must be used. But this make light contrast ratio (the light contrast ratio is the ratio between the while and black parts in an image) became weak and bubble boundary became unclear.

2) For front side-lighting scheme 3-5, all images have clear bubble boundaries and difficult to be distinguished by human intuitions. Nevertheless, the calculated results showed that the dual light source with an inclination of 30^0 scheme can obtain the most clear bubble images.

3) For normal-lighting scheme 6 and 7, reflected lights from the PMMA of experiment setup enter into the lens directly that makes bubbles unclear.

4) To stress with, different experiments with different lighting conditions, such as lighting direction, light intensity and colors, therefore the best light scheme depend on actual experiment conditions.

CONCLUSION

Taking clear images that can be recognized easily by computer has being the key and difficulty to design image measurement system. In this paper we put forward several evaluation methods according to the broadness and slopes of image boundaries, which enable quantitative evaluation for the quality of images in different lighting conditions. Compared with traditional methods with human intuitions, these methods are more rational and reliable.

REFERENCES

1. Chen Xilin, Shan Shiguang, Gao Wen. Pose-Varied Face Recognition. Journal of Image and Graphics. 2002,5(3):

252-255.
2. HE Jia-feng, YE Hu-nian, YE Miao-yuan. A Study on Iris Image Quality Evaluation. Journal of Image and Graphics. 2003(4):387-391.
3. A.Tokuhiro，M.Maekawa，K.Iizuka，K.Hishida，M.Maeda. Turbulent flow past a bubble and an ellipsoid using shadow-image and PIV techniques. International Journal of Multiphase Flow 24(1998) 1383-1406.

Experimental Investigations of Two-phase Boiling Flow in a Complicated Channel

Guangyao LU, Jing WANG, and Qilin TANG

School of Mechanical Engineering, Shanghai Jiaotong University, 200240, Shanghai, China

Abstract. Experiments were carried out to investigate the characteristics of two-phase boiling flow and heat transfer characteristics in a vertical complicated channel. The results showed that boiling heat transfer characteristics for R113 flow in the complicated channel were different from those in tubes with normal size, and flow boiling in a complicated channel can enhance the efficiency of convective heat transfer. The influences of flow rate, heat flux and temperature difference were analyzed. It was found the flow rate had small influence on the convective heat transfer coefficient in low heat flux area and in small temperature difference region. In addition, the flow visualization experiments were conducted to observe flow regimes and their transitions. It was found that flow regimes and transitions in the complicated channel were different from other normal tubes. In the present experiments, there were only four main flow patterns, which were bubble flow, slug flow, churn flow, and annular flow. In slug flow region, the gas plug was segmented by the U-type electrical heater. And the plug was broken immediately after it was formed owing to the geometric influence of complicated channel.

Keywords: Complicated channel; Flow boiling; Flow regime; Visualization

INTRODUCTION

As a means of heat transfer augmentation, boiling is widely used in power, energy sources, chemical industry and other industrial fields. Flow boiling could enhance the efficiency of heat exchange, reduce energy loss, and minify the volume and mass of heat exchanger et al. All these merits are important in the realm of navigation and military affairs [1]. Researches of flow boiling have always been one of the most important branches in the field of multiphase flow. In nuclear industry, two-phase boiling flow and heat transfer occur in most facility, for example, the steamer, gas generator, transporting pipelines and so on. The characteristics of boiling flow and heat transfer are very important for the security and safety of nuclear instruments.

The present study is carried out to investigate two-phase flow and heat transfer in a complicated channel with small dimension in nuclear gas generator. Bergles and Dormer [2] performed one of the earliest studies of flow boiling in small tubes. Later, many experiments were carried out to investigate heat transfer coefficient, pressure drop, critical heat flux and flow regimes of refrigerant boiling in small channels [3~8]. In their experiments, most hydraulic diameters were in the range of 1.0~5.0mm. Huo et al. [8] investigated R134a boiling heat transfer in small diameter tubes. Their heat transfer experiments were conducted with two stainless steel tubes of internal diameter 4.26 and 2.01mm respectively. It was found that the nucleate boiling was dominant when the vapour quality was less than about 40–50% for the 4.26 mm tube and 20–30% for the 2.01mm tube. They also conducted flow visualization experiments in Pyrex glass tubes. Six typical flow patterns were observed, i.e. dispersed bubble, bubbly, slug, churn, annular and mist. Warrier et al. [9] conducted experiments to study heat transfer and pressure drop in narrow rectangular channels. Two new correlations were proposed – one for subcooled flow boiling heat transfer and the other for saturated flow boiling heat transfer. On the point of dimension influence, some researchers reported that the existing models or empirical maps for normal size tubes could not well predict flow patterns in channels with small dimension [10-12]. There are some contradictory conclusions [13, 14]. And some other contradictory points in two-phase flow and boiling heat transfer still need to be solved.

On the basis of the results obtained before, experiments were carried out to investigate the characteristics of two-phase flow and boiling heat transfer in a vertical complicated channel. The influences of flow rate, heat flux and temperature difference were also analyzed. A new correlation was proposed to calculate boiling heat transfer for

CP914, *Multiphase Flow: The Ultimate Measurement Challenge,* edited by X. Cai, Y. Wu, Z. Huang, S. Wang, and M. Wang

R113 flow in a complicated channel. Furthermore, flow visualization experiments were carried out with high-speed camera. Through the comparison between the experimental results and other works, it was found flow regimes and their transitions in the complicated channel were different from other normal tubes.

EXPERIMENTAL

The test section is composed of two electrical heaters and a thimble, which is shown in Fig. 1. Two electrical heaters are manufactured as "U" type, and the thimble is made of organic glass for the purpose of flow visualization. The dimension of thimble is Φ35mm and 2100mm long. The dimension of the electrical heater is Φ8mm and 1500mm long; the gap inside electrical heater is 10mm; the width between electrical heater and thimble is 4.5mm; and the pressure measuring length is 650mm and 750mm, respectively.

The fluid temperature and wall temperature are measured by copper constantan thermocouples. The placements of the thermocouples at axial distance are 100, 300, 500, 700, 900, 1100, and 1300mm part from the inlet of test section. The thermocouples are placed into holes drilled into the electrical heater surface and secured in place by high conductivity thermocouple cement to measure the wall temperature. At each measuring place, two thermocouples are used to measure the wall temperature and one thermocouple to measure the fluid temperature. At the inlet and outlet of test section, thermocouples are also installed to measure the fluid temperature.

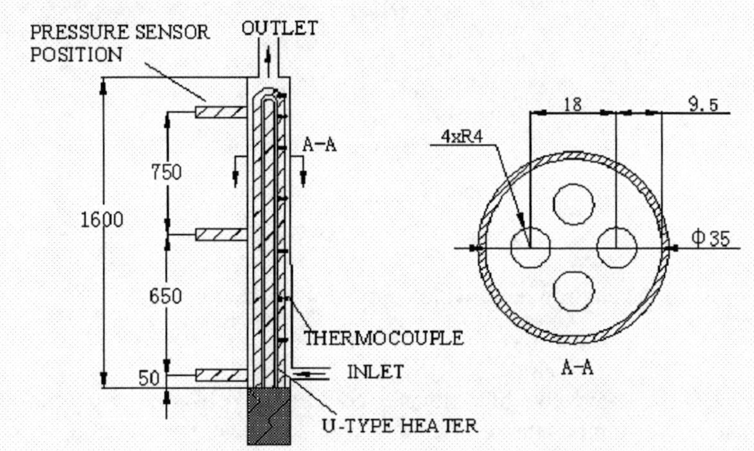

FIGURE 1. Structure of test section

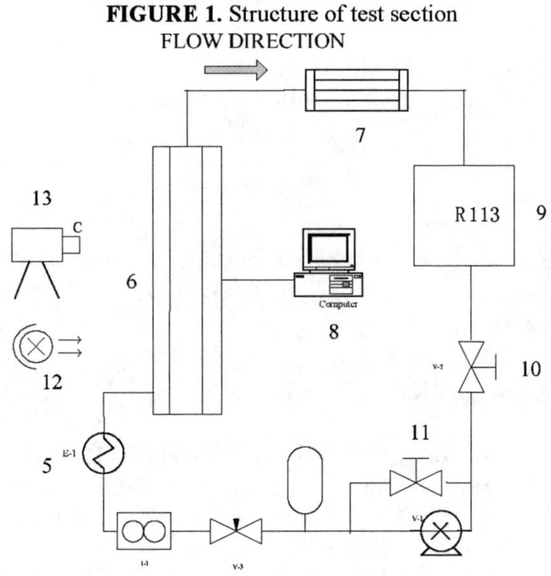

FIGURE 2. Schematic diagram of experimental apparatus

1.Magnetic pump 2.Manostat 3.Diaphragm valve 4.Flowmeter 5.Preheater 6.Test section 7.Condenser 8.Computer 9.Liquid tank 10,11.Ball valve 12.Lamp-house 13.High-speed camera

The schematic diagram of experimental apparatus is shown in Fig. 2. In the experiments, refrigerant 113 is adopted as working fluid. R113 liquid, driven by magnetic pump, flows through flowmeter, preheater, test section and condenser, and then returns back to liquid tank. The liquid temperature at the inlet of test section is maintained at steady values, by regulating the heat input of preheater. At the outlet of test section, two-phase R113, cooled in the condenser, becomes single-phase liquid and then flows into liquid tank. The flow rate is measured by float-type flowmeter, and the heat flux is calculated by the electrical current and voltage. The error in the heat flux measurement is less than ±3%. The error in flow rate measurement is less than ±3%, and for temperature measurement is less than ±1%.、

RESULTS AND DISCUSSION

In the experiments, the temperature range of R113 is 40~78℃; the flow rate ranges from 50 to 350L/h; and the heat flux range is 0~60kw/m2.

Heat Transfer

In the experiments, the total power supplied to the heaters is,

$$Q_{supply} = VI \tag{1}$$

Where, V=Voltage, I=Current.

And the net power to the complicated channel ($Q_{channel}$) can be calculated by,

$$Q_{channel} = Q_{supply} - Q_{loss} \tag{2}$$

Where, Q_{loss} is the heat loss to the environment, which can be calculated by,

$$Q_{loss} = h_{environment}(T_{outer} - T_{environment}) \tag{3}$$

Where, henvironment is the convective heat transfer coefficient between the outer walls and the environment; Touter and Tenvironment are the temperature of the outer wall and the environment, respectively.

The wall heat flux is assumed to be uniform and is calculated as,

$$q = Q_{channel} / A_{heater} \tag{4}$$

Where, Aheater is the total surface area of the electrical heater.

The heat transfer coefficient is determined from the following equation,

$$h = q/(T_w - T_f) \tag{5}$$

Where, T_w is the inner wall temperature; T_f is the fluid temperature.

The vapour quality, x, is determined based on the heat transferred to the fluid.

$$x = (H - H_l)/(H_g - H_l) \tag{6}$$

Where, H_l and Hg are the specific enthalpy of saturated liquid and vapor, respectively. H is the local specific enthalpy of the fluid, which is determined from the enthalpy of the previous section and the heat transferred to the fluid.

The relationship between the convective heat transfer coefficient and the heat flux is shown in Fig. 3. The convective heat transfer coefficients increase with the augmentation of the heat flux. When the heat flux is lower than 10kw/m2, there is small difference in the convective heat transfer coefficients for different flow rates. It is shown that the flow rate has small influence on the convective heat transfer coefficient in low heat flux region. This result is similar to Tran [5]. Corresponding to the same heat flux which is larger than 10kw/m2, the heat transfer coefficient of small flow rate is bigger than that of large flow rate. The heat transfer efficiency is better when the flow rate is small in the present experiment bound. KYTEOB [16] carried out experiments to investigate the heat transfer characteristics of R113 flowing in tubes, and proposed the correlation to calculate the convective heat transfer coefficient, which is shown as,

$$h = 0.7574q^{0.75} \tag{7}$$

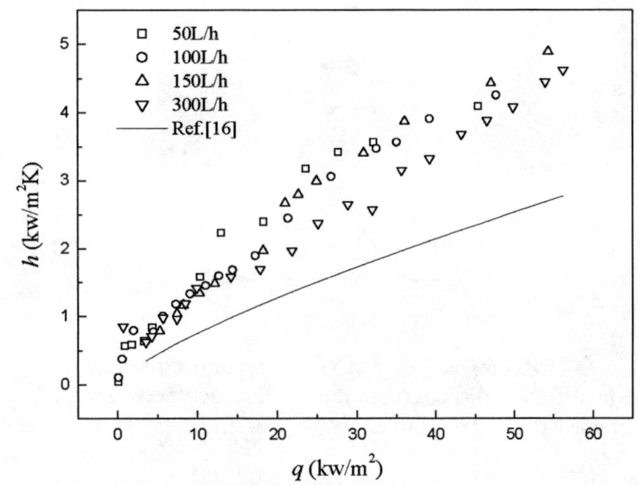

FIGURE 3. Convective heat transfer coefficient *vs.* heat flux

From Fig. 3, it is shown the convective heat transfer coefficients of the present experiments are larger than those in normal tubes [16]. The flow boiling in the complicated channel can enhance the efficiency of convective heat transfer more than tubes with normal size.

The relationship between the convective heat transfer coefficient and the temperature difference between the heater wall and the fluid is shown in Fig.4. The convective heat transfer coefficients increase with the augmentation of the temperature difference. Similar to the relationship between the convective heat transfer coefficient and the heat flux, there is small difference in the convective heat transfer coefficients for different flow rates when the temperature difference is lower than 8℃. In the region of temperature difference larger than 8℃, the convective heat transfer coefficient of 100L/h is the largest in these four flow rates. It is shown that for one specific temperature difference ($T_w - T_f > 8$ ℃), there is a optimal flow rate to gain the best heat transfer efficiency. In the present experimental bound, 100L/h is this optimal flow rate.

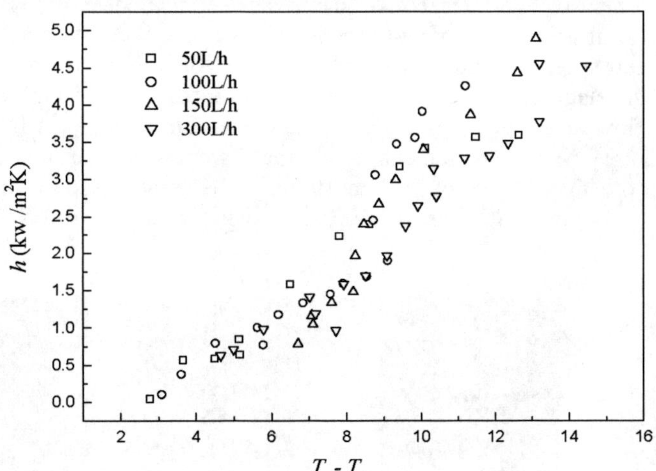

FIGURE 4. Convective heat transfer coefficient *vs.* temperature difference

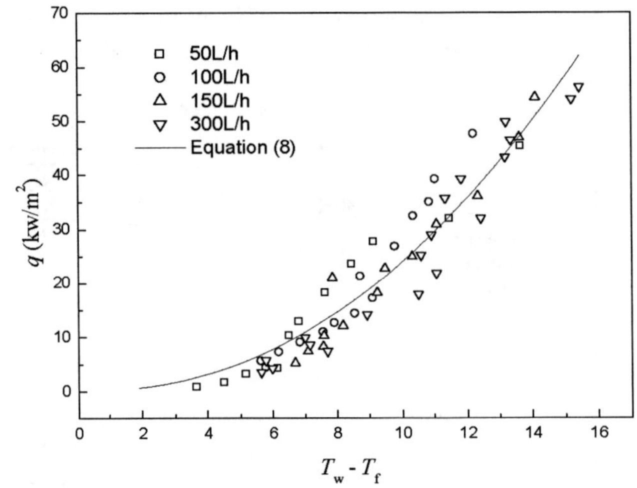

FIGURE 5. Heat flux *vs.* temperature difference

The relationship between the heat flux and the temperature difference between the heater wall and fluid is shown in Fig.5. In the experimental bound, the regressing equation for the heat flux, q, as a function of temperature difference is gained as,

$$q = 152.33(T_w - T_f)^{2.19} \tag{8}$$

Flow Pattern

Most researchers agreed to categorize their observations into four main flow patterns: stratified flow, intermittent flow, annular flow and bubble flow. Each main class could be subdivided into subclasses [8]. In the flow visualization experiments, the flow patterns and their transitions are obtained by the digital high-speed camera. From the experimental results, it is shown that there are only four main flow patterns in the present experiment, which are bubble flow, slug flow (intermittent flow), churn flow, and annular flow. Figure 6 shows the above flow patterns obtained in the experiments. In bubble flow region, numerous small bubbles float in a continuous liquid phase. The quantity and volume of bubble begin to increase with the augmentation of heat input. The gas plug, which takes shape as the increment of bubble size and quantity, is segmented by the U-type electrical heater. The plug is broken immediately after it is formed owing to the geometric influence of complicated channel. Then, slug flow is transformed to churn flow soon after slug is formed. In churn flow region, the gas phase and liquid phase stir violently in the test section. The electrical heater and thimble are in intense scour, and there are some dry spots on the heater wall. The annular flow takes its shape with the augmentation of the void fraction. When the heat flux continues to increase, there might be the phenomena of alternate wetting and drying cycles upon the top of the U-type electrical heater. Then the incensement of the heat flux input is ceased here for the experimental safety. It is shown that the flow patterns and their transitions in the present experiment are different from other works [8, 12, 15].

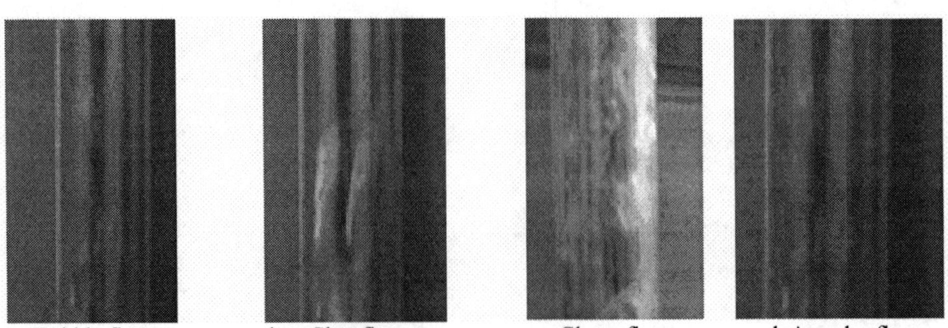

a. Bubble flow b. Slug flow c. Churn flow d. Annular flow

FIGURE 6. Flow patterns

CONCLUSION

Experiments were carried out to investigate the characteristics of two-phase boiling flow and heat transfer in a vertical complicated channel. The influences of flow rate, heat flux and temperature difference were analyzed. A new correlation was proposed to calculate boiling heat transfer for R113 flow in a complicated channel. The experiments showed that the boiling heat transfer characteristics for R113 flow in the complicated channel were different from those in tubes with normal size. And flow boiling in complicated channels can enhance the convective heat transfer more than the normal tubes. The experimental results showed that in low heat flux area and in small temperature difference region, the flow rate had small influence on the convective heat transfer coefficient. For one specific temperature difference, which was larger than 8℃, there was a optimal flow rate to gain the best heat transfer efficiency.

Furthermore, the visualization experiments were conducted to observe the flow regimes and their transitions. Through the comparison between the experimental results and those of other channels, it was found that flow regimes and their transitions in the complicated channel were different from other normal tubes. There were only three main flow patterns in the present experiments, which were bubble flow, slug flow, churn flow and annular flow. In slug flow region, the gas plug was segmented by the U-type electrical heater. And the plug was broken immediately after it was formed owing to the geometric influence of complicated channel.

ACKNOWLEDGEMENTS

The authors would like to acknowledge the financial support from the National Natural Science Foundation of China (Grant No. 50476015).

REFERENCES

1. Kemal O. Pasamehmetoglu, Padmanabha R. Chappidi, Cetin Unal, Saturated pool nucleate boiling mechanisms at high heat fluxes, Int. J. Heat Mass Transfer, Vo. 36, (15) , 1993, pp.3859-3868.
2. A.E. Bergles, T. Dormer Jr., Subcooled boiling pressure drop with water at low pressure, Int. J. Heat Mass Transfer, Vol.12, 1969, pp.459–470.
3. Z.Y. Bao, D.F. Fletcher, B.S. Haynes, Flow boiling heat transfer of Freon R11 and HCFC123 in narrow passages, Int. J. Heat Mass Transfer, Vol.43, 2000, pp.3347–3358.
4. G.M. Lazarek, S.H. Black, Evaporative heat transfer, pressure drop and critical heat flux in a small vertical tube with R-113, Int. J. Heat Mass Transfer, Vol.25 (7) , 1982, pp.945–960.
5. T.N. Tran, M.W. Wambsganss, D.M. France, Small circular- and rectangular-channel boiling with two refrigerants, Int. J. Multiphase Flow, Vol.22 (3) , 1996, pp.485–498.
6. M.W. Wambsganss, D.M. France, J.A. Jendraejczyk, T.N. Tran, Boiling heat transfer in a horizontal small diameter tube, J. Heat Transfer, Vol.115, 1993, pp.963–972.
7. Y.Y. Yan, T.F. Lin, Evaporation heat transfer and pressure drop of refrigerant R-134a in a small pipe, Int. J. Heat Mass Transfer, Vol.41, 1998, pp.4183–4194.
8. X. Huo, L. Chen, Y.S. Tian, T.G. Karayiannis, Flow boiling and flow regimes in small diameter tubes, Applied Thermal Engineering, Vol.24, 2004, pp.1225-1239.
9. Gopinath R. Warrier, Vijay K. Dhir, Leslie A. Momoda, Heat transfer and pressure drop in narrow rectangular channels, Experimental Thermal and Fluid Science, 2002, Vol.26, pp.53–64.
10. D.A. Damianides, J.W. Westwater, Two-phase flow patterns in a compact heat exchanger and in small tubes, 2nd UK Nat. Conf. on Heat Transfer, vol. 2, sessions 4a–6c, 1988, pp.1257–1268.
11. S. Lin, P.A. Kew, K. Cornwell, Characteristics of air/water flow in small tubes, Heat Technol. Vol.17 (2), pp.63–70, 1999.
12. K. Mishima, M. Ishii, Flow regime transition criteria for upward two-phase flow in vertical tubes, Int. J. Heat Mass Transfer, Vol.27 (5) , 1984, pp.723–737.
13. J.W. Coleman, S. Garimella, Characterization of two-phase flow patterns in small diameter round and rectangular tubes, Int. J. Heat Mass Transfer, Vol.42,1999, pp.2869–2881.
14. T. Oya, Upward liquid flow in small tube into which air streams (1st report, experimental apparatus and flow pattern), Bull. JSME, Vol.14 (78) , 1971, pp.1320–1329.
15. Y. Taitel, Flow pattern transition in two phase flow, Keynote lecture, 9th International Heat Transfer Conference, Jerusalem, Israel, 1990, pp. 237–254.

16. A.M.KYTEOB, Dynamics and heat exchange in the process of forming steam, translated in Chinese by Fan Cong-zhen and Sa Ying-lu, Beijing: Water Conservancy and Electric Power Publishing House, 1983.

Determination of Polychlorinated Diben-p-dioxins and Dibenzofurans in Flue Gas by High Resolution Gas Chromatography Coupled with High Resolution Spectrometry

T. Chen*, J.H. Yan, X.D. Li, K.F. Cen

*State Key Laboratory of Clean Energy Utilization , Zhejiang University, 38 Zheda Road 310027,Hangzhou, China

Abstract. In 1988 the first municipal solid waste (MSW) incinerator plant was built in Shenzhen, China. Since then Beijing, Shanghai and other big cities have built a few MSW incineration plants. MSW incineration has gradually been used in some cities of China due to its advantages such as significant volume reduction (about 90%), mass reduction (about 70%), and toxicity reduction of the waste and energy recovery. However, MSW incineration is sometimes considered to the general public as the secondary pollution source, because of concerns about is toxic combustion byproducts (TCBs). Polychlorinated dibenzo-p-dioxins and dibenzofurans (PCDD/Fs) are the most toxic compounds among the TCBs. In this paper, PCDD/Fs in flue gas produced from a MSW incinerator were isokinetically withdrawn and collected in a multi component sampling train. Then the PCDD/Fs samples were extracted, concentrated and cleaned up step by step. Finally, the determination of PCDD/Fs was carried out by high-resolution gas chromatography coupled with high-resolution mass spectrometry (HRGC/HRMS). The two PCDD/Fs emission levels were the similar, which was 0.115 and 0.096ng TEQ/Nm^3 although different sampling time. The PCDD/Fs emission levels were lower than the PCDD/Fs emission regulation in China (1.0 ng TEQ / Nm^3) and close to the developed countries' regulations, i.e., 0.1ng TEQ/Nm^3. 2,3,4,7,8-Pentachlorodibenzofuran (2,3,4,7,8-PeCDF) was the dominant congener of the total TEQ.

Keywords: municipal solid waste incinerator (MSWI), flue gas, polychlorinated dibenzo-p-dioxins and dibenzofurans (PCDD/Fs), high-resolution gas chromatography coupled with high-resolution mass spectrometry (HRGC/HRMS)

INTRODUCTION

Landfill is the dominant municipal solid waste (MSW) disposal method in China, which accounts for more than 80% the MSW disposal [1]. However, the number of large-scale MSW incinerator plants has gradually increased in developed cities in China recently. In 1988, the first municipal solid waste (MSW) incinerator plant was built in Shenzhen, China. Since then Beijing, Shanghai and other big cities have built a few MSW incineration plants. For example, 9 MSW incineration plants have been constructed in Zhejiang Province, which can dispose 4400 MSW tons per day, and 12 others under construction, which are designed to dispose 6600 tons per day [2]. MSW incineration has gradually been used in some cities of China due to its advantages such as significant volume reduction (about 90%), mass reduction (about 70%), and toxicity reduction of the waste and energy recovery. However, MSW incineration is sometimes considered to the general public as the secondary pollution source, because of concerns about is toxic combustion byproducts (TCBs). Polychlorinated dibenzo-p-dioxins and dibenzofurans (PCDD/Fs) are the most toxic compounds among the TCBs.

In this paper, PCDD/Fs in flue gas produced from a MSW incinerator plant were isokinetically withdrawn and collected in a multi component sampling train. Then the PCDD/Fs samples were extracted, concentrated and cleaned up step by step. Finally, the determination of PCDD/Fs was carried out by high-resolution gas chromatography coupled with high-resolution mass spectrometry (HRGC/HRMS).

CP914, *Multiphase Flow: The Ultimate Measurement Challenge,* edited by X. Cai, Y. Wu, Z. Huang, S. Wang, and M. Wang

EXPERIMENTAL

The MSW incinerator plant consists of three Martin stoke grate, which number is 1, 2 and 3. Air pollutant control devices (APCDs) are composed of bag filter and semi-dry scrubber. Activated carbon also used to remove PCDD/Fs from the flue gas. The sampling location lies after APCDs of the number 1. Schematic of PCDD/Fs sampling system is shown in Fig.1. This sampling system was based on EPA method 0023A [3]. Sampling volume is approximately $2Nm^3$. The sampling time is two in different days. The PCDD/Fs surrogate standards were added to the XAD-2 resin for check the PCDD/Fs sampling efficiency. After each sampling, the system was rinsed with acetone, toluene respectively and the rinse was stored in brown glass bottle. Filter and XAD-2 resin also stored and maintained in the dark at <4℃ from the field until receipt at the laboratory.

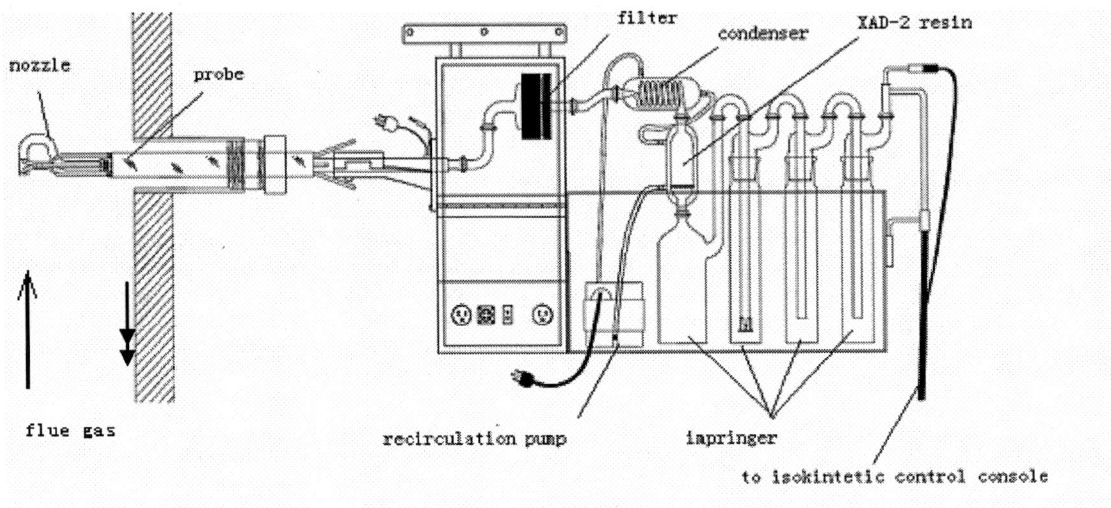

FIGURE 1. Schematic of method for PCDD/Fs sampling from flue gas

Analysis of PCDD/Fs in the collected flue gas was conducted according to the EPA method 1613 B [4]. The analysis flow chart for PCDD/Fs is shown in Fig.2. The analysis was performed by HRGC/HRMS on a 6890 Series gas chromatograph (Agilent, USA) and coupled to a JMS-800D mass spectrometer (JEOL, Japan). A DB-5ms (60m×0.25mm I.D., 0.25μm film thickness) capillary column

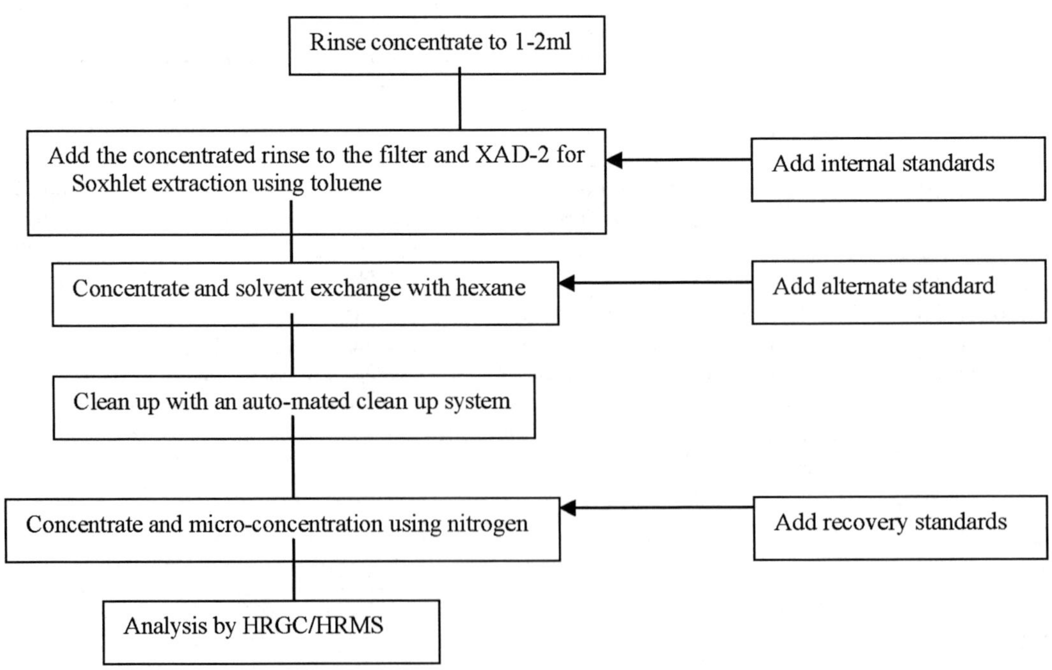

FIGURE 2. The analysis flow chart for PCDD/Fs

was used for separation of PCDD/Fs congeners. The GC temperature program was optimized as follows: splitless injection of 2μl at 150℃, initial oven temperature of 150℃for 1min, then increased at 25℃/min to 190℃, finally increased at 3 ℃/min to 280℃ and held for 20min. He was used as the carrier gas. The mass spectrometer was operated in the electron impact ionization mode using selected ion monitoring (SIM). Electron energy was set to 38eV. Source temperature was 280℃. The mass system was tuned to a minimum resolution of 10000 (10% valley) using PFK (perfluorokerosene) as lock mass. The detailed quantitative determination of PCDD/Fs was referred to US EPA method 1613 B.

DISCUSSION AND RESULTS

Fig.3 shows the 17 2,3,7,8-substituted PCDD/Fs distribution from flue gas. The PCDD/Fs congeners distribution was similar although different sampling time. 2,3,4,7,8-PeCDF, 1,2,3,7,8-PeCDD, 1,2,3,6,7,8-HxCDF and 1,2,3,4,7,8-HxCDF were the major congeners among the 17 2,3,7,8-substituted PCDD/Fs. Especially 2,3,4,7,8-PeCDF, which contributed to over 50% in the total TEQ. Several researchers found there is a linear correlation between the 2,3,4,7,8-PeCDF with the TEQ value [5]. Kato et al summarized the PCDD/Fs data of 64 facilities and concluded the following Equation (1) [6].

[I-TEQ of PCDD/Fs]=1.4[2,3,4,7,8-PeCDF], r=0.99 (1)

So 2,3,4,7,8-PeCDF could be used as an indicator for TEQ prediction.

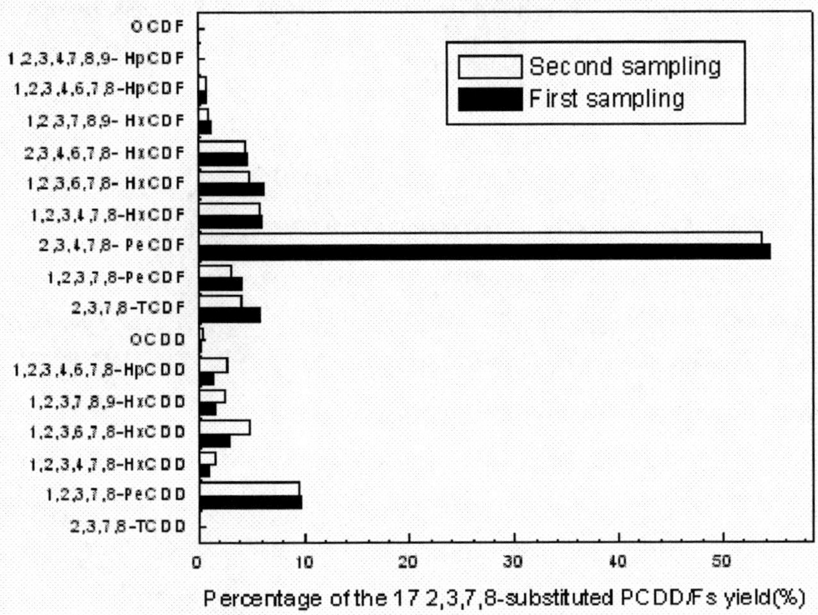

FIGURE 3. Congener profiles of 17 2,3,7,8-substituted PCDD/F in flue gas from two samplings

Table 1 shows the 2,3,7,8-substituted PCDDs and PCDFs levels in flue gas from two samplings. It could be seen the 2,3,7,8-PCDDs levels were much lower than that of 2,3,7,8-PCDFs, which means de novo mechanism may be more important than precursor routes. The permitted emission level for PCDD/Fs from MSW incineration in China is 1.0 ng TEQ/Nm^3, which is based on standard dry flue gas corrected to 11% O_2 condition. So O_2 content is an important parameter for PCDD/Fs control. The two PCDD/Fs emission level corrected to 11% O_2 was 0.115, 0.096 respectively, which was more lower than the PCDD/Fs emission regulation in China, i.e., 1.0 ng TEQ / Nm^3 and close to the developed countries' regulations, i.e., 0.1 ng TEQ / Nm^3.

TABLE 1. 2,3,7,8-substituted PCDD/Fs levels in flue gas

Concentration (ng TEQ/Nm^3)	First sampling	Second sampling
2,3,7,8-PCDDs	0.025	0.025

2,3,7,8-PCDFs	0.123	0.088
Total PCDD/Fs	0.148	0.113
Total PCDD/Fs (11% O_2)	0.115	0.096

CONCLUSION

The PCDD/Fs emission from the MSW incinerator was 0.115 and 0.096ng TEQ/Nm^3 respectively during different sampling time, which was lower than the PCDD/Fs emission regulation in China, i.e., 1.0 ng TEQ / Nm^3 and close to the developed countries' regulations, i.e., 0.1 ng TEQ / Nm^3.

2,3,4,7,8-PeCDF was the dominant congener of the total TEQ, which could be used as a indicator for PCDD/Fs prediction.

O_2 content is an important parameter for PCDD/Fs control from MSW incineration.

REFERENCE

1. M. Li, J. Xiang, S. Hu, L. S. Sun, S. Su, P. S. Li, X. X. Sun, *Fuel*, 83,1397-1405 (2004).
2. P.L. Gu, The 2nd China-Korea joint symposium on incineration/pyrolysis and environmental management, Hangzhou, China, Mar. 2006.
3. US EPA Method 0023 A, Sampling Method for Polychlorinated Dibenzo-p-dioxins and Polychlorinated Dibenzofurans Emissions from Stationary Sources, Dec.1996.
4. US EPA Method 1613, Revision B, Tetra- through Octa-chlorinated Dioxins and Furans by Isotope Dilution HRGC/HRMS, Oct. 1994.
5. F.Iino, T. Takasuga, A. Touati, B.K.Gullett, *Waste Management*, 23, 729-736 (2003).
6. M.Kato, K.Urano, *Waste Management*, 21, 55-62 (2001).

Mercury Emission Measurement in Coal-Fired Boilers by Continuous Mercury Monitor and Ontario Hydro Method

Yanqun Zhu[*,+], Jinsong Zhou[+], Sheng He[+], Xiaoshu Cai[*], Changxin Hu[+], Jianming Zheng[+], Le Zhang[+], Zhongyang Luo[+], Kefa Cen[+]

**:College of Power Engineering, University of Shanghai for Science &Technology, Shanghai 20093, China,*
+:State Key Laboratory of Clean Energy Utilization, Zhejiang University, Hangzhou 310027, China
E-mail: yqzhu@zju.edu.cn

Abstract. The mercury emission control approach attaches more importance. The accurate measurement of mercury speciation is a first step. Because OH method (accepted method) can't provide the real-time data and 2-week time for results attained, it's high time to seek on line mercury continuous emission monitors(Hg-CEM).Firstly, the gaseous elemental and oxidized mercury were conducted to measure using OH and CEM method under normal operation conditions of PC boiler after ESP, the results between two methods show good consistency. Secondly, through ESP, gaseous oxidized mercury decrease a little and particulate mercury reduce a little bit, but the elemental mercury is just the opposite. Besides, the WFGD system achieved to gaseous oxidized mercury removal of 53.4%, gaseous overall mercury and elemental mercury are 37.1% and 22.1%, respectively.

Keywords: Mercury, On line CEM method , OH method, Measurement of mercury;

INTRODUCTION

Mercury (Hg) is a high-priority regulatory concern because of its persistence and bioaccumulation in the environmental and evidence of serious adverse effects on the neurological development of human being. Statistically, the mercury form coal-fired utility is now one of the largest single source mercury in China with an annual ever-increasing rate of 4.8%[1]. Definitely, Hg emission standard for coal-fired boilers would be carried out in the future. Hereby, it's imperative to seek a new, effective technology approach to control mercury emissions.

Ordinary, mercury exists in three primary forms, gaseous elemental mercury(Hg^0), gaseous oxidized mercury (Hg^{2+}), and particulate-bound mercury (Hg_p) .The mercury that is absorbed onto the solid surfaces, such as fly ash or unburned carbon, is known as particulate-bound mercury, which can be captured by downstream particulate mater (PM) control devices. Due to its soluble, oxidized mercury (mainly $HgCl_2$) is liable to be captured in wet flue gas desulfurization(WFGD) system. While, elemental mercury is volatile, less soluble in water, so little of effect by the WFGD. Injection of activated carbon(AC) upstream of either an ESP or fabric filter (FF) has been investigated as a potential control method. Thus the specific form of mercury varies with the control technology configuration.

In order to achieve mercury emission control properly, it needs to better understand the transformation and speciation of mercury in coal combustion flue gas. They are strongly dependent on serious of factors such as the type of coal, operating conditions, flue gas temperature, the residence time, load changing, existing air pollution control (Electrostatic Precipitation(ESP), WFGD, Low-NO_x burner, etc) and so on. The measurement of mercury in flue gas is the basic of the mechanism studies. The traditional measurement method-Ontario Hydro(OH) method is suggested by U.S.EPA for the mercury emission[2]. But the OH method, based on chemical absorption, needs lots of procedure and several hours for one sample. Nowadays, several mercury Continuous Emission Monitoring (CEM) systems are proposed by different manufactures, but the accuracy and stability need to be investigated.

In this paper, 300MW and 200MW utility tangential coal fired PC boiler were chosen to carry out the field test measurement and measure the mercury and its speciation emission by Ontario Hydro method (OH) and Hg Continuous Emission Monitoring system (Hg-CEM).

CP914, *Multiphase Flow: The Ultimate Measurement Challenge,* edited by X. Cai, Y. Wu, Z. Huang, S. Wang, and M. Wang

EXPERIMENTAL

Ontario Hydro method (OH method)

Ontario Hydro method (OH) which is concerned as a American Standard testing method of measuring elemental, oxidized, particle-bound, and total Mercury in flue gas generated from Coal-Fired Stationary. Based on EPA method 5, the isokinetic sampling is adopted, made by the KNJ Engineering, INC. Figure 1. is the isokinetic sampling system. Isokinetic sampling is defined as the uniform sampling of particles and gases moving within a stack. It is achieved when the velocity of gas entering the sampling nozzle is exactly equal to the velocity of the approaching gas stream. This provides a uniform, unbiased sample of the particulate matter being emitted by the source.

A sample is withdrawn from the flue gas stream isokinetically through a probe/filter system, maintained at 120℃, followed by a series of impingers in an ice bath. Particle-bound mercury is collected in the front half of the sampling train. Oxidized mercury is collected in impingers containing a chilled aqueous potassium chloride(1mol/L KCl) solution. Elemental mercury is collected in subsequent impingers (one impinger containing a chilled aqueous acidic solution of hydrogen peroxide(5% v/v HNO_3 /10%v/v H_2O_2) and three impingers containing chilled aqueous acidic solutions of potassium permanganate(4%w/v H_2SO_4 /10%v/v $KMnO_4$). Samples are recovered, digested, and then analyzed for mercury using Atomic Fluorescence Spectroscopy (AFS), whose detection limit is 1ppb.

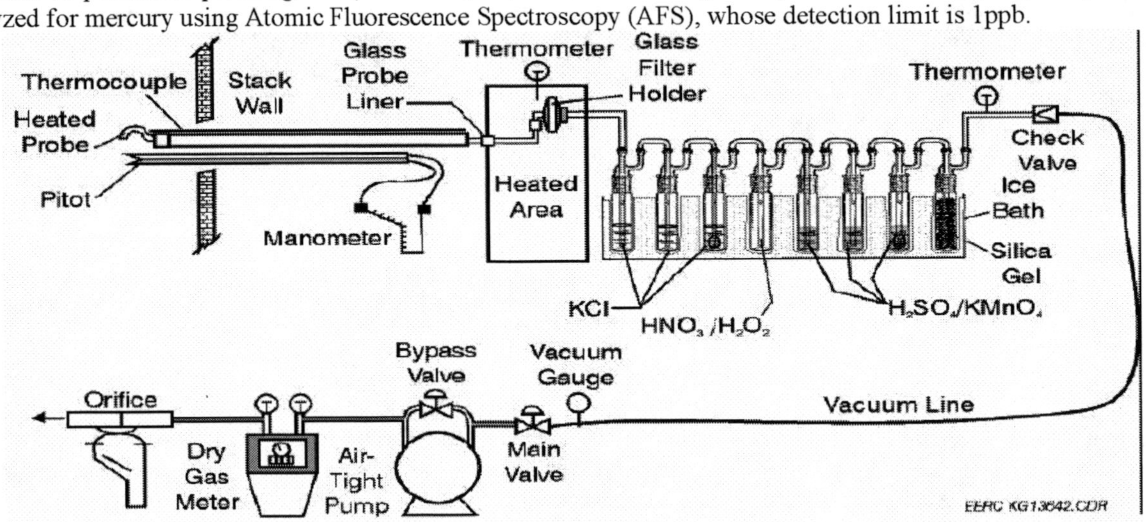

FIGURE 1. OH method sampling system

On-line Hg Continuous Emission Monitoring system (Hg-CEM)

Hg Continuous Emission Monitoring system (Hg-CEM) uesd in this study, Nippon DM-6B/MS-1A, manufactured by NIC (Nippon Instrument Corporation).

The flow diagram of the CEMs system is shown in Figure 2. The system consists of four parts: ①Sampling system includes a probe, filter, pump and heating Teflon sampling lines,20 meters long, being kept at 200℃. ② Pretreatment/conversion unit (MS-1A here): contains two streams. One stream is to get the elemental mercury through KCl solution which removes oxidized mercury. That is, flue gas and 1mol/L KCl are mixed in a gas-liquid contact tube(reaction tube), and any water-soluble Hg^{2+} are removed. Gas and liquid are then separated in a gas-liquid separating tube. The gas is cleaned in a gas scrubber tube filled with 1mol/L KOH solution and transfered through a dehumidification tube to detection unit, where measurement of gaseous elemental mercury(Hg^{o}) is carried out. Another is to get oxidized mercury, then reduce hem to Hg^0 through a stannous chloride (1%w/v $SnCl_2$+10%v/v H_2SO_4) solution by reaction (1-2). That is, solution containing Hg^{2+} mercury are pumped up to the 2nd reaction tube(gas-liquid contact tube),where the solution is mixed with a reducing agent gaseous elemental mercury in this tube. The gas separated in the gas-liquid separating tube is similarly transferred through a gas scrubber tube and a dehumidification tube to detection unit, where measurements for Hg^{2+} mercury are carried out. ③ Detection unit(DM-6B detector): two same detectors(based on cold-vapor atomic absorption(CVVA) spectroscopy)to monitor Hg^0 and Hg^{2+} (Hg^{2+} is reduced to Hg^0 firstly) individually, but CVVA just analyses Hg^0 directly.④ Calibration unit: Before analysis, it needs to calibrate the system. The calibration system has a Hg vapor generator, which supplies constant stream of Hg vapor (about 1L/min), air gas (10L/min), as calculated, the Hg concentration is about 14.9ug/m^3. But the data varies with the system bias as a result of probe and sampling line[3]. Hg-CEM method provides direct measurement, real or near-real time results, fast response to mercury control. The technical parameters: detection range(0-1mg/m^3), resolutions: (0.1μg/Nm^3: 0-99.9μg/Nm^3), (1g/m^3:100-1000μg/Nm^3); reproducibility: an indicated value ±(3%+0.5)μg/Nm^3 ; response characteristics: reach 90% of targeted values in 1 min or less (when distance between detection probes is set at 20 m).

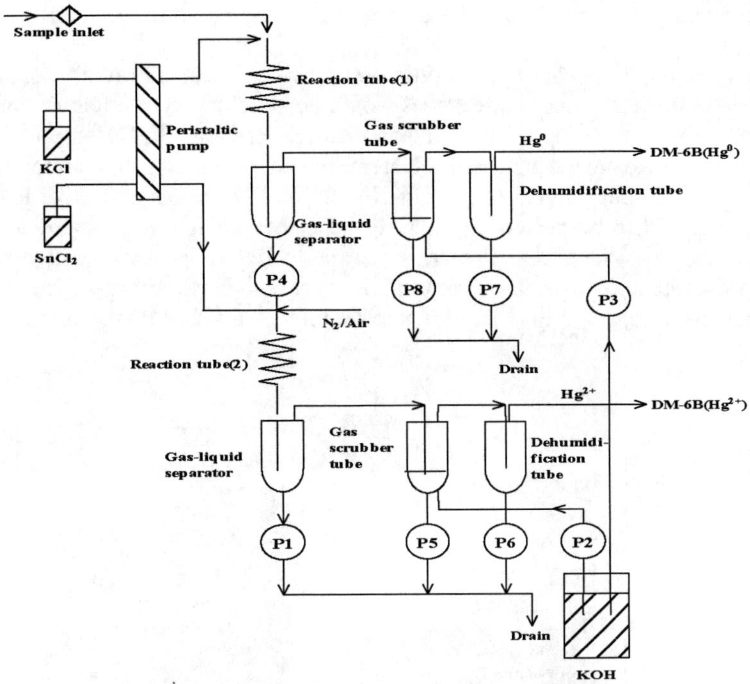

FIGURE 2. Schematic of the CEM's pretreatment/conversion system

$$Hg_2^{2+} + Sn^{2+} = 2Hg + Sn^{4+} \quad (1)$$

$$2Hg^{2+} + Sn^{2+} = Hg_2^{2+} + Sn^{4+} \quad (2)$$

RESULTS AND DISCUSSION

OH and CEM results

In this paper, a utility 300MW tangential coal-fired PC boiler was chosen to carry out for the comparison of OH and CEMs system. During the field test, the boiler is at the normal operating conditions. It should be noted that a kind of blended coals were burned for this unit. Table 1 is analysis data of the feeding coal.

TABLE 1. Analysis data of the feeding coal (dry basis)

Approximate analysis / %				Ultimate analysis / %					Trace element / 10^{-6}
M	A	V	F	C	H	N	S	O	Hg
1.31	30	21.04	47.65	58.54	3.94	1.07	0.75	4.39	14

Mercury measurements by the CEM method and OH method were made after ESP during the different periods of one day. In Figure3., it is examined that the results of gaseous elemental mercury and gaseous oxidized mercury at the post-ESP point of 140℃ during 2 hours continuously.

In the coal combustion flue gas, gaseous elemental mercury(Hg^0) is usually the predominant form, with the flue gas cooling, gaseous elemental mercury would be oxidized. Clearly, the same behaviour is observed here. Average concentrations of elemental mercury and oxidized mercury measured by CEM method were 9.1μg/Nm^3 and 8.9 μg/Nm^3, respectively. Concentrations of gaseous elemental mercury and oxidized mercury by OH method were 5.9μg/Nm^3 and 8.0 μg/Nm^3, which the percentage of elemental form arranges from 42% to 50%, 50% to 58% of oxidized mercury. It is indicated that oxidized mercury is slightly more than the elemental form. However, it shows a bias between the gaseous elemental mercury. Firstly, bias of the results may be partly attributed to different measurement time. And some research reported that biases caused by the fly ash accumulated on the filter in the OH. That's because such fly ash acts as sorbent which has the potential to capture mercury in the flue gas and make the mercury speciation varied[3].

Although the results of CEM method were slightly more than those of Ontario-Hydro method to some degree, there is still good consistency between the results of both methods, so as to confirm the accuracy of continuous mercury monitoring.

Comparison the both results, it appears to be a little bias between them. In order to achieve the long-time CEM on-line measurement results more reliable, some issues would be attached more attention. Sampling line tube should be less absorbed by mercury as possible. Hence, making the experiment of Teflon and Silicon tube on different length, 1m, 2m, 3m, 4m, to make sure which is the better choice. As a result, the silicon tube can absorb mercury nearly 0.6-1.0μg/Nm^3 per meter, but only0.1-0.2μg/Nm^3 for Teflon. Teflon, as expected, is the perfect one. Besides, Teflon sampling lines are ought to be heated at 200℃ to keep the speciation the same as that in the stack and to reduce the condensation of acidic gases[4].Because lower temperature will convert the elemental mercury to oxidized mercury. The constant flow rate and pressure of sampling line are also needed. What's more, with the requirement of cleaned or ash-free sampling flue gas, a tube filter is installed necessary and compressed air is used to help to suck the flue gas.

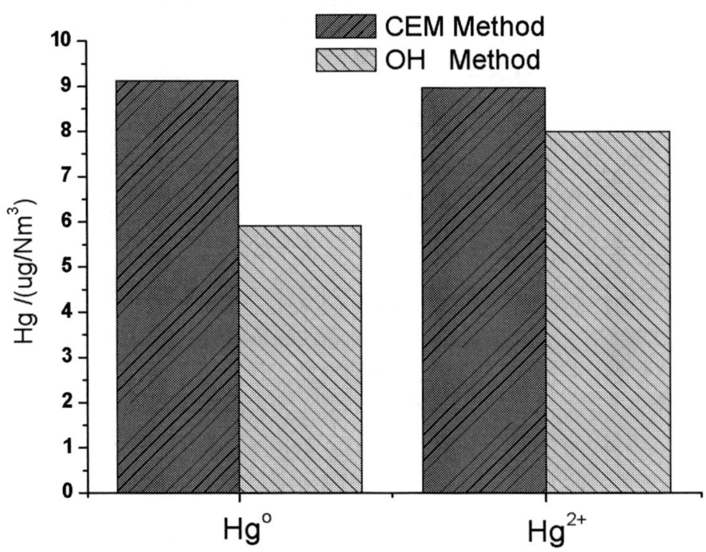

FIGURE 3. Gaseou elemental (Hg^0) and gaseous oxidized (Hg^{2+}) mercury after ESP

ESP effects on mercury measurement

Usually the existing ESP is installed in the coal fired power plant, it's necessary to investigate the influence of ESP on the emission of mercury. Figure 4. describes the measured data of the gaseous elemental mercury and gaseous oxidized mercury by OH method at the locations of pre-ESP and post-ESP respectively. Because the measurements were made at different two days, little difference from true fielding testing existed. In Fig 4. average concentrations of gaseous elemental mercury and oxidized mercury are 2.4 μg/Nm^3 and 8.1μg/Nm3 before ESP, 5.9μg/Nm^3 and 8.0 μg/Nm^3 after ESP. As mentioned before in Figure 3., the concentrations of Hg^{2+} are higher a little bit than Hg^0 at the exit of ESP, so does before the ESP, and even more higher. A possible explanation would be before the ESP, high levels of fly ash on the OHM filter would convert elemental mercury to oxidized mercury. It's reported by Zhou Jinsong[5] that fly ash, fly ash components and flue gas appears to have a well effect on Hg^0 oxidization via homogeneous(gas-gas) or heterogeneous (gas-solid, surface catalyzed) reactions. By comparison, the specific surface area of fly ash is key factor on Hg^0 oxidization, the second impact is nonmagnetic phase derived from fly ash.

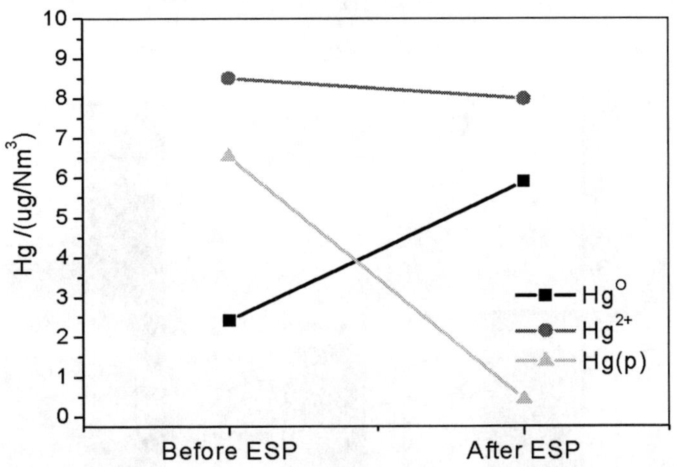

FIGURE 4. Gaseous elemental (Hg^0) , oxidized (Hg^{2+}) mercury and particulate mercury(Hg_p)before and after ESP

From Fig 4., it is clearly that gaseous oxidized mercury is reduced a little through ESP. More evidently is that the results of Hg_p though ESP decreases dramatically. The data of particulate mercury at the points of pre-ESP and post-ESP are 6.5μg/Nm^3 and 0.4μg/Nm^3 individually. It is explained that before ESP, the high levels of fly ash contribute to absorption of mercury too much and form particulate mercury (Hg_p). A large quantity of the bigger fly ash particle is removed as the flue gas passes through ESP, the fine particles are left after ESP. Even though usually the concentrations of mercury in fine particles are higher than those of larger particles, but the absolute amounts of particles are less, evidently, as compared to those of before ESP, so that the particulate mercury (Hg_p) decreases swiftly at the same volume of flue gas.

On the other hand, the trend of gaseous elemental mercury is just the opposite. At the inlet and outlet of ESP are 2.4 μg/Nm^3 and 5.9 μg/Nm^3 respectively. Surprisingly, vapor-phase mercury in elemental form increase relatively fast. A possible explanation is that there is a link with high levels of particulate mercury before ESP, gaseous elemental mercury are absorbed onto the solid surface, such as fly ash and unburned carbon through heterogeneous reaction[6]. The absorbed elemental mercury may be reemitted out through the discharge process of ESP. And some oxidized mercury may be reduced into elemental mercury in the ESP. The effect of discharge to the absorption of mercury by fly ash and the conversion of oxidized and elemental mercury needs further detail investigation in the future.

WFGD effects on mercury measurement

It's suggested by U.S.EPA that mercury can be removed in the existing air pollution control equipments [7]. Currently, the Wet FGD systems have been mostly installed. The approach is being tested to enhance the co-benefit mercury capture in WFGD, coupled with SO_2 reduction. The WFGD system includes the common used lime and limestone forced oxidizaion (LSFO) wet scrubber. It's accepted that oxidized form(Hg^{2+}) is highly water soluble, so wet scrubber can capture 80-90% of the oxidized mercury, with little impact on elemental mercury[8].

This paper is conducted that long-term continuous monitoring the mercury speciation in another 200MW coal-fired boiler, which burned the bituminous coal. Because it had only one on-line CEM system, measurement couldn't finished in same day, but the boiler ran steadily. Even though different measurements resulting in the bias from the true condition, the trend will be similar. Figure 5. shows the average measured data of gaseous elemental mercury, gaseous oxidized mercury and total gaseous mercury at the locations of inlet and outlet of wet scrubber. It's obvious that the flue gas passes though wet scrubber, the gaseous oxidized and total mercury are reduced a little bit. As calculated, the wet scrubber is capable of highly removal of the gaseous oxidized mercury at 53.4%, with removal of elemental form at 22.1%, and that removal of the total mercury is approximately 37.3%. Because oxidized mercury (Hg^{2+}) is highly water soluble, it's easy to captured in the WFGD system. But the figure of mercury removal is somewhat lower than average, which does occur is ascribed to the absorption of the oxidized species. Maybe the sulfite present in the scrubber reduce the oxidized mercury back to the elemental form, or other factors may come into play, such as the lower liquid-gas ratio (L/G)[9].The last reason may be the different measurement time. Despite the lower removal of overall mercury, the WFGD system has a substantial effect on the mercury removal.

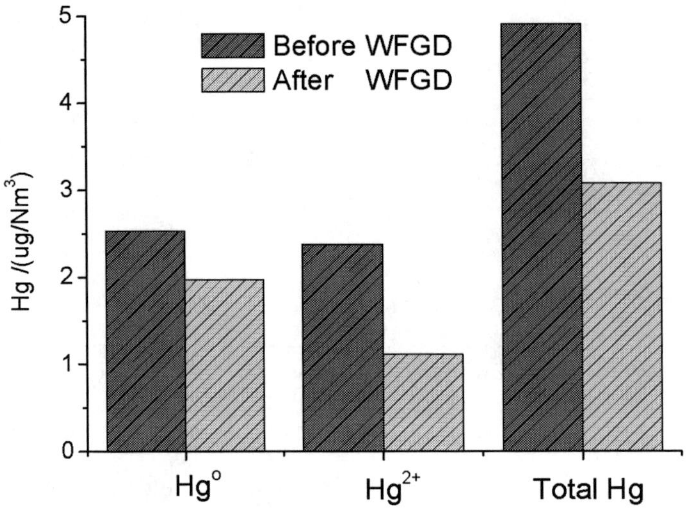

FIGURE 5. Gaseous elemental (Hg^0) , oxidized (Hg^{2+}) and total gaseous mercury before and after WFGD

CONCLUSION

1) OH and CEM produce the similar results for the measurement of gaseous elemental mercury and oxidized mercury in order to confirm the accuracy of CEM. Because the slight bias between two method, some modifications should be made.

2) Because CEM system is on-line measurement, it's a promising and effective mercury measurement.

3) Through ESP, particulate mercury (Hg_p) reduced sharp higher than gaseous oxidized mercury, gaseous elemental mercury is just the opposite.

4) Average removal of overall mercury across the WFGD system is 37.3% with the removal of gaseous oxidized mercury and elemental form 53.4% and 22.1% respectively.

ACKNOWLEGEMENT

The research work was supported by the Chinese Special Funds for High Tech Research and development (863) Program (No.2005AA520080), Doctorate Foundation of the State Education Ministry of China (No.20050335057) and the National Natural Science Foundation (No.50476056).

REFERENCES

1. Wang Q C, Ma R L. Mercury in Coals and Ashes[J].Chinese Journal of Environmental Sicence,1997, 17(1),pp.76-79.(in Chinese).
2. Shawn Kellie, Yufeng Duan, Mercury emissions from a 100-MW wall-fired boiler as measured by semicontinuous mercury monitor and Ontario Hydro Method, Fuel Proceeding Technology 85, pp.487-499,2004.
3. Sjostrorm S, Bustard J, Mercury removal trends in full-scale ESP's and Fabric Filters [A].In: Proceeding of the Mega Symposium[C], Chicago:Air & Waste Management Association,2001,Session 6, No.235.
4. Yufeng Duan, Yan Cao, In-situ Measurement and Distribution of Flue Gas Mercury for a Utility PC Boiler System, Journal of Southeast University(English Edition), Vol21, No.1, pp.53-57,2005.
5. Jinsong Zhou, Zhongyang Luo, Effects of Fly ash on Mercury Transformation in the Coal-fired Fule Gas, Proceedings of the 4th International Conference on Combustion, Incineration/Pyrolysis and Emission Control (4th i-CIPEC), pp.327-330, 2006.
6. Robert Hall, Chun Wai Lee, Mercury Control Technolgy-A Review, Proceedings of the 4th International Conference on Combustion, Incineration/Pyrolysis and Emission Control (4th i-CIPEC), pp.13-16, 2006.
7. John H.Pavlish, Everett A. Status Review of Mercury Control Options for Coal-fired Power Plants, Fuel Processing

Technology 82, pp. 89-165,2003.
8. Dennis L.Laudal, Jeffrey S.Thompson, Use of continuous mercury monitors at coal-fired utilities, Fuel Processing Technology 85, pp. 501-511, 2004.
9. Paul S. Nolan, kevein E. redinger, Demnonstration of additive use for enhanced mercury emissions control in wet FGD systems, Fuel Processing Technology 85, pp.687-600,2004.

Measurement of Particle Size Distribution and Volume Concentration based on Ultrasonic Attenuation Spectrum in Fat Emulsion

Lili Dong, Mingxu Su, Minghua Xue, Xiaoshu Cai, Zhitao Shang

Institute of Particle and Two Phase Flow Measurement,
University of Shanghai for Science and Technology, Shanghai 200093, China

Abstract. The determination of particle size distribution in concentrated polydisperse fat emulsions measured by ultrasonic attenuation spectrum is studied in this paper. Based on theoretical analysis, ultrasonic attenuation spectrum of fat emulsion samples containing different droplet concentrations (1~20%) is measured over the range of 2~13MHz. Then the droplet size and distribution are determined by processing the experiment data with inversion arithmetic. The particle size distribution of original sample determined by ultrasonic spectrum gives excellent agreement with that of diluted sample measured by an optical instrument TSM(Totally Scattering Measurement) . This indicates that ultrasonic spectrum is capable of determining the particle size distribution and dispersed-phase volume fraction of concentrated fat emulsions in a non-destructive approach.

Keywords: applied acoustics; ultrasonic; attenuation; fat emulsion; particle size distribution

INTRODUCTION

Fat emulsion injection is a homogeneous and stable bactericidal fat emulsion consisting of soybean oil droplets, purified lecithin and glyceride. It can provide high energy and essential fatty acids to patients needing parenteral nutrition. The deliquescence and absorbency of fat emulsion have a very close relationship with particle size distribution. Chinese Pharmacopoeia prescribes severely the particle size range. The measurement of particle size distribution of fat emulsions is very important in clinical medicine[1].

Many technologies have been developed into commercial instruments which are now used routinely to analyze fat emulsions, such as microscopy, electrical sensing zone method (viz. Coulter method) and light scattering[2]. Nevertheless, each technique has its own advantages and limitations. Microscopic techniques can provide the most direct size information about the droplets of emulsions, However, the preparation of samples for analysis is often time consuming and laborious. Moreover, it's prone to human error. Optical microscope can measure the emulsions only containing bigger droplets (diameter>1μm). Electron microscope can measure much smaller droplets than optical microscope, but it costs much. The instruments based on electrical sensing zone method which are rapid and precise have been widely used. However the lower limit of Coulter Counter is about 0.5μm, which is not fit for fat emulsions containing small droplets. Microscopy and electrical sensing zone method are not capable of measuring on-line because of their principles. Particle-sizing instruments based on light scattering are simple to operate and give rapid measurements over a wide range of droplet size distributions, besides, this method is easy for automation of data processing. But the concentrated sample must be diluted considerably prior to analysis because of limited penetrability of light. The same as before, light scattering couldn't be applied on-line.

Ultrasonic spectrometry utilizes measurements of the frequency dependence of the ultrasonic attenuation coefficient of dispersions to obtain information on the concentration and size distribution of particles. Instruments based on ultrasonic spectrometry are capable of analyzing emulsions which are concentrated and optically opaque because ultrasonic wave has powerful penetrability. This technique can give measurements over a wide rang of droplet size distributions (typically 0.1～1000μm), which enables ultrasonic spectrometry to measure emulsions quickly and on-line without dilute and disturbing the samples[3]. In this paper experiment is carried out to study the measurement of particle size distribution of fat emulsions using ultrasonic spectroscopy.

CP914, *Multiphase Flow: The Ultimate Measurement Challenge,* edited by X. Cai, Y. Wu, Z. Huang, S. Wang, and M. Wang

ULTRASONIC SCATTERING THEORY

When an ultrasonic wave passes though an emulsion, many phenomena may occur such as absorption, transmission, scattering, reflection and resonance which result in the decrease of its amplitude.

ECAH model [4,5] studies the effect of four most important mechanisms (viscosity, thermal dissipation, scattering and intrinsic absorption)on ultrasonic attenuation and velocity. It's a comprehensive model assuming that:

1) droplets are spherical;
2) droplets do not physically interact with their neighbors. And there is only single scattering,etc.

This theory gives the following formula of complex propagation constant:

$$\left(\frac{\kappa}{k_c}\right)^2 = 1 + \frac{3\phi}{jk_c^3 R^3}\sum_{n=0}^{\infty}(2n+1)A_n \tag{1}$$

Where $k = \omega / c_s(\omega) + j\alpha_s(\omega)$, α_s and c_s are the equivalent to ultrasonic attenuation coefficient and ultrasonic velocity of two-phase flow, respectively. R is the droplet radius, $j=(-1)^{1/2}$, ϕ is the disperse phase volume fraction. The subscript c refers to the continuous phase.

In highly concentrated emulsions, since the ultrasonic scattering mechanism is more complex , multiple scattering has more effect on the ultrasonic scattering. One of the most comprehensive and widely used multiple scattering theories is that derived by Waterman and Truell[6]

$$\left(\frac{\kappa}{k_c}\right)^2 = 1 + \frac{3\phi}{jk_c^3 R^3}\sum_{n=0}^{\infty}(2n+1)A_n - \frac{9\phi^2}{4k_c^6 R^6}\times[(\sum_{n=0}^{\infty}(2n+1)A_n)^2 - (\sum_{n=0}^{\infty}(-1)^n(2n+1)A_n)^2] \tag{2}$$

It is clearly that the first and second terms in the right of equation(2) are the same as that of single scattering. The last term represents the influence of multiple scattering, which shows concentration has stronger effect on the transmission of ultrasonic wave in emulsions.

ECAH model are applicable to a wide range of particle sizes and material types. However, it's usually hard to know ultrasonic parameters of many materials which this model need. Its calculation is time consuming and its linear equations are fairly morbid which make it difficult to apply[7].

David J McClements[8] believes that the mathematical description of ultrasonic propagation can be simplified significantly in the long wavelength regime(LWR:$R<<\lambda$). In the LWR only three types of interaction between ultrasonic waves and emulsions are important: intrinsic absorption, droplet pulsation, and droplet oscillation. The effects of the pulsation and oscillation of droplets in an ultrasonic field on the ultrasonic properties of an emulsion are contained in the scattering coefficients A_0 and A_1. All higher-order scattering coefficients are negligible in the LWR. So the equation (1) or (2) can only refer to A_0 and A_1, which is called McClements model. At the same time McClements gives the suitable relationship of droplets and medium density for these two expressions. It is $0.9<\rho_p/\rho_c<1.2$ which most emulsions are suitable for. Explicit expressions for the A_0 and A_1 terms have been derived as follows:

$$A_0 = \frac{i(kR)^3 \cdot [(\rho k'^2 / \rho' k^2) - 1]}{3} - k^2 RCT\rho\tau H \cdot (\beta_0 / \rho C_p - \beta_0' / \rho' C_p')^2 \tag{3}$$

$$A_1 = \frac{-i(kR)^3}{3} \cdot \frac{(\rho - \rho')}{3\rho + 2(\rho' - \rho)/[1 + 3(1+i)\delta_v / 2R + 3i\delta_v^2 / 2R^2]} \tag{4}$$

Where, subscript "ʹ" refers to the properties of droplets. T is the absolute temperature, β_0 is the coefficient of volume expansion, c is the ultrasonic velocity, C_p is the specific heat capacity. And

$$H = \{1/(1-iz) - \tau/\tau' \cdot \tan(z')/[\tan(z') - z']\}^{-1} \tag{5}$$

Where, $z=(1+i)\cdot R/\delta_t$. The terms δ_v and δ_t are the thermal and viscous skin depths, they represent the heat flow and viscosity flow across the interface between the droplet and surrounding liquid, respectively.

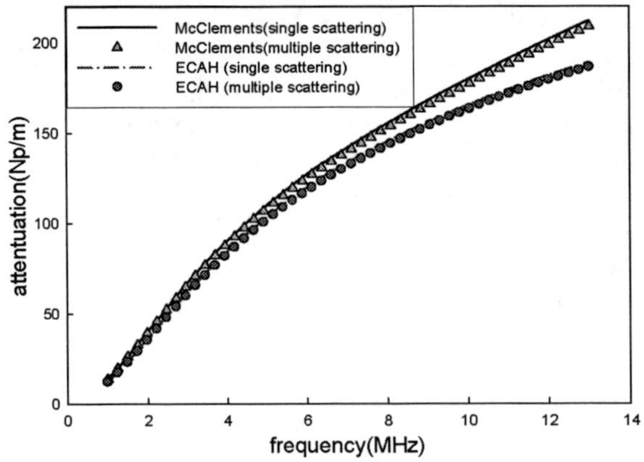

FIGURE 1. Theoretical numerical results of ultrasonic attenuation spectrum in fat emulsion with mean droplet radius of 0.20μm and concentration of 20%.

Figure 1 shows the frequency dependence of ultrasonic attenuation coefficient in fat emulsion with mean droplet radius of 0.20μm and concentration of 20%. The dashed and real line are numerical simulation of ECAH model and McClements simplified model, respectively. Obviously, in the LWR the effect of multiple scattering on ECAH model and McClements model are all small and can be neglected in practice. Meanwhile, in the case of single scattering the two theories give closed numerical simulation results. In order to improve calculation efficiency the McClements model can be applied in stead of ECAH model in many cases.

APPARATUS STRUCTURE

The sample provided by Shanghai Institute of Pharmaceutical is used without dilution. The initial droplets volume fraction is 20%. Figure 2 shows the Experimental setup which measures the attenuation spectrum of a series of emulsions with different droplet concentrations (1~20 vol %). To avoid effect of temperature fluctuation the sample temperature should be stabilized at 25℃.

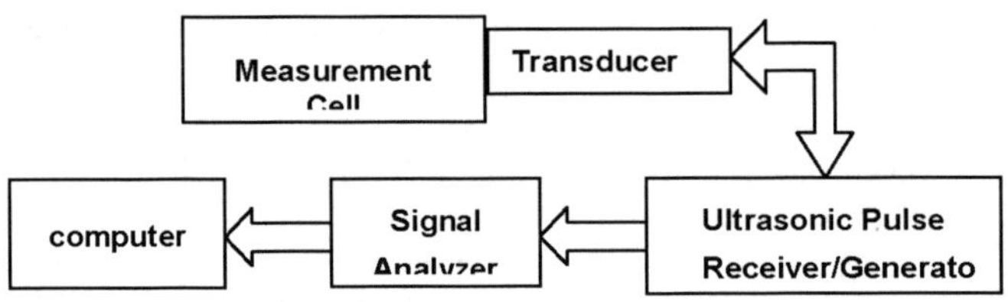

FIGURE 2. Schematic of the experimental setup.

The measuring system contains ultrasonic pulse receiver/generator (Panametrics-NDT, Model 5800), high speed signal analyzer and computer. A single transducer is used to both generate and receive the ultrasonic pulses in the experiment. Two broad-band transducers are used in turn to acquire ultrasonic attenuation spectrum over the range of 2~13MHz., the center frequencies of them are 5MHz and 10MHz respectively. Figure 3 shows a variable sound path length device. The reflect distance between the transducer and moveable reflector plate is adjustable(4mm~8mm), which make it easy to measure at different points.

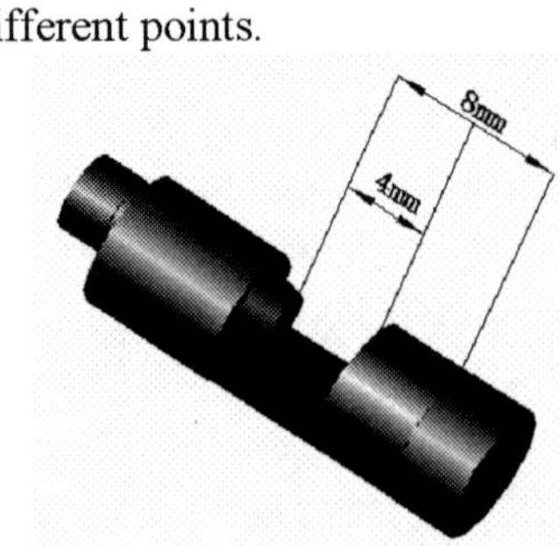

FIGURE 3. Three-dimensional view of ultrasonic probe transducer.

Experiment is performed in the following steps. Ultrasonic pulse receiver/generator produces an broad-band electrical pulse. This signal is applied to the ultrasonic transducer which converts it into an ultrasonic wave which propagates through a sample in measurement cell. After reflected wave passing through the sample, ultrasonic signal is detected by the same transducer and converted back into an electrical signal which, then, is received by the ultrasonic pulse receiver. The electrical signal is amplified and digitized by signal analyzer and transmitted into computer. A software analyzes this data to determine the ultrasonic attenuation coefficient and velocity of the sample.

EXPERIMENTAL STUDY AND INVERSION ARITHMETIC

The frequency dependence of the amplitude and phase of the pulse are determined using a Fast Fourier Transform(FFT) technique over a wide range of frequency. According to the measurement detector shown in figure 3, the attenuation coefficient is calculated by comparing the reduction in the amplitude of the pulse that is before traveling with that of after traveling through the sample:

$$\alpha = \frac{\ln\left(A_2/A_1\right)}{2\left(L_2 - L_1\right)} \tag{6}$$

Where A_1 and A_2 are amplitude of ultrasonic wave in the distance L_2 and L_1, respectively.

The attenuation coefficient of a series of fat emulsions with different disperse phase volume fractions (1–20 vol%) are shown in Figure 4. The droplets per unit volume are more and more with the droplets volume fraction increasing. Thus the higher wave energy dissipation leads to the increase of attenuation coefficient. And the upward trend of this increase by degrees is strengthened with increasing frequency in this figure.

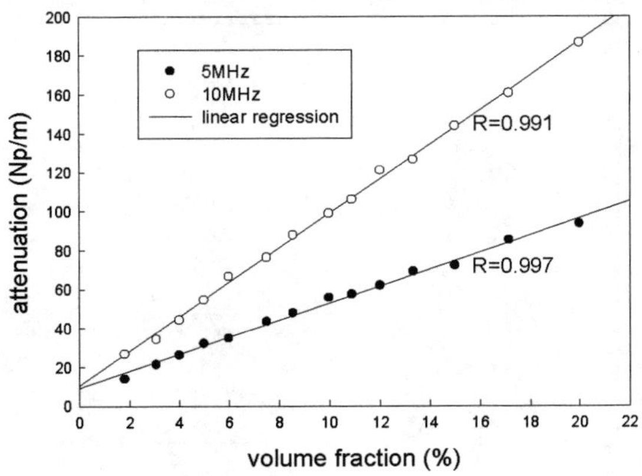

FIGURE 4. The attenuation coefficient of a series of fat emulsions with different disperse phase volume fractions(1~20 vol%). The circle and crisscross represent attenuation spectrum at 10MHz and 5MHz, respectively.

When ultrasonic wave passes though an emulsion, absorption and reflection may occur. Assume the effect of absorption and reflection are independent, then the attenuation of ten droplets should equal to ten times the value of that of one droplet, which means attenuation effect can be superposed linearly. on the other hand, if particle-particle interaction exists, linear superposition effect could be wrong. And the curve will present non-linear one because of multiple scattering effect. In figure 4, both particle-particle interaction and multiple scattering don't have main effect on attenuation. Also in figure 1,it is clear the multiple scattering effect is small.

A series of fat emulsions with different droplet concentrations (1~20 vol %) is prepared by diluting the initial emulsion with different amounts of distilled water. Figure 5 shows the attenuation spectra of a series of fat emulsions with different disperse phase volume fractions.

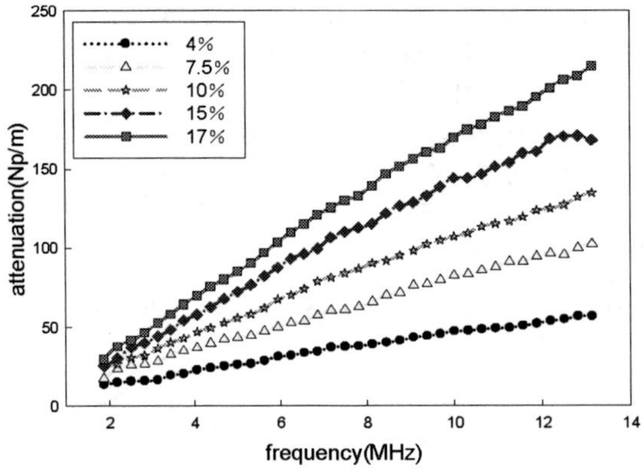

FIGURE 5. Ultrasonic attenuation spectra of fat emulsions containing different droplet concentrations.

Ultrasonic attenuation coefficient of any emulsion over a wide range of frequency (2~13MHz) can be calculated using McClements model. The predicted values are compared with the experimental values, and then the sum of the squares of the differences (the Sum of the Squares of the Differences, SSD)[9] is calculated.

$$SSD = \sum_i [\alpha_{theory}(f_i) - \alpha_{meas}(f_i)]^2 \tag{7}$$

Where $\alpha_{theory}(f_i)$ and $\alpha_{expt}(f_i)$ are the predicted ultrasonic attenuation coefficient and the experimentally measured values at frequency f_i. The mean and standard deviation are optimized until the computer searches the minimum in SSD. Then the particle size distribution can be determined.

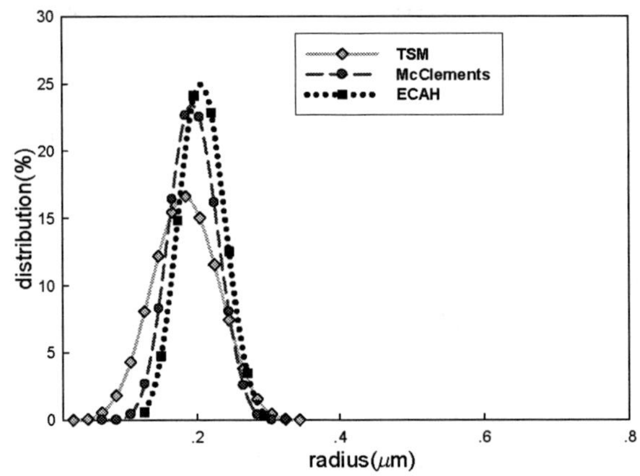

FIGURE 6. Comparison of particle size distribution determined by ultrasonic spectrum and TSM(Totally Scattering Measurement) . The solid and dashed represent the results of ultrasonic method and TSM method, respectively.

In the Figure 6 the two dashed represent particle size distribution derived from experimental data by Optimal Regularization Algorithm[3] and the real line is that determined by TSM(Totally Scattering Measurement) based on light extinction principle[10] in dilute fat emulsion. The results based on two ultrasonic models are very similar. The particle size is bigger using ECAH theory and particle size distribution is wider measured by TSM.

Table 1 shows the comparison of characteristic size determined by ultrasonic method and light extinction theory. The values of R_{10}, R_{50}, R_{90} represent the size by number below which 10%, 50% and 90% of a sample particle falls, respectively.

TABLE 1. Comparison of characteristic size determined by ultrasonic method and extinction theory (μm).

	Extinction TSM	Ultrasonic Method
R_{10}	0.11	0.13
R_{50}	0.18	0.21
R_{90}	0.25	0.25

Table 1 illuminates the particle size distribution determined by ultrasonic method gives good agreement with that by light extinction theory in the respect of particle size and width of distribution of sample.

Figure 7 shows the experimental values of R_{10}, R_{50}, R_{90}[2] when the droplet concentration changes. Clearly, though the droplet concentration of fat emulsions has great effect on the value of ultrasonic attenuation, it has little effect on the experimental results of particle size distribution of fat emulsion.

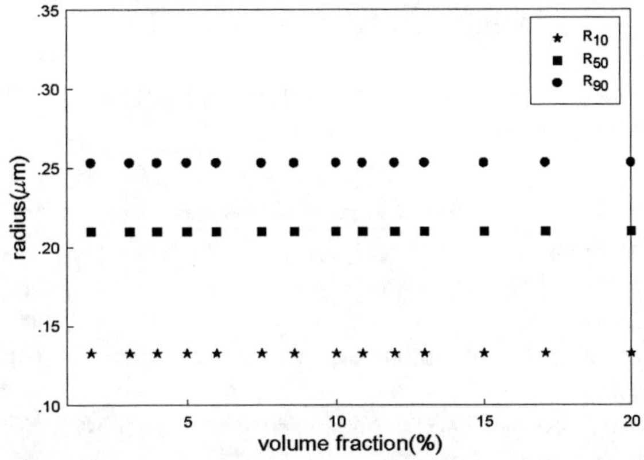

FIGURE 7. The value of mean particle size of different droplet concentration.

Meanwhile, the volume fraction of sample also can be calculated by inversion arithmetic simultaneity. It is 20.6% and droplet concentration can be calculated as a uncertain parameter using Optimization Algorithm. The simplified solution which contains two steps can be applied assuming that the droplet concentration dependence of ultrasonic attenuation is nearly linear trend. First, particle size distribution of the sample is calculated neglecting the term of droplet concentration. Second, droplet concentration is available by finding the minimum difference between experimental and predicted attenuation spectrum. This method is applied in this paper. Compared with the initial droplet concentration 20%, the deviation is about 3%. Figure 8 shows the retrieval spectrum. SSD calculated by equation 7 is 2.5E3. Also the *Err* of retrieval spectrum and experimental spectrum could be calculated as follow:

$$Err = \sqrt{\frac{\sum_{i=1}^{N_f} \left(\frac{\alpha_{meas}(f_i) - \alpha_{theory}(f_i)}{\alpha_{theory,i}}\right)^2}{N_f}} \quad (8)$$

Where N_f is number of frequency. According to this formula the deviation *Err* of retrieval spectrum simulated by McClements model and experimental spectrum is 9.37%.

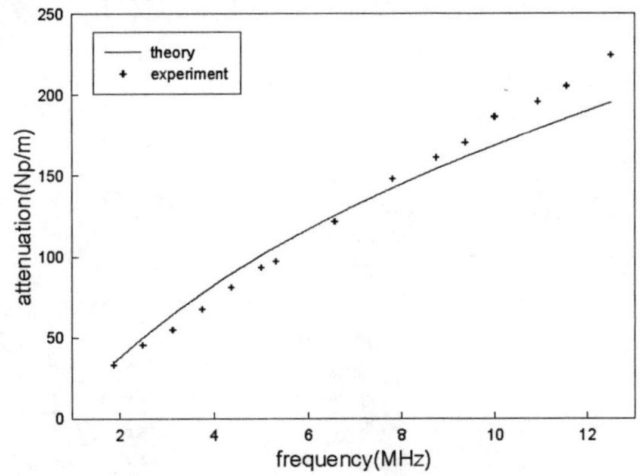

FIGURE 8. Comparison of experimental and retrieval attenuation spectrum based on McClements model.

CONCLUSIONS

The particle size distribution and droplet concentration in concentrated polydisperse fat emulsions are measured by ultrasonic attenuation spectrum and compared with the particle size distribution determined by light extinction method. Since this method is capable of measuring the concentrated fat emulsion (20 vol %) quickly and determining the particle size distribution and dispersed-phase volume fraction, It will enable manufacturers of instruments that utilize ultrasonic spectrometry to apply the technique to monitoring the fat emulsions with higher droplet concentrations.

ACKNOWLEDGEMENTS

This material is based upon work supported by National 863 project (No. 2006AA03Z349) and Shanghai Education Committee Youth Foundation project (04EC30).

REFERENCES

1. Chinese Pharmacopoeia.2000. the second volume：Appendix Ⅸ C
2. N.-N. Wang, Optical Particle Sizing and its Applications. Atomic Energy Press, Beijing 2000 (in Chinese).
3. Su Mingxu, Cai Xiaoshu. The Measurement of Particle Size and Concentration in Suspensions by Ultrasonic Attenuation. Acta Acoustics, 29(5), 2004, 440-444.
4. Epstein P S, Carhart R R. The Absorption of Sound in Suspensions and Emulsions: I. Water Fog in Air. J. Acoust. Soc. Am. 1953, 25: 553-565.
5. Allegra J R, Hawley S A. Attenuation of Sound in Suspensions and Emulsions: Theory and Experiments. J. Acoust. Soc. Am. 1972,51: 1545-1564.
6. Waterman P C, Truell R. Multiple Scattering of Waves. J.Math. Phys. 1961,2:512-540.
7. Su Mingxu Cai Xiaoshu. The numerical study of acoustical attenuation and velocity in the suspension of superfine particles. Acta Acoustics, 2002, 27(3):218-222.
8. D.J.McClements, Comparison of multiple scattering theories with experimental measurements in emusions[J]. J.Acoust.Soc.Am. 1992,91(2):849-854.
9. D.J.McClements, Principles of Ultrasonic Droplet Size Determination in Emulsions, Langmuir 1996,12, 3454-3461.
10. Cai Xiao-Shu, A New Dependent Model for Particle Sizing with Light Extinction, J.Aerosol Sci.1995,Vol. 26,No.4,685-688.

Measurement of Three-Dimensional Temperature Distribution in an Absorbing, Emitting, and Anisotropically Scattering Medium

Fei WANG *, Dong LIU, Qunxing HUANG, Jianhua YAN, Kefa CEN

State Key Laboratory of Clean Energy Utilization, Zhejiang University, No.38 Zheda road, Hangzhou, 310027, China
**Tel: 86-571-87952628, Email: wangfei@cmee.zju.edu.cn*

Abstract. Many advances in the thermal sciences depend on accurate determination of three-dimensional temperature distribution. Previous measurements by thermocouples and pyrometers were mainly restricted to time-average "point" measurements of temperature. In this paper, a model to measure the three-dimensional temperature distribution according to the flame image captured by CCD camera is studied numerically. The formation of the flame image is considered as two stages: radiation transfer in the participating medium and the optical imaging process through the lens of the camera. The inverse analysis based on Monte Carlo method and the optical model of the flame image are introduced to describe these two stages respectively. The flame of the pulverized coal is considered as a three-dimensional, absorbing, emitting, anisotropically scattering medium and full of CO_2, N_2 and carbon particles. In the temperature reconstruction case, eight CCD cameras are located on the front, back, left and right surfaces of the cube with 0.4m×0.4m×0.4m size. The temperature reconstruction equations are ill-posed linear system of equations. LSQR (Least Square QR-factorization) and damping LSQR have been used for solving the ill-posed equations. For an assumed temperature profile, the numerical simulation results show the reconstruction errors are maintained at the lower levels under different measurement errors and temperature near the center of the cube can be reconstructed well even under large measurement errors.

Keywords: Diode laser; Absorption line; lineshape function; Gas concentration measurement

INTRODUCTION

Temperature distribution measurement is essential for the study on combustion process. Traditional contact method for temperature measurement can only obtain temperature in some local positions and more measuring points must be laid in combustion space to get the whole temperature distribution, which obviously will bring inconvenient and difficult problems. So contact method for temperature measurement has limitation for furnace temperature measurement in the modern pulverized coal-fire plant.

Recently, investigations have focused on the reconstruction of temperature distribution in a target space according to its boundary heat flux obtained by detectors, which is a kind of inverse problem of radiation. Among these investigations, Li HY et al. [1-3] and Liu LH et al. [4,5] have reconstructed the temperature distributions in the plane-parallel, spherical, cylindrical, semitransparent, and gray media by the inverse analysis from the data of the radiation intensities exiting the boundaries. Zhou HC et al. [6-8] have reconstructed the temperature distribution in two-dimensional (2-D) area without considering scattering and improved a modified Tikhonov regularization method for solving the inverse problem of three-dimensional (3-D) temperature. A.J. SILVA NETO et al. [9] presented an inverse radiation analysis for simultaneous estimation of optical thickness, single scattering albedo and the coefficients of phase function for an anisotropic scattering plane-parallel medium. R.G. Keanini et al. [10] described a method for measuring time-varying temperature distribution using high frame rate visible imaging CCD cameras. An imaging-based instrumentation system for 3-D temperature measurement of a combustion flame was presented by Yong Yan et al. [11]. However, until now, few studies on the reconstruction of temperature distribution in an absorbing, emitting, and anisotropically scattering medium have been reported.

The formation of the flame radiative energy image is considered as two stages: radiation transfer in the participating medium and the optical imaging process through the lens of the camera. The inverse analysis based on

CP914, *Multiphase Flow: The Ultimate Measurement Challenge,* edited by X. Cai, Y. Wu, Z. Huang, S. Wang, and M. Wang

Monte Carlo method and the new optical model of the flame image are introduced to describe these two stages respectively. After radiative energy images are obtained, temperature reconstruction equations will be gained according to temperature profile reconstruction model. But temperature reconstruction equations are ill-posed large linear system of equations and normal methods for solving linear system of equations can not get satisfying solutions. Therefore LSQR and damping LSQR methods, which are suit for dealing with ill-posed problems, are adopted for solving the ill-posed linear system of equations and then three-dimensional temperature distribution will be obtained.

MONTE CARLO METHOD

Simple Method description

System description

A three-dimensional medium full of CO_2, N_2 and carbon particles in a rectangular, cold, black enclosure, with dimensions, $W = 2, L = 5, H = 3$ m has been analyzed, in Fig.1. The total mixture pressure is specified to be 1atm, with a volume fraction of carbon dioxide of 0.21. The carbon particles are specified to be of uniform diameter, 30 μ m. The temperature and particle concentrations varied over the medium according to the following two equations, respectively:

$$T(x,y,z) = T_0\left[\left(1-2\frac{|x|}{W}\right)\left(1-2\frac{|y|}{L}\right)\left(1-2\frac{|z|}{H}\right)+1\right] \tag{1}$$

$$N(x,y,z) = N_0\left[3\left(1-2\frac{|x|}{W}\right)\left(1-2\frac{|z|}{H}\right)+1\right] \tag{2}$$

Where, T_0 and N_0 are chosen to be 500K and either 5×10^7 or 5×10^8 particles per cubic meter, respectively [12, 13]. The radiative properties of particles are referred to literature [13]. The spectral absorption coefficient of carbon dioxide is calculated according to literature [14].

The medium is divided into many small subregions in which the temperature, pressure and particle concentration can be considered as homogeneous, respectively. The enclosure walls are divided into homogeneous area elements. Initial launching locations and directions of rays are random.

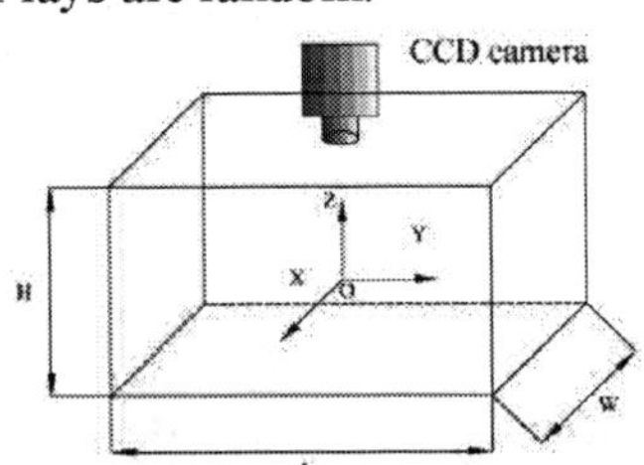

FIGURE 1. Coordinate system and dimensions for three-dimensional rectangular media

Rays tracing process

The ray travels through the volume full of the participating media. The ray's energy is continually absorbed by the gas with absorbability.

Before the ray collides with the particle or the boundary, total energy absorbed during traveling by the gas is calculated. If the absorbed energy is larger than the ray's energy, it is considered that the ray be absorbed completely and a new ray be emitted.

When the ray collides with the particle it is should be judged if the ray's energy is absorbed or scattered by the particle. If the energy is absolutely absorbed by particles, a new ray is emitted and goes into the new loop. If the ray is scattered by the particle, the direction of scattering is then determined and the direction of the ray is renewed.

When the ray transmits through the boundary, the boundary is judged if it is the whole system boundary or it is the volume element inside boundary. If the ray collides with the whole system boundary, it is considered to be absorbed completely.

The details of rays tracing process can be referred to literature [12].

Results and comparisons

From comparisons shown in Fig.2, the results from Monte Carlo method presented in this paper are similar to the results in literature [12]. So we can use this method to describe energy bundles propagation in the medium which is a part of three-dimensional temperature profile reconstruction.

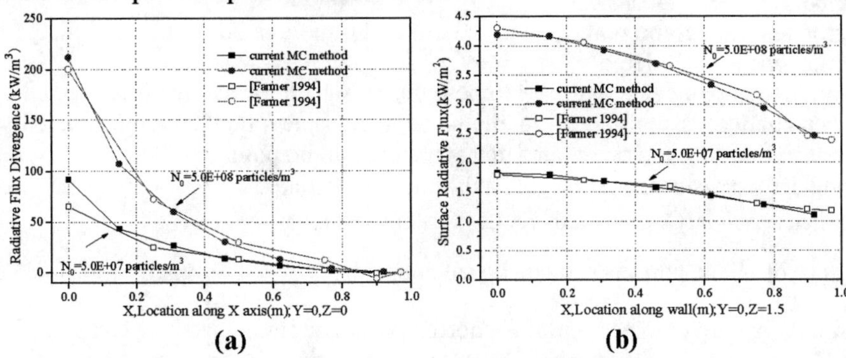

FIGURE 2. a) Comparison of radiative flux divergence at Y=0, Z=0, b) comparison of Surface Radiative flux at Y=0, Z=1.5m, for inhomogeneous medium, with T_0 =500K, P =1atm (21% carbon dioxide, by volume), and N_0=5×10^7 or 5×10^8 particles/m^3

CCD LENS IMAGE-FORMING OPTICAL MODEL

CCD camera is mounted in the center of top surface of system, shown in Fig.1.

Judging if angle of incidence is within the range of field angle of CCD camera

The direction of vector of energy bundle can be obtained through Monte Carlo process above when the bundle emitted in sub-volume collides with the lens of CCD camera on the surface. The included angle between the bundle direction and Z axis is then gained by direction of vector. We can judge if the incidence bundle is within the range of field angle according to this included angle.

Judging if incidence bundle collides with lens of CCD camera

When incidence bundle collides with surface where CCD camera is mounted, the collision position $P(x_p, y_p, z_p)$ can be determined by Monte Carlo process above. The distance between point P and point $O(x_O, y_O, z_O)$ (the center of lens of CCD camera) is then calculated, that is $|PO| = \sqrt{(x_p - x_o)^2 + (y_p - y_o)^2 + (z_p - z_o)^2}$. Let lens radius be r . If $|PO| > r$, the collision position is out of lens area range. If $|PO| \le r$, the collision position is within the lens area range.

Image formation calculation after energy bundle into lens

The coordinate systems are divided into space coordinate system XYZ and image coordinate system $X'Y'Z$, shown in Fig.3. The CCD target area is on the plane $X'O'Y'$. Let $|OO'| = L$.

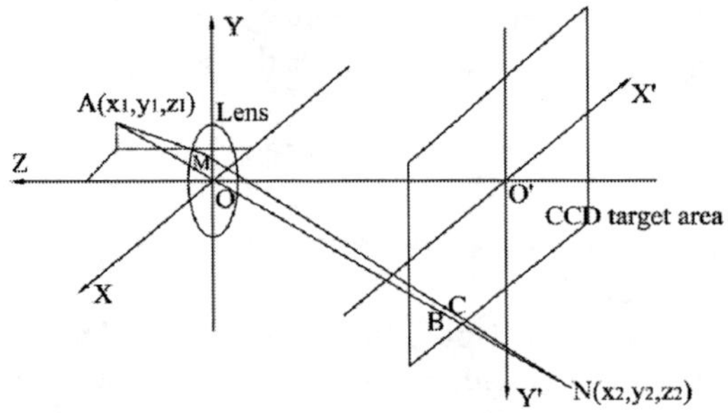

FIGURE 3. Imaging process in the CCD canmera

1. Luminous point is assumed to be point A and the emitted bundle is AM. Here the image formation of single energy bundle into lens is condidered.
2. Image point of luminous point A is assumed to be point N, which means that image point of A is located after target area (other conditions considered in the same way, here other conditions not listed). The point of intersection between line MN and target area and is assumed to be point C.
3. The included angle α between line AO and Z axis can be calculated according to point A coordinate. Let lens focus be f and then secondary optical axis focus is calculated by $f' = f / \cos\alpha$;
4. d is image distance of image point N and can be calculated by $\frac{1}{f'} = \frac{1}{d} + \frac{1}{s}$ .Where, s is object distance and can be obtained by point A. After d obtained, point N coordinate can be immediately calculated.
5. Incidence point M coordinate is determined by Monte Carlo process above and then the line MN equation is obtained. So the intersection point C can be calculated through line MN equation and plane $X'O'Y'$ equation according to space analytic geometry.
6. The pixel position where emitted bundle AM reaches is point C and image formation condition is then obtained.

TEMPERATURE DISTRIBUTION RECONSTRUCTION

Reconstruction model description

Three-dimensional furnace is divided into N sub-volumes. In the actual furnace the walls temperature values are much lower than those in the furnace so the radiative energy emitted by walls is ignored. There are N unknown temperature values to solve. CCD target area is assumed to be divided to M pixels so M equations can be obtained:

$$\begin{array}{lcl} f_1(T_1,T_2,......,T_N) & = & P_1 \\ f_2(T_1,T_2,......,T_N) & = & P_2 \\ \vdots & \vdots & \vdots \\ f_M(T_1,T_2,......,T_N) & = & P_M \end{array} \tag{3}$$

Where, P_i is radiative energy received by each pixel on the CCD target area and f_i represents Monte Carlo process. j=1, 2,…, M.

The energy emitted by sub-volume i in pulverized coal flame is calculated by:

$$E_i = 4(K_{gco2,i} + K_{p,i})\sigma_0 dV \cdot T_i^4 \tag{4}$$

The radiative energy emitted by sub-volume i arriving to pixel j on the CCD target area:

$$E_{i\to j} = E_i \cdot \alpha_{i\to j} \tag{5}$$

The following equation is obtained according to (3) and (4):

$$E_{i\to j} = A_{i\to j} \cdot T_i^4 \tag{6}$$

Where, $A_{i\to j} = 4(K_{gco2,i} + K_{p,i})\sigma_0 dV \cdot \alpha_{i\to j}$ 。

And then we can get the following equations from (3) and (6):

$$\begin{aligned}\sum_{i=1}^{N}(A_{i\to 1}\cdot T_i^4) &= P_1\\ \sum_{i=1}^{N}(A_{i\to 2}\cdot T_i^4) &= P_2\\ \vdots\qquad &\ \ \vdots\ \vdots\\ \sum_{i=1}^{N}(A_{i\to M}\cdot T_i^4) &= P_M\end{aligned} \tag{7}$$

The equations above are transferred into a matrix equation as:

$$A_{M\times N}\times T^4_{N\times 1}=P_{M\times 1} \tag{8}$$

Change the symbols in equation (8) and get:

$$\mathbf{E} = \mathbf{A}_{\mathbf{E}}\mathbf{T} \tag{9}$$

The equation (9) is a reconstruction equation.

Reconstruction case

System description

A three-dimensional furnace with 0.4m×0.4m×0.4m size is divided into 7×7×7 volume elements, shown in Fig.4. The origin of coordinate system is in the center of system. Eight CCD cameras are chosen to obtain radiative energy images. Four CCD cameras are mounted on the left and right surfaces, respectively. Another four CCD cameras are mounted on the front and back surfaces (not marked out in Fig.4). Here eight CCD cameras are used mainly to satisfy future much larger space. It has been proved through calculation that temperature field can be obtained by only one image with space size above.

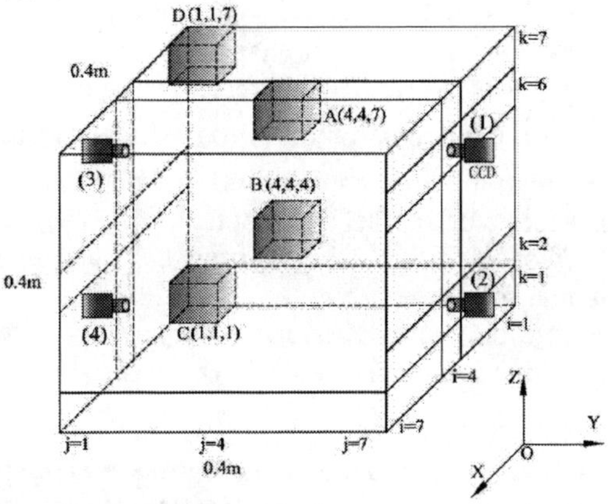

FIGURE 4. Temperature profile reconstruction system

In this reconstruction case the response wavelength range of CCD camera is 0.4～0.7 μ m which is wavelength range of visible light. Spectral absorption coefficients of CO_2 are considered to be zero in visible light wavelength range. In each volume element average scattering coefficient of particles is $0.15m^{-1}$ and extinction coefficient is 0.32 m^{-1}(simulating carbon particles concentration to be 2.0×10^8 per cubic meter) for simplicity.

A temperature distribution in the case of literature [12] is adopted here to be original assumption temperature profile and T_0 is adjusted according to actual furnace combustion temperature.

$$T(x,y,z)=T_0\left[\left(1-2\frac{|x|}{W}\right)\left(1-2\frac{|y|}{L}\right)\left(1-2\frac{|z|}{H}\right)+1\right] \tag{10}$$

Where, T_0 =750K。

Temperature distribution on the typical section k=4 is shown in Fig.5.

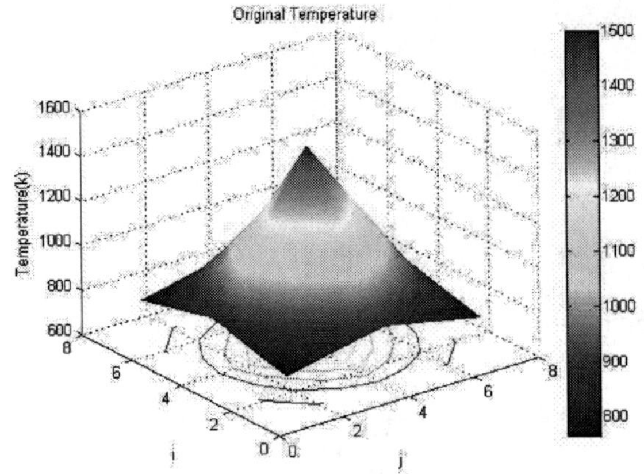

FIGURE 5. k=4 section original assumption temperature distribution

Results

Here radiative energy image captured by four CCD in Fig.4 is shown in Fig.6.

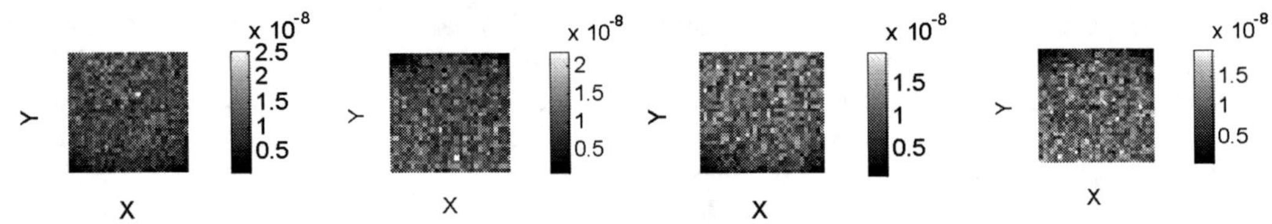

CCD(1) CCD(2) CCD(3) CCD(4)

FIGURE 6. Radiative energy images

The coefficient matrix $\mathbf{A_E}$ in equation (9) is a seriously ill-posed matrix due to its very large condition number which is 1.4351e+006 so the matrix equation (9) is an ill-posed overdetermined equation and general methods can not obtained satisfactory solutions. Moreover in the real applications, the obtained radiative energy consists of measurement errors which greatly affect solutions using normal calculation methods. At this time solutions we obtained is greatly different from the real solutions.

In order to simulate measurement errors, energy E in matrix equation (9) is added by random errors of normal distribution with zero average value and mean square deviation σ :

$$\mathbf{E_{err}} = (\mu + \sigma\xi)\mathbf{E_{orignal}} + \mathbf{E_{orignal}} \tag{11}$$

Where, average value μ =0， ξ is a random variable of standard normal distribution， $-3 < \xi < 3$.

LSQR method and damping method which can effectively deal with large ill-posed problems (more details referred to [15~18]) are adopted here to solve the ill-posed matrix equation (9).

LSQR algorithm for solving

$$\mathrm{Ax} = \mathrm{b} \tag{12}$$

Initial process:

$$\beta_1\mathbf{u}_1 = \mathbf{b}, \alpha_1\mathbf{v}_1 = \mathbf{A^T u_1}, \mathbf{w_1} = \mathbf{v_1}, \mathbf{x_0} = \mathbf{0}, \overline{\phi_1} = \beta_1, \overline{\rho_1} = \alpha_1, i = 0 \tag{13}$$

$i = i + 1$, go to 3)

Two diagonalizations:

$$\begin{aligned} \beta_{i+1}\mathbf{u_{i+1}} &= \mathbf{Av_i} - \alpha_i\mathbf{u_i} \\ \alpha_{i+1}\mathbf{v_{i+1}} &= \mathbf{A^T u_{i+1}} - \beta_{i+1}\mathbf{v_i} \end{aligned} \tag{14}$$

QR-factorization

$$\begin{aligned} &\rho_i = \left(\overline{\rho_i}^2 + \beta_{i+1}^2\right)^{1/2}; c_i = \overline{\rho_i} / \rho_i; s_i = \beta_{i+1} / \rho_i \\ &\theta_{i+1} = s_i\alpha_{i+1}; \overline{\rho_{i+1}} = -c_i\alpha_{i+1}; \phi_i = c_i\overline{\phi_i}; \overline{\phi_{i+1}} = s_i\overline{\phi_i} \end{aligned} \tag{15}$$

Renew solutions

$$\mathbf{x_i} = \mathbf{x_{i-1}} + (\phi_i / \rho_i)\mathbf{w_i}$$
$$\mathbf{w_{i+1}} = \mathbf{v_{i+1}} - (\phi_{i+1} / \rho_i)\mathbf{w_i} \tag{16}$$

Check convergence criterion, if it is satisfied, stop iteration or go to 2).

Some 4th power of the temperature values are negative in the calculated temperature values which is not permitted in the furnace. These negative temperatures almost appear in the lower temperature regions, which denotes that ill-posed characteristic affects lower temperature reconstruction more easily. It is considered that lower temperature should be higher than a limited lowest value which is assumed to be 750K here and the modifying method is given below:

$$\begin{cases} T_i^{'} = T_i & T_i \geq 750 \\ T_i^{'} = T_i + T_{\min} & 0 < T_i < 750 \\ T_i^{'} = T_{\min} & T_i \leq 0 \\ T_{\min} = 750K \end{cases} \tag{17}$$

Where, T_i is the obtained reconstruction temperature and T_i' is the modified temperature.

Here the temperature distribution after modified could not be optimal so long as the reconstruction errors can be maintained at a lower level which satisfies errors requirement.

$$\sigma = 0.01$$

The average relative error between modified energy Eerr and original Eoriginal is 0.79% and maximum relative error is 3.75%.

Here LSQR method and another general method to solve overdetermined linear system of equations which can be found in software Matlab are both presented to solve equation (9) for comparison.

In soft Matlab matrix division is useful primarily for solving equations, and especially for solving simultaneous linear equations. For example, you want to solve for x in Ax = B. x = A\B solves for x in Ax = B. If coefficient matrix A is an m×n matrix and m>n, x = A\B is least square solutions for the overdetermined equation Ax = B. This method is based on orthogonal transformation and called method LSOT (Least Square based on Orthogonal Transformation) in the following text.

First x = A\B is used to solve (9) and solutions which have been modified by (17) are shown in Fig.7.

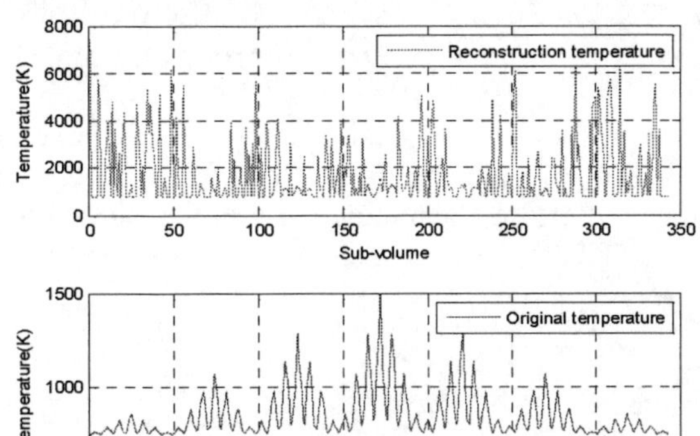

FIGURE 7. LSOT reconstruction profile and original profile

From Fig.7 we can see the obtained solutions are greatly differently from original assumption temperature values due to ill-posed characteristic in matrix equation (4-7) and measurement errors. So in the following calculations LSQR method is used. Fig.8 shows reconstruction profile and original assumption temperature profile. Typical section k=4 temperature distribution is shown in Fig.9. Calculation time is 3.278523 seconds. The average relative error of reconstruction temperature distribution is 8.19%. The relative error at the maximum temperature is 0.0217%.

It is obviously concluded that LSQR method has much better stability through comparisons between results shown in Fig.7 and Fig.8.

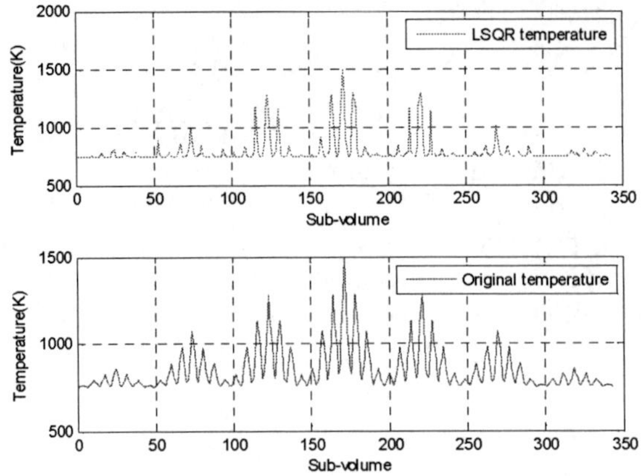

FIGURE 8. Reconstruction and original assumption temperature distribution

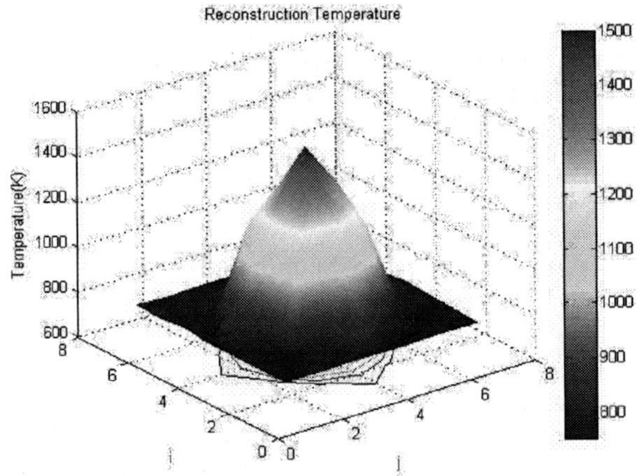

FIGURE 9. Typical section k=4 temperature distribution

$$\sigma = 0.05$$

The average relative error between modified energy Eerr and original Eoriginal is 4.02% and maximum relative error is 18.58%.

Figure.10 shows reconstruction profile and original assumption temperature profile. Typical section k=4 temperature distribution is shown in Fig.11. Calculation time is 2.580159 seconds. The average relative error of reconstruction temperature distribution is 8.01%. The relative error at the maximum temperature is 0.0044%.

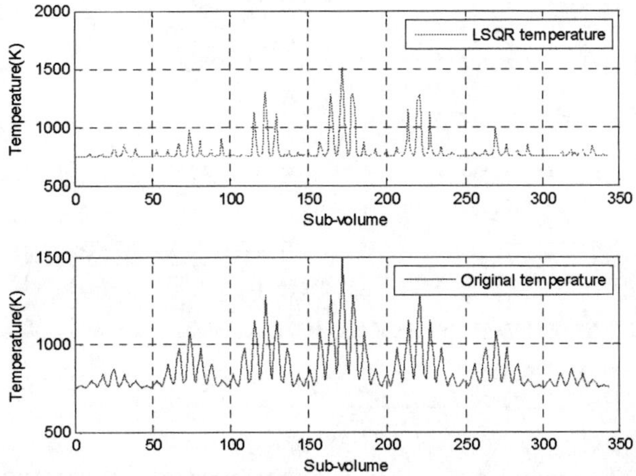

FIGURE 10. Reconstruction and original assumption temperature distribution

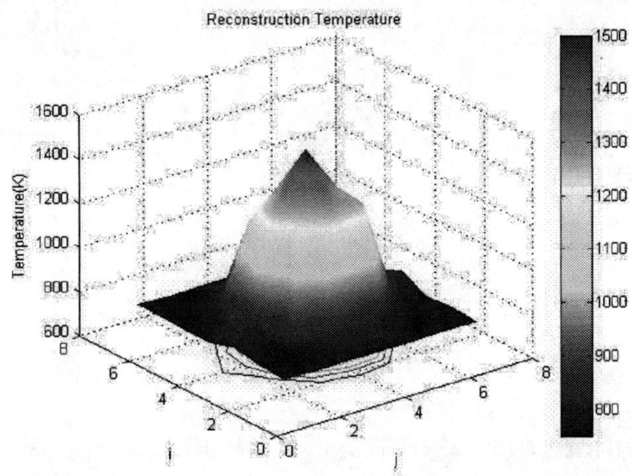

FIGURE 11. Typical section k=4 temperature distribution

$$\sigma = 0.50$$

The average relative error between modified energy Eerr and original Eoriginal is 40.22% and maximum relative error is 209.91%.

Figure.12 shows reconstruction profile and original assumption temperature profile. Typical section k=4 temperature distribution is shown in Fig.13. Calculation time is 2.233718 seconds. Here damping LSQR is used here because of the larger error and damping coefficient is 3.0e-21. The average relative error of reconstruction temperature distribution is 10.09%. The relative error at the maximum temperature is 1.31%.

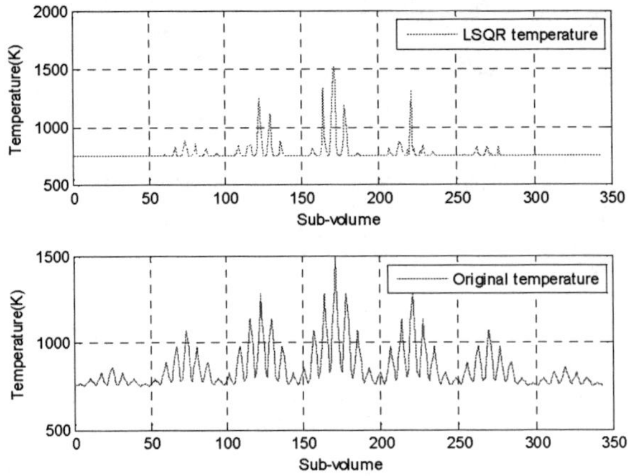

FIGURE 12. Reconstruction and original assumption temperature distribution

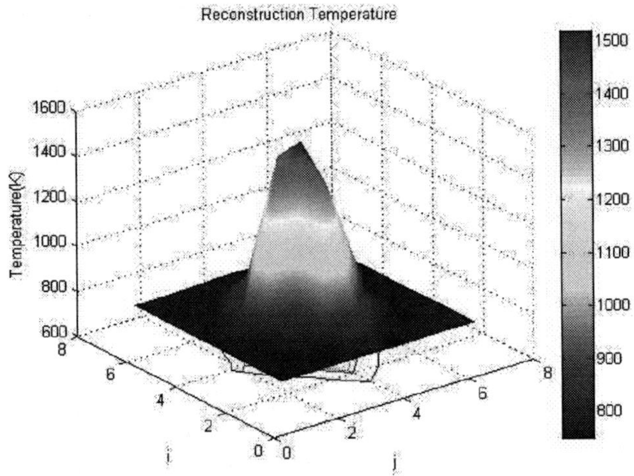

FIGURE 13. Typical section k=4 temperature distribution

Discussions

Reconstruction temperature distribution is similar to the original temperature distribution when measurement errors are lower. But when measurement errors increase, the low temperature region reconstruction effect becomes worse and under very high measurement errors, three of original seven sections temperature profiles can be only reconstructed out and lower temperature regions have not been able to be obtained according to Fig.12. This is mainly due to ill-posed characteristic of the reconstruction equation (9).

But people generally concern temperature profile in the high temperature regions of the furnace in large coal-fire power plant, especially temperature level of the flame center in furnace. From results above it can be concluded that high temperature region reconstruction effect is well and even when mean square deviation of measurement error is up to 0.50, the relative error at the maximum temperature is only 1.31%. The whole average relative errors under different measurement errors can be maintained at a lower level. This demonstrates that reconstruction method has anti-measurement error characteristic which has good practical application prospect.

The reconstruction time is shorter, which is generally between 2~4s. This satisfies real-time reconstruction requirement.

CONCLUSIONS

This paper develops a Method Carlo method in the three-dimensional absorbing, emitting and anisotropically scattering medium. The results from Monte Carlo method presented in this paper are similar to the results in literature [12].

Temperature reconstruction equations are ill-posed large linear system of equations and normal methods for solving linear system of equations can not get satisfying solutions. Therefore LSQR and damping LSQR methods, which are suit for dealing with ill-posed problems, are adopted for solving the ill-posed linear system of equations and then three-dimensional temperature distribution has been obtained.

Reconstruction temperature distribution is similar to the original temperature distribution when measurement errors are lower.

When measurement errors increase, the low temperature region reconstruction effect becomes worse and under very high measurement errors, three of original seven sections temperature profiles can be only reconstructed out and lower temperature regions have not been able to be obtained.

High temperature region reconstruction effect is well and even when mean square deviation of measurement error is up to 0.50, the relative error at the maximum temperature is only 1.31%. The whole average relative errors under different measurement errors can be maintained at a lower level.

The reconstruction time is shorter, which is generally between 2~4s. This satisfies real-time reconstruction requirement.

ACKNOWLEDGEMENTS

This work has been supported by National Natural Science Foundation of China (60534030) and Program for Changjiang Scholars and Innovative Research Team in University (IRT0434).

REFERENCES

1. Li HY and Ozisik MN. Identification of the temperature profile in an absorbing, emitting, and isotropically scattering medium by inverse analysis. Journal of Heat Transfer, 1992; 114: 1060-1063.
2. Li HY. Estimation of the temperature profile in a cylindrical medium by inverse analysis. Journal of Quantitative Spectroscopy & Radiative Transfer, 1994; 52: 755-764.
3. Li HY. An inverse source problem in radiative transfer for spherical media. Numerical Heat Transfer Part B, 1997; 31: 251-260.
4. Liu LH, Tan HP and Yu QZ. Inverse radiation problem of sources and emissivities in one-dimensional semitransparent media. International Journal of Heat and Mass Transfer, 2001; 44: 63-72.
5. Liu LH and Tan HP. Inverse radiation problem in three-dimensional complicated geometric systems with opaque boundaries. Journal of Quantitative Spectroscopy & Radiative Transfer, 2001; 68: 559-573.
6. Zhou HC, Han SD, Sheng F et al. Visualization of three-dimensional temperature distributions in a large-scale furnace via regularized reconstruction from radiative energy images : numerical studies. Journal of Quantitative Spectroscopy & Radiative Transfer, 2002; 72: 361-383.
7. Zhou HC, Yuan P, Sheng F et al. Simultaneous estimation of the profiles of the temperature and the scattering albedo in an absorbing, emitting, and isotropically scattering medium by inverse analysis. International Journal of Heat and Mass Transfer, 2000; 43(23): 4361-4364.
8. Chun Lou, Huai-Chun Zhou. Deduction of the Two-Dimensional Distribution of Temperature in a Cross Section of a Boiler Furnace from Images of Flame Radiation. Combustion and Flame, 2005; 143: 97-105.
9. A.J. SILVA NETO and M. N. OZISIK. An inverse problem of simultaneous estimation of radiation phase function, albedo and optical thickness. Journal of Quantitative Spectroscopy & Radiative Transfer, 1995; 53(4): 397-409.
10. R.G. Keanini and C.A. Allgood. Measurement of time varying temperature fields using visible imaging CCD cameras. International Communication of Heat and Mass Transfer, 1996; 23(3): 305-314.
11. Phillip M. Brisley, Gang Lu and Yong Yan, et. al. Three-dimensional Temperature Measurement of Combustion Flames Using a Single Monochromatic CCD Camera. IEEE Transaction Instrumentation and Measurement, 2005; 54(4): 1417-1421.
12. J.T.Farmer, J.R.Howell. Monte Carlo prediction of radiative heat transfer in inhomogeneous, anisotropic, nongray media. Journal of Thermophysics and Heat Transfer, Vol.8, No.1, pp.133-139, 1994.

13. Tong W. T., Skocypec R.D. Summary on comparison of radiative heat transfer solutions for a specified problem. Developments in Radiative Heat Transfer, ASME HTD 203, pp.253-258, 1992.
14. Weiming Li, Timothy W. Tong, Dean Dobranich, Louis A.Gritzo. A combined narrow- and wide-band model for computing the spectral absorption coefficient of CO2, CO, H2O, CH4, CH2, and NO. J.Quant Spectrosc.Radiat.Transfer, Vol.54, No.6, pp.961-970, 1995.
15. Paige C C, Saunders M A. LSQR: Sparse linear equations and least squares problems. AMC Transactions . Math , Vol.8, No.2, pp.195-209, 1982.
16. Yang Wen-cai, Du Jian-yuan, A New Algorithm of Seismic Tomography with Application to Engineering Detection, ACTH GEOPHYSICA SINICA, Vol.37, No.2, pp.239-244, 1994.
17. James Baglama, Lothar Reichel, Decomposition methods for large linear discrete ill-posed problems, Journal of Computational and Applied Mathematics Vol.198, pp.332-343, 2007.
18. S. Morigi, L. Reichel, F. Sgallari, F. Zama, Iterative methods for ill-posed problems and semiconvergent sequences, Journal of Computational and Applied Mathematics, Vol.193, pp.157-167, 2006.

NOMENCLATURE

f_i	Monte Carlo process defined in Eq. (4-1)
P_i	radiative energy received by each pixel on the CCD target area defined in Eq. (4-1)
N	total number of volume elements
M	total number of pixel on the CCD camera target area
E_i	energy emitted by sub-volume i
T_i	temperature in volume element i
K_{gco2}	absorption coefficients of CO2
K_p	absorption coefficients of particles
σ_0	Stefan-Boltzmann constant
dV	volume of sub-volume
$\alpha_{i\to j}$	the radiative energy fraction of the total energy emitted by volume element i, received by the image element j
$E_{i\to j}$	the radiative energy emitted by sub-volume i arriving to pixel j on the CCD target area
E	M×1 energy vector defined in Eq. (4-7)
A_E	M×N coefficient matrix defined in Eq.(4-7)
T	N×1 4th power of the temperature vector defined in Eq.(4-7)
A	M×N coefficient matrix defined in Eq.(4-8)
b	M×1 vector defined in Eq. (4-8)
x	N×1 solutions defined in Eq.(4-8)
u_1	M×1 vector defined in Eq. (4-9)
v_1,w_1	N×1 vector defined in Eq.(4-9)
x_0	N×1 initial solution vector defined in Eq.(4-9)
$\overline{\phi_1}$, β_1, $\overline{\rho_1}$, α_1	initial parameters defined in Eq.(4-9)
u_i	M×1 vector at the i generation iteration
v_i,w_i	N×1 vector at the i generation iteration
ρ_i, $\overline{\rho_i}$, c_i, β_i	parameters at the i generation iteration
θ_i, s_i, α_i, $\overline{\phi_i}$	parameters at the i generation iteration
x_i	N×1 solution vector at the i generation iteration

Subscripts

i	volume element , i= 1, 2,…, N
j	pixel on the CCD camera target area, j=1,2,…,M
i	iteration defined in Eq.(4-9)

On-line Measurement of Soot Concentration in Duct Using Scattering-Transmission Method

Yanjun Zhao*, Yongtao Zhang*, Weiying Ding*† and Shimin Wang¶

*College of Computer and Automatic Control, Hebei Polytechnic University, China,
†Department of Education, Tangshan Normal College, China
¶The Key Laboratory of Clean Coal Power Generation and Combustion Technology Ministry of Education, Southeast University, China

Abstract. A method for the soot concentration measurement based on Scattering integration and the Extinction theory is presented. The soot concentration can be determined by measuring the light intensity of forward small angel and transmission light. The method is related to the range of scattered light collection angel and parameters of measurement system, and is independent of the mean size and size distribution of soot particle. The numerical simulation results based on Mie scattering theory confirmed the validity of the method, and also gave its the measurement range of soot concentration. Additionally, the sample calibration equipment based on the Scattering-Transmission method is also introduced. Experimental results show the method has advantages of good reliability and high solution.

INTRODUCTION

In China, dust pollution increases rapidly, and dust emission from all kinds of coal-fired boilers is one of major sources of atmospheric dust pollution. Over one hundred years ago, Ringelmann, a french engineer, proposed the eyeballing dust concentration measurement called Ringelmann blacken method [1,2]. This method is simple, but the measurement results are influenced largely by the operators' subjective factors and weather objective environment etc. In order to satisfy the need of dust pollution source monitoring and control, people present more stringent requirements for the on-line measurement of soot concentration. With the development of modern science and technology, people have developed several measuring methods for soot particles concentration based on different sensing principle, such as sampling [1,2], Optics [3,4], electrostatic induction [5], ß-rays method [6] and ultrasonic method [7], etc. The methods have their own characteristics and adaptability fields. For sampling method, a certain number typical flue gas is extracted by Iso-kinetic sampler and pretreated. Then the separated dust particles are weighed and the dust concentration can be calculated. But it is very difficult to realize the Iso-speed sampling when implementing the method and the fractional losses of dust sampled in the separation course will cause the measurement error. Additionally, a fetal shortcoming of sampling method is that it can't realize the on-line and continuous measurement of soot concentration. But if sampling method is used in proper mode, the higher measurement accuracy can also be obtained. Therefore up to date, several countries including Europe, the United States and other developed countries, still regard Iso-kinetic sampling method as a standard of no-online measure particle concentration. In recent years, optical methods are widely applied to measuring soot concentration. In the 1970s, optical opacity method had been used to measure dust concentrations, which is based on Lambert-Beer's theorem. When a bunch of light passes through the smoke, its light intensity will be weakened due to absorption and scatter by particles. Transmission light intensity is related to the absorption coefficient, concentration of soot and light length. For given medium and light wavelength, the absorption coefficient of soot particles can be regarded as constant, and for given stack, light length can be also regarded as a constant. Therefore, the concentration of soot particles by measuring the original and transmission light intensities can be determined. In recent years, in-depth study found that the absorption coefficient of soot particles is not constant but is a complex function of particle size distribution and particle matter characteristics [8]. So optical opacity does not exactly reflect the dust concentration, and the average extinction coefficient and average particle size must be known when using extinction method. Extinction coefficient is a function of light Wavelength, its value is approximately 2 in a certain particle range [9]. But in most industrial environments, particle average size is a random parameter. If the unknown particle size can't

CP914, *Multiphase Flow: The Ultimate Measurement Challenge,* edited by X. Cai, Y. Wu, Z. Huang, S. Wang, and M. Wang

be obtained, measurement error will be introduced when using optical opacity. Another article presents use Opacity method to online measure particle concentration at the meantime online measurement particle size, but it needs complex light system [10].

A new method for measuring dust concentration based on Scattering and Transmission theory is presented in this paper, whose validity is verified in the lab. And its main advantage is not necessary to know the average diameter of soot in the dust flow.

PRINCIPLE OF THE SCATTERING-TRANSMISSION METHOD

Scattering integration method for soot concentration measurement can be expressed by the following equation [11]:

$$C_V = \frac{V_P}{sA_L} = \frac{4\pi}{ksA_L}\int_0^{2\pi}\frac{I(\theta)}{I_0}\mathrm{d}\theta = K\int_0^{2\pi}\frac{I(\theta)}{I_0}\mathrm{d}\theta \tag{1}$$

Where C_V is the soot's volume concentration, V_p is the sampling area, S is the light length, A_l is the section area of the laser, θ is scattering angle, $K=2\pi/\lambda$, $I(\theta)$ is the forward scattered light intensity at θ.

The formula of soot concentration based on Extinction theory is

$$I_t = I_0 \exp[-\beta_{\mathrm{ext}} \cdot s] \tag{2}$$

Where I_t is the extinction light intensity, I_0 is the incidence light intensity, β_{ext} the average extinction coefficient.

The denominator of the Scattering Integration method equation I_0 is large, so the sensitivity of the method is not high. If the denominator changes to the extinction light intensity, the sensitivity should be increased. Considered that the incidence light intensity is constant and its value is large, and the extinction light intensity decreases gradually with the increase of soot concentration, I_0 can be regarded as constant. Then replacing I_0 with I_t, we can obtain a new measurement formula termed scattering-transmission method

$$C_V = K\frac{\int_0^{2\pi} I(\theta)d\theta}{I_0 \cdot \exp[-\beta_{\mathrm{ext}}(\lambda)\cdot s]} = K_1\frac{\int_0^{2\pi} I(\theta)d\theta}{\exp[-\beta_{\mathrm{ext}}(\lambda)\cdot s]} \tag{3}$$

Average value method measuring volume concentration is based on the scattering theory of a single particle, Therefore, it is not applicable to the measurement of high particle concentration. This is because of the multi-particle scattering phenomena between the particles [12].

NUMERICAL SIMULATION BASED ON MIE SCATTERING THEORY

The scattering-transmission method is simulated based on Mie scattering theory. In simulations, the incidence light wavelength is 0.67 μm and the range of the measuring angle is $0° \square 5°$. The different average diameter can be obtained by adjusting the particles diameter distribution function and the largest diameter. The different indexes of 1.80 and 1.57-0.56i are considered respectively because the Mie scattering calculation is related to the refractive index m of the soot. Three types of the soot size distribution based on Johnson_SB function [13] are chosen, and their distribution parameters are $\sigma=2.5$, $\mu=-2.0$; $\sigma=3.8$, $\mu=2.0$; $\sigma=4.8$, $\mu=0.0$.

Figure. 1 show the relation between particle concentration of diffusion particle and the ratio of the scattering and extinction light intensity and the average diameter when integration region of the scattering light intensity are $0° \square 2°$ and $0° \square 5°$ respectively. Figure. 2 and Figure.3 show the relation between particle concentration of non-diffusion particle and the ratio of the scattering and extinction light intensity and the average size. From the figures, in can be concluded: for diffusion particle and non- diffusion particle, in the small particle diameter range, the concentration and the ratio of the scattering and extinction light intensity has a certain dependent relations with the particle average size. With the particle size increasing, this kind of dependent relations vanishes gradually. When the integration area of scattering light increases, the dependent relations of the ratio and particle average size gradually decrease. As the particle size increases, regardless of how the scattering angle range changes, the dependent relations of ratio and particle average diameter gradually reduce. The ratio of non-diffusion particle exist obviously fluctuations, with the particle average diameter size increases, the wave character gradually weakens; diffusion particle doesn't exist fluctuations, it mainly because it has a certain relationship with the extinction coefficient of the particles [14].

EXPERIMENTAL AND RESULT DISCUSSIONS

In order to simulate the actual situation in two-phase flow, a dust experimental device as shown in Figure.4 was designed [15]. Concentration of soot calibration is based on the experiment system. The dust particles from electrical precipitator in power plant were used to simulate the on-line measurement of concentration in the experiment. And the Iso-Kinetic sampling method was regarded as the real value. To verify the accuracy of the measuring instrument, the measuring result with the calibration value were compared.

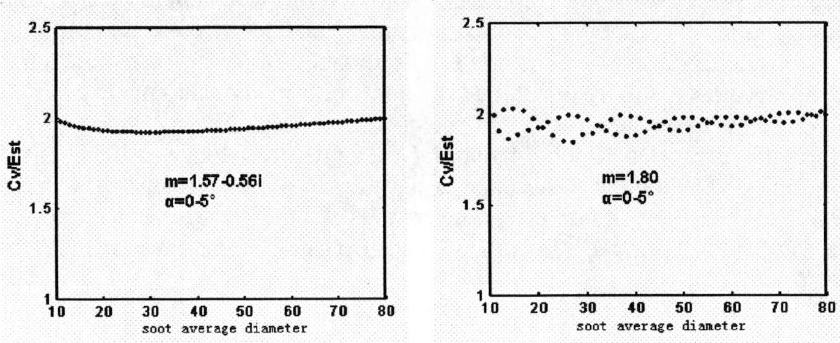

(a) (b)

FIGURE 1. Numerical Simulation1 of the different *m*

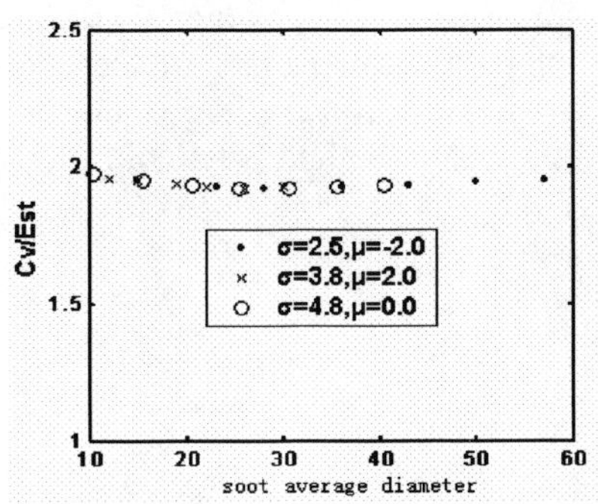

FIGURE 2. Numerical Simulation1 of the different diameter distribution. m=1.57-0.56i.

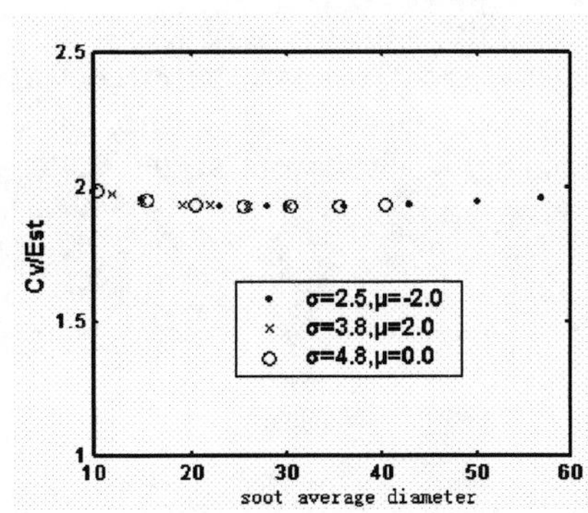

FIGURE 3. Numerical Simulation1 of the different diameter distribution. m=1.80.

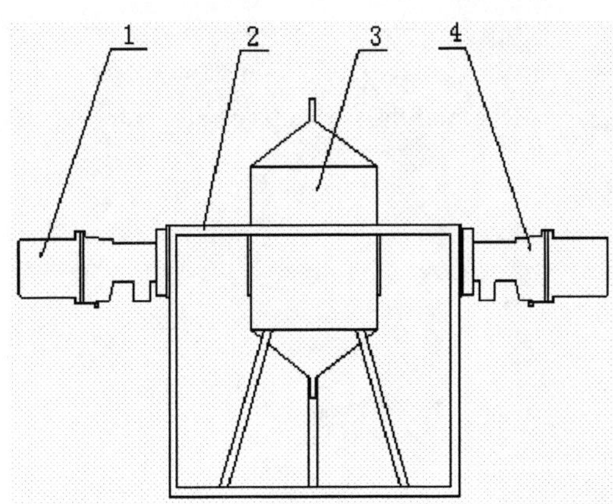

FIGURE 4. Experiment system of soot measurement

1.transmission of measurement; 2.bracket; 3. occurrence instrument of soot concentration; 4, receiver of measurement

The particles with two different size distributions were used in the calibration experiments, which come from second and third stages of electrical precipitator in power station. Their average sizes are 15.0 μm and 11.2μm respectively. The calibration results are shown in Figure 5 and Figure 6. Horizontal ordinate shows the non-calibration result measured by scattering-transmission method, and vertical ordinate shows the real result measured by Iso-Kinetic Sampling method.

As is shown from Figure 5 and Figure 6:

(1) With the dust concentration increase, the indication value measured by Scattering -Transmission measurement apparatus correspondingly rises, which shows that the relationship between the indication value and dust concentration is positive correlation

(2) The correlation coefficient of the Scattering-Extinction method and Iso-Kinetic Sampling method is 0.93, which shows that the two group data is significantly correlative. The relative error calculated by simple regression analysis is 2% and it can completely satisfy the requirements for the measurement of dust concentration in the field.

(3) The two fitted equations of Scattering-Extinction method and Iso-Kinetic Sampling method consistent. It sticks out a mile that the Scattering-Extinction method is independence of the average diameter.

(4) Obviously, the results are not strictly linear, and the non-linear relationship may be because of the stability of the device itself.

(5) Measurement results using Iso-Kinetic Sampling method are average values, but Measurement result adopting scattering-transmission method is an instantaneous value. In the experiment, the average value is compared with the instantaneous value, which may cause error, and so it is difficult to estimate this error.

(6) In addition, the sampling measurement uncertainty can also cause a certain errors.

CONCLUSION

A novel technique for measuring soot concentration based on the Scattering integration method and the Extinction theory, has been developed. Numerical simulation by using the Mie theory has also infirmed the validity of this technique. According to the measurement principle, a measurement apparatus is designed and calibrated by Iso-Kinetic Sampling method. The correlation coefficient of calibration results in the laboratory is 0.93. The approximation of the fitted equation verifies that the Scattering-Extinction method is independent of the average particle size. The laboratory result shows that the Scattering-integration method is reliable and feasible.

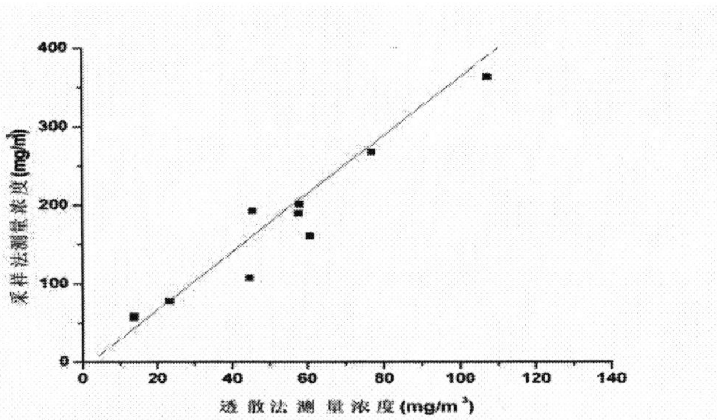

FIGURE 5. Result of the Iso-Kinetic Sampling（11.64 μm ）

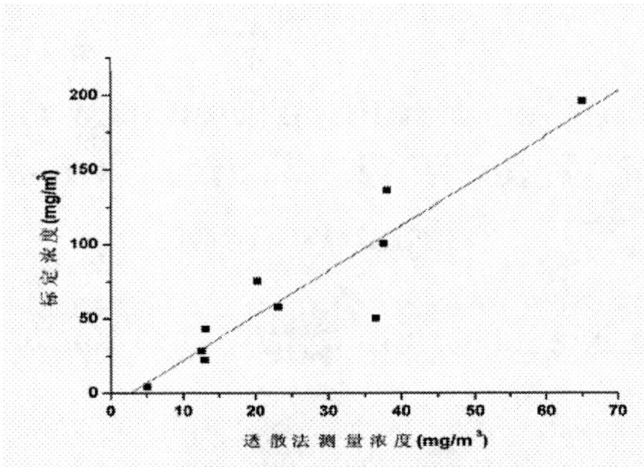

FIGURE 6. Result of the Iso-Kinetic Sampling（15.06 μ m ）

ACKNOWLEDGEMENTS

This research was supported by TangShan Application Basic Research Projects (06234501A-1). The authors wish to express their gratitude.

REFERENCES

1. Zhao Yanjun, Wang Shimin, Mu Ning etc, The Study Of Light Scattering Average Value Method OnlineMonitoring Exhaust Dust Particle Concentration, *Journal Of Electrical Engineering Of China*, Volume 11, No. 24, pp217-221, 2004.11
2. Yu Zhengran etc, Dust Testing Practical Skills [M], Beijing: China Environmental Science Press, 1990
3. Mu Ning, Continuous Measurement of Soot Concentration in the Exhaust Duct, Southeast University, 2004, 3
4. Cai Xiaoshu, Pan Yongzhi, Ouyang Xin, The Study of Diagonosing the Running Condition of Pulverized Coal in Pipe, Proceedings of the CSEE Vol. 21, No.7 July 2001, pp83-86, China
5. Juliuse B.Gajewski, Non-intrusive Solids Charge and Mass Flow Measurements with an electrostatic Flow Probe[J],Journal of Electrostatics,46,pp317-330,1999
6. Li Qiping, the Current Situation and Problem of Continuous On-line Testing Technology for Stack Dust in China [J],Sensor World,pp25-29,2001,6
7. Su Mingxu, the Ultrasonic Measurement Theory of the Particle Size Distribution and Concentration in the Gas-Solid Flow [D], University of Shanghai Science and Technology,2002,5
8. Van de Hulst, H. C., 1957, Light scattering by small particles.[M] John Wiley & Sons, New York, U.S.A
9. Dick, V. P., Applicability limits of Beer's law for dispersion media with a high concentration of particles. [J] Appl. Optics, 37, pp4998-5004, (1998)
10. Wang Shimin, Xu Yiqian and Lu Yong, On-line Monitoring Particulate Mass Concentration Emitted by Coal-fired Boiler [J]. Journal of Combustion and Technology, Vol. 3 No. 1, 1997.
11. Ye Mao,Light scattering methods for in-situ measuring particle size Distribution and Concentration in two-phase flow,[D],Southeast University, 2000
12. Hodkinson J. R., Particle sizing by means of the forward scattering lobe. Appl. Optics, **5**, pp839-844, (1966)
13. Wang Naining etc, Optics Measurement Technology And The Application Of Particle Size, [M] IAEA Press, January 2000, p. 1
14. Zhao Yanjun, Light Scattering Particle Concentration Online Monitoring Methods And CEM System[D], Doctoral Thesis, Southeast University, 2004.12
15. Xu Wei, The Experiment Study Of online measuring exhaust soot concentration From Electric Power Plant and average size, master's degree thesis, Southeast University. Nanjing 1996

Experimental Study on Performance of Turbine Flowmeter and Venturi Meter in Oil-Water Two-Phase Flow Measurement

Zhiyao Huang, Xia Li, Yian Liu, Baoliang Wang and Haiqing Li

National Laboratory of Industrial Control Technology, Department of Control Science and Engineering, Zhejiang University, Hangzhou, 310027, P. R. China, zyhuang@iipc.zju.edu.cn

Abstract. The performance of turbine flowmeter and Venturi meter in oil-water two-phase flow measurement was investigated. Experiments were carried out on horizontal pipelines of 0.5-inch, 1.0-inch and 1.5-inch diameters, with the total flowrate range of 0.9~4.5m^3/h and the oil volume fraction range of 15% ~ 85%. Experimental results show that the measurement errors of the turbine flowmeter and the Venturi meter obviously increase, whether the static mixer is installed on the experimental loop or not. Also, the non-homogeneity of the oil-water two-phase flow and the swirl flow produced by the static mixer have negative influence on the performance of turbine flowmeter and Venturi meter. Research work further indicates that the oil fraction has significant influence on the measurement results of Venturi meter.

Keywords: Turbine flowmeter, Venturi meter, Oil-water two-phase flow, Static mixer
PACS: 47.61.Jd 47.80.-v 47.27.nf

INTRODUCTION

Oil-water two-phase flow is frequently encountered in such domains as chemical and petroleum engineering[1-3]. There is a great need for the on-line measurement of oil-water two-phase flow. However, despite the industrial importance of this problem, compared with the study on the measurement of other kinds of two-phase flow (e.g., gas-liquid two-phase flow), this topic has received relatively little attention. More research work should be undertaken[1,3].

As conventional flowmeters, turbine flowmeter and Venturi meter are widely used for the flowrate measurement of single-phase flow[4]. Due to their advantages of low cost and simplicity, turbine flowmeter and Venturi meter have also been widely studied for the measurement of gas-liquid two-phase flow or gas-solid two-phase flow for many years [3,5-9]. Recently, the mentioned two kinds of single-phase flowmeters have been studied in the research field of oil-water two-phase flow measurement. Skea and Hall report their research result concerning the performance of turbine flowmeter in 2-inch and 4-inch pipelines with the second component fraction range of 3%~15% [10]. Pal publishes their research work concerning the performance of Venturi meter in 26.92 mm pipeline with the oil fraction range of 0~84.32%[1,11]. However, the published information is limited and there is still some lack of knowledge about the performance of turbine flowmeter and Venturi meter in the oil-water two-phase flow.

The aim of this work is to investigate the performance of turbine flowmeter and Venturi meter in oil-water two-phase flow in 0.5-inch, 1.0-inch and 1.5-inch pipelines, with the total flowrate range of 0.9~4.5m^3/h and the oil volume fraction range of 15%~85%. Meanwhile, the static mixer is introduced to investigate the influence of the fluid homogeneity and the swirl flow produced by the static mixer on the measurement result of the oil-water two-phase flow.

EXPERIMENTAL FACILITY

Experiments were carried out on a horizontal oil-water two-phase flow loop of the National Laboratory of Industrial Control Technology, Zhejiang University. The experimental facility was schematically shown in Fig. 1.

The experimental materials were diesel oil and tap water. The diameters of the pipelines were 0.5 inch, 1.0 inch

CP914, *Multiphase Flow: The Ultimate Measurement Challenge,* edited by X. Cai, Y. Wu, Z. Huang, S. Wang, and M. Wang

and 1.5 inches. The total volume flowrate of the oil-water two-phase flow was in the range of 0.9~4.5m^3/h and the oil volume fraction varied from 15% to 85%. The operation pressure of the experiment was around 0.4Mpa and the temperature varied from 15℃ to 25℃.

Three turbine flowmeters (Flat-bladed turbine flowmeter) and three Venturi meters were tested. The original accuracy of each turbine flowmeter was ±0.5% for single-phase flow. The original accuracy of each Venturi meter was ±1.0% for single-phase flow and their diameter ratios were 0.74 (0.5-inch Venturi meter), 0.68(1.0-inch Venturi meter) and 0.58 (1.5-inch Venturi meter) respectively. Meanwhile, for each flowmeter, sufficient upstream straight length and downstream straight length were installed to meet the requirement of the standard measurement.

A precise Oval gear flowmeter and a precise Roots flowmeter were used to obtain the pure oil flowrate. Meanwhile, a precise Electromagnetic flowmeter was used to obtain the pure water flowrate. Based on the pure oil flowrate and the pure water flowrate, the reference total flowrate of oil-water two-phase flow could be obtained and the equivalent density of the oil-water two-phase flow could be further determined (assuming the oil-water two-phase flow was the homogenous two-phase flow). The change of the oil fraction could be accomplished by adjusting the pure oil flowrate and the pure water flowrate.

To investigate the influence of the fluid homogeneity and the swirl flow produced by the static mixer on the measurement result of oil-water two-phase flow, three helical static mixers were adopted in our research work. The experiments were divided into two groups. One group were carried out under such experimental state that the static

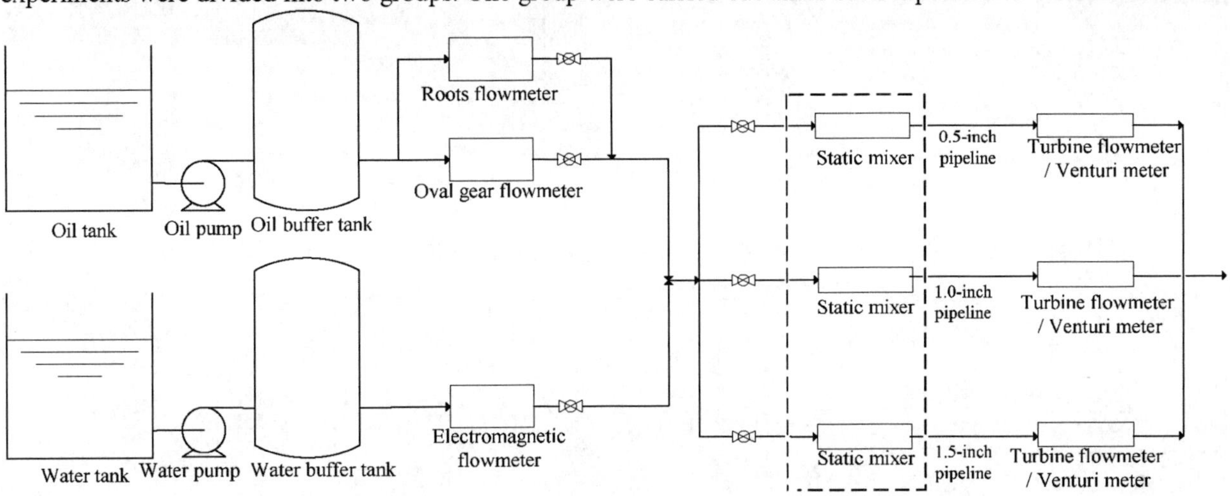

FIGURE 1. Experimental facility of oil-water two-phase flow.

mixer was installed on the upstream pipeline of the tested flowmeter. The other group were carried out without the static mixer.

The measurement model of turbine flowmeter could be briefly described as[4]:

$$q_v = f / K_T \tag{1}$$

Where q_v was the volume flowrate, f was the rotation frequency, , K_T was the meter coefficient of the turbine flowmeter.

The measurement model of Venturi meter could be briefly described as[4]:

$$q_v = K_V \sqrt{\Delta p / \rho} \tag{2}$$

Where K_V was the meter coefficient of the Venturi meter, Δp was the pressure difference, ρ was the density.

In this work, the meter coefficients of the turbine flowmeters and the Venturi meters were calibrated from the single-phase flow (pure water flow).

EXPERIMENTAL RESULTS AND DISCUSSIONS

Turbine Flowmeter

Figure 2 and Fig. 3 illustrate the experimental results of the tested turbine flowmeters in measuring the oil-water two-phase flow. The relative measurement errors increase up to ±5% while the results vary slightly with the diameter of the pipeline. Compared with the original accuracies of the flowmeters, the measurement errors obviously increase whether the static mixer is introduced into our experimental loop or not. This experiment result is not in accord with Skea and Hall's research result obtained from 2-inch and 4-inch turbine flowmeters with the second component fraction range of 3~15%. Skea and Hall indicate that the second phase has little influence on the flat-bladed turbine flowmeter, both for water-in-oil and for oil-in-water. Our research work indicates that the second phase has significant influence on the small-middle diameter (0.5-inch, 1.0-inch or 1.5-inch) flat-bladed turbine flowmeter if the component fraction is relatively high (15%~85%). Experimental results also show that the static mixer contributes little to the improvement of the measurement results. Whether the static mixer is installed on our experimental loop or not, the maximum measurement errors are both less than 5%. Meanwhile, due to the

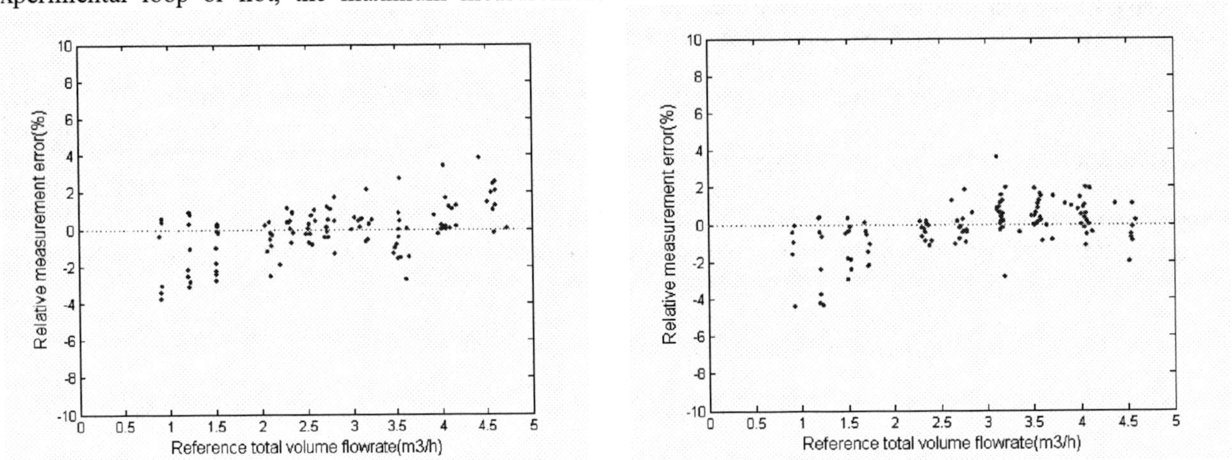

(a) (b)

FIGURE 2. Measurement result of turbine flowmeter. (a) with static mixer, (b) without static mixer.

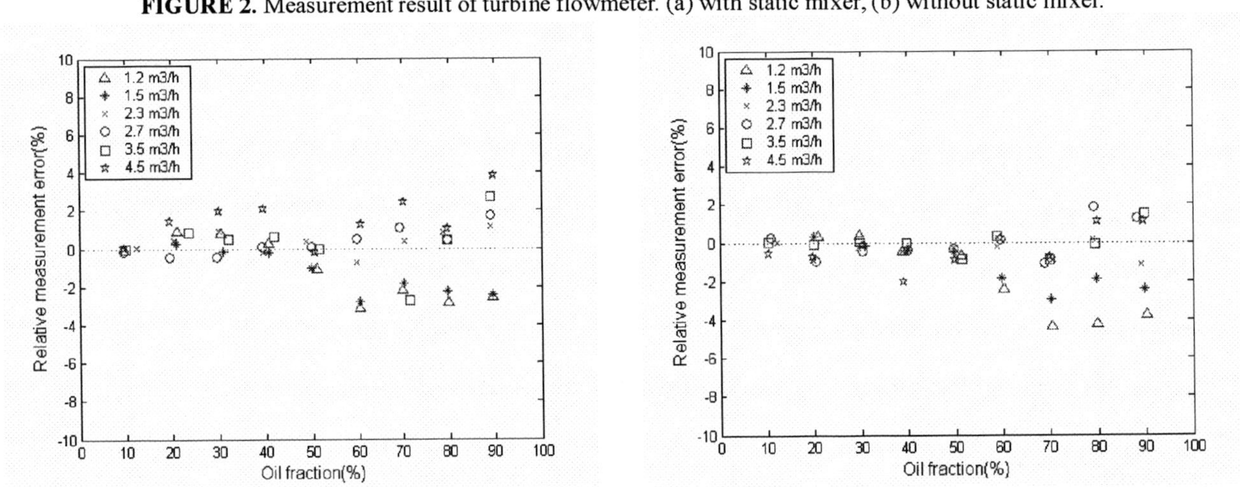

(a) (b)

FIGURE 3. Influence of oil fraction on the result of turbine flowmeter. (a) with static mixer, (b) without static mixer.

complexity of the liquid-liquid two-phase flow, no obvious regular relationship between the component fraction and the measurement result is observed from our experiments though a trend is found from our experiment results that the measurement errors of turbine flowmeter increase a little at high oil fraction.

Venturi Meter

The experimental results of the Venturi meter in measuring the oil-water two-phase flow are shown in Fig. 4 and Fig. 5. Generally, the experimental results of the Venturi meter are similar to that of the turbine flowmeter. However, there are still some differences between the results of the turbine flowmeter and the Venturi meter. The measurement results of the Venturi meter are strongly affected by the oil fraction. At low oil fraction (< 40%), almost all the relative measurement errors are less than 2.5% whether the static mixer is introduced or not. But at high oil fraction the measurement errors increase up to 5.5% when the static mixer was not adopted and 6% when the static mixer is installed on the experimental loop. This experiment result is generally in accord with Pal's research result obtained

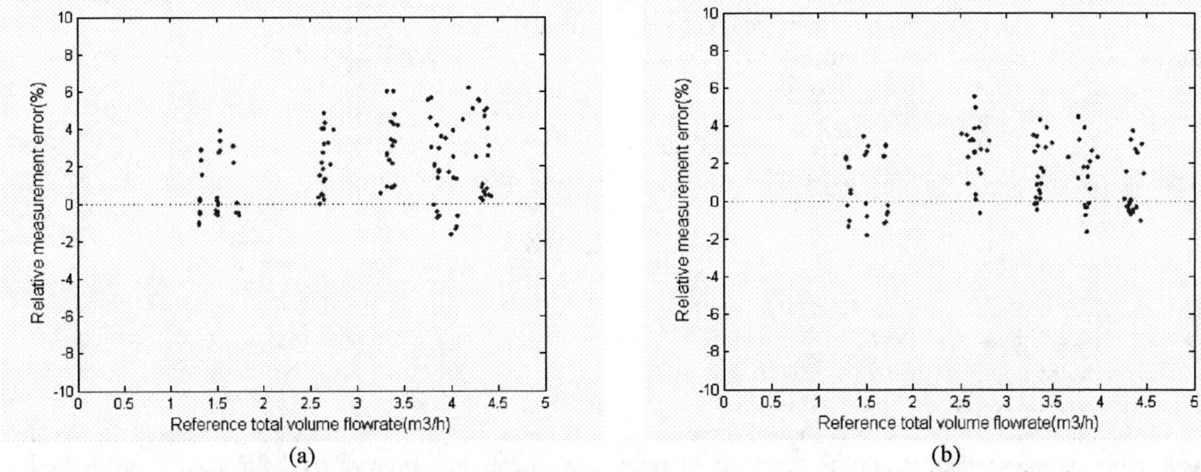

FIGURE 4. Measurement result of Venturi meter. (a) with static mixer, (b) without static mixer.

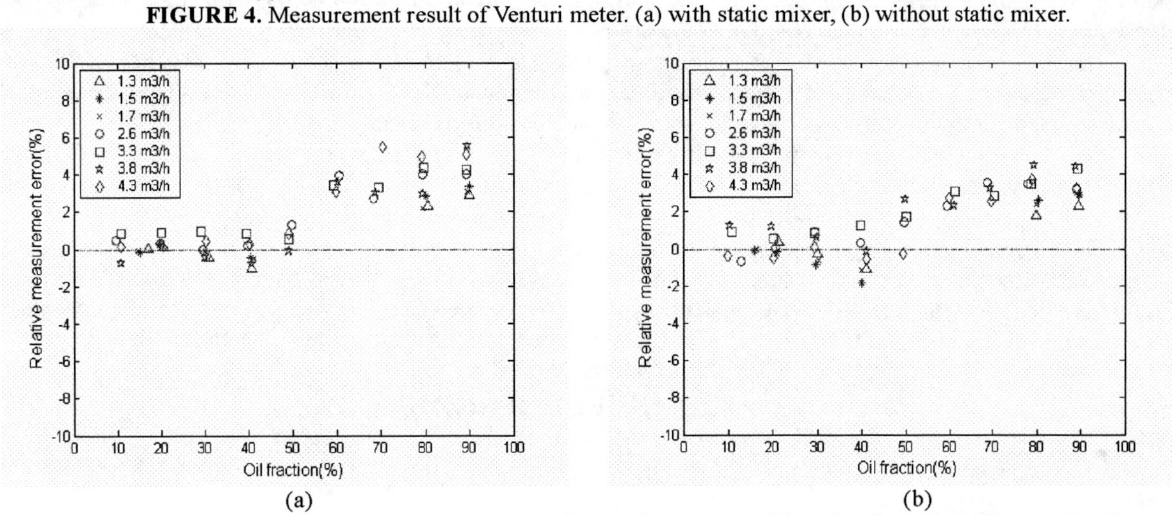

FIGURE 5. Influence of oil fraction on the result of Venturi meter. (a) with static mixer, (b) without static mixer.

from a Venturi meter (the inner diameter of the straight pipe section is 26.92 mm and the diameter of the throat is 16.0mm) with the oil fraction range of 0~ 84.32%. Pal indicates that Venturi meter is a feasible flow metering device for oil/water emulsions, with accuracy of $\pm 5\%$. Our research work further indicates that the oil fraction has significant influence on the measurement results of Venturi meter. The Venturi meter obviously overestimates the total volume flowrate of oil-water two-phase flow at high oil fraction.

Discussions

Our experimental results indicate that the measurement error will increase if we use conventional single-phase flowmeters (the turbine flowmeter and the Venturi meter) in the oil-water two-phase flow measurement. The increase of the measurement error can be analyzed from two aspects: the fluid homogeneity of the oil-water two-phase flow and the swirl flow produced by the static mixer.

If the static mixer is not introduced, the oil and the water are directly mixed and the oil-water two-phase flow is non-homogeneous flow. The non-homogeneity of the oil-water two-phase flow may be the main reason for the increase of the measurement error under such experimental state that the static mixer was not adopted, especially for the experimental results of the Venturi meter (in this work, the equivalent density of the oil-water two-phase flow is determined on the basis of the homogenous two-phase flow assumption, that will cause some additional measurement errors).

If the static mixer is introduced, the introduction of the static mixer can keep the homogeneity of the oil-water two-phase flow. However, the swirl flow will be produced by the static mixer and cause negative influence on the measurement result [12]. The swirl flow may be the main reason for the increase of the measurement error under such experimental state that the static mixer is installed on the experimental loop.

From our experimental results, the measurement errors of the Venturi meter are significantly affected by the oil fraction and the Venturi meter obviously overestimates the total volume flowrate of oil-water two-phase flow at high oil fraction. A possible reason for this difference is that: at the low oil fraction, water is the continuous phase of the oil-water two-phase flow, but at high oil fraction, oil becomes the continuous phase of the oil-water two-phase flow. Which phase is the continuous phase may be the main factor that influences the final measurement results.

CONCLUSIONS

The performance of turbine flowmeter and Venturi meter in oil-water two-phase flow has been investigated by experiments. The experiments have been carried out in 0.5-inch, 1.0-inch and 1.5-inch horizontal pipelines. The total flowrate of oil-water two-phase flow is in the range of 0.9～4.5m^3/h and the oil volume fraction varies from 15% to 85%.The experimental results are listed as follows:

(1) The measurement errors of the turbine flowmeter and the Venturi meter obviously increase, whether the static mixer is introduced into our experimental loop or not. The contribution of the helical static mixer to the improvement of the measurement result is not manifest.

(2) The non-homogeneity of the oil-water two-phase flow has negative influence on the measurement performance of turbine flowmeter and Venturi meter. It may be the main reason for the increase of the measurement error under such experimental state that the static mixer is not adopted.

(3) The swirl flow produced by the static mixer also has negative influence on the measurement performance of turbine flowmeter and Venturi meter. It may be the main reason for the increase of the measurement error under such experimental state that the static mixer is installed on the experimental loop.

(4) The oil fraction has significant influence on the measurement results of Venturi meter. The Venturi meter obviously overestimates the total volume flowrate of oil-water two-phase flow at high oil fraction.

Useful knowledge for the oil-water two-phase flow measurement has been obtained in this work. However, due to the complexity of liquid-liquid two-phase flow, further research work is required in this area.

ACKNOWLEDGEMENT

The project is supported by National Natural Science Foundation of China under Grant No.50576084 and No.60532020.

REFERENCES

1. R. Pal, *Industrial & Engineering Chemistry Research* **33**, 1413-1435 (1994).
2. P. Vigneaux, P. Chenais, and J.P. Hulin, et al., *AIChE Journal* **6**, 781-789 (1988).
3. R. Thorn, G.A. Johansen, and E.A. Hammer, *Measurement Science & Technology* **8**, 691-701 (1997).
4. R.C. Baker, *Flow Measurement Handbook: Industrial Designs, Operating Principles, Performance, and Applications*, New York: Cambridge University Press, 2000, pp.130-138 and 215-234.
5. Z.H. Lin, *Gas-liquid-solid Multiphase Flow Measurement*, Beijing: China Metrology Press, 1988, pp. 75-76, 82-111 and 120-123.
6. H.Q. Li, *Measurement of Two-phase Flow Parameter and its Applications*, Hangzhou: Zhejiang University Press, 1991, pp.123-130.
7. M.W. Johnson and S. Farroll, *Flow Measurement and Instrumentation*, **6**, 279-282 (1995).
8. G. Oddie and J.R.Anthony Pearson, *Annual Review of Fluid Mechanics*, **36**, 149-172 (2004).
9. Z.Y. Huang, D.L. Xie and H.J. Zhang, et al., *Flow Measurement and Instrumentation*, **16**, 177-182 (2005).
10. A.F. Skea and A.W.R. Hall, *Flow Measurement and Instrumentation*, **10**, 151-157 (1999).
11. R. Pal, *Industrial & Engineering Chemistry Research*, **32**, 1212-1217 (1993).
12. A. Soleimani, C.J. Lawrence and G.F. Hewitt, *Trans IChemE, Part A*, **80**, 145-153 (2002).

Analysis on Dynamic Differential Pressures of Multi-Loop Flowmeter for the Measurement of Gas-Liquid Two-Phase Flow

Dailiang Xie, Guowei Liang, Fang Wang

College of Metrological Technology and Engineering, China Jiliang University, Xueyuan Road, Xiasha Higher Education District, Hangzhou, 310018
E-Mail: dlxie@cjlu.edu.cn

Abstract. A multi-loop flow sensor with new structure is designed for the parameters measurement of gas-liquid two-phase flow. The dynamical differential pressure signals of gas-liquid two-phase flow captured from the rising section at the horizontal diameter of the multi-loop sensor are analyzed. A characteristic value related to the root-mean-square of the dynamical differential pressure signals is obtained. The relationship between the characteristic value of differential pressures and the volumetric void fraction is established. The gas density is introduced for the correction of this mode. The experiments were carried out on the horizontal multi-phase flow loop. The non-dimensional method was applied to analyze the data. The relationship between the volumetric void fraction and the non-dimensional value are achieved. The online measurement of the volumetric void fraction of gas-liquid two-phase is realized through the multi-loop flow sensor. The experiments showed that the results are satisfactory.

Keywords: two-phase flow; multi-loop meter; dynamical differential pressure signal; volumetric void fraction.
PACS: 47.27.ed

INTRODUCTION

The gas-liquid two-phase flow widely exists in many industrial processes, such as petroleum, chemical engineering, metallurgy and energy production. The operation of industrial devices is influenced the parameters including voidage fraction, flow patterns and flowrate of this two-phase flow. So the improvement of measurement technology for two-phase flow will bring significant benefits to the industry, particularly, to the petroleum industry. However, it is difficult to measure the parameters of this flow because of the complexity of the two-phase flow [1,2,3]. In resent years, many efforts have been made and the new methods, for example, optical fibers[4], tomography system[5] and correlation technique[6], are employed to measure the gas-liquid two-phase flow. Fluctuation is one of the inherent natures of two-phase flow when it flowing in the pipe. Researches indicate that some parameters are related to the characteristic values of the fluctuation signals obtained from the two-phase flow [7,8,9]. These attempts provide useful information for the parameters measurement through analysis of fluctuations.

The loop flowmeter is one of the centrifugal meters. A new flowrate measurement method for gas-liquid two-phase flow is proposed in this work based on the loop flowmeter. A multi-loop flowmeter, which, to our knowledge, is a new extension of elbow flowmeter or traditional loop flowmeter, is developed. On the presented multi-loop flowmeter, four different positions are selected as the place to measure differential pressure, in which two positions locate at the horizontal diameter and two locate at the vertical diameter. In this paper, the details are provided on the establishment of the volumetric void fraction measurement models using fluctuation signal serials. The volumetric void fraction of the two-phase flow is calculated through dynamic differential pressure analysis. Experiments were carried out on the horizontal multiphase flow loop and the results show that this method is effective.

EXPERIMENTS SETUP

A loop flowmeter with new structure is designed for parameters measurement of gas-liquid two-phase flow (as shown in Figure 1 and Figure 2). The sensor bends into two loops, on which four groups of pressure ports are settled.

CP914, *Multiphase Flow: The Ultimate Measurement Challenge,* edited by X. Cai, Y. Wu, Z. Huang, S. Wang, and M. Wang

The two groups at the horizontal diameter are set to measure the differential pressure on the wall of the inner circle and the outer circle (as show in Figure 1). The other two groups at the vertical diameter are set to measure the differential pressure between the top and the bottom of the multi-circle flowmeter (as show in Figure 2). The diameter of the pipeline is 50mm, and the radius of the circle is 250mm. The ratio of the elbow is $R/D = 5$.

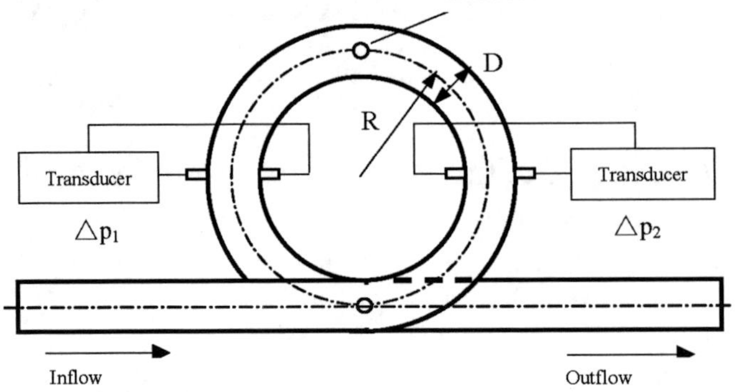

FIGURE 1. Diagram of pressure ports at the horizontal diameter.

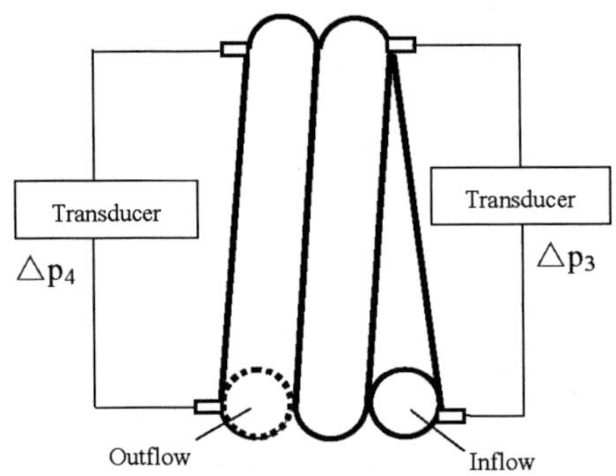

FIGURE 2. Diagram of pressure ports at the vertical diameter.

The experiments of flowrate measurement were carried out in a horizontal multiphase flow loop at China Jiliang University. The water and the air were selected as the experimental materials. The facilities are shown in Figure 3. The water was given by a pump and went to a water tank (large as $3m^3$) that kept the pressure stable. Upstream the mixter, an EMF, calibrated to within ±0.2% of full range, was installed to provide the accurate water flowrate. The gas flux was provided by a compressor. The pressure of the gas maintained 0.7MPa by the air tank. The accurate gas flowrate was measured by a roots type flowmeter ($\pm 1.5\%$). After the mixer there were 6m straight pipelines to ensure the full development of two-phase flow. The straight pipeline included 4m transparent sections (for flow patterns observation, not used in these experiments) and the removable test section that allowed easy installation of instrumentations. The measurement range of the sensor is 50 m^3/h. The experimental materials were water and air. During the experiments, the water flowrate was in the range of $4m^3/h$ to 45 m^3/h under the pressure of 0.2 Mpa and the air flowrate was in the range of 0.21 m^3/h to 2.2 m^3/h under the pressure of 0.7Mpa. The KELLER PD23 was used as the differential pressure transducer. A high-speed sampling card was used to collect the time serials of the differential pressure values.

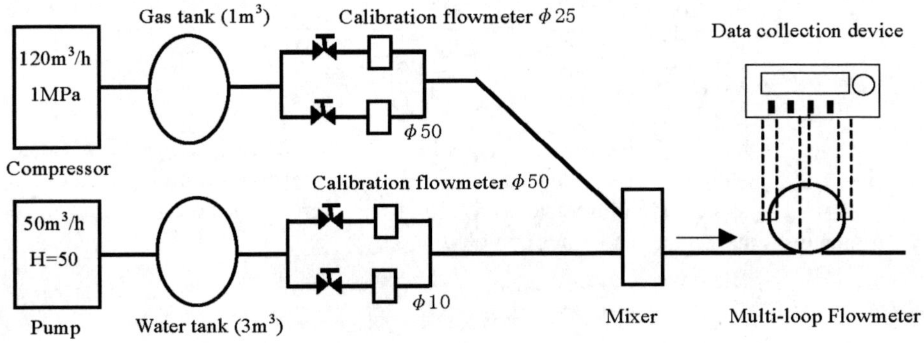

FIGURE 3. Experimental facilities

DYNAMICAL ANALYSIS METHOD AND RESULTS

Theoretical Model

The loop flowmeter is one of the centrifugal meters. The flowrate measurement principle of the loop flowmeter is similar to that of the conventional elbow flowmeter. According to the forced swirl flow theory, forced swirl flow occurs when the fluid goes through the annuli because of the bend of the pipe. The angular acceleration of the fluid results in the centrifugal force, which brings the differential pressure between the wall of the inner loop and outer loop of the pipe. The differential pressure values are proportion to the average velocities of the fluid. So the flowrate can be measured if the geometrical structure of the loop flowmeter is determined.

The volume flowrate can be given by[10]:

$$q_v = C\sqrt{\frac{R}{2D}}(\frac{\pi}{4}D^2)\sqrt{\frac{2\Delta p}{\rho_m}} \tag{1}$$

where q_v is the volume flowrate, C is the correction factor generally determined by the position of the pressure port and can be obtained by experiments, R denotes the radius of curvature of the annulus tube, D is diameter of the pipe, Δp is the value of differential pressure between the wall of the inner loop and outer loop of the pipe and ρ_m is the density of the fluid.

For gas-liquid two-phase flow, the density of the mixture ρ_m is expressed as:

$$\rho_m = \rho_g \alpha + \rho_l (1-\alpha) \tag{2}$$

where α is the voidage fraction, ρ_g is the gas density and ρ_l is the liquid density.

Combining equation (1) and equation (2), the measurement model of total flowrate can be deduced as:

$$q_v = C\sqrt{\frac{R}{2D}}(\frac{\pi}{4}D^2)\sqrt{\frac{2\Delta p}{\rho_g \alpha + \rho_l (1-\alpha)}} \tag{3}$$

Because the relationship of voidage α, volumetric void fraction β and slip ratio between gas phase and liquid phase S is:

$$\alpha = \frac{1}{1+S(\frac{1}{\beta}-1)} \tag{4}$$

for homogeneous flow(S=1, $\alpha = \beta$), equation (3) can be modified as:

$$\frac{\Delta p}{q_v^2} = k\beta + b \tag{5}$$

where

$$k=\frac{\rho_g-\rho_l}{\left[C\sqrt{\frac{R}{D}}(\frac{\pi}{4}D^2)\right]^2},\quad b=\frac{\rho_l}{\left[C\sqrt{\frac{R}{D}}(\frac{\pi}{4}D^2)\right]^2} \tag{6}$$

where Δp, q_v and β are average values of the dynamical signal serial, k and b are determined by the geometrical structure of loop flowmeter and the density of gas phase and liquid phase. Researches indicate that the instantaneous values of differential pressure, flowrate and volumetric void fraction are agreed with the relationship of average values [8, 9]. The relationship of instantaneous values can be given by:

$$\frac{\Delta p_i}{q_{vi}^2}=k\beta_i+b \tag{7}$$

RMS (root-mean-square) of the differential pressure signals can be calculated as

$$\delta=[\frac{1}{n}\sum_{i=1}^{n}(\Delta p_i)^2-\Delta p^2]^{1/2} \tag{8}$$

where n is the sampling number and δ denotes the vibration amplitude of differential pressure signals.

Because

$$q_v^2=\frac{1}{n}\sum_{i=1}^{n}q_{vi}^2, \tag{9}$$

combing equation (5), equation (8) and equation (9), we will obtain:

$$\frac{\delta}{q_v^2}=[\frac{1}{n}\sum_{i=1}^{n}(k\beta_i+b)^2-(k\beta+b)^2]^{1/2} \tag{10}$$

A nondimensional value can be achieved by:

$$I=\frac{\delta}{\Delta p}=\frac{[\frac{1}{n}\sum_{i=1}^{n}(k\beta_i+b)^2-(k\beta+b)^2]^{1/2}}{k\beta+b}=f(\beta) \tag{11}$$

Equation (11) indicates that I is the single-valued function of β.

Experimental Results

TABLE 1. Experimental Data.

q_v (m^3/h)	β	I	q_v (m^3/h)	β	I
12.624	0.4043	3.2809	44.079	0.0401	0.6696
20.657	0.5353	5.1913	28.551	0.2575	3.3071
24.686	0.3790	3.6817	31.509	0.2123	2.3309
17.310	0.5124	4.1602	35.821	0.1631	2.3097
22.580	0.6187	3.5111	40.007	0.1239	2.3101
14.540	0.4120	3.3684	44.956	0.0949	2.7714
19.335	0.2532	2.5451	25.751	0.2357	2.7165
17.336	0.2738	1.9695	30.117	0.1782	2.4280
21.675	0.3103	2.3353	34.648	0.1191	1.9944
26.289	0.4404	2.2644	39.748	0.0893	1.1653
24.398	0.1745	2.7203	44.535	0.0693	1.3887
23.494	0.1483	1.7400	33.183	0.2587	2.8391
28.038	0.1012	1.3453	36.839	0.2014	3.3071
31.618	0.0834	1.4316	41.200	0.1597	1.3704
38.645	0.0571	0.5457	45.708	0.1216	3.1607
28.833	0.1288	1.5113	39.158	0.0635	1.8729
34.086	0.0932	0.9080	42.982	0.0512	0.9642

A data collection device with the sampling rate of 1000 values per second was employed to capture the differential pressure signals at the position 2 (Δp_2, as shown in Figure 1). 8000 values were collected for each measurement point. The differential pressure signals were obtained under various flowrate and volumetric void fraction. The nondimensional value I was calculated according to the method mentioned in Section 3.1. The results were listed in Table 1.

The relationships between the volumetric void fraction β and I, β and $1/I$ are shown in Figure 4 and Figure 5 respectively.

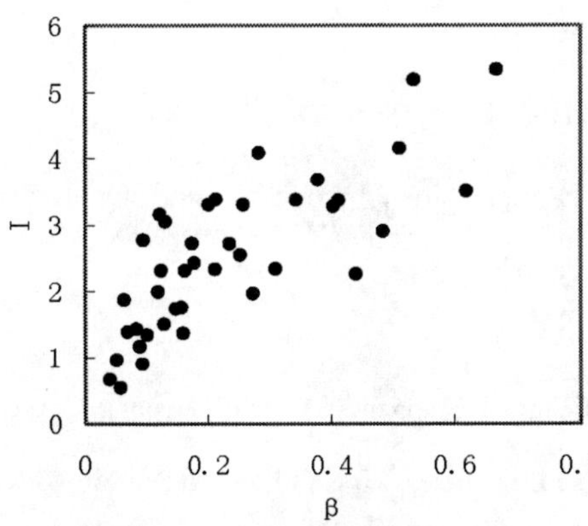

FIGURE 4. Relationship of I and β

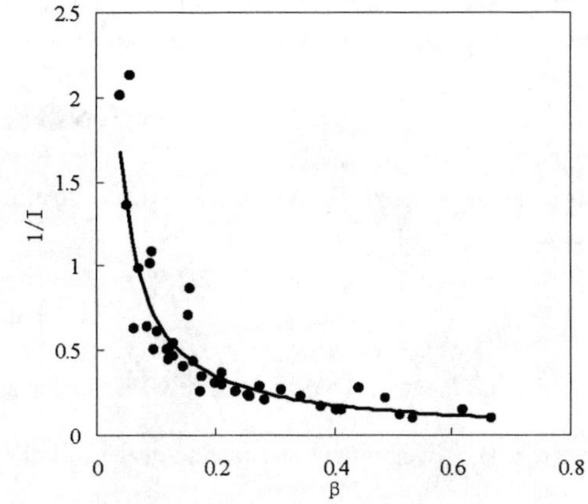

FIGURE 5. Relationship of $1/I$ and β

Because of the variation of the gas density during the experiments, the variable I is corrected according to the gas density. A nondimensional value ρ_g / ρ_l is employed for the correction. The model can be modified as:

$$I' = I\left(\frac{\rho_g}{\rho_l}\right)^{\xi} \tag{12}$$

where ρ_l, ρ_g is the density of the liquid phase and gas phase respectively, and ξ is the coefficient determined according to the experimental condition.

The relationship between I' and β can be determined after ξ is selected. Figure 6 shows the relationship of I' - β when $\xi = -1$, which is satisfied the linear relationship and can be given by:

$$I' = K\beta + B \tag{13}$$

where $K = 14.465, B = 0.1182$, that are calculated by least square regression.

Figure 7 shows the measurement result of the volumetric void fraction β.

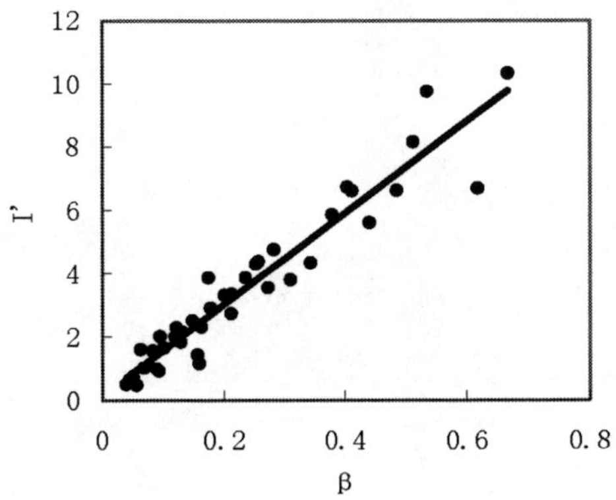

FIGURE 6. Relationship between β and I'

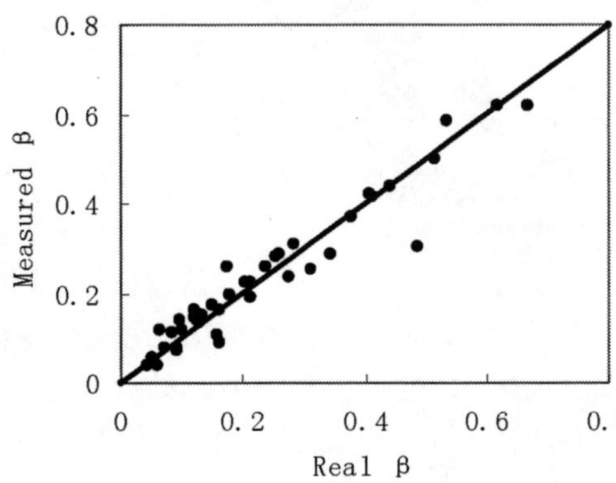

FIGURE 7. Measurement result of β

CONCLUSION

A multi-loop flowmeter is developed for parameters measurement of gas-liquid two-phase flow. The volumetric fraction of the two-phase flow is calculated through dynamic differential pressure analysis. The details on the establishment of the volumetric void fraction measurement models are provided. Experiments were carried out on the horizontal multiphase flow loop and the results show that this method is effective.

ACKNOWLEDGMENT

This work is supported by General Administration Quality Supervision, Inspection and Quarantine of the People's Republic of China Grant #2006QK23 and China Jiliang University Grant #XZ0514.

REFERENCES

1. H. Q. Li. Measurement of Two-Phase Flow Parameter and Its Applications. Hangzhou, China: Zhejiang University Press, 1991.
2. G. F. Hewitt et al. Development and testing of the mixmeter multiphase flow meter, Proc.13th North Sea Flow Measurement, 1995.
3. Z. H. Lin. Gas-liquid Two-phase Flow and Boiling Heat Transfer. Xi'an, China, Xi'an Jiaotong University Press, 2004.
4. C. Alain. Measurement of gas phase characteristics using new monofiber optical probes and real-time signal processing. Nuclear Engineering and Design, 1998, 184(2-3): 393-408.
5. Z. Y. Huang, D. L. Xie, H. J. Zhang, H. Q. Li. Gas–oil two-phase flow measurement using an electrical capacitance tomography system and a Venturi meter. Flow Measurement and Instrumentation, 2005, 16(2-3):177-182.
6. B. Gurau, P. Vassallo, K. Keller. Measurement of gas and liquid velocities in an air-water two-phase flow suing cross-correlation of signals from a double sensor hot-film probe. Experimental Thermal and Fluid Science, 2004, 28(6): 495-504.
7. V. C. S. Ferreira. Differential pressure spectral analysis for two-phase flow through an orifice plate. The International Journal of Pressure Vessels and Piping, 1997, 73(1): 19-23.
8. H. J. Zhang, W. T. Yue, L. B. Ma, H. L. Zhou. Relationship between Fluctuating Differential Pressure and Void Fraction of gas-liquid two-phase Flow in Venturi Tube. Journal of Chemical Industry and Engineering (China), 2005, 56(11): 2102-2107.
9. L. J. Xu, J. Xu, F. Dong, T. Zhang. On fluctuation of the dynamic differential pressure signal of Venturi meter for wet gasmetering . Flow measurement and Instrumentation, 2005, 14(4-5): 177-182.
10. G. W. Liang, W. C. Cai, et al. Flowrate measurement technology and instruments (Chinese). Beijing, China: China Machine Press, 2002.

Numerical Simulation of Optic-pneumatic Combined Probe in On-line Measurement of Pulverized Coal

Li Ye, Xiaoshu Cai, and Bingtao Zhao

Department of Power Engineering, University of Shanghai for Science and Technology, Shanghai 200093, China

Abstract. Optical-pneumatic combined probe apparatus invented by authors institute are capable of measuring size, velocity and concentration of pulverized coal simultaneously. To analyse the accuracy of measured values, numerical simulation for the probe in on-line measurement was carried out with CFD software. Simulation results show that the intrusion of the probe brings negative effects on the measured velocity values, making them relatively higher than the actual ones; and the flow of purge air brings negative effects both on the velocity measurement and concentration measurement, making the measured values relatively higher than original ones. Such influences can't be avoided but can be controlled.

Keywords: optical-pneumatic combined probe; pulverized coal; two-phase flow; on-line measurement; light fluctuation; correlation method

INTRUDUCTION

Size, velocity and concentration of pulverized coal are of great influence on the efficiency, emission and safety of boilers in power plants[1]. To detect transportation conditions of pulverized coal real time and guarantee the high-performance and trouble-free running of boilers, it is necessary to carry out on-line measurement and on-line diagnosis on those three parameters. Different from most current on-line measurement technologies which can only measure the concentration values[2~4], optical-pneumatic combined probe apparatus invented by authors institute are capable of measuring size, velocity and concentration of pulverized coal simultaneously, easing the operation of on-line measurement and on-line diagnosis at a great degree. Developed on the basis of light fluctuation and correlation method, this apparatus owns attributes of small dimension, simple structure, rapid response and convenient installation etc. Used in practice for a long time, it has been proved to be an effective on-line instrument to monitor the transportation conditions of pulverized coal[5]. While it is estimated that the intrusion of probe and the flow of purge air will bring some influence on the flow field of air-coal stream inevitably and affect the accuracy of measured values accordingly. The objective of this work was just to find out how the measured values were affected by those affecting factors through simulating the flow fields within and around the probe's measurement region with Computational Fluid Dynamics (CFD) software. Air-coal stream velocities and purge air flow flux were chosen as variable parameters during the simulation process.

MEASUREMENT THEORY AND APPARATUS STRUCTURE

Optical-pneumatic combined probe apparatus is developed on the basis of light fluctuation and correlation method[6]. When particles moving across a small bundle of laser light, the number and size of particles contained in the light-region are variable at all time, causing random changes of the intensity of transmission light. Having measured such changes of intensity of transmission light, the size, velocity and concentration of particles can be calculated with relevant formulas. The measurement principle is illustrated in fig. 1.

CP914, *Multiphase Flow: The Ultimate Measurement Challenge,* edited by X. Cai, Y. Wu, Z. Huang, S. Wang, and M. Wang

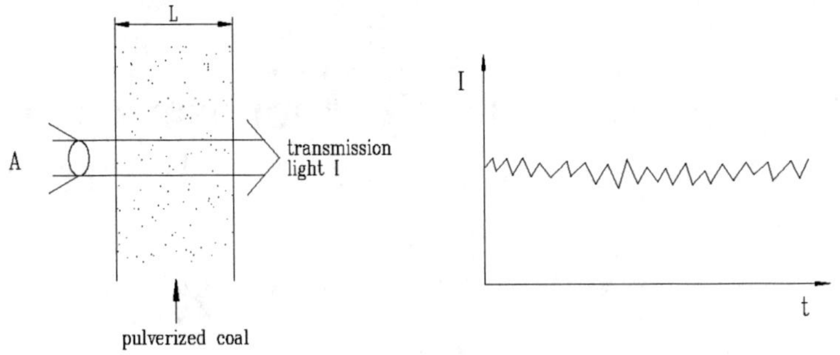

FIGURE 1. Measurement principle of light fluctuation method

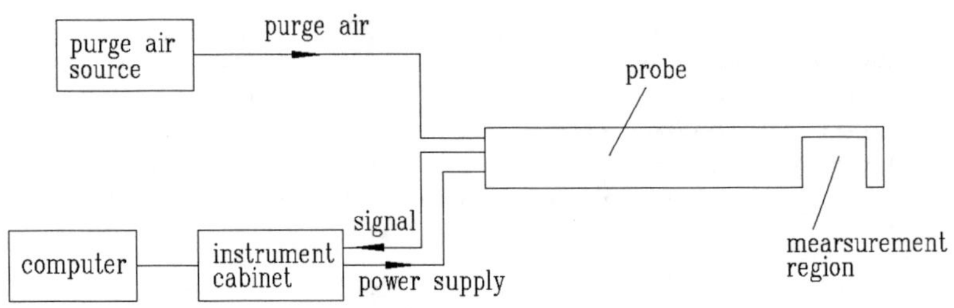

FIGURE 2. Schematic diagram of the optical-pneumatic combined probe apparatus

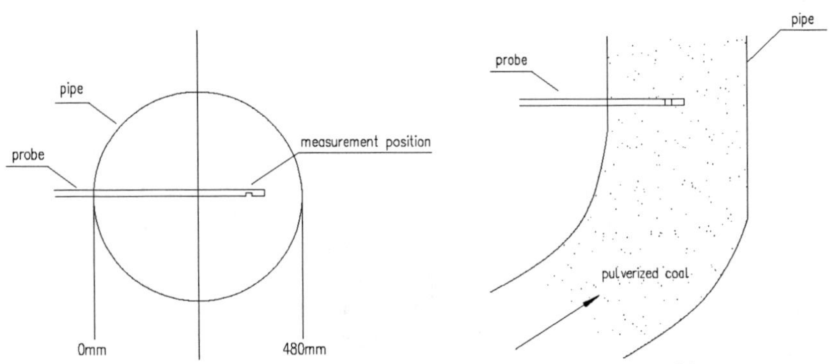

FIGURE 3. Schematic diagram of the probe's measurement position

Optical-pneumatic combined probe apparatus mainly consists of a probe, an instrument cabinet, a purge air source and a computer. Other components such as laser emitter, optical fibers, photoelectric signal detector and 0~20mA signal convertor are all integrated into the probe, so it is very easy to install and operate this apparatus on the spot. At the front part of the probe, lies a measurement region with a length of 12 mm. Optic signals detected by the measurement region will be transformed into electrical signals through semiconductor elements, and then transformed into readable values which corresponding to different size, concentration and velocity of pulverized coal after further processed by computers. The structure of optical-pneumatic combined probe apparatus is shown in fig. 2 and the measurement position of the probe on the spot is shown in fig. 3.

NUMERICAL SIMULATION

Creation of Simulation Model and Meshes

The columnar probe has an outer diameter of 25mm. At the front part of the probe, locates a measurement region; and at the inner surfaces of the measurement region, locates 4 measurement holes which emit and accept laser light bundles. To prevent those measurement holes from being jammed by coal-particles, purge air is sent into the hollow probe and blown through them, so the measurement holes also acts as purge air holes. The measurement region has a

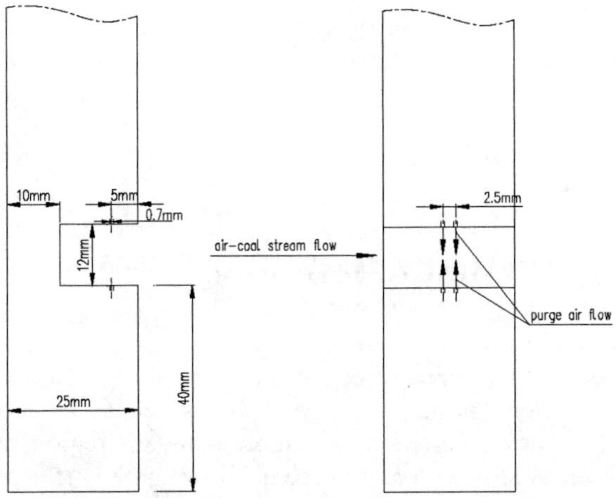

FIGURE 4. Structure and dimensions of the probe

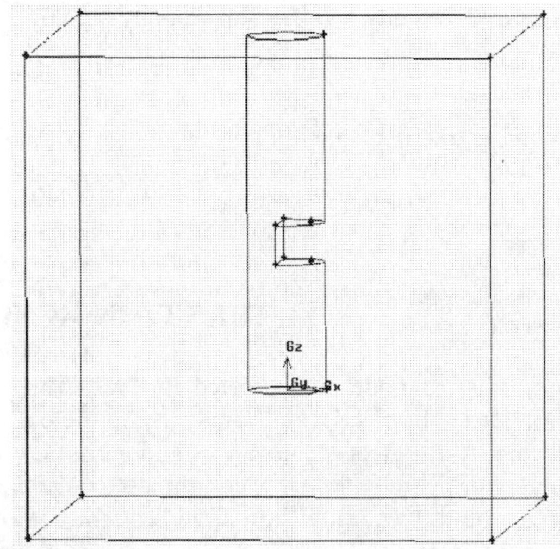

FIGURE 5. Schematic diagram of the simulation model

length of 12mm, and each measurement holes (purge air holes) has an aperture of 0.7mm. The structure and dimensions of probe as well as the flow directions of air-coal stream and purge air are shown in fig. 4.

Compared to 480mm inner diameter of the transportation pipe, the dimension of measurement region is so small that the influence of pipe's wall on this region can be considered negligible. So it just needs to pick out a finite space around the measurement region for this simulation and a three-dimensional model was created as shown in fig. 5. Air-coal stream velocity and purge air flow flux constitute the major affecting factors which influence the accuracy of measured values, and according to measurement results in practice, typical air-coal stream velocity from 10m/s to 40m/s and purge air flow flux from 2L/min to 10L/min were selected as variable parameters during this simulation process.

TGrid were created to divide the flow field herein. And to get more accurate simulation results and shorten calculation time at the same time, different sizes of meshes were adopted within and around the measurement region.

Governing Equations

Mixture model was selected to solve this gas-solid two-phase flow problem. In mixture model, air was treated as primary phase and pulverized coal was treated as secondary phase. This model allows the phases to move at Different velocities, using the concept of slip velocities.

(1) Continuity equation for the mixture：

$$\frac{\partial}{\partial t}(\rho m)+\nabla\cdot(\rho_m\vec{v}_m)=\dot{m} \tag{1}$$

where $\vec{v}_m$ is the mass-averaged velocity:

$$\vec{v}_m=\frac{\sum_{k=1}^{n}\alpha_k\rho_k\vec{v}_k}{\rho_m} \tag{2}$$

and ρ_m is the mixture density:

$$\rho_m=\sum_{k=1}^{n}\alpha_k\rho_k \tag{3}$$

(2) Momentum equation for the mixture:

$$\frac{\partial}{\partial t}(\rho_m\vec{v}_m)+\nabla\cdot(\rho_m\vec{v}_m\vec{v}_m)=-\nabla p+\nabla\cdot[\mu_m(\nabla\vec{v}_m+\nabla\vec{v}_m^{T})]+\rho_m\vec{g}$$
$$+\vec{F}+\nabla\cdot(\sum_{k=1}^{n}\alpha_k\rho_k\vec{v}_{dr,k}\vec{v}_{dr,k}) \tag{4}$$

(3) Slip velocity equation:

$$\vec{v}_{qp}=\vec{v}_p-\vec{v}_q \tag{5}$$

The drift velocity and the Slip velocity are connected by the following expression:

$$\vec{v}_{dr,p} = \vec{v}_{qp} - \sum_{k=1}^{n} \frac{\alpha_k \rho_k}{\rho_m} \vec{v}_{qk} \tag{6}$$

(4) Volume fraction equation for the secondary phases

$$\frac{\partial}{\partial t}(\alpha_p \rho_p) + \nabla \cdot (\alpha_p \rho_p \vec{v}_m) = -\nabla \cdot (\alpha_p \rho_p \vec{v}_{dr,p}) \tag{7}$$

SIMULATION RESULTS AND DISCUSSION

The simulation model was created in three-dimensions, while illustrations of flow fields on two-dimensional planes are needed to make results analysis more clearly. Two planes were selected for this purpose. The first one is perpendicular to the X coordinate-axis and across the four measurement holes (purge air holes), called as plane-1; the second one is parallel to the X coordinate-axis and near to the lower inner face of the measurement region, called as plane-2. The influences on flow field brought both by the air-coal stream and purge air flow can be reflected on either plane, so they are considered as the optimal planes to make the analysis.

Discussion on the Accuracy of Velocity Measurement

Influence of air-coal stream on the accuracy of velocity measurement

Figure 6 shows the velocity distributions of pulverized coal on plane-1 within and around the measurement region of the probe under the same purge air flow flux (2L/min). From both (a) and (b), it can be found that the coal particles are accelerated partially within the passage of measurement region. Such acceleration of coal particles is brought by the air-coal stream flow, i.e. caused by the shrinkage of streamline when stream flows from a larger space into a smaller space. Since the velocity values detected by the probe are average velocities of particles within the light region, the measured velocity values of pulverized coal will be relatively higher than the actual ones. Observed from simulation results, such deviations of velocity values are limited. The largest deviation within desired range of air-coal stream flow rate will not exceed 3m/s, just as shown in table 1.

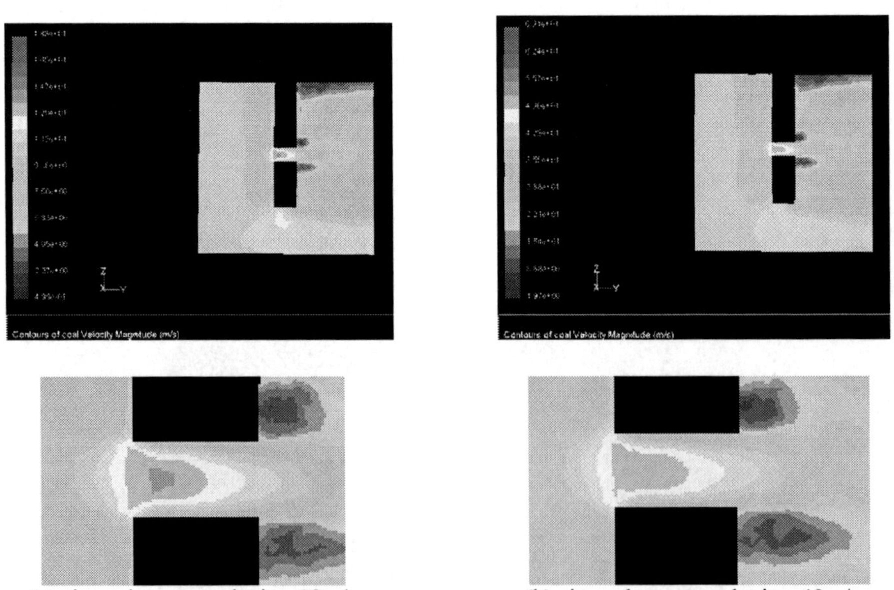

(a) air-coal stream velocity: 10m/s
purge air flow flux: 2L/min
(plane-1)

(b) air-coal stream velocity: 40m/s
purge air flow flux: 2L/min
(plane-1)

FIGURE 6. Contours of coal velocity magnitude under different air-coal stream velocities

TABLE 1. Deviation of velocity values within desired range of air-coal stream flow rate

Actual velocity of coal particles(m/s)	Measured velocity of coal particles (m/s)	Deviation (m/s)
10	11.0	1.0
15	16.5	1.5
20	21.7	1.7
25	27.0	2.0
30	31.4	1.4
35	37.6	1.6
40	43.0	3.0

Influence of purge air flow on the accuracy of velocity measurement

Figure 7 shows the velocity distributions of pulverized coal within and around the measurement region of the probe under the same air-coal stream velocity (20m/s). It is hard to find the purge air flow with lower flux (2L/min) has any influence on the pulverized coal's velocity from (a) of fig. 7--the faint influence only can be observed on the corresponding plane-2 in (b). On the contrary, purge air flow with higher flux (10L/min) has quite obvious influence on the measured velocity values, as shown in (c) of fig. 7, in which it can be observed the particles' velocity around the four measurement holes is increased as much as 4 times than their original values, causing quite big deviation of velocity measurement. The simulation results guide us to find an optimum flux of purge air flow which can prevent the measurement holes from being jammed and bring as small deviation of measured values as possible at the same time. The optimum flux will be got investigated in our further studies.

Discussion on the Accuracy of Concentration Measurement

According to light fluctuation theory, the concentration of pulverized coal is calculated by the following expression:

$$Ln(I/I_0) = -\frac{\pi}{4}LC \tag{8}$$

where C represents the concentration of pulverized coal, and L represents the travel length of transmission light. In the calculations, the travel length of light is often assumed to be equal to the length of measurement region (12mm). Obviously, if the moving tracks of pulverized coal are changed by foreign forces, that presumption will not be correct any more and the calculated concentration values will be different from the actual ones.

From the analysis in above 4.1 section, it can be concluded that the air-coal stream flow will not bring much influences on the accuracy of measured concentration values, while purge air flow will influence the concentration accuracy in different degree depending on its flux. As the purge air flow will shorten the actual travel length of light, the calculated concentration values will be relatively higher than the original ones.

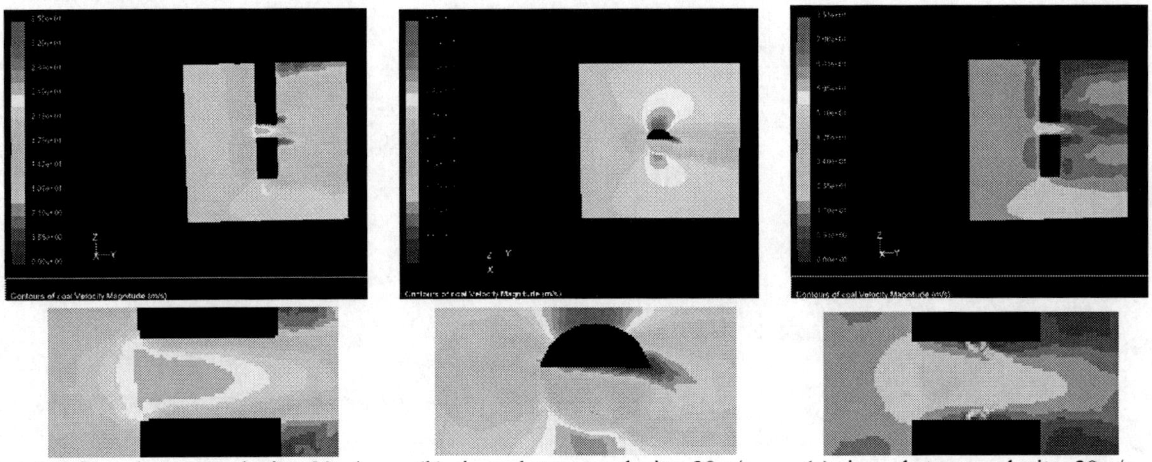

(a) air-coal stream velocity: 20m/s purge air flow flux: 2L/min (plane-1) (b) air-coal stream velocity: 20m/s purge air flow flux: 2L/min (plane-2) (c) air-coal stream velocity: 20m/s purge air flow flux: 10L/min (plane-1)

FIGURE 7. Contours of coal velocity magnitude under different purge air flow fluxes

CONCLUSIONS

Based on the simulation results and theoretical analysis, the following conclusions can be drawn:

The intrusion of probe brings negative effects on the measured velocity values, making them relatively higher than the actual ones. The deviation of measured velocity caused by intrusion of the probe can't be avoided but is limited.

The flow of purge air brings negative effects on both of velocity measurement and concentration measurement, making measured values relatively higher than the actual ones. The influence brought by purge air flow with high flux is quite remarkable and can't be ignored, so it is urgent to find an optimum flux of purge air flow which can prevent the measurement holes from being jammed and bring as small deviation of measured values as possible at the same time.

The simulation results of CFD software confirm the applicability of optical-pneumatic combined probe in on-line measurements of pulverized coal and provide us guidance on our future work. The optimum flux of purge air will be got investigated in our further studies.

REFERENCES

1. Chang Ruili and Wang Fei et al, Study on the Flame Monitoring and Combustion Diagnosis Control Optimizing inside Pulverized Coal Boiler of Power Plant. Power Engineering, Vol.4, pp.10-13, 2006
2. Wang Kai, Zhao Haishen, Chen Zenghong, Pulverized Coal Concentration Monitoring System. Power Engineering, Vol.14, pp. 45-48, 1994
3. Wang Guoxiong, Pulverized Coal Injection in Modern Blast Furnace. Beijing: Metallurgical Industry Press, 1977
4. Laux S, Grusha J, McCarthy K, Coal and Air Flow Measurement for Reduced NOx and UBC. Internet information, 2002
5. Cai Xiaoshu, Pan Yongzhi, et al, The Study of On-line Measurement Technology for Size, Concentration and Velocity of Pulverized Coal. Power Engineering, Vol.19, pp. 466-470, 1999
6. Wu Weilian, The Study of Measurement Technology for Gas and Solid Two Phase Flow and its Application in Pulverized Coal Monitoring in Power Station. Ph.D dissertation, University of Shanghai for Science and Technology, 1999

NOMENCLATURE

ρ	density, kg/m^3
m	mass, kg
α	volume fraction
v	velocity, m/s
p	pressure, pa
μ	viscosity, cP
g	gravity acceleration
F	body force, N
I	intensity of transmission light
L	travel length of light, m
C	concentration, kg/m^3

SUBSCRIPTS

m	the mixture
n	the number of phases,
k	phase k
dr	drift velocity
p	the secondary phase
q	the primary phase

Ultrasonic Method for Concentration and Particle Size Analysis in Dense Coal-water Slurry

Xue Minghua , Su Mingxu , Dong Lili , Shang Zhitao , Cai Xiaoshu

Institute of Particle & Two-phase Flow Measurement, University of Shanghai for Science & Technology, Shanghai, China

Abstract. The concentration and particle size distribution in particulate two-phase flow are the important parameters in a wide variety of industrial areas. For the purpose of on-line characterizing dense coal-water slurry, ultrasonic methods have many advantages, such as no-dilution, real-time and no-invasion, while light-based techniques are not capable of providing information because optical methods normally require the sample to be diluted. In this paper, the modified Urick equation including temperature modification, which can be used to determine concentration from the measurement of velocity in the coal-water slurry, is evaluated on the basis of model analysis and experiments. Combined with the optimum regulation technique inverse algorithm, the particle size distribution of the coal-water slurry is obtained according to the attenuation prediction of Coupled-phase model plus Bouguer-Lambert-Beer-Law scattering model and experimental attenuation at frequencies ranges from 3MHz to 12MHz. This technique brings the possibility of using ultrasound for on-line measurement of dense slurry.

Keywords: ultrasonic method; velocity; concentration; attenuation; particle size distribution; on-line.

INTRODUCTION

Dense particulate two-phase flow is very common in different industrial areas and production processes. It is often necessary to know the concentration and particle size distribution (PSD) of the dispersed phase. For example, the dense Coal-water slurry (CWS) is an effective substitute fuel for oil and is commonly used in boilers of power plant. Several literature [1,2] have reported that the concentration and PSD are of big influence on the property of the CWS. In order to improve the burning efficiency and stability, it is very important to determine the concentration and PSD of the CWS.

For dense CWS two-phase flow measurements several techniques have been developed in literature [3,4]. Optical scattering is currently the most widely used technique for characterizing the two-phase flow. However, since the principle requires transmission of light and assumes single particle scattering, the optical method is limited to making measurement on dilute slurry. Ultrasound waves interact with particles in a similar manner to light but have the advantage that they can travel through concentrated particulate two-phase flow. Therefore ultrasonic method is more suitable to on-line characterize the original dense slurry.

There has been increasing interest in concentrated suspension using Coupled-phase model to describe the acoustic propagation in recent decades[5]. Urick and Hampton measured the ultrasonic attenuation and velocity for suspensions at high-volume fractions and the acoustic velocity was calculated with averaged values of density and compressibility of the medium components. Harker and Temple analysed the acoustic propagation in suspension and extended the model including irreversible heat transfer between the phases. Pinfield and Povey improved the Urick equation by taking into the effect of ultrasound scattering. This modified Urick's equation can determine slurry concentration from the measurement of velocity in two-phase flow.

The attenuation spectrum of ultrasonic waves through the sample may provide the information of the PSD[6]. In this paper, the total attenuation of the dense CWS is predicted with Coupled-phase model plus Bouguer-Lambert-Beer-Law (BLBL) scattering model. Combined with the optimum regularization technique (ORT) analysis algorithm, the PSD of CWS is obtained according to the model prediction and experimental data.

CP914, *Multiphase Flow: The Ultimate Measurement Challenge,* edited by X. Cai, Y. Wu, Z. Huang, S. Wang, and M. Wang

THEORY

Ultrasonic propagation model in two-phase flow

The Coupled-phase theory is commonly used to model acoustic propagation in dense particulate two-phase flow. This theory has generally been addressed from the view of hydrodynamic point, which consists of the basic formulation of the wave problem involves the elasticity of each phase, conservation of mass, and conservation of momentum. The aim of this model is to arrive at the dispersion relationship which relates the complex wavenumber k for compression waves in the particulate mixture to the physical properties of the components of the two-phase flow.

Urick[7,8] measured the velocity in 1947 for suspensions of kaolin particles in water. This is one of the earliest theoretical works. Velocity is determined from an effective density and compressibility. Both magnitudes depend on the density and the compressibility of solid and liquid phases, and on the concentration of particles:

$$c = \sqrt{\frac{1}{\rho_{ef}\beta_{ef}}} \tag{1}$$

Where

$$\rho_{ef} = \rho_p\varphi + \rho_l(1-\varphi) \tag{2}$$

$$\beta_{ef} = \beta_p\varphi + \beta_l(1-\varphi) \tag{3}$$

The subindex p refers to the particles, l refers to the liquid and φ is volume fraction of the particles. The most important character of Eq.(1) is to establish the relationship between velocity and volume fraction with the conception of "effective volume-averaging". However, Urick model supposes that particles size is much smaller than wavelength ($kR << 1$) and not considering the effect of scattering, thermal and viscous losses in particulate two-phase flow[9].

Harker and Temple[10] analyzed the acoustic propagation in suspension and extended the model including irreversible heat transfer between the phases. They used Vand's view for viscosity and obtained an expression for the complex wave number:

$$k^2 = \omega^2\beta_{ef}\rho_{ef}(\omega) \tag{4}$$

Where

$$k = \omega / c(\omega) + i\alpha(\omega) \tag{5}$$

$$\rho_{ef}(\omega) = \frac{\rho_l[\rho_p(1-\varphi+\varphi S)+\rho_l S(1-\varphi)]}{\rho_p(1-\varphi)^2+\rho_l[S+\varphi(1-\varphi)]} \tag{6}$$

The imaginary part of k gives the attenuation and the ratio of ω to the real part of k gives the velocity. Parameter S is the function of the density, viscosity of continuous phase and the particle size, volume fraction of particulate phase. For using the conception of "effective volume-averaging", the Eq.(4) is self-consistant[11,12], which means that it is valid for all values of volume fraction.

Compared to acoustic velocity, which is mainly dependent on the density and compressibility of the medium, the acoustic attenuation is caused by many factors, such as scattering, absorption, thermal loss, viscous loss and so on. Because Coupled-phase model assumes the particulate phase is a continuum, there is no scattering description. It is necessary to combine Coupled-phase model with BLBL scattering model[13,14] for the purpose of available attenuation prediction. The attenuation coefficient α_s in BLBL scattering model is given by

$$\alpha_s = \frac{3\varphi}{8R}K_{\text{Ext}} \tag{7}$$

Where

$$K_{\text{Ext}} = -\frac{4}{\sigma^2}\sum_{n=0}^{\infty}(2n+1)\text{Re}\{A_n\} \tag{8}$$

Here σ is defined as the dimensionless particle diameter, given by $\sigma = \omega R / c$, ω is the frequency, R is the particle radius, c is the acoustic velocity in the slurry. A_n is scattering coefficient of a single particle.

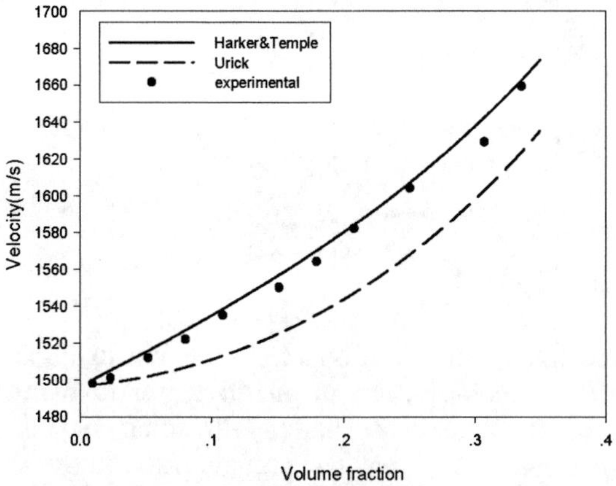

FIGURE.1 Comparison between theory velocity and experimental velocity

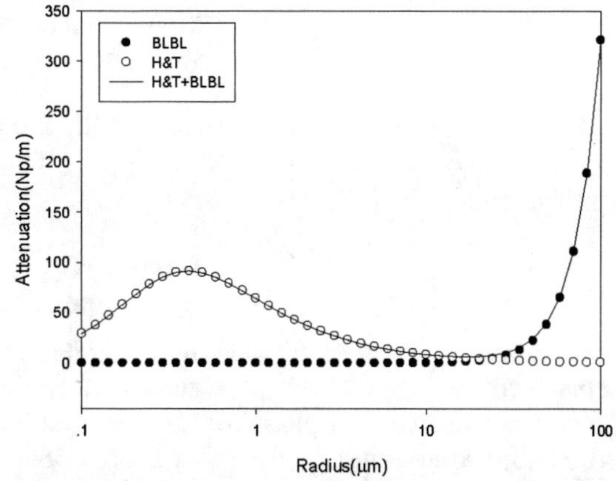

FIGURE.2 The relationship between attenuation with particle size

The theoretical velocity predictions of the Urick and Harker & Temple model in CWS were presented in Fig.1. The results of Harker & Temple model were closer to experimental data than the results of Urick model. The theoretical attenuation predictions of Harker & Temple and BLBL model were presented in Fig.2. For the reason of the viscous loss region (from 0.1 μm to 10 μm) and scattering loss region(from 10 μm to 100 μm), the total attenuation in the CWS was described with Harker & Temple plus BLBL scattering model from the particle size ranges of 0.1 μm to 100 μm.

Inverse problem

Dense slurry is analyzed using ultrasonic spectrum by measuring its attenuation coefficient as a function of frequency. The PSD of slurry is then determined by finding the PSD which gives the best fit between theory and experimental data. There are two main approaches to solving this inversion problem: dependent model inversion and independent model inversion. In the dependent model inversion it is assumed that the PSD of slurry follows some common form which can simply be modeled mathematically, e.g. Rosin-Rammler distribution[15].

$$V(D) = 1 - \exp(-(D/\bar{D})^k) \tag{9}$$

Here $\bar{D}$ is the characteristic diameter, D is the diameter of the particle, k is the narrowness index of the distribution. $V(D)$ is the accumulative volume distribution. The PSD can be calculated from experimental ultrasonic measurement using a computer. A assumption at the parameters $\bar{D}$ and k is used to calculate the ultrasonic prediction of the slurry using the Coupled-phase plus BLBL scattering theory. The predicted values are then compared with experimental data, and the sum of the squares of the difference (SSD) is calculated. The SSD for attenuation measurements is given by[16]

$$\min(SSD) = \min(\sum_i [\alpha_{theory}(f_i) - \alpha_{\exp t}(f_i)]^2) \tag{10}$$

Where $\alpha_{theory}(f_i)$ is the predicted ultrasonic attenuation coefficient, and $\alpha_{\exp t}(f_i)$ is the experimental values at frequency f_i. The parameters $\bar{D}$ and k are then calculated until the computer finds the minimum in SSD. Once the parameters are determined, the PSD $V(D)$ can also be calculated using Eq.(9).

For the case that the assumed distribution form is not a reasonable representation of the slurry, it is necessary to use the independent model inversion. The Eq.(7) may be more generally written as

$$\alpha_s = \int_{D_{\min}}^{D_{\max}} \frac{3\varphi}{8R} \cdot \frac{dV}{dR} \cdot K_{ext} dR \tag{11}$$

For the inversion calculation, the first-kind of Fredholm integral equation, Eq.(11), is transferred into a linear equation $\boldsymbol{AF} = \boldsymbol{G}$ after the discretization.

$$\alpha_s = \frac{3\varphi}{8} \sum V(c) \int_{\Delta R} \frac{K_{ext}(R)}{R} dR \tag{12}$$

Here $\boldsymbol{A}$ is the result of $\int_{\Delta R} \frac{K_{ext}(R)}{R} dR$. $\boldsymbol{F}$ is the accumulative volume distribution $V(c)$. $\boldsymbol{G}$ is the experimental attenuation spectrum.

EXPERIMENTAL METHOD

Measurements of the acoustic velocity and attenuation were taken using the apparatus shown in Fig.3. The experiments were conducted in a plexiglass cell equipped with a variable speed magnetic stirrer to maintain a homogeneous particulate two-phase flow. The system was used in pulse-echo mode. Ultrasonic signals were generated by ultrasonic transducer using a Model 5800 ultrasonic pulser/receiver. The signals through the sample were acquired by high-speed data acquisition card, which was controlled with a personal computer. The ultrasonic transducer was designed as the style of probe in order to be dipped into different regions of the test particulate two-phase flow.

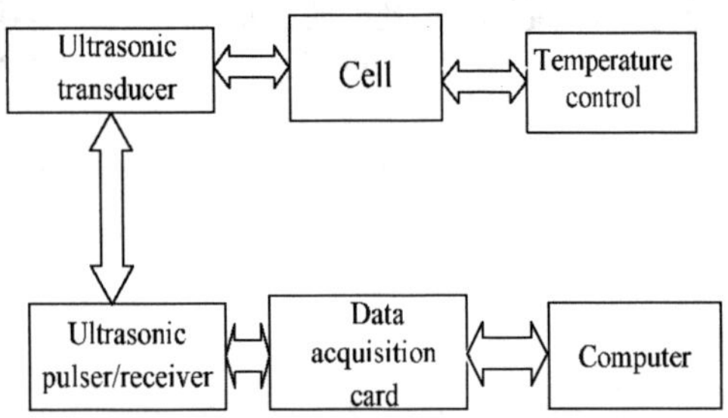

FIGURE.3 Schematic of ultrasonic measurement system

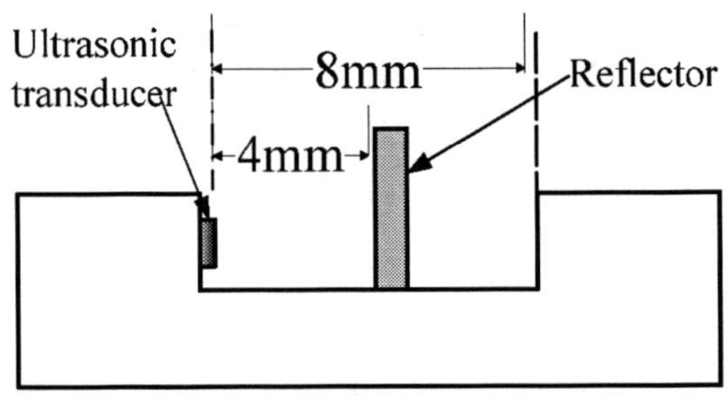

FIGURE.4 Schematic diagram of ultrasonic probe

The velocity is measured based on the distance (L) between the transducer and the reflector and the transit time (Δt). The velocity is given by

$$c = L/\Delta t \tag{13}$$

The attenuation coefficient α is determined as a function of frequency through the simple relation:

$$\alpha = \ln(V_2/V_1)/(R_2 - R_1) \tag{14}$$

Here V_2/V_1 is the ratio of amplitudes measured at two positions R_2 and R_1.

RESULTS AND DISCUSSION

Concentration measurement

The relationship of velocity and particle size in the CWS with different volume fractions by numerical simulation was presented in Fig.5. The velocity was sensitive to the particle size from the ranges of 0.1 μm to 1 μm. In contrast, the velocity was not sensitive from the ranges of 10 μm to 100 μm. However, several literature [17] have reported that the range of the PSD of CWS is from 1 μm to 10 μm .Hence the difference of particle size is of no influence on the velocity in dense CWS.

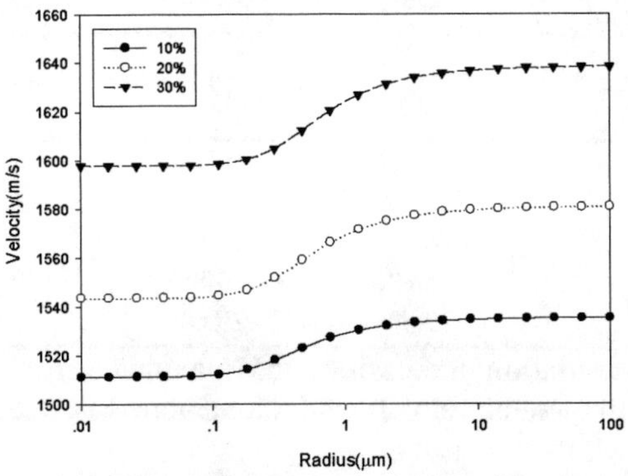

FIGURE.5 The relationship of velocity and particle size under different volume fractions

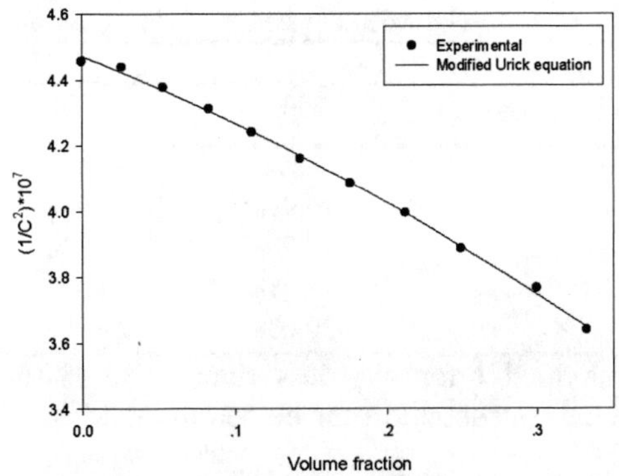

FIGURE.6 The curve of velocity and volume fraction

Although the results of the Coupled-phase model were close to the experimental data (Fig.1), it is not suitable for using this model in engineering application because of its complex properties of different phases. For practical purposes, the relationship of acoustic velocity with volume fraction can be correlated by modified form of Urick's equation as proposed by Pinfield and Povey[18]:

$$1/C^2 = 1/C_l^2 + \beta_1\varphi + \beta_2\varphi^2 \tag{15}$$

Where C_l is the velocity in the continuous phase. β_1 and β_2 are experimentally determined constants which are functions of particle size. Compared to Harker & Temple model, Eq.(15) can obtain the relationship of velocity and volume fraction only to determine the coefficient β_1 and β_2 by the experimental method.

Fig.6 shows calibration lines for the Coal-water slurry used in the present study at 25℃. Adopting the Marquardt-Levenberg analysis algorithm[19], β_1 and β_2 can be determined as follows:

$$1/C^2 \times 10^7 = 4.4733 - 1.8593\varphi - 1.8507\varphi^2 \tag{16}$$

For it has been proved that the difference of particle size in CWS was no influence on velocity, Eq.(16) can be used to determined the concentration of CWS from the measurement of acoustic velocity.

The influence of temperature on the different volume fractions of the velocity in the Coal-water slurry was shown in Fig.7. The velocity increased with increasing temperature from the range of 25℃ to 45℃ at all volume fractions.

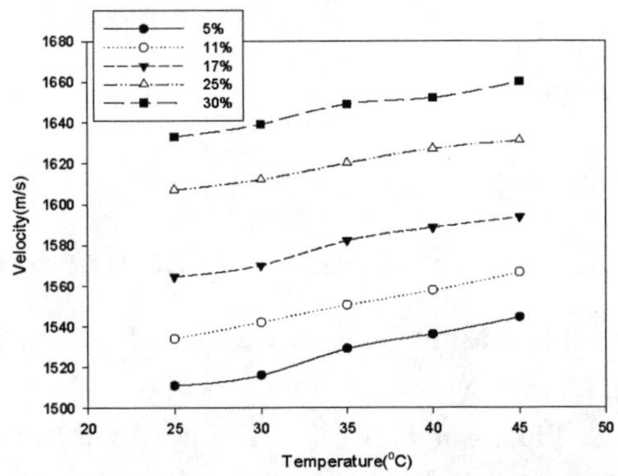

FIGURE.7 The relationship of velocity and temperature with different volume fractions

Based on the velocity with different volume fractions at 25℃, adopting the linear analysis algorithm, the modified Urick's equation considering temperature influence was given as follows:

$$1/(C - 1.4791\Delta t)^2 \times 10^7 = 4.4733 - 1.8593\varphi - 1.8507\varphi^2 \tag{17}$$

Where $\Delta t = t - 25°C$. The comparison between the standard concentration and the results of Eq.(17) was shown in Table.1.

TABLE.1 Comparison between the standard concentration and the results of Eq.(17)

Standard Concentration	The results of Eq.(17)					Maximal Error/%
	25℃	30℃	35℃	40℃	45℃	
5%	4.8%	4.4%	5.7%	5.5%	5.7%	14%
11%	10.9%	11.0%	11.3%	11.2%	11.6%	5.4%
17%	17.7%	17.3%	18.1%	18.1%	17.6%	3.5%
25%	25.7%	25.3%	25.4%	25.3%	24.8%	2.8%
30%	30.0%	29.8%	30.1%	29.5%	29.6%	1.6%

The maximal error was less than 15%. With the concentration increasing, the maximal error decreased. Therefore the concentration can be determined from the measurement of velocity and temperature in CWS.

Particle size distribution measurement

The principle of PSD measurement is based on the properties of attenuation spectrum generated by different particle sizes. Combined with the optimization analysis algorithm, the PSD of CWS was obtained according to the experimental attenuation spectrum. The data of experimental attenuation spectrum from 3MHz to 12MHz was shown in Tab.2.

TABLE.2 Experimental attenuation spectrum from 3MHz to 12MHz(Units Np/m)

Volume fraction \ Frequency(MHz)	3	4	5	6	7	8	9	10	11	12
10%	16.4	21.7	26.5	35.8	50.0	48.9	58.4	69.4	87.3	108.7
20%	30.1	39.5	50.2	63.9	87.1	97.9	121.1	146.4	194.7	223.2
30%	47.3	63.6	80.4	100.6	137.8	155.5	187.9	237.9	305.2	344.0

Adopting Optimum Regulation Technique(ORT), the results of PSD in CWS with different volume fractions were shown in Fig.8.

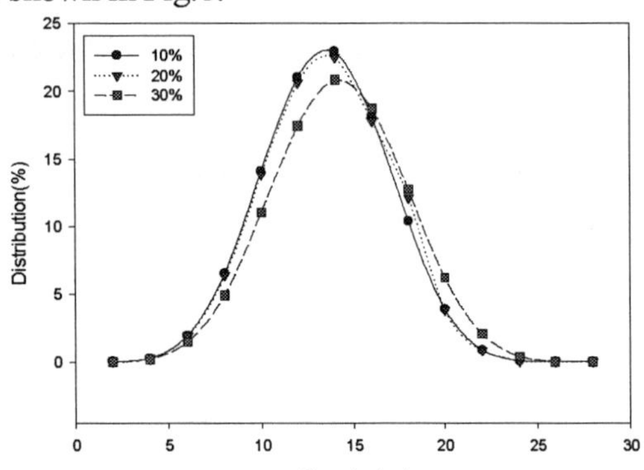

FIGURE 8 The curves of PSD of Coal-water Slurry with three volume fractions

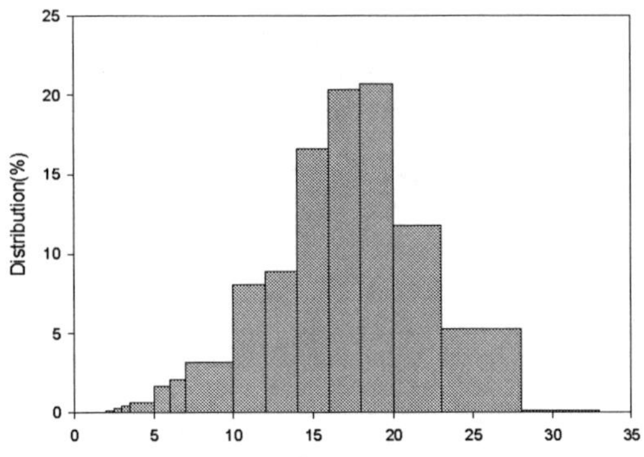

FIGURE.9 Measurement PSD of CWS with optical method

The range of the measured PSD with ultrasonic method was from 5 μm to 25 μm. The measured number mean diameter were 13.2 μm,13.4 μm,14.0 μm with the volume fractions of 10%,20%,30%. Fig.9 shows the PSD of CWS measured on a Picture Image Processor (PIP)[20]. The measured number mean diameter was 17.5 μm. Considering the reason that the particles of CWS were not dispersed with PIP method, it was understood that value of number mean diameter with PIP method was bigger. The range of PSD with ultrasonic method was similar to the results with PIP method.

CONCLUSIONS

By measuring the acoustic velocity and attenuation in the CWS, the concentration and PSD of the dispersed phase can be obtained. Numerical simulations show that particle size of CWS is of no influence on velocity. The concentration was measured by the velocity and temperature in the slurry using the modified Urick's equation considering temperature influence. The PSD was obtained by an inversion calculation which including the prediction of Harker & Temple plus BLBL scattering, experimental data, and Optimum Regulation Technique inverse algorithm. The measurement results of ultrasonic method agree well with the optical image measurement results

ACKNOWLEDGMENTS

This project is supported by the Shanghai Education Committee Youth Foundation (04EC30) and 863 Program, China (2006AA03Z349).

REFERENCES

1. Cao Yingyu, Wen Weiguo, Huang Zhengyu, et al. Industrial Test for Burning Water-Coal Slurry in 230t/h Boiler of Baiyanghe Power Plant[J]. Electric Power, 2001, 34(7): 16-18 (in Chinese).
2. Li Shanshan, Cheng Jun, Li Yanchang, et al. Analysis of some factors affecting CWS Viscosity[J]. Coal Conversion, 2006, 29(1): 23-26 (in Chinese).
3. Wang Renzhe, Zhang Rongzeng, Xu Zhiqiang. Experiment and Analysis of Sizing Particles in Coal-Water Slurry Fuel[J]. Coal Engineering, 2004, 59(7): 54-56 (in Chinese).
4. Huang Zheng, Zhang Gongpeng, Zhu Entong, et al. The Technology of Particle Measurement in Coal-Water Fuel[J]. China Measurement Technology, 2004, 30(7): 21-24 (in Chinese).
5. Su Mingxu, Cai Xiaoshu. The Numerical Study of Acoustical Attenuation and Velocity in the Suspension of Superfine Particles[J]. Acta Acustica, 2002, 27(3): 218-222 (in Chinese).
6. Su Mingxu, Cai Xiaoshu, Xu Feng, et al. The Measurement of Size and Concentration of Glass Beads by Using Ultrasonic Attenuation in Three Frequency[J]. Acta Acustica, 2004, 5(9): 440-444 (in Chinese).
7. Urick R J. The Absorption of Sound in Suspension of Irrgular Particles[J]. The Journal of the Acoustical Society of America, 1948, 20: 283-289.
8. Urick R J. A Sound Velocity Method for Determining the Compressibility of Finely Divided Substance[J]. The Journal of the Acoustical Society of America, 1947, 18: 983-987.
9. V. Stolojanu, A. Prakash, Chrarcterization of Slurry Systems by Ultrasonic Techniques, Chemical Engineering Journal, 2001, 84: 215-222.
10. Harker A. H, Temple J. A. G.. Velocity and Attenuation of Ultrasound in Suspensions of Particles in Fluids[J]. The Journal of the Acoustical Society of America, 1988, 21: 1576-1588.
11. Evans J M, Attenborough K. Coupled Phase Theory for Sound Propagation in Emulsions[J]. The Journal of the Acoustical Society of America, 1997, 102: 278–282.
12. Evans J M, Attenborough K. Sound Propagation in Concentrated Emulsions: Comparison of Coupled Phase Model and Core-shell Model[J]. The Journal of the Acoustical Society of America, 2002, 112 : 1911–1917.
13. Riebel U, Loeffler F. The Fundamentals of Particle Size Analysis by Means of Ultrasonic Spectrometry[J]. Part. Part ,Syst. Charact. 1989, 6 : 135-143.
14. Khatchikian P, Riebel U, Kräuter U. Ultrasonic Particle Sizing[J]. KONA Powder and Particle. 1995, 13: 31-43.
15. Wang Naining, et al. The Optical Measurement Technology and Application of Particle Size[M]. Shanghai, China: University of Shanghai for Science and Technology, 1999.
16. McClements. Principles of Ultrasonic Droplet Size Determination in Emulsions[J]. Langmuir, 1996, 12: 3454-3461.
17. Wu Kuntai. The Research of Dense CWM Manufacture Technology[J]. Coal Engineering, 2002, 3: 38-41 (in Chinese).
18. V.J. Pinfield, M.J.W. Povey. The Application of Modified Forms of The Urick Equation to the Interpretation of Ultrasound Velocity in Scattering Systems[J]. Ultrasonics, 1995, 33: 243-251.

19. Press, W. H., Flannery, B. P., Teukolsky, S. A., et al. Numerical Recipes[D]. Cambridge: Cambridge University Press, 1986.
20. OMEC Technology Co, Ltd. The Particle Image Processor Handbook. 2006 (in Chinese).

SESSION PT-1

DEVELOPMENTS OF PT TECHNOLOGY

MEASUREMENT OF WATER FRACTION IN LIQUID HYDRO-CARBONS USING ECT SENSOR

Syed Faisal Ahmed Bukhari, Idris Ismail and Wuqiang Yang

School of Electrical and Electronic Engineering, The University of Manchester, PO Box 88, Manchester M60 1QD, United Kingdom, sfaisal_bukhari@yahoo.com

ABSTRACT. In the oil industry, it is important to measure the water fraction in liquid hydrocarbons in oil pipelines for optimising operation and transportation, e.g. loading/unloading of oil tankers, where the flow rate is variable. As a feasibility study an electrical capacitance tomography (ECT) sensor has been used for on-line water fraction measurement. Data from different electrode combinations, e.g. adjacent electrodes and opposite electrodes are combined. Preliminary results indicate that ECT can be used for measuring the fraction of water in oil with 0.05% error in a range of water-in-oil up to 10%.

Keywords: Multi-phase flow, on-line measurement, electrical capacitance tomography (ECT), water fraction

INTRODUCTION

Water-cut is defined as the volume flow rate of water relative to the volume flow rate of the total liquid. Usually, water-cut is different from water hold-up because hydrocarbons travel faster than water in a production line. Water-cut in an oil production well may be estimated by capacitance or impedance of the fluid. Water-cut meter and flow meter are often combined together to estimate water-cut.

While it is important to perform on-line measurement of water fraction in liquid hydrocarbons during their transfer in pipelines or crude oil tankers (Hammer and Johansen 1997; Gamio et al. 2004), the following issues must be addressed:

Variable hydrocarbon compositions in crude oil
Different components flow at different velocity
Water content varies from 0 to 10%
It is likely to form slug flow.

A single-electrode capacitance probe was suggested by Schüller et al. (2003). It is based on the change in frequency of an oscillator by the medium adjacent to a single electrode. The drawback of this method is that the assumption of linear model results in under-estimate of water volume fraction.

For accurate water fraction measurement in industry, ideally a device can be installed on an existing bypass loop with sufficiently fast response. This work aims to measure water fraction using an electrical capacitance tomography (ECT) sensor.

ECT SENSOR

Considering the industrial requirements, an ECT sensor is suitable for on-line water fraction measurement. Processes involving dielectric materials or mixtures can benefit from ECT (Dyakowski et al. 1997), which provides appropriate sensitivity and stability that are regarded as key issues in sensor selection. In a conventional ECT system, the sensor takes the form of an array of electrodes placed around an insulating pipe and surrounded by a grounded screen. The data acquisition unit measures the capacitance of all possible electrode pairs, thus producing a total of n(n-1)/2 measurements, where n is the number of electrodes. The capacitance sensor selected for this work has 8 electrodes. The length of each electrode is 15 cm. These electrodes are placed on the outside of a Perspex pipe of 127 mm in diameter and 2 mm in thickness, which protects from the fluids from shortening electrodes. The pipe is thin and thus its effect on the measured inter-electrode capacitance is small. The sensor is electrically shielded with a 3 mm aluminium screen, which also provides a robust mechanical support.

CP914, *Multiphase Flow: The Ultimate Measurement Challenge,* edited by X. Cai, Y. Wu, Z. Huang, S. Wang, and M. Wang

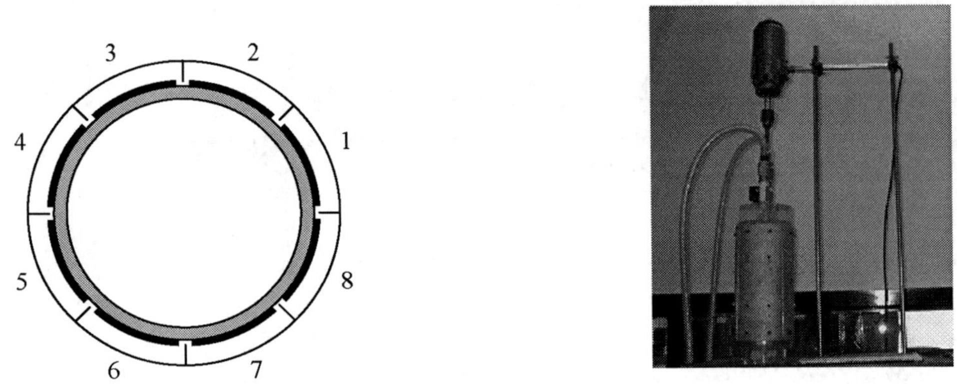

(a) Cross-sectional view

(b) ECT sensor with stirrer

FIGURE.1 ECT sensor test set-up

DATA ANALYSIS METHOD

As suggested by Romanowski (Andrzej et al. 2005), the data analysis methods applied in process tomography can be divided into two main categories: (1) raw measurement data and (2) visualisation of 2D or 3D images. While many algorithms, most of which are time-consuming, have been developed (Yang and Peng 2003) for visualisation, the quantitative errors associated with the image reconstruction techniques may not be smaller than direct analysis of raw measurement data (Grudzien et al. 2005). Currently most researchers working on tomography have not used direct analysis of raw data, but in favour of more sophisticated and challenging approaches requiring solving the inverse problem.

According to Andrzej et al. (2005) and Grudzien et al. (2005), useful information about phase distribution in a sensor can be derived using raw data. For example, principal component analysis has been deployed to raw data for determining material concentration (Zhou et al. 2001).

In this work raw capacitance measurements have been used to estimate the water fraction in liquid hydrocarbons. Capacitance measurements are taken from four groups of electrodes as below.

- Group A – Adjacent electrodes (1-2, 1-8, 2-3, 3-4, 4-5, 5-6, 6-7, 7-8)
- Group B – One electrode in between electrode pair (1-3, 1-7, 2-4, 2-8, 3-5, 4-6, 5-7, 6-8)
- Group C – Two electrode in between electrode pair (1-4, 1-6, 2-5, 2-7, 3-6, 3-8, 4-7, 5-8)
- Group D – Opposite electrodes (1-5, 2-6, 3-7, 4-8)

where 1-2, 1-8, 2-3 ... 4-8 etc. represent electrode pairs as shown in Fig. 1 (a). Fig. 1 (b) shows the test facility, where oil and water are mixed together using a stirrer.

The parallel, series and Maxwell models can be applied to ECT measurements for calculation of normalised data (Chaplin et al. 2005), which can be expressed in terms of measured capacitance (C_m) and high and low capacitance readings at calibration (C_h, C_l) by the following equations.

Parallel model: $$C_{nP} = \frac{C_m - C_l}{C_h - C_l} \tag{1}$$

Series model: $$C_{nS} = \frac{\frac{1}{C_m} - \frac{1}{C_l}}{\frac{1}{C_h} - \frac{1}{C_l}} \tag{2}$$

Maxwell model: $$C_{nM} = \frac{C_{nP}(k+2)}{3 + C_{nP}(k-1)} \tag{3}$$

In equation (3), k is the high to low permittivity ratio (C_h / C_l).

EXPERIMENTAL RESULTS AND DISCUSSION

AMSOIL synthetic 10W-40 motor oil was used for experiment. The results of the measured capacitance values are shown in Fig. 2 (a). For all groups, it indicates a nearly linear relationship between the measured capacitance and water percentage. As expected, Group A is the most sensitive and Group D is the least sensitive. The results for the normalised values using the parallel, series and Maxwell model are shown in Fig. 2 (b), (c) and (d). It indicates that for all groups, the Maxwell model gives more accurate measurement than the parallel and series models.

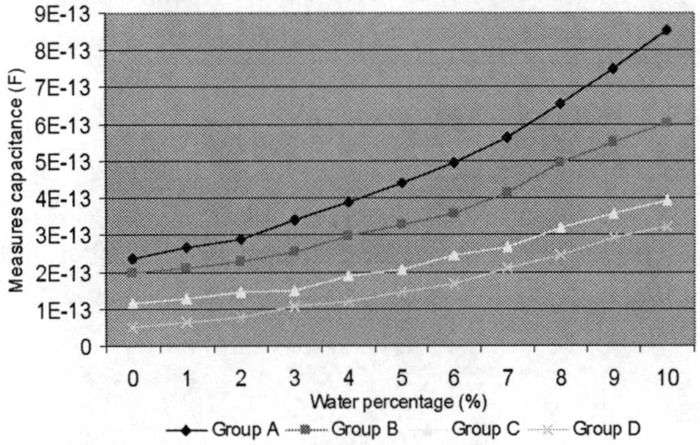

(a) Measured capacitance

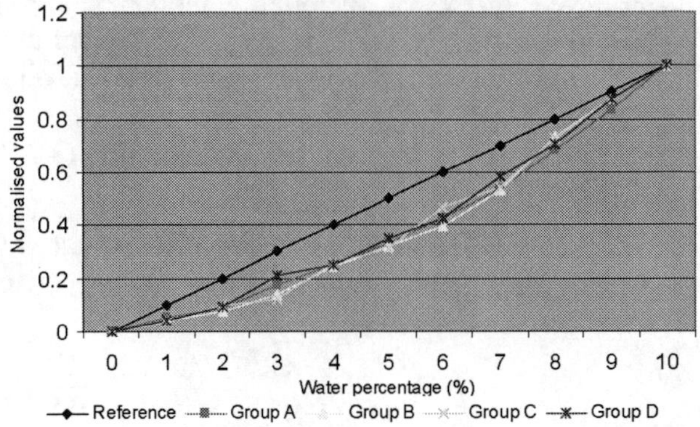

(b) Normalised value using parallel model

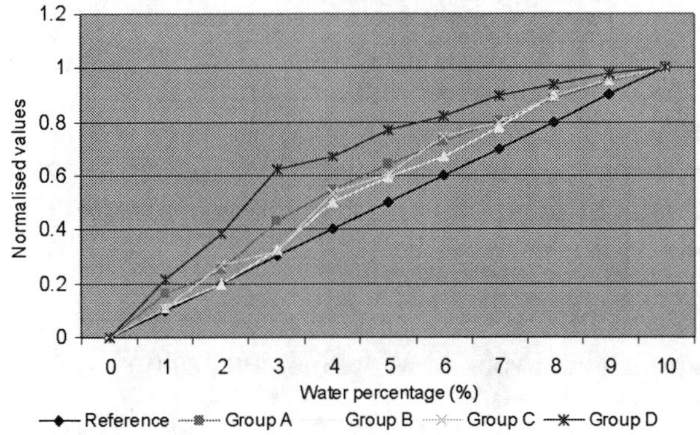

(c) Normalised value using series model

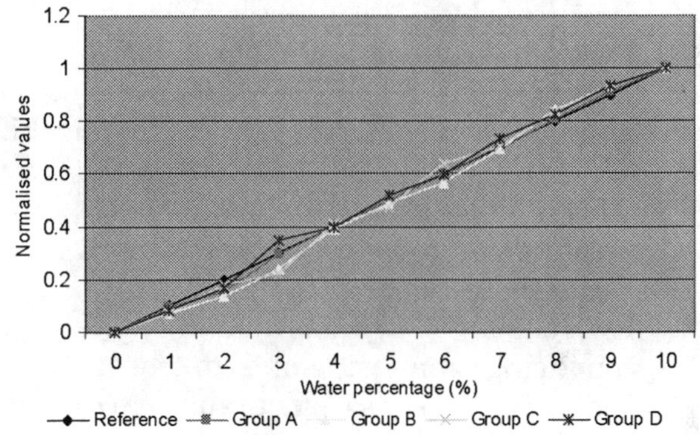

(d) Normalised value using Maxwell model

FIGURE 2 Measured and normalised capacitance vs. water percentage

The measurement error using Maxwell model is shown in Fig. 3. The maximum absolute errors for Groups A, B, C and D are 0.04%, 0.06%, 0.08% and 0.07% respectively. The sum of squared-difference (SSD) has been used to select the best curve for each group of electrode pairs. It is concluded that the Maxwell model, which assumes that the material is evenly distributed throughout the volume, gives the most accurate results with maximum absolute average error of 0.04% (see Fig. 3 (b)).

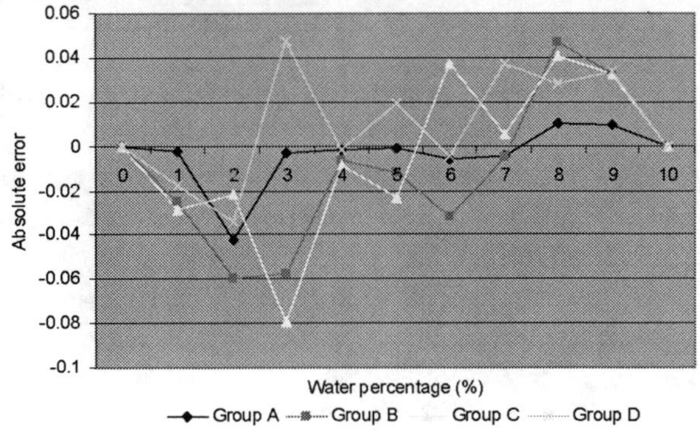

(a) All groups

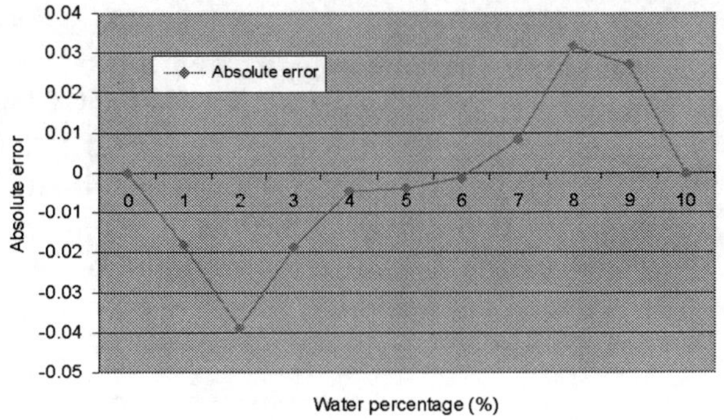

(b) Average

FIGURE.3 Measurement error using Maxwell model

CONCLUSION

An ECT-based test rig has been used to measure homogeneous water-in-oil mixture. The results indicate that ECT can be used to provide on-line measurement of water fraction provided the oil and water are properly mixed. Calculation using the Maxwell approach shows that the maximum average measurement error is 0.04% for homogeneous water-in-oil mixture up to 10%. The results obtained with the ECT sensor are in good accordance with the recommended measurement error of less than 0.5% as required for water fraction meter by the Norwegian Society for Oil and Gas Measurement. This would be suitable for the case of hydrocarbon sales metering where typical operating condition with water content of less than 10%. However, further tests are needed to evaluate the performance of ECT under dynamic and salinity conditions and validation of the dynamic measurement range for the homogenous water-in-oil mixture.

ACKNOWLEDGEMENTS

Syed Faisal Ahmed Bukhari and Idris Ismail would like to thank the Ministry of Education, Pakistan and Universiti Teknologi Petronas, Malaysia respectively for supporting their PhD study at the University of Manchester.

REFERENCES

1. G.Chaplin, T.Pugsley, L.V.Lee, A.Kantzas and C.Winters, The dynamic calibration of an electrical capacitance tomography sensor applied to the fluidized bed drying of pharmaceutical granule, Meas. Sci. Technol., Vol.16, pp.1281-1290, 2005.
2. T.Dyakowski, R.B.Edwards, C.G.Xie and R.A.Williams, Application of capacitance tomography to gas-solid flows, Chemical Engineering Science, Vol.52, pp.2099-2110, 1997.
3. C.Gamio, J.Castro, F.Garcia-Nocetti, L.Aguilar, L.Rivera and H.Balleza, High pressure gas-oil two-phase flow visualisation using electrical capacitance tomography, Proc. 7th Biennial ASME Conference on Engineering Systems Design and Analysis ESDA 2004, Manchester, United Kingdom, pp.1-5, 2004.
4. K.Grudzien, A.Romanowski, R.G.Aykroyd and R.A.Williams, A novel approach to pneumatic conveying monitoring and control strategy development, Proc. 4th World Congress on Industrial Process Tomography, Aizu, Japan, pp.886-891, 2005.
5. E.A.Hammer and G.A.Johansen, Process tomography in the oil industry: state of the art and future possibilities, Measurement + Control, Vol.30, pp.212-216, 1997.
6. T.R.McKeen and T.S.Pugsley, The influence of permittivity models on phantom images obtained from electrical capacitance tomography, Meas. Sci. Technol., Vol.13, pp.1822-1830, 2002.
7. A.Romanowski, K.Grudzien, and R.A.Williams, A review of data analysis methods for electrical industrial process tomography applications, Proc. 4th World Congress on Industrial Process Tomography, Aizu, Japan, pp.916-921, 2005.
8. R.B.Schüller, B.Engebretsen and M.Halleraker, Measurement of water concentration in oil/water dispersions by a single-electrode capacitance probe, Proc. Instrumentation and Measurement Technology Conference, Vail, CO, USA, pp.635-639, 2003.
9. W.Q.Yang and L.Peng, Image reconstruction algorithms for electrical capacitance tomography, Meas. Sci. Technol., Vol.14, pp.R1-R13, 2003.
10. L.Zhou, Y.C.Zhang, D.Y.Yao, L.H.Peng and B.F.Zhang, Using principal component analysis to measure two-phase flow concentration, Proc. 2nd World Congress on Industrial Process Tomography, Hannover, Germany, pp.35-40, 2001.

NOMENCLATURE

C_{nP}	Normalised capacitance for parallel model
C_{nS}	Normalised capacitance for series model
C_{nM}	Normalised capacitance for Maxwell model
C_m	Measured capacitance
C_l	Relative permittivity of lower permittivity material
C_h	Relative permittivity of higher permittivity material
k	High to low permittivity ratio

Wire-Mesh Capacitance Tomography in Gas-Liquid Flows

Huang Shan-Fang Zhang Xiu-Gang Wang Dong Lin Zong-Hu

State Key Lab. of Multiphase Flow in Power Engineering, Xi'an Jiaotong University, Xi'an, 710049, China, Shanfang_huang@mail.xjtu.edu.cn

Abstract. A new intrusive, wire-mesh capacitance tomography for phase distributions in horizontal pipes is presented. A wire with an insulating film is used as a capacitor. The capacitance is in direct proportion to water film thickness, independent of water distribution and the salinity. We can obtain film thickness by measurement of capacitance. The sensor consists of four parallel wire capacitors, vertically placed over the cross section of the pipe. The interface is the result of linking the discrete film thicknesses with smooth curve, which are the tomographic result only when both phases are continuous. Static and dynamic experiments were carried out in air-water two-phase flows in horizontal pipes, from which we analyze the disturbance factors of the sensor to the flow field and measurement error of film thickness. The experimental results show that the tomographic results are consistent with expectation. The method has several advantages of no reconstructed algorithms, high signal-to-noise ratio, high sensitivity, and high spatio-temporal resolution. The disturbance and error can be neglected in the range of this paper.

Keywords: Wire-mesh capacitance tomography; Gas-liquid flow; Film thickness
PACS: 47.61Jd, 47.80.-v, 47.80.Jk

INTRODUCTION

Process Tomography (PT) has been one of the focuses in the field of two-phase flow measurement in the past 20 years, among which electrical PT, including electrical resistance tomography (ERT) and electrical capacitance tomography (ECT), is very popular because it's convenient, cheap and safe compared with other methods based on nuclear magnetic resonance(NMR), X-ray or γ-ray, and ultrasound etc.. However, there are some defects in electrical PT such as low signal-to-noise ratio, low sensitivity, low spatio-temporal resolution, owing to the inherent characteristic of soft field. An intrusive method[1], namely, wire-mesh resistance tomography, is proposed, which is able to conquer the above shortcomings because the new method has a feature of hard field, similar to that of X-ray PT. A sensor of two or three grids with parallel and equispaced wires is placed over the cross-section of the pipeline, contacting with the fluids directly. The conductance output of adjacent wires varies when two phase fluids pass through the wires alternately. Much research work[2-4] has been performed to confirm the feasibility to reconstruct phase distributions in the pipe. Furthermore, disturbance of wire-mesh to flow field and comparisons with X-ray CT have been comprehensively analyzed, which shows that the intrusive wire-mesh tomography is comprehensively better. However, any method that measures the fluid impedance is influenced by the salinity and the distribution of fluids. Besides, almost all the PT techniques regard two-phase flow system as a "black body", for they seldom consider flow laws and try to reconstruct the phase distributions only depending on the measurement signals.

The present paper proposes an intrusive wire-mesh capacitance tomography. The aim is to reconstruct the phase distribution without the influence of water salinity and its distributions. Only several capacitance data are needed for reconstruction without reconstruction algorithms in air-water two-phase flows in horizontal pipes. Several typical flow patterns were tested under the conditions that air and water are both continuous.

THE MEASUREMENT PRINCIPLE

The principle of wire-mesh capacitance tomography is based on the measurement of water thickness. We use a wire with an insulation film as a capacitor, depicted schematically in Fig. 1. The dielectric of the capacitor is the insulating film and the two electrodes are the metal core of the wire and water film covering it, respectively. We may

CP914, *Multiphase Flow: The Ultimate Measurement Challenge,* edited by X. Cai, Y. Wu, Z. Huang, S. Wang, and M. Wang

obtain the capacitance:

$$C = \frac{2\pi\varepsilon l_1}{\ln\frac{d_2}{d_1}} + \frac{2\pi\varepsilon l_2}{\ln\frac{d_2}{d_1}} + \ldots + \frac{2\pi\varepsilon l_n}{\ln\frac{d_2}{d_1}} = \frac{2\pi\varepsilon}{\ln\frac{d_2}{d_1}}\sum_{i=1}^{n} l_i \tag{1}$$

where $L_i (i=1,n)$ are the discrete film thicknesses of water covering the capacitor, ε is the dielectric constant, and d_2 an d_1 are the diameters of the wire with and without the film, respectively. We suppose that ε, d_1, and d_2 are constant, so the capacitance is proportional to the water film lengths $\sum_{i=1}^{n} L_i$, and is independent of the distribution and the salinity of water. The capacitance is very sensitive to the variation of film because the denominator is close to zero for $d_1 \approx d_2$.

Static experiment was carried out to confirm Eq. (1) with the result shown in Fig. 2. The water salt content range was from 0 to 5 percent, and the temperatures were at 17° C and 50° C. The result shows that all the data under different conditions are in one line. The capacitance is in proportion to the film thickness and the capacitance coefficient is 4.8 pF per mm length for a wire with the diameter of 0.5mm, the core of copper and the film of Teflon. Fagundes Netto[5] measured the film thickness of long bubble flow using this method and obtained good results. It's a pity that he conducted no experiments of other flow patterns.

If several such capacitors are placed over the cross section of the pipe, we will obtain the phase interface by linking the discrete film thicknesses by linking the discrete film thicknesses based on Eq.1, which is right the aim of wire-mesh capacitance tomography.

Since the proposed tomographic technique is intrusive, it is a hard-field method. The principle is similar to that of X-ray tomography and the wires are like X-ray bundles. The measured capacitance of each capacitor is similar to X-ray output after attenuation. The phase interface can be decided by the film thicknesses at different positions on the basis of flow regime knowledge. The authors succeeded in reconstructing the phase distributions of several typical static flow patterns in horizontal pipes by such idea [6].

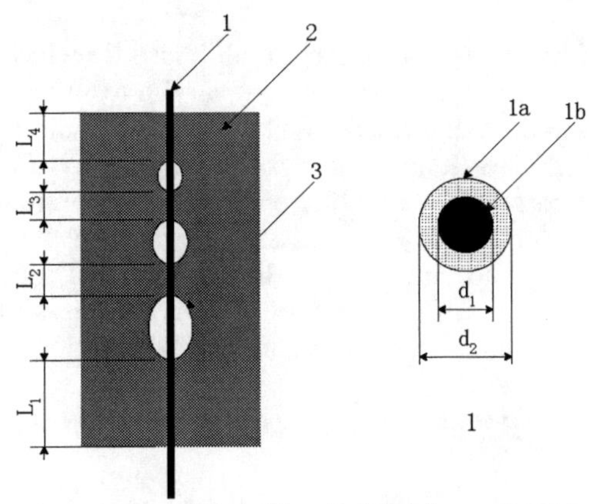

FIGURE1. 1: Wire capacitor; 1a: Insulating film; 1b: Metal core;2: Water film; 3: Bubble

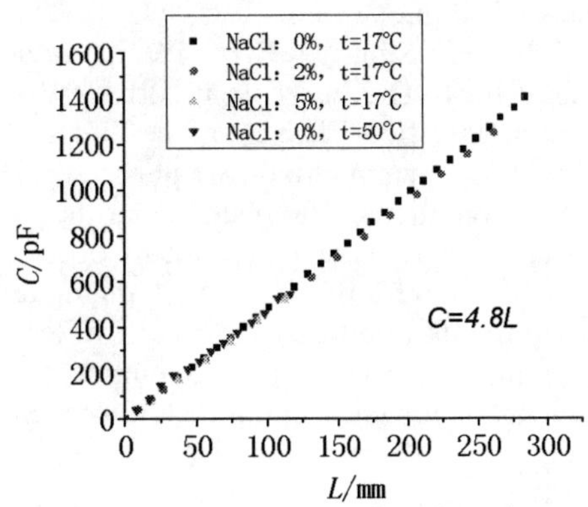

FIGURE2. Static output of capacitor as a function of water film thickness

EXPERIMENTAL SETUP

Sensor Structure

The wire used here is made of copper with an insulating film of Teflon. The diameter of the wire is 0.2mm and the output coefficient is 1.5pF/mm. The sensor is made up of two planes, Plane A-A and Plane B. Plane A-A includes four parallel wire capacitors, vertically placed over the cross section of the pipe, from No.1 to No.4, and No.2 is across the center of the pipe. The pitches between the wires are 10mm, 6mm, and 12mm respectively. Plane B is a lead plane from water including four leads. The leads located at the bottom of the pipe. Each lead is 5mm high from the bottom and is 5mm from the corresponding wire in Plane A-A. The sensor is shown schematically in Fig.3. The diameter of the pipe is 50mm and the test section is 1m long with the sensor located 800mm from the entrance of the test section.

Measurement Circuit

Capacitance is measured by integrate circuit of CAV414[7] which is special for capacitance-to-voltage transforming. The circuit would work as long as several resistors and capacitors are provided according to the requirement of the capacitance range, from 50pF to 100pF in this study. The four capacitances can be sampled simultaneously by four independent circuits. There would be no disturbance between adjacent capacitors when the pitch is bigger than 10mm. The disturbance can be neglected when the pitch is bigger than 4mm. After the capacitance is transformed into voltage, all the data are sampled by NI product, NI-PCI6225, and stored in computer automatically.

The working frequencies of each integrate circuits is 1000Hz. It is much lower than the sampling rate of computer, so imaging rate is 1000 frames/s It can be regarded as continuous sampling. For convenience, only three typical frames are given for each condition. The sampling time for each experiment is 10 seconds long.

Experimental Setup

The gas-liquid flow experiments were carried out in a multiphase flow loop with the pipe diameter of 50mm. All the horizontal part of the flow loop is made up of transparent acrylic resin and it is easy to observe the flow process. The fluids were air and tap water and the experiments were carried out under ambient conditions. The horizontal part before the test section is $l/d=200$, which is long enough for the phases to fully mix. Air flow rate was measured by orifice flow meter with the maximum relative error of 1.5%. Water flow rate is measured by electric magnetic flow meter with the maximum relative error of 0.5%. Phases are separated in the separator at the outlet of the pipeline. Water was recirculated in a closed circuit and air was directly rejected into air. We used flow photos, taken by SpeedCam Pro-LT digital high-speed camera, as standard of film thickness measurement. The camera was located 0.5m upstream from the sensor. The running parameters and flow descriptions were summarized in Tab.1.

RESULTS AND DISCUSSION

Observation Description

Several flow patterns with both phases continuous were observed: (1) Static interface. The interface is completely 2-D, corresponding to Condition 1 in Tab.1; (2) Stratified flow. The interface is still 2-D, but there exist ripples on the interface, Condition 2; (3) Semi-slug flow. The interface is already 3-D. There is one water peak within the sampling time 10s, Condition 3; (4) Semi-slug flow. The interface is completely 3-D.There are four water peaks within the sampling time 10s, Condition 4. Ta.1 summarizes the phase superficial velocities and the descriptions to the flow structures under 4 different flow conditions. The phenomena that the sensor disturbs the flow field, water film stick to the wire, and water climbs up the wire were not obviously observed in the experiments. There is no error at all when some drops of water stick to the wire, for the drops are hanging and discrete to the continuous film at the bottom of the pipe. The four sampling circuits interfered with each other when the capacitances were measured simultaneously, which would improve the disturbance to the output signals, but this would not change the measurement results very much. The disturbance would completely disappear if the distance between the capacitors is great enough, about 10mm in this study.

TABLE1. Parameters and descriptions (m/s)

Conditions	v_{gs}	v_{ls}	Descriptions
1	0	0	Static interface
2	0.4	0.0042	Wavy stratified flow
3	1.83	0.014	One peak
4	1.81	0.09	Four peaks

Tomographic Process

Two steps to reconstruct the phase distributions over the cross-section are necessary: First, flow pattern type is told by the central capacitor, namely No.2 capacitor, by analyzing the output trend with time of film thickness based on the flow regime knowledge; (2) The discrete interface positions are decided by the film thicknesses. The whole interface is decided by connecting the discrete positions by smooth curve. The tomographic method mentioned above can only work when both phases are continuous. It is impossible to reconstruct the complete interface because the number of the measured capacitors is limited. The tomographic results should be regarded as the mean results of phase interface. It's easy to reconstruct the images up to 1000 frames per second. As an example, six groups of tomographic results are shown in figure 4 under different conditions summarized in Tab. 1. There are three parts in one experiment. On the top left of each group, there is a film thickness curve with time of No.2 unit, and there are three frames of reconstructed results of three different moments on the top right, marked under each frame, according to the curve on the left. It is the side view of the flow view at the below part. We gave only the front part of the structure for the scope of the camera lens is limited in some cases.

Error Analysis

Since it is an intrusive method, disturbance of wires to flow field is inevitable. The process for the fluids to flow over the wire is very complicated. The extent of the disturbance depend on many factors such as size and characteristics of wire surface, characteristics of fluids, surface tension, fluids' velocities, etc. It is almost impossible to make the process clear only by analytic method.

As to the precision of film thickness measurement, it is even more difficult to analyze. All the parameters seem to have influence on the measurement precision. What's more, the coupling of the factors is serious.

Let's first consider the process of water film going up and down the wire surface. When water is rising, the water surface surrounding the wire is able to keep up with the water level. There is no obvious measurement error of film thickness. But when the water is falling, the water surface surrounding the wire will stick to the surface of the wire and it will fall at a slower speed v_f. For a given wire and given fluids, the falling speed v_f, can be seen constant. At last, the film sticking to the surface of the wire will shrink into bubbles because of surface tension. The process of water falling down the wire is schematically shown in Fig. 5. The discrete bubbles would not bring error on the film thickness measurement because they do not contact with the water at the bottom. The error only appears at the falling stage. When the falling velocity of water v_w is low, the measurement error can be neglected. But when v_w is big enough, the measured film thickness would be obvious.

The flow pattern of condition 4 is semi-slug flow. The flow structure can be divided into three parts, stratified flow at the front and at the rear, and semi-slug flow at the middle. The front part and the rear part of the semi-slug flow are basically symmetrical from the photos taken by high-speed camera. From the measurement curve of film thickness of condition 4, we can see that the curve rises sharply, but falls much more gently. There is little error of measurement of film thickness at the front and obvious error at the rear. The bigger the speed of the phase interface, the larger the error. The largest error in this study, in condition 4, is estimated to be less than 0.05D by comparison the measured curve with photos.

As to the disturbance of the wires to the flow field, Prasser [4] analyzes the disturbance process in bubbly flow regime by taking photos using high-speed camera. The results show that the intrusive wires really destroy the flow structure and the shape of bubbles is deformed, but the output is not influenced seriously and the reconstructions are still able to reflect the undisturbed flow structure. The disturbance effect in this paper is estimated to be much slighter compared with Prasser's, because the number of wires is much less.

CONCLUSION

In this paper, we present an intrusive, wire-mesh capacitance tomography, based on the measurement of film thickness. Static and dynamic experiments are carried out to validate the method in air-water two-phase flows in horizontal pipes.

The capacitance is proportional to the film thickness and is independent of water salinity and its distribution. The error of film thickness is estimated to be below 0.05D. Only four film thicknesses are enough to reconstruct the phase distribution when both air and water are continuous. The reconstructions of phase distribution in the pipe can be achieved based on the measurements of 4 film thicknesses and on the knowledge of flow regimes, without reconstruction algorithms. The agreement between the reconstructed results and what were observed is very good.

The phase interface is the averaged result of the real interface. The method has several advantages of quick response, no reconstructed algorithm, high signal-to-noise ratio, etc.

However, there are still details missing in analyzing quantitatively the influence factors on the measurement of film thickness; the method should be extended to other flow patterns. Further study needed.

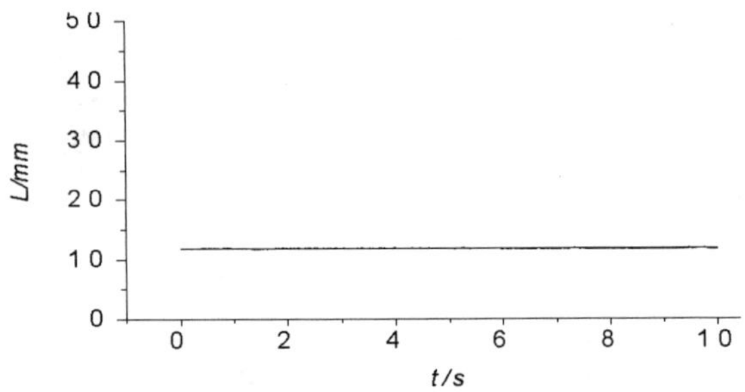

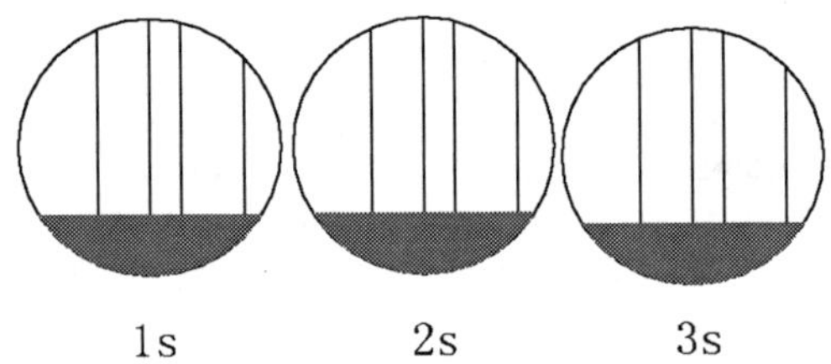

CONDITION1

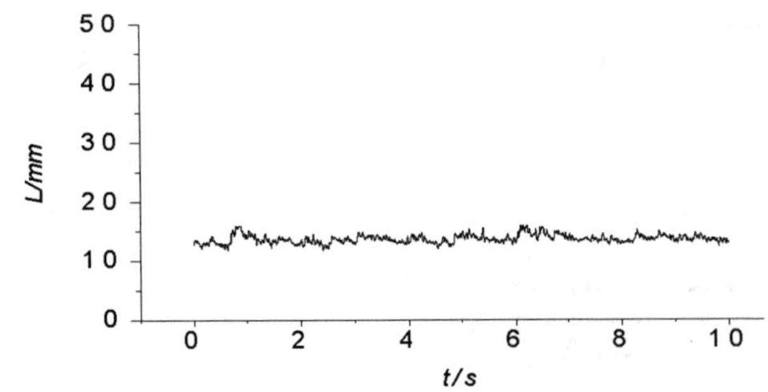

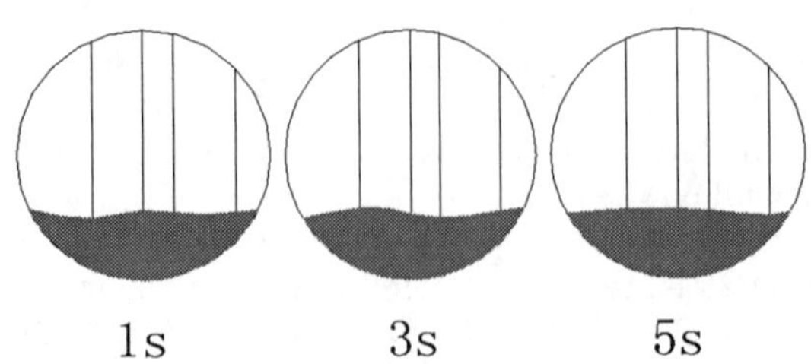

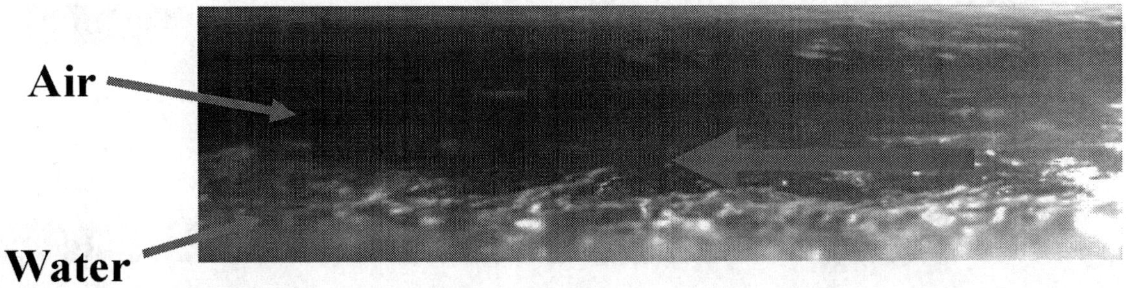

CONDITION2

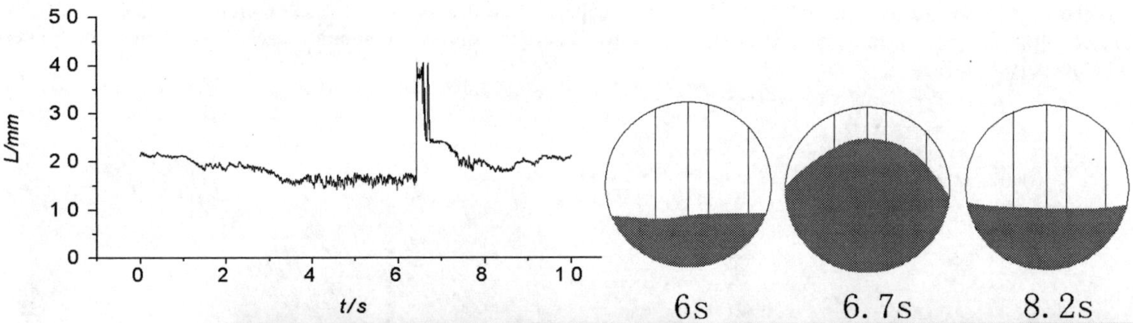

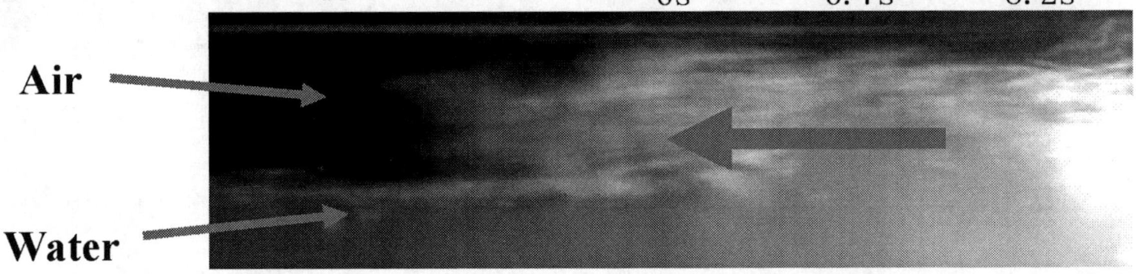

CONDITION3

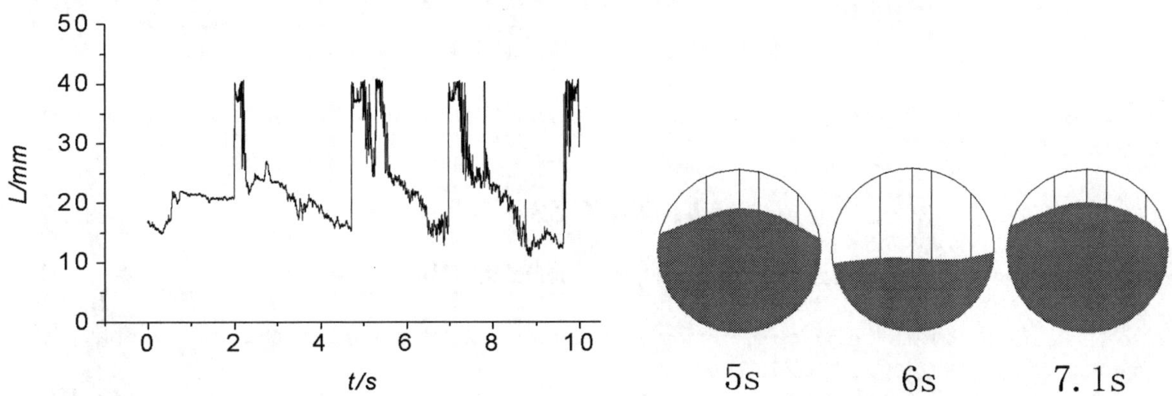

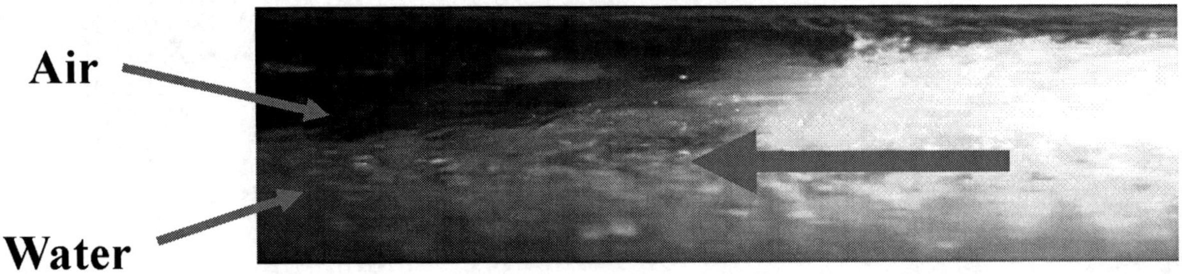

CONDITION4

Notes: There is the film thickness curve measured by No.2 wire capacitor on the top left for each condition and there are three tomographic results at different moments marked under the pictures. There is a side view photo of each flow taken by high-speed camera at the below part.

FIGURE4. Tomographic results

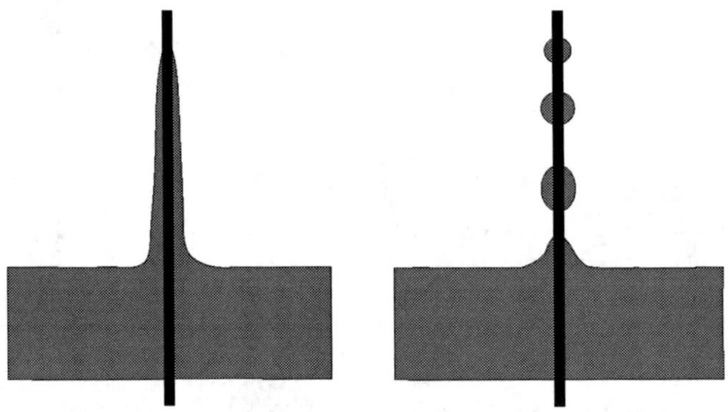

FIGURE5. Process of water film falling down the wire

ACKNOWLEDGEMENTS

The authors express their great thanks to National Excellent Doctoral Thesis Fund of Department of Education of China for their finantial support (No.200337).

REFERENCES

1. Johnson I D. Method and apparatus for measuring water in crude oil. United States Patent, No. 4,644,263, Date of patent: Feb. 17, 1987.
2. Reinecke N, Petritsch G, Boddem M, et al. Tomographic imaging of the phase distribution in two-phase slug flow[J]. Int. J. Multiphase Flow, 1998, 24(4):617-634.
3. Prasser H M, Bttger A, Zschau J. A new electrode-mesh tomography for gas-liquid flows[J]. Flow Meas. and Instru, 1998, 9(2):111-119.
4. Prasser H M, Scholz D, Zippe C. Bubble size measurement using wire-mesh sensor[J]. Flow Meas. and Instru. 2001,12(4):299-312.
5. Fagundes Netto J. R, Fabre J, Peresson L. Shape of long bubbles In horizontal slug flow[J]. Int. J. Multiphase Flow. 1999, 25(6-7): 1129-1160.
6. Huang S F, Zhang X G, Wang D, et al. Wire-mesh capacitance tomography in horizontal two-phase flow[J]. Thermal Energy and power engineering. 2006, 21(4) :414-417.
7. Liu C S, Wang Y M, Wang S P, et al. Modern integrated circuits instruction manual and their application examples[M]. Xidian University Press, Xi'an, 2003:179-184.

Design of an Electrical Impedance Tomography Sensor for Flow Measurement in an Oscillatory Baffled Reactor.

G. Vilar, R.A.Williams, M.Wang

Institute of Particle Science & Engineering, Houldsworth Building, University of Leeds, Clarendon Road, Leeds, West Yorkshire, LS2 9JT, United Kingdom. Email: g.vilaruso04@leeds.ac.uk

Abstract. In this paper, a new application of electrical impedance tomography (EIT) for an advanced on-line measurement is presented. This application involves the design, manufacture and adaptation of an EIT sensor for the measurement in an oscillatory baffled reactor (OBR). The main goal is to develop of a novel measurement and modeling method for control of the OBR. The reactor itself enables the production of water-in-oil/oil-in water emulsions along with the use of chemical reagents for a variety of manufacturing processes. Use of electrical tomography facilitates detailed measurement of the concentration and flow of components in the reactor. The paper reports on design philosophy of the EIT for this application that has not, to our knowledge, been reported previously.

Keywords: tomography sensor, oscillatory baffled reactor.

INTRODUCTION

Electrical Impedance Tomography (EIT) is a powerful tool for mapping the concentration and velocity distributions in two phase flows, where electrical impedance differences between the two-phase fluids exist [1]. The chemical process consists of two immiscible fluids. The mineral oil (dispersed phase) is dispersed as liquid droplets in water (continuous phase). The measurement of the different impedance of the two-phase fluid in the process, the analysis of the data and the conversion to different variables (e.g. velocity, concentration, etc) are the main steps in the control of the process by the EIT Technique [2].

The oscillatory baffled reactors (OBR) are a new type of continuous reactor built of equally spaced orifice plate baffles. In these reactors, the process fluid has an oscillatory movement that is superimposed upon all the fluid in the reactor. The addition of this sinusoidal imposed movement enables the radial mixing unlike the traditional plug flow reactors (Fig.1). The oscillatory movement and the baffles enable the formation of several vortex cycles in the inter-baffle cavities of the reactor that make the fluid to loop and therefore improve the mixing [3]. The combination of baffles and the oscillatory movement creates a flow pattern that is leading to highly efficient heat and mass transfer while maintaining plug flow [4].

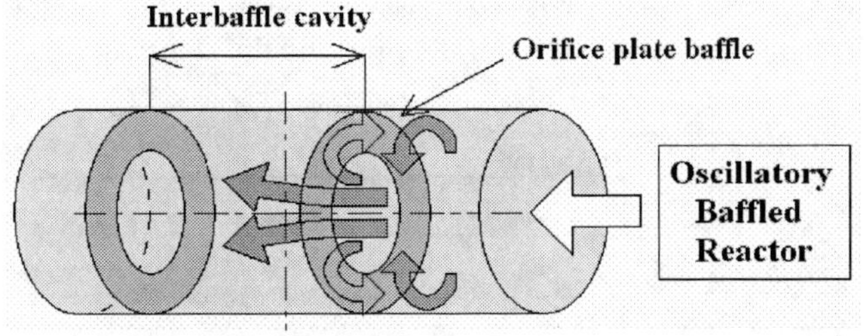

FIGURE 1. Mixing mechanism in the OBR.

CP914, *Multiphase Flow: The Ultimate Measurement Challenge,* edited by X. Cai, Y. Wu, Z. Huang, S. Wang, and M. Wang

EXPERIMENTAL ARRANGEMENT

EIT Measurement Equipment

An EIT system is built on a sensor, a data collection system and an image reconstruction algorithm, which are known as the three key elements of a tomography system [2]. The objective of the EIT sensor is the detection of differences of electrical impedance then they are converted in concentration. The measurement of the impedance in the dispersed phase in a multiphase flow is transformed into an electrical signal on electrodes that is the main input of the data collection system.

The data collection system reported here is the ITS P2000 System (Industrial Tomography Systems). The data collection system transforms the electrical signal into raw data that is processed by an image reconstruction algorithm. The resulted image shows the mapping of the distribution of concentration of the two-phase flow in the chemical process. The block diagram for the control of the emulsion process with the EIT Sensor is shown in Fig. 2.

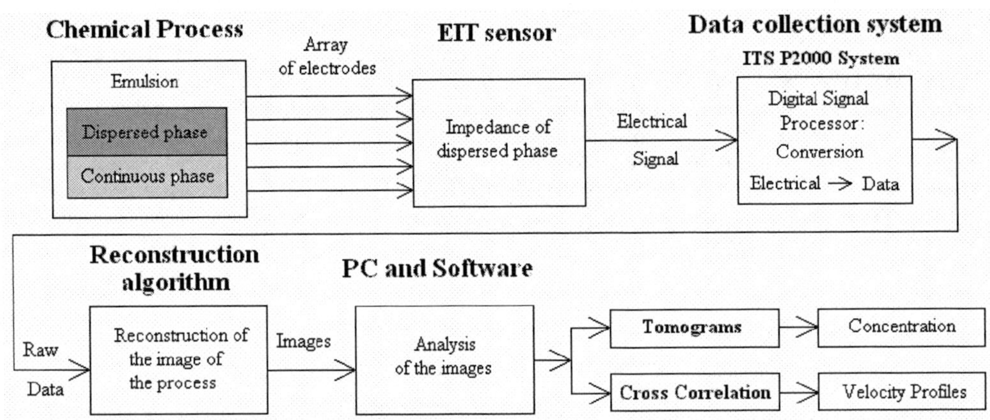

FIGURE 2. Block diagram of the control of the process with an EIT System.

The use of cross correlation method (CCS) enables the velocity profiles of the components of the oil-in-water emulsion in the longitudinal and transverse axis of the OBR to be obtained [5,6]. The application of this software is based on the comparison of the conductivity data (tomograms) obtained from one plane of electrodes with the data obtained from another plane of electrodes where the distance between planes of electrodes is known. [7]

OBR Characterization

The OBR consist of a continuous chemical reactor where equally spaced orifice plate baffles are firmly positioned inside the main pipe (Fig. 3). The water flow (continuous phase) is supplied by an external pump (Table 1). The mineral oil (dispersed phase) is added from an insertion point of the reactor pipe in order to create the emulsion. The EIT sensor is installed between the end of the pipe and the starting of the bend.

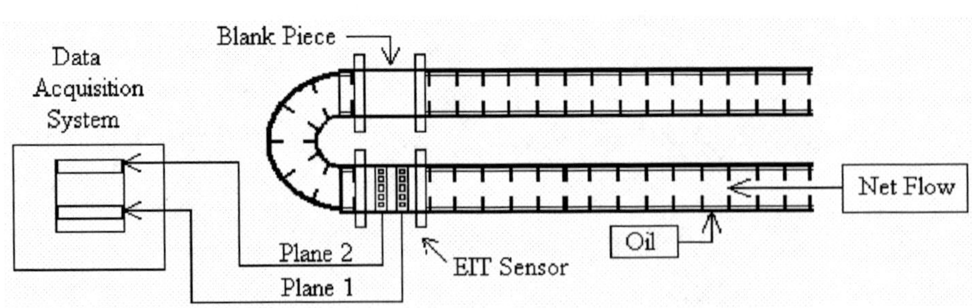

FIGURE 3. Experimental set-up of the OBR.

TABLE 1. Characteristics of the OBR.

Properties	Experimental Values
Flow rate (liter/min)	0.25 to 4
Diameter of the reactor (m)	0.05
Diameter of the orifice plates (m)	0.011

DESIGN OF THE EIT SENSOR

Here we consider the procedures to design the dimensions and measurement on the OBR system, based on a discussion of the stages summarized in Fig. 4.

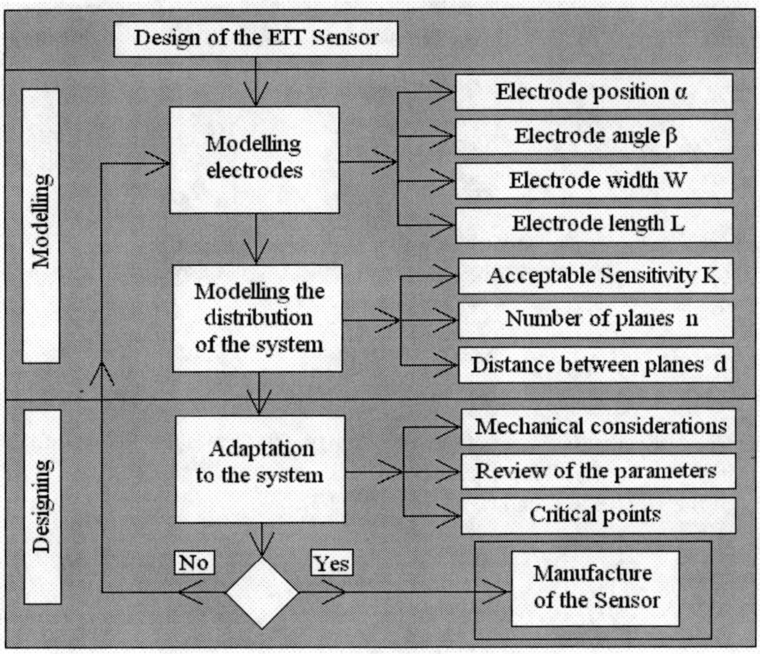

FIGURE 4. Experimental set-up of the OBR.

Introduction

The philosophy behind design of the EIT sensor involves a calculation of the minimum requirements for the sensor in order to obtain a sufficiently accurate measurement of key properties in the OBR, notably flow rate and velocity. The main parameters of design of the EIT Sensor are the electrode position angle, electrode width angle, electrode width and length. These parameters determine the shape and the position of the electrodes in the sensor.

Modeling of the Electrodes

In electrical impedance tomography typically 16 or more electrodes are used in the measurement. One group of the total number electrodes are used for inject currents or voltages into the flow, the other group of electrodes are used for measure these currents or voltages. Normally current injection is used [7], because it avoids the transient time to reach the desired value that occurs when voltage is injected. Electrical current is affected by the electrical conductivity differences between the two-phase fluids. The electrode position angle (α) is determined by the number of electrodes (N_E):

$$\alpha = \frac{2\pi}{N_E} \tag{1}$$

The estimation of the electrode angle enables the distribution of the electrodes inside the pipe. Previous works form Wang et al. showed the challenges involved in simulating a physical electrode angle [8] when the number of electrodes is limited and the effect of different electrode configurations depending on number of nodes considered in each electrode in the finite element method mesh (single node, grouped 2-nodes, grouped 4-nodes). The strategy of measurement determines the calculation of the width of the electrode. The electrode width angle (β) depends on the number of boundary nodes (N_B) and the number of nodes per electrode (N_D). It can be obtained by [9]:

$$\beta = \frac{2 \cdot \pi \cdot N_D}{N_B + N_E} \tag{2}$$

Consequently, the minimum required width (W) of the electrode for the EIT sensor can be obtained from the diameter of the pipe of the reactor (D) as [9]:

$$W = \frac{D}{2} \cdot \beta \tag{3}$$

The distance (S) of separation between electrodes is one of the most important factors in the manufacture of the EIT sensor. It will determine the way the electrodes are attached to the bore surface of the sensor. It can be obtained as:

$$S = (\alpha - \beta) \cdot \frac{D}{2} \tag{4}$$

The design of the electrode length depends on the velocity of the flow and the velocity of the data acquisition system. Faster flow measurements require longer electrodes and faster systems in order to obtain accurate data. The minimum required length of the electrode L can be obtained as [9]:

$$L = V \cdot \sigma_{frame} \tag{5}$$

The minimum length of electrode is obtained considering the worst conditions of operation. It occurs when the flow rate V and the time required per frame σ_{frame} are the highest.

Modeling of the Sensor Configuration

The modeling of the distribution of the system involves the determination of the configuration of the planes of electrodes in the EIT sensor. These parameters are very important because they affect the sensitivity and accuracy of the measurement with the EIT Sensor. The fractional discrimination or sensitivity of the measurement system K can be defined as the ability of detecting the smaller changes of the values of a velocity variable in measurement. It is a percentage that describes the minimum difference in a measured value that the system is able to detect compared to the resolution of the instrument. It depends on the velocity of the data acquisition system and the transit time between planes of electrodes τ. The instrument will be able to obtain more accurate measurements when the velocity of the flow is slower and the distance between planes of electrodes is higher. Theoretically, an infinite distance between planes of electrodes means that the instrument would be able to detect every change in the measurement. The fractional discrimination K of the EIT measurement system can be calculated by:

$$K = \frac{\sigma_{frame}}{2 \cdot \tau} \tag{6}$$

A lower value of K means that the EIT system is able to see small changes in the values of velocity of the dispersed phase; consequently the instrument has a higher precision on velocity measurement.

The measurement of the impedance of the dispersed phase with the plane of electrodes enables to obtain the mapping of the distribution of concentration. The use of various planes of enables the mapping in different surfaces and the comparison of the data between planes. In this case, the important data obtained from the tomograms can be improved with the calculation of parameters of velocity of the flow with cross correlation software between two planes of electrodes. The use of cross correlation method enables the velocity profiles of the flow in the longitudinal and transverse axis of the OBR to be obtained. The assumption of the value of K enables the determination of the minimum required transit time between planes for a particular value of the velocity of the flow. The transit time between planes of electrodes can be defined as:

$$\tau = \frac{d}{V} \tag{7}$$

Consequently, the substitution of the transient time in equation 6 leads to the calculation of the minimum distance required between planes of electrodes:

$$d = V \cdot \frac{\sigma_{frame}}{2 \cdot K} = \frac{L}{2 \cdot K} \tag{8}$$

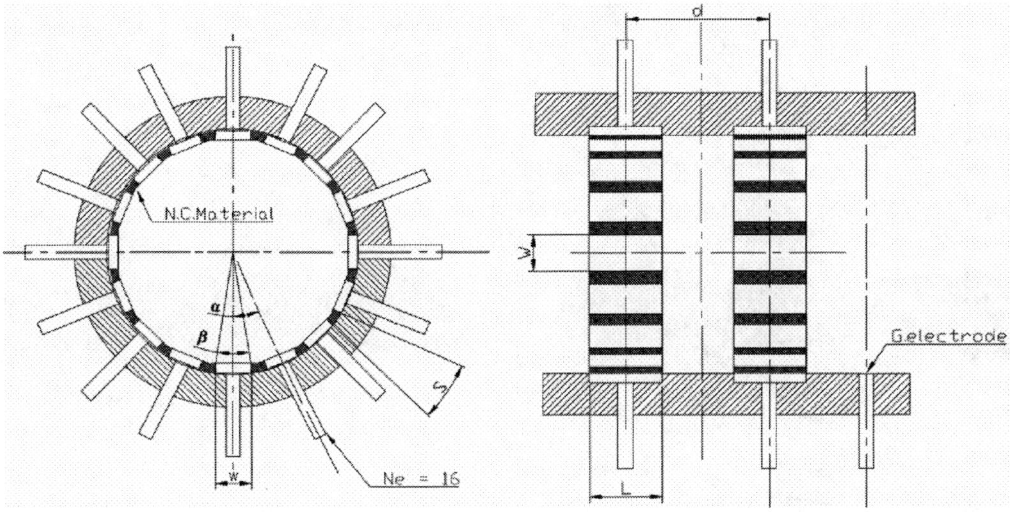

FIGURE 5. Parameters of design of the electrodes.

RESULTS

Parameters of Design of the Sensor

The results of the calculation of the parameters are shown in the following tables. The sensor is made of 16 electrodes, every electrode is considered as a single node in the design. Applying equations 1 to 5 the results are:

TABLE 2. Electrode properties.

Properties	Values
Electrode position angle α	0.393
Electrode width angle β	0.169
Electrode Width W (mm)	4.9
Separation between electrodes S (mm)	4.925

The velocity of acquisition of data collection system is one of the most important factors in the calculation of the minimum length of the electrode.

TABLE 3. Calculated Values of the electrode length required (mm).

Flow rate (liter/min)	**t_{frame} (frame/s)**			
	10	20	50	100
0.25	0.2122	0.1061	0.04244	0.02122
2	1.698	0.8488	0.3395	0.1698
4	3.395	1.698	0.6791	0.3395

The minimum length of electrode is obtained considering the worst conditions of operation. It occurs when the flow rate and the time required per frame are the highest. The minimum length of the electrode required in this system is 3.395 millimeters. This small result is due to the slow velocity of the flow compared to the velocity of the data acquisition system, the conclusion is that the flow parameters can be measured in this system without length limitations from the design of the electrodes. The minimum distance required between planes of electrodes in order to use the cross correlation software can be obtained from the length of the electrode. The following graph represents the values of the minimum required distance between planes depending on the assumption of different values of fractional discrimination K according to equation 8.

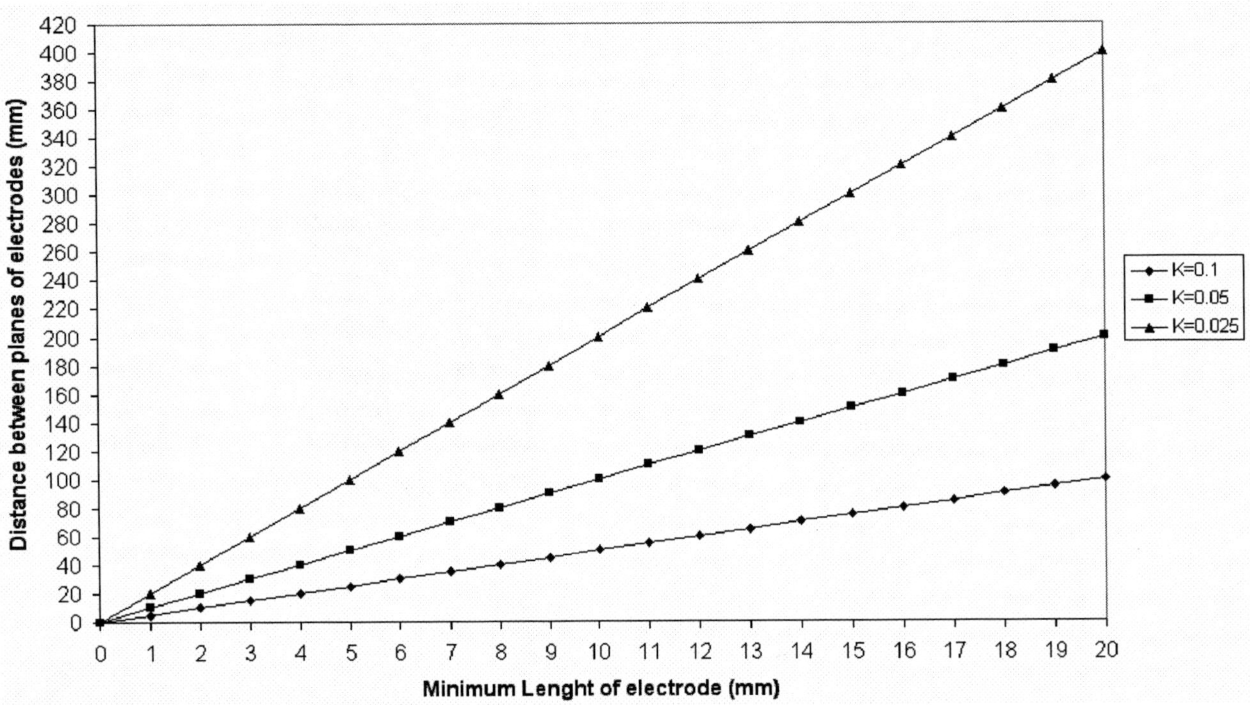

FIGURE 6. Minimum required distance between electrodes depending on K.

The maximum value of K that will be considered in this design is K = 0.05 (5%). This assumption enables the determination of the minimum required transit time between planes and consequently the minimum distance between planes in order to achieve the assumed value of sensitivity:

TABLE 4. Calculated Values of the distance between electrodes required (mm).

Flow rate (liter/min)	**t_{frame} (frame/s)**			
	10	20	50	100
0.25	2.122	1.061	0.4244	0.2122
2	16.98	8.488	3.395	1.698
4	33.95	16.98	6.791	3.395

The minimum distance required between two planes of electrodes in this system is 33.95 millimeters. If the considered distance between the planes of electrodes is increased then the transit time τ between the two planes of electrodes is higher and consequently the sensitivity K is reduced.

Mechanical considerations

The mechanical design is one of the most important factors in the manufacture of the EIT sensor. Depending on the chemical process that is going to be measured by the sensor, the calculated parameters of design can vary in order to make the sensor available to work correctly in the process. The ideal way should be to build the sensor and install it with a minimal influence in the process itself. It means giving the sensor the correct parameters of design in order to avoid the modification of the flow pattern and the velocity profiles. According to the mechanical suggestions and the calculation of previous parameters, the fixed parameters of the sensor are the following:

TABLE 5. Fixed parameters of design of the sensor.

Properties	Values
Number of electrodes	16
Number of planes of electrodes	2
Distance between planes of electrodes (mm)	30
Width of electrode (mm)	7
Length of electrode (mm)	15

TABLE 6. New Electrode properties.

Properties	Values
Electrode position angle α	0.393
Electrode width angle β	0.28
Electrode Width W (mm)	7
Separation between electrodes S (mm)	2.825

Critical Points in the Operation

According to the calculation the minimum distance required between two planes of electrodes in this system is 33.95 millimeters. The mechanical design suggests a lower distance of 30 millimeters in order to maintain the flow pattern. The sensitivity K for the worst conditions (maximum velocity of the fluid and maximum time required per frame) can be obtained as:

$$K = \frac{V \cdot \sigma_{frame}}{2 \cdot d} = 0.057 \tag{9}$$

The sensitivity in the worst case is 5.7 % that is higher than the maximum sensitivity considered in the design (5%). It means that the EIT system will not be able to detect smaller changes accurately in the variable in this particular situation. This is a critical point that must be taken into account in the operation of the system.

Application of the Design

The measurement with the EIT sensor enables to map the oil volume fraction of the mineral oil in the emulsion. The test was carried out at water flow rate of 4.2 liter/min and mineral oil flow rate of 0.125 liter/min. The application of cross correlation method enables to obtain the axial velocity profiles of the mineral oil.

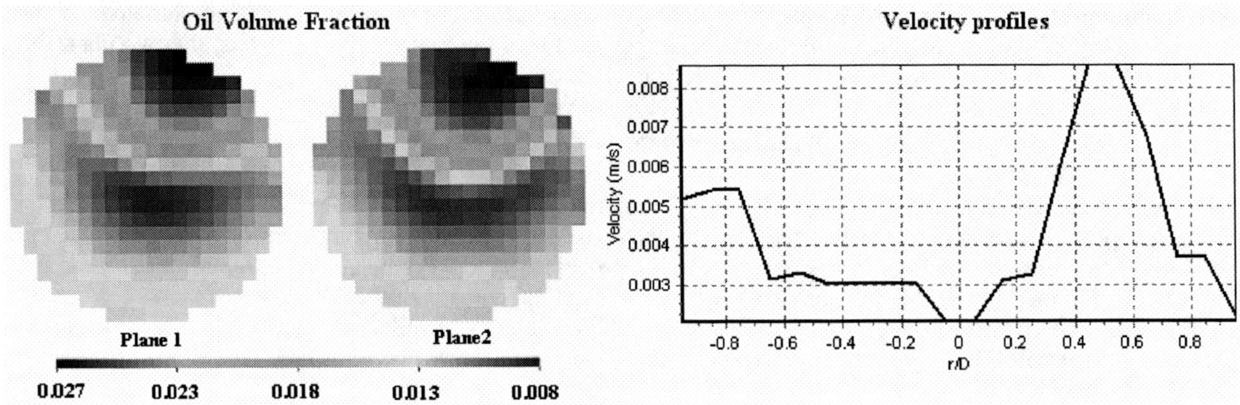

FIGURE 7. Experimental results from the measurement with the EIT sensor.

CONCLUSIONS

In this paper we have reported for the first time a methodology of design of an EIT Sensor for a new application in chemical engineering. The main issues of the design and adaptation of the EIT Sensor to the OBR reactor are discussed. The future work with the EIT Sensor will be the application of CCS Software to obtain the velocity profiles and develop a mathematical modeling of the flow based on the experimental data.

REFERENCES

ITOMS Industrial Tomography Systems. 47 Newton Street, Manchester, M1 1FT, United Kingdom. Tel: +44 (0) 161 237 5779, Fax: +44 (0) 161 237 5669.

CCS Cross Correlation Software version 2.0. Institute of Mechanics, Chinese Academy of Sciences and Institute of Particle Science and Engineering, University of Leeds. Copyright 2004. All rights reserved.

1. M.Wang and F.J.Dickin, "Electrical resistance tomography for process applications" in Measurement Science and Technology Vol. 7 (1996) pp. 247-260.
2. M.Wang, "Impedance mapping of particulate multiphase flows" in Flow Measurement and Instrumentation Vol. 16 (2005) pp. 183-189.
3. T.Howes, M.R.Mackley and E.P.L.Roberts, "The simulation of chaotic mixing and dispersion of periodic flows in baffled channels" in Chemical Engineering Science Vol. 48 (1991) pp. 799-803.
4. C.R.Brunold, J.C.B.Hunns, M.R.Mackley and J.W.Thompson, "Experimental observations on flow patterns and energy losses for oscillatory flow in ducts containing sharp edges" in Chemical Engineering Science 44 (1998) 1227-1244.
5. Y.Dai, M.Wang, N.Panayotopoulos, G.Lucas and R.A.Williams, "3-D Visualisation of a Swirling Flow Using Electrical Resistance Tomography", Keynote Lecture: 4th World Congress on Industrial Process Tomography, Aizu, Japan
6. H.Li, M.Wang, Y.Wu, Y.Ma and R.A.Williams, "Characterization of Air-water Two-phase Characterization of Air-water Two-phase Vertical Flow by Using Electrical Resistance Vertical Flow by Using Electrical Resistance Imaging" in The Canadian Journal of Chemical Engineering Vol. 83, February 2005.
7. H.Li, M.Wang, Y.Wu, Y.Ma and R.A.Williams, "Measurement of oil volume fraction and velocity distributions in vertical oil-in-water flows using ERT and a local probe" in Journal of Zhejiang University SCI 2005 6A(12): pp. 1412-1415. ISSN 1009-3095.
8. M.Wang, F.J.Dickin, R.A.Williams and G.Lucas, "The grouped node technique as a means of handling large electrode surfaces in electrical impedance tomography"in Physio. Measurement Vol. 16 (1995) pp. 219-226.
9. M.Wang, R.Mann and F.J. Dickin, "Electrical Resistance Tomographic sensing systems for industrial applications" in Chemical Engineering Communications (1999) Vol. 175 pp.49-70.

AC-based Capacitance Tomography System With Small-diameter And High-pressure Pipe

Yang Daoye, Xu Chuanlong, Zhou Bin, Wang Shimin,Wu Xin

Key Laboratory of Clean Coal Power Generation and Combustion Technology of Ministry of Education, Southeast University, Nanjing,210096, China

Abstract. Electrical capacitance tomography (ECT) is a technology which can visualize the internal behavior of industrial process comprising dielectric components. An AC-based capacitance tomography, which is used in dense-phase transportation of pulverized-coal with small-diameter pipe, is introduced in this paper. The inside diameter of ECT pipeline sensor is 10 mm, the working pressure in pipeline is as much as 6 MPa. Wide range capacitance value can be measured by AC-based and DC-based programmable gain amplifiers. It could collect tomographic image data at more than 100 frames per second.

Keywords: capacitance measurement, AC-based, tomography
PACS: 47.61.Jd Multiphase flows

INTRODUCTION

Two-phase(multi-phase) flow is a commix flow mode which is widely existed, and its parameter measurement technology takes a more and more important role in science research and industrial production, thus process tomography comes out with this trend. Electrical capacitance tomography (ECT) is one of the most mature and promising methods, and it has many characteristics such as low cost, fast response, non-intrusive method, broad application, safety, so it will be widely used in industrial process.

At present, most of the ECT sensors researched is run in lower pressure (0.1 to 1MPa), and the sensor in higher-pressure condition is seldom researched. The ECT system introduced in this paper is applied to gas/solid flow in pneumatic conveyors, and the pressure in sensor pipeline is as much as 6 MPa, so much more rigorous condition is needed in this system than in the system with normal pressure parameter. The other characteristic of the system is that the ECT sensor diameter is only 10 mm, this produces a possibility to apply ECT in mini measurement. Therefore, the application range of ECT can be widened.

ECT SENSOR

Sensor Configuration Argumentation

As mentioned above, the ECT sensor pipeline works in high pressure, so vigorous demand is needed to sensor pipeline material: On the one hand, the material of sensor pipeline should have good tensile strength; On the other hand, definite thickness of pipeline wall is necessary in adequate tensile strength circumstance. Capacitance sensor sensitive area is near the electrode [1], but pipeline wall occupies this sensitive area in the sensor. The standing capacitance enlarges with the increase of the pipeline thickness because the permittivity of pipeline is much larger than the flow medium, thus the linearity between capacitance value and dielectric constant will degrade with the increase of the pipeline thickness[2]. So it is advantageous to use thin pipeline sensor to improve system sensitivity in the condition of adequate mechanical strength.

CP914, *Multiphase Flow: The Ultimate Measurement Challenge,* edited by X. Cai, Y. Wu, Z. Huang, S. Wang, and M. Wang

The pipeline material of the sensor is quartz glass, which has outstanding tensile strength. By calculation, the pipeline thickness of 3 mm is adequate to ensure the mechanical strength demand in high pressure.

Although the pipeline thickness is 3 mm, the inner diameter is only 10 mm, 61 percent of area surrounded the electrodes is occupied by the pipeline wall, and this area is high sensitive to electrode capacitance change, so the sensitivity of the capacitance sensor is badly degrade. For the purpose of reducing the difficulty of capacitance measurement, the only way is to decrease the number the measure so that the electrodes capacitance value is big enough to be measured by ECT measuring circuit. In addition, there has been a higher demand for dimension accuracy in the small-diameter capacitance sensor, and less electrode number can reduce the demand for dimension accuracy and the difficulty of machining the sensor.

As discussed above, when the demand of basic image quality is ensured, it is more reasonable to choose 8 electrodes sensor. As a result, the small-diameter and high-pressure ECT sensor is shown in Fig.1.

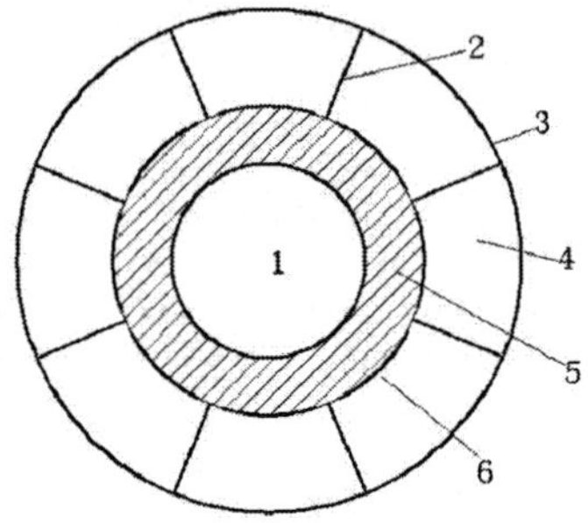

1. Imaging area 2. Radial electrode 3. Earthed screen 4. Filling layer 5. Insulating pipeline 6. Measurement electrode

FIGURE 1. Small-diameter and high-pressure ECT sensor with 8 electrodes

For the purpose of decreasing the dimension error of sensor and facilitating its manufacture, the technique of copperizing on the surface of glass is adopted. The electrodes are no longer sheet coppers mounted around the circumference of the insulating pipeline, they are directly plating on the outside of pipeline. Comparing with the normal mounted electrodes, the plating electrode has several advantages: Firstly, the plating electrodes are thin and they are felted outside the pipeline compactly, so the sensor structure is nearly perfect, and the measurement error introduced for sensor dimension will be decreased greatly. Secondly, copperizing on the pipeline can leave out some cockamamie working procedure such as mounting the electrode, thus the manufacture of ECT sensor becomes easier. Lastly, the integration of pipeline and electrodes can increase the longevity of sensor, and the sensor performance deterioration produced by external conditions such as temperature can be avoided.

Capacitance Simulation Of Sensor

To a certain sensor pipeline, the capacitance values between electrodes are mainly determined by electrode area. For a given electrode number and angle, the capacitance between electrodes is proportional to the axial length of electrode [3], so to make capacitance measurement more easily, long electrode is preferable. However, too long electrode can result in spatial filter effect, and make the sensor loss the response ability to high frequency signal of spatial structure. Therefore, it is necessary to calculate the sensor capacitances so that both of the requirements are satisfied.

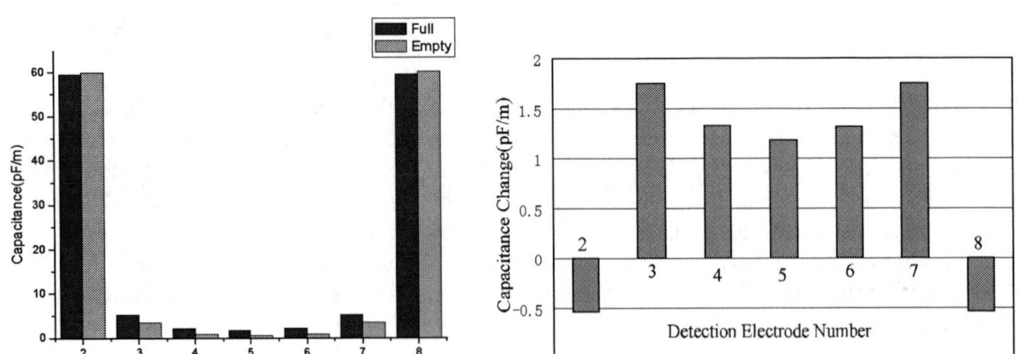

(a) Capacitance contrast between empty and full (b) Changes in capacitance

FIGURE 2. Capacitance simulation of the sensor electrode pairs

The result of ECT sensor capacitances simulated by ANSYS 9.0 is shown in Fig.2. For the 8-electrode system, sensor electrode 1 is selected as excitation electrode, and the others as detect electrodes, so 7 electrode pair capacitances can be calculated. The result includes the empty sensor capacitances when the sensor is filled with air and the full sensor capacitances when it's filled with coal powder. The relative dielectric constant of coal powder, air, pipeline and filling layer is 3.5, 1, 8 and 2.5 respectively.

From the simulation result, the standing capacitance of adjacent electrodes is about 60 pF per meter, and they are 10 to 100 times larger than the other electrodes. However, the capacitance change of adjacent electrodes is about 0.53 pF/m, smaller than the other ones (1.3 to 1.75pF/m). Further more, their changes is negative because the pipeline wall is too thick compared with the sensor inner diameter so that the high sensitivity area is occupied by sensor pipeline wall [4].

The above capacitance simulation gives a powerful reference of ascertaining the axial length of electrode. For the purpose of enabling the measurement circuit to measure the capacitance changes and keeping adequate frequency response ability, the adopted electrode length is 10 cm [5]. Thus, the capacitance changes between empty and full sensor range from 0.05 pF to 0.18 pF.

CAPACITANCE MEASURING CIRCUIT

At present, there are many different kinds of weak capacitance measuring circuits, and two of them are most suitable for use in ECT system because of their stray–immunity[6], which are DC charge/discharge circuit and AC-based circuit. DC charge/discharge circuit suffers from several disadvantages such as CMOS switch charge injection and zero point drift, so it's measuring resolution is limited at 0.3fF. And the AC-based measuring circuit has much more high resolution but it is complex. Therefore, to overcome the disadvantage of low sensor sensitivity in the ECT system, the AC-based circuit is adopted.

AC-based Capacitance Measuring Circuit

The AC-based measuring circuit is shown in Fig. 3. Signal generator chip generates a 500 kHz sine wave and a same frequency square wave synchronously. The sine wave is used as an excitation voltage signal and the square wave as a reference signal. The amplitude of sine wave generated by signal generator is only 2 V peak to peak. To increase the intensity of excitation signal, the sine wave is enlarged nine times at first, and then excites the excitation electrode of the unknown capacitance C_x, the detection electrode is connected to C/V conversion circuit. The C/V conversion circuit is composed of a high-speed and high-precision operational amplifier, feedback capacitance C_f and feedback resistance R_f. The weak current signal on the detection electrode, produced by the excitation electrode, is converted to AC voltage V_o by the C/V conversion circuit. There are two stray capacitances between the two measurement electrodes and earth, which are marked as Cs_1 and $Cs2$. Cs_1 is directly driven by the excitation signal, so it does not affect the voltage applied to the unknown capacitance C_x and hence has no effect on the capacitance measurement. Because the invert input of the operational amplifier is held at virtual earth by feedback and the potential difference across C_{s2} is close to zero, there is nearly zero current passing through C_{s2}, and hence it does not affect the capacitance measuring circuit either. Therefore, the capacitance measuring circuit is not sensitive to the stray capacitance, i.e. it has the capability of stray-immune.

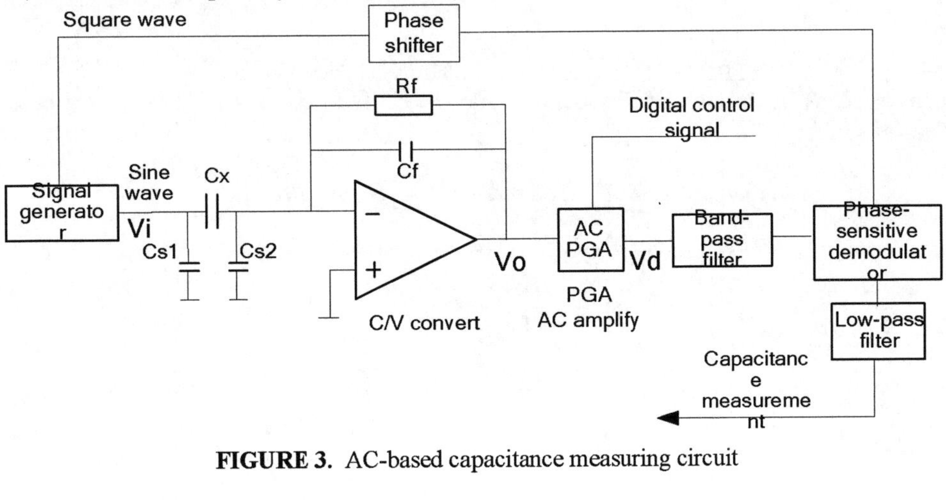

FIGURE 3. AC-based capacitance measuring circuit

The main effect factors in weak signal amplify and transmission are the white noise and 1/f low-frequency noise caused by the thermal noise and shot noise from electrical components[7], and these noise can not be screened. Hence a band-pass filter (BPF) is adopted after the AC amplifier to pre-filter the noise introduced by former class operational amplifier so that the noise out of the band-pass filter bandwidth can be reduced largely, then the signal-to-noise ratio (SNR) of the measuring system will be improved. The experiment indicates that the band-pass filter can improve the distortion and stability of the sine wave signal effectively.

In ideal stance, the current via C_x is equal to the current via C_f and R_f. So the relationship between input and output of C/V convert can be expressed as follows:

$$V_o = -\frac{j\omega C_x R_f}{j\omega C_f R_f + 1} V_i \tag{1}$$

Where ω is the angular frequency of the excitation sine wave.

When capacitance feedback is selected to be dominant, i.e. $|1/j\omega C| << R$, thus $|j\omega CR| >> 1$, Equation.(1) becomes

$$V_o = -\frac{C_x}{C_f} V_i \tag{2}$$

Because C_x is much smaller than C_f, V_o is a weak signal and easy to be interfered by noise. It is necessary to amplify the weak signal to a suitable value. In Fig. 2(a), the change range of C_x is large, from 0.005pF to 6pF, hence the AC amplify circuit should have the capability of gain adjustable so that its output V_d can work in a suitable voltage range. The AC gain adjustable amplify circuit is mainly composed of high-speed and high-precision operational amplifier and switch. The amplifier multiple is determined by adjusting the feedback resistance.

The phase-sensitive demodulator can turn the amplified modulated signal to DC or slow-change signal. Normally there are two kinds of phase-sensitive demodulator circuit: analogue multiplier and electronic switch mode. The phase-sensitive demodulator of electronic switch mode is equivalent to the analogue multiplier when the reference signal is a square wave. In this ECT system, the demodulator of electronic switch is adopted. Comparing with the analogue multiplier, the electronic switch mode has several advantages such as low cost, fast speed, excellent stability and credibility [8].

The input excitation signal is

$$x(t) = V_s \cos(\omega t + \theta) \tag{3}$$

Where V_s is the amplitude of sine wave signal, θ is the angle dispersion between sine wave signal and square wave signal. The input waves are shown in Fig. 4.

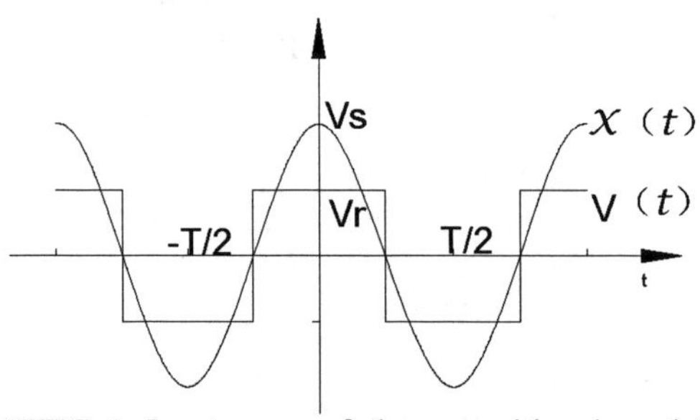

FIGURE 4. Input waves of phase-sensitive demodulator

The reference signal is a square wave, whose amplitude is $\pm V_r$ and period is T. Thus the periodic function can be conversed to Fourier series:

$$r(t) = a_0 + \sum_{m=1}^{\infty} a_m \cos m\omega\, t + \sum_{m=1}^{\infty} b_m \sin m\omega\, t \tag{4}$$

Where

$$a_0 = \frac{1}{T}\int_{-T/2}^{T/2} r(t)dt \tag{5}$$

$$a_m = \frac{2}{T}\int_{-T/2}^{T/2} r(t)\cos(m\omega t)dt \tag{6}$$

$$b_m = \frac{2}{T}\int_{-T/2}^{T/2} r(t)\sin(m\omega t)dt \quad (7)$$

From equation (4) to equation (7), the Fourier series of *r(t)* can be expressed as

$$r(t) = \frac{4V_r}{\pi}\sum_{n=1}^{\infty}\frac{(-1)^{n+1}}{2n-1}\cos[(2n-1)\omega t] \quad (8)$$

Thus the output of phase-sensitive demodulator becomes

$$V_d(t) = x(t)\bullet r(t) = \frac{2V_sV_r}{\pi}\sum_{n=1}^{\infty}\frac{(-1)^{n+1}}{2n-1}\cos[(2n-2)\omega t-\theta] + \frac{2V_sV_r}{\pi}\sum_{n=1}^{\infty}\frac{(-1)^{n+1}}{2n-1}\cos[2n\omega t+\theta] \quad (9)$$

The first part of Equation (9) is the frequency dispersion item and the second part is the frequency addition item. After low-pass filtering, only the first item of frequency dispersion item of $n=1$ is remained:

$$V_d(t) = \frac{2V_sV_r}{\pi}\cos\theta \quad (10)$$

The electronic switch mode demodulator does not change the amplitude of sine wave signal, so $V_r=1$, thus the final demodulator result can be expressed as:

$$V_d(t) = \frac{2V_s}{\pi}\cos\theta \quad (11)$$

Realization of electronic switch mode demodulator: First, the sine wave output of band-pass filter is amplified with gain of +1 and -1 respectively, the two sine waves are switched alternately by a SPDT switch and the SPDT switch is controlled by the reference square wave generated by the signal generator. Thus a full-wave rectify signal is obtained. The demodulated signal consists of DC signal and high frequency signal components.

A 4th-order Butterworth low-pass filter with a cut-off frequency of 20 kHz is used to reject the AC signal, and the resultant DC voltage after low-pass filtering represents the unknown capacitance value.

The signal generator produces sine wave and reference square wave synchronously. The sine wave enters the demodulator after a series of conversion, thus there exists time dispersion between *x(t)* and *r(t)*. The time delay is small but the measuring circuit works at high frequency, so the phase dispersion angle θ produced by time delay can not be ignored. As can be known from Equation (11), the larger θ becomes, the weaker the measured signal will be. Thus it is necessary to delay the square wave signal so that the two waves can synchronize again in the demodulator. As a result from the experiment, the time delay is between 50 ns and 350 ns. So a time delay circuit composed of a programmable 8-bit silicon delay line is adopted, and the delay value, which can be adjusted from 10 ns to 522ns in step of 2ns, is controlled by 8-bit digital signal. Therefore, the phase angle θ can be adjusted from 0°to 360°in step of 0.36°.

Data Acquisition System

Operation Theory

Normally each electrode except the first one follows by a capacitance measuring circuit in AC-based capacitance measuring system. However, for the complexity of AC-based capacitance measuring circuit, the whole circuit of ECT system will be much complex. On the other hand, the AC-based measuring circuit is expensive, so the whole cost of the ECT hardware system will be very high in multiple capacitance measuring circuit. Therefore, single capacitance measuring circuit is adopted in this system, and the whole hardware part of the ECT system is shown in Fig. 5. One electrode is chose as excitation electrode by 8-to-1 multiplexers, another one as detection electrode, and the others are connected to earth by CMOS switches. The output of the capacitance measuring circuit (Fig.3) is connected to non-invert input of differential amplifier and subtract from the D/A feedback voltage so that the standing capacitance can be eliminated. The DC amplify gain is determined by this difference value. The amplified DC voltage is converted to digital value by analog-digital converter, and then transferred to MCU.

Each electrode of the ECT sensor has three operation modes: excitation, detection and earthing. In the process of data acquisition, electrode 1 is set as excitation mode via CMOS switches, electrode 2 is set as detection mode and the others as earthing; and then the electrodes from 3 to 8 are set as detection respectively. By the same token, the electrode 2 is set as excitation, the following electrodes are set as detection respectively. When electrode 7 is set as excitation and 8 as detection, all the capacitances of a frame are measured. To n-electrode sensor, the number of obtained capacitance is $N=n(n-1)/2$. In 8-electrode sensor, the value of N is 28.

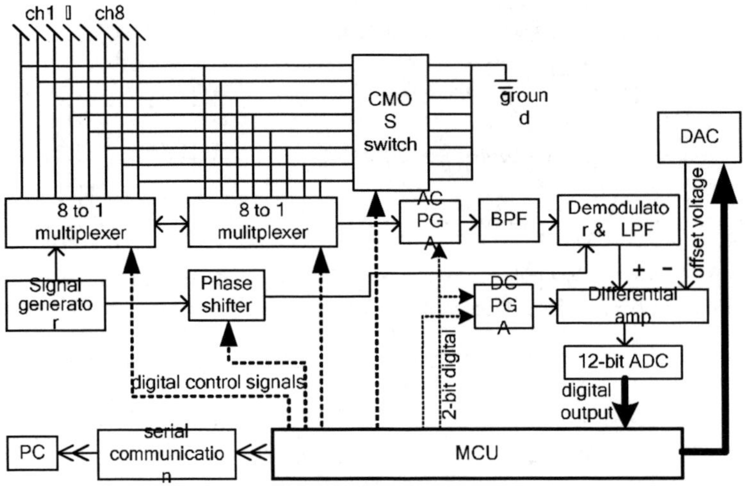

FIGURE 5. Configuration of ECT system.

The electrode will be at the mode of excitation, detection and earthing if the switch 1, 2, 3 in Fig.6(a) is closed by themselves respectively. When switch 2 is closed and switch1, 3 is open, the electrode is at detection mode, then there exists potential between the two sides of switch. Thus a coupling capacitance of switch 1, which is in parallel with the measured interelectrode capacitance, is added to the measured capacitance. The coupling capacitance is very small, about 0.5 pF [9], but it can not be ignored when compared with the measured capacitance, also the capacitance value is not constant. Therefore, the interference of coupling capacitance must be eliminated. In practice, a T-style switch arrangement is necessary as shown in Fig.6(b). In the detection mode, switch 1 and 2 are closed while switch 3, 4 and 5 are open, connecting the coupling capacitance of switch 4 to earth and thus eliminating its effect on the measuring capacitance.

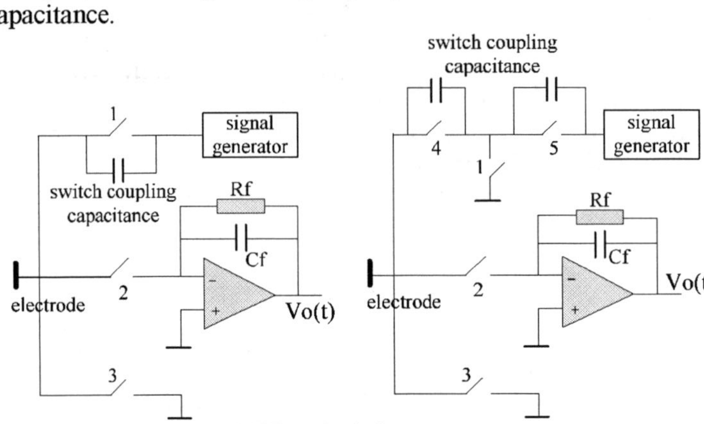

FIGURE 6. Switch arrangement

The ECT system is under the control of MCU, its main tasks include:

(1) Electrodes arrangement control

Correct electrode pair is chose via 8-to-1 multiplexers and CMOS switches so that all the electrodes work at the corresponding modes: excitation, detection and earthing.

(2) Gain control

2-bit digital signal is used to control the AC amplify gain and the DC differential amplify gain, one of the digital signal controls both the AC gain and the DC gain, so that the two amplify stages can be associated with each other.

(3)Time delay control

The 8-bit silicon delay line can adjust the delay time by 256 steps. MCU gives corresponding digital signal according to the different delay time of different electrode capacitances.

(4) Analog-Digital Converter data acquisition control

The analog-digital converter converts the output voltage of DC-PGA, and transfers the digital signal to MCU via interface circuit.

(5) Offer digital-analog Converter compensation

In the process of calibration, the standing capacitances of each electrode pairs are saved in MCU. And corresponding signals are fed back in data acquisition procedure so that the standing capacitance can be cancelled, thus the changes of capacitances are amplified effectively.

(6) Serial communication

The obtained data is transferred to computer with the help of serial communication circuit. The whole system operational process is shown in Fig. 7. The information of channel selecting, gain and the delay time of each channel are saved in the ROM of MCU. At the beginning of operation, the empty pipeline capacitance is measured then saved in the RAM of MCU as the feedback of D/A converter. Several hundred microseconds is necessary to make the measuring circuit stabilified after selecting a pair of electrodes. However, the serial communication is also a bottleneck of system. For the purpose of improving the speed of system, the serial communication operates after channel select, thus the two time-consuming processes is running synchronously so that the acquisition speed is improved effectively.

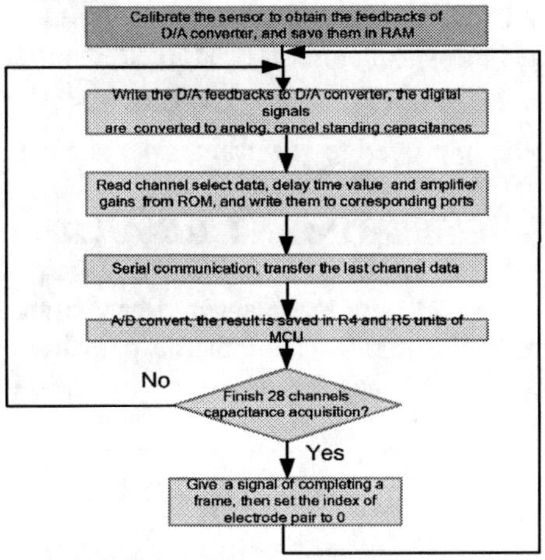

FIGURE 7. Flow chart of hardware system operation

Data Acquisition

The data acquisition system converts the analog signals of capacitance measuring circuit to digital signals, and then transfers them to MCU. The analog signals are converted to digital ones using a 12-bit A/D converter. Its conversion time is 22 μs, operation mode is bipolar and input voltage is at the range of -10 V ~ 10 V. And 12-bit D/A converter is adopted, whose operation mode is unipolar, output voltage range is from 0 to 4.096V and conversion time is less than 15μs.

As can be seen from Fig .2, the standing capacitances and the capacitance changes of the sensor are different from each other. For the purpose of adjusting the capacitance signal to an appropriate value, both AC amplify and DC amplify are adopted. The MCU provides two bits digital signal (A and B) to control the choice of gains (Fig .5). Only one bit (B) is used in AC PGA to provide optional gains of 2 and 20. And the DC PGA has two control signals to provide gains of 5, 10, 20 and 50. The control bit B is a common signal in AC PGA and DC PGA, thus the two amplifies can be associated with each other; also the number of control signals is decreased. The relationship between gain and sensitivity of the measuring circuit is shown in Table 1. There are four optional general gains which are determined by the capacitance range. The capacitance range in the table is only correlated to the operation voltage range of demodulator, and the DC amplify is not associated with the standing capacitance because the standing capacitance is cancelled before DC amplify.

TABLE 1. Relationship between gains and sensitivity

Control signal		AC gain	DC gain	General gain	Sensitivity(V/pF)	Measuring range (pF)
B	A					
0	0	2	20	40	4.584	0~33
0	1	2	50	100	11.459	0~33
1	0	20	5	100	11.459	0~3.3
1	1	20	10	200	22.918	0~3.3

System Dynamic Characteristic

As discussed above, this measuring system is different from normal one. There is only one capacitance measuring circuit in the whole system, thus the circuit is predigested and the cost is decreased, but the speed of measuring system is decreased. To obtain a frame of capacitance data, the channel arrangement should switch for 28 times, and a long time is necessary to make the output analog signal stabilized, so the system speed is limited.

The total time of a frame is calculated by

$$T_0 = 28(\mathrm{Max}\ (T_1,\ T_2) + T_{adc}) \tag{12}$$

Where T_1 is the switch time of channel and corresponding stable time, about 300 μs, T_2 is the serial communication time of transmitting 2 bytes. When the baud rate is 115.2 kB/s, T_2 will be 140 μs. T_{adc} is the data acquisition time, it is less than 25μs. The action of T_1 and T_2 happen synchronously, so the larger is adopted. Therefore, the total time of a frame is 9.1 ms, and the data acquisition speed is more than 100 frames per second.

If the system of multiple capacitances measuring circuit is adopted, the time of a frame will be calculated:

$$T_0 = 7T_1 + 28(T_2 + T_{adc}) = 6.02\ \mathrm{ms}$$

Thus the data acquisition speed can be improved to 160 frames per second.

EXPERIMENT RESULT

The parasitic capacitance, which is parallel with the measured capacitance, is caused by the PCB layout and it can not be eliminated by the stray-immune circuit. Though the parasitic capacitances are different in different capacitances, they are nearly constant. So they can be measured by the measuring circuit, and cancelled in the procedure of data processing.

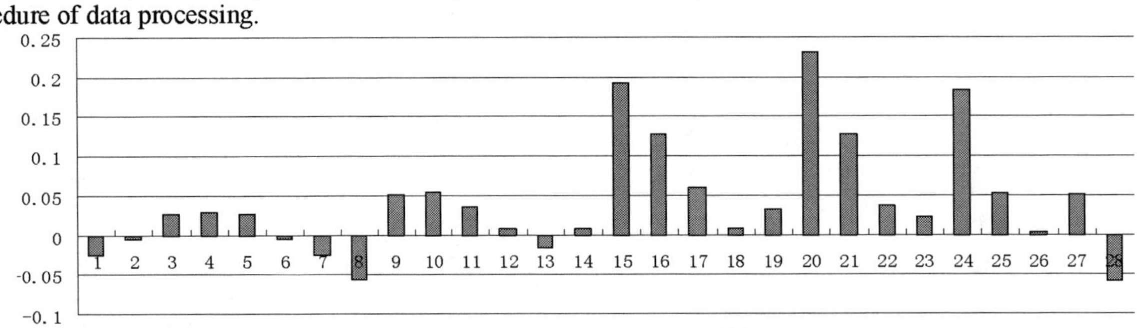

FIGURE 8-1. Capacitance changes of stratified flow

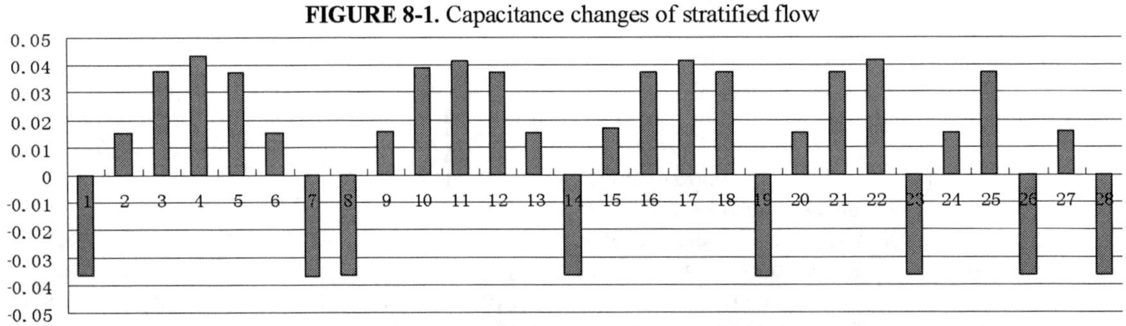

FIGURE 8-2. Capacitance changes of core flow

To evaluate the ECT system, experiments were carried out. The stratified flow and core flow (Shown in Fig. 9-1) are measured respectively, and the measured capacitance changes are displayed in Fig .8. There are 28 capacitances in one frame, and the corresponding capacitance from left to right in the Figure are the measuring result between electrodes1-2, 1-3, … ,1-8, 2-3, 2-4, … ,2-8, … up to 7-8. Some electrode capacitances especially the adjacent ones are particularly affected by the pipeline wall, thus their capacitance changes are negative.

A stochastic inverse technique based on genetic algorithms (GA-ECT) has been developed for image reconstruction in this ECT system. This reconstruction algorithm isn't based on sensitivity field model, which is the main error source of ECT image reconstruction, so it's suitable for solving non-linear problem in ECT image reconstruction. The reconstructed images of stratified flow and core flow are shown in Fig .9-2.

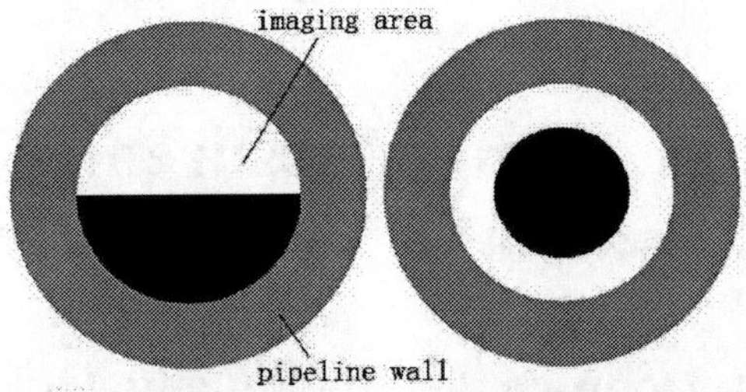

FIGURE 9-2. Original images of stratified flow and core flow.

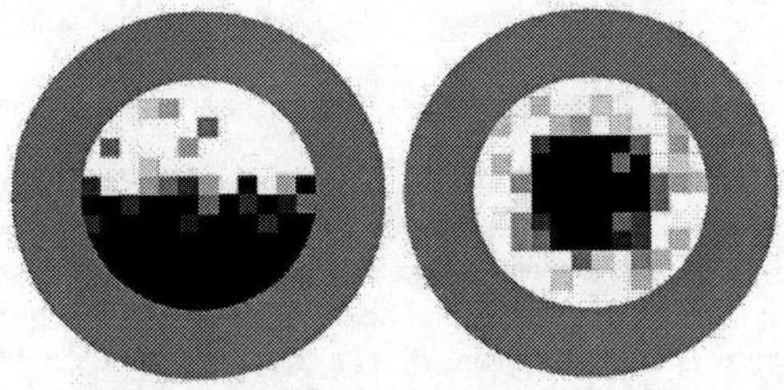

FIGURE 9-2. Reconstructed images of stratified flow and core flow

CONCLUSIONS

As mentioned above, the ECT sensor has special characteristics, so the whole ECT system has its features:

(1)There are only 8 electrodes in the sensor, so the capacitance change value increase, thus the capacitance measurement is easier;

(2)The capacitance changes of adjacent electrodes are negative, so that the image reconstruction becomes more difficult;

(3)A band-pass filter is adopted before the phase-sensitive demodulator, which can greatly improve system SNR and capability of anti-jamming.

(4)Single capacitance measuring circuit is used to predigest the circuit, which can reduce the cost. At the same time it also decreases the measuring speed, but the request of high-speed data acquisition is still satisfied.

From the above result of flow experiment, a conclusion can be drawn that this ECT system, which is used in small-diameter and high-pressure pneumatic conveyor, can obtain the real-time flow information with high speed, and reconstruct the inner pipeline image.

ACKNOWLEDGMENTS

This research was supported by Special Funds for Major State Basic Research Projects (2004CB17702-04). The authors wish to express their gratitude.

REFERENCES

1. LI Haiqing and HUANG Zhiyao, *Special measurement technology and application*, Hangzhou: Zhejiang University Press, 2000,pp.59-73
2. Jaworski A J,Bolton G T, *Measurement of science and technology*,**11**, 743-757(2000)
3. W. Q. Yang, T. A. York, *IEE Proc.-Sci. Meas. Technol*, **146**, 47-53(1999)
4. Wang Baoliang, "Electrical capacitance tomography and its application in two phase parameter measurement", Ph.D. Thesis, Zhejiang University, 1998
5. W. Q. Yang, Lihui Peng, *MEASUREMENT SCIENCE AND TECHNOLOGY*, **14**, R1-R13(2003)
6. S. M. Huang, C. G.. Xie, R.Thorn, "Design of sensor electronics for electrical capacitance tomography," in *IEE PROCEEDINGS-G*, Vol, 139, No.1, 1992, pp.83-88
7. Chen Lingli, Feng Zhihua, *Measurement technique and system*. Beijing: Tsinghua University Press, 2005, pp:153-156
8. Gao Jinzhan, *Detection of Weak Signals*. Beijing: Tsinghua University Press. 2004, pp.154-161
9. Data conversion product databook (Analog Devices, Inc. 2005)

Gas-Liquid Two-Phase Flow Measurement with Dual-Plane ERT System in Vertical Pipes

Tan Chao, Dong Feng, Hua Shuang, Qi Guohua

School of Electrical Engineering and Automation, Tianjin University, Tianjin, 300072, China
Email: fdong@tju.edu.cn

Abstract. The real-time parameters of gas-liquid two-phase flow in vertical pipe are measured by dual-plane electrical resistance tomography (ERT) system. With the discrete wavelet transform the flow regime feature is extracted from the measured data, and the Support Vector Machines (SVM) is applied to the flow regime recognition. Based on the flow regime recognition results, the flowrate is measured by cross-correlation method with the flow model of each flow regime, and cross-correlation velocity and kinematic wave propagation speed are compared and analyzed.

Keywords: two-phase flow, electrical resistance tomography, wavelet, support vector machine, flow regime recognition, flowrate, cross-correlation, kinematic theory

INTRUDUCTION

The accurate measurement of two-phase flow parameters has always been a key issue and tough work for decades. Due to the complex behavior of multiphase flow, it presents a great challenge to the study of the flow mechanisms and the measurement of multiphase flow [1-4].

By reason of the complex relative interaction between phases, precise measurement on flow parameters is difficult to achieve. To solve the issues, a great deal of flow models have been brought out, such as the drift-flux model, presented by Zuber and Findlay, is widely used in one-dimensional two-phase flow with taking the volumetric void fraction and relative velocity between phases into account [5]. This kind of models is based on the flow regime, because the flow status varies with it. So the accurate measurement of two-phase flow depends on the precise recognition of flow regime. General way of identifying flow regime is to extract feature value from the measured data of two-phase flow, which contains flow regime information.

Electrical resistance tomography (ERT) is a kind of process tomography that can be widely used in the measurement of multi-phase flow parameters [6]. In this paper, a method with wavelet transform and support vector machine (SVM) is brought forward for recognizing the flow regime of two-phase flow. The cross-correlation velocity is derived by dual-plane ERT system, and basing on the recognition results, the drift-flux models are adopted to measure the flowrate. The cross-correlation velocity and kinematic wave propagation speed are compared and analyzed.

EXPERIMENT DESCRIPTION

Dual-Plane ERT System

ERT, which is based on the principle that the conductivity of different medium varies, is one of the process tomography. By using ERT technique, the medium distribution of measured field can be identified if the conductivity distribution of sensing field is acquired.

The operation mode of ERT system is current-exciting and voltage-measured. The system in this experiment works in adjacent exciting mode. The procedure of this strategy is: the exciting current firstly injects into two electrodes and the voltages measured from successive pairs of neighboring electrodes, then the current is applied through the next pair electrodes and the voltage measurements repeated until all the independent measurement have

CP914, *Multiphase Flow: The Ultimate Measurement Challenge,* edited by X. Cai, Y. Wu, Z. Huang, S. Wang, and M. Wang

been made.

In this experiment, the dual-plane ERT system measures 208 voltage data from one cross-section and the two planes are excited alternately. The data acquirement rate of the system is 15 frames/sec·plane.

Experiment Facility

The flow regime can be observed through the transparent organic-glass-made pipe (diameter is 0.125m), and can be varied by controlling liquid and gas valve. Two planes of 16 alloy electrodes are installed uniformly around the pipe as sensors of dual-plane ERT system. The experiments herein are presented in gas-liquid two-phase slug flow at two kinds of separation of sensor planes, 0.125m and 0.25m. The structure is shown in Fig. 1 and Fig. 2.

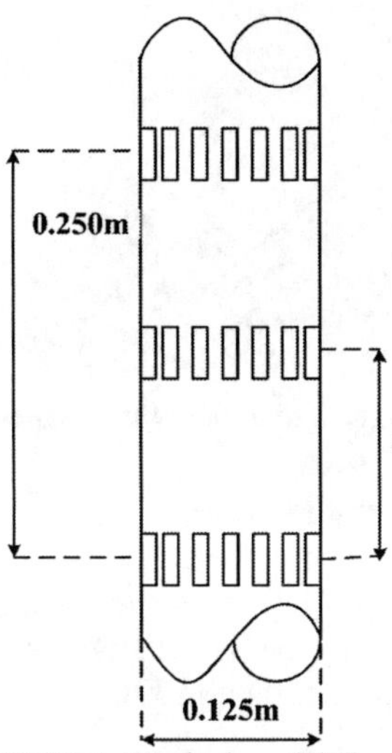

FIGURE 1. Dual-plane ERT sensors in the pipe

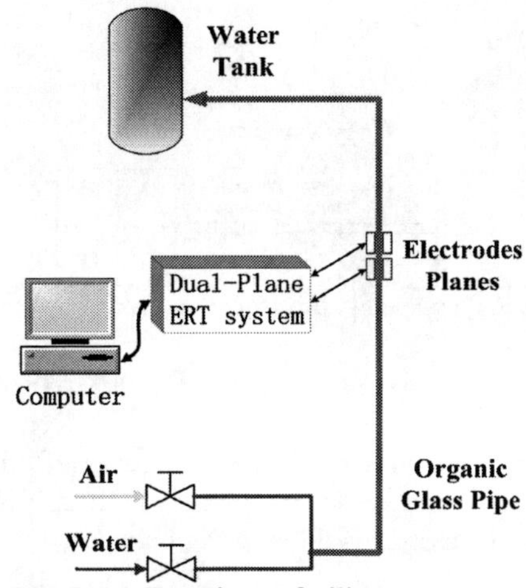

FIGURE 2. Experiment facility structure

FLOW REGIME RECOGNITION

Feature Extraction with Two-Dimension Discrete Wavelet Transform

Wavelet is a function or a signal in space $L^2(R)$, for any function or signal $f(x)$, wavelet can be defined as:

$$W_f(a,b) = \int_R f(x)\overline{\psi_{(a,b)}(x)}\,dx = \frac{1}{\sqrt{|a|}}\int_R f(x)\overline{\psi\left(\frac{x-b}{a}\right)}\,dx \tag{1}$$

where $\overline{\psi(x)}$ represents the complex conjugate of function $\psi(x)$ which is called wavelet generating function. In this research, the Harr wavelet theory in discrete wavelet theory is applied to carry out the wavelet transform on discrete time series, and obtained the corresponding wavelet parameters. Then the flow regime is identified by extracting the regime information from these parameters.

Three typical regimes: bubbly flow, slug flow and churn flow in vertical pipes have been measured continuously. Because of the particular working procedure of ERT system, the quantity and dimension of the measurement are too high for the data feature to be distinctive. So the feature must be extracted before it can be directly used:

$$y_i = \frac{1}{13}\sum_{j=1}^{13}\left(V_{ij} - V_{ij0}\right) \tag{2}$$

where V_{ij} is the measured value of jth boundary voltage in ith excitation, and V_{ij0} is the measured value of jth boundary voltage in ith excitation when pipe is full of water. After the data processing, the characteristic vector y_i is formed; it contains the flow regime information of the data from each frame. 50 frames data is used to illustrate the entire state of the flow regime, so the flow status can be expressed by a matrix:

$$
\begin{bmatrix} y_{1,1} & y_{1,2} & \cdots & y_{1,16} \\ y_{2,1} & y_{2,2} & \cdots & y_{2,16} \\ \vdots & \vdots & & \vdots \\ y_{50,1} & y_{50,2} & \cdots & y_{50,16} \end{bmatrix} \tag{3}
$$

Two-dimension wavelet transform is used for this matrix. Its application in the flow regime identification is only a trial, so only one-level two-dimension wavelet transform is carried out, as shown in Fig. 3. After the wavelet transform on the data of each flow regime, four power values E_A, E_H, E_V and E_D in each flow regime make up of a characteristic vector [7].

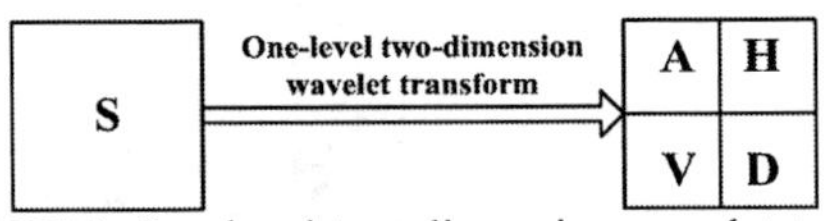

FIGURE 3. One-level two-dimension wavelet transform

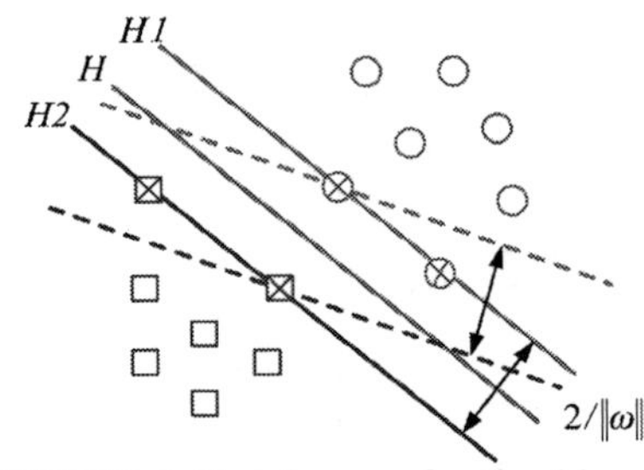

FIGURE 4. Sketch map of optimal hyperplane

Flow Regime Recognition with Support Vector Machine

Support vector machine (SVM) is brought out from the optimal hyperplane on condition of linear separable. The optimal hyperplane is a separating plane which in high dimensional space not only exactly separates two classes of samples, but also maximizes the interval between the two classes. The sketch map is shown in Fig. 4.

Recognition Results

Three flow regimes is studied and 30 groups experiments are made for each regime, therefore Directed Acyclic Graph SVM (DAGSVM) arithmetic [8] presented by Platt etc can be applied herein. All the data in experiments are trained by Radial Basis Function (RBF) and Sequential Minimal Optimization (SMO) arithmetic [9]. The cross validation [10] is adopted: divide the samples into three subclasses; two of them constitute the training samples, and the rest one for testing. Two examples of each flow regime of the first experimental results are shown in table 1:

TABLE 1. Flow regime recognition results in the first experiment

Sequence number	Apriority flow regime	Characteristic vector (E_A E_H E_V E_D)	Recognition result
1	Bubbly flow	(3.6895 0.2492 0.1058 0.1109)	Slug flow
2	Bubbly flow	(2.5506 0.1447 0.0968 0.0908)	Bubbly flow
3	Slug flow	(4.4251 0.5374 0.1552 0.1844)	Slug flow
4	Slug flow	(3.5773 0.5439 0.1711 0.1756)	Slug flow
5	Churn flow	(355.6366 44.4783 4.7922 4.8619)	Churn flow
6	Churn flow	(343.5400 36.3794 5.9158 5.5291)	Churn flow

There are only 5 groups of bubbly flow are wrong judged in the three experiments, the average identification rate is 94.4%, from which conclude that by extracting the flow regime feature with two-dimension Haar wavelet transform, the internal information of flow regime can be furthest presented and the recognition results are relatively better.

FLOWRATE MEASURED BY DUAL-PLANE ERT SYSTEM

Cross-Correlation Technique in Dual-Plane ERT System

The cross-correlation technique used in dual-plane ERT system is characteristic value correlation which correlates the characteristic value extracted from the measured data of each cross section. Defined V_R, the mean value of the relative change of measured boundary voltage, as the characteristic value of cross-correlation:

$$V_R = \frac{1}{N}\sum_{i,j}\frac{V_{ij} - V_{ij0}}{V_{ij0}} \tag{4}$$

With the two series of V_R values from upstream and downstream measured data, characteristic velocity can be calculated by normalized cross-correlation method.

One-Dimensional Drift-Flux Model

Based on Zuber's study [5], the continuity equation of disperse phase in gas-liquid two-phase flows is

$$\frac{\partial(\alpha\rho_g)}{\partial t} + div\left[\rho_g \alpha v_g\right] = 0 \tag{5}$$

It is considered that (1) the superficial velocity of each phase, average over an appropriate time interval, is constant at all axial positions; (2) the homogeneous velocity has the same value at all axial positions in the flow, at all times; (3) the cross-correlation average void fraction, average over an appropriate time interval, is constant for all axial positions [11].

$$\frac{\partial \alpha}{\partial t} + \frac{dj_g}{d\alpha}\frac{\partial \alpha}{\partial Z} = 0 \tag{6}$$

Zuber presented $\frac{dj_g}{d\alpha}$ as kinematic wave propagation velocity, a new equation is obtained:

$$u_{cc} = u_{kw} = \frac{dj_g}{d\alpha} = C_0\langle j\rangle + \overline{v_{gj}} \tag{7}$$

As the analysis above, the flow model of cross-correlation velocity and kinematic wave propagation speed is determined by flow regime, different flow regime has different flow model. Therefore, basing on the flow regime recognized in chapter 3, the cross-correlation velocity will be derived by choosing the model of each regime.

Cross-Correlation Results

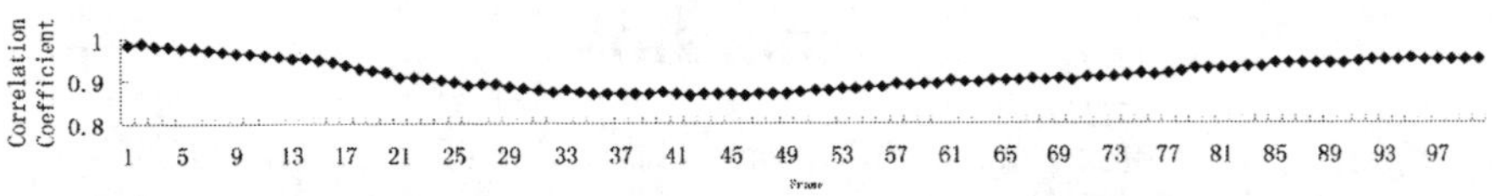

(a) Bubbly flow

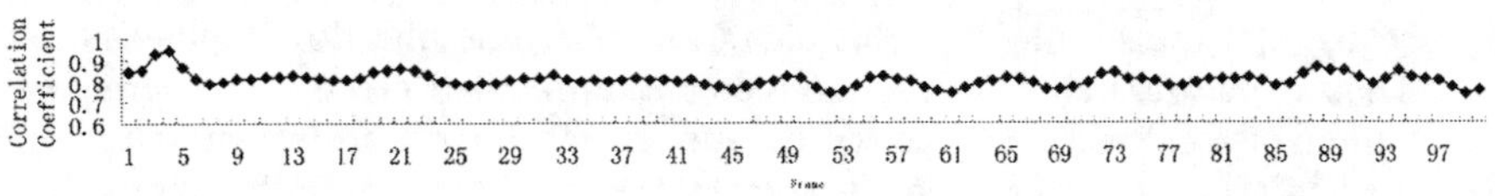

(b) Slug flow

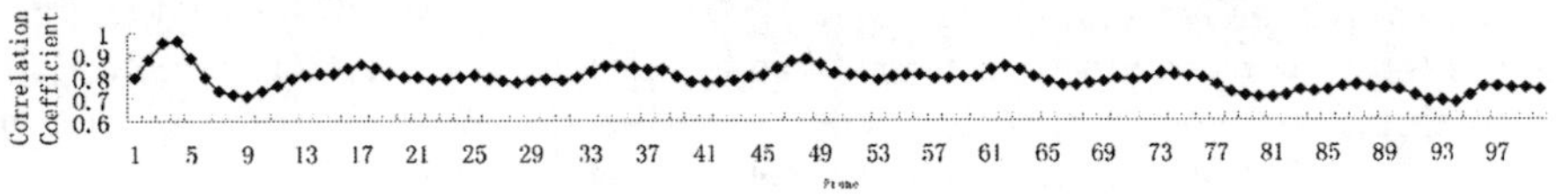

(c) Churn flow

FIGURE 5. Correlation coefficient of typical flow regimes

Correlating the V_R value of the two sensor planes to derive the transit time of the flow, and then calculate the cross-correlation velocity in each flow regime. Figure 5 shows the correlation coefficient of typical flow regimes, from which clearly seen that the correlation coefficient varies with the flow regimes' fluctuation. Tiny fluctuation in bubbly flow does not vary the correlation coefficient as great as churn flow. In slug flow, the correlation coefficient series has a distinct peak. From the position of peak value of the correlation coefficient series, the transit time can be derived.

Comparison of the Velocities Based on Kinematic Wave Theory

It is believed that the cross-correlation velocity can be used to measure the propagation of kinematic wave, and each flow regime has its flow model. Table 2 lists the comparison of cross-correlation velocity with the speed of kinematic wave propagation at different flow regime.

It can be seen from table 2 that the deviation of cross-correlation velocity is different in each regime. When the separation is 0.125m, the deviation is smaller than at 0.25m, it is because the separation is too long for the characteristic values of the two sensor planes to maintain correlation. Comparing the deviation between flow regimes, the measurement precision of slug flow is higher than the other two regimes, the reason is the influence are notable than other flow regimes when the big air bubble passing the two sensor planes. In bubbly flow, if the separation is overlength the air bubbles can not meet the frozen assumption, and result in measurement error. In churn flow, the flow status is too drastic to maintain correlation between sensor planes, so the deviation is larger than in slug flow.

TABLE 2. Comparison of cross-correlation velocity with kinematic propagation

Separation of the Planes (m)	Flow Regime	Total volumetric flux density (m/s)	Kinematic wave velocity (m/s)	Cross-correlation velocity (m/s)	Deviation of cross-correlation velocity
0.125	Bubbly Flow	0.544972	0.888377	>0.936330	>5.12%
0.125	Slug Flow	0.455307	0.933768	0.936330	0.27%
0.125	Slug Flow	0.063098	0.463118	0.561798	17.57%
0.125	Churn Flow	0.053257	0.451308	0.561798	19.67%
0.25	Bubbly Flow	0.340154	0.639648	1.872659	65.84%
0.25	Slug Flow	0.616282	1.126938	1.123596	-0.30%
0.25	Slug Flow	0.137807	0.552768	0.802568	31.13%
0.25	Churn Flow	0.109266	0.518519	0.802568	35.39%

The results also give a clarification that the cross-correlation velocity of two-phase flow is not the mixed velocity or even the disperse phase velocity but fits the kinematic wave theory.

CONCLUSION

The parameters of gas-liquid two-phase upward flow are measured in the experiments, including the pressure and the temperature in the pipe, the average void fraction, etc. The feature of the data acquired by dual-plane ERT system is analyzed, from which extracted the four power values as the characteristic value for each flow regime with two-dimension discrete wavelet transform. With this characteristic value, the flow regime is recognized by SVM technique. The results show that the new flow regime recognition method meets the requirement of identification rate for real-time measurement, and manifest that the two-dimension matrix or characteristic vectors can lead to a better identification rate; and also can represent the flow regime precisely for the further research.

By using the dual-plane ERT system, the cross-correlation velocity is derived with normalized cross-correlation method. Based on the flow regime recognition results, the flow model is selected for each flow regime. And in view of the kinematic wave theory, cross-correlation velocity and kinematic wave velocity are nearly equal in different flow model, it means that the cross-correlation velocity is not just the disperse phase velocity or even the mixed velocity, but has close relation to the kinematic wave velocity.

ACKNOWLEDGEMENTS

The author appreciates the support by National Natural Science Foundation of China (No. 60532020) and Natural Science Foundation of Tianjin (No.05YFJMJC11600).

REFERENCES

1. C. L. Gu and S. P. Dong, "The Wavelet Noise-Immune Disposal for the Signal of Gas/Liquid Two-Phase Flow Regime", *J. Univ. Sci. Technol. B.* **24,** 76-78 (2002).
2. Z. X. Jiang, F. Dong, L. J. Xu and L. A. Xu, "Gas/Liquid Two-Phase Flow Regime Identification Based on Neural Network in Horizontal Pipes", *Chinese. J. Sci. Instrum.* **23**, 511-512 (2002).
3. F. Dong, X. T. Qiao, Z. X. Jiang and L. A. Xu, "Application of Electrical Resistance Tomography to Two-Phase Pipe Flow Parameters Measurement" in *World Congress on Industrial Process Tomography-2003*, VCIPT Proceedings, Banff, Canada, 2003, pp.299-304.
4. D. A. Goodwin, R. G. Aykroyd and S. Barber, "Modelling and Predicting Flow Regimes Using Wavelet Representations of ERT Data" in *World Congress on Industrial Process Tomography-2005*, edited by A. de Jong etc, VCIPT Proceedings, Aizu, Japan, 2005, pp.904-909.
5. N. Zuber and J. A. Findlay, "Average Volumetric Concentration in Two-Phase Flow Systems", *J. Heat. Trans.* **87**, 453-468 (1965).
6. F. Dong, X. P. Liu, X. Deng, L. J. Xu and L.A. Xu, "Identification of Two-phase Flow Regimes in Horizontal, Inclined and Vertical pipes",*Meas. Sci. Technol.* **12**, 1069-1075 (2001).
7. L. Q. Zhang and B. H. Wang, "Feature Extraction Methods for Pulse Signal Based on Wavelet Transform", *J. Data Acqui. Process.* **19**, pp.323-328 (2004).
8. J. Platt., N. Cristianini and N. Taylor., "Large Margin DAG's for Multiclass Classification," in *Advances in Neural Information Processing Systems*, Cambridge: MIT Press, 2000, pp. 547-553.
9. G. W. Flake, S. Lawrence, "Efficient SVM Regression Training with SMO",*Mach. Learn.* **41**, 271-290 (2002).
10. B. G. Wei, A. M. Wang and L. S. Shen, "A New Multi-classification Pattern Recognition Support Vector Machine", *Pattern Recogn. Artif. Intel.* **15**, 178-181 (2002).
11. G. P. Lucas and I. C. Walton, "Flow Rate Measurement by Kinematic Wave Detection in Vertically Upward, Bubbly Two-phase Flows", *Flow. Meas. Instrum.* **8,** 133-143 (1997).

Optimization Design of Sensor in Electromagnetic Tomography System

Dong Feng*, Xue Yixuan*, Liu Ze+

** School of Electrical Engineering and Automation, Tianjin University, Tianjin, 300072, China,*
Email: fdong@tju.edu.cn
+ Department of Automation, Beijing Jiaotong University, Beijing, 100044 China,
zliu@center.njtu.edu.cn

Abstract. The sensor design of the Electromagnetic tomography (EMT) system is always the most important part of the whole system structure, and it also is an important research direction in the EMT system study. This paper based on the fundamental principle and the positive question model of the EMT system, analyzed the factors which affect the sensor performance, and proposed the improving design scheme of the system sensor. The novel design will simplify the structure of system sensor, and therefore improve the stability of the system.

Keywords: Electromagnetic tomography, sensor design, inductive coil, excitation

INTRUDUCTION

Electromagnetic tomography (EMT) is a non-intrusive and non-hazardous visualization technique, which uses rugged and noninvasive sensors. As a relatively new tomography technique, EMT employs magnetic coupling between sensors to provide images that represent the distributions of electrically conductive and magnetically permeable material within object space. Potential applications for EMT techniques are where the target materials of interest are characterized by contrasts in their electrical conductivity or magnetic permeability and in particular in harsh industrial production environments [1].

The purpose of EMT technique is to do research for the spatial distributions of electrically conductive and magnetically permeable material within the object space. The EMT system adds an alternating excitation field to the measured field which has electrically conductive material or magnetically permeable material or both of them. Basing on the electromagnetic induction principle, the magnetically permeable material will form an induction magnetic field, and the electrically conductive material will form the electric eddy flow. They both will modulate the initial excitation field, their superposition will generate induction voltages with different amplitude and phase on the detection coils which located in the boundary of the field. Therefore, the distribution of the magnetic permeability μ and the electric conductivity σ in the detected field can be constructed.

Development of an EMT system requires a custom data acquisition system and physical inductive sensors. Much effort has been made in developing a wide variety of EMT systems for industrial process applications over the past decade. For example, Korjenevsky, a Russia scholar, developed a similar model instrument of EMT experiment system, which planed to apply in the imaging of biologic tissue. The experiment system adopting the method of single coil excitation separately, only reconstructs the image of electrically conductive material in the induction field. And its image reconstruction algorithm adopts the filtering linear back projection algorithm [2]. In 1994, Yu Zhizhong in UMIST University proposed an EMT system which adopted parallel exciting mode [3] [4]. The main advantage of parallel exciting mode is that it can get higher sensitivity in the centre of the induction field. In 1996, Peyton reported a multiple pole system with 16 coils operating at 100 kHz and showed it was possible to distinguish target materials with high conductivity but low permeability from those non-conductive materials with high permeability.

The design of sensor in the system is an important part of whole system structure design. The design of sensor array must fully consider the suitable selection of the number of coils, the materials used to construct the sensor and the operating frequency. Normally more coils can increase spatial resolution of the system due to the increased

CP914, *Multiphase Flow: The Ultimate Measurement Challenge,* edited by X. Cai, Y. Wu, Z. Huang, S. Wang, and M. Wang

number of measurements. However, the increase of the coil number will increase the complexity and uncertainty of the system. So it's necessary to find the optimum point of all the conditions. And this paper will mainly concern on this problem.

THE PRINCIPLE OF EMT SYSTEM

The positive question model of EMT can be founded from the Maxwell functions. In fact, the sensing field of EMT system is a three-dimensional (3-D) uninformed field. We must gain the answer of the positive question under the 3-D condition for simulating the conductive and permeable distribution of sensing field exactly. When we want to gain the image reconstruction in iteration method, the 2-D field should approximately replace the 3-D field for increasing the imaging speed.

We can deduce the mathematic description of EMT from the Maxwell functions. We suppose the conditions as following:

Firstly, the magnetic sensing field of EMT will follow the discipline of the quasi-stable field. In order words, the time lagging effect of the electromagnetic field which transmits in the space will be ignored. In the typical EMT system, when the excitation frequency of the electromagnetic field is less than 1MHz, the wavelength of the excitation signal is longer than 300m which is far longer than the pipe diameter. So, the assumption of the quasi-stable field is reasonable.

Secondly, in the space where we investigate, the EMT electromagnetic sensor field accord with the distribution of the 2-D field. The theory analysis and simulation study tell us: when the length of the electromagnetic shield layer along the pipe's axial direction is longer than three times of the pipe diameter, we can ignore the impact of the edge effect, and consider the magnetic filed as the 2-D distribution field.

Finally, the measured objects are linear and in different directions their characteristics are the same. In other words, the measured objects would satisfy the functions: $B = \mu$, $D = \varepsilon E$, $J = \sigma E$. Base on the principle of electromagnetic induction [5], in the condition of sine excitation, all the vector quantity of the field which moves with the time simple harmonically could be represented by $e^{j\omega t}$, $j = \sqrt{-1}$. Under the conditions which supposed before, the EMT system's electromagnetic sensing field will fit the simple harmonica form of the Maxwell functions as following:

$$\begin{cases} \nabla \times \mathbf{H} = \sigma \mathbf{E} + j\omega\varepsilon \mathbf{E} \\ \nabla \times \mathbf{E} = -j\omega \mathbf{B} \\ \nabla \cdot \mathbf{B} = 0 \\ \nabla \cdot \mathbf{D} = 0 \end{cases} \tag{2-1}$$

Define the distribution function of magnetic permeable μ on the measured section as $\mu = \mu(x, y)$. And define the magnetic potential vector as $\mathbf{A}$, it satisfies the function $\nabla \times \mathbf{A} = \mathbf{B}$. Substitute this function into the equation set (2-1), and then gets the function (2-2):

$$\nabla\left(\mu^{-1}(x, y) \cdot \nabla \cdot \mathbf{A}\right) = j\omega\sigma(x, y)\mathbf{A} \tag{2-2}$$

Spread the function (2-2) in the circular cylindrical coordinates, we can get:

$$\frac{1}{\rho}\frac{\partial}{\partial \rho}\left(\rho \cdot \mu^{-1}(\rho,\phi)\frac{\partial \mathbf{A}}{\partial \rho}\right) + \frac{1}{\rho^2}\mu^{-1}(\rho,\phi)\frac{\partial^2 \mathbf{A}}{\partial \phi^2} = j\omega\sigma(\rho,\phi) \cdot \mathbf{A} \tag{2-3}$$

Suppose the excitation current distribution is area current along the circle outside the pipe, and on the circle between the excitation field space and the inner magnetic shield which radius is R, it will satisfy the interface edge condition as following:

$$n \times \left(\mathbf{H}_2 - \mathbf{H}_1\right)\Big|_{\rho=R} = \mathbf{I}(\phi) \tag{2-4}$$

where n is the outer normal direction of the interface, $\mathbf{H}_1$ is the magnetic intensity in the magnetic shield itself, $\mathbf{H}_2$ is the magnetic intensity in field edge, $\mathbf{I}(\phi)$ is the excitation current, and R is inner circular radius of the shield.

Suppose magnetic shield is combining with the non-conductive material whose magnetic permeable is infinite, the magnetic-curve is considered access perpendicularly into the magnetic shield. So the interface condition could be

$$\left.\frac{\partial \mathbf{A}(\rho,\phi)}{\partial \rho}\right|_{\rho=R} = \mu_0 \mathbf{I}(\phi) \tag{2-5}$$

Therefore, the system can be described by the functions as following:

$$\begin{cases} \dfrac{1}{\rho}\dfrac{\partial}{\partial \rho}\left(\rho \cdot \mu^{-1}(\rho,\phi)\dfrac{\partial \mathbf{A}}{\partial \rho}\right) + \dfrac{1}{\rho^2}\mu^{-1}(\rho,\phi)\dfrac{\partial^2 \mathbf{A}}{\partial \phi^2} = j\omega\sigma(\rho,\phi)\cdot \mathbf{A} \\ \left.\dfrac{\partial \mathbf{A}(\rho,\phi)}{\partial \rho}\right|_{\rho=R} = \mu_0 \mathbf{I}(\phi) \end{cases} \tag{2-6}$$

The EMT positive question could be solved by the analysis method, the numerical value method or experiment investigation method. Through the analysis method, the characteristic of measured spatial field could be described accurately by the analysis function. And then, the detected value of the detection coils which distribute on the field edge could be calculated. Furthermore, through the analysis of the solution's expression, we can see the general rule interacting between the EMT excitation field and the acting field of measured material, and lead the design of the EMT sensor. So, adopting the direct analysis method is important for the research of EMT positive question. However, when there're electrical inductive, magnetic permeable or the both materials distributing in the EMT excitation field, the solution conditions of the question will quite complexity, it's difficult to gain the analytic solution. Fortunately, the numerical value method provides the effective approach to solve the problem [6].

THE SYSTEM STRUCTURE

The EMT system is mostly composed by several parts. It contains the sensor array, data acquirement and processing unit, image reconstruction and information extraction unit. The structure of the system is showed in the Fig.1.

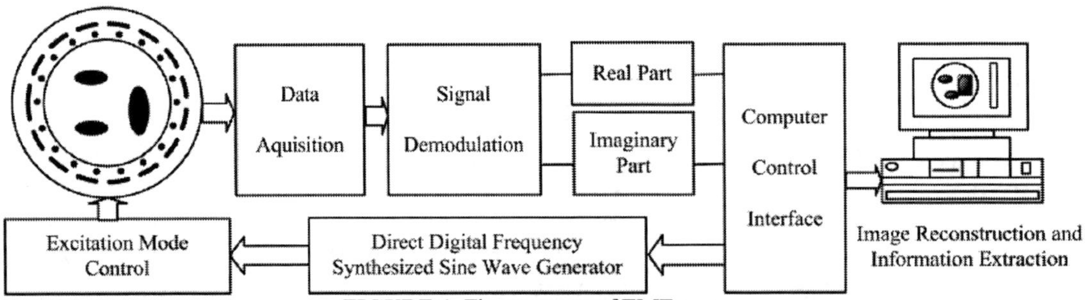

FIGURE 1. The structure of EMT system

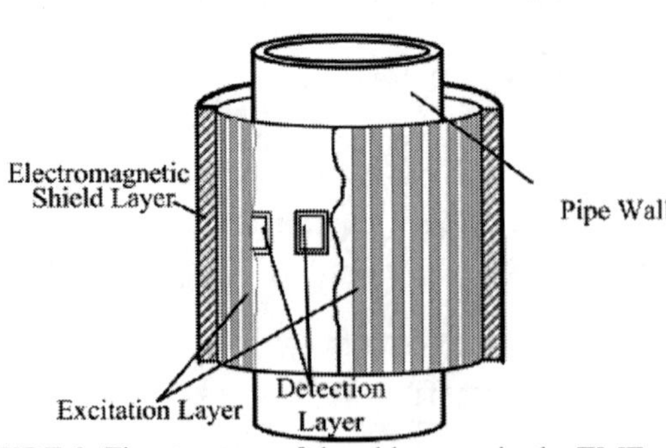

FIGURE 2. The structure of the old sensor in the EMT system

Figure 2 shows the old sensor structure of the EMT system. The diameter of detected pipe is 70mm; the sensor includes pipe wall, excitation layer, detection layer and electromagnetic shield layer from inside to outside in turn. The detection layer consists of 8 precise coils which distribute on the pipe wall surface equidistantly; they will accomplish the measurement of the boundary conditions. The excitation layer is made-up by the flexural circuit boards. And the electromagnetic shield layer is composed of ferrite and Permalloy. The excitation layer consists of 32 polar plates which are symmetrical distribution, and different exciting modes could be realized by changing the current distribution of the polar plates.

Using flexural circuit boards as exciting coils, and controlling the switch array and signal sources, the system's exciting current distribution and object fields scanning modes can be controlled smartly. So, the parallel exciting mode, the quasi-parallel exciting mode and the single coil exciting mode can be easily realized in the system respectively. But it will increase the system's complexity in a great degree, and will decrease the reliability of the system. The sensor structure maybe suit for the experimental condition in the lab, but it isn't suit for the practice application environment.

RESULTS

The design of EMT sensor includes the selection of the material and number of the inductive coils, the selection of the excitation mode and the determination of the excitation current. Actually, more coils can increase spatial resolution of the system due to the increased number of measurements. But the coil number can't be too large because the pipe circle is limited. Furthermore, the increase of the coil number will increase the complexity and uncertainty of the system. It also set a limit to the coil number in another aspect. The actual design schemes mainly contain the eight, twelve, sixteen coils design, and so on. In this design, we will adopt eight coils which distributed outside the pipe equidistantly. The material of the coils will be the copper winding. The excitation mode adopts the single coil excitation, and the other seven coils use for detection.

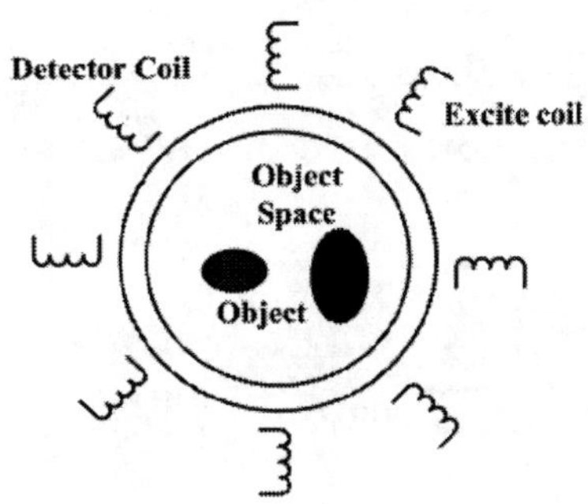

FIGURE 3. Structure of single coil excitation with eight coils design

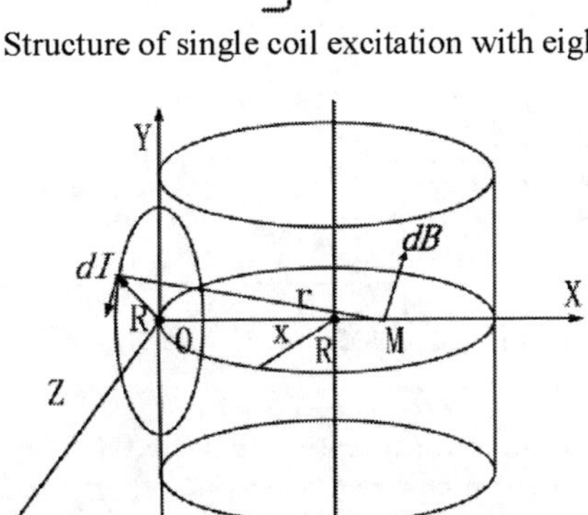

FIGURE 4. Calculation diagram of single coil excitation field

Figure 3 shows the single coil excitation EMT sensor system with eight coils design. When the i-th coil ('i' is the sequence number of the coils) excited by the current $I(i)$, the other coils are played as detectors. The induced voltage of the detection coil will be $U(i, j)(j = 1, 2, \cdots, N;\ j \neq i)$. So when the '$i$' changes from one to eight, we can get eight independent conditions of exciting magnetic field distribution. And seven measured values of boundary will be gained from each condition. Therefore, the independent data number of each measurement will be 56.

The excitation field's distribution could be calculated from the method showed in Fig.4. The excitation current in the system adopt the sinusoidal current signal $I(i) = I_m Sin(\omega t)$.

As the Fig.4 shown, R_e is the radius of the coils; x is the distance between the coil center and the point M which located on X axial line; R_p is the pipe radius; dI is the excitation in the exciting coil; dB is the magnetic induction intensity in the point M which generated by the point current dI; r is the distance between point current dI and point M.

Suppose the circle number of excitation coils is n, the current enter the excitation coil is I_e, according to Biot-Savart rule, the expression of exciting magnetic field intensity on the coil axes would satisfied the function as follows.

$$\mathbf{B} = n\int d\mathbf{B}_x = n\int_0^{2\pi R} \frac{\mu_0}{4\pi}\frac{IR}{r^3}dl = \frac{\mu_0}{2}\frac{nIR}{(R^2+x^2)^{3/2}} \tag{3-1}$$

When $R_e = 15mm$, $n = 100$, $I_e = 10mA, 20mA, 30mA$, the magnetic field intensity change with the distance between the object space and coil center from 0mm to 70mm could be expressed by the curve in Fig.5.

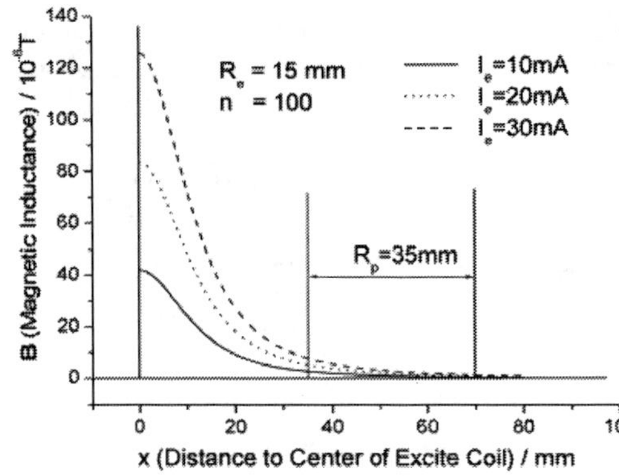

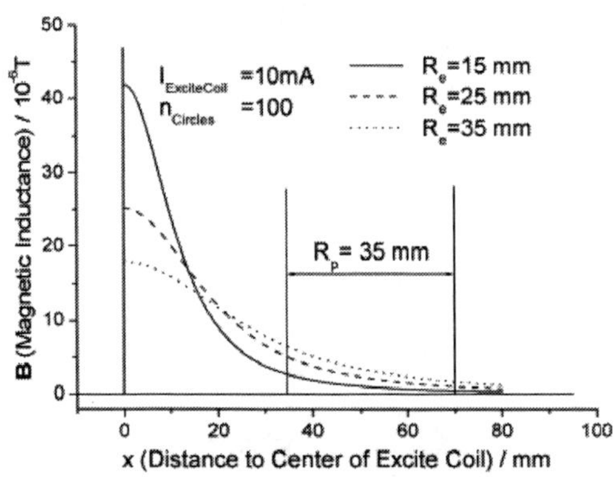

FIGURE 5. Magnetic Induction in Different Excite Current **FIGURE 6.** Magnetic Induction of in Different Coil Radius

In Fig.5, R_p is the radius of measured pipe. When $x = 2R_p = 70mm$ and the exciting current $I_e = 10mA, 20mA, 30mA$, the values of magnetic induction intensity B will be $0.38532\times10^{-6}T$, $0.77064\times10^{-6}T$, $1.15596\times10^{-6}T$ respectively. These values is about 0.92% of the value when $x = 0mm$. It indicated that the increase of excitation current could increase the excitation field's intensity, but it can't improve the uniformity of the excitation field. The increase of coil circles will obtain same effect as the increase of excitation current.

When $I_e = 10mA, n = 100, R = 15mm, 25mm, 35mm$ and the distance between coil center and measured space is no more than $70mm$, the variation curve of magnetic induction intensity was shown in Fig.6, which the values of magnetic induction intensity B will be $0.38532\times10^{-6}T$, $0.95623\times10^{-6}T$, $1.60567\times10^{-6}T$ respectively. These values are about $0.92\%, 3.80\%, 8.90\%$ of the value when $x = 0mm$ respectively. It indicated that the increase of excitation coils' radius could improve the uniformity of the excitation magnetic field, but the field still asymmetry in a certain degree. Furthermore, because the limit of pipe circle's length, the increase of excitation coils' radius will decrease the number of independent measure point.

CONCLUSION

Because the non-homogeneous of the sensing field exciting by the single coil mode, the magnetic intensity which generated by the excitation coil will attenuate with the distance increase between the space and the excitation center. Therefore, the gain compensation is necessary for the detection coils, which can balance the effect which brought by the attenuation of magnetic intensity. For example, when one coil is excited, suppose the selected gain is 1 for the coil being excited, the gain will be 128 for the coil near the excited one, 384 for the next coil and 512 for the coil opposite to the excited one; this programmable gain function will amend each measured value, and compensate the asymmetry of excitation field.

The initiative simulation results indicate that, this design scheme will reduce the complexity of the system, increase the stability of the system and at the same time guarantee the detection quality. However, the determination of the excitation current is also the important part of the senor design. The increase of the exciting current's frequency will make more detection data in the same time, but it's not good for the stable of the sensing field. The problems on the selection of excitation current require the further investigation.

ACKNOWLEDGEMENTS

The author appreciates the support by National Natural Science Foundation of China (No. 60302011) and Natural Science Foundation of Tianjin (No.05YFJMJC11600).

REFERENCES

1. X. Ma, S. R. Higson, A. Lyons and A. J. Peyton, "Development of a Fast Electromagnetic Induction Tomography System for Metal Process Applications" in *World Congress on Industrial Process Tomography-2005*, edited by A. de jong etc, VCIPT Proceedings, Aizu, Japan, 2005, pp. 196-202.
2. A Korjenevsky, V Cherepenin and Sapetsky, " Magnetic Induction Tomography: Experimental Realization" , *Physiol. Meas.* , **21,** 89-94 (2000).
3. Z.Z. Yu, "Non-intrusive Inductance based Electromagnetic Tomography (EMT) Techniques for Process Imaging", Ph.D. Thesis, UMIST, 1994.
4. Z.Z. Yu, A. J. Peyton, L. A. Xu, M. S. Beck, "Electromagnetic Inductance Tomography (EMT): Sensor, Electronics and Image Reconstruction Algorithm for a System with a Rotatable Parallel Excitation Field" in *IEE P-Sci. Meas. Tech.* **145,** 20-25 (1998).
5. C. Z. Feng, *Electromagnetic Field*, Chap. 1, Beijing: Higher Education Press, 1983.
6. Z. Liu, "The Investigation of Electromagnetic Tomography Technology and System", Ph.D. Thesis, Tianjin University, 2002.

Dynamic Flow Behavior Measurements in Gas-Solid Fluidized Beds Using Different Non-Intrusive Techniques and Polyethylene Powder

Bangyou Wu*†, Guang Yu†, Celine Bellehumeur†, Apostolos Kantzas*†

*Tomographic Imaging and Porous Media Laboratory
†Department of Chemical and Petroleum Engineering
Schulich School of Engineering
University of Calgary, Calgary, AB, Canada T2N 1N4

Abstract. Pressure fluctuations and X-ray computed tomography (CT) measurements were utilized to characterize the flow behavior of gas-solid fluidized beds using polyethylene particles in three Plexiglas columns with diameters of 10cm, 20cm, and 30cm. Air was used as the gas phase. Gas-solids flow dynamic under ambient conditions was characterized from statistical analysis of pressure fluctuation data and CT images. The time-averaged voidage distribution, bubble phase area fraction, bubble diameter and bubble number distribution varying with the bed heights were extracted from all the three columns. Bed scales had significant effect on the hydrodynamics. Scale up effects on the gas-solids two-phase flow behavior were discussed.

Keywords: X-ray CT; Pressure fluctuation; Flow dynamics; Fluidized bed; Polyethylene powder

INTRUDUCTION

Fluidized bed reactors are used in a wide range of industrial applications. The production of polyolefins is one important example of such applications. However, the hydrodynamic modeling of fluidized bed polymerization reactors is often based on data obtained from different gas-solid systems, which need to be improved and verified. This work aims to fill this gap.

For gas-solids fluidized bed systems, various measurement techniques have been used to characterize the flow behavior, to monitor or control the fluidization quality, and to detect the gas or solids phase properties. Basically, these techniques can be categorized into intrusive measurement techniques, such as solids concentration probes, and non-intrusive techniques, such as pressure measurements. As non-intrusive methods do not affect the gas-solids flow behavior inside the fluidized bed, these techniques were used in this study. Pressure measurements are widely used in both laboratory and industry. This technique is convenient and simple, and often the only possible way to get the flow information inside the industrial reactors. Pressure fluctuations generally reflect the global flow behavior, which is the behavior of the full reactor scale. X-ray Computed Tomography (CT) has been used to visualize gas-solid fluidized beds since the 1990's. He and Kantzas [1] summarized the application of CT scanning in fluidized beds. CT analysis can provide high-resolution and detail information on density distribution inside the fluidized bed, and may be useful for investigation of bubbling behavior. However, it is difficult to apply the CT technique directly in industry. Thus combining CT scanning with pressure measurement may help in interpreting plant scale pressure measurements.

Also, scale up is the ultimate purpose for laboratory research work on fluidized bed systems. Therefore, in this work, three Plexiglas columns with diameters of 10cm, 20cm, and 30cm were used to investigate the scale effect on gas-solids flow behavior.

CP914, *Multiphase Flow: The Ultimate Measurement Challenge,* edited by X. Cai, Y. Wu, Z. Huang, S. Wang, and M. Wang

EXPERIMENTAL

The fluidized bed experimental systems are shown in Fig. 1. Plexiglas columns have diameters of 10cm, 20cm, and 30cm and the same height of 1.0m. The fluidized beds have a wind box filled with plastic balls to disperse the gas, a porous plate distributor, and an expansion section. Air enters the fluidized bed from the inlet at the bottom, goes through the wind box, the distributor, the bed section, and flows out from the outlet at the top of the expansion section. A cyclone is connected to the top of the fluidized bed to collect the particles that are entrained out of the reactor. A rotameter is used to control the operating pressure and flow rate of the incoming compressed air.

FIGURE 1. Experimental setup of fluidized bed columns

Measurement ports at different heights on the wall of the column are designed for pressure measurement. The locations are at 0cm, 17cm, 34cm and 50cm from the distributor. The particles are linear-low-density polyethylene resins with same mean particle size (677μm), particle size distribution (140-1500μm), and a plaque density of 918kg/m^3. The static bed height is 40cm for the three columns. Four pressure transducers (Schlumberger Solartron, model 8000 DPD) are connected to the measurement ports with plastic tubes. A data acquisition board is use to convent analog signal from the transducers into digital signal. A computer is used to collect and record the digital signals. The pressure data was collected at a rate of 500Hz for 60 seconds at each flow rate.

The x-ray Computed Tomography (CT) scanner is a GE-9800 Highlight Advantage™ system, modified from its original design for medical applications to suit chemical and petroleum engineering applications. The scanner was operated at 120keV voltage and 100mA current. The 10cm and 20cm columns were scanned from bottom to top in 3mm increments, with a slice thickness of 3mm. The 30cm column was scanned in 5cm intervals because of limitations with our lifting device. Each scan took 2 seconds for a cross-section. The scanning resolution was 490μm by 490μm. More detail on the CT scanner can be found elsewhere [1].

RESULTS AND DISCUSSION

Pressure Fluctuation Measurements

Minimum fluidization velocity (U_{mf}) was determined experimentally from the profiles of mean pressure versus superficial gas velocity (U_g). The values are 10.3 cm/s, 9.52 cm/s, and 7.96cm/s for the 10cm, 20cm, and 30cm column, respectively. The minimum fluidization velocity decreases with increasing diameter of fluidized beds. This may result from the higher friction forces in the small-scale beds. Average absolute deviation (*AAD*) and average cyclic time (*ACT*) were calculated from the pressure fluctuation series. *AAD* is the average of the absolute difference between each pressure measurement and the mean value of a pressure measurement series. AAD can be calculated using the equation (1) below.

$$AAD = \frac{\sum_{i}^{n} |P_i - \overline{P}|}{n} \quad (1)$$

where P_i is the ith pressure measurement in a series of n pressure measurement and $\overline{P}$ is the average value of the n measurements. *AAD* of pressure signal reflects the amplitude of fluctuation within the fluidized bed at a certain gas flow rate. Average cyclic time is defined as the average time that a complete wave takes. A wave is recognized as a complete one if the signal crosses its average twice. *ACT* is calculated using equation (2) as follows [2]:

$$ACT = \frac{total \quad time}{(number \quad of \quad crossing \quad with \quad the \quad average)/2} \quad (2)$$

ACT of pressure signal reflects how fast the fluctuation is in the fluidized bed propagate at a certain gas flow rate.

Pressure measurement positions did not affect the statistical analysis results (*AAD* and *ACT*) noticeably. In other words, the amplitude and the period of pressure fluctuation in a fluidized bed are independent of the height of measurement position. Only one position at bed height (*h*) of 17cm from the distributor is shown in this paper. *AAD* and *ACT* varying with U_g and bed scales are shown in Fig. 2 and 3. *AAD* and *ACT* increase with an increase of U_g significantly at $U_g > 1.25U_{mf}$. These trends reflect the forming and passage of bubbles in the fluidized bed. For $U_g > 2U_{mf}$, *AAD* and *ACT* are much larger and increase faster with U_g. Pressure fluctuation was mainly caused by formation, coalescence, split, and passage of bubbles in the bed and bubble bursting at the bed surface. Larger *AAD* and *ACT* indicate that large bubbles were formed in the bed. For the 10cm column, slugs were observed for $U_g > 2U_{mf}$. That is why *AAD* and *ACT* are significantly higher in the 10cm column than those in the 20cm and 30cm columns. The slugging effect decreases with an increase in the diameter of fluidized beds, so the AAD and ACT in the 30cm bed are expected to be less than those in the 20cm bed. However, the experimental results show that *AAD* and *ACT* values of the 30cm bed are much larger than those of the 20cm bed. It is suspected that gas distribution in the 30cm column was not uniform therefore larger bubbles were formed compared to 20cm column. It was observed that bubbles always came up from one side of the bed. The mal-distribution of gas will be further confirmed from X-ray CT measurement.

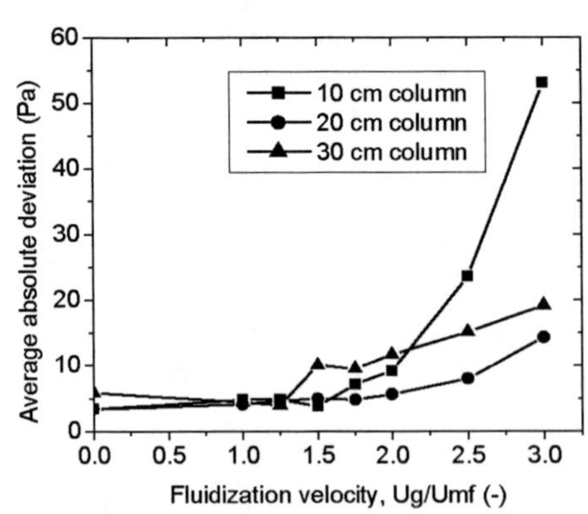

FIGURE 2. *AAD* versus U_g and bed scale at bed height h = 17cm

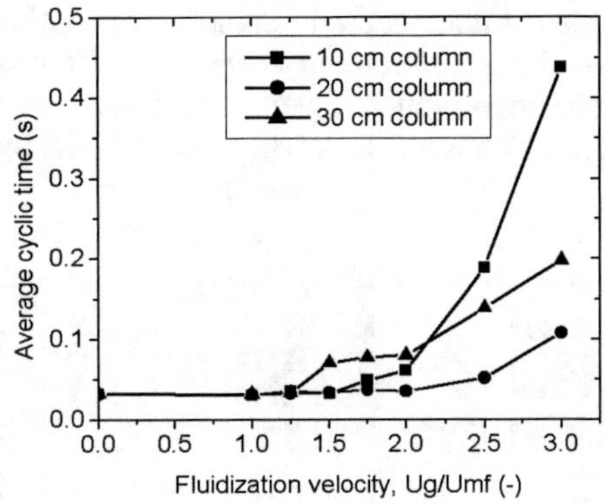

FIGURE 3. *ACT* versus U_g and bed scale at bed height h = 17cm

X-ray CT Measurements

CT images from the three columns were processed using similar methods to those by He and Kantzas [1]. After preprocess and artifact removal, CT images were converted into voidage maps using a calibrated equation. They were then binarized using a threshold value of maximum voidage at U_{mf} for statistical analysis of bubble phase properties. Examples of voidage maps and frequency distributions for the 20cm column are shown in Fig. 4 and 5. From the voidage map, average voidage for each bed height can be calculated. With the increase of U_g, average voidage also increases. The fraction of high voidage, the voidage value with highest frequency (Fig. 5) and area of bubbles with high voidage (>0.61) (Fig. 4) also increased with an increase in U_g.

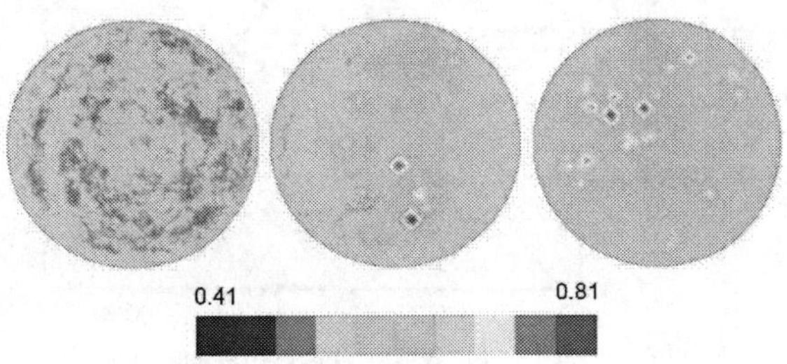

FIGURE 4. Voidage map for 20cm column at bed height of 17cm and U_g of 1.0, 1.5, 2.0 U_{mf} (from left to right)

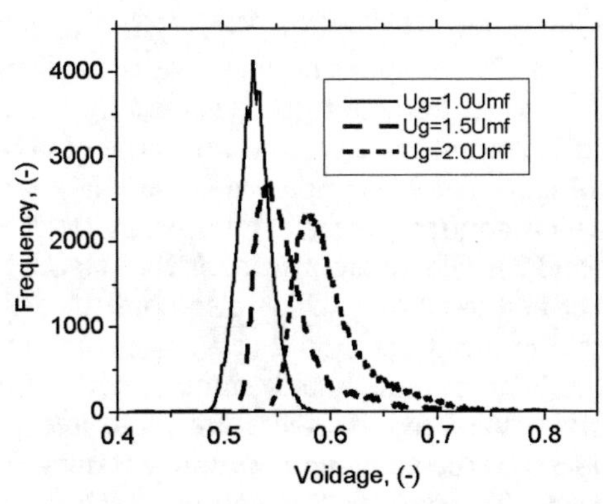

FIGURE 5. Frequency distribution for 20cm column at bed height of 17cm

Figures 6-9 show the statistical analysis results of the bubble properties in the 3 columns and $U_g = 2U_{mf}$. At other U_g, these profiles are of different trends, but $2U_{mf}$ is a typical velocity in a bubbling bed. U_g effect on the results will be examined later. Average voidage (Fig. 6) does not vary significantly with bed height at certain U_g, which means that the volumetric distribution of gas along the column keeps almost constant. It is worth to point out that the average voidage values in the 10cm column are much higher than those in the larger columns. It is likely due to the strong slugging effect in the small diameter column. This will be confirmed further from the bubble diameter profile in Fig. 8.

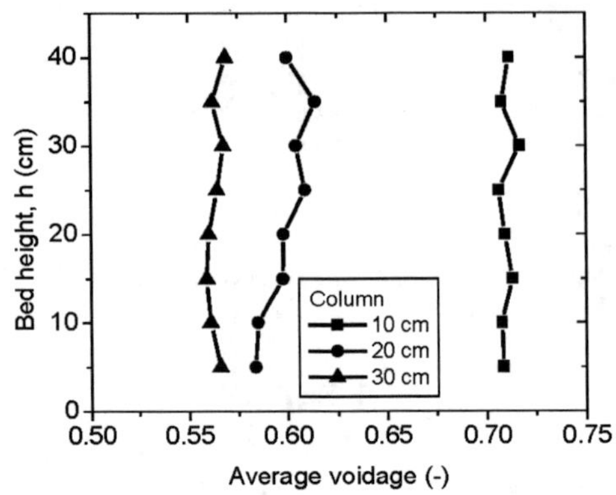

FIGURE 6. Average voidage along the column at $U_g=2U_{mf}$

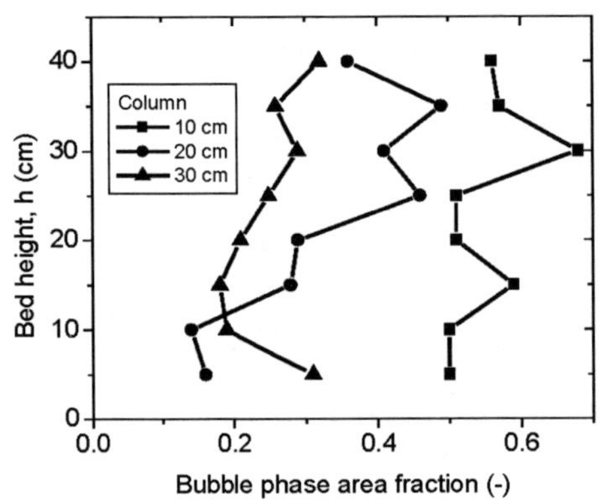

FIGURE 7. Bubble phase area fraction along the column at $U_g=2U_{mf}$

From the profiles of bubble area fraction distribution (Fig. 7) in the columns, bubble area fraction generally increases with an increase in bed height. There might be many very small bubbles or small volumetric gas phase fraction at lower bed heights, which may not be able to be detected by the CT scanner. Also, bubble coalescence must be very significant at higher bed locations, which also means larger bubble diameter and less bubble number at higher bed height as confirmed from Fig. 8 and 9. There are relatively larger bubbles and less number of bubbles in the 10cm column (Fig. 8 and 9), which confirms the slugging behavior. This is consistent with the results from pressure fluctuation analysis (Fig. 2 and 3). There are also relatively larger bubbles and less number of bubbles in the 30cm column, especially at higher bed locations, which also confirmed the mal-distribution of gas close to the distributor region from pressure fluctuation analysis (Fig. 2 and 3). At the lower bed height (h < 10 cm), the bubble number decreases dramatically. The reduction in the bubble number with an increase in bed height further indicates bubble coalescence in the upper part of the bed. As there are more and relatively smaller bubbles in the 20cm column, the gas solids mixing should be the best among the three columns. From pressure fluctuation analysis, the *AAD* and *ACT* are relatively low, which also confirms this uniform bubbling behavior. This is favorable for the gas phase polymerization process. Therefore, in the scale up of fluidized bed reactors, careful design of distributor to ensure uniform distribution of gas and uniform bubbling behavior inside the reactor is also a critical issue.

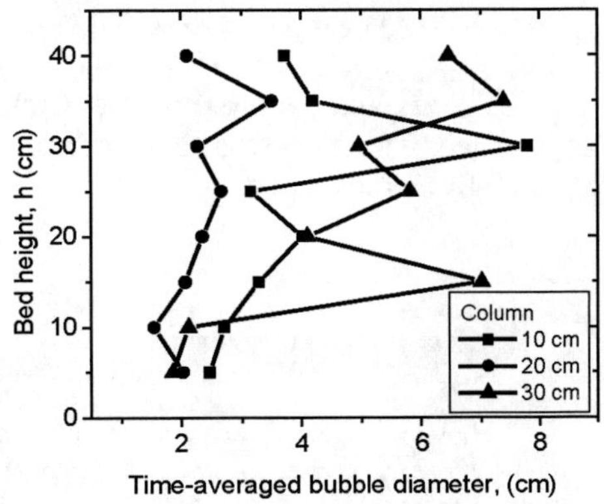

FIGURE 8. Time-averaged bubble diameter along the column at U_g=2U_{mf}

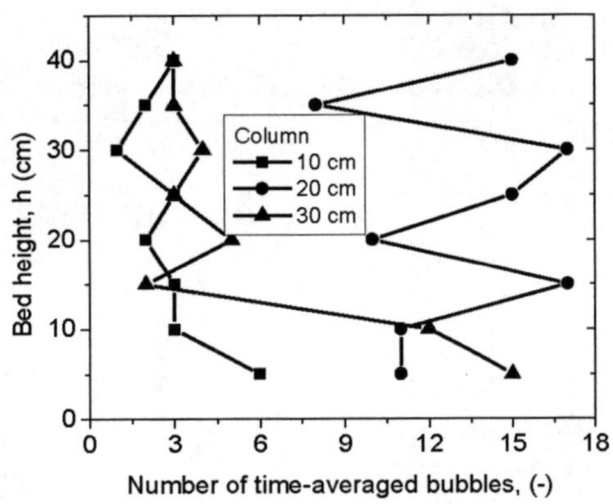

FIGURE 9. Number of time-averaged bubbles along the column at U_g=2U_{mf}

Although the general trends of the profiles in Fig. 6-9 could be identified, the profiles show very irregular variation along the column, which may demonstrate the chaotic characteristics of the fluidized bed systems. The chaotic behavior can also be extracted from pressure fluctuations, and examples can be found elsewhere [3]. The consistent results from pressure fluctuation measurement and CT scanning ensure the accurate investigation of the hydrodynamics, especially bubble properties in the fluidized bed systems. The results extracted from our analysis will be used to validate models and will facilitate implementation of changes for further improvement of hydrodynamic modeling and simulation of gas-solid fluidized bed system. The results can also be used for design, scale-up, operation optimization and assessment of new processes of gas-solids fluidized bed reactors.

CONCLUSIOINS

The pressure fluctuation in the fluidized beds reflected primarily the gas phase or bubble phase flow behavior. *AAD* and *ACT* significantly varied with U_g, especially at higher U_g (> 2U_{mf}). Based on the processed CT images, the time-averaged voidage distribution, bubble area fraction, bubble diameter and bubble number along the bed height were obtained. Both pressure fluctuation and X-ray CT measurements can consistently reveal the bubble phase properties. Bed scales were found to greatly affect the hydrodynamics in fluidized bed systems. The small column may exhibit a slugging behavior, and the largest column may have mal-distribution of the gas phase. Pressure fluctuation measurements can be used for monitor or control of the flow dynamics in gas-phase polymerization reactors, as it is a simple, fast and global measurement.

ACKNOWLEDGEMENTS

The financial support of the National Sciences and Engineering Research Council of Canada and the Canada Research Chairs Program is greatly appreciated. The help from Mr. Yubin Li and Mr. Quanzhong Zhang for the experiments and image analysis is also gratefully acknowledged.

REFERENCES

1. Z.X. He and A. Kantzas, "Comparison of the Behavior of Glass-beads and Polyethylene Resin Fluidized Beds through Flow Visualization with X-ray Tomography", In: *4th World Congress on Industrial Process Tomography*, edited by A.de Jong, M. Takei, R.A. Williams, M. Zeidan, Aizu, Japan, 2005, pp. 770-775.
2. M.L.M. van der Stappen, "Chaotic Hydrodynamics of Fluidized Beds", Ph.D. Thesis, Delft University of Technology, Netherlands, 1996.
3. B.Y. Wu, L. Briens, and J-X. Zhu, *Chemical Engineering Journal*, **117**, 187-195 (2006).

High Resolution Gamma Ray Tomography and its Application to the Measurement of Phase Fractions in Chemical Reactors

Uwe Hampel[1], Andre Bieberle[1], Eckhard Schleicher[1], Günther Hessel[1], Cornelius Zippe[1], Hans-Jürgen Friedrich[2]

[1]*Institute of Safety Research, Forschungszentrum Rossendorf e.V., Bautzner Landstraße 128, 01328 Dresden, Germany, U.Hampel@fzd.de*
[2]*Verein für Kernverfahrenstechnik und Analytik Rossendorf e.V. (VKTA), Bautzner Landstraße 128, 01328 Dresden, Germany*

Abstract. We applied gamma ray tomography to the problem of phase fraction measurement in chemical reactors. Therefore, we used a new tomography device that is operated with a Cs-137 source and a high resolution gamma ray detector. One application example is the reconstruction of the fluid distribution and the measurement of radial gas fraction profiles in a laboratory scale stirred vessel. The tomograph was used to obtain radiographic projections of the averaged gamma ray attenuation for different stirrer speeds along the height of the vessel. With tomographic reconstruction techniques we calculated the angularly averaged radial distribution of the attenuation coefficient for as many as 150 single cross-sectional planes and synthesised from this data set the axial and radial fluid distribution pattern. Further, we exemplarily reconstructed the radial gas fraction distributions induced by the stirrer in the area of the stirrer blades. In a second application the gamma ray measurement system was used to visualise gas inclusions in a water cleaning column that is used to remove hazardous heavy metal species from water.

Keywords: Gamma ray tomography, Chemical reactor, Flow measurement, Gas hold-up, Liquid hold-up.
PACS: 47.80.Jk

INTRODUCTION

The measurement of the hydrodynamics in chemical reactors is a challenging problem. Gamma ray tomography is one potential method that may be applied to measure the distribution of phases respectively the hold-up of liquid or gas in metallic vessels. It is, however, a rather slow measurement technique that provides time integral information on the phase distribution. There are some reports in literature describing the application of gamma ray densitometry or tomography to flow measurement problems [1]-[4]. Most of the introduced systems have a comparatively low spatial resolution due to the use of photomultiplier technology. Due to considerable interest in gamma ray tomography in the field of nuclear thermal hydraulics we recently developed a high resolution gamma ray tomography device. The detector uses a new photon detection technology, namely scintillator coupled large area avalanche photodiodes, which enables the construction of a larger detectors with small and robust detector elements. With this technology the gamma ray detector is also less susceptible to electromagnetic noise in non-laboratory environments, such as thermal hydraulic test facilities or chemical processing plants.

THE GAMMA RAY CT SYSTEM

The gamma ray tomography system (Figure 1) consists of an isotopic source and a detector arc. Both are placed opposite to each other about the object that is to be investigated. The isotopic source that we use is a Cs-137 capsule with 662 keV gamma photon energy, 165 GBq activity and 5.6 mm source spot size. The capsule resides within a shielding container and can be manually drawn into working position by means of a bowden cable. The detector arc

CP914, *Multiphase Flow: The Ultimate Measurement Challenge,* edited by X. Cai, Y. Wu, Z. Huang, S. Wang, and M. Wang

is made of 320 single lutetium yttrium orthosilicate scintillation crystals coupled to avalanche photodiodes and associated detector electronics [5]. The crystals have an active area of 2 mm width, 8 mm height and 22 mm length. Stopping efficiency is roughly 75% whereas energy resolution is about 25%. The in-plane resolution of the gamma ray CT setup has been measured as 3.5 mm. The axial resolution is about 7 mm for the uncollimated system but may be increased by applying further spatial collimation to source and detector. Thus, in the presented study we have used additional tungsten collimators to reduce the axial extension of the fan to 2 mm. The detector is operated in pulse counting mode, that is, each gamma photon that interacts in the scintillation crystal is electronically converted to a voltage pulse. Each detector channel comprises a preamplifier stage followed by a Gaussian pulse shaper, a pulse height detector with dual window comparator and a 24 bit counter stage with digitally multiplexed output. Tomography requires the rotation of either the source-detector system or the object and the acquisition of radiographic projections from many different viewing angles, usually equally spaced along a full 360° rotation. In the laboratory setup the object is therefore placed between detector and source on a translational-rotational stage where it can be rotated and vertically moved.

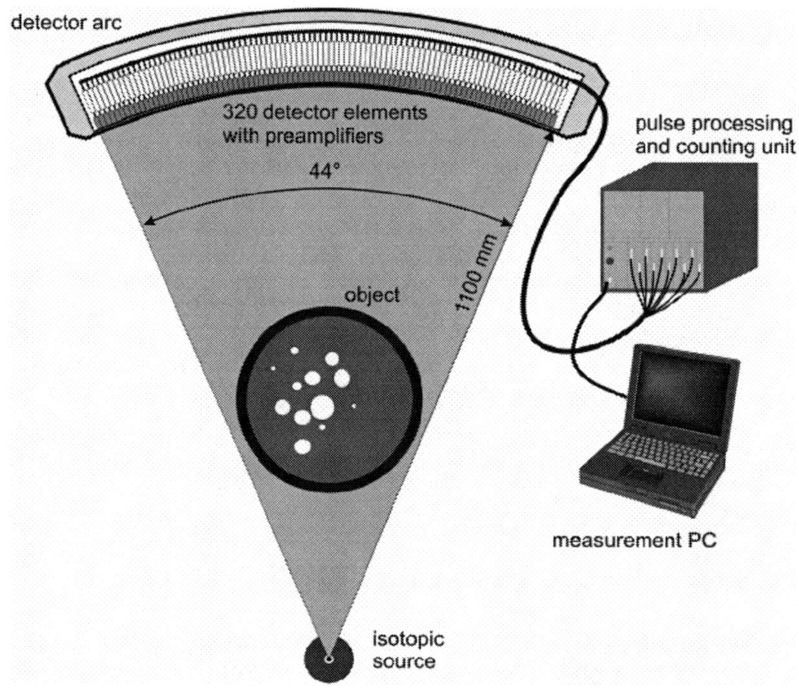

FIGURE 1. Sketch of the gamma ray CT setup.

MEASUREMENTS ON A STIRRED REACTOR

Gamma ray CT has been applied to the visualisation of the fluid distribution in a stirred tank reactor at different stirrer speeds. The aim of this study was the provision of data for CFD simulations. Thus, special parameters of interest are the shape of the fluid-gas boundary around the stirrer shaft and the gas distribution in the impeller region. The vessel is a laboratory scale cylindrical borosilicate glass autoclave reactor Ecoclave 150 (Büchi AG, Switzerland) with a spherically shaped bottom. The reactor inner diameter is 80 mm and its walls are of 8 mm thickness. As a stirrer we used a hollow shaft gas inducing turbine (Figure 3). During the experiments the stirrer speed was precisely computer controlled. The reactor was filled with 1.5 l isopropanol as a model fluid.

Due to the rapidly rotating fluid distribution inside the reactor we may assume a cylindrical symmetry of the flow with respect to the non-rotating laboratory frame. It is clear, that there are angular structures of the flow field within the frame of rotation, such as gas vanes at the impeller tips and waves on the fluid-gas boundary. However, such structures can at the moment not be captured by the measurement system due to integration times which are much longer than a single period of the stirrer rotation. During the experiments the reactor was placed in the centre of the gamma CT setup and scanned in 150 successive axial slices of 2 mm separation. Due to the assumed cylinder symmetry we did not rotate the vessel. In each plane photon counting was done for 10 seconds giving an average number of approximately $2 \cdot 10^4$ counts in the fluid region and $1.2 \cdot 10^4$ counts at the stirrer shaft, that is, a statistical noise/signal ratio of less than 1%. The first step of data processing consists in the calculation of the linear extinction value for each measurement point. Let i denote the index of the detector element (the radial coordinate) and j the index of the plane (the axial coordinate), then the extinction values are calculated according to

$$E(i,j) = -\log \frac{N(i,j) - N_d(i)}{N_0(i) - N_d(i)}, \qquad (1)$$

where N(i,j) denotes the count rate for a given measurement channel (plane and detector), N_d the dark count rate and N_0 the reference count rate. The dark count rate is recorded once with the isotopic source not being in operation. The reference count rate could be measured on the empty vessel, which would then give the extinction from the fluid only. But in practice it is difficult to fill or empty the vessel between two measurements without causing some displacement that would result in artefacts. Thus, we used the free space reference, that is, the detector readings without any object in the fan. This data is present in each data set just in the planes below the bottom of the vessel.

Since the measurement data is still a projection of the fluid distribution we need to apply tomographic inversion algorithms to reconstruct the fluid distribution in details. For this purpose we use the filtered backprojection technique. First, we synthesise a backprojection image $\mathbf{b}_j$ by applying

$$b^j_{m,n} = \sum_{p=0}^{N_P-1} \sum_{i=0}^{N_D-1} a_{p,i,m,n} E_{i,j}, \qquad (2)$$

where $a_{p,i,m,n}$ denote the pixel weights, i. e. the geometric share of pixel (m,n) to the finite width fan beam ray (p,i). The backprojection operation is carried out for a sufficient number of projections which secures an artefact free image although there is only data from a single projection. An alternative here is using the Abel inversion technique [6] which should lead to similar results. After backprojection the two-dimensional image is restored applying a suitable inverse filter, where in our case we chose a Hamming modified ramp filter. Having done this subsequently for all the 150 scanned slices we synthesise an axial cross-section image through the very centre of the vessel. The results for scans at different stirrer speeds are shown in Figure 2. It can be seen, that the liquid-gas boundary can be clearly restored at a high spatial accuracy. However, it turned out, that a scan time of 10 seconds is not sufficient to measure the gas fraction in the impeller region. The critical stirrer speed in this particular case was at about 970 rpm. From earlier experiments with X-ray cone beam tomography [7] we have determined a gas fraction below 5% in the outer region of the impeller plane. This low gas fraction value requires a higher integration time, therefore we performed measurements at 120 s integration time only in the plane in the very axial centre position of the impeller structure (Figure 3). Figure 4 shows the radial gas fraction values for different stirrer speeds which have been obtained with this procedure.

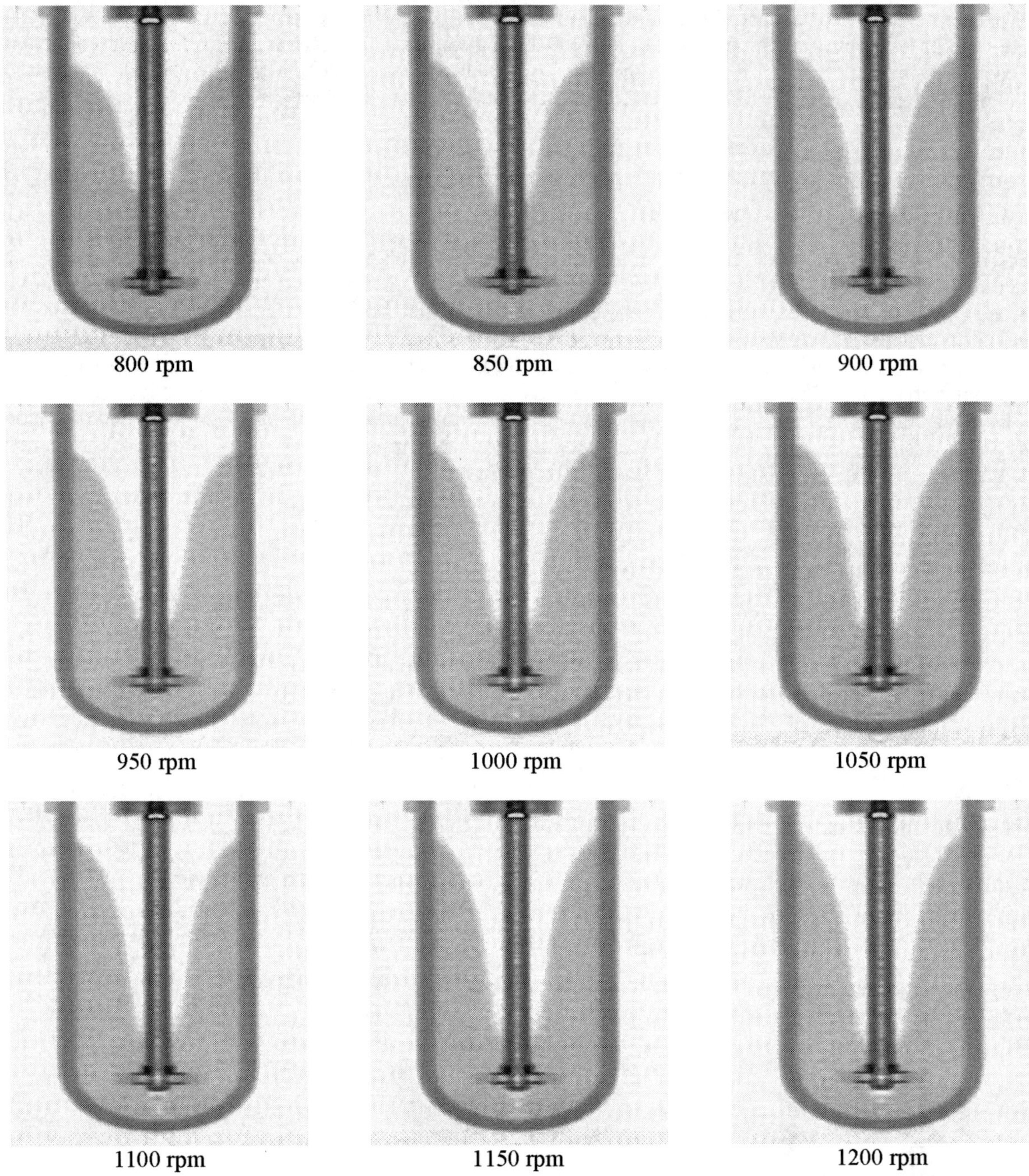

FIGURE 2. Results of the fluid phase reconstruction for a stirred vessel at different stirrer speeds.

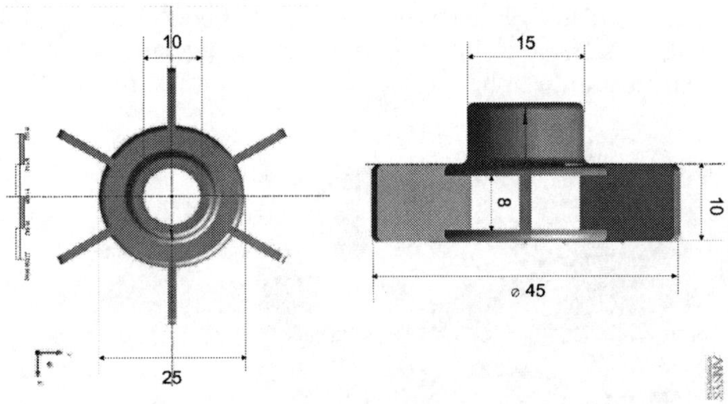

FIGURE 3. Impeller geometry.

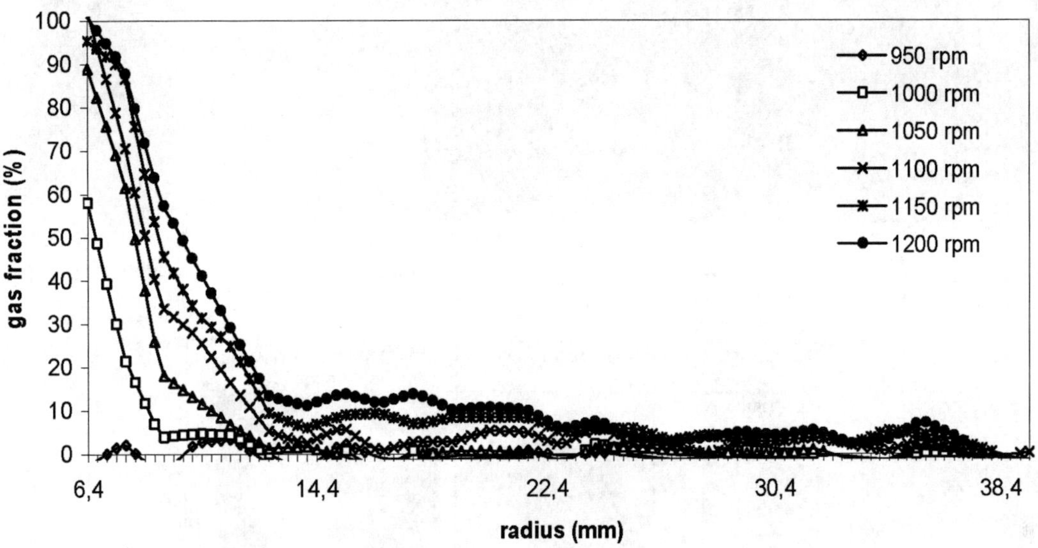

FIGURE 4. Measured radial gas fraction profiles in the impeller plane.

MEASUREMENTS ON A WATER CLEANING COLUMN

The hydrodynamics in trickle bed reactors and fluid processing columns may also be assessed by gamma ray tomography. In this particular case gamma ray tomography has been used to detect gas inclusions in a column for mineral water cleaning. The column is a laboratory scale plastic pipe of 80 mm inner diameter and 500 mm length, filled with gravel with a density of 1.49 g/cm^3 and a porosity of 0.272. Within the gravel bed there are four layers of surface modified iron particles which adsorb certain unwanted trace heavy metals, such as uranium. The iron itself acts as a catalyst for hydrogen production. The hydrogen may in turn accumulate in the pores of the bed and this again may lead to local gas clogging in the column with a corresponding flow obstruction. To assess the amount of gas accumulated in the column we performed a gamma CT measurement. The experimental setup is shown in Figure 5. The column is mounted upright on the rotational table. A pump produces the required water flow from a reservoir through the column. The water flow goes bottom up through the column and is measured by a rotameter. The flow rate has been varied between 5 l/h and 25 l/h. An initial radiographic scan has been performed to detect the iron particle layers (figure 5, right). Tomographic planes for gas measurement have been chosen in the gravel regions in between the iron layers where gas clogging is to be expected. Again two scans have been made for each measurement, one for the reference data and one for the actual measurement. As reference data we used scans which were taken immediately after starting up the flow when negligible gas fraction is suspected. After one hour of continuous flow we have performed the gas measurement scans. Figure 6 shows the results. It can clearly be recognised, that in this particular case there is only very little gas accumulation especially in the peripheral areas of the column with an overall gas fraction well below 3%. This result proved to be valid for all mass flow rates and scanned planes.

FIGURE 5. Gamma CT measurement on a water cleaning column: experimental setup (left) and radiography of the column with indicated CT planes (right).

Plane	Tomography image	Binary phase distribution image
160	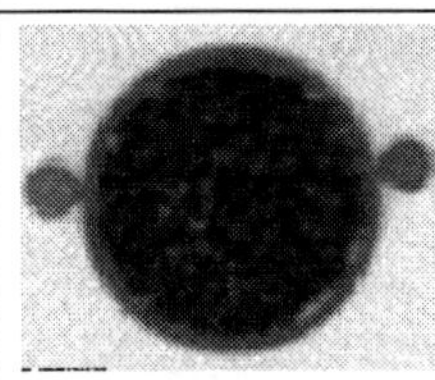	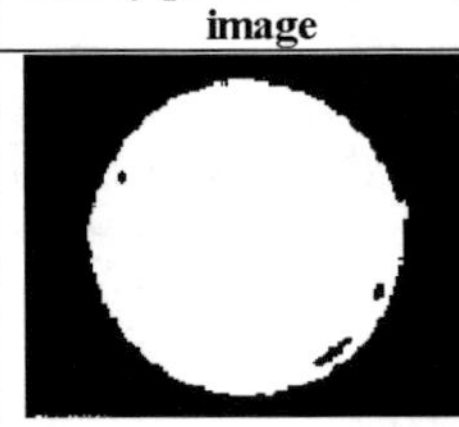
255	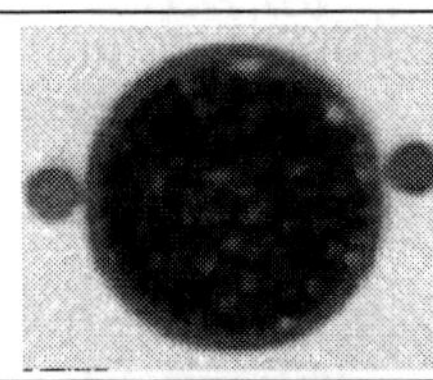	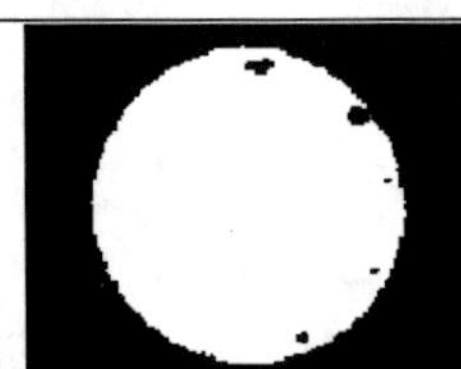
340	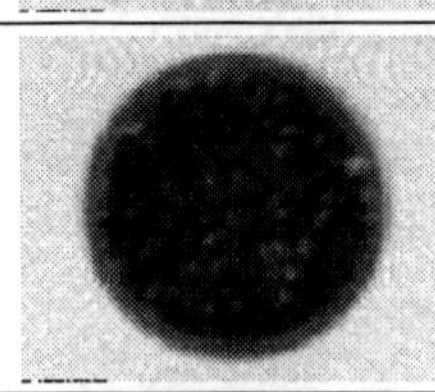	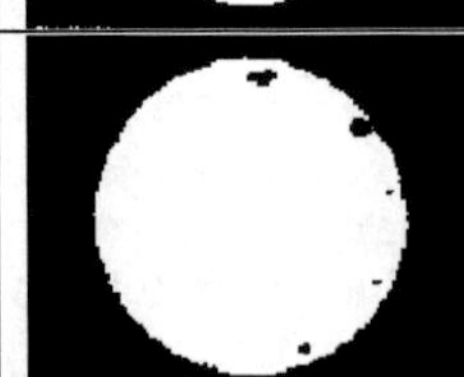
425	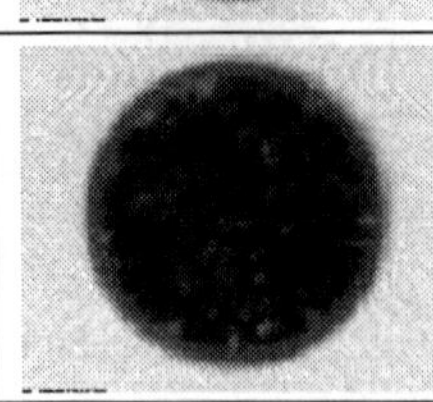	

FIGURE 6. Results of the gamma CT measurement with detected gas inclusions in bed pores (left: original tomography images, right: binary images that show the gas filling).

CONCLUSIONS

After the successful development of a high resolution gamma ray tomography system it has been used in first experimental studies of the hydrodynamics in a stirred chemical reactor and a water processing column. Advantages of the gamma ray CT technique are the small effort spent for preparation of the setup and vessel, the non-intrusiveness of the measurement and its ability to provide three-dimensional information on the phase distributions inside the investigated objects within a comparatively short time. For the stirred vessel we were able to obtain the fluid-gas boundary at high spatial resolution for the whole vessel height assuming cylinder symmetry of the flow and to measure radial gas fraction profiles in the impeller region. This data may now be used for evaluation of CFD codes. For the water cleaning column we were able to detect local gas inclusions in the bed of the column and to estimate the overall gas fraction. These first experiments show the potential of gamma ray tomography for such studies.

REFERENCES

1. A. C. Devuono, P. A. Schlosser, F. A. Kulacki, P. Munshi, *IEEE T. Nucl. Sci.*, **NS-27**, 814-820 (1980).
2. S. B. Kumar, D. Moslemian, M. P. Dudukovic, Flow Meas. Instrum. **6**, 61-73 (1995).
3. K. A. Shollenberger, J. R. Torczynski, D. R. Adkins, T. J. O'Hern, N. B. Jackson, Chem. Eng. Sci. **52**, 2037-2048 (1997).
4. U. P. Veera, Chem. Eng. Journal **81**, 251-260 (2001).
5. A. Bieberle, U. Hampel, E. Schleicher, D. Hoppe, H.-M. Prasser, T. Sühnel, C. Zippe, "Design and performance of a high-resolution gamma tomography device" in Proceedings of the 4th World Congress on Industrial Process Tomography, Aizu, Japan, Sept. 2005, 2005, pp. 18-23.
6. A. C. Kak, M. Slaney, *Principles of Computerized Tomographic Imaging*, IEEE Press, New York, 1987.
7. S. Boden, U. Hampel, M. Bieberle, Chem. Eng. J, 2006 (submitted).

INVESTIGATING THE LIQUID DISTRIBUTION IN A REACTIVE DISTILLATION PACKING USING HIGH ENERGY X-RAY TOMOGRAPHY

J. Abdullah[2], S. Aferka[1], M. Crine [1], A.K. Saroha [3], D. Toye[1] and Pierre Marchot[1],

[1]*Chemical Engineering, University of Liège, Institut de Chimie - B6c, Liège, 4000, Belgium,*
[2]*Malaysian Institute for Nuclear Technology Research, (MINT) Bangi 43000, Kajang, Malaysia*
[3]*Chemical Engineering Department, IIT, Delhi, Hauz Khas, 110016 New Delhi, India*

Abstract.
Liquid distribution in a 0.1 m diameter column packed with Katapak-SP12 is investigated with a 420 kV X-ray tomograph. The distribution of the various liquid hold up between baskets and corrugated sheets is quantified..

Keywords: Tomography, Imaging, Reactive Distillation, Hydrodynamics.
PACS: 87.59.Fm; 89.20.Bb; 89.20.Kk.

1. INTRODUCTION

Catalytic distillation is a type of process intensification that utilises close coupling of separation and chemical reaction systems. The combination of heterogeneously catalysed chemical reactions and distillation in one single apparatus has several advantages. Such a hybrid process offers the potential for large savings of capital, energy, materials and significant reduction of waste products that satisfies the increasing demand for sustainable technology. Various catalyst packing configurations based on a regular alternation of zones with weak permeability (catalytic zones) and of high permeability (liquid and vapour transport zones) have been reported in the literature. These sandwich structures allow postponing the flooding beyond usual operating flow rates. However, designing catalytic distillation columns remains a challenge. The introduction of separation and reaction zones in one single apparatus leads to complex interactions between hydrodynamics, vapour-liquid equilibrium, vapour-liquid mass transfer, internal diffusion and chemical kinetics. Different models of the process, as well as new structured catalytic packings have been developed. However, to take advantage of these potentialities requires a better understanding of complex multiphase flow phenomena occurring in catalytic distillation packings. X-ray tomography has been shown to be an efficient non-intrusive tool to see inside and to adequately image the liquid and vapour flow distribution in packed columns. In this paper, we report on the use of a high energy and high resolution X-ray tomograph able to penetrate through stainless steel packing and dense catalytic zones and to achieve a spatial resolution of a few hundreds of microns.

2. DESCRIPTION OF THE X-RAY TOMOGRAPH SET-UP

The generator is a Baltograph CSD450 constant potential generator (Balteau NDT, BE). It may be operated between 30 and 420 kV. The X-ray source is an oil cooled, bipolar TSD420/3 tube (Comet, CH and Balteau NDT, BE), producing a 40° aperture fan beam. Its minimum focus size is 0.8x0.8 mm following norm IEC336 - EN12543. Its intensity may be varied between 2 and 8 mA. The detector is a X-Scan 0.4f2-512-HE (Detection Technology, FI). It is constituted by a linear array of 1280 photodiodes each coupled with a $CdWO_4$ scintillator. The detector is 512 mm long with a 0.4 pixel pitch, which corresponds to a spatial resolution at the centre of the object of approx. 360 µm. Reconstruction of object cross sections are obtained by a classical linear back projection algorithm adapted to

CP914, *Multiphase Flow: The Ultimate Measurement Challenge,* edited by X. Cai, Y. Wu, Z. Huang, S. Wang, and M. Wang

the fan beam geometry implemented in the Fourier domain. The test rig allows manipulation of rather large objects (max. height: 4m, max. diam: 0.45 m). During the tomographic measurements, the column rotates slowly around its vertical axis, while the X-ray source (S on fig 1) and the detector (D) remain fixed. The entire scanning unit is located inside a room surrounded by 0.6 m thick high density concrete wall to prevent exposing operating personnel to X-ray radiation [1].

The column used in this study has a height of 4 m. and an inner diameter of 0.1 m. It is made of transparent PVC. The packing (1.8 m high - 0.1 m diameter) is constituted by the superposition of, from the bottom to the top, one Mellapak 752Y element, four Katapak-SP12 elements and three Mellapak 752Y used to get a initial uniform liquid distribution. These packings are manufactured by Sulzer Chemtech, CH. They are made of stainless steel. Baskets in the Katapak SP12 elements are filled with 1 mm glass spheres [2]. Both types of packings are 0.10 m diameter and 0.2 m high. An air-water system is used. The superficial velocities of the liquid range between 10 and 40 $m^3/m^2.h$.

3. EXPERIMENTAL RESULTS

Tomographic measurements are realised in packing cross sections situated at different heights. To distinguish between the liquid and the solid contributions to the attenuation, the projection data obtained on the dry column are subtracted, before reconstruction, from those obtained at the same height on the irrigated column.

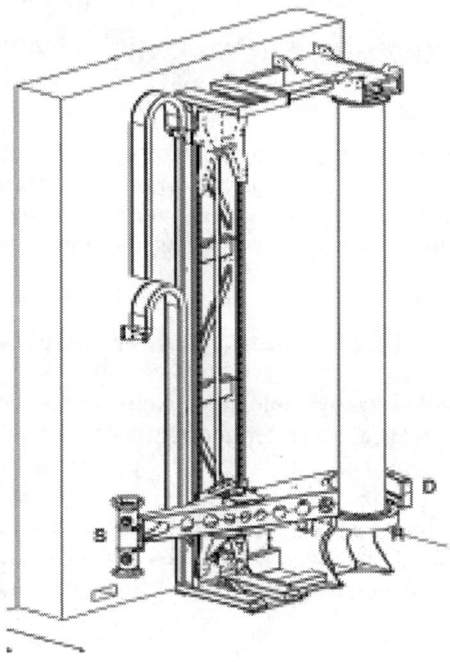

FIGURE 1: X-ray tomography. (S): source; (D): linear detector; (R): rotating table.

To eliminate the background noise affecting reconstructed images, they are thresholded using the method of Otsu (implemented in the Matlab 7 Image Processing Toolbox 4.2). This automatic thresholding is applied only to the pixels located inside the external wall of the column obtained using image masking. Reconstructed sections are squares of 499*499 pixels, the spatial resolution in the plane is around 0.36 mm.

3.1 Packing Geometry

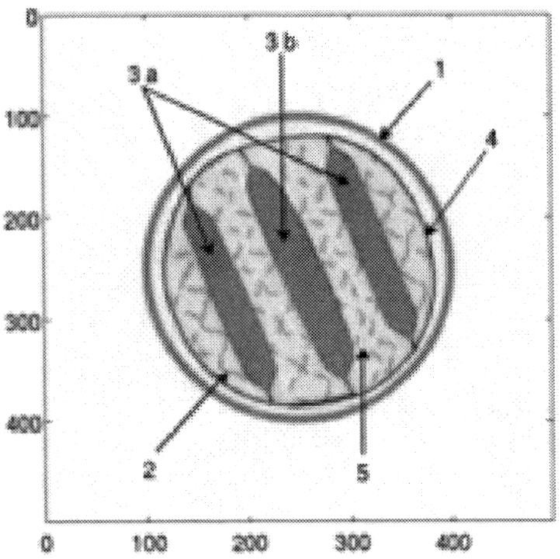

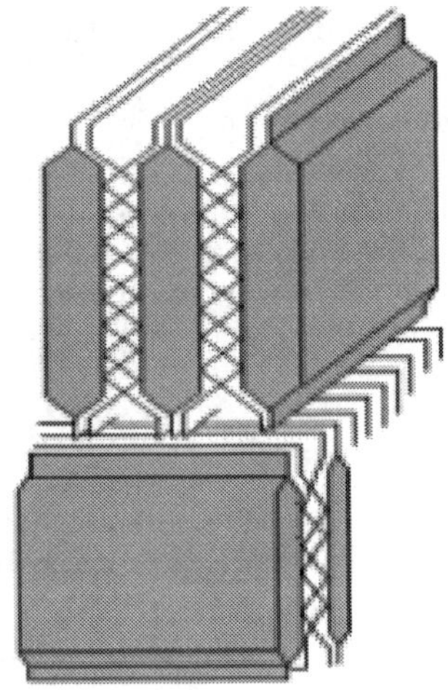

FIGURE 2: image in grayscale of a cross-section of Katapak-SP12. **FIGURE 3: arrangement of the baskets and sheets of Katapak-SP12**

Figure 2 (left) shows a reconstructed cross section of a Katapak SP2 element, scanned at a height corresponding around the middle of an element. Column wall (1), corrugated sheets (2), outermost baskets (3a), central basket (3b), wall wiper (4), perforations (5) and embossings are clearly visible. However, the small glass beads, within the baskets, are not resolved. Figure 3 (right) is a sketch showing the arrangement of the baskets and sheets in the column.

3.2 Irrigated Packing Results

The total liquid hold up is composed of the dynamic, pore, capillary and residual hold up [3].The dynamic liquid hold up is the liquid which drains from the column after interrupting the liquid and gas flows. The porous hold up is located in the pores of the catalyst (here it is zero, because we experiment on non porous glass spheres). The capillary hold up results from the capillary rise of the liquid in the packing of spheres and from the presence of the gauze wire mesh. The residual liquid hold up corresponds to liquid menisci trapped by capillarity at the points of contact of the spheres (at higher locations than those reached by capillary rise) or at points of contact between the corrugated sheets or to liquid drops remaining on the sheets after that the liquid film covering the sheets is broken by drainage (see also § 3.6).

3.3 Total Liquid Hold up Results Images

Figures 4 a-e illustrate the total liquid holdup distribution in the packing (dynamic + static hold up). To illustrate as clearly as possible the liquid and the solid phases, we proceed as follows. First, dry packing images are recorded, reconstructed and thresholded following Otsu's method, leading to the gray part of the images. Then corresponding irrigated cross sections are reconstructed (on the difference of the attenuation data, as explain before) and thresholded following the same method leading to the blue part of the images. Finally, gray and blue parts are superimposed [4].

Figure 4 presents a series of irrigated packing cross sections at various heights for a liquid flow rate of 38 m^3/m^2.h. showing that the major part of the liquid, flows within the baskets. The wall wiper seems to perform well in redistributing the liquid towards the inside of the bed, since the wall flow is limited. When the liquid flows from one Katapak element to the next , the basket located in the middle is not immediately filled (compare fig 4a and 4b, as well as 4c and 4d). This effect results from the irregular shape of the bottom part of the baskets.

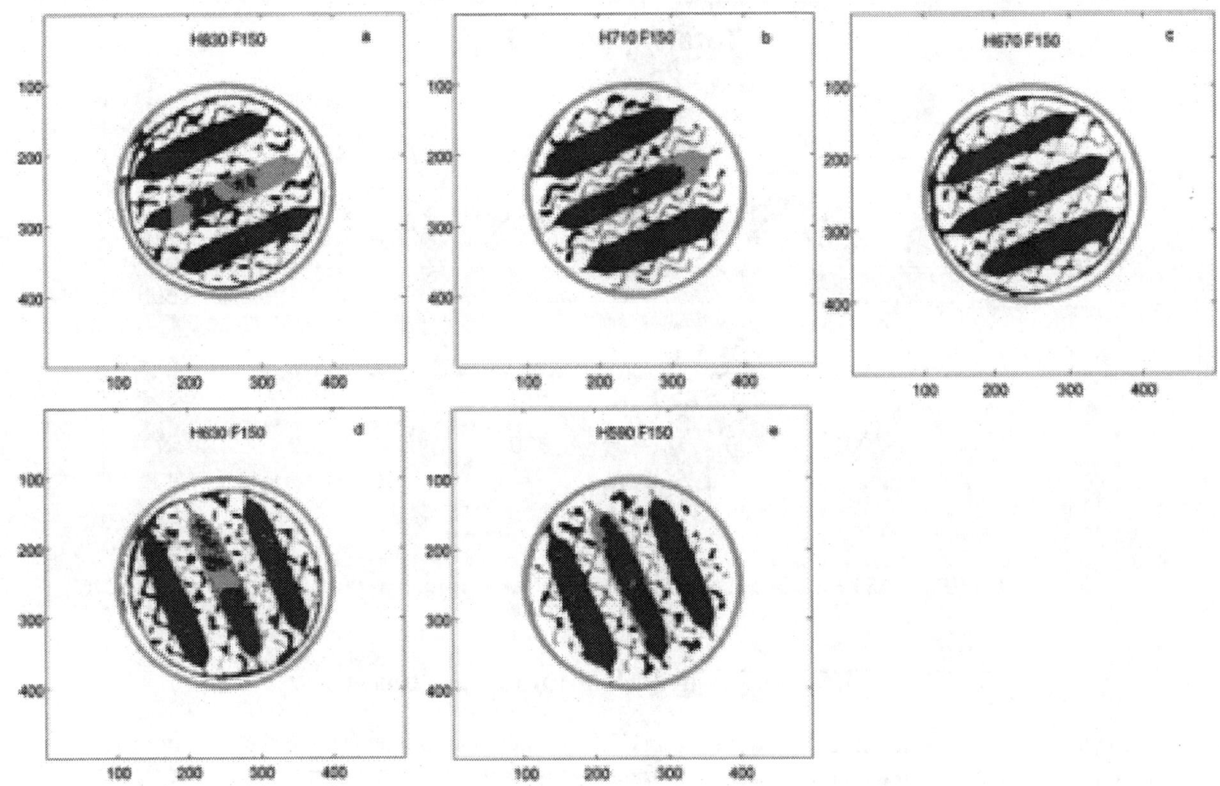

FIGURE 4 a,b,c,d,e : Evolution of liquid distribution inside Katapak-SP12.

To get quantitative values for the liquid hold up, a mask corresponding to the pixels located inside the column internal diameter is determined. Applying this mask to each of the images of the total liquid hold up, we obtain a maximum pixel value on each cross sections. These maxima differ by less than 10 % relative. Assuming that their average corresponds to a pixel totally filled by the liquid, the liquid hold up may be calibrated at the pixel level.

3.4 Total Liquid Hold up as a function of bed height

Measurements are performed using 9.5, 19, 38 $m^3/m^2.h$ flow rates, on a pre wetted packing. Figure 5 plots the liquid hold-up at the cross section level, for sections located at different distances from the liquid distributor, as a function of liquid velocity. Points labeled from a to e correspond to sections illustrated on figure 4. Lines joining the experimental points intend to facilitate the reading of the graph. Vertical bars represent the beginning / end of the four Katapak-SP12 elements. All experimental points are located within the elements and not in between. As expected, the total liquid hold up increases with liquid velocity. Moving from the bottom of the bed (left size of the graph), the hold up within each Katapak element increases first, then decreases as the end of the element is reached. The maximum hold up for each elements is located around its middle height. For the highest velocity investigated, the liquid hold up increases with the distance from the distributor.

Contrary to the total liquid mass flow rate, the total liquid hold up (as well as its components) needs not to be conserved from sections to sections. Actually, a decrease in liquid hold up must go with an increase of the corresponding liquid velocity. Between two Katapak elements the surface area of the solid decreases nearly to zero. There, the liquid flows as rivulets of small cross sections with high velocities. This drives the liquid filling the baskets to leave them and contributes to increase the liquid solid transfer.

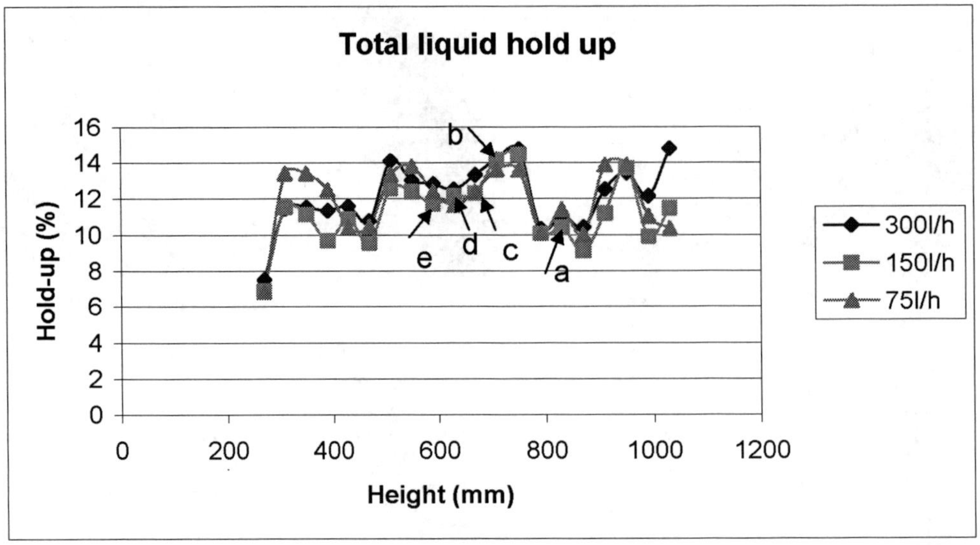

FIGURE 5 : F75, F150 and F300 are in l/h and corresponds respectively to 12, 23 and 47 $m^3/m^2.h$

3.5 Total Liquid Hold up of the baskets

Applying a mask corresponding to the cross section of the baskets, we can determine the total liquid hold up for each basket and consequently the total liquid hold up fraction inside the baskets. Figure 6 plots this fraction, which ranges about 70% , as a function of the height in the packing for three liquid flowrates.

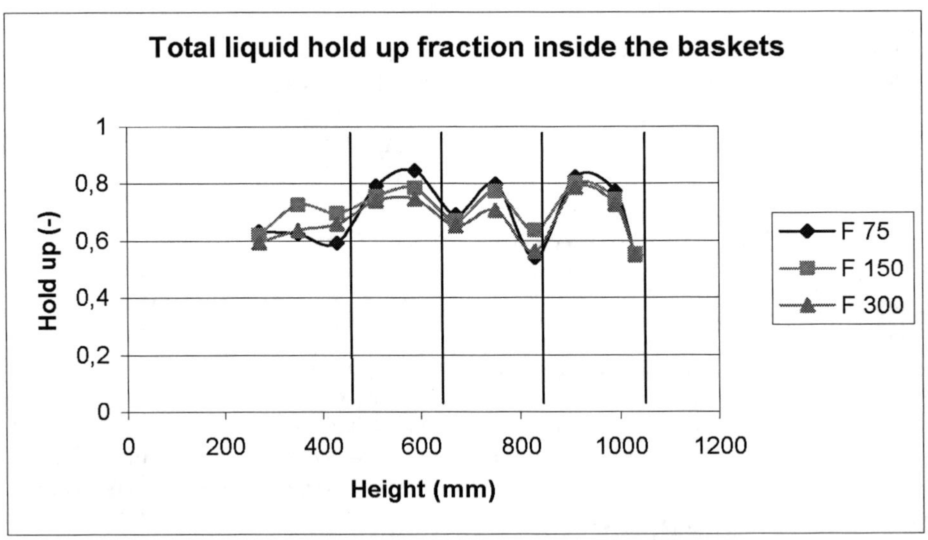

FIGURE 6 : F75, F150 and F300 are in l/h and corresponds respectively to 12, 23 and 47 $m^3/m^2.h$

3.6 Static Liquid Hold up

The column is totally filled up by the liquid, then drained for 15 hours. Figures 7 are obtained in the same way as figures 4. Figures 7 a b show the static liquid hold up obtained near the bottom (7a) and near the top (7b) of a Katapak-SP12 element. The static liquid is mainly trapped at the bottom of the baskets due to capillary rise and constitutes the capillary hold up. Indeed, because of the gravity, the liquid drains from the baskets until it is balanced by the capillary forces. The residual hold up, i.e. the static liquid remaining trapped on the corrugated sheets and within the baskets at heights greater than the capillary rise, is evidenced on figure 7b.

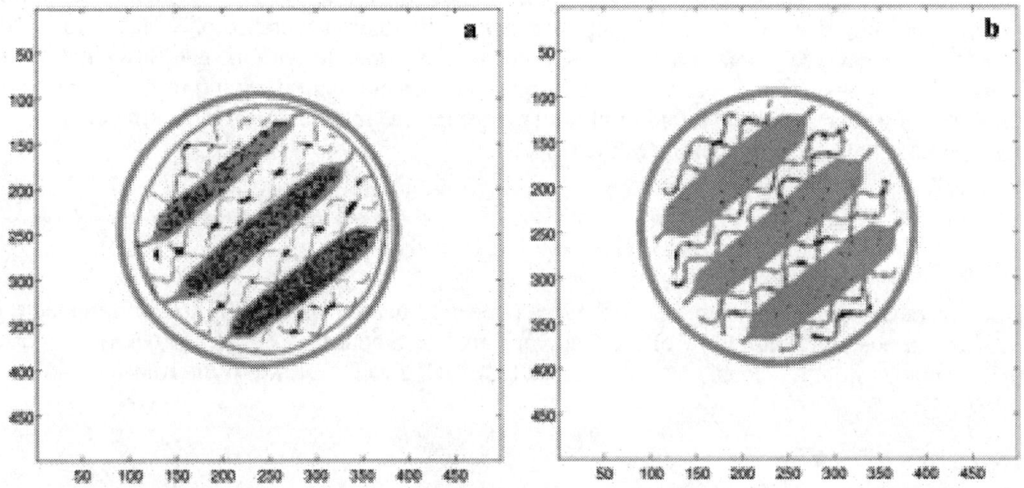

FIGURE 7 : bottom (a) and top (b) cross sections showing the static hold up in a Katapak-SP12

Figure 8 plots the static hold up (% of total cross section) within the baskets as a function of the height in a Katapack SP12 element. Measurements are performed every 2 mm. On the left side of this figure (which corresponds to the lower part of the baskets), the static hold up is high and fluctuates around 13 % : it constitutes the capillary hold up. Fluctuations result from fluctuation of the baskets width. On the right side, from around 4.8 cm, the static hold up decreases markedly and forms the residual hold up.

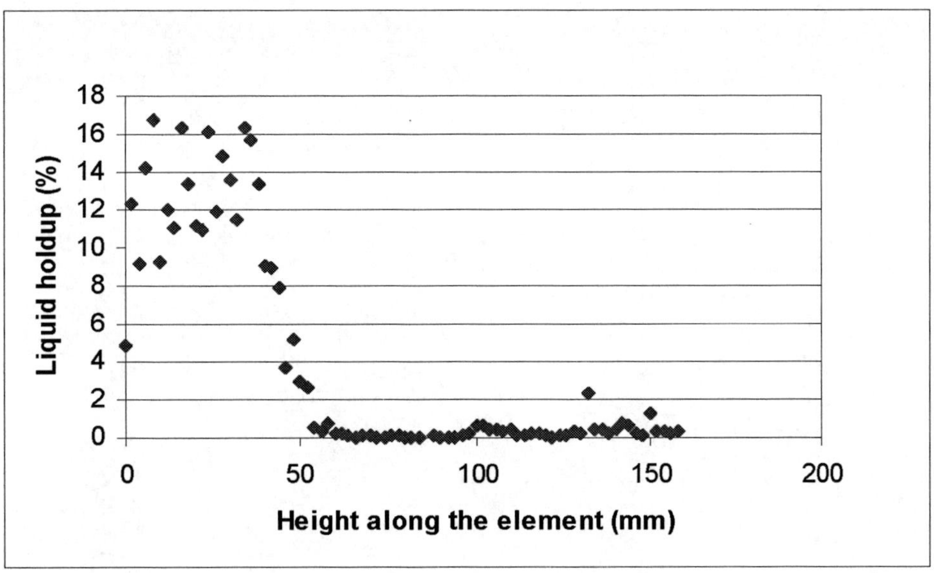

FIGURE 8: Static liquid hold up

4. CONCLUSIONS

These are preliminary results, nevertheless they are extremely encouraging. They evidence the possibility of the method to investigate liquid distribution in complex reactive distillation metallic packings. From the distillation engineering side, they may seem somewhat academic because we operate without gas flow, and we use water as liquid. However, the various components of the total liquid hold up are quantified and useful observations are done. In particular, the strong decrease in the total liquid hold up around the junction of two packing elements is evidenced as well as the capillary hold up at the bottom of the baskets

ACKNOWLEDGMENTS

This work was performed in the frame of a research concerted action financed by the Communauté Française de Belgique. A.K.S is a post-doctoral fellow of the Belgian Fund for Scientific Research (FNRS). Mellapak 752Y and Katapak-SP12 elements were graciously provided by SULZER CHEMTECH AG Winterthur.

REFERENCES

1. D. Toye, M. Crine, P. Marchot, ''Imaging of liquid distribution in reactive distillation packings with a new high energy x-ray tomograph'', *Measurement Science and Technology*, 16 (2005), pp.2213-2220.
2. S. Ratheesh, A. Kannan, ''Holdup and pressure drop studies in structured packings with catalysts'', *Chemical Engineering journal*, 104 (2004), pp. 45-54.
3. M. Behrens, '' Hydrodynamics and Mass Transfer Performance of Modular Catalytic Structured Packing '', PhD Thesis, ISBN 13:978-90-9020347-8, TU Delft, (2006).
4. P. Marchot, D. Toye, M. Crine, G. L'Homme, A.-M. Pelsser, Z. Olujic, '' Liquid distribution images on structured packing by X-ray computed tomography'', *AIChE Journal*, 47 (2001),pp.1471-1476.

Particle Field Characterization by Digital In-line Holography

Shiliang Pu, Xingguo Pu, Kefa Cen and Zhenfu Yuan

State Key Laboratory of Clean Energy Utilization, Zhejiang University, Hangzhou 310027,China E-mail:pushiliang@yahoo.com.cn , Tel: 0086-571-87951293, Fax:0086-571-87951616

Abstract. Recent developments have shown the potential of digital in-line holography for diagnostics in fluids. This new method provides a low-cost and an easy access method for measuring both size and spatial distribution of particles in a volume. Here, it is shown that, by applying traditional image processing tools on the particle images digitally reconstructed, statistical reliable results on particles size and spatial distribution can be obtained. The particle sizing method is experimentally illustrated by glass microspheres that are moving in a turbulent flow generated by a annular jet. A comparison with the histogram diameters provided by a common diffraction particle sizer are presented.

1. INTRODUCTION

Recently, Digital In-line Holography (DIH) has been successfully applied on several domains. 3D location of small fibers and particles by numerical reconstruction of holograms was reported by Belaid et al. (1997), Buraga-Lefebvre et al. (2000) and Lebrun et al. (1999, 2003). Biological applications like cell biology, imaging and tracking submicrometer particles in seawater have been published by Xu et al. (2001,2002,2003), Owen et al. (2000) and Milgram et al.(2002) . The use of DIH has also been demonstrated by Sun et al. (2002) for observation of erosion processes. In the field of fluid mechanics Coëtmellec et al. (2001), Pan and Meng (2001), Sheng et al. (2003) and Malek et al. (2004) have shown that velocity vector maps could be determined in several slices of interest within a sample volume by this technique. The main advantage of DIH is the simplicity of the recording apparatus : a collimated beam illuminates the objects and the far-field diffraction pattern is directly recorded by a digital camera. In fact this system could be seen as a lens-less imaging setup where the focusing stage is realized numerically. Furthermore, by analyzing the diffraction pattern recorded on the camera (hologram) several slices of the volume can be digitally reconstructed and each particle image can be extracted before being processed. The aim of this paper is to show that such a method can give reliable information on the statistical distribution of particle diameters and velocity within a sample volume. The paper is organized as follows: at first, the method used for hologram recording and reconstruction is recalled. The second paragraph is devoted to present the image processing method used for particle field characterization. Experimental results are reported on section 4.

2. DIGITAL IN-LINE HOLOGRAPHY

The reconstruction of Digital in-line particle holograms has been widely described in previous papers (Buraga-Lefebvre et al. (2000), Malek et al. (2003)). The objective of this section is to recall briefly the background theory used in this study. Let us consider a small object (particle, fiber…) illuminated by a monochromatic collimated plane wave (see Fig.1).

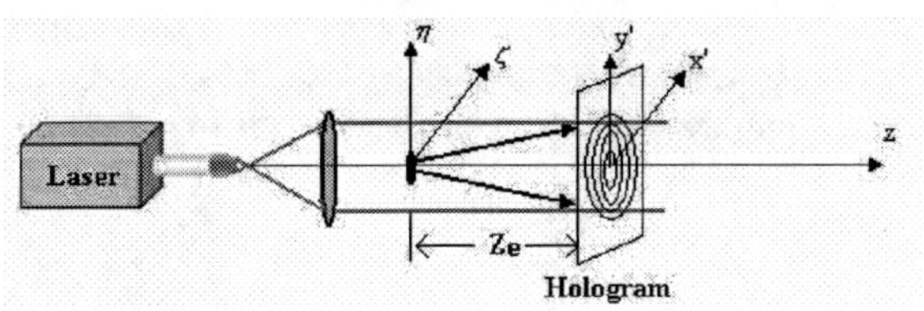

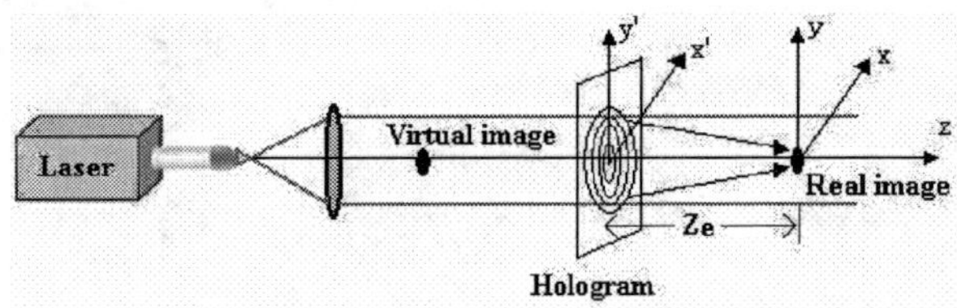

CP914, *Multiphase Flow: The Ultimate Measurement Challenge,* edited by X. Cai, Y. Wu, Z. Huang, S. Wang, and M. Wang

FIGURE 1. Representation of the diffraction pattern recording

FIGURE 2. Representation of the diffraction pattern reconstruction

The transmission function at object plane can be described by $1 - O(\xi,\eta)$ where :

$$O(\xi,\eta) \begin{cases} = 1 \text{ if } (\xi,\eta) \text{ belongs to the object} \\ = 0 \text{ elsewhere} \end{cases} \quad (1)$$

The intensity recorded by a quadratic sensor located at a distance z_e from the object plane can be approximated under Fraunhofer conditions (i.e. object size negligible as regards to z_e) by the following convolution:

$$I_{z_e}(x,y) = 1 - \left[O ** \left(H_{z_e} + \overline{H}_{z_e}\right)\right](x,y) \quad (2)$$

H_{ze}, also called the Fresnel Kernel, is defined from the wavelength λ by:

$$H_z(x,y) = \frac{i}{\lambda z} exp\left[\frac{i\pi}{\lambda z}\left(x^2 + y^2\right)\right] \quad (3)$$

When a given diffraction pattern $I_{z_e}(x,y)$ is illuminated by the same laser source (fig. 2), the intensity of a reconstructed image $WT(\alpha,x,y)$ at a distance z_r can be computed by the convolution operation $WT(\alpha,x,y) = \left[I_{ze} ** \psi_{z_r}\right](x,y)$. The reconstruction function (a wavelet function) is given by :

$$\psi_z(x,y) = \frac{\pi}{\lambda z}\left[\sin\left(\pi \frac{x^2+y^2}{\lambda z}\right) - M_\psi\right] \times \exp\left(-\pi \frac{x^2+y^2}{\lambda z \sigma^2}\right) \quad (4)$$

M_ψ is used to ensure a zero mean value of $\psi_z(x,y)$ which is defined as $M_\psi = \frac{\sigma^2}{1+\sigma^4} \cdot \exp\left(-\pi \frac{x^2+y^2}{\lambda z \sigma^2}\right)$ is the aperture of the reconstruction function. σ depends on the recording system (CCD camera) specifications. At first the wavelet function should be approximately confined within the dimension of CCD sensor. σ should respect equation (5) in order to ensure that the value of the wavelet function falls down under a small value at the edge of the CCD sensor.

$$\sigma_1 \leq \frac{N\delta_{CCD}}{2\alpha} \sqrt{\frac{1}{\ln(\varepsilon^{-1})}} \quad (5)$$

Here, N^2 is the number of pixels of the recording device, δ_{CCD} is the size of pixel and ε is a small constant used as a tolerance parameter. Here ε is set to 0.01.

According the Shannon's sampling limitation, σ should be fixed so that under sampling effect is avoided, it leads to:

$$\sigma_2 \leq \frac{\pi\alpha}{2\delta_{CCD}} \sqrt{\frac{1}{\ln(\varepsilon^{-1})}} \quad (6)$$

By the above analysis, in order to check both the spatial and frequency limitation of digital sensor, σ should satisfy following condition:

$$\sigma \leq \min(\sigma_1, \sigma_2) \tag{7}$$

Fig.3 illustrated the reconstruction steps. At first the reconstruction plane is chosen, then the wavelet function is generated by equation (4). The image at the plane is reconstructed by the convolution between the diffraction pattern and the wavelet function. The convolution step is performed by fast Fourier transform (FFT) in order to speed up calculation.

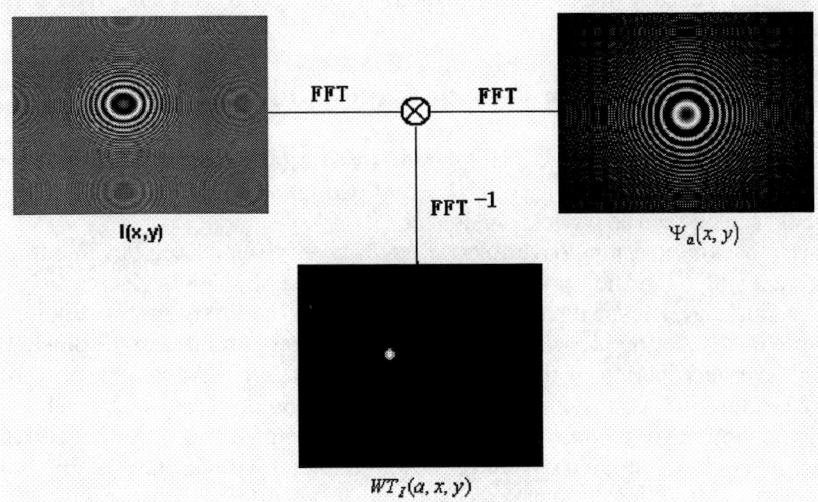

FIGURE 3. The steps of digital reconstruction by wavelet transform

3. IMAGE PROCESSING METHOD

In this section the processing method for extracting particles' 3D coordinates and diameters from the reconstructed volume is presented. The image processing steps are summarized on the following diagram (fig.4):

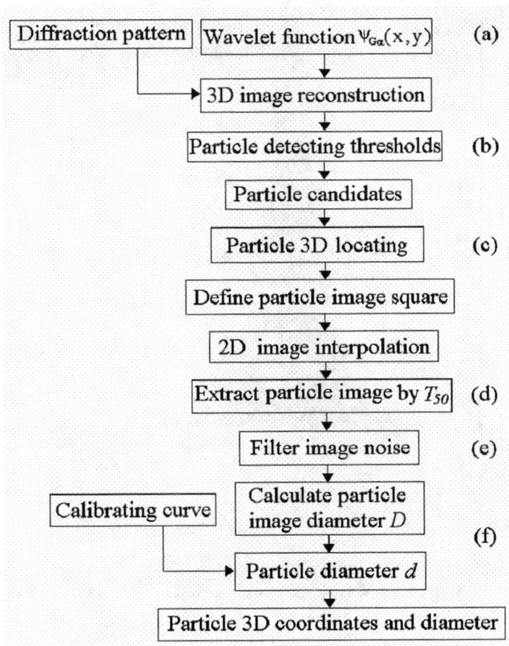

FIGURE 4. Image processing steps

Step A: Three-dimensional Image Reconstruction

At first the depth of the analyzing volume z_1-z_2 is defined. Here, z_1 is the minimum recording distance of the analyzing volume and z_2 is the maximum recording distance of the analyzing volume. Then the axial sampling period Δz (the axial distance between two successive reconstructed planes) is chosen according to the required axial location accuracy. A family of wavelet function $\psi_{Ga}(x,y)$ is previously generated in accordance with each reconstruction plane. Then, each plane is reconstructed by calculating the convolution between the diffraction pattern and the wavelet function.

Step B: Particle Candidate Selecting

Since the intensity distribution of image background is essentially a Gaussian function with zero mean value. Therefore the gray level threshold can be used to detected particle images. The detected regions with higher gray level than the threshold can be seen as particle candidates.

However the image background may have uneven intensity in different regions due to particles uneven spatial distribution. Therefore a single threshold may miscalculate a group of spatial close particles as one. Consequently, instead of applying a single gray level threshold to detect particle, we firstly use a serial of increasing gray levels from the lowest threshold T_{min} to the highest threshold T_{max} to detect particles. Fig 5 illustrates the definition of the thresholds. The gray level distribution of the background can be approximated as a zero-mean Gaussian function. Therefore the image background has a symmetrical intensity distribution around zero value. It is interesting to note that the intensity distribution of the reconstructed image in negative region is only constituted by image background whereas the positive region is constituted by image background and particle images. By assuming its symmetry to zero value, the intensity distribution of image background in positive region can be estimated by reverting the intensity distribution in negative region. By comparing the intensity distribution of reconstructed image and the intensity distribution of image background, T_{min} is defined by the gray level where the pixel number of reconstructed image begins to be more than the pixel number of image background. By this way, theoretically T_{min} is the minimum gray level of the particle images. T_{max} is defined as the maximum gray level of the image background.

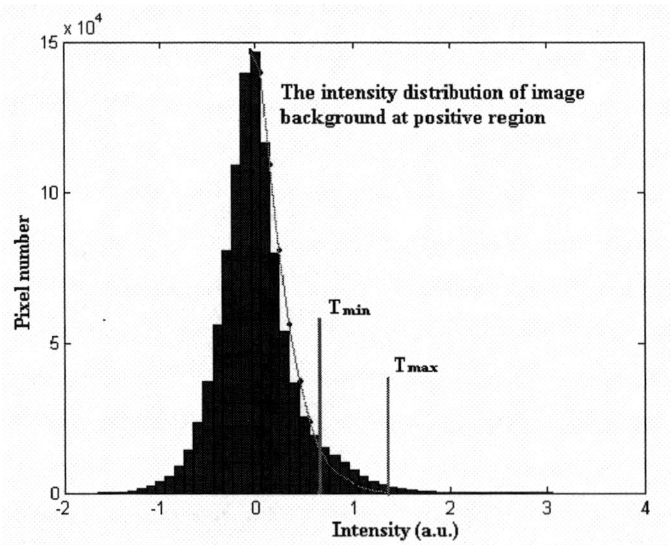

FIGURE 5. Example of gray level histogram for definition of T_{min} and T_{max}

Step C: Three-dimensional Particle Locating

Assuming that the reconstructed particle image in its focused plane can be seen as a convolution between the object function and a Gaussian PSF, the particle maximum gray level should be found at center of the particle image. As a result the particle coordinates are given by the maximum intensity of the detected particle candidates in the reconstructed volume.

Step D: Particle Image Extracting

A small image window surrounding the particle coordinates that includes particle image is selected. Then two-dimensional interpolation is performed on the image window in order to increase particle sizing accuracy. The particle sizing threshold T_{50} is defined at mid level between the background intensity and the maximum intensity of the particle image. By applying T_{50}, particle image is extracted from the interpolated image. The particle image diameter D is obtained by counting pixel number of the separated particle image.

Step E: Image Noise Filtering

In order to remove noise from the particle candidates, an image noise filter is applied. This filter uses the statistical property of image noise to identify it from particle candidates. Here, we use the relationship between the mean intensity of particle candidate image, intensity standard deviation and image diameter D to calculate the probability of a candidate being a particle.

Step F: Particle Sizing

After image noise is removed from the particle candidates, the correction factor that expresses the relationship between particle image diameter D and particle diameter d is obtain by simulation way. Then the particle diameter d is calculated from the particle image diameter D by a correction factor.

4. EXPERIMENTAL RESULTS

The Experiment Of Particle Sizing

In order to check the validity of the method, calibrated glass spheres have been inserted in a flow produced by an annular jet. The inner and outer diameters of the annular nozzle are $\phi_i = 48.75\ mm$ and $\phi_o = 53.88\ mm$. As a result, the thickness of the annular jet near exit is $e=2.56\ mm$. Figure 6 shows the geometry of the experimental setup. A beam coming from a Nd:Yag Laser illuminates the flow and the diffraction patterns produced by glass particles are directly recorded onto the photosensitive plane of a CCD sensor.

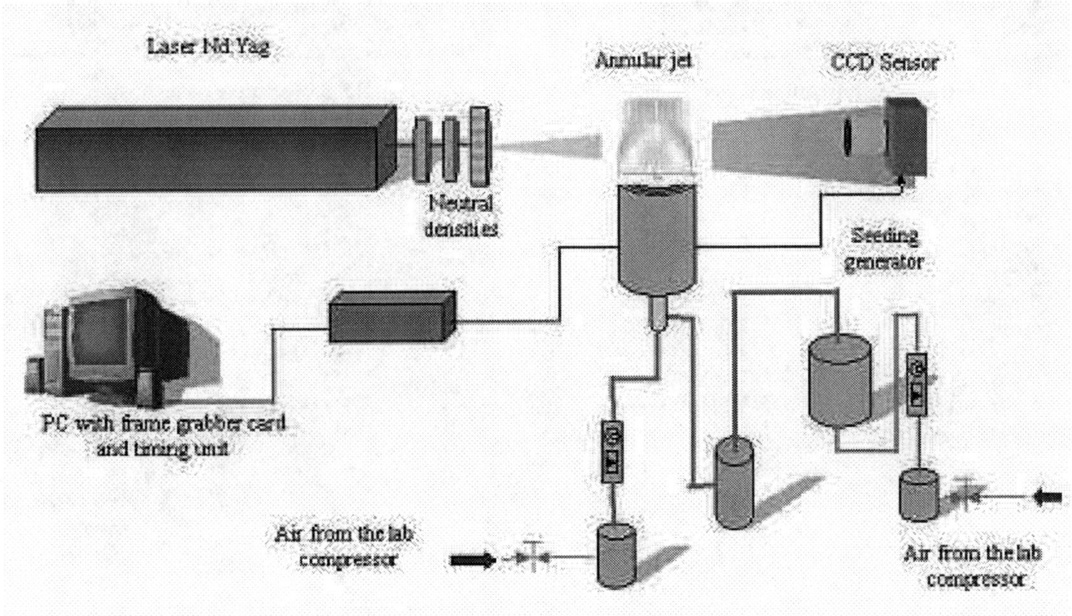

FIGURE 6. Geometry of the experimental setup

Fig.7 is an example of a diffraction pattern recorded by the CCD camera. Fig. 8 shows the image reconstructed by our process (z-plane) at a distance z_r=85mm from the CCD plane. Fig. 9 presents an example of particle diameter distributions calculated by Digital in-line holography (circle dots) from the diffraction pattern. The diameter range obtained here confirm the technical specifications delivered by the manufacturer (90-100% of the particle-diameters lower than 90 µm). In order to compare with an another sizing apparatus, the glass particles which are used here have been also calibrated by a diffraction based particle sizer (Beckman Coulter). As we can see on Fig. 9, the two methods give approximately the same diameter range (20-100 µm). Note that both size distributions are bimodal. However, small differences can be observed, particularly for big particles. In fact the automatic procedure used for the DIH method may select a group of particles and process it as a single particle image. Moreover,1455 particle images have been extracted and measured. From this result, the beam obscuration can be roughly calculated : here we found F=3.7%.

Another advantage of DIH over diffraction based methods is that, knowing the 3D coordinates of each particle image, the spatial particle density distribution as well as the spatial size distribution are available at a given time, this advantage is very important for studying of the characteristic of two-phase flow. Fig. 10 illustrates the 3D particle spatial distribution of the measuring region from *z=40mm* to *z=160mm*. Each particle is displayed by a circular dot with located three-dimensional coordinates and a gray level corresponding to the estimated diameter. It is noticeable that the highest concentration of particles are observed near the exit of the annular jet: z=80 mm and z=125 mm. These results confirm previous experimental studies that were realized by Patte-Rouland (2001) on the same annular jet. The variations of particle density observed by DIH are in good agreement with the particle concentration observed by this author with Laser Doppler Velocimetry.

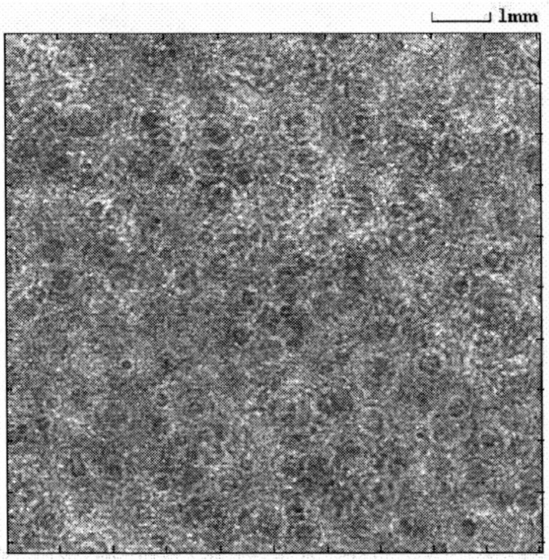

FIGURE 7. The diffraction pattern

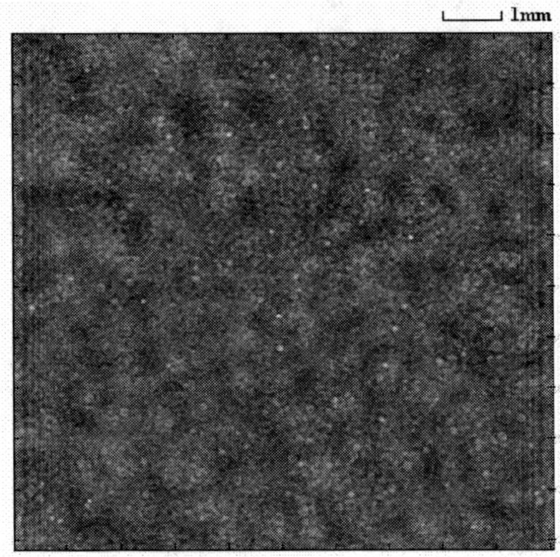

FIGURE 8. Example of reconstructed image

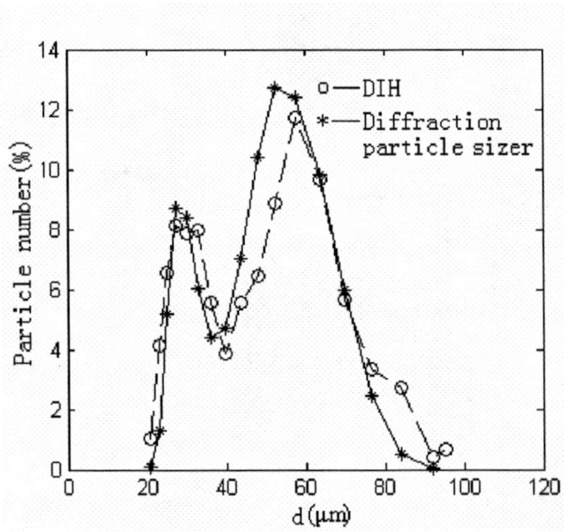

FIGURE 9. Estimation of particle diameter distribution

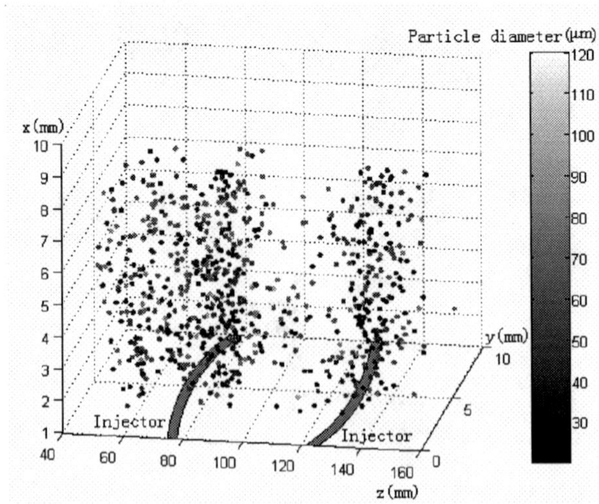

FIGURE 10. Particle density and diameter spatial distribution

5. CONCLUSION

We have developed a new particle sizing instrument for the characterization of particle field. As this method is based on image processing of particle images reconstructed by digital in-line holography, the 3D coordinates can be known at the same time. A particle image character filter is applied to separate the background noise (speckle) from particle image. A correction factor, which is deduced from a simulation curve, is applied to estimate the particle diameter from the measured image diameter. An experimental comparison between particle diameter distributions obtained by DIH and those delivered by a standard diffraction particle sizer approved that the particle diameter distribution obtained by DIH is robust and reliable.

REFERENCES

1. Belaïd S; Lebrun D; Özkul C (1997). Application of two dimensional wavelet transform to hologram analysis : visualization of glass fibers in a turbulent flame. Opt. Eng. 36: 1947-1951.
2. Buraga-Lefebvre C; Coëtmellec S; Lebrun D; Özkul C (2000). Application of wavelet transform to hologram analysis: three dimensional location of particles. Opt. Lasers Eng. 33: 409-421.
3. Coetmellec S; Buraga-Lefebvre C; Lebrun D; Özkul C (2001). Application of in-line digital holography to multiple plane velocimetry. Meas. Sci. Technol. 12: 1392-1397
4. Fournier C. ; Ducottet C. ; Fournel T.(2004) Digital in-line holography : influence of the reconstruction function on the axial profile of a reconstructed particle image. Meas.Sci.Technol. 15:686-693
5. Lebrun D; Touil C.E.; Özkul C (1996) Methods for the deconvolution of defocused-images pairs recorded separately on two CCD cameras : application to particle sizing. App. Opt. 35(32): 6375-6381
6. Lebrun D; Belaïd S; Özkul C (1999). Hologram reconstruction by use of optical wavelet transform. App. Opt. 38: 3730-3734.
7. Malek M; Coëtmellec S; Allano D; Lebrun D (2003). Formulation of in-line holography process by a linear shift invariant system: application of the measurement of fiber diameter, Opt. Comm 223: 263-271
8. Malek M; Allano D; Coëtmellec S; Özkul C; Lebrun D (2004). Digital in-line holography for three-dimensional-two components particle tracking velocimetry. Meas. Sci. Tech. To be published
9. Malek M; Allano D; Coëtmellec S; Özkul C; Lebrun D (2004). Digital in-line holography for three-dimensional-two components particle tracking velocimetry. Meas. Sci. Tech. To be published
10. Pan G; Meng H (2001) Digital in-line Holographic PIV for 3D particulate Flow Diagnostics. 4th International Symposium on Particle Image Velocimetry, Göttingen, Germany, September 17-19, 2001.
11. Sheng J; Malkiel E; Katz J (2003) Single beam two views holographic particle image velocimetry, App. Opt. 42: 235-249
12. Sun H; Dong H; Player MA; Watson J; Paterson DM and Perkins R (2002) Inline digital video holography for the study of eronsion processes in sediments, Meas. Sci. Technol. 13: L7-L12
13. Xu W; Jericho MH; Meinertzhagen IA; Kreuzer HJ (2001) Digital in-line holography for biological applications. PNAS 98: 11301-11305
14. Pu Shi-liang, D. Lebrun, Wang Qin-hui *et al* Application of digital holography to circle flow bed boiler measurement[J]. Proceedings of the Chinese society of electrical engineering, 2005，25（15）：111-115

NOMENCLATURE

$O(\xi,\eta)$	object function
$I_z(x,y)$	Intensity distribution at a distance z
λ	Wavelength of the illuminating source
$H_z(x,y)$	Fresnel Kernel
$\psi_z(x,y)$	Reconstruction wavelet function
$R(x,y)$	Reconstructed image
PSF(x,y)	Point Spread Function
δ_{CCD}	Pixel size
d	Diameter of the theoretical image
D	Diameter of the measured image
F	Beam obscuration

ε	Tolerance parameter for sampling condition
δz	Precision on axial coordinate measurement

Visualization of Flow Pattern in Thermosyphon by ECT

Shi Liu[1*], Jingtao Li[1], and Qi Chen[2]

1. Institute of Engineering Thermophysics, Chinese Academy of Sciences, Beijing 100080, China, Tel. +86 10 82543062, Fax. +86 10 62575913, email: s.liu@mail.etp.ac.cn
2. Beijing Jiaotong University, Beijing 100044, China

Abstract. An investigation on the visualization of the flow of the liquid in a two-phase closed thermosyphon (TPCT) is carried out. Electrical capacitance tomography (ECT) with optimal step lengths for iterative image reconstruction algorithm is applied in the study. Measurements are made under a range of heating rates of the TPCT. As against the common assumption of "low spatial resolution of ECT", the experimental results have demonstrated ECT's capability of reconstructing images of thin films on the order of 100 μm in thickness. Results for film thickness by ECT are compared with the established theory. Agreements and discrepancies are observed among the reconstructed images corresponding to difference thresholds. ECT images of the liquid distributions in the TPCT show clear dependence on the variation of working conditions, which proves the potential of ECT as an advantageous technique for monitoring the operation of TPCT.

Keywords: flow visualization, thermosyphon, electrical capacitance tomography.

INTRODUCTION

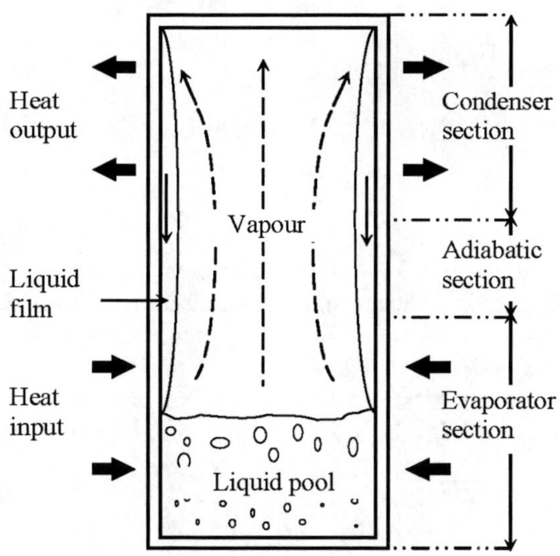

FIGURE 1. Working principle of TPCT

A TPCT is a sealed wickless heat pipe. The working principle of a TPCT is depicted in figure 1. The TPCT is s partially filled with a working fluid. The bottom is a heating section, the middle an adiabatic section, and the upper part a condensation section. From the bottom vapour flows upwards, cooled and condensed into liquid at the condensation section. Then the liquid flows downwards along the wall as a very thin film. However, when abnormal working conditions happen, liquid plugs may also form, which should be avoided.

CP914, *Multiphase Flow: The Ultimate Measurement Challenge,* edited by X. Cai, Y. Wu, Z. Huang, S. Wang, and M. Wang

The film thickness is an important parameter in determining the resistance to heat transfer, but very difficult to measure due to its very small size, often on the order of μm. The curvature of the tube tends to cause measurement errors in common optical visualization methods and the required high level vacuum inside the TPCT prohibits intrusive methods (such as electrical wire probe). Up to now, successful online measurement of the thickness of the film in a tubular TPCT has not been reported in the literature.

Relevant reference may be directed to the Nusselt model[1] for a condensed film on a flat surface, assuming that the liquid film is stable with no shear on the vapour–liquid interface. For the film thickness on an flat adiabatic plate, the Nusselt's equation is:

$$\delta = \left[\frac{4\mu_l k_l L_c (T_v - T_w)}{g h_{fg} \rho_l (\rho_l - \rho_v)} \right]^{1/4} \tag{1}$$

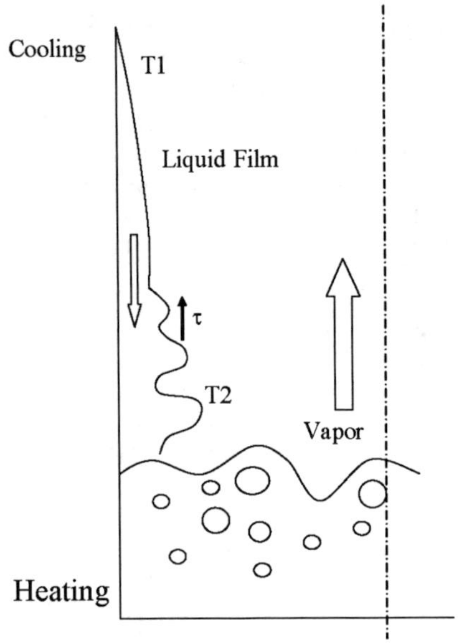

FIGURE 2. Factors affecting δ

Heat transfer inside TPCTs has been investigated both experimentally and theoretically[2,3,4]. It was discovered that the condition inside a TPCT is more complex than the Nusselt's model. This can be demonstrated by examining a more realistic description of a falling liquid film in a TPCT as in figure 2.

First, there exists a temperature gradient, i.e. T2>T1,which tends to create a thermocapillary motion under the effect of surface tension. This tangential stress on the film surface is directed from the regions with the highest surface temperature to the regions with the lowest temperature and amounts to [5, 6]

$$\tau = \nabla\sigma = \left(\frac{\partial\sigma}{\partial T}\right)\nabla T \tag{2}$$

Although at current stage it is difficult to find an exact model for the thermocapillary effect in TPCTs, however, for a shallow wide cuvette with different temperature on the side walls, the rate of thermocapillary convection on the film surface may be estimated by[5, 6]:

$$u_T = \frac{1}{4\pi}\frac{\partial\sigma}{\partial T}\delta\frac{dT}{dx} \tag{3}$$

As against the downward flow of the liquid film, the thermocapillary effect tends to reduce the flow speed of the film, which would consequently increase the film thickness.

Second, the counter-current flow of the vapour exerts a shear force on the film surface in the form:

$$\tau = f\left(u_v + u_i\right) \tag{4}$$

This shear may deter the downward flow of the film, thus also cause a thicker film.

Last, under certain conditions, waves tend to appear on the film surface, causing uneven thickness of the film that can be modelled by a sinusoidal function:

$$\delta = \delta_{avg} + f\left(\sin(x)\right) \tag{5}$$

The above highlights the fact that, although Nusselt's equation is a handy reference, it would almost certainly under predict the film thickness. Therefore, data from real measurements will be of great value.

ECT SYSTEM AND IMAGE RECONSTRUCTION

ECT is a merging technique for visualizing phase distribution according to the permittivity variations in the sensing area. Data acquisition of ECT follows a sequence similar to a medical CT: each electrode takes turns to be charged, and when one electrode is charged, signals are taken from the other electrodes. A typical ECT system is shown in Figure 3. The electrodes encircle the sensing area enabling non-intrusive measurement, the

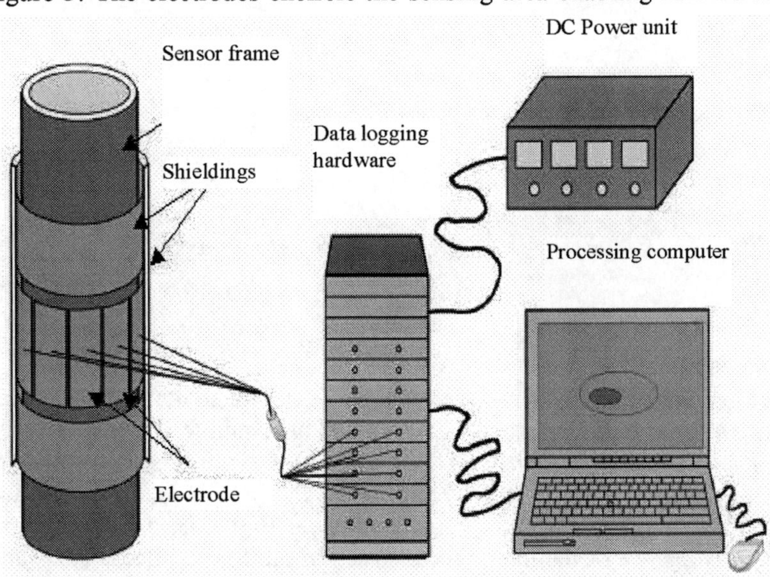

FIGURE 3. ECT system

logging electronics collect data at a high speed, and images are produced by the computer via certain image reconstruction algorithms. More details can be found in Liu et al.[7]. With the advance in hardware design, in sensor optimization and in image reconstruction method, ECT is becoming a very promising technique to solve the problems on hand. It is therefore the purpose of this study to examine the feasibility of ECT to visualize the flow inside a TPCT, as well as to find a suitable image reconstruction algorithm for such a task.

Despite the successful applications of ECT to many multiphase systems, real-time visualizing thin film is a special challenge for ECT due to its limited spatial resolution. Among available image reconstruction algorithms, the Landweber iterative method[8] is a most frequently used one and in many cases produces good images. However, the step length and the number of iteration require careful considerations for the current case. It is known that inappropriate step length will cause either prolonged image reconstruction time or even divergence of the iterations. Also, insufficient number of iterations will not produce much improved images, but an excessive number of iterations will in tern deteriorate the quality of the images[9].

The above problems are tackled by adopting an optimal step length. The Landweber iteration is expressed as:

$$\mathbf{G}^{(k+1)} = \mathbf{G}^{(k)} + \alpha^{(k)}\mathbf{S}^T\mathbf{e}^{(k)}: \qquad \mathbf{e}^{(k)} = \mathbf{C}^{(k)} - \mathbf{S}\mathbf{G}^{(k)}, \quad (k = 0, 1, 2, 3, \ldots), \tag{6}$$

where vector **e** is the error, or discrepancy, between the measured capacitance **C** and the image-converted capacitance **SG**. A fixed step length α is not optimized for iteration efficiency or convergence. Alternatively, a dynamically auto-adjusting step length, i.e. an optimal step length, will both accelerate the speed of image reconstruction and reduce the tendency of divergence. The optimal step length is derived by minimizing the 2-norm of the error vector **e**:

$$\mathbf{e}_{k+1} = \mathbf{C} - \mathbf{S}\mathbf{G}_{k+1} = \mathbf{C} - \mathbf{S}\left(\mathbf{G}_k + \alpha_k\mathbf{S}^{\mathrm{T}}(\mathbf{C} - \mathbf{S}\mathbf{G}_k)\right) = \mathbf{e}_k - \alpha_k\mathbf{S}\mathbf{S}^{\mathrm{T}}(\mathbf{C} - \mathbf{S}\mathbf{G}_k) \tag{7}$$

Let $f_k = \| \mathbf{e}_k \|^2$ ($k = 0, 1, 2, 3, \ldots$), then

$$f_{k+1} = \| \mathbf{e}_{k+1} \|^2 = \| \mathbf{e}_k - \alpha_k\mathbf{S}\mathbf{S}^{\mathrm{T}}(\mathbf{C} - \mathbf{S}\mathbf{G}_k) \|^2 \tag{8}$$

The aim is to minimize f_{k+1} at each step by letting

$$\frac{df_{k+1}}{d\alpha_k} = 0. \tag{9}$$

By solving the above equation, α_k can be found:

$$\alpha_k = \left\|\mathbf{S}^T\mathbf{e}_k\right\|^2 / \left\|\mathbf{S}\mathbf{S}^T\mathbf{e}_k\right\|^2. \tag{10}$$

Further details can be found in Liu et al[10].

Many tests have been conducted. In general, the optimal α can reduce the number of iterations and produce a series of consistent images in the iteration process. Figure 4 is an example where (a) shows the distribution of a rod above a layer; (b) is the image by LBP, in which the rod is not reconstructed; (c) and (d) show the Landweber iteration with a step length 1.2, and (e) and (f) are the images reconstructed by the optimal α. After 30 iterations, the image in (e) matches that in (c), However, after 10,000 iterations, (d) has many artefacts in the images, an indication of divergence of the image reconstruction process, while (f) still shows a similar image to (e), indicating a consistent image reconstruction process. Further information can be fund in Liu et al.[10].

The above method is suitable for high quality off-line image reconstruction for the collected data. Meanwhile, for fastest online monitoring of the flow, other methods can be used, such as the Linear Back Projection (LBP) method that is simply $\mathbf{G} = \mathbf{S}^T\mathbf{C}$[11], or the recently developed OIOR algorithm that offers images with the same quality as the Landweber's method, yet with a fastest reconstruction speed as LBP[12].

The significance of this optimal step length is that it enables unlimited iterations without deteriorating the images, thus the quality of the images can be improved to approach the best possible, which is a unique advantage for the quality-demanding liquid film visualization.

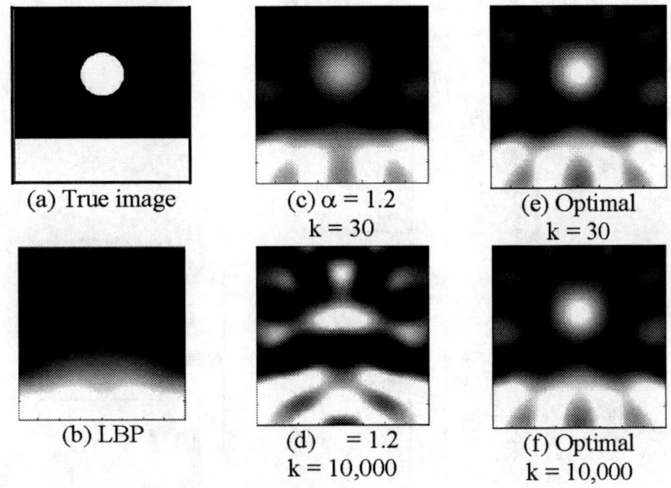

FIGURE 4. Effect of optimal α

The application of certain thresholding will enhance the ability to distinguish the water-vapour interface. If the range of the grey level of the elements in **G** is 0 to 1, a commonly used global (or fixed) threshold is:

$$\mathbf{G}=\mathbf{F}\mathbf{G}:\quad F_{i,i}=\begin{cases}0, & if\ G_i\le Lo\\ 1, & if\ 0\le G_i\le 1\\ 1/G_i, & if\ G_i\ge 1\end{cases} \tag{11}$$

where Lo is the low threshold according to particular requirement, **F** is a diagonal matrix, $F_{i,i}$ and G_i are elements in **F** and in **G** respectively.

If the histogram of an image is bimodal, which is apparently a reasonable assumption for the liquid-vapour flow in the TPCT, then a very often effective method is the Otsu's thresholding that selects the threshold by minimizing the within-class variance of two groups of pixels separated by the thresholding operator. Details of Otsu's method can be found in the literature[13] and will not be elaborated here. Effects of thresholding are compared in the subsequent section of this paper for film thickness measurement.

EXPERIMENTAL

The experimental system is sketched in Figure 5. The TPCT is a glass sequentially followed by an adiabatic section and an evaporator section. All the three sections are of an equal length of 200 mm. The condenser section is encircled by a water jacket to cool the vapour into water film. Cold water flows through the water jacket with controllable flow rates to maintain required cooling rates. An electric heating wire is wound round the evaporator to provide controlled heating rates by adjusting the voltage of the power supply. The TPCT is wrapped with glass wool as an insulating material to minimize heat loss. Ethanol is chosen as the working fluid for its appropriate electrical permittivity for ECT measurements. Data acquisition speed is adjustable.

The sensor has been depicted in figure 1. Eight electrodes are evenly spaced round the frame that is actually a part of the adiabatic section of the TPCT. The length of the electrodes is 10 mm and the gap between each pair of electrodes is 1 mm. To eliminate environmental interferences, a ring of metal shields is fixed both below and above the electrodes, and the whole sensor is again enclosed by another metal shield. The ECT's data logging electronics are based on AC excitation and acquisition circuits. Data collecting rate can be adjusted up to 140 frames per second. The measured capacitance data are processed by the computer and images are reconstructed.

Before the experiments, the working fluid was charged into the TPCT through a vacuum sealed valve. The amount of filled liquid was 15 ml. The vacuum level inside the TPCT was maintained at 10^{-2} torr after filling.tube, with an outer diameter of 15 mm, a wall thickness of 1.5 mm and a length of 700 mm. 100 mm below its top starts the condenser section,

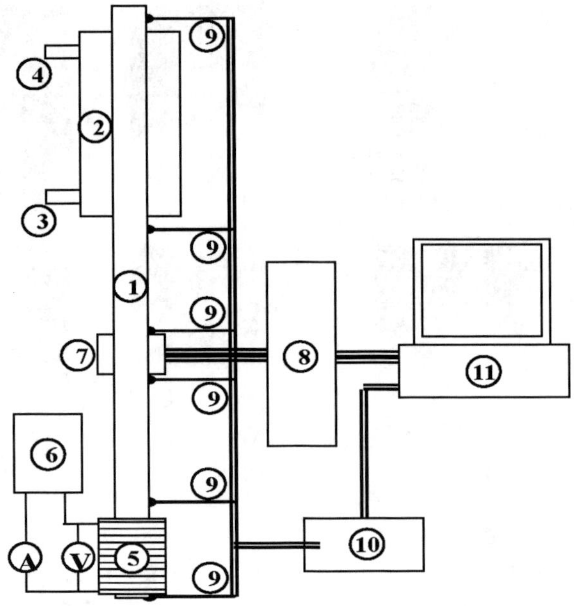

FIGURE 5. Experimental apparatus

1. TPCT, 2. water jacket, 3. cooling water inlet, 4. cooling water outlet, 5. heater, 6. power unit, 7. ECT sensor, 8. ECT data acquisition unit, 9. thermal couples, 10, temperature measurement unit, 11. computer

FLOW PATTERN VISUALIZATION

A series of tests were performed covering a range of heating voltages from 30 V, at an increment of 10 V, to the upper limit of TPCT's capacity of heat transfer, which is a point when the TPCT is no longer able to convey all the heat input, marked by the constantly rising temperature of the evaporator.

TABLE 1. Comparison of images reconstructed by three algorithms

V	40	60	80	100	120	140
LBP	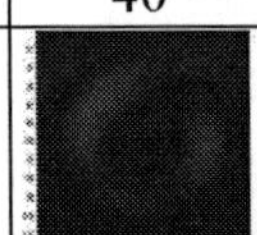	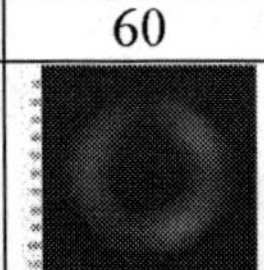	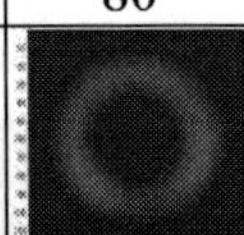	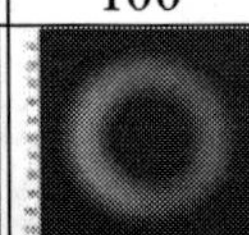	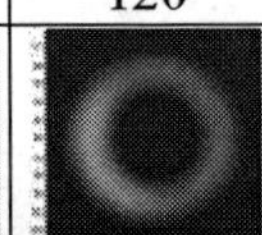	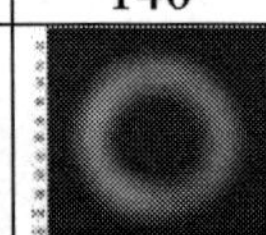
Land	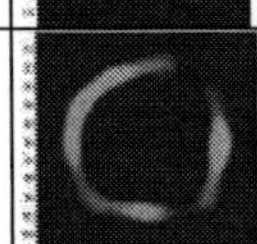	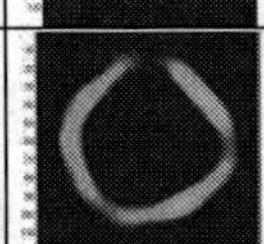	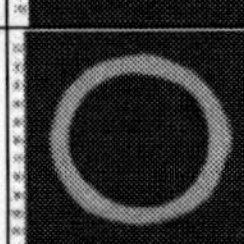	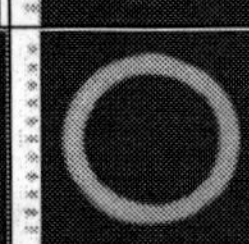	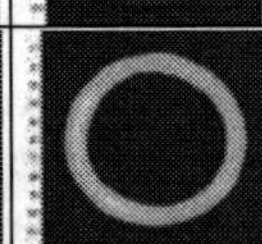	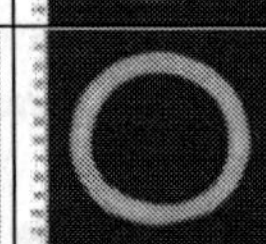
Opt α	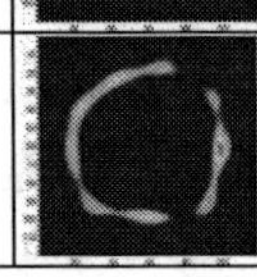	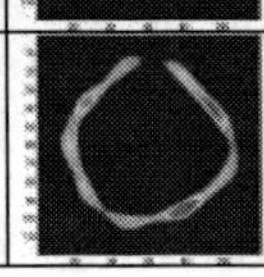	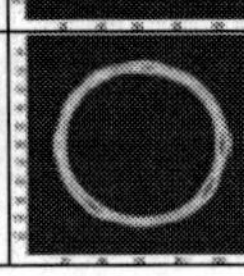	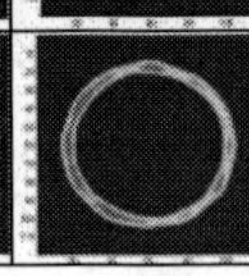	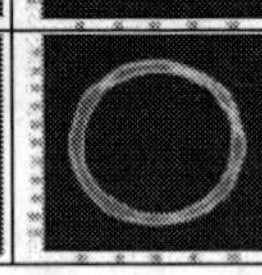	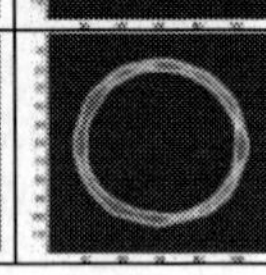

To compare the effects of different image reconstruction methods, LBP, the Landweber iteration (20 steps) and the Landweber iteration with the optimal step length (also 20 steps) are applied and the images of an annular flows are listed in Table 1, indicated by LBP, Lan, and Opt α respectively. It can be seen that the Landweber iteration with the optimal step length yields the finest thickness and highest grave level of the images.

Based on the data for the images in Table 1, figure 6 is the characteristics of the variation of the voidage (a) and the film thickness (b) in the tube, corresponding to a 60 volts heating rate (heat flux of 5.5 kW/m^2). The x coordinate is the frame number, at a ratio of 10 frames per second. From the figure it can be seen that the fluctuations are relatively small, implying a constant annulus flow pattern.

TABLE 2. Flow pattern for an increased heating rate

Frame 2834	Frame 2835	Frame 2836	Frame 2837
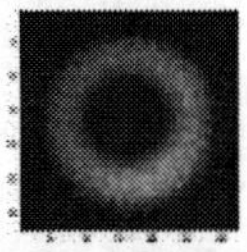	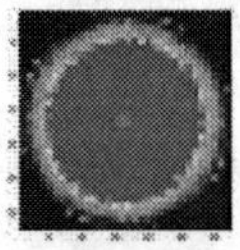	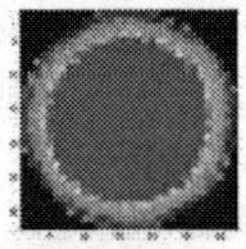	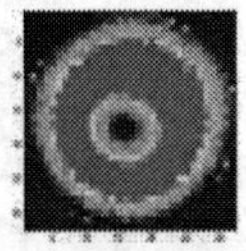
Frame 2838	Frame 2839	Frame 2840	Frame 2841
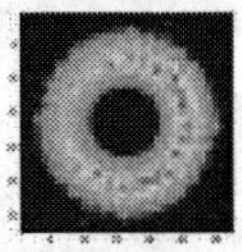	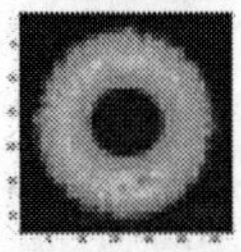	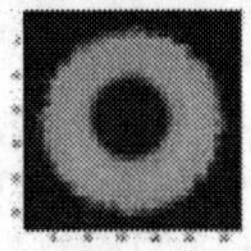	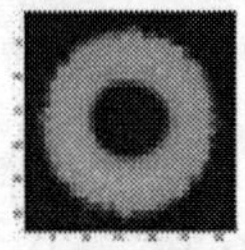

As demonstrated in Table 2, when heating rate increases, the flow become unstable with alternating annular flow (frames 2834, 2838-2841) and liquid plugs (i.e. frame 2835, 2836, 2837). For the same sampling rate,

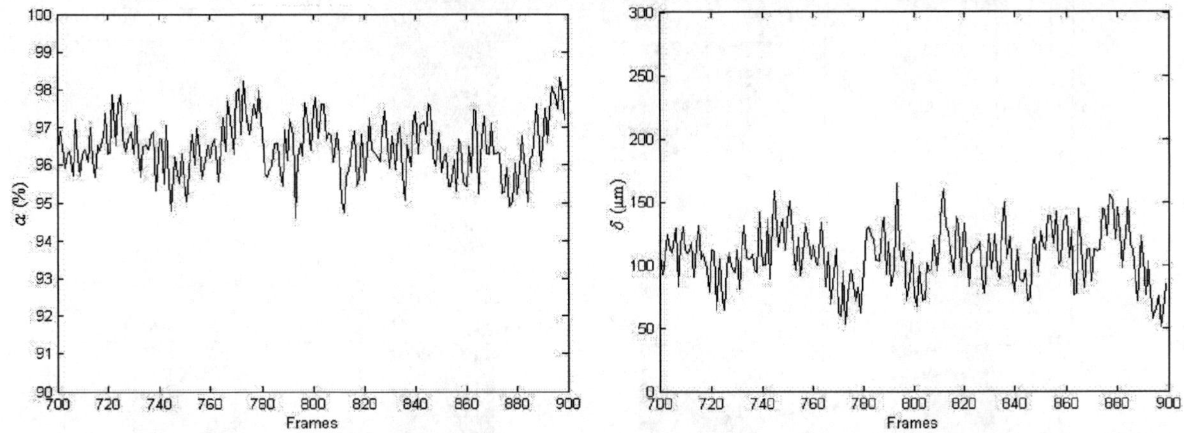

(a) Variation of voidage (b) Variation of film thickness

FIGURE 6. Variations of voidage and film thickness in TPCT for moderate heating rates

figure 7 shows the characteristics of the variation of the voidage (a) and film thickness (b), corresponding to a

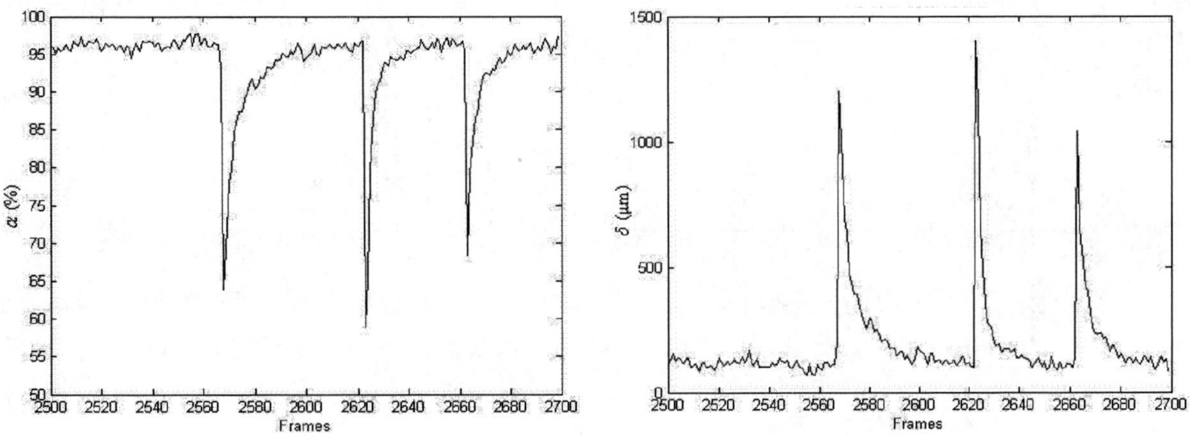

(a) Variation of voidage (b) Variation of film thickness

FIGURE 7. Variations of voidage and film thickness in TPCT for stronger heating rates

100 volts heating rate (heat flux of 15.2 kW/m^2). It can be seen that within about 100 frames, i.e. equivalent to 10 seconds, three pulses occur in voidage and in film thickness, indicating an unstable flow pattern.

For very large heating rates, the flow pattern in the TPCT will be very unstable with frequent pulses in voidage and in film thickness. This can be seen in Table 3, where the working fluid frequently fill the whole pipe cross section as liquid plugs.

Figure 8 is the characteristics of the variation of the voidage and film thickness in the tube, corresponding to a 150 volts heating rate (heat flux of 34.1 kW/m^2). It can be seen that with the increase in heating rate, the flow becomes increasingly unstable. Frequent pulses occur in voidage and in film thickness. In this case, the flow pattern has changed totally from annular flow to a kind of "foaming" flow, an unfavourable flow condition for TPCT.

Table 3. High frequency pulses in voidage and film thickness

Frame 2999	Frame 3000	Frame 3001	Frame 3002
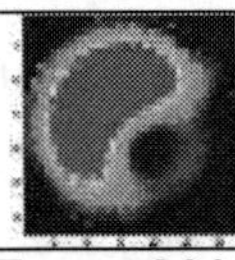	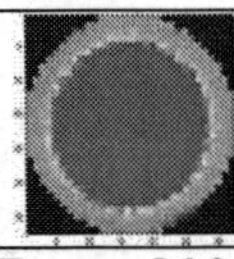	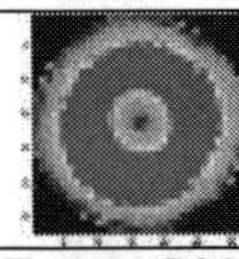	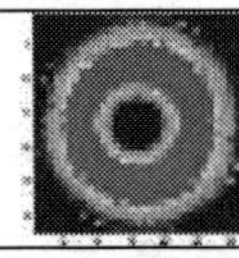
Frame 3003	Frame 3004	Frame 3005	Frame 3006
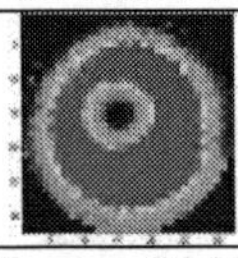	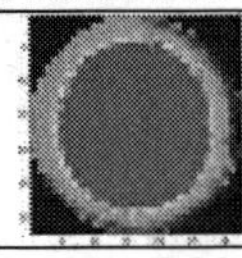	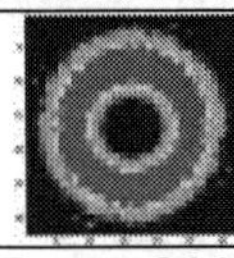	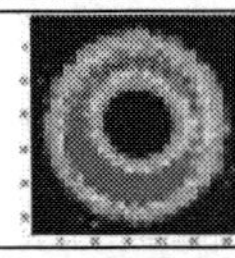
Frame 3007	Frame 3008	Frame 3016	Frame 3017
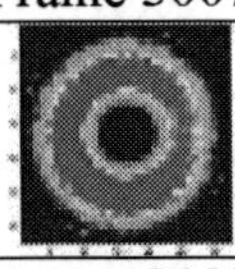	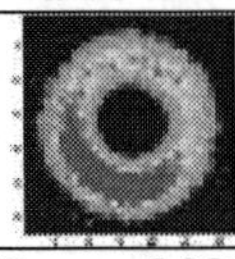	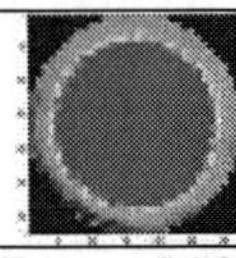	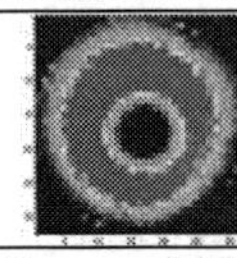
Frame 3027	Frame 3028	Frame 3029	Frame 3030
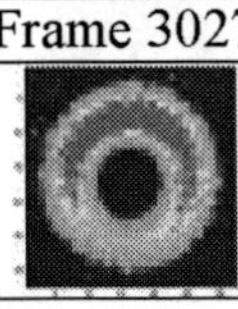	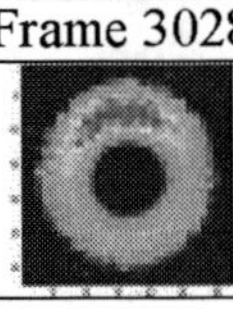	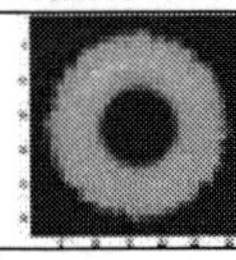	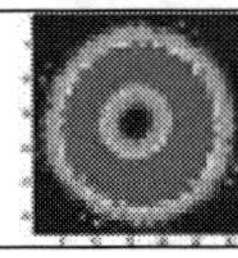

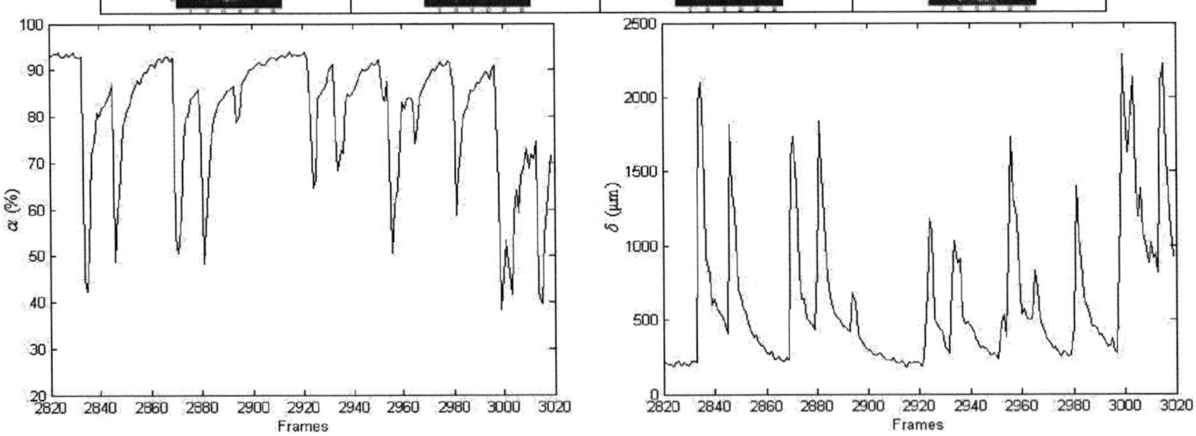

(a) Variation of voidage

(b) Variation of film thickness

FIGURE 8. Variations of voidage and film thickness in TPCT for high heating rates

FILM THICKNESS MEASUREMENT

For the case of stable annular flows, figure 9 provides a comparison among several sets of results for the peripherally averaged film thickness, including (A) Nusselt's predictions, (B) CFD simulations in which the vapour shear and surface wave effect are modelled[14], (C) ECT data with global threshold (Lo= 0.19), (D) ECT

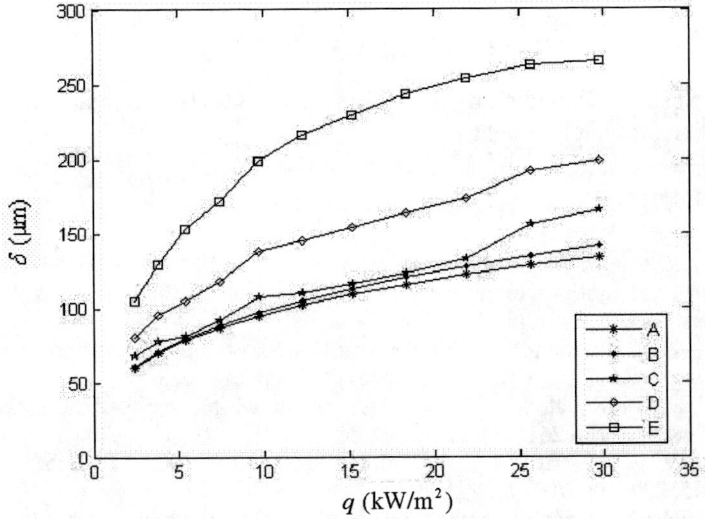

FIGURE 9. Comparison between film thickness by ECT with Nusselt's model
A: Nusselt model, B: CFD simulation, C: global threshold, D: Otsu's threshold, E: Landwber

results with the Otsu's threshold, (E) ECT results by the Landwber method without thresholding. The x-coordinate is the heating rate of the TPCT.

As discussed before, the interfacial shear, the wave of the film surface and the thermocapillary effect all tend to yield higher film thickness than predicted by the Nusselt's equation. From figure 9, it can be seen that CFD model gives slightly higher values than the Nusselt's model. However, at the current stage, the thermocapillary effect has not been included in the CFD model. Thus the true thickness could be higher than this CFD result. From ECT data, increasing film thickness are reconstructed by the methods, in the order of (i) the Landweber iterations with a 19% global threshold, (ii) the Landweber iterations with the Otsu's thresholding, and (iii) Landweber iterations without thresholding; all with 20 steps of iteration. From the above it is seen that, apart from the Landweber iteration without thresholding, all the other results deviate by only small amounts.

It is also seen that the image reconstruction methods have apparent influences on the accuracy of the measured film thickness. Therefore, further studies on this subject are needed to identify the most appropriate methods. More advanced pixel-division methods would also be desired for finer definition of the shape of the liquid film.

On the other hand, the characteristics of condensation inside a TPCT are more complex than predicted by the Nusselt's model. This problem has been investigated by investigators such as Yang and Pan[15], with a conclusion that the inter-facial shear does have a strong effect on the condensation and reflux film inside the thermosyphon. The experimental results reported in this paper seem to agree with the above analysis.

CONCLUSIONS

Various flow patterns of the liquid phase in a TPCT are visualized by ECT, which demonstrates the ECT's capability of monitoring the working condition of TPCTs, and presents a novel technique for the investigation of various heat pipes. The μm-scale film thickness measurement by ECT also proves the potential of high spatial resolution of ECT when proper image reconstruction algorithms are used.

At low heating rates, annular flow predominate in TPCT. As heating rate increases, more liquid plugs tend to form and the flow pattern becomes increasingly unstable.

Film thickness measured by ECT agrees quantitatively with the theoretical predictions. However, differences exist showing thicker film than the Nusselt's predications. The Landweber's algorithm with the optimal step length enhances the convergence and speed of the iterations, but different thresholds will reconstruct different film thicknesses, suggesting that further studies are needed on the most proper image reconstruction algorithms for further improved accuracy. Also, more studies on the special characteristics of the film flow in TPCTs should be conducted.

ACKNOWLEDGEMENTS

The National Natural Science Foundation of China (No. 60672151 and 60532020) is gratefully acknowledged for sponsoring the research project.

REFERENCES

1. Nusselt W., *Die Oberflachenkondensation des Wasserdampfes*, VDIZ., Vol 60, 1916, pp. 541-544.
2. Spendel T., "Laminar film condensation heat transfer in closed two-phase thermosyphons", *Advances in heat pipe technology*, Pergamon Press, Oxford, 1981, pp. 163-173.
3. Gross U. and Hahne E., "Condensation heat transfer inside a closed thermosyphon–generalized correlation of experimental data", *Proc. of the 6th IHPC*, Grenoble, France, 1987, pp. 618–623.
4. Ma T. Z., Lin X. and Wu J. p., "Flow patterns and operating limits in two-phase closed thermosyphon", *Proc. of the 6th IHPC*, Grenoble, France, 1987, pp. 256-261.
5. Kabov , O. A. and Chinnov, E. A., "Heat Transfer from a Local Heat Source to Subcooled Liquid Film", *High Temperature*, Vol. 39, No. 5, 2001, pp. 703–713.
6. Levich, V.G., "Fiziko-khimicheskaya gidrodinamika (Physicochemical Hydrodynamics)", Moscow: Gos. Izd. Fiz.-Mat. Literatury (Fizmatgiz), 1959.
7. Liu S., Wang H. G., Jiang F. and Yang W. Q., "A new image reconstruction method for tomographic investigation of fluidized beds", *AIChE J.*, V48, N8, 2002, pp. 1631-1638.
8. Landweber L., "An iterative formula for Fredholm integral equations of the first kind", *Amer. J. Math*, Vol 73, 1951, pp. 615-624.
9. Liu S., Fu L., and Yang W. Q., "Comparison of three image reconstruction algorithms for electrical capacitance tomography", *2nd World Congress on Industrial Process Tomography*, Hannover, Germany, 2001, pp. 29-34.
10. Liu S., Fu L., and Yang W. Q., "Optimization of an iterative image reconstruction algorithm for electrical capacitance", *Meas. Sci. Technol.*, 1999, 10, pp. 37-39.
11. Xie, C.G., Huang, S.M., Hoyle, B.S., Thorn, R., Lenn, C., Snowden, D. and Beck, M.S., "Electrical capacitance tomography for flow imaging: system model for development of image reconstruction algorithms and design of primary sensors", *IEE Proceedings-G*, 1992, 139 (1), pp. 89-97.
12. Liu, S., L. Fu, W.Q. Yang, H.G. Wang and F. Jiang, "Prior-online iteration for image reconstruction with electrical capacitance tomography", *IEE Proc.-Sci. Meas, Technol.* 2004, 151(3), pp. 195-200.
13. Otsu, N., "A threshold selection method from gray-level histogram", *IEEE Transactions on System Man Cybernetics*, Vol. SMC-9, No. 1, 1979, pp. 62-66.
14. Li J., "Investigation on ECT flow pattern recognition and heat transfer characteristics in pulsed heat tube", *PhD Thesis*, Institute of Engineering Thermophysics, Chinese Academy of Sciences, 2007.
15. Yang P., "Condensation heat transfer characteristics and concept of sub-flooding limit in a two-phase closed thermosyphon", *Int. Comm. Heat Mass Transfer*, Vol 28, N3, 2001, pp. 311-322.

NOMENCLATURE

C	Measured capacitance vector by ECT
e	Error vector
G	Image vector
g	Gravitational acceleration (m^2/s)
h_{fg}	Latent heat of vaporization (J/kg)
k	Thermal conductivity ($W/m^2 \cdot ℃$)
L	Length (m)
q	Heat flux (W/m^2)
S	Sensitivity map for image reconstruction
T	Temperature (K)
u	Velocity (m/s)
V	Voltage of power supply (V)
x	Coordinate alone the pipe wall, positive downwards (m)

Greek letters:

α	Step length in Landweber iteration
δ	Liquid film thickness in adiabatic section (m)
μ	Dynamic viscosity (Pa·s)
ρ	Density (kg/m^3)
σ	Surface tension coefficient (N/m)
τ	Tangential stress on the film surface (N/m^2)

Subscripts

avg	average
c	Condenser
i	Surface of the liquid film
l	Liquid
T	Thermocapillary effect
w	Wall
v	Vapour

Investigating Jet Mixing Using Electrical Resistance Tomography

Stephenson DR*, Cooke M*, Kowalski, A+ and York TA**

*School of Chemical Engineering and Analytical Science, University of Manchester, UK,
david.stephenson@postgrad.manchester.ac.uk
**School of Electrical and Electronic Engineering, University of Manchester, UK,
t.york@manchester.ac.uk
+Unilever Research and Development, Port Sunlight, Wirral, UK,
adam.kowalski@unilever.com

Abstract. Coaxial jet and side entry mixers are used in a wide range of industries for a variety of processes including precipitation, polymerization and neutralization duties [1,2,3]. Jet mixers are characterized by short contact times between the fluids and can be operated continuously or semi-batch. Coaxial and side entry jets can be designed in order to deliver rapid turbulent mixing using short sections of pipeline. As the energy required for mixing is provided by the addition stream, the process-side pressure drop required for homogeneity is very low. A key design parameter for jet mixers is the mixing length, the length of pipe downstream of the injection point required to achieve a given degree of homogeneity. The mixing length can be affected by the addition geometry (for example coaxial or side entry), orifice size and shape, operating conditions and material properties. This paper presents the use of Electrical Resistance Tomography (ERT) to monitor jet mixing via the addition of a conductivity tracer through coaxial and side entry jets. Multiple ERT sensors are fitted along the pipe downstream of the jet addition point. The ERT sensors enable real time, non-invasive measurement of conductivity within the pipe, furnishing approximately 2500 conductivity measurements per ERT sensor when modeled independently in 3D. The effect of secondary (main pipe) flow rate and jet configuration on the nature of the tracer plume evolution and axial mixing is determined using this technique.

Keywords: Electrical Resistance Tomography Jet Mixing.

INTRODUCTION

Previous studies have employed single point measurement techniques, empirical relationships and computational fluid dynamics [4,5,1] in order to better understand jet mixing phenomena. Electrical tomography has been used to study many processes including mixing in stirred tank reactors [6], gas-liquid mixing [7] and separation processes [8, 9]. This paper will present a study to explore the use of electrical resistance tomography (ERT) to yield information on key parameters and attempt to determine if ERT is an appropriate tool with which to study jet mixing within pipes. Results using a circular orifice in a coaxial jet configuration will be compared to those using a side entry jet under various operating conditions. The work has been performed on pilot scale plant equipment, through 3.4 cm stainless steel pipe, with key process parameters monitored via a DeltaV control system. The mixing rig is equipped with 3 rings of 16 electrodes that are interrogated using an adjacent pair drive strategy with the ITS P2000 instrument (Industrial Tomography Systems, www.itoms.com).

EXPERIMENTAL METHODOLOGY

Flow Rig

The flow rig used in the experiments is shown in Figure 1(a) below. Mains water is fed to a 3.4 cm internal diameter (ID) stainless steel pipe. The pipe extends for approximately 1.2 m before reducing to 2.2 cm ID and turning a U bend. On exiting the U bend, the fluid passes through an expansion back up to 1.2 m of 3.4 cm ID

CP914, *Multiphase Flow: The Ultimate Measurement Challenge,* edited by X. Cai, Y. Wu, Z. Huang, S. Wang, and M. Wang

before exiting to drains. The flow rig is controlled by a DeltaV control system (Emerson), enabling online measurement of the secondary (main pipe) flow rate through the rig. The conductivity tracer, 5 g l^{-1} NaCl (Sigma-Aldrich) in mains water, for the jet addition stream was fed from a centrifugal pump, flows measured using a float-type flow meter.

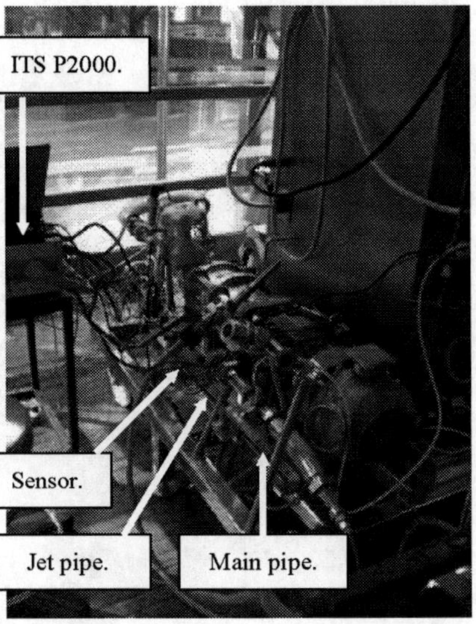

(a)

(b)

FIGURE 1. Experimental flow rig (a) and ERT sensor array (b).

The ERT sensor arrays are constructed of Perspex and are cylindrical of 3.4 cm ID. Figure 1(b) shows the sensors consisting of 16 circular electrodes of diameter 3.4 mm, sitting flush with the inside wall of the sensor and are therefore in contact with the process fluid. The sensors are connected to the ITS P2000 data acquisition system via 1 meter lengths of coaxial cable; all cable shields are grounded to the metal pipe work to reduce electrical noise. The circular orifice used for the coaxial and side entry jets has an internal diameter of 2 mm. The ERT sensors were arranged in two configurations along the main pipe to enable measurements for several experimental objectives. The first objective aims to visualize the plume of tracer formed by jet addition, under various secondary flow rates and jet configuration. For this objective, the sensors were aligned adjacent to the jet entry. The second objective aims to monitor mixing along the axial length of the main pipe, hence the sensors were positioned axially along the pipe. Each experimental objective was carried out at three secondary water flow rates, 10 liters min^{-1}, 20 liters min^{-1} and 35 liters min^{-1} whilst the jet flow was held constant at 0.3 liters min^{-1}. The experimental programme is summarized in Table 1 below.

TABLE 1. Overview of Experimental Programme.

Experiment	Secondary flow, Q_s, liters min^{-1}.	Jet flow, Q_j, liters min^{-1}	Jet type	Objective
Vis_10_c	10	0.3	Coaxial	Visualisation
Vis_20_c	20	0.3	Coaxial	Visualisation
Vis_35_c	35	0.3	Coaxial	Visualisation
Vis_10_s	10	0.3	Side entry	Visualisation
Vis_20_s	20	0.3	Side entry	Visualisation
Vis_35_s	35	0.3	Side entry	Visualisation
Mix_10_c	10	0.3	Coaxial	Mixing analysis
Mix_20_c	20	0.3	Coaxial	Mixing analysis
Mix_35_c	35	0.3	Coaxial	Mixing analysis
Mix_10_s	10	0.3	Side entry	Mixing analysis
Mix_20_s	20	0.3	Side entry	Mixing analysis
Mix_35_s	35	0.3	Side entry	Mixing analysis

Image Reconstruction

EIT belongs to a family of non-invasive, soft field electrical tomographic modalities. In EIT, the internal complex admittivity (electrical conductivity and permittivity) distribution of a domain is determined by injecting low frequency (typically below 100kHz) AC current and measuring the amplitude and phase of the potentials on electrodes placed on the periphery of the domain. In forward EIT calculations, an estimation of the interior scalar potential field is made given the internal admittivity distribution, under mixed Neumann and Dirichlet boundary conditions, typically using a finite element discretisation of the domain. The inverse EIT calculation involves estimating an admittivity distribution given the measured amplitude and phase on periphery electrodes.

In order to solve the inverse admittivity boundary value problem, one requires a forward operator that predicts voltage observations for a given admittivity distribution. For EIT, the forward operator is required to calculate boundary voltages (and the scalar potential distribution for determining the Jacobian, a matrix of partial derivatives), given the current stimuli and an admittivity distribution. For this operator, a mathematical model is derived from Maxwell's equations. Underlying assumptions are that the admittivity is linear and isotropic, and also that the electric and magnetic fields described by Maxwell's equations are slowly varying. The application of the above assumptions to Maxwell's equations results in the derivation of the Laplacian elliptic partial differential equation shown in (1), where γ is the complex admittivity and u is the scalar potential distribution. The forward problem is formulated when (1) is solved (often by finite element discretisation) using a set of boundary conditions, a common choice is the complete electrode model [10].

$$\nabla \cdot (\gamma \nabla u) = 0 \tag{1}$$

Given the complex vectors $x \in C^n$ and $b \in C^m$, and a complex matrix $A \in C^{mxn}$, the aim of the inverse problem is to find x, given b and A ($Ax = b$). In the case of EIT, b is the measurement vector, x the admittivity (conductivity in ERT) vector and A the Jacobian matrix. Ideally one would solve a system of equations of this type by determining the inverse of the matrix A, however, due to its compact nature, it has an unbounded (discontinuous) inverse. This causes A^{-1} to be unstable against variations in the data, hence violating Hadamard's third criterion for well posedness. Several techniques exist for updating the admittivity distribution once the Jacobian matrix has been computed, many using regularization techniques in order to make the problem less ill-posed [11].

RESULTS

Results from the two experimental objectives are presented in the following sections. Various image reconstruction algorithms were investigated and the Generalized Singular Value Decomposition (GSVD) [11] was determined to be the most appropriate algorithm for this study. A further discussion of algorithm choice and modeling approach can be found below.

Jet Visualization

The results from the ERT measurements visualizing the jet plume structure close to the jet addition points are shown in Figure 2 below. The size of the reconstructed domain is approximately 15 cm in axial length and 3.4 cm diameter; encompassing the three ERT sensors. Figure 2 illustrates the results by visualizing cross-sectional slices through the domain, however, it is important to note that the domain has been reconstructed in a 3D model consisting of approximately 9000 elements. The reconstructed images in Figure 2 are time-averaged over the tracer injection period. Inversion of voltage data into a solution space of 9000 elements requires computational ability above that of standard desktop computers. The computer used for reconstructing the results in Figure 2 was a high performance Linux workstation, running Matlab 64 under Scientific Linux 3 with a dual core AMD processor and 8 GB of RAM.

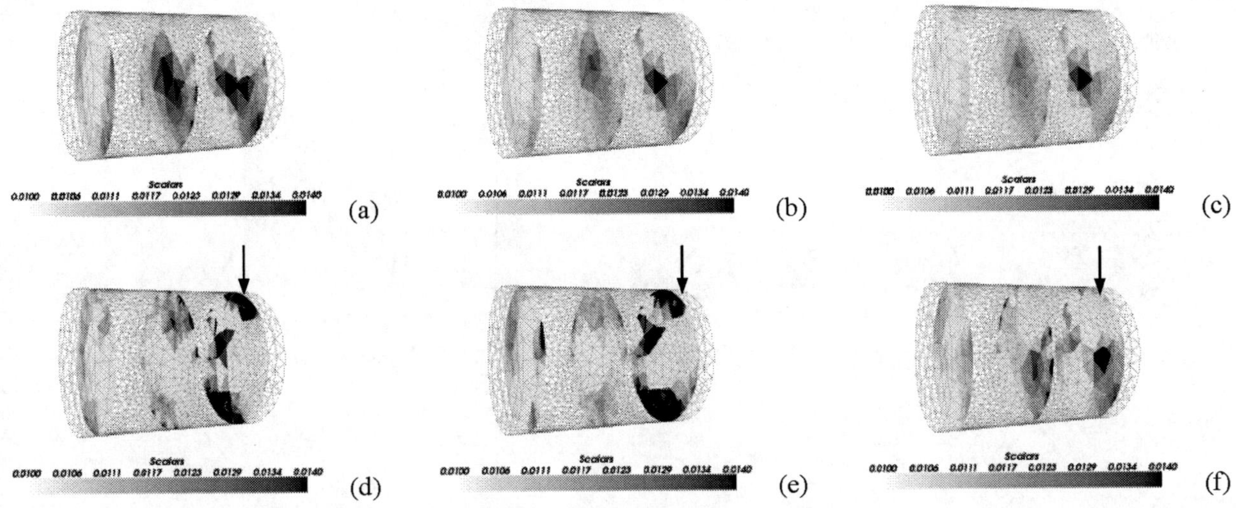

FIGURE 2. Tracer plume visualization experiments for Vis_10_c (a), Vis_20_c (b), Vis_35_c (c), Vis_10_s (d), Vis_20_s (e) and Vis_35_s (f). Grey scales are constant, darker regions indicating higher tracer concentration.

It can be seen from Figure 2(a) to Figure 2(c) that as the secondary flow rate increases, the coaxial jet (areas of higher conductivity shown darker) is better defined and dynamic results show that the jet is more stable. The ERT results illustrate that at lower secondary flow (Figure 2a), increased radial mixing occurs. ERT detects significant differences between the coaxial and side entry jet configurations; comparing vertically adjacent results in Figure 2 confirms this. The results from the side entry jet, Figure 2(d) to Figure 2(f), indicates that the side entry jet is fully penetrating the secondary flow until the secondary flow is increased to 35 liters min^{-1}. The arrows indicate the position of the side entry jet, several centimeters upstream of the reconstructed domain.

Axial Mixing

The results from the study of axial mixing along the pipe for the two jet configurations are presented in Figure 3 below. Plane 1 is at an axial distance of 5 cm from the jet, plane 2 at 55 cm and plane 3 at 80 cm. Coefficient of Variation of the reconstructed conductivity, COV, (100 x standard deviation / mean) is plotted against frame number (at constant frame rate) in Figure 3; a zero value of COV indicates a homogeneous distribution, hence well mixed. Analysis of the rows of Figure 3 enables a comparison of the jet configuration at constant flows and analysis of the columns of Figure 3 enables a comparison of secondary flow rates within the same jet configuration. Figure 3 shows the COV as a function of frame number (~2 fps), during an addition of conductivity tracer lasting between 100 and 200 frames. Initially the COV is low (no tracer has been injected), and then rises sharply as the tracer is added. Once the tracer injection has ceased, the COV values return to initial pre-injection values. It is important to note that the duration of the COV increase is a function of tracer injection time and not primarily influenced by the respective jet and secondary flows.

Figure 3(a), (c) and (e) illustrate that increasing the secondary flow past the coaxial jet from 10 to 35 liters min^{-1} lowers the COV of plane 2 towards the COV value of plane 3, and hence improves the axial mixing along the pipe. Of particular interest is plane 2; illustrating that as the secondary flow increases, plane 2 becomes almost as well mixed as plane 3. The same result is not replicated for the side entry jet, Figure 3(b), (d) and (f), although there is a reduction in the plane 3 COV with increasing secondary flow. A comparison of the rows of Figure 3 (i.e. a & b) indicates that the coaxial jet has lower COV values (more homogeneous) and delivers superior axial mixing at constant jet to secondary flow ratios.

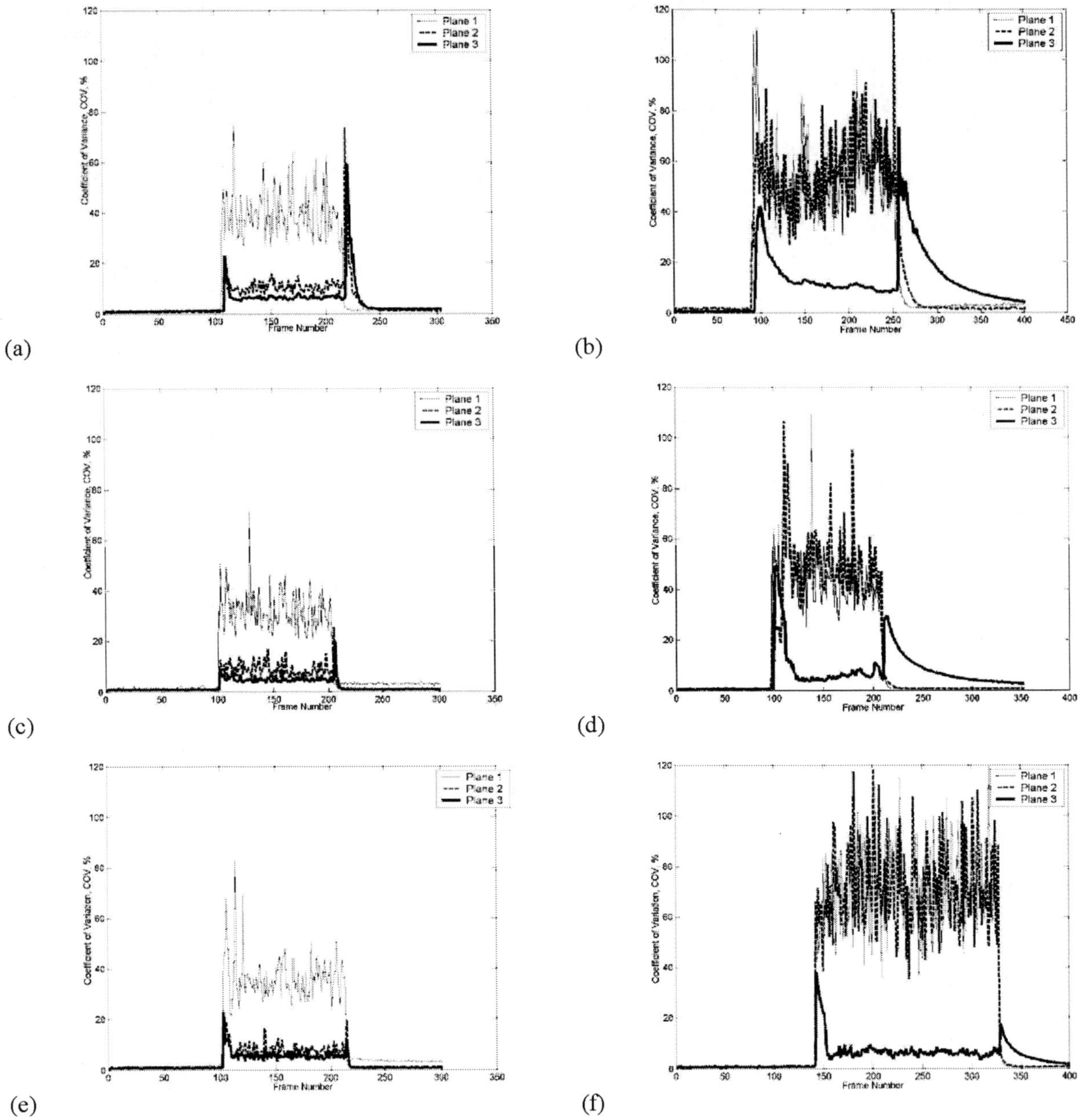

FIGURE 3. Mixing analysis experiments for Mix_10_c (a), Mix_10_s (b), Mix_20_c (c), Mix_20_s (d), Mix_35_c (e) and Mix_35_s (f).

Choice of Algorithm, Parameters and Modelling Approach

It is known that the choice of reconstruction algorithm strongly affects reconstructed images, and the choice of algorithm should be based upon some prior knowledge of the system under investigation [12]. The data presented above have been analyzed using various 2 and 3D reconstruction algorithms using the finite element forward solver and Jacobian algorithms in the Eidors toolkit [13]. It was found that the real data could not be fitted to enough precision to warrant the use of non-linear algorithms, therefore the choice of algorithms was limited to Linear Back-Projection, Landweber, Conjugate Gradients, Linear Tikhonov Regularized Gauss-Newton, Total Variation Regularization and Generalized Singular Value Decomposition (GSVD). It was found through visual comparison of the reconstructed images with video images of the experiments that the GSVD algorithm had superior performance with a regularization parameter of 0.001. The choice of reconstruction algorithms, parameters and modeling approach (i.e. 2D or 3D) strongly impacts on the reconstructed conductivity, both quantitatively and spatially. It is therefore important to note that ultimately, the results derived are dependent upon these choices available to ERT practitioners. The impact of choices available is a subject of ongoing research and forms the basis of future work on confidence in ERT reconstruction and derived process measures.

CONCLUSIONS AND FUTURE WORK

It has been demonstrated that electrical resistance tomography is a useful tool in the analysis of jet mixing within pipelines. ERT is capable of visualizing tracer plume development exiting a coaxial or side entry jet. It is possible to monitor mixing along the axial length of the pipe using series of ERT sensors, furnishing approximately 2500 conductivity measures per ERT sensor. Analysis of the variance of the conductivity along the length of the pipe allows a measure of mixing progress under various operating conditions. Whilst ERT cannot yet produce quantitative data on conductivity levels within domains it remains a useful tool for comparative studies. The road to quantitative conductivity data necessitates the use of non-linear algorithms, more accurate measurements and improved regularized inverse methods. Future work on jet mixing will study industrially relevant process chemistry, improved quantitative analysis of conductivity and methods to measure mixing lengths within pipes.

REFERENCES

1. Wei, H., and Garside, J. Application of CFD modelling to precipitation systems. *Chemical Engineering Research and Design*, 75(2), 219-227, 1995.
2. Kolhapure, N.H., Tilton, J. N., Pereira, C., J. Integration of CFD and condensation polymerization chemistry for a commercial multi-jet tubular reactor. *Chem. Eng. Sci.,* 59, 5177–5184, 2004.
3. Pipino, M., and Fox, R. O. Reactive mixing in a tubular jet reactor: a comparison of PDF simulations with experimental data. *Chem. Eng. Sci.,* 49(24B), 5229-5241, 1994.
4. Smith, J. V., and Perona, J. J. Axial mixing of fluids in turbulent flow through concentric annuli. *Industrial & Engineering Chemistry Fundamentals*, 8(4), 621, 1969.
5. Revill, B. K. Jet Mixing. *In* Harnby, N., Edwards, M.F. and Nienow, A.W. (Eds.), *Mixing in the Process Industries*, 2nd Ed., Butterworth-Heinemann, Oxford, 1992.
6. Mann, R., Dickin, F. J., Wang, M., Dyakowski, T., Williams, R. A., Edwards, R. B., Forrest, A. E., and Holden, P. J. Application of ERT to interrogate mixing processes at plant scale, *Chem. Eng. Sci*., 52(13):2087-2097, 1997.
7. Wang, M., Dorward, A., Vlaev, D., and Mann, R. Measurements of gas-liquid mixing in a stirred vessel using electrical resistance tomography (ERT), *Chemical Engineering Journal*, 77:93-98, 2000.
8. Gutierrez J.A., Dyakowski T., Beck M.S., and Williams R.A. Using electrical impedance tomography for controlling hydrocyclone underflow discharge, *Powder Technology*, 108:180-184, 2000.
9. Grieve, B.D., Davidson, J.L., Mann, R., Lionheart, W.R.B. and York, T.A., Process compliant electrical impedance instrumentation for wide scale exploitation on industrial vessels. *Proc. 3rd World Congress on Industrial Process Tomography*, Banff, Canada (Eds. Dyakowski, T., Kanzas, A. and Ronson, P.), 806-812, 2003.
10. Somersalo, E., Cheney, M., Isaacson, D. Existence and uniqueness for electrode models for electric current computed tomography. *SIAM J. Appl. Math.*, 52:1023-1040, 1992.
11. Hansen, P.C., *Rank-deficient and discrete ill-posed problems: numerical aspects of linear inversion*, SIAM, Philadelphia, 1998.
12. Stephenson D.R., Davidson J.L., Lionheart W.R.B., Grieve B.D. and York T.A. Comparison of 3-D Image Reconstruction Techniques Using Real Electrical Impedance Measurement Data. *4th World Congress on Industrial Process Tomography*, Aizu, Japan, Sept. 5-8 2005, 643-650.
13. Polydorides, N., Lionheart, W.R.B. A Matlab toolkit for 3-dimensional electrical impedance tomography: a contribution to the Electrical Impedance and Diffuse Optical Reconstruction Software project. *Meas. Sci. Technol.*, 13:1871-1883, 2002.

Electrical impedance tomography with square sensor

Zhang, Cao and Huaxiang, Wang

School of Electrical Engineering and Automation, Tianjin University, Tianjin 300072, China
Email: zhang.cao.tju@gmail.com, hxwang@tju.edu.cn

Abstract. A square electrical impedance sensor has been designed for circulating fluidized beds(CEBs). Based on the conformal transformation theory and precise integration method for elliptic function, a fast image reconstruction algorithm using an analytical model has been introduced. And the results validate the feasibility of the algorithm.

Keywords: square sensor, electrical impedance tomography, analytical model, precise integration method, conformal mapping
PACS: 41.20.Gz

INTRODUCTION

Circulating fluidized beds (CFBsx) have been used for coal combustion in the power industry. The continuing development of electrical impedance tomography (EIT) provides an opportunity to investigate the fluidization phenomena. And ECT has been used to investigate pneumatic conveyors and fluidized beds. In the past, circular ECT sensors have been used. Since square or rectangular CFBs are commonly used in the combustion industry, there is a demand on ECT to image such apparatus. For this purpose, a square sensor, together with the corresponding sensitivity maps, was developed by Yang and Liu [1]. And a preliminary experimental result is given by Liu, etc [2].

Usually, EIT has been mostly treated in a single mode [3, 4, 5], either as ERT [6] or ECT [7], or a simple combination of the two single modes [8] and this single mode treatment limits the range of multi-phase flow visible measurement, since ERT and ECT usually deal with electrical conductivity distribution and electrical permittivity distribution, respectively. The information of both capacitive and resistive mode is not acquired from the same electromagnetic filed, either at different locations of the pipe/vessel or at different time intervals, and this restricts the capability of EIT to capture the details of the multi-phase flow. In order to extract the maximum information of the material distribution of the pipe/vessel cross-section and minimize hardware cost, the EIT sensor is expected to obtain the resistive and capacitive information on the same cross section simultaneously. In this paper, a simple model for EIT information fusion impedance square sensor is established and a new strategy for EIT voltage excitation is introduced, and a fast image reconstruction algorithm based on mathematical formulas [9] for the case of the homogeneous distribution, is introduced using Tikhonov regularization based on the sensitivity theorem given by Geselowitz [10] and Lehr [11].

CP914, *Multiphase Flow: The Ultimate Measurement Challenge,* edited by X. Cai, Y. Wu, Z. Huang, S. Wang, and M. Wang

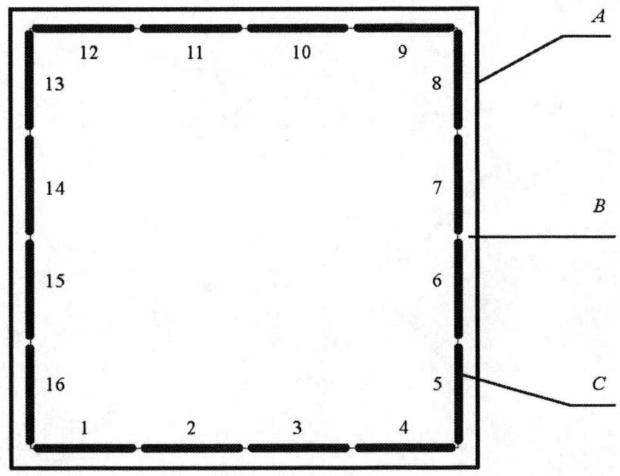

FIGURE 1. The cross-sectional view of the square EIT sensor with 16 electrodes and a copper shielding. *A*. Copper shielding *B*. Plastic frame *C*. Measurement electrode

MODEL OF THE SQUARE SENSOR

FIGURE 1 shows a cross sectional view of the square EIT sensor equipped with 16 electrodes. In this design, the electrodes are installed at equal intervals at each side of the square sensor. The electrodes are embedded into the inner surface of the pipe, in contact with the media. There is a copper shielding surrounding the sensor, in order to minimize the disturbance from outside and intra-electrode. The electrodes are much bigger than the traditional ERT electrodes, and are in contact with the media, other than the traditional ECT electrodes, which are usually installed in a noninvasive way, i.e. outside the pipe made of dielectric material. Since in most EIT systems the frequency of electrical signal which imposes electrical field in the measurement domain is of the order of 100 kHz-1 MHz, the corresponding wavelength of electromagnetic radiation is therefore typically of the order of hundreds of meters. This wavelength exceeds the sensor size by several orders of magnitude as typical sensor diameter is in the 1cm-1m range. And the electrical potential distribution obeys the Laplace equation.

Physical Model

An analytical model has been established by Cao and Wang[10] in the case of the circular sensor. According to the conformal mapping theory, there exists an elliptic function that maps the square area to the unit circular area, and the size of electrodes do not equal with each other after the conformal transformation, see FIGURE 2. Here, we use the precise integration method [12, 13] to perform the elliptic function calculation, and the precise integration method gives the results in the machine precision. The elliptic function used for the conformal transformation from the complex plane with the shape of a square to the complex plane with the shape of the unit disk, has the form as in equation (1).

$$w = \frac{sn(\sqrt{2}z, \frac{1}{\sqrt{2}})}{\sqrt{2}dn(\sqrt{2}z, \frac{1}{\sqrt{2}})} \tag{1}$$

Where $sn(\cdot,\cdot)$ and $dn(\cdot,\cdot)$ are the elliptic functions.

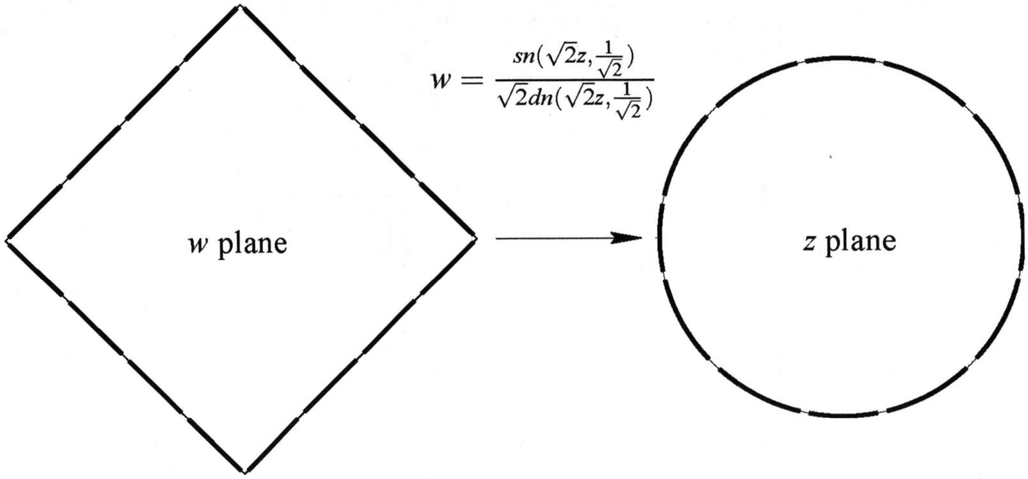

FIGURE 2. TThe diagram for the conformal transformation from a square to the unit circle

In the case of circular sensor [9], the electrical potential in the homogeneous material distribution satisfies equation (3),

$$\varphi(z) = \frac{1}{2\pi}\int_0^{2\pi}\frac{R^2-|z|^2}{|\xi-z|^2}\varphi(\xi)d\theta \tag{2}$$

$$= \mathbf{Re}[\frac{1}{2\pi i}\int_{|\xi|=R}\frac{\xi+z}{\xi-z}\varphi(\xi)\frac{d\xi}{\xi}] \tag{3}$$

The bracketed expression is an analytic function of z for $|z| < R$ [14]. It follows that $\varphi(z)$ is the real part of

$$f(z) = \frac{1}{2\pi i}\int_{|\xi|=R}\frac{\xi+z}{\xi-z}\varphi(\xi)\frac{d\xi}{\xi} = \varphi(z)+i\psi(z) \tag{4}$$

Where $\psi(z)$ is the imaginary part of $f(z)$, and a conjugate function of $\varphi(z)$. And

$$\frac{df(z)}{dz} = \frac{\partial\varphi(z)}{\partial x} - i\frac{\partial\varphi(z)}{\partial y} = \overline{E(z)} \tag{5}$$

Where $\overline{E(z)}$ is the conjugate of the electric field intensity $E(z)$ at z.

Use the 1-by-1 excitation mode [9], as only electrical potential at one electrode is excited to V_0 and all the others are grounded, then equation (5) has a simplification as equation (6).

$$\overline{E(z)} = \frac{V_0}{\pi i}(\frac{1}{z-Re^{i\theta_1}} - \frac{1}{z-Re^{i\theta_0}}) \tag{6}$$

where θ_0 and θ_1 are the starting angle and ending angle of the arc associated with the excitation electrode, respectively.

Based on equation (6) and elliptic function of the associated conformal transformation, the electrical field intensity in a square sensor can be easily calculated.

For the capacitive mode with unit axial length, in the sense of circular region, the capacitance measured between an electrode with a starting angle α and ending angle β and the excitation electrodes [9] is expressed by the following expression

$$C = \frac{\varepsilon}{\pi}\mathbf{Re}[\log(\frac{e^{i\alpha} - e^{i\theta_1}}{e^{i\alpha} - e^{i\theta_0}}) - \log(\frac{e^{i\beta} - e^{i\theta_1}}{e^{i\beta} - e^{i\theta_0}})] \tag{7}$$

For the resistive/conductive mode with unit axial length, in the sense of circular region, the conductance measured between an electrode with a starting angle α and ending angle β and the excitation electrodes [9] is expressed by the following expression

$$G = \frac{\sigma}{\pi}\mathbf{Re}[\log(\frac{e^{i\alpha} - e^{i\theta_1}}{e^{i\alpha} - e^{i\theta_0}}) - \log(\frac{e^{i\beta} - e^{i\theta_1}}{e^{i\beta} - e^{i\theta_0}})] \tag{8}$$

Since the conformal transformation holds the values of mutual conductances and capacitances between electrode pairs, the mutual conductance and capacitance between an electrode pair for the circular sensor can be calculated. So the square sensor here is self-calibrated.

Voltage excitation mode

Generally, for the N-electrode system, there are naturally $\frac{N}{2}$ excitation modes, which could be termed as 1-by-1 mode, 2-by-2 mode, 3-by-3 mode, $\cdots$, $\frac{N}{2}$-by-$\frac{N}{2}$ mode. In the case of the k-by-k mode, the voltages on the neighboring k electrodes are excited to V_0, while all the other N-k electrodes are grounded. In the k-by-k mode $(1 < k < N)$, there are $N(N-k)$ independent measurements. However, only $\frac{N^2}{4}$ independent measurements exist in $\frac{N}{2}$-by-$\frac{N}{2}$ mode, and $\frac{N(N-1)}{2}$ independent measurements exist in 1-by-1 mode, with respect to the symmetry configuration. In the excitation strategy with 1-by-1 mode, the electrode 1,2,3,4,5,...,15,16 is excited subsequently, with all the other electrodes grounded, to generate a different electric field distribution from the previous excitation, also not the same with that generated by the other modes. Although, there exist only $\frac{N(N-1)}{2}$ independent measurements in a N-electrode sensor, different excitation mode provide a flexible measurement strategy.

RESULTS

In this section, only 120 independent measurements in the 1-by-1 excitation mode are used. In the numerical simulation, the gap between two neighboring electrodes is 1.5% of the length of the sensor, the data are obtained by using ANSYS.

Based on the sensitivity theorem [10, 11] and formula (6), the sensitivity map for electrode pair $m(1 \leq m \leq 16)$ and $n(1 \leq n \leq 16)$ is given by the equation (9).

$$S_{m,n}(z) = \frac{E_m(z) \cdot E_n(z)}{V_0^2} \tag{9}$$

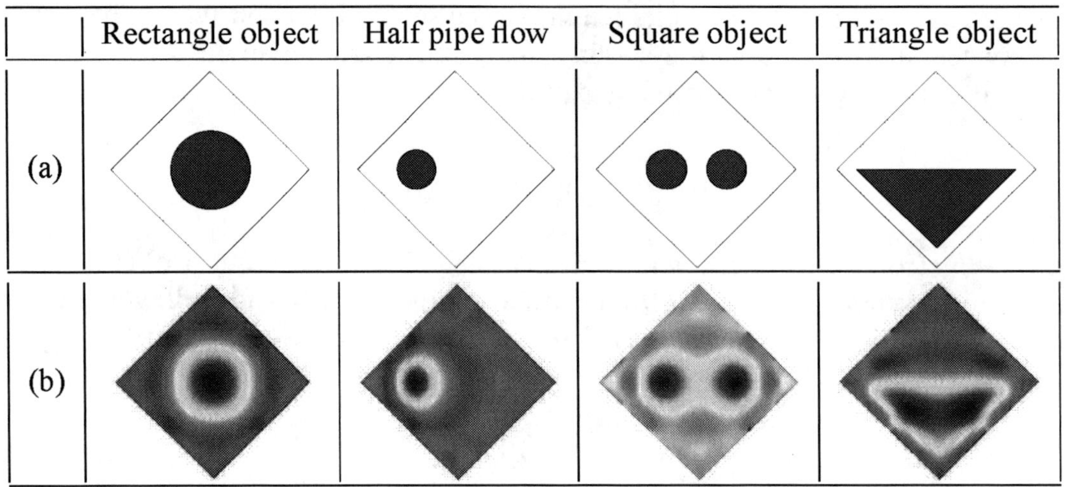

FIGURE 3. The results of reconstructed images using Tikhonov regularization based on the model in comparison to the Test profiles. Row (a): Test profiles, Row (b): Images reconstructed using TR.

Where $E_m(z)$ and $E_n(z)$ are the electric field intensity at z when a voltage V_0 is applied to m-th electrode and n-th electrode, respectively.

Using the standard Tikhonov regularization(TR) procedure, the inverse solution of EIT is expressed as

$$\Delta v = (S^T S + \mu I) S^T \Delta M \tag{10}$$

Where I is an identity matrix, and μ is the regularization parameter, which must be positive, in this paper $\mu = 50$. Δv is the changed distribution and ΔM is the measurement difference between a disturbed distribution and a homogeneous distribution. The results of image reconstructions of TR are shown in row (b) of FIGURE 3 and the quality of reconstructed images is good, both in position and size.

CONCLUSIONS

In this paper, an analytical model of a square impedance sensor of EIT system has been established. It derives a fast algorithm using Tikhonov regularization. This model is suitable for ECT and ERT information fusion in dual mode, and provides a flexible voltage excitation strategy. For the real time performance, this new strategy has the following advantages:

- fast response, since only one step multiplication of the matrix is needed, so it is suitable for real time image reconstruction;
- simple and easy to implement on the hardware system;
- the sensor has a set of calculable standard value and is self-calibrated.

ACKNOWLEDGMENTS

The authors gratefully acknowledge the financial support from the High Technique R&D Program of China (20001AA 413210), the National Natural Science Foundation of China (60532020, 60301008, 60472077, 50337020), and the supporting program of the Ministry of Science and Technology of China (2006BAI03A00).

REFERENCES

1. W. Q. Yang, and S. Liu, *IEEE., Electron. Lett.* **35**, 295–296.
2. S. Liu, and W. Q. Yang, *Meas. Sci. Technol.* **12**, 1120–1125 (2001).
3. T. Dyakowski, L. F. Jeanmeure, and A. J. Jaworski, *Powder Technology.* **112**, 174–192 (2000).
4. T. York, *J. Electronic Imag.* **10**, 608–619 (2001).
5. H. S. Tapp, A. J. Peyton, E. K. Kemsley, and R. H. Wilson, *Sensors and Actuators B: Chemical* **92**, 17–24 (2003).
6. M. Wang, *Meas. Sci. Tech.* **13**, 101–117 (2002).
7. I. Ismail, J. C. Gamio, S. F. A. Bukhari, and W. Q. Yang, *Flow Meas. Instrum.* **16**, 145–155 (2005).
8. B. S. Hoyle, X. Jia, F. J. W. Podd, and etc, *Meas. Sci. Technol.* **12**, 1157–1165 (2001).
9. Z. Cao, and H. X. Wang, "The Study of a 2D Model and Image Reconstruction Algorithms Based on EIT System," in *IMTC2006-Instrumentation and Measurement Technology Conference*, Sorrento, Italy, 2006, pp. 2323–2326.
10. D. B. Geselowitz, *IEEE Trans., Biomed. Eng.* **18**, 38–41 (1971).
11. J. Lehr, *IEEE Trans., Biomed. Eng.* **19**, 156–157 (1972).
12. W. X. Zhong, *Journal of Computational and Applied Mathematics* **163**, 59–78 (2004).
13. W. X. Zhong, and Z. Yao, *Advaces in applied mechanics(Ying Yong Li Xue Jin Zhan)(in chinese)* pp. 106–111 (2004).
14. L. V. Ahlfors, *Complex analysis*, China Machine Press, Beijing, China, 2004, 3rd edn.

SESSION PT-2

IMAGE RECONSTRUCTION AND APPLICATIONS OF PT TECHNOLOGY

Simulation Studies on the On-line Concurrent Measurement of Velocity and Size Distribution of Pneumatically Conveyed Particles through Digital Imaging

Ding Song, Lihui Peng, Geng Lu, Yan Zhang, and Shiyuan Yang

Department of Automation, Tsinghua University, Beijing 100084, China

Abstract. Pneumatic conveyance of particulate materials is very important in many industries, particularly in the coal fired electrical power generation. In order to improve plant control and operational efficiency, it is desirable to know the parameters of the particle flow. This paper presents the principle of an on-line digital imaging based method for measuring the velocity and size distribution of particles simultaneously. By controlling the time of exposure of a CCD camera, blurred images of moving particles are captured, which contain the information of moving particles. The motion blur parameters (include motion length and motion direction) are first estimated from the acquired images and then used to infer the velocity of particles in the pipeline by estimating the PSF (Point Spread Function) of the blurred image. The relationship between a 2-D motion blur image and the velocity of moving particles has been established. As to particle sizing, the image has to be deblurred so that the original image can be retrieved and ultimately the particle size distribution can be obtained. The Wiener deconvolution is applied to restore the image degraded due to motion blur. Simulation results indicate that the proposed method is feasible for the on-line concurrent measurement of particle velocity and size distribution.

Keywords: Particle velocity measurement; Particle size distribution; Motion blur; Wiener deconvolution.

INTRODUCTION

Pneumatic conveying systems are widely seen in many industries, particularly in the coal fired power generation industry. It has been recognized that combustion efficiency can be improved if the parameters of the gas/solids two-phase flow are known [1]. Solids velocity is important because it is the key parameter to estimate the mass flow rate, which is a common requirement in many pneumatic conveyance processes. Moreover, the particle velocity needs to be maintained around a safe value to achieve optimum conveying conditions. Particle size distribution determines the interface area between particles and atmosphere, consequently, affect combustion efficiency and pollutant emissions in coal fired processes.

It has long been recognized that the on-line measurement of mass flow rate of particulate solids in pipelines is a technically challenging area, since the two-phase flow is a random process, compared to single-phase flow [2]. Much work has been carried out in the field of gas/solids two-phase flow measurement over the past three decades. In addition to thermal, electrical [3][4][5] and acoustical methods [6], almost all regions across the electromagnetic spectrum from gamma-rays to microwaves, have been applied to develop suitable sensors for this application [1]. To date, there are very few commercial systems available for on-line continuous measurement of particle velocity and size distribution.

Substantial efforts have been made to develop instruments for particle velocity measurement. A cross-correlation velocity measurement system based on electrostatic sensors has been studied by Yan et al [5]. LDV (Laser Doppler Velocimetry) is also applied to particle velocity measurement [7], which has the potential to measure a wide range of solids velocities. However, the LDV is only applicable to point velocity measurements, making it suitable for profiling the spread of velocities within a conveyor for research purposes. Particle image velocimetry (PIV) [8] is a technique which enables instantaneous measurement of the flow velocity at several positions in a plane, but it's very complicated and not cost-effective for industry installations. R M Carter and Y Yan [9] proposed a novel approach to on-line particle size measurement through digital imaging. The general principle is, when particles are flowing through the cross section of a pipeline, it is possible to acquire an image of the particles using a CCD camera.

CP914, *Multiphase Flow: The Ultimate Measurement Challenge,* edited by X. Cai, Y. Wu, Z. Huang, S. Wang, and M. Wang

Digital image processing and statistical analysis method are then applied to determine the particle size distribution. Motion blur occurs because of the motion of particle at high speed, which can lead to image degradation and subsequent inaccuracies in the measurement of particle size distribution. However, a blurred image due to particle motion contains useful information that may be used for particle velocity measurement. In this paper a new approach to particle velocity measurement based on motion blurred images is presented. The relationship between the 2D images of moving particles and the particle velocity and size distribution is established. A critical issue here is to estimate the motion blur parameters which can be used to deduce particle velocity from the images.

MEASUREMENT PRINCIPLE

The principle of the proposed method is using a CCD camera with exposure time adjustable to obtain the images of the moving particles. Because the relative motion between the camera and the particles when the camera shutter remains open, the captured image is blurred. The motion blur parameters are first estimated from the acquired images and then the velocity of particles is then inferred by estimating the PSF (Point Spread Function) of the blurred image. Meanwhile, Wiener deconvolution is applied to restore the original image, which is used for the measurement of particle size distribution.

Estimation of the PSF

The observed image $g(m,n)$ is modeled as the output of a 2-D linear space-invariant system, which is characterized by its point spread function (PSF) $h(m,n)$. Equation (1) shows the relation between the observed image $g(m,n)$ and its original image $f(m,n)$ [10]:

$$g(m,n) = h(m,n) \otimes f(m,n) \tag{1}$$

where $\otimes$ stands for convolution. In this paper, $h(m,n)$ is restricted to be the PSF of a uniform motion system [11][12][13].

$$h(m,n) = \begin{cases} \frac{1}{L}, & |m| \le L\cos\theta/2, n = m\tan\theta \\ 0, & otherwise \end{cases} \tag{2}$$

where θ is the angle between the motion direction and the m axis, L is the length of the motion blur, in other words, it's the number blurred pixels during the period of exposure time. Taking Fourier transform on both sides of (1) gives

$$G(u,v) = H(u,v)F(u,v) \tag{3}$$

where G,H,F are the frequency responses of g,h,f, respectively. For the sake of technical clarity, we assume that motion is along the horizontal direction ($\theta = 0$). In this condition, equation (3) can be rewritten as

$$G(u,v) = H(u)F(u,v) \tag{4}$$

The frequency response of motion blur PSF is given by

$$\begin{aligned} H(u) &= \sum_{m=0}^{N} h(m) e^{-j2\pi(\frac{um}{N})} \\ &= \sum_{m=0}^{L-1} \frac{1}{L} e^{-j2\pi(\frac{um}{N})}, \quad 0 \le u \le N-1 \end{aligned} \tag{5}$$

If $u \neq 0$, $H(u)$ can be rearranged so that

$$H(u) = \frac{1}{L}\frac{1-e^{-j2\pi\frac{uL}{N}}}{1-e^{-j2\pi\frac{u}{N}}} \tag{6}$$

Clearly zero-crossings will periodically appear in $H(u)$ when u satisfies the following equation:

$$Lu/N = k, \quad k = \pm 1, \pm 2, \cdots, \pm L/2 \tag{7}$$

In other words, parallel dark lines would appear in $G(u)$, when $u = N/L,\ 2N/L, \cdots (L-1)N/L$, $H(u) = 0$. Hence, there are $(L-1)$ zero points in $G(u)$.

Once the "length" of the motion blur L is identified, a pinhole camera model is used to estimate the velocity of the moving particles.

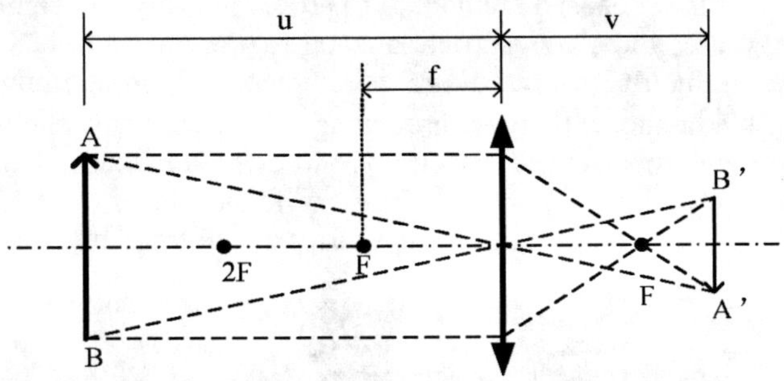

FIGURE 1. Model of pinhole camera.

The pinhole camera model can be described as:

$$\frac{1}{u} + \frac{1}{v} = \frac{1}{f} \tag{8}$$

where, u, v and f stand for the object distance, image distance and focus distance of the lens system, respectively. As shown in Figure 1, the relationship between the real distance of the particle motion D and the number of blurred pixels L during a period of time is given by:

$$\frac{v}{u} = \frac{L}{D} = \frac{Le}{V \cdot t} \tag{9}$$

where, t stands for the exposure time of the CCD camera and e is the size of the CCD pixel. From equations (8) and (9), the particle velocity can be described as:

$$V = \frac{(u-f)Le}{f \cdot t} \tag{10}$$

where, V represents the velocity of particles. For a real system, the focus distance f and pixel size e are given, object distance u depends on the configuration of the sensor, exposure time t can be adjusted manually, and L can be obtained from estimating the PSF of the blurred image as demonstrate above.

Image Restoration

The image has to be deblurred so that the original image can be retrieved and the particle size distribution estimated. Once the exact nature of the blur has been determined, many approaches can be deployed to restore the image. Wiener deconvolution is deemed suitable, at least to some degree, when the frequency characteristics of the image and additive noise are known [14]. It minimizes the mean-square error between the ideal image f and the restored image $\hat{f}$ [13].

$$F(u,v) = \frac{H^*(u,v)}{|H(u,v)|^2 + \gamma} G(u,v) \tag{11}$$

where, $H(u,v)^*$ is the conjugate of $H(u,v)$, γ stands for the signal noise ratio (SNR), which is given by $\gamma = P_n(u,v)/P_f(u,v)$, $P_n(u,v)$ and $P_f(u,v)$ are the power spectra of the noise image and ideal image, respectively.

SIMULATION RESULTS

To evaluate the effectiveness of the proposed methods for particle velocity and sizing measurements, a series of simulation tests was conducted. Carter and Yan's particle image [9] was chosen to be our original image, as shown in Figure 2. The blurred images in Figure 3(a), (b) and (c) were generated through image synthesis as proposed by Potmesil and Chakravarty [15]. It is assumed that the three images have the same shutter speed, consequently, the difference of the blurred length corresponds to the velocity of particles. The blurred length of the three images is 9, 29 and 59 pixels, respectively. Figure 4(a), (b) and (c) show the logarithm of the Fourier Transform of the three images. The logarithm here helps to bring out details of the Fourier transform in regions where it is very close to 0.

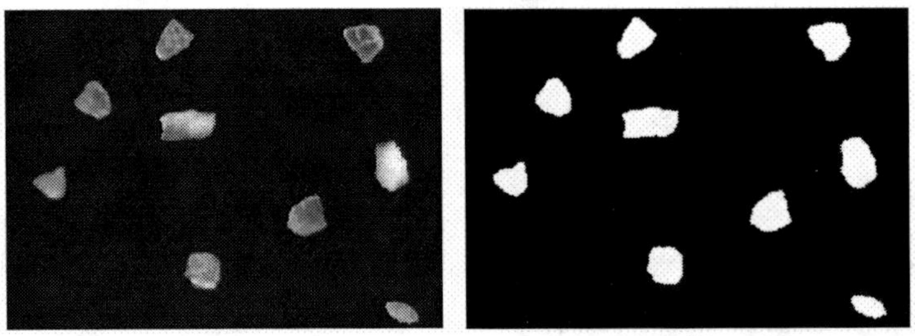

(a) (b)

FIGURE 2. Original image and thresholded image. (a) Original image. (b) A thresholded image.

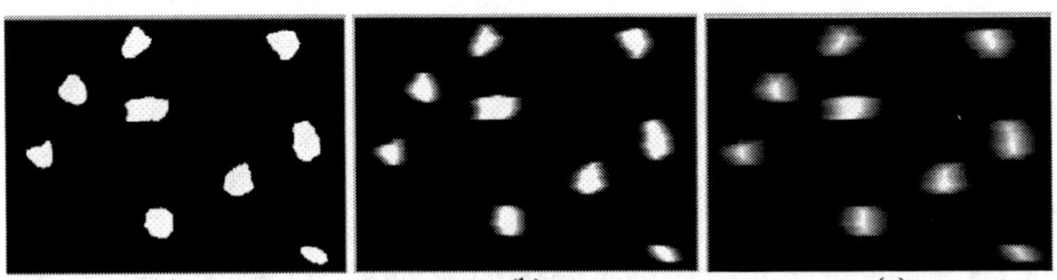

(a) (b) (c)

FIGURE 3. Motion blurred image. (a) Motion length=9 pixels. (b) Motion length=29 pixels. (c) Motion length=59 pixels.

(a) (b) (c)

FIGURE 4. Frequency responses of motion blurred images. (a) Motion length=9 pixels. (b) Motion length=29 pixels. (c) Motion length=59 pixels.

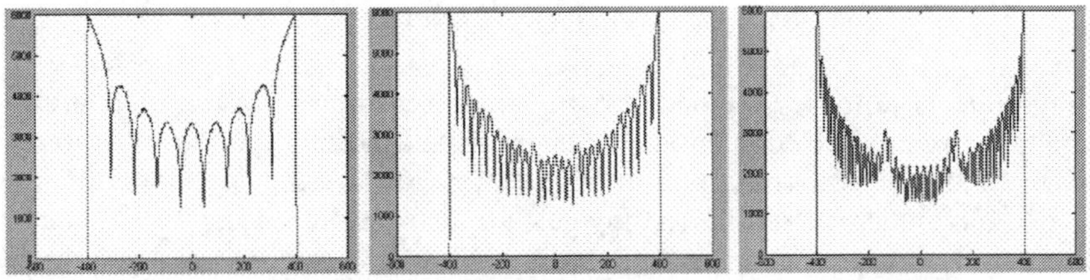

(a) (b) (c)

FIGURE 5. Radon transform of the frequency responses. (a) Motion length=9 pixels. (b) Motion length=29 pixels. (c) Motion length=59 pixels.

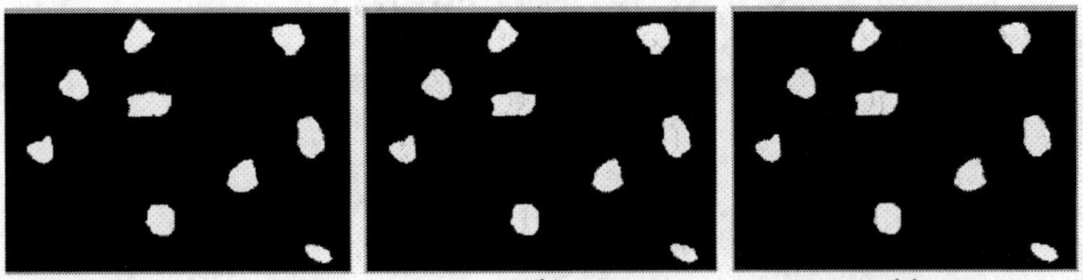

(a) (b) (c)

FIGURE 6. Restored images through Wiener deconvolution. (a) Motion length=9 pixels. (b) Motion length=29 pixels. (c) Motion length=59 pixels.

Looking carefully at these lines in Figure 4(a), (b) and (c), one can see that increasing the length of motion blur yields more lines in the frequency response. Radon transform can help to estimate the motion length $\hat{L}$. Let K stands for the number of minimum points in Figure 5(a), (b) and (c), then $K_a = 8, K_b = 28, K_c = 58$, $\hat{L}_a = 9, \hat{L}_b = 29, \hat{L}_c = 59$, they are the same as the initialization. The velocity of particle can be calculated from equation (10).

Wiener deconvolution can be used here to restore the original image and ultimately the particle size distribution. The more information in the frequency domain we know about the motion blurred image, the more accurate original image would be. Hence, it's very helpful that motion length is known in advance. Figure 6 illustrates the restored images from the motion blurred versions in Figure 3. The particle sizing data obtained from the original image and the restored image for motion length=59 pixels are shown in Figure 7.

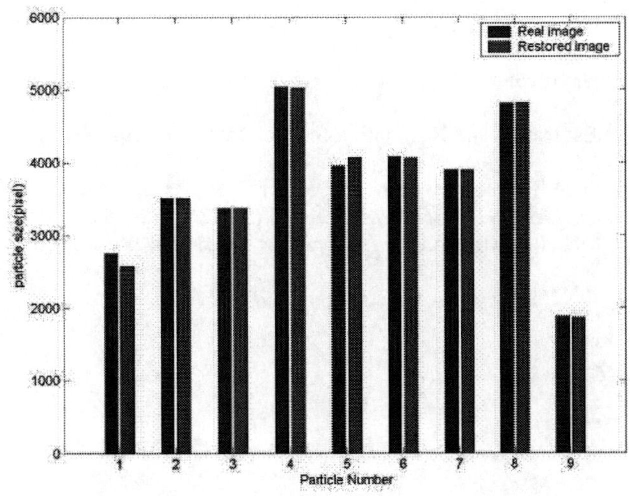

FIGURE 7. Particle sizing data from the images.

CONCLUSION

In this paper we presented a proposed method for the concurrent measurement of particle velocity and size distribution through digital imaging. The simulation studies presented in this paper have demonstrated, in principle, that one can build a mathematical model of the blurred particle images to estimate the motion length, which is then used to determine the particle velocity. The blurred images are also restored using Wiener deconvolution, so that particle size distribution can be obtained. The simulation results have also demonstrated that the proposed method is feasible for concurrent measurement of particle velocity and size distribution. Future work will be focused on the practical design and implementation of a prototype system based on the simulation studies.

ACKNOWLEDGEMENTS

The authors wish to thank the National Natural Science Foundation of China for providing financial support for a two-base research project between Tsinghua University, China and University of Kent, U. K. under the contract No. 60440430641.

REFERENCES

1. Y. Yan, Mass flow measurement of bulk solid in pneumatic pipelines, *Meas. Sci. Technol.* **7(12)**, 1687-1760 (1996).
2. H. L. Hu, T. M. Xu, S. E. Hui, Q. L. Zhou, A novel capacitive system for the concentration measurement of pneumatically conveyed pulverized fuel at power stations, *Flow Measurement and Instrumentation* **17(2)**, 87–92 (2006).
3. H. Amano, M. Sakakibara, Y. Doi, Y. Morimatu, H. Kawahara and S. Matsushita, Two-phase flow meter for pulverized coal/air flow, *Proc. Int. Conf. on Fluid Control and Measurement*, Sep, Tokyo, Japan, 1985, pp. 965-971.
4. G. A. Irons and J. S. Chang,. Particle fraction and velocity measurements in gas–powder streams by capacitance transducers. *Int. J. Multiphase Flow*. **9 (3)**, 289–297 (1983).
5. Y. Yan, B.Byrne, S. Woodhead and J. Coulthard, Velocity measurement of pneumatically conveyed solids using electrodynamic sensors, *Meas. Sci. Technol.* **6(5)**, 515–537 (1995).
6. S. H. Sheen and A. C. Raptis, Active ultrasonic cross-correlation flowmeters for mixed-phase pipe flows, *ISA Trans* **24 (1)** 53–58 (1985).
7. A. Birchenough and J. S. Mason, Local particle velocity measurement with a laser anemometer in an upwardflowing gas–solids suspension, *Powder Technol.* **14 (1)**, 139–152 (1976).
8. J. Westerweel, Fundamentals of digital particle image velocimetry, *Meas. Sci. Technol.* **8 (12)**, 1379-1392 (1997).
9. R. M. Carter and Y. Yan, On-line particle sizing of pulverized and granular fuels using digital imaging techniques. *Meas. Sci. Technol.* **14 (7)**, 1099-1109 (2003).
10. M. Cannon, Blind Deconvolution of Spatially Invariant Image Blurs with Phase, IEEE Trans. *Acoust. Speech & Signal Process* **24 (1)**, 56-63 (1976).
11. H. Y. Lin and K. J. Li, Motion Removal and Its Application to Vehicle Speed Detection, *International Conference on Image Processing (ICIP)*,Oct, Singapore 2004, pp.3407-3410.
12. Y. Yitzhaky and N. S. Kopeika, Identification of blur parameters from motion blurred images. *Graphical Models and linage Processing* **59 (5)**, 310-320 (1997).
13. Q. Li and Y. Yoshida, Parameter Estimation and Restoration for motion blurred Images, *IEICE Trans. Fundamentals* **80 (8)**, 1430-1437 (1997).
14. H. Lim, K. C. Tan, T. G. Tanb, Edge Errors in Inverse and Wiener Filter Restoration of Motion Blurred Images and Their Windowing Treatment, *Graphical Models and Image Processing* **53 (2)**, 186-195 (1991).
15. M. Potmesil and L. Chakravarty, Modeling Motion Blur in Computer-Generated Images, *Computer Graphics* **17 (3)**, 389-399 (1983).

A Hybrid Image Reconstruction Algorithm for Electrical Capacitance Tomography

Lihui Peng, Geng Lu, and Nin Sun

Department of Automation, Tsinghua University, Beijing 100084, China

Abstract. Image reconstruction for electrical capacitance tomography is to retrieve the permittivity distribution of processes with different dielectric materials inside from the measured capacitances outside. It is a typical inverse problem and has been always a challenge for its nonlinearity and ill-posedness. This paper explores the possibility of applying Tikhonov regularization, widely used for ill-posed problems, to the image reconstruction for electrical capacitance tomography. Two methods using different regularization matrices are investigated by using the simulated data, which are the standard Tikhonov regularization and the Tikhonov regularization based on the second order derivative operator. Results show that both the standard Tikhonov regularization and the Tikhonov regularization with the second order derivative operator do not provide reconstructions with qualities good enough. In order to improve this, a hybrid image reconstruction algorithm, which is based on the Tikhonov regularization with the reconstruction from the linear back projection as a constraint, is developed. Numerical results show that this hybrid image reconstruction algorithm takes advantages from both the linear back projection and the Tikhonov regularization and provide reconstructions better than those from the LBP and the Tikhonov regularization.

Keywords: Electrical capacitance tomography; Image reconstruction; Linear back projection; Tikhonov regularization.

INTRODUCTION

Electrical capacitance tomography is a measurement technique for visualizing dielectric processes, such as pneumatic conveying systems and fluidized beds, by using tomographic imaging [1][2][3]. The investigation about electrical capacitance tomography started in the late 1980s [1][2]. Remarkable progresses have been made in the past two decades [3][4][5][6]. However, challenges still need to be overcome in the further development and application of electrical capacitance tomography. One of them is the promotion of new image reconstruction algorithms.

The task of image reconstruction for electrical capacitance tomography is to find the permittivity distribution inside the sensor by using capacitance measurements. It is a typical inverse problem, which involves uncovering the cause e.g. the permittivity distribution leading to the measured capacitances.

There are some difficulties with image reconstruction related to electrical capacitance tomography. Firstly, image reconstruction of electrical capacitance tomography is a non-linear problem, which means that electrical field in the imaging area strongly depends on the permittivity distribution inside the sensor. This is the so-called "soft field effect", which distorts reconstructed images. Secondly, it is an under-determined problem. A typical capacitance tomography sensor only has 8 or 12 electrodes, which provide 28 or 66 measurements. However, the imaging area is usually meshed into several hundreds pixels for a good looking tomogram. For this reason, the number of unknown variables is much larger than the number of equations. Finally, image reconstruction for electrical capacitance tomograhpy has to solve integral equations, which are ill-posed. In other words, the solution is very sensitive to the error in measurements.

It was reported by Yang [6] that two categories of image reconstruction algorithms for electrical capacitance tomography exist. One is the direct method. The other is the iterative method. The direct method is fast, which makes on-line imaging possible, although it only provides qualitative reconstruction. In contrast to the direct method, the iterative method takes some time but provides quantitative reconstruction.

CP914, *Multiphase Flow: The Ultimate Measurement Challenge,* edited by X. Cai, Y. Wu, Z. Huang, S. Wang, and M. Wang

From the point view of application, the direct method is more attractive for the purpose of on-line imaging. Several direct methods have been developed and applied to image reconstruction for electrical capacitance tomography, such as the linear back projection [7], the least squares method base on singular value decomposition [6], and the Tikhonov regularization [8][9]. However, certain weaknesses exist in these algorithms, which make them only suitable for certain permittivity distributions. The paper will focus on the possibility of developing more adaptable direct image reconstruction algorithms for electrical capacitance tomography.

HYBRID IMAGE RECONSTRUCTION ALGORITHM

The discrete and linearized mathematical model for electrical capacitance tomography image reconstruction is

$$\boldsymbol{\lambda} = \mathbf{A} \cdot \mathbf{g} + \mathbf{e} \tag{1}$$

where, $\boldsymbol{\lambda}$, $\mathbf{A}$, and $\mathbf{g}$ represent normalized capacitance, sensitivity matrix, and normalized permittivity respectively. $\mathbf{e}$ is the error in measurements. The direct method for electrical capacitance tomography image reconstruction is related to solving the equation (1) directly.

Linear back projection was the first direct method for electrical capacitance tomography image reconstruction. The solution based on linear back project can be expressed as

$$\hat{\mathbf{g}}_{LBP} = \mathbf{A}^{\mathrm{T}} \cdot \boldsymbol{\lambda} \tag{2}$$

Mathematically, it's very difficult to explain equation (2), in which $\mathbf{A}^{\mathrm{T}}$ is simply treated as the inverse of $\mathbf{A}$. Therefore, mathematicians did not accept it. However, the equation does work in practice. A related explanation is if $\mathbf{A}$ is considered as a map from the permittivity vector space to the capacitance vector space, then $\mathbf{A}^{\mathrm{T}}$ can be considered as the adjoint map from the capacitance vector space to the permittivity vector space.

Tikhonov regularization is widely used to solve ill-posed problems [10]. The essence of Tikhonov regularization is using some extra information as constraints. Mathematically, it is the result of minimizing the following objective function:

$$\|\mathbf{A}\mathbf{g} - \boldsymbol{\lambda}\| + \mu \|\mathbf{L}(\mathbf{g} - \mathbf{g}_0)\| \tag{3}$$

where, μ is a positive number, which is called regularization parameter. $\mathbf{L}$ is the matrix corresponding to certain linear operator. $\mathbf{g}_0$ is the assumed true solution. Minimizing the above function means keeping the solution close to the true solution as possible as it can be while the discrepancy is minimized. This is actually the basic principle behind regularization. In practice, the true solution $\mathbf{g}_0$ is exactly what image reconstruction what to retrieve and is unknown before reconstruction. A common way is choosing $\mathbf{g}_0$ as zero, which leads to the solution as

$$\hat{\mathbf{g}}_{Tik} = \left(\mathbf{A}^{\mathrm{T}} \cdot \mathbf{A} + \mu \mathbf{L}^{\mathrm{T}} \cdot \mathbf{L}\right)^{-1} \mathbf{A}^{\mathrm{T}} \cdot \boldsymbol{\lambda} \tag{4}$$

Choosing $\mathbf{L}$ as identity matrix, which means finding the solution with minimum energy, gives the solution of the standard Tikhonov regularization written as

$$\hat{\mathbf{g}}_{Tik\_I} = \left(\mathbf{A}^{\mathrm{T}} \cdot \mathbf{A} + \mu \mathbf{I}\right)^{-1} \mathbf{A}^{\mathrm{T}} \cdot \boldsymbol{\lambda} \tag{5}$$

If $\mathbf{L}$ is chosen as $\mathbf{L}_2$ as

$$\mathbf{L}_2 = \begin{bmatrix} -2 & 1 & 0 & \cdots & \cdots & \cdots & 0 \\ 1 & -2 & 1 & 0 & \cdots & \cdots & 0 \\ 0 & 1 & -2 & 1 & 0 & \cdots & 0 \\ \vdots & \ddots & \ddots & \ddots & \ddots & \ddots & 0 \\ 0 & \cdots & 0 & 1 & -2 & 1 & 0 \\ 0 & \cdots & \cdots & 0 & 1 & -2 & 1 \\ 0 & \cdots & \cdots & \cdots & 0 & 1 & -2 \end{bmatrix} \tag{6}$$

which is a discrete form for the second order derivative operator, the corresponding Tikhonov solution with minimum curvature can be written as

$$\hat{\mathbf{g}}_{Tik\_L_2} = \left(\mathbf{A}^{\mathrm{T}} \cdot \mathbf{A} + \mu \mathbf{L}_2{}^{\mathrm{T}} \cdot \mathbf{L}_2\right)^{-1} \mathbf{A}^{\mathrm{T}} \cdot \boldsymbol{\lambda} \tag{7}$$

In practice, it is found that the reconstruction quality using Tikhonov regularization strongly depend on the permittivity distributions. For certain permittivity distributions, Tikhonov regularization provides good reconstructions. But for some other permittivity distributions, Tikhonov regularization gives the reconstructions even worse than the reconstructions from the LBP. For this reason, we propose a hybrid method using the Tikhonov regularization with the second order derivative operator and the LBP result as the constraints, which corresponding solution can be described as

$$\hat{\mathbf{g}}_{Hybrid} = \left(\mathbf{A}^{\mathrm{T}} \cdot \mathbf{A} + \mu \mathbf{L}_2{}^{\mathrm{T}} \cdot \mathbf{L}_2\right)^{-1} \left(\mathbf{A}^{\mathrm{T}} \cdot \boldsymbol{\lambda} + \mu \mathbf{L}_2{}^{\mathrm{T}} \cdot \mathbf{L}_2 \cdot \hat{\mathbf{g}}_{LBP}\right) \tag{8}$$

Furthermore, the quality of reconstructions based on this hybrid algorithm can be improved by thresholding as

$$\begin{cases} \bar{\lambda} = \dfrac{\sum_{m=1}^{M} \lambda_m}{M}, \quad \alpha = \dfrac{\sum_{k=1}^{N} \hat{g}_k}{N}, \quad g_t = \alpha\left(1 - \bar{\lambda}\right) \\[2ex] \hat{g}_k = \begin{cases} 0 & \hat{g}_k < g_t \\ \hat{g}_k & g_t \le \hat{g}_k < 1 \\ 1 & \hat{g}_k \ge 1 \end{cases} \end{cases} \tag{9}$$

where, λ_m is the mth component of the normalized capacitance vector $\boldsymbol{\lambda}$, $\hat{g}_k$ is the kth component of the reconstructed permittivity distribution vector $\hat{\mathbf{g}}$.

RESULTS AND DISCUSSION

In order to assess this hybrid image reconstruction algorithm, numerical simulation with respect to a square-shaped ECT sensor with 8 electrodes is carried out. Four typical permittivity distributions are chosen for obtaining the simulated data. The capacitances between all possible electrode combinations are calculated by using the finite element method.

The normalized image error and correlation coefficient between the test distributions and reconstructions described in equation (10) and equation (11) were used to compare reconstructions based on different algorithms, including linear back projection, the standard Tikhonov regularization, and the hybrid algorithm.

$$error = \left\|\mathbf{g} - \hat{\mathbf{g}}\right\| / \left\|\mathbf{g}\right\| \tag{10}$$

$$r = \frac{\sum_{i=1}^{N} \left(\hat{g}_i - \bar{\hat{g}}\right)\left(g_i - \bar{g}\right)}{\sqrt{\sum_{i=1}^{N} \left(\hat{g}_i - \bar{\hat{g}}\right)^2 \sum_{i=1}^{N} \left(g_i - \bar{g}\right)^2}} \tag{11}$$

where, $\mathbf{g}$ is test distribution and g_i is the ith component of $\mathbf{g}$. $\hat{\mathbf{g}}$ is reconstructed result and $\hat{g}_i$ is the ith component of $\hat{\mathbf{g}}$. $\bar{g}$ and $\bar{\hat{g}}$ are mean values of $\mathbf{g}$ and $\hat{\mathbf{g}}$, respectively.

Figure 1 depicts the numerical results. The related image errors, correlation coefficients and the regularization parameters are demonstrated in table 1.

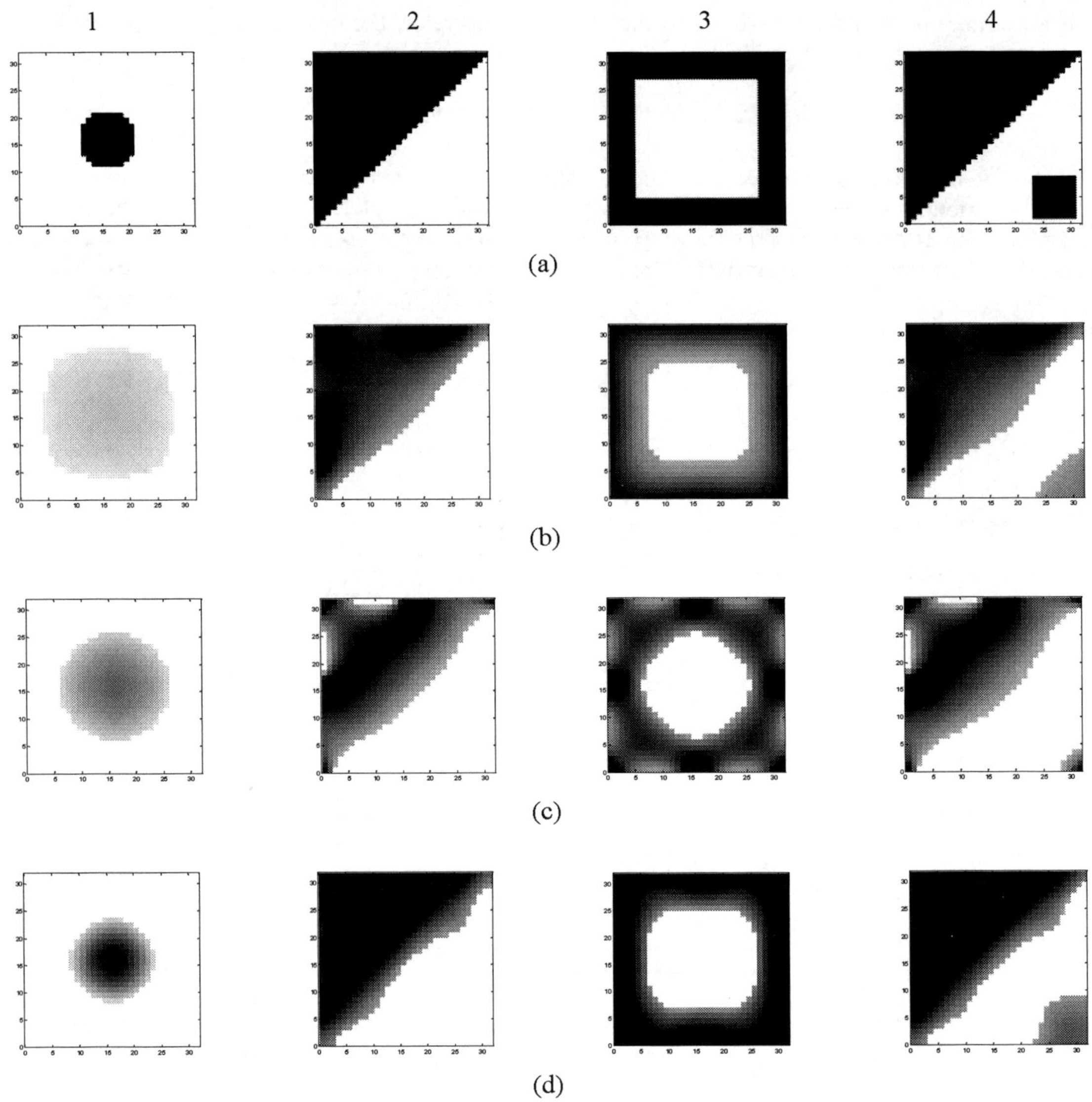

FIGURE 1. Numerical results. (a) test phantoms; (b) reconstructions based on the LBP; (c) reconstructions based on the standard Tikhonov regularization; (d) reconstructions based on the hybrid algorithm.

TABLE 1. Image errors and correlation coefficients related to the reconstructions in figure 1.

Test phantom	**1**	**2**	**3**	**4**
Reconstructions based on the LBP	error = 0.876 r = 0.541	error = 0.294 r = 0.926	error = 0.372 r = 0.871	error = 0.398 r = 0.828
Reconstructions based on the standard Tikhonov regularization	error = 0.579 r = 0.812 μ = 0.0001	error = 0.410 r = 0.857 μ = 0.005	error = 0.511 r = 0.671 μ = 0.0001	error = 0.507 r = 0.738 μ = 0.005
Reconstructions based on the hybrid algorithm	error = 0.413 r = 0.907 μ = 0.0001	error = 0.207 r = 0.956 μ = 0.1	error = 0.309 r = 0.903 μ = 0.0005	error = 0.301 r = 0.891 μ = 0.0005

It is shown in figure 1 that the reconstruction based on LBP for test phantom 1 is of poor resolution because of low sensitivity in the central imaging area, while the standard Tikhonov regularization produces much better result. For the test phantoms 2, 3 and 4, all reconstructions from the standard Tikhonov regularization are a little bit worse than those from the LBP. No matter what, an obvious fact is that all reconstructions based on the hybrid method for anyone test phantom are better than those from both the LBP and the standard Tikhonov regularization. Simply speaking, the hybrid algorithm takes advantages from both the linear back projection and the Tikhonov regularization and provides reconstructions better than the LBP and the Tikhonov regularization.

It is worthy of notice that the reconstruction qualities from both the standard Tikhonov regularization and the hybrid method depend on the regularization parameter. In fact, how to choose the optimal regularization factor has been always crucial and even harder than the reconstruction itself. The regularization parameters in our simulation are chosen tentatively. Further investigation is needed for choosing the optimal regularization parameters by using the related methods such as the L-Curve method, the GCV method and so on [11].

CONCLUSION

This paper demonstrates a hybrid image reconstruction algorithm for electrical capacitance tomography, which principle is using the Tikhonov regularization with second order derivative operator and the linear back projection reconstruction as prior constraint. The hybrid image reconstruction algorithm takes advantages from both the linear back projection and the Tikhonov regularization and provides reconstructions better than LBP and the standard Tikhonov regularization.

Future work will focus on the adaptability of the hybrid algorithm by applying it to more permittivity distribution cases.

ACKNOWLEDGEMENTS

The authors wish to thank the National Natural Science Foundation of China for supporting this research under the contracts No.60532020 and No.60204003.

REFERENCES

1. S. M. Huang, A. B. Plaskowski, C. G. Xie and M. S. Beck, Capacitance-based tomographic flow imaging system, *Electron. Lett.* **24 (7)**, 418-419 (1988).
2. G. E. Fasching and N. S. Smith, A capacitive system for 3-dimentional imaging of fluidized-beds, Rev. Sci. Instrum. **62 (9)**, 2243-2251 (1991).
3. M. S. Beck and R. A. Williams, Process tomography: a European innovation and its applications, Meas. Sci. Technol. **7 (3)**, 214-224 (1996).
4. O. Isaksen, A review of reconstruction techniques for capacitance tomography, Meas. Sci. Technol. 7 (3), 325-337 (1996).
5. W. Q. Yang, Hardware design of electrical capacitance tomography systems, Meas. Sci. Technol. 7 (3), 233-246 (1996).
6. W. Q. Yang and L. H. Peng, Image reconstruction algorithm for electrical capacitance tomography, Meas. Sci. Technol. 14 (1), R1-13 (2003).
7. C. G. Xie, S. M. Huang, B. S. Hoyle, R. Thorn, C. Lenn, D. Snowden and M. S. Beck, Electrical capacitance tomography for flow imaging-System model for development of image reconstruction algorithms and design of primary sensors, *IEE Proc.-G* **139 (1),** 89-98 (1992).
8. L. H. Peng, H. Merkus and B. Scarlett, Using regularization methods for image reconstruction of electrical capacitance tomography, *Part. Part. Syst. Char.* **17 (3),** 96-104 (2000).
9. W. R. B. Lionheart, Reconstruction algorithms for permittivity and conductivity imaging, in *Proc. of 2nd World Congress on Industrial Process Tomography*, 29th-31st Aug, Hannover, 2001, pp. 4-11.
10. A. N. Tikhonov and V. Y. Arsenin, *Solutions of ill-posed problems*, Washington D C: Winston & Sons, 1977.
11. P. C. Hansen, *Rank-deficient and discrete ill-posed problems: numerical aspects of linear inversion*, Philadelphia : SIAM, 1998.

Characterization of Window Functions for Regularization of Electrical Capacitance Tomography Image Reconstruction

Peng Jiang, Lihui Peng, and Deyun Xiao

Department of Automation, Tsinghua University, Beijing 100084, China

Abstract. This paper presents a regularization method by using different window functions as regularization for electrical capacitance tomography (ECT) image reconstruction. Image reconstruction for ECT is a typical ill-posed inverse problem. Because of the small singular values of the sensitivity matrix, the solution is sensitive to the measurement noise. The proposed method uses the spectral filtering properties of different window functions to make the solution stable by suppressing the noise in measurements. The window functions, such as the Hanning window, the cosine window and so on, are modified for ECT image reconstruction. Simulations with respect to five typical permittivity distributions are carried out. The reconstructions are better and some of the contours are clearer than the results from the Tikhonov regularization. Numerical results show that the feasibility of the image reconstruction algorithm using different window functions as regularization.

Keywords: Electrical Capacitance Tomography; Image Reconstruction; Window Function; Regularization.

INTRODUCTION

Electrical Capacitance Tomography (ECT) is a technique for two-component flow measuring, which has been under development for more than a decade. ECT has become increasingly popular and widely used because it is non-intrusive, robust and economical [1][2][3].

An ECT system is often made up of eight or twelve electrodes that are mounted around pipes or vessels to be investigated. The inside permittivity distribution can be reconstructed from the measurements of capacitance between all possible electrode combinations.

Generally, the mathematical model for ECT is simplified as a linear form as below

$$\mathbf{Sg} = \boldsymbol{\lambda} \tag{1}$$

where $\boldsymbol{\lambda}$ is an M dimensional vector indicating the normalized capacitance, $\mathbf{g}$ is an N dimensional vector standing for the normalized permittivity distribution, and $\mathbf{S}$ is an M by N dimensional sensitivity matrix. The purpose of ECT image reconstruction is to solve $\mathbf{g}$ from equation (1).

Existing ECT image reconstruction algorithms [4] mainly consist of two categories: direct methods and iterative methods. Direct methods, such as linear back projection (LBP) algorithm [5], truncated singular value decomposition (TSVD) algorithm [6][7], Tikhonov regularization and so on [8][9], need less computation and cost less time. Direct methods can be used for on-line imaging, but their reconstruction qualities are low. Iterative methods, such as Newton-Raphson method [10], simultaneous iterative technique (SIRT) algorithm [11], Landweber iteration [12][13], etc., usually provide reconstructions with higher qualities, but they need intensive computation and are hardly used for on-line imaging.

Image reconstruction for electrical capacitance tomography is a typical ill-posed inverse problem. Because of the small singular values of sensitivity matrix, the solution is sensitive to the measurement noise. To solve this problem, many regularization algorithms have been proposed. The principle of regularization is to reduce the weighting for the data associated with the small singular values. Meanwhile, the theory on filtering which uses different window functions in signal processing has been well established. For the filtering characteristic of window function, a

CP914, *Multiphase Flow: The Ultimate Measurement Challenge,* edited by X. Cai, Y. Wu, Z. Huang, S. Wang, and M. Wang

method using different window functions as regularization, which is capable of suppressing the instability caused by the small singular values of sensitivity matrix and the errors in measurement data, is proposed for ECT image reconstruction.

PRINCIPLE OF REGULARIZATION USING WINDOW FUNCTION

There are two main problems in ECT image reconstruction. Firstly, equation (1) is underdetermined because the number of unknowns is much larger than the number of equations. Secondly, the solution of equation (1) is sensitive to measurement noise because it is ill-conditioned.

Tikhonov regularization, which is perhaps the most popular regularization method, has been wildly used for solving ill-posed inverse problems. Its standard form is generally expressed as

$$\hat{\mathbf{g}} = \arg\min\left\{\left\|\mathbf{S}\mathbf{g}-\boldsymbol{\lambda}\right\|^2 + \mu\left\|\mathbf{g}-\mathbf{g}^{\infty}\right\|^2\right\}. \tag{2}$$

where, μ is the regularization parameter describing the extent of regularization when necessarily adjusted, $\mathbf{g}^{\infty}$ is the assumed true solution, and $\hat{\mathbf{g}}$ is the regularized solution.

The solution to equation (2) is written as

$$\hat{\mathbf{g}} = \left(\mathbf{S}^{\mathrm{T}}\mathbf{S}+\mu\mathbf{I}\right)^{-1}\left(\mu\mathbf{g}^{\infty}+\mathbf{S}^{\mathrm{T}}\boldsymbol{\lambda}\right). \tag{3}$$

Writing the matrix $\mathbf{S}$ using the singular value decomposition as

$$\mathbf{S} = \mathbf{U}\boldsymbol{\Sigma}\mathbf{V}^{\mathrm{T}} = \sum_{k=1}^{m}\sigma_k\mathbf{u}_k\mathbf{v}_k^{\mathrm{T}}. \tag{4}$$

By substituting the matrix $\mathbf{S}$ in equation (3) with (4) [14], the solution can be rewritten as

$$\hat{\mathbf{g}} = \sum_{k=1}^{N} f_k \cdot \mathbf{v}_k \tag{5}$$

where,

$$\hat{\mathbf{f}}_l = \begin{cases} \dfrac{\mu}{\sigma_l^2+\mu}\mathbf{v}_l^{\mathrm{T}}\mathbf{g}^{\infty} + \dfrac{\sigma_l^2}{\sigma_l^2+\mu}\left(\dfrac{\mathbf{u}_l^{\mathrm{T}}\boldsymbol{\lambda}}{\sigma_l}\right) & for\ l=1,2,\cdots,m \\ \mathbf{v}_l^{\mathrm{T}}\mathbf{g}^{\infty} & for\ l=m+1,\cdots,n \end{cases} \tag{6}$$

The quantity of $\sigma_l^2/(\sigma_l^2+\mu)$, which multiply the data coefficient $\mathbf{u}_l^{\mathrm{T}}\boldsymbol{\lambda}/\sigma_l$, are called filter factors [14], which show how different measurement data components corresponding to different singular values are weighted.

A vector marked as $\mathbf{W}(\boldsymbol{\sigma})$, made up of a series of filter factors $\mathbf{w}(\sigma_k)$ corresponding to different singular values, constructs a window function described as

$$\begin{aligned}\mathbf{W}(\boldsymbol{\sigma}) &= \left[\mathbf{w}(\sigma_1),\mathbf{w}(\sigma_2),\cdots,\mathbf{w}(\sigma_m)\right] \\ &= \left[\frac{\sigma_1^2}{\sigma_1^2+\mu},\frac{\sigma_2^2}{\sigma_2^2+\mu},\cdots\frac{\sigma_m^2}{\sigma_m^2+\mu}\right]^T \quad \left(\sigma_1\ge\sigma_2\ge\cdots\ge\sigma_m\right)\end{aligned} \tag{7}$$

and satisfies the following conditions [16]

- $\mathbf{w}(\sigma_k) \le 1$, for any $\mu > 0$;
- $\lim_{\mu \to 0} \mathbf{w}(\sigma_k) = 1$, for any σ_k;
- $\frac{\mathbf{w}(\sigma_k)}{\sigma_k}$ is bounded for any $\mu > 0$ and σ_k.

The above analysis indicates that regularization is characterized by a corresponding window function. Hence some other window functions such as Hanning, Blackman, and Gauss window functions developed for spectral filtering can be used for ECT image reconstruction. The weights corresponding to the small singular values need to be reduced in order that these window functions are adapted to the ECT image reconstruction.

Taking Hanning window as an example, the window function is modified as

$$\mathbf{w}(\sigma_i) = \frac{1}{2} + \frac{1}{2}\cos\left(\frac{\pi(\log\sigma_i - \log\sigma_1)}{\log\sigma_m - \log\sigma_1}\right) \tag{8}$$

where, σ_1 is the maximum singular value of the sensitive matrix, σ_m is the minimum singular value of the sensitive matrix, and σ_i is the i-th singular value of the sensitivity matrix. Similarly, the Blackman window function, the cosine window function and the Gauss window function are modified in equation (9), (10) and (11), respectively.

$$\mathbf{w}(\sigma_i) = 0.42 + 0.5\cos\left(\frac{\pi(\log\sigma_i - \log\sigma_1)}{\log\sigma_m - \log\sigma_1}\right) + 0.08\cos\left(\frac{2\pi(\log\sigma_i - \log\sigma_1)}{\log\sigma_m - \log\sigma_1}\right) \tag{9}$$

$$\mathbf{w}(\sigma_i) = \cos\left(\frac{\pi(\log\sigma_i - \log\sigma_1)}{2(\log\sigma_m - \log\sigma_1)}\right) \tag{10}$$

$$\mathbf{w}(\sigma_i) = \exp\left[-\left(\frac{2(\log\sigma_i - \log\sigma_1)}{\log\sigma_m - \log\sigma_1}\right)^n\right] \tag{11}$$

where, n is the power order of the Gauss window.

Substituting the filter factors in equation (6) with $\mathbf{w}(\sigma_i)$ gives equation (12) so that a family of regularization methods is obtained.

$$\hat{\mathbf{f}}_l = \begin{cases} (1 - \mathbf{w}(\sigma_l))\mathbf{v}_l^{\mathrm{T}}\mathbf{g}^{\infty} + \mathbf{w}(\sigma_l)\left(\frac{\mathbf{u}_l^{\mathrm{T}}\boldsymbol{\lambda}}{\sigma_l}\right) & \textit{for } l = 1, 2, \cdots, m \\ \mathbf{v}_l^{\mathrm{T}}\mathbf{g}^{\infty} & \textit{for } l = m+1, \cdots, n \end{cases} \tag{12}$$

Furthermore, a projection operator in equation (13) is applied to the solution to make sure that the reconstructed image is non-negative and upper-bounded.

$$\mathbf{P}[\mathbf{f}(x)] = \begin{cases} 0 & \textit{if} \ \ \mathbf{f}(x) < 0 \\ \mathbf{f}(x) & \textit{if} \ \ 0 \le \mathbf{f}(x) \le 1 \\ 1 & \textit{if} \ \ \mathbf{f}(x) > 1 \end{cases} \tag{13}$$

RESULTS AND DISCUSSION

To evaluate the algorithm, simulations are carried out and the results are compared with those from the LBP and the Tikhonov regularization. A square sensor with 8 electrodes is used. The imaging area is meshed using 32×32

grids, which gives 1024 pixels for one reconstruction. The regularization parameter for the Tikhonov regularization is chosen as 0.003. The power of Gauss window is taken as 2. g^{∞} is chosen as the results from the LBP reconstruction.

Image error and correlation coefficient between the test phantom and the reconstruction, as defined by equation (14) and (15) respectively [15], are calculated to assess the reconstruction quality.

$$image\ error = \frac{\|\hat{\mathbf{g}} - \mathbf{g}\|}{\|\mathbf{g}\|} \tag{14}$$

$$r = \frac{\sum_{i=1}^{N} \left(\hat{g}_i - \bar{\hat{g}}\right)\left(g_i - \bar{g}\right)}{\sqrt{\sum_{i=1}^{N} \left(\hat{g}_i - \bar{\hat{g}}\right)^2 \sum_{i=1}^{N} \left(g_i - \bar{g}\right)^2}} \tag{15}$$

where, $\hat{\mathbf{g}}$ is the reconstructed permittivity distribution, $\mathbf{g}$ is the permittivity distribution used as test phantom, and N is the dimension of vector $\mathbf{g}$. $\hat{g}_i$ and g_i are the i-th components of vector $\hat{\mathbf{g}}$ and $\mathbf{g}$.

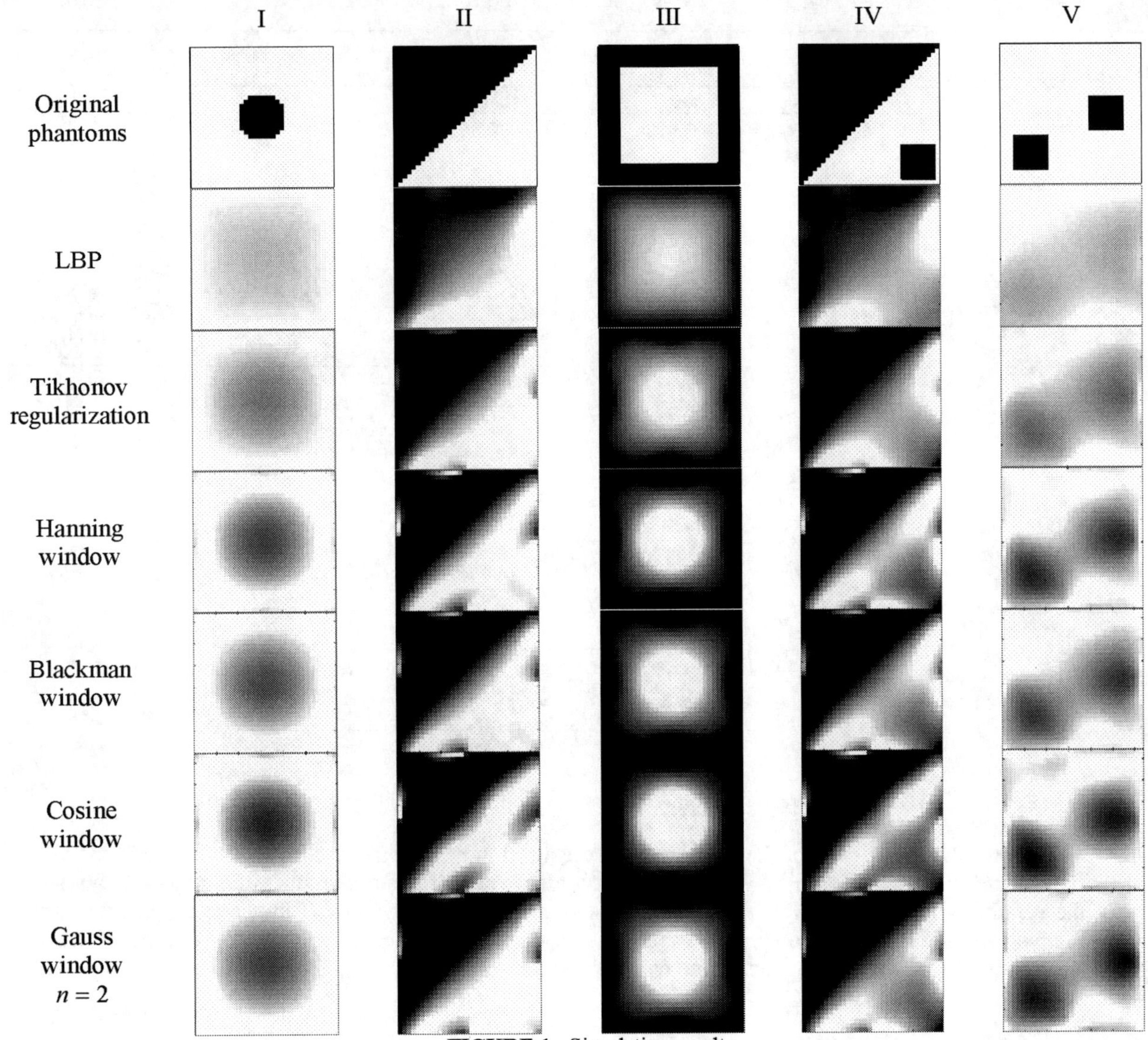

FIGURE 1. Simulation results

Figure 1 shows the reconstructions respectively corresponding to LBP, Tikhonov regularization, the Hanning window, the Blackman window, the cosine window and the Gauss window. It manifests that these regularization methods, which are obtained from the window functions introduced above, are feasible and efficient for the five permittivity distributions. The results of image reconstructions are favorable and much better than LBP algorithm, and are comparable with Tikhonov regularization.

Table 1 and 2 show the image errors and the correlation coefficients. The results using the Blackman window and the Gauss window are similar and better correspondingly. The results using the cosine window are better in sparse permittivity distributions such as I and V.

TABLE 1. Image Errors (%)

	I	II	III	IV	V
LBP	87.85	30.07	38.68	38.52	85.09
Tikhonov regularization	75.60	21.50	33.72	32.18	74.62
Hanning window	66.66	25.55	38.49	32.46	65.41
Blackman window	69.30	22.88	36.22	31.20	68.91
Cosine window	64.98	31.06	41.66	37.46	61.56
Gauss window $n = 2$	68.03	23.04	36.16	31.80	67.76

TABLE 2. Correlation Coefficients

	I	II	III	IV	V
LBP	0.5507	0.9184	0.8635	0.8292	0.5463
Tikhonov regularization	0.6581	0.9533	0.8918	0.8704	0.6441
Hanning window	0.7324	0.9310	0.8653	0.8680	0.7397
Blackman window	0.7079	0.9457	0.8794	0.8796	0.7032
Cosine window	0.7518	0.8963	0.8433	0.8179	0.7815
Gauss window $n = 2$	0.7217	0.9448	0.8797	0.8743	0.7162

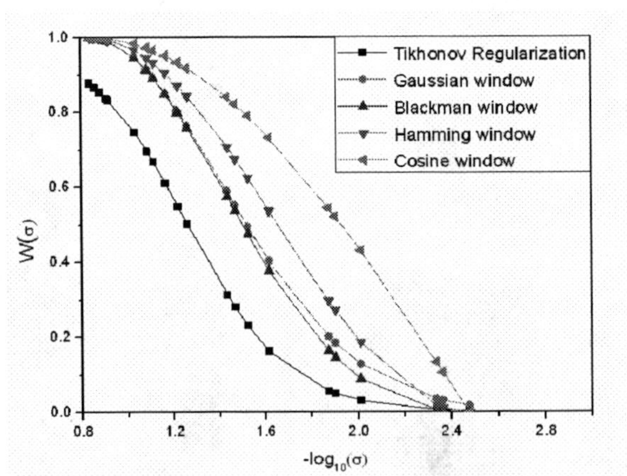

FIGURE 2. Comparison of the window functions used in ECT image reconstruction.

Figure 2 shows the comparison of the window functions used in ECT image reconstruction. The Gauss window and the Blackman window are similar, which are in accordance with the simulation results above. Obviously, to reduce the weighting for the data which is associated with small singular values, Tikhonove regularization is the most effective one, while the cosine window is the least correspondingly. However, the image reconstructions of sparse permittivity distributions using the cosine window are the best.

CONCLUSION

A method using discrete window functions as regularization for electrical capacitance tomography image reconstruction is presented in this paper. From the viewpoint of spectral filtering, the filter factors in Tikhonov regularization show how it reduces the weighting for the data associated with the small singular values, which suggests that other window functions including the cosine window, the Blackman window and the Gauss window can be used as regularization for ECT image reconstruction. The numerical results show the feasibility of the image reconstruction method using window functions as regularization. Future work will focus on the robustness of this method.

ACKNOWLEDGMENTS

The authors would like to express their gratitude to the National Natural Science Foundation of China for supporting the project under the contract (No.60532020, No.60204003).

REFERENCES

1. D. M. Scott, Introduction to process tomography, in: D.M. Scott, R.A. Williams(Eds.), *Proceeding of Frontiers in Industrial Process Tomography I*, Engineering Foundation, California, 1995, pp.1-7.
2. M. S. Beck, T. Dyakowski and R. A. Williams, Process tomography-the state of the art, *Meas. Control* **20 (4),** 163-177 (1998).
3. M. S. Beck and R. A. Williams, Process tomography: a European innovation and its applications, *Meas. Sci. Technol.* **7 (3)** 215-224 (1996).
4. W. Q. Yang and L. H. Peng, Image reconstruction algorithms for electrical capacitance tomography, *Meas. Sci. Technol.* **14 (1),** 1-13 (2003).
5. C. G. Xie, S. M. Huang, B. S. Hoyle, R. Thorn, C. Lenn, D. Snowden and M. S Beck, Electrical capacitance tomography for flow imaging system model for development of image reconstruction algorithms and design of primary sensors, *IEE Proc.-G* **139 (1),** 89-98 (1992).
6. B. W. Rust, Truncating the singular value decomposition for ill-posed problems, Tech. Report NISTIR 6131, National Institute of Standards and Technology, U.S. Dept. of Commerce, July 1998.
7. P. C. Hansen, *The truncated SVD as a method for regularization*, BIT **27**, 534-553(1987).
8. L. H. Peng, H. Merkus and B. Scarlett, Using regularization methods to do image reconstruction of electrical capacitance tomography, *Part. Part. Syst. Charact.* **17 (3),** 96-104 (2000).
9. W. R. B. Lionheart, Reconstruction algorithms for permittivity and conductivity imaging, in *Proc. of 2nd World Congress on Industrial Process Tomography*, 29th-31st Aug, Hannover, 2001, pp.4-11.
10. A. S. Householder, *Principles of Numerical Analysis*, New York: McGraw-Hill, 1953, pp. 135-138.
11. B. L. Su, Y.H. Zhang, L.H. Peng, D.Y. Yao and B.F. Zhang, Use of simultaneous iterative reconstruction technique for electrical capacitance tomography, *Chen. Eng. J.* **77 (1),** 37-41 (2000).
12. W. Q. Yang, D .M. Spink, T. York and H. McCann, An image reconstruction algorithm based on Landweber's iteration method for electrical capacitance tomography, *Meas. Sci. Technol.* **10(11),** 1065-1069 (1999).
13. S. Liu, L. Fu and W. Q. Yang, Optimization of an iterative image reconstruction algorithm for electrical capacitance tomography, *Meas. Sci. Technol.* **10 (7)** L37-L39 (1999).
14. S. M. Tan and Colin Fox, Inverse Problems, Lecture Notes of the University of Auckland, pp.6-4.
15. C. G. Xie and S. M. Huang, Experimental Evaluation of Capacitance Tomographic Flow Imaging Systems Using Physical Models, *IEE Proc-Circuits, Devices & Systems* **141 (5),** 357-368 (1994).
16. M. Bertero and P. Boccacci, *Introduction to Inverse Problem in Imaging*, Bristol, Philadelphia, Institute of Physics Publishing, 1998.

New Image Reconstruction Methods for ECT Measurement in Gas-Solids Two-Phase Flow

Cuiping Wang[1] Dingkai Li[2] Haiying Qi[2]

1. Electromechanical Engineering College, Qingdao University, Shandong Province, China, 266071
2. Department of Thermal Engineering, Tsinghua University, Beijing ,China, 100084
[1]whom correspondence should be addressed E-mail: wangcuiping@tsinghua.org.cn;

Abstract: For measuring the transient gas-solids two-phase flow with ECT (Electrical Capacitance Tomography), an adaptive image reconstruction method is needed especially for the diluted gas-solid flow. In this study, two new image reconstruction methods are introduced into ECT measurement. The higher space resolution can represent the diluted flow regimes, as the concentration distribution in a flow region. GVSPM is a method to reduce the corner dimension between calculated capacitances with measured capacitance vector till to a satisfied convergence error. Semi-step Landweber iterative method use the frontier two steps iterative information to do next iteration, the convergence is speed-up. Reconstructing images with this algorithm, the calculation is rapid without divergence till to satisfy the required error. The designed flow and measured capacitance data of layer flow are reconstructed by Landweber, Semi-Landweber and GVSPM iterative methods in order to test these new algorithms. For a required error, the calculation by GVSPM lasts shortest time, and the widest applied method for all flow regime is semi-Landweber iterative way.

Keywords: Electrical Capacitance Tomography (ECT); image reconstruction method; semi-step Landweber iterative method; GVSPM

INTRODUCTION

Electrical capacitance tomography (ECT) has been developed for over a decade and the measurement technique has been used for mine detection, wildcat and grease detection and gas-solids flow measurement[1-2]. With the new developments in the hardware and image reconstruction algorithms, ECT has become one of the favorite tools for measuring multi-phase flows. In the previous studies on gas-solids flow measurement, ECT has been used mostly to deal with dense flows. Because the high solids concentration causes the distinctive changes in dielectric constant field, the measured capacitance signal is relatively strong and the noise is relatively low. For a dilute gas-solids flow in a large CFB, however, the signal-to-noise ratio (SNR) may be too small to achieve meaningful results.

Therefore, it is necessary to optimize the sensor design and develop new image reconstruction algorithms (that is the method to solve the inverse problem to deduce concentration distribution from measured capacitance data and then represent the solution in an image), so that ECT can be applied to measure a large CFB and represent dilute gas-solids flows.

The simplest and rapidest image reconstruction method is linear back projection (LBP) with no iteration, and the popular recognized method is Landweber interative method[3]. New methods which can improve the space solution, especially to distinguish the distribution of center flow region, are studied in this paper.

NEW IMAGE RECONSTRUCTION METHODS

Semi-step Landweber Iterative Method

When the inverse problem is solved with Landweber method, "half-convergence" always occurred during the iterative procedure, that is, steady convergence in the early solving period but suddenly divergence after a threshold. In order to avoid this phenomenon and speed the solving process, the very first two steps iterative results by Landweber method are used to do the next iterative calculation, with weighted coefficient of correction. So this

CP914, *Multiphase Flow: The Ultimate Measurement Challenge,* edited by X. Cai, Y. Wu, Z. Huang, S. Wang, and M. Wang

method is called as Semi-step Landweber (or Semi-Landweber) iterative algorithm[4-6].

Semi-Landweber iterative method is adopted as equation (1).

$$\mathbf{g}_k = \mathbf{g}_{k-1} + \mu_k(\mathbf{g}_{k-1} - \mathbf{g}_{k-2}) + \omega_k \mathbf{S}^*(\boldsymbol{\lambda} - \mathbf{S}\mathbf{g}_{k-1}) \tag{1}$$

where μ_k and ω_k are dynamic functions of parameters v, and $v>0$; k is the iteration number; $\boldsymbol{\lambda}$ is the normalised capacitance, which consists of 28 independent capacitance values from an 8-electrode ECT system measurement in a scanning period; $\mathbf{S}$ is a 28×1024 matrix, called sensitivity map; $\mathbf{g}_k$ is the kth solution of solids distribution, can be presented as gray level. With 32×32 pixels in a cross-sectional field, the calculated solution $\mathbf{g}_k$ has 1024×1 elements, arranged into 32×32 to represent an image.

The initial values can be obtained after the choice of v, as the equations (2).

$$\begin{cases} \mu_1 = 0 \\ \omega_1 = \dfrac{4v+2}{4v+1} \\ \mu_k = \dfrac{(k-1)(2k-3)(2k+2v-1)}{(k+2v-1)(2k+4v-1)(2k+2v-3)}, k>1 \\ \omega_k = \dfrac{4(2k+2v-1)(k+v-1)}{(k+2v-1)(2k+4v-1)}, k>1 \end{cases} \tag{2}$$

The solving procedure can be carried out as the following steps:

(1) Initiate the values. The measured capacitance vector $\boldsymbol{\lambda}$ is known and the solids distribution gray level $\mathbf{g}$ is to be calculated by LBP method of $\mathbf{g} = \boldsymbol{\lambda}^T\mathbf{S}$. The $\mathbf{g}_0$ is the initial value.

(2) Calculate the next step results. $\lambda_1 = \dfrac{4v+1}{4v+2}\lambda$ and $g_1 = S^*\lambda_1$.

(3) Begin the iteration. For $k \geq 2$, the coefficient is calculated by equation (2) and the iteration is by equation (3) till to a required convergence error.

$$\begin{cases} \lambda_k = \lambda_{k-1} + \mu_k(\lambda_{k-1} - \lambda_{k-2}) + \omega_k(\lambda - Sg_{k-1}) \\ g_k = S^*\lambda_k \end{cases} \tag{3}$$

The choice of v is very important to speeder the calculation, for the ECT image reconstruction, $0 < v < 1$ should be the best range.

GVSPM (Generalized Vector Sample Pattern Matching Method)

Generalized Vector Sample Pattern Matching Method is developed from Sample Pattern Matching Method for solving ill-posed equations[7-9]. GVSPM is studied applying for image reconstruction of EIT in medical measurement[9]. For the gas-solids distribution by ECT measurement, the image reconstruction principle is similar to be applied. The solving and convergence process can be expressed by the example of two-dimension non-orthogonal vector in Figure.1 (from the $\boldsymbol{\lambda}^{(0)}$、$\boldsymbol{\lambda}^{(1)}$ approaching to $\boldsymbol{\lambda}$, the angle is reduced）:

(1) Normalization, the $\boldsymbol{\lambda} = \mathbf{S}\mathbf{g}$ becomes $\underline{\boldsymbol{\lambda}} = \underline{\mathbf{S}}\,\underline{\mathbf{g}}$;

where $\underline{\boldsymbol{\lambda}} = \dfrac{\boldsymbol{\lambda}}{|\boldsymbol{\lambda}|}$, $\underline{\mathbf{S}}_i = \dfrac{\mathbf{S}_i}{|\mathbf{S}_i|}$, then $\underline{\mathbf{g}}_i = \dfrac{\mathbf{g}_i|\mathbf{S}_i|}{|\boldsymbol{\lambda}|}$ (In Figure.1，i is the iterative time)

(2)Initiation, the projection on each coordinate axis of $\underline{\boldsymbol{\lambda}}$ as the initiation solution, then the $\underline{\mathbf{g}}^{(0)}$ can be obtained, $\underline{g}^{(0)} = \underline{S}^{T}\underline{\lambda}$;

(3) Iterative process:

I. When the iteration is beginning, k=0, $\lambda^{(k)} = \underline{S}\underline{g}^{(k)}$, do the normalization $\underline{\lambda}^{(k)} = \frac{\lambda^{(k)}}{|\lambda^{(k)}|}$;

II. Calculate $\underline{\Delta\lambda}^{(k)} = \underline{\lambda} - \underline{\lambda}^{(k)}$, the angle $\theta^{(k)}$ between $\underline{\lambda}$ and $\underline{\lambda}^{(k)}$ is the difference, after once iteration, the angle reduce $\Delta\vartheta$ and the vector changes $\underline{\Delta\vec{\lambda}}$.

III. Iteration continuously to reduce the angle and $\underline{\Delta\vec{\lambda}}$: $\underline{g}^{(k+1)} = \underline{g}^{(k)} + \underline{S}^{T}\underline{\Delta\lambda}^{(k)}$.

IV. Until the angle is lower enough:

$$\underline{\vec{\lambda}} \bullet \underline{\vec{\lambda}}^{(k)} = \cos\vartheta \to 1 \tag{4}$$

V. Conversion: $\underline{g}^{(k+1)}$ should be converted into $\underline{g}_i = \frac{g_i|S_i|}{|\lambda|}$, the vector g_i is the last optimum solution。

Because GVSPM method is based on the corner dimension reduced to iterate without inversion matrix, the solution error is very fine. Like semi-Landweber method, with the required error and threshold, the calculation will be convergence and stop automatically.

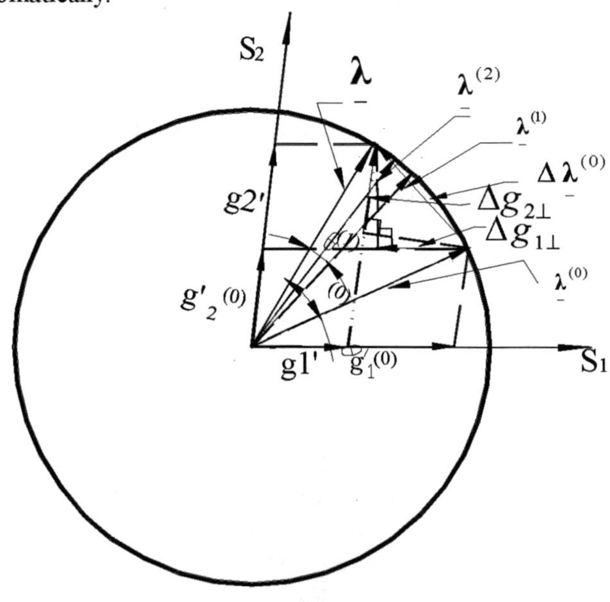

FIGURE 1. Presentation of convergence process by GVSPM iteration method

IMAGE EFFECTIVENESS

The image effectiveness is evaluated comparing the reconstructed images of simulated flow regimes and experimental flow regimes, as shown in Figure.2 and Figure.3.

Figure.2 represents that the four typical flow regimes are designed for simulation, the images reconstructed by

LBP, Landweber iterative, semi-Landweber and GVSPM iterative methods. The image error is listed below each image, it is the average difference between the designed and the reconstructed gray distribution in the image. The image reconstructed time is iterative number.

For a required convergence error, Landweber, semi-Landweber and GVSPM methods iterate different times and reach different accuracy, the number in bracket after the image error is the iterative time (shortened as Its) responding to each iterative method.

For the designed flow models and judged by the reconstructed images in Figure.2, only semi-Landweber iterative method is adaptive to all types of flow regime to reconstruct the images with the less error and less iterative times. GVSPM method also can obtain better quality images with rapid convergence velocity for some flow regimes, such as the points flow, the image by GVSPM reaches to the smallest error in pipe center. It is clear that semi-Landweber iterative method can improve the space solution of reconstructed pipe field while GVSPM is more suitable to center flow regime. For the annular flow, images by GVSPM are distorted, there produced a false circle in the pipe centre, this is because of half-divergence in GVSPM iteration. The gas-solids interface in each designed flow regime is more clearly reconstructed by semi-Landweber method than LBP, Landweber methods, and GVSPM method is only suitable to reconstruct the point or center flow regime.

The experimental layer flow in a square pipe was measured by ECT and reconstructed all images by above four algorithms. Figure.3 represents the images reconstructed from the measured capacitance vector at one time of the flows. The actual flow distribution at this time may be concluded by the image of semi-Landweber method, the gas-solids interface is clearer than other three methods though some errors should be exist. As for the "cavity" in solids and "feather" in gas phase (not means the feather on

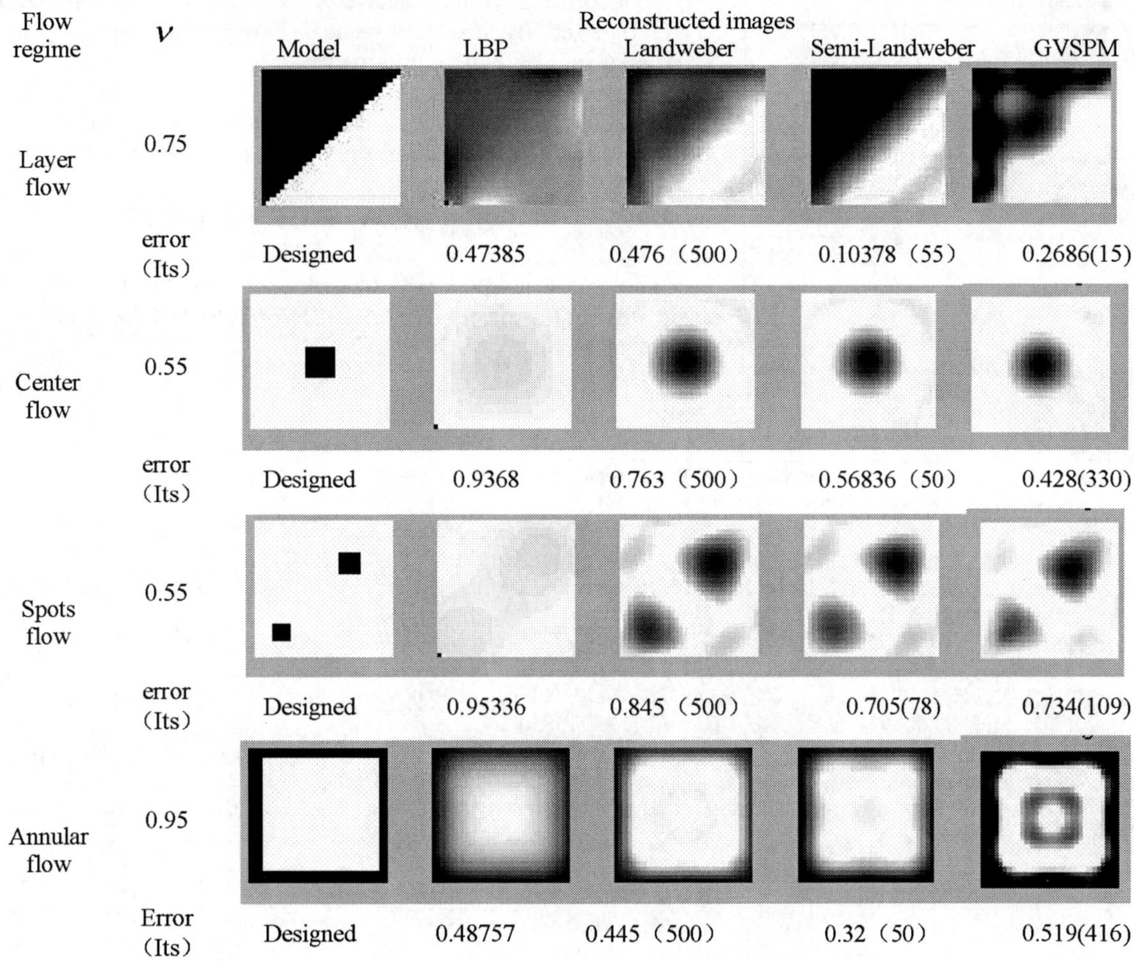

FIGURE 2. The effectiveness of image reconstruction on the simulation distribution

the gas-solid boundary) shown in Figure.3, it is not sure if they were exists, perhaps the flow has local high void rate and non-uniform in the bulk, local loosening will cause changes of the measured capacitance vector, then the reconstructed cavity may be right. Because the operation of layer flow is slow and steady, the "feather" in the gas phase space should not be exist. It is reconstructed from the spray capacitance that always is caused by the screen between electrodes is not completely. It is the superior for an image reconstruction method to reduce the feather on the gas-solid boundary, so the semi-Landweber iterative method is an excellent one.

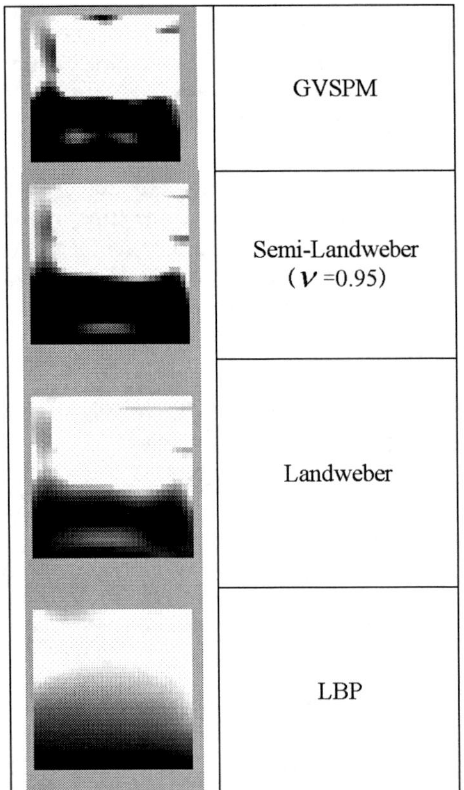

FIGURE 3. Reconstructed image of layer flow in a square pipe

CONCLUSIONS

a. The new introduced methods of GVSPM and semi-Landweber iterative one all improved the image quality, the image errors are less and the iterative times are reduced than Landweber method.
b. GVSPM is more suitable to reconstruct the center point flow. But it is not suitable to an annual flow.
c. Semi-Landweber has the widest applicability to reconstruct all designed flow regimes. Especially for the annual flow and layer flow, the reconstructed images have the least error. So the diluted flow in a CFB riser (typical core-annual flow) can use this method to reconstruct ECT measured images.
d. The reconstructed image error is still not weak enough to be neglected. More works need to be carried out to utilize ECT images to present the transient information in gas-solids flow

ACKNOWLEDGEMENTS

This study is sponsored by the Special funds for Major National Basic Research Projects (No. 2006CB200305) and the Laboratory Fund of Tsinghua University.

REFERENCES

1. W. Q. Yang and S. Liu, *Flow Measurement and Instrumentation*, 11, 237-244 (2000).
2. S Liu, W Q Yang and H Wang, *Meas. Sci. Technol.*, 12, 1120-1125(2001).
3. Lihui Peng, Henk Merkus, and Brian Scarlett, *Part. Part. Syst. Charact.* 17, 96-104(2000).
4. Tingyan Xiao, Shengen Yu and Yanfei Wang, *Numberical Calculation of Inverse Problem,* Publisher City:Beijing, Science Press (China), 2003, pp. 157-176.
5. Guangzheng Ni, Xiuying Qian, *Numberical Calculation of Electromagetic Field*, Publisher City: Beijing, High Education Press, 1996, pp.75-81.
6. Jianni Sheng, *Numberical Analysis of Electromagetic Field*, Publisher City: Beijing, Science Press, 2003, pp.45-49.
7. Y. Saito, E. Itagaki, and S. Hayano, *J. Appl. Phy.* 67 (1990) 5830-5832.
8. H. Endo, S. Hayano, Y. Saito, et al, "Generalized vector sampled pattern matching method -Theory and applications" in *Electromagnetic Nondestructive Evaluation* (VI), edited by F. Kojima, Netherlands, 23 (2002) 285-292.
9. Guoya Dong, "Study of EIT and electrical conductivity measurement methods of head tissues", Ph.D. Thesis, Beijing, Tsinghua University, 2003.

Visualization Of Water Vaporization During Electrical Heating Process With X-ray Tomography

Jie Wang[1], Yubin Li[1,2], Michael Erath[1,2], Apostolos Kantzas[1,2*]

1Department of Chemical and Petroleum Engineering, Schulich School of Engineering
2Tomographic Imaging and Porous Media Laboratory
University of Calgary, 2500 University Drive, Calgary, AB, Canada T2N 1N4.
akantzas@ucalgary.ca

Abstract: This paper deals with the study of electrical heating as a potential method for recovery of heavy oil and bitumen from unconsolidated sand formations of Northern Alberta. Electrical heating is a preferred energy transfer method for heating relatively shallow formations containing bitumen with limited support for high-pressure steam injection. Simulation results demonstrated that although electrical heating can contribute to the recovery of oil, substantially more oil would be recovered if the formation or injected water were allowed to boil. Thus the in-situ steam creation has been the target of simulation and experimental work presented here.

Keywords: X-ray CT; Heavy oil and Bitumen; Electrical heating; Computer Simulation; Thermal recovery

INTRUDUCTION

Electrical heating is a thermal process, which is used for heavy oil or bitumen production. In electrical heating, the formation water acts as an electrical circuit so electrical energy is dissipated into the reservoir. During this process, the formation water can be vaporized near the well bore (electrode) due to the overheating of the near well bore area. This could result in cutting off the electrical circuit and process termination. Therefore water vaporization is often considered as a catastrophic phenomenon for electrical heating processes. On the other hand, additional oil production can be expected by converting some water into steam.

In our work we consider only Ohmic heating processes where low frequency current is applied. The process has been applied in environmental clean-up of heavy organic pollutants in soil and is patented for use in bitumen recovery [1]. The work is split into two parallel procedures. One deals with the simulation of the process while the other deals with experimental measurements at different scales. Progress on both fronts is presented next.

COMPUTER SIMULATION

Computer simulation of typical heavy oil and bitumen formations, using state-of-the-art commercial reservoir simulators, revealed that given adequate energy, boiling water could expand and displace bitumen efficiently. The coordinated injection of water and injection of energy through electrical heating can also further improve oil recovery [2]. In order to investigate the heavy oil or bitumen recovery numerically, we performed simulations in a prototype 1m3 Cartesian model. All simulations were performed with the commercial simulator STARS [3]. The reservoir simulation model mimics some aspects of a typical Athabasca reservoir. The reservoir properties are tabulated in Table 1. The oil is modeled as dead oil. The viscosity data for the oil are those of typical Athabasca bitumen and are tabulated in Table 2. The electrical and thermal properties of the rock as well as relative permeability data are taken from electrical heating template files for STARS [3].

CP914, *Multiphase Flow: The Ultimate Measurement Challenge,* edited by X. Cai, Y. Wu, Z. Huang, S. Wang, and M. Wang

Table 1. Reservoir Properties [2]

Properties	**Values**
Porosity	0.4
Water saturation	0.4
Oil saturation	0.6
Initial reservoir temperature	10 °C
Initial reservoir pressure	1.5 MPa
Permeability in I, J, K directions	2000 md

Table 2. Viscosity data, Athabasca bitumen [2]

Temperature, °C	**Viscosity, mPas**
10	3000000
40	30000
60	4500
80	1000
100	300

3D Cartesian model was set up to simulate 1m3 rock in order to determine the magnitude of incremental oil recovery by water vaporization. The reservoir model is divided into 15 (I) x 15 (J) x 5 (K) grid blocks, with each grid block size 0.1m x 0.1m x 0.2m. The producer/electrode is located at the center of the reservoir, and ground is placed around four side surfaces of the Cartesian reservoir. Grounding in this manner mimics infinite grounding, which is common in field electrical heating applications. The model is insulated so no heat loss is taken into account at this stage.

Numerous simulation runs were performed to study two specific cases in order to demonstrate the effect of electrical heating in bitumen recovery. Specifically, two sets of parameters were studied:

The well produces at the saturation temperature (197°C) under the reservoir pressure 1.5MPa (case I).

The well produces at a temperature 8°C lower than the saturation temperature (i.e., 189°C, case II).

The objective of the above simulation runs is to find the difference between two oil recoveries, and establish whether the simulator can offer an effect of boiling. The magnitude of incremental oil recovery can be estimated numerically and it can be compared to simple analytical calculations. Setting the minimum bottom hole pressure of 1.5MPa is to ensure that the production of the well can only attribute to thermal expansion of the fluids and water vaporization.

The electrical operating constraint is maximum voltage and set at 220 Volts for both cases. In order to make the well produce at the saturation temperature, the maximum voltage constraint works all the way through the production life (120 days) for case I. For case II, the initial electrical operating constraint is maximum voltage (220V). At day 32, the operating constraint switches to maximum current (0.05A). Therefore the well can operate at 8°C lower (189°C) than the saturation temperature (197°C).

In order to study the effect of water injection after water boiling on the incremental oil production, the third case (case III) was built. The water injection was set to start from day 120 and run for 4 months, then the injection well was shut in and the producer was open.

EXPERIMENTAL

Simulation alone cannot provide answers on the pore level physics of the problem. In order to actually observe these phenomena, preliminary laboratory experiments were designed. A cylindrical geometry sand pack with 8.9cm in diameter and 30cm in height was designed and constructed. The holder is made out of x-ray transparent materials (Plexiglas and Aluminum). In order to keep matters simple, the pore pressure and temperature are maintained ambient. The holder is packed with the sand saturated with synthetic heavy oil and brine and the electrode will be inserted in the central axis of the sand pack along with the capability to inject water and produce oil. CT scanning is used to experimentally visualize water vaporization during the application of electrical heating in the sand pack. The sand pack is subjected to electrical heating and CT scanning simultaneously. The amount of oil produced is monitored by a data acquisition system, at the same time the sand pack is scanned vertically. The CT images show the resulted distribution of oil, water and steam during oil production in the sand pack.

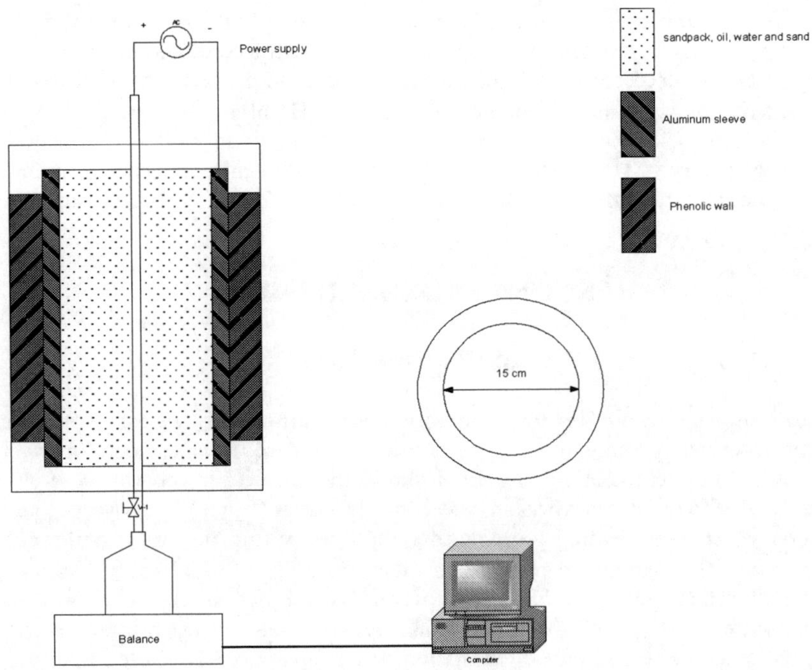

FIGURE 1. Experimental setup

FIGURE 2. Vertical CAT scanner assembly

The thickness of the aluminum grounding-cylinder (see Figure 1) is 5mm. The sand pack volume is 1466cm3 and it has porosity of 0.3 and initial water saturation of 20%. The oil used in the preliminary experiment was synthetic oil that has a viscosity of 10,000cS at a temperature 25oC. The sand was originally mixed with water (3%wt NaCl brine) to form a paste and then this paste was mixed with oil and was pressed into the core holder. A single-phase power supply with maximum output of 140V was used. Fiberglass insulation was used to prevent heat losses as much as possible.

The CAT scanner used in this study is a General Electric (GE) Highlight Advantage™, which is based on the GE9800 third generation design. The scanner has been extensively modified so that it can be flipped to a vertical scanning position (horizontal gantry). Thus vertical objects of up to 1.8m in length can be scanned with positioning precision of 0.01mm. The scanning parameters for this experiment were 120keV, 170mA, 3mm thickness and 0.49mm x 0.49mm pixel resolution.

RESULTS AND DISCUSSION

Computer Simulation

Figure 3 shows typical temperature profiles for simulation case I and case II. In case I, temperature is allowed to reach the saturation point (reservoir pressure maintains at 1.5 MPa). In case II, the temperature was increased to 189˚C (8 degree lower than the saturation temperature). Figure 4 shows the oil recovery curves of these two cases. It can be seen that case I has more than 40% OOIP recovery but case II has less than 5% OOIP recovery. The simulation confirms that heating the formation water to its boiling point, rather than not letting the water boil can result in significant incremental oil recovery. Since the bottomhole pressure is maintained at its initial value, viscous forces (ΔP) are not considered in both cases. The production mechanisms that play a role are thermal expansion of fluids (oil and water) as well as water vaporization (internal steam drive). In case I, all these mechanisms are present because the temperature is increased to the saturation temperature and part of formation water is vaporized. While in case II, the temperature is lower than the saturation temperature; therefore the production is only attributed to thermal expansion of fluids. The difference in thermal expansion brought by 8°C accounts for an insignificant incremental recovery. Therefore it is highly likely that the additional 30% oil recovery in case I came from water vaporization alone.

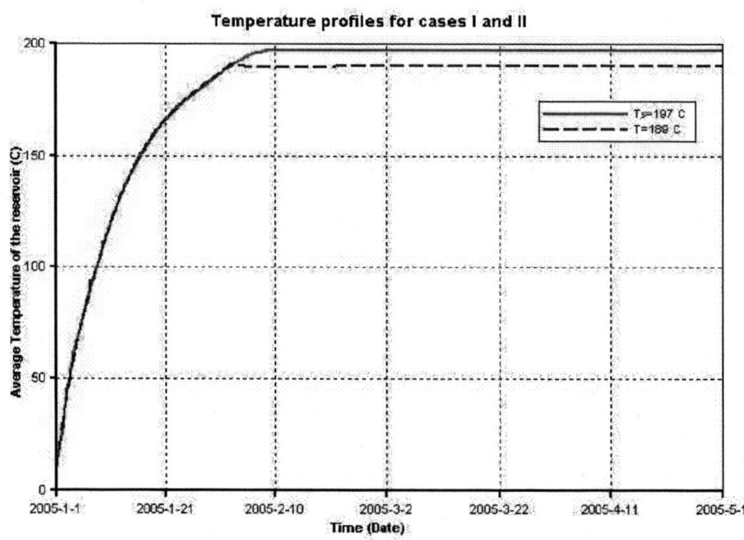

FIGURE 3. The temperature profiles for cases I and II

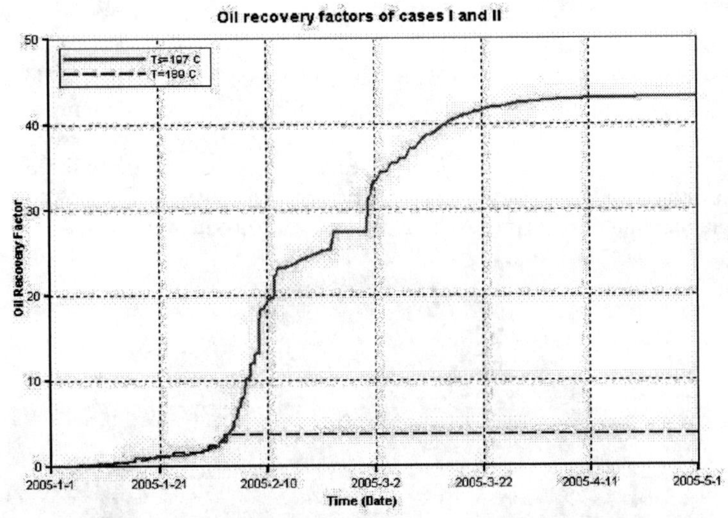

FIGURE 4. The oil recovery curves for cases I and II

The results for case III are shown in Figure 5. The red curve in Figure 5 is the temperature profile of the reservoir. Water injection at day 120 quenched the reservoir from 197˚C to approximately 150˚C. Water injection continues and the producer is shut-in. The reservoir pressure increases and the saturation temperature also increases to the new saturation temperature corresponding to that pressure. After 4 months, the producer opens and the injector is shut in. The saturation temperature is back to 197˚C and a jump in oil recovery factor is observed. An additional 10% OOIP is recovered (blue curve) was obtained by injecting water at 197˚C.

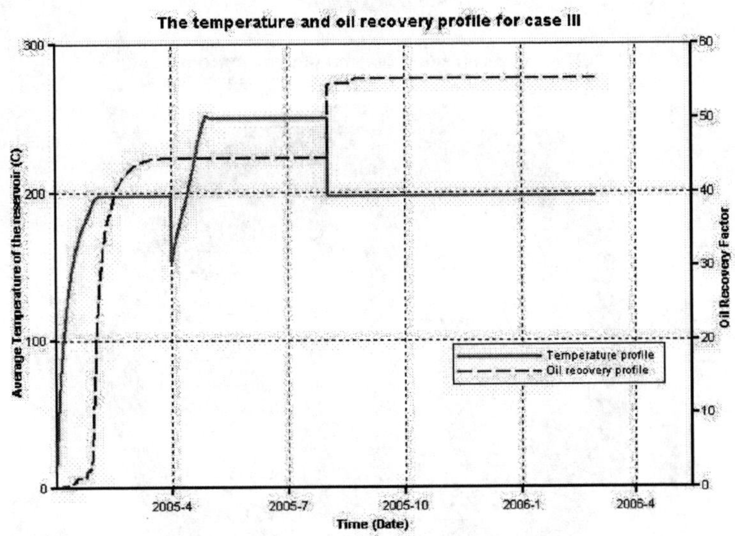

FIGURE 5. The temperature and oil recovery profiles for case III

X-ray CT Measurements

The sand pack was exposed to heating cycles that span over one week. Initially thermal losses dominated the process and heating was minimal. Better insulation and higher heat flux allowed us to reach temperatures of up to 99oC by the time this manuscript was compiled. Production started after the sand pack reached temperature of 90oC. The production trickled through the well. After one week of heating 14.6g of oil were produced. X-ray CAT images were collected on a daily basis and the results were analyzed for saturation and sweep efficiency calculations. Figures 6 and 7 show two images on a single location of the sand pack before and after production started.

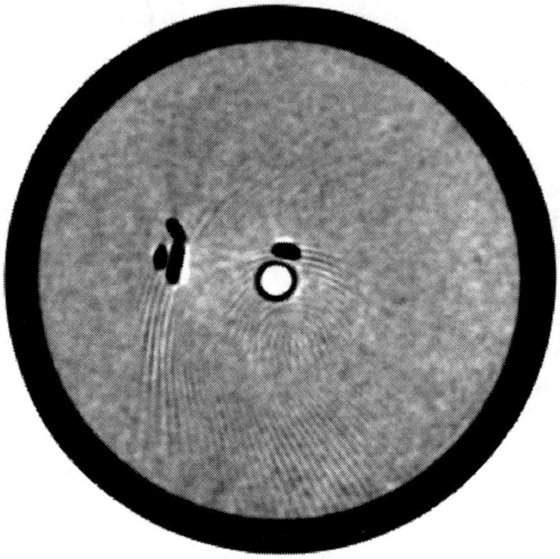

FIGURE 6. Cross-section of sand pack prior to production

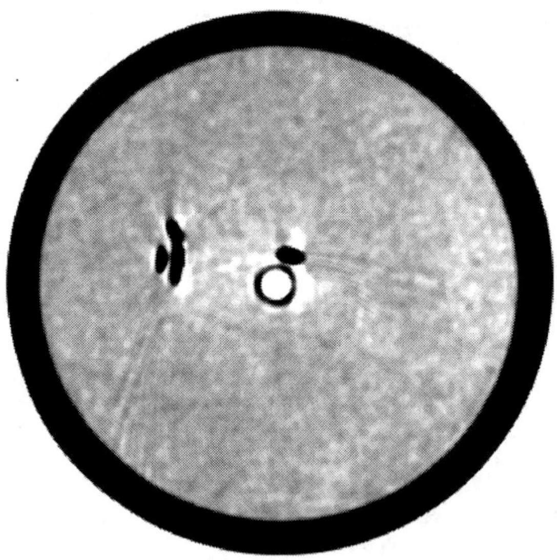

FIGURE 7. Cross-section of sand pack after production

Figure 8 shows the saturation evolution for the first week of production as a function of sample length. It can be clearly seen that the sand pack is draining at the top of the model (although very slowly). About 6.8% by weight of the total oil in place was recovered this first week of production. Although we have reached approximately the boiling point for water (96.2oC at 662mmHg pressure) we have not seen yet any explosive production that was observed in beaker tests and simulation [2]. The work now continues with incremental injection of water and further heating of the sand pack.

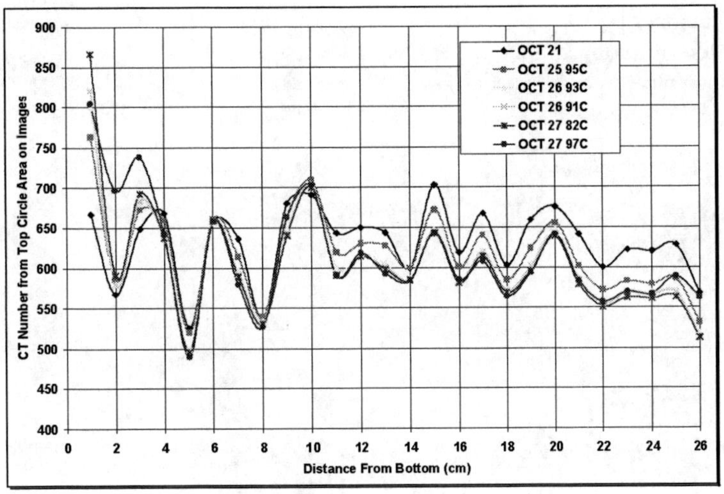

FIGURE 8. Saturation profile along the sand pack

CONCLUSIOINS

The simulation results indicated that if the reservoir is heated at its boiling temperature with various grounding options, the recovery factors could range up to 40+% of the original oil in place (OOIP). It is also expected that addition of water to replace voidage and reestablish the electrical circuit could give much higher recovery factors. The additional oil recovery brought by water injection can reach about 10% of OOIP.

Experimental work demonstrated that oil indeed drains with electrical heating and recoveries similar to what was observed with simulation were observed. However, the process is very slow, the control of heat losses is critical and the potential of water boiling to increase recovery has not been as apparent as it was in the simulation.

ACKNOWLEDGEMENTS

The financial support of the National Sciences and Engineering Research Council of Canada, the Canada Research Chairs Program and its industrial affiliates (PetroCanada, Canadian Natural, Devon Canada, Shell, Suncor Energy, Nexen, Paramount, Laricina, E-T Energy, Shclumberger) is greatly appreciated.

REFERENCES

1. Vermulan, F. and McGee, B., In Situ Electromagnetic Heating for Hydrocarbon Recovery and Environmental Remediation. JCPT. 39(8), 25-29, 2000.

2. Wang, J. McGee, B. and Kantzas, A., Calculations of the Effect of Boiling Water in Bitumen Production. CIPC2006-046, presented at The Petroleum Society's 7th Canadian International Petroleum Conference (57th Annual Technical Meeting), Calgary, Alberta, Canada, June 13 – 15, 2006.
3. CMG STARS user's Guide, version 2004.

NOMENCLATURE

CT_n	CT number [-]
S_w	Water saturation [fraction]
S_b	Bitumen saturation [fraction]

FPGA Based Data Acquisition and Processing for Gamma Ray Tomography

H. Inaki Schlaberg[¶], Donghui Li[*], Yingxiang Wu[*] and Mi Wang[¶]

[¶]*Institute of Particle Science and Engineering, University of Leeds,Leeds,LS2 9JT, UK*
[*]*Institute of Mechanics, Chinese Academy of Sciences, Beijing, China*
Email: h.i.schlaberg@leeds.ac.uk

Abstract. Data acquisition and processing for gamma ray tomography has traditionally been performed with analogue electronic circuitry. Detectors convert the received photons into electrical signals which are then shaped and conditioned for the next counting stage. An approach of using a FPGA (Field programmable gate array) based data acquisition and processing system for gamma ray tomography is presented in this paper. With recently introduced low cost high speed analogue to digital converters and digital signal processors the electrical output of the detectors can be converted into the digital domain with only simple analogue signal conditioning. This step can significantly reduce the amount of components and the size of the instrument as much of the analogue processing circuitry is eliminated. To count the number of incident photons from the converted electrical signal, a peak detection algorithm can be developed for the DSP (Digital Signal Processor). However due to the relatively high sample rate the consequently low number of available of processor cycles to process the sample makes it more effective to implement a peak detection algorithm on the FPGA. This paper presents the development of the acquisition system hardware and simulation results of the peak detection with previously recorded experimental data on a flow loop.

Keywords: gamma ray, tomography, FPGA, data acquisition

INTRODUCTION

Data acquisition and processing for gamma ray tomography has traditionally been performed with analogue electronic circuitry. Using scintilators with their associated photomultiplier tubes as detectors for gamma rays, these generate electrical signals in the form of pulses for each incident photon. Analogue circuits shape the pulses making their width uniform and adjust the baseline voltage which drifts over time. In the case of dual energy gamma rays, i.e. with two distinct energy sources, additional amplitude discrimination circuitry is added to differentiate signals originating from each of the sources. The output of these circuits then feeds a digital pulse counter which adds the number of photons in a given time frame.

An approach of using a FPGA based data acquisition and processing system for gamma ray tomography is presented in this paper. With the availability of more lost cost high speed analogue to digital converters and digital signal processors, the electrical output of the detectors can be converted into the digital domain with only simple analogue signal conditioning. This step can significantly reduce the amount of components and size of the instrument as the bulk of the analogue processing circuitry is eliminated. In typical tomography applications an instrument will consist of 16, 32 or more identical channels. Once the pulse signals have been converted into a digital signal it can be more easily used for other operations such as calculating the distribution of the peak amplitudes which is presented in the form of a histogram.

The sampling frequency of the A/D converter was chosen at 20MHz which results in an interval between samples of 50 ns. This time determines the accuracy with which the peak amplitude can be measured and can easily be increased in the future by using pin compatible devices with higher sampling rates. Operations on the digital signal were designed to be performed by a DSP and one of the first tasks of the processor is to extract the peak amplitudes using a peak detection algorithm. The DSP chosen for this work operates at 200Mhz. While this is fast

CP914, *Multiphase Flow: The Ultimate Measurement Challenge,* edited by X. Cai, Y. Wu, Z. Huang, S. Wang, and M. Wang

enough for many operations, it only allows 10 processor clock cycles to complete the peak detection algorithm between samples provided by the ADC.

A field programmable gate array chip is also used in the circuitry. This component provides all the glue logic to interface the ADC to the DSP, has a buffer memory for samples and handles the communication with other parts of the system as well as implementing a serial link to a host computer for monitoring during the development of the system.

Only a small fraction of the chip's resources are used for these functions leaving spare capacity that can be employed for processing of the samples. As opposed to the sequential nature of the program on the DSP, the FPGA can use resources for circuitry that operates in parallel. Taking advantage of this property, a peak detection algorithm and counter was developed that processes each sample from the ADC in just two clock cycles. The code is written in VHDL (Very high speed Hardware Description Language) and a simulation of the code with data captured during other experiments was run. The results in the form of a histogram showing the peak distribution of the data, are very similar to those using a sequential algorithm developed in Matlab and which used the same data set.

The physical implementation of the system consists of a number of acquisition cards which connect to the detectors and one controller card which communicates with a host computer and has controlling functions such as orchestrating the simultaneous start of the acquisition on all channels and retrieving the results for transmission to the host. Each board is implemented on a standard Eurocard size PCB (100x160mm). Every acquisition card has two channels, i.e. two detectors, but further channels can be added as the computing load of the DSP has been transferred to the FPGA which still has ample resources available.

This work describes the development and first results of a FPGA based data acquisition system for the application of a gamma ray tomography. It has the advantages over analogue based signal processing systems of reduced component numbers, potentially smaller size and increased flexibility.

SYSTEM OVERVIEW

The data acquisition system communicates with a host computer which sends parameters, such as the acquisition time frame, sends commands, such as the start signal of an acquisition cycle and collects the acquired results. One processor board of the acquisition system is dedicated to perform the communication with the host and interfaces with the rest of the channels of the system. This controller board distributes the commands for the host to each of the channels via a backplane that connects all the boards of the system.

Each channel of the tomography system can operate independently and report the results to a controlling processor which will communicate with a host computer. This architecture provides a highly parallel setup in which the addition of further channels adds little time to the total acquisition process which only would be increased due to the transmission of the results to main processor while all the processing is integral to the channel circuitry.

A similar parallel architecture based on DSP and FPGA has been developed for optical tomography (Garcia-Castillo and Ozanyana, 2005). However optical tomography presents different processing requisites than gamma ray tomography and features such as a digital lock-in amplifier are not required for this application. Nevertheless, the basic principle can be applied to other tomography modalities such as ERT or ECT.

After simple signal conditioning to bring the voltages within the range of the A/D converter specification a dual channel converter chip takes the data of two detectors to the FPGA chip. The FPGA is connected to a DSP which can operate on the sample data or instruct the FPGA to perform operation. A backplane connector provides the bus for communication between the channels and the main controlling board. Although the connector is physically attached to the FPGA, it is the DSP which controls the communication. See Figure 1.

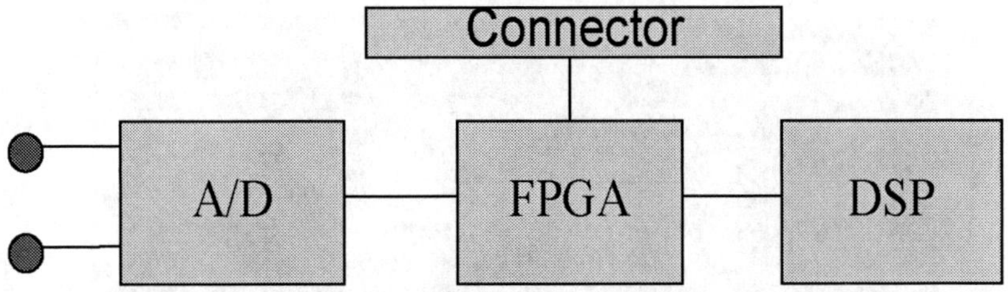

FIGURE 1. Block diagram of a processing board with two channels

COMPONENT SELECTION

The main components of each channel are the A/D converter that brings the signals into the digital domain, the DSP that performs processing and communication functions and the FPGA which connects these and other components of the channel and can also perform operations on the incoming signals.

The component that will influence the system specification most is the A/D converter. The main parameters are the sampling speed in samples per second and the resolution in number of bits. Previous experiments with a dedicated single channel commercial acquisition board showed that a 20MSPS (mega samples per second) with 12 bits resolution provided a sufficiently accurate data for flow loop measurements. Another system described by Bruneau et al. (2005) uses a 10MSPS with 8 bit resolution for a fluidised bed application using gamma rays.

The A/D converter chip selected operates at 20MSPS with 12 bits resolution. It is pin compatible with 14 bit devices and devices that can sample at rates up to 65MSPS. In this way it provides an easy upgrade path if higher specifications are required in future, without having to redesign the hardware. A further feature of the device is that it houses two converters on the same chip, reducing the chip count of the final system. The part is manufactured by Analog Devices.

The DSP chip is a low cost 200MHz 32-bit Sharc device with 2Mbit on chip memory. This DSP has a at first inspection sufficient processing capability and on chip peripherals to handle the tasks of the acquisition, data processing and communications.

Connecting the A/D converter and DSP is the FPGA which also interfaces to the backplane that interconnects all the channels to the controller board. The is a great variety of devices to choose from and the selection criteria is based on:

a) cost,

b) the number of pins necessary to connect the A/D converter to the DSP and to the backplane connector as well as other peripheral such as a simple serial RS232 link for the debugging stage and a USB link for connection to the host,

c) available on chip memory to hold the samples of an acquisition frame and the resulting histogram without requiring external memory

d) sufficient available logic to implement processing and communication functions without further external components.

The two main FPGA manufacturers (Altera and Xilinx) offer devices with very similar capabilities although their internal structure is different. The part that was finally selected was a Xilinx Spartan 3 device with 208 pins and 400,000 internal gates.

Figure 2 shows the initial tests with a signal generator connected to the two channels. The converted data is that of a 800kHz triangular waveform sampled at 20MHz. The two channels overlap very well indicating a balanced buffer amplifier. The triangular wave is also clearly reproduced despite that there are only around 20 data points per cycle.

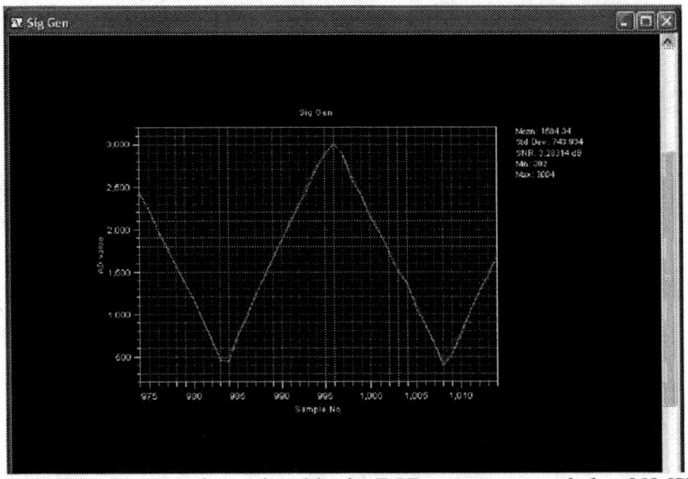

FIGURE 2. Converted test signal in the DSP memory sampled at 20MSPS

IMPLEMENTATION OF PEAK DETECTION ALGORITHM

Each time an incident photon interacts with the detector. An electric signal in the form of a pulse is generated. The shape of the pulse has a fast rising time and a slower, exponential decay. The amplitude of the signal is dependent on the energy of the source. The initial application of multiphase flow will be in a flow loop that employs two sources, ^{241}Am and ^{137}Cs. The electric signal will thus consist of pulses with two predominant amplitudes. Pulses with other amplitudes, due to scattering and cosmic rays will also be present in the received signal but need to be filtered out. Therefore only pulses with amplitudes that lie within two narrow bands are of interest to the application.

The number of received pulses is mainly dependent on the activity level of the source. An object between the source and the detector absorbs part of the photons emitted and thus reveals some aspects of the nature of the object.

The task of the acquisition channel is to count the number of pulses at each energy level and transmit these two values to the host. Combined with the results from other channels these will be used in an image reconstruction algorithm. The procedure for obtaining the information is to first detect the peaks from the data stream of the A/D converter and record its maximum amplitude.

An array A[N] of all the possible values of amplitudes ($N=2^{12}= 4096$ for the 12bit case) is reserved to hold this information. Each time a peak value is detected in the data, the corresponding location in the array is incremented. For a peak of value p, A[p] = A[p]+1. At the end of the pre-selected frame duration , the contents of the array will represent a histogram with the count of the number of peaks at different amplitudes.

From this histogram the count of the peaks in the band of interest can be added and transmitted as the result of the pulse count from each energy level. If p_1 and p_2 are the central peak amplitudes and m_1 and m_2 the width of the peak that will be considered, the total pulse count for each energy level will be:

$$C_1 = \sum_{i=p_1-\frac{m_1}{2}}^{p_1+\frac{m_1}{2}} A[i] \quad C_2 = \sum_{i=p_2-\frac{m_2}{2}}^{p_2+\frac{m_2}{2}} A[i] \tag{1}$$

To detect the peak of a pulse, an algorithm was developed using Matlab. Data recorded during previous experiments on a flow loop was used to test the algorithm. The acquisition of this data was performed with a commercial PCI card that sampled the electrical signal at 20MSPS with 12 bits resolution. A compressed view of $2x10^6$ samples can be seen in Figure 3. Here a large number of peaks with the same amplitude can be observed at a level just under the value of 250.

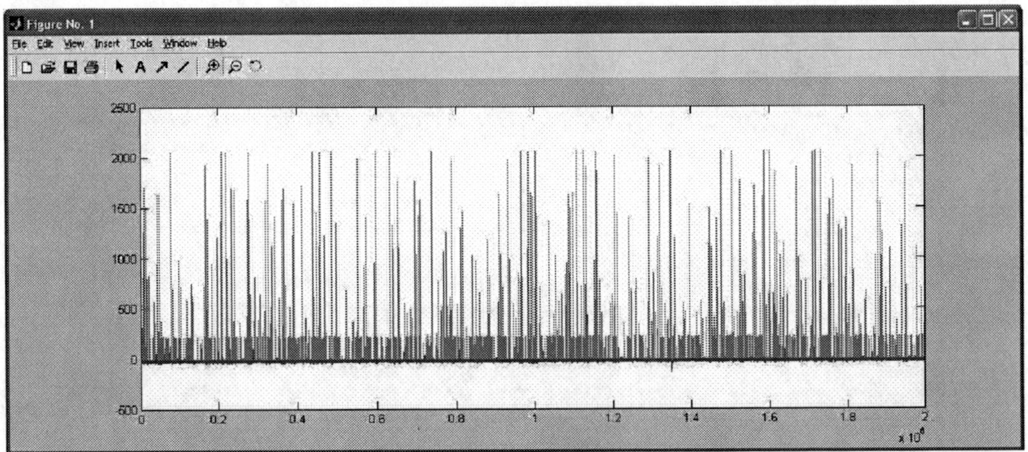

FIGURE 3. Compressed view of peaks from flow loop experiment

FIGURE 4 shows a more detailed view of an arbitrary pulse where the rise time and the slower decay time can be clearly seen. Superimposed on the pulse are the different stages of the peak detection algorithm. When the signal first exceeds a predetermined threshold value, the algorithm enters a state where it will look for a peak. When a peak is found, the detected position is indicated with a spike pointing to the peak. After the signal starts to decay, the algorithm enters a different state and finally when the signal falls below the threshold the algorithm returns to its initial state.

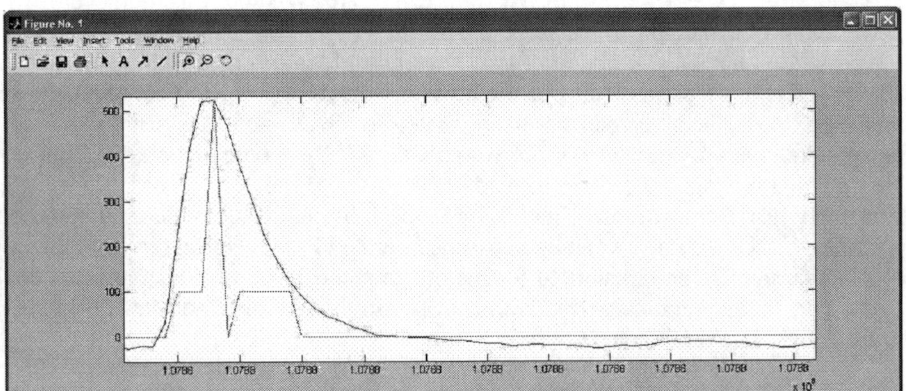

FIGURE 4. Detailed view of an arbitrary peak with superimposed stages of peak detection algorithm

The algorithm can be further refined to deal with overlapping pulses and for pulses that are not as monotonic as those typically found in this data set.

Figure 5 shows a section of the histogram with an accumulation of peaks around the value of 210. This corresponds to the multiple peaks under the value of 250 that were previously observed in Figure 3. The deviation from the central of 210 is around +/- 15 which at the full scale resolution of 12 bits, amounts to a spread of less than 1% of the total width of the histogram. This demonstrates that a simple peak detection algorithm can work with the expected signals from the sensors.

As the data stream enters the FPGA at 20MSPS, the DSP which operates at 200MIPS only has 10 clock cycles to implement the peak detection algorithm between incoming samples. This may prove challenging and also limits the functionality of the algorithm. A further demand on the resources is the fact that each A/D converter has dual inputs and delivers data to the FPGA (and the DSP) from two separate detectors.

The dual data stream may however be compensated by the fact that the DSP has a dual core and can simultaneously operate on two separate data operands albeit with the same instruction. In other words the DSP uses a SIMD (Single instruction, multiple data) architecture that could be exploited by the dual data stream of the A/D converter.

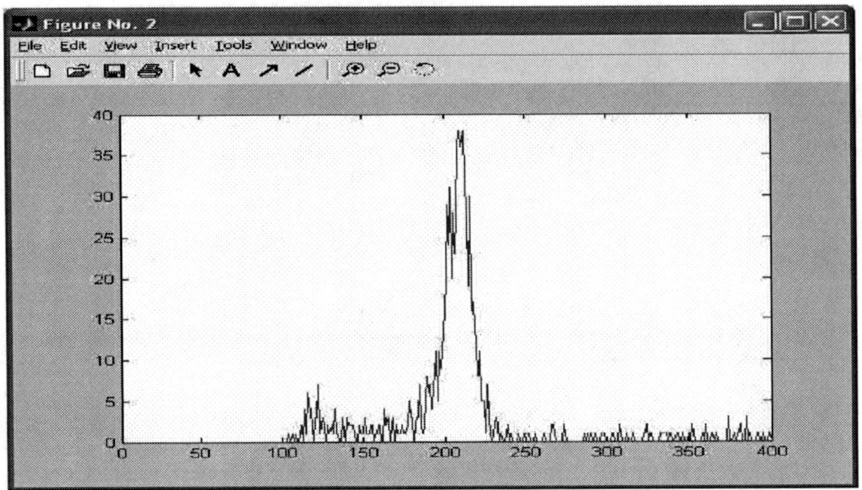

FIGURE 5 . Detail of histogram where maximum values accumulate, using Matlab simulation

An alternative method to implement the peak detection algorithm makes use of the FPGA. In this way the resources of the device can be employed to operate on the data in parallel rather than the sequential nature of the DSP program. The number of clock cycles can be reduced by employing a pipelined design. However this is at the expense of increased use of internal resources and a longer latency depending on the number of stages in the pipeline. For example, if the design has a five stage pipeline, it would take five cycles for the first data item to emerge. However, a delay of a few clock cycles is not significant in the whole acquisition cycle.

Bolic et al (2001) employ a method of pulse amplitude determination also using FPGA technology. In their method the pulse is shaped with a simple operation and the amplitude is determined by the integration of the digital samples that belong to a single shaped pulse. For the present work however, a direct measurement method of the peak value has been employed. The algorithm has been written in VHDL and uses just 2 clock cycles to detect and increment the corresponding memory contents of the array that holds the whole histogram of pulse amplitudes. After the frame period has elapsed it is only necessary to read the contents of the array in order to transfer to histogram to the host PC. Alternatively just the positions of the expected peak amplitudes can be read and then added together. Even the addition of the peak values can be can be performed on the FPGA resulting in just two values, one for each of the energy sources, that need to be transmitted for the image reconstruction. The algorithm developed in VHDL was tested with same data set as previously used with the algorithm developed in Matlab, by writing a VHDL test bench with file IO functions. The result can be seen in Figure 6.

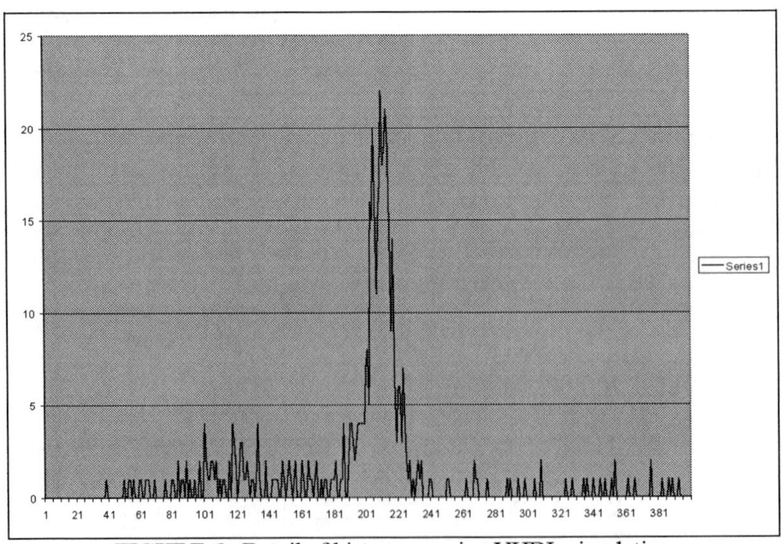

FIGURE 6 . Detail of histogram using VHDL simulation

The shape of the peak in the histogram and it's position is very similar to that obtained with the sequential Matlab algorithm. However further work is necessary since the amplitude if the histogram peaks is different.

CONCLUSIONS AND FURTHER WORK

This paper has presented the development of a data acquisition system for gamma ray tomography based on FPGA technology and combined with DSP processing and control. The system uses dual channel A/D converter chips that operate at 20MSPS and 12 bit resolution, but can be upgraded to 65MSPS and 14bits without hardware modifications other than replacing the pin compatible converter chip. The components used are widely available and result in a low cost per channel. The channel hardware has been successfully tested with a standard signal generator. The data processing for the peak counting has been implemented on the FPGA resulting in a fast and flexible processing of the sensor information. The algorithm has been simulated on the FPGA platform with data obtained during previous tests on a flow loop and the results show the peaks of energy levels that were expected. Further work on this systems needs to be performed by integrating the data of multiple channels and performing experimental evaluation with the gamma ray sources on the flow loop.

ACKNOWLEDGEMENTS

This work was a collaboration between the Institute of Mechanics, Chinese Academy of Sciences in Beijing and the University of Leeds and partly supported by the Royal Society.

REFERENCES

1. M. Bolic, V. Drndarevic, Digital gamma-ray spectroscopy based on FPGA technology, *Nuclear Instruments and Methods in Physics Research A* **482** (2002) pp. 761–766
2. S. Garcia-Castillo and K.B. Ozanyana, Field-programmable data acquisition and processing channel for optical -tomography systems, *Review of Scientific Instruments* **76**(2005), 095109
3. M. Ma, A. van Genderen and P. Beukelman, Developing and Implementing Peak Detection for Real-Time Image Registration, *ProRISC*, 2005, pp. 647-652
4. P.R.P. Bruneau, R.F. Mudde, T.H.J.J. van der Hagen, High-speed flow imaging using gamma-ray computed tomography, *4th WCIPT congress*, 2005

Interpretation of CT-Scan Sinograms to Study the Dynamic Flow Properties of a Fluidized Bed

Bennett MA[*+], Kryuchkov S[*+] and Kantzas A[*+]

*Department of Chemical and Petroleum Engineering, University of Calgary, Calgary, Alberta, Canada, T2N 1N4
+Tomographic Imaging and Porous Media Laboratory, Calgary, Alberta, Canada, Akantzas@ucalgary.ca.

Abstract. Fast moving objects typically produce poor quality images and motion-artifacts in Computed Tomography (CT) scans. The imaging of bubbling gas-solid fluidized beds, where many bubbles may pass through the sensing cross-section in a tomographic measurement, gives such a problem. Previously, one method of dealing with this problem was to remove artifacts from the reconstructed image itself through low pass filtering. However, this method leads to lost information from the images. When the concentration of bubbles is low, and their shape is close to cylindrically symmetric, analysis of consecutive scans in the raw data sinogram can also provide valuable information about the frequency of bubbles passing, and about their individual size, shape and position. Then, utilizing the inverse Abel transform, this information can then also be used to contribute to time averaged properties of the bed's behaviour such as voidage distribution, bubble phase area fraction and spatial bubble number distribution.

Keywords: sinogram, CT scanner, bubbling fluidized bed, Abel transform.
PACS: 07.85.-m, 47.55.Lm, 42.30.Wb

INTRODUCTION

CT scan images are well known to have problems with motion artifacts [1]. In the case of fluidized beds this is chiefly because of the presence of moving bubbles and density fluctuations caused by particle motion, which are inconsistent with algorithms that are used for image reconstruction. Although removing artifacts from the reconstructed image itself has been previously attempted [2], the easiest way to try to overcome this problem is to go directly to the projection data collected during the scan before they are used for image reconstruction. When represented as an image the projection data are normally called a sinogram (Figure 1) and this term is also used in this work. It has been previously postulated that valuable information such as bubbling rate, and the position, size and shape of individual bubbles can be extracted [3]. In this further work, an algorithm based on the application of the inverse Abel transform has been initially applied to give a more detailed analysis of the density distribution both within the bubbles and the emulsion in the rest of the fluidized bed. These density distributions can then be put together to give an 'image' based on an individual projection. Information from all the 'images' produced from a complete sinogram can then be used to give important time averaged parameters such as bubble phase area fraction, spatial bubble number distribution and bubble diameter distribution, and with superimposing of the projection "images" from the complete sinogram, a voidage distribution can be produced, representing an artifact-free reconstructed tomographic image.

CP914, *Multiphase Flow: The Ultimate Measurement Challenge,* edited by X. Cai, Y. Wu, Z. Huang, S. Wang, and M. Wang

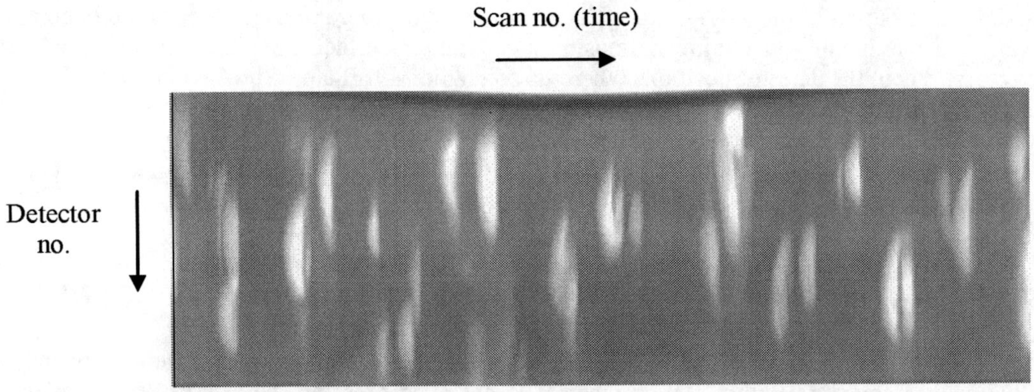

FIGURE 1. A sinogram of a bubbling fluidized bed (bubbling rate ~ 17 bubbles per second) [3].

EXPERIMENTAL METHOD

The analyzed sinograms were obtained from experiments with a 100 cm height, 20 cm diameter Plexiglas column. A GE9800 Highlight Advantage™ X-ray CT scanner with a scan time of 2 seconds and a cross-section of 3 mm was used for measurement. Individual projections were separated by about 2 milliseconds in time and each projection was taken at approximately 0.4°, so that from the first projection to the last projection 360° was covered. Measurements of typical flow regimes were taken using polyethylene resin.

ANALYSIS OF AN INDIVIDUAL BUBBLE

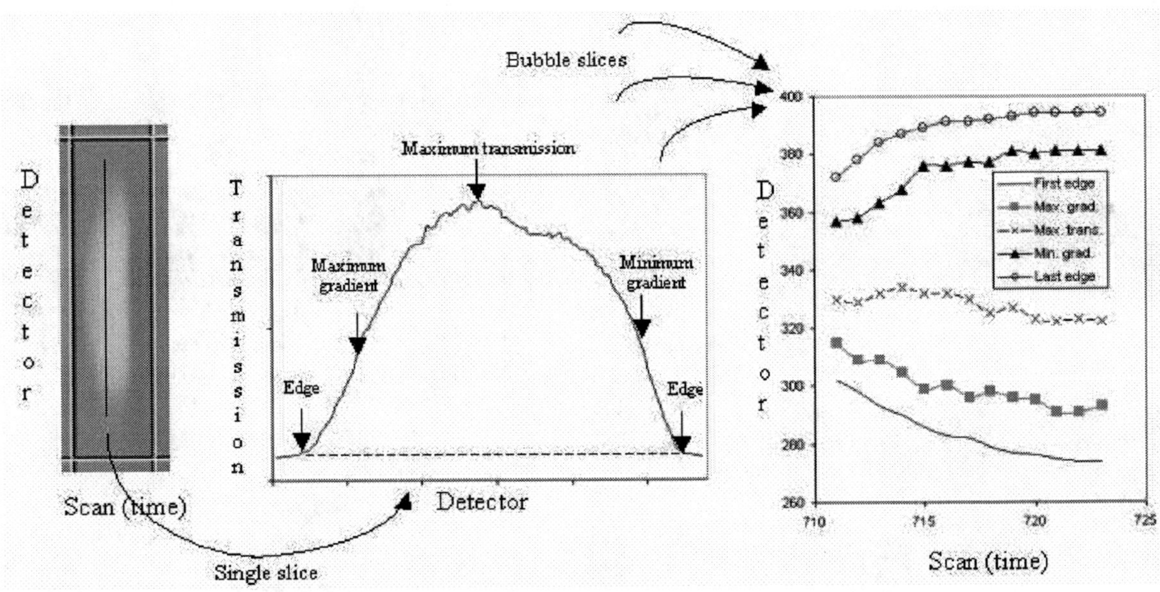

FIGURE 2. Analysis of an individual bubble.

Upon examining a sinogram, individual bubbles could be identified as a series of transmission increases across detector measurements in consecutive scans/projections of the sinogram forming a 'white patch', and one such bubble is represented in Figure 2, left. Using software, the area of the patch could be selected and each consecutive fraction of a scan within the patch (representing a slice of the bubble passing through the X-ray cross section), could be analyzed to give the edges of the bubble 'slice' as detector positions (Figure 2, middle). These edges were found by first finding the detector position of maximum transmission increase (the approximate center of the bubble slice) and from that finding the maximum and minimum transmission gradients either side (the steepest transmission changes across detectors). Then, the detector positions where these gradients had tapered off sufficiently were taken to represent the 'edges' of the bubble slice. On taking detector positions and transmission values for a series of slices, a bubble passing through the X-ray cross-section could be effectively represented (Figure 2, right).

From the obtained analysis in Figure 2 it can be observed that this individual bubble is not spherical, but looks more like a semi sphere. This is consistent with both CFD models and X-ray fluoroscopy measurements obtained from our group [4].

CALCULATING POSITION AND DIAMETER OF AN INDIVIDUAL BUBBLE

For the position and horizontal diameter of the bubble to be calculated, an assumption is made that the bubble is approximately cylindrically symmetric within a slice when it is observed crossing the X-ray beam, and that the detector array and X-ray source are mounted on the same frame and are rotating as a single unit. Calculations demonstrated with Figure 3 show that from the known direction and angular velocity of the X-ray source, and from the rate of change of the angle of the detector corresponding to the center of the bubble, it is possible to find the position of the bubble center using principles based on stereoscopy, and then using the source distance from the bubble the diameter of each bubble slice can be found.

In the case of single bubbles, this was a straight forward process, but occasionally overlapping of bubbles occurred in the projections, and a consistent trend could not be discerned. An example of this is demonstrated in Figure 4, where two different trends for the detector positions for the center of the bubble can be seen, most probably corresponding to the influence of two separate, but angularly overlapping bubbles (which can also be observed in the analysed portion of sinogram shown). This can be counterracted by lowering the flowrate.

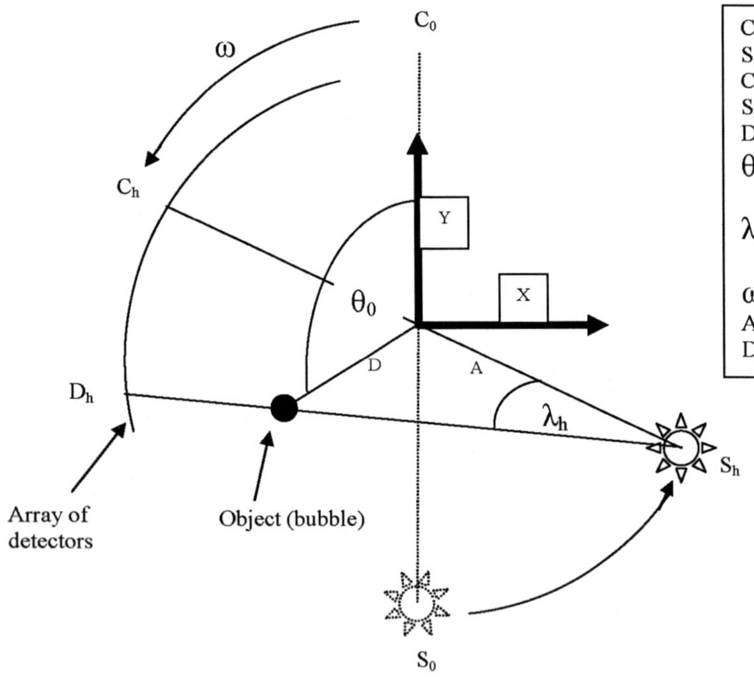

C_0	Central detector at T_0.
S_0	Source at T_0.
C_h	Central detector at T_h.
S_h	Source at T_h.
D_h	Detector observing object at T_h.
θ_0	Angle of object from central detector starting point (C_0).
λ_h	Angle of detector observing object from central detector at T_h.
ω	Angular velocity.
A	Distance of source to center.
D	Object distance from center.

At any moment (h):

$$\operatorname{Tan} \lambda_h = \frac{\operatorname{Sin}(\theta_0 - \omega T_h)}{\frac{A}{D} + \operatorname{Cos}(\theta_0 - \omega T_h)}$$

Using several λ_h values over time the D and θ_0 values can be found by the steepest descent curve fitting procedure [5] to give the object coordinates.

FIGURE 3. Calculating the position of an individual bubble.

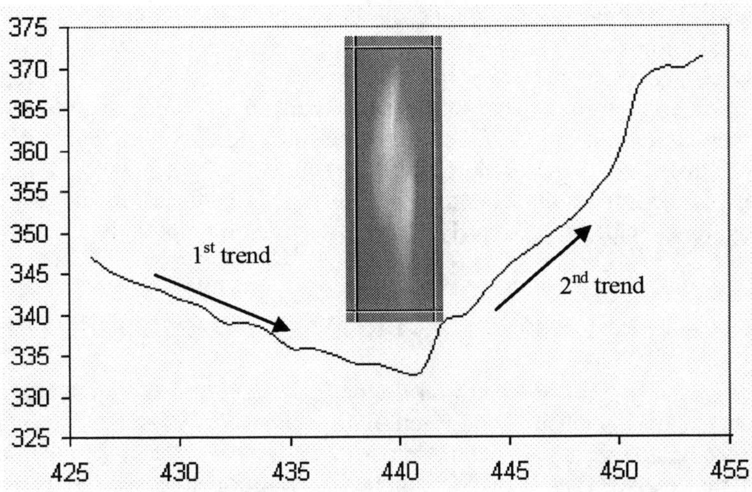

FIGURE 4. Different trends from bubbles overlapping in projections.

APPLYING THE ABEL TRANSFORM TO INDIVIDUAL PROJECTIONS

With the known position and diameter of the bubble, further steps in obtaining additional information from sinograms are possible and we can reconstruct the density distributions both within a bubble and the emulsion in the rest of the column, by separately applying the Abel transform [6] to each before superimposing the density distributions produced to give a projection image.

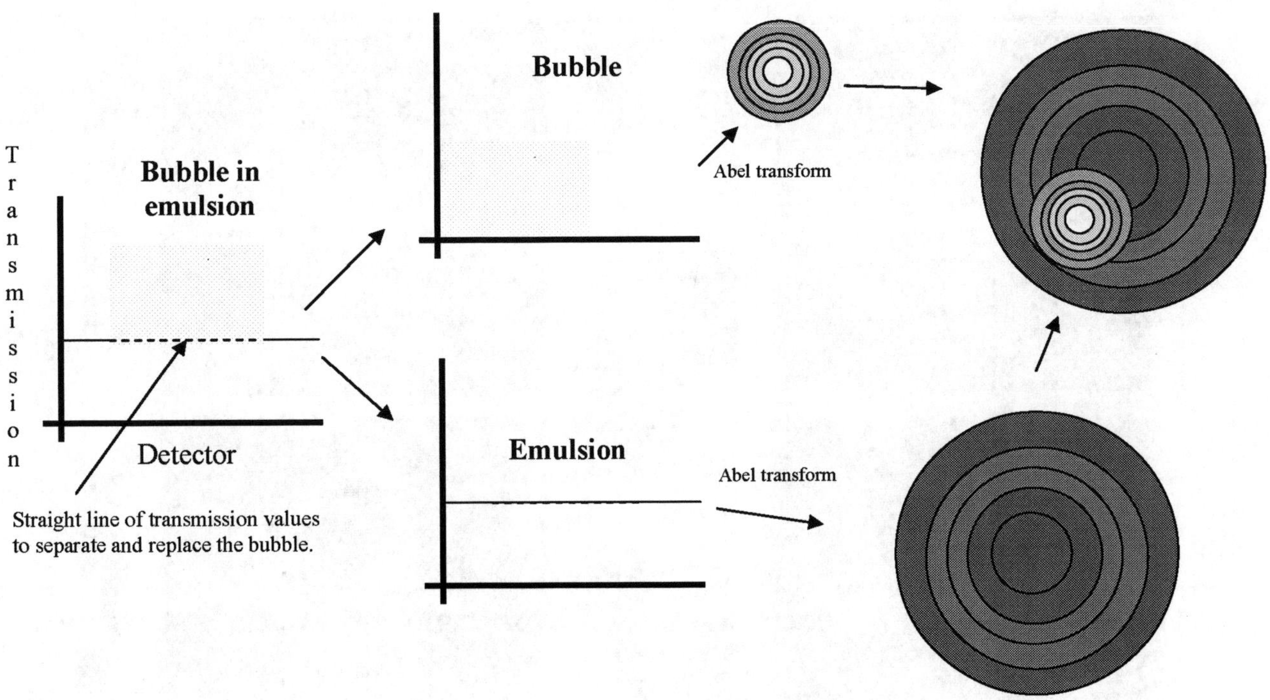

FIGURE 5. Applying the Abel transform to both bubble and emulsion.

In brief, this is achieved by firstly producing a straight line of 'virtual' transmission values for each 'active detector' (detectors between the bubble edges) to represent the expected transmission values obtained if the bubble did not exist (Figure 5). This line is then used to separate off the bubble, and to replace the active detector values in the emulsion. The solid fraction distribution of the bubble is then calculated by calculating the pathlengths of gas that would be measured by each active detector between the bubble edges if the bubble was a circle of pure gas, and then subtracting the actual pathlength of gas measured by each active detector due to the bubble (calculated from the increase in transmission caused in each detector). This gives the pathlength of solid measured by each active detector due solely to the bubble. The distribution of pathlengths for the active detectors must then be separated by a line at the central active detector to give two halves, and then each active detector value is averaged with it's 'mirror image' partner of equal distance on the other side of the separating line. This symmetric distribution is then inverse Abel transformed to give the radially symmetric solid fraction distribution of the bubble. A similar process is then performed on the emulsion distribution in the column, and knowing the position of the bubble, the two are superimposed to give an image for a single projection. The basic principle of the process is summarised in Figure 5.

TIME AVERAGED PARAMETERS FROM FLUIDIZED BED FLOWS

Applying this Abel transform analysis technique to all the projections in the sinogram can give important time averaged statistical parameters, such as emulsion solid fraction distribution (or voidage distribution – essentially an artifact-free "reconstructed tomographic image), bubble phase area fraction (the average fraction of bubbles in different spatial positions of the image), spatial bubble number distribution (the number of bubbles occurring in different spatial positions) and bubble diameter distribution and bubbling rate.

In treating the whole sinogram, software was firstly used to select 'white patches' corresponding to individual bubbles. The detector representing the center of the bubble was then calculated for each bubble slice and the results graphed. Where a consistent trend in these values could be observed, the contributing bubble slices were selected to give the position of the bubble and their information was separately stored away from the sinogram, while their effect was 'smoothed' out in the sinogram by the straight line method (discussed earlier). Those slices not contributing to the position calculation were left in the sinogram, and after this process was performed on all available bubbles the sinogram projections effectively represented the emulsion (while the bubble information was stored separately). Shown in Figure 6 are some initial semiquantitative sample images before and after separation of bubble information from the sinogram, and it is hoped to obtain statistically significant results with the analysis of multiple images.

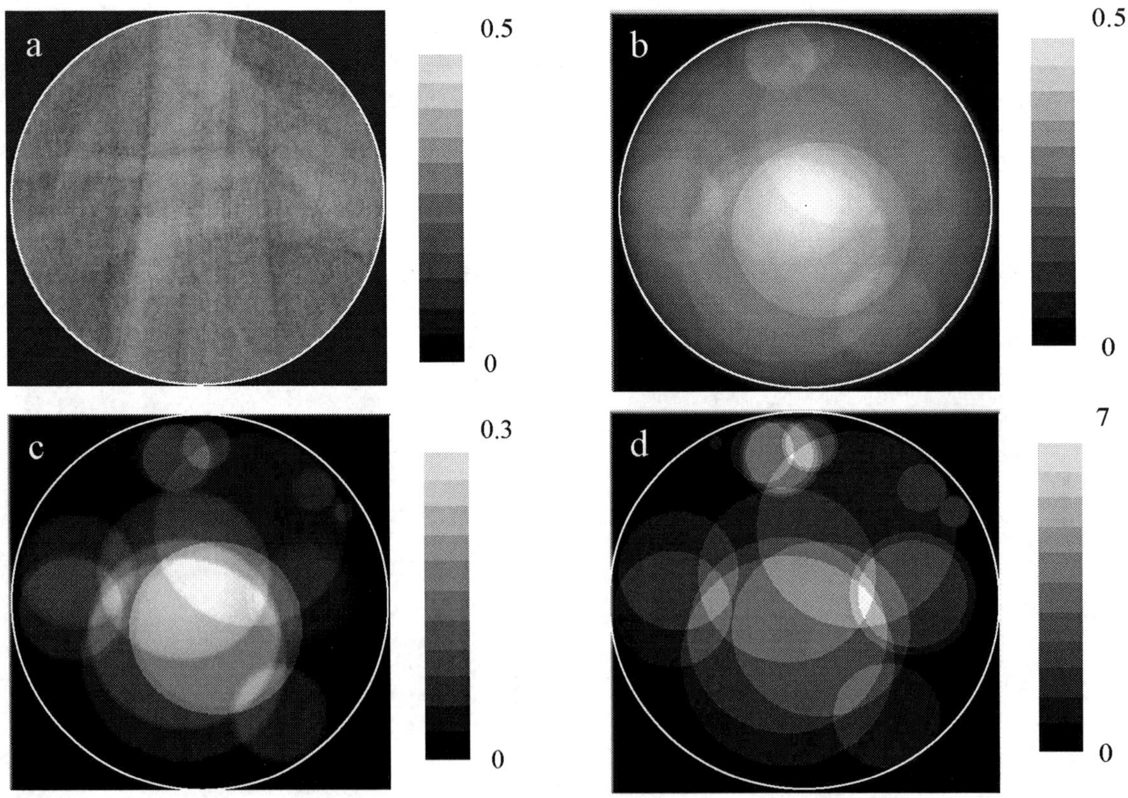

FIGURE 6. Sample images (a) a voidage distribution before bubble analysis, (b) voidage distribution after bubble analysis, (c) bubble phase area fraction, (d) spatial bubble number distribution.

CONCLUSIONS

Sinograms can be used to provide valuable information on dynamic objects passing through the sensing zone, and in an application like bubbling fluidized beds this can be parameters such as the size and position of individual bubbles, as well as time averaged parameters such as voidage distribution, bubble phase area fraction and spatial bubble number distribution. An algorithm based on the application of the Abel transform is now in the initial stages and it is hoped that it will provide a more detailed analysis of density distribution within the bubbles and in the rest of the fluidized bed.

REFERENCES

1. J. Hsieh, "Computed Tomography: Principles, Design, Artifacts, and Recent Advances", The Society of Photo-Optical Instrumentation Engineers, Bellingham, 2003.
2. Z. He, "Comparison of the Behaviour of Glass-Beads and Polyethylene Resin Fluidized Beds using X-ray Imaging Experiments and CFD Simulation", M. Sc. Thesis, University of Calgary, Calgary, 2005.
3. S. Kryuchkov and A. Kantzas, "Non-conventional Use of Conventional CT-scanner in Imaging of Bubbling Fluidized Beds", 4th International Symposium on Process Tomography in Poland. Warsaw 14- 15 September 2006.
4. B. Chandrasekaran, "A Validation Study of the Computed Hydrodynamics of a Gas-Solid Fluidized Bed", M. Sc. Thesis, University of Calgary, 2005.
5. M. Avriel, *Nonlinear Programming: Analysis and Methods*, Mineola:Dover Publishing,, 2003.
6. I.N. Sneddon, *The Use of Integral Transforms*, New York:McGraw-Hill, 1972.

Electrical Resistance Imaging of Two-Phase Flow Through Rod Bundles

Umer Zeeshan Ijaz♣, Anil Kumar Khambampati♣, Jeong Seong Lee♦, Kyung Youn Kim♣, Min Chan Kim• and Sin Kim♦+

♣Department of Electronic Engineering, Cheju National University, Cheju 690-756, South Korea
•Department of Chemical Engineering, Cheju National University, Cheju 690-756, South Korea
♦Department of Nuclear and Energy Engineering, Cheju National University, Cheju 690-756, South Korea
+Corresponding Author(sinkim@cheju.ac.kr)

Abstract. This paper considers the dynamic electrical resistance tomography (ERT) for the visualization of phase distribution in heat transfer systems with rod bundles. Typical measurement techniques like conductivity probes, ultrasonic and radiological imaging suffer from difficulties in the measurements due to these internal structures. On the other hand ERT imaging does not get affected by the presence of internal structures. These internal structures can be used as internal electrodes to further improve the dynamic ERT performance. This paper discusses the current injection protocols for such systems. The main goal is to find the optimal current patterns to reconstruct the conductivity distribution. We provide extensive simulation results to evaluate the performance of different current patterns.

Keywords: Electrical Impedance Tomography, Two-Phase Flow, Rod Bundles, Extended Kalman Filter.
PACS: 87.63.Pn

INTRODUCTION

A rod bundle is a common engineering structure used in many heat transfer systems. In nuclear power plants, for example, the fuel assembly is composed of a bundle of cylindrical fuel rods and the steam generator is made up of tube banks. For a pressurized water reactor, normally the water coolant in the reactor coolant system is maintained at the subcooled condition. The coolant entering into the reactor flows through the spaces called the subchannels surrounded by fuel rods and is heated up by the heat emitted from the fuel rods. Under the pressurized condition, the coolant leaving the reactor remains still at the subcooled condition. The heat acquired from the fuel rods in the reactor should be released from the reactor coolant system to the secondary system through the steam generator. However, in an abnormal condition, the system pressure may decrease or the cooling capability may deteriorate. In that case, the coolant can evaporate and exists in a two-phase state. The vapor thus affects the heat transfer from the fuel rods to the coolant resulting in an undesirable heating-up of the fuel rods. Under such a condition, the knowledge on the two-phase flow is of great importance in the safety analysis of the reactor coolant system because the phase distribution in two-phase flow or the flow regime has a great influence on the heat transfer and hydrodynamics. In practice, the two-phase flow analysis highly relies on the experiments since the phenomena in the two-phase flow are very complex and analytical approaches are often limited.

The phase distribution is of major concern in the analysis of two-phase flow fields and there have been many attempts to measure two-phase flow fields. Recently, the electrical resistance tomography is highlighted as a new modality to measure multi-phase flows as it is non-intrusive and has a high temporal resolution [1]. The electrical resistance tomography (ERT) reconstructs the conductivity distribution, namely the phase distribution, in the domain of interest based on the relationship between the injected currents through the electrodes placed on the domain boundary and the resultant electrical potentials measured at the electrodes. Especially for two-phase flows through rod bundles, conventional methods like conductivity probe, ultrasound imaging, and X-ray imaging often face

CP914, *Multiphase Flow: The Ultimate Measurement Challenge,* edited by X. Cai, Y. Wu, Z. Huang, S. Wang, and M. Wang

difficulties in the measurements because of the presence of rods, which can hinder the insertion of probes, transmission of ultrasound waves and X-rays. However, the ERT is free from such limitations and offers a promising application to the visualization of two-phase flows. For the application of the ERT to two-phase flows through rod bundles, the conventional ERT algorithm should be improved to cope with the fast transient of the flow field and the geometrical complexity of the flow domain.

The conventional ERT also assumes the conductivity distribution to be stationary during the data collection to obtain a single frame of image in a classical sense. In the application of ERT to visualize two-phase flows, however, such an assumption may be invalid since the phase distribution or the conductivity distribution may change during the time taken to acquire the full set of measurement data to reconstruct a single frame of image. Hence, the present work introduces a dynamic ERT algorithm, where the estimation of the conductivity distribution is considered as a state estimation problem. In order to solve the state estimation problem, the extended Kalman filter approach is adopted.

The ERT of the object containing high conductive material suffers from so called masking effect, which indicates that high resistive targets near the high conductive material are very difficult to reconstruct since near the high conductive material, the currents hardly flow through the high resistive targets. This study considers the use of rods as internal electrodes, which will enhance the reconstruction performance significantly. If the rods are used as the electrodes, then the proper design of electrical current injection protocol may play an important role to improve the reconstruction performance further. With the combined current patterns of the boundary and the internal electrodes, several protocols including trigonometric, adjacent, and opposite protocols are tested to find an optimal pattern. For the verification of the proposed dynamic ERT with internal electrodes, extensive numerical experiments are successfully conducted.

MATHEMATICAL MODEL

Dynamic ERT Model

ERT reconstructs the conductivity distribution based on the relationship between the injected current and the induced boundary voltages. The procedure to find the solution consists of: the forward problem where the voltage distribution is calculated with the assumed conductivity distribution; and the inverse problem where the conductivity distribution is updated to minimize the difference between the measured and calculated boundary voltages. The numerical solution of the forward problem is obtained by FEM with triangular elements. The complete electrode model is used to include the effect of the contact impedance and the shunting effect. For the inverse solution, this work uses the extended Kalman filter (EKF) algorithm [2] in which the conductivity distribution is regarded as the state variable to be estimated and the boundary voltages at a certain step are linearized about the previous step, while the linearized Kalman filter (LKF) [3] linearizes the boundary voltages about the initial or known fixed step. As for the process model, the random walk model is used since the physical characteristics of the process may not be known *a priori*. Also, two different mesh structures are used for the forward (2904 mesh elements) and the inverse (726 mesh elements) solutions to avoid the inverse crime. The backbone of the numerical model is based on EIDORS [4]. The detailed description is omitted here since the full description of the mathematical model of ERT and the extended Kalman filter algorithm can be found in many publications, e.g. [2] and [3].

Current Injection Methods

The adjacent method, the opposite method, the cross method and the trigonometric method are the most common current injection protocols used in ERT. Given the protocol, in the static ERT, all the independent current patterns are injected to reconstruct a single frame of image and the number of independent current patterns is determined by the current pattern and the number of electrodes. Recalling that in the dynamic ERT only one current injection is allowed for each image reconstruction, each current pattern comprising a current injection protocol can be an independent current pattern. For example, in the trigonometric pattern, each mode of cosine and sine pattern can be an independent pattern. Hence, a variety of current patterns is possible in the dynamic ERT.

Isaacson [5] investigated the optimal current pattern in a two-dimensional cylindrical object with concentric inhomogeneity and showed that the trigonometric pattern is optimal when the L2 norm of the current is kept constant. However, if the maximum injected current is kept constant, Köksal and Eyüboğlu [6] found that the opposite method gives the best distinguishability. Kim *et al.* [7] reported a qualitative comparison of the reconstructed dynamic images obtained with various current patterns. The comparison shows that the trigonometric and the opposite methods generate the best reconstructed images.

This paper will focus on the adjacent, trigonometric and the opposite methods. As for the trigonometric pattern, only the first mode is considered because the higher modes tend to deteriorate the reconstruction performance [7]. With a 16-electrode system, the injected current of the l-th electrode, I_l, for the cosine and the sine patterns are defined as

$$I_l = \begin{cases} I_0 \cos(2\pi l/16) \\ I_0 \sin(2\pi l/16) \end{cases} \quad l = 1, 2, \cdots, 16 \tag{1}$$

While injecting the designed currents, the induced voltages at the electrodes, V_l, are measured. A number of current injection protocols were tested; however some of them are reported in the paper and they are listed in Table 1. The configuration of electrodes is shown in Fig. 1.

TABLE 1. Description of current patterns.

Conventions used	Description	Measurements	Repeat
ADJ-EXCL-IE	Conventional adjacent protocol excluding internal electrode	1/frame	0
OPP90DEG-EXCL-IE	Choose I_1=I_0, I_9=-I_0 in one frame and I_5=I_0 and I_{13}=-I_0 in next frame, consecutively	1/frame	0
TRICOS1-EXCL-IE	Use first cosine current pattern	1/frame	
ADJNEAREST-INCL-IE	I_1=$I_0/2$, I_2= $I_0/2$, nearest internal electrode=-I_0 in one frame and then I_2=$I_0/2$, I_3= $I_0/2$, nearest internal electrode=-I_0 in next frame and so on	1/frame	0
OPP90DEG-INCL-IE	Choose I_1=$I_0/2$, I_9= $I_0/2$, I_{21}=-I_0 in one frame and I_5=$I_0/2$, I_{13}= $I_0/2$, I_{21}=-I_0 in next frame, consecutively	1/frame	0
OPP90DEG2-INCL-IE	Choose I_1=$I_0/2$, I_9= $I_0/2$, I_5=$I_0/2$, I_{13}=$I_0/2$, I_{21}=-2I_0 in each frame, consecutively	1/frame	1
~-R5	Repeat current patterns in inverse solver	1/frame	5
~-FRAME	Use whole frame	frame	0

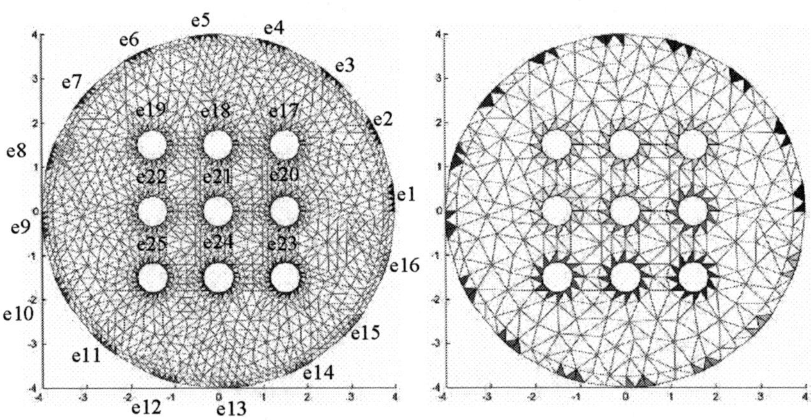

FIGURE 1. Type of mesh structure used and electrode numbers

1. Forward mesh 2. Inverse mesh

NUMERICAL RESULTS

We have considered a transient scenario in which a single anomaly moves through the domain. A total of *15* frames are used. Simulations are performed for two cases. In the first case, the contrast of the anomaly to the background is set to *2*. In the second case the contrast is set to *100:1* and *3%* relative measurement noise is added.

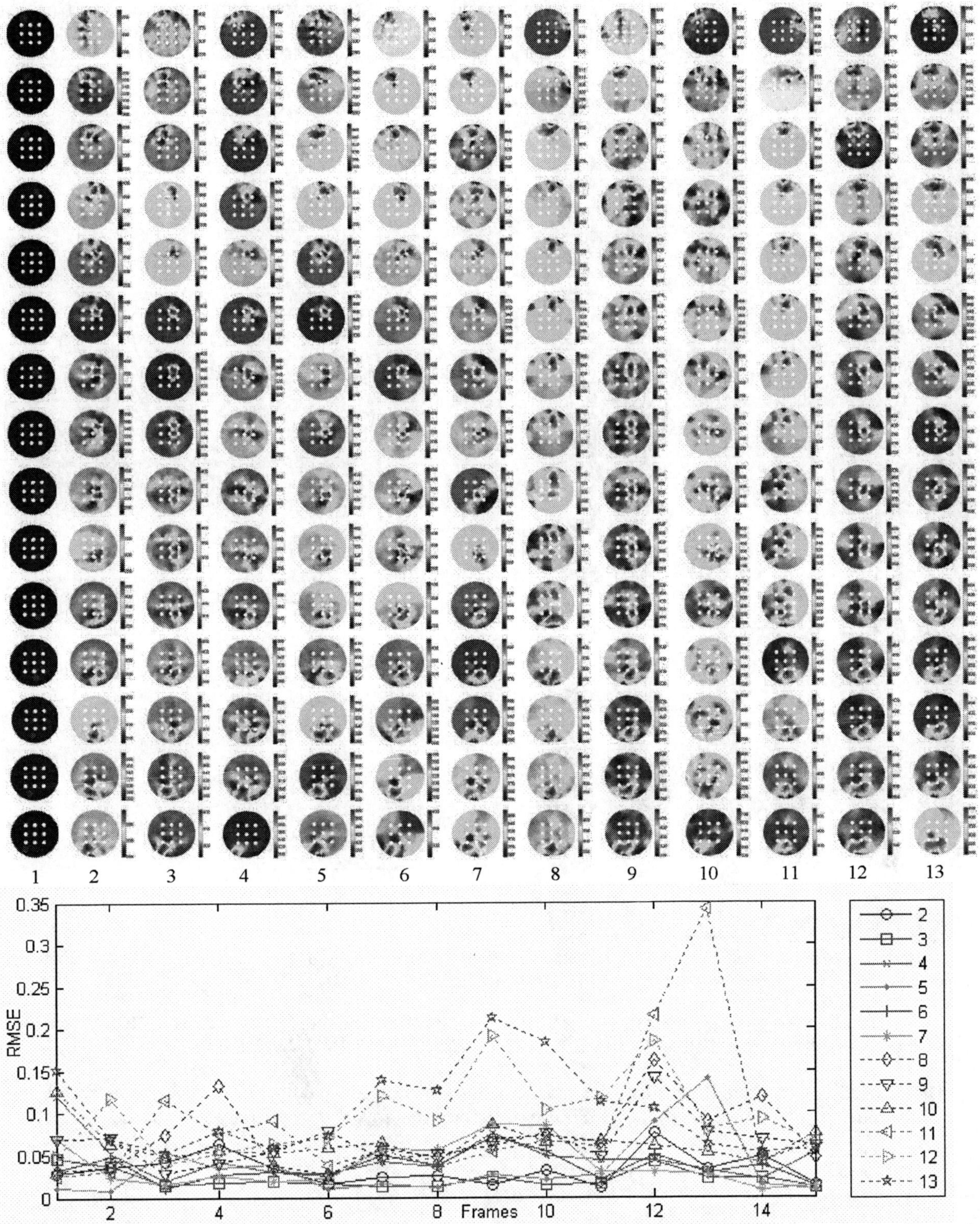

FIGURE 2. Reconstructed images and RMSE with one current pattern per frame. 2-7 are the results for contrast ratio of 1:2 and no measurement noise whereas 8-13 are the results for contrast ratio of 1:100 with 3% measurement noise.

1. True movement of target 2. ADJ-EXCL-IE 3. OPP90DEG-EXCL-IE 4. TRICOS1-EXCL-IE 5. ADJNEAREST-INCL-IE 6. OPP90DEG-INCL-IE 7. OPP90DEG2-INCL-IE 8. ADJ-EXCL-IE 9. OPP90DEG-EXCL-IE 10. TRICOS1-EXCL-IE 11. ADJNEAREST-INCL-IE 12. OPP90DEG-INCL-IE 13. OPP90DEG2-INCL-IE

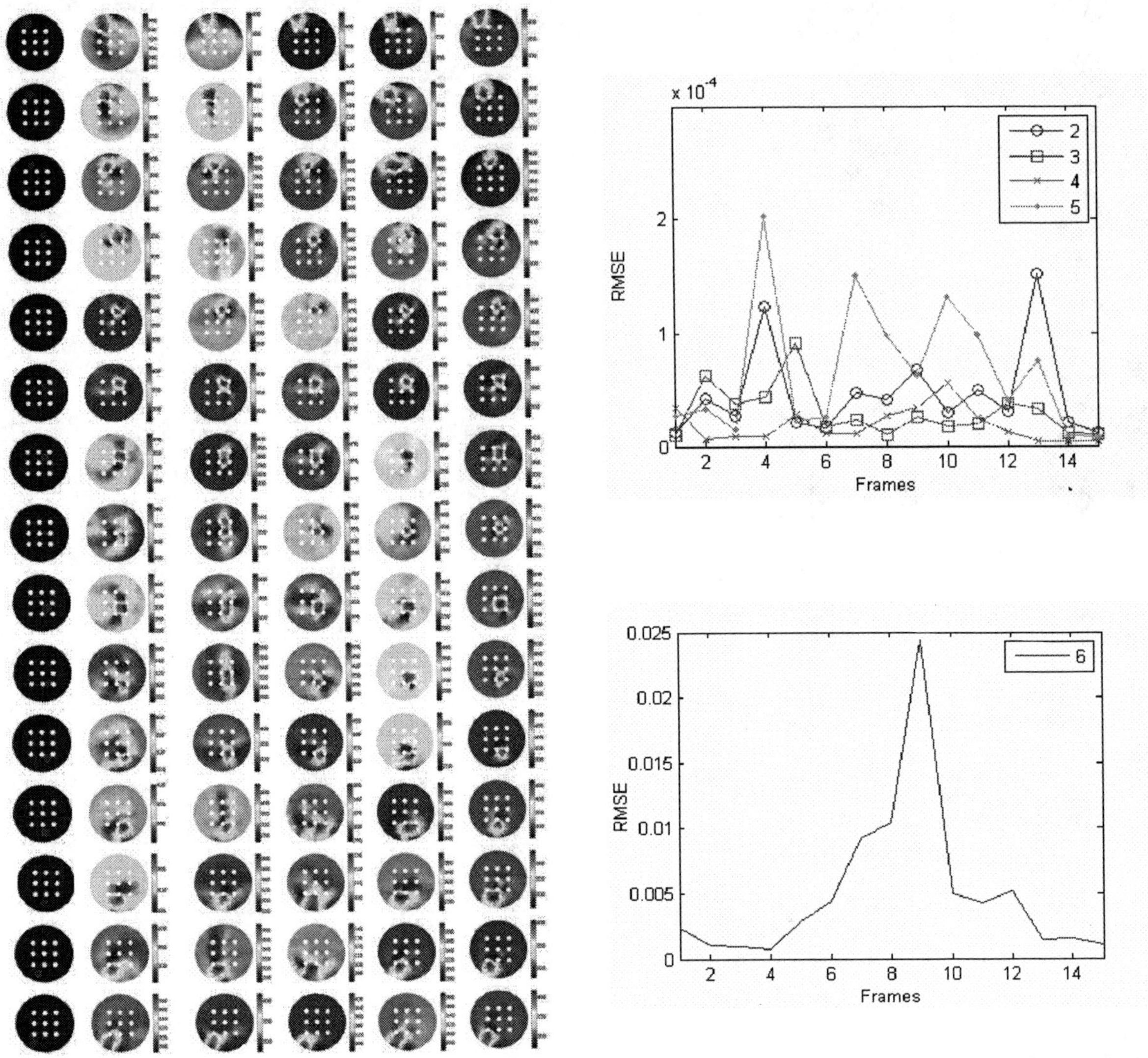

FIGURE 3. Reconstructed images and RMSE for multiple usages of the same measurement data obtained from a single current injection per frame for a contrast ratio of 1:2 and no measurement noise.

1. True movement of target 2. ADJ-EXCL-IE-R5 3. OPP90DEG-EXCL-IE-R5 4. TRICOS1-EXCL-IE-R5 5. ADJNEAREST-INCL-IE –R5 6. ADJNEAREST-INCL-IE –FRAME

There are several measures to evaluate the reconstruction performance and the image quality. The RMSE will be one of the most important measures. The RMSE can be defined in terms of the boundary voltages:

$$RMSE = \sqrt{\sum_{l=1}^{L}\left(V_{meas,l} - V_{cal,l}\right)^2 \Big/ \sum_{l=1}^{L}\left(V_{meas,l}\right)^2} \tag{2}$$

The reconstruction results are shown in Figs. 2 and 3. From Fig. 2, it can be noticed that the with the contrast ratio of *1:2* between background and the anomaly and no measurement noise, the visual performance of current patterns excluding and including the internal electrodes is almost similar. However, lower RMSE values can be observed for opposite current patterns (OPP90DEG-EXCL-IE, OPP90DEG-INCL-IE and OPP90DEG2-INCL-IE). For higher contrast ratio *1:100* and *3%* measurement noise, the opposite current patterns that include internal electrodes perform better than the ones that don't use internal electrodes. For *3%* noise, the best current pattern is OPP90DEG2-INCL-IE as it has the lowest RMSE. It should be noted that the simulation results shown in Fig. 2 are obtained with only one current pattern per frame.

In Fig. 3, the simulation results are shown for repeated usage of the same measurement data obtained from a single current injection per frame in the inverse solver. Namely, no more measurement is needed, so that the temporal resolution is unchanged while the computational cost will increase as the number of repetitions. However, it has been found that additional iterations in the EKF help reduce the error covariance matrices and the RMSE is much less than the single usage of the measurement data. Also in Fig. 3, to compare with other single-current-injection protocols, we reports ADJNEAREST-INCL-IE-FRAME in which *15* current patterns are used per frame. Even though multiple current injections are used, the RMSE seems to be slightly bigger than the repeated usage of the same measurement data obtained from a single current pattern per frame. However the visualization is quite good as compared to other current patterns.

CONCLUSIONS

In the application of imaging heat transfer systems with rod bundles, dynamic ERT offers a promising technique to visualize the phase distribution inside heat transfer systems. In most cases of two-phase flows, current injections used in static ERT will not be possible and only one current injection may be allowed. Furthermore, the internal structures can be used as internal electrodes to make the visualization better. Hence the design of current injection becomes important in the current context. Various current injection patterns are tested to compare their reconstruction performance. It is concluded that:

- The repeated use of the same measurement data obtained from a single current injection per frame has an obvious advantage as the additional iterations tend to reduce the error covariance in EKF.
- OPP90DEG-INCL-IE and OPP90DEG2-INCL-IE have better image resolution when the anomaly moves through the rod bundles.
- ADJNEAREST-INCL-IE –FRAME can be used when more current patterns are allowed in the duration when the target remains static.
- For higher noise level, the current patterns that use internal electrodes offer better visualization when the anomaly moves through the rod bundles.

ACKNOWLEDGMENTS

The part of the researchers participating in this study is supported by the grant from "the 2nd phase BK21 project".

REFERENCES

1. M. Wang, A. Dorward, D. Vlaev, R. Mann, "Measurement of gas-liquid mixing in a stirred vessel using electrical resistance tomography (ERT)", *Chem. Eng. J.* 77, 93-98 (2000).
2. K. Y. Kim, B. S. Kim, M. C. Kim, Y. J. Lee and M. Vauhkonen, "Image reconstruction in time-varying electrical impedance tomography based on the extended Kalman filter", *Meas. Sci. Technol.* **12**, 1032-1039 (2001).
3. M. Vauhkonen, "Electrical impedance tomography and prior information", Ph.D. thesis, University of Kuopio, 1997.
4. M. Vauhkonen, W. R. B. Lionheart, L. M. Heikkinen, P. J. Vauhkonen, and J. P. Kaipio, "A MATLAB package for the EIDORS project to reconstruct two-dimensional EIT images", *Physiol. Meas.* **22**, 107-111 (2001).
5. D. Isaacson, "Distinguishability of conductivities by electric current computed tomography", *IEEE Trans. Medical Imaging* **5**, 91-95 (1986).
6. A. Köksal and B. M Eyüboğlu, "Determination of optimum injected current patterns in electrical impedance tomography", *Physiol. Meas.* **16**, A99-A109 (1995).
7. K. Y. Kim, B. S. Kim, M. C. Kim, K. J. Lee, Y. J. Ko and S. Kim, "Electrical impedance imaging of two-phase flows undergoing rapid transient. Part II. Effect of input current pattern", *Int. Comm. Heat Mass Transfer* **32**, 649-657 (2005).

Nonlinear Image Reconstruction Using a GA-ECT Technique in Electrical Capacitance Tomography

Zhou Bin, Xu Chuanlong, Yang Daoye, Wang Shimin and Wu Xin

Key Laboratory of Clean Coal Power Generation and Combustion Technology of Ministry of Education, Southeast University, Nanjing, 210096, China

Abstract. Electrical capacitance tomography (ECT) is a non-invasive measurement technique that estimates the dielectric permittivity distribution of an inhomogeneous object from the boundary potentials at floating electrodes [1, 2]/or mutual capacitances. In this paper, a stochastic inverse technique based on genetic algorithm (GA-ECT) is developed, which is adapted to the two different methods, i.e. potential measurement and capacitance measurement. Numerical simulation results are presented to evaluate the inverse technique both for noise free and noisy data and the results show that quantitative image can be reconstructed not only with the low permittivity contrast but also with the high contrast. Furthermore, the influence of a priori knowledge to image reconstruction is discussed.

Keywords: Electrical capacitance tomography, Inverse problem, Nonlinear reconstruction algorithm GA-ECT

INTRODUCTION

Electrical capacitance tomography (ECT) is a measurement technique that infers the permittivity distribution $\varepsilon(x, y)$ of object using boundary measurements from an array of electrodes attached to the body interested [3, 4]. With development of this technique worldwide over almost two decades, potential industrial and medical applications such as flow pattern recognition in pneumatic conveying [5, 6] and reconstruction of the temperature change field in brain [7] are promising.

The estimation of permittivity distribution is usually called image reconstruction. From the mathematical point of view, the image reconstruction task is to recover a coefficient (permittivity) of elliptic partial differential equation under given boundary conditions, which is a non-linear, ill-posed inverse problem. Due to the complexity of the problem, the analytical solution of the equation can be achieved only in some simple situations. Up to now, various types of numerical algorithms have been developed to solve this problem. The most frequently used algorithms are based on sensitivity field model. In principle, the sensitivity field of a non-linear model needs to be updated after each round of iteration to overcome the soft-field problem. However in practice as for gas-oil two phase flow, if the differences between sensitivity fields for different flow patterns are not significant and the scope and contrast of material permittivity are also small (nearly 3:1) then conditions for supposition theorem and linear approximation can be satisfied, so the initialized sensitivity field can be used without updating to save computation time cost. The algorithms based on non-iterative sensitivity field include linear back-projection (LBP), singular value decomposition, Tikhonov regularization, Landerweber iteration etc. (refer to [8, 9] in detail). Currently, some algorithms such as regularized Gauss-Newton scheme based on iterative sensitivity field are also investigated ([10, 11]), the quantity of the reconstructed image has been improved, and the images with comparatively high contrast permittivity (3:80) can be recovered accordingly. However, the simplified assumption $\partial u / \partial n = 0$ in the gap between the electrodes is introduced in implementing the algorithm, which means decoupling the relationship between the interior region of the electrode ring and the region between the screen and the electrodes. This is not adaptable to all cases.

All the algorithms mentioned above have their own merits and drawbacks in practical application and research investigation. In fact, the effectiveness of an algorithms' application depends on certain measurement environment (instrument environment). For example, what measurements can be realized and what are resolutions of them. As far as the ECT system is concerned, there exist two different measurement methods: (1) Detection of mutual

CP914, *Multiphase Flow: The Ultimate Measurement Challenge,* edited by X. Cai, Y. Wu, Z. Huang, S. Wang, and M. Wang

capacitance. (2) Measurement of potentials at floating electrodes (the potential includes the information of mutual capacitance indirectly). Accordingly, signal conditioning circuits are also different [2]. In this paper, a stochastic inverse technique based on genetic algorithm (GA-ECT) is developed, which is adapted to the two different measurement methods. The traditional sensitivity field model is avoided; consequently the soft-field problem does not exist. Numerical examples are presented to evaluate the technique both for noise free and noisy data. The results show that quality image can be reconstructed with not only low contrast permittivity but also with high contrast. It is well known that three sources provide information in inverse problem: physical laws, measurements and a priori information, Therefore errors of inversion results also come from these three sources. The image reconstruction results based on GA-ECT related to these three sources are discussed.

IMPLEMENTATION OF GA-ECT TECHNIQUE

Model Formulation

Firstly, a simple four-electrode ECT square sensor without radical screen, depicted in Fig.1, is modeled in order to clarify the implementation performance of GA-ECT technique.

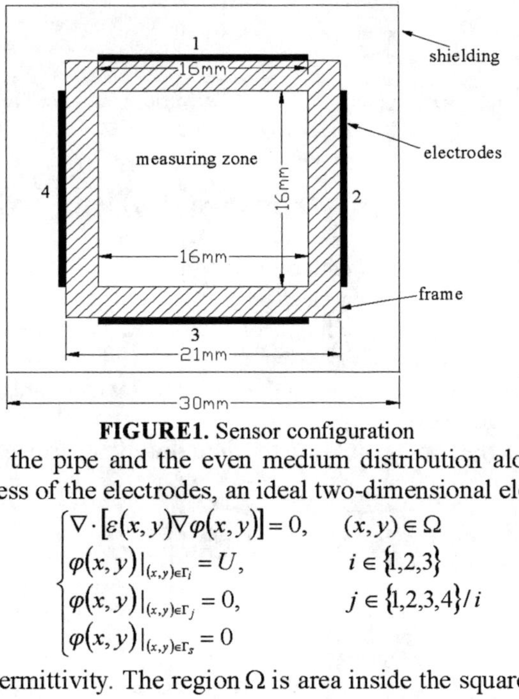

FIGURE1. Sensor configuration

Assuming zero free charge in the pipe and the even medium distribution along the pipe axis, ignoring wave propagation effects and the thickness of the electrodes, an ideal two-dimensional electrostatic model is as follows:

$$\begin{cases} \nabla \cdot [\varepsilon(x,y)\nabla\varphi(x,y)] = 0, & (x,y) \in \Omega \\ \varphi(x,y)|_{(x,y)\in\Gamma_i} = U, & i \in \{1,2,3\} \\ \varphi(x,y)|_{(x,y)\in\Gamma_j} = 0, & j \in \{1,2,3,4\}/i \\ \varphi(x,y)|_{(x,y)\in\Gamma_s} = 0 \end{cases} \tag{1}$$

Where $\varepsilon(x,y)$ is the dielectric permittivity. The region Ω is area inside the square screen excluding the electrodes. $\varphi(x,y)$ is the voltage distribution on Ω. Γ_i denotes the excitation electrode with voltage U. Γ_j is the sensing electrodes with zero voltage. Γ_s is the earthed screen. When one electrode is chosen as the excitation electrode, the mutual capacitances between it and other electrodes are detected with sequence using the first measuring method. According to the reciprocal theorem, a maximum of six independent capacitances can be acquired for a four-electrode system.

The mathematical model based on the potentials measuring is slightly different from the capacitance method in boundary conditions.

$$\begin{cases} \nabla \cdot [\varepsilon(x,y)\nabla\varphi(x,y)] = 0, & (x,y) \in \Omega \\ \varphi(x,y)|_{(x,y)\in\Gamma_i} = U, & i \in \{1,2,3\} \\ \varphi(x,y) = constj, & j \in \{1,2,3,4\}/i \\ \varphi(x,y)|_{(x,y)\in\Gamma_s} = 0 \end{cases} \tag{2}$$

All the symbols have the same meaning as those mentioned above. The term $\varphi(x,y)|_{(x,y)\in\Gamma_j} = 0$ is substituted with $\varphi(x,y) = constj$, where $constj$ **denotes the potential on the** j_{th} **electrode**. It means that the electrode is an equal potential body and the net charge at the floating electrode is zero although there are charges distributed on it. When one electrode is chosen as the excitation electrode, the voltage between the floated electrodes and the earth are detected in turn. Fig. 2 shows the electric fields are different under the two different measuring methods for the same dielectric permittivity distribution but different boundary conditions.

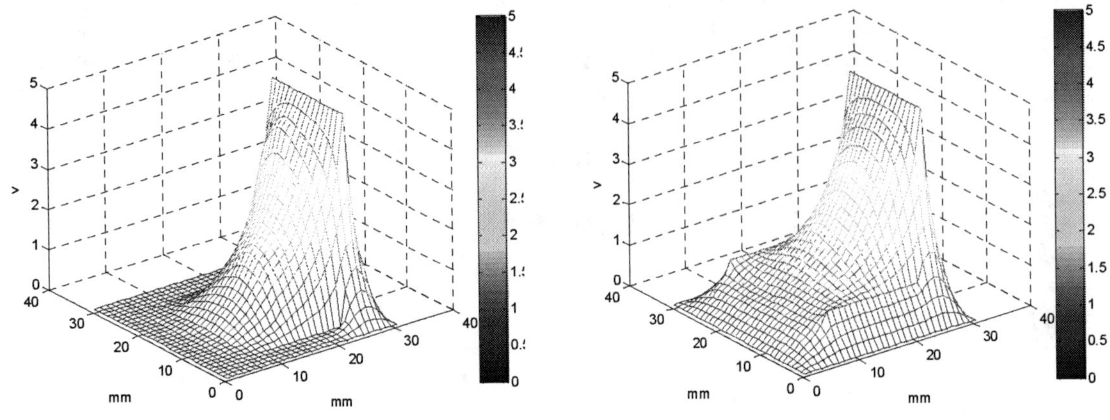

(a) electric field under capacitance measuring method **(b)** electric field under the second measuring method

FIGURE 2.

Solution of Forward Problem

The aim of solving the forward problem is to determine the electric field under certain permittivity distribution, then to calculate the mutual capacitances between electrodes using capacitances detecting method or the potential measurement method. We look for a numerical solution of the boundary problem (1) and (2). In this setting, the finite difference method is employed under rectangular coordinates. The domain Ω is discretized into square meshes. Here the nodes equations are not constructed in the traditional way based on Taylor series theory as discrete partial differential equation but in a way based on physical meaning.

Figure 3 shows the situation that the nodal point is in the non-uniform distributed permittivity region.

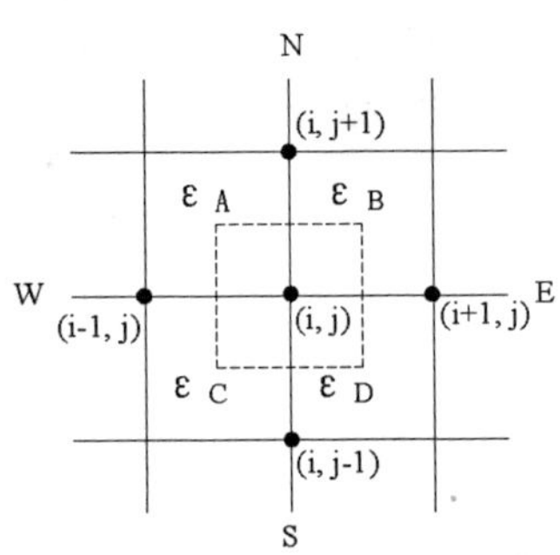

FIGURE 3. Mesh Discretization in non-uniform distributed permittivity region

The grid depicted with solid lines denotes the mesh used to discretize the dielectric permittivity region and the grid depicted with dashed line is the control volume centered in node (i, j). Supposing no free charge distribution in the medium, the Gauss-flux theorem can be applied to the control volume,

$$\int_l Dds = \int_l \varepsilon Eds = q = 0 \tag{3}$$

where l is the boundary around the control volume, E is the intensity of electric field. The discrete equation (3) can be expressed as:

$$\left(\frac{\varepsilon_A+\varepsilon_C}{2}\right)\frac{(\varphi(i-1,j)-\varphi(i,j))}{\Delta x}\cdot\Delta y+\left(\frac{\varepsilon_A+\varepsilon_B}{2}\right)\frac{(\varphi(i,j+1)-\varphi(i,j))}{\Delta y}\cdot\Delta x$$
$$+\left(\frac{\varepsilon_B+\varepsilon_D}{2}\right)\frac{(\varphi(i+1,j)-\varphi(i,j))}{\Delta x}\cdot\Delta y+\left(\frac{\varepsilon_C+\varepsilon_D}{2}\right)\frac{(\varphi(i,j-1)-\varphi(i,j))}{\Delta y}\cdot\Delta x=0 \quad (4)$$

The four terms on the left side of the above equation denote the sum of the electric flux entering the boundary of the control volume which should be equal to the free charge in the volume, it is zero $q=0$, i.e. the right term in the above equation is zero. Let $\Delta x=\Delta y$ .we have:

$$(\varepsilon_A+\varepsilon_B+\varepsilon_C+\varepsilon_D)\varphi(i,j)=\frac{(\varepsilon_A+\varepsilon_C)}{2}\varphi(i-1,j)+\frac{(\varepsilon_A+\varepsilon_B)}{2}\varphi(i,j+1)$$
$$+\frac{(\varepsilon_B+\varepsilon_D)}{2}\varphi(i+1,j)+\frac{(\varepsilon_C+\varepsilon_D)}{2}\varphi(i,j-1) \quad (5)$$

The format of the above equation is comparable to the five-point differential equation laid out by Taylor series. In particular for nodal points in uniform permittivity; the following equation can be obtained:

$$4\varphi(i,j)=\varphi(i-1,j)+\varphi(i,j+1)+\varphi(i+1,j)+\varphi(i,j-1) \quad (6)$$

Boundary conditions are discretized accordingly as follows. For the capacitance measuring method, the potentials on the electrodes and the screen are known ($\varphi(i,j)=b(i,j)$), where (i,j) is the node on the boundary, $b(i,j)$ being the known voltage. For the potentials measuring method, the potentials at the floating electrodes are unknown, and then Gauss-flux formula $\oint_{l_i} Dds=0$ is used again, where l_i is the boundary around the electrode. So equations can be established corresponding to each node. Rearranging the unknowns $\{\varphi(i,j)\}$ into an N-vector φ, the linear algebraic equations are formed:

$$A_{N\times N}(\varepsilon)\cdot\varphi=B_i \quad (7)$$

where N is the number of the total nodes, the N×N coefficient matrix A is the discrete representation of the operator. N-vector B_i is determined by the boundary values $\{b(i,j)\}$. The corresponding boundary condition terms $B_i(i=1,2,3)$ change with excitation modes. The relationship between capacitance and permittivity distribution is governed by the following equation:

$$C=\frac{Q}{V}=\frac{1}{V}\oint_{\Gamma_k}\varepsilon\frac{\partial\varphi}{\partial n}ds \quad (8)$$

Where V is the potential difference between two electrodes forming the capacitance, n is the inward normal on the k_{th} electrode. However, for the potentials measuring method, the coefficient matrix $A_i^{'}$ is different because the nodes equations on the electrodes are not the same for different excitation models. The potentials at the floating electrodes can be acquired after the solution of the equations (9):

$$A_i^{'}(\varepsilon)\cdot\varphi=B_i^{'} \quad (i=1,2,3) \quad (9)$$

Inverse Problem

Generally, inversion in ECT is a process that allows the reconstructed permittivity distribution to approach its true distribution gradually, and to minimize the difference between the observation values and the calculation values. A great amount of forward calculation is needed in this process.

It is well known that the four-electrode ECT sensor can provide a maximum of $C_4^2=6$ independent mutual capacitances. For the potentials measuring method, supposing the number of independent measurements is not more than 6, the potentials on the floated electrodes include the mutual capacitances information. Fig. 4 shows the inner region of the pipe depicted in Fig. 1 which is divided evenly into four parts and the permittivity of each part is considered as a constant. They are denoted as ε_A ε_B ε_C ε_D .

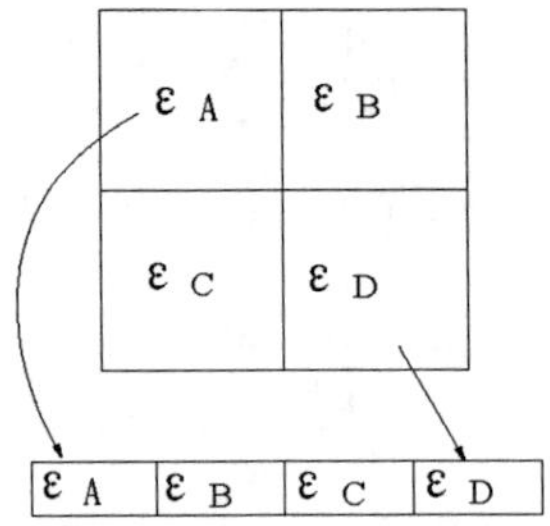

FIGURE 4. Region separation of the square pipe inner

The aim of modeling the sensor is to test whether the inversion of pre-set four permittivity constants $\varepsilon_A \;\; \varepsilon_B \;\; \varepsilon_C \;\; \varepsilon_D$ can be obtained accurately from the six capacitance or voltage measurements simulated from the forward calculation. In most cases, the number of the independent measurements contaminated by noise is far less than the number of the estimated parameters, and some a priori knowledge is introduced by regularization to overcome the ill-posed problem. However the error contribution caused by algorithm itself is not known. For example even if the sufficient measurements free of noise are given, ('sufficient' means the number of independent measurement is no less than the number of the estimated parameters.), the true solution can still not be found using LBP algorithm . The insufficiency of common measurements mentioned above will be discussed later.

The nonlinear multi-variable optimization problem is formed for the four electrodes ECT problem:

For the first measuring method,

$$\begin{cases} Min \;\; F(\varepsilon) = \|C - C_m\|_2^2 \\ s.t. \;\; A_k(\varepsilon)\varphi = B_k, k = 1,2,3 \\ \varepsilon = \{\varepsilon_A \;\; \varepsilon_B \;\; \varepsilon_C \;\; \varepsilon_D\}, \varepsilon_i \geq 1, i = A,B,C,D \end{cases} \tag{10}$$

where C is the mutual capacitance from calculation and C_m is the measurement.

For the second measuring method,

$$\begin{cases} Min \;\; F'(\varepsilon) = \|\varphi - \varphi_m\|_2^2 \\ s.t. \;\; A'_k(\varepsilon)\varphi = B'_k, k = 1,2,3 \\ \varepsilon = \{\varepsilon_A \;\; \varepsilon_B \;\; \varepsilon_C \;\; \varepsilon_D\}, \varepsilon_i \geq 1, i = A,B,C,D \end{cases} \tag{11}$$

where φ is the calculated potential and φ_m its measurement value.

Considering the practical situation where the medium components in the pipe are known, the a priori information $l \leq \varepsilon \leq u$ can be added, where l,u denote the lower and upper boundary of ε respectively. Furthermore other a priori information such as smoothness of the reconstructed image or minimization of total variation (TV) can also be included in the object function.

GA can be used to solve this nonlinear multi-variable optimization problem [12, 13, and 14]. GA's borrow much of their terminology from biological genetics. We choose GA to solve this problem based on its following characteristic:

1) GA search the solution directly based on the object function, no evaluation of the derivatives is needed. For the object function $F(\varepsilon) = \|C - C_m\|_2^2$, it is difficult to find the derivative expression of $C(\varepsilon)$ or $\varphi(\varepsilon)$, the obstacle can be avoided using GA.
2) GA searches the solution from many points while the traditional algorithm start the iteration process from only one point with less information, which is easily trapped in the local optimization.

Figure 5 shows the implementation flow chart of the GA-ECT.

GA starts with the random creation of an initial population $p(t)$. One evaluates the fitness of the individuals in the population by substituting each individual $\{\varepsilon_A \;\; \varepsilon_B \;\; \varepsilon_C \;\; \varepsilon_D\}$ into the coefficient matrix and solving the forward problem and thus determining their survival potential. Then highly fit individuals are selected as part of the crossover pool. The individuals in the cross pool will combine and crossover to produce new individuals $p'(t)$. Mutation according to a predetermined probability is often encouraged so that new feature can be introduced at the gene level. After such genetic operations as selection, crossover, and mutation of current population have been performed, a new population $p''(t)$ will reproduce and replace the old one. Fitness will again be evaluated, and the population will begin a new cycle of reproduction. Because the new population usually has higher fitness than the old one, the population is improved from generation to generation. A reasonable stop criterion is often chosen as the point at which the prescribed maximum generation is reached or convergence is detected.

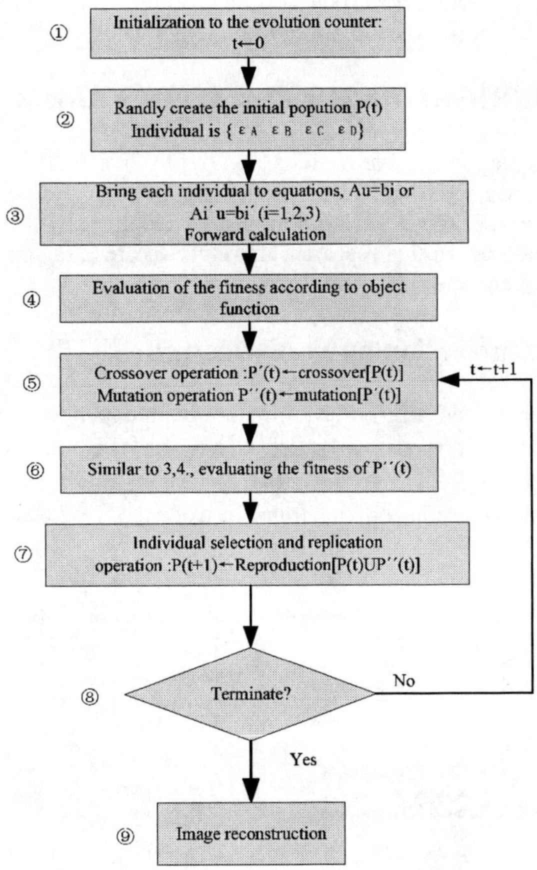

FIGURE5. The implementation flow chart of the GA-ECT

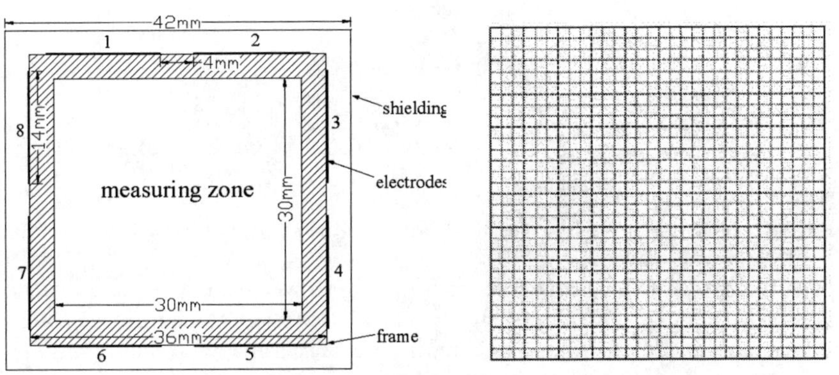

(a) Eight-electrode square ECT sensor. **(b)**Two meshes used in the calculation

FIGURE.6

The sixth part in the flow chart is a time consuming operation. $m \times (n-1)$ linear algebraic equations need to be solved in each iteration, where n is the number of electrodes and m is the number of individuals in population(let m =30). The computation time of each step in the iteration depends mainly on the size of the forward mesh. For meshes used in the four-electrode ECT, the computation time for each individual is 0.03s. The computer used has an Intel Pentium 4 processor with 2.66GHz clock rate with 512MB of RAM. The total computation time consumed in the sixth part at each iteration is 30×0.03=0.9s. It is clear that the effectiveness of the algorithm will be significantly improved if the technology of parallel computation is used.

Similar to the four-electrode ECT sensor, an eight-electrode ECT sensor which is more accorded with practical situation is constructed. Fig. 6 shows the eight-electrode ECT sensor and the two meshes used in calculation.

The most subdivided mesh is used for forward solver and the larger grid is used for inverse solver.

NUMERICAL RESULTS AND DISCUSSIONS

In this section, Numerical simulation tests have been carried out to investigate the performance of the GA-based inverse technique with noisy and noise free data. This is the first attempt at nonlinear ECT reconstruction using GA to the author's knowledge. Because GAs are random-search techniques, the results obtained from a single numerical test may have unique features. In this research, therefore the results were averaged by several tests with different random-number series to overcome such drawback.

Inversion on Four-electrode ECT

First of all, four measurements chosen arbitrarily from the six independent capacitances are used under the condition that $\varepsilon_i \geq 1, i = A, B, C, D$ which is always satisfied because the relative permittivity of any dielectric material is not less than 1.

Figure7 shows nonlinear inversion results on the four-electrode ECT sensor given the condition mentioned above.

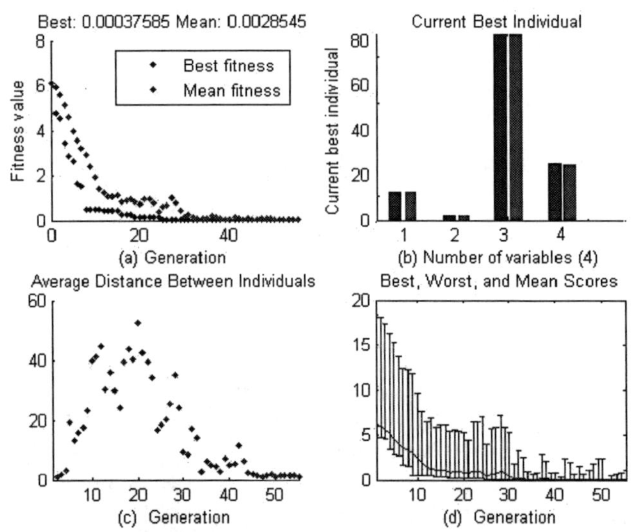

FIGURE.7 (b) Inversion results from noise-free measured data; **(a), (c)** and **(d)** Implementation performance of GA-ECT

Figure7 (a) show that the speed of GA's convergence is getting slower with the generation and the fitness value is almost unchanged from generation 30. If the criterion is chosen reasonably, significant computation time can be saved. (b) shows the inversion results, the blue bars are the pre-set permittivity values and the brown bars are the inversion outcomes. If the measurements are sufficient and not contaminated by noise, the inversion is almost 100% accurate. (c) shows the change of the average distance between individuals with generation. (d) shows the best , worst, and mean scores of the individuals and the diversity of the population decreasing with evolution.

The combination of the mutual capacitance and the potentials can also be used to inversion successfully. In fact it is a two-objective nonlinear optimization problem now. However, when the inner region of the pipe is subdivided evenly into nine square parts, inversion result using the twelve measurements (six capacitances and six potentials) is no better than the one with only six capacitances. It can be explained that the potential information included in the capacitance information can not provide new independent information any more.

In practical situation the medium component permittivity can be obtained as a priori information. Inversion results in many cases are still satisfying only using two or three capacitance or potential values if the information $l \le \varepsilon \le u$ is supplemented. Fig. 8 shows the examples with supplementary a priori information. In our case $l=1$ and $u=3$.

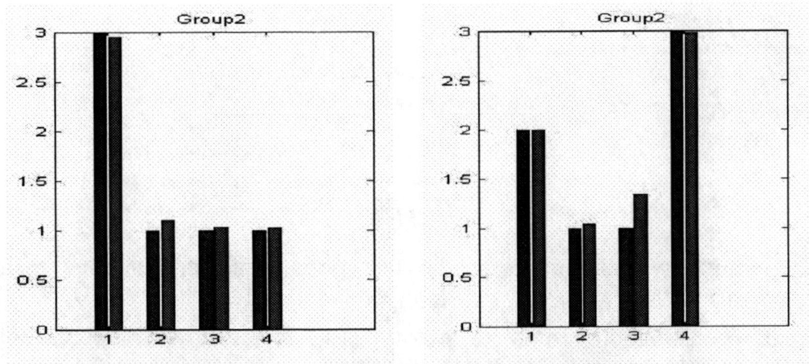

(a)Inversion with two measurements **(b)Inversion with three measurements.**

FIGURE 8.

$l \le \varepsilon \le u$ is considered both for (a) and (b). It can be illustrated that a priori knowledge $l \le \varepsilon \le u$ is indeed helpful for the inversion calculation when the measurement data are insufficient.

Inversion on Eight-electrode ECT

To simulate the real conditions, a specific amount of random noise is added to the noise-free measured data I_0, using the follow equation:

$$I = I_0(1 + k \times RND) \tag{12}$$

Where I_0 is the capacitance or potential vector which is calculated using FDM given the test distribution. I is the capacitance or potential vector contaminated by noise. RND is a random scale value drawn from a normal distribution with mean 0 and standard deviation 1. Defined in this way the noise level k is equal to $\|I - I_0\| / \|I_0\|$. Fig. 9 shows the potential and noise vectors.

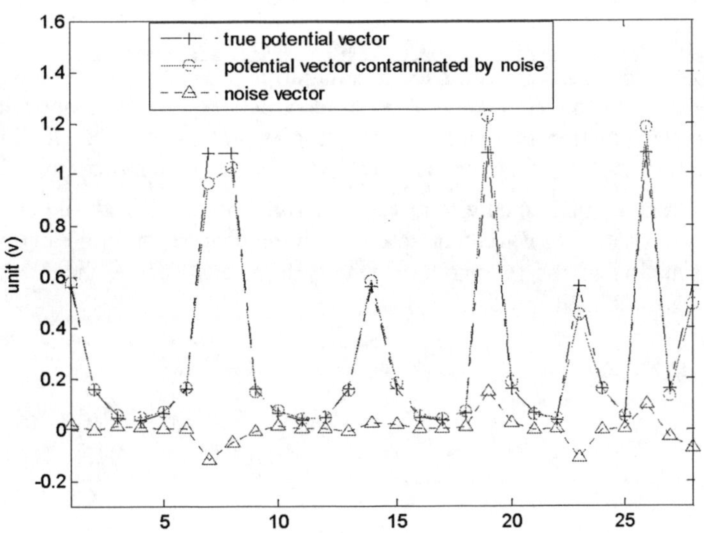

FIGURE 9. Potential and noise vectors

To quantitatively evaluate the quality of reconstructed images, three criteria [9] were used:
Relative image error:

$$\frac{\left\|\varepsilon^{*}-\varepsilon\right\|}{\left\|\varepsilon\right\|} \tag{13}$$

Relative capacitance or potential residual:

$$\frac{\left\|I(\varepsilon)-I(\varepsilon^{*})\right\|}{\left\|I(\varepsilon)\right\|} \tag{14}$$

Correlation coefficient:

$$\frac{\left\langle \varepsilon-\bar{\varepsilon},\varepsilon^{*}-\bar{\varepsilon}^{*}\right\rangle}{\left\|\varepsilon-\bar{\varepsilon}\right\|\cdot\left\|\varepsilon^{*}-\bar{\varepsilon}^{*}\right\|} \tag{15}$$

Where ε is the true permittivity distribution of the test permittivity distribution, ε^{*} is the reconstructed permittivity distribution, $\bar{\varepsilon}$ and $\bar{\varepsilon}^{*}$ are the mean value of ε and ε^{*} respectively.

In Fig. 10, we present reconstruction results for two types of low contrast of permittivity (3:1) profiles: a stratified pattern and a centered object using potentials data contaminated by 1% noise without the usual regularizing penalty term. The color bar on the right of each image indicates the change of permittivity value from 1 to 3. (Similar results can also be achieved with capacitance values.)

Test profile

Reconstructed image

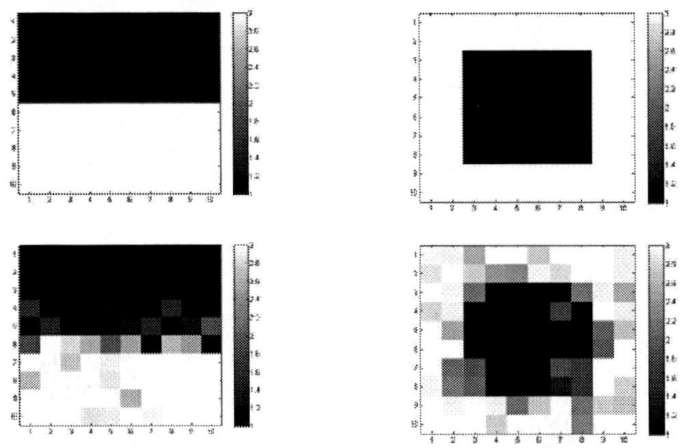

Image error (%)	16.14	27.85
Poten. residual	0.0046	0.0079
Corr .coeff	0.9343	0.7149

FIGURE.10 Reconstruction results for stratified and centered objects (%1 noise)

Figure 11 shows the image reconstruction results for an annular pattern and two adjacent square objects. For each case, the noise-free capacitance data as well as three sets of noisy data with relative noise level at 2%, 4% and 6% were used respectively, and the function of regularization term is also considered. In this case the regularizing penalty term $\alpha\|\varepsilon\|^2$ is added to the object function. Its physical meaning is minimization of the image energy. One can try to use other forms of regularizing penalty term such as: smooth generalized Tikhonov regularization for the smoothly varying case or total variation (TV) regularization for jump changes and find an optimal value for α . Here our purpose is to test the function of the regularizing penalty term and choose the value of α from experience. (Similar results can also be achieved using potentials.)

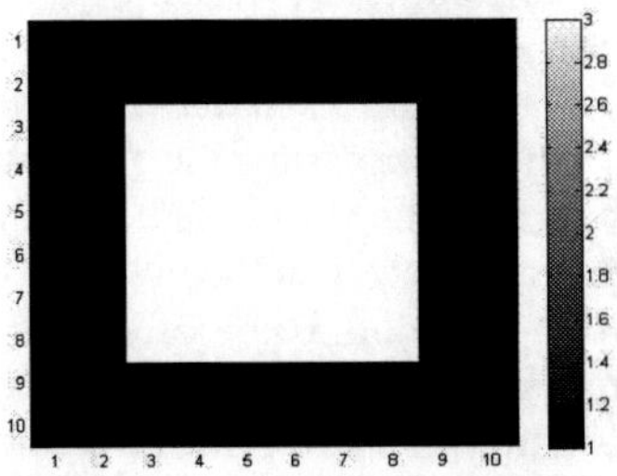

(a) An annular object

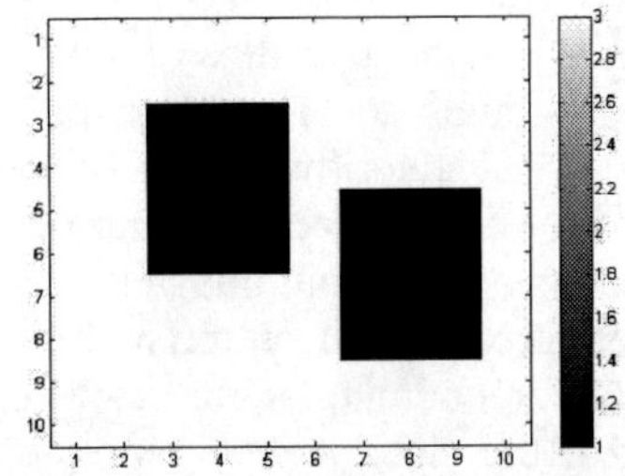

(b) Two rectangular objects

Data noise	0%	2%	4%	6%

Reconstructed image without regularization

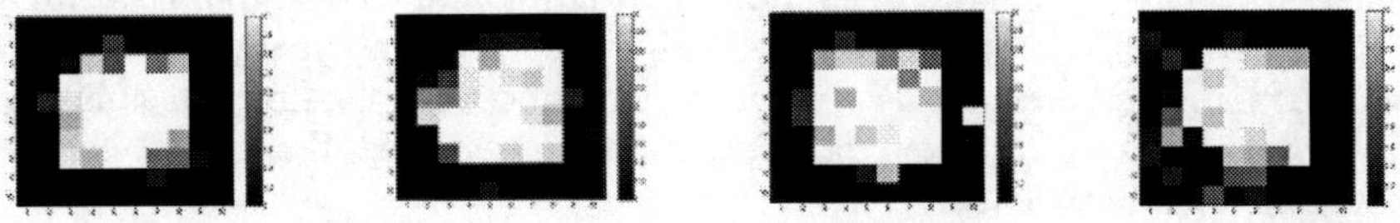

Image error (%)	27.86	36.56	38.12	35.04
Cap. residual	0.0079	0.0036	0.0091	0.0125
Corr .coeff	0.7138	0.7016	0.6783	0.7186

Reconstructed image with regularization

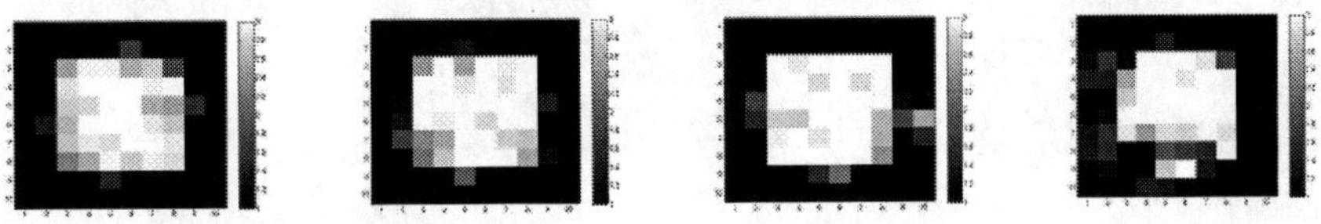

Image error (%)	33.22	33.88	36.58	33.03
Cap. residual	0.0018	0.0023	0.0099	0.0124
Corr .coeff	0.7495	0.7435	0.7062	0.7488

(c) Image reconstruction results for the object in (a)

Data noise	0%	2%	4%	6%

Reconstructed image without regularization

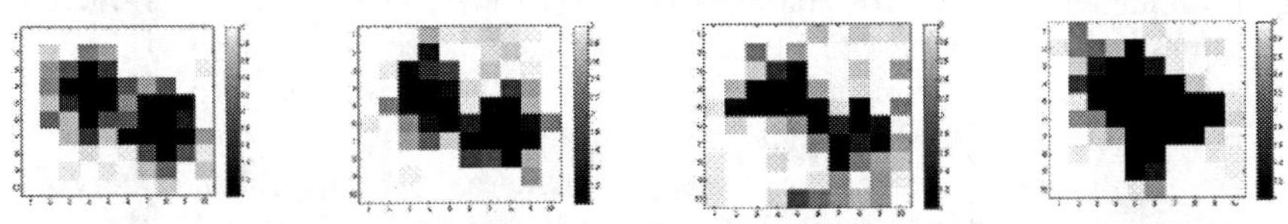

Image error (%)	23.78	24.60	28.02	30.24
Cap. residual	0.0022	0.0050	0.0146	0.0194
Corr .coeff	0.6999	0.6844	0.6463	0.5946

Reconstructed image with regularization

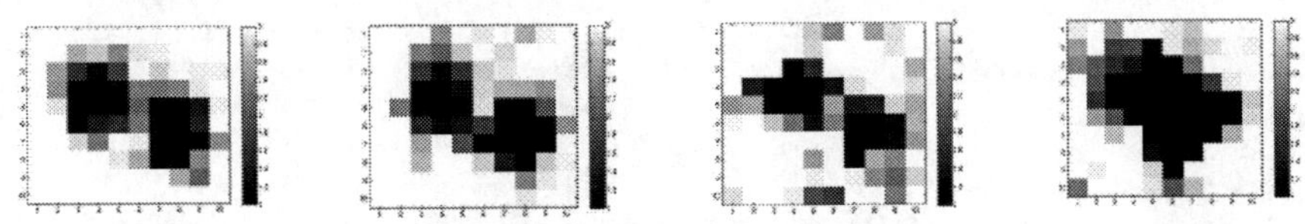

Image error (%)	23.06	23.62	26.96	30.10
Cap. residual	0.0015	0.0051	0.0102	0.0120
Corr .coeff	0.7184	0.7080	0.6619	0.6144

(d) Image reconstruction results for the object in (b)

FIGURE 11.

The images reconstructed in Fig.11 demonstrate that the interference of noise to the reconstructed image can be alleviated through regularization. The relative image errors are smaller and the correlation coefficients are higher with a regularizing penalty term than those reconstructed without regularization in the same noise level. It also shows the capacitance residuals are near zero for most cases in Fig. 11, which are much smaller than the ones reconstructed with traditional algorithm [9]. However, this is not always true when comparing the reconstructed free-noise results with and without a regularizing penalty term in Fig. 11 (c). Generally, the former is better than the latter in reconstructed image effect, and this is to say a regularizing penalty term sometimes has the negative effects for some dielectric permittivity distribution with very low noise level. (In fact, we could not offer absolutely accurate simulation data, for example the mesh size used in the forward calculation is reduced to the half and quarter of the current one and if the simulated measurements of the last are taken as true, then the relative errors from forward calculation are 6% for the current and 3.2% for the half.) This can be explained that a priori knowledge offered by the regularization is not consistent to the true situation. In most cases mentioned above, the regularization penalizes the extreme permittivity values' variation and stabilizes the reconstruction process. With only 28 independent measurements and 100 pixels low spatial resolution is expected. Certainly, more pixels can be used in the inversion mesh which need the more fined mesh used in forward calculation. It is still a non-linear optimization problem with more variables and the Pareto satisfying solution is anticipated. However the time costs increases intensively with the more fined mesh, so a compromise exists between the time consumption and spatial resolution.

Figure12 gives the results with a square object (permittivity 80) near the top left corner of the square pipe filled with dielectric material (permittivity 3).

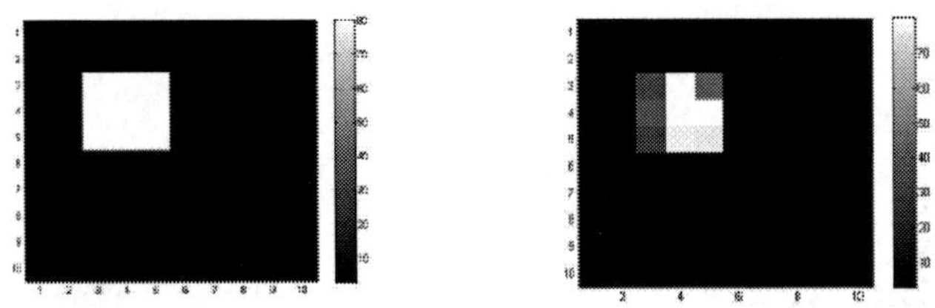

Image error (%)	46.06
Cap. residual	0.0640
Corr .coeff	0.8739

(a) True phantom **(b)** Results with 1% noise and a regularization term

FIGURE 12.

It can be seen that the location of the high permittivity is found comparatively right but the relative image error is large. This is because that the accuracy of high permittivity value calculation is low, which dominates the image error. The capacitance residual is still very small in this case. Three factors contribute to the large image error: (1) Insufficient measurements. (2)The error coming from the forward calculation mentioned above. (3)The error due to measurements saturation. (As depicted in Fig.13. This makes the reconstruction of high values of permittivity very difficult as the difference in measurements is very small). Furthermore considering the noise disturbance, the heavily nonlinear problem is more difficult in practical situation than in simulation.

The saturation phenomenon of capacitance measurements have already been observed by [11]. In our numerical test, the saturation phenomenon of potential measurements is found. The permittivity of the object in Fig. 12 is varied from 3 to 80 and the accordingly voltage changes of the floated electrodes 3 and 5 to the earth respectively are shown in Fig. 13.

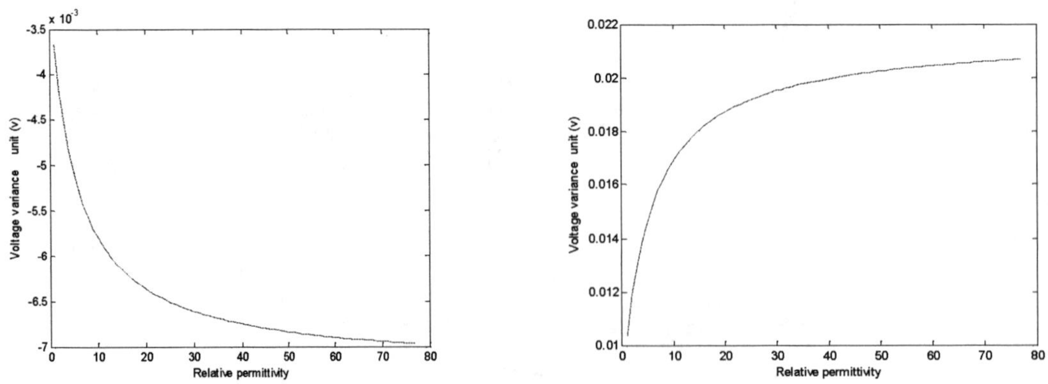

(a) **(b)**

FIGURE 13. Measurements saturation phenomenon. (a) voltage change with permittivity between electrode 3 and the earth. (b) voltage change with permittivity between electrode 5 and the earth

It can be seen that all the numerical inversion results presented above are obtained using the raw simulated measurement values without normalization.

CONCLUSIONS

A stochastic inverse technique based on genetic algorithms (GA-ECT) has been developed for image reconstruction in electrical capacitance tomography. Different from many traditional methods based on sensitivity field model, raw measurements such as potentials and capacitances are used and more forward calculation is needed in implementation without sensitivity conception. Numerical tests were performed on the four-electrode and eight-electrode ECT sensors. The results show that with sufficient numbers of measurements of free noise the inversion technique can be applied satisfactorily and a priori knowledge of the materials' permittivity is helpful to image reconstruction. The deficiency of independent measurements and noise cause the blurring of the reconstructed image. A regularizing penalty term added to the object function can penalize the variation of extreme permittivity value and stabilizes the reconstruction. However some times negative effects occur when a priori knowledge is deviated from the true situation. Due to the error from forward calculation and saturation in measurements, the relative image error of the high contrast permittivity object is large, but its location recognition is found comparatively accurate.

It has been shown that the GA-ECT can be an effective and stable technique for inverting permittivity distribution with measurements containing random noise, although it is expensive in terms of computing time and resources. With further development of fast computer processor, the author is confident that the real-time application of GA-ECT will be realized in the near future.

ACKNOWLEDGEMENTS

This research was supported by Special Funds for Major State Basic Research Projects (2004CB17702-04). The authors wish to express their gratitude. We also thank Dr Jianyong Zhang, now at Teesside University, for many useful discussions.

REFERENCES

1. Daniel Watzenig, Markus Brandner and Gerald Steiner Dynamic Image Reconstruction in Electrical Capacitance Tomography using Particle Filters, 2005 IEEE Sensors (IEEE Cat. No.05CH37665C)
2. Hannes Wegleiter, Anton Fuchs,Gert Holler, and Bernhard Kortschak, Analysis of Hardware Concepts for Electrical Capacitance Tomography Applications, 2005 IEEE Sensors (IEEE Cat. No.05CH37665C), 2005, p 4
3. Huang S M, Plaskowski A B, Xie C G and Beck M S 1988 Capacitance-based tomographic flow imaging system Electron.Lett
4. Beck M A, Dyakowski T and Williams R A 1998 Process tomography-the state of the art Meas Control 20 163-77
5. Jaworski A J and Dyakowski T 2001 Application of electrical capacitance tomography for measurement of gas-solids flow characteristics in a pneumatic conveying system Meas.Sci.Technol.12 1109-19
6. Xie Dailiang, Ji HaiFeng, Huang Zhiyao, Wang Baoliang and Li Haiqing A New Void Fraction Measurement Method for Gas-oil Two-phase Flow Based on Electrical Capacitance Tomography System and Otsu Algorithm. Proceedings of the 5th World Congress on Intelligent Control and Automation 6.2004
7. Akira Kimoto, Takayuki Nakatani, Yu-Ichiro Matsuoka and Katsunori Shida Reconstruction of Temperature Patterns in the Cylindrical head Model From Electrical Capacitance Tomography IEEE Transaction on instrumentation and meas. Vol.54, No.6,2005
8. Isaksen O. A review of reconstruction technique for capacitance tomography, Meas Sci Technol, 1996, 7:325-337
9. Yang W Q and Peng L H 2003 Image reconstruction algorithms for electrical capacitance tomography Meas.Sci.Technol. 14 R1-13
10. Fang W 2004 A nonlinear image reconstruction algorithm for electrical capacitance tomography Meas.Sci.Technol.15 2124-32
11. Soleimani M and lionheart W R B2005 Nonlinear image reconstruction for electrical capacitance tomography using experimental data Meas.Sci.Technol. 16 1987-96
12. R.Olmi, M.Bini, and S.Priori A Genetic Algorithm Approach to Image Reconstruction in Electrical Impedence Tomography IEEE TRANSACTIONS ON EVOLUTIONARY COMPUTATION. VOL.4,NO.1,2000
13. Mou Changhua, Peng Lihui, Yao Danya, Xiao Deyun, Image reconstruction using a genetic algorithm for electrical capacitance tomography,Tsinghua Science and Technology, v 10, n 5, Oct. 2005, p 587-92

14. Chen De-yun, Zheng Gui-bin, Yu Xiao-Yang, Sun LI-quan Image reconstruction algorithm based on genetic algorithms for two-phase flow electrical capacitance tomography system, Electric Machines and Control, v 7, n 3, Sept. 2003, p 207-11

NOMENCLATURE

$\varepsilon(x,y)$	permittivity distributionof object
Ω	everything inside the square screen excluding the electrodes
$\varphi(x,y)$	voltage distribution on Ω
Γ_i	excitation electrode
Γ_j	the sensing electrodes
Γ_s	earthed screen
D	electric displacement vector
(i,j)	node
l	boundary around the control volume
E	intensity of electric field
V	potential difference between two electrodes
A_i'	coefficient matrix
ε_A ε_B ε_C ε_D	permittivity of each part
C	mutual capacitance from calculation
C_m	measurement capacitance
φ	potential from calculation
φ_m	potential measurement value
l	lower boundary of ε
u	upper boundary of ε
$F(\varepsilon)$	object function
$p'(t)$	population reproduced by crossover
$p^{*}(t)$	population reproduced by mutation
I_0	noise-free measured data
I	capacitance or potential vector contaminated by noise
RND	a random scale value
k	noise level
ε^*	the reconstructed permittivity distribution
$\bar{\varepsilon}$	mean value of ε
$\bar{\varepsilon}^*$	mean value of ε^*
α	regularization parameter

X-Ray Visualization Measurement of Slurry Flow in Solid Propellant Casting

Toru Shimada*, Hiroto Habu*, Yoshio Seike†, Seiji Ooya§, Hideo Miyachi¶, and Masaaki Ishikawa**

*Department of Space Transportation Engineering, ISAS/JAXA, 3-1-1 Yoshinodai, Sagamihara, 229-8510, Japan
†R&D Department, NOF Corporation, 61-1 Kitakomatsudani Taketoyo, Chita, 470-2398, Japan
§M&D Section, Hitachi Engineering & Services Co., Ltd., 2-2, Saiwai-cho 3, Hitachi, 317-0073, Japan
¶Visualization Division, KGT Inc., 2-8-8 Shinjuku, 160-0022, Japan
**Dept. of Mechanical System Engineering, University of Ryukyus, 1 Senbaru, Nishihara-cho, 903-0213, Japan

Abstract. Simulated solid propellant slurry containing lead sphere tracers is experimentally cast into a double circular cylinder container. During the casting, the temperature and the pressure environment has been mimicked to an actual composite solid propellant casting of solid rocket motors. X-rays are projected on to the slurry flow from two directions and penetration images are recorded by a flat panel detector and an X-ray image intensifier. By suppressing the mutual interference of the X-rays, the two-directional X-ray photography has been successfully conducted. Using the time series data of the X-ray images from the two directions, the identification of each tracer particle in space and time has been done and their three-dimensional paths have been computed. From these procedures, the flow field or the velocity field inside the slurry flow, which is invisible usually, has been estimated.

Keywords: Solid propellant casting, Slurry flow field, X-ray penetration photography, Particle tracking velocimetry

INTRODUCTION

Although advanced technology of active thrust-amount control of solid rocket motor (SRM) is being developed, it has been basically conducted passively by designing a grain shape of slid propellant, i.e., burning-surface-area evolution. Therefore, to improve the performance and the reliability of SRM, it is important to improve the precision of the estimation of the ballistic performance. In manufacturing solid propellant, slurry consisting of fine particles of solid-oxidizer and the metal fuel mixed in liquid polymer are first made in a mixer. The slurry is then cast into a motor case with a core inserted and, finally, the propellant grain is complete via the curing process. Without being sufficiently quantified, recent researches show impacts on the ballistic performance of the casting process of solid propellant slurry[1].

One important aspect that the casting process influencing the performance is the thrust variation due to the multiple-batch casting for large-sized solid rocket motors. This appears specifically as the thrust imbalance problem during the combustion tail-off of multiple solid rocket boosters. In order to predict this characteristic of such systems, it is demanded that both the shape of each batch after cure and the ballistic performance reflecting the local burning rate variation of each batch be evaluated. Another important aspect is related to the problems which occur in the casting process, such as segregation and orientation of particles and inclusion of air bubbles. Quantitative investigation of the cause and the characteristic of these phenomena and moreover evaluation of the degree of influence of them on the ballistic performance and the structural stability of SRM are required.

To answer such needs, research and development of the casting simulation and the ballistic performance analysis have been conducted using the CFD technology with free interface capturing technique[2]. By the practical use of such analysis tools, the precision and the degree of freedom of the propellant grain design will improve, and a better optimization of SRM design will be possible. Hence, there is possibility to contribute for the development-cost reduction.

To verify the numerical simulation of the casting, the measured data of the propellant slurry flow in casting is necessary. However, since the slurry flow in casting is unsteady phenomenon with free interfaces and since

CP914, *Multiphase Flow: The Ultimate Measurement Challenge,* edited by X. Cai, Y. Wu, Z. Huang, S. Wang, and M. Wang

propellant slurry is opaque to the visible light, the measurement of the flow quantity is not an easy task. As far as the author knows, a well-organized database does not exist on the solid-propellant slurry flow in the casting process.

In this paper, a flow-visualization and velocity-measurement experiment of casting simulated propellant slurry is described. In order to visualize the slurry flow, two X-ray penetration and video photography systems, which are placed in the orthogonal direction each other, are activated simultaneously. The inner velocity field is then restored by image analysis of the obtained digital video data. The objective of the study is to verify the feasibility of the present visualization and data deduction schemes for a slurry flow in solid propellant casting.

EXPERIMENTAL

A simulated propellant is used in the experiment for the safety reasons. It is because metallic spheres as the tracer must be kneaded into the propellant. Simulated propellant slurry is prepared using glass beads, vinyl chloride powder, aluminum powders, and liquid polymer (binder). Instead of the oxidizer particles, the glass beads and the vinyl chloride powders are used with the same grain sizes and the same grain-size combination of the real propellant. The quantity of the binder is adjusted for the density and the viscosity to become equal to the real propellant.

The experimental setup for the casting experiment is illustrated in Fig. 1. The casting facilities which have been made for the experiment are equipped with the container to heat the slurry and the casting hopper with a hot water jacket to keep the heat. The stand and the hot water circulating equipments are arranged not to interfere with the X-ray photography. The case into which slurry is poured is a double-circular cylinder made of acrylic glass and is equipped with a core. The ratio of the case caliber and the outer diameter of the core is adjusted to a practical rocket motor. The size of the case is selected to be suitable for the X-ray photography. Moreover, the case is surrounded by a hot-water bath and endures a vacuum-casting having thickness with the degree the X-ray photography is also possible. The temperature of the water is set to a value which is general with the practical propellant casting.

The simulated-propellant slurry with lead ball tracers mixed in is dropped from the slurry container and accumulated in the case. Being divided into two batches, it is also examined how the interface shape of the first batch will change by casting the second batch. For this purpose, two batches of slurry are cast sequentially into the case. The slurry of the first batch and the second batch is made be visually distinguishable by coloring one of the two.

Two X-ray generators of non-destructive inspection equipments are used with two X-ray detectors, a flat-panel detector (FPD) and an X-ray image intensifier (XII). The two X-ray equipments are arranged in the mutually orthogonal direction (Fig. 2). Videos are taken simultaneously from the two directions. The X-ray with the tube voltage from 160 kV to 170 kV is used. The distance from the generator to the sample is from 1000 to 1200 mm and is 300 mm from the sample to the detector. The X-ray image intensifier is Toshiba 9-inch E5764HV-P1K and the flat-panel detector is VARIAN PaxScan2520.

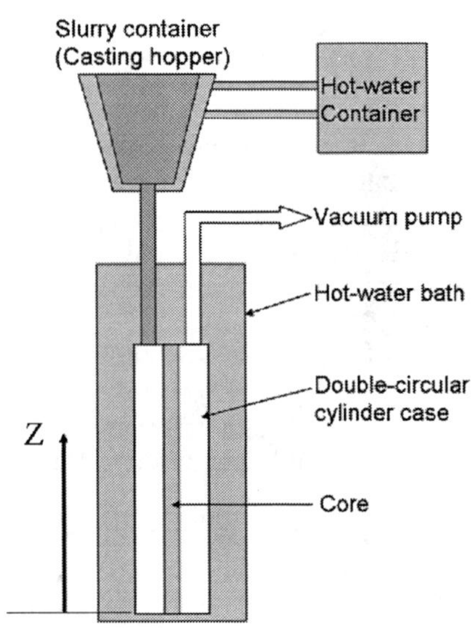

FIGURE 1. Experimental setup of casting

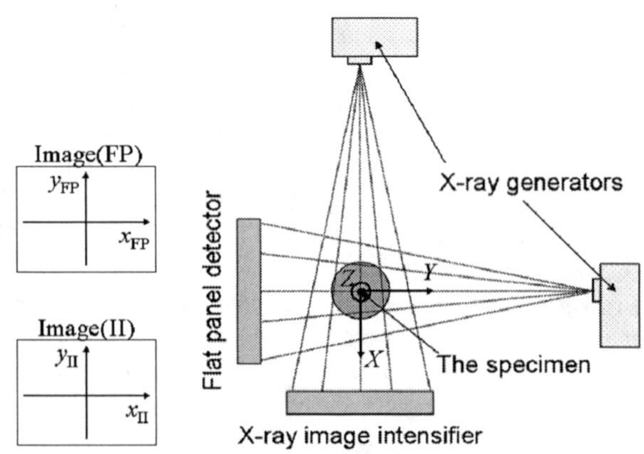

FIGURE 2. X-ray photography setup

IMAGE ANALYSIS

The outline of the image analysis is shown in Fig. 3. Serial-numbered still images (bitmaps) are obtained by resolving the video obtained as the output of FPD and XII. Using the sequential static images (bitmap), the position coordinates (pixel) of tracer particles are detected with the visual observation and also the particle identity in the time-wise direction is recognized. The time-sequential position coordinates (pixel) of tracer particles are transformed to real coordinates (mm) by using 8 calibration marker particles. It is assumed that the X-ray is irradiated in parallel. The transformation includes the aspect ratio of the image in the unit of mm/pixel. The identification of particle during two images of FPD and XII is done by the pair-matching among those which have been poured at the same time. Using the position coordinates of FPD-particle and XII-particle identified by the pair-matching, the three-dimensional position coordinates (mm) of tracer particle is determined. The inner velocity field is reconstructed by the obtained three-dimensional tracks data of trace particles. Detailed description on the image analysis will be available in our other paper[3].

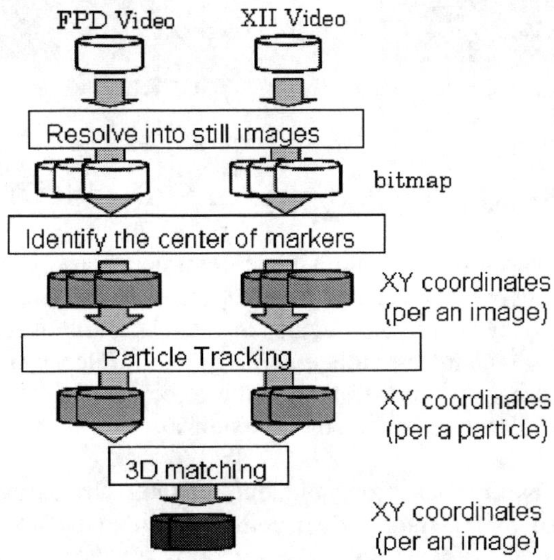

FIGURE 3. Image analysis outline

RESULTS AND DISCUSSION

Casting of the simulated propellant

Two casting experiments have been conducted, and in each casting, two batches of slurry are poured sequentially. As for the first casting, the temperature of the slurry has been cooled down in both the flow tube and the valve to the extent that the slurry has lost the fluidity. In order to refrain from cooling too much in the flow tube, the valve has been removed and a heat insulating material has been wound outside the flow tube. The second casting has been successfully done with these measures. The picture in Fig. 4 shows the state of slurry during the second casting. Poured slurry piles up spirally in the container. The air caught during casting has expanded in the vacuum container. Then, it generates air bubbles just like water boils up. The same phenomenon happens in the casting of a practical rocket motor. However, it becomes the cause of the tracer's unnatural motion. It is observed that the piling slurry spreads to the either side of the core and unites again. The slurry in the second batch, too, has piled up in the same way. However, the second batch has pushed downward the first batch. The second batch also has spread around the core and reunited at the opposite position to the pouring port. Then, it has become a mountain-like peculiar shape. The picture in Fig. 5 shows the situation.

FIGURE 4. State of slurry during the second casting

FIGURE 5. Mountain-like shape of the batch interface

X-ray photography with FPD and XII

As for the X-ray photography, various preparations have been necessary. First, it has been confirmed that the photograph could be taken by using two X-ray penetration equipments at the same time, and that the detection of the lead ball is easy from the picture. At this time, the number of the lead balls in the slurry is 300 / 5 kg. During the preparation, partial halation has been recognized in the picture by the influence from the scattered rays when taking X-ray photographs from two directions. In order to reduce the effect of the scattered X-rays, a partition has been provided at the side of the X-ray detectors and also the extension collimator of the X-ray generator has been improved.

Various tracer candidates such as, iron ball, iron fragment, lead ball, lead fragment, wire, and so on, have been tested to result in the best contrast of the tracer when being irradiated by an X-ray. As the result, it has been confirmed that a good contrast and, at the same time, slurry-motion traceability could be acquired with 1 mm lead balls. Immediately after the slurry drips, a calico pattern is formed in the container due to the air bubbles entrained into the slurry. This degrades the contrast of the lead ball depending on the position. However, the contrast of the overall lead ball has been good. A typical example of X-ray images is shown in Fig. 6.

It is possible to confirm what degree of difference between the digital value (FPD) and the analog value (XII) exists.

FIGURE 6. Typical X-ray image. Larger black dots on both sides are marker particles for calibration. Smaller black dots in the central region are the tracer particles.

Image Analysis

A snapshot of the three-dimensionalized track of tracer particle is shown in Fig. 7. In this figure, the tracks of 29 particles (of the 2nd batch) which could be chased three-dimensionally are displayed. A sphere is placed at the end of the track. The color is for the purpose of identification and does not have any physical meaning. It can be observed that the tracks are turning around the core.

The velocity vectors and particle paths of the second-batch slurry after the first batch's casting are expressed in Fig. 8 by a three-dimensional image. The slurry which has fallen in the back of the core at the beginning flows as it surrounds both sides of the core. First the slurry flows downward and then moves upward in the near side of the core due to surrounding flow collision. Such qualitative flow characteristics are depicted in Fig. 8. In order to realize it, the speed information is interpolated onto the lattice points in the container from the position and the speed data at the 2114 traced positions of all sampled timings. Therefore, the speed information at each lattice point is an average of total time period of the measurement.

To select the size of the lead ball tracer, the sedimentation velocity of a lead ball placed in the simulated propellant has been measured. By this measurement it is proved that the sedimentation amount over the whole casting time can be suppressed below the ball diameter. The validity of using a lead ball for a tracer has thus been confirmed.

Because the extraction and the chase of particle are done by the visual observation, the precision of the center of gravity coordinates of particle is the unit of the pixel. Therefore, the center of gravity coordinates contains the error of about 2 pixels, i.e., about 1 mm. Due to the moderate number of tracer particles, the chase of particle in the time-wise direction has been successfully done. Because the number of the particle poured at the same time is at most five, the pair matching procedure has been almost correctly conducted. However, it will be wanted in the future to increase the number of tracer particles in order to visualize a detailed flow field. For this purpose, it is thought necessary that the extraction and the chase of particle will be automated as well as be conducted with higher degree of accuracy.

Because a simple transformation is used in the three-dimensional restoration, the positioning error becomes large. The error which is estimated from the tracer position this time is about 5 diameters at the edges away from the center of X-ray axis. In the future, a technique for the high-precision three-dimensional coordinate computation must be developed.

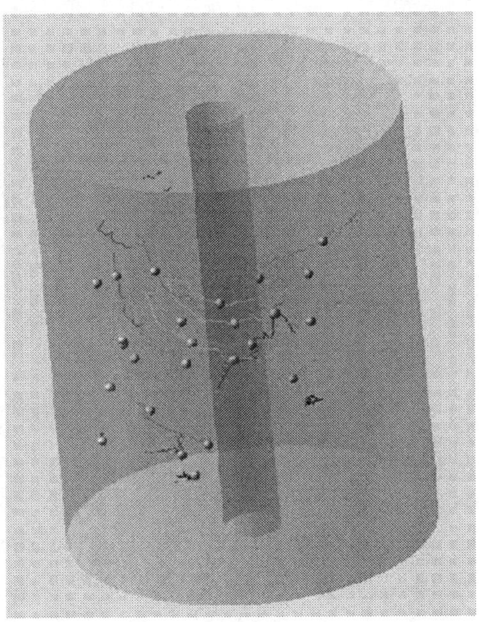

FIGURE 7. Snapshot of 3D track of tracer particles

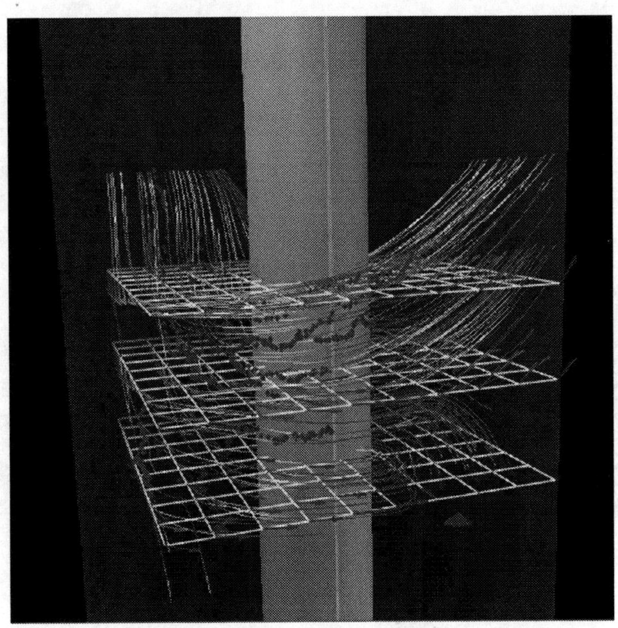

FIGURE 8. Velocity and particle paths of the second-batch slurry flow

CONCLUSION

In this paper it has been shown that it is possible to visualize and measure a three-dimensional non-stationary flow in casting of simulated-propellant slurry by the use of bi-directional real-time X-ray penetration measurements and digital image analysis. There are several technical issues recognized during the course of this study. As future studies, development of a real-time X-ray image acquirement of tracer motion, that of automated particle identification with high accuracy, and that of the technique of a high-precision three-dimensional reconstruction are considered.

ACKNOWLEDGEMENTS

This research is supported by JAXA Grant U4-4K2-USRPJ-1003.

REFERENCES

1 C.W. Beckman and R.L. Geisler, "Ballistic Anomaly Trends in Subscale Solid Rocket Motors", AIAA-82-1092, AIAA/SAE/ASME 18th Joint Propulsion Conference, Cleveland, Ohio, June 1982.

2 D. Ribéreau, P. Le Breton, and E. Giraud, "Solid Rocket Motor Three-Dimensional Surface Burnback Computation Using Mixes Stratification Deduced from Three-Dimensional Grain Filling Simulation", AIAA Paper 99-2802, 35th AIAA/SAE/ASME/ASEE Joint Propulsion Conference, Los Angeles, CA, June 1999.

3 M. Ishikawa, H. Miyachi, T. Shimada, H. Habu, S. Ooya, and Y. Seike, Y., "Flow visualization of slurry fluid using two-directional X-ray photograph" in Proceedings of Kobe Visualization Symposium 2006, Kobe, Japan, Sept. 2006.

Measurement of Gas Holdup Profiles in A Gas Liquid Cocurrent Bubble Column Using Electrical Resistance Tomography

Haibo Jin*, Suohe Yang*, Mi Wang+ and R. A. Williams+

*Department of Chemical Engineering, Beijing Institute of Petrochemical Technology, Beijing 102617, P. R. China. Email:jinhaibo@bipt.edu.cn.
+Institute of Particle Science & Engineering, School of Process, Environmental and Materials Engineering, University of Leeds, Leeds LS2 9JT, UK. Email: m.wang@leeds.ac.uk.

Abstract. Radial variation of the gas holdup and mean holdups were investigated in 0.160m i.d. bubble column using electrical resistance tomography with two axial locations (Plane 1 and Plane 2). In all the experiments tap water was liquid phase and air was gas phase. Superficial gas velocity was varied from 0.02 to 0.25 m/s, and the liquid velocity varied from 0 to 0.01 m/s. The effect of liquid velocity on the mean holdups and radial gas holdup distribution was discussed. The experimental results showed the liquid velocity influence slightly the mean holdup and radial hold-ups distribution in the operating condition, and the liquid flow can improve the trasition gas velocity for homogeneous regime to heterogeneous regime. Meanwhile the average column gas holdups as a function of gas velocity derived from differential pressure method and electrical resistance tomography method. The agreement between results obtained by two methods is generally very good in homogeneous regime. But in transition regime and heterogeneous regime, results with ERT are slightly larger than one with the differential pressure method. According to the experimental results, a correlation for the centreline holdup is obtained.

Keywords: gas liquid cocurrent bubble column, ERT, radial gas holdup profile

INTRODUTION

Bubble columns are widely used as gas-liquid-solid contacting devices due to the ease of their operation and ability to enable good mixing and high heat transfer rates in a controlled manner [1]. During the past few years, special attention has been paid to local and global bubble flow characteristics as a key parameter that determines the overall reactor performance owing to the complexity of fluid mechanics in these systems [2, 3]. It is essential to develop qualities as well as quantitative understanding of the axial and radial holdup distributions in bubble columns [4].

In recent years, applications of process tomography as a robust non-invasive tool for direct analysis of the characteristics of multiphase flows have increased in number [5]. Some tomographic techniques have been applied to measure the axial and radial gas holdup distributions in bubble columns, e.g. gamma-ray computed tomography for measuring cross-sectional gas holdup distributions in bubble columns, gamma-densitometry tomography combined with electrical impedance tomography for measuring gas and solid holdup distributions in gas liquid solid three phase bubble columns, and multi-modal computed tomography for measuring simultaneously cross-sectional gas and solid holdups in three-phase bubble columns [6]. However, these data were captured in time-averaged mode therefore, all the tomographic techniques presented in these reports were too slow to capture the real-time image of highly fluctuating multiphase flows and eventually, some important information could be lost. Owing to its non-intrusive and fast characters, electrical resistance tomography might provide very useful tools for measuring and monitoring bubble column operation on line. The speed of the technique to capture real-time data of highly fluctuating flow systems may be one of the most important concerns for the purpose of imaging multiphase flows in process industries. In this regard, electrical resistance tomography is considered to be the most powerful tool among other available tomographic techniques due to its high-speed capability, low construction cost, high safety and suitability for small or large vessels.

CP914, *Multiphase Flow: The Ultimate Measurement Challenge,* edited by X. Cai, Y. Wu, Z. Huang, S. Wang, and M. Wang

Electrical resistance tomography (ERT) has been used in various industrial investigations for visualisation of the concentration profiles and characteristics of the fluid dynamics in gas-liquid two-phase systems [7-9]. The aim of this study is to determine the gas holdup profiled in a gas liquid cocurrent bubble clolumn size by the electrical resistance tomography. Based on results of the overall gas holdup in the cocurrent gas liquid operation, the effect of liquid velocity on the gas liquid flow regime is discussed in this paper, and a correlation for the centreline holdup is obtained.

EXPERIMENTAL

Experimental Setup

The experimental setup was a transparent Perspex acrylic column with 2.5 m heigh and an inner diameter of 0.160 m operated in gas liquid cocurrent mode as shown in Figure 1. A thermometer was used to provide a continuous monitoring of the water temperature. Two differential pressure sensors for measuring the differential pressure drop were placed along the column at 0.16 m and 0.42 m above the distributor. The differential pressure sampling frequency was 100 Hz. Two neighboring pressure-sampling ports were connected to the high end and low end of each different-pressure transducer, respectively.

Two rings of ERT sensors, each composed of 16 rectangular electrodes, were mounted in the inner wall of the column in a non-invasive fashion. The electrodes were made of stainless steel with a contact area of 8 mm (w) by 16 mm (h). The bottom ERT sensor ring was located 0.30m bove the air distributor. The details of the column configuration are shown in Figure 1. The column was equipped with a perforated plate distributor of 0.003 in thickness consisting of 55 holes in 1mm diameter, which was arranged with an equilateral triangular pitch of 14mm. A commercially electrical resistance tomography system was used for data collection. The data collection rate was 2 frames per second with an excitation signal frequency of 9.6 kHz. No conductivity-temperature compensation was applied since the water temperature was monitored and maintained as about 22 °C throughout the experiment.

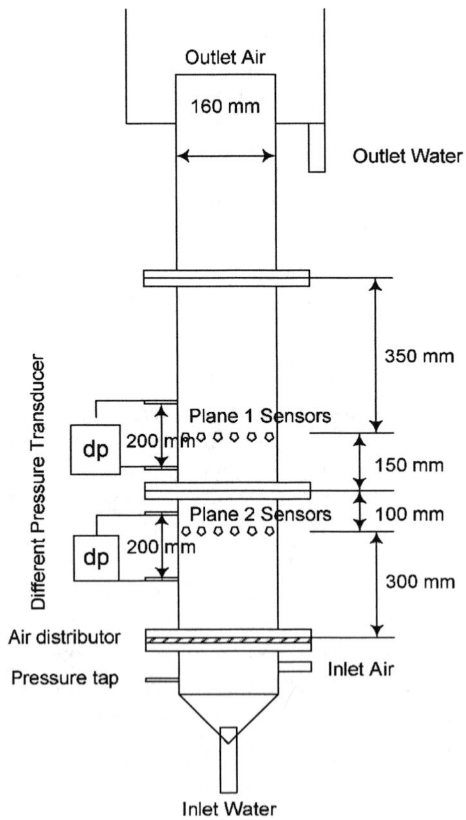

FIGURE 1. Schematic diagram of bubble column

Experimental Systems And Operating Conditions

All experiments were carried out at a ambient temperature of approximately 22°C. Tap-water was used as the liquid phase, and air, which were measured by two flow-meter, was introduced into the bubble column. The superficial gas velocity varied from 0.02 m/s to 0.25 m/s and the liquid velocity varied from 0 to 0.011 m/s. The water temperature was maintained under ambient conditions, at about 22.0°C.

Measurements and Estimation of Hold-up Profile

The adjacent electrode pair strategy was adopted using 10 mA injection current at 9.6 kHz for parameter measurement. Data collection rates were 50 ms, with time for image reconstruction and other overheads, yielding approximately 2 images per second.

With conductivity data from ERT, the local gas volume fraction (ε_g) can be determined by applying the Maxwell equation [10]:

$$\varepsilon_g = \frac{2\sigma_1 + \sigma_2 - 2\sigma_{mc} - \sigma_{mc}\sigma_2/\sigma_1}{\sigma_{mc} - \sigma_2/\sigma_1 + 2(\sigma_1 - \sigma_2)} \tag{1}$$

where σ_1 is the conductivity of the first phase, σ_2 is the conductivity of the second phase, and σ_{mc} is the local value of mixture conductivity distribution.

If the second phase is assumed to be non-conductive material, such as air in this study, the above equation can be simplified as follows:

$$\varepsilon_g = \frac{2\sigma_1 - 2\sigma_{mc}}{2\sigma_1 + \sigma_{mc}} \tag{2}$$

the local mixture conductivity (σ_{mc}) is determined from the pixel conductivity of ERT image.

The axial hold-up can be measured from the two differential pressure signals, the calculated equation showed as follow [11].

$$\varepsilon_g = \frac{\Delta P}{\rho_l g \Delta H} \tag{3}$$

where ΔP is the differential pressure between two pressure sensor points, ΔH the vertical distance between two pressure sensor points, and ρ_l liquid density.

RESULTS AND DISCUSSION

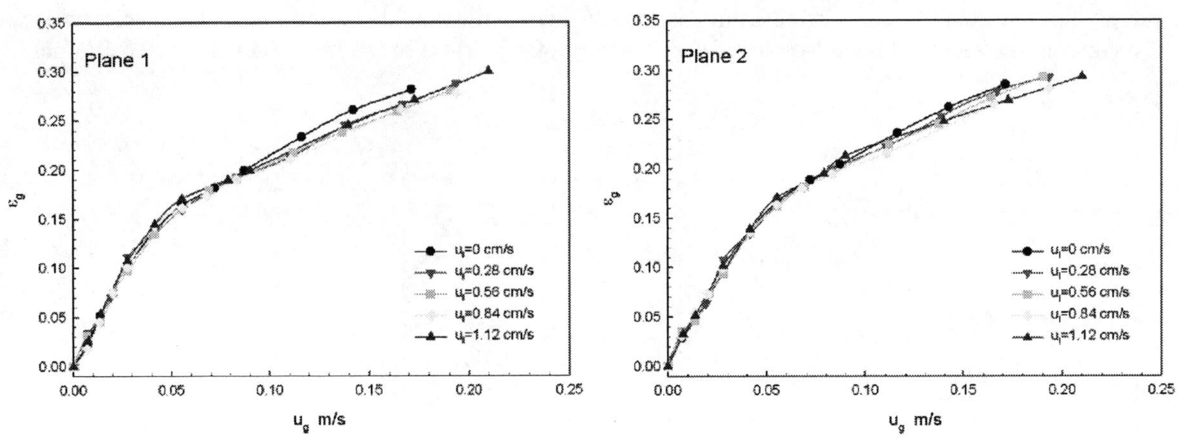

FIGURE 2. The average gas hold-ups as a function of gas velocity at different liquid velocity

Figure 2 shows the relationship between average gas holdups and gas velocity using electrical resistance tomography method with water/air in gas liquid cocurrent operation. The agreement between results obtained in different operating condition is generally very good. But there is a little error in hetergeneous regime, namely the overall holdups slightly decrease with an increase of liquid velocity, which is similar to information reported in the literature [1]. The reason can be interpreted that the rising velocity of gas bubble increases with increasing the liquid velocity in cocurrent operation, and results in reducing the resistance time of gas bubble in the column and the overall holdup.

The homogeneous regime is characterised by the linearity of the gas holdup versus gas velocity at lower gas velocity, and the fully developed heterogeneous regime is observed at higher gas velocity from Figure 2. But the transition regime can not be obtained directly from the relationship between the gas holdup and gas velocity, such as Figure 2.

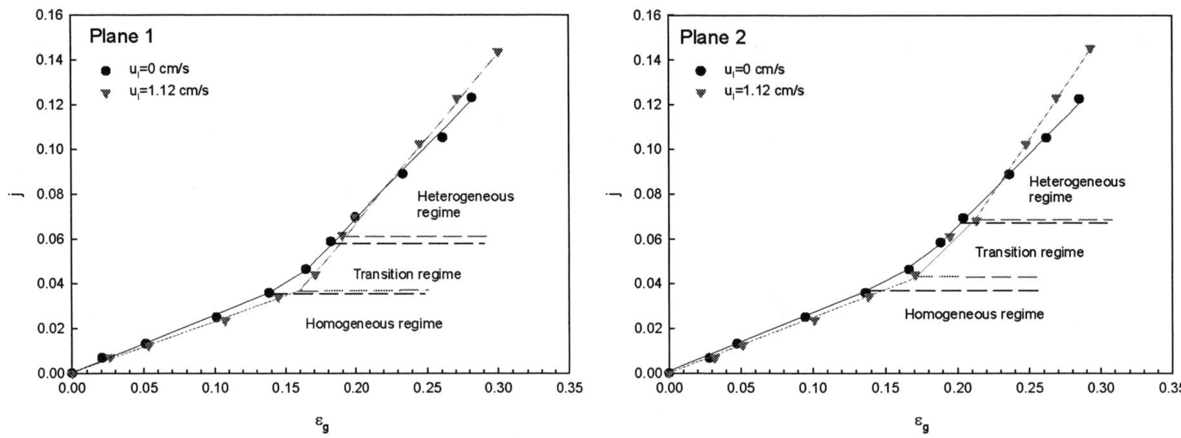

FIGURE 3. The effect of liquid velocity on flow regime

Numerous studies and methods have given the indentification of the flow regime transition [12]. The classical method like drift-flux model is a good technique to analysis to the flow regime. In this paper, the regimes and transition regimes were determined using the classical methods with data of electrical resistance tomography. The drift flux can be expressed as follow [13]:

$$j = \varepsilon_g (1-\varepsilon_g)(\frac{u_g}{\varepsilon_g} - \frac{u_l}{\varepsilon_l}) \tag{4}$$

The relationship between drift flux (j) and hold-ups (ε_g) was shown in Figure 3. From Figure 3, we can see that the liquid velocity slightly influences the gas-liquid flow regime, such as transition gas velocity point. The transition gas velocity at a certain liquid vecolity condition is larger than one at batch operation. This reason can be interpreted that the liquid velocity affects the gas liquid flow stability. Futhermore when liquid velocity is larger than 10 cms^{-1} in reference [1], there is a significant effect on gas hold-up in bubble columns with directional liquid flow.

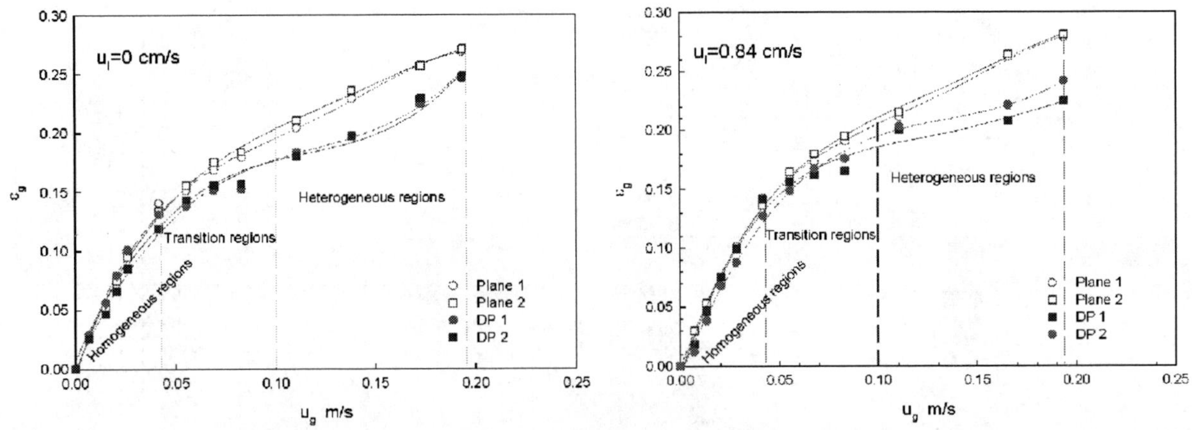

FIGURE 4. Comparison of average gas hold-ups using two measurement methods

The average column gas hold-ups as a function of gas velocity obtained from differential pressure method and electrical resistance tomography method with water/air in the gas liquid cocurrent operation are shown in Figure 4. The agreement between results obtained by two methods is generally very good in homogeneous regime. But in transition regime and heterogeneous regime, results with ERT are slightly larger than one with the differential pressure method. Moreover, the errors increase with an increase of gas velocity. The reason for this discrepancy is due to the effect of gas liquid churn-turbulent flow on the conductivity of mixture, and furthermore affect on the values of local hold-up. However, the local hold-up obtained by the differential pressure method is unaffected by gas liquid flow. The electrical resistance tomography has verified the validity and effectiveness of the measurement technique for bubbly flow regime.

Radial holdup profiles for different liquid velocity at two axial locations as mentioned above are shown in Figure 5. It is seen from the Figure 5 that the radial gas holdup increases with an increase of gas velocity. Meanwhile, the radial gas holdup profiles are steeper at the central region of the column with an increase of gas velocity, namely the higher gas hold-ups at the center of column and lower gas hold-ups at the wall of column. Meanwhile the gas liquid belongs to the homogenous regime at u_g=0.015 and u_g=0.041 m/s. In this range, there is a narrow bubble size distribution, which results in a radially uniform gas holups. With increasing gas velocity, bubble coalescence and breakup phenomena can be neglected, which a wide bubble size distribution and a marked radial gas holdup profiles have appeared. But the liquid velocity at such lower range influences slightly the gas holdup distribution from Figure 5.

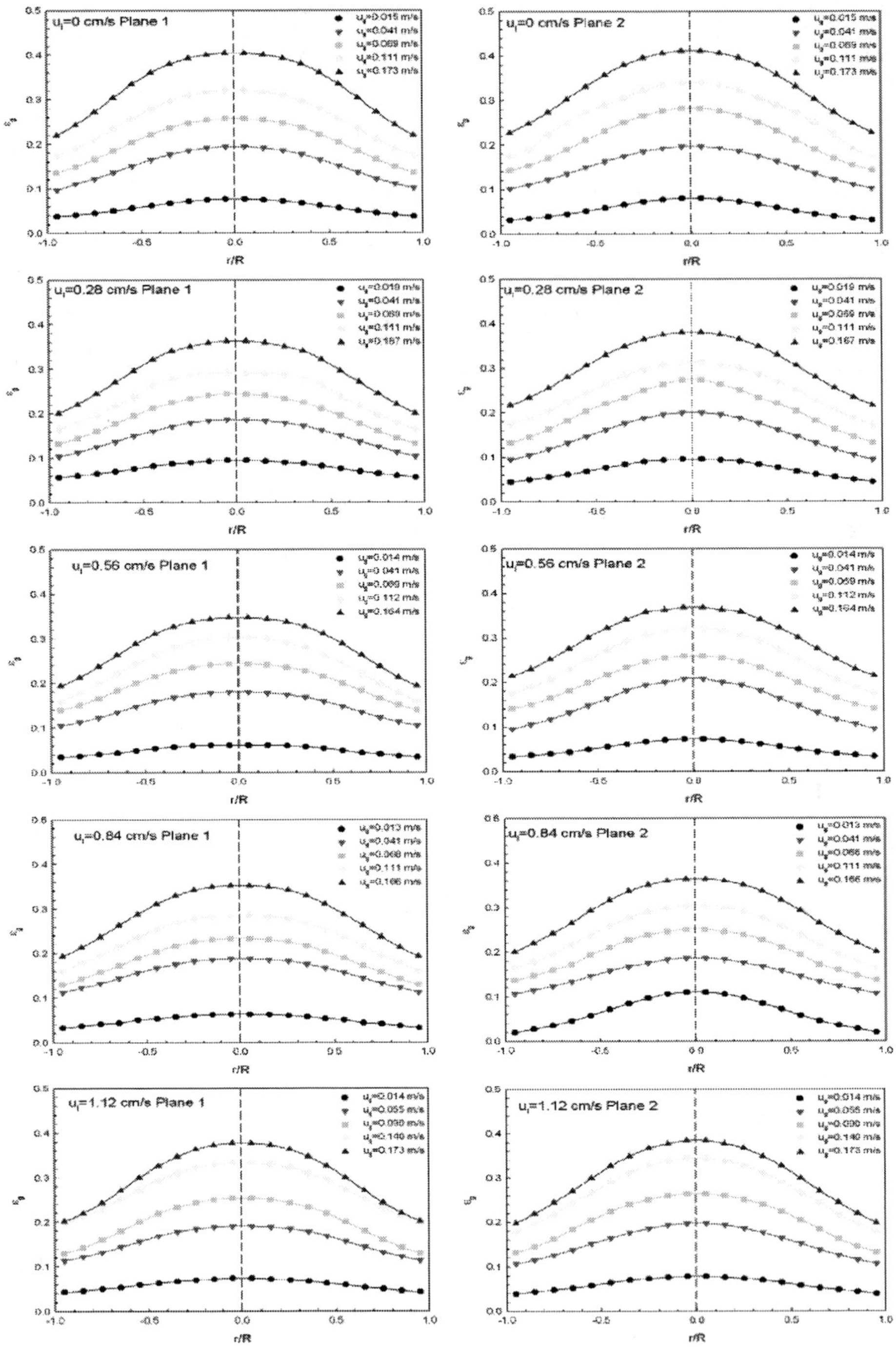

FIGURE 5. The radial gas holdup profiles at different liquid velocity

According to these results, we can obtain that the lower liquid velocity cannot effect the radial gas holdup distribution. From Figure 5, a correlation is developed for the centreline holdup with respect to gas velocity (u_g) and height to diameter ratio (H/D) as shown Eq. (5) and Figure 6 shows comparison calculated $\varepsilon_{g,\ center}$ values with experimental $\varepsilon_{g,\ center}$ values. The centreline holdup increases with an increase of gas velocity, however, it slightly decreases with an increase in distance from sparger.

$$\varepsilon_{g,center} = 1.042 u_g^{0.523} \left(\frac{H}{D} \right)^{-0.096} \tag{5}$$

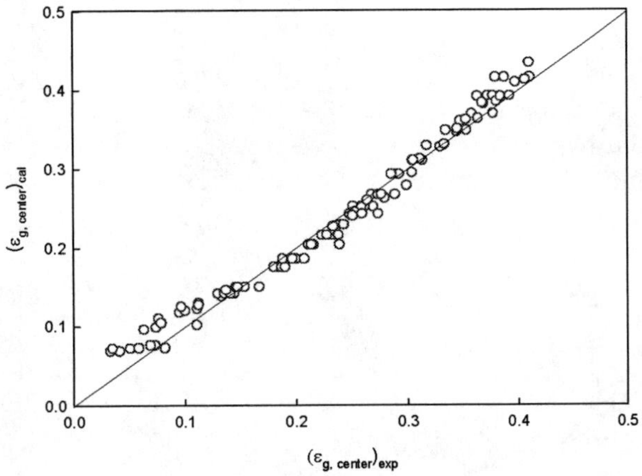

FIGURE 6. Comparison calculated $\varepsilon_{g,\ center}$ values with experimental $\varepsilon_{g,\ center}$ values

CONCLUSIONS

Electrical resistance tomography measurements are used to study the overall and radial holdup in a gas liquid cocurrent bubble column. The effect of liquid velocity on the mean holdups and radial gas holdups distribution was discussed. The experimental results showed the liquid velocity influence slightly the mean holdup and radial hold-ups distribution in the operating condition, and the liquid flow improve the transition gas velocity for homogeneous regime to heterogeneous regime. Meanwhile the average column gas hold-ups as a function of gas velocity obtained from differential pressure method and electrical resistance tomography method. The agreement between results derived by two methods is generally very good in homogeneous regime. But in transition regime and heterogeneous regime, results with ERT are slightly larger than one with the differential pressure method. So the electrical resistance tomography has verified the validity and effectiveness of the measurement technique for bubbly flow regime in this operating condition.

REFERENCES

1. W. D. Deckwer, *Bubble Column Reactors,* New York, John Wiley and Sons, 1991.
2. M. P. Dukovic, *Oil & Gas Sci. and Technology* 55,135-158(2000).
3. L. S. Fan, *Gas-Liquid-Solid Fluidization Engineering,* Boston, Buttorworth, 1989.
4. H. Jin, M. Wang and R. A. Williams, *Chinese J. Chem. Eng.* 14, 127-131(2006).
5. R. A. Williams and M. S. Beck, *Process Tomography Priciples, Techniques and applications*, Oxford, UK: Butterworth-Heinemann, 1995.
6. W. Warsito and L S. Fan, *Chem. Eng. Sci.* 56, 6455-6462(2001).
7. M. Wang, R. Mann and F. J. Dickin, *Chem. Eng. Comm.* 175, 49-70(1999).
8. E. Fransolet, M. Crine, P. Marchot and D. Toye, *Chem. Eng. Sci.* 60, 6118–6123(2005).
9. E. Fransolet, M. Crine, G. L'Homme, D. Toye, *Meas. Sci. Technol.* 21, 1055–1060(2001).
10. J. C. Maxwell, *A Treatise on Electricity and Magnetism*, Clarendon Press, Oxford, 1873.
11. H. Jin, S. Yang, T. Zhang and Z. Tong, *Chem. Eng. Technol.* 27(9), 1007-1013(2004).
12. C. Vial, S. pncin, G. Wild, and N. Midoux, *Chem. Eng. Process.* 40, 135-151(2001).
13. R. C. Darton and D. Harrison, *Chem. Eng. Sci.* 30, 581-586(1975).

NOMENCLATURE

D	Diameter of the column, m
G	Gravitational constant, m/s^2
H	Height of measurement location from the sparger, m
$\triangle H$	The vertical distance between two pressure sensor points, m

$\triangle P$	The differential pressure between two pressure sensor points, Pa,
u_g	Gas superficial velocity, m/s
u_l	Liquid phase velocity, cm/s

Greek Letters

ε_g	Gas hold-up
$\varepsilon_{g, center}$	The centreline gas hold-up
σ_1	The conductivity of the first phase
σ_2	The conductivity of the second phase
σ_{mc}	The local mixture conductivity
ρ_L	The liquid density, kg/m^3

Gas-Liquid Two-Phase Flows Inverse Analysis in An Aeration Tank

Wen CHENG1,2, Zhang-bin WANG 2, Yuichi MURAI 3, Fujio YAMAMOTO4

1 State Key Laboratory of Multiphase Flow in Power Engineerig; Xi'ann Jiaotong University, Xi'an 710049, China, Email wencheng@xaut.edu.cn
2Environment Research Institute, Xian University of Technology, Xian 710048, China
3Div. of Mec. Science, Graduate School of Engineering, Hokkaido University, Japan, murai@eng.hokudai.ac.jp
4Graduate School of Engineering, University of Fukui, Bunkyo 3-9-1, Fukui 910-8507, Japan, yamamoto@fv.mech.fukui-u.ac.jp

Abstract. In bubbly two-phase flow, the gas phase and liquid phase have different flow fields. The mathematical expression for the motion of a small bubble at low Re number has been already established, the liquid velocity along the trajectory of the bubble is calculated inversely by using the motion equation. Whole field liquid flow structure has also estimated using a spatial and/or temporal interpolation method. This paper proposes an algorithm for estimating the liquid phase flow field based on measurement data on bubble data on bubble motion. The applicability of the algorithm is examined with Taylor-Green vortex flow as an analytical test case. Meanwhile, The result, based on the theory, has been applied to reconstruct liquid phase velocity field by using the data of bubble velocity in an aeration tank.

Keywords: Gas-liquid two-phase flows; Particle Image Velocimetry; inverse analysis; aeration tank

INTRODUCTION

Gas-liquid two-phase flows are widely applied in engineering and environmental processes. However, so far there have been many problems for human beings to know clearly, such as overfall, nonsteady flow, mechanism of interaction, heat and mass transfer between the two phases and so on. Particle image velocimetry (PIV) and particle tracking velocimetry (PTV) have rapidly advanced and spread in last decade [1]. The method can be used to measure time-dependent multi-dimensional bubble velocity vectors, so it is expectable that the carrier phase flow is reconstructed with certain accuracy. So they have been applied to disperse multi-phase flows in order to understand the physics of dispersion motion and the interaction between two phases. Through the theoretical research on bubble dynamics, the motion equation for a low Re number spherical bubble (Re<200) has been established nowadays. Using the equation, the liquid flow can be estimated on the data of bubble motion in principle [2]. This paper concerns such an algorithm based on recent knowledge of bubble dynamics and the application of the technique of PIV in an aeration tank.

The motion equation for a bubble ordinarily consists of seven force components, i.e., inertia, added inertia, drag, lift, surface, gravity, and history forces. Among these components, lift force is the one of unclear components. Sridhar-Katz (1995) measured the lift force acting on a bubble using three-step PTV. Kitagawa et al. (2002) measured a bubble-bubble interaction force in wall-sliding bubbles based on the PTV data of bubble motions. On one hand, their researches based on PIV and/or PTV provided useful information for improving the CFD model. On the other hand, the interpolation schemes for discrete data of velocity vectors were proposed recently (Ido, 2002 and Murai, 2002). Such the post-processing schemes are also applicable to the reconstruction of carrier phase in bubbly flow.

The motivation of developing the method is to test an algorithm to obtain the carrier phase velocity under the conditions that tracer particles cannot be used, or that tracer particles may distort the velocity of the carrier phase. Hoping the method can give a theoretic reference.

CP914, *Multiphase Flow: The Ultimate Measurement Challenge,* edited by X. Cai, Y. Wu, Z. Huang, S. Wang, and M. Wang

BASIC EQUATION FOR INVERSE ANALYSIS

Translation Motion Equation Of Single Bubble

The equation of motion for a sphere in an unsteady viscous fluid is expressed as follows:

$$\vec{F}_g+\vec{F}_i+\vec{F}_d+\vec{F}_a+\vec{F}_s+\vec{F}_l+\vec{F}_h=0 \quad (1)$$

This equation consists of gravity force $\vec{F}_g$, inertia force $\vec{F}_i$, drag force $\vec{F}_d$, surface force $\vec{F}_s$, added inertia force $\vec{F}_a$, lift force $\vec{F}_l$, and history force $\vec{F}_h$. Mathematical formulation of each force component has been nowadays well established for a low Re number spherical bubble in incompressible fluid. When the detailed expression for each force is substituted [1], Eq. (2) becomes.

$$\rho_g V_g g-\rho_g V_g \frac{dv}{dt}-C_D\frac{\pi r_g^2}{2}\rho_l|v-u|(v-u)+C_M\rho_l V_g\left(\frac{Du}{Dt}-\frac{Dv}{Dt}\right)+\rho_l V_g\left(\frac{Du}{Dt}-g\right)$$
$$-C_L\rho_l V_g(v-u)\times\omega=0 \quad (2)$$

Where, r, u, v and g stand for density, liquid velocity vector, bubble velocity vector and acceleration of gravity, respectively. Subscripts g and l stand for gas and liquid phases, respectively. Vg stands for volume of a bubble. The surface force in Eq. (4) equals to the force due to pressure gradient and viscosity in liquid replaced by Navier-Stokes equation. The added inertia force in Eq. (5) is expressed by Lagrangian acceleration components for two phases with added mass coefficient X_M. C_D and rg in Eq. (6) are drag coefficient and radius of bubble, respectively. The lift force in Eq. (7) is inertia lift force derived by Auton [7] with a lift coefficient C_L. And ω ($\omega=\nabla\times v$)stand for fluid vorticity. History force is ignored in this study because it has a form of temporal integration that makes inverse analysis impossible. Nevertheless, it is actually negligible for a clean surface bubble at Re > 50.

Denoting the density ratio of gas to liquid by γ, Eq. (9) is simply expressed by

$$u=v-\left[(C_M+1)\frac{Du}{Dt}-(C_M+\gamma)\frac{Dv}{Dt}-C_L(v-u)\times\omega+(\gamma-1)g\right]\times\left[C_D\frac{3}{8r_g}|v-u|\right]^{-1} \quad (3)$$

At the process of the Eq. (3), we found that the inverse method is only be used to calculate the liquid-phase velocity field. When four qualifications following are satisfied as following [1]

1) The air bubble is approximately spherical.
2) No self-excited behavior of the bubble such as zigzag or spiral motion emerges, when Re<200. Also, the Reynolds number should be higher than 50 so that one can ignore the history force.
3) The interaction force of bubbles can be ignored for interstice rate between bubbles are low.
4) The size of typical flow liquid is bigger than the bubble.

The Inverse Analysis Equation

Providing that use (u_g,v_g), (u_l,v_l) and $\frac{Dv}{Dt}=a_g$ replace bubble velocity(v), liquid-phase velocity(u), and($\frac{du_g}{dt},\frac{dv_g}{dt}$)in Eq.(3).

$$u_l=u_g-\left[(C_M+1)\left(\frac{\partial u_l}{\partial x}u_g+\frac{\partial u_l}{\partial y}v_g\right)-(C_M+\gamma)\frac{du_g}{dt}-C_L(v_g-v_l)\left(\frac{\partial v_l}{\partial x}-\frac{\partial u_l}{\partial y}\right)+(\gamma-1)g_y\right]$$
$$\times\left[C_D\frac{3}{8r_g}|v-u|\right]^{-1} \quad (4)$$
$$v_l=v_g-\left[(C_M+1)\left(\frac{\partial v_l}{\partial x}u_g+\frac{\partial v_l}{\partial y}v_g\right)-(C_M+\gamma)\frac{dv_g}{dt}+C_L(u_g-u_l)\left(\frac{\partial v_l}{\partial x}-\frac{\partial u_l}{\partial y}\right)+(\gamma-1)g_y\right]$$

$$\times\left[C_D \frac{3}{8r_g}|v-u|\right]^{-1} \quad (5)$$

Bubble velocity at anytime in Eq.(5) can be calculated by using the PIV program. So This corresponds liquid phase speed can also be calculated by the method of iterative using the Eqs.(4) and (5). If lift coefficient $C_L = 0$ in Eq.(5),the liquid-phase velocity of the air bubble position is very easy calculated according to the eddy motion result. If lift coefficient $C_L \neq 0$, the liquid-phase velocity of the air bubble position must be calculated using the LER method.

The carrier phase field is calculated by LER (Laplace Equation rearrangement). The LER method rearranges scattered data on velocity vectors on a regular grid system using the Laplace equation. The measurement area of the objective flow field is divided into small Cartesian coordinate grid, and the discrete velocity vectors obtained by PIV are initially arranged to the nearest grid point by parallel shifting. Refining the grid division can be reduce the error due to the parallel shifting. The basic equation of LER is

$$\nabla^2 \mathrm{u} = 0 \quad (6)$$

The original data on velocity vectors are protected from modification using the following switching function:

$$u_{i,j}^{**} = u_{i,j}^{*} * \alpha_{i,j} + u_{i,j}(1-\alpha_{i,j})$$

$$v_{i,j}^{**} = v_{i,j}^{*} * \alpha_{i,j} + v_{i,j}(1-\alpha_{i,j}) \quad (7)$$

Where α is a switching parameter which take a value of 0 at PIV data points, and a value of 1 at the other points. $u_{i,j}^{*}$ and $v_{i,j}^{*}$ are velocity vectors which are to be interpolated, while $u_{i,j}^{*}$ and $v_{i,j}^{*}$ are the rearranged velocity vectors. the carrier phase field are calculated by using Eq. (7).

Parameters Of The Eq.(12): C_D, C_M and C_L

Drag Coefficient C_D

The value of drag coefficient C_D is changed with respect to Reynolds number (Re)

If $\mathrm{Re} << 1$, drag coefficient C_D can be express by

$$C_D = \frac{16}{\mathrm{Re}}\left(\frac{1+(3\mu_p)/(2\mu_f)}{1+\mu_p/\mu_f}\right) \quad (8)$$

The " μ_f " here is hydrodynamics viscosity coefficient. The value of Shear stress between superficial activation bubble and liquid equals to zero. So $\mu_f >> \mu_p \approx 0$, drag coefficient C_D is given by $C_D = 16/\mathrm{Re}$.

If $1 < \mathrm{Re} < 60$, drag coefficient C_D can be established by Mei [4]

$$C_D = \frac{16}{\mathrm{Re}}\left\{1 + 0.15\,\mathrm{Re}^{0.5}\right\} \quad (9)$$

If $\mathrm{Re} > 60$, drag coefficient C_D can be established by Magnaudet [5]

$$C_D = \frac{48}{\mathrm{Re}}\left(1 - \frac{2.2}{\mathrm{Re}^{1/2}}\right) \quad (10)$$

the curve of drag coefficient C_D with **Re** is given by Fig 1

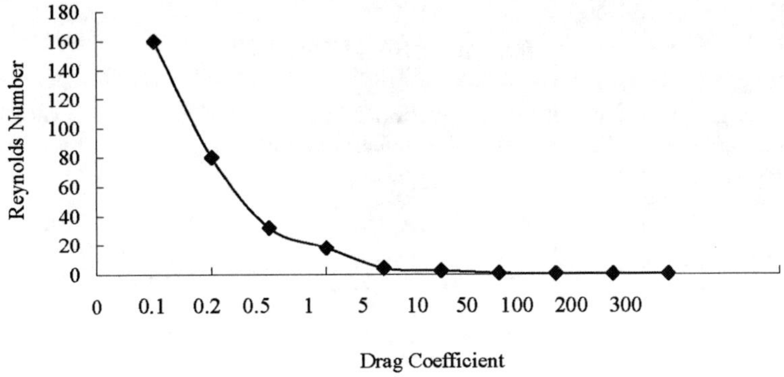

FIGURE. 1: Drag coefficients corresponding to Reynolds

2.3.2 Added Inertia Force Coefficient C_M

Because of the globosity of bubble, added inertia force coefficient C_M has nothing to do with Re and the flow of the carrier-phase. so added inertia force coefficient is a constant [5].

C_M=1/2

Lift Coefficient C_L

So far, the value of lift coefficient C_L has not been confirmed This paper adopt an coefficient that is provided by D.Legendre and J.Magnaudet [6]. They had developed the variety of the lift coefficient with Reynolds number(0.1< Re < 500= and shear rate (Sr) ,and they had provided an empirical formula,

The curve in Fig.2 show the relation among C_L Sr and Re, when shear rate is 0.2 and 0.02,respectively.

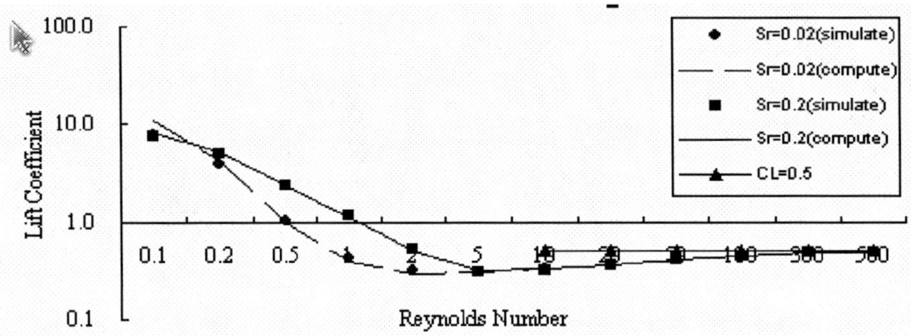

FIGURE 2 Lift coefficients corresponding to Reynolds and shear rate [6]

They had found : (1) lift coefficient will goes down when Re goes up in the range of 0.1<Re≤5. (2) when Re equals to 3, the value of lift coefficient is minimum 0.3. In case of Re = 0.3, both C_L and Sr appear go up, C_L is expressed by empirical formula,

$$\varepsilon = (Sr / \mathrm{Re})^{\frac{1}{2}} \qquad J'(\varepsilon) = \frac{J(\infty)}{(1+0.2\varepsilon^{-2})^{3/2}}$$

$$C_{L(\mathrm{I})}(\mathrm{Re}, Sr) = \frac{6}{\pi^2}(\mathrm{Re}\, Sr)^{-1/2} J'(\varepsilon) \tag{11}$$

The "$J(\infty)$" here equals to 2.255.

In case of Re>5 ,lift coefficient goes up with the raise of Re , and approach quickly 0.5 which is found by Auton [7] . C_L has nothing with Sr in the range of Re>5 。 C_L is given by the Eq.(12):

$$C_{L(\mathrm{II})}(\mathrm{Re}) = \frac{1}{2}\frac{1+16\,\mathrm{Re}^{-1}}{1+29\,\mathrm{Re}^{-1}} \tag{12}$$

In summary, C_L is given by empirical formula in case of 0.1<Re<500

$$C_L(\mathrm{Re}, Sr) = ([C_{L(\mathrm{I})}(\mathrm{Re})]^2 + [C_{L(\mathrm{II})}(\mathrm{Re})]^2)^{1/2} \tag{13}$$

In this paper, the Eq. (13) has been used to calculate lift coefficient for Re in the range (50-200) [8].

Examination Flow Fifld

A vortex flow which is given by the following equation in this section, is used to evaluate the performance of the combination of the inverse analysis and PIV with the post-processing algorithm:

$$u = W \sin\left(\frac{2\pi kx}{L}\right)\cos\left(\frac{2\pi ky}{L}\right)$$

$$v = -W \sin\left(\frac{2\pi ky}{L}\right)\cos\left(\frac{2\pi kx}{L}\right) \tag{14}$$

Here k stand for the wave-number of a vortex existing in measurement size L. W is the amplitude of the velocity. Eq.(14) satisfy simultaneously the equation of the continuity condition and the Navier-Stokes equation, known as an instantaneous analytical solution for the Taylor-Green vortex flow field. This vortex flow is suitable for comprehensively evaluating the performance of the algorithm because it is a periodic flow field in 2D space with nonzero vortices. Fig. 3 (a) gives the Taylor-Green vortex flow field. assuming $W=1$, $k=0.25$, $L=1$.

hows the liquid velocity distribution obtained by the Eulerian-Lagrangian model. Fig. 3 (b) show the rearrangement results for the liquid phase of the combination the inverse analysis method with the LER method .from the figure, It is obvious that the result based on time-series data is better rearranged than those based on instantaneous data. In order to quantitatively compare the results with each other, the cross correlation coefficient, which is defined as the inner product of two functions, is used as an evaluation parameter:

$$C = \frac{\sum(\phi_{th}\cdot\phi_R)}{\sqrt{\sum\phi_{th}^{\;2}\cdot\sum\phi_R^{\;2}}} \tag{15}$$

Where ϕ represents the physical quantities of the original flow, and th that of the rearranged flow which is to be evaluated.. In this paper, the cross correlation coefficient for the velocity has been compared. The correlation coefficient for the velocity is here defined as the average of two values for the two velocity components.

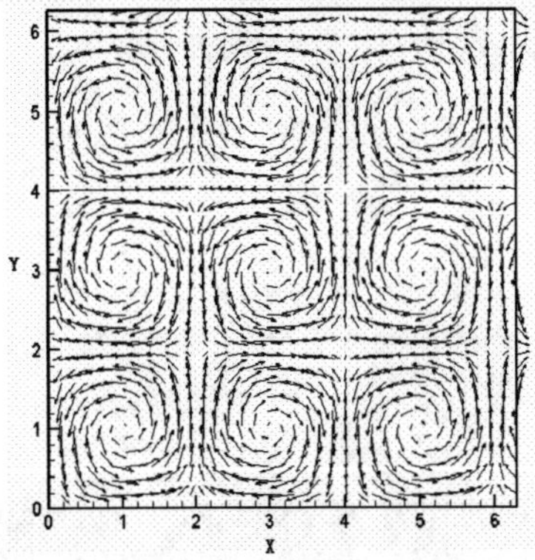

Fig. 3(a): liquid velocity field of the Taylor-Green vortexflow by the E-L model of gas velocity fieldequals 200.

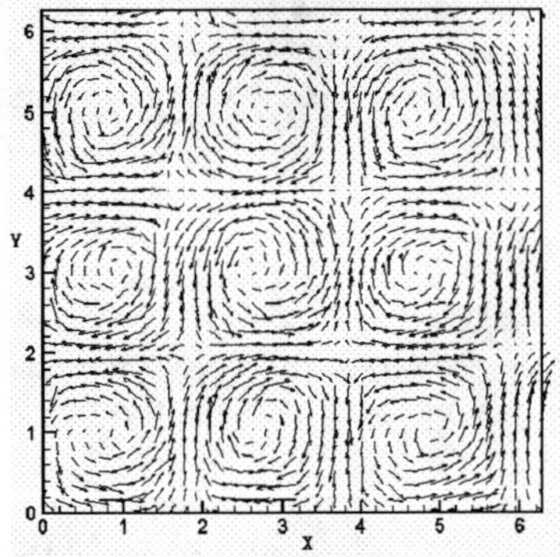

Fig. 3(b): liquid velocity field of the Taylor-Green vortex flow by inverse analysis while n of gas velocity field equals 200

RESULT AND DISCUSSION

The rearrangement performance is quantitatively evaluated by using the coefficient of correlation between the original flow and the inverse analysis of the rearrangement flow. The number of bubble velocity can be obtained by using the Eq.(15). The cross-correlation coefficients between liquid-phase velocity field obtaining the inverse analysis method and real velocity field in the u direction and v direction are shown in figure 4 and 5.

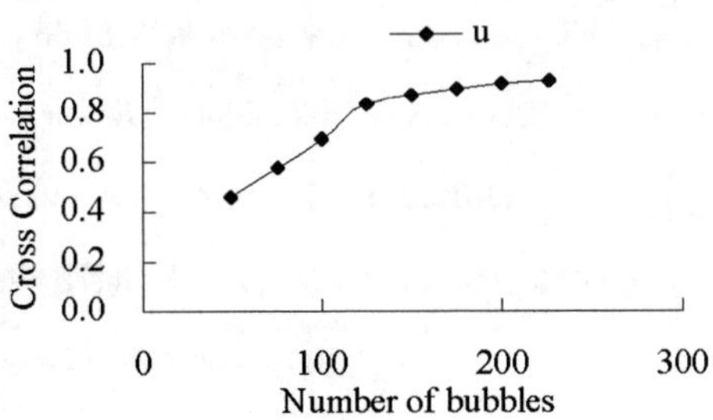

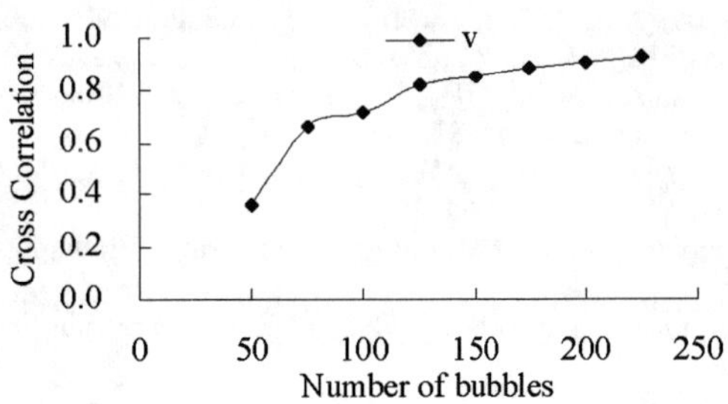

Fig. 4: Correlation coefficient in u direction

corresponding to the number of gas velocity

corresponding to the number of gas velocity

Fig. 5: Correlation coefficient in v direction

Figs.4-5 shows the conclusion as follow: The more increasing of gas velocity we known, the higher effect of inverse analysis we will get. However, the effect will not be evident when the number of gas velocity increase to some extent. If we want to obtain the value of the cross correlation more than 0.9, we must know more than 1/6 number of bubble full-field velocities.

Table 1： Experimental conditions

Liquid: water; Gas: air(ρ :1.28 kg/m3, T: 13-15℃)
Coefficient of kinematics viscosity （υ)： 10^{-6} m2/s，

Case	Q/ ($10^{-5}m^3/s$)	H/m
Case 1	0.99	0.451
Case 2	7.04	0.617
Case 3	8.88	0.31

At the experimental aeration tank, air flux and aspect ratio of the tank were two important factors of affecting the result of the experiment. Three aspect ratio cases were introduced in the aeration tank as following table 1. In case 1 and 2 (here H/B>2, H/B—the ratio between height and width of water), the bubble in the tank had shorter time and weaker turbulent intensity than the case 3(here H/B=1). In case 3, the bubble in the equipment has longest rest time and stronger turbulent intensity. In case 3, the bubble in the equipment has the longest rest time and stronger turbulent intensity, so oxygen transferring speed and efficiency will be advanced greatly.

CONCLUSION

This paper mentioned the inverse analysis approach for estimation the carrier phase flow field from the measurement results for dispersion motion. The following points have been clarified from this study.

（1） Because two parameters (C_D & C_L) of the equation can change with the difference of Re, the more the number of the bubble velocities, the higher value of the cross correlation of inverse analysis we could get. However, when the number of bubble velocity increased to some extent, the effect will not be evident. If we want to obtain the value of the cross correlation more than 0.9, we must know more than 1/6 number of bubble full-field velocities.

（2） In the aeration tank equipment, gas-liquid two-phase flows will behave three flowing cases. When flows are in case 3, the bubbles in the equipment will have longest rest time and stronger turbulent intensity, so oxygen transferring speed and efficiency will be advanced greatly. When we design aeration tank, we must make gas-liquid two-phase flows in case 3.

REFERENCES

1. Wen Cheng, Yuichi Murai, Toshio Sasaki, Fujio Yamamoto, Bubble Velocity Measurement with Recursive Cross Correlation PIV Technique, Fluid Measurement and Instrumentation, 16: 35-46, 2005.
2. Kitagawa A., Ashihara M., Mural Y., Yamamoto F, Bubble-Bubble Interaction Observed in a Swarm of Wall-Sliding Bubbles, Trans. JSME, B, Vol. 67, No. 657, pp. 1111- 1118, 2002.
3. Kiger K T, Pan C, PIV Technique for the Simultaneous Measurement of Dilute Two-Phase Flows, Journal of Fluids Engineering, 122: 811-818, 2000.
4. Mei R, Klausner J F, LAWRENCE C J, A Note on the History Force on a Spherical Bubble at Finite Reynolds number, Phys Fluids, 6: 418-420, 1994.
5. Magnaudet J, Rivero M, Fabre J, Accelerated Flows Past a Rigid Sphere or a Spherical Bubble, J Fluid Mech [J], 284: 97-135, 1995.
6. Legendre D, Magnaudet J, The Lift Force on a Spherical Bubble in a Viscous Linear Shear Flow, J Fluid Mech, 368: 81-126, 1998.
7. Auton T R, The lift force on a spherical body in a rotational flow, J Fluid Mech, 183: 199-218, 1987.
8. Wen Cheng, Ce Song, Xiao-de Zhou, Numerical Simulation and Experimental Study Gas-liquid Two-phases Flow in Aeration Tank, Journal of Hydrodynamics, 12: 32-35, 2001.

Visualization Study of the Effect of the Prolonged Vertical Tube on the Separation Performance of a Cyclone

Qian Fuping *, Zhang Mingyao

*School of Civil Engineering and Architecture, Anhui University of Technology, Ma'anshan 243002, China
+Key Laboratory on Clean Coal Power Generation and Combustion Technology of Ministry of Education, Thermo-energy Engineering Research Institute, Southeast University, Nanjing 210096, China

Abstract. Cyclones have been widely used in different industrial processes. The most widely used experimen techniques for the flow pattern include Pitot tubes; hot-wire anemometers, as well as the non-intrusive techniques such as laser Doppler anemometry (LDA). However, all these available techniques have the limitation of only one-point measurement at a time, by which the whole-field fluid characterization in cyclones cannot be obtained simultaneously. Particle image velocimetry (PIV) is a novel non-intrusive visualization experimental technique to measure the whole-field three-dimensional fluid velocity instantaneously. However, only a few papers that examined the flow pattern in cyclones with PIV have been found. The objective of this work is to carry out an elementary experimental investigation of the effect of the prolonged vertical tube on the separation performance of a cyclone by PIV. Since the tangential velocity results in the centrifugal force for particle separation, and is regarded as the most important component of the gas flow in a cyclone, this paper only presents this velocity component. To facilitate visual observation all cyclone parts are constructed of perspex. The lengths of the prolonged vertical tube (attaching a vertical tube at the bottom of the dust outlet) are 0.30m, 0.40m and 0.50m, respectively. Experiments were conducted at 18m/s inlet gas velocity. The model cyclone is set horizontally in order to take images by charge coupled device (CCD) easily. Particles of about 1 μm in diameter are added as tracer particles. They are generated by the atomizer. Air with tracer particles is sucked into the cyclone through the entry, subsequently travels through outlet tube of the cyclone and the gate valve, and is discharged into the atmosphere. The experimental results show that a typical Rankine vortex with inner quasi-forced vortex and outer quasi-free vortex in the vertical tube and dustbin which are the same as that in the body of the cyclone; the tangential velocity in the dustbin reduces1 greatly when the prolonged vertical tube attaching into the dust outlet, which can avoid the re-entrainment of already separated dust effectively. The results also indicate that PIV technique can be used to measure the strong swirl gas flow such as cyclones, and the prolon。ged vertical tube has practical significance for improving the separation performance of a cyclone. tal observation

Keywords: Cyclone; PIV; Separation performance; Prolonged vertical tube

INTRODUCTION

Cyclones have been regarded as one of the simplest and cheapest type of deduster on account of their high efficiency, adaptability, and relative economy in power. Cyclones have been widely used in different industrial processes. Since the flow pattern plays a very important role in understanding its operating principle and separation performance. The most widely used experimental observation techniques for the flow pattern in cyclones include Pitot tubes [1]; hot-wire anemometers [2], as well as the non-intrusive techniques such as laser Doppler anemometry (LDA)[3-6]. However, all these available techniques have the limitation of only one-point measurement at a time, by which the whole-field fluid characterization in cyclones cannot be obtained simultaneously. Particle image velocimetry (PIV) is a novel non-intrusive visualization experimental technique to measure the whole-field three-dimensional fluid velocity instantaneously. However, only a few papers [7, 8] that examined the flow pattern in cyclones with PIV have been found. The objective of this work is to carry out an experimental investigation of the effect of the prolonged vertical tube on the separation performance of a cyclone by PIV technology.

* Corresponding author. Tel.: +86 555 2311862; fax: +86 555 2311862. *E-mail address:* qfpqd@sohu.com (Qian F. P.).

CP914, *Multiphase Flow: The Ultimate Measurement Challenge,* edited by X. Cai, Y. Wu, Z. Huang, S. Wang, and M. Wang

EXPERIMENTAL SET-UP

Cyclone Used

The geometrical lay-out of the cyclone used in this study is shown in Fig. 1. To facilitate visual observation most cyclone parts were constructed of acrylate (Perspex). The diameter of the cylindrical body of the cyclones is equal to 0.2m, and the diameter and height of the dustbin are also 0.2m. Experiments are conducted at 18m/s inlet gas velocity. Additional characteristic values for the cyclone are presented in Table 1.

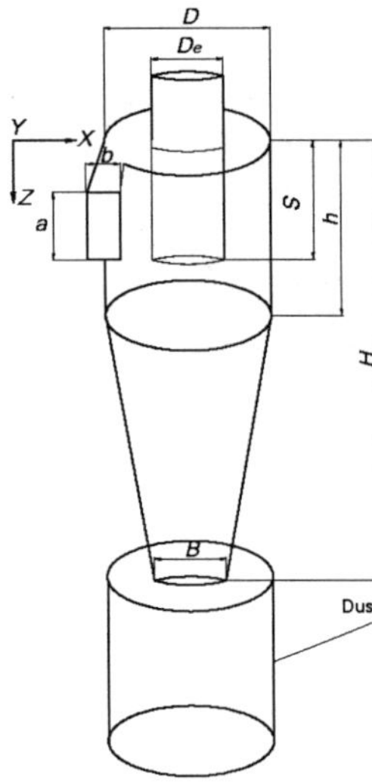

FIGURE 1. Schematic Representation of the Cyclone

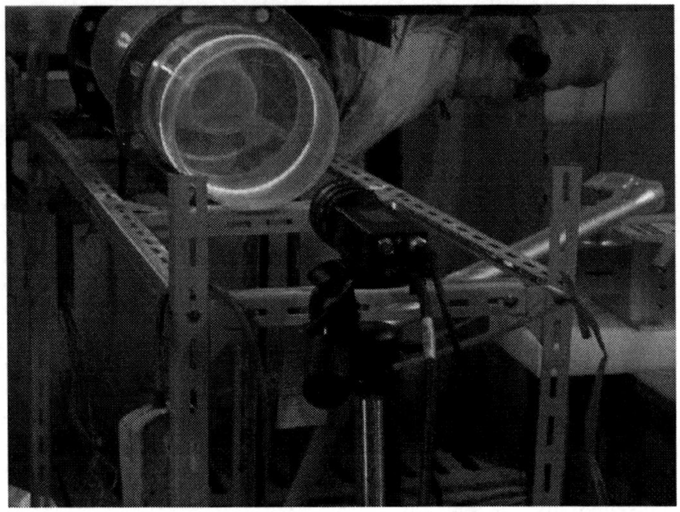

FIGURE 2. Impression of Experimental Set-up Used for PIV Measurements

TABLE 1. Characteristic Values for the Cyclone Used in This Measurement, Unit: m

a	*b*	D_e	*S*	*h*	*H*	*B*	*D*
0.091	0.04	0.065	0.178	0.261	0.756	0.08	0.2

Experimental Set-up

Experiments were carried out in an experimental set-up as shown in Fig. 2. The detailed description of the set-up can be referred to Qian and Zhang [9]. With the negative generated by the blower, air with tracer particles is sucked into the cyclone through the entry, subsequently travels through outlet tube of separator and gate valve, and is the discharged into the atmosphere. The model cyclone is set horizontally in order to take images by charge coupled device (CCD) easily. There is a flow meter which is used to measure the air flow rate in the entry. A gate valve is used to adjust the gas flow rate. Laser beam travels through a set of lens and is changed into a laser sheet to light up the measured flow field. Images of the tracer particles, which carry the information of the field, are recorder by CCD and converted into digital signals, and finally stored in a computer.

The all test conditions of this study are shown in Fig. 3. Owing to the limitation of the experimental condition, this study only presents the flow field of the vertical tube and dustbin. Additionally, since the tangential velocity results in the centrifugal force for particle separation, and is regarded as the most important component of the gas flow in a cyclone. Therefore, this paper only gives this velocity component.

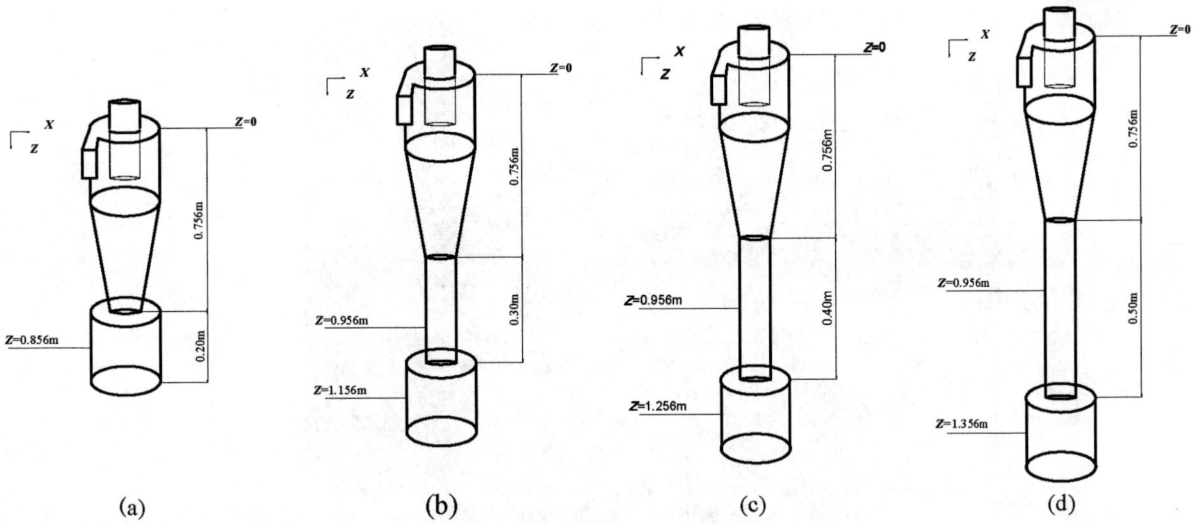

FIGURE 3. Cyclone Geometries and All Test Conditions
(a) Conventional cyclone; (b) The length of the prolonged vertical tube is 0.3m; (c) The length of the prolonged vertical tube is 0.4m; (d) The length of the prolonged vertical tube is0.5m

PIV System

PIV system mainly included a CCD camera, an image-gathering card, synchronizing controls and two pulsed lasers serving as light sources (As shown in Fig. 2). The adopted CCD camera, whose pixel array is 1600*1600; works under the model of two channels and double exposure. The speed of image acquisition amounts to 30 frames per second or 15 measurements per second and the adjustable time interval between every pair of double frame images is below 1μs. The image acquisition card supports the working mode of two channels and the transport rate of images is 40Mb/s. Two ND: YAG pulsed laser as light sources provide up to 50mJ laser pulse at 15Hz. The pulsed laser finally transfers into green light with 532nm wavelength to light up the flow field. A time-delay device can send out delay signal, which serves as the external triggering controlling signal for the xenon lamp and Q switch of the two lasers and CCD camera. The discrepancy of triggering time is less than 10 nanoseconds.

Tracer Particles

In general, liquid flows are naturally seeded with sufficient amounts of scattering particles, whereas for aerodynamic applications additional seeding of the flow is necessary. The choice of tracer particles for PIV measurements is a compromise between their light scattering properties, and their ability to track the flow accurately. It is also important that a seeding system is capable of producing clean particles at sufficient quantities in a cheap and easy way.

The size of the tracer particles has a large effect on the signal quality. A considerable reduction of the Signal-to-Noise-Ratio (SNR) occurs if the particle has a size comparable to or larger than the fringe spacing. Hence, it is important to remove over-sized particles to reduce noise. The scattering power of a particle strongly depends on its diameter. The scattering properties of particles of diameters greater than the wavelength of the incident light obey Mie's theory [10].

The way of seeding in a strong swirling flow field is critical. The difficulty is that the diameter of particles should be controlled to less than 1μm for illustrating the main flow regime. The stronger the swirling flow is, the more difficult to seed the center of the vortex core. Oil drops produced by one kind of particle generators are used in this study. Fig. 4 shows this particle generator (AB-1500), which mainly includes a high-pressure gas flows through valve and filter, and enter into the high-pressure nozzle. The outlet of nozzle is an annulus with gradual tapering and adjustable nozzle area. The wider the area is, the larger the diameter of oil drops. Impacted by the high-pressure stream, the oil changes into granular oil mist with higher internal pressure on the surface of liquid, which is then introduced into the cyclone as seeding.

The test shows that generated mist particles are spherical, distributed uniformly in the flow field, and scatter sufficient amount of light. The average diameter of particles is below 1μm, which can meet the seeding requirement.

FIGURE 4. The generator of the tracer particles

RESULTS AND DISCCUSION

Instantaneous Flow Field

The original image of the tracer particles in some section in the cyclone are shown in Fig. 5(a) and (b). The time interval between (a) and (b) is 10μs. This Figure also gives the velocity and vorticity obtained by MicroVec V2.0 based on the autocorrelation. After obtaining the velocity vector of this section, the tangential and radial velocity can be calculated. Owing to the radial velocity is much lower than the tangential one; hence, this paper doesn't give the radial velocity. Additionally, because of the limitation of the experimental condition, the data of the section parallel to the axial line can't be obtained. Therefore, this paper doesn't present the axial velocity, too. From Fig. 5, the measurement result of the section located in the center of the dustbin and vertical tube is not satisfied, which may be induced by the strongly swirling in these regions.

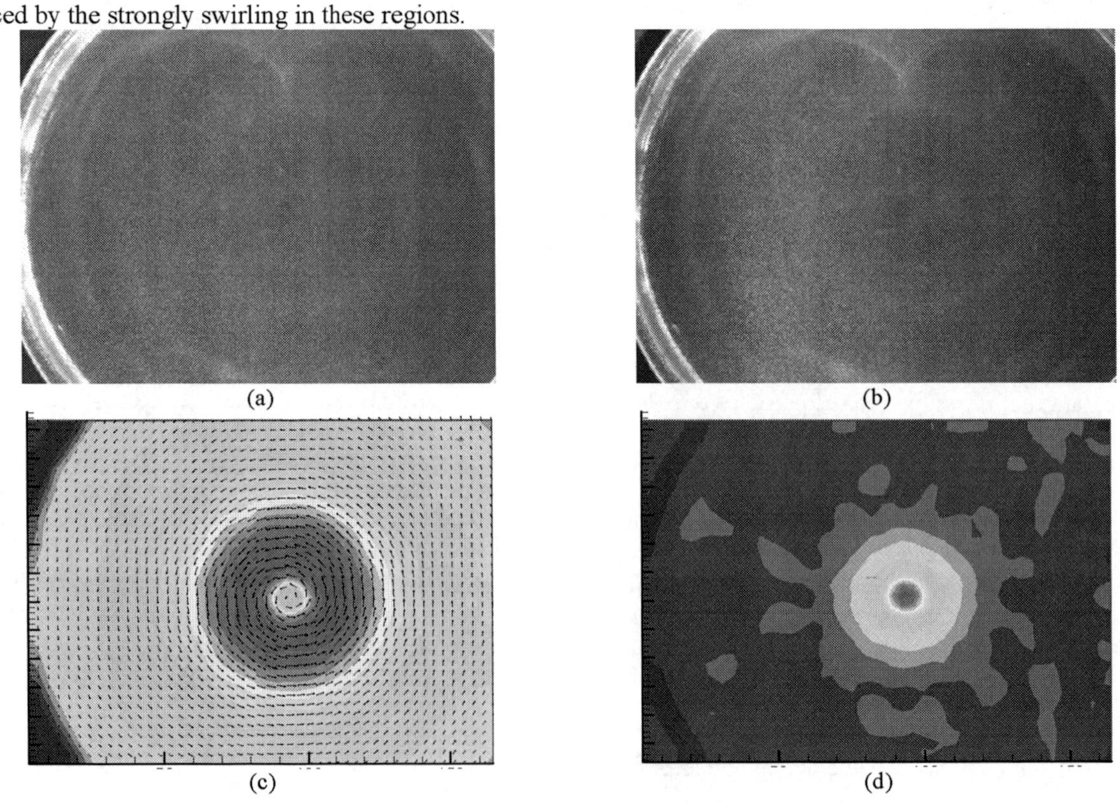

FIGURE 5. Original Image of the Tracer Particles, Velocity Vector and the Vorticity Distribution
(a) and (b): Original image of the tracer particles; (c) Velocity vector by autocorrelation; (d): Vorticity distribution

Effect of the Prolonged Vertical Tube on the Flow Field of the Cyclone

Fig. 6 presents that the velocity distribution along the radius at the same section for the cyclone with different vertical tube lengths. Fig. 6 shows that the structure of the flow field in the dustbin in this study is the same as other's conclusion [11]. Fig. 7 gives the comparison of the tangential velocities in the dustbin and vertical tube of a

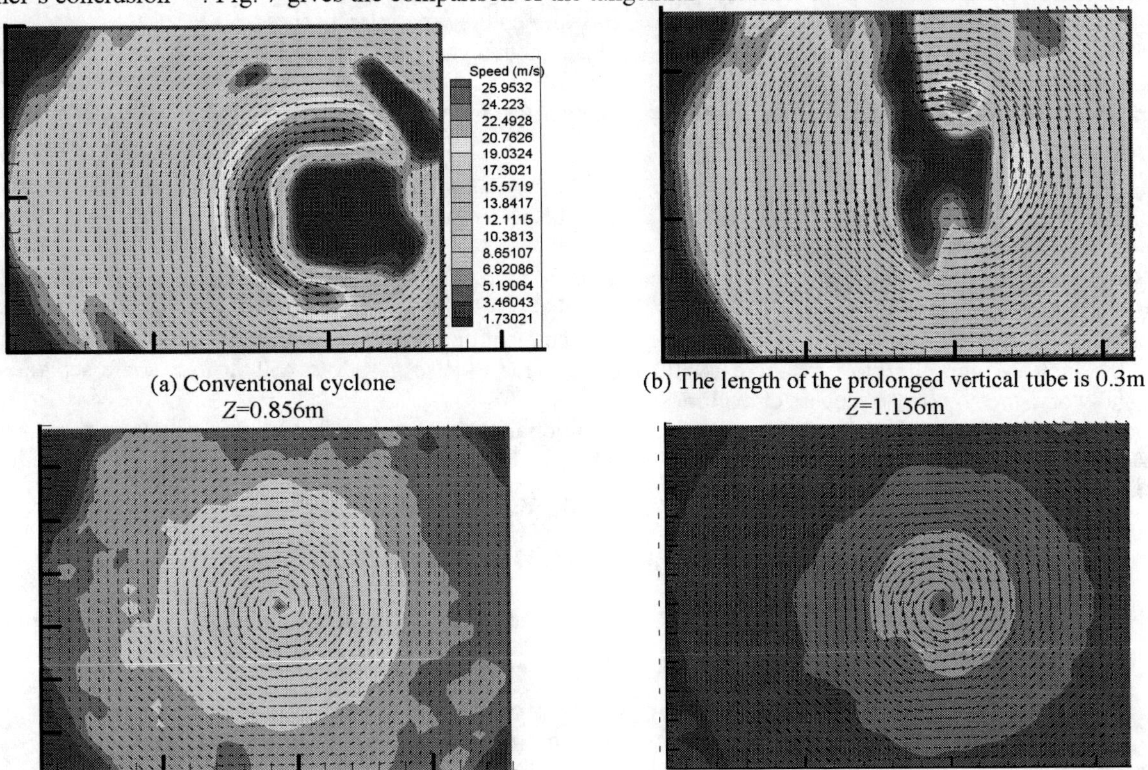

(a) Conventional cyclone
Z=0.856m

(b) The length of the prolonged vertical tube is 0.3m
Z=1.156m

(c) The length of the prolonged vertical tube is 0.4m
Z=1.256m

(d) The length of the prolonged vertical tube is 0.5m
Z=1.356m

FIGURE 6. Comparison of the Velocity in the Dustbin of a Cyclone at the Same Section with Different Vertical Tube Lengths

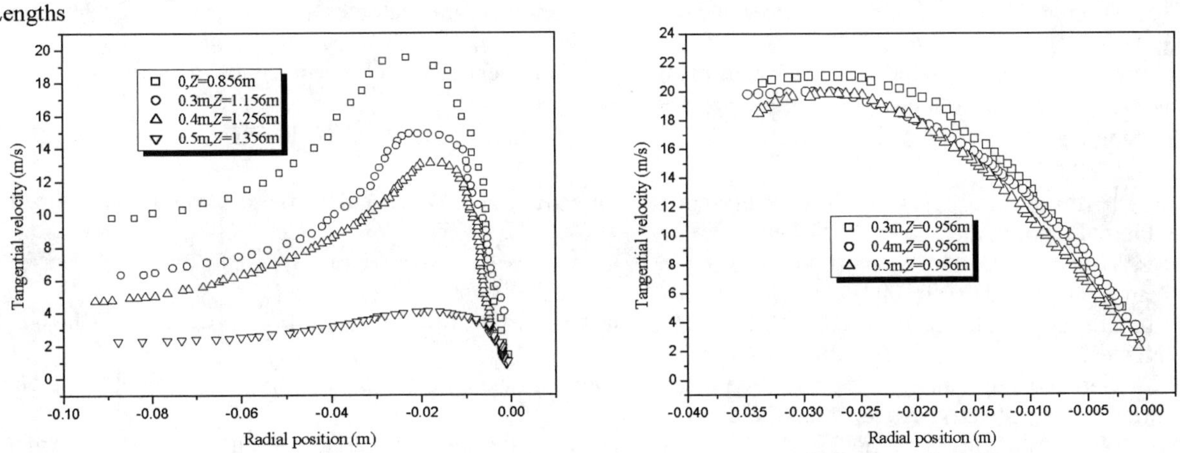

(a) In the dustbin

(b) In the vertical tube

FIGURE 7. Comparison of the Tangential Velocities in the Dustbin and Vertical tube of a Cyclone with Different Vertical Tube Lengths

cyclone with different vertical tube lengths. It is noted that the tangential velocity is the time-averaged value. Figs. 6 and 7 indicate that the flow fields in the vertical tube and in the dustbin are similar to that in the cyclone, i.e. the tangential velocity here also exhibits a combined vortex structure, i.e. the tangential velocity increases with increasing radius in the neighborhood of the axis, which is called quasi forced vortex, reaches a maximum approximately at the vortex finder radius, and decreases thereafter with increasing radius, which is called quasi free vortex. The maximum tangential velocity is located at 0.7-0.8 radius of the vortex finder. Fig.7 indicates that the gas tangential velocity is large in the dustbin for the conventional cyclone, and the maximum is up to 21m/s. With a dustbin only, the potential vortex continues from the cyclone cone into the bin. Due to the strong flow in the dustbin, a cyclone with such a bin will not result in good separation efficiency. On the other hand, the gas tangential velocity of prolonged cyclone in the dustbin is reduced greatly. Large gas tangential velocity and turbulent kinetic energy [9] will produce the re-entrainment of already separated dust, and worsen the separation process.

Fig. 7 also indicates that the tangential velocity is reduced in the vertical tube of the cyclone, owing to the increased friction of the larger wall surface. Reduced tangential velocity will lead to worse separation efficiency, which can be offset by the additional separation process in the vertical tube and the reduced dustbin re-entrainment.

CONCLUSIONS

(1) The strong swirling turbulent flow in the dustbin and vertical tube of a cyclone has been successfully examined using particle image velocimetry technology.

(2) The tangential velocity is still large in the dustbin of the conventional cyclone, and the average value is about 21m/s, which indicates that the gas eddy here is still intense. This intense gas flow will disturb some separated particles, and worse the separation process of cyclones.

(3) When a prolonged vertical tube attached into the conventional cyclone, the swirl will enter into the tube, and make dust separate further. The gas tangential velocity of prolonged cyclone in the dustbin is reduced greatly, which is useful to improve the separation performance of the cyclone.

ACKNOWLEDGEMENTS

The authors would like to thank Dr. Wei runjie and Yu jian (Beijing Lifang Tiandi technological Ld. Inc., Beijing, China) for his guidance and support.

REFERENCES

1. A.J.Ter Linder, Investigation into cyclone dust collectors, Proc. Inst. Mech. Eng.,**160**, 233-240 (1949).
2. A.Gil, and C.Cortes, Gas-particle flow inside cyclone diplegs with pneumatic extraction, Powder Technology, **128**, 78-91(2002).
3. L. X. Zhou and S. L. Soo, Gas-solid flow and collection of solids in a cyclone separator, Powder Technology, **63**, 45-53(1990).
4. A. J. Hoekstra and J. J.Derksen, An Experimental and Numerical Study of Turbulent Swirling Flow in Gas Cyclones, Chemical Engineering Science, **54**, 2055–2065(1999).
5. W. Peng and A.C. Hoffmann, Flow pattern in reverse-flow centrifugal separators, Powder Technology, **127**, 2 12 – 222(2002).
6. 6 S. Obermair and C. Gutschi, Flow pattern and agglomeration in the dust outlet of a gas cyclone investigated by Phase Doppler Anemometry, Powder Technology, **156**, 34-42(2005).
7. X. L. Wu and M. X. Shi, Visualization of the processing vortex core in a cyclone separator by PIV, Chinese J. Chem. Eng., **11**, 633-637(2003).
8. Z. L. Liu and Y. Zheng, Stereoscopic PIV studies on the swirling flow structure in a gas cyclone, Chemical Engineering Science, **61**, 4252-4261(2006).
9. F. P. Qian and J. G. Zhang, Effects of the prolonged vertical tube on the separation performance of a cyclone. Journal of Hazardous Materials, **136**, 822-829 (2006).
10. A. J. Hoekstra, "Gas flow field and collection efficiency of cyclone separators", Ph. D thesis, Technical University of Delft, 2000.
11. S. Obermair and J. Woisetschlager, Investigation of the flow pattern in different dust outlet geometries of a gas cyclone by laser Doppler anemometry, Powder Technology, **138**, 239-251(2003).

12. Nomenclature

a height of cyclone inlet, m
b width of cyclone inlet, m
D diameter of cyclone body, m
D_e diameter of cyclone vortex finder, m
H length of cyclone , m
h length of cyclone cylinder, m
S deepness of vortex finder insertion , m
B dust outlet diameter , m

In-situ Measurement of Two-Phase Fraction in Jet-Pump Using the ERT Technique

J. Y. XU+, M. WANG*, Y. X. WU+, Z. C. ZHENG+, H.I. SCHLABERG*, and R. A. WILLIAMS*

*Institute of Particle Science and Engineering, University of Leeds, Leeds, UK
+Institute of Mechanics, Chinese Academy of Sciences, Beijing, China

Abstract. Due to the complexity of two-phase flow field in jet-pump, it is of difficult to accurately measure in-situ phase fraction in the pump. Electrical resistance tomography (ERT) has been successfully used in predicting gas concentration, dispersed phase velocity and flow regimes in both vertical and horizontal flows. Thus, the present study carried out an experimental investigation of in-situ the phase fraction in a jet-pump using the ERT technique. A new jet-pump with ERT sensors was designed to measure in-situ phase fraction and flow regime. The study was based on laboratory experiments that were carried out on a 50 mm vertical flow rig for various gas and liquid phase superficial velocities. A total of 6 experimental tests were conducted. The flow pattern of gas-liquid in the jet-pump was studied using the ERT technique. Results suggest the flow patterns recognition using ERT systems can not only be applied for most two-phase flow regimes, but also for the visualisation of a complex system. Furthermore, the visualizations of a rapid mixing process in the throat of a jet pump obtained in the present work provide a reliable basis for theoretical study and optimal design of jet pumps.

Keywords: Jet-pump, Electrical resistance tomography, In-situ phase fraction, Flow pattern, Vertical flow.

INTRODUCTION

The jet-pump transfers energy from a liquid or gas primary fluid to a secondary fluid. The latter may be a liquid, a gas, a two-phase gas-in-liquid mixture, or solid particles transported in a gas or a liquid. All these combinations occur in a wide range of practical applications in the chemical, oil and process industries. The main components of a jet pump, as shown in Fig.1, are a primary nozzle, secondary fluid injectors, a mixing chamber, a throat, and a diffuser. Due to the high reliability of these devices which have no moving parts, considerable interest has been shown in theoretical and experimental studies of the Jet-pump (Folsom, 1948; Takashima, 1952; Higgins, 1964; Witte, 1962; Jiao, 1985; Zheng, 1985). These studies, in general, have been similar in applying momentum-continuity modelling of the throat process. More recent work for predicting the performance of hydraulic jet-pump handling two-phase fluids was proposed by Noronha et al. (1997). It was a model that, as the one developed by Cunningham (1974), considered the gas compressibility, and used the data obtained by Jiao (1998) to evaluate his model and to determine two constant energy-dissipation factors for the flow inside the throat and through the nozzle.

With the development of particle image velocimetry (PIV), some authors are also using the PIV technique to study the flow field of jet-pump (He, 2004). However it is of difficulty to accurately measure the in-situ phase fraction and flow pattern in jet-pump by PIV alone. Due to the characteristics of its structure, the jet-pump should have an important influence on flow pattern when it is used to transport multi-phase flow, especially for liquid-liquid flow (Xu et al., 2005). Electrical resistance tomography (ERT) has been successfully applied in predicting gas concentration, disperse phase velocity and flow regimes in both vertical and horizontal flows (Wang, 2001, 2004). To understand the performance of a two-phase jet pump which influences on the flow pattern, the present study carried out an experimental investigation of in-situ phase fraction and flow pattern in a jet-pump on a vertical pipe with the ERT technique.

CP914, *Multiphase Flow: The Ultimate Measurement Challenge,* edited by X. Cai, Y. Wu, Z. Huang, S. Wang, and M. Wang

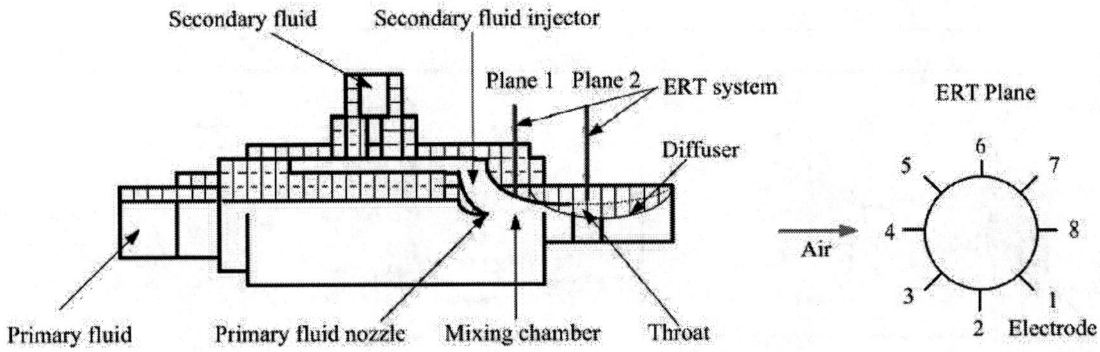

FIGURE 1. Schematic of a new jet-pump with ERT sensor

EXPERIMENTAL SETUP AND PROCEDURE

The experiments reported in this paper were performed in the multiphase flow facilities at the Institute of Particle Science and Engineering at University of Leeds, which has a gas-liquid flow loop of about 12 m length with an inner diameter of 50mm. The experimental investigations are conducted in the setup shown in figure 2. The flow loop can run at a maximum superficial liquid velocity of 1 m/s and a superficial gas velocity greater than 40 m/s. By controlling flow-rates of the gas phase and the liquid phase, many flow patterns, such as bubble slow, bubbly flow, slug flow, churn flow, can be created in this device. Tomographic sensors with four sensing planes were installed in the flow loop in order to obtain the in-situ phase fraction and flow patterns. Tap-water (conductivity 0.284 mS/cm and temperature 21℃) was used as the liquid phase which was scaled with an accumulating tank during the experiments to get water flow-rate and superficial velocity, and air was introduced into the flow loop, which was measured by a gas flowmeter. Liquid phase and gas phase were fed into the pipeline via the jet-pump with dual-plane ERT sensors. In this work, we will focus our attention on a particular jet-pump, the liquid-jet-gas pump (LJGP). This device uses liquid as the primary fluid and will use compressed air as a secondary fluid. The schematic of the new jet-pump with ERT sensors is shown in Fig.1. Photographs were recorded as visual presentations of these different flow patterns.

Experimental results presented in this paper were obtained from an ITS P2000 ERT system (Industrial Tomography Systems Ltd. Manchester, U.K) in monitoring gas-liquid flow in the jet-pump and vertical pipes at various superficial velocities. In each plane, eight stainless steel electrodes are mounted flush to the surface at equal intervals. Since the current data collection speed was still limited, we set the distance between the two sensing planes in vertical pipe to 50mm and the two sensing planes in jet-pump to 30mm apart, respectively. The voltage potential differences for tomography images were collected based using the normal adjacent protocol, with a data collection speed of 29 frames per second at an AC current injection frequency of 38400Hz and a current value of 15mA for flow pattern recognition, and with a 5 frames per second at the signal frequency of 9600 Hz for average fraction calculation. The reconstruction of the image was carried out by the use of the Linear Back Protection (LBP) algorithm. The average void fraction was determined from Maxwell equations (Maxwell, 1881; Lucas et al. 1999; George et al. 2000).

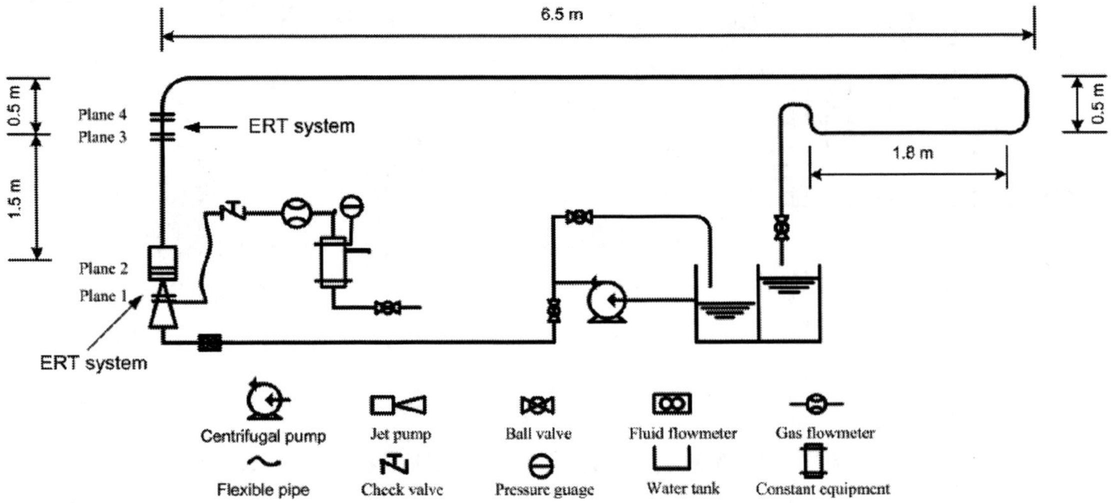

FIIGURE 2. Schematic of the test flow loop

RESULTS AND DISCUSSION

Two-dimensional (2-D) slice images were sequentially reconstructed using ITS 2000 ERT system. Fig.3 shows that typical in-situ air fraction in jet-pump and the vertical pipe under slug flow regime. Where the blue area represent the air cavities or low conductivity regions, and the red area represent the water or high conductivity regions. These slice images were accumulated to show the flow pattern. It can be seen from plane 3 in Fig.3 that a long-bubble is moving via the vertical test section. Furthermore, it can be also observed from a comparison between planes 1 and 2 that the flow of gas-phase in jet-pump may be rotational round the axes. This will be investigated in detail in a future study using a new fast ERT system. Flow patterns in three different planes in both the vertical test section (Plane 3) and jet-pump (Plane 1 and 2), namely bubble, bubbly and slug, are shown in Fig.4. The flow patterns were stacked within three different planes with 10 continuous images from a data set of 400 images for bubble and bubbly flow, and 20 continuous images from 500 images for slug flow. Different flow patterns between in the jet-pump and in the vertical section can be clearly imaged by ERT using the conductive ring sensing strategy under the same superficial velocities.

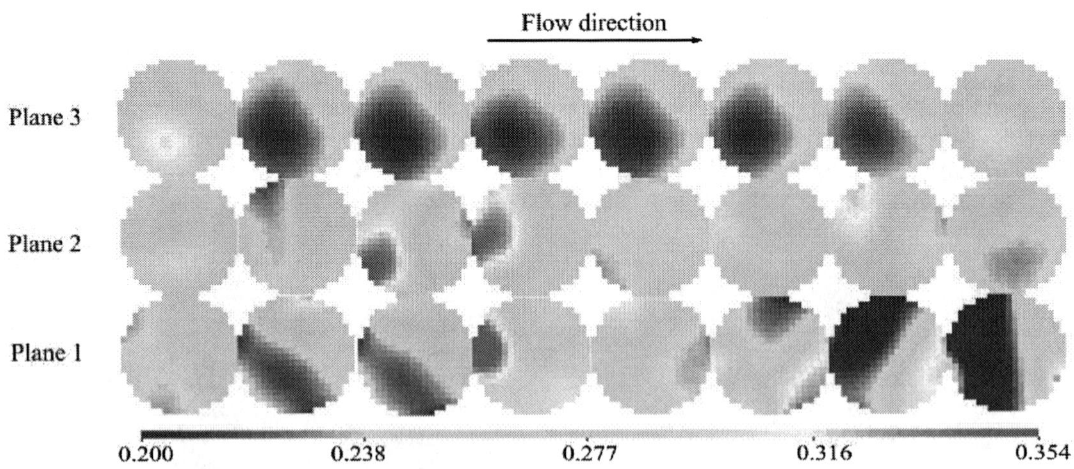

FIIGURE 3. Reconstructed 2-D images in respect to typical in-situ phase fraction in jet-pump and the vertical pipe, respectively, at superficial water velocity, 0.008 m/s and superficial gas velocity, 0.170~0.212 m/s.

It is shown from the comparison between the planes 1 and 2 in Figs.4 that the liquid jet leaves the nozzle at the position of plane 1 simultaneously with the air phase, it is clear that gas phase is continuous, most of air is moving along the boundary near the injector. After entering the throat, the phases start to mix strongly as can be seen from plane 2 in Fig. 4. In the mixing process in the throat region, in which the liquid jet entrains, accelerates and compresses the gas, the transfer of momentum from the liquid serves largely to compress the gas in which significant momentum transfer is involved in increasing the kinetic energy of air phase. The pressure recovery in the LJGP diffuser is significantly reduced because the liquid performs most of the work in compressing the entrained gas bubbles. It can be seen that both gas and liquid phases distribution and their dynamical varying processes (flow processes) in the jet-pump and the vertical pipe can be clearly demonstrated using the ERT technique.

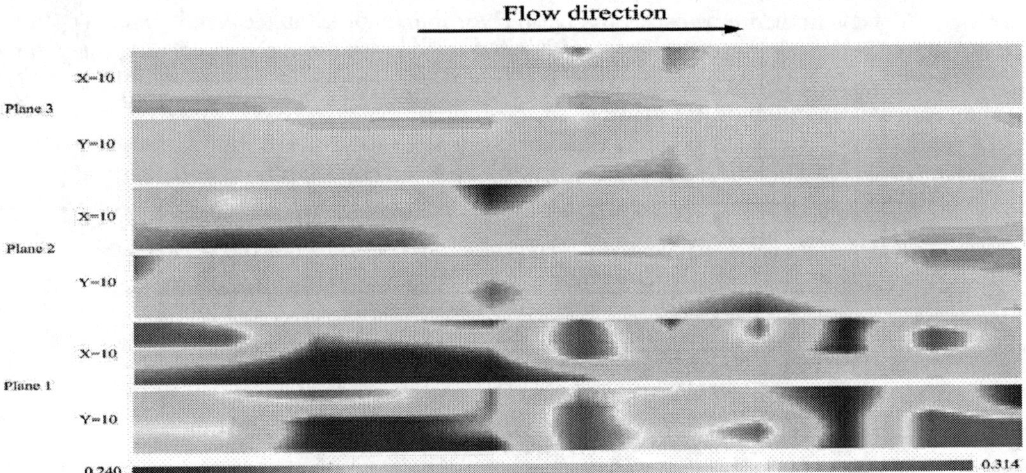

(a) Bubble flow: Frames from 250 to 260 at superficial water velocity, 0.398 m/s and superficial gas velocity, 0.008~0.024 m/s.

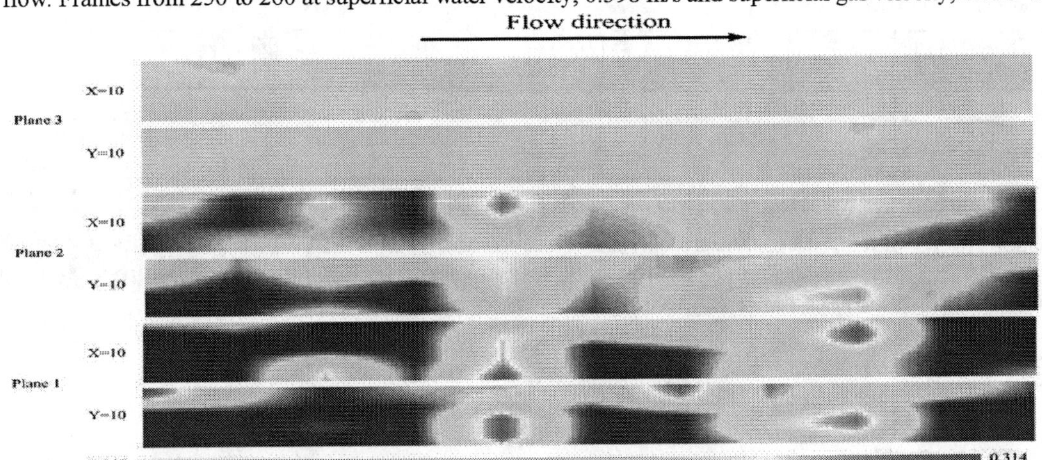

(b) Bubbly flow: Frames from 250 to 260 at superficial water velocity, 0.785 m/s and superficial gas velocity, 0.127~0.153 m/s.

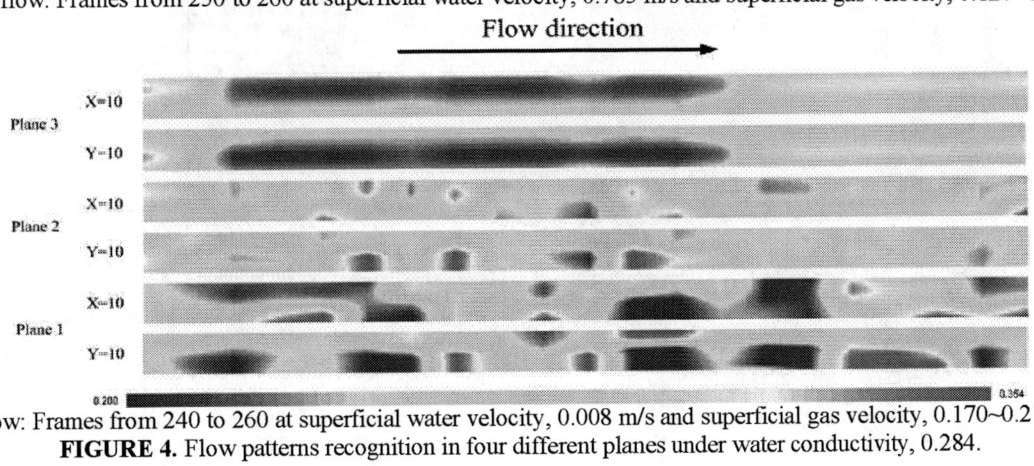

(c) Slug flow: Frames from 240 to 260 at superficial water velocity, 0.008 m/s and superficial gas velocity, 0.170~0.212 m/s.

FIGURE 4. Flow patterns recognition in four different planes under water conductivity, 0.284.

For each superficial velocity, 350 images were collected for all loading concentration. The images were averaged and merged over the recorded voltage values. From the merged images produced for each plane, the concentration profile was obtained from the Maxwell equation. The mean concentration was calculated from

$$\overline{C_V} = \frac{\sum_{i=1,j=1}^{i=n,J=n} (C_{i,j}, A_{pixel})}{A_{total}} \quad (1)$$

Since there are 316 pixels in the linear back projection algorithm, and the area of each pixel is 6.2mm2, the area of the pipe is 1963mm2 according to the 50mm diameter. Table 1 shows the mean concentration of three different planes under bubble flow, bubbly flow and slug flow, respectively. It can be observed from the concentration data obtained that the mean concentration of gas-phase in the position of plane 2 (i.e. the throat) is smaller than those of the position of planes 1 and 3 due to the compressible characteristic of gas phase. Furthermore, owing to the effect of the slip velocity and gas condensability on phase fraction, the mean concentration in the vertical pipe will also be lower than those in the position of plane 1 as shown in table 1.

TABLE 1 Mean concentration of different planes under the same superficial velocities.

Table 1.1 Superficial water velocity =0.785 m/s

Superficial gas Velocity	Plane 1	Plane 2	Plane 3	Plane 4	Flow pattern in vertical pipe
0.008~0.024 m/s	0.51%	0.13%	0.24%	0.28%	Bubble
0.051~0.076 m/s	1.63%	0.32%	0.56%	0.70%	Bubbly
0.127~0.153m/s	4.04%	0.99%	0.88%	0.92%	Bubbly

TABLE 1.2 Superficial water velocity =0.398 m/s

Superficial gas Velocity	Plane 1	Plane 2	Plane 3	Plane 4	Flow pattern in vertical pipe
0.008~0.024 m/s	3.70%	0.65%	0.54%	0.71%	Bubble
0.051~0.076 m/s	7.62%	2.86%	1.26%	1.36%	Bubbly

TABLE 1.3 Superficial water velocity =0.008 m/s

Superficial gas Velocity	Plane 1	Plane 2	Plane 3	Plane 4	Flow pattern in vertical pipe
0.170~0.254 m/s	10.9%	6.99%	8.39%	7.98%	Slug

Fig.5 shows the plot of the calculated mean concentration at the different plane position against the axial position with slug regime. As shown above, the void fraction near the left pipe well (electrode 5) is higher than those near the right as a result of the position of air injector.

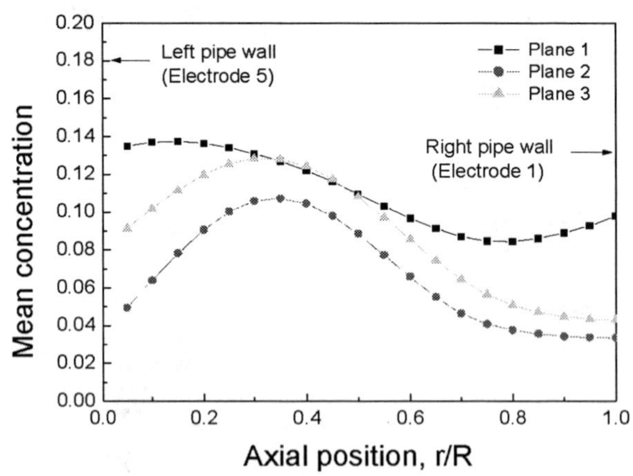

FIIGURE 5. The dependence of the mean in-situ concentration on the axial position within slug regime under superficial gas velocity,0.170~0.254 m/s and superficial water velocity, 0.008 m/s.

CONCLUSIONS

To understand the performance of a two-phase jet pump which influences on the flow pattern, the presented study was carried on an experimental investigation of in-situ fraction and the flow pattern in jet-pump and vertical pipe with ERT technique. A series of experiments was carried out, and an ITS 2000 ERT system was used to monitor gas-liquid flow in jet-pump and vertical pipes at different superficial velocities. Results suggest the flow patterns recognition using ERT system can not only be applied for most two-phase flow regimes, but also for the visualisation of the complex system. Furthermore, the visualizations of a rapid mixing process in the throat of a jet pump obtained in the present work provide a reliable basis for theoretical study and optimal design of jet pumps.

ACKNOWLEDGEMENTS

The authors gratefully acknowledge the work described here was financially supported by the China-UK joint project (Royal Society Grant Number 15933).

REFERENCES

1. Cunningham, R.G., Dopkin, R.J., Jet Breakup, and Mixing Throat Lengths for the Liquid-Jet Gas Pumps, Journal of Fluids Engineering 94, Trans. ASME, 1974.
2. Folsom, H.G., Jet Pump with Liquid Drive, Chemical Engineering Progress 44, 19487, 65-770
3. George, D.L., et al., Validation of Electrical-Impedance Tomography for Measurements of Material Distribution in Two-Phase Flows, International Journal of Multiphase Flow 26, 549–581, 2000.
4. He, P.J., et al., PIV Measurement of Flow Field in Jet Pump, Advances in Water Science 15, 2004, 296-299.
5. Higgins, H.W., Water Jet Air Pump Theory and Performance, MS thesis, The Pennsylvania state university, U.S.A., 1964.
6. Jiao, B., Behavior of Hydraulic Jet Pumps When Handing A Gas-Liquid Mixture, MS thesis, University of Tulsa, Tulsa, 1985.
7. Jiao, B., Performance Model for Hydraulic Jet Pumping of Two-Phase Fluids, Ph.D. dissertation, University of Tulsa, Tulsa, 1988.
8. Lucas, G.P., et al., Measurement of The Solids Volume Fraction and Velocity Distributions in Solids–Liquid Flows Using Dual-Plane Electrical Resistance Tomography, Flow Measurement and Instrumentation 10, 1999, 249–258.
9. Maxwell, J.C., A Treatise on Electricity and Magnetism, Clarendon Press, Oxford, UK, 1881.
10. Noronha, F.A.F., France, F.A., Alhanati, F.J.S., Improved Two-Phase Model for Hydraulic Jet Pumps. SPE 37427, SPE Production operations symposium, Oklahoma , 1997
11. Takashima, Y., Studies on Liquid Jet Gas Pumps. Journal Scientific Research Institute 46, 1952, 230-246.
12. Wang, M., et al., Gas-Water Flow Interface Characterization Using Electrical Resistance Tomography, 2nd World Congress on Industrial Process Tomography, Hannover, Germany, 2001.
13. Wang, M., Electrical Mapping of Particulate Multiphase Flows, The 4th International Symposium on Measurement Techniques for Multiphase Flows, Huangzhou, China, 2004.
14. Witte, J.H., Mixing Shocks and Their Influence on the Design of Liquid-Gas Ejectors, Ph.D. Dissertation, Delft, 1962.
15. Xu, J.Y., Wu, Y.X., Li, D.H., Experimental Studies for Oil-Water Two-Phase Pipeline Transportation with Jet Pump. Journal of Experiments in Fluid Mechanics 19, 2005, 49-55.
16. Zheng, Z.C., Energy Conservation of Jet-Pump. Mechanics and Practice 1, 1985, 46-50.

Particle Swarm Optimization Technique For Elliptic Region Boundary Estimation In Electrical Impedance Tomography

Umer Zeeshan Ijaz♣, Anil Kumar Khambampati♣, Min Chan Kim•, Sin Kim♦, Jeong Seong Lee♦ and Kyung Youn Kim♣○

♣*Department of Electronic Engineering, Cheju National University, Cheju 690-756, South Korea*
•*Department of Chemical Engineering, Cheju National University, Cheju 690-756, South Korea*
♦*Department of Nuclear and Energy Engineering, Cheju National University, Cheju 690-756, South Korea*
○*Corresponding Author(kyungyk@cheju.ac.kr)*

Abstract. In this study we consider the recovery of smooth elliptic region boundary in electrical impedance tomography (EIT). The assumption made is that the resistivity profile is known *a priori* but some of the geometric information is missing. This missing information may for example be shape, size and location. This leads to a nonlinear ill-posed inverse problem. In this study we propose a population based parallel evolutionary computation technique, particle swarm optimization (PSO). We formulate the forward problem as mapping of a set of Fourier coefficients representing boundary shapes to the resistivity profile in the domain. Then PSO is used to iteratively seek the boundary configuration minimizing a cost functional. The final goal of this study is to recover the region boundary with only one measurement data (i.e., one current pattern) contaminated with measurement noise along with a very high contrast ratio. The simulation results are also provided to assess the merit of PSO technique.

Keywords: Electrical Impedance Tomography, Particle Swarm Optimization
PACS: 02.60.Pn, 87.63.Pn

INTRODUCTION

Image reconstruction in EIT is a kind of nonlinear optimization problem in which the solution is obtained iteratively through the forward and inverse solvers. The physical relationship between the internal resistivity and the surface voltage is governed by a partial differential equation with appropriate boundary conditions. It is in most cases impossible to obtain an analytical solution for the forward problem so that the numerical techniques such as finite element method (FEM) is often employed. The inverse problem in EIT is a highly nonlinear, underdetermined and ill-posed problem. There are many different types of current injection protocols used. For example, most common one being trigonometric, opposite and adjacent current patterns. For a 32 electrodes configuration, there are 31 (16 cosine, 15 sine) trigonometric, 16 opposite, and 32 adjacent current patterns in a single image frame. Usually two types of scenarios exist: static in which the resistivity profile does not change within the time taken to acquire full set of voltage measurements and; dynamic in which the resistivity profile changes dynamically during the application of consecutive current patterns. For static case, modified Newton Raphson (mNR) is applied in which the image is reconstructed after every image frame. In dynamic case, extended Kalman filter (EKF) and recently interacting multiple model (IMM) scheme is used. The elliptic region boundary estimation through the use of IMM is already covered in [5] in which at least 8 trigonometric current patterns were used before the target changes its position.

With regard to optimal current pattern, Isaacson *et al.* in [8] showed that the best current patterns to distinguish a central concentric inhomogeniety inside an otherwise homogeneous circular conductor are trigonometric current patterns. Gisser *et al.* in [9] compared the distinguishabilities by using adjacent, opposite and cosine current patterns on a circular conductor model without a centered circular target and showed that the maximum cosine current equal to the current injected with opposite or the adjacent electrode is the optimum current pattern. Newell *et al.* in [10]

CP914, *Multiphase Flow: The Ultimate Measurement Challenge,* edited by X. Cai, Y. Wu, Z. Huang, S. Wang, and M. Wang

showed that cosine current patterns can distinguish smaller inhomogeneities than the use of the opposite and adjacent current patterns. It is generally known that the opposite current patterns are optimal if the total current (sum of the amplitude of the injected current) is kept constant and trigonometric current patterns are optimum if the maximum of the injected current is kept constant.

The PSO is a parallel evolutionary computation technique based on the social behavior metaphor developed by Kennedy and Eberhart [1]. PSO was inspired by the sociological behavior associated with birds flocking and fish schooling. The PSO algorithm is initialized with a population of random candidate solutions, conceptualized as particles. Each particle is assigned a randomized velocity and is iteratively moved through the problem space. It is attracted towards the location of the best fitness achieved so far by the particle itself (Pbest) and by the location of the best fitness achieved so far across the whole population (Gbest). They also have stochastic exploration capability via the use of the random multipliers.

The PSO algorithm includes some tuning parameters that greatly influence the algorithm performance. There are a number of considerations in using PSO in practice; one might wish to restrict the velocities to a certain maximum amount, for instance. The considerable adaptability of PSO to variations and hybrids is seen as a strength over other robust evolutionary optimization mechanisms, such as genetic algorithms. In order to exploit the reconstruction capabilities of a single current pattern, in this paper we use PSO as an inverse algorithm for reconstruction of elliptic region boundary. A single opposite current pattern is used while the sum of the amplitude of the injected current is kept constant. The succeeding sections elucidate the proposed methodology.

MATHEMATICAL MODEL AND FORWARD SOLVER

When electrical currents $I_l\,(l=1,2,...,L)$ are injected into a body $\Omega \in \mathbb{R}^2$ through the electrodes $e_l\,(l=1,2,...,L)$ attached on the boundary $\partial\Omega$, and the resistivity distribution $\rho(x,y)$ is known for Ω, the corresponding electrical potential $u(x,y)$ on Ω can be determined uniquely from a partial differential equation, which can be derived from the Maxwell equations:

$$\nabla.(\rho^{-1}\nabla u)=0 \text{ in } \Omega \tag{1}$$

with the following boundary conditions based on the complete electrode model:

$$u+z_l\rho^{-1}\frac{\partial u}{\partial n}=\overline{u}_l \text{ on } e_l, l=1,2,...,L \tag{2}$$

$$\int_{e_l}\rho^{-1}\frac{\partial u}{\partial n}dS=I_l, \quad l=1,2,...,L \tag{3}$$

$$\rho^{-1}\frac{\partial u}{\partial n}=0 \text{ on } \partial\Omega \backslash \bigcup_{l=1}^{L} e_l \tag{4}$$

where z_l is the effective contact impedance between the lth electrode and electrolyte, $\overline{u}_l$ is the potential on the lth electrode, e_l, n is the outward unit normal and L is the number of electrodes. Various forms of boundary conditions may be used in the forward model. We have chosen the complete electrode model (CEM) which takes into account the discrete electrodes, effects of the contact impedance and the shunting effect of the electrodes. In addition, the following two constraints for the injected currents and measured voltages ensure the existence and uniqueness of the solution:

$$\sum_{l=1}^{L} I_l=0 \tag{5}$$

$$\sum_{l=1}^{L} \overline{u}_l=0 \tag{6}$$

The computation of the potential $u(x,y)$ on Ω and the voltages $\bar{u}_l$ on the electrodes for the given resistivity distribution $\rho(x,y)$ and the boundary conditions is called the forward problem. There are different numerical methods such as the finite difference method (FDM), boundary element method (BEM) and finite element method (FEM). In this paper, we used the FEM to obtain a numerical solution. In the FEM, the object area is discretized into sufficiently small elements having a node at each corner, while it is assumed that the resistivity distribution is constant within each element. The potential at each node and the 'referenced' electrode voltages, defined by the vector $\upsilon \in \mathbb{R}^{N_n+L-1}$, are calculated by discretizing (1) into $Y\upsilon = c$, where $Y \in \mathbb{R}^{(N_n+L-1)\times(N_n+L-1)}$ is the so-called stiffness matrix, N_n is the number of FEM nodes. Y and c are functions of the resistivity distribution in the object and the injected currents through the electrodes, respectively. For more details on the forward solution and the FEM approach, see [2],[3],[6] and [7].

PHASE BOUNDARY REPRESENTATION AND INVERSE SOLVER

We assume that the outer boundary of the body, that is, $\partial\Omega$ is known. If phase boundaries of the objects are sufficiently smooth, they can be approximated in the form ([2],[6],[7]).

$$C_l(s) = \begin{pmatrix} x_l(s) \\ y_l(s) \end{pmatrix} = \sum_{n=1}^{N_\theta} \begin{pmatrix} \gamma_n^{x_l}\theta_n^x(s) \\ \gamma_n^{y_l}\theta_n^y(s) \end{pmatrix} \tag{7}$$

where $C_l(s)(l = 1,2,...,S)$ is the boundary of the lth object, S is the number of the objects in the body, $\theta_n(s)$ are period and differentiable basis functions and N_θ is the number of basis functions. In this paper, we express the phase boundaries as Fourier series in two-dimensional coordinates with respect to parameter s, that is, we use the basis functions of the form

$$\theta_1^\alpha(s) = 1 \tag{8}$$

$$\theta_n^x(s) = \sin(2\pi \frac{n}{2} s), \qquad n=2,4,6,\ldots,N_\theta-1 \tag{9}$$

$$\theta_n^y(s) = \cos(2\pi \frac{(n-1)}{2} s), \qquad n=1,3,5,\ldots,N_\theta \tag{10}$$

where $s \in [0,1]$ and α denotes either x or y. Furthermore, using (7), the boundaries are identified with the vector γ of the shape coefficients, that is,

$$\gamma = (\gamma_1^{x_1},\ldots,\gamma_{N_\theta}^{x_1},\gamma_1^{y_1},\ldots,\gamma_{N_\theta}^{y_1},\ldots,\ldots,\gamma_1^{x_s},\ldots,\gamma_{N_\theta}^{x_s},\gamma_1^{y_s},\ldots,\gamma_{N_\theta}^{y_s})^T \tag{11}$$

where $\gamma \in \mathbb{R}^{2SN_\theta}$. The next step is to use the PSO algorithm to estimate the Fourier coefficients.

The basic PSO algorithm in vector notation is

$$v_{k+1} = a \otimes v_k + b_1 \otimes r_1 \otimes (p_1 - \gamma_k) + b_2 \otimes r_2 \otimes (p_2 - \gamma_k) \tag{12}$$

$$\gamma_{k+1} = c \otimes \gamma_k + d \otimes v_{k+1} \tag{13}$$

The symbol $\otimes$ denotes element-by-element vector multiplication. At iteration k, the velocity v_k is updated based on its current value affected by a momentum factor a and on a term which attracts the particle towards previously found best positions: its own previous best position p_1 and globally best position in the whole swarm p_2. The strength of the attraction is given by the coefficients b_1 and b_2. The Fourier coefficients γ_k (representing the boundary in [2],[6],[7]) are updated using its current value and the newly computed velocity v_{k+1}, affected by coefficients c and d. Randomness for good state space exploration is introduced through the vectors of random numbers r_1 and r_2 where $r_1, r_2 \in Uniform[0,1]$. Trelea in [4] has done the convergence analysis of the PSO and formulated a deterministic algorithm in one-dimension in which the parameters were set as $c = 1$, $d = 1$, $b = \frac{b_1 + b_2}{2}$, $p = \frac{b_1}{b_1 + b_2} p_1 + \frac{b_2}{b_1 + b_2} p_2$, $r_1 = r_2 = \frac{1}{2}$ and found out that the PSO is convergent if $a < 1, b > 0, 2a - b + 2 > 0$ is satisfied. Based on their work, they also formulated two properties of PSO i.e., harmonic oscillations when the property $a^2 + b^2 - 2ab - 2a - 2b + 1 < 0$ is met, zigzagging when the property $a < 0$ or $a - b + 1 < 0$ is met. A balance of these two properties is essential in exploring and exploiting the problem domain. Eventually two sets of parameters were proposed, Trelea Method 1 (*a=0.6* and *b=1.7*) and Trelea Method 2 (*a=0.729* and *b=1.494*) giving the best published results so far [4]. Applying PSO to EIT, we need to formulate the cost functional to obtain Gbest and Pbest fitnesses at each iteration. The cost functional is then defined as

$$\Phi(\gamma_k) = \frac{1}{2}(V(\gamma_k) - \overline{u})^T (V(\gamma_k) - \overline{u}) \tag{14}$$

where $V(\gamma_k)$ is the boundary voltages obtained through FEM in the iteration k.

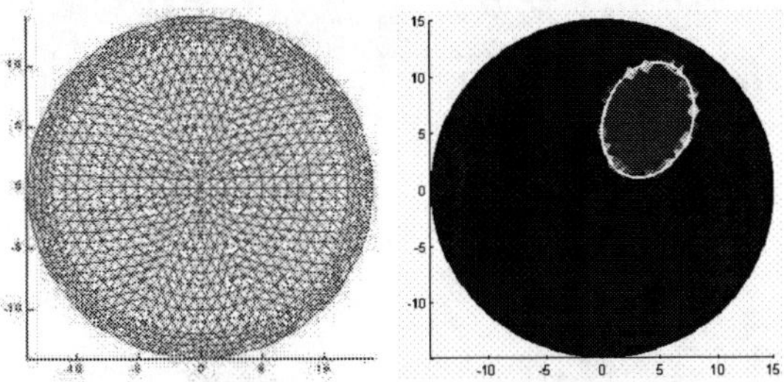

1 2

FIGURE 1. Mesh structure and mapping between resistivity profile and boundary represented by truncated Fourier series. 1. Mesh structure used 2. Resistivity profile and boundary representation

NUMERICAL RESULTS

In order to test the PSO algorithm, a circular domain of 30cm diameter was considered. 32 electrodes are mounted on the boundary of the circular domain. The mesh structure used in the PSO is shown in Fig. 1 with 1497 nodes and 2793 elements, respectively. Here it is assumed that the object and background resistivities are known *a priori* and the values are 250×10^6 and 250, respectively so that the contrast ratio between the object and background is $10^6:1$. Furthermore the contact impedance was set to 0.005. A total of 6 Fourier coefficients are used to represent the elliptic region boundary i.e., $N_\theta = 3$. For simulation purposes, only one measurement data is obtained by applying a single opposite current pattern between e1 and e17. The measurement data thus obtained is contaminated with 0%, 1%, 2% and 3% relative white Gaussian noise. As discussed in [4], Trelea method 1 and Trelea method 2 are used giving the best published results so far. A total of 30 particles were generated per scenario initialized randomly with the range of each Fourier coefficients between -8 and 8. In [4] it is recommended that a total of 30 particles give good convergence.

The reconstructed boundaries are shown in Fig. 2 along with Gbest value which represents the global best value of the cost functional (14) in the swarm. From the reconstructed images in Fig. 2 it can be seen that for 0% relative noise, the PSO technique estimates the boundary quite well. For both Trelea method 1 and Trelea method 2, the PSO converges fast in the start and slow convergence is observed at the end. Here it can be noticed that a total of 100 epochs were considered for simulation purposes. The reason behind choosing more epochs is that not all of the Fourier coefficients are equally sensitive and hence the higher modes of Fourier coefficients take more time to converge. It was also found that the higher modes are also sensitive to measurement noise. As seen in Fig. 2 for a measurement noise of 1%, 2% and 3% even though the location of the target is estimated well, however there is

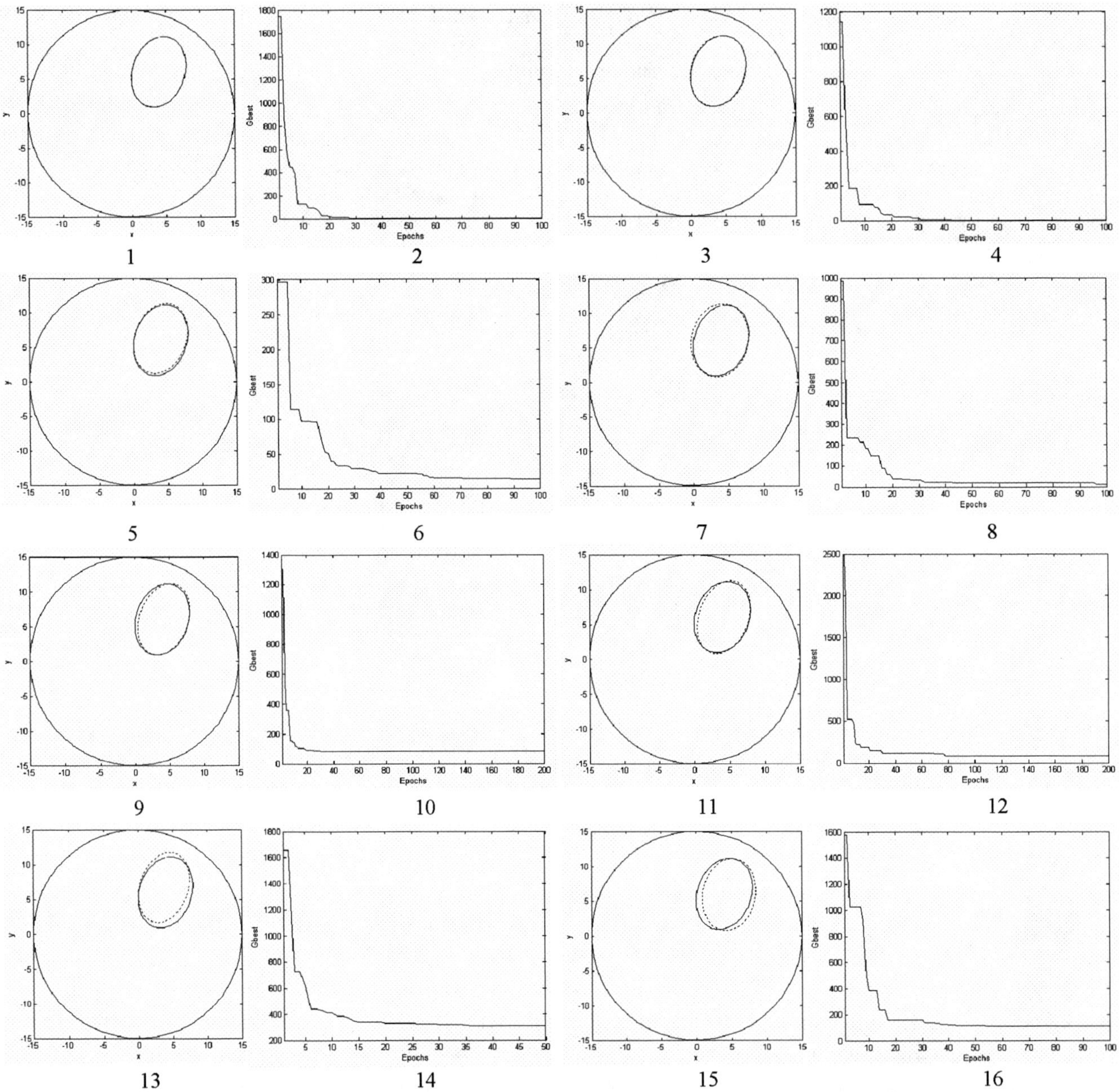

FIGURE 2. Reconstruction results. True boundary is represented by solid line and estimated boundary is represented by dotted line in 1, 3, 5, 7, 9, 11, 13 and 15.

1. Reconstructed image for Trelea method 1 with 0% relative measurement noise. 2. Gbest for Trelea method 1 with 0% relative measurement noise. 3. Reconstructed image for Trelea method 2 with 0% relative measurement noise. 4. Gbest for Trelea method 2 with 0% relative measurement noise. 5. Reconstructed image for Trelea method 1 with 1% relative measurement noise. 6. Gbest for Trelea method 1 with 1% relative measurement noise. 7. Reconstructed image for Trelea method 2 with 1% relative measurement noise. 8. Gbest for Trelea method 2 with 1% relative measurement noise. 9. Reconstructed image for Trelea method 1 with 2% relative measurement noise. 10. Gbest for Trelea method 1 with 2% relative measurement noise. 11. Reconstructed image for Trelea method 2 with 2% relative measurement noise. 12. Gbest for Trelea method 2 with 2% relative measurement noise. 13. Reconstructed image for Trelea method 1 with 3% relative measurement noise. 14. Gbest for Trelea method 1 with 3% relative measurement noise. 15. Reconstructed image for Trelea method 2 with 3% relative measurement noise. 16. Gbest for Trelea method 2 with 3% relative measurement noise.

some discrepancy in the shape. In all cases, it can be seen that the PSO converges fast in the first 20 iterations. Furthermore, Trelea method 1 and Trelea method 2 have similar performance, however based on our observations, we recommend Trelea method 1 as in most cases it has lower values of cost functional. And so for 3% relative measurement noise, Trelea method 1 was used for a total of 50 epochs which gives almost similar reconstructed image as compared to Trelea method 2 with 100 epochs. From the simulation results, it is established, that given a single measurement data it is possible to estimate the region boundary which was done in the previous attempts using partial or whole image frame. However, since it is an evolutionary computational technique, it is recommended to be used offline. The vector, floating-point nature of the algorithm suggests that high-performance implementations could be created that take advantage of modern hardware extensions pertaining to vectorization and so there is a possibility of online usage as well.

CONCLUSIONS

In this paper, we propose a new numerical method for the recovery of elliptic boundary region . The proposed region boundary is represented by Fourier coefficients. Numerical results were given with synthetic data for particle swarm optimization (PSO) technique. Optimal parameters, Trelea method 1 and Trelea method 2 are used. Based on the performance it seems that the overall performance is promising considering the usage of only one current pattern to obtain measurement data and that too is contaminated by noise. The higher contrast ratio and elliptic region boundary representation can be used to recover the boundary of air bubbles in conducting medium. The approach can be used offline as it is an evolutionary computational technique. The PSO technique does not require the complex mathematical derivation of Jacobian and is helpful in those cases where a complex boundary representation is desirable. The algorithm shows a very fast convergence in the start. But with Fourier coefficients representation, some of the coefficients are more sensitive and require more iterations to converge even though the cost functional does not change significantly.

ACKNOWLEDGMENTS

The part of the researchers participating in this study are supported by the grant from "the 2nd phase BK21 project".

REFERENCES

1. J. Kennedy, R. C. Eberhart and Y. Shi, *Swarm Intelligence*, San Francisco, CA: Morgan Kaugmann Publishers, 2001.
2. V. Kolehmainen, A. Voutilainent and J. P. Kaipio, "Estimation of non-stationary region boundaries in EIT-state estimation approach", *Inverse Problems* **17**, 1937-1956 (2001).
3. M. Vauhkonen, W. R. B. Lionhart, L. M. Heikkinen, P. J. Vauhkonen and J. P. Kaipio, "A MATLAB package for EIDORS project to reconstruct two-dimensional EIT images", *Physiol. Meas.* **22**, 107-111 (2001).
4. I. C. Trelea, "The particle swarm optimization algorithm: convergence analysis and parameter selection", *Information Processing Letters* **85**, 317-325 (2003).
5. R. C. Eberhart and Y. Shi, "Comparing intertia weights and constriction factors in particle swarm optimization" in *Congress of Evolutionary Computation-2000*, San Diego, CA, 2000, pp. 84-88.
6. B. S. Kim, U. Z. Ijaz, J. H. Kim, M. C. Kim, S. Kim and K. Y. Kim, "Nonstationary phase boundary estimation in lectrical impedance tomography based on the interacting multiple model scheme", *Meas. Sci. Technol.* **18**, 62-70 (2007).
7. V. Kolehmainen, S. R. Arridge, W. R. B. Lionhart, M. Vauhkonen and J. P. Kaipio, "Recovery of region boundaries of piecewise constant coefficients of an elliptic PDE from boundary data, *Inverse Problems* **15**, 1375-1391 (1999).
8. D. Isaacson, "Distinguishability of conductivities by electric current computed tomography", *IEEE Trans. Medical Imaging* **M1-5**, 91-95 (1986).
9. D. C. Gisser, D. Isaacson, and J. C. Newell, "Theory and performance of adaptive current tomography system", *Clinical Physics and Physiological Meas.* **9**, 35-42 (1988).
10. J. C. Newell, D. C. Gisser, and D. Isaacson, "An electric current tomograph", *IEEE Trans. Biomed. Eng.* **35**, 823-833 (1988).
11. V. Kolehmainen, M. Vauhkonen, P. A. Karjalainen and J. P. Kaipio, "Assessment of errors in static electrical impedance tomography with adjacent and trigonometric current patterns", Report No. 2/97, University of Kuopio, 1997.

3-D turbulent Premixed Flamelet Stretch Visualization Based On High Speed Stereoscopic Image

Qunxing Huang, Fei Wang, Qiang Zhang, Zengyi Ma, Jianhua Yan and Kefa Cen

State Key Laboratory of Clean Energy Utilization, Hangzhou/China hqx@zju.edu.cn

Abstract. According to flamelet model hypothesis, the turbulent premixed flame was assumed to be consisting with series of laminar flamelet with deferent scales and structures. Due to the complexity in turbulent flow mechanism and chemical kinetics, the most prevalent numeric simulation techniques only available with simplified cases. For deep understanding of the model and premixed combustion process, high performance 3-D visualization diagnostic method should be developed. In this paper, a modified stereoscopic technique based on high speed images was provided for flamelet stretch characters measurement. The stereoscopic flame images were captured simultaneously with one high resolution CCD chip which was formerly calibrated, and a novel algorithm was developed for flamelet matching and the 3-D flame front stretch rate and stretch speed were discussed. Comparing with normal multi-camera 3-D visualization systems, the method provided shows great advantages and potential for flame visualization, such as simple structure, low cost, and without camera synchronization, high robust and accuracy.

Keywords: Premixed flame; stereoscopic; camera calibration; image pattern matching.
PACS:

1. INTRODUCTION

Due to the complexity of the chemical kinetic and physical motion, the premixed turbulent combustion has not been good understood over centuries. Flamelet hypostasis is one of the most popular models for flame analysis during the last decade which treat with the flame to be consists of series laminar flamelets under different scales and orientations[1][2][3]. The surface structure and stretch character are two kernel parameters for flamelet statistic. However the inherently 3D dynamic motion character has made the flame presentation and measurement to be extremely difficult. Most experiment setups are only available for 1D or 2D quantitatively measure. Tomography reconstruction, holographic imaging and are two predominant 3D structure visualization methods widely used as non-intrude techniques. But only last one has been discussed for flame surface topology measurement because the spatial resolution of tomography has been limited by the projection number and system is too much complexity and expensive [4][5]. Holographic imaging is reasonable choice for instantaneous 3D spatial object surface reconstruction, but many prior research results shows the precious has been restricted by image noise from soot particles scattering and thermophoresis force [6].

Resent years, stereoscopic imaging system has been widely discussed for object surface reconstruction which simulate human eye's mechanism and recover the in-deep information from two different views. Zhang et al. (2003) has reconstructed the impinging flame surface by a single Olympus digital camera assembled with stereo adapter and got very good precious [7]. Limei Song (2006) has developed a grating match method to deal with the left-right images to visualize a state cup surface from grating stereo cameras [8]. Andrea Stella (2001) combined the stereoscopic image with traditional PIV instruments to form a 3D velocimetry system [9] and got reasonable results. The great advantage of stereoscopic imaging system is its' simplicity which made it easy for operating and possible for in-situ measurement outside laboratory.

In this paper, the Zhang's [7] system has been implemented and also been improved with high speed camera for premixed methane flame surface stretch speed measurement under 3D orientation. Why we chose Zhang's system is that by adopting a stereo adapter before normal camera lens, we can capture left-right or up-down stereo images instantaneous through the same CCD chip. So there is no synchronization device needed. The 3D structure of

CP914, *Multiphase Flow: The Ultimate Measurement Challenge,* edited by X. Cai, Y. Wu, Z. Huang, S. Wang, and M. Wang

methane Bunsen flame surface has been recovered under different oxygen ratio. And the flamelet in-deep stretch speed has also been calculated based on coordination disparities at different time. The results show the potential of stereo imaging for 3D unsteady flame surface motion visualization and quantitatively measurement.

2. EXPERIMENTAL SET-UP

The flame discussed in this paper comes from a methane Bunsen burner. Oxygen and methane were premixed and ignited at the burner nozzle. The ratio between oxygen and methane could be changed by two gas flow gauges (Fig.1). The high speed camera supplied by RedLake Company was located on a tripod which made the vertical height and horizontal orientation of camera all tunable. The shuttering speed of the camera ranges from 1fps to 10000fps which covering flame flicker frequency. The stereo adapter which was made up of four delicate reflection mirrors with 45 parallel degrees was fixed to the front of camera lens. The captured image was separated into left and right parts representing two symmetry projections of the object. The image data captured by the camera was transfer to the computer hard disk by net cable.

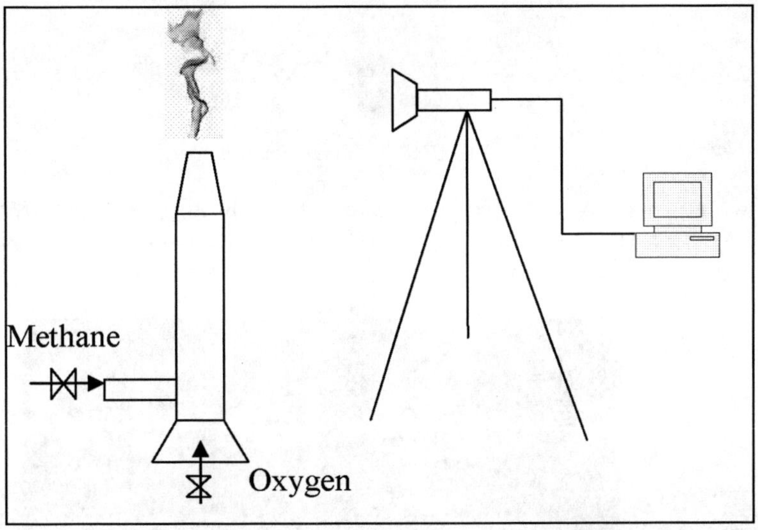

FIGURE 1. The burner and stereo imaging system

The most notable advantage of above arrangement is that it avoids the synchronization task if two or more cameras were implemented. However the stereo adapter has restricted the reconstruction geometry range. The surface that appeared only in one side image could not be recovered and also unavailable in dealing with transparent flame. In a word, the stereoscopic imaging system just quantitatively simulated human eye and brain mechanism.

3. STEREO ADAPTER OPTICAL MODEL

The kernel mechanism of a stereo adapter is combining object image under different view orientation into a single image as illustrated in Fig 2.

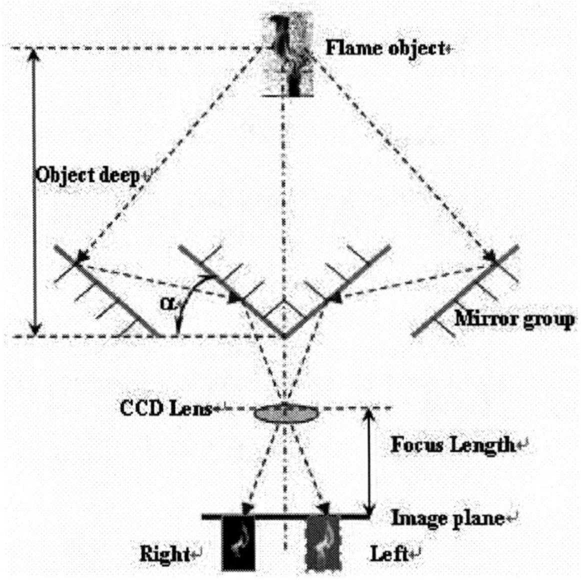

FIGURE 2. Stereo adapter model

The same side two mirrors should be arranged parallel to each other and symmetric to .another side mirrors respectively. Otherwise there will be extremely distortion between left and right part images. Typically flame image capture by above set-up is like Fig.3 showed.

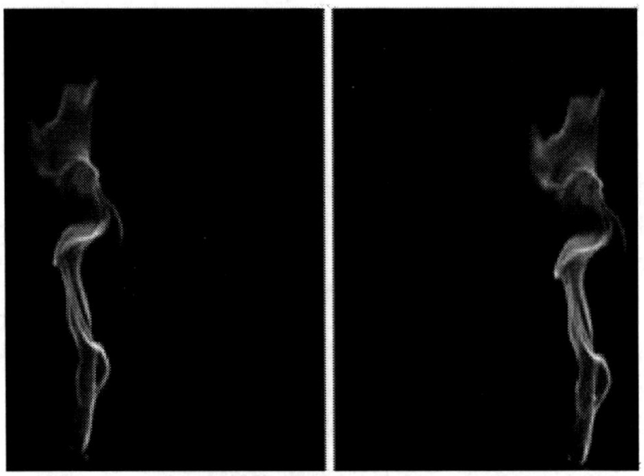

FIGURE 3. Stereoscopic flame image

4. CAMERA CALIBRATION

Each side imaging system complies with pinhole camera principle (Fig.4) and must be calibrated to figure out the intrinsic and extrinsic parameters. Intrinsic parameters include camera principle point location (C_x,C_y), and focus length (f_x,f_y) which were expressed in pixel unit, and the skew coefficient (α_c) defining the angle between the x and y pixel axes. All these parameters are only depending on camera self features and once estimated, can be re-used as long as the focal length is fixed. The extrinsic parameters are used to describe the camera motion around a static scene, or rigid motion of an object in front of the still camera, the matrix format of the extrinsic parameters likes:

$$\begin{bmatrix} r_{11} & r_{12} & r_{13} & t_1 \\ r_{21} & r_{22} & r_{23} & t_2 \\ r_{31} & r_{32} & r_{33} & t_3 \end{bmatrix} \text{ or } [R|t]$$

Where $\boldsymbol{R}$ indicates the rotating vector and $\boldsymbol{t}$ indicates the translation vector respectively. By concatenating the intrinsic and extrinsic parameters, the relationship between the image pixel and object surface points spatial position can be expressed as:

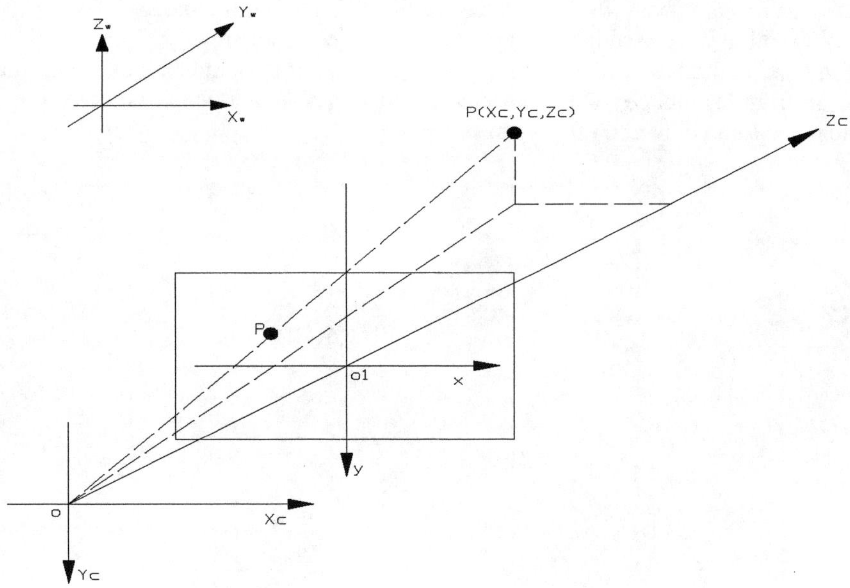

FIGURE 4. pinhole camera imaging model

$$\begin{bmatrix} u \\ v \\ 0 \end{bmatrix} = \begin{bmatrix} f_x & \alpha_c & c_x \\ 0 & f_y & c_y \\ 0 & 0 & 1 \end{bmatrix} \begin{bmatrix} r_{11} & r_{12} & r_{13} & t_1 \\ r_{21} & r_{22} & r_{23} & t_2 \\ r_{31} & r_{32} & r_{33} & t_3 \end{bmatrix} \begin{bmatrix} X \\ Y \\ Z \\ 1 \end{bmatrix}$$

In this paper, the self-calibration model [10][11][12] has been implemented. According to this technique several chessboard pictures with different orientation are taken by the same camera with fixed intrinsic parameters and all upper unknowns are recovered by image post-processing algorithm. The typical stereo calibration chessboard pattern was illustrated in Fig.5

FIGURE 5 stereoscopic chessboard calibration pattern

After the left and right imaging camera have been calibrated by the same pattern, a Matlab based stereo calibration tool [13] could be used to figure out the coordinate transfer matrix between them.

$$X_R = R * X_L + T$$

5. FLAME SURFACE MATCHING MODEL

Another important thing before in-deep information recover is to determine the correspondence facet which was the projection of the same part from flame surface in the left and right images. The epipolar constraint and maximum like hood algorithm were used in this paper to solver such problem.

As showed in Fig.6, ***p1*** and ***p2*** are called correspondence points, if they indicate the same flame surface point ***p***. When p1 is picked up from left image, ***p2*** is restricted in the epipolar line which is intersection between the right image plane and the plane formed by C_1C_2 and point ***p***.

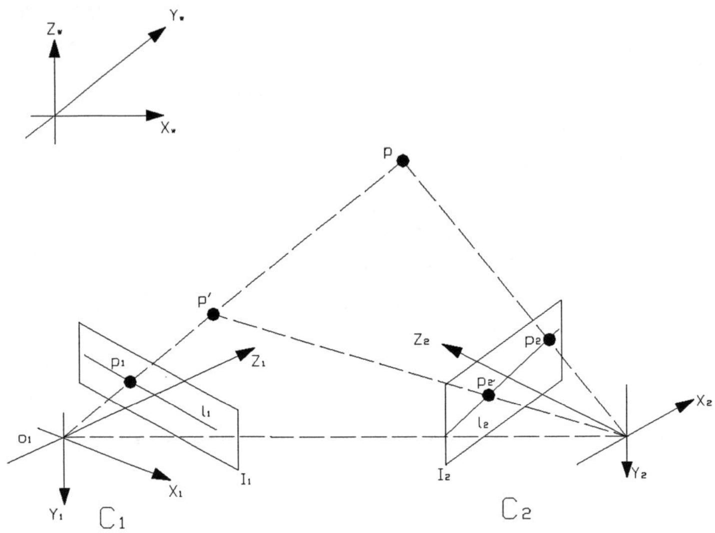

FIGURE 6. Epipoar geometry

So the mathematic description of the epipolar line can be expressed as:

$$p_2^T * F * P_1 = 0$$

Where ***F*** is the fundamental matrix determined by stereo camera calibration. And after the epipolar line is determined, a maximum like hood method can be used to search the matching point ***p2*** along the line. For flamelet matching, we choose a searching window of pixel size 16×16 to improve the matching precious, not just a single point. The most likelihood results can be obtained by minimizing the following functional:

$$\sum_{i=1}^{16}\sum_{j=1}^{16}\left\| p_{1,ij} - p_{2,ij}(p_{1,ij}, F)\right\|^2$$

After the match points have been figured out, the depth can be calculated easily. In this paper, the energy–based disparity estimation algorithm developed by Alvarrez (2002) has been implemented [15].

$$z = \frac{f \times (p_1 - p_2)}{D}$$

Where ***D*** is the distance between the camera baseline and the surface point ***p***. Reader can refer to Alvarrez 's work for any further information.

As long as the 3D coordinates of flame surface have been recovered from each stereoscopic image, the surface flamelet stretch speed can be calculated at once based on the coordinate disparity at different time.

6. 3-D RECONSTRUCTION RESULTS

In this paper, five different position and orientation calibration pictures were captured, Fig.7 showed the calibration results.

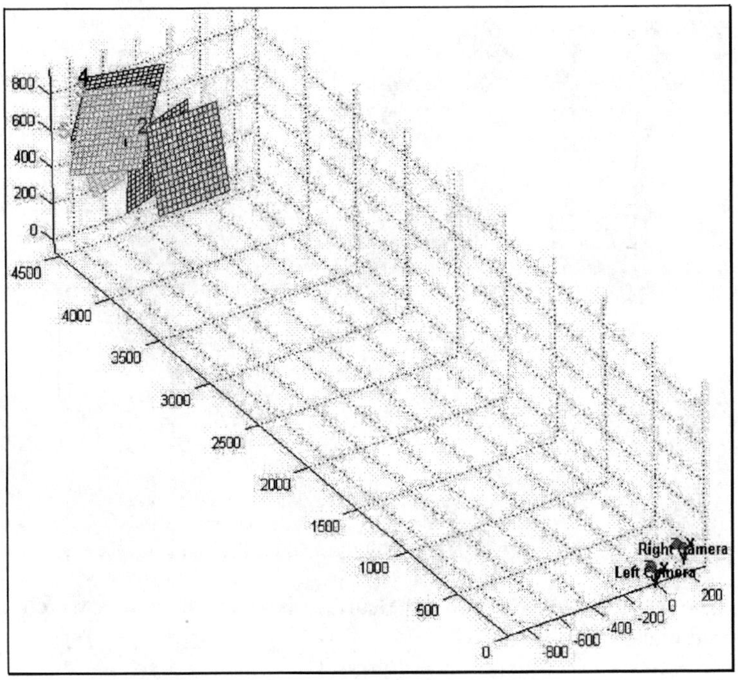

FIGURE 7. stereoscopic calibration results

The left and right camera indicates the left and right observing scope of the stereo adapter. And there is about 5% difference between the measured distance from the chessboard to the camera center with the recovered valued. The distortions of each camera in radial and tangential direction also have been calculated to make the results more reasonable.

Fig.8 shows the reconstructed 3D plotting from stereo image illustrated in Fig.3. The result still looks like 2D image. It is because that the stereoscopic angle is less than 15 degree. The flame surface edge also turned out to be obscured by noise. Zhang [7](2003) has separated some 'island' flamelet pieces to observe the wrinkle effect from the impinging flame surface. But in this paper, no such phenomenon has been captured due to the low gas flow rate. The shadow part was meshed in Fig.9.

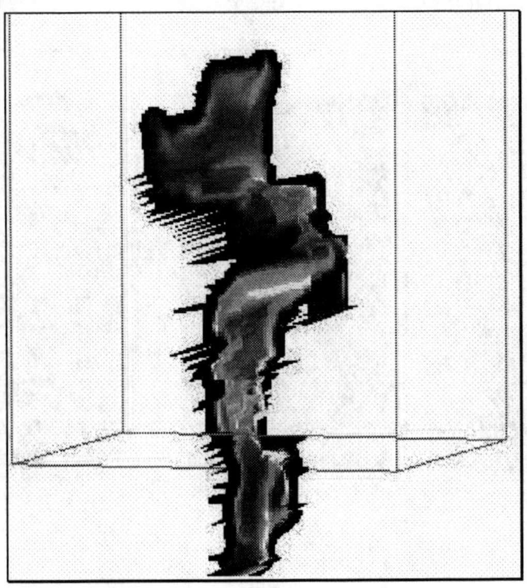

FIGURE 8. 3D flame surface.

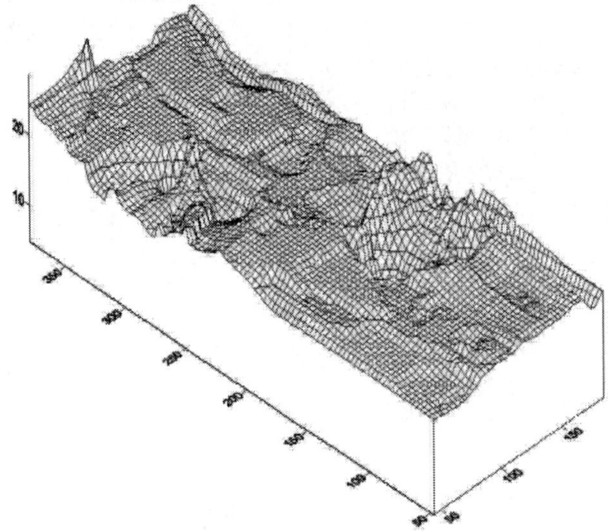

FIGURE 9. 3D meshed surface.

From the meshed surface, we can see clearly that the surface is concave-convex which indicating the combustion process to be roughness and unsteady.

7. 3-D VELOCIMETRY RESULTS

The largest camera shutter speed is up to 10000fps, which means maximum 10000 pictures could be captured in one second. But the actual picture number depends on the picture size and camera inner memory storage. In this paper, the speed is set to 2500fps which is about 10 times the fire flickering frequency. The time interval between two successive images is 250 microseconds.

Based on the reconstructed surface position and orientation, the three-components of the stretch vector are calculated. Fig.10 shows the X-Y speed components which were expressed with arrow vectors. The molar Oxygen and fuel ratio of all cases are less than 1.0 to keep the flame unburned out which will great increase the flame brightness through the soot particles thermal emission. Otherwise a flame brightness intensifier must be applied to make the flame visible in the picture under such a short exposure time.

From the speed distribution and average values, we can clearly observe the eddy motion in the central.

α	**0.3**	**0.4**	**0.5**	**0.6**
Stereo image	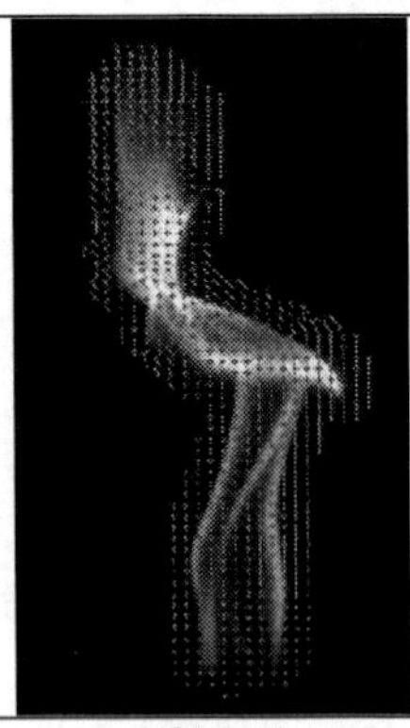	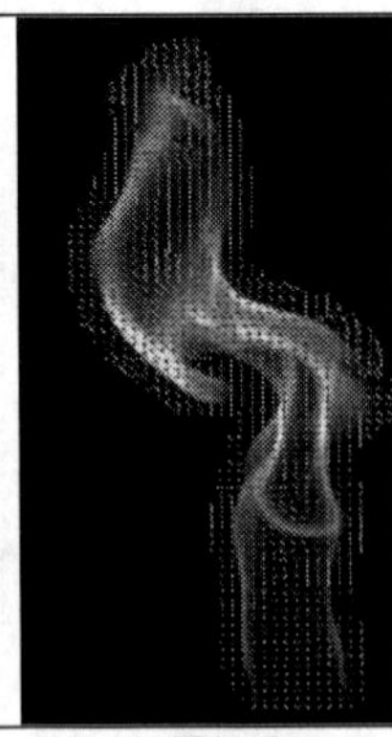	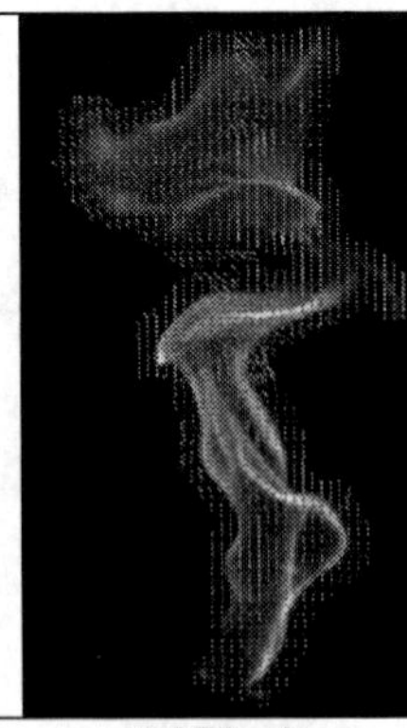	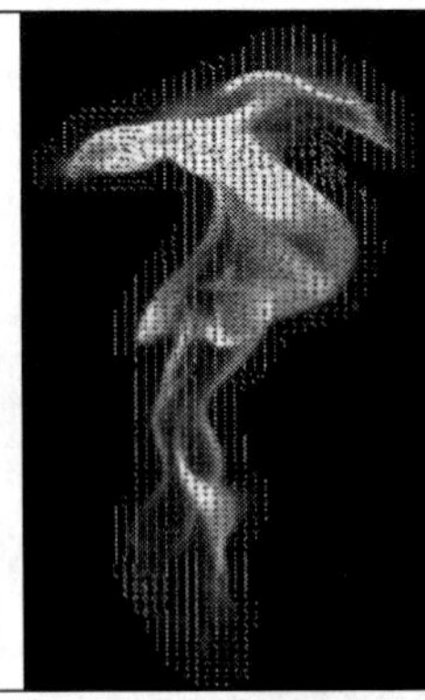
Average Vx	**1.311m/s**	**1.671m/s**	**1.971m/s**	**2.367m/s**
Average Vy	**6.48m/s**	**7.521m/s**	**9.645m/s**	**10.225m/s**

FIGURE 10. X-Y flame stretch

The third component vector which indicating in-deep stretch speed was plotted by pseudo-color contour. From the mean speed we can see that stronger disturbance occurred as more oxygen supplied.

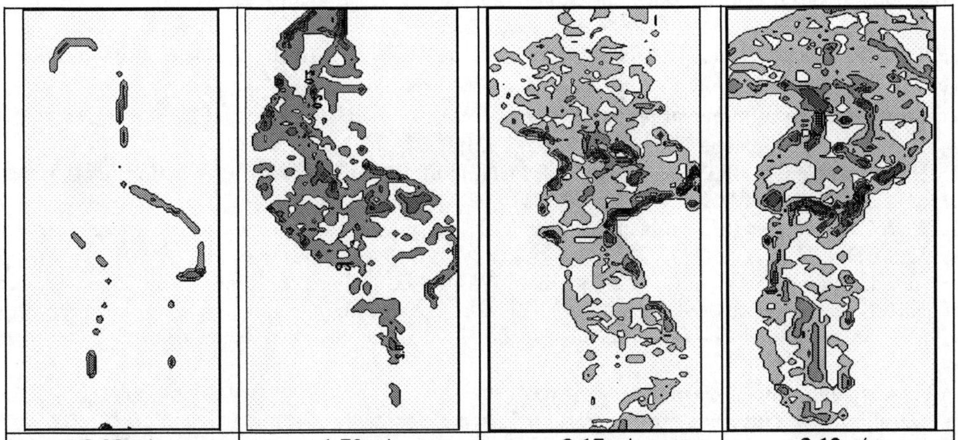

0.98m/s	1.72m/s	2.17m/s	3.12m/s

FIGURE 11. In-deep flame stretch

8. DISCUSSION

The classic stereoscopic technique has been widely used in movie industry over decades. It simulated human eyes mechanism, which simplified imaging system and providing 3-D information through a single picture or film. The implementation of stereo imaging in flame or other motion objects surface reconstruction owes to the development of image processing technique and electronic sensors.

The flame discussed in this paper is the most classic case which has been studied for lots researcher over decades, but still not enough because the combustion mechanism is so complicated. In a rigorous manner, the reconstructed results are not so reasonable, because the gas flame is inherently semitransparent. By feeding tracers can make the results more precious.

The experiment set-up and results show the capability of stereo imaging technique for flame visualization and quantitative measurement. The stereo adapter not only saves one more camera, but also avoids synchronization work making it possible for in-situ measurement.

After the flame surface structure has been reconstruct, many other flame characters can be deduced, such as wrinkling or flamlet scale and vortex motion rate. It also provides a promising tool for LES or DNS simulation test.

9. CONCLUSION

In this paper, a stereoscopic experiment set-up for turbulent premixed flamelet 3-D structure reconstruction and stretch speed measurement was provided. By adopting a stereo adapter, only a single camera chip was need for in-deep information recovery. The adapter was made up of four delicately arranged mirrors which reflecting two flame scenes into left and right part of the same CCD sensor. Self-calibration algorithm was applied based on a chessboard pattern. After calibration, the intrinsic and extrinsic parameters of left and right imaging geometry were recovered. The 3-D surface structure and flamelet stretch speed were reconstructed of a Bunsen gas flame under different oxygen ratio. The results show the stereoscopic technique is very capable for 3-D visualization and quantitative measurement along with the development of image processing method and CCD device.

ACKNOWLEDGMENTS

The work is supported by the Natural Science Foundation of China under Grant Nos. 50606031 and 60534030. It also supported by Program for Changjiang Scholars and Innovative Research Team in University (IRT0434). The authors would like to thank Dr. Zhang Yang for providing the stereo adapter.

REFERENCES

1. Petes N.laminar Diffusion Flamelet Models in Non-premixed Turbulent Combustion Prog. , Energy Combust Sci. 1984,10:319~339[J]
2. P. J. COELHO, N. PETERS, Unsteady Modelling of a Piloted Methane/Air Jet Flame Based on the Eulerian Particle Flamelet Model, COMBUSTION AND FLAME 124:444–465 (2001) [J]
3. Sergei A. Measured properties of turbulent premixed flames for model assessment, including burning velocities,stretch rates, and surface densities, Combustion and Flame 141 (2005) 1–21[J]
4. Pelliccia-Kraft BJ, Watt DW (2000) Three-dimensional imaging of a turbulent jet using shearing interferometry and optical tomography. Exp Fluids 29:573–581
5. McMackin L, Hugo RJ, Pierson RE, Truman CR (1997) High speed optical tomography system for imaging dynamic transparent media. Opt Express 1:302–311
6. Meng H, Estevadeirdal J, Gogineni S, Goss L, Roquemore WM (1998) Holographic flow visualization as a tool for studying three-dimensional coherent structures and instabilities. J Visualiz 2:133–144
7. W.B. Ng, Y. Zhang, Stereoscopic imaging and reconstruction of the 3D geometry of flame surfaces, Experiments in Fluids 34 (2003) 484–493
8. LiMei Song,A novel grating matching method for 3D reconstruction, NDT&E International 39 (2006) 282–288
9. Andrea Stella a, Giulio Guj b, Three-components Particle Image Velocimetry measurements in premixed flames, Aerosp. Sci. Technol. 5 (2001) 357–364
10. Z. Zhang. A Flexible New Technique for Camera Calibration. Technical Report MSRTR-98-71, Microsoft Research, December 1998. Available together with the software at http://research.microsoft.com/~zhang/Calib/
11. S. J. Maybank and O. D. Faugeras,. A theory of self-calibration of a moving camera, the International Journal of Computer Vision, 8(2):123–152, Aug. 1992.
12. Heikkila, J. Silven, O. A four-step camera calibration procedure with implicit imagecorrection, Proceedings., 1997 IEEE Computer Society Conference ,(1997) 1106-1112
13. R. Y. Tsai, A versatile camera calibration technique for high accuracy 3D machine vision metrology using off-the-shelf TV cameras and lenses, IEEE J. Robotics Automat., p323-344, Vol. RA-3, No. 4 1987
14. Jean-Yves Bouguet and Pietro Perona. Camera Calibration from Points and Lines in Dual-Space Geometry. Submitted to European Conference on Computer Vision, 1998
15. Alvarez L, Deriche R, Sa´nchez J, Weickert J (2002) Dense disparity map estimation respecting image discontinuities: a PDE and scalespace based approach. J Vis Commun Image Represent 13:3–21

Total Variation Regularization Used in Electrical Capacitance Tomography

Huaxiang Wang, Lei Tang

School of Electrical Engineering and Automation, Tianjin University, Tianjin 300072, P. R. China.
Email: hxwang@tju.edu.cn, lei.tang@gmail.com

Abstract. To solve ill-posed problem and poor resolution in electrical capacitance tomography (ECT), a new image reconstruction algorithm based on total variation (TV) regularization is proposed and a new self-adaptive mesh refinement strategy is put forward. Compared with the conventional Tikhonov regularization, this new algorithm not only stabilizes the reconstruction, but also enhances the distinguishability of the reconstruction image in areas with discontinuous medium distribution. It possesses a good edge-preserving property. The self-adaptive mesh generation technique based on this algorithm can refine the mesh automatically in specific areas according to medium distribution. This strategy keeps high resolution as refining all elements over the region but reduces calculation loads, thereby speeds up the reconstruction. Both simulation and experimental results show that this algorithm has advantages in terms of the resolution and real-time performance.

Keywords: Electrical Capacitance Tomography (ECT) image reconstruction algorithm Total Variation (TV) Edge-preserving property Self-adaptive mesh refinement

PACS: 41.20.Gz

INTRODUCTION

Electrical capacitance tomography (ECT) is a tomographic measurement technique which was developed during the late 1980s [1, 2, 3]. Since then it has been improved in many ways and applied in various areas of industrial processes for non-invasive monitoring and measurement, such as imaging gas/liquid and gas/solid flows in pipelines, reactors, fluidized beds and the visualization of flames [4, 5, 6]. A typical set-up of an ECT sensor consists of a ring of several thin rectangular metal electrodes around the pipe or vessel, on either the exterior or interior wall of the pipe/vessel. The entire sensor is usually enclosed by a screen to shield off other stray electromagnetic fields. Then, in the first step, an excitation voltage is applied on one electrode (source) while the remaining electrodes (detectors) are held at zero potential, and the charge on each of the detector electrodes is measured, giving the source/detector capacitance distribution until each electrode has served once as a source electrode. Thus, a set of independent integral measurement values are obtained. In the second step, the distribution of the electrical

CP914, *Multiphase Flow: The Ultimate Measurement Challenge,* edited by X. Cai, Y. Wu, Z. Huang, S. Wang, and M. Wang

permittivity $\varepsilon(x,y)$ and hence material distribution over the cross-section is mathematically reconstructed from the capacitance measurements C.

The relationship between capacitance and permittivity distribution is governed by the following equation:

$$C = \frac{Q}{V} = -\frac{1}{V}\iint_{\Gamma} \varepsilon(x,y)\nabla\phi(x,y)d\Gamma \tag{1}$$

where V is the potential difference between two electrodes forming the capacitance, $\phi(x,y)$ is the potential distribution and Γ is the electrode surface.

Based on FEM, the linearized and discrete form of the forward problem can be expressed as

$$\underset{M\times 1}{\Delta C} = \underset{M\times N}{J} \cdot \underset{N\times 1}{\Delta\varepsilon} \tag{2}$$

where J is a Jacobian matrix, i.e. the sensitivity distribution matrix, giving a sensitivity map for each electrode pair, M and N are the number of measurements and elements, respectively.

Equation (2) is written in a normalized form[7]

$$\lambda = Sg \tag{3}$$

where λ is the normalized capacitance vector, S is the Jacobian matrix of normalized capacitance with respect to normalized permittivity, i.e. the transducer sensitivity matrix and g is the normalized permittivity vector, i.e. the grey level of pixels for visualization.

In the discrete form, the aim of image reconstruction for ECT is to find the unknown g from the known λ by equation (3).

Many papers about reconstruction in electrical capacitance tomography have been published [8, 9]. Most of them can be summarized as implementations of linear back projection algorithms (LBP) and iterative reconstruction algorithms. Other reconstruction techniques are model-based reconstruction by neural networks and pattern recognition.

There are three principal difficulties with image reconstruction for ECT: (1) the relationship between the permittivity distribution and capacitance is non-linear and the electric field is distorted by the material present, which is called 'soft-field' effect; (2) the number of independent measurements is limited, leading to an under-determined problem; (3) the inverse problem is ill posed and ill conditioned, making the solution sensitive to measurement errors and noise. Regularization methods are needed to treat this ill-posedness[9], and the common one used is Tikhonov and iterative types.

In this paper, a new image reconstruction algorithm based on total variation regularization is proposed and a new self-adaptive mesh refinement strategy is put forward. Compared with conventional Tikhonov regularization, this new method not only stabilizes the reconstruction, but also enhances the distinguishability of the reconstruction images, which possesses a good edge-preserving property.

TIKHONOV REGULARIZATION

Regularization techniques have been adopted in order to stabilize the inversion. They are used to determine a set of solutions using prior constraint information and then one solution from this set is chosen. Tikhonov method is one of the most commonly used universal regularization tools for solving ill posed inverse problems and has been applied to electrical tomography[13].

The reconstruction is commonly stabilized using the Tikhonov regularization by minimization of the follow function:

$$J(g) = \frac{1}{2}\|Sg - \lambda\|^2 + \mu R(g) \tag{4}$$

$R(g)$ is the regularization functional, and μ is the regularization parameter. The functional is often expressed in L^2 form as:

$$R(g) = \|L(g - \overline{g})\|^2 \tag{5}$$

where L is an appropriate regularization matrix and $\overline{g}$ is a prior estimate of the permittivity distribution.

In practice, it is difficult to obtain $\overline{g}$ when prior information is not available. In general, choosing $\overline{g}$ as zero and L as an identity matrix I give the standard Tikhonov regularization, which is equivalent to solving the following equation:

$$(S^T S + \mu I)g = S^T \lambda \tag{6}$$

It has been proved that the inverse of $(S^T S + \mu I)$ always exists, so standard Tikhonov method can be solved by (7).

$$\hat{g} = \left(SS^T + \mu I\right)^{-1} S^T \lambda \tag{7}$$

Iterative Tikhonov regularization based on Newton-Raphson method is expressed as:

$$\hat{g}_{k+1} = \hat{g}_k - \left(S^T S + \mu I\right)^{-1} S^T \left(S\hat{g}_k - \lambda\right) \tag{8}$$

There are situations in almost every field of application of ECT where the imaged permittivity has discontinuities. For example, in process tomography a multiphasic fluid will give rise to discontinuities at each phase interface. It is therefore important to be able to reconstruct these situations correctly.

Standard L^2 regularization, because of its simplicity, has been the common framework for solving several inverse problems. But this kind of methods tends to smooth the reconstructed images. They impose a certain degree of smoothness in order to obtain stability in the reconstruction process. So they penalize non-smooth images in ECT, leading to the reconstruction algorithm incapable of describing step permittivity distributions.

TOTAL VARIATION REGULARIZATION

To solve the poor distinguishability of L^2 regularization, a different approach is represented by the choice of the total variation functional, which is still a differential operator but leads to a L^1 regularization.

$$R(g) = TV(g) = \int_{\Omega} |\nabla g| d\Omega \tag{9}$$

where Ω is the region to be imaged.

The TV functional was firstly employed by Rudin, Osher and Fatemi[10] for regularizing the restoration of noisy images. The technique is particularly effective for recovering 'blocky' images, and has become well known to the image restoration community[11]. TV regularization allows a much broader class of functions to be the solution of the inverse problem, including functions with discontinuities[12].

Suppose that the permittivity is described by piecewise constant elements, the discretized version of $TV(g)$ can be expressed as:

$$TV(g) = \sum_i l_i \left| g_{m(i)} - g_{n(i)} \right| \tag{10}$$

where l_i is the length of the ith edge in the mesh, $m(i)$ and $n(i)$ are the indices of the two elements on opposite sides of the ith edge, and the index i ranges over all the edges. Equation (10) can be expressed in terms of matrices as:

$$TV(g) = \sum_i \left\| L_i g \right\| \tag{11}$$

where L is a sparse matrix, with one row per each edge in the mesh. Every row L_i has two non-zero elements l_i and - l_i in the columns $m(i)$ and $n(i)$, respectively .

Commonly, a smooth approximation is used to solve the non-differentiability of $|\nabla g|$ in the TV regularization functional. Then the minimization function becomes

$$J(g) = \frac{1}{2}\left\| Sg - \lambda \right\|^2 + \mu TV_{\beta}(g) = \frac{1}{2}\left\| Sg - \lambda \right\|^2 + \mu \sum_i \sqrt{\left\| L_i g \right\|^2 + \beta} \tag{12}$$

where $\beta > 0$ is a pre-determined small constant.

The gradient and Hessian of J can be calculated as follows:

$$J'(g) = S^T(Sg - \lambda) + \mu L_\beta(g)g \tag{13}$$

$$H(g) = S^T S + \mu L_\beta(g) \tag{14}$$

where

$$L_\beta(g) = L^T(diag(\sqrt{\|L_i g\|^2 + \beta}))^{-1} L \tag{15}$$

Then the Newton-Raphson iteration method based on total variation regularization can be stated as follows:

$$g_{k+1} = G(g_k - t_k H_k^{-1} J'_k) \tag{16}$$

where t is step length and G is a gate operator defined by

$$G(g) = \min\{\max\{g, g_{\min}\}, g_{\max}\} \tag{17}$$

where g_{min}, g_{max} are based on *priori* knowledge of the high and low permittivity of the multiphase flow.

This new algorithm not only stabilizes the reconstruction, but also enhances the distinguishability of the reconstruction image especially in areas with discontinuous medium distribution. It possesses edge-preserving and scale-dependent properties[12].

SELF-ADAPTIVE MESH REFINEMENT

In order to show the different properties between the L^2 Tikhonov and TV regularization, the values of $\|L_i g\|$, which is proportional to the permittivity difference between two neighboring elements sharing the ith mesh edge, have been analyzed by simulation test. In some cases of 'blocky' image, the TV reconstruction presents values of $\|L_i g\|$ as high as several times as for the Tikhonov reconstruction. The plots of $\|L_i g\|$ for the TV reconstruction, sorted in increasing order, presents a slowly rising trend followed by a steep bend to high values on the right of the graph as in Fig. 1.

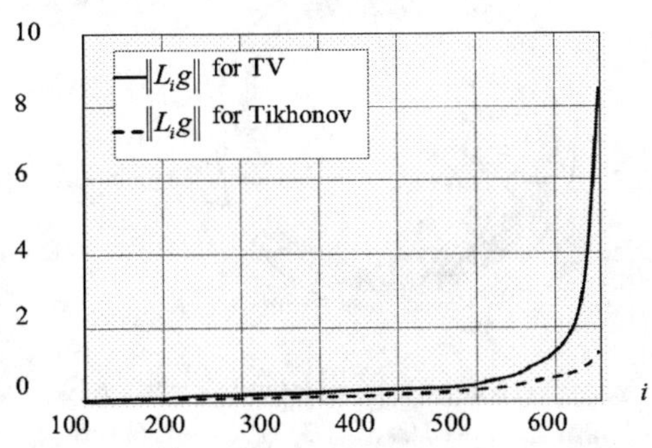

FIGURE 1. Plots of $\|L_i g\|$ for TV and Tikhonov regularization

By setting a threshold it is possible to identify a small number of edges as belonging to the boundary of the reconstructed targets. For each edge, select the neighboring triangles for refinement, and then the refinement is carried out as Fig. 2 (The two black circles in the figure stand for supposed phase interface).

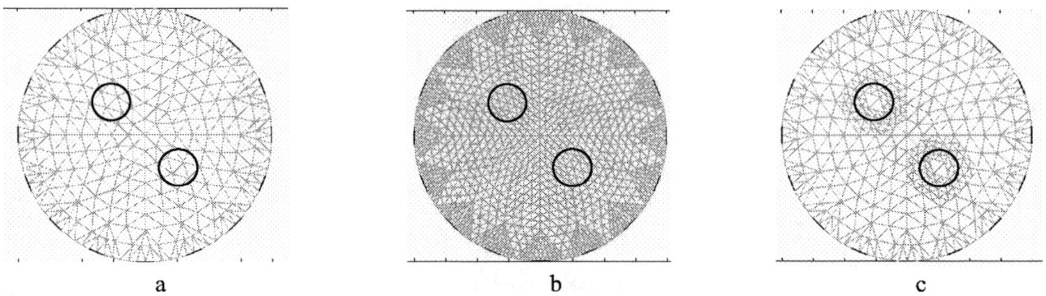

FIGURE 2. a. mesh before refinement b. mesh after all refined c. mesh after refinement by $||L_i g||$

In Fig.2.b, the mesh contains 1920 elements, but in 2.c the number decreases to 654. This self-adaptive mesh generation technique can refine the mesh automatically in specific areas according to medium distribution. It keeps high resolution as mesh refinement all over the region but reduces calculation load.

RESULTS

In order to prove the efficiency of TV regularization method and the self-adaptive refinement technique, both simulation and real-time experiments are carried out. And images reconstructed by standard Tikhonov regularization, TV method without refinement and TV method with adaptive refinement are evaluated, respectively.

In our work, a 16-electrode ECT model is used and the configuration is shown in Fig.3.

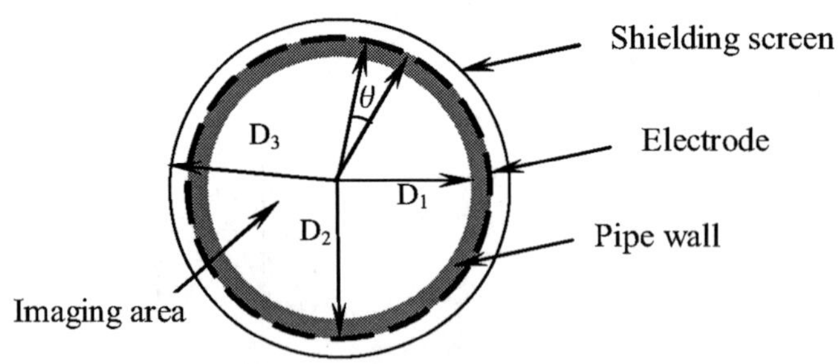

FIGURE 3. Cross section of a 16-electrode circular ECT sensor
D_1=100mm, D_2=110mm, D_3=120mm, θ=18°

In simulations, 1% noise is added and relative image error defined in equation (18) is chosen as the evaluation criterion.

$$Image \quad error = \frac{\left\| g - g^{*} \right\|}{\left\| g^{*} \right\|} \tag{18}$$

where g^{*} is the true permittivity distribution of the test object.

Simulation results obtained by MATLAB are shown in Table 1.

TABLE 1. Simulation results obtained by three methods

Phantoms	**Standard Tikhonov**	**TV without refinement**	**TV with refinement**
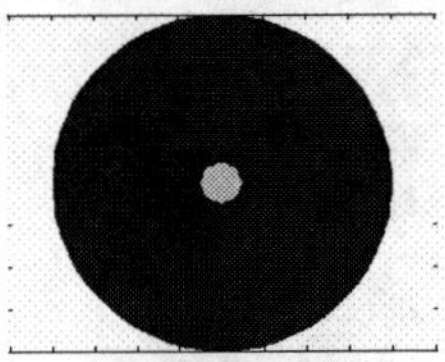	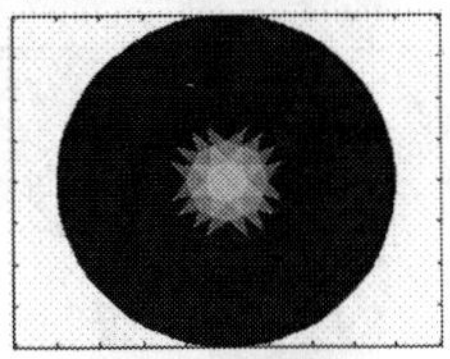	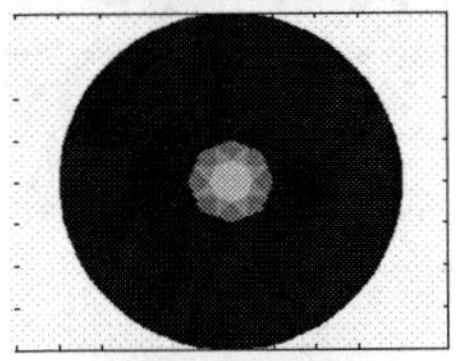	
Image error	7.504%	4.165%	3.571%
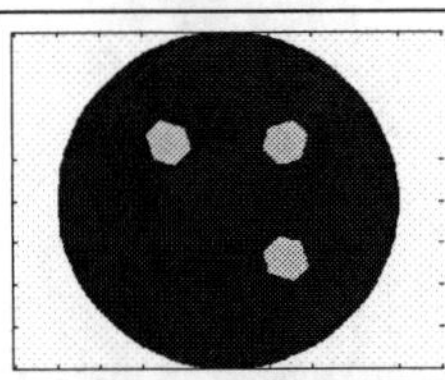	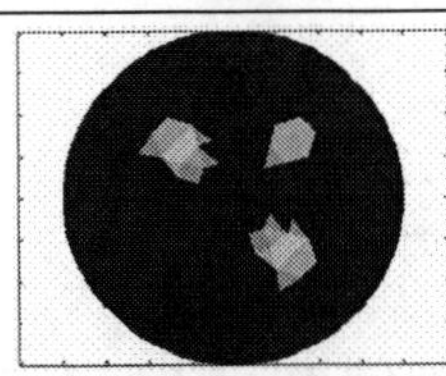	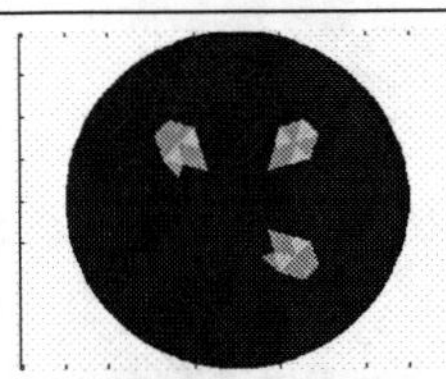	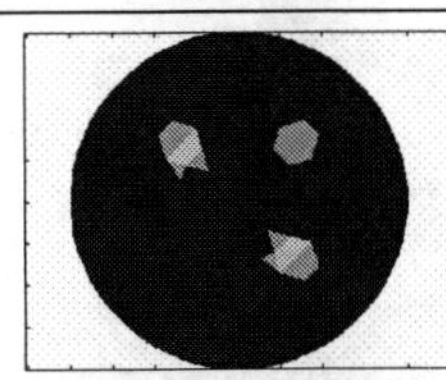
Image error	6.707%	3.443%	3.026%
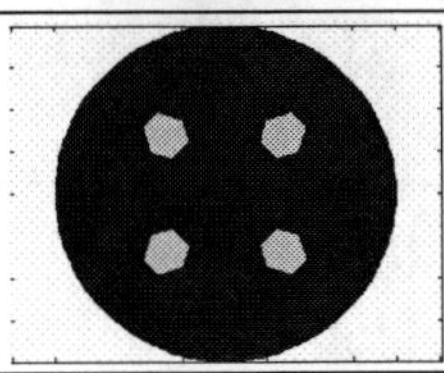	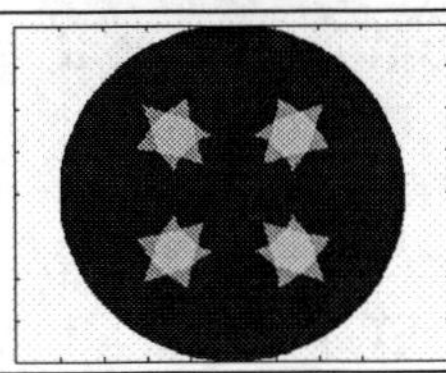	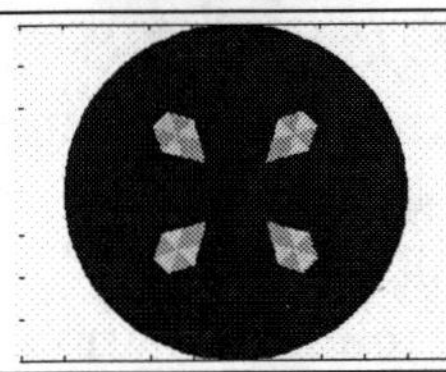	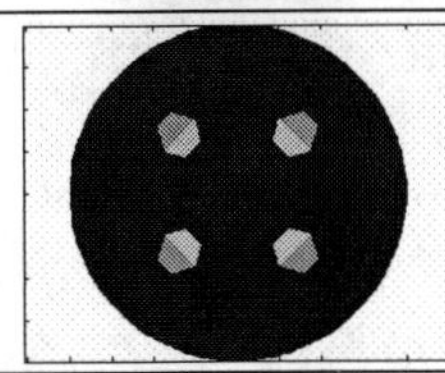
Image error	6.987%	3.102%	2.877%

All three algorithms are applied to TJU_MMT_System (a Multi-Mode Tomography system which can also be used as a ECT system separately. The sensor configuration is as Fig.3, and the excitation frequency is 500 kHz) . Experimental results for three typical medium distributions are shown in Table 2.

On a PC with a Pentium^{TM} 4 1.0GHz CPU and 256M memory, the reconstruction speed of the three methods are respectively as follows, standard Tikhonov regularization: 195 f/s; TV regularization without refinement: 150 f/s; TV method with self-adaptive meshing: 80 f/s. Experimental results show that, not only the resolution of the reconstructed images, but also the real-time performance of the methods based on TV regularization meet the demand of application.

TABLE 2. Images reconstructed by three methods for typical medium distributions

Phantoms	Standard Tikhonov	TV without refinement	TV with refinement
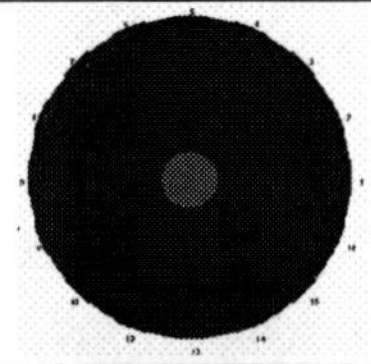	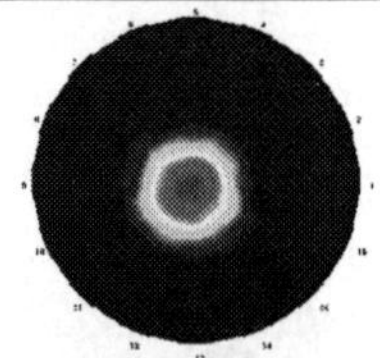	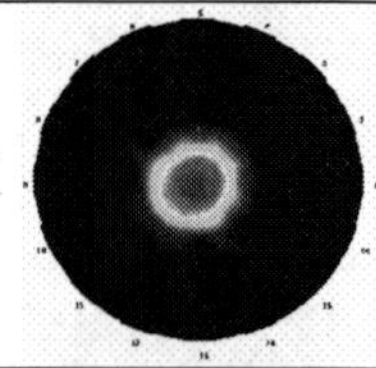	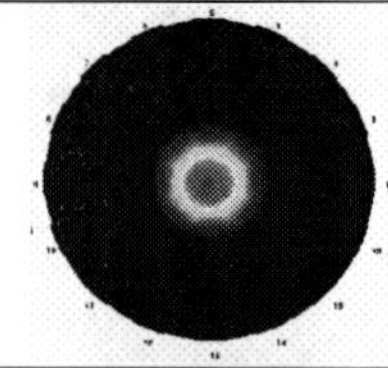
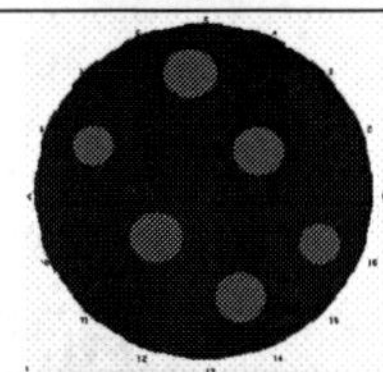	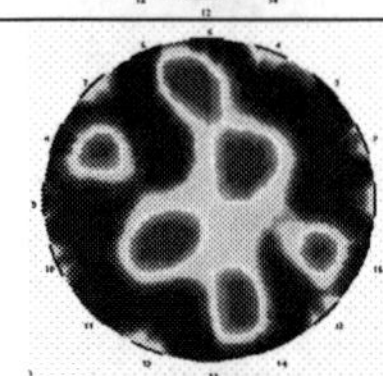	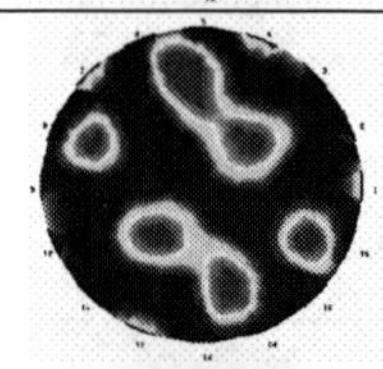	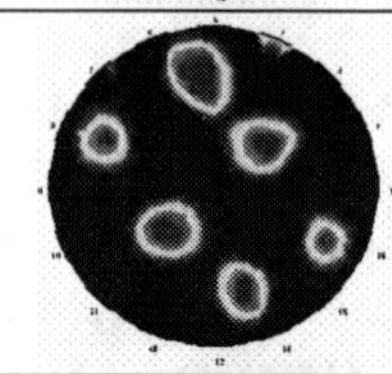
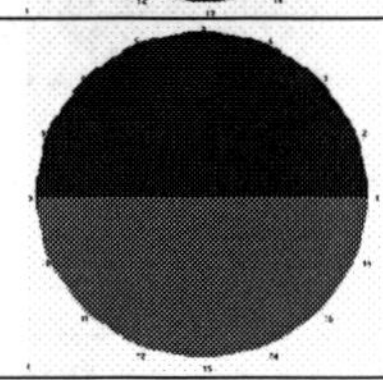	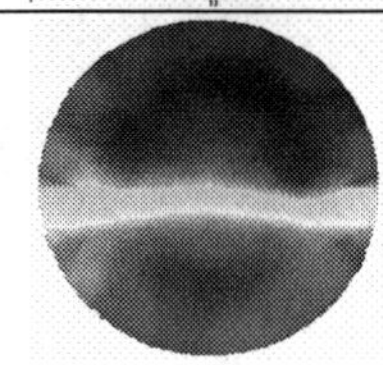	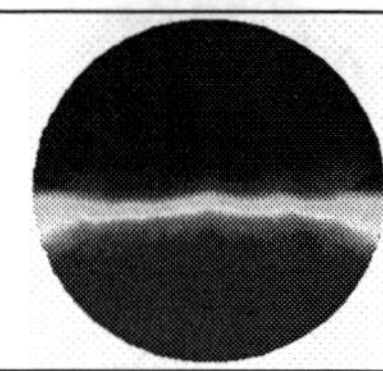	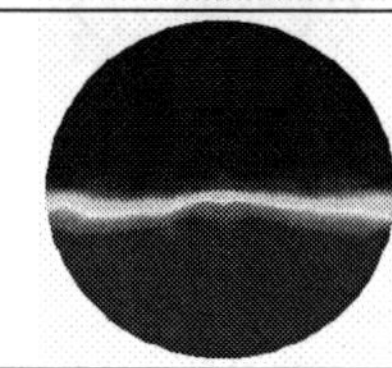

DISCUSSION

Both simulation and experimental results show that TV regularization has advantages in terms of the resolution and real-time performance, especially the distinguishability of several objects in the region. The new mesh refinement strategy can improve the resolution of the image with less computing cost than conventional refinement.

Some other algorithms based on TV regularization were put forward, such as Primal-Dual Method[11] which is called an 'efficient method' in resolution. But it doesn't perform well in real-time implementation. In ECT applications which require both fine resolution and high imaging speed, we should find a better strategy to derive a mixed method of several algorithms.

ACKNOWLEDGEMENTS

The authors gratefully acknowledge the financial support from the High Technique R&D Program of China (20001AA 413210), the National Natural Science Foundation of China (60532020, 60301008, 60472077,50337020), and the supporting program of the Ministry of Science and Technology of China (2006BAI03A00).

REFERENCES

1. M.S. Beck, A.B. Plaskowski, and R.B. Green, Imaging for measurements of two-phase flow, Proc. Flow Visualisation IV, Paris, pp.585-588, Hemisphere, London, Aug. 1986.
2. Huang S M, Plaskowski A, Xie C G and Beck M S, Electron. Lett., 24, pp.418-419, 1988.
3. Xie C G, Plaskowski A and Beck M S, Proc. IEE, 136, pp.173-83, 1989.
4. Reinecke N and Mewes D, Meas. Sci. Technol., **7,** pp.325-37, 1996.
5. Beck M S, Dyakowski T and Williams R A, Meas. Control, 20, pp.163-77, 1998.
6. Dyakowski T, Jeanmeure L F C and Jaworski A J, Powder Technol, 112, pp.174-92, 2000.
7. Xie C G, Huang S M, Hoyle B S, et a1, IEEE Proc G, 139(1), pp.89-98, 1992.
8. Isaksen Ø, Meas. Sci. Technol., **7,** pp.325-37, 1996.
9. Yang W Q, Peng L H, Measurement Science and Technology, 14, pp.1-13, 2003.
10. Rudin L. I., Osher S., Fatemi E., Physica D,60, pp. 259-268, 1992.
11. Chan T. F., Golub G. H., Mulet P., A Nonlinear Primal-Dual Method for Total Variation-Based Image Restoration, UCLA Math Department CAM Report, pp.95-43, 1995.
12. D. Strong, T. Chan, Inverse Problems, 19, pp.165-187, 2003.
13. Dobson D. C., Santosa F., Inverse Problems, 10, pp.317-334, 1994.
14. Fang W, Cumberbatch E., J. Eng. Math., 51, pp. 127-146, 2005.

Author Index

A

B

C

D

E

F

G

H

I

J

K

L

M

N

O

P

Q

R

S

T

U

V

W

X

Y

Z